Dictionary of Plant Lore

Dictionary of Plant Lore

By Donald Watts BA MIL
Bath, United Kingdom

AMSTERDAM • BOSTON • HEIDELBERG • LONDON
NEW YORK • OXFORD • PARIS • SAN DIEGO
SAN FRANCISCO • SINGAPORE • SYDNEY • TOKYO
Academic Press is an imprint of Elsevier

ELSEVIER

Academic Press is an imprint of Elsevier
30 Corporate Drive, Suite 400, Burlington, MA 01803, USA
525 B Street, Suite 1900, San Diego, California 92101-4495, USA
84 Theobald's Road, London WC1X 8RR. UK

Library of Congress Cataloging-in-Publication Data
Application Submitted

British Library Cataloguing-in-Publication Data
A catalogue record for this book is available from the British Library.

ISBN 13: 978-0-12-374086-1
ISBN 10: 0-12-374086-X

For information on all Academic Press publications
visit our Web site at www.books.elsevier.com

Printed and bound by CPI Group (UK) Ltd, Croydon, CR0 4YY
Transferred to Digital Print 2011

In Memory of Megan and Don

PREFACE

Sadly, this book is published posthumously. It results from the author's interest in plant and tree lore and ethnobotanical studies developed over several decades of research and teaching. Although the author was checking the first print at the time of his death, any inaccuracies or omissions would undoubtedly have been rectified if he had lived a few months longer, and I can only apologise for these.

Throughout the book there are references to recipes of the past use of plants for a variety of uses—some of which would be considered highly imaginative or mythical today. It goes without saying that such recipes should not be attempted by the lay person, but should be seen in the light of man's evolutionary understanding of the potential of plants within the culture and environment that he lived.

The author had a very considerable collection of antiquarian books, but I would like to acknowledge the excellent support he received from the London Library and West Wiltshire Libraries. I would also like to thank Sheila Tomkinson for her generous help with proof-reading.

S. Watts

CONTENTS

Preface   vii

- A   1
- B   21
- C   53
- D   97
- E   127
- F   143
- G   159
- H   179
- I   206
- J   213
- K   217
- L   220
- M   235
- N   263
- O   269
- P   283
- Q   311
- R   312
- S   335
- T   376
- U   395
- V   400
- W   408
- X   439
- Y   440
- Z   448

Bibliography   449

A

Abies alba > SILVER FIR

Abies balsamea > BALSAM FIR

ABORTION

At the same time as being an aid to conception, TANSY leaves were chewed by unmarried pregnant girls to procure an abortion (Porter. 1969). Indeed, the poisonous oil of tansy has long been taken to induce abortion (Grigson). Exactly the same ambivalence is shown in beliefs concerning PARSLEY. It is an aid to conception on the one hand, and a contraceptive as well as an abortifacient on the other. Cambridgeshire girls would eat it three times a day to get an abortion, but the belief is actually widespread (Waring). Eating HORSERADISH leaves three times a day was a valued means in the Fen country of causing abortion (Porter. 1968), knowledge apparently not confined to East Anglia, for Whitlock mentions it as a Wiltshire remedy, if that is the right word to use.

LADY LAUREL (*Daphne mezereum*) is a poisonous plant, dangerously violent in its action. Presumably that is the reason for its inclusion in a list of abortifacients used in Dutch folk medicine (van Andel). According to Dodonaeus, it is so strong that it had only to be applied on the belly to kill the child.. OLEANDER, too, is extremely poisonous, and has been used for abortions in India (P A Simpson). Hungarian gypsies recommended oleander leaves in wine, together with those of peony, as well as ergot, for abortions (Erdos. 1958). A certain Granny Gray, of Littleport, in Cambridgeshire, used to make up pills from the very poisonous HEMLOCK, pennyroyal and rue. They were famous in the Fen country for abortions (Porter. 1969) in the mid-19th century. SAVIN, probably the most notorious of the poisonous abortifacients, had been in use, as a matter of common knowledge, since ancient times, and was certainly well known in the 16th and 17th centuries, for instance, the scurrilous lines in Middleton's play, *A game of chess*, act 1; sc.2:

> To gather fruit, find nothing but the savin-tree,
> Too frequent in nuns' orchards, and there planted
> By all conjecture, to destroy fruit rather.

Some of the common names for the tree bear witness to this usage, names like Cover-shame and Bastard Killer. And its contraceptive properties were well-known, too. It was said that a stallion would never cover a mare if there was any savin in the stable (G E Evans. 1966). Even ARNICA, dangerous as it is, has been used in folk medicine (Schauenberg & Paris). MARSH ROSEMARY (*Ledum palustre*) is another traditional abortifacient (Schauenberg & Paris), and even the dried flowers of FEVERFEW

have been used in Europe (Lewis & Elvin-Lewis). A volatile oil made from the wood of VIRGINIAN JUNIPER, a close relative of Savin, has been used for abortions (Weiner); if this is the substance known as Red Cedar Wood Oil, then it must have been a thoroughly dangerous process (Usher). It is claimed that the Blackfoot Indians used the rhizome of SWEET FLAG to cause abortions (Johnston). That sounds highly unlikely, considering the plant's universal usage as a mild carminative.

RUE should never be taken if the patient is pregnant (Gordon. 1977) – with good reason, for it has often been used for abortions (Clair). French folklore insists that there was a law forbidding its cultivation in ordinary gardens. It was said that the specimen in the Paris Botanical Garden had to be enclosed to prevent pregnant girls from stealing it. In the Deux-Savres region of France it was believed that it caused any woman who merely touched it with the hem of her dress to miscarry (Sebillot). HEMP leaves were recommended in Cambridgeshire, the aim being to cause severe vomiting, often enough to result in a miscarriage. In the Scottish Highlands, it seems that the MOUNTAIN CLUBMOSS was used for the purpose. James Robertson, who toured the West Highlands and Islands in 1768, noted that "the Lycopodium selago is said to be such a strong purge that it will bring on an abortion" (quoted in Beith).

NUTMEGS were used at one time – the women in London who were the practitioners were actually known as "nutmeg ladies" (Emboden. 1979). BITING STONECROP is traditionally known as abortive (Schauenberg & Paris). 'A Middle English Rimed Medical Treatise' has:

> She that drynkes fumiter and the stoncrope
> Schal neuere yong childe in cradell roke.

In other words, they caused sterility or abortion (I B Jones). A surprising report involves HOUSELEEK. Some of the plants would be boiled, and the water given to the girl to drink. Later on, she would be told to climb a high wall and jump down, and that would do the trick (Vickery). One assumes that the second part of the treatment would have been the only operative one, yet the plant appears in a list of abortifacients used in Dutch folk medicine (Van Andel). The dried flowers of SAFFLOWER have been used in Chinese medicine as a "blood invigorator", whatever that means, though one meaning is certainly to promote menstruation, which is presumably why the flowers have the reputation in China of causing abortion (F P Smith). PENNYROYAL, too, is a known abortifacient (V G Hatfield. 1994), surprising, perhaps, for a plant that is almost a panacea. But it is also an emmenagogue (Cameron), which would explain the usage.

It is said that a root decoction of the African tree CATCHTHORN (*Zizyphus abyssinica*) will cause abortion (Palgrave & Palgrave). BUSY LIZZIE was an African plant originally, and some groups used its root as an abortifacient (Watt & Breyer-Brandwijk). SCARLET LEADWORT (*Plumbago indica*) is a well known abortifacient in India and Malaysia, but it is a thoroughly dangerous practice, and could be lethal.

(P A Simpson). It is the root that is used, apparently by introduction into the vagina, and that would cause violent local inflammation (Gimlette). Its close relative, CEYLON LEADWORT (*Plumbago zeylanica*) is used in the same way in the Philippines (Watt & Breyer-Brandwijk), and in Guyana, where it is called Gully-root, the leaf decoction of GUINEA-HEN WEED (*Petiveria alliacea*) is used (Laguerre).

Abrus precatorius > PRECATORY BEAN

ABSCESS
GROUNDSEL, MALLOW or MARSH MALLOW poultices were quite common for boils and abscesses (Hampshire FWI, Flück, Tongue. 1965), and a hot compress made from FENUGREEK seeds was used in the same way (Flück). The inner leaves of CABBAGE could be used, too (V G Hatfield. 1994), and in Ireland, a favourite treatment was to make a tea from **BROOM** tops, and bathe the place with this (Maloney). MADONNA LILY petals, macerated in alcohol, usually brandy, were bound to abscesses, boils and ulcers (Porter), and to all sorts of other skin eruptions. A poultice of WILD SORREL leaves was used in parts of South Africa to treat an abscess (Watt & Breyer-Brandwijk). Similarly, a poultice of chickweed is still prescribed by herbalists (Warren-Davis).

ABSINTHE
Thr drink known as Absinthe was actually taken as a "tonic drink"; it became very popular by the end of the 19th century. Made from oils of WORMWOOD, combined with anise, coriander and hyssop, it is actually a narcotic alcoholic drink, banned now that it is realised that it causes permanent neural damage (Emboden. 1979). Besides upsetting the nervous system, it irritates the stomach and increases heart action, and could cause disorientation, delirium and hallucination (Le Strange). At one time, wormwood was used in the preparation of all sorts of medicated wines and ales. Nowadays, extract of aniseed has replaced wormwood in aromatic liqueurs, in Pernod for instance, though small amounts of wormwood are still added to vermouth, which is a fortified white wine (Le Strange).

Acacia catechu > CUTCH TREE

Acacia dealbata > SILVER WATTLE

Acacia karroo > MIMOSA THORN

Acacia seyal > SHITTAH

Acanthus mollis > BEAR'S BREECH

Acer campestre > FIELD MAPLE

Acer negundo > BOX ELDER

Acer pseudo-platanus > SYCAMORE

Acer saccharum > SUGAR MAPLE

Achillea lanulosa > WOOLLY YARROW

Achillea millefolium > YARROW

Achillea ptarmica > SNEEZEWORT

Aconitum napellus > MONKSHOOD

ACORNS
have their own folklore. In some parts of the Continent they were put into the hands of the dead (Friend). Their cups and stems are the pipes smoked by the leprechauns (Ô Súilleabháin), and the cups are fairies' shelter, as Shakespeare knew : "All their elves, for fear, creep into acorn-cups, and hide them there" (*Midsummer Night's Dream act 1. sc 1*). Carrying one around in your pocket or purse is a way to keep yourself youthful, and to preserve health and vitality (Waring), or to prevent rheumatism (Thomas & Thomas). Dreaming of then is a good sign – it shows that health, strength and wordly wealth will be the dreamer's (Raphael), for acorns were in ancient times the symbol of fecundity – the acorn in its cup was one of the earliest phallic emblems (the acorn is the masculine, and the cup the feminine) (Wellcome). But over most of Europe, and in America, a plentiful crop of acorns augurs a poor corn crop next year. There was a form of marriage divination connected with them, or rather, their cups – two of them were taken, one named for the lover, and the other for one'sself. Then they were set to float in a bowl of water; watch them – if they sailed together, there would be marriage, but if they drifted apart then it was obvious what the result would be (Trevelyan). Of course, acorns have been of economic importance since prehistoric times as a substitute food. The offical hunger bread of Tsarist Russia consisted of ten parts of acorn flour, two parts of rye flour and two parts of rye bran. In parts of Poland, acorn flour was quite usual as an ingredient of bread – in fact, loaves made purely from cereal flour were virtually unknown. In Norway, too, acorn flour was used for bread right up to the 19th century (Clark). Another way of using them was to grind them up as flour with peas and beans (Ablett). Very important, too, was the practice of releasing pigs into oak woods – Gerard said,"swine are fatted [on acorns], and by feeding thereon have their flesh hardy and sound". Evelyn followed with "A peck of acorns a day, with a little bran, will make an hog ('tis said) increase a pound-weight per diem for two months together".

Acorus calamus > SWEET FLAG

ADAM-AND-EVE

(Boland), or **ADAM -AND-EVE-IN-THE-BOWER** (Macmillan) are names given to WHITE DEADNETTLE, "because in the flower, if you hold it upside down, the stamens can be seen lying for all the world like two people side by side in a curtained bed".

Adansonia digitata > BAOBAB

ADDER'S TONGUE

(*Ophioglossum vulgatum*) A plant with an evil reputation, destroying the grass in which it grows, and injuring the cattle that graze there (Addison). Nevertheless, the leaves, when pressed, produce a green oil, sometimes known as Green Oil for Charity (Leyel. 1937), considered by the older herbalists as a balsam for green wounds. An ointment, often called Green Adder ointment, is also used. This dates from ancient times, and is still available in the 20th century (Savage). The doctrine of signatures ensured that it would be used for snake-bites.

ADENOIDS

In Russian folk medicine, a gargle of BEET juice is a recognized remedy for adenoids. (Kourennoff).

Adhatoda vasica > MALABAR NUT

Adiantum capillus-veneris > MAIDENHAIR FERN

Aegopodium podagraria > GOUTWEED

Aesculus hippocastanum > HORSE CHESTNUT

Aesculus octandra > YELLOW BUCKEYE

Aframomum melegueta > MELEGUETA PEPPER

AFRICAN BLACKWOOD

(*Dalbergia melanoxylon*) Among some of the peoples of the Sudan, the causes of illness are divined through burning the young growths of this tree. The wood will burn "bright and sharp", so giving information about the activities of witches, which can then be interpreted by diviners (Rival). More empiricially, decoctions of the roots are given throughout East Africa in the treatment of sickness ascribed to the effect of an evil spirit, and the bark is one of the ingredients of the steam bath used for the same purpose (Koritschoner).

AFRICAN CORN LILY

(*Ixia viridiflora*). *Ixia*, the generic name, comes from the Greek word ixos, meaning bird lime, for these plants have a sticky sap, and can be used for the purpose.

AFRICAN MAHOGANY

(*Khaya ivorensis*) A sacred tree in Yoruba belief, venerated because of its usefulness, but the foot of this tree was believed to be the favourite haunt of wizards. It is, too, the symbol of vengeance, and must not be cut down until the tree spirit that lives in its branches has been propitiated by the offering of a fowl, or some palm oil (J O Lucas).

AFRICAN MARIGOLD

(*Tagetes erecta*) Not African at all, as is well-known, but a native of Mexico, brought back to Spain, and then naturalised in North Africa, where the soldiers of the Emperor Charles V, in the 1535 expedition to free Tunis from the Moors, found it growing, and assumed it was native (Fisher). It is claimed that this plant, or indeed any of the species, will reduce eelworm damage if planted among valued plants. The root secretion will kill eelworms at three feet distance (M Baker. 1977). They will, too, kill nematodes in the soil, as well as whitefly, and are really great to grow near potatoes and tomatoes (Boland & Boland), and also carrots and onions (Vickery. 1995), to keep pests away.

Ageratum conyzoides > GOATWEED

Agrimonia eupatoria > AGRIMONY

AGRIMONY

(*Agrimonia eupatoria*) A bitter herb, one of those once used to make beer keep (Bottrell). But:

> Kirn milk and agrimony
> Mak' the lasses fair and bonny (Simpkins).

A Guernsey preservative against spells, and a typically complicated one, to be hung round the neck: take nine bits of green broom, and two sprigs of the same, which you must tie together in the form of a cross; nine morsels of elder, nine leaves of Betony, nine of Agrimony, and a little bay salt, salammoniac, new wax, barley, leaven, camphor and quicksilver. The quicksilver must be enclosed in cobbler's wax. Put the whole into a new linen cloth that has never been used, and sew it well up so that nothing will fall out. Hang this round your neck. It is a sure preservative against the power of witches (MacCulloch). Agrimony was used in Guernsey for divinations too; one charm was to put two fronds of agrimony, each bearing nine leaflets, crosswise under the pillow, securing them by two new pins, also crossed. The future husband would appear in a dream (MacCulloch). But actually dreaming of agrimony foretells sickness in the house (Mackay).

Agrimony tea can be drunk as an ordinary beverage tea, popular once with French peasants (Cullum). It was known as Tea Plant in Somerset (Macmillan), and it is thé des bois in France, too (Clair). But it can be used as a gargle for a dry cough (Conway), or for a whole variety of ailments, lumbago among them (Vickery. 1995). It had a reputation for curing jaundice, and Culpeper assured us that "it openeth and cleanseth the liver", and in fact that was what it was used for in Gaelic folk tradition (Beith). Gerard had already recommended the leaf decoction as "good for them that have naughty livers, and for such as pisse blould upon the diseases of the kidneys". It is still used by herbalists as a liver tonic (it is sometimes known as

Liverwort), as well as for some kinds of rheumatism (Conway), for which it was a popular medicine in Cumbria (Newman & Wilson). The tea was a gypsy remedy for curing coughs, too (Vesey-Fitzgerald), and it is good for sore throats (A W Hatfield). It is still used for digestive disorders, and as a "blood purifier" (Clair), the latter recorded as a folk medicine in Dorset (Dacombe).

Agrimony was an ingredient in genuine arquebuscade water, as prepared against wounds inflicted by an arquebus, and mentioned by Philip de Comines in his account of the Battle of Morat, 1476. Eau d'arquebuscade was still supplied in 1897 for sprains and bruises (Fernie). "Agrimony sod in red wine, wherewith if wounds be washed, it cleanseth all filth and corruption from it …"(Lupton). It was an ingredient of a plaster to get thorns or splinters out of a wound (Dawson. 1934), according to a 15th century leechdom. Agrimony and black sheep's grease were used for a Scottish ointment (Dalyell), and another ointment, for backache, was made with our plant and mugwort, in the 14th and 15th centuries (Henslow Dawson,).

Agrimony wine is made, typically, to drink when one has a cold (Grigson. 1955), and the infusion was once used as an eye lotion for the cure and prevention of cataract. This kind of use is ancient; the Anglo-Saxon version of Apuleius, for example, had a remedy for "sore of eyes", as well as for a number of other ailments, including snake-bite, for which there were other remedies in that period. Storms quoted another Anglo-Saxon magical cure, in which he recommended the use of agrimony to "make one ring about the bite, it (the poison) will not pass any further…". Lupton, too, noted that agrimony by itself was enough: "… with a wonderful facility (it) healeth the bites of serpents and other venomous beasts".

Agropyron repens > COUCHGRASS

Agrostemma githago > CORN COCKLE

AGUE
A charm, using GROUNDSEL. A woman suffering from the disease was recommended to tell her husband to tie a handful of groundsel to her bare bosom while the charmer spoke the necessary incantation (Black). It had to remain there, and as the herb withered, the ague would go away. Wesley prescribed another charm for the same sickness – "For an ague … take a Handful of Groundsell, shred it small, put it in a Paper Bag, four Inches square, pricking that Side which is to [go] next the Skin full of Holes. Cover this with thin Linnen, and wear it on the Pit of the Stomach, renewing it two Hours before the Fit : Tried". That charm was being used in Cornwall long after Wesley's time. PANSY leaves in the shoe were said to cure ague, but Gerard was more conventional. "It is good, as the late Physitions

write, for such as are sicke of an ague, especially children and infants …". In Sussex, it used to be said that eating nine SAGE leaves on nine consecutive mornings (or some say seven leaves on seven mornings (Thompson. 1947)) is a cure for the ague (Fernie).

A Suffolk cure for ague was a mixture of beer, gin and ACORNS (*Gentleman's Magazine; 1867*). Surely the strangest, and one of the more disgusting, remedies for the ailment was the Scottish one that prescribed "a little bit of ox-dung drunk with half a scruple of masterwort" (Graham). Wearing round the neck a bag filled with grated HORSERADISH was a Cambridgeshire ague preventive. TEASEL was used in some parts (Black). The *Gentleman's Magazine* in 1867 mentions a remedy where the patient had to gather teasels and carry them about with him. But the remedy lay apparently in what was found inside the teasel – in Lyte's translation of Dodoens, 1586, he says "the small wormes that are founde within the knops of teasels do cure and heale the quartaine ague, to be worne or tied about the necke or arme" (see HULME. 1895). In the Fen country of England, OPIUM **POPPY** was grown so as to provide the medicine to treat the endemic ague or malaria rife there (see MALARIA), and the doctrine of signatures would ensure that WILLOWS would feature in ague medicine; it grows in wet conditions, and the disease is caused by damp. A charm used in the Lincolnshire marshes to get rid of the ague was to take a sprig of ROWAN over a stile, and then return home by another route. In that way, the disease would leave the patient, and the next person who passed over the stile would then take it (Gutch & Peacock).

AKEE
(*Blighia sapinda*) An African tree, but cultivated in tropical regions and southern Florida for its fruit, known as akee, which is from an Ashanti word. Canning and exporting them is now an industry in Jamaica. The aril is the edible part, but if they are under-ripe, they are poisonous, and risky if over-ripe (Menninger). African women grind the bark and mix it with locally made black soap to wash with during pregnancy. This is supposed to ensure an easy delivery when the time comes (Soforowa).

Alcea officinalis > MARSH MALLOW

ALDER BUCKTHORN
(*Frangula alnus*) The berries, bark and leaves are toxic, for they are a violent purgation. As Gerard pointed out, "the inner bark hereof is used by divers countrymen, who drinke the infusion thereof when they would be purged. It is… a medicine fit for clownes rather … than for dainty people". Hill noted that in Yorkshire "they bruise the bark with vinegar, and use it outwards [externally] for the itch, which it cures very safely". And in the Balkans, a rash would

be bathed in a decoction, which had to be left under the eaves overnight (Kemp).

ALEXANDERS

(*Smyrnium olusatrum*) A Mediterranean plant, naturalised now in Britain, chiefly near the sea, probably as a relic of old cultivation as a potherb (Grigson). There have been a number of medicinal uses in past times, notably for dropsy, for which Dioscorides recommended it. A 15th century leechdom also prescribed it, and other herbs, "for all manner of dropsies: take sage and betony, crop and root, even portions, and seed of alexanders, and seed of sow thistle, and make them into a powder, of each equally much; and powder half an ounce of spikenard of Spain, put it thereto, and then put all these together in a cake of white dough and put it in a stewpan full of good ale, and stop it well; and give it the sick to drink all day …" (W M Dawson). Alexanders also used to be prescribed for bladder problems.

ALGAROBA BEANS

CAROB pods are known in commerce as Algaroba-beans. The word itself is exactly the same as 'carob'.

ALKANET

(*Anchusa officinalis*) "The gentlewomen of France do paint their faces with these roots. as it is said" (Gerard). *Anchusa* is from the Greek meaning to paint or dye (another species is Dyer's Bugloss (*A tinctoria*). Alkanet seems to be one of the most ancient of face cosmetics (Clair). This use of the roots for making rouge led to the plant becoming known as a symbol of falsehood (Folkard).

Allium ascalonium > SHALLOT

Allium cepa > ONION

Allium porrum > LEEK

Allium sativum > GARLIC

Allium schoenoprasum > CHIVES

Allium ursinum > RAMSONS

ALLSPICE

(*Pimento dioica*) It is called Allspice because it seems to combine all the flavours of nutmeg, clove, cinnamon, and juniper berries, making it a favourite ingredient for mulled wine (Leyel.1937). The ripe berries have been used in Jamaica for flavouring a special drink based on rum, and known as pimento dram (Brouk). There used to be something called a spice plaster to put on parts affected by rheumatism or neuralgia. The way to make it was to crush an ounce or so of whole allspice, and boil it down to a thick liquor, which was then spread on linen ready to be applied (A W Hatfield). One sometimes comes across the belief in America that a necklace of allspice worn round a baby's neck will help teething (H M Hyatt).

ALMOND

(*Prunus amygdalus*, or sometimes *Amygdalus communis*). The two varieties are *dulcis* and *amara*, sweet and bitter almonds. The former are often known as Jordan Almonds, a name that has nothing to do with the Middle East. Jordan is the same as the modern French *jardin*, in other words, it is the cultivated kind. A crazy tree, according to beliefs on the Greek islands. Crazy, because it is apt to bloom as early as January; if frosts occur afterwards, as they often do, then there will be no crop (Argenti & Rose). These early flowering Prunuses have always been a source of comment – witness the name Precocious Tree for the apricot. Such a habit accounts for its symbolism of hope.

To dream of eating almonds, according to the old dream books, signifies a journey, its success or otherwise depending on whether they were the sweet or bitter kind you were eating (Dyer). Some superstitions from the Mediterranean area show the almond to be a protective influence. In Morocco, for example, they would say that a jinn does not come near a person who carries a stick cut from a bitter almond tree, or who is carrying a charm made from its wood (Westermarck).

There is sexual symbolism, too. Greek mythology has the Phrygian tale of Attis. In one version he is castrated by the gods, and dies. His testicles fall to the ground and sprout new life in the form of an almond tree (Edwards). Not that this symbolism is entirely Greek, for at gypsy weddings in parts of Spain everyone heaps pink almond blossoms over the bride's head as she dances. Such throwing over the bride's head is known in Greece as, significantly, "pouring", as if the almond blossom were life-giving water. An almond tree will always fruit well near the home of a bride, it is said. Almond paste, called 'matrimony', since it blends sweet and bitter flavours, usually appears on wedding cakes. In Italy, pink and white sugared almonds are distributed in ornamental boxes at weddings, like bread cake elsewhere. So universal are almonds at Greek weddings that the question "when will we eat sugared almonds?" is asked instead of when will the wedding be (Edwards). They are equally prominent at Indian weddings, and some are put at every table setting at the reception; they may indicate both prosperity and children, i.e., money, fertility and sexual fertility. Sugared almonds may, in certain circumstances, be given at a Greek funeral, but the symbolism is the same. The funeral would have to be that of a spinster, and the almonds are for a parody of a wedding that did not take place in life, and also to mark her wedding to Christ (Edwards). In another manifestation of the symbolism, unmarried girls often keep some of the almonds for divination. Three under the pillow will inspire dreams of the future husband.

Aloe chinensis > BURN PLANT

Aloe vera > MEDITERRANEAN ALOE

ALOPECIA

MAIDENHAIR FERN has been used to stop the hair falling out, a use stemming from the legend that the hair of Venus (*capillus-veneris*) was dry when the goddess came out of the sea, since when the fern has been used in hair lotions, particularly for lotions to prevent the hair from going out of curl on damp days. From there it is but a short step for the doctrine of signatures to ensure that it should be used for alopecia. It is the ashes of the fern, mixed with olive oil and vinegar, that are used (Leyel. 1937). SOUTHERNWOOD had a similar reputation. See Gerard's prescription "the ashes of burnt Southernwood, with some kind of oyle that is of thin parts … cure the pilling of the hairs of the head, and make the beard to grow quickly".

A cap of **IVY**-leaves worn on the head was supposed to stop the hair falling out (Leather), or to make it grow again when illness had caused it to fall. Gerard claimed that a gall from a **DOG ROSE**, stamped with honey and ashes "causeth haires to grow which are fallen through the disease called Alopecia, or the Foxes Evill".

Alopecurus myosuroides > HUNGRY GRASS

Alopecurus pratensis > FOXTAIL, i.e., MEADOW FOXTAIL

Althaea rosea > HOLLYHOCK

ALUM-ROOT

(*Heuchera americana*) The root of this plant would be used when a powerful astringent was needed, for example, a wash can be made for wounds and obstinate ulcers (Lindley). The Meskwaki Indians used the leaves in a similar way for sores (H H Smith. 1928), and a tea is an American country remedy for diarrhoea (H M Hyatt).

AMBOYNA WOOD

(*Pterocarpus indicus*) The red resin of this tree is known as dragon's blood, and was once much used in medicine, but now seems to be confined to use as an aphrodisiac, or at least as a man-attracting charm (C J S Thompson. 1897).

AMERICAN BLACKBERRY

(*Rubus villosus*) Like most of the brambles, this was used as an astringent tonic for diarrhoea, dysentery, and the like (Grieve.1931). The astringency probably helps the throat membranes, for the Cherokee people made a habit of chewing the roots and swallowing the saliva when they had a cough. They also used it as a poultice for piles. A syrup of blackberry root is an American country remedy for dysentery, and the juice of the fruit, spiced and laced with whisky, is a well-valued carminative drink in Kentucky, the original of the well-known "Blackberry Cordial" (Lloyd).

AMERICAN CHESTNUT

(*Castanea dentata*) Very rare in the United States now, the result of chestnut blight disease. While the tree was plentiful, it was used in domestic medicine, particularly an infusion of the dried leaves, given for whooping cough, and carrying one in the pocket was said in Iowa to bring good luck (Stout).

AMERICAN COWBANE

(*Cicuta maculata*) Cherokee women, so Weiner reported, chewed and swallowed the highly toxic root for four consecutive days to induce permanent sterility!

AMERICAN PENNYROYAL

(*Hedeoma pulegioides*) It was used medicinally by the American Indians. The Iroquois, for example, steeped the leaves to make a tea for headaches (Weiner), and they also used it to keep off insects (a volatile oil is still made from it for just this purpose). Another use, known by the Penobscot, was for suppressed menstruation (Youngken). This is still in use, often with sage (Corlett), and is a Hoosier remedy for the condition, as a form of birth control (Tyler).

AMERICAN WHITE HELLEBORE

(*Veratrum viride*) Poisonous, of course, like the true hellebores. People like the Blackfoot Indians have been known to use the plant for suicidal purposes (Johnston). The alkaloids in it lower the blood pressure, and so reduce the heart rate (cf WHITE HELLEBORE (*Veratrum album*). Owing to its emetic qualities it is seldom fatal to man (in accidental circumstances), but overdoses are distressingly energetic (Lloyd). The Cree Indians used the powdered roots for a snuff in the treatment to reduce hernia. The patient would be raised on a platform to a horizontal position. He would take a good pinch of the snuff, and during the violent sneezing that followed someone would be standing ready to push the hernia back with his fist! (Corlett). The Quinault Indians boiled the whole plant, and drank it in very small doses, for rheumatism. Others, who know all about its poisonous character, just tie a leaf round the patient's arm, or wherever, to relieve pain (E Gunther). Because it is so poisonous, it is no surprise that the root was carried as a charm to ward off evil spirits, or to kill sea monsters, by the Salish Indians of Vancouver Island (Turner & Bell).

AMERICAN WORMSEED

(*Chenopodium ambrosioides*) As the name implies, this plant is a famous anthelmintic. But care has to be taken, for an overdose can be dangerous, even fatal (Tampion). Brazilian curanderos use it as an ingredient in the ritual baths that form part of healing ceremonies (P V A Williams). It must grow in Africa, too, for the leaves are one of the ingredients used in the steambath used to treat sickness ascribed to the effect of evil spirits (Koritschoner). To prevent nightmares, Amazonian peoples like the Ka'apor bathe their heads with the leaves (Balée).

AMULETS

Usually used for prophylactic purposes rather than therapeutic. For example, it used to be said in Warwickshire that a child wearing a cross made of the white pith of an ELDER would never have whooping cough (Palmer). Necklaces made of small twigs from a churchyard elder for the same purpose are recorded too (Lewis). That same necklace would have been used to prevent teething fits (Friend). Similarly, in Ireland, nine pieces of a young twig were used for epilepsy; should the necklace fall and touch the ground, it had to be burned, and a new one made (Wilde). That was probably because it was believed that the really efficacious elders themselves would never have touched the ground, for the preferred amulet tree would have grown in the stump of some other tree, a willow for preference, where birds had dropped the seed. PEONY roots were, in the same way, used as amulets. A necklace turned from the roots was worn by Sussex children to prevent convulsions, and to help teething (Latham), and a necklace made from the seeds was also used in parts of France to keep children free from fits (Sebillot). Gerard also mentioned a necklace made from the roots "tied about the neckes of children" as an effective "remedy against the falling sicknesse". Langham went beyond the falling sickness; he claimed that it protected against the "haunting of the fairies and goblins". These necklaces were known as Anodyne necklaces in the 17th and 18th centuries. The galls on a DOG ROSE were once treated as amulets, by wearing them round the neck to cure whooping cough (Grigson. 1955), or even merely hanging them in the house (Rolleston); they were worn against rheumatism (Bloom), or piles (Savage). Putting one under the pillow was a Norfolk way to cure cramp (Taylor), and carrying one in the pocket would prevent toothache (Leather). Occasionally, whole plants of BUCK'S HORN PLANTAIN were worn round the neck as amulets against the ague, a practice recorded in Sussex (Allen), but it seems to have come from Gerard, but he warned, "unto men, nine plants, roots and all; and unto women and children seven". Yorkshire schoolboys carried them as a charm against flogging (Gutch); that is why they were known there as Savelick, or Save-whallop (Gutch, Robinson).

In Italy, amulets for the prevention of insomnia were made by binding OAK twigs into the form of a cross (Leland. 1898). Amulets of FENNEL were sometimes made – the seeds were hung round a child's neck to protect from the evil eye (W Jones), and in Haiti it protects from loupgarous (F Huxley). A medieval Jewish protective amulet used a sprig of fennel over which incantations had been recited, wrapped in silk along with some wheat and coins, and the whole lot encased in wax (Trachtenberg). CAMPHOR was once used to keep off evil spirits (Maddox), while in more recent times it was worn to protect the wearer from epidemics (the strong smell would be enough to repel evil, and so would be antiseptic in some way). In the Balkans, VALERIAN was sewn into a child's clothes, as an amulet to ward off witches (Vukanovic), and WOODY NIGHTSHADE was hung round a baby's neck to help teething, and around the neck of cattle that had the staggers (Brand/Hazlitt).

Welsh tradition in particular valued VERVAIN as an amulet, the dried and powdered roots to be worn in a sachet round the neck (Trevelyan). But similar practices were recorded elsewhere. In the Isle of Man, neither the mother nor a new-born baby were let out of the house before christening day, and then both had a piece of vervain sewn into their underclothes for protection (Gill. 1963). In Sussex, too, the practice was to dry the leaves and put them in a black silk bag, to be worn round the neck of a sickly child (Latham), most probably to avert witchcraft rather than to effect a cure, and it was sewn into children's clothing to keep fairies away. ST JOHN'S WORT is just as good, at least at its proper season, which would be St John's Eve and Day. It will keep out all evil spirits and witches. Some say it should be found accidentally. In the Western Isles it had to be sewn into the neck of a coat, and left there. Interfering with it in any way would rob it of his powers (Spence. 1959).

Figures cut from a MANDRAKE root were worn as amulets by both men and women in Palestine as fertility charms (Budge). The plant was believed to be an aphrodisiac, and Palestinian women quite often used to bind a piece of the root to their arm, the belief being that it could only exert its magical influence if worn in contact with the skin (G E Smith).

Amygdalus communis > ALMOND

Anacyclus pyrethrum > PELLITORY-OF-SPAIN

Anadenanthera peregrina > COHOBA

ANAEMIA

NETTLE tea is an Irish remedy for anaemia (Ó Súilleabháinn), and there seem to be safe grounds for considering it so. Young nettle leaves have the effect of increasing the haemoglobin in the blood (Bircher), hence the various "spring tonic" uses one finds in folk medicine, whether it is called that or "blood purifier". The medicine usually takes the form of a tea, and sometimes dandelion leaves are included, as in a Dorset "blood tonic" recipe (Dacombe). SWEET CICELY is said to be a great help in curing, or avoiding, anaemia (A W Hatfield), and **WOOD SORREL** formed part of an Irish cure for anaemia (Ó Súilleabháin). BLACKBERRIES, too, are good for the condition (Conway). ONIONS will reduce blood pressure, and improve the blood in cases of anaemia (Schauenberg & Paris). CHICKWEED contains both copper and iron, so it is useful for anaemia. In what seems to be an example of the doctrine of signatures, Cherokee medicine men used BLOODROOT in remedies for regulating menstruation, and also for anaemia (B L Bolton).

ANAESTHETIC

MANDRAKE was probably the first anaesthetic, and could be prepared simply by steeping the juice, or boiling the root, in wine; then a draught could be given before surgery, and the pain possibly dulled (H F Clark). A 5th century work has the observation (in translation) "if anyone is to have a member amputated, cauterized, or sawed, let him drink an ounce and a half in wine; he will sleep until the member is taken off, without either pain or sensation". Mixing it with lettuce seed and mulberry leaves would make it more potent. It has been said that the first volatile anaesthetic was a sponge boiled in such a mixture and held under the face of the patient. But of course, it is a powerful poison, and too much of it could bring on madness, paralysis or death. HEMP, too, has been used as an anaesthetic. It was in AD 220 that a Chinese physician and surgeon used cannabis resins mixed with wine as an anaesthetic, rather like the early Greek use of Mandrake. Both are pain-relievers (Emboden. 1972).

Mexican surgeons have used an alkaloid from the bark of TREE CELANDINE (*Bocconia frutescens*) as an anaesthetic (Perry. 1972). In central Africa, the leaves of CHINESE LANTERN (*Dichrostachys glomerata*), when used as a treatment for snakebite, are said to act as a local anaesthetic (Palgrave).

Anagallis arvensis > SCARLET PIMPERNEL.

Anaphalis margaretacea > PEARLY IMMORTELLE

Anastatica hierochuntica > ROSE OF JERICHO

Anchusa officinalis > ALKANET

Anemone coronaria > **POPPY** ANEMONE

Anemone nemerosa > **WOOD ANEMONE**

Angelica archangelica > ARCHANGEL

Angelica sylvestris > WILD ANGELICA

ANGINA

Russian folk practitioners have always treated angina pectoris with an infusion of HAWS (a glassful three times a day at meals). In Germany, the infusion in alcohol is considered to be the only effective cure for angina. LADY'S SMOCK is sometimes combined with the infusion of haws. To what purpose, one wonders? (Kourennoff). Here in Wiltshire a tea made from haws is said to be good for heart disorders, for it helps to prevent arteriosclerosis (Leete). Herbalists warn, though, that the effect is only noticeable after a prolonged course of treatment. In China, CASSIA twigs have been used to improve blood circulation,

and also to treat angina, as did the heartwood of SANDALWOOD (Geng Junying), from which Oil of Santal is also made.

ANISE (ANISEED)

(*Pimpinella anisum*) It is not a popular flavouring in Britain, though apparently it was cooked as a potherb in England in the mid-16th century (Lloyd). Anisette is one of the cordial liqueurs made by mixing the oil from the seeds with spirits of wine, added to cold water on a hot day for a refreshing drink (Grieve. 1933). But it is also used in ouzo, raki, Pernod, etc., (Brouk), and, in South America, aguardiente (Swahn). Another use for the oil has been to cover up the bad taste of medicines (Swahn), but it is also said to be a good mice bait, if smeared in traps. It is poisonous to pigeons, and will destroy lice.

This is one of the herbs supposed to avert the evil eye (Grieve. 1931), and it was used in a Greek cure for impotence; ointments were made of the root of narcissus mixed with the seeds of nettle or anise (Simons). Gerard claimed, among other things, that "it helpeth the yeoxing or hicket, [hiccup, that is] both when it is drunken or eaten dry; the smell thereof doth also pre-vaile very much". It is best known as an indigestion remedy. The Romans offered an anise-flavoured cake at the end of rich meals to ease indigestion.

ANODYNE NECKLACES

The name given in the 17th and 18th centuries to necklaces of PEONY roots or seeds, used as amulets. They were worn by children to prevent convulsions, or to help teething (Latham). Gerard also mentioned a necklace made from the roots, and "tied about the neckes of children" as an effective "remedy against the falling sicknesse". The use went further : it "heals such as are thought to be bewicht…". Langham, too, went beyond the falling sickness; he claimed that it protected against "the haunting of the fairies and goblins".

Antennaria dioica > CAT'S FOOT

Antennaria plantaginifolia > PLANTAIN-LEAVED EVERLASTING

Anthemis cotula > MAYDWEED

Anthoxanthum odoratum > SWEET VERNAL GRASS

ANTHRAX

An African treatment for anthrax is reported using WATERCRESS. It is apparently a Xhosa remedy (Watt & Breyer-Brandwijk).

Anthyllis vulneraria > KIDNEY VETCH

ANTI-DEPRESSANTS

Chewing MILK THISTLE leaves can relieve depression (Boland. 1977), a fact known by Gerard's sources.

Antiaris toxicaria > UPAS TREE

Antirrhinum majus > SNAPDRAGON

Aphanes arvensis > PARSLEY PIERT

APHRODISIACS

Many plants have been claimed as such, upon what grounds beggars the imagination. Who, for instance, would have thought that PURSLANE (Haining), or NETTLE ever enjoyed such a reputation, even as a flagellant? (Leyel. 1937). The seeds, so it was claimed, powerfully stimulate the sexual functions, and they figured, too, in a Greek remedy for impotence, when an ointment was made from the roots of narcissus with the seeds of nettle or anise (Simons). On the other hand, "to avoid lechery, take nettle-seed and bray it in a mortar with pepper and temper it with honey or with wine, and it shall destroy it …" (Dawson). In other words, exactly the opposite of the aphrodisiac claim. Another unlikely claimant, also ambivalent, is LETTUCE. The Romans certainly thought of it as promoting sexual potency (R L Brown), and the Akan belief, from West Africa, was that Min, a sky fertility god, was associated with a plant assumed to be some kind of lettuce, believed to stimulate procreation. The reason is that the juice of some of the lettuces is milky, resembling either, in the female aspect, the flow of milk, or in the male aspect, semen (Meyerowitz). By Gerard's time, he asserts that the juice "cooleth and quencheth the naturall seed if it be too much used …". Women were wary of lettuce, for it would cause barrennness, so an old superstition runs. It probably arose because it was thought that the plant itself was sterile (M Baker.1980). It is recorded that women in Richmond, Surrey, would carefully count the lettuce in the garden, for too many would make them sterile (R L Brown), but what the maximum acceptable number was is not revealed.

CYCLAMEN was reckoned aphrodisiac, a reputation that it enjoyed since ancient times. In fact, it became the very symbol of voluptuousness (Haig). Gerard repeated the belief, and recommended that the root should be "beaten and made up into trochisches, or little flat cakes", when "it is reported to be a good amorous medicine to make one in love, if it be inwardly taken". GARLIC in this category is difficult to understand. Chaucer's Somnour, who was "lecherous as a sparwe", was particularly fond of it: "Wel loved he garleek, onyons and eek lekes". And it had the same reputation in Jewish folklore (Rappoport). PARSLEY wine had this reputation, too (Baker. 1977), but a good many of the superstitions pertaining to this herb are connected with conception and childbirth – "sow parsley, sow babies" and so on. Surely it was nothing more than sympathetic magic that led Gerard to recommend ASH seeds to "… stirre up bodily luste specially being poudered with nutmegs and drunke". WALNUT is mentioned as an aphrodi-

siac in *Piers Plowman*, probably on the strength of its being an ancient symbol of marriage, the nuts being of two halves (I B Jones). NUTMEGS were reckoned to be aphrodisiac at one time, standard ingredients in love potions, and widely used. They still are, apparently, for Yemeni men take them even now to enhance their potency (Furst). Even TOBACCO leaves were thought at one time to be aphrodisiac (Brongers), and in 16th and 17th century Europe, potions for perennial youth were made from it, and in medieval times DEADLY NIGHTSHADE was included, for hallucinations caused by drugs derived from this very poisonous plant could take on a sexual tone. Large doses are liable to result in irresponsible sexual behaviour, hence the aphrodisiac tag (Rawcliffe).

At least with CUCKOO-PINT the reason is obvious enough. Its method of growth, the spadix in the spathe, stood for copulation. This is the reason for all the male + female names, and for the sexual overtones in a lot of others. The 'pint' of Cuckoo-pint is a shortening of pintel, meaning penis; a glance at the plant will show why. Recent name coinage carries on the theme, for Mabey.1998 has recorded Willy Lily, as ribald as any of the older ones. Even SUMMER SAVORY (or JASMINE (Haining)), was claimed as an aphrodisiac, but that belief rested on the derivation of the generic name, *Satureia*, which some thought was from 'satyr' (Palaiseul). Leland said that VERVAIN was a plant of Venus. In other words, it was used as an aphrodisiac, or as an ingredient in some kind of love philtre (Folkard). Lyte recommended WILD SAGE seeds drunk with wine, and so did Culpeper. HOGWEED is another unlikely candidate for inclusion here, but, so it is claimed, it has been shown to have a distinct aphrodisiac effect (Gerard). Even LOVAGE had this reputation, surely only as a result of misunderstanding the name, for Lovage has nothing to do with love. TOMATOES, too, owed a one-time reputation of being aphrodisiac to etymological confusion. The original Italian name was pomo dei mori (apple of the Moors), and this later became pomo d'ore (hence Gerard's Gold-apples). It was introduced to France as an aphrodisiac, and the French mis-spelled its name as pomme d'amour. So the tomato eventually reached England under the name pome amoris – love-apple, which name went back to America with the colonists (Lehner & Lehner). VALERIAN also was supposed to be aphrodisiac (Haining), and there is a record of Welsh girls hiding a piece of it in their girdles, or inside their bodices, to hold a man's attention (Trevelyan).

PANSIES were once thought to be aphrodisiac. Shakespeare, of course, knew this. Oberon's instructions to Puck were to put a pansy on the eyes of Titania. And the plant was dedicated to St Valentine; all this accounts for the numerous "love" names, of the Jump-up-and-kiss-me type (see Watts. 2000), including the one given by Shakespeare – Cupid's Flower. On the priciple of homeopathic magic, that which causes love will also cure it, or the result of it. That was why it was prescribed for venereal disease. Gerard noted the belief, and prescribed "the distilled water of the herbe or floures given to drinke for ten or more daies together … (it) doth wonderfully ease the paines of the French disease, and cureth the same…". Culpeper too regarded it as "an excellent cure for the French disease, the herb being a gallant Antivener-ean", the latter remark being contrary to the accepted belief of his time. But such a hopeless idea as pansy being aphrodisiac must be reflected in the best-known of the "love" names – Love-in-idleness, for that can only mean Love-in-vain, a name that is actually recorded in Somerset (Grigson. 1955).

Who would ever have thought of POTATOES as aphrodisiacs? But Shakespeare was only echoing popular belief when he had Falstaff say: "Let the sky rain Potatoes … and hail Kissing-comfits, and snow Eringoes". Almost certainly he was referring to sweet potatoes, but no matter, for the idea lingered after the introduction of our potato, and all because of a fundamental error. Being a tuber, it was mistaken by the Spanish who first came across both the potato (papa) and sweet potato (batata), for a truffle, and the truffle was the trufa, eventually meaning testicle, and so an aphrodisiac (Wasson). The other Spanish term for the truffle was turma de tierra, even more explicitly 'earth testicle'. In the same way, the testicle-suggesting tubers of EARLY PURPLE ORCHID ensured that the root would be regarded as aphrodisiac, the old tuber being discarded, and the new one used. It would be dried, ground, and secretly administered as a potion (Anson). Another orchid with the same reputation, among the American Indians, was FROG ORCHID (Yarnell). Similarly, a root with that reputation was that of SEA HOLLY, preserved in sugar, and known as Kissing Comfits, as mentioned above, in Falstaff's speech (see KISSING COMFITS). Even WILLOW was once credited with being an aphrodisiac – "spring water in which willow seeds have been steeped was strongly recommended in England as an aphrodisiac, but with the caveat that he who drinks it will have no sons, and only barren daughters" (Boland. 1977). GLOBE ARTICHOKE has to be included. As Andrew Boorde had it, "they doth increase nature, and dothe provoke a man to veneryous actes".

Among African examples, the Zezuru chewed the roots of MIMOSA THORN (*Acacia karroo*) as an aphrodisiac (Palgrave & Palgrave), and in Malawi, the leaves of CATCHTHORN (*Zizyphus abyssinica*) are chewed for the effect (Palgrave & Palgrave).

CORIANDER seed was one of the many plants supposed to be aphrodisiac. It is mentioned as such in the Thousand and one Nights. Albertus Magnus (*De virtutabis herbarum*) includes it among the ingredients

of a love potion. SESAME seed, soaked in sparrow's eggs, and cooked in milk, also bore this reputation, and so did GINSENG. The name is Chinese, Jin-chen, meaning man-like, a reference to the root, which, like those of mandrake, was taken to be a representation of the human form, and it was this supposed resemblance that resulted in the doctrine of signatures stating that the plant healed all parts of the body (W A R Thomson. 1976). The more closely the root resembled the human body, the more valuable it was considered, and well-formed examples were literally worth their weight in gold as an aphrodisiac (Schery; Simons). It was the the Dutch who brought the root to Europe, in 1610, and its reputation as an aphrodisiac came with it. The court of Louis XIV in particular seemed to value this reputation (Hohn). AMBOYNA WOOD (*Pterocarpus indicus*) once had this sort of reputation, or at least was used as a man-attracting charm (C J S Thompson. 1897), as was PATCHOULI perfume, too (Schery). MANDRAKE was held to have aphrodisiac as well as narcotic virtues. Theophrastus, in the 4th century BC, recommended the root, scraped and soaked in vinegar, for the purpose (Simons). But the plant was perhaps better known as an aid to conception, and to put an end to barren-ness, even independently of sexual intercourse. And see Genesis 30. 14-16, in which it is said that Rachel bargained for the mandrake with her sister Leah (by giving up her husband to her). She sunsequently bore her first-born, Joseph, though she had previously been barren (see Hartland. 1909). Mandrake's associates in British flora, BLACK BRYONY and WHITE BRYONY, have inherited the aphrodisiac beliefs, the former, according to East Anglian farm horsemen, benefiting both man and horse (G E Evans. 1966). CARDAMOM has long been famous as an aphrodisiac, and it has been suggested that the practice of blending coffee with cardamom, still current, it seems, in Saudi Arabia, is that the cardamom would eliminate the bad effects of drinking the coffee (Swahn).

Apparently SAFFRON, like coca, enjoyed in the Aztec court the reputation of being an aphrodisiac (De Ropp). However unlikely that may sound, there are comparable beliefs in the Old World – see Leland. 1891: "Eos. the goddess of the Aurora, was called the one with the saffron garment. Therefore the public women wore a yellow robe". There is a doubtful looking observation that Rorie made, when he claimed that an infusion of *Deutzia gracilis* was taken as an aphrodisiac in Scotland (Rorie. 1994).

Apium graveolens > CELERY

Apium nodiflorum > FOOL'S WATERCRESS

APPLE

(*Malus domestica*) There seems to be an archetypal belief that the apple can show man the way to a god-like state of wisdom and immortality (cf. the story of Newton's discovery of gravity – the apple was the instrument of his enlightment). In Avalon, apples grow in profusion, and it has been suggested that the controlled eating of apples was once a mystical practice. This would explain the belief that the apple was the fruit of the Garden of Eden, through which knowledge came (though the story has made it the symbol of temptation (Leyel. 1937), for a quite different reason). The apple is also the emblem of Christ, as the new Adam, so when it appears in the hands of the original Adam it means sin, but with Christ, it symbolizes the fruit of salvation (Ferguson). It was also used to symbolize the secret of immortality. Perhaps that is why apples feature so prominently in divination games, in various ways. In what remains the best-known of these divinations, young giels would peel the apple skin off in one continuous piece and throw it over the left shoulder, whence it was hoped it would fall in the shape of a letter which would be the initial of the man they would marry. Or, as in parts of America, the peel had to be put over the door, and the first man to enter through the door would be the husband (Stout). If the peel broke, she would not marry at all (Waring). Or, as in Lancashire for instance, the peel should hang on a nail behind the door. The initials of the first man to enter the house afterwards would be the same as that of the future husband (Opie & Tatem).

The pips too can be used in these divinations. If, for instance, a girl cannot choose among several of her suitors, she should take a pip and recite one of the men's names, then drop the pip on the fire. If it pops, well and good, for it shows the man is "bursting with love for her". Of course, if it is consumed without making any sound at all, she will know the man is no good for her (Waring). The rhyme to be spoken is:

> If you love me, pop and fly,
> If you hate me, lay and die (Halliwell. 1869).

Another divination game involving apple pips was to take one of them between finger and thumb and to flip it into the air, while reciting "North, south, east, west, tell me where my love doth rest". You had to watch the direction in which it fell, and then draw your own conclusions (Courtney. 1887). Another way of doing it was to stick two pips on the cheek or forehead, one for the girl's choice and the other for another man who was not. The one named for the man she really wanted would stick longest, not all that difficult to manage, or to make sure the unwanted one fell first (Opie & Tatem). A Kentucky version requires five seeds on the face, named. Then the first to fall off shows the one that the girl will marry (Thomas & Thomas). Another American children's game merely involves counting the seeds to predict the future:

> One I Love
> Two I love
> Three I love I say;
> Four I love with all my heart
> And five I cast away;
> Six he loves
> Seven she loves
> Eight they both love;
> Nine he comes
> And ten he tarries,
> Eleven he courts
> And twelve he marries (Stout).

Similarly, the number of seeds found indicates the number of children you will have. Even the stalks can be used; the girl has to twist the stalk to find whom she will marry. The game is to twist while going through the alphabet, a letter for each twist. The letter she has reached when the stalk comes off is the initial of the first name of the man she will marry (Opie & Tatem). An Austrian divination involved cutting an apple in two on St Thomas's Eve (20 December) and counting the number of pips. If it was an even number, then she was soon to marry. But if she had cut one of the pips, she would have a troubled life and end up a widow (Waring).

To dream of apples means long life, success in trade, and a lover's faithfulness (Gordon. 1985). But there is a darker side of apple belief. True, it was used as a sanctuary in catching games in Somerset, but one children's rhyme is open to quite a different interpretation:

> Bogey, Bogey, don't catch me!
> Catch that girl in the apple tree! (Tongue).

And there are a number of death omens connected with apples. Out of season blossom is sinister. If it happens when there is fruit on the tree, it is a sign of death in the family, put into rhyme in Northamptonshire as:

> A bloom upon the apple tree when apples are ripe,
> Is a sure termination of somebody's life (Baker. 1980).

The "somebody" being a member of the owner's family, it must be understood. Never leave a last apple on the tree, for that too would mean a death in the family. Not in Yorkshire, though, for there one *must* be left on, as a gift for the fairies (Baker. 1980). But if one stayed on, that was a sign of a death in the family (Gutch. 1911). It is also very unlucky to burn apple wood, in spite of its undoubted fragrance. As the tree is an ancient symbol of plenty, to destroy it might disturb the household's prosperity (Baker. 1974). Never eat an apple without first rubbing it clean, for that would constitute a challenge to the devil (Waring). Never eat an apple until they have been christened. i e not until after St Swithin's Day (15 July), when rain can be traditionally expected, to perform their christening. In Cornwall the injunction covers St James's Day (25 July), when they say they get their final blessing (Vickery. 1995), and apples should only be picked at the "shrinking of the moon" (*Notes and Queries. 1st series. vol 10; 1854 p156*). And never plant a rowan near an apple tree, as one will kill the other (Tongue). But there are some lucky omens; for instance, it is lucky to see the sun through the apple branches on Christmas Day (Baker. 1980). But it is really Old Christmas Day (6 January) that is the important time, for then it means a good apple crop to come (Vickery. 1995). Put into rhyme on Dorset, the belief is:

> If Old Christmas Day be fair and bright,
> You'll have apples to your heart's delight.

Old Christmas Day is the great apple wassailing time. Wassail literally means 'good health', and to wassail a tree implies going through some ceremony that will ensure its health and ability to produce an abundant crop.

> Wassail the trees, that they may bear
> You many a plum and many a pear;
> For more or less fruit they will bring,
> As you do them wassailing.

There was nearly always a rhyme to be said or a song to be sung, the best known being:

> Here's to thee, old apple tree;
> Whence you may bud, and whence you may blow.
> And whence you may bear apples enow
> Hats full, caps full,
> Bushel-bushel sacks full,
> And my pockets full, too (Brand).

If the parson happened to be popular, the line:

> Old parson's breeches full

was added (Ditchfield.1891).

In the areas of east Cornwall and west Devonshire the custom was to take a milkpanful of cider, into which roasted apples had been chopped, into the orchard. This was put as near as possible dead centre of the orchard, and each person (and it was important that everyone partook; the children were brought out, and so were the sick and invalids, for if anyone were missing, the charm would not be effective (Whitlock. 1977)) would take a cup of the drink, and go to each separate tree, and say the ritual formula. Part of the cupful of cider was drunk as a health to the tree. But the rest was thrown at it (Hunt. 1881). The throwing is deliberate, and acts as a warning to the tree. Guns are actually fired into the branches, and before guns were common, the trees were beaten with sticks (Whitlock. 1977). The Hampshire advice to knock a rusty nail into the tree if it is not bearing (Hampshire FWI) is another example of this threatening or warning behaviour towards the tree. After all, a good

apple crop is important enough in the cider-producing areas (Leather) to take whatever steps are necessary as insurance (see also WASSAIL).

"An apple a day keeps the doctor away" is a very well known prescription, varying slightly in different regions. In Herefordshire it was:

> Take an apple going to bed,
> 'Twill make the doctors beg their bread (Leather).

However, it is not often that they are linked with any particular ailment, (Galen prescribed apple wine as a cure-all (Krymow)), although advice from the Highlands of Scotland enjoins a decoction of apples and rowan, sweetened with brown sugar to be taken for whooping cough (Beith). A Yorkshire practice was to use a poultice of rotten apples for what were known as botches, described as small boils (Gutch. 1911). American opinion suggested that apples would relieve rheumatism (Thomas & Thomas), and another American domestic remedy is a lotion to cure dandruff, made of one part of apple juice to three parts of water (H M Hyatt).

There are one or two charms recorded, like this Devonshire wart cure: cut an apple in two, rub one half on the wart. Give it to the pig to eat, and eat the other half yourself (Choape). More widespread was a similar one for warts, which were rubbed nine times with an apple cut in two. The sections were re-united and buried where no human foot was likely to tread. In Northumberland the warts were opened to the quick, or until they bled, and then they were rubbed well with the juice of a sour apple. The apple was then buried (Drury. 1991). These are all simple transference charms, but there is one more, for rheumatism this time, that merely involved carrying half an apple in the pocket (Foster). Half a potato is more usual than an apple for this purpose, but a hazel nut is sometimes used instead.

APPLE-RINGIE
A Scottish name for SOUTHERNWOOD. Simpson's explanation is that apple is from the old word aplen, a church, and ringie is Saint Rin's, or St Ninian's, wood (Ringan was the Scots form of Ninian). Aitken offers another derivation. Appelez Ringan, pray to Ringan, became first Appleringan, then Appleringie. But Jamieson says that it is from the French 'apile', strong, and 'auronne', southernwood, which derives from *abrotanum*.

ARABIAN JASMINE
(*Jasminum sambac*) In India, the flowers, sacred to Vishnu (Pandey), are strung together as neck garlands for honoured guests, and in Borneo, women roll up jasmine blossoms in their hair at night (Grieve. 1931), as Indian women do, too. It is a symbol of chastity and conjugal fidelity (Gupta), but the flowers are supposed to form one of the darts of Kama Deva, the Hindu god of love (Pandey).

Arachis hypogaea > GROUND NUT

Araucaria araucana > MONKEY PUZZLE TREE

ARCHANGEL

(*Angelica archangelica*) It blooms around the Archangel Michael's Day, 8 May in the earlier tradition (Emboden), hence the comon name, though it is more likely that the name was given because Tradescant found it near the Russian town of that name (Fisher). According to Grimm, the name is given because its efficacy against such epidemic diseases as cholera and the plague was revealed by an angel in a dream. Then there was the name Holy Ghost – "some call this an herb of the Holy Ghost; others more moderate call it Angelica, because of its angelic virtues, and that name it retains still, and all nations follow it" (Culpeper). The root was the special part (radix Sancti Spiriti), chewed during the Great Plague in an attempt to avoid the infection. With this background it is hardly surprising that it was used as a protection from other things, evil spirits, witchcraft, for instance, and against the cattle disease called elf-shot (Prior). Cornish folklore still regards it as a strong witch repellent (Deane & Shaw). The belief is at least as old as Gerard, who said "it is reported that the root is availeable against witchcraft and inchantments, if a man carry the same about him …". The Lapps believed it prolonged life, and they chew and smoke it in the same way as tobacco (Leyel. 1937). Sometimes one comes across mention of a Holy Ghost pie, apparently used in the Black Mass. It is suggested, by Rhodes, that this was an angelica-flavoured cake, and therefore a host.

The name ensures wonders in medical treatment, but there are genuine usages, too. A Cornish cold cure requires that elder flowers and angelica leaves be steeped in boiling water for ten minutes, strained and sweetened to taste (Deane & Shaw), while a good gargle for a sore throat can be made with an infusion of the leaves and stems (Conway). An ointment made from the roots can soothe rheumatic pains and skin disorders, a use that was already known in medieval times, as the following prescription from the Welsh text known as the Physicians of Myddfai shows: "for scabies. Take the roots of archangel, boil well, and boil a portion of garlic in another water. Take a good draught of the decoction, and wash your whole body therewith every morning. Boil the residue of the archangel and garlic in unsalted butter, make into an ointment and anoint your whole body therewith for nine mornings".

There is a piece of pure fantasy, from "A booke of Phisicke and Chirurgery", written in 1610, but obviously of much earlier date, and offering a receipt "for one that hath loste his mind - take and shave off the hayre of the mouilde [apparently the dent in the upper part of the head] of his heade, then take archangel and stampe it, and binde it to his heade where it is shaven, and let him take a sleep therewithall, and when he awaketh he shall be right weake and sober enoughe".

Arctostaphylos uva-ursi > BEARBERRY

Arenaria serpyllifolia > THYME-LEAVED SAND-WORT

Argemone mexicana > MEXICAN POPPY

Arisaema triphyllum > JACK-IN-THE-PULPIT

Aristolochia clematitis > BIRTHWORT

Aristolochia serpentaria > VIRGINIAN SNAKE-ROOT

Armeria maritima > THRIFT

Armoracia rusticana > HORSERADISH

ARNICA

(*Arnica montana*) The tincture, in use till recently, but now replaced by a much safer cream, is applied to whole chilblains, and to sprains and bruises, hence its name "tumbler's heal-all" (Thomson. 1978). Internal use of the tincture would almost certainly be lethal, but there are a number of homeopathic uses, in minute doses, for shock, for example (M Evans). In folk medicine, it has even been used as an abortive (Schauenberg & Paris), and a decoction of ivy and arnica is used in the Balkans for skin diseases (Kemp). One of the names for the plant is Mountain Tobacco. The leaves, or indeed all parts, can be used to make a tobacco substitute, known in France as tabac des savoyards, tabac des Vosges, or herbe aux prêcheurs (Sanecki). One of the French names can be translated as Sneezewort, for the flowers, if smelt when freshly crushed, will certainly cause a sneezing fit (Palaiseul).

ARQUEBUSCADE WATER

AGRIMONY was an ingredient of genuine arquebuscade water, as prepared against wounds inflicted by an arquebus, and mentioned by Philip de Comines in his account of the Battle of Morat, 1476. Eau d'arquebuscade was still being supplied in 1897 for sprains and bruises (Fernie).

ARROW TREE

(*Sapium bilocurare*) Natives of Lower California used to warn that this is a dangerous tree; travellers were told that if they fell asleep under it they would wake up blind (Menninger).

Artemisia abrotanum > SOUTHERNWOOD

Artemisia absinthium > WORMWOOD

Artemisia arborescens > SHRUBBY WORMWOOD

Artemisia cina > LEVANT WORMSEED

Artemisia dracunculoides > RUSSIAN TARRAGON

Artemisia dracunculus > TARRAGON

Artemisia gnaphalodes > WHITE MUGWORT

Artemisia ludoviciana > LOBED CUDWEED

Artemisia maritima > SEA WORMWOOD

Artemisia pontica > ROMAN WORMWOOD

Artemisia tridentata > SAGEBRUSH

Artemisia vulgaris > MUGWORT

ARTERIOSCLEROSIS

GARLIC is still prescribed for the condition, though its virtues seem to lie in reducing blood pressure, reports of which are recorded from Ireland, and Alabama. A decoction of HAWS , taken instead of tea or coffee is used for high blood pressure (Kourennoff) for it helps to prevent arteriosclerosis. In any case, haws in various preparations have been prescribed for angina pectoris, particularly in Russian folk medicine (Kourennoff), and in Germany it is claimed to be the only effective cure for the condition. Herbalists, though, still maintain that HAZEL nuts improve the condition of the heart, and prevent hardening of the arteries (Conway). Like true garlic, the wild garlic (or RAMSONS) is prescribed by herbalists as a tea made from the dried leaves, or by eating the fresh leaves, for this complaint (Flück). Herbalists still use DAISIES for improving the circulation. They will keep the artery walls soft and flexible (Conway), and GLOBE ARTICHOKE has been used, too (Schauenberg & Paris). LIME-FLOWER tea, good for many conditions, is said to be good for arteriosclerosis, too, for it thins the blood, and so improves the circulation (M Evans). The dose is given as one cupful, four times a day, between meals. See also HYPERTENSION.

ARTHRITIS

A Kentish village remedy for arthritis required NETTLES to be cooked and eaten, and then the water in which they had been boiled had to be drunk (Hatfield. 1994). **PRIMROSE** leaves and flowers used in salads will, so it is claimed, help to keep off arthritis (Page. 1978). COMFREY root tea has been taken, too (Painter). A decoction of BIRTHWORT has been used to soothe the pain of arthritis (Schauenberg & Paris), and in Russian folk medicine. The much-vaunted Oil of EVENING PRIMROSE has also been recommended to help arthritis (M Evans).

WHORTLEBERRY leaves or young shoots were used– one part to one part of alcohol, kept in a warm place for 24 hours, and then strained. The dose would be a tablespoonful in warm water, twice a day (Kourennoff). The Indiana remedy for the condition is to eat fresh or dried POKEBERRIES each day (Tyler). In Barbados, the leaf decoction of GUINEA-HEN WEED (*Petiveria alliacea*) is taken for the conplaint (Laguerre).

Artocarpus altilis > BREADFRUIT

Artocarpus heterophyllus > JACKFRUIT

ARUM LILY

(*Zantedeschia aethiopica*) A thoroughly unlucky plant, not to be taken indoors, and *never* brought into a hospital (Deane & Shaw). The reason is its association, in the 19th century, with funerals (Vickery. 1985).

Arum maculatum > CUCKOO-PINT

ASAFOETIDA

(*Ferula assa-foetida*) To prevent colds, tie a small bag of it round the neck. Sometimes the asafoetida would be soaked in camphor first (Stout). Tied round a baby's neck, it will help it to cut teeth without pain. "Wear asafoetida to keep the itch away" (Stout) – or to keep diphtheria away – or cure whooping cough – or, in Maryland, for hysteria (Whitney & Bullock).

German Hexenbänner used to advise people who thought they were bewitched to burn asafoetida all night in every room of the house, with doors and windows shut. The witch would be bound to visit the house within three days (J Simpson. 1996).

ASH

(*Fraxinus excelsior*) Yggdrasil, the tree of the universe of Scandinavian mythology, is generally supposed to have been an ash (see Yggdrasil), the tree upon which Odin hanged himself in his quest for wisdom (Davidson, Turville-Petrie). According to Hesiod, the men of the third age of the world (the Bronze Age) grew from the ash tree, and Teutonic mythology has it that the first men came from this tree (Rydberg). Ash and human birth are linked in many ways. In the Highlands, at the birth of a child, the midwife used to put a green ash stick into the fire, and while it was burning, let the sap drop into a spoon. This was given as the first spoonful of liquor to the newborn baby (Ramsay). It is said that it was given as a guard against witches, or against the evil eye. The mythology claimed that the fruit of Yggrdrasil ensures safe childbirth. When Ragnarok draws near, it was said the ash tree will tremble, and a man and woman who hide in it, Lif and Lifthrasir, will survive the ensuing holocaust and flood. They stand alone at the end of one cycle and the beginning of another. From these two, the earth will be re-peopled, and Yggrdrasil itself will survive Ragnarok. In other words, Yggdrasil is the source of all new life (Crossley-Holland).

The Irish tree, Bile Tortan, one of the five ancient sacred trees of Ireland, is said to have been an ash. It was an enormous tree, said in the later literature to have been 300 cubits high, and 50 cubits thick. "When the men of Tortu used to meet together round the huge conspicuous tree, the pelting of the storms did not reach them, until the day when it was decayed". It fell somewhere around AD 699, so the legend says (Lucas). Another of the Irish sacred ashes was still growing in Borrisokane parish, County Tipperary, where it was called a bellow-tree. Another account gives its name as Big Bell Tree. Both these versions are actually the Irish bile, a sacred tree. Water that lodged in a hollow between the branches of this ash was regarded as holy, and no part of it would ever be used as fuel, for the belief was that if that were done, the house itself would burn down (Lucas).

Yggdrasil itself was sacred to Odin (Graves), and that is enough to make ash a lightning tree. Elton has a note that the ash, together with the houseleek and hawthorn (all thought to avert the lightning) were all sacred to Taramis, the northern Jupiter, who was worshipped by the Britons under titles derived from words for fire and thunder, In this connection, note the belief that it is unlucky to break a bough off an ash:

> Avoid an ash,
> It courts a flash (Northall).

On the other hand, ash for the fire is, in Evelyn's words, "fittest for ladies' chambers":

> Burn ash wood green
> 'Tis fire for a queen.
> Burn ash wood sear,
> 'Twill make a man swear.

In Ireland, ash wood is burned to banish the devil (Ó Súilleabháin), and in Devonshire ash faggots are burned at Christmas, probably for the same reason, though the Christmas Ashen Faggot has an extensive folklore of its own (see ASHEN FAGGOT). Ash was certainly regarded as all-powerful against witchcraft – in fact, it was anathema to witches. In Lincolnshire, the female ash, called Sheder, would defeat a male witch, while the male tree, Heder, was useful against a female one (M Baker.1980). Eating ash buds provided invulnerability to witchcraft (Banks). The Witches' Well at Pandlestone, in Somerset, is no longer dreaded – now that ash trees grow round it, it is safe (Tongue). Ashwood sticks were preferred to any other, as they would protect the cattle from witchcraft. A beast struck with one could never be harmed, as it would never strike a vital part (Wiltshire), and an ash twig (from a tree that had a horseshoe buried among its roots) stroked upward over cattle that had been overlooked would soon charm away the evil (Pavitt). Branches of it were wreathed around a cow's horns, and round a cradle, too (Wilde). English mothers rigged little hammocks to ash trees, where their children might sleep while field work was going on, believing that the wood and leaves were a sure protection against dangerous animals and spirits. A bunch of the leaves guarded any bed from harm, and a house that was surrounded by an ash grove would always be secure (Skinner); a bunch of ash leaves in the hand would preserve the bearer from witchcraft (Denham). Norman peasants used to sew a little piece of ash (with a piece of elm bark) into their waistcoats, for protection (Sebilllot). For a different reason, Cornish people used to carry a piece of ash wood

in their pockets as, in this case, a rheumatism cure (Deane & Shaw).

There is a long-standing belief (dating from Pliny's time) in the power of ash to repel serpents. Pliny said snakes would rather creep into a fire than come into contact with it. In Cornwall, people used to carry an ashwood stick with this in mind. A single blow from an ash stick was enough to kill an adder; struck by any other wood, the adder is said to remain alive till the sun goes down (Deane & Shaw). The belief is as widespread as it is ancient (Fiske recorded it in America in 1892, referring to the White Ash, *F americana*). (Harper). Cowley uses the superstition in one of his poems:

> But that which gave more wonder than the rest
> Within an ash a serpent built her nest
> And laid her eggs, when once to come beneath
> The very shadow of an ash was death.

Evelyn knew about the "old imposture of Pliny's, who either took it upon trust, or we mistake the tree", and Gerard also repeated the belief, less critically: "The leaves of this tree are of so great virtue against serpents that they do not so much as touch the morning and evening shadows of the trees, but shun them afar off". "And if a serpent be set between between a fire and Ash-leaves, he will flee into the fire sooner than into the leaves" (Bartholomew Anglicus/Seager). The trees were actually planted round houses, just to keep adders away (Bottrell). In Devonshire, they said it only needed a circle drawn round an adder with an ash stick to kill it (Whitlock). In west Somerset, a wreath of flowers was hung on the ash tree nearest the farm to protect both men and cattle against snake bite for the year (Tongue). On Dartmoor until very recently, and perhaps to the present day, if a dog is bitten by an adder, fresh green ash-tips are gathered and boiled, and the liquid given to the dog to drink (St Leger-Gordon), and "the juyce of Ash-leaves, with pleasant white wine" was a mid-16th century remedy for snakebite. You could also apply the fresh leaves to the place bitten (T Hill). Most lightning plants (hazel, fern, etc.,) had similar anti-snake properties. See the Somerset charm for adder's bite:

> Ashing-tree, ashing tree
> Take this bite away from me.

Suck the wound and spit, then say the charm. Do this three times. If you can make it bleed, so much the better (Tongue). There is too a belief that "a few ash-boughs, thrown into any pond where there are plenty of toads and frogs will undoubtedly destroy them in two or three days (Atkyns). Another connected belief was that the shade of an ash tree was destructive not only to snakes but to all vegetation over which it extended. That is the source of a saying current at one time in Guernsey: "it is better for a man to have a lazy fellow in his service than an ash-tree on his estate" (MacCulloch), an opinion at odds with the general. Of course, the belief in the snake's antipathy to the ash gave rise to a number of pseudo-medicinal uses against snake-bite. "The juice of the leaves themselves being applied, or taken with wine, cure the bitings of vipers, as Dioscorides saith" (Gerard). A Welsh practice was to keep a piece of bark in the pocket, or to rub on the hands, to scare snakes away (Trevelyan).

A very well known belief connected with the tree is that a failure in the crop of ash keys portended a death in the royal family (or at the very least it was a sign of some great disaster (Hunter)). This actually happened, so it is said, in 1648, and so was connected with the execution of Charles I in January 1649 (Leather). The even ash beliefs are just as well known, for the leaf is often used for invoking good luck ("luck and a lover" (Leather)), and there is always a simple rhyme to accompany it. One from Cornwall runs:

> Even ash, I do thee pluck,
> Hoping thus to meet good luck;
> If no good luck I get from thee
> I shall wish thee in that tree.

Another is:

> With a four-leaved clover a double-leaved ash and
> a green-topped seave
> You may go before the queen's daughter without
> asking leave (Friend).

Perhaps better known than the good luck charms are those recited when the even ash is used for divination purposes:

> Even, even ash
> I pull thee off the tree;
> The first young man I do meet,
> My lover he shall be.

The leaf is then put in the shoe. That is from Northumberland (Denham), but a Buckinghamshire charm simply required an ash leaf to be put in the right shoe – " … the first man you meet you have to marry" (Heather). Another North country rhyme shows how the even ash was carried, unless, that is, the evidence is merely assonance:

> The even ash in my bosom
> The first man I meet shall be my husband
> (Denham).

Slight variations in the rhyme occur over the country, but it would be tedious to quote all of them here.

Charms for a different purpose are typical of other lightning plants. Ash rods are used in many parts of England to cure cattle, and even more widespread is the custom of passing children through holes in ash trees as a remedy for hernia. In Cornwall, the ceremony had to be performed before sunrise, and a further Cornish belief was that the child would recover only if he were washed in dew collected from the branches on three successive mornings (Deane & Shaw). Gilbert White reported that it was customary to split an ash, and to pass ruptured children through. Evelyn, too, knew all about the belief. The Herefordshire practice was for the child's father to pass him through to another man. The father said, "The Lord giveth", and the other man replied "The Lord receiveth" (Leather). In Suffolk, apparently, the charm was used for epilepsy, and in places as far apart as Norfolk and Jersey, for rickets (Le Bas). If any injury should happen to the split tree, the child would suffer accordingly. The practice of planting a tree to commemorate the birth of a child may be a relic of this belief that the life of an individual is bound up in that of the tree. Perhaps this is one of the reasons why it is always so unlucky to break a branch off an ash (see above). In west Sussex, the child had to be attended by nine persons, each of whom passed him through, west to east (Black). The rules given for the split ash in Suffolk are:

> Must be early in the spring before the leaves come
> Split the ash as near east and west as possible
> Split exactly at sunrise
> The child must be naked
> The child must be put through the tree feet first
> The child must be turned round with the sun
> The child must be put through the tree three times
> (Gurdon)

The Somerset rules include 2,4 and 6, but go on further to say that the child must be handed in by a maiden, and received by a boy (Mathews).

Sebillot says that children with coqueluche, which must be whooping cough, were passed through split ashes. He quoted an ancient ash in Richmond Park, Surrey which was visited in 1853 by mothers "dont les enfants étaient ensorcelés, malade de la coqueluche ou d'autres affections". It had to be done before sunrise, and no stranger could be present. The child was passed nine times under and over. It seems, too, that whooping cough could be cured by pinning a lock of the patient's hair to an ash tree (Addison & Hillhouse). A feature of a lot of these charms is that illnesses would be handed over to the tree. So too with warts:

> Ashen tree, ashen tree
> Pray buy these warts of me (Northall).

That is a Leicestershire rhyme to accompany the charm, which was to take the patient to an ash tree, and to stick a pin into the bark. Then that pin would be pulled out and a wart transfixed with it until pain was felt, after that the pin would be pushed back into the ash, and the charm spoken. Each wart was treated, a separate pin being used for each (Billson).

An East Anglian cure was to cut the initial letters of both one's Christian and surnames on the bark of an ash that has its keys. Count the exact number of your warts, and cut the same number of notches in the bark. Then, as the bark grows, so will the warts go away (Glyde). Another method was to cross the wart with a pin three times, and then stick the pin into the tree (Northall), and recite the appropriate rhyme. The Cheshire cure was to steal a piece of bacon, and to rub the warts with it, then to cut a slot in the bark and slip the bacon underneath. The warts would disappear from the hand, but would make their appearance as rough excrescences on the bark of the tree (Black). An Irish cure for jaundice operated in a similar way (see Wood-Martin). A most unlikely sounding charm is one from Sussex, to stop a child from bed-wetting. The child had to go alone to an ash, then going another day to gather a handful of keys, which have to be laid with the left hand in the hollow of the right arm. They are carried like this, and then burned to ashes (Latham).

A shrew ash is one in which a hole has been bored in the trunk, and a shrew-mouse put inside and left there. At one time, almost every country village had its shrew-ash. The point was that cattle and horses, when suffering from any sickness that seems to cause a numbness of the legs, were thought to have been bitten by a shrew, and the only cure for this was thought to be the application of a branch or twig from a shrew ash (Clair). Such a tree was known in the Black Country as a "nursrow" tree, and was not necessarily confined to ash – oak and elm could be treated in the same way (Raven). But inserting something into an ash could have other results. In Wiltshire, for instance, finger- and toe-nail clippings used to be put in a hole in a maiden ash and the hole then stopped up; this was a neuralgia preventive. A maiden ash is one that has never been pollarded or topped (Clark).

Ash has its share of weather lore, the best known being the comparison with the oak to foretell a good or bad summer:

> If the oak before the ash come out,
> There has been, or will be, a drought.

There are quite a number of jingles of the same import, the most succinct of which is, from Surrey:

> Oak, smoke
> Ash, squash

(Northall). Or sometimes

> Oak, choke,
> Ash, splash

i.e., if the oak leafed first, there would be dry, dusty weather (M Baker.1980).

Ash provides the toughest and most elastic of British timbers, hence its use for spear shafts; indeed aesc

in OE came to mean spear, and aesc-plega the game of spears, or battle. Then it was further extended to the man who carried the spear. The handles of most garden tools are best made of the wood – some rakes are still made entirely of ash (Freethy). Clothes posts, billiard cues (Wilkinson), hockey and hurley sticks, cricket bat handles and police truncheons were all traditionally made of ash timber. It was tough enough for windmill cogwheels, and boats also were made of it – OE aesc, Norse aske came to mean a vessel as well as a spear. In ancient Wales and Ireland all oars and coracle-slats were made of it (Graves). Evelyn mentions that the inner bark was used as paper, before the invention of the latter, and he also mentions that the keys are edible, and often pickled – "being pickled tender, [they] afford a delicate sallading"; Sir Robert Atkyns, a number of years later, spoke of them as "an excellent wholesome sauce, and a great expeller of venom". Recipes are still given; a recent one suggested that one should boil the keys in salt water for ten or fifteen minutes, then strain and put into warmed jars. Cover with boiling spiced vinegar. The keys should be picked while they are still green and soft (Cullum). Yorkshire carters used a spray of ash in the head stall of their horses, to keep off the flies (Nicholson), and medicinal uses for man or beast were many indeed. The bark is good for agues and fevers (Atkyns), and is still used in herbal medicine as a substitute for quinine. In Vermont, USA, a story used to be told of a man who cured himself of fever by tying himself (and the fever) to an ash tree, and then crawling out and leaving the disease tied there (Bergen). Burnt ash bark was a Highland remedy for toothache (Beith), and in Ireland the sap of a young tree was used to cure earache. This is actually a very old remedy, for there are recorded leechdoms from the fifteenth century, as well as similar usages in the early Welsh text known as the Physicians of Myddfai. Evelyn had heard of it, but misunderstood the malady, for he claimed that the "oyl from the ash … is excellent to recover the hearing …"

Apparently, there was a belief that the wood, provided it was cut at certain holy seasons, was incorruptible, and so would heal wounds (Kelly); hence Aubrey, even if the moment of cutting does not agree with "holy seasons": "To staunch bleeding, cut an ash of one, two or three years' growth, at the very hour and minute of the sun's entering Taurus: a chip of this applied will stop it". James II's nosebleed, so it is said, was staunched in this way in 1688. There is a veterinary usage of some interest – Devonshire farmers were quite convinced that feeding infected cattle with ash leaves was a cure for foot and mouth disease (*Devonshire Association. Transactions. Vol 65; 1933 p127*).

ASHEN FAGGOT

A Devonshire Christmas Eve custom, forming a part of more general Yule Log celebrations around

Christmas. The faggot is made up entirely of ash timber, with green ash bands round it, and as big as possible. A quart of cider was served each time a hoop round the faggot burst, which it does usually with a loud bang, being green wood. The faggot used on Christmas Eve, 1952 at the inn at Northleigh, Devon, was described as weighing over a thousand kilos, was 7 meters long and about 22 centimeters thick. The largest stick of green ash was 7 centimeters diameter, and this example was bound with five strips of hazel rather than ash (Coxhead). Christmas Eve is still called Ashen Faggot Night in the next county, Somerset (Rogers), when divinations are made according to the bursting of the, in this case, willow bands round the log as the fire grows. (see also YULE LOG)

Asimina triloba > PAPAW

ASPARAGUS

(*Asparagus officinalis*) Always leave at least one stem in the asparagus bed to blossom – for luck (Igglesden), though actually it is no more than common sense – that is the way to get seed for next year's sowing. Like parsley, asparagus must not be transplanted. Someone in the family would die if that were done. That, at least, was once the belief in Devonshire (Read). Another strange belief was that if asparagus root were worn as an amulet (for what purpose?) the wearer became barren (Simons) (then why wear it?). Dreaming of it, gathered and tied up in bundles, is an omen of tears. On the other hand, dreaming of it growing, is a sign of good fortune (Mackay). The roots were once used on the Continent for "falsifying sarsaparilla" (Lindley).

There have been some medicinal uses. It is diuretic, "a powerful diuretic", Hill (1754) would have it, and it is used in homeopathy for dropsy and rheumatism (Schauenberg & Paris), the latter complaint was also treated with this plant in Ireland, as was gout (Moloney). Indiana folk medicine also advised eating lots of asparagus, which, they claimed, brought relief in just a few days (Tyler). Thomas Hill (1577) listed the ailments to be treated with "sperage" as "the Palsie, King's Evil [scrofula], Strangurie, a hard Milt [spleen], and stopping of the Liver".

Asparagus officinalis > ASPARAGUS

ASPEN

(*Populus tremula*) 'Tremula' describes it well. It is a shivering, quaking tree, the symbol of fear, and of scandal, the latter the result of comparing the constant motion of the leaves to the wagging of women's tongues. " … it is the matter where of women's tongues were made… which seldome cease wagging" (Gerard). It was actually called Women's Tongues in some places, or Old Wives' Tongues (Lowsley, Grigson). That constant quivering of the leaves accounts for one of its medicinal uses. The doctrine of signatures claims its use for the ague – a shivering

tree, to make a medicine for the shivering disease. In one region of France, such a fever could be transferred to the tree, simply by tying a ribbon to it (Sebillot).

In the Scottish Isles, aspen was a cursed tree, since it held up its head when other trees bowed down during the procession to Calvary, and also since the Cross was made of it. Curses and stones used to be flung at aspens, and crofters and fishermen would avoid using its wood for their gear (Grigson). In Somerset, too, they would say that the Cross was made from aspen wood, and that is why the tree shivers incessantly (Tongue), and Welsh folklore has the same belief. The belief spread to America, too. There it is the American Aspen (*Populus tremuloides*) that is at fault. The legend in Brittany was that not only did it refuse to bow, but declared that it was free of sin, and had no cause to tremble and weep, whereupon it immediately began to tremble, and will go on doing so until the last day (Grigson). All sorts of reason are given for this incessant quaking. In the Forez district of France, they confirm that it was the aspen's pride that causes it to shake now, but it was St Pardoux before whom it refused to bow (Sebillot). German legend has it that it was cursed by Jesus on the flight into Egypt, because it refused to help him, while in Russia the cause is stated to be that it was the tree of Judas. By a kind of inverted reasoning, teething rings used to be made of aspen wood in the Highlands. The argument was that since the Cross was made of it, far from being a cursed tree, it was a blessed one, and no harm could possibly come to the child (Rorie). Perhaps, too, this accounts for a traditional Russian use of the wood to pierce the buried body of a witch through the heart, or to lay on her grave, to prevent it rising to the surface again (Warner, J Mason).

Asperula cynanchica > SQUINANCYWORT

Asperula odorata> WOODRUFF

Asperula tinctoria > DYER'S WOODRUFF

ASPHODEL BREAD
The dried root of WHITE ASPHODEL, boiled, yields a mucilage which is mixed with grain or potato to make Asphodel bread.

Asphodelus ramosus > WHITE ASPHODEL

ASPIDISTRA
(*Aspidistra elatior*) The Victorians coined the name Cannonball Plant for aspidistra, on account of its tolerance of shade, fluctuating temperatures, dust, smoke, and general neglect (F Perry. 1972). The same applies to other names given to it – Cast-Iron Plant, for example (Bonar), and Bar-room Plant (Hyam & Pankhurst).

Asplenium ruta-muraria > WALL RUE

Asplenium trichomanes > SPLEENWORT

Aster tripolium > SEA ASTER

ASTHMA

was treated by "a spoonful of NETTLE-juice mixt with clarified Honey, every night and morning" (Wesley). Domestic medicine agrees on nettle's efficacy in chest complaints, from coughs to tuberculosis, but COMFREY root tea, taken for a variety of ailments, is not so well known (Painter). An infusion of ELECAMPANE roots has been used for asthma (Forey), as well as for coughs and whooping cough. Cockayne quotes a Saxon leechdom "ad pectoris dolorem" in which elecampane played its part along with many other herbs. Mulled ELDER berry wine is good for asthma (Hatfield). Another recipe is to make a conserve of HONEYSUCKLE flowers, beaten up with three times their weight of honey; a tablespoonful dose is to be taken night and morning, to relieve the condition (A W Hatfield).

An Irish remedy was to use GORSE flowers. They would be packed tightly in a crock, and brown sugar put on top. The crock would be covered, and put in a saucepan to stew slowly (Lucas). Also from an Irish source, sufferers were advised to drink of a potion of GROUND IVY (or dandelion), with a prayer said over it before drinking (Wilde.1890). Watercress too was used in Ireland for the complaint, and so was SEA HOLLY (Ô Súilleabháin), or DANDELION tea (P Logan), while a tea made from **OX-EYE DAISY** was used in Scotland, and so was PENNYROYAL tea (Beith). HOLY THISTLE was another Irish source of an asthma treatment (Maloney). HORSERADISH was used in Russian folk medicine – half a pound of fresh root, grated, mixed with the juice of two or three lemons. The dose would be half a teaspoonful twice a day. Another Russian remedy used GINGER. The recipe given is a pound of ginger grated, put in a quart bottle, which was filled with alcohol. This was kept warm for two weeks, shaken occasionally, until the infusion was the colour of weak tea. Then it was strained, and the sediment allowed to settle. Then the clear liquid was poured into another bottle, and the infusion taken twice a day (Kourennoff). A root decoction of ROSE-BAY WILLOWHERB has been used too (Leyel. 1937). **BILBERRY** tea (up to ten cups a day, very hot) was a Russian remedy for asthma (Kourennoff). RIBWORT PLANTAIN leaf tea is used for bronchitis and asthma (Conway), and a tincture of LARKSPUR seeds has also been used (Lindley), as has syrup made from the roots of BLACK BRYONY (Brownlow). The leaf decoction of CHILE PEPPER is used as a treatment for asthma in Trinidad (Laguerre), and the root and bark of LESSER EVENING **PRIMROSE** have in recent times been used for asthma and whooping cough (Grieve. 1931).

The best known treatment has been the smoking of dried COLTSFOOT, which is still an ingredient in all herbal tobaccos (Grigson. 1955), as it is also in Chinese medicine (F P Smith), for asthma, and even for lung cancer. There was, too, an Irish usage that involved boiling the fresh leaves in milk, and then eating the lot (Ó Súilleabháin). Another Irish remedy was smoking the dried leaves of MULLEIN for asthma and bronchitis (Ô Súilleabháin). Similarly, a cigarette used to be made from THORN-APPLE leaves and smoked to ease the condition, or it may have been treated with the infusion (Kingsbury. 1967). Another mixture smoked has INDIAN TOBACCO (*Lobelia inflata*) as its base. The leaves and flowers of Thorn-apple are mixed in to make "asthma powders", which can be bought as such. A little nitre is included to make it burn, and the smoke is inhaled. The mixture is often made up into cigarettes, for convenience (Hutchinson & Melville). In Ghana, the leaves of NEVER-DIE (*Kalanchoe crenata*), either boiled or macerated in water, are used as a sedative for asthma sufferers (Dalziel).

ATAMASCO LILY

(*Zephyranthes atamasco*) Equally well-known under the names Swamp Lily, or Zephyr Lily, this is an American plant from the southern states of the USA. From Alabama: "For boils, take equal parts of sumac, sage and swamp-lily root, and boil into a strong infusion, strain, and put in a cupful of lard and fry until the water is out. Apply on a flannel cloth" (R B Browne).

Atropa belladonna > DEADLY NIGHTSHADE

AUGSBURG ALE

is said to owe its peculiar flavour to the addition of a small bag of AVENS in each cask (Grieve. 1931).

AURICULA

(*Primula auricula*) Dreaming of auriculas seemed to have some significance at one time. If they were bedded out, then it was a good luck sign; if they were growing in pots, then it was a promise of marriage. But if you were picking the flowers in your dream, that was apparently a portent of widowhood (Mackay). It is said that pieces of meat were once put about the root, and that "a good part of its bloom is actually owing, like an alderman's, to this consumption of flesh" (Ingram). They are used in Russian folk medicine – a tea made from the whole herb is said to keep the kidneys working well and to prevent the formation of kidney stones (Kourennoff). In Britain, the leaves were once used as a styptic to heal wounds (Tynan & Maitland).

Bear's Ears is a very common name for this plant (that is what is suggested by auricula, anyway). It is the shape of the leaf that accounts for the name, but as is very often the case, it has got corrupted into odd forms. In the north of Scotland it has become Boar's Ears, or Bore's Ears (Britten & Holland). Interestingly, Jamieson said that a bear is called a boar in northern Scotland. However that may be, bear's ears becomes a little more unrecognizable as it goes south. In Lancashire, it is Baziers, or Basiers (Nodal & Milner). There is a May song that had its origin in south Lancashire, and its refrain is "The baziers are sweet in the morning of May" (Chambers). In Gloucestershire, the name is further changed to Bezors (Britten & Holland).

AUTUMN GENTIAN

(*Gentianella amarella*) It is a herb of St John in Russia, according to Gubernatis. And they used to say that whoever carried it about with him would never incur the wrath of the Czar.

Avena sativa > OATS

AVENS

(*Geum urbanum*) Avens used to be hung over a door to keep the devil from crossing the threshold (Boland. 1977), for this is herba benedicta, the blessed herb, or perhaps, as Prior suggested, benedicta is from St Benedict, founder of the Benedictine order, who was a hard ruler. His monks plotted to murder him by poisoning his wine. St Benedict made the sign of the cross over the glass, and it flew into pieces (he doesn't say what this has to do with avens, but goes on to say that the plant became known as an antidote to poison). In fact, the plant, which was called St Benedict's Herb, and Herb Bennett, etc., has nothing to do with the saint, but is the blessed herb, which stops the devil from entering. Even having it growing near the house is enough to deter the devil (Tongue. 1965). No venomous beast would come near it. It is herbe de St Benoit in France, and Benediktenkraut in German, still keeping the error in etymology, but it is Erba benedetta in Italian (Barton & Castle).

Avens was grown as a potherb in the 16th century (Grigson. 1955), and the young leaves were sometimes used in salads (Barton & Castle). A small amount of the root put in ale gives it a flavour and perfume, popular in Culpeper's day, and prevents it from turning sour. Augsburg Ale is said to owe its peculiar flavour to the addition of a small bag of avens in each cask (Grieve. 1931). The roots, tied in small bundles and put in an apple tart, will give it a clove flavour (Genders. 1971), hence the name Clove-root, or Clovewort. The roots were used to tan leather, and to dye wool a permanent dark yellow. They were also believed to repel moths – "the roots taken up in autumn and dried, do keepe garments from being eaten with moths …" (Gerard).

Gypsies would use the crushed root as a cure for diarrhoea (Vesey-Fitzgerald), and not only gypsies, for it is quite common as a herbal cure for the condition and similar ailments. The Maoris chewed the leaves as a dysentery remedy (Goldie). It is an old febrifuge, and was recommended in the 19th century as a quinine substitute (Thornton). In Ireland, it was given for a chill (Moloney). In some French country regions, the root, gathered before sunrise, is put in a linen bag, to be worn round the neck, as an amulet to stop all bleeding, particularly haemorrhoids, and to strengthen the sight (Palaiseul). Wiltshire people used the powdered roots in boiling water as a spring pick-me-up. It is said that these roots should be dug on 25 March, from dry ground (Wiltshire).

B

BALDNESS

NETTLE juice combed through the hair to prevent baldness has been a common folk practice (Baker). The Wiltshire cure for dandruff was to massage the scalp with a nettle infusion each day (Wiltshire). PARSLEY was recommended for baldness as far back as Pliny's time (Bazin), repeated a long time afterwards as "powder your head with powdered parsley seed three nights every year, and the hair will never fall off" (Leyel. 1926). Actually, it really does make a good lotion for getting rid of dandruff, and helps to stave off baldness (A W Hatfield). ROSEMARY, besides providing the base of various hair rinses (see COSMETICS), was also used for the more serious purposes of preventing baldness. A manuscript from 1610 claims that "if thou wash thy head with [rosemary water] and let it drye on agayne by itselfe, it causeth hayre to growe if thou be balde" (*Gentleman's Magazine Library ; Popular superstitions p162*). Equally optimistic was "a wash to prevent the hair from falling off", noted in the *Housekeeper's & Butler's Assistant* for 1862. It required "a quarter of an ounce of unprepared tobacco leaves, two ounces of rosemary, two ounces of box leaves, boiled in a quart of water in an earthen pipkin with a lid, for twenty minutes …". Sniffing HORSERADISH juice will cure baldness, so it was believed (Page. 1978), and the Anglo-Saxon version of Apuleius recommended WATERCRESS "in case that a man's hair fall off, take juice; put it on the nose; the hair shall wax" (Cockayne). Watercress actually is a good hair tonic. There is a saying in French that a bald man "n'a pas de cresson sous le caillou" – loosely, has no watercress on his head (Palaiseul). Gypsies use ST JOHN'S WORT as a hair dressing, to make it grow (Vesey-Fitzgerald).

WALNUT, by the doctrine once current bearing the signature of the head (see DOCTRINE OF SIGNATURES), was used for all maladies relating to the head and brain, from madness back to baldness. Even in Evelyn's time, the distillation from walnut leaves "with honey and wine", was being used hopefully to "make hair spring on bald-heads". HAZEL leaves had that reputation, too (Anderson), as well as providing a dark hair dye, just as CASSIA oil, mixed with olive oil, is used in Palestine to keep the hair dark, and also to prevent baldness (Genders. 1972). ONION juice "anointed upon a pild bald head in the sun", will bring "the hair again very speedily" (Gerard). The tuber of a JERUSALEM ARTICHOKE, cut in half, and the cut side rubbed on the roots of the hair, was an old country remedy for baldness (Quelch).

SOAPWEED (*Yucca glauca*) roots were widely used by native Americans as a soap substitute, and by Pueblo Indians as a hair wash shampoo as part of the ritual in initiation ceremonies (La Fontaine), though people like the Kiowa claimed it was an effective cure for baldness and dandruff (Vestal & Schultes). One of the more extraordinary remedies for baldness involved SPEAR PLUME THISTLE – see Gerard: "… being stamped before the floure appeareth, and the juice pressed forth, causeth the haire to grow where it is pilled off, if the place be bathed with the juyce". Langham was equally optimistic about BEETROOT – "the asches of the root with hony, resoreth haire, and keepeth the rest from falling". Aqua Mellis, which Burton took to be a decoction of BALM, was much used in 17th century England against baldness.

Early settlers in New Zealand rubbed the juice from cut stems of the tree called by the Maori RIMU over bald heads; they found it an excellent hair restorer (C Macdonald). INDIAN TOBACCO (*Lobelia inflata*) is used in Indiana, where the practice is to fill a bottle with the pulverized herb, add equal parts of brandy or whisky, and olive oil. Let it stand for a few days, then bathe the head once a day with the liquid (Tyler).

BALM
see BEE BALM (*Melissa officinalis*)

BALSAM FIR
(*Abies balsamea*) A fir from the eastern side of North America. Oil of fir is distilled from the bark and needles, the latter aromatic, and often made up into balsam pillows (Schery). Blisters on the bark are the source of Canada Balsam, used in American domestic medicine as an application to sore nipples (Weiner). Native Americans, such as the Menomini, would press the liquid balsam from the trunk and use it for colds and lung troubles (H H Smith. 1923). Another use by the same people was to steep the inner bark, and drink the subsequent tea for chest pains. The Ojibwe used it for sore eyes (H H Smith. 1945) and for gonorrhea (Weiner).

BALSAMINT
(*Tanacetum balsamita*) Introduced into Britain, and naturalized, as escapes from old physic gardens. In Elizabethan times it was used as a strewing herb for floors, shelves and closets (Macleod). At one time it was taken as a symbol of impatience. The plant went out of fashion (Leyel. 1937), and even at the beginning of the 19th century, the past tense had to be used when describing its virtues "for strengthening the stomach and curing headaches" (Hill). Gerard, among other conditions, recommended the seed, that "expelleth all manner of worms out of the belly", or "wormes both small and great", in Langham's words. It is still occasionally used, mainly in making an ointment for burns, bruises and skin troubles; more immediately, bruising a leaf and putting it on a bee sting will give relief (Brownlow).

Costmary is another name for Balsamint. It is from costus, an aromatic plant used for making perfumes in the east. There is too a reference to French coste amère, Latin costus amarus. So the "mary" part of the name seems to be a simple mistake. Nevertheless it did not stop the plant being dedicated to the Virgin Mary in most west European countries, and receiving names like Herbe Sainte Marie in France (Rohde), and Herb Mary or St Mary etc., in this country (Mabey. 1977, Dawson).

BAMBOO

A very important motif in Chinese art, being the symbol of longevity in both Chinese and Japanese systems, and the most frequently portrayed plant form on Chinese porcelain. The bamboo, prunus, and pine together are the emblems of Buddha, Confucius and Lao Tzu, the Three Friends (Savage. 1964). In some parts of Japan, there is a superstition that bamboo will bring death within three years of its planting (Bownas), which is odd, because Japanese mythology has it as generally a lucky symbol, representing tenacity and courage (J Piggott), and, as mentioned, long life, as is the pine. They usually combine the two to decorate gateposts at New Year (Seki).

BANANA

(*Musa x paradisiaca*) A complex group of hybrids, all sterile and so seedless, so cultivated by planting out side shoots that develop on the old growth. In the 16th and 17th centuries, the banana was the favourite candidate for the tree of the knowledge of good and evil (Prest). Gerard reported that the Jews "suppose it to be that tree, of whose fruit Adam did taste", and he named it Adam's Apple Tree in consequence.

In fertility cults in West Africa, the banana is an obvious male symbol, with its phallic-shaped fruit. It is often used in the symbolism in combination with the terminal bud, which, with its oval form, stands for the female reproductive organs (Talbot). In Uttar Pradesh, the image of the goddess Nanda Devi is carved from the trunk of a banana tree, and the fruit is a symbol of fecundity – a newly married bride is given the fruit. If a baby is born prematurely, the new-born child is made to sleep each day on a fresh banana leaf (Upadhyaya). Another kind of symbolism is used by some Negrito groups in Malaysia, who use the banana plant in stories to explain man's mortality. When the deity gave to one of the superhumans some "water life-soul" to give to the humans they had made, it was inadvertently lost. So the superhuman borrowed some from a banana plant. This was "wind life-soul" that he then gave to the inert bodies. "Wind life-soul" is a "short" life-soul, whereas what had been lost (the "water life-soul" was a "long" life-soul, and that would have made man immortal, whereas the one that the human beings eventually received was merely borrowed and this provides only temporary life). Some say that not only the life-soul but also the heart and blood were borrowed from a banana plant, and this is supposed to account for the resemblance, in colour and viscosity, between coagulated banana plant sap, which dries to a dark brown colour, and dried human blood (Endicott).

There are one or two superstitions to notice. An American idea, recorded in Illinois, is that dreaming of them is a good sign (Dorson. 1964). Another American belief, if that is the right word, is that you must eat bananas to grow tall (HM Hyatt), which must be homeopathic in origin. And from Britain, there is a divination game, which must be modern, that children play with the fruit. To find out whether a boy is being faithful, the question is put, and the lower tip of the fruit is cut off. The answer is found in the centre of the flesh, either a Y, meaning yes, or a dark blob, meaning no (Opie & Opie. 1959). Clearly, the system can be used to predict the outcome of many other activities, or to solve a problem that requires a simple yes or no answer (Vickery. 1995). There is one extraordinary medicinal use. It comes from Norfolk, from a man who had facial skin cancer. While he was waiting for treatment, a gypsy advised him to rub the cancer with the pith of a banana. It seems that the cancer was cleared up entirely by this means alone (V G Hatfield. 1994).

Banisteriopsis caapi is a tropical American hallucinogen. The narcotic drink made from the bark of this liana is variously known as caapi (or kahpi, which is apparently nearer the original (Furst)), ayahuasca, yaje, natema, or pinde, according to the area and Indian group using it. In the westernmost part of its range, the bark is prepared in a cold water infusion; elsewhere it is boiled, sometimes for a long time. In parts of the Orinoco region, the fresh bark may be chewed, and perhaps also a snuff may be taken.

Ayahuasca, a Quechua name meaning 'vine of the dead', or 'vine of the souls', is its Peruvian name, and the narcotic has become submerged in the total culture of the people who take it. Partakers often experience a kind of "death", and the separation of body and soul. To some Colombian Indians, drinking the preparation represents a return to the womb; the drinkers see all the gods, the ancestors and the animals. Those who take it "die", only to be reborn in a state of greater wisdom. It serves, too, for prophecy, divination, etc., and to fortify the bravery of male adolescents at initiation. But it may be taken at funeral ceremonies, and, in other contexts, by a shaman to diagnose an illness or divine its cure, or to establish the identity of an enemy (for a description of the proceedings, see Reichel-Dolmatoff).

The effects may be violent and with unpleasant after-effects, especially when the bark is boiled, and certainly when some other toxic plants are mixed in. Nausea and vomiting are almost always early characteristics; this is followed by pleasant euphoria and visual hallucinations. The men take it ceremonially after nightfall, and they maintain that they do not just experience visual hallucinations, but also hear music and see dances. In fact, they say that both their present-day dances and their music are based on hallucinations, and certainly all their visual art forms stem from the same source. The most frequent hallucinations are highly coloured: large snakes, jaguars, spirits, trees, often falling, and lakes filled with anacondas. All participants speak of a sense of motion and rapid change or "transformation", and few have ever admitted that they find it a pleasant experience, for they drink it to learn about things, persons or events which could affect the society as a whole, or its individuals (Kensinger).

BANYAN

(*Ficus benghalensis*) A tree that is sacred to Kali, that is, to time (O'Neill), and it was under this tree that Vishnu was born (Gordon. 1985). According to Hindu mythology, the banyan is the male to the Peapul, the female (Pandey). A silver coin should be put under the roots of a young banyan. In some places, Celebes, for instance, banyan is still a sacred tree. People will not cut one down, and will not plant crops in a field where one is growing (Mabuchi). In India, too, they say it should never be felled, otherwise the woodman who cut it would have no son (Gupta). In some of the Pacific islands, the banyan is a sort of world-tree, for on its branches live the soul-birds of people. As often as a man is born, a leaf sprouts on the tree. There is a branch for each village, and of course the leaf appears on the appropriate branch. As long as the leaf is there the man to whom it belongs continues to live, and if he is sick, the leaf withers. If it falls to the ground, it is a human being who is dead, but the leaf will only be pulled by the soul-bird (Roheim). The Santals and other Indian peoples wind the young aerial roots of this tree around the neck, to ensure conception (Pandey).

The fruit is very small, not much larger than a hazel nut, and of no use. So there used to be a saying among sailors of a "banyan dinner", when they were put on short commons (Ablett).

BAOBAB

(*Adansonia digitata*) A remarkable tree, remarkable enough for Kenyans to say that the devil planted it upside down (F Perry. 1972). But Shona-speaking people of southern Africa revered it, and one of them was often adopted as their land shrine (Bucher), while it would usually mark a Yoruba sacred site (Awolalu).

Baobabs are extremely useful trees. Rope, strong enough to give rise to the Swahili saying that translates : "secure as an elephant bound with baobab rope" (Prance), is made from the bark, as well as all sorts of plaited cordage (H G Baker). Nets and sacks are made from it, too, and in parts of West Africa, it is woven into a coarse material for clothing (Prance).

The wood makes a strong paper, and a red dye is obtainable from it (Dale & Greenway). Even the pollen is useful, for it can make a good glue (Palgrave & Palgrave). The inside of the tree is pulpy, and it is often hollowed out for water storage, though it flourishes in the Kalahari Desert, and as much as a thousand gallons of water can be tapped naturally from the tree (Emboden. 1974). The young leaves can be cooked as a potherb (Palgrave & Palgrave), and the dry leaves can be used either as a medicine or to thicken stews (H G Baker). In Sierra Leone, the leaf is used as a prophylactic against malaria (Emboden. 1974). The Bushmen in particular value the seed as a winter food (Lee). The pods, which can be up to a foot long, and look like sausages, fall to the ground when ripe, and are collected daily. And so on – every part of the tree has a use, it seems. Perhaps the strangest thing about this tree is the way that the trunks of old specimens can be hollowed out, and then used, not only to hold water, but also for dry storage of materials. They have even been put to use in storing a corpse, presumably indefinitely, for mummification has been practiced in these cases (H G Baker). Perhaps that is why witches are said to meet in baobab (and iroko) trees (Parrinder. 1963).

BAR-ROOM PLANT

A name given to the ASPIDISTRA (Hyam & Pankhurst). Nothing, it seems, could harm this plant, not even the atmosphere of a Victorian bar-room. Cf, too, Cannon-ball Plant, and Cast-Iron Plant, both stressing the seemingly invincible nature of this plant.

BARBADOS NUT

see PHYSIC NUT (*Jatropha curcas*)

BARBERRY

(*Berberis vulgaris*) A hedgerow shrub, scarce now in Britain, for most of it has been eradicated by farmers owing to the belief that its proximity to wheat caused fungus rust, though the fungus on barberry leaves is entirely different from wheat rust (Quelch). The belief, though, seems to underlie the story quoted by Bottrell, from an old Cornish droll, which speaks of a farmer cutting down and uprooting all barberries around his property, in an attempt to break a spell. Or has the plant a connection with witchcraft?

According to Culpeper, the ley of barberry ash and water as a hair wash will turn it yellow (Wykes-Joyce), a use also reorded by Langham. From this, it was a small step to believing that it actually made the hair grow. You had to wash the head with the water in which barberry had been boiled – but "take care that the wash does not touch any part where the hair should not grow" (Leyel. 1926). "To cause the hair to grow. take the barberry, and fill an iron pot therewith, fill it up with as much water as it will contain, then boil on a slow fire to the half. With this water, wash your head morning and evening" (Physicians of Myddfai).

The doctrine of signatures shows in a remedy for jaundice. The bark is yellow, and a decoction taken in ale or white wine was often used for the condition (Dyer. 1889). Irish folk medicine recommended the bark in stout, with sulphur, the whole cooked together (Moloney). Another Irish remedy involved brewing the root bark to a strong decoction that had to be taken every morning, fasting, for nine successive mornings (Wilde. 1890). In Lincolnshire, a tea was made from the twigs and bark for gallstones and jaundice (Gutch & Peacock), and gypsies use a weak infusion of the berries for kidney trouble (Vesey-Fitzgerald).

BARK FABRIC

The bast (inner bark) of LIME trees has served many purposes. J Taylor. 1812 talks of "the bark macerated in water, is made into cordage, ropes, and fishing nets; and mats and rustic garments are also made of the inner rind". Shoes of plaited bast were still worn in very recent times in eastern Europe, particularly in the Volga district (Buhler. 1940).

BARLEY

(*Hordeum sativum*) An Irish charm for warts was to get ten knots of barley straw (though it was more usual to use ten slices of potato), count out nine and throw away the tenth. Rub the wart with the nine, then roll them up in a piece of paper, and throw them before a funeral. Then the wart would gradually disappear (Haddon). Large amounts of boiled barley juice were recommended in Scotland to be drunk for kidney disease, and Jewish folklore has a recipe for retention of urine, i. e. water in which barley, eggshells and parsley had been boiled (Rappoport).

In seventeenth century Skye a mixture of barley meal and white of egg was applied as a first aid measure for broken bones. After that splints were used (Beith). Martin gave an example of a cure used in Harris for drawing "worms" out of the flesh. It involved applying a "Plaister of warm Barley-dough to the place affected". Eventually the swelling went down, and it drew out "a little Worm, about half an inch in length, and about the bigness of a Goose-quill, having a pointed head, and many little feet on each side". They called this creature, whatever it was, a Fillan.

Barley is required for many Hindu religious ceremonies, and is particularly associated with the god Indra. It is important at ceremonies attending the birth of a child, at weddings, funerals, and some sacrifices (Pandey).

BARWOOD

(*Pterocarpus angolensis*) A Central African tree, and the best of African timbers, very durable. Both the bark and roots are used medicinally, the bark by hot infusion mixed with figs, and used as a breast massage to induce lactation. The bark on its own is used as a cure for nettle rash, and the infusion for stomach upsets, headaches and mouth ulcers (Palgrave). One of the Mashona witch doctor's medicine for a persistent cough in adults is to take a piece of bark from the east side of the tree, and a piece from the west side, and crush them with the inner part of a waterlily (*Nymphaea caerulea*) root. The result is soaked in water and given in thin porridge (Gelfand). Or they cook the roots with a chicken, and take the resultant soup as a cough medicine (Gelfand). Elsewhere, a root decoction is used to cure malaria and blackwater fever. The sap of this tree is red and sticky, hence the name Bloodwood (Palgrave), or Bleedwood (Howes). It is this red gum that many people compare with blood. So it (or the wood) is used in rituals in which blood flows – boys' circumcision rites, hunting rituals, and in rites concerned with menstrual disorders, and childbirth (V W Turner). After a boy's circumcision, he is made to sit on a log of this wood, the belief being that this will cause the cuts to stop bleeding, in sympathy with its coagulative qualities. In all the hunting cults, this wood represents blood, in particular the blood of animals. So it also stands for their meat (V W Turner).

BASIL

i.e., Sweet Basil (*Ocimum basilicum*) A plant that is quite important in Greek folklore, used in divination practices and also to dismiss the evil Karkantzari at the proper season (Abbott). "A belief there is that basil comes into flower when the heavens burst apart at dawn on Epiphany" (Megas). Basilikos means royal (the plant is herbe royale in French (Grieve. 1933)), and it is looked on by the Greek peasant as a prince among plants. It is a holy plant on the Greek island of Chios, connected with the True Cross, for when St Helena was seeking the Cross, she came to a place where there was a lot of basil growing, and the plant's scent guided her to the right place to find the relic. At the feast of the Invention of the Cross (14 September), women bring basil plants to the church, and the priests distribute twigs to the congregation (Argenti & Rose). It is burnt on mainland Greece to discover the worker of witchcraft. While it is burning, a number of names are repeated in succession. A loud pop or crackle denotes that the name of the offender has been reached (Lawson). In the Balkans, basil in vinegar is recommended to "drive the snake and any other creature out of a man" (Kemp).

In the Middle East, it was the herb of grief, and was put on graves. Jews carried sprigs of it to give them strength and endurance (A W Hatfield). Similarly, in Crete, it is a symbol of mourning. Perhaps better known is the plant's symbolising hatred (Leyel. 1937). The Romans used to sow the seeds with curses through the belief that the more it was abused the better it would prosper. When they wanted a good crop they trod it down with their feet, and prayed the gods it might not grow. The Greeks too supposed basil to thrive best when sown with cursing – this explains the

French saying "semer le basilic", as signifying slander (Fernie). It also probably explains why in Italian folklore, basil always stands for hatred, although it had the opposite meaning in eastern countries (H N Webster). In India, where it is known as tulasi (Hemphill) (see HOLY BASIL, rather), it is sacred to Vishnu and Krishna (Clair). It is kept in every Hindu home as a disinfectant, and to protect the family from evil (A W Hatfield).

It is said that basil will wither in the hands of the impure (Barraclough). A prospective husband could test a girl's chastity by making her hold a sprig of basil in her hands. If it quickly withered it was taken as a sure sign that she was not a virgin (Higgins). If a young man accepted a sprig of basil from a girl, he was instantly in love with her, or so it was believed (A W Hatfield). Another belief was that it was the smell of basil that would attract a lover, hence one of its names in Italian, bacia Nicola (kiss me Nicholas). That was why Italian girls would pick basil and put it in their bosoms; married women put it in their hair (Gubernatis). In Smyrna, if a girl wanted to get married within the coming year, she would plant a pot of basil in May the year before. She would tend it carefully until Epiphany, when she would break off a small sprig and give it to the priest during his round, and was given in return the sprig with which he had blessed the waters. This sprig was then put in the frame of one of the family icons, and the girl waited patiently for the husband, who could not fail to come (Megas). In Sicily and parts of southern Italy, a pot of basil on the balcony signals that the family has a daughter of marriageable age for whom they seek a suitor (Simoons).

In Tudor times, little pots of basil were often given as compliments by farmers' wives to their landladies and to visitors (Grieve. 1933). In Mediterranean countries a pot of basil is kept on windowsills to keep flies out of the room (G B Foster), and a sprig in the wardrobe will keep moths and insects away (Conway). A strange early belief that Browne counts as one of the Vulgar Errors, was that "there is a property in Basil to propagate scorpions, and that by the smell thereof they are bred in the brains of men". He says that one Hollerius "… found this insect in the brain of a man that delighted much in the smell", also "whosoever hath eaten basil, although he be stung with a scorpion, shall feel no paine thereby". Gerard had already mentioned the superstition: "there be that shun Basill and will not eat thereof, because that if it be chewed and laid in the sun, it engendreth wormes. They of Africke do also affirm, that they who are stung of the scorpions and have eaten of it, shall feele no paine at all". Those "wormes" engendered in the sun are, of course, serpents (Hulme. 1893).

Basil is an embalming herb, already used as such in ancient Egypt. This tradition is also met in Keats'

poem called *Isabella, or the pot of basil*, originally told by Boccaccio. Isabella laid the head of her murdered lover in a pot of basil, which kept it "fairly unspoilt" (Swahn). It is used in cooking, of course, but only a tiny pinch is needed in soups (particularly turtle soup). It was said by Parkinson "to procure a cheerful and merrie heart", and Gerard also says that "the seeds drunken is a remedy for melancholy people", but on the other hand, notes that "Dioscorides saith that if Basill be much eaten, it dulleth the sight, it mollifieth the belly, breedeth winde, provoketh urine, drieth up milke, and is of hard digestion". Evelyn also warned that it was "sometimes offensive to the eyes; and therefore the tender tops to be very sparingly us'd in our Sallet" (Evelyn. 1699). It is said to have been the characteristic taste in the famous Fetter Lane sausages, a 17th century invention. The sausage-maker made a fortune by spicing his sausage with basil (A W Hatfield).

It was used as a strewing herb (Brownlow), and it counters headaches and colds, either by an infusion, taken hot at night (Quelch), or by taking it as snuff. Dried basil leaves in that form have been used for nervous headaches and head-colds for centuries (Hemphill). In Britain, basil, mixed with blacking, has been used to get rid of warts (Leyel. 1926).

BASKETRY

THREE-LEAF SUMACH (*Rhus trilobata*) was used more extensively for basketry than any other plant except willow. American Indian groups like the Navajo and Apache always used the twigs, while the Zuñi reserved them for the very best baskets, while the Navajo made their sacred baskets from them. The peeled branches were used for both warp and weft; for sewing materials the branch was usually split into three strips. The bark and brittle tissue next to the pith would be removed, leaving a flat, tough strand. It was used, too, to produce a black dye, both for baskets, and for leather.

Bassia latifolia > MAHUA

BASTARD MYROBALAN

(*Terminalia bellirica*) An Indian species, with pale greenish-yellow, bad-smelling, flowers. The fruit is used for dyeing and tanning. Unripe, it is astringent, while the ripe ones are purgative. The kernel is said to produce intoxication if a lot is eaten, the symptoms being nausea and vomiting, followed by narcosis. Perhaps this is why Hindus in northern India look on the tree as inhabited by demons. So they avoid it, never sitting in its shade (Pandey). (see also MYROBALAN (*Terminalia catalpa*).

BAY, or LAUREL (*Laurus nobilis*) Originally from the Mediteranean area, but widely cultivated now. The name 'Bay' is French baie, a contraction of Latin baccae, berries, or better, baccae lauri.

In Greek mythology, Daphne fled from Apollo, and was changed into a bay tree, which from that time became sacred to him (Clair). Did the priestess chew bay leaves before delivering the oracle? Palaiseul suggests that the leaves chewed would put them in a favourable state. Every sanctuary to Apollo had a bay tree, and none could be found where the soil was unfavourable to the tree's growth. No worshipper could share in the rites who did not have a crown of laurel on his head or a branch in his hand (Philpot). Since Apollo was the god of poetry, it follows that the crown of bay leaves became the customary award in the universities to graduates in rhetoric and poetry (Clair); we still speak of the "Poet Laureate" as the highest award for a poet in this country. Bachelor is from French bachilier, and Latin baccalaureus – laurel berry. Students who took their degree were not expected to marry, so single men are still bachelors (Wilks). The staff of bay of a reciting poet was assumed to assist his inspiration, just as the bay rod in the hand of a prophet or diviner was assumed to help him to see hidden things. That is why the use of bay played an essential part in the oracular ceremony at Delphi, to name but the most famous (Philpot).

Bay was used at weddings in a similar way to rosemary (Andrews. 1898) (see *Rosmarinus officinalis*). It featured in weddings in Burgundy, when, decorated with ribbons, a bay used to be hoisted to the highest chimney of the wedding house by the best man and six assistants. Then a bottle of brandy would be broken over it, and healths drunk, as guests sang:

Il est planté, le laurier.
Le bon vin l'arrose
Qu'il amème aux mariés
Ménage tout rose,
Tout rose (Baker.1977).

This is a lightning tree, and a protector from lightning, which was believed powerless to hurt a man standing by one (Dyer. 1889), one of the "vulgar errors" listed by Aubrey (Aubrey. 1686). But people have been known to carry branches of it over their heads in a storm (Waring). "He who carrieth a bay leaf shall never take harm from thunder" (Browne. 1646), and Culpepper added to the belief – " … neither Witch nor Devil, Thunder nor Lightning, will hurt a Man in the Place where a Bay-tree is". As garlic protected the boats from storms and the evil eye, so laurel protected them from lightning (Bassett). It was said (by Pliny) that the emperor Tiberius wore a laurel chaplet during thunderstorms for this reason. In the New Forest, the bay was planted because of the protection it gave from lightning and forest fires, but also because it averted evil (Boase), and in East Anglia, a bay (or holly) growing near a house has the same effect (G E Evans. 1966).

There are many more superstitions attached to the bay. The crackling of the leaves in the fire was a good omen. But if they just smouldered, the signs were not so good (M Baker. 1980). It used to be said that the decay of the tree was an omen of disaster, just as oaks were. Every Roman emperor solemnly planted one by the Capitol, and it was said to wither when he was about to die. Before the death of Nero, though the winter was very mild, all these trees withered at the roots; a great pestilence in Padua was preceded by the same phenomenon (Evelyn. 1678). It was the custom, too, for a successful general to plant a laurel at his triumph in a shrubbery originally set by Livia. Hence, bay is a symbol of glory (Leyel. 1937), or triumph, and as it is evergreen, of eternity (Ferguson). Shakespeare speaks of this superstition :

'Tis thought the king is dead; we will not stay,
The bay-trees in our country all are wither'd
(*Richard II. 2. 4.7*).

See also Holinshed : "in this year 1399 in a maner throughout all the realme of England, old baie-trees withered, and contrary to all men's thinking grew greene againe, a strange sight, and supposed to import some unknowne event".

It was believed in ancient Greece that spirits could be cast out by the laurel, and a bough was often fixed over the door in cases of illness (Philpot). That is why in ancient times a man would put a bay leaf in his mouth when he got up in the morning (Durham). That, though, can be quite rational, for a bay leaf has antiseptic properties, so that chewing one first thing was a good cleanser (like toothpaste) for a furry tongue. The practice on Chios of bathing in water to which bay and hazel leaves have been added (Argenti & Rose), must surely be another protective measure. Similarly, if a baby is born feet first, it will be lamed in an accident while still young, unless bay leaves are immediately rubbed on its legs (Waring). Aubrey mentions that branches of bay were strewn on coffins at 17th century funerals, and Jersey burial customs required the coffin to be covered with laurel and ivy (L'Amy). It used to be the custom in some parts of Wales for a funeral to be preceded by a woman carrying bay. She sprinkled the leaves on the road at intervals (J Mason). It is a symbol of resurrection, for seemingly dead trees often revive from the roots (Drury. 1994).

Cornelius Agrippa said that a sick magpie puts a bay leaf into her nest to cure herself (Berdoe), and, according to Aelian (*De nature anim.*), the pigeons put laurel sprigs in their nests to protect their young against the evil eye. The same use was noted in Morocco, where people would insure their ploughs against the evil eye by making some part of them in laurel wood.

It was used for love divination charms in this country. A St Valentine's Eve charm was to put two bay leaves across the pillow, after having sprinkled them with rose water, and saying:

Good Valentine, be kind to me,
In dream let me my true love see (Dyer. 1889).

Another charm from Devonshire called for five leaves, one pinned at each corner and the fifth in the middle of the pillow. The operator of this charm had to say the rhyme seven times, and count seven, seven times over at each interval:

Sweet guardian angels, let me have
What I most earnestly do crave –
A Valentine endued with love,
Who will both true and constant prove.

The future husband would appear in a dream (Vickery. 1995). The same number of leaves, to be disposed in the same pattern, was the rule at Gainsborough, in Lincolnshire (Rudkin). Usually, the girl had to put on a clean nightgown for the operation, often inside out (Drury. 1986). If a girl writes her lover's name with a pin on a laurel leaf, and puts it in her bosom, the writing will turn red if he is true to her (Leather). That comes from Herefordshire, but surely no-one actually tried the charm.

It is said that lovers should pick a twig from a laurel bush, break it in two, and each keep a piece (Igglesden). Why? The dream books had something to say of the meaning of bay-tree dreams. For a man to dream of one, it is a sign that he will marry a rich and beautiful wife, but have no success in his business undertakings. It is a good thing for physicians and poets to dream of it (Raphael). Wasn't a bay chaplet the proper accolade for a poet? (see above)

Bay had its ordinary uses, in addition to the medicinal and folkloric. Lupton tells us that "if there be branches of bay wrapt up or laid among cloths or books, it will keep the same safe from moths, worms and other corruptions". It was said too, in the 10th century collection called the Geopontica, that if a water supply was bad, it could be made wholesome by steeping laurel in it (Rose). Bay Rum has nothing to do with this tree; oil of bay is distilled from the fresh leaves of *Pimenta acris*, and is used solely for Bay Rum and Florida Water for toilet articles (Grieve. 1933).

Both the leaves and berries have been used in medicine. Pomet described the berries as "cephalick, nemotick, alexipharmick, and anti-colick; they mollifie, discuss, expel Wind, open Obstructions, provoke Urine and the Terms, facilitate the travel of Women in Labour, and help Crudities in the Stomach. They are good for the Nerves in Convulsions and Palsies, give ease in the most extream Colicks, and take away the After-Pains of Women in Child-Bed". Evelyn, earlier, had called them emollient, sovereign in affectioins of the nerves, collics, gargarisms, baths, salves, and perfumes …" "… taken in wine [they] are good against all venom and poison … [and] the juice pressed out [of the leaves] is a remedy for pain in the eares, and

deafnesse, if it be dropped in with old wine and oile of Roses …"(Evelyn, Gerard). We are told, too, that "pigeons and blackbirds when suffering from loss of appetite, eat bay leaves as a tonic" (Hulme. 1895), and they "heal stingings of bees and wasps, and do away all swellings" (Bartholomew Anglicus). A few bay leaves soaked in brandy formed a cure for colic in Illinois. One to four teaspoonsfuls of the result would be given (Hyatt). We have already seen that the berries, too, were used for the complaint. And "it is reported that common drunkards were accustomed to eat in the morning fasting two leaves thereof against drunkennesse" (Gerard), but that, as we have already seen, may very well be to get rid of a hangover and furry tongue. The most engaging of the early leechdoms is one from Langham, aimed at "one that is stricken with the Fayrie". The treatment was to "spread oyle de Bay on a linnen cloth, and lay it above the sore, for that will drive it into every part of the body; but if the sore be above the heart, apply it beneath the sore, and to the nape of the necke".

BAY RUM
Nothing to do with the Bay tree (*Laurus nobilis*). Oil of bay is distilled from the fresh leaves of *Pimento acris*, the Bay Rum Tree, and is used solely for Bay Rum and Florida Water for toilet articles (Grieve. 1933).

BEAR'S BREECH
(*Acanthus mollis*) A strange name. Halliwell quotes as an archaism a verb to breech, meaning to flog or whip, and this is made interesting by another old name for the plant, Brank Ursin. Branks, it seems, was a word for a kind of halter or bridle. Ursin, of course, is the bear (Watts. 2000). The leaves of this and other *Acanthus* species (notably *A spinosissimus*) were the motifs for the designs of the capitals of Corinthian columns of the Greeks and Romans, and were much imitated in the architecture of the Middle Ages. According to one legend, Callimachus,a Greek architect, was visiting the tomb of a young girl who had died on the eve of her wedding. There, standing on an Acanthus plant and left by a previous visitor, was a basket covered with a tile. Callimachus noticed that the leaves had been forced back by the tile into a decorative shape, and he adopted the motif to fit the pillars of the temple he was building at Corinth (Perry. 1972). Acanthus patterns often figure in old needlework. See, for instance, the 10th century St Cuthbert's stole in Durham cathedral. In some legends it was one of the plants used for Christ's Crown of Thorns. In the language of flowers, the acanthus symbolises the cult of fine arts (Rambosson), and in Christian symbolism, it was used to indicate Heaven (in the Ravenna mosaics). The trees of Jesse and the Trees of Life in early art are also founded on the Acanthus. After the 13th century, the use in symbolism ceased, and the use as a purely decorative motif took its place (Haig).

The roots, boiled and mashed up into a poultice, have been used to treat abscesses (Barton & Castle). Culpeper even suggested that "they [the leaves] are excellent good to unite broken bones, and strengthen joints that have been put out".

BEAR'S EARS is a very common name for AURICULA (*Primula auricula*) (that is what is suggested by the word 'auricula' itself). It is the shape of the leaf that gave rise to the name. It appears as Boar's Ears, or Bore's Ears in northern Scotland. Jamieson said that a bear is called a boar in the north of Scotland. However that may be, the name changes as one goes south. In Lancashire it is Baziers, or Basiers (Nodal & Milner). There is a May song from Lancashire that has as its refrain "The baziers are sweet in the morning of May". Further south still, in Gloucestershire, the name is further changed to Bezors (Britten & Holland).

BEARBERRY
(*Arctostaphylos uva-ursi*) Dried, the leaves would be smoked by the American Indians as a substitute for tobacco (Sanford); the Keres Indians seem to have mixed them with tobacco in the ordinary way (L A White), while the Chippewa claimed they smoked it "to attract game" (Densmore). The North-west coast Indians also used the leaves for the smoking substance kinnikinnick (Enboden. 1979), which is an Algonquin word meaning "that which is mixed", usually tobacco (Johnston).

BED-WETTING
It is said that a preparation of MULLEIN flowers in olive oil, made for earache (see EARACHE), can be used to cure children of bed-wetting. The dose would be a few drops in warm water (Genders. 1976). HORSETAIL tea, being rich in silicic acid, can be used for urinary problems, including bed-wetting (M Evans). Alabama children were given a tea made from the berries of THREE-LEAF SUMACH to cure them of bed-wetting, and PUMPKIN seeds were also used there for the same purpose (R B Browne). This is strange, because the same tea is known as an efficient diuretic. Similarly, DANDELION, the best known of all diuretics, the recipient of vernacular names like Pissabed, was also used to stop the misfortune. The flowers were given to Fenland children to smell on May Day to inhibit bed-wetting for the next twelve months (Porter. 1969). Perhaps these are examples of homeopathic magic. An infusion of ST JOHN'S WORT also was used. In Russian folk medicine, CENTAURY and St John's Wort were mixed in equal amounts for this (Kourennoff).

BEE BALM
(*Melissa officinalis*) The leaves have a lemon fragrance, and because of this it was used as a strewing herb (Clair), with a "quasi-medicinal" effect, as one writer put it (Fletcher). The stems were woven into chaplets for ladies to wear (Genders. 1972), and even the juice was used as a furniture polish which also gave the wood a sweet perfume. Not surprisingly, balm is the symbol of pleasantry (Leyel. 1937).

Melissa means a bee, and has the reputation of keeping bees in their hive. Gerard said :" The hives of bees being rubbed with the leaves of Bawme, causeth the bees to keep together, and causeth others to come unto them", a belief still current in East Anglia, where they say that if this grows in the garden, the bees will not leave the hive (G E Evans. 1966). Wiltshire beekeepers agree; they rub the inside of the skeps with it (Wiltshire) after hiving a new swarm, to encourage them to stay.

But it is in the sphere of popular medicine that balm is important. A tale from Staffordshire tells how Ahasuerus, the Wandering Jew, knocked at the door of a cottage, and found the occupant ill. The Jew was asked in and offered a glass of ale. In return, the patient was told to gather three balm leaves and to put them in a cup of ale, and to drink it, refilling the cup when it was empty, and adding fresh leaves every fourth day. He was cured in twelve days (M Baker. 1980). Aubrey. 1696 mentions a story that is probably the same as the Staffordshire legend, about an old man who was cured of his lameness by taking balm leaves in beer. But balm tea is the most widely used medicine, for stomach upsets or colic in Gloucestershire, but more commonly elsewhere for colds, especially if feverish, for it has the effect of promoting sweating (Conway). It makes a pleasant drink for influenza patients (A W Hatfield. 1973), and has even been recommended for bronchitis (Flück). Fresh leaves are best, and the usual recipe is an ounce of leaves to a pint of boiling water, when lemon juice or sugar can be added when cool, if the patient prefers it (Rohde. 1936).

Oil of balm is useful for drying sores and wounds (Gordon. 1977). It is a wound plant in the Balkans – balm, the leaves of centaury and the dust of a live coal, pounded (Kemp). From now on, its uses become more and more esoteric. We are told that "... eius decoctio in aqua menstrua provocat et matrica mundificat et confortat et conceptum aduivat" (Circa Instans/ Rufinus, quoted in Thorndike), and Gerard, taking his lead from Dioscorides, maintained that the leaves "drunke in wine, or applied outwardly, are good against the stingings of venomous beasts, and the bitings of mad dogs ...". He was down-to-earth enough to prescribe a mouthwash of the decoction for toothache, but went on to claim that it is "likewise good for those that cannot take breath unless they hold their neckes upright"! Not only that, but it "comforts the heart, and driveth away all melancholy and sadnesse ..." (it was still in use in the 20th century for nervous complaints and depression (Boland. 1979). We even hear that "essence of balm", drunk daily, will preserve youth. Llewellen, prince of Glamorgan, who lived to 108, attributed his long life to it (M Baker. 1980). Indeed, there was a once popular "restorative cordial", supposed to confer longevity, called Carmelite Water, apparently still made in France, under the name Eau de mélisse des Carmes, by macerating the fresh flowers and tops in fortified white wine, together with a variety of spices (Clair). There was also an Aqua Mellis, taken to be a decoction of balm, that was much used in 17th century England against baldness (Burton).

After all this it should come as a surprise to find balm used as a remedy for illness caused by witchcraft. It comes from a deposition made to the Assizes in Leicester in 1717, and was described as "used and prescribed by the cunning men", who put rosemary, balm, "and many gold flowers in a bagge to the patients brest as a charm and to give them inwardly a decoction of the same in a quart of ale and their own blood …" (Ewen. 1929). As far off as South America, there is a similar belief, for balm was used as an ingredient in the ritual bath that is part of a Brazilian healing ceremony (Williams).

BEE-KEEPING

Beekeepers in the East Riding of Yorkshire used to sprinkle the hive with an elder branch dipped in sugar and water when the bees were ready to swarm (Addy). In Cornwall, too, they say that the inside of hives should be scrubbed with elder flowers to prevent a new swarm from leaving (Courtney). But bees do not seem to like the smell. When they swarm, a sprig of elder is often held about nine inches above them. The idea is that the elder will drive them out of the tree in which they are swarming. In any case, elder is a well-known insect repellant. SWEET CICELY is very attractive to bees, and was often rubbed over the inside of hives to induce swarms to enter (Northcote). So was THYME, which was always grown near the hives (Gordon. 1977). BEE BALM, too, could be used, for it had the reputation of keeping bees in their hive, as Gerard said, and this is a belief still current, at least in East Anglia, where they say that if this grows in the garden, the bees will not leave the hive (G E Evans. 1966). Wiltshire bee-keepers agree; they rub the inside of the skeps with it (Wiltshire) after hiving a swarm, to encourage them to stay. One could hang a piece of JUNIPER inside the hive, to protect the bees from adverse magic (Boland. 1977).

BEE STINGS

Bruise a leaf of BALSAMINT and put it on a bee sting – it will bring relief (Brownlow). So will SUMMER SAVORY (and Winter Savory, *Satureia montana*) leaves (Clair), and Gerard advised using MALLOW leaves. He went on "if a man be first anointed with the leaves stamped with a little oile, he shal not be stung at all". ONION juice rubbed on a bee or wasp sting was an old Wiltshire beekeepers' remedy (Wiltshire), but there is nothing either local or esoteric about that;

the remedy is recorded in Norfolk, too (Randell), and in Ireland, and in America (H M Hyatt). It appeared, too, in the *Gardeners' receipt book*, 1861. Lawrence Durrell also mentioned it as standard practice in Corfu. CHICKWEED can be applied to a sting, to get the swelling down (Vickery. 1995), or rub MARIGOLD petals on them (Rohde. 1936). WATER MINT is "… good against the stingings of bees and wasps, if the place be rubbed therewith" (Gerard).

BEECH

(*Fagus sylvatica*) Apparently a British native only on chalk and limestone in south-east England. Elsewhere it was planted, albeit a very long time ago, as witnessed by place names (see examples in Cameron). In spite of the early spread of the tree, there is very little folklore attached to it, but what there is shows that it is held in high regard. In Somerset, for instance, they say that if you are lost in a wood at night nothing can harm you if you sleep under a beech tree, and if you say your prayers under a beech, they will go straight to heaven. On the other hand, the tree will be avenged if you use bad language under it, for then the leaves will rustle, and a bough may even drop on you (Tongue). Like a number of other trees, beech was supposed to be proof against lightning (Dyer), or that lightning never strikes it (Sebillot). Again like other trees, beech has its protectors, as with an enormous specimen known as "The Lady's Beech", a few miles from Copenhagen. Legend says this tree was once part of a large forest. While a girl was sheltering under this tree, a white-clad figure appeared to her and prophesied that the girl would become mistress of Kokke-dal. She also made the girl give her her promise never to consent to the tree's being felled. It was said that each owner of Kokke-dal was bound to leave the tree standing (Craigie).

A potent drink, probably originating from the Chilterns, called leaf noyau, can be made from beech leaves. A recipe given requires one to pack an earthenware or glass jar about 90% full of young, clean leaves. Pour gin into the jar, pressing the leaves down all the time, until they are just covered. Leave to steep for about a fortnight, and then strain off the gin, which will now be a brilliant green. Add sugar dissolved in boiling water, and add a dash of brandy. Mix well, and bottle when cold (Mabey. 1972).

"We must not omit to praise the mast, which fats our swine and deer, and hath in some families even supported men with bread …" (Evelyn). But one has to be careful with them, for they are toxic to some people. "The nuts, when eaten by the human species, occasion giddiness and head-ache: but when well-dried and powdered, they make wholesome bread" (Taylor). Beechmast oil can be used like any other cooking oil (Mabey. 1972), and beechnut butter is still made in some country districts of America. It was even claimed that they could be substituted for coffee.

Medicinal uses for beech are few, in fact the only recipe involving the leaves dates from the 15th century, and is for deafness : "take the juice of leaves of a beech-tree, and good vinegar, even portions, and put thereto powder of quick-lime; and then clear it through a cloth; and of this, when it is cleansed, put hot into the sick ear" (Dawson). More attention was paid to the water that collected in hollow parts of the tree. It cures, according to Evelyn, "the most obsinate tetters, scabs, and scurfs, in man or beast, fomenting the part with it", and it also prevents baldness, according to a report from Devonshire (*Devonshire Association. Transactions. vol 103; 1971 p103*).

BEET (ROOT)

(*Beta vulgaris 'Maritima'*) Very popular in Russia, where it is mainly used as a base for the soup borsch. The only note there is as to the use of anything but the root comes from the Isle of Wight, where apparently it was the custom to eat the leaves, under the name Wild, or Sea, Spinach (Grigson. 1955).

There is not much in the way of superstition recorded, though the Pennsylvania Germans say that if beets run to seed the first year, it foretells a funeral, or some-one in the family will die (Fogel), a belief that must presumably have travelled from Europe. A distinct oddity from Kentucky insists that eating beet is a sign that you are in love! (Thomas & Thomas). The dream books say that if you dream you are eating beetroot, it is a sign that your troubles will disappear, and that prosperity will follow (Raphael). Certainly, one or two of the so-called medicinal usages look more like superstitions, or charms, at least. For instance, from Corfu – a small bottle of beetroot juice is corked and put in the heart of an uncooked loaf of bread. The bread is baked, and the bottle removed, and then the medicine is drunk in small doses for dysentery or diarrhoea (Durrell). And Langham's optimistic statement must come under the same heading – "the asches of the root with hony, restoreth haire, and keepeth the rest from falling".

Prescriptions involving beetroot have been used for fevers since ancient times. One "confection for the fevers" is included in a 15th century collection of medical recipes, and reads "take centaury a hand-ful; of the root and of the leaves of the earthbeet a handful; of the root of clover a handful; of ambrose a handful; and make powder of them, then mix honey therewith. And make thereof balls of the greatness of half a walnut. And give the sick each day one of them fasting, and serve him nine days. For this is a good confection for fevers and the mother and for the rising of the heart" [flatulence] (Dawson). That may have worked, especially as nine days may have been enough to see the fever off naturally, but one would have to question the Balkan practice of treating a fever by laying beet leaves on the skin round the waist, and changing it morning and evening, for three days (Kemp). The earliest reference to the use of beet, for headache this time, comes from Anglo-Saxon times. The Lacnunga has this prescription: "Roots of beet, pound with honey; wring out. Apply the juice over the nose. Let him (the patient) be face upward toward the hot sun and lay the head downward until the brain be reached. Before that, he should have butter or oil in the mouth, the mucus to run from the nose. Let him do that often until it be clean" (Grattan & Singer). Gerard endorsed the practice, in fewer words, and even in the 18th century it was still repeated by Hill. There are still one or two more odd-sounding medical practices. In Russian domestic medicine, for example, a gargle of beet juice is a rec-ognized remedy for adenoids (Kourennoff). Another is the Pennsylvania German claim that crushed beet leaves put in a rag and bound on a wound will cure lockjaw (Fogel).

BEGGARWEED

Names like this usually mean either that too much of it will make the soil fit for nothing else, and so beggar the farmer, or that the presence of the plant is a sure indication that the soil is no good anyway, and will still beggar one. So the name is given to plants like CORN SPURREY (*Spergula arvensis*), or Knotgrass (*Polygonum aviculare*), and so on. Cf Poverty-weed, Pickpocket, which carry the same message.

BELLADONNA

(*Atropa belladonna*) see DEADLY NIGHTSHADE. The name itself, or its English translation, Fair Lady (Friend. 1883), which looks strangely out of place, refers to an ancient belief that the nightshade is the form of a fatal witch called Atropai, who in fact was the eldest of the Fates, the one whose duty it was to cut the thread of life. There is an old superstition that at certain times the plant takes the form of an exceedingly beautiful enchantress, dangerous to look upon (Skinner). Certainly, it was reckoned to have been grown in Hecate's garden (Clair). But the usual explanation of the name refers to the custom on the Continent for women to use it as a cosmetic to make the eyes sparkle (atropine is still used by oculists to dilate the pupils (Brownlow)).

Bellis perennis > DAISY

Berberis vulgaris > BARBERRY

BERRY-TREE

A Yorkshire name for a GOOSEBERRY bush, for gooseberries are taken there as berries par excellence (Hunter). Gooseberry-pies are berry-pies. Gooseberries in the north of England have long been the subject of esteem and competition.

Beta vulgaris 'Altissima' > SUGAR BEET

Beta vulgaris 'Maritima' > BEET(ROOT)

BETONY

(*Stachys officinalis*) i.e., WOOD BETONY Sown round the house, it protects it from witchcraft. "The house where Herba Betonica is sowne, is free from all mischeefes" (Scot). The Anglo-Saxon Herbal mentions it as a shield against "frightful goblins that go by night and terrible sights and dreams" (Bonser). "For phantasma and delusions: Make a garland of betony and hang it about thy neck when thou goest to bed. that thou mayest have the savour thereof all night, and it will help thee" (Dawson). The first item on Apuleius' list is "for monstrous nocturnal vistiors and frightful sights and dreams" (Cockayne). A Welsh charm to prevent dreaming was to "take the leaves of betony, and hang them about your neck or else the juice on going to bed" (Bonser).

Roy Vickery was quite right when he said that there was "little evidence for betony being much used in British and Irish folk medicine (Vickery. 1995). It is surprising in view of the number of prescriptions in early and classical herbalism, and also in view of the proverb "Sell your coat and buy betony" (Dyer. 1889; Whitlock. 1992). There is a Cumbrian recommendation to drink betony tea for indigestion (Newman & Wilson), and in Somerset a cure for headache is to drink the tea hot (Tongue. 1965). Gypsies, too, take an infusion of the fresh leaves to relieve stomach trouble, and they make an ointment from the juice of fresh leaves and unsalted lard to remove the poison from stings and bites (Vesey-Fitzgerald). There is, too, an injunction to chew a fresh betony leaf to prevent drunkenness before a party (Conway). But that is all, yet next to vervain, betony was the most esteemed of all plants by the early herbalists. To sum it all up: "A medicine against alle maner of infirmities. Take and drinke a cupful of the juyce of betonye, the first Thursday in May, and he shalle be delivered from alle maner of diseases for that year" (*Gentleman's Magazine Library: popular superstitions*). It was the first to be dealt with by Apuleius, and the Anglo-Saxon translation has no less than twenty-nine prescriptions (Cockayne), and there are at least that number of leechdoms independently written in Anglo-Saxon times. A Middle English rimed medical treatise prescribed either betony or fennel for the digestion, to be taken "in drage after mete". "Dragges" were a kind of digestive powder for weak stomachs, and were used by Chaucer's Doctor. The term was applied to such things as sweetmeats served in the last course as stomach closers, according to Cotgrave's Dictionary, 1660 (I B Jones).

Betula pendula (*verrucosa*) (*alba*) > BIRCH

BHANG

according to Burton, the Arab Banj and the Hindu Bhang, is the most frequently used word referring to the drug Cannabis (HEMP).

BIBLE LEAF

One of a series of similar names given to TUTSAN (*Hypericum androsaemum*). When dried, its leaves have a very sweet smell, likened to ambergris. Picking the leaves and pressing them in books used to be a favourite pastime, hence the name, and others like Sweet Leaf (Devonshire – Grigson. 1955), as well as Book Leaf (Dorset – Macmillan).

BILBERRY

see WHORTLEBERRY. Bilberry is as common as Whortleberry for *Vaccinium myrtillus*. It seems to be from a Danish word, boelle-baer, meaning ball berry (Britten & Holland).

BINDWEED

see (a) FIELD BINDWEED (*Convolvulus arvensis*) (b) GREAT BINDWEED (*Calystegia sepium*)

BIRCH

Birch, the Lady of the Woods, is another of the trees looked on as having some protective power over attack by witches, or even the fairies. Over quite a large part of the Continent, it used to be said that if a witch were struck with a birch broom, she would lose all power (Lea). The Irish still say that fairies do not like the birch, and in the west of England, birch crosses would be hung over cottage doors to repel enchantment. Sometimes that other powerful protector, rowan, would be coupled with the birch on particular days (May Eve and Day, for example) when the risk was greatest. Presumably this is the reason why throughout Europe, birch twigs are used for beating the bounds (Graves). In Worcestershire, birch branches, hung with cowslip balls, were put on the door at May Day (Chamberlain, E L). Connected with this must be the Breton custom of taking birch leaves if a child is sickly, drying them in an oven, and then putting them in the baby's cradle to give it strength, so it was claimed (Gubernatis), but probably to keep away the evil that was causing the sickness.
But there is a sinister side to birch mythology. Take, for instance, the ballad, The Wife of Usher's Well:

> It fell about the Martinmas,
> When nights are lang and mirk,
> The carlin wife's three sons came hame
> And their hats were of the birk
> It neither grew in dyke nor ditch
> Nor yet in one sleugh;
> But at the gates o' Paradise,
> That birk grew fair enough.

In other words, the birch grew at the gates of paradise, and it furnished the ghosts with their "hats o' birk". It is a sacred tree, and traditionally the tree of death. To dream of pulling the "birk sae green" portends death in the Braes o Yarrow. A "wand o' bonny birk" is laid on the breast of the dead in Sweet William's Ghost (Wimberley). Birch is traditionally the tree of death in Somerset legend, too. "The one with the white hand" (a birch, so it would seem) was a spirit that haunted the moorland near Taunton. It would rise up at twilight out of a scrub of birch and oak and come drifting across to lonely travellers so fast that they had no time to escape. She was deadly pale, with clothes that rustled like dead leaves, and her hand looked like a blasted branch. Sometimes she pointed a finger at a man's head, and then he ran mad, but more often she put her hand over his heart, and he fell dead. The spirit was laid by a farmer who carried salt with him, and he put his own hand with the salt into that of the spirit (Tongue).

Delinquents, and formerly lunatics, were flogged with birch boughs, "for which anciently the cudgels were us'd by the lictor, for lighter faults, as now the gentler rods by our tyrranical pedagogues", in Evelyn's words. "Gentler", because "birchen twigs break no ribs" (Hazlitt). But the flogging, especially in the case of lunatics, was more likely to be the means of driving out the evil that was causing the condition.

Birch branches were often used, along with milkwort, in the Rogationtide processions. Whit Sunday, too, was marked as a day to use birch for church decoration. But May Day is the special feast day associated with the tree. The Welsh maypole was always made of it (Trevelyan) – in fact the Welsh word bedwen serves both for birch and maypole. And in Herefordshire a birch tree was brought into the farmyard on May Day, and decorated with red and white rags, then propped against the stable door to protect the horses within from being hag-ridden (Leather).

There is a saying, "as bare as the birch at Yule eve". It is spoken of anyone in extreme poverty (Denham), and birch was used as a symbol of meekness (Leyel). It also seems to have been a symbol of fertility (Hartland). One of the names for a Germanic rune was berkana, birch twig, and this must have been connected with fertility rites. Birch saplings were put in houses and stables, and men and women, as well as the cattle, were struck with birch twigs, with the avowed intention of increasing fertility (Elliott). Birch twigs were fixed over the lover's door on May morning in Cheshire (Wimberley). At one time, when a Welsh girl accepted a proposal of marriage, she presented her lover with a wreath of birch leaves (if she refused him, she sent hazel (Trevelyan)). There is an extensive body of customary lore in Wales on this subject. The love poems of medieval Welsh poets contain many references to seeking to court girls under the green leaves of the birch tree, for what is in effect a "daytime date", when a comfortable bed of birch leaves and twigs would be provided. Things made with birch leaves and twigs served as emblems of love when exchanged between lovers, the *cae bedw*, the birchen garland, being the most popular. Hats made of birch twigs were common, too (Stevens).

Birch, like most trees, has its practical uses; the timber is good enough for clog-making in Ireland (Ô Súilleabháin), and the inner bark was used for writing long before the invention of paper (Gerard). John Clare tried his hand at paper-making from birch bark, and found that it "is easily parted in thin lairs & one shred of bark round the tree would split into 10 or a dozen sheets. I have tryd it & find it receives the ink very readily" (Clare). Birch bark had found a use even in Mesolithic times, for it produces a sticky substance like pitch. Rolls of birch bark have been recovered from the site at Star Carr, which was occupied about 7500 BC. It seems that they were used for glue to hold microliths in their wooden seating, for flints with birch glue still on them have been found (Helm).

In the Highlands, birch used to be employed at one time for a great deal of purposes – the timber made beds, chairs, tables, spoons and the like, and the outer rind, called *meilleag* in Gaelic, was used for making candles (Fairweather). Rope was made from it as well, far more durable than hempen ones. The wood was used too for building, toolmaking and for constructing carts and fences, and it furnished thatch and brooms (Ablett). There is an Irish record of the use of the bark for tanning (Ô Súilleabháin), and the leaves were used for dyeing (Jenkins). An aromatic oil is obtained from it, known as Russian leather, because bookbinders used it to rub on leather, to give it the characteristic perfume, and also to preserve it. That same oil, which contains methyl salicylate, mixed with alcohol, is rubbed on the skin as a protection against midges and mosquitoes (Hutchinson & Melville).

Birch tea is used for urinary complaints, especially dropsy. But it is useful also for gout, and has even been recommended as being helpful for the heart (Schaunberg & Paris). It has also been given for rheumatism (Grieve), and another folk remedy for the complaint, from Russia in this case, involved boiling birch leaves in water for half an hour, and putting that water into a hot bath. One bath daily before going to bed, for 30 days at least, is prescribed (Kourennoff). But there were some strange claims made in times gone by, perhaps the most hopeful being from the Physicians of Myddfai – "for impotency. Take some birch, digest it in water, and drink". A pure transference charm is recorded in Suffolk, for toothache, by clasping the tree in one's arms, and then cutting a slit in it. Cut a piece of hair with one's left hand from behind the ear. That has to be buried in the slit, and when the hair has disappeared so will the toothache (Burn).

BIRD CHERRY

(*Prunus padus*) Bird Cherry fruit is sometimes used to flavour home-made wines, and was even used to flavour gin and whisky once (C P Johnson), while the bark was used sometimes to dye wool a light brown colour (Fairweather). It is associated with witches in Germany, and called Hexenbaum, a name which is too near such words as Heckberry to carry much conviction as to its witch connection. However, there is a saying from Roxburghshire, in Scotland, that may help, in a negative way:

> Hagberry, hagberry, hang the de'il,
> Rowan-tree, rowan-tree, help it weel.

But, at least in Wester Ross, bird cherry in some cases played the protective role taken by rowan. A walking stick made from it prevented the bearer from getting lost in a mist (Denham).

BIRD-LIME

MISTLETOE berries, "of a clammy or viscous moisture", are such "whereof the best Bird-lime is made, far exceeding that which is made of the Holm or HOLLY bark" (Gerard). It was made by drying and pounding the berries soaking them in water for twelve days, and pulverizing them again. Bird-lime was used up to medieval times for taking small birds in the branches of trees, and also for catching hawks, whch were decoyed by a bird tethered between the arches of a stick coated with the stuff (J G D Clark. 1948). "It is reported that the bark of the root of the (WAYFARING TREE) buried a certaine time in the earth, and afterwards boyled and stamped according to Art, maketh a good Bird-lime for Fowlers to catch birds with" (Gerard). In Africa, the latex of the CANDELABRA TREE (*Euphorbia ingens*), which becomes sticky when partially dry, is used for the purpose (Palgrave), and another African plant that is used is AFRICAN CORN LILY (*Ixia viridflora)*, which has a sticky sap. The generic name, *Ixia*, comes from a Greek word ixos, meaning bird-lime.

BIRD'S EYE is a name given to a number of plants (see Watts. 2000). Descriptive, of course in most cases, but there was a belief, particularly applying to GERMANDER SPEEDWELL, that if you pick the flowers, birds would come and pick your eyes out, or your mother will suffer that fate (cf MOTHER-DIE).

BIRTHWORT

(*Aristolochia clematitis*) The name *Aristolochia* comes from two Greek words meaning 'best birth'. The greenish-yellow flower constricts into a tube that opens into a globular swelling at the base. The swelling was interpreted as the womb, the tube as the birth passage. Thus by the doctrine of signatures, it was used to help delivery, to encourage conception, and to "purge the womb". Oddly enough, it seems that the plant does have an abortive effect (Grigson). Pliny said it was prescribed for securing a male child (Bonser). Another strange ancient usage was described by Guainino (c1500). He prescribed a kind of medicated pessary made of honey and birthwort, to be introduced before sleeping. The woman was judged to be fertile if on awakening she had a sweet taste in her mouth! (T R Forbes). Hippocrates recommended the plant as a purgative for women, or for uterine complaints, and it seems that Theophratus even prescribed it for prolapsed uterus (Dawson).

In addition to the doctrine of signatures practices, we find that in ancient times it was used for other ailments. Hippocrates, again, recommended it for pain in the side, and for dressing ulcers (Dawson), the latter remedy is still in use today (Schauenberg & Paris). Early medieval texts from the Byzantine school prescribed it for epilepsy, and it was also recommended for asthma (Dawson). The Anglo-Saxon version of Apuleius quoted birthwort "against strength of poison", and also "for bite of adder" (Cockayne), which had already appeared in Dioscorides/Rufinus ("… et ad vipera morsem et ad alia venena intrinsicus facit") (Thorndike). The Anglo-Saxon text also prescribed it for fevers, as did the Welsh text known as the Physicians of Myddfai ("For intermittent fever. Take the mugwort, the purple dead nettle, and the round birthwort, as much as you like of each, bruising them well in stale goat's milk whey, and boiling them afterwards. Let the patient drink some thereof every morning, and it will cure him"). There were some very strange conditions mentioned in Apuleius for which birthwort was a cure, like "for sore of nostrils", or "if any child be vexed", and, strangest of all, "in case that to anyone an ulcer grow on his nose"!

The plant is still used by herbalists. For instance, a decoction of the fresh plant is recommended to treat infected wounds. Used internally, it is emmenagogic, and is also taken to soothe arthritis and rheumatism pains (Schauenberg & Paris). A tincture is used in Russian folk medicine for gout (Kourennoff).

BISTORT

(*Polygonum bistorta*) "Herb Pudding", or "Yarby Pudding" (Vickery. 1995), was made from Bistort leaves on Easter Day, or more properly, at Passiontide. The leaves were boiled in broth, with barley, chives, etc., and served to accompany veal and bacon. Easter Giants, or Easter Mangiants, both from the French manger, to eat, are other names, and there is Ledger Pudding, as well. Mabey. 1972 says that the last two weeks of Lent was the proper time to eat this pudding, but many people make enough to freeze, to have for breakfast on Christmas Day (J Smith. 1989). The Yorkshire and Cumbria Dock Pudding is not apparently connected with Easter, but is simply a cheap meal. It contains bistort, young nettles, onions and oatmeal. The mixture would be simmered till cooked, strained and allowed to go cold. Then slices would be fried, with bacon. The Cumbrian version was more elaborate, with a lot more different spring leaves (Schofield). Bistort leaves contain starch, and have been used as a marginal food (Browning), in fact they can provide a form of flour once the tannin has been steeped out (Dimbleby).

In country medicine, the root is the only part still used, being rich in tannin (20%). It is said to be the best herbal medicine for a sore throat (Conway), and, of course, being so rich in tannin, the root is a strong astringent ("one of the best astringents in the world" was Hill's opinion), great for diarrhoea (Flück), and it was used for staunching wounds, and internal haemorrhages. "The root doth glew wounds together..." (Langham). As an astringent, it is still used as a mouthwash (V G Hatfield. 1994). The infusion of the dried herb is used in Russian folk medicine for jaundice (Kourennoff). And it was used in Scotland, too, for urinary complaints (Beith). The root is regarded as a sure cure for incontinence (Mitton & Mitton). Gypsies use it for diphtheria (Vesey-Fitzgerald), and bistort tea is drunk in Cumbria to get rid of a headache (Newman & Wilson). There were old recipes for snakebite, but they were doctrine of signatures, from the twisted roots. Bistort is from the Latin bis, twice, and torta, twisted. It is the roots that are twice-twisted, or writhed, like a nest of snakes, which explains names like Snakeweed (Gerard) and Snakeroot (Clapham, Tutin & Warburg), and which also explains the use for snakebite. One veterinary use, from East Anglia, has been recorded. The juice from the leaves is rubbed round horses' teeth to prevent decay (V G Hatfield. 1994).

BITING STONECROP

(*Sedum acre*) It is traditionally known as an abortive (Schauenberg & Paris). 'A Middle English Rimed Medical Treatise' has:

> She that drynkes fumiter and the stoncrope
> Schal neuere yong childe in cradell roke.

In other words, they caused sterility or abortion (I B Jones). Stonecrop "hath the signature of the gums", and so was used for scurvy (Berdoe), and Hill, in the 18th century, independently advised that "the juice ... is excellent against the scurvy and all other diseases arising from what is called foulness of the blood".

The bruised plant is applied to wounds to help heal them, and also to cure warts and corns (Flück). It had also a use in skin complaints; in East Anglia, the juice has been used for dermatitis, and in Scotland an infusion was made to treat erysipelas (Beith).

BITTER GOURD

(*Colocynthus vulgaris*) Colocynth means "bitter gourd" – it *is* exceedingly bitter (indeed it is "of an intolerable bitterness" (Pomet)). The gall of the Bible often refers to this (Moldenke). In minimal doses it is a violent purgative; in larger doses it is lethal, which makes the practice of rubbing it on the nipples to wean a child surprising, to say the least (Van Andel). A piece of root set in a gold or silver case, was hung round a baby's neck as a teething amulet, a practice recommended by a Roman physician, Actius of Amida, in the 6th century AD. Wasson, in a footnote,

says that colocynth was the base for "general issue" purgative pills in the British army in the first World War. Pomet mentions the practice of confectioners who "cover these Seeds with Sugar, and sell them to catch or delude Children with, and People of Quality upon extraordinary Occasions ..."

It was known in Britain as early as the 11th century (Thompson. 1897), and was certainly known to Shakespeare – Iago, in *Othello*, says "the food that to him now is as luscious as locusts, shall be to him shortly as bitter as coloquinrida". Topsell, writing in 1697, seemed to know about it, too – "it is said that he who will go safely through the mountains or places of [the hyena's] abode must carry in his hand a root of coloquintida".

BITTERSWEET

An alternative name for WOODY NIGHTSHADE (*Solanum dulcamara*), and a translation (in reverse) of the specific name, dulcamara, which is really amara dulcis : "faire berries ... of a sweet taste at the first, but after, very unpleasant". Gerard, obviously studying dulcamara, got it the wrong way round, and should have paid attention, if he were still alive, to Shakespeare, who knew the real sequence:

> I should not think it strange, for 'tis a physic
> That's bitter to sweet end.

BITTERVETCH

(*Lathyrus montanus*) The tubers can be eaten like potatoes, and in Holland, they were roasted like chestnuts, and they taste rather like chestnuts, too (G M Taylor). It is also used in the Hebrides for flavouring whisky (Murdoch McNeill), and the roots were chewed as a tobacco substitute in the Scottish Highlands (G M Taylor). Pennant observed that the roots dried "are the support of the highlanders in long journies; and a small quantity ... will for a long time repel the attack of hunger..." (Pennant. 1772).

BLACK BRYONY

(*Tamus communis*) Like a piece of silver, the root of back bryony was used when a hare suspected of being a metamorphosised witch was to be shot. Powder was put in the barrel and a piece of this root, the whole rammed down as if it were real shot. "And they say you could cut a hole through a door using this root" (Evans & Thomson). East Anglian farm horsemen used to put black bryony root, shredded, into their horses' feed to bring up the gloss on their coats (G E Evans. 1966). But they believed it had supernatural powers as well – the association with mandrake (see **WHITE BRYONY**) was evident here, for they said it had aphrodisiac qualities for both man and horse.

A syrup made from the root was once used against asthma (Brownlow), and they (the roots) used to be applied as a plaster for rheumatism (Hulme) and gout (Whitlock. 1992). But these roots are irritant and acid, so dangerous to experiment with. Lindley described the applications as "stimulating plasters". Nevertheless, they continued to be used through the centuries. Pomet, for example: "the Root … apply'd fresh upon Contusions oer Wounds, stops the Bleeding, and heals the Part; so that it has obtain'd the Name of Wound-root". It was called Chilblain-berry, too. The berries and roots steeped in gin were often applied to chilblains (Wiltshire; Vickery. 1995). Another name was Blackeye-root (North). Gerard reported that the roots "do very quickly waste away and consume away blacke and blew marks that come of bruises and dry-beatings" (a French name for the plant is Herbe aux femmes battues"! (Baumann)). Tetterberry, or Tetterwort would indicate another medicinal use, a tetter being some kind of skin eruption. Then there are Murrain-berry (Britten. 1880), or Murren-berry (W H Long), showing some connection between the plant and the cattle disease of that name, probably foot-and-mouth disease in modern terms.

BLACK HAW

(*Viburnum prunifolium*) An American species, from the eastern side of the continent. The root was given for female troubles in Alabama (R B Browne), or rather, as a permanent contraceptive. It was quite a complicated procedure, for the roots had to be dug from the north and south sides of the plant. In a separate bottle, some tea from *Ceanothus americanus* roots, mixed with red pepper and a teaspoonful of gunpowder, had to be kept. Every time the moon changed, a little from each bottle had to be taken (Puckett).

BLACK MUSTARD

(*Brassica nigra*) This is the mustard used as a condiment. It first appeared in Britain in 1720, but it is apparent from an edict of Diocletian of AD301 that it was already regarded as a condiment by that time, at least in the eastern parts of the Roman empire (Fluckiger & Hanbury); further west, though, it seems to have been used more as a medicine than a condiment (Lloyd). It was apparently used as a symbol of indifference (Leyel. 1937).

Mustard seeds were chewed for toothache, taken internally for epilepsy, lethargy, stomach ache and as a blood purifier (Lehner). Externally it was used in the form of a poultice as a powerful stimulant, though it was rarely used in the pure state, and would usually be found to contain some white mustard. Anyway, it would be dangeroius to leave a mustard plaster on too long, as it is such an irritant. They would be used for the treatment of rheumatism, sciatica, etc. The American Indians used it medicinally, too, in spite of the fact that it is not a native plant there. Some groups ground the seed to use as a snuff for a cold in the head (H H Smith. 1928). Mexican Indians have used it for a children's cough remedy, by heating the oil from the seed and rubbing it on the chest, which was

then covered with a flannel cloth (Kelly & Palerm), a remedy that sounds very like similar practices in Europe. But people like the Totonac (Mexico) use the seed to cope with something they call "malviento", almost like illness caused by an evil spirit, or an evil eye. For this, they burn the seed, and blow the smoke on the victim, or the patient has to bathe in water in which the plant has been rubbed (Kelly & Palerm). The seed was used in the southern states of America in black "conjure" ceremonies to break up a home, or to protect one from "conjure" (Hurston).

BLACK NIGHTSHADE

(*Solanum nigrum*) In some parts of Europe, the leaves used to be put in babies' cradles, the idea being that they would soothe them to sleep (Grieve. 1931). There may be some justification for this, for the generic name, Solanum, comes from a word meaning 'to soothe' (Young). Some South American Indian peoples use this plant for insomnia, by steeping a small quantity of the leaves in a large amount of water (Weiner).

The main folk medicinal use is for skin complaints, an ancient practice. "Dioscorides writeth, that Nightshade is good against S Anthonies fire, the shingles, …" Gerard wrote, while still warning his readers of the dangers of using such a toxic plant. We find this use against erysipelas, for that is what St Anthony's fire is, in America, too. In Mexico, for instance, the Totonac grind the whole plant, add salt and lime juice to it, and apply it as a plaster (Kelly & Palerm). In South Africa, too, a paste made of the unripe berries is in general use as an application to ringworm (Watt & Breyer-Brandwijk). Sunburn is treated in Indiana by crushing the leaves and stirring them in a cup of cream. When ready, put the cream on the sunburned area (Tyler).

BLACK WALNUT

(*Juglans nigra*) Or American Walnut. It is said in America that no plants will grow in the shade of a Black Walnut. Some research has been done on this, and it seems that the roots do secrete some substance harmful to other plants (Baker. 1977). The leaves, too, will keep away house-flies (Bergen. 1899). A dozen or so of these leaves boiled in a quart of water, with a teaspooon of sulphur added, is an Alabama eczema cure (R B Browne), and rubbing ringworm with the inside of a green walnut will cure it, according to Illinois practice (Hyatt). One other belief, once common in Missouri, is that a walnut carried in the pocket will prevent rheumatism (Bergen. 1899).

BLACKBERRY

(*Rubus fruticosus*) It is said that blackberries were a taboo food in Celtic countries, and that when the taboo was only a dim memory, explanations were thought up - that the fruits are poisonous, that they belong to the fairies, or that the bush was chosen for the Crown of Thorns, and the berries are Christ's blood. There is still a remnant of the taboo left in the superstition found all over the southern counties of England that blackberries have a sell-by date, after which the devil is supposed to put his foot on them (Graves), or that he "has been on them", or "spat on them" (Widdowson). The devil bears this grudge against blackberries because when he was expelled from heaven (on 10 September, apparently), he landed in a blackberry bush on his way to hell (Briggs. 1980). That cut-off date is usually Michaelmas, so that it is unlucky to gather them after this date, for in some places they used to say that October blackberries are actually poisonous. But the date is rather variable – it could be interpreted as Old Michaelmas Day (10 October), or in some places the day is given as SS Simon and Jude (28 October) (Folkard). The devil arrives earlier in Devonshire, though – 20 September is the day he spits on the blackberries (Whitlock. 1977). In Ireland, it used to be said that it was the fairies, or pooca, who passed over both blackberries and sloes, on November Eve, and made them unfit to eat. Eat them after that and there is a risk of serious illness (Gregory. 1970). All this is reasonably justifiable, for after the first frosts, blackberries are often tasteless and watery, and so not worth the picking. When they get to this state, the devil is said to have "cast his club over them" in Derbyshire (Addy. 1895), and in Scotland, "thrown his cloak over them" (Folkard). Another explanation is that the green bug that infests bramble-bushes in late autumn is called pisky in Cornwall, and it is pisky that spoils the fruits (Courtney. 1887).

Shakespeare used blackberries as symbols of worthlessness, and the way it grows makes it an emblem of lowliness. Yet another of its symbolic attributions is that of remorse, from the "fierceness with which it grips the passer-by" (Dyer. 1889). A propos of that quality, there is a legend to explain it: the cormorant was a wool merchant. He entered into partnership with the bramble and the bat, and they freighted a large ship with wool. She was wrecked, and the firm became bankrupt. Since then, the bat skulks about till midnight to avoid his creditors, the cormorant is for ever diving into the sea to discover the sunken vessel, while the bramble seizes hold of every passing sheep to make up his loss by stealing the wool (Dyer. 1889). Another version has the heron, cat and bramble as partners who bought the tithe of a certain parish. The heron bought the hay, but lost it all in a storm, the cat bought the oats, but the mice got them, and the bramble bought the wheat, but never got the money from the person to whom it sold the crop. That is why bramble takes hold of everyone and says "Pay me my tithe", for it forgot to whom the wheat was sold (Owen).

There are many superstitions connected with brambles. The dream books were full of them. To dream of passing through places covered with brambles means trouble; if they prick you, secret enemies will do you an injury with your friends; if they draw blood, expect heavy losses in trade. Or the latter can mean many difficulties, poverty and privation all your life (Raphael). To dream of passing through brambles unhurt means a triumph over one's enemies (Folkard), or troubles, but only short-lived ones (Raphael). Gathering blackberries is a sign of approaching sickness. If you see others gathering them, you have enemies where you least expect it.

In Guernsey, wreaths of brambles were hung from the rafters to drive off witches, the idea being that they would get scratched while flying through the air (MacCulloch). Sometimes, a sprig of bramble was used like rowan to put in or under the milking pail to ensure that the substance of the milk could not be taken by some evil agency – even burning them offers protection, for it is said that a bridal bedchamber ought to be fumigated as a safeguard against ill-wishing (Boland. 1977). When a girl's dress gets caught by a bramble while she is out walking with her boy-friend, it is taken as a sign that he will be faithful to her (this from Lancashire) (Harland & Wilkinson). There must have been some death associations, for there is a record of the young shoots being used to bind down the sods on newly-made graves (Folkard), and presumably this is not just a utilitarian measure. They used to say in Cornwall that blackberry stains could only be washed away when their season was over, just as, in Portuguese folklore, a scratch from a bramble will only heal when the bush is cut down and burned (Gallop). But a good blackberry crop foretold a good herring season (Deane & Shaw). There is a belief that when a bramble blooms early in June, an early harvest could be expected (Swainson. 1873). A Yorkshire tradition tells that an abundance of blackberries in autumn foretells a hard winter to come (Gutch. 1901), on the "many haws, many snows" basis. But at least, the weather is usually good when the blackberries ripen, and that period at the end of September and beginning of October is quite often called the blackberry summer (Denham. 1846).

Welsh children who were late learning to walk were made to creep under a bramble (Trevelyan), but this was by no means just a Welsh custom, even if the reason for carrying it out differs. In Devonshire, they say to cure boils, you have to find a bramble "growing on two men's lands", that is, roots on one man's land, grown over the hedge, and rooted on the other side, on someone else's land. The patient had to creep under it three times (Crossing); similarly in Cornwall, creeping under a bramble was done for boils (Grieve. 1931). Or in Zennor, blackheads were treated by crawling nine times round a bramble (Deane & Shaw). To cure boils in Dorset, you had to creep under a bramble three mornings running, against the sun, while for whooping cough, a child is passed nine times under and over (Udal), while this rhyme is recited :

> Under the briar, and over the briar
> I wish to leave the chin cough here (Raven).

That usage is mentioned by Aubrey, too (Aubrey. 1686/7). The remedy in Warwickshire was to pass the child three times beneath a "moocher", as it was called – a bramble that had bent back to root at both ends (Palmer. 1976). The Essex whooping cough remedy was to draw the child under "the wrong way", presumably, that is, by the ankles (Newman & Wilson). In the Midlands, the child had merely to walk under the bramble arch "a certain number of times" to cure his whooping cough (*Notes and Queries; 1853*). In Somerset, a child was passed through, apparently for hernia (Mathews). As far away as the Balkans, the blackberry arch was negotiated for illness – jaundice in this case (Kemp), and the custom was known in America, too, for, of all things, colic (H M Hyatt). On the Welsh border, an offering of bread and butter was put under the arch after the child had passed through, and sometimes, the patient had to eat the bread and butter while the adults present recited the Lord's Prayer. The rest of the food was given to an animal or bird on the way home – this would die, the disease dying with it (Baker. 1980). There were other, even stranger beliefs connected with a blackberry arch. For instance, if a shrewmouse ran over a horse it "enfeebled his hindquarters and made him unable to go". The cure was to drag the horse backwards through a bramble arch (Jacob), a procedure that must have been worth watching! From Ireland – if a man on Hallowe'en creeps under the long, trailing branches of the bramble, he will see the shadow of the girl he is to marry. If a gambler hides under a bramble, and invokes the devil's help, he will have luck at cards, no matter what colour he bets on. But there were genuine uses for the long shoots – binding thatch was one of them, and making beehives was another (C P Johnson. 1862). Its rate of growth is utilised in a Portuguese belief. A woman who wants her hair to grow longer cuts the tips off and puts them on a bramble shoot. As the bush grows, so will her hair. There is a snag, though. This is sympathetic magic, and it works both ways. If perchance the bush is cut down, her hair, in sympathy, will wither at the roots (Gallop).

There are a lot of medicinal uses for bramble leaves, some perfectly rational in view of the high tannin content, others no more than charms, like this Cornish use in cases of scalds and burns. Nine leaves are moistened in spring water, and these are applied to the affected part. While this is being done, the following charm has to be recited three times:

> There came three angels out of the west.
> One brought fire,and two brought frost;

> Out fire, in frost;
> In the name of the Father, Son and Holy Ghost
> (Hawke. 1981).

In Somerset bramble tips are used for bronchitis simply by peeling a shoot and nibbling it if the cough starts (Tongue.1965). Bramble vinegar, (made with the fruit) used to be made in Lincolnshire for coughs and sore throats (Gutch & Peacock), and the decoction of the tips with honey was an old sore throat remedy (Hill. 1754) (so is blackberry jam (Page. 1978)). Langham's *The garden of health* … was written in 1578, and we can find something very similar there: "the new sprigs … doe cure the hote and evill ulcers of the mouth and throat and the swellings of the gums, uvula and almonds of the throat, being iften chewed …". Equally efficacious is the use for diarrhoea, for both leaves and the root bark contain a lot of tannin, and so are astringent enough to be useful. Is that why bramble leaves are chewed to stop toothache? (Hatfield. 1994).

They say in Dorset that an ointment made from bramble tips and primroses is excellent for curing spots and pimples on the face (Dacombe), an interesting remedy in view of the fact that as far away as the Balkans, blackberry roots boiled slowly are the remedy for skin diseases. So are leaves in olive oil (Kemp), and on Chios it is bramble leaves that are used to treat a suppurating wound (Argenti & Rose), while in Scotland a poultice made from the leaves is a recognised cure for erysipelas (Beith). Gerard had: "the leaves of the bramble boyled in water, with honey, allum, and a little white wine added thereto, make a most excellent lotion or washing water …" The same preparation, he went on, was a "present remedy against the stone" – so, according to Pliny, are the berries and flowers as a decoction in wine. But what did Gerard mean by "they heale the eies that hang out … if the leaves be laid thereunto"? Something else that must be taken with a degree of scepticism is the use against heart trouble – according to Apuleius, in a 12th century manuscript, the treatment was simply laying the leaves on the left breast. Finally, and something that no herbalist would quarrel with - the berries are good for anaemia (Conway).

BLACKCURRANT

(*Ribes nigrum*) Used in folk medicine against sore throat (a quinsy, or squinancy), long before cultivation, and it is still so used, even for whooping cough, for which blackcurrant tea is taken in Cumbria (Newman & Wilson). A wine or jelly usd to be made in Yorkshire from the fruit, and set aside for sore throats (Nicholson), while the juice from the berries is a diarrhoea remedy (W A R Thomson. 1987). Other medicinal uses are more esoteric. What, for example, lies behind the statement that "their medicinal value lies in their ability to improve the eyesight …"?

(Schauenberg & Paris). In Ireland, it was used as a remedy for hydrophobia (Wood-Martin). Wesley was able to recommend it for similar misfortune: "a venomous sting, …, take inwardly, one Dram of black Currant leaves powder'd. It is an excellent counterpoison".

BLACKTHORN

(*Prunus spinosa*) Blackthorn is the first of the hedgerow shrubs to come into bloom in the spring, the white flowers appearing, according to where in the country it is growing, any time between mid-March to mid-April. One would think it would be a welcome and auspicious plant, but it is a thoroughly unlucky one. Even its early blooming brings talk of a "blackthorn winter", but evidently not from those who believed it would bloom on old Christmas Eve (*Folklore. vol 39; 1928*). There are often some warm days at the end of March and beginning of April which are enough to bring it into flower, and they are nearly always followed by a cold spell, the Blackthorn Winter. "Beware the Blackthorn Winter", or "blackthorn hatch" as it is sometimes called (Clinch & Kershaw)), is a well-known admonitory saying. The north-east winds that seem to prevail in the spring, about the time the blackthorn is in flower, were known as "blackthorn winds". A blackthorn winter means a spoiled summer, they say in Somerset (Tongue. 1965). Sometimes, it seems, there is a second blackthorn winter, which is said to fall in the second week in May. This may be just coincidence, for the festivals of the Ice Saints (Mamertus, etc) fall then (Jones-Baker. 1974). Even when the sloes themselves appear, there is foreboding. Is it not said that:

> Many haws, many sloes,
> Many cold toes ? (Denham).

The more berries there are, the worse the coming winter is said to be, and the more sinister the result. Note the Devonshire rhyme:

> Many nits [nuts]
> Many pits;
> Many slones,
> Many groans (Choape).

The flowers are extremely unlucky things to bring indoors; if it comes to that, virtually any white flowers in the house being dire results. But more fuss seems to be made about the blackthorn than anything else. It is just as bad to wear it as a buttonhole. Sussex people looked on it as a death token (Latham); in Suffolk, too, they used to say that it would foretell the death of some member of the family (Gurdon). And in Somerset, it would mean you would hear of a death (Tongue. 1975). Of course, all this might possibly amount to preventive superstition – Vickery pointed out that a scratch from the fierce thorns could very well cause blood poisoning (Vickery. 1985). Against this is a Lincolnshire belief that is not

quite a death token – at Alford, in that county, they used to say that blackthorn flowers indoors would result in the relatively lesser misfortune of a broken arm or leg (Gutch & Peacock). With all this ill-luck attached to the blackthorn, it seems almost logical that whitethorn, which is almost universally regarded as a good influence, should be thought the more dominant. If the two grow near each other, it would be the whitethorn that would destroy the blackthorn, so it used to be said (Wiltshire). It was once the custom in some parts to put May morning garlands of various plants over girls' doors. Each plant conveyed a message – the opinion of the villagers about any girl's behaviour was made explicit. Whitethorn was the most complimentary, but blackthorn, according to one authority (Tynan & Maitland) was reserved for a shrew (nettle was the worst of insults). But of course, blackthorn is a fairy tree, under the protection of a special band of them, said by Irish people to guard them specially on 11 November, which is Samhain old style, and 11 May, Beltane O S. They would let no-one cut a stick from it on those days. If anyone tried to, then he would be bound to suffer misfortune (Wentz). An extension of the fairy belief in Ireland is that they are supposed to blight the sloes at Samhain, just as the devil spits on blackberries at some time usually a little before that. So the sloe that was put in a County Roscommon Hallowe'en cake was the last eatable sloe of the year; the recipient of that was reckoned to be the longest liver (Byrne), by some convoluted reasoning.

The thorns were the ones used to stick into wax images made for black magic, and blackthorn wood was often used for a witch's walking stick (Wiltshire). This "black rod" carried by witches caused miscarriages, and when Major Weir was burned in Edinburgh in 1670, a blackthorn staff was burned with him as the chief instrument of his sorceries (Graves). The Lay of Runzifal makes a blackthorn shoot out of the bodies of slain heathens (and a white flower from fallen Christians) (Dyer. 1889) – the latter being logically the whitethorn, presumably. A blackthorn stake is to be used to impale a vampire (Kemp); this is a widespread belief in the Balkans. From the same area comes the corollary – the tree itself should never be injured or cut down (Kemp). It seems to have been buried with corpses in Ireland (Ô Súilleabháin). In Normandy, they used to say that fleas could be acquired by the ill-wishing of a witch. The only way to break the spell was to go down to the river before sunrise, and to beat one's shirt for an hour with a branch of blackthorn (W B Johnson), an action that seems to suggest that a blackthorn stick is a good thing after all. So it was in Slav folklore – it was a protection, and bits of the plant were carried sewn into the clothing (Lea). In Irish folklore, too, a blackthorn stick was used to overcome evil spirits (Ô Súilleabháin); this is the traditional knobbly walking stick known as the shillelagh (Hart), although it is claimed that the shillelagh was always an oak cudgel (Sheehy & Mott). Gypsy men often carry a blackthorn walking stick as a protection against any kind of danger; it is a traditional charm against ill-wishers (Boase), and, apparently, against mildew in a wheat crop. Part of a blackthorn branch would be burned in a large fire in the field, and the remainder hung up in the house (Drury. 1992). Blackthorn is the accepted timber with which Irish tinkers fight at fairs. Robert Graves said that not for nothing was it called bellicum in Latin. This is an extremely tough wood, tough enough to make flail swingels from, holly being equally tough, and equally popular for the job (the handles were usually made of ash).

The Crown of Thorns was sometimes said to have been made from the blackthorn (Graves). A Worcestershire New Year morning custom used to be to make a blackthorn crown, which was baked in an oven until calcined. This ash was then taken to a cornfield and scattered, to ensure a plentiful crop. In Herefordshire, scorched blackthorn was mixed with mistletoe as a Christmas decoration to bring good luck (R L Brown). Like the Glastonbury Thorn, blackthorn was said to bloom at midnight on old Christmas Eve (M Baker. 1980).

It is still known as the wishing thorn, as it is the tree from which wishing rods were cut (Philpot). This is from Germany, but there is a similar idea that was current in Wales. Ffynon Saethan (Caernarvonshire) was visited (on Easter Monday?), and blackthorn points thrown into the water. If they floated, the lover was faithful, but if they sank, matters were rather more doubtful. Similarly, at the Silver Well, Llanbethian, in Glamorgan – if the blackthorn floated, the lover was faithful; if it whirled round, he had a cheerful disposition; if it sank a little, he was stubborn, and if it sank out of sight, he was unfaithful. If a number of thorn points slipped into the well from the visitor's hand, then it showed that the lover was a great flirt, and so thoroughly unreliable (F Jones).

Getting rid of warts by rubbing a snail on them and then impaling it on a blackthorn used to be common practice; or, from East Anglia, you could rub the wart with a green sloe, and then throw the slow over your left shoulder (Glyde). They are both transference charms; cattle doctors in Worcestershire used to cure footrot by cutting a sod of turf from the spot on which the animal was seen to tread with its bad foot, and then to hang the turf on a blackthorn. As the sod dried out, so would the hoof heal (Drury. 1985). To rise above the level of charms, there were some quite genuine folk remedies involving sloes – in North Wales they were used for a cough cure (Friend. 1883); so they were in the Highlands, too, for sloe jelly was reckoned the best cure for relaxed throat (Grant), while the juice of boiled sloes was an East Anglian

gargle for a sore throat (V G Hatfield. 1994). Sucking a sloe is said to cure gumboils (Addison & Hillhouse). And a gypsy remedy for bronchitis involves peeling the bark, boiling it in a saucepan of water, and then allowing it to cool. Add sugar, and then drink it when needed (Page. 1978). In Sussex, the inner bark is scraped off and made into a tea to be taken for various ailments. Equally varied and unspecified are the disorders for which sloe wine used to be taken in Northamptonshire (Friend. 1883). Blackthorn leaves were used in Ireland as an indigestion remedy, or to cure "summer fever" (Ô Súilleabháin), while Thornton said that ague could be cured sometimes with the powdered bark. He also reckoned that an infusion of a handful of the flowers "is a safe and easy purge", but the Welsh belief that if a person ate the first three blackthorn blossoms he saw, he would not have heartburn all through the year (Trevelyan), can only be classed as superstition, not even a charm.

Sloes will give a slate-blue dye with no mordant, and sloe juice is indelible, as careless handling during the making of sloe gin will prove! Juice squeezed out of the unripe fruit was sold at one time under the trade name German Acacia, and used to mark linen – an ideal laundry marking, in fact. Sloe gin is the best-known use of the fruit, but they have been used in a mixture of various ingredients that was sold as choice old port! (Hulme). The leaves have been used as an adulterant of tea, notoriously so, in fact, in Victorian times ('sloe poison', Punch was moved to call tea in the 1870's). C P Johnson said that at one time four million pounds of blackthorn leaves were packaged up in a fraudulent attempt to sell them off as genuine China tea! They were also an Irish substitute for tobacco (Ô Súilleabháin). Sloes were used for fevers at one time in the Highlands, and the flowers as a laxative (Beith). Can that be feasible? Blackthorn thorns, in infusion, is an Irish cure for diarrhoea (Buckley). Tusser recommended the berries for veterinary use:

> Keepe sloes upon bow,
> For flixe of thy cow.

Later in his five hundred points, he gave instructions to:

> "seeith water and plump therein plenty of sloes,
> Mix chalke that is dried in powder with thoes,
> Which so, if ye give, with the water and chalke,
> Thou makest the laxe fro thy cow away walke".

Norfolk pig-farmers used to hang the afterbirth on a blackthorn tree after a sow had farrowed, "so that the pigs do well" (Norfolk FWI).

The thorns have been used as fish-hooks – in fact they survived in Wales until fairly recent times. They were put in an oven for some days to harden the points, and were fitted to hand-lines or long-lines that were particularly effective for catching flatfish.

The lugworm that was used as bait was threaded from the bottom upwards and over the point of the thorn (Jenkins).

And:

> "When the sloe is white as a sheet,
> Sow your barley, whether it be dry or wet" (M E S Wright)

BLADDER CAMPION
(*Silene vulgaris*) The "bladder" is the inflated calyx, which snaps when suddenly compressed, and so could be used as a sort of love-charm, according to Coles. The degree of success depended on the loudness of the pop. There was a strange name for the plant in Dorset – White Flowers of Hell (Macmillan). Apparently, there was a superstition that the leaves and "bladders" were poisonous. Actually, children often eat the young leaves, which are supposed to taste like green peas (Jordan). They have even been used as a substitute for asparagus.

Gerard indulged a fantasy about this plant – he said that this plant, which he called Behen, was a protection against the "stinging of Scorpions and such like venomous beasts. Insomuch that whoso doth hold the same in his hand can receive no damage or hurt by any venomous beast". On a more rational level, he prescribed the root decoction to help strangury and "paines about the neck and hucklebone". Far more practical is the gypsy use of the plant – the leaves were applied externally as a poultice to cure erysipelas (Vesey-Fitzgerald).

BLADDER COMPLAINTS
By the doctrine of signatures, BLADDERWORT, of course, would be used to treat bladder complaints (B L Bolton). Gerard recommended BROOKLIME leaves, "stamped, strained, and given to drinke in wine" to help with strangury, "and griefes of the bladder". So did Langham, and, much later, Thornton. In Irish folk medicine, the decoction, either alone, or mixed with watercress, was taken for gravel, and urinary disases generally (Egan).

BLADDER SENNA
(*Colutea arborescens*) The pods of this leguminous plant are inflated, hence "bladder", of the common name. A laxative, of course, but Sir John Hill's words have to be heeded, for "… they are very rough : they work both upwards and downwards, and are only fit for very robust constitutions …".

BLADDERWORT
i.e., GREATER, or COMMON, BLADDERWORT (*Utricularia vulgaris*) By the doctrine of signatures, these can, of course, be used for bladder complaints (B L Bolton).

Blighia sapinda > AKEE

BLOODROOT
(*Sanguinaria canadensis*) A North American plant whose rootstock produces a reddish juice, used by the Indians as a dye, for basketry decoration (Speck), for clothes, for weapons (Sanford), or for staining their face or bodies, by which it served not only as a decoration, but kept away insects, too (Lloyd). One other use – with various other plants, they were used to dye porcupine quills a brilliant scarlet (Densmore).

It was also used as a love charm. A man would rub some of the root on the palm of his hand, and then contrive to shake hands with the girl he wanted, in the belief that after five or six days she would be willing to marry him (Corlett).

Small doses of the rootstock were given as a tonic, mild stimulant, and to induce sneezing and vomiting (Sanford). An overdose must have been dangerous (Lindley), as it has been described as "an arterial sedative" (O P Brown). Leighton mentions it as a powerful emetic. The Indians of the Mississippi region drank a root tea for their rheumatism (Weiner). In what seems to be an example of the doctrine of signatures, Cherokee medicine men used bloodroot in remedies for regulating menstruation (Cunningham & Côté), and for anaemia (B L Bolton). There is a recipe for cramp from Alabama: mix one teaspoonful of crushed bloodroot with half a cup of vinegar and four teaspoons of sugar. Heat to boiling, strain, and give up to one teaspoonful every half an hour (R B Browne). The Cherokee used a preparation of bloodroot for treating breast cancer. This was taken up in a big way in the 19th century by the cancer quacks in the USA. It has been tested, but there was insufficient evidence to prove the genuine case, but it became quite popular for the treatment of warts, and nasal polyps (Thomson. 1976).

BLUE FLAG
(*Iris germanica*) Bits of the dried rhizome were once given to babies to help their teething (Schauenberg & Paris), and the sliced roots, like those of Yellow Iris, were once applied to the skin for cosmetic effects, mainly to get rid of freckles (Le Strange). The Anglo-Saxon version of Dioskorides recommended the root for snake bite, and "for kernels and all evil lumps" (Cockayne). It is still being used, with camphor, for swellings, in the Balkans (Kemp).

BLUE LOTUS
(*Nymphaea caerulea*) The ancient Egyptians believed that the creator had sprung from a lotus flower, so from very early times it was a symbol of reincarnation. It has been found in garlands laid on the bodies of the dead (Tutankhamun included) (Grigson. 1976). It was sacred, too, because of its perfume (Genders. 1972).

Both rhizomes and the seed are eaten by the Africans on the Guinea coast, and the seed has also been used as a remedy for diabetes (Watt & Breyer-Brandwijk). One of the Mashona witch doctor's remedies for a persistent cough in adults consisted in powdering the bark of *Pterocarpus angolensis* with the inner bark of this waterlily root. The powder is soaked and given to the patient in his porridge (Gelfand. 1956).

BLUE PASSION FLOWER

(*Passiflora caerulea*) A South American plant that was introduced into England in 1629 (Gordon. 1988). It eventually became the ecclesiastical symbol for Holy Rood Day, whether it was the spring celebration of the festival, 3 May, or the autumn day, 14 September, for this is the flower that stood for the emblem of Christ's Passion. It is said that when the Spaniards, obviously gifted with extraordinary imagination, first saw it in the New World, they took it as an omen that the Indians would be converted to Christianity. The descriptions vary somewhat. One of them is that the three styles represent the three nails, and the ovary is a sponge soaked in vinegar. The stamens are the wounds of Christ, and the crown (located above the petals) stands for the Crown of Thorns. The petals and sepals indicate the Apostles (Bianchini & Corbetta). Another description has it that the ten white petals show the Lord's innocence; the outer circlet of purple filaments symbolise his countless disciples; the inner brown circlets the Crown of Thorns; the ovary is either the chalice he used at the Last Supper or the column to which he was tied, or the head of the hammer that drove in the nails; the five anthers are his wounds; the three divisons of the stigma the nails with which he was fastened to the Cross, and the tendrils are the lashes of the scourging, just as the leaves are the hands of those who reached out to crucify him (Whittle & Cook). The difficulty is that virtually every description of the symbolism is different.

BLUEBELL

(*Hyacinthoides nonscripta*) Bluebells are fairy flowers. In Somerset they used to say that you should never go into a wood to pick them. If you were a child you may never come out again, and if you were an adult you would be pixy-led until someone met you and took you out (Briggs. 1967), and in Devonshire it is one of the flowers thought to be unlucky to bring indoors (*Devonshire Association. Transactions. vol 65; 1933*). The bluebell is appropriated to St George (23 April) (Geldart), and it was once the custom, according to some, to wear bluebells on that day, and to decorate churches with them (Gordon. 1985). Surely they were not talking about this bluebell? There are other flowers with the name, Harebell for example, the Bluebell of Scotland:

On St George's Day, when blue is worn,
The blue harebells the fields adorn (Friend. 1883).

That is all very well, but harebells are not naturally in flower in April. Even Early Purple Orchids are called

Bluebells in Dorset, and so are Periwinkles. So what did they have in mind for St George's Day?

Without compromising dock's pre-eminence, they used bluebell sap to put on nettle stings in the Wye valley (Waters. 1982). Gerard listed Sea Onion as a name for this Bluebell, not because they grew by the sea, but because in his time sailors took the bulbs to sea, to be eaten like onions (Genders. 1976) (but fresh bluebell bulbs are poisonous (North).

BO-TREE

See PEAPUL (*Ficus religiosa*)

Bocconia frutescens > TREE CELANDINE

BOG ASPHODEL

(*Narthecium ossifragum*) It is said in the Highlands that this plant (what Otta Swire called the little golden bog orchid, if that is correct identification, for she may have had in mind the rare Bog Orchid, *Hammarbya paludosa*) first grew by the Falls of Ragie, Strathpeffer, when the Virgin Mary, walking over the moor, stopped at some marshy ground to tuck up her robe. As she undid her golden girdle it slipped through her fingers into the peat bog and sank out of sight. Where it had fallen, the orchids grew (Swire. 1963).

The specific name, *ossifragum*, means "bone-breaker", because as it grows on wet moors and mountains, sheep pasturing there often suffered from foot-rot, and this was attributed to their browsing on these plants (Grieve. 1931); it "softened their bones", it was said in the Lake District (Vickery. 1985). In Norway they went further – not only that sheep and cattle that fed on the plant got their bones soft, but it was also said they could be rolled up and moulded into any desired shape! (Coats. 1975). The plant is called Cruppany Grass in Donegal, cruppany being the disease in sheep that the plant was believed to cause – "bone stiffness", seemingly quite different to the English and Scottish descriptions of the disease. But it is not all just superstition, for it is known to be toxic to both sheep and cattle.

One of the names for this plant is Lancashire Asphodel (Prior), and the classification at one time was *Asphodelus lancastriae*. In the same county it was called Maiden's Hair, for bog asphodel was, in the 17th century, the basis of a popular Lancashire hair dye (Putnam). After the yellow flowers have faded, the stems change to a saffron colour, and it is from these that the dye is extracted.

BOG COTTON

– see COTTON-GRASS

BOG ST JOHN'S WORT

(*Hypericum elodes*) John the Conqueror Root was a love charm favoured by men. It is a dried root with a prong or spike growing out of it, an obvious piece of phallic sybolism. It was carried in a little chamois leather bag, or one made of red cloth. What the root was was kept secret by those who sold the charm, but it is said to be Bog St John's Wort.

BOGBEAN
see BUCKBEAN. Bogbean is often preferred as the common name, but must presumably be a transformation of buckbean to suit its habitat (Grigson., 1974)), but it is a very widespread local name just the same.

BOILS
A LEEK poultice was used to "mature" a boil, according to the Welsh medical text known as the Physicians of Myddfai, and of course ONIONS have always been used too. A poultice (with or without black treacle) is a widespread treatment for a boil, not only in Britain, but in the Aegean (Argenti & Rose), and in America (Sackett & Koch). Irish country people used cooked NETTLES to cure boils (O'Farrell). GROUNDSEL poultices were commonly used for boils; there are records from Dorset (Dacombe), Norfolk (Randell), the Highlands (Grant) and northern Ireland (Foster), and doubtless many other areas. Just as commonly used were DOCK leaves, either those of Broad-leaved Dock or of Curled Dock. The Blackfoot Indians used the latter as a poultice (Johnston), and dock tea was another way of dealing with them (Fernie). COMFREY, at least in County Kerry, has been used, too, the roots being made into a poultice. The belief was that the comfrey drove away the worms in the boil. They could not stand the smell of the stuff (Logan). DWARF ELDER was used in the Balkans; the leaf decoction was taken internally as well as being used externally (Kemp). RIBWORT PLANTAIN is mentioned as a Highland remedy for boils and bruises (Grant), and it is common knowledge that a MALLOW or MARSH MALLOW compress should be used to deal with them. A compress made by cooking FENUGREEK seeds until they were like a paste or porridge, was used hot on boils, abscesses and the like (Flück). A hot poultice of CATMINT leaves was used in Alabama (R B Browne).

OATMEAL poultices were part of Scottish domestic medicine for the cure of minor boils and suppurations. The oatmeal would be prepared with water and a slight dressing of salt butter. For "difficult" boils urine would take the place of butter (Beith). Another possibility is to use the infusion of SCENTED MAYWEED for such a poultice (Flück), and in Somerset a poultice of GOOSE-GRASS was the remedy (Tongue. 1965).

VIOLET leaf plasters were used for ulcers and boils – there is a recipe from the 15th century for "hot botches", described as inflamed boils: "Take violet, and stamp it with honey and vinegar, and make

thereof a plaster; and anoint the head [of the botch] on the beginning of its growing with the juice of violet, and then lay on the plaster" (Dawson). The condition used to be treated by Irish country people by applying a MULLEIN leaf roasted between dock leaves, and moistened with spittle – as long as the spittle be that of an Irishman! (Egan). The inner leaves of a CABBAGE formed the ideal medicament to draw a boil, according to belief in Shropshire and Cheshire (Barbour). An Irish cure was to use a pain-killing poultice made by mixing HEMLOCK leaves with linseed meal, and this could double up as a preparation to put on boils (Maloney). A CHICKWEED poultice used to be a common Dorset remedy for boils (Dacombe), as it was in Ireland, too (Barbour).

In Devonshire, they say that to cure boils, you have to find a BRAMBLE "growing on two men's lands", that is, roots on one man's land, grown over the hedge, and rooted on the other side, on someone else's land. The patient had to creep under it three times (Crossing); similarly in Cornwall, creeping under a bramble was believed to cure boils (Grieve. 1931). In Dorset, the cure was to creep under a bramble three mornings running, against the sun (Udal). Even amulets have been used to combat boils. A CAMPHOR bag round the neck was used in Maine (Beck), for example, and so was a NUTMEG, but it had to be nibbled nine mornings fasting (Hawke). In Devonshire, a little more ritual was needed. The patient must be given the nutmeg by a member of the opposite sex. He carries it in his pocket, and nibbles it from time to time. Only when the nutmeg has quite disappeared will the boils have gone (*Devonshire Association. Report. vol 91; 1959 p199*). SLIPPERY ELM bark could be used to put on a boil. For example, mixed with lard, it can be put straight on the boil (R B Browne), or a paste could be made by pouring boiling water over the bark, and that could be put on (R B Browne, Beck), and splitting a fig and applying that as a poultice was the practice in Indiana (Tyler). From Alabama: "For boils, take equal parts sumac, sage and swamp-lily (that is, ATAMASCO LILY) root, and boil into a strong infusion, strain, and put in a cupful of lard and fry until the water is out. Apply on a flannel cloth" (R B Browne). The Kiowa Indians treated them simply by rubbing the leaves of POISON IVY over the boil. The resultant dermatitis would last about as long as the boil. So the disappearance of the two afflictions together may explain this strange practice (Vestal & Schultes).

Shona witch doctors were reported to use a medicine involving the CANDELABRA TREE (*Euphorbia ingens*) to treat boils. One way was to to crush the shell of a particular snail into a powder and mix it with the milky sap, and apply that (Palgrave).

Bombax ceiba > RED SILKCOTTON TREE

BOMBWEED

A name given to Rosebay Willowherb (*Chamaenerion angustifolium*) in south-east England, from the way it so quickly colonised bomb sites during World War 2. (Mabey. 1998). Its seeds are usually the first to germinate in areas cleared of vegetation by fire, etc. That is why it is known as Fireweed in the USA.

Botrychium lunaria > MOONWORT

BOUNDARY OAKS

Boundaries were defined by oak trees from very ancient times in England, even from the days of mythology. The Cadnam Oak, a few miles from Lyndhurst, in the New Forest, was a famous "boundary tree" of the Forest.

BOX

(*Buxus sempervirens*) An evergreen, with many of the associations shared with other evergreens, notably its use at funerals. In the north of England, a basin full of box sprigs was often put at the door of the house, or sometimes at the church door, before a funeral, and everyone who attended was expected to take one to carry in the procession and then to throw it into the grave (Ditchfield. 1891). Or a table was put at the door, with sprigs of rosemary, or sometimes lavender, and box, for each mourner to pick up as he came into the house (Vaux). Box grown in Lancashire gardens used to be known as Burying Box. It was thrown into the grave in Lincolnshire, too, as a symbol of life everlasting. Small sprigs are sometimes found when old graves are disturbed; though dry and brittle, they are usually quite green (Gutch & Peacock). They have even been found associated with Romano-British burials in Cambridgeshire and Berkshire (Vickery. 1984). Presumably, this practice led to the Dorset superstition that a sprig of box in flower brought indoors meant that death would soon cross the threshold (Udal). Again, the Bavarian custom that Frazer mentions of putting a piece of boxwood in the holy water basin stems directly from the association with the dead. Box was said to have been one of the timbers used to make the Cross, the others being cypress, cedar and pine. In the Eastern tradition, olive and palm are substituted for box and pine (Child & Colles). Chaucer writes of it as a dismal tree, and described Palaman in his misery as:

> "Like was he to byholde,
> The Boxe Tree or the Asschen dead and colde"
> (Knight's Tale).

At Yport, Normandy, a branch of box that had been blessed by the priest was thrown into the sea on Palm Sunday, in a blessing of the sea ceremony (Salle). The Palm Sunday box also appeared in a Breton folk tale. When put in a baby's cradle, it protected the child, and stopped it being changed (Wentz). Box was used, too, to dress wells at Llanishen in Gwent, on New Year's

Eve (Baker. 1980). Another example of its worth as an evergreen is the fact that it was used to replace the Christmas decorations, from Candlemas to Easter (Dallimore). Herrick, *Hesperides*, notes the custom under Candlemas Day:

> Down with the rosemary and bayes,
> Down with the mistletoe:
> In stead of holly now upraise
> The greener box for show;
> The holly hitherto did sway,
> Let box now domineer;
> Until the dancing Easter Day,
> Or Easter's Eve appear.

In France, it had many of the same associations as the mistletoe (Salle), and there is an American love divination played with box leaves. Name some and lay them on a hot hearth. The one that swells and whirls towards you will be your future husband or wife. If one turns in the opposite direction, he or she will shun you (Whitney & Bullock).

In spite of the death connection, box is generally a plant that brings good luck, and charms away sorcery (Sebillot). To dream of it "augurs well for love affairs" (Dyer. 1889), and foretells long life and prosperity, with a happy marriage and large family (Raphael).

Box hedges used to be planted as a plague preventive, particularly in Dorset. It is said that traces of these borders planted in the 16th century can still be seen in Netherbury (Dacombe). Box is a febrifuge, still prescribed by herbalists and homeopathic doctors, who treat it as a substitute for quinine, in malaria. In the form of a tincture, it is given for rheumatism, and diarrhoea, as well as fever (Palaiseul). A volatile oil distilled from the wood has been prescribed for epilepsy, and has been used too for piles and toothache (Grieve. 1931). The latter was particularly mentioned by Hill, while Evelyn said it was used in the treatment of venereal diseases. In Ireland, the leaves were used as a remedy for hydrophobia (Wood-Martin); compare this with the 14th century recipe: "For bytynge of a wood hound. Take the seed of box, and stampe it with holy watyr, and gif it hym to drynke" (Henslow).

In the 16th century, it was being claimed (Langham) that "the lee wherein the leaves have been sodden or steeped maketh the haire yelowe (or auburn, according to Parkinson) being often washed therewith". It is quite startling to find, some three hundred years later, that box leaves still appear in a recipe for a hair restorer : Ash to prevent the hair from falling off; a quarter of an ounce of unprepared tobacco leaves, two ounces of rosemary, two ounces of box leaves, boiled in a quart of water …"(*Housekeeper's and Butler's Assistant; 1862*). Boxwood turners' chips were used by herbalists as the basis of "hysteric ale", which also contained iron filings, and a variety of herbs. It was recommended to be "taken constantly by vaporous women" (Wilkinson. 1973).

There is one veterinary usage recorded – the leaves were fed to horses at one time to cure them of the bots (Drury. 1985).

BOX ELDER

(*Acer negundo*) A North American tree, and a favourite source for a divining rod, at least in Iowa (Stout). Some native American peoples used the wood of this tree for charcoal for ceremonial body decoration and tattooing (Gilmore).

BOXTY BREAD

Irish boxty bread was made from POTATO flour, and cooked on the griddle, just as Shetland tattie bannocks were (Nicolson).

BRACKEN

(*Pteridium aquilinum*) It will protect the house from lightning if hung up inside, but if it is cut or burnt, it will bring on rain. And if you tread on the plant it will cause you to become confused, and to lose your way (Waring). Cut in two, the root was supposed to show the initial letters of a lover's name, a quite widespread superstition (see Leather, Courtney, Forby). And initial letters are evident in the Guernsey belief that if you want to win at cards, gather bracken in the early morning of St John's Day, and make a bracelet of it in the form of the letters MUTY, the bracelet to be hidden under the sleeve while playing (Garis). MUTY must be the first letters of some forgotten cryptogram. A lover's faithfulness could be tested by taking a stem and plucking off the fronds one by one. If the result was an odd number, everything was OK (Stevens). It used to be said in Shropshire that bracken flowers only on Michaelmas Eve at midnight, when it puts forth a little blue flower, which vanishes at dawn. But in Germany it was said that fern seed shines like gold on St John's night (J Mason).

Once established, bracken is very difficult to remove, and in the green stage is actually poisonous, though perfectly harmless once it has turned brown, when it makes an invaluable bedding for farm stock (Mabey. 1977). There are a few 'snake' names for bracken, to reinforce the toxic qualities. Perhaps it is only toxic to stock, for Tynan & Maitland insisted that the young fronds are edible, with a flavour that has been compared to asparagus. The Japanese certainly eat it (Rackham. 1986). Perhaps the toxicity is eliminated by cooking.

BRAMBLE

see BLACKBERRY. The word bramble is used as frequently as blackberry for the bush; the two words seem to be interchangeable, though it would be better to reserve blackberry for the fruit. There are occasions when bramble is used for the fruit as well – there may be blackberry jam, but there can never be anything but bramble jelly. The word originally meant anything

thorny (Ellacombe) (Shakespeare often used 'thorn' for bramble – *Midsummer Night's Dream*, iii, 2 - 'For briers and thorns at their apparel snatch', for example (Grindon). In Chaucer's time it had not yet been fixed on this shrub – he used it for Dog Rose, and there is to this day confusion between the words bramble and briar.

BRANDY BOTTLES

A quite widespread name for the YELLOW WATERLILY (Dartnell & Goddard; Grigson. 1955; Nall). Perhaps it arises from the shape of the fruits, which are like bottles, or because they are said to smell like the stale dregs of brandy (Grigson. 1955). Genders. 1971 prefers to liken the smell to that of ripe plums. It is due to a combination of acetic acid and ethyl alcohol.

Brassica campestris var. rapa (Brassica rapa)
> TURNIP
Brassica nigra > BLACK MUSTARD
Brassica oleracea 'capitata' > CABBAGE

BREADFRUIT

(*Artocarpus altilis*) A Polynesian tree originally, it is cultivated widely throughout Asia, but it is wild in the Andamans, where the islanders roast the fruits in hot ashes till the hard pulp and seeds are eatable (Cipriani), the flesh apparently tasting exactly like freshly baked bread. The tree was first discovered in Tahiti by Captain Cook's expedition, and they introduced it into the West Indies as a cheap diet for slaves (the *Bounty* voyage was for this purpose, and in fact Bligh actually did take the seedlings later to the Caribbean) (Brouk).

Several Polynesian islands have stories of the origin of breadfruit, all quite similar. The version from Hawaii tells how a man called Ulu died during a famine, and his body was buried near a spring. During the night, his family, who had stayed indoors, could hear the sound of dropping leaves and flowers, and then heavy fruit. In the morning, they found a breadfruit tree growing from his grave, and the famine was over (Poignant). Ulu in fact signifies an upright, i.e., male, breadfruit, which is called ulu-ku. The low, spreading tree whose branches lean over, is ulu-ha-papa, and is regarded as female. There are, of course, perfectly rational stories in Hawaii of the introduction of the breadfruit from other islands, though some of them suggest that it was taken from an island inhabited only by gods, and preserved for human use (Beckwith. 1940).

From the West Indies, where the tree was taken, there are one or two medicinal uses recorded; for example, a leaf decoction is taken in Trinidad to treat diabetes (Laguerre), and from Guyana there is a report of the decoction of yellow leaves being used to treat high blood pressure (Laguerre).

BREWING

GROUND IVY, for its tonic bitterness, was at one time much used in brewing, as a series of names, starting with Alehoof, testify. Alehoof means literally that which causes ale to heave, or work (Britten & Holland), though there have been other attempts at its etymology which may safely be ignored. There are variations on Alehoof, as with Allhoove, or Hove, and there is Tunhoof, tun being a cask (of ale). The place where such medicated beer was sold was known as a gill-house (Barton & Castle), and Gill, in many different forms, is still one of the names for Ground Ivy.

BRIAR

see DOG ROSE. Briar is a general name for the species, elaborated perhaps, as with the North-country Briar-rose or the Cheshire Bird-briar (Grigson. 1955), and varying into Breer or Brere, and there is a shortened form, Bree.

BRIGHT'S DISEASE

used to be treated in Irish folk medicine with MALLOW juice (Maloney).

BRISTLY OX-TONGUE

(*Picris echioides*) In the Gironde district of France, they used to say that to cure a prick made by the bristles of this plant (if the identification is correct), one had to get up and cut a head of the same plant, and throw it behind one, without looking round (Sebillot). The leaf of Bristly Ox-tongue is used as a vegetable in southern Europe (Watt & Breyer-Brandwijk), and was cooked in England as a potherb in medieval times (Harvey). Binding the roots to the affected part in varicose veins was recommended as a cure at one time (see Tusser).

Briza media > QUAKING GRASS

BROAD BEAN

(*Vicia faba*) Beans are fairy food. The Green Children captured near Wolfpits in Suffolk would only eat beans, but gradually became used to mortal food. The boy pined and died, but the girl lived, and eventually married a local man. By the same token beans are the food of the dead, or, according to another tradition, they contain the souls of the dead (Waring). At any rate, they were sacred to them. That is why they were prohibited in the Vedas as ordinary food. They were a favourite offering to the departed in ancient Greece, and for this reason they were forbidden to his followers by Pythagoras. Pliny says that beans were used in sacrificing to the dead because the souls of the dead were in them, and Ovid said that the witch put beans in her mouth when she tried to call up spirits. On the other hand, spitting beans at a witch was a Wiltshire way of rendering her spells ineffective (Whitlock. 1992). Eventually, they were used at funerals in classical times, a use that found its way to the north of England, where the tradition used to be (at least in the 1890s) that broad beans should always be buried with the coffin (Pope). Children used to recite:

> God save your soul,
> Beans and all (Tongue).

Presumably all this accounts for the Somerset saying that to sniff the scent of a beanfield is fatal. Similarly, in Leicestershire, it is said that to sleep overnight in a beanfield can cause insanity, or at least horrifying dreams. It is a fact that the heavy perfume has been known to cause fantasies during sleep (Genders). Dreaming of beans is reckoned to bring trouble of some kind (Gordon.1985), quite apart from the belief that they "cause vain dreams and dreadful" (Bartholomew Anglicus). Both North country and Midlands (Davies) miners insisted that colliery accidents were more frequent when the beans were in flower (Waring). And another superstition said that when beans grew upside down in the pod, it was an omen of some kind. They did so in the summer of 1918 apparently, and then it was remembered that the last time they did that was the year the Crimean war ended. Unfortunately the correspondent of *Notes and Queries* who reported it dated his letter in 1941, when once more they were growing upside down, but that war took another four years to end.

There were prescribed times for sowing beans, although they show wide variations. The favourite seems to have been St Thomas's Day (21 December) (Baker. 1974, Wright), but equally well known is the Somerset rule that they should be set in the Candlemas Waddle, that is, the waning of the February moon, or on the February new moon, as some say (Watson). Otherwise they would not flourish. Elsewhere a date a little later than this is preferred, according to the rhyme:

> Sow beans or peas on David or Chad,
> Be the weather good or bad;
> Then comes Benedict,
> If you ain't sown your beans –
> Keep 'em in the rick (Baker. 1977).

Warwickshire custom required bean-planting to start on St Valentine's Day, and they agree it must be finished by St Benedict. St Valentine's Day is 14 February, St David's Day 1 March, and Chad the day after. St Benedict is celebrated on 21 March. Usually three beans are put in each hole, though the Warwickshire rhyme gives four:

> One for the pigeon, one for the crow,
> One to go rotten, and one to grow (Savage).

There is quite a widespread belief in the efficacy of rubbing warts gently with the furry inside of a bean pod as a way to get rid of them – it has been recorded from East Anglia to Wiltshire and Somerset, (Randell, G E Evans. 1966, Whitlock. 1988) as well as from Cumbria. Sometimes one finds relics of a charm attached to this. For instance, in Essex, there is the injunction to throw the pod down a drain after rubbing the warts (Newman & Wilson):

> As this bean shell rots away
> So my warts shall soon decay (Hardy. 1878).

Just the smell of the flowers is enough to cure whooping cough, so it is still claimed in Norfolk (Hatfield. 1994). Burns are treated with broad beans on the Greek island of Chios. The beans are roasted, powdered and mixed with oil (Argenti & Rose).

A manuscript of 1610, published in the *Gentleman's Magazine*, prescribes bean flowers, distilled, as a lotion with which to wash the face, "and it will be fair". Gerard had already said that "the meale of Beanes clenseth the filth of the skin", and much later, Hill wrote that "it has been customary to distil a water from bean-flowers, and use it to soften the skin …". In China, they used the shoots, boiled in oil and salt, as a medicine to arouse a drunkard from his stupor (F P Smith).

BRONCHITIS

A Somerset bronchitis remedy was to make an ointment of lard and GARLIC, and rub it on to the soles of the feet at night (Tongue), a recipe that was recorded as recently as 1957. Garlic has always been in great demand for chest complaints, and there are similar prescriptions for whooping cough, coughs and colds, asthma, and even tuberculosis. Gypsies in this country have used CUCKOO-PINT, either by root decoction or by an infusion of the dry, powdered flowers, to cure croup and bronchitis (Vesey-Fitzgerald). Another gypsy remedy for this complaint involved peeling the bark, boiling it in a saucepan of water, and then allowing it to cool. After sugar was added, the liquor could be drunk as needed (Page. 1978). FENNEL tea is still taken sometimes for the complaint, and BALM tea has been recommended as an expectorant, as has a tea from dried HOLLYHOCK flowers (both Fluck), and PENNYROYAL tea is often taken (Beith). SWEET CICELY too, is taken for chest complaints and bronchial colds in popular medicine (Gibson), and so is a leaf tea of RIBWORT PLANTAIN (Conway), or GROUND IVY, which was always known as "Gill-tea", Gill being one of the names given to that herb (Clair). WATERCRESS, boiled with whisky and sugar, is an Irish cure for bronchitis (Wood-Martin), as is smoking the dried leaves of MULLEIN, both for bronchitis and asthma (Ô Súilleabháin). Smoking dried COLTSFOOT is the best known remedy for coughs, asthma and bronchitis, and a tea made from the flowers or leaves can be taken, too (Thomson. 1978). YARROW tea, either made from the dried herb or the fresh plant is used for a bad cold, and is also taken for bronchitis (V G Hatfield. 1994). Some of the American Indian peoples used to boil the whole flower head of a SUNFLOWER for lung trouble (Gilmore), which is interesting, for Russian folk medicine used sunflowers for the same complaint, listed as bronchitis, laryngitis and other pulmonary disorders (Kourennoff). JUJUBE (*Zizyphus jubajuba*) fruits have been famous since ancient times for colds and bronchitis. They would be made up into lozenges, still called jujubes (Mitton).

BROOKLIME

(*Veronica beccabunga*) The specific name, *beccabunga*, is O Norse bekh, a brook, plus bung, which is the name of a plant. Brooklime seems to be the exact Anglo-Saxon equivalent of the Norse (which is still retained as a common name in Shetland as Bekkabung (Grigson. 1955)). It is edible as a spring salad plant, and it is used as such all over northern Europe. It often grows with watercress, and the two were gathered and eaten together (Barton & Castle). "Spring Juice" was fresh brooklime and scurvy-grass, cut and beaten in a mortar, and left to steep for twelve hours. Then it was strained, and the juice of Seville oranges to an equal amount was added. A wineglassful taken fasting each morning for a week was a spring tonic (Quelch).

It is said to be excellent for skin diseases. In the Balkans they make a poultice by boiling it with onions and wheat chaff in sour milk (Kemp). Gypsies use the leaves for a poultice for piles, boils, etc., (Vesey-Fitzgerald), a use Wesley knew: "the Piles (to cure): a Poultis of boil'd Brook-lime", and the poultice was prescribed for whitlows and burns as well (Barton & Castle). In Irish folk medicine, the decoction, either alone, or mixed with watercress, used to be taken for gravel, and urinary diseases generally (Egan). Boiled and sweetened, it was used around Belfast as an expectorant (Barbour). In Wicklow, too, the water in which it had been boiled was taken to be an excellent cold cure. One should stay in bed, however, as "it opens all the pores" (O Cleirigh).

BROOM

(*Sarothamnus scoparius*) The original broom, whether for domestic or magical (witch) uses, was a stalk of the broom plant with a tuft of leaves at the end, made by "broom-dashers" (Sargent). The specific name *scoparius* is from Latin scopae, a broom or besom. Broomstick marriages were once fairly common, and jumping over the broomstick was said to be part of the gypsy marriage ceremony (Vesey-Fitzgerald). But "jumping the broomstick" is a Scottish expression for an irregular marriage (Cheviot), and in Somerset it was said you should never step over a broom if you are unmarried, for if you do you will bear a bastard child (Tobgue. 1965). Note the folk song known as Green Besoms:

One day as I was trudging
Down by my native cot,
I saw a jolly farmer,
O happy is his lot.
He ploughs his furrows deep,
The seeds he layeth low,
And there it bides asleep
Until the green broom blow.
   O come and buy my besoms,
   Bonny green besoms,
   Besoms fine and new,

Bonny green-broom besoms,
Better never grew.

The song gives the adventures of the broom gatherer with first a farmer, then a miller, a squire and a parson. The significance of broom gathering is quite clear in the final stanza:

O when the yellow broom is ripe
Upon its native soil,
It's like a pretty baby bright
With sweet and wavily smile (Reeves).

"Going to the broom", in ballads and folk songs, as David Atkinson pointed out, is always associated with sex. The Breton matchmaker carried a rod of broom as a symbol of his office when he came to a house on serious business (Tynan & Maitland), and dreaming of them means an increase in the family (Mackay).

As you would expect of a witch plant, broom has its blasting properties, too:

If you sweep the house with blossomed broom in May, You're sure to sweep the head of the house away.

It is still believed to be an unlucky flower to bring indoors (Widdowson), as, for instance, in the Isle of Man (Gill. 1932), or in Sussex, where they say "it sweeps someone out of the house" (Vickery. 1985). But the belief is much more widespread than the two instances quoted here. In some cases, the prohibition is only in force during the month of May, but again, the mere picking of the flowers is enough to cause the father or mother to die in the course of the year. In this context, it is worth remembering that the Welsh used it as a charm – if waved over a restless person, it induced sleep (Trevelyan). The ballad known as the Broomfield Hill also tells of the sleep-inducing effects of broom. A kind of circle magic coupled with the strewing of broom flowers causes the men to fall asleep on the hillside:

Ye'll pu' the bloom frae off the broom,
Strew't at his head and feet,
And aye the thicker that ye do strew,
The sounder he will sleep (Atkinson).

It is one of the plants that should never be used to beat a child, for the belief was that the child would stop growing, for the broom never attains the height of a tree (Burne. 1883).

But broom could be used as a prophylactic, too, especially, as in Italy, when burning it was regarded as an anti-witch charm (Leland. 1898). Riding a broom on Good Friday round the meadows was the Bohemian farmer's way of keeping them free of moles, and in East Prussia riding a broom was a means of preventing a fever, in popular imagination (Runeberg). It has its place in Celtic mythology, for in the Mabinogion story of Math, the son of Mathonwy, Gwydion and

Math made Blodeuedd from the "flowers of the oak, and the flowers of the broom, and the flowers of the meadowsweet". The colour of the flowers was used in Celtic languages as descriptive of golden hair – goddesses with hair like gold or the flowers of the broom (Elton).

Broom is the symbol of humility (Friend. 1883), quite why is difficult to determine, unless it is simply because of its use as a humble domestic implement. Indeed, it is also the emblem of the housewife in a negative sort of way. The display of one at the house door showed she was away from home, and her husband would welcome visits from his male friends. Conversely, no respectable housewife would dream of leaving a broom on view if she *were* at home, for it showed that a man's company would be welcome (M Baker. 1974).

Sprays of broom flowers were a traditional feature of Whitsuntide decorations (A W Hatfield). In Scotland, where the broom and the whin (gorse) were rich in blossom, it was looked on as a sign of a good harvest to come (Gregor) (the fuller the flowers on the bush, the better the harvest (Inwards)). It has been used as a dye plant – it will dye wool a moss-green colour (Jenkins. 1966), and there is another odd usage – broom juice mixed with oil of radishes is an old recipe for killing lice and other pests (Drury. 1992).

It has many medicinal uses, not least for heart conditions – broom tops are a very old popular remedy for it, perfectly justified, too, it seems, for the active principle is sparteine (Thomson. 1976), an alkaloid contained in the seeds, also used to induce labour and to hasten birth. It is also used as an antidote to many poisons (Schauenberg & Paris). It is interesting that Auvergne shepherds say that sheep that have grazed on broom show an immunity to adder bites. A tea made from broom tops was an Irish cure for an abscess, by bathing it (Maloney). Broom tea, a known diuretic, was used to cure dropsy, and there are widespread reports of its use in domestic medicine in Britain (it was taken for rheumatism, too, in Cumbria (Newman & Wilson)). Another remedy for dropsy is to soak the feet in a bath of the hot liquid (quoted by Gibbings. 1942). There are, too, records of its use for jaundice, probably doctrine of signatures (yellow flowers for the yellow disease), but not necessarily so, given the diuretic properties. An infusion of the flowers, stems and root has been used in Norfolk to treat this ailment (*Folk-lore. vol 36; 1925*).

An ointment made from the flowers has been used for gout (C P Johnson), and gypsies use an infusion for kidney troubles (Vesey-Fitzgerald), by boiling a few sprigs in water, and drinking a wineglassful in the morning. It has even enjoyed a reputation of being a stauncher of blood, from the 14th century onwards. Note the Scottish belief that a bunch of broom wrapped round the neck will stem a nosebleed (Beith).

There is one piece of veterinary information concerning broom. Aubrey noted its use in Hampshire and Wiltshire to prevent rot in sheep (Aubrey. 1847). He knew of "careful husbandmen" who cleared their land of broom, "and afterwards their sheep died of the rott, from which they were free before the broom was cutt down". So then they made sure of leaving some plants of it round the edges of their land, just for the sheep to browse on, "to keep them sound".

BROOMSTICK MARRIAGES

Fairly common at one time, though still irregular, and jumping over the broomstick was said to be part of the gypsy marriage ceremony (Vesey-Fitzgerald). But "jumping the broomstick" is a Scottish expression for an irregular marriage (Cheviot), and in Somerset they say you should never step over a broom if you are unmarried, for if you do you will bear a bastard child (Tongue. 1965).

BROOMWEED

(1) (*Gutierrezia microcephala*) An American plant, used by the Navajo for a yellow dye (Kluckhohn *et al*), and they also recognised its medicinal properties. Women made a tea of the whole plant to drink in order to promote expulsion of the placenta, and they also chewed the plant and then applied it to insect stings. The Hopi sometimes used the tea for stomach upsets (Weiner), while Plains tribes gave a decoction to their horses for diarrhoea (Youngken). There was a very strange usage further east, for this is a "root-doctor" prescription for "loss of mind", used by African Americans from Florida : "Sheepweed (another name for our plant leaves), bay leaf, sarsaparilla roots. Take the bark and cut it all up fine. Make a tea. Take one tablespoonful and put in two cups of water and strain and sweeten. You drink some and give some to patients" (Hurston).

BROOMWEED

(2) (*Malvastrum coromandelianum*) This has been used in Jamaican obeah practice to detect a thief. The ceremony was described as holding two broomweed sticks that have first been dipped in lye, one on each side of a man's neck, and saying:

By Saint Peter, by Saint Paul,
By the living God of all,

and if he is a thief, the weed will grow around his neck and choke him to death. Another way of doing it is to make a "switch" of broomweed, rosemary and willow, by virtue of which a person suspected of being the thief will be struck with pain when his name is called out, and will never be free of pain again (Beckwith. 1929). It is not very clear in the report how this charm is supposed to work.

BRUISES

RIBWORT PLANTAIN is mentioned as a Highland remedy for boils and bruises (Grant). A Cornish ointment for bruises is made from MALLOW leaves pounded and mixed with pig's lard (Deane & Shaw), and they were one of the ingredients in a medieval prescription "for swelling of a stroke" (Henslow), which presumably means a bruise. They used to be treated with WORMWOOD – "the plant steeped in boiling water, and repeatedly applied to a bruise will remove the pain in a short time, and prevent the swelling and discolouration of the part" (Thornton). Earler herbalists repeatedly prescribed a similar treatment with wormwood (Henslow, Dawson, Cockayne). A liniment, using the root of RUSSIAN TARRAGON, was used by some native American peoples, the Mescalero, and Lipan, for instance. The liniment was made by pounding the root and soaking it in cold water (Youngken), and LESSER EVENING PRIMROSE was made into a poultice by peoples like the Ojibwe, by soaking the whole plant in warm Water (H H Smith; 1945).

Gypsies use an ointment made from the leaves or roots of SOLOMON'S SEAL in lard to apply to a bruise or black eye (Vesey – Fitzgerald). So did Fenland people (Porter. 1974). It is said that a certain Sussex pub landlord grew beds of the plant specially to put on the black eyes received after a pub fight (Sargent). The treatment was described in Scotland as "the root of Solomon's Seal, grated, and sprinkled on a bread poultice" to remove "bruise discolouration" (Gibson). Gerard mentioned their use for bruises "gotten by falls or womens wilfulness, in stumbling upon their hasty husbands fists, or such like". Gypsies also used a decoction of the root of SOAPWORT to apply to a bruise or black eye (Vesey-Fitzgerald), or a freshly dug root put on the black eye would do (Campbell-Culver), though that may take longer to achieve the desired effect. Similarly, BLACK BRYONY got the name "herbe aux femmes battues" in French (Baumann), for this is another plant that has the reputation of healing bruises – the roots "do very quickly waste away and consume away blacke and blew marks that come of bruises and dry-beatings" (Gerard). The crushed leaves of HOUSELEEK are another good remedy (Randell; Grant).

Gerard called DAISY Bruisewort, with good reason, for he advised daisy leaves for "bruises and swellings". Variations on this name include Briswort and Brisewort (Britten & Holland). MARIGOLD is used as a tincture in much the same way as Arnica. If the tea is taken after an accident, it brings out the bruises and prevents internal complications. A lotion would be applied to sprains and bruises as well (Moloney). MELILOT is another useful plant. Sir J Hill gave the advice: "the fresh plant is excellent to mix in pultices to be applied to swellings", and that advice can apply equally well to bruises (Thomson. 1978).

Bryonia dioica > WHITE BRYONY

BUCK'S HORN PLANTAIN

(*Plantago coronopus*) In Glamorgan, where it appears to have been very common, the roots and leaves were made into a decoction, sweetened with honey, and given as a cure for hydrophobia (Trevelyan). Hill had heard of the cure, but gave it little credit: "it is said also to be a remedy against the bite of a mad dog, but this is idle and groundless", but he did recommend the leaves, "bruised, and applied to a fresh wound [to] stop the bleeding and effect a cure". Sometimes the whole plant was hung round the neck as an amulet against the ague (Allen), but this seems to come from Gerard, although he wanted for wearing round the neck: "unto men, nine plants, roots, and all; and unto women and children seven".

Early prescriptions include one for toothache "Shave hartshorn and seethe it well in water, and with the water wash the teeth, and hold it hot in thy mouth a good while. And thou shalt never have the toothache again" (!) (Dawson. 1934), hartshorn being another name for Buck's Horn Plantain.

BUCKBEAN

(*Menyanthes trifoliata*) The roots used to be pounded up to provide an edible meal (J G D Clark), while the Chinese pickled them to use either as a food or as a sedative (F P Smith). There was one other use, in brewing, and the leaves this time, not the roots – "one ounce will go as far as half a pound of hops" (Thornton). The point is that this is an intensely bitter plant, and that is the reason for its use as a substitute for hops in brewing (C P Johnson). Bitter or not, the leaves used to be shredded and smoked in the Faeroes in times of tobacco scarcity (Williamson). Where bitter tonics were needed medicinally, the root, gathered in August, and the seeds, were used. Gypsies made a leaf infusion to take as a blood purifier (Vesey-Fitzgerald).

Scurvy is the disease with which Buckbean is most often associated. Other skin diseases were treated with it, boils, for example, in Ireland (Ô Súilleabháin), while in Orkney, the crushed leaves would be applied for scrofula (Leask). A decoction of the seeds was used to treat rheumatism (V G Hatfield. 1994), or to prevent it (Sargent), it was also used for gout and dropsy – there is a recipe from South Uist, in the Outer Hebrides, that involved cleaning and boiling the whole plant, putting the juice in a bottle, to be drunk daily (Shaw). This may be doctrine of signatures, of course, given the plant's preference for wet, marshy ground. Perhaps the same argument could be used to explain its use for malaria. Hill, in the mid 18th century, mentions this use for the dried leaves, and it also crops up in Russian domestic medicine for the same complaint. Four or five tablespoonfuls of the dried herb in a gallon of vodka, kept for two weeks, and one small wineglassful to be taken daily (Kourennoff).

It used to be a Highland remedy for all kinds of stomach pains, and herbalists still give the leaf decoction for stomach disorders, loss of appetite and the like (Schauenberg & Paris), an interesting usage, for the Kwakiutl Indians of the Pacific north-west of America use the water from the boiled roots for some kind of stomach trouble, briefly translated as "when the pit of our stomach is sick" (Boas). Even stomach ulcers were treated, in Scotland, successfully apparently, with infusions from this plant (Beith). They even used it for constipation on South Uist. They took the root, cleaned it and boiled it in water all day until the juice was dark and thick. It was strained, and a teaspoonful given to the patient; it was even given to calves for the same complaint (Shaw), though the dose must have been increased.

BUCKTHORN

(*Rhamnus cathartica*) The black berries are a powerful purgative, but are dangerous to children (P North); in large doses they could cause intestinal haemorrhage (Flück). "It is a rough purge", said Hill. 1754, "but a very good one", and they have been known as such since Anglo-Saxon times. Syrup of blackthorn, made from the juice of the berries, can still be bought (Flück), but a hundred years or more ago it was used more for animals than for men (Fluckiger & Hanbury).

The berries have got other uses, for dyeing and making pigments, Sap Green among others, provided the proper mordant is used. As a dye, glovers would use them to give a yellow colour to their leather (Aubrey. 1867), and the yellow colour was used in the 19th century to tint paper (CP Johnson).

As with other thorns, this one could be used as a counter to witchcraft. There are records of this from Germany, for with a buckthorn stick a man can strike witches and demons, and no witch dared approach any vessel made of it (Lea). There are examples as far back as Ovid, who described a ceremony for countering a vampire witch. The final act was to put branches of buckthorn in the window (Halliday). Following the belief, it is easy to see how it could be used to protect one from the dead – chewing it was deemed protection enough. (Beza). Buckthorn appears also in the Cromarty legend of Willie Miller, who went to explore the Dripping-cave. "He sewed sprigs of rowan and wych-elm in the hem of his waistcoat, thrust a Bible into one pocket and a bottle of gin into the other, and providing himself with a torch, and a staff of buckthorn which had been cut at the full of moon … he set out for the cave …"(H Miller). Is there a misreading in this case, for one would have expected his stick to be of the belligerent blackthorn rather than of buckthorn?

BUCKWHEAT

(*Fagopyrum esculentum*) Both the growing plant and the straw are poisonous to livestock, and in man an allergic rash may occur from eating foods made from buckwheat flour (P North). That, though, is rare, and it was widely cultivated once for cattle and hen food, as well as for a poor man's flour. Bread made from this was described by Gerard as "of easie digestion, and speedily passeth through the belly, but yeeldeth little nourishment". Cobbett, too, disliked buckwheat bread, finding it heavy, and meriting its French name, Pain Noir. But he liked buckwheat cakes enough to give the recipe (see Cobbett. 1822). They are like pancakes, and these pancakes are a national dish in America in the winter, made more or less as Cobbett suggested, and served with maple syrup as breakfast cakes. This meal is also made into crumpets in Holland (Grieve. 1931), and the porridge known as Kasha in Russia is also made from it (Brouk). The flour is eaten by Hindus during their fast days, being one of the grains lawful for fasts (Pandey).

Apart from all this, buckwheat grain was used at one time in making gin (C P Johnson), and one can even get a blue dye from the straw (Schery). The Japanese are fond of buckwheat noodles, called Soba, and Japanese goldsmiths have long used buckwheat dough to collect gold dust in their shops. So the grain is considered a potent charm for collecting riches. Every Japanese family eats buckwheat noodles on New Year's Eve to get the luck for making money in the coming year, and every household serves soba at feast times, as, for instance, when there is a move to a new house, when a Japanese will give a present of soba to all his neighbours (Schery). There is a folk tale from Japan that tells how an ogress's blood flowed on to buckwheat plants, whose roots are even now red because of what happened then (Seki).

Strangely, there is very little in the way of folk belief attached to it, though Turgenev (*Fathers and sons*) must have known at least one superstition: "Arina Vlasyevna … believed that if on Easter Sunday the lights did not go out at vespers, then there would be a good crop of buckwheat". The only other belief recorded refers to the time of sowing – according to Pennsylvania Germans the proper day is the Seven Sleepers (27 June) (Fogel).

BUFFALO GOURD

(*Cucurbita foetidissima*) American Indian tribes in the west of the country valued this, as they believed it simulated the form of the human body, and was either male or female. So it was a heal-all, that portion of the root being used that corresponded to the affected part of the patient's body (Youngken).

BUILDING TIMBER

The "witch-post" was found in Yorkshire as a protective amulet – it was always made of ROWAN wood, and marked with an inscribed cross. It always formed part of the structure of a house, usually under the cross-beams of the timber framework (Davidson). The old Scots word for a cross-beam in the chimney is rantree, because it was so often made of rowan (F M McNeill). SWEET CHESTNUT timber is unsafe. Taylor. 1812 said that "it is very apt to be shaken, and there is a deceitful brittleness in it which renders it unsafe to be used as beams…". Theophrastus seems to agree, for he reported a case where chestnut beams in a public bath gave warning before they broke, so that all the people in the building were able to get out before the roof collapsed (Meiggs). Evelyn, though, was enthusiastic about the quality of the timber. "The chestnut is (next the oak) one of the most sought after by the carpenter and joyner : it hath formerly built a good part of our ancient houses in the city of London …".

BULLACE

(*Prunus domestica 'institia'*) "As black, as sour, as bright, as a bullace" are all Yorkshire similes (Gutch. 1909). Bullace, probably from O French buloce, which meant sloe. The Welsh name bwlas, which also signifies sloe, must be the same word. Damson is the cultivated form of this shrub. An infusion of the flowers, sweetened with sugar, has been used as a mild purgative for children (Grieve. 1931).

BULRUSH

(*Scirpus lacustris*) This is the true Bulrush, the plant usually meant by that name is in fact the False Bulrush, *Typha latifolia*. In aboriginal America, bulrush provided the sacred yellow pollen of the San Carlos Apache. The medicine-man rubbed it on the affected part of his patients, and then sang, after which he would extract the disease in the usual sucking way of shamans. Sometimes he would put a little pollen in the patient's mouth (Youngken). Other Indian groups sometimes made the pollen into cake. It was collected by beating it off into a cloth (Havard).

BURIAL

ELM timber endures well under water, and so was the first material for water pipes (Grigsin, 1955), and for piles, as in London and Rochester bridges (Rackham. 1976). For the same reason, elm wood was used for coffins, a practice that Vaux wholeheartedly condemned – "the evil practice of making coffins of elm in order to keep the body from the corrupting effects of contact with the earth for as long as possible, the very opposite to which is what all sensible people desire".

It is said that when MEADOW SAFFRON (*Colchicum autumnale*) blooms on a grave, it is a good sign for the deceased (Friend. 1883).

BURN PLANT

(*Aloe chinensis*) The common name says it all; the juice is used to treat burns, particularly X-ray burns, apparently. When evaporated to a semi-solid, it can be used as a laxative (Schery).

BURNET ROSE

(*Rosa pimpinellifolia*) There used to be a belief around the Bristol Channel that if the Burnet Rose should bloom out of season, it would be an omen of shipwreck (Radford & Radford). It can be used as a dye plant, more especially in the Hebrides; a brown dye can be made from it, with copperas (Murdoch McNeill). Elsewhere, using alum as the mordant, the hips were used to dye silks violet (C P Johnson).

BURNET SAXIFRAGE

(*Pimpinella saxifraga*) It is called Saxifraga "for the propertye that it hath in breaking of the stone in a man's bodye" (Turner. 1551). Gerard elaborated on this, and went on : "the juice of the leaves ... doth cleanse and take away all spots and freckles of the face, and leaveth a good colour ...". In Italy, they used to say that if a woman eats burnet saxifrage, her beauty will increase (Skinner). The distilled water, too, was used as a cosmetic (Leyel. 1937). In plant symbolism, this stood for affection (Leyel. 1937).

The root infusion is taken for catarrh, and sometimes as a diuretic. Externally, this infusion can be used as an application for healing a wound (Flück). Another application from the roots, a tincture this time, is given for sore throats (Schauenberg & Paris), and the fresh root is chewed for toothache (Grieve. 1931).

BURNS

Using the petals of MADONNA LILIES, usually macerated in alcohol, was said to leave no scar (Wiltshire). As far back as the 13th century the lily was prescribed by a monk as a sovereign remedy for burns, for "it is a figure of the Madonna, who also cures burns, that is, the vices or burns of the soul" (Haig). It is said that MARIGOLDS should be used for burns. Not only does it cure, and help to relieve the pain, but it will also prevent the formation of scars. And it is even claimed that it will take away existing scars (Leyel. 1937). BROAD-LEAVED DOCK has been used for all sorts of skin complaints for a very long time. The leaves would be used as a poultice to apply to the spot, and in the Hebrides, the roots were boiled with a little butter and applied on a bandage to the burn (Shaw). Or apply a GREAT PLANTAIN leaf (A W Hatfield), or a CABBAGE leaf, fried, according to belief in County Cavan (Maloney).

A Cornish charm for scalds and burns required nine BRAMBLE leaves, moistened in spring water, and these are applied to the affected part. While this is being done, the following charm has to be recited three times:

> There came three angels out of the west.
> One brought fire, and two brought frost;
> Out fire, in frost;
> In the name of the Father, Son, and Holy Ghost
> (Hawke).

HOUSELEEK, being a protector from fire or lightning, would naturally be taken as a cure for burns. As Gerard put it, "they take away the fire of burning and scaldings ...". The leaves would be put on the burn, like a plaster (Jamieson ; O P Brown), or they could be bruised with cream and laid on (Morris). An ointment prepared from the flowers and leaves of ST JOHN'S WORT, mixed with olive oil is good for burns (Toingue. 1965).

Slices of POTATO could be put on the burn, provided it was a small one. A Sussex treatment for a larger one was to scrape the insides out of a potato, mash it, and put that on (Sargent). At least according to Gerard, burns used to be treated with ONIONS: "the juice taketh away the heat of scalding with water or oile as also burning with fire and gunpowder". "Pollypodden", presumably the fern POLYPODY, is used in Ireland for burns. The procedure is to boil the stems with butter. The green juice sets to a jelly, and this is put on the burn (Maloney). Another fern, HARTSTONGUE, was made into ointment in Scotland for a burn cure (Beith). BURN PLANT (*Aloe chinensis*) juice is excellent for burns, and so is the juice of its Old World relative, MEDITERRANEAN ALOE (*Aloe vera*). The Maoris would pulp the bark of RIMU (*Dacrydium cupressinum*) to apply to burns (Leathart).

BUSY LIZZIE

(*Impatiens walleriana*) Originally from tropical Africa, where the root has been used as an abortifacient (Watt & Breyer-Brandwijk).

BUTCHER'S BROOM

(*Ruscus aculeatus*) The strange-looking common name came about because, so it is said, butchers once used the shrub to sweep their blocks. It suggests that this shrub was commoner in those times than is so today. It is a British native, but is very local, in the south. They do make hard brooms, for the shoots are stiff and spiny, stiff enough to be used as household besoms in Italy (Folkard). There is too a suggestion that butchers used them to tenderize their meat. They are certainly part of the crest of the Butchers' Company (Billington).

The young shoots are edible, and were once eaten "as sperage in sallads". It was, too, quite important as a medicinal plant, and has been so in quite recent times. Maud Grieve, for instance, was able to report that a root decoction was still given for jaundice and gravel in her day, and it was once used for dropsy too (Barton & Castle). It still is a favourite medicine for stone in Ireland. There the roots are boiled for eight hours, and a pint of whiskey added to a quart of the water, and the whole strained and bottled. The patient takes three glasses daily (Logan). But this is a venerable remedy, for Gerard wrote that the root decoction made in wine "and drunken, provoketh urine, breaketh the stone, driveth forth gravell and sand …". There are a few other conditions that were treated with Butcher's Broom, one being fractures, when a decoction of the berries and leaves was made into a poultice (Leyel. 1937). And chilblains were flogged with the branches (Grieve), just as holly was used (see HOLLY).

BUTTER BEAN
(*Phaseolus lunatus*) see LIMA BEAN

BUTTERCUP
i.e., *Ranunculus acris* (Meadow Buttercup).
R bulbosus (Bulbous Buttercup) and *R repens* (Creeping Buttercup). Meadow buttercup was rubbed on cows' udders with some ceremony on May Day in Ireland. One wonders why, though it is because of the long-lived tradition, flying in the face of all the facts, that buttercups impart a good colour to butter, or that they improve the quality of the milk. In fact, of course, the milk of cows that eat them becomes tainted, and the same holds true for the butter. But usually, cows will not eat buttercups at all, for they are all extremely acrid. The point is that they grow in good soil, and such soil produces a lot of good grass. That is what cows eat to improve their milk, not the buttercups.

The acrid principle made the plants notorious at one time, for beggars used them to raise blisters on their feet. As Gerard said, "cunning beggars do use it to stampe the leaves, and lay it unto their legs and arms, which causeth such filthy ulcers as we day by day see (among such wicked vagabonds) to move people the more to pitie". As a contrast, what could be more innocent than the pastime small children indulge in,

of holding buttercups under each others' chins to see if they like butter? It must be the golden colour that makes this a symbol of riches (Leyel. 1937).

The principle of the counter-irritant made the medicinal use of buttercups possible. Chinese medicine certainly used it as such (F P Smith). Thornton's remedy for rheumatism must come under that heading – he recommended pounding the leaves and applying it as a poultice, "when it produces a vesication like a blister". Gerard recognized the idea, and in a burst of humour offered this: "Many do use to tie a little of the herbe stamped with salt unto any of the fingers, against the pain in the teeth; which medicine seldome faileth; for it causeth greater paine in the finger than was in the tooth …".

Thornton actually said that they have been used internally for worms. Certainly, it would do the worms no good at all, but surely their host would suffer equally! Ô Súilleabháin quotes an Irish use of the juice for jaundice. This must surely be our doctrine of signatures. Any yellow plant, or anything with yellow juice, would automatically recommend itself to treat the yellow disease. A homeopathetic tincture of buttercup is taken, internally, for shingles (Leyel. 1937), and there are early records from Ireland of its use for St Anthony's Fire, i.e., erysipelas. Earlier still, Apuleius seemed confident with a recipe using one of the buttercups: "For a lunatic, take this wort, and wreathe it with red thread about the man's swere [neck], when the moon is on the wane, in the month which is called April, or in the early part of October, soon will he be healed" (Cockayne). Perhaps we are still talking about counter-irritants!

Buxus sempervirens > BOX

C

CABBAGE

(*Brassica oleracea 'capitata'*) "If thou desirest to die, eat cabbages in August". That was one of the medical maxims from the Book of Iago ab Dewi (Berdoe), and shows that cabbages were regarded with some suspicion at the time. Now it is difficult to see, or even think of, anything uncanny about a cabbage, but the Pennsylvania Germans used to say that a cabbage plant running to seed the first year, or one with two heads on one stalk, is a sign of death (Whitney & Bullock, Fogel). If one of them had white leaves, it meant a funeral. But it is good luck if you find one growing "double", that is, with two shoots from a single root (Waring) (different from two heads on one stalk). Equally odd is a superstition from Illinois that if a woman eats cabbage during her confinement, she will die. Similarly, mothers are advised not to eat it while breast-feeding a baby – it will give the child colic (H M Hyatt). In Normandy, they say that you should never eat or pull a cabbage on St Stephen's Day (26 December), for the saint, according to their tradtion, was stoned to death in a cabbage patch (W B Johnson). Similarly, American belief holds that it is most unlucky to have a head of cabbage in the house on New Year's Day (H M Hyatt), but in Kentucky they say that if you eat cabbage on New Year's Day, you will have money all the year (Thomas & Thomas).

To ensure a good crop, make a cross on the stump every time you cut one (Tongue. 1965). There is a medieval Jewish tale which is very similar to the myth of Baldur in Scandinavian mythology. Every tree had sworn it would not bear Christ's body, except for a cabbage stalk, and it was on that he was crucified. Another twist brings the Baldur myth even closer, for there were English traditions according to which Christ was crucified on a mistletoe (Turville-Petrie). Next to a broomstick, ragwort is the likeliest thing to serve as a witch's horse, but another tradition says they use cabbage stalks on which to ride through the air (Wood-Martin, Simpkins). Because green is an unlucky colour at Scottish weddings, care used to be taken that no cabbage or any other green vegetable should be served up on the occasion. All over France, children are found under a cabbage, not a gooseberry as in England. Parents, speaking of the time before their child was born, would tell them "c'était du temps où ils étaient encore dans les choux" (Sebillot). So they are in Maryland, too, though not under the cabbage, but in the cabbage head (Whitney & Bullock). To dream of cutting cabbage is a sign that your wife, or husband, or lover as the case may be, is very jealous. If you are actually eating a cabbage, then it is a sign of sickness for said wife, husband or lover (Raphael). The American interpretation of dreaming of them is that you will experience sorrow if the dream is of eating it, or if it is a growing plant, good fortune is coming to you (H M Hyatt). Sow them on St Gertrude's Day, which is 17 March, for the best results, say the Germans, and the Pennsylvania Germans, too (Fogel), but in Kentucky, the favoured day is 9 May, and scatter elder leaves over them to keep insects away. Another piece of American wisdom advises sprinkling flour over the plants while the dew is on them, to drive away worms. Pennyroyal leaves can be used to the same effect (H M Hyatt).

Cabbage is very popular in eastern Europe and in Russia, where it is used for making soups, and also as a stuffing for cakes. Fermented cabbage is known in English by its German name, Sauerkraut. Even in this country, there were other uses for it than the almost inevitable boiled greens. In the 19th century, it was the custom to bake bread rolls wrapped in cabbage leaves; it was said to give them a nice flavour (Fernie), though the reason given in Warwickshire was that doing so would give the bread a thinner crust (Bloom). Cabbage leaves are used medicinally, too. Wrapping a leaf round the affected part is a Cornish remedy for rheumatism (Hawke), and holding them to the affected side will cure a stitch, they say (Page. 1978). Fried cabbage leaves used to be the Irish method of deadening the pain caused by a burn (Wilde. 1890). If they are fried, then it must be the grease that is doing the work, and the leaf is merely the vehicle for it. They are spoken of as "roasted" in co Cavan, and used to put on a cut (Maloney). Logan also records the Irish use of cabbage leaves for a burn, but he says they are macerated thoroughly, and applied along with the juice. In much the same way, he says, green cabbage leaves used to be a favourite Irish remedy to treat ulcers; he must be talking of a poultice, though not necessarily, for the leaves could simply be applied, and herbalists prescribe cabbage juice for stomach ulcers (Thomson. 1978). Put a cabbage leaf to "draw the cold out", in other words, to cure a headache, they used to say in Essex, and they would cure a toothache by putting a hot cabbage leaf sprinkled with pepper to the cheek (Newman & Wilson). Not only that, but the inner leaves are the ideal medicament to draw a boil, at least that is the practice in Shropshire and Cheshire (Barbour), and in East Anglia, too, where abscesses are similarly treated (V G Hatfield. 1994). And it is even recorded in co Cavan that a boil could be treated simply by eating cabbage leaves (Maloney). In Scotland, they were made into a cap to put on a child's head for eczema (Rorie). And eminently practical is the Kentucky practice of putting a wet cabbage leaf on the head to keep off sunstroke (Thomas & Thomas). There is, too, from Illinois, a recommendation that raw cabbage chopped fine and mixed with salt, pepper, vinegar and sugar, and eaten once a day, is the best thing to bring blood pressure down (H M Hyatt).

The water in which cabbage has been cooked (with no salt!) is good when cooled to bathe sore eyes (Quelch), and Gerard reported its use for dim eyes. He went on to prescribe the juice for "the shaking palsie", and snake-bite; the seed for worms, freckles and sunburn, and the like. Even the Maori, it seems, use it in the shape of a hot decoction, for colic (Goldie). There were, too, the usual fantastic remedies – in some way, it was used in Ireland as a cure for hydrophobia (Wood-Martin). Gerard hopefully said that "it is reported that the raw Colewort beeing eaten before meate, doth preserveth a man from drunkennesse …", and claimed that the juice, mixed with wine, and "dropped into the eares, is a remedy against deafeness".

Cajanus indicus > PIGEON PEA

CAJUPUT

(*Melalauca leucadendron*) Green Cajuput oil, distilled from the fresh leaves and twigs, is often used as a counter-irritant, as for instance, in rubbing it on the gums to relieve toothache (Mitton). It is also a good mosquito repellent (Chopra, Badhwar & Ghosh), but the best known use is as a lotion or ointment to use for rheumatism and bruises. A few drops on sugar quickly ends hiccoughs.

Calendula officinalis > MARIGOLD

CALIFORNIAN BUCKTHORN

(*Rhamnus purshiana*) Famous for its product, the Cascara Sagrada, the sacred bark that is the best known of all laxatives, no matter under what name it is marketed (see Weiner). The name cascara sagrada was given by the Spanish pioneers, who took notice that the Californian Indians (Schenk & Gifford; Spier) used the bark infusion as a physic. There is also a brown dye to be obtained from the bark. In some parts of California in the early days, the scriptural term Shittim Bark was applied to this. It was said locally to be the Shittim wood of which the Hebrew ark was made (Maddox). It was used by the Californian Indians for toothache; the root was heated as hot as could be borne, put in the mouth against the aching tooth, and tightly gripped between the teeth (Powers).

CALIFORNIAN JUNIPER

(*Juniperus californica*) The pulp of the fruit used to be a staple article of Indian diet; they could be eaten either fresh or dried, and in the latter state they were ground and made into bread (Yanovsky). Another use was as a dyestuff – red could be made from the ashes (Jaeger).

The Hopi Indians made use of it as medicine. A leaf decoction was taken as a laxative, for example, and the same decoction was taken by women who wanted a female child. A piece of chewed juniper or a tea made from the leaves was given to women during childbirth. During the lying-in period, all of the mother's food had to be prepared in some degree with a decoction of juniper leaves. Her clothes, too, had to be washed in water that had some of the leaves in it. The newborn baby itself was rubbed with ashes from burned juniper, and if later on in its life the child misbehaved, recourse was made again to the juniper. The child was taken, at the mother's request, and held by some other woman in a blanket over a smouldering fire of juniper. He soon escaped, of course, half suffocated, supposedly a better and wiser child. Another ritual was recorded; when the men returned from burying a corpse, they washed themselves outside the house with water in which a branch had been boiled (Whiting).

CALIFORNIAN LAUREL

(*Umbellularia californica*) There is something odd about the leaves of this tree, for it is well-known that the scent of them has a depressing effect on those who inhale it (G B Foster), and handling the foliage can cause skin irritation and running of the eyes in some people (Hora).

Calla palustris > WATER ARUM

Calluna vulgaris > HEATHER

Caltha palustris > MARSH MARIGOLD

CAMMOCK

is the proper English name for Rest Harrow. When dairy produce is tainted by this plant, the result would be "cammocky", as Wiltshire dialect had it (Dartnell & Goddard). (see REST HARROW*)*

CAMOMILE

(*Chamaemelum nobile*) Camomile is rarely used for lawns these days, but was in Elizabethan times. Even turfed seats made of it were garden features (Rohde. 1936). Shakespeare has Falstaff say (*Henry VI, pt 1*) "For though camomile, the more it is trodden, the faster it grows; yet youth, the more it is wasted, the faster it wears". But by his time that had become proverbial. Dyer's rendering was:

> Like a camomile bed,
> The more it is trodden,
> The more it will spread.

That must be the reason why it symbolizes energy in adversity (Leyel. 1937). Gardeners looked on camomile as a "plant physician", restoring to health any sickly plant near which it grew (Bardswell). Of course, one had to be judicious in its use. It should only be put near the ailing plant for a short time; if the camomile clump grew too big, it had to be moved, otherwise the other plant would weaken again, "as though the patient had become over-dependent on the doctor and tiresomely hypochondriac" (Leyel. 1937). Camomile was the principal ingredient (along with ground ivy and pellitory-of-the-wall) in Elizabethan snuffs (Genders. 1972). But there is one old usage that is still current, that of using the dried flowers in shampoos for fair hair (it is the double variety (*C nobile pl*) that is commercially grown). Or camomile water can be made at home. Simply steep the dried flowers in boiling water and strain off when cool (Hawke). On the Greek island of Chios, camomile is used to dye the hair a light chestnut colour, almost gold (Argenti & Rose).

It is, though, in the field of medicine that camomile was, and is, used so much, a use that goes back to the beginning of records (Lloyd). The Egyptians held it sacred, and the Romans believed it cured snakebite. One of the medical maxims from the Book of Iago ab Dewi was "if a snake should crawl into a man's mouth, the patient was to take camomile powder in wine" (Berdoe).

Camomile tea is a virtual panacea, used for a remarkable number of often unrelated ailments. It is a great standby for an upset stomach (Hawke), or as a laxative (V G Hatfield. 1994). Gypsies in Britain use it for flatulence (Vesey-Fitzgerald), and it is very popular in France (Clair), but even more so in Italy (Thomson. 1976). On Chios, it is drunk "for the good of the stomach", which is more or less what Gerard said : it is good "against coldnesse in the stomack, soure Belchings, voideth winde, and mightily bringeth downe the monethly courses", this last being interesting in view of the German belief that camomile tea is good for women in labour (Thonger). It is listed as a bitter stomachic and tonic (Fluckiger & Hanbury), and as such is recorded in Hampshire as good for "clearing the blood" (Hampshire FWI), by which a spring tonic is probably meant. And it is also recommended for neuralgia and migraine (Schauenberg & Paris), as well as for a simpler kind of headache (Newman & Wilson), and for insomnia as well (Brownlow), even for a cold. It can, too, be used as a lotion for external use, to treat ulcers, wounds, etc.,

Even the dew shaken from the flowers was used (in Wales) for consumption (Trevelyan), and the real juice from the plant had its uses, too, for sore eyes, to take one example. It had to be gathered before dawn, and the gatherer had to say why he was taking it; "let him next take the ooze, and smear the eyes therewith" (Cockayne). The root is another part used as a toothache cure in Ireland. The instructions are simple – just put a piece on the aching tooth (Vickery. 1995). Ointments made from the plant were quite widely used, too. An example from Skye shows that camomile and fresh butter made into an ointment was used for, of all things, a stitch! (Martin). There is a 14th century example of the ointment being used for cramp (Henslow).

Camomile poultices, made with flowers or leaves, were often used, also. The flowers were used in

Orkney, particularly to prevent a boil coming to a head, in other words to allay inflammation and swelling (Leask). That poultice was used in Scotland for gumboils (Gibson. 1959). "Warty eruptions" were dealt with by using bunches of camomile, according to an old leechdom (Cockayne), and the flowers were used in an old treatment for deafness, the recipe telling the patient to "take camomile and seethe it in a pot, and put it in the ear that is deaf, and wash the ear; and so do for four days or five, and he shall be whole" (Dawson. 1934). Something similar used to be the custom in Wiltshire domestic medicine for earache. People made a flannel bag, and filled it with camomile heads. This was warmed by the fire and held against the ear (Wiltshire). One of the most engaging of all these prescriptions was the way Wiltshire mothers were advised to deal with fractious children – the flowers, picked when the sun was on them, dried in the sun and kept in a close stoppered jar for use when needed. A draught containing ten heads, over which a pint of barley water was poured, and sweetened by a large tablespoonful of honey, was the dose given, hot at night, cold during the day (Wiltshire).

Campanula glomerata > CLUSTERED BELL-FLOWER

Campanula medium > CANTERBURY BELL

Campanula rapunculus > RAMPION BELL-FLOWER

Campanula rotundifolia > HAREBELL

Campanula trachelium > NETTLE-LEAVED BELL-FLOWER

CAMPHOR

(*Cinnamonium camphora*) Camphor, though solid, is the essential oil from this tree, made by distillation of the shoots. Aboriginal people observed sanctions in the gathering of camphor. The Malayan Jokuns, for example, used to believe that there is a "bisan", or spirit, that looks after the camphor trees, and that without propitiating the spirit, it would not be possible to get the camphor. The bisan makes a shrill noise at night, and that is a sign that there are camphor trees about (actually the noise is made by a cicada). The offering to the bisan was simply a part of the food being taken into the jungle. No prayers were offered, but all food had to be eaten dry, without condiments. Salt must not be pounded fine, or the camphor would only be found in fine grains. Conversely, the coarser the salt, the larger the camphor grains (Skeat).

The drug use probably emanates from the earlier belief that the smell keeps off evil spirits. It was used as an amulet for this purpose (Maddox), and there are still signs of it in recent times – in Maine, for instance, they used to say that the way to cure boils was to hang a camphor bag round the neck (Beck). As recently as the end of the 19th century, pieces of camphor were carried around on the person in the belief that it was in some way antiseptic, quite fallacious, of course. Another practice was to saturate a piece of brown paper with spirits of camphor and then burn the paper in a sick room. This was supposed to disinfect and "freshen the air" (Ackermann). The wood was once burned during epidemics, too (Genders. 1972). And Pomet claimed that "the Oil is very valuable for the Cure of Fevers, being hung about the Neck, in which scarlet Cloth has been dipped …".

It has certainly been claimed that there are some stimulant properties attached to camphor, and it has been used quite a lot in India that way. According to Lewin, there were in the earlier part of the 20th century many camphor-eaters in England and America, the drug being taken solely for its stimulating and exciting properties, though women claimed that it was good for the complexion. Maddox says, too, that it has been prescribed as a nerve stimulant, while in the Far East it has been in use for heart disease for centuries (Tosco). But to show with what grave suspicion such a strong-smelling drug was regarded, Browne. 1646 found it necessary to refute the belief that "camphor eunuchates, or begets in men an impotency unto venery". As he points out, "observation will hardly confirm (it); and we have found it to fail in cocks and hens, though given for many days…". But black mothers in the southern states of America used to rub camphor oil on their breasts to dry up the milk after weaning (Puckett. 1926) (and it was used for that in Dutch folk medicine, too) (van Andel).

CANADA BALSAM

Blisters on the bark of BALSAM FIR (*Abies balsamea*) are the source of this balsam, used once in American domestic medicine as an application to sore nipples (Weiner). Native Americans such as the Menomini pressed the balsam from the trunk to use for colds and lung troubles (H H Smith. 1923). It is better known because it has the same refractory index as glass, so has been used for cementing glass in optical instruments (Wit).

CANCER

One is never sure whether cancer or canker is meant in older documents, but evidently cancer is meant in this advice from Alabama: "for cancer of the breast three or more quarts of RED-CLOVER blossom tea a day" (R B Browne). Gypsies used to steep VIOLET leaves in boiling water and use the result as a poultice for cancerous growths. An infusion of the leaves, they said, would help internal cancers (Vesey-Fitzgerald). This is also found in Welsh folklore (Trevelyan), and as a Dorset herbal cure (Dacombe). An Irish remedy was to drink a decoction of the dried flowers of WILD SORREL (Egan), and an ointment used to be made there from the leaves, for cancer (Egan), who also said that the leaves were eaten for stomach cancer.

The bark of SPURGE LAUREL (*Daphne laureola*) was used in the treatment of cancer (Grigson. 1955), though this was only a cottage remedy, and there seems no record of how it was done, nor whether it was really cancer (and not canker) that was being treated. The use of MISTLETOE for a tumour seems to be quite genuine. The juice is applied on the tumour as a well-known form of treatment (Thomson. 1976). The rhizomes of ZEDOARY (*Curcuma zedoaria*), a close relative of Turmeric, have been used in Chinese medicine to treat cervical cancer (*Chinese medicinal herbs of Hong Kong. 1978*), and another remedy, for cancer of the stomach, in Chinese medicine, is the use of MELONS, in some way (F P Smith). COLTSFOOT is well-known as a treatment for chest complaints, from coughs to asthma, and it seems that the smoked leaves have been used in Chinese medicine for treating lung cancer. (Perry & Metzger). The Cherokee Indians of North America used a preparation of BLOODROOT for treating breast cancer. This was taken up in a big way in the 19th century by the cancer quacks in the USA. It has been tested, but there was insufficient evidence to prove the genuine case (Thomson. 1976).

There is an extraordinary report from Norfolk about a man who had facial skin cancer. While waiting for treatment, a gypsy advised him to rub the cancer with the pith of a BANANA. It seems that the cancer was cleared up entirely by this means alone (V G Hatfield. 1994). Another strange case was that of a Cornish blacksmith, Ralph Barnes, in 1790. He was supposed to have cured himself of a cancer by taking immense quantities of HEMLOCK juice (Deane & Shaw) (primitive chemotherapy?). There is another East Anglian report that GREATER CELANDINE has been used to treat liver cancer there (V G Hatfield. 1994), and HERB ROBERT is still used by herbalists to treat skin cancer (Beith). Gypsies claim that GOOSE-GRASS is a very ancient remedy for the condition (Vesey-Fitzgerald). Thornton does record its use for tumours in the breast.

The drug colchicine, obtained from MEADOW SAFFRON, can, it seems, bring acute leukemia under control (Thomson. 1976). But other authorities say it is far too toxic to be of use (e.g., Schauenberg & Paris). Another red herring was EAGLE VINE (*Marsdenia condurango*), which became well-known in the latter half of the 19th century as a "cure" for cancer. Legend had it that a South American Indian woman administered the bark decoction to her husband, hoping to kill him because he was in such agony from a growth. Instead of killing him, it cured him. Unfortunately, controlled tests have shown that it is not so valuable as expected (Le Strange).

The Sotho in South Africa are reported as using the CANDELABRA TREE for cancer, in some unspecified way (Palgrave). Another *Euphorbia*, PETTY

SPURGE, has been clinically investigated for use in cancer treatment (Beith).

CANDELABRA TREE

(*Euphorbia ingens*) This is a tree-like spurge from central Africa. The latex becomes sticky when partially dry, and is used as bird lime. It is said to be poisonous, and can cause skin irritation and blistering. It has been used for homicidal purposes, too, in which case the "milk" would be evaporated to dryness, and the residue carefully collected and put in beer or food, which is then offered to the victim (Palgrave). The Zulus use it in very small doses as a drastic purgative, and the Sotho administer the latex as a cure for dipsomania, and also use the plant in some way as a cancer remedy. Shona witch doctors use a medicine involving the candelabra tree to treat boils. Another way to do it is to mix a certain snail shell with the milky sap, and apply that to the boil. An earache remedy is to crush a certain caterpillar with the roots of this plant in a cloth, and express the resultant fluid through the cloth into the patient's ear (Gelfand).

CANDLEMAS

(2 February), the feast of the Purification (of the Virgin Mary), called Candlemas because the tapers and candles to be used in church during the coming year were consecrated at this time. SNOWDROPS are dedicated to this feast (the plant is sometimes known as Candlemas Bells). A more general name for it was Fair Maids of February, for it was once the custom for young women dressed in white to walk in procession to the Festival (Prior). A stage from this would mean that the plant was sacred to virgins in general. In Shropshire, in the Hereford Beacon area, where the plant may possibly be native, a bowl of snowdrops was brought into the house (evidently the only day on which such a thing could be done with safety). They were thought to purify the house (Coats), in keeping with the name of the festival.

CANDYTUFT

(*Iberis amara*) Candytuft seeds have long been taken in the treatment of gout and rheumatism (O P Brown), and Hill, in the 18th century, wrote that "the leaves are recommended greatly in the sciatica; they are to be applied externally, and repeated as they grow dry".

Canna indica > INDIAN SHOT

Cannabis sativa > HEMP

CANNON-BALL PLANT

A name coined by the Victorians for the ASPIDISTRA. Nothing can harm a cannon-ball, and it seemed that the same applied to the plant, which would put up with any amount of neglect without coming to any harm. Cf Cast-Iron Plant, and Bar-room Plant.

CANTERBURY BELL

(*Campanula medium*) The name, of course, is taken from the bells of Canterbury Cathedral, so this plant was used as their emblem by Chaucer's pilgrims (Genders. 1976). Or perhaps the name came about from the bells carried by their horses (Radford & Radford). But it is also called Coventry Bell. Strangely, it is an unlucky plant in the Weald of Kent. If you pick them from your garden, it will toll the death-bell in the village. On the other hand, if it grows well in the garden, it will be a lucky one (Radford & Radford).

CAPER SPURGE

(*Euphorbia lathyris*) The fruits are quite often used green as a caper substitute (Browning), hence the common name, but it can be dangerous to eat them, poisonous as they are. It is the purging quality that most spurges have that causes trouble, and they have been known to be fatal (Salisbury. 1964). Goats are quite liable to eat quantities of it – then, it is said, their milk had the poisonous properties of the plant (Long. 1924). The toxin is in the milky latex, causing blistering and ulceration on the skin. Ingestion causes severe abdominal pain and nausea, leading to vomiting and diarrhoea, and possibly internal haemorrhage, though death is unlikely (Jordan). Nevertheless, the seeds are emetic and laxative, and some herbals have even recommended them for rheumatism (Schauenberg & Paris), though that sounds rather dangerous. In Chinese medicine, the flowers, seeds and leaves are all prescribed for diarrhoea (F P Smith), but it seems to have been most widely used, if the ascription is correct, in Anglo-Saxon times; in Cockayne's translation of Apuleius there are listed leechdoms for "sore of the inwards", warts, and even against leprosy ("take heads … sodden with tar, smear therewith"). Other separate leechdoms quoted are for "thick eyelids", a swelling, fever, and snake-bite.

Cockayne refers to this plant as Springwort. Grimm also associated *Euphorbia lathyris* with the mystical Springwort, which has the power of opening doors and locks, even medicinal ones - "if anyone is bound by herbs, give him springwort to eat and let him sip holy water …" (Meaney). (see also **SPRINGWORT**).

If you plant caper spurge in the garden, it will keep moles away. That is a widespread belief, and may very well be true. Anyway, the practice spread to America, and there they call this plant Mole-plant, or Mole-tree (G B Foster ; Bergen. 1899).

Capsella bursa-pastoris > SHEPHERD'S PURSE

Capsicum annuum > CHILE PEPPER

Capsicum frutescens > TABASCO PEPPER

CARAWAY

(*Carum carvi*) So ubiquitous was the use of caraway seeds in this country in the nineteenth century and into the twentieth that "Seeds" alone always identified them. Seed cake is a very old recipe – it used to be an essential institution at the feasts given by farmers to their labourers at the end of wheat-sowing (Grieve.1931). In Wiltshire, seed cake was always given at funerals, and in Lincolnshire, seed bread, made with either caraway or tansy seeds, is still traditional at funerals (Widdowson). Lozenge-shaped buns with caraway seeds were called shittles in Leicestershire; they were given to children and old people on St Valentine's Day. The village of Hawkshead, in Lancashire, used to be famous for its Seed Whigs – oblong buns like teacakes in consistency, and flavoured with seeds (Lancs FWI). Caraway comfits are the usual flavouring added to cabbage as it is being salted down to make sauerkraut (Mabey.1972), and Germans use them to flavour cheese, cabbage soup and household bread (Grieve.1933), and one finds them with meat, and in sausages, too (Usher). They have even been put in beer (Johnson).

Caraways are often mentioned by old writers as an accompaniment to apples (Ellacombe). The custom of serving roast apples with a little saucerful of caraway seeds, well known in Shakespeare's time, is still kept up at Trinity College, Cambridge, and at some London livery dinners. And in Scotland a saucerful is put down at tea to dip the buttered side of bread into, and called "salt water jelly" (Grieve.1933). All in all, it is difficult to understand why "caraway seed" entered Lincolnshire dialect as something quite worthless. "I wouldn't give a caraway seed to have it one way or the other" (Peacock). An essential oil from the seed is used in perfumery, but consumption of it in Europe is far more important as a spice, or in the form of oil as an ingredient of alcoholic liquors (Fluckiger & Hanbury), Kummel for instance. A spice wine used to be made from the seeds, too – it was called Aqua compositis. Henry VIII was apparently very fond of it (Genders).

Unlikely as it sounds, caraway was an essential ingredient in love philtres, because it was believed to induce constancy (Clair). It was thought, too, that it conferred the gift of retention (including retention of a husband – even a few seeds in his pocket would prevent the theft of a husband! (Baker)); and it prevented the theft of anything that had some seeds in it, at the same time holding the would-be thief in custody within the house (Grieve.1931). The idea that you could make tame pigeons stay quietly in their lofts if you gave them a piece of dough with the seeds in it (Baker), obviously belongs to the same "retention" superstition.

The medicinal use of the seeds dates back to early times – the Arabian origin of the name caraway points to their knowledge of its values, and by the 12th century that knowledge had spread at least as far as Germany, for two medicine books, of the 12th and 13th centuries, mention the word Cumich, which is still the popular name of caraway in southern Germany. It was certainly in use in England by the end of the 14th century (Fluckiger & Hanbury). Its main use, in the form of an essential oil, has always been as a carminative. "It consumeth winde", Gerard wrote. The prescription appears again in American domestic medicine. Gerard went on with his list of virtues, "it helpeth conception … and is mixed with counterpoysons …" There are just as fantastic prescriptions much later than in his day – for earache, as an example – the patient was advised to pound up a hot loaf with a handful of bruised seeds, and clap this to his ear (Fernie). And Culpeper said the seed "helpeth to sharpen the Eye-sight", and the seed was used in Tibetan medicine to treat eye diseases (Kletter & Kriechbaum). A Cambridgeshire cough remedy sounds more realistic: two ounces of caraway seeds boiled in a quart of water down to a pint, half strain off, sweeten with sugar, add a glass and a half of rum. Take a wineglassful every night on going to bed (Porter.1969)

CARBUNCLE

An ancient cure for the condition used KNAPWEED. The Welsh medical text known as the Physicians of Myddfai has: "For a carbuncle … take the flowers of the Knapweed or the leaves, pounding with the yolk of an egg and fine salt, then applying thereto, and this will disperse it". They have been dealt with too, by using a TOBACCO leaf as a poultice (Thomas & Thomas). Another American remedy was by the use of SWEET FLAG - the roots would be mashed to a similar consistency as mashed potatoes, and that would be spread on a bandage, and used as a poultice (Indiana) (Tyler). One can assume the use of TUTSAN for a carbuncle by one of its Welsh names, Dail fyddigad, carbuncle leaves (Awbery). Its close relative, ST JOHN'S WORT, could also be used, by direct application (Physicians of Myddfai). A poultice of CHICKWEED is still prescribed by herbalists (Warren-Davis).

Cardamine pratensis > LADY'S SMOCK

CARDAMOM

(*Elettaria cardamomum*) An ingredient of curry powder, and a seasoning in many kinds of sausage (Schery), and a flavouring agent in many medicines (Soforowa). The seeds are sometimes burned, too, to produce an incense-like atmosphere (Valiente). In the Near East, coffee was blended with cardamom by the 16th century, and by the 17th the practice had reached Italy. It seems it still survives in Saudi Arabia. It is suggested that the reason was that cardamom was famous as an aphrodisiac. So mixing it with coffee would eliminate the bad effects of drinking the latter (Swahn).

CARDINAL FLOWER

(*Lobelia cardinalis*) An American plant. The Indians had some ritual use for it; the Meskwaki, for example, would throw the "tobacco" made from it into the air when a storm was brewing, to try and dispel the bad weather. They used to throw it into the grave at a funeral, too (H H Smith. 1928). Both the roots and flowers were used by some Plains Indians in love charms (Gilmore), preferably mixed with ginseng and angelica, according to Sanford.

Carduus benedictus > HOLY THISTLE

Carduus nutans > MUSK THISTLE

CARMELITE WATER
Made with BEE BALM, it was once popular as a "restorative cordial" supposed to confer longevity. Probably still made in France under the name Eau de mélisse des Carmes. It is made by macerating the fresh flowers and tops in fortified white wine, together with lemon peel, cinnamon, cloves, nutmeg and coriander (Clair).

CARNATION
(*Dianthus caryophyllus*) Loudon said that we owe the introduction of the carnation into Britain to the Flemings in the 16th century, but actually the plant was well-known under the name of gillofres, for centuries before that (Friend. 1882). It was certainly well in cultivation in Edward II's time, and it was used to give a spicy flavour to ale and wine, particularly the sweet wine presented to brides after the wedding ceremony (Grindon). Hence the name Sops-in-wine. See Shakespeare, *Taming of the shrew*, act 3, sc 2 :

> Quaff'd off the muscadel,
> And threw the sops all in the sexton's face.

Chaucer called it the Clove Gilliflower, an odd combination, for both names mean 'clove', gilliflower being a corruption of *caryophyllus* (clove), the specific name, more properly, 'nut-leaved'. 'Carnation' itself is really 'coronation', from Latin corona, a chaplet, or garland. The flowers must have been used to make chaplets, though Gerard speaks of the "pleasant Carnation colour, whereof it tooke his name". He must have been thinking of incarnardine, the flesh, or crimson, colour.

This is a good luck plant, and figured as such quite a lot in oriental carpet symbolism (Bouisson). It was also a symbol of fascination (Leyel. 1937), and of conjugal felicity, or of marriage itself (Ferguson). In early German painting, it was occasionally used as the emblem of the Virgin Mary, and in Italy too it was put with the lily in a vase by the Virgin (Haig). These days, red carnations are the emblems of workers' movements in most European countries (Brouk).

Lands and tenements in Ham, Surrey, were once held by John of Handloo of the men of Kingston on condition of rendering them three clove gilliflowers at the king's coronation (Friend. 1883). Red roses as rent are perhaps more familiar, but the Ham record of the use of carnations is not the only example. The manor of Mardley, in Welwyn parish (Hertfordshire) was held for the annual "rent of a clove gilliflower", while two of them plus 3s. 6d. paid the yearly rent of 100 acres and a 40-acre wood in Stevenage. And at Berkhamstead a tenant of the royal manor provided one clove gilliflower "at such times as anie King or Queen shall be crowned in the Castle" (Jones-Baker. 1974).

The good luck of a red carnation was once extended to protection from witchcraft in Italy. On St John's Eve, when they were always specially active, all you had to do was to give them a flower. For any witch had to stop and count the petals, and long before she had done that, you were well out of their reach (Abbott).

CAROB
(*Ceratonia siliqua*) "Now John wore a garment of camel's hair, and a leather girdle around his waist; and his food was locusts and wild honey" (Matthew 3; 4). Were Carob pods the locusts that John the Baptist ate in the desert? In any case, they are certainly edible, and known as Algaroba-beans in commerce (F G Savage). The name Locust Bean is applied to the pod (Wit) – this is the name under which they are imported into Britain, and names like St John's Sweetbread (Kourennoff), and St John's Bread (Potter & Sargent) are given in deference to the Biblical association with John. They were the "husks" of Jesus' parable of the prodigal son: "And he would fain have filled his belly with the husks that the swine did eat: and no man gave unto him" (Luke 15; 16). The pods are still used, as in Biblical times, for feeding cattle, horses and pigs, and in times of scarcity they are used for human consumption. When they are completely ripe they are full of a honey-like syrup (Moldenke & Moldenke). Palestinian children will eat them raw, but they are generally boiled down into a kind of molasses (Crowfoot & Baldensperger).

The seeds were once used as weights by apothecaries and jewellers, because of their uniformity (Hyam & Pankhurst). The word carat may come from keration, the Greek name for carob (Potter & Sargent).

There is a cough medicine called carob molasses, which is very popular (Bianchini & Corbetta). Carob is used in Russian folk medicine (half a pound, chopped, in a quart of vodka, taken twice a day), for piles. The patient is warned that no other alcohol should be taken! (Kourennoff).

CAROLINA NIGHTSHADE
(*Solanum carolinense*) Assuming that the name Horse Nettle is meant for this plant, there are a number of uses that R B Browne lists from Alabama. It is recommended for 1) retention of virility in old age (eat a quarter-inch of the stalk), 2) a cough medicine, 3) neuralgia (leaf tea and leaf poultice), 4) toothache (chew it, or put it in the cavity). It was good for teething, too, as well as for toothache – they were strung on a thread and left round a baby's neck until they wore out (Puckett).

Carpobrotus edulis > HOTTENTOT FIG

Carpobrotus equilaterale > FIG MARIGOLD

CARRON OIL

So named because it was first introduced in the Carron Iron Works, is a mixture of equal parts of linseed oil and lime water, and is an embrocation for small injuries, burns, and for rheumatism and gout (Wickham).

Carthamus tinctorius > SAFFLOWER

Carum carvi > CARAWAY

Carya ovata > SHAGBARK HICKORY

Carya tomentosa > WHITE HICKORY

CASCARA SAGRADA

The bark of the CALIFORNIAN BUCKTHORN (*Rhamnus purshiana*). The sacred bark is the best known of all laxatives, no matter under what name it is marketed. The name was given by the Spanish pioneers, who took notice that the Californian Indians used the bark infusion as a physic (Weiner; Schenk & Gifford; Spier).

CASSAVA

(*Manihot esculenta*) Widely cultivated throughout the whole of tropical America as a food plant. It is the tubers that are used, and they are an important carbohydrate source. Tapioca is made from them by peeling, expressing the juice, grating and soaking the pulp, then heating it. This causes the starch to form the small lumps characteristic of tapioca (Kingsbury. 1964). There are two kinds of *M esculenta*, the "sweet" (often described as a separate species, *M dulcis*) and "bitter", differentiated by the hydrocyanic acid content, which is low in the "sweet" kind, so that they can be eaten raw if necessary. Varieties that have higher amounts of the acid are "bitter", and the toxin must be eliminated. It can be lethal, but rendered innocuous simply by peeling the roots and heating them (Kingsbury. 1964).

The Spanish adopted it as a staple, and a substitute for wheat bread. From very early in the colonial period, Jamaica was the important centre for its production and export (Sturtevant). It still is, but there is a ritual use of the poisonous properties of cassava there. It is a valuable food plant, but in spite of that some small farmers will not grow it, simply because they may be suspected, or even accused, of poisoning by obeaj practice. The juice of the bitter cassava caught under the finger nail is said there to be sufficient to cause death, and the record shows all sorts of unlikely practices were believed in, like introducing maggots bred in bitter cassava, or soaking a victim's underclothes in it (Beckwith. 1929).

Cassava has some ritual significance, too, in parts of Africa. In his account of Ndembu symbolism, V W Turner quotes one of his informants: "Cassava is used instead of powdered white clay to purify. Indeed, if people have no white clay to invoke with, they should use cassava meal … Cassava is important at birth and death. When a child is born it is given thin porridge, made from cassava meal. If a sick person is nearly dead, before he dies he asks for thin cassava porridge. He drinks it and dies … Again, when women pass a grave they throw down cassava roots for the dead. They are food for the dead …".

CASSIA

(*Cinnamonium cassia*) In Palestine, cassia oil was mixed with olive oil, and this was used to rub on the feet, and also to massage into the scalp, as it kept hair dark and greasy, and prevented baldness (Genders. 1972). The twigs are used in China to improve blood circulation, and also to treat angina (Geng Junying). But Cassia oil was one of the precious perfumes, an ingredient of the holy oil of the Old Testament. It was also used as part of the incense burnt in the Temple (Zohary). In much the same way as the oak in European mythology, some of the earlier peoples of India regarded the Cassia as the origin of human life (Porteous. 1928).

Castanea dentata > AMERICAN CHESTNUT

Castanea sativa > SWEET CHESTNUT

CAST-IRON PLANT

A name given to the ASPIDISTRA (Bonar), in the same vein as Cannon-ball Plant. It seemed to the Victorians that nothing could harm this plant. Cf too Bar-room Plant.

CASTOR OIL PLANT

(*Ricinus communis*) It once had some very odd properties ascribed to it, affecting the very name of the plant and its by-product. It was thought to be proficient in assuaging the natural heat of the body, and it had the reputed power to "soothe the passions", so it was called by the French Agnus Castus, Spanish agno casto, a name now reserved for *Vitex agnus-castus*, the Chaste Tree (Maddox). So 'castor' was originally 'casto', solely because of a dubious reputation, which, given its undoubted laxative virtues, may very well have been justified.

The plant's dissemination must have taken place very early, for it was cultivated as a drug plant by the ancient Egyptians. It was sacred enough to have been put in sarcophagi by 4000BC, so that the dead would have use for the bean in the other world. There is a whole list of medicinal uses in the Ebers Papyrus and other lesser known ones. They used the seeds as a purge, taken with beer, or as a painkiller, particularly for sores, in the form of an ointment, which was also used for a disease tentatively identified as alopecia. Again, the seeds, mixed with beans, were used as a fumigant, and employed "to drive away the influence of a god or goddess, a male poison or a female poison, a dead man or a dead woman" (Dawson. 1929).

Castor oil is a vermifuge, and, of course, a laxative, and a particularly valuable one, though the smell and taste have made it "a by-word for offensiveness" (Maddox). In Haiti it is used for colic, for eye trouble and headaches, or, in America, to put on a wart (Thomas & Thomas). African peoples like the Mano, of Liberia, use it for headaches, too, by rubbing the leaves in water, and bathing the head with the infusion (Harley), while in the southern states of America a similar practice merely involves wrapping the forehead with the leaves, which will treat a fever as well (Puckett). Kentucky practice was to carry a castor-bean about on the person, for indigestion (Thomas & Thomas). There is, too a certain amount of ritual use, which probably includes the Shona (Zimbabwe) habit of smoking the leaves like a cigarette, ostensibly to cure hiccups, it is said (Gelfand. 1956). Certainly, the Brazilian curanderos used it ritually, as a fumigant, and as an ingredient in ritual baths in their healing ceremonies (Williams. 1979). Palma Christi is a name sometimes used for this plant. It is at least as old as Turner (1548), and means Christ's hand (palma is the palm of the hand). It is a reference to the palmately divided leaves. The open hand, as well as the fist, is a potent instrument for dispelling ill-wishing or the evil eye, and it is interesting to find that castor oil plant leaves have been worn round the neck to ward off devils, because the leaf is like that open hand (C J S Thompson. 1897).

Catalpa bignonioides > LOCUST BEAN

CATARRH

was cured at one time by taking the powder made from CONKERS as a snuff. The Pennsylvania Germans used it that way (Fogel), but this was quite an early habit (Thornton), and the idea was to grate them up and use the powder to make one sneeze. Apparently it was recommended not only as a powder, but also as an infusion or decoction to take up the nostrils. SANICLE can be used to treat catarrh (an infusion of the astringent leaves) (Conway). Smoking the crushed berries of VIRGINIAN JUNIPER is an American domestic remedy for catarrh (H M Hyatt). The fern known as POLYPODY was made in Scotland into a medicine for catarrh (Beith), and FENUGREEK is also used as a traditional treatment for the condition (Schauenberg & Paris), who also suggest that the infusion of the flowers of SMALL-LEAVED LIME was used.

CAT'S FOOT

(*Antennaria dioica*) Hungarian gypsy girls, according to a correspondent of the *Journal of the Gypsy Lore Society. Vol 37; 1958*, used to gather large bunches of these flowers at the time of their betrothal. They were kept hidden for a year with her marriage clothes, for they kept her free from sickness and harm. During the week before the wedding, she took them to the nearest crossroads and burnt them, so that no other girl, not even her own sister, could steal them, and by so doing, alienate her bridegroom. There have been a few medicinal uses. Herbalists prescribe it for treating bronchitis and bilious conditions (Schauenberg & Paris). And a tea made from the dried plant is drunk for diarrhoea (Flück). The astringent properties implicit in that treatment will also explain its use for quinsy and mumps (Grieve. 1931). One of the names given to it in America, is Ladies' Tobacco (Leighton), which must mean that experiments have been made in smoking the dried herb, and that it was found to be very mild in character – hence "Ladies'".

CATCHTHORN

(*Zizyphus abyssinica*) In Malawi, an extremely potent alcoholic drink called kachaso is made from the berries (Palgrave & Palgrave). The leaves are chewed as an aphrodisiac, and a root decoction is used as an abortifacient, while the root infusion is taken for dysentery. There is a lot of tannin in both the bark and leaves, so its use for dysentery probably depends on this. A powder prepared by drying and pounding is used to rub into incisions made on the chest in cases of pneumonia (Palgrave & Palgrave).

Catheranthus roseus > MADAGASCAR PERIWINKLE

CATMINT, or CATNIP

(*Nepeta cataria*) Cats "are much delighted with catmint, for the smell of it is so pleasant unto them, that they rub themselves upon it, and wallow and tumble into it, and also feed on the branches very greedily" (Gerard).

> If you set it
> The cats will eat it;
> If you sow it
> The cats will not know it (Barton & Castle).

Rats, on the other hand, hate it. Plant it thickly round the walls of a rat-infested house, and they will soon be cleared (Quelch). It was once believed that chewing the plant created quarrelsomeness (Dyer. 1889), or rather, courage. There was also a legend that the hangman could never pereform his duty until he had chewed a root of catmint (Genders. 1971).

Pillows are stuffed with the dried leaves, for the smell is supposed to help the sleepless (Sanford), just as catmint tea, infused from the dried herb, is helpful in fevers, for producing quiet sleep, and for nervous headaches (Brownlow). The tea was also given for stomach-ache, especially for babies with colic (R B Browne, Stout, H M Hyatt). They say in Alabama that "catnip tea is good for baby's head" (R B Browne). This tea has even been given in cases of pneumonia (Sanford), herbalists still recommend it for chronic bronchitis (Schauenberg & Paris), and it is taken for a cold in Ireland, the instructions being to boil catmint and drink the water (Maloney). It has been used as a jaundice remedy (Barton & Castle), and a hot poultice of catmint leaves is applied to a boil (R B Browne).

Nip, or Nep, are contractions of Catnip; there is a saying in Suffolk, "as white as Nip" (Gurdon). For the wild plant (*Nepeta cataria*), has white flowers. The more familiar garden catmint, with blue flowers, is *Nepeta faassenii*.

CATSTAIL
(*Phleum nodosum*) In the Faroes, the flower heads of this grass are treated as plantain would be, on St John's Eve. The stamens would be pulled off, and the flower put under the pillow, while making a wish before going to bed. If the stamens had grown again next morning, the wish would come true (K Williamson).

CAYENNE PEPPER
Cayenne pepper is "an indiscriminate mixture of the powder of the dried pods of many species of capsicum, but especially of the *capsicum frutescens* [see TABASCO PEPPER]…, which is the hottest of all" (Thornton).

Ceiba pentandra > KAPOK TREE

CELANDINE
i.e., GREATER CELANDINE (*Chelidonium majus*), or LESSER CELANDINE (*Ranunculus ficaria*)

CELERY
(*Apium graveolens*, which is Wild Celery; Garden Celery is *A graveolens 'Dulce'*). Poisonous in its wild state, at least to cattle (Kingsbury. 1964) but blanching is the simple process that turns a poisonous plant into an edible one, though eating it will bring bad luck, at least according to Kentucky belief (Thomas & Thomas), but to dream of it is said to be a sign of robust good health (Raphael). On the other hand, it did appear, according to Graves, in an English formula for witches' flying ointment. Jacob says it was used by witches to prevent their getting cramp while flying. Why not? Celery was quite a common medieval remedy for cramp, at least, the earthbound kind.

The seeds are a food, as well as the stems (celery spice, celery salt, etc.,), and the seed has been used in medicine, too, since medieval times. The pre-scription for "aching of the hollow tooth", given by Dawson, dates from the 15th century: "Take the seed of apium, a scruple, that is to say, a pennyweight, and of the leaves of henbane and of the seeds of avens two scruples, and grind it small with aqua vitae and make pellets thereof of the size of a vetch and lay on the tooth that acheth and it shall cease anon for it be proved that it hath ceased the ache in half an hour". It is still being used; a decoction of the seeds relieves lumbago and rheumatism, so it is claimed (Newman & Wilson), or the medicine is sometimes taken in the form of a tea (Browning). An American rheumatism cure was to use celery seed worn in a bag round the

neck. A wedding ring in the bag would clinch the cure (Whitney & Bullock). Russian folk medicine agrees that celery is good for rheumatism, but here it is the stem that is used. Half a pound of celery, without the roots, is simmered in a quart of water, strained, and this provides a day's dose divided into three, and taken hot (Kourennoff). Indiana home remedies agree, too, that lots of celery, boiled in milk or water, will give relief (Tyler).

Medieval medical receipts were quite keen on celery, as it was prescribed for widely differing complaints – "ache of wound" (Dawson), for example. Topsell suggested that "apium seed" was used for snakebite, even. Other conditions included a stroke (a blow in this case), heart disease, or "who that spitteth bloud", backache, and so on. Gerard was just as keen, as were other herbalists of his generation, recommending it for agues, jaundice, mouth ulcers, whitlows, etc., It has well-known diuretic effects, and would certainly help anyone suffering from kidney stone, strangury, etc., Lastly, sufferers in Norfolk would cure a hangover by the simple expedient of chewing celery (V G Hatfield. 1994), and eating it is a simple Irish remedy for indigestion (Maloney).

CELERY-LEAVED BUTTERCUP
(*Ranunculus sceleratus*) This is the most poisonous of the buttercups. It is, as Rambosson described it, "…une espèce très dangéreuse, dont les seules émanations excitent l'éternuement et des armes. Prise à l'intérieur, elle produit la contraction de la bouche et des joues, un rire sardonique…", which is probably what Apuleius was talking about in the Anglo-Saxon version of his herbarium, translated by Cockayne as "…whatsoever man fasting eats this wort, leaves his life laughing". (There is a French name, Herbe sardonique). Nevertheless, he goes on to list a few medicinal uses, for wounds and running sores (which can be caused anyway by contact with the extremely acrid juice, for which the plant had to be prepared as some kind of poultice, and "against swellings and against warts" (that is: strumae et furunculi), when it had to be pounded with swine's dung, and laid on – "within a few hours it will drive away the evil, and draw out the pus". Celery-leaved Buttercup crops up in Chinese medicine, the seeds being used for colds, rheumatism, etc., (F P Smith), and the whole herb is sued for ailments as diverse as malaria, boils and snake-bite (*Chinese medicinal herbs of Hong Kong*).

Centaurea cyanus > CORNFLOWER

Centaurea nigra > KNAPWEED

Centaurea solstitialis > ST BARNABY'S THISTLE

Centaurium erythraea > CENTAURY

CENTAURY
(*Centaurium erythraea*) The name is taken as Centaur (Chiron the Centaur is said to have healed a wound with

this plant). Another strange belief comes from the pen of Albertus Magnus : "Witches say that this herb hath a marvellous virtue, for if it be joined with the blood of a female lapwing … and be put with oil in a lamp, all they that compass it about shall believe themselves to be witches, so that one shall believe of another that his hand is in heaven and his feet in the earth …"

Centaury earned a reputation as a tonic (Vickery. 1995). Gypsies used it as such (Vesey-Fitzgerald). The juice (in whisky?) is described on South Uist as excellent for the one in need of a nerve tonic or for weakness following an illness (Shaw). From an Irish recipe: a cordial was made of one part gentian, two parts of centaury, bruised well together, and mixed with distilled water for a drink (Wilde. 1890). It is, too an Irish remedy for asthma (Ô Súilleabháin), and it is taken internally for muscular rheumatism (Moloney).The flower or leaf infusion makes a good wash for wounds and sores, for it is strongly antiseptic (Conway). It is also used as a wound salve in the Balkans, and is drunk in decoction to check menstruation. or when blood is found in urine or the spittle, and thus for coughs or chest pains (Kemp). And in early times it had a great reputation for healing snake-bites (see Cockayne, and Dawson. 1934). Topsell [*The history of four-footed beasts* (1607), and *The history of serpents* (1608)] added his weight. For snake-bite, he wrote " … after cupping glasses and scarifications, there is nothing that can be more profitably applied than centaury, myrrh, and opium …". Later, "for the trial of the party's recovery, give him the powder of centaury in wine to drink; and if he keeps the medicine, he will live; but if he vomits or casts it up, he will die". Gerard noted that "the juice is good in medicines for the eies …", and long before his time the Anglo-Saxon version of Apuleius had "for sore of eyes, take this wort his juice; smear the eyes therewith; it heals the thinness of sight …".

In the 16th century, Lupton offered the advice, "drink the juice of centaury, once every morning four days together, and it will thee sing clear, and speak with a good voice…".

Ceratonia siliqua > CAROB

Cercis siliquastrum > JUDAS TREE

CEYLON LEADWORT
(*Plumbago zeylanica*) It is popular throughout Africa as a remedy for parasitic skin diseases, especially leprosy; it is also used as a root decoction to treat scabies and ulcers (Sofowora). It is used in the Philippines as an abortifacient (Watt & Breyer-Brandwijk), but this is just as toxic as its relative, *P indica*. The sliced root and leaves are acrid enough to produce blisters, deliberately done sometimes. That acrid juice was the means of tattooing among some of the southern African peoples like the Gwembe Tonga.

After several hours contact with the flesh, it is said to leave an indelible mark (Scudder).

Chaerophyllum temulentum > ROUGH CHERVIL

Chamaemelum nobile > CAMOMILE

Chamaenerion angustifolium > ROSEBAY WIL-LOWHERB

CHANGELINGS
Part of fairy lore, changelings are the fairy substitutes for human babies that they have abducted. There are standard procedures, often violent, for forcing the fairy abductors to return the true child and take away the changeling. Sometimes the procedure involves plants, as with FOXGLOVE, for instance. The suspected changeling is bathed in the juice, and fairy-struck children had to be given the juice of twelve (or ten, some say), leaves of foxglove (Wilde. 1902). Or a piece of the plant could be put under the bed. If it is a changeling, the fairies would be forced to restore the true child (Mooney). Simpler still, put some leaves on the child itself, and the result would be immediate (Gregory). And instructions from County Leintrim advised a suspicious parent to "take lusmore [an Irish name for foxglove] and squeeze the juice out. Give the child three drops on the tongue, and three in each ear. Then place it [the suspected changeling] at the door of the house on a shovel (on which it should be held by someone) and swing it out of the door on the shovel three times, saying "If you're a fairy, away with you". If it is a fairy child, it will die that night; but if not it will surely begin to mend" (Spence. 1949).

A Scottish way to get rid of a suspected changeling was to build the fire with ROWAN branches, and to hold the child in the thick of the smoke. The brat would disappear up the chimney, and the true child would be returned (Aitken).

CHAPPED HANDS
Rubbing on the leaf juice of REST HARROW is an Irish means of dealing with chapped and rough hands (Moloney). The leaves of WATER BETONY were used in the Fen country to treat chapped hands and feet, sore heels, and the like (Marshall). PRIMROSE flowers and young leaves boiled in lard make an ointment for healing cuts and chapped hands, and an ointment made from the crushed roots of MARSH MALLOW is good for the condition (Page.1978).

CHARMS
(against the evil eye) GARLIC – see EVIL EYE
"(for the cure of warts) see WART CHARMS
"(to transfer an illness to another person, or tree, etc.,) see TRANSFERENCE CHARMS
"(to discover the identity of a future partner) see DIVINATIONS
"(for good luck) see GOOD LUCK CHARMS

CHASTE TREE

(*Vitex agnus-castus*) The "chaste" of the common name is the Latin castus, Sometimes the shrub is called Chaste Willow (Pomet), "willow" because of the pliant branches, used in basketry, and particularly for bee-skeps (Mabberley).

There is a suggestion that the leaves may be contraceptive; certainly Dioscorides said so, and Pomet's comment on the name *agnus-castus* is equally explicit: "and the name of Agnus Castus, because the Athenian ladies who were willing to preserve their Chastity, when there were Places consecrated to the Goddess Ceres, made their beds of the Leaves of this Shrub, on which they lay; but it is by way of ridicule that the name of Agnus Castus is now given to this seed, since it is commonly made use of in the Cure of venereal Cases, or to assist those who have violated, instead of preserv'd, their Chastity". But it was still used by Athenian women as a symbol of chastity in religious rites. And for that reason it became also a symbol of indifference (Pomet).

Chelidonium majus > GREATER CELANDINE

Chenopodium album > FAT HEN

Chenopodium ambrosioides > AMERICAN WORM-SEED

Chenopodium bonus-henricus > GOOD KING HENRY

CHERRY LAUREL

(*Prunus laurocerasus*) This is the shrub, usually known simply as Laurel, so beloved of Victorian shrubbery owners, and so typical of them that the shrub is known in America as English Laurel (Cunningham & Côté). It is difficult to see why it achieved such popularity. Certainly it more or less looks after itself, but it is highly poisonous. As early as 1731, Madden of Dublin drew the attention of the Royal Society to some cases of poisoning that had occurred by the use of a distilled water of the leaves. This water had been used in Ireland for flavouring puddings and creams, and also as an addition to brandy. This is actually hydrocyanic, that is, prussic acid! O P Brown recommended the leaves as "an excellent sedative", and in fact a tincture made from the leaves is still in use in homeopathy as a sedative (Schauenberg & Paris).

A Witney (Oxfordshire) love divination was for a girl to prick her lover's name on a laurel leaf. If the prick marks turned pink, it was a sign that he would marry her, but if not, he would desert her (*Oxfordshire and District Folklore Society. Annual Record; 1952*).

CHESTNUT SUNDAY

Chestnut Sunday is (or rather was) the Sunday before Ascension Day, when crowds flocked to Kew Gardens and Bushy Park (Surrey), in particular, to see the chestnuts in full flower. The custom was revived in 1977, the year of Queen Elizabeth's jubilee, and is now held on the nearest Sunday to 11 May (Mabey. 1998).

CHICK PEA

(*Cicer arietinum*) After a birth, every relation of the family on the Aegean island of Chios brings with him a plate of chick peas on his visit of congratulation (Argenti & Rose). Why this and no other pulse is not clear.

CHICKENPOX

Herbalists still prescribe marigold flowers to lay on the rash for both chickenpox and shingles (Warren-Davis).

CHICKWEED

(*Stellaria media*) A lucky plant, at least to the people of the Fen country of England, where it used to be grown in pots, to bring good luck to the house. If the plant was gathered when the dew was still on it, it was thought to turn the plainest woman into a beauty (Porter. 1969). Apparently, chickweed in the language of flowers symbolized rendezvous (Leyel. 1937), quite why is not clear. There is a little weather lore associated with the plant – it expands its leaves when fine weather is to follow. But if it should shut up, then the weather will be foul. As a compromise, half-opening flowers are a sign that the wet will not last long.

Chickweed was used extensively in the medicinal field. It contains copper and iron, and so it is useful for anaemia, and kidney and skin disorders can be treated with it. The last-named was known to Gerard, too – "the leaves boiled in vinegar and salt are good against manginesse of the hands and legs, if they be bathed therewith". Chickweed was used to make poultices for boils in Irish folk medicine (Barbour), and the treatment was known in Newfoundland, too (Bergen. 1899). In Somerset, the poultice is used for abscesses and ulcers, too (Tongue. 1965), and it was a common Dorset remedy for gatherings and boils (Dacombe), and in Norfolk for quite severe cases of dermatitis or eczema (V G Hatfield. 1994). The practice in Hampshire was to mix groundsel with chickweed in making this poultice (Hampshire FWI), but this kind of use was well-known in the Highlands of Scotland, too, for carbuncles and abscesses. The traditional way of preparing it there was to bruise the plant between a flat and round stone kept for the purpose (Beith). They can be used for warts, too, by rubbing the fresh juice on them. But it seems the warts had to be pared to their quick first, then they would fall off (Fernie). It can be applied to bee stings, too, to get the swelling down (Vickery. 1995). Half genuine and half charm is one cure from Ireland: the healer rolls up some chickweed into a small ball. This he rubs "upwards" on the rash (i.e., towards the heart), at the same time saying "In the name of the Father, and of the son, and of the Holy Ghost" (Buckley).

Chickweed can be used for rheumatism with some effect. Homeopathic doctors prescribe what is described as an essence of the fresh plant, to be taken to relieve rheumatic pains, and it can be used externally as a rub (Schauenberg & Paris), or, as in Sussex, it can be crushed and laid on as a poultice (Allen). In Scotland an ointment made from it is used to like effect, and that ointment is also used on children's chilblains and rashes (Vickery. 1995). Herbalists still prescribe it for carbuncles, abscesses, etc., (Warren-Davis). In Ulster, it was used for sprains, either boiled and made into a poultice, and applied very hot, or "roasted", and made into an ornament (Maloney).

Gypsies use a leaf infusion to cure coughs and colds (Vesey-Fitzgerald), and it is an old Irish remedy for whooping cough (Ô Súilleabháin), while in Skye the feet and ankles of a fever patient were at one time washed in warm water in which chickweed had been put, as a means of getting the patient to sleep (Martin). The plant is used by herbalists in the treatment of stomach ulcers, and as an aid to digestion (Conway). A lotion made of chickweed and rose leaves is used in Somerset for sore eyes (Tongue. 1965), or it could be used with unsalted lard as an ointment. There is a strange usage from Japan: an infusion of the leafy shoot with sugar is given internally to stop a nosebleed (Perry & Metzger).
The name Chickweed has a fairly obvious derivation. Chickens like it - "chickens and birds love to pick the seed thereof" (Coles). It is, in fact, a rich iron tonic, long given to cage birds, too.

CHICORY

(*Cichorium intybus*) It was cultivated on the Continent up to World War II, for the root, which is used as a substitute for coffee, and is sometimes mixed with real coffee as an adulterant. It is very bitter, and contains no caffeine or tannin (Sanford). It was introduced as a coffee substitute in the 18th century, but it was actually banned by a law of 1832, repealed in 1840, though it has now virtually died out as a coffee substitute (Brouk). Another use for the foliage, which is edible once the bitter principle is removed by twice boiling, was for a blue dye (Hemphill).

The seed was at one time used as a love potion. In a German story, a girl whose lover had gone away on a voyage devoted her life thereafter to sitting at the wayside and looking for him. She waited so long that she finally took root and became the blue chicory, which is known as Wegenwarte (Philpot), the watcher of the road. One version of the story attributes the woman's desertion to good cause, and where that idea prevails, the plant is known as the accursed maid (J M Skinner). It is a symbol of frugality (Leyel. 1937). Country people used to tell the time by chicory flowers, and Linnaeus used them in his "floral clock" (Krymow).

Chicory has the power of conferring invisibilty, and it was once hung on the banners of those going on a crusade or exploring new lands. Many a prospector in the Californian gold rush had a chicory root in his pocket (R L Brown), for the plants that conferred invisibility almost invariably found treasure and opened locks. Like Moonwort, it could open a locked box – a leaf had to be held against the lock. For the magic to work, it had to be picked on St James's Day, 25 July, and had to be cut in silence with a golden knife. If the cutter spoke during the work, he would die (M Baker. 1980). Another of the magical uses has to do with crystal gazing. An exponent said that good results could be had by drinking an infusion of mugwort, or else chicory (Kunz). One story about the woodpecker concerns chicory. It was said that the bird got its strength by rubbing its beak against a plant that flowered only at midnight on Midsummer Eve. Chicory was one of the candidates for this identification (Hare).

It was very much a liver (it was known as "liver's friend) and jaundice plant (see Gerard, and Thornton). Gypsies use a root decoction for jaundice (Vesey-Fitzgerald), and chicory flowers were used for liver complaints (Pollock). Distilled water of the flowers has been recommended as a good eyewash (Genders. 1976); this may be a cosmetic, too, for an infusion of the whole plant has long been used as such – applied at bedtime, it will remove blemishes from the skin. There is nothing new about that, for Thomas Hill in 1577 wrote that "Ciccorie cureth scabbed places, causeth a faire skin …". He went on to claim that "it helpeth … the kings evill, the plague, burning agues … and cureth the shingles". It has been recommended for easy childbirth (V G Hatfield. 1984), and Lupton quoted Mizaldus in making the extraordinary claim that "if a woman anoint often her dugs or paps with the juice of succory, it will make them little, round and hard; or if they be hanging or bagging, it will draw them together, whereby they shall seem as the dugs of a maid". Succory, incidentally, is, most probably, another version of chicory.

CHILBLAINS

In some places, the juice of LEEKS mixed with cream was used (Dyer). Gerard recommended TURNIPS for chilblains, for which "oil of Roses boiled in a hollow turnep under the hot embers" is also good. They can be treated with HORSERADISH, too, by wrapping the grated root round the finger/toe, and keeping it in place with a piece of lint (Rohde. 1926). But the best known treatment was to thrash them with HOLLY till they bled, some say, to let the chilled blood out. This is a logical enough cure, for chilblains are caused by poor circulation, and thrashing them deals with this. But there is a variant to the belief – the feet should be crossed while it is being done (Igglesden), and that must surely be illogical. Another chilblain remedy from Essex also enjoined the use of holly, but this time as an ointment. The berries are to be powdered and mixed with lard (V G Hatfield). Similarly, from Wiltshire, with the added proviso to use the ointment liberally, wrap the part with an old stocking, and then toast it in front of the fire until the heat becomes unbearable (Whitlock. 1988). The prickly stems of BUTCHER'S BROOM were used also to thrash the chilblains (Grieve), just as holly was.

WOODY NIGHTSHADE's red berries were used to deal with whitlows (felons, as they used to be called). They were bruised and applied directly to the place. Warwickshire people did exactly the same for their chilblains, and made a habit of preserving the berries in bottles for just that purpose in winter time (Bloom). A cure from Essex was an ointment made from lard and the juice of DEADLY NIGHTSHADE (V G Hatfield. 1994), a logical remedy, for a decoction of the plant, used externally, was prescribed as a means of keeping up the blood circulation (Brownlow). TUTSAN roots boiled, and the liquid poured "upon curds. Pound the same with old lard, and apply as a plaster" were used in Wales (Physicians of Myddfai). The infused leaves of SANICLE were once used as a lotion for chilblains (Mitton), and a decoction of QUINCE pips is still used sometimes, to apply to skin complaints like chilblains (Schauenberg & Paris). An Essex treatment of chilblains was to take WHITE BRYONY berries, crushed and rubbed on (V G Hatfield. 1994), and BLACK BRYONY, entirely unrelated, berries were used, too (they and the roots were steeped in gin and then applied (Wiltshire, Vickery. 1995). They used to be treated in Ireland simply by rubbing ONION juice on them (Maloney), or putting a roasted onion on them as hot as could be borne. In Gerard's opinion, "the inner part of SQUILLS boiled with oile and turpentine, is with great profit applied unto the chaps or chilblains of the feet or heeles ..." In Scotland, an ointment made from CHICKWEED was used on children's chilblains and rashes (Vickery. 1995). WALL PENNYWORT, too, was used at one time (Tynan & Maitland), and so was the tincture of ARNICA, provided the chilblains were whole (Grieve. 1931). An ointment made from the crushed bulbs of SNOWDROPS has been used on chilblains (and to treat frostbite!) (Conway).

CHILDBIRTH

BIRTHWORT (*Aristolochia clematitis*) must be mentioned first in this connection. *Aristolochia* itself comes from two Greek words meaning 'best birth', and the association stems from the shape of the flower, which constricts into a tube that opens into a globular swelling at the base. The swelling was interpreted as the womb, the tube as the birth passage. So by the doctrine of signatures, it was used to help delivery, to encourage conception, and to "purge the womb". Oddly enough, it seems that the plant apparently does have an abortive effect (Grigson). Pliny said it was prescribed for securing a male child (Bonser). That prime fertility agent, MISTLETOE, was used in parts of France in cases of difficult labour

(Salle). The shoots of HOLLYHOCK, too, have been used in Chinese medicine to make labour easy (F P Smith). Sudeten women used to put LADY'S BEDSTRAW in their beds to make childbirth easier and safer, and since women who have just had a child are susceptible to attack by demons, they would not go out unless they had some lady's bedstraw with them in their shoes (Grigson. 1955).

RASPBERRY leaf tea is a general country drink taken to ensure easy childbirth. They say it should be started three months before the birth is due, and taken two or three times a week (Page. 1978). Highland women used it too as a means of strengthening the womb muscles (Beith). The tea was an old remedy for relieving morning sickness, and powdered leaves can be bought in tablet form (Addison. 1985); they are said to ensure relaxation in childbirth, a function that the fresh fruit will perform just as well. African women grind the bark of AKEE and mix it with locally made black soap to wash with during pregnancy. This is supposed to ensure easy delivery when the time comes (Soforowa), while other African peoples recommend that seven seeds of MELEGUETA PEPPER, with a piece of paw-paw root, should be chewed during labour. It is supposed to cause immediate delivery (Soforowa). Gypsy women used to drink LINSEED tea during pregnancy to ensure an easy birth (Vesey-Fitzgerald), and a Swedish belief had it that HEATH SPOTTED ORCHID (*Dactylorchis maculata*), known there as Maria's Keys, used to be put in the pregnant woman's bed as an amulet for easy delivery. A prayer was said at the same time in which the Virgin's keys were referred to, and the loan of them asked during childbirth (Kvideland & Sehmsdorf).

Greek midwives made sure that, at the birth of a child, the whole room smelled of GARLIC, and a few cloves had to be fastened round the baby's neck either at birth or immediately after baptism (Lawson). Palestinian mothers and new-born babies must be protected from Lilith with garlic cloves, for she would otherwise strangle the babies, and frighten the mother into madness (Hanauer). All this highlights garlic as a protector from evil influences. Another Greek, or rather Cretan, practice was to use DITTANY in difficult childbirth. That plant was dedicated to the goddess Lucina, who watched over women in childbirth (Gubernatis). But the earlier belief was "a hind ... eateth this herb that she may calve easilier and sooner ..." (Bartholomew Anglicus), and observation of this led women to adopt the practice. PEARLWORT, put under the right knee of a woman in labour, soothed her mind and protected her child and herself from the fairies (J G Campbell. 1902), for this is the mystical Mothan, very important in Gaelic communities.

In the Highlands, at the birth of a child, the midwife used to put a green ASH stick into the fire, and while it was burning, let the sap drop into a spoon. This was given as the first spoonful of liquor to the newborn baby (Ramsay). The universe tree in Scandinavian mythology, Yggdrasil, is taken to be an ash. The cooked fruit of this tree ensures safe childbirth. Yggdrasil itself is the source of all new life (Crossley-Holland).

HONEYSUCKLE. In the Scottish ballad of Willie's Lady, the witch tries various means of preventing the birth of the Lady's child, including a "bush o' woodbine" planted between her bower and the girl's. Once this "restricting, constricting, plant" has been removed, the birth proceeds normally (Grigson).

The Physicians of Myddfai made the extraordinary claim, "if a woman be unable to give birth to her child, let the MUGWORT be bound to her left thigh. Let it be instantly removed when she has been delivered, lest there should be haemorrhage". Clearly this is a garbled version of an older, perhaps genuine, usage of the plant. It is a theme taken up by Gerard, for he said "that it bringeth down the termes, the birth, and the afterbirth…". Perhaps the answer lies in the old belief that Artemis helped women in childbirth (the generic name for this plant is *Artemisia*). FAIRY FLAX (*Linum catharticum*) has a similar use. Just putting it under the soles of the feet, so it was believed until quite recently in the Hebrides, was an aid to easy childbirth (V G Hatfield. 1984). But this plant, as its specific name implies, is an effective purge (Purging Flax is another name in English), and it was well known in the Highlands for gynaecological and menstrual problems (Beith). CAMOMILE tea is good for women in labour (Thonger) – it is good for virtually anything, and a regular panacea.

PARSLEY superstitions include many connected with conception and childbirth, summed up in the saying "sow parsley, sow babies" (Waring). The parsley bed, like the gooseberry bush, was once "the euphemistic breeding grounds of babies" (Gordon. 1977), or at least girl babies were found there (Baker. 1977). But many of these parsley beliefs are confused, ranging as they do from aphrodisiacs to abortifacients. WILD PARSNIP was used in Anglo-Saxon times for a difficult labour (M L Cameron).

The Hopi Indians made ritual use of CALIFORNIAN JUNIPER during childbirth, either by a chewed piece, or a tea made from the leaves. During the lying-in period, all of the mother's food had to be prepared in some degree with a decoction of juniper leaves. Her clothes, too, had to be washed in water that had some of the leaves in it. The newborn baby itself was rubbed with ashes from burned juniper, and if later on in its life the child misbehaved, recourse was made again to the juniper. The child was taken, at his mother's request, and held by some other woman in a blanket over a smouldering fire of juniper. He soon escaped, of course, half suffocated, supposedly a

better and wiser child (Whiting). Some Indian peoples made a tea from the leaves of PLANTAIN-LEAVED EVERLASTING for mothers to drink for two weeks after giving birth (H H Smith. 1928). Menomini Indian women used the tea made from DUTCH RUSH, which is a Horsetail, to clean up the system after childbirth (Youngken).

Fenland midwives used to give a "pain-killing cake" to women in labour. Apparently it was made from wholemeal flour, hemp seed crushed with a rolling pin, crushed rhubarb root, and grated DANDELION root. These were mixed to a batter with egg yolk, milk and gin(!), turned into a tin and baked in a hot oven. At the woman's first groan, a slice of cake would be handed to her (Porter. 1969). Hempseed, rhubarb and gin would have quite an effect, but it is not clear how dandelion fits into the pattern.

An example of sympathetic magic at work is the use of ROSE OF JERICHO (*Anastatica hierochuntica*) when there is a difficult birth. As the seeds ripen on the plant, during the dry season, the leaves fall off, and the branches curve inwards to make a ball, which is blown about the desert until the rainy season begins. So, when birth is difficult, they put the ball in water, so that it will slowly open, thus sympathetically bringing about the same result for the women in labour.

CHILDREN'S GAMES

Children still make "arrows" of the stems of RUSHES, described as half-peeling a strip of the outer skin away from the pith, balancing the stem across the top of the hand, then pulling sharply on the half-peeled strip, which propels the arrow at a target. Girls make "Lady's Hand-mirrors", by bending the stem of a rush sharply at the middle at two points about a quarter of an inch apart. The stem is then plaited by bending each side in turn sharply over the other at right angles, and so on (Mabey. 1998). REEDS, too, have their uses. Children would put the leaf between their palms and blow, to make a piercing whistle, and another pastime in the Fen country was to make boats out of the leaves. "You took the leaf with its hard little stalk still on it and folded each end back. Then you split the folded ends into three and tucked one of the outside ones through the other outside ones, leaving the middle one flat for the little boat to sail on, and the stalk would stick up in the middle like a real little mast" (Marshall).

COUCHGRASS earns the name Grandmother Grass from a game children play with it. It involves cutting the head of a piece of the grass, and sticking it in another head, still on its stem. A flip of the hand holding the stem, and "Grandmother, grandmother, get out of bed" is recited as the first head springs out (Mabey. 1998).

The best known of all is CONKERS, the nuts of HORSE CHESTNUT, that are Conquerors or shortened to Conks. The reference, of course, is to the game children play with the nut on a string, to be struck by the opponent's similarly strung conker in an effort to establish a conquering, or champion, nut. So popular has it been that there is now a World Conker Championship, held at Ashton, Northamptonshire, on the second Sunday in October (Mabey. 1998). A variation on the game was described by George Bourne. It was called Mounters, and consisted in whirling a conker on its string round and round, and letting go so that it "mounted" up into the sky. Not much of a game, perhaps, but the tree was actually called Mounter-tree at the time of year the game was played. The petioles of Horse Chestnut are known as Knuckle-bleeders in Norfolk, evidence of another boys' game.

The flower heads of RIBWORT PLANTAIN are used as "soldiers" or "fighting cocks" in a children's game. "Soldiers" is described in Kintyre : one child holds out a duine dubh (black man), and his opponent tries to decapitate it with another. If one soldier knocks the head of another, it is called a Bully of one, Bully of two and so on (MacLagan). Hebridean children knew the ribwort stalks as "Giants", but the game was played in exactly the same way. Girls find the "giants" useful for making daisy chains – they pick the daisy tops and string them on the tallest giants they can find (Duncan). The "soldiers" game is also known as Black Man (the duine dubh already mentioned), Cocks-and-Hens, Hard Heads, or Knights (Opie & Opie. 1969). "Cocks", for the plant as well as the game, is the name in northern England (Brockett), and in Kent it is apparently called "Dongers" (Mabey. 1998). The game had another significance in Somerset, where the plant is given the name (among many others) of Tinker-tailor Grass. The blow that knocks the head off marks the profession of the future husband, in the Tinker-tailor-soldier-sailor tradition (Elworthy. 1888), a divination game is also played with RYE-GRASS. The "Soldiers" game is also known as Kemps in the northern counties and Scotland. To kemp is to fight (OE campe, soldier, etc.,). In some parts of Scotland, the game is called Carldoddie, probably from the names of Charles Stuart and King George (MacLagan) – carl is Charles (the Prince) and doddies were the supporters of King George, doddie being the local name for George.

A children's divination game once played with COWSLIPS was called Tisty-tosty, or Tosty-tosty. Blossoms, picked on Whit Sunday for preference, were tied into a ball, and strictly, the balls were the tisty-tosties, though the growing flowers got the name too. Lady Gomme mentioned the game as belonging to Somerset, but it had a much wider spread than that. The cowslip ball is tossed about while the names of various boys and girls are called till it drops. The name called at that moment is taken to be the "one

indicated by the oracle", as Udal puts it, for the rhyme spoken at the beginning is:

> Tisty-tosty tell me true
> Who shall I be married to?

That is quoted as the Dorset rhyme, but the same is recorded in Herefordshire (Leather). The game was also known in Wales, where the purpose was different, for the rhyme there was:

> Pistey, Postey, four and twenty,
> How many years shall I live?
> One, two, three, four … (Trevelyan).

There is, too, a game described by Northall, which mentions the name Cobin-tree, by which WAYFARING TREE is meant. The rhyme is:

> Keppy-ball, keppy ball, Cobin-tree,
> Come down and tell me
> How many years old
> Our (Jenny) will be …

The number of keps or catches before the ball falls is the age. It seems that the game was played under this tree (the rhyme is said by Northall to be a Tyneside one, so presumably the name Cobin-tree was local there, too). Another of these divinatory games involved RYE GRASS. One kind of the Love-me, love-me-not variety is played by pulling off the alternating spikelets, hence the Somerset name Yes-or-no, and there is also Aye-n- Bent, from Gloucestershire. A different version of the game accounts for Does-my-mother-want-me, yet another Somerset name (Grigson. 1959). Another game that girls used to play involved striking the heads of the grass together, and at each blow saying Tinker, tailor, soldier, sailor, etc., The blow that knocks the head off marks the profession of the future husband (Elworthy; Leather). Hence Tinker-tailor Grass, from Wiltshire.

Children at one time liked to strip the stems of WATER BETONY (and FIGWORT) of their leaves, and scrape them across each other, when they will produce a squeaking sound, hance the name Squeakers (Macmillan), and more imaginatively, Fiddles, and also Fiddlewood, Fiddlesticks (Grigson. 1955, Goddard & Goddard. 1934), and Fiddlestrings (Macmillan). Crowdy-kit is another relevant name for these plants – crowd is a fiddle or similar instrument, the Welsh crwth.

MARBLES VINE is a plant that grows in West Africa on sandbanks or seashore, and the seeds are used by children all over the area in a game played like marbles (Dalziel), hence the common name.

Churn is a name given to DAFFODILS in Lancashire (Britten & Holland), where children played a game that involved separating the corolla from the stem bearing the pistil, and then working it up and down with a churning motion while repeating the rhyme:

Churn, churn, chop,
Butter come to the top (J B Smith).

Children play a sort of game with the seed pod of SHEPHERD'S PURSE. They hold it out to their companions, inviting them to "take a haud o' that". It immediately cracks, and there follows a triumphant shout, "You've broken your mother's heart". That is why the plant has the name Mother-die (Vickery. 1985), or Mother's Heart (J D Robertson), a shortened version of Pick-your-mother's-heart-out (Grigson. 1955).

Some of the many names given to GOOSE-GRASS, Bleedy-tongues, Tongue-bleed for example, are references to a children's game, if it can be called that. "Children with the leaves practise phlebotomy upon the tongue of those playmates who are simple enough to endure it" (Dyer. 1899). It simply involved getting someone to draw a bristly leaf or stem across the tongue, and it inevitably would draw blood.

CHILE PEPPER

(*Capsicum annuum*) Slices of green pepper put under the fingernails are claimed to keep rheumatism at bay. It is the smell, they say, that drives the "spirit" of rheumatism away (Waring). The leaf decoction is used as a treatment for asthma in Trinidad (Laguerre).

CHILE PINE

(*Araucaria araucana*) see MONKEY PUZZLE TREE

CHILEAN FIREBUSH

(*Embothreum coccineum*) When a Yaghan (Chile) woman was pregnant, she would put the flowers of this plant over her genitalia, after which she held her thighs tightly together. The idea was to make her child as beautiful as these flowers (Coon).

CHINABERRY TREE

(*Melia azedarach*) A small tree, originally from Afghanistan and northern India, now planted as an ornamental in all warm countries. This is the lien tree in China, the leaves of which were said to frighten dragons (Hogarth). In India, it is often associated with the pipal tree (*Ficus religiosa*), as the symbolic female to the male of the pipal. The two are planted together in "marriage", so that the vines of the pipal can twine round the limbs of its mate (Lincoln).

The berries were often used as beads in America (La Barre), hence the name Bead Tree (Barber & Phillips), also as rosary beads (it is Paternosterbaum in German). In India, too, they make beads. During epidemics of smallpox, etc., they are hung as a charm over doors and verandahs to keep the infection out (Pandey). But on a more practical level, there are these instructions from Alabama on how to make shoe blacking: get about half a gallon of chinaberries. Cover them with warm water, boil until tender, then rub the pulp through a sieve. Put in a clean vessel, add a small lump of goat tallow and a pint of sifted soot (R B Browne). They have been used as an insecticide,

and, during the American Civil War, as a source of alcohol (Hora). Another traditional American use is to put the seeds in with dry fruit, to keep the worms out, and to keep them away from turnips and other greens (R B Browne). Flies and fleas can be driven off by using the leaves, too.

The bark is used for rheumatism in Indian domestic medicine (Codrington), and in America the roots too had their medicinal value. There is a record from the southern states of a cure for scrofula involving chinaberry roots and poke-root. They were boiled together, with a piece of bluestone in the water, and then strained off. The sores had to be salved with this mixture, and then anointed with a feather dipped in pure hog lard. This would bring the sores to a head, so that the core could be pressed out (Puckett).

CHINESE DATE PLUM

(*Diospyros kaki*) 'Chinese' it was originally – now it can be found both in China and Japan. In a practice reminiscent of the treatment of walnut trees in Europe, (and apple trees – see **WASSAILING**), Japanese householders at the New Year ceremony, in the custom known as narikizome (torturing the fruit tree), maltreat the Date Plum tree by wounding it slightly with a hatchet. As he does so, he asks, "Fruitful or not?", and a man hiding behind the tree replies, "I'll be fruitful". This, they say, forces the tree to bear abundant fruit (Oto).

CHINESE HAIRCUT

The name given to a particularly painful prank played by one child on another. (MEADOW) FOXTAIL (or TIMOTHY *Phleum pratense*) is the grass used. The flowers would be stripped off the stalk, and this stalk used to twiddle into the hair of the child sitting in the desk in front. A swift tug would quickly remove all the hair attached (Vickery. 1995).

CHINESE LANTERN

(*Dichrostachys glomerata*) A shrub from central Africa, whose roots are chewed and macerated, then put on snake or scorpion bites to remove the poison. Leaves, also used for the purpose, are said to produce local anaesthesia. An extract of the leaves, mixed with salt, is applied to sore eyes. In southern Malawi, the roots and leaves are used as a toothache cure (Palgrave & Palgrave). A related species, *D. mutans*, is used in Ambo (Zambia) boys' puberty medicines. Three incisions are made on the abdomen and on the back, and a compound made of burnt wool of he-goat, burnt penis of a particular lemur, and the scraped roots of this shrub, is inserted. The he-goat is a symbol of strong sexual powers; the lemur is believed to have a strong penis, and the wood of this shrub is exceptionally hard (Stefaniszyn).

CHIVES

(*Allium schoernoprasum*) The word Chives came into the language through the French cive which is ultimately from Latin cepa, onion. The medieval versions of the word are siethes, sieves or sithes, still present in local forms. Predictably, Gerard was not impressed with chives as food, for, he said they "attenuate or make thin, open, provoke urine, ingender hot and grosse vapors, and are hurtfull to the eyes and brain. They cause troublesome dreames …", "yet of them", Lupton recorded a few decades later, "prepared by the art of the Alchymist,, may be made an excellent Remedy for the stoppage of urine". One cannot help wondering what he had in mind in requiring the services of an alchemist in making a herbal medicine.

There is a record from Devonshire that chives were looked upon as a fairy musical instrument, but apart from that, there is no indication of any folklore association. It has its uses, though, apart from well-known culinary processes, for chive leaves infused in boiling water, and the resultant liquid diluted with twice the amount of plain water, is often used to combat gooseberry mildew and apple scab. Even planting chives beneath apple trees will, it appears, reduce scab.

Chlorophora excelsa see IROKO

CHOCOLATE ROOT

A name given to WATER AVENS in America. A decoction of the rootstock of this plant was a favourite beverage among the American Indians (Indian Chocolate is another name for it (Leighton), and it is still used as a winter hot drink (Genders. 1976), boiled in milk and sweetened, giving a beverage not too different from chocolate (Sanford).

CHOKE CHERRY

(*Prunus virginiana*) An American species that is cultivated in Mexico and central America. The cherry is small, black and bitter (hence Choke Cherry, presumably). Birds often get drunk eating them. However, the cherries are quite useful – country people infuse them in brandy as a flavouring (Lloyd), and native Americans used them as food; the Ojibwe used to pound them, stones and all, and dried them to store as food (Densmore). The bark is slightly narcotic, making the user a little drowsy, and its sedative qualities gave it quite a reputation in America, in dyspepsia and tuberculosis (Lloyd). The Indians made a tea from this bark for diarrhoea (HH Smith. 1923), or any stomach ailment. Apparently, the bark was also used in the treatment of syphilis (Lloyd). However the kernels are as poisonous as those of the rest of the genus, and children have been known to die after eating them (Tampion) – it is the cyanide content that causes the damage. The root, too, has been used – Blackfoot Indians chewed the dried root, which was then put into a wound to stop the bleeding (Johnston). Eating half a cupful of these cherries each day was reckoned to cure gout (Tyler).

CHOKE PEAR

(*Pyrus communis*) See PEAR

CHOLERA

It used to be said in Alabama that WORMWOOD tea was good for cholera (R B Browne).

CHRIST-THORN

(*Paliurus spina-christi*) A Mediterranean plant, which may very well be the genuine Crown of Thorns of the New Testament. The young stems are pliant, and could be woven into a crown (Moldenke & Moldenke). The thorns will account for the statement, originally from Dioscorides, that "it is said that the branches thereof, being layd in gates or windows, doe drive away the enchantments of witches" (Dioscorides, *edited by* R T Gunther).

CHRISTMAS DECORATIONS

"Against the feast of Christmas every man's house, as also the parish churches, were decked with holm, ivy, bays and whatever the season of the year afforded to be green", Stow reported in 1598. Not only the houses and churches, but "the conduits and standards in the streets were likewise". So the custom was well-established in the 16th century, but it owes its origins to practices that are a great deal older. The use of HOLLY, always mentioned first in any account of Christmas decorations, is apparently a survival of an ancient Roman custom occurring during the Saturnalia, for holly was dedicated to Saturn. While the Romans were celebrating the feast, they decked the outside of their houses with sprays of holly; at the same time the early Christians were quietly celebrating the birth of Christ, and to avoid detection, they outwardly followed the custom (Napier). Since then, holly has become almost the embodiment of Christmas, actually called Christmas in many places, or, as in Suffolk, Christmas-tree, and Prickly Christmas in Cornwall (Grigson). In the Scandinavian countries it is known as Christ-thorn (Hadfield & Hadfield). All the evergreens used were called collectively the Christmas in East Anglia (Forby).

Parson Woodforde recorded in his diary for 24 December 1788 that "this being Christmas Eve I had my Parlour Windows dressed of as usual with Hulver-boughs well seeded with red Berries, and likewise in Kitchen". He had got the timing right, for it is bad luck to take holly into the house before Christmas Eve (or hawthorn, blackthorn or gorse at any time, possibly because of the connection with the Crown of Thorns) (Palmer). Only a man could bring it in (Baker. 1974) (for holly is a male emblem), and it could not remain in the house after 6 January. If the rules were not kept, it was considered in Oxfordshire, "you will have the devil in the house", or, from Cardiganshire, and referring to the MISTLETOE, "a ghost will sit on every bough" (Winstanley & Rose). Wherever else it was put, and in America it was bad luck to put them upstairs (Whitley & Bullock), it was always the custom to put a spray of berried holly in the window (Crippen), possibly as a visible protection, for houses were hung with holly in the Highlands to keep mischievous fairies away (McNeill. Vol 3). In Wiltshire, they say that the holly and bay garland hung outside the front door at Christmas is to keep the witches outside. They have to stay and count the holly berries – indefinitely, for witches can only count to four, and then have to start again (Wiltshire).

In Derbyshire, it was thought unlucky not to have both holly and MISTLETOE in the house at Christmas; they should be taken in together, and part of the holly had to be of the smooth, and the remainder of the prickly kind (Wright). At Burford, in Shropshire, only "free", or smooth, holly is used to decorate houses (Burne. 1883). IVY, though used in its own right, was sometimes made to simulate holly, by having its berries reddened with ruddle left over from sheep marking (Baker. 1974). But the two plants usually accompanied each other, and an old carol gives the impression that it was once the custom to set up a long pole decorated with holly and ivy, after the fashion of a maypole (Friend). Ivy was the natural companion of holly. And holly is the male emblem, so ivy is the female. Aubrey mentions an Oxfordshire custom, or at least a custom kept at Launton in that county – "it is the custom for the Maid Servant to ask the Man for Ivy to dress the House, and if the Man denies or neglects to fetch in Ivy, the maid steals away a pair of his Breeches and nails them up to the gate in the yard or highway". So the man had to bring the ivy in, just as he had to with holly, but it was at the behest of the female, who was recognised by custom to be in control of her own plant. Not surprisngly, care had to be taken not to let ivy predominate in the decorations. It is far too unlucky a plant to risk, unlucky for men, no doubt (Sternberg). Just to emphasise the point, it used to be the custom for church ivy saved from the Christmas decorations to be fed to ewes, and it said to induce the conception of twin lambs (Baker. 1974).

Nowadays, mistletoe has taken the place of ivy as the female element, even though the berries contain the male essence. But, as with ivy elsewhere, Worcestershire farmers gave their Christmas mistletoe to the first cow to calve in the New Year (Baker. 1974). The mistletoe is still used as a kind of love token at Christmas. Girls are ritually kissed under the mistletoe. It was said that each time a girl was thus kissed she had to be given one of the berries, which implies that these Yule fertility ceremonies had a strictly limited life (Sandys), just as there was a strictly limited period in which the mistletoe could be cut. In the Worcestershire apple-growing areas, where mistletoe grows so well in the orchards, they say it is very unlucky to cut it at any time other than at midnight on Christmas Eve. Elsewhere a little more leeway was allowed, but

Christmas was the only time permitted (Drury. 1987). In Herefordshire, mistletoe was not allowed in the house till New Year's morning, halfway through the twelve days of Christmas (Briggs. 1974).

Not surprisingly, given its reputation, mistletoe was hung in houses (and it is unlucky *not* to have it at Christmas), but never, with a very few exceptions, in churches. The most notable exception was York Minster, to which the mistletoe was ceremonially carried on Christmas Eve, and laid on the high altar, after which a universal pardon was proclaimed at the four gates of the city – liberty for all as long as the branch lay on the altar (Hole. 1941). There are a few more exceptions in the Midlands, but it must have been used in Welsh churches, for it is recorded that a sprig of mistletoe that had been used in the Christmas church decorations would bring good luck for the coming year to whoever had it. Again, there was a belief that if a girl put a sprig taken from the church under her pillow, she would dream of her future husband (Trevelyan).

Sometimes, one finds that mistletoe used in house decoration was kept an abnormally long time. The Staffordshire custom was to keep it to burn under next year's Christmas pudding, or else to hang it round the neck as a witch repellent. The Devonshire custom was also to keep it till next Christmas. It would prevent the house being struck by lightning; or, as at Ottery St Mary, it would ensure that the house would never be without bread. In Herefordshire, scorched BLACKTHORN was mixed with mistletoe as a Christmas decoration to bring good luck (R L Brown).

The importance of the mistletoe is emphasised by default, as it were, for in those areas such as the Lincolnshire marshes, where it does not grow, a bunch of some other evergreen has been used, but it had the same functions and privileges as the real thing (Gutch and Peacock. JUNIPER serves in Italy (Elworthy. 1895), but then it has many protective uses in its own right, and in France branches of it serve as a Christmas tree, and presents are hung from it (Salle). ROSEMARY was used at Ripon in the 18th century. The choirboys brought baskets of red apples, in each of which a sprig of rosemary was stuck. One of these apples was offered to every member of the congregation for a small payment (Crippen). POINSETTIA has become in recent years an accepted Christmas emblem in America, and has spread to this country. Evidently it was considered acceptable before Dr Poinsett discovered it in 1828, for its Mexican name means Flower of the Holy Night.

YEW is an evergreen seldom if ever used in the decorations. In Suffolk, they said it was unlucky enough to bring the death of a member of the family within a year (Forby). The Christmas evergreens should never be hung in a bedroom (Gutch & Peacock), and this sanction really applies at any time to any flowers or greenery, and has a superstition all of its own. "Set" Christmas decorations appeared quite early, for Barnaschone mentions the "quaint devices" appearing in the windows of 18th century Tenby houses, and business places too. They were made up of evergreens, usually box, myrtle or holly. The most widespread of these "quaint devices" was the one known variously as the Kissing Bough, Kissing Bush, or, as in Worcestershire, Kissing Boss (L M Jones), usually hung from cottage ceilings. The simplest way to make one is to fix two iron hoops together at right angles, and then decorate it, but tradition usually demands something more elaborate. We read, for instance, of the hoops being "bent in the form of a crown" (Wright), while Laurence Whistler's description (quoted in Hadfield & Hadfield) required five equal circles of thickish wire, bound together so that one became the horizontal "equator", with the four others crossing at the "poles". Whatever the framework, it had to be covered with greenery, and decorated with apples and lighted candles. Often a bunch of mistletoe was fixed to the underside. In the Cleveland district the kissing-bush had "roses" cut from coloured paper, and was hung with apples and oranges. When the railways came, mistletoe was added. That is interesting, for obviously mistletoe was remembered, but had become unobtainable locally (Spence. 1949). Devonshire kissing boughs were much simpler affairs, for a small FURZE bush served, dipped in water and then powdered with flour, after which it was studded all over with holly berries (Wright).

The last question is what to do with the decorations after Christmas has passed. Generally speaking, the accepted dictum is "all your Christmas should be burnt on Twelfth Day morning" (*Notes and Queries; 1853*). It would be unlucky to throw them away carelessly with the other household rubbish; then the luck of the house went with them (Drury. 1987), or a death would occur before next Christmas (Opie & Tatem). But in eastern Yorkshire it was specifically said they should be thrown out to decay (Nicholson). In Lincolnshire, too, it was reckoned to be very bad luck to burn the evergreens (Gutch & Peacock), the exact opposite of most considered opinion. In Somerset, they said that if you did not burn them, they would turn into pixies and plague you for a year (Tongue). Perhaps, as Drury suggested, the injunction was against burning them indoors, as in Norfolk, where they must always be put on a fire outdoors (Randell), but Opie & Tatem quoted a woman in Redhill, Surrey, who always burned holly in the grate on Twelfth Night.

After the Christmas decorations came down, BOX replaced them from Candlemas to Easter, or more correctly, Candlemas Day to either Palm Sunday or Easter Eve (Dallimore). Herrick, *Hesperides*, notes the custom under Candlemas Day:

Down with the rosemary and bayes,
Down with the mistletoe;
In stead of holly now upraise
The greener box for show;
The holly hitherto did sway,
Let box now domineer;
Until the dancing Easter Day,
Or Easter's Eve appear.

CHRISTMAS ROSE

(*Helleborus niger*) These pure white flowers are symbols of purity, so were used to purify houses, and to drive away evil spirits (Gordon. 1985). This symbolism also accounts for its use as the emblem of St Agnes, the patroness of purity (Hadfield &Hadfield). The time of its blooming emphasises this, for her feast day is 21 January, when the plant should be well in flower.

Christmas Rose was planted near a house because it was believed that no evil spirit would enter a dwelling near which these plants were grown (Rohde. 1936). Cattle were blessed with it, and were treated for cough with pieces of the root, which would be put in a hole made in the cow's ear or dewlap (Drury. 1985), just as Stinking Hellebore would be used. In the Lozère district of France, it was hung in stables and mangers to drive away "les serpents suceurs et les salamandes" (Sebillot). Wonderful enough, but there is an even greater wonder: it is said that scattering the powdered remains of the plant was a way of conferring invisibility! (Emboden. 1974). In some parts of west Wales, the blossoming of the Christmas Rose late in spring was said to indicate "unexpected events" (Trevelyan), as does unseasonal blooming with a lot of other plants. But then, this is one of the plants that blooms on Christmas night, whatever the weather, between 11 and 12, according to the Pennsylvania Germans (Fogel).

The plant had its uses, dubious as they might be, though one in Albertus Magnus is mundane enough: "... when thou wilt that Flies come not nigh thy house, then put Condisum [identified as *Helleborus niger*] et Opium [*Papaver somniferum*] in white Lime, and after make thy house white with it, then Flies shall in no wise enter". But it is difficult to know what to make of the statement that the ancient Gauls rubbed their arrow-heads with Hellebore, believing that it made the game more tender (Coiats. 1968). At least the medical uses are realistic enough, though Pliny made a marvel of its gathering, when the person doing so had to be clad in white, and go bare-footed, plucking (not cutting) the flower with the right hand, and then, covering it with his robe secretly, it had to be conveyed to the left hand. Afterwards, the gatherer had to offer a sacrifice of bread and wine (Pettigrew). After that, it seems extremely mundane to record the fact that its use was as a drastic purgative (Fluckiger

& Hanbury). The drug Radix Hellebori Nigri was still being imported from Germany in the 19th century and sold for the use of domestic animals. Much earlier, Topsell had reported that "the small roots of hellebore, which are like to onions, have power in them to purge the belly of dogs". Just like Stinking Hellebore, this species was used for expelling worms in children, and was even recommended by experts like Buchan, but of course this is a highly dangerous practice. Gerard recommended a similar treatment for "mad and furious men, …, and for all those that are troubled with blacke choler, and molested with melancholy", i.e., for nervous disorders and hysteria. Thornton, much later, more or less repeated this. There is an old legend that may have contributed to this notion that a purge with hellebore cured madness, to the effect that the daughters of Prae-tus, king of Argos, were cured of their madness by the soothsayer and physician Melampus, who had apparently noticed its effect on goats. The Greeks certainly believed it to be a remedy for madness.

The roots are also used as a heart stimulant, and to control menstruation (Usher), and they have also been used to deal with a number of skin diseases, listed by Gerard as the "morphew and blacke spots in the skin, tetters, ring-wormes, leprosie, and scabs". Even its bad smell, so it was believed, was the result of its absorbing sickness from the patient (Skinner).

CHRISTMAS TREE
Christmas trees are of German origin. One legend connects them with St Boniface, the 8th century English missionary to Germany. He cut down a sacred oak at Geismar one Christmas Eve, and is said to have offered the outraged pagans a young FIR-tree in its place, to be an emblem of the new faith he preached. A later story says Martin Luther introduced the custom by using a candlelit tree as an image of the starry heavens (Tille). But the fact is that the first record of a Christmas tree occurs in 1604, at Strassburg, where the adorning of houses with fir branches at New Year is witnessed at the end of the 15th century and the beginning of the 16th century. Fir trees were put up in rooms, decorated with "roses" cut out of many-coloured papers, and with apples, gold-leaf, sweets, etc., and fixed in a rectangular frame. It is interesting to compare this with the legend of flowers blooming at Christmas (usually Old Style) (see GLASTONBURY THORN), for often boughs of cherry trees and hawthorn were, a fortnight before New Year, put into water in a warm place, in the hope that they would bloom at New Year or Christmas. The blossoms, if any, were used as an oracle; if there were a lot of them, well shaped, it meant luck; if the opposite, the omen was not at all good (Tille). Are we to assume that modern Christmas trees are just a substitute for trees and bushes actually in bloom?

The cult of the Christmas tree was familiar in the United Sates long before it was known in much of Europe. There is evidence of one being in use by German settlers in Pennsylvania as early as 1746. It spread to France from Germany in the early 19th century, and German merchants and court officials brought it to England in the late 1820s (Hadfield & Hadfield). Prince Albert really established the custom here, but apparently Queen Caroline had one in her court in 1821, and by 1840 it was well-known in Manchester, where it had been introduced by German merchants who had settled there (Hole. 1941). Technically though, there were much earlier English examples. For instance, Twelfth Night used to be celebrated at Brough, in Cumbria, by carrying through the town a holly-tree with torches attached to its branches. And the Wassail-bob, also called Wesley-bob, was really another Christmas tree example. It was still being carried round on a stick in Yorkshire, particularly Huddersfield, Leeds and Aberford, in the mid-19th century (J O'Neill).

A common substitute for the Christmas tree in Saxony last century was the "pyramid", a wooden structure decorated with coloured papers and lights. In Berlin, the pyramid had green twigs as well as candles and papers, and so looked more like the genuine thing. Pyramids, without lights apparently, were known in England before 1840, and in the Faeroes, too, before the relatively recent introduction of the real thing (Williamson). Pyramids in Hertfordshire were made of gilt evergreens, apples and nuts, and were carried about just before Christmas for presents. They appeared in Herefordshire at the New Year (Miles). Trees of various kinds are known as Christmas trees in other lands. New Zealand has one that goes under that name because of the red flowers that open in December and January; its native name is Pohutukawa (Hadfield & Hadfield, Leathart).

Other trees have been the recipient of the name in this country. HOLLY, of course, already associated with Christmas, and SPRUCE FIR or SILVER FIR, the "standard" Christmas trees, even the MONKEY PUZZLE, or Chilean Pine (*Araucaria araucana*). The oddest example must be the HORSE CHESTNUT, which, although having no association with Christmas, seems to be a reminder when it is in full bloom. Each flower spike is a Candle, then a Christmas Candle.

In France, branches of JUNIPER act as a Christmas tree; they are put round the chimney breast, and presents for the children are hung from them (Salle). In Italy, too, juniper is hung up at Christmas time (Elworthy. 1895).

Chrysanthemum segetum > CORN MARIGOLD

Cicer arietinum > CHICK PEA

Cichorium intybus > CHICORY

Cicuta maculata > AMERICAN COWBANE

Cicuta virosa > COWBANE

CIMARUTA

i.e., cima di ruta, sprig of RUE. An amulet to preserve the wearer from the evil eye, and one of the most potent of charms in use in Naples. A representation of the plant these days, but obviously the real thing originally. It is absolutely necessary that it is of silver and carries the silversmith's hallmark, otherwise there is no virtue in it (Gifford). It contains a compound charm as well as representations of the plant, comprising a hand, moon, key, flower (interpreted variously as vervain (Valiente) or orange blossom (Gunther. 1905), horn or fish (as phallic emblems), cock or eagle). Additionally one sometimes finds a heart, serpent, cornucopia and cherub, all contained in an amulet that is typically about 7 x 5cm. The charm must have originated in an ancient practice of holding in the hand a sprig of rue taken from the plant, and later a dried sprig may have been attached by a mount to a charm or ribbon round the neck, which was how it was worn in the time of Aristototle, who recorded its efficacy (W Jones. 1880). Elworthy. 1895 records an early Etruscan amulet that is very similar to the modern charm, so the cimarute must be one of the very oldest of existing amulets. This one was made of bronze, with no subsidiary amulets (Gifford), and it has to be said that there is no real evidence that it was used against the evil eye, though the modern charm has three main branches, each branch ending in a clenched fist, the thumb pushed between the index and middle finger, the classic protection against the evil eye.

CINNAMON

(*Cinnamon zeylanicum*) According to medieval legend, the spice was found in birds' nests, particularly that of the phoenix, "and may not be found, but what falleth by its own weight, or is smitten down with lead arrows. But these men do feign, to make things dear and of great price …" (Bartholomew Anglicus). It is often used as incense, when a little is burned on charcoal, which is good for "aiding meditation and clairvoyance" (Valiente). There is a ritual use in Rhodes; at wedding ceremonies the bride's hands were anointed with it (Rodd) (as an oil, presumably).

Pomanders were once carried as a preservative against infection, and they usually contained cinnamon bark amongst other aromatic substances (Clair). It has even been used, in Spain, to cure rabies (H W Howes). A spoonful of cinnamon with water was an Alabama domestic remedy for a headache (R B Browne). Cinnamon powder taken in milk is given as a cure for dysentery, and strong cinnamon tea taken at the beginning of mumps will, it is claimed, reduce the violence of the complaint and prevent complications. It can even be used for warts. The practice in Illinois was simply to cover the wart with cinnamon in order to get rid of it (H M Hyatt). Cinnamon brandy is

good to take at the beginning of a cold or influenza (Leyel. 1937). But it appeared very early in medical receipts: Dawson, for instance, quoted a "powder to gather flesh. Take powder of mastic and frankincense, canell [cinnamon, that is], and coral, of each equally much, and make powder thereof". That word 'canell' is the Italian cannella, French cannelle, and was given because the sticks of rolled bark resemble pieces of reed, which is what cannella means. It is mentioned as early as the 13th century, in a poem on the Land of Cockaigne – "pudrid with gilofre and canel", and is still in use – l'eau de cannelle is cinnamon water, and it used to be essential in Jersey to celebrate a christening (Lemprière).

Cinnamonium camphora > CAMPHOR

Cinnamonium cassia > CASSIA

Cinnamonium zeylanicum > CINNAMON

CINQUEFOIL

(*Potentilla reptans*) Cinquefoil means five leaves, and that features in many of the names of this plant, whether they are called leaves, or fingers, as in Five-finger Grass (Turner), or Five-finger Blossom (Grigson. 1955), etc., It is good for fevers, so it was said, and at a time when fevers were considered to be the work of witches, the plant came to be an antidote to witchcraft. Reginald Scot refers to the custom of those "who hang in their entries an herb called Pentaphyllon, Cinquefoil", with hawthorn gathered on May Day, in order "to be delievered from witches" (Scot). In Wales they would dig up a root on May Eve and wear it in their coats, for luck (Trevelyan). According to Graves, it was used by medieval French witches as the chief ingredient of flying ointment. Montague Summers gave two recipes for flying ointment, taken from Scot again, in which cinquefoil is mentioned as an ingredient, along with other poisonous or revolting substances. Very similar is the recipe in Bacon, *Sylva Sylvarum* (1627), who says that witches' ointments were made "of the fat of children digged out of their graves, of the juices of smallage, wolfbane, and cinquefoil, mingled with the meal of fine wheat". But what is cinquefoil doing in all this? It appears again in a witch philtre for love or hate, composed of adders, spiders, cinquefoil, the brains of an unbaptised baby, and so on (Summers). That philtre, with only cinquefoil, makes an unlikely appearance with the Pennsylvania Germans. "To gain the admiration of girls, carry cinquefoil in your pocket" (Fogel). There is one piece of weather lore to notice. It comes from Devonshire, and states that if the flowers of cinquefoil expand, you can expect rain; if they close, fine weather (Hewett). It sounds the wrong way round, but that is what the author, Sarah Hewett, wrote.

Cinquefoil is used as a tea for diarrhoea (but tormentil is much more effective). Externally, the decoction is sued to bathe wounds, and as a mouthwash for throat and gum irritants (Flück). In Alabama, a tea is made from it to be used for fevers (R B Browne). Aubrey. 1696 quoted a recipe for ague, tertian or Quartan : "Gather cinquefoil in a good aspect of…" [here follows astrological instructions] "and take… of the powder of it in white wine … With this receipt, one Bradley, a Quaker at Kingston Wick upon Thames (near the bridge end), hath cured above a hundred". Earlier herbalists seem to be agreed that Cinquefoil offers a cure for toothache. (see Gerard, Dawson).

Cirsium acaulon > PICNIC THISTLE

Cirsium heterophyllum > MELANCHOLY THISTLE

Cirsium vulgare > SPEAR PLUME THISTLE

Citrullus vulgaris > WATER MELON

CLAIRVOYANCE

An infusion of MUGWORT tea is quite seriously believed to be an aid in the development of clairvoyance. The young leaves are used, sweetened with a little honey; but the herb has to be gathered at full moon. It is clear from elsewhere in the world that the plant itself is the agent for clairvoyance – in Japan, for example, if a house were robbed in the night, and the burglar's footmarks were visible next morning, the householder will burn mugwort in them, so hurting the robber's feet and making his capture easy (Radford). Perhaps it is rather a question of protection; mugwort will hold evil at bay, or make it easier to overcome.

CLAPPEDEPOUCH

This name for SHEPHERD'S PURSE (Prior) is an allusion to the licenced begging of lepers, who stood at crossroads, with a bell and clapper. They could receive their alms in a cup or basin at the end of a long pole, the money-receiving end of which is likened to the purses or pouches of Shepherd's Purse. Rattlepouch is a variant name (Grieve. 1931).

CLARY

(*Salvia sclarea*) Whatever Clary means, it is not "clear-eye", though that has been used as a name for it, with justification, for the seeds swell up when put into water, and are mucilaginous. These can then be put like drops into the eye, to cleanse it (Grigson. 1955). As Gerard said, "the seed of Clarie poudered, finely scarced [sieved, that is] and mixed with hony, taketh away the dimnesse of the eies, and cleareth the sight". Long before his time, though, the seed was being used as an eye salve, and not only the seed, for the leaves are prescribed in a 15th century leechdom, which runs "for the pearl in the eye [cataract?], and the web: take the leaf of Oculus Christi [a common medieval name for clary], peeled downward, and hyssop, with a leaf of sage: and drink the juice of these three days first and last [that is, morning and evening] (Dawson. 1934). The editor does not say how

this, or anything else taken internally, could possibly work. Gerard even reported it to be an aphrodisiac!

Clary is cultivated these days for its oil, which is used in the cosmetics industry as a perfume fixer. It is a highly aromatic oil, with a scent resembling ambergris (Clair). Clary was once used apparently by German wine merchants as an adulterant. It was infused with elder flowers and then added to Rhenish wine, so converting it to the likeness of Muscatel. It is still called in Germany Muskateller Salbei (Grieve. 1931).

CLOVE

(*Syzygium aromaticum*) 'Clove', which is from the French clou, a nail, and ultimately from Latin clavus, is the dried, unopened flower bud of this evergreen tree. The fruit is called "Mother of Cloves" (J W Parry). They were in use in China in the third century BC, and they were also known to the Romans. But it was not until the Middle Ages that they were introduced into the rest of Europe (H G Baker), and then they must have been extremely expensive, valuable enough to pay for a year's rent on a manor, for the manor of Pokerley, County Durham, was held by the provision of one clove on St Cuthbert's Day, annually (Blount). Oil of cloves is used in medicine. It is a disinfectant and dental analgesic (Schauenberg & Paris), much used in domestic medicine, and even as a laxative, according to a Suffolk record. They had to be boiled in water and steeped overnight (V G Hatfield. 1994). It is a magical spice, too. Not all that long ago, people in Indonesia wore cloves stuck into their nostrils and lips, so that demons could not enter the body there (Swahn).

CLUBMOSS

(*Lycopodium inundatum*) Bretons thought it had occult powers. So did people in Cornwall, where it was considered good against all diseases of the eyes. It was said that if anybody revealed this secret, the virtues of the moss would disappear (Courtney. 1890) (that must have happened, as we are recording it here). Anyway, there was a proper way of gathering it. It had to be done on the third day of the moon, when the new moon was seen for the first time. Show the moon the knife with which the moss for the charm was to be cut, and repeat:

> As Christ ealed the issue of blood,
> So I bid thee begone;
> In the name of …

In Scotland, clubmoss is something that brings luck to a house. As long as a piece of it is in the house, bad luck cannot enter (Gregor. 1888).

CLUSTERED BELLFLOWER

(*Campanula glomerata*) Tradition has it that this plant sprang up wherever the Danes' blood was shed in battle (Cf Dwarf Elder). It is actually known as Danes' Blood in Cambridgeshire (Dyer). In Bartlow, in that county, there are four mounds supposed to have been thrown up by the Danes as monuments of the battle fought between Canute and Edmund Ironside in 1006. This plant apparently grows prolifically on those mounds.

COB-NUTS

HAZEL nuts are so-called, though the name actually refers to a game played with the nuts, much like conkers, though there was more than one version. Halliwell considered the one like conkers was the most recent, but he says the older consisted of pitching at a row of nuts piled up in heaps of four, three at the bottom and one at the top of each heap. All the nuts knocked down became the property of the pitcher, and the nut used for pitching was the one called the Cob. Hunter's description of the other game as it was played in his time (1829) runs "Numerous hazel-nuts are strung like the beads of a rosary. The game is played by two persons, each of whom has one of these strings, and consists in each party striking alternately with one of the nuts on his own string, a nut of his adversary's. The field of combat is usually the crown of a hat. The object of each party is to crush the nuts of his opponent. A nut which has broken many of those of the adversary is a cob-nut. Hence the Cornish Victor-nut and the Devonshire Crack-nut.

COBIN-TREE

is a name given to WAYFARING TREE (*Viburnum lantana*), Coven-tree is the variation in some areas (Grigson. 1955; Dartnell & Goddard). The name appears in a rhyme spoken in a children's game:

> Keppy-ball, keppy-ball, Cobin-tree,
> Come down and tell me,
> How many years old
> Our (Jenny) will be ...

The number of keps, or catches before the ball falls is the age. But these names were given in Scotland to any tree that stood before a Scottish mansion house; it was under this tree that the laird always went out to greet his visitors. So the word is covyne, a meeting place. Keppy in the song may be the Scottish kep, to meet (Rowling).

COCKLEBUR

(*Xanthium strumarium*) In China, the leaves are used for dyeing yellow (F P Smith). Dioscorides had a recipe for making the hair yellow – "the fruit being gathered before it be perfectly dry, and beaten, and put up into an earthen vessel, is of force to make hair yellow". It is a poisonous plant, but nevertheless, it is used medicinally. Cocklebur tea has been used to reduce fevers (H M Hyatt), and a sore throat remedy from Indiana uses the leaves and root, powdered. Mix with a little flour and water, and put a little on the back of the tongue, so that it drips into the throat (Tyler). An Alabama ringworm cure uses the juice and cream, applied (R B Browne), and crushing cocklebur leaves over a snakebite is recommended there (R B Browne).

COCONUT

(*Cocos nucifera*) The name comes from the Portuguese coco, meaning a grimace, for the nuts are said to resemble a human face (Lehner). In India, the coconut was offered to the gods instead of a human head (Upadhyaya), and there are tales in Malaysia of the coconut originating from a human head. Mabuchi has this story from Simalut : "Once there were two brothers, Rabin and Rachman. Their father died and left a lot of debt. The people deprived the brothers of all the property, but the brothers still had to get out of the remaining debt. Thus, they fled from the village to the jungle, where they made a patch of field. The elder brother said to the younger: 'Let us dig a hole in the ground. Cut off my head and bury it in the hole'. The younger brother was reluctant to do so. But at last he could not refuse the request of his elder brother. He saw a tree growing there. He took fruit of the tree and tasted the juice and the pulp. He planted many of these fruits there, while he brought a number of them to the village where his father had lived. The villagers had never seen this kind of fruit, and they found it delicious. This was the coconut. Because the coconut originated from the head of the elder brother for the benefit of the people, this could well counterbalance the debt left by the father". There are many other origin myths showing how the coconut grew from a head, human or otherwise (see J Campbell. 1960 for quotes), from the Pacific islands. This one from Tonga is typical: 'A male child, Eel, was born to a human couple, who had also a pair of human daughters. Eel, living in a pool, sprang towards his sisters in eager affection, but they fled, and when he pursued them they jumped into the sea and became two rocks that may be seen to this day off the shore of Tonga-tabu. Eel went on swimming, to Samoa, when he again took up life in a pool. But when a virgin, bathing there, became pregnant because of his presence, the people decided to kill him. He told the girl to ask the people to give her his head when he had been killed, and to plant it, which she did. And it grew into a new sort of tree, the coconut tree', which is a symbol of fecundity in India. Women who want a son are given a coconut by their priest, and in Gujarat the bride gives a coconut to the bridegroom (Upadhyaya). The myth of the coconut deriving from an eel lover is common throughout the south Pacific. Another typical story is one from Tahiti: 'Hina, whose gods are sun and moon, is betrothed to a chief who has an eel body. She flees to the god Maui for help, He baits his fishhook, the eel swallows it, Maui cuts up the body and gives the head to Hina to take home and plant. Hina forgets and puts the head down while she bathes at Paui, and the head sprouts into a coconut ...' (Beckwith. 1940).

The "eyes" of the coconut feature in Trobriand myths telling of how the spirits became invisible to humans. Malinowski quotes two examples : when the spirit's feelings were hurt by some human act, she (the spirit) decided to go and live in Tuna, the underworld. "She then took up a coconut, cut it in half, kept the half with the three eyes, and gave her daughter (the human) the other. 'I am giving you the half that is blind, and therefore you will not see me. I am taking the half with the eyes, and I shall see you when I come back with other spirits'. This is the reason why spirits are invisible, though they themselves can see human beings". The "eyes" appear again, even though in a negative way, in a Malayan belief that when the nut lacks the usual three "eyes" it could act as a protection in warfare against the enemy's bullets, an obviously magical charm (Skeat).

Malayan peasants used to say that coconuts should only be planted with a full stomach. The gardener had to run quickly and throw the coconut into the hole made for it without straightening his arm; if he did straighten it the fruit stalk would break. They should be planted in the evening, so that they will bear fruit while still near the ground. When you pick seed coconuts off the tree someone should stand below to watch and see if the "monkey face" of each nut turns either towards himself or to the base of the tree, in which case the seed will be good, or looking away from both, in which case it is not worth planting (Skeat).

The many virtues of the tree made it a favourite candidate for identification as the Tree of Life in the 17th century, for, as Raleigh said, "the Earth yeeldeth no plant comparable to this", such a tree as "giveth unto man whatsoever his life beggeth at Nature's hand" (Prest). Almost every part of the tree is useful: when the nuts are green, the milk makes a refreshing drink; later, when they are ripe, they are covered with a husk of coir which can be used to make rope and such things as doormats, as well as providing a pest-free compost material for gardeners. When the nuts are ripe, they can provide copra, which is the dried fruit, used for soap making and many other products. Cooking oil can be extracted from it. The fronds of the tree have many uses, mats to make a hut wall, or thatching, basket making. Fronds can also be stuck upright in the ground to make a fencing around a courtyard (Caplan). In India, they are a major source of sugar-rich sap. It is the main flower spike that is tapped. A few hours after collection, this sap will have fermented to an alcoholic drink called palm wine, or toddy, which must be drunk the same day, or it will ferment again to become a sour vinegar. Palm wine can be distilled to a spirit called arrak (Edlin. 1976).

Cocos nucifera > COCONUT

Coeloglossum viride > FROG ORCHID

COHOBA
(*Anadenanthera peregrina*) A South American plant, particularly in the Orinoco basin. A snuff, usually known as yopo, or partica, made from the powdered seeds, is one of the most famous of New World hallucinogens, extremely powerful and rapid. The original inhabitants of Haiti made this narcotic stuff, which they took through a bifurcated tube, and in fact it was much used by the aboriginal population over most of the area in religious ceremonies (Hostos). Lewin described the use among one of the Brazilian tribes: "...begin to take Parica snuff ... they assemble in pairs, everyone with a (bamboo) tube containing Parica in his hand, and ... everyone blows the contents of his tube with all his strength into the nostrils of his partner. The effects produced in these generally dull and silent people are extraordinary. They become very garrulous and sing, scream and jump about in wild excitement ...".

Cola nitida > KOLA

Colchicum autumnale > MEADOW SAFFRON

COLD CURES
GARLIC, boiled, and the juice drunk (Ireland) (Maloney). They have a way of treating a cold in Alabama, either just by eating a spoonful of it, or if the sufferer does not like garlic, take a baked potato, open it, and put the garlic in it; then butter it, or put gravy on it, and eat the potato without chewing it (Browne). Dried TURNIP, grated and mixed with honey is an American cold cure (Stout), and a gypsy remedy is to take a leaf infusion of CUCKOO-PINT to treat their chills and colds (Boase). HORSERADISH can be used, too, by smelling it, or rather inhaling the vapour from the grated root (Vickery. 1995). In a similar way, American Indians would burn twigs of VIRGINIAN JUNIPER and inhale the smoke to clear up a cold (Gilmore).

ELDER flower tea is good for colds, coughs, etc., as well as for sore throats (so is mulled elderberry wine, which is also said to be good for asthma (Hatfield)). A concoction of elderberries was a Highland cold cure (Thornton), but what better than the wine? A Cornish cold cure involved picking elder flowers and angelica leaves, steeping in boiling water for ten minutes and straining, adding sugar to taste (Deane & Shaw). Elder flower is still infused in Cornwall and used as a tisane for hay fever and catarrh. TANSY flower tea was also given for colds (Palmer. 1976) and so was DOG ROSE hips tea (Thomson. 1978), and ROSEMARY tea is recommended, too – it helps clear the head (Rohde). BEE BALM tea is given, as well, particularly if it is a feverish cold, for this medicine has the effect of promoting sweating (Conway). Dried HOLLYHOCK flower tea is good, too (H M Hyatt). WATERCRESS tea is drunk for a cold in Trinidad (Laguerre), and an American remedy is to drink hot, sweetened SAGE tea in bed, after soaking the feet in hot water (Stout). The leaf infusion of WHITE HOREHOUND is probably the best known of cough or cold remedies, and this dates back at least to Anglo-Saxon times (Cockayne). It is still available in the form of lozenges (Cameron). A GROUND IVY infusion was an Irish cure for a cold (Moloney). Actually, "Gill-tea", as it was called (Gill being one of the names for Ground Ivy), mixed with honey or sugar to take away the bitterness, has always been a favourite remedy for coughs and colds (Clair). It is also mixed with wood sage in a tea, and given for colds. This is a New Forest gypsy remedy (Boase). It is said that the Basuto treat colds by stuffing the fresh leaves of HORSE MINT up their nostrils (Ashton).

The flannelly leaves of MULLEIN could be used like flannel, and wrapped round the throat to help relieve coughs and colds (Sanford). Hot mullein tea was also drunk for a cold (R B Browne), and so was YARROW tea, made either from the dried herb, or from the fresh plant (a handful of the whole herb to a pint of boiling water (Jones-Baker. 1974). WOOD SAGE was similarly used in Ireland for colds.(Ô Súilleabháin). Some native American groups would grind up the seeds of BLACK MUSTARD to use as a snuff for a cold in the head (H H Smith. 1928). Taking BASIL as a snuff is another example of a cold cure, or an infusion, taken hot at night, is another way of curing a head cold (Quelch). GINGER is a fairly obvious choice for a cold. Parihar & Dutt pointed out that ginger jam was a favourite for coughs and colds. JUJUBE (*Zizyphus jubajuba*) fruits have been famous since ancient times as a cold cure. They used to be made up into lozenges; such lozenges are still called jujubes (Mitton). AGRIMONY wine used to be made just to take to cure a cold (Grigson. 1955).

COLIC
CARAWAY, in the form of the essential oil from the seeds, has always been used as a carminative. "It consumeth winde", Gerard wrote. It appears in American domestic medicine, too. Hill preferred his patients to chew the seeds, for colic and the like. Gypsies were using GROUNDSEL, i.e., the bruised leaf and stem, to relieve the condition until very recently (Vesey-Fitzgerald). OATMEAL is reckoned good for the complaint, too (Beith), and so are BAY leaves and berries. A traditional cure in Illinois was to soak a few bay leaves in brandy, and to give one to four teaspoonfuls of the result (H M Hyatt). Pomet claimed that the berries "give ease in the most extream Colicks". CATMINT tea is another remedy, especially for babies with the complaint (R B Browne). WOOD SAGE was used in Ireland (Moloney), and BEE BALM tea in Gloucestershire, while in the Fen country the roots of PERIWINKLE were a popular remedy for the condition (Porter. 1969). In parts of France, bilberries were said to be "souveraines contre la colique des enfants" (Loux).

CASTOR OIL is used in Haiti for the complaint, and a baby's colic was dealt with in Alabama by roasting an onion in hot ashes, and squeezing the juice out so that the baby receives a few drops at a time, and the use of GINSENG (*Panax quinquefolium*) in the form of a tea was another medicine that mothers gave to babies who had colic (R B Browne), and it is, in fact, used in the southern states as a remedy for all kinds of stomach troubles (W A R Thompson. 1976). The seeds of MELEGUETA PEPPER, soaked in gin or palm wine, were taken by African people like the Mano, of Liberia, for colic. (Harley).

TORMENTIL is a famous remedy for colic. Its very name seems to proclaim, for this is French tormentille. And that in its turn comes from Latin tormentum. Tormentil survived as a medicine for colic until comparatively recent times in isolated parts, notably Northumberland and the Hebrides (Grigson. 1955), the recipe on South Uist being "boil the entire plant and drink the juice" (Shaw). Pennant. 1772 records its use on Rum – "if they are attacked they make use of a decoction of the roots … in milk".

Colocasia antiquorum > TARO

Colocynthus vulgaris > BITTER GOURD

COLTSFOOT
(*Tussilago farfara*) "If the down flyeth off colt's foot, dandelyon, and thistles, when there is no winde, it is a sign of rain" (1656, quoted in M E S Wright) (probably Coles, W. *Knowledge of plants*). The only other instance of folklore is recorded by Dyer. 1889, who said that "on Easter Day the Bavarian peasants make garlands of coltsfoot and throw them into the fire". What was the ritual? Coltsfoot had its uses, apart from the medical. We are told that the downy seed covering was used in the Highlands of Scotland for stuffing mattresses (Pratt). However much of this would be necessary to fill a mattress? Another account confines itself, a little more reasonably, to stuffing pillows (M Evans).

But it is far better known for its medicinal value, so valuable that it was apparently used as the sign of an apothecary's shop, in Paris (Grieve. 1931). It was on coughs and all chest ailments that it built its reputation, as a tea (often with honey (Thomson. 1976)), as a piece of coltsfoot rock, to chew, or as coltsfoot wine (M Evans), or as pectoral beans or jelly (Grigson. 1955). And it can be smoked, like tobacco. Bechion, the plant in Dioscorides taken to be coltsfoot, was smoked against a dry cough, and it is still smoked in all herbal tobacco (Grigson. 1955), as it is also in Chinese medicine (F P Smith), for asthma and bronchitis, and even for lung cancer (Perry & Metzger). Gypsies smoke the dried leaves, and the tea is taken for coughs (Vesey-Fitzgerald). Ô Súilleabháin quotes its use in Ireland for asthma; the fresh leaves would be boiled in milk, and the lot eaten. It was smoked by

Cornish miners as a precaution against lung disease (Deane & Shaw). There is nothing new about all this. Gerard was advising the same treatment in his day, in different words: "A decoction made of the greene leaves and roots, or else a syrrup thereof, is good for the cough that proceedeth from a thin rheume … The fume of the dried leaves taken through a Funnell or tunnell, burned upon oles, effectually those that are troubled with the shortnesse of breath… Being taken in manner as they take Tobacco, it mightily prevaileth against the diseases aforesaid".

Gypsies use the juice from the fresh leaves in making an ointment for the treatment of ulcers and the like (Vesey-Fitzgerald), and the juice was used in Ireland for earache (Ó Súilleabháin). The fresh leaves put on the chest and changed frequently were said in Russian folk medicine to bring relief for heartburn in tubercular cases (Kourennoff). A flower decoction has been used in China as an eyewash (F P Smith), but the oddest use must be Topsell's "for the drawing forth of a thorn or splinter out of a dog's foot, take coltsfoot and lard, or the powder thereof burned in a new earthen pot either of these applied to the foot draws forth the thorn and cures the sore".

Colutea arborescens > BLADDER SENNA

COMFREY

(*Symphytum officinale*) The name comfrey comes, through Anglo-Norman cumfirie, from Latin confervere, to heal. Another line, through confirma and conserva, produced consolida, and eventually consound, a name much applied to this and to plants with a similar reputation. Comfrey too, although now confined to this particular genus, once had a more general significance. The reputation rests on the interpretation of the Greek word which gave the botanical name *Symphytum*; it comes from sumphuo, to grow together. Whether this really is the plant that the Greeks named and described as useful for knitting bones is doubtful, but it was certainly taken to be such in medieval times, and has been used for the purpose ever since. It was the glutinous matter of the root that was used, grated, for a plaster that set hard over a fracture. Sometimes a charm had to be spoken at the same time. One from Exmoor ran:

> Our Lord rade,
> The foal slade,
> Sinew to sinew and bone to bone
> In the name of the Father, Son and Holy Spirit

said three times (Tongue). In Ireland it was taken internally as well, as part of the process of knitting fractures. Gypsies used it in the usual way, and probably still do (Vesey-Fitzgerald). It is said that soldiers carried the herb with them in medieval times (Tongue), and that it was known to the Crusaders as a wound herb. As an extension of all this, the Pennsylvania Germans even say it will reduce hernia,

But this is a magical cure – you have to hold it on the hernia until it is warm. Then the comfrey should be planted. If it grows, the hernia will be cured (Fogel). And surely the next must rank as the ultimate repair job, for it was even said to be able to hide the fact that a girl had lost her virginity. She was advised to take a long bath on the eve of her wedding in hot water and comfrey (As it was said, that "would do up whatever has been undone") (Page).

Comfrey leaves are edible (in Ireland they were used like spinach (Vickery.1995)), and they make a good animal feed; chopped leaves in with the feed will increase hens' laying production (Painter). In Bavaria, they have a special way of treating the leaves – they are dipped in batter and fried (see German cookery books under Schwarzwurz). The flowers too have been used, for giving flavour to cakes, and comfrey wine is quite well known.

The rest of the uses are purely medicinal. The leaves and root made quite a well-known cough remedy. The use is an ancient one, for an Anglo-Saxon leechdom prescribed for the dry cough elecampane and comfrey eaten in virgin honey (Cockayne). In the 14th century we find: For the quinsie: take colymbyn and fethernoyge (feverfew, perhaps?) and levys of comfrey and stampe hem to-gedre and drynke the ius with stale ale" (Henslow). A hundred years later, there is "for the dry cough: take horehound and comfrey, and eat [it] with honey three mornings and three evenings" (Dawson). The root tea is an Alabama remedy for dysentery (Browne), and elsewhere it is drunk for rheumatism, arthritis and asthma (Painter). It is also a good thing to put on cuts and bruises. The standard Irish country method to ease the pain is simply to apply a poultice of comfrey roots (Logan). Elsewhere, it was not so simple, for there is a widespread superstition that the purple-flowered comfrey is for a man, and the white one for a woman (Burne, Foster). Gypsies too recognize the difference, and it has to be male flowers for a woman, and female ones for a man (Boswell). In the Isle of Man, it was said that the leaves, one side rough and the other smooth, would heal a wound if put in the right order, by first drawing and cleaning it, and then healing (Killip). It has been appearing in medical texts as a wound herb for centuries.

Comfrey root is used in Ireland to improve the complexion, by macerating the root and pressing out the juice as needed (Logan), and an ointment made from it is still in use for skin complaints and burns (Painter), and also for ulcers (Vickery. 1995). Boils, too, were treated with comfrey, at least in County Kerry, where they made a poultice from the boiled root. In this case, the belief was that the comfrey drove away the worms in the boil. They could not stand the smell of the comfrey, so it was said (Logan).

It has veterinary uses, too. In some parts of Ireland, swine fever is treated by boiling comfrey roots in milk, and adding everything, roots and all, to the pig's food. This has to be kept up for some weeks (Logan). But Norfolk pigkeepers added comfrey leaves to the pig's feed just to keep them in good health (Randell), but also to make sure they could not be bewitched. Pigweed is a Wiltshire name for the plant (Dartnell & Goddard), though it must have had a much wider significance.

COMPASS PLANT

(*Silphium laciniatum*) A plant from the North American prairies. In an exposed position, its leaves turn their edges north to south, and avoid the mid-day radiation (Willis). Their directional accuracy was relied on by the early hunters on the prairies (A Huxley). This will account for another name, Pilot-weed (Youngken).

CONCEPTION, aids to

Women who wanted children were told to eat LEEKS, and not only in Wales, though Evelyn noted that "the Welch, who eat them much, are observed to be very fruitful", and the Welsh medical text known as the Physicians of Myddfai lists among the virtues of leeks that "it is good for women who desire children to eat (them)". Perhaps the reason for such claims lies in the Germanic peoples' belief that leeks contribute to "manly vigour" (Wimberley), clearly derived from their upstanding growth. CARAWAY, at least according to Gerard, "helpeth conception", but he did not go into details. BIRTHWORT must obviously be included here. The doctrine of signatures, from the shape of the flowers which constrict into a tube that opens into a globular swelling at the base, interpreted as the womb and the birth passage, ensured its use to help delivery, etc., The generic name is *Aristolochia*, which comes from two Greek words meaning 'best birth'. As well as helping delivery, it was said to encourage conception – of a male child, according to Pliny. Another strange usage was described by Guainino (c 1500). He encouraged the use of a kind of medicated pessary made of honey and birthwort, to be introduced before sleeping. The woman was judged to be fertile if on awakening she had a sweet taste in her mouth! (T R Forbes). Just as obvious in this regard is EARLY PURPLE ORCHID, whose twin tubers, suggesting testicles, are its signature for use for impotence and as an aphrodisiac. An old belief, which Gerard ascribed to Dioscorides, was that "if men do eat of the great or fat roots … they cause them to beget male children; and if women eat of the lesser dry or barren root, which is withered or shrivelled, they shall bring forth females". He disclaimed the belief, though - "these are some Doctours opinions only". WHITE BRYONY is another plant with a reputation for ensuring conception, but only because this was taken to be the English Mandrake (see MANDRAKE). A childless woman in East Anglia would drink "mandrake tea", and it was even given

to mares as an aid to conception (Drury. 1985). True MANDRAKE is *Mandragora officinalis,* the most famous of all aphrodisiacs; it was the fruit (actually called Love-apples) that was to be used, and it had the power, so it was said, to put an end to barrenness, quite independently of sexual interourse. See Genesis 30; 14-16, for example. Palestinian women used to bind a piece of mandrake root to their arm (for it could only exert its magical influence if worn in contact with the skin (G E Smith), to promote their fertility, and figures cut from the roots were worn as amulets by both men and women (Budge).

Eating a TANSY salad with a view to procuring a baby was undertaken by childless women at one time. No less an authority than Culpeper advised it: "Let those Women that desire Children love this herb, 'tis their best Companion, their Husband excepted". He recommended it either bruised and laid on the navel, or boiled in beer, and drunk to stay miscarriages. But the real authority as far as Fenland couples were concerned was rabbits. They used to say that where there were wild rabbits, there was sure to be tansy. And everybody knows what large families they produce, so the plant must have the same effect on humans. There are a lot of PARSLEY superstitions connected with conception and childbirth, summed up in the saying "Sow parsley, sow babies" (Waring). In Guernsey folklore it is the man who should wear parsley under the arm, though both should drink large quantities of parsley tea as an aid to conception (Garis). SAGE, too, has a reputation for increasing women's fertility, quite deservedly, apparently, and it has been used for that purpose since ancient Egyptian times (Schauenberg & Paris). "Sauge is good to helpe a woman to conceyve", as Andrew Boorde said, and it seems to do so by arresting lactation. So a strong infusion has been used to dry up the breast milk for weaning (Fernie). Given all this, it is difficult to understand the old wives' tale that drinking salvia cooked in wine would ensure that a woman would never conceive (Boland. 1977). Even MUGWORT, steeped in wine, has been used for the purpose, according to a 17th century prescription (K Knight).

Given the multitude of seeds in a POMEGRANATE, it is natural that it should be linked with conception and pregnancy. In a Persian story, Khodaded was one of fifty children begotten by a childless nomad upon his fifty wives, after eating as many pomegranate seeds. He had incessantly prayed for offspring, and was commanded in a dream to rise at dawn, and to go, after saying certain prayers, to his chief gardener, who was to give him a pomegranate. He had to take from it as many seeds as seemed best to him (Hartland. 1909). KOLA nuts are a sexual stimulant, so it is said, and act as aids to conception in women (Lewin).

Russian folklore contains a belief that WILLOW branches put under the marriage bed would ensure a pregnancy (Kourennoff). But this is a fruitless tree, and it would be used for contraception.

Surely one of the strangest practices under this heading was the use of DARNEL as a fumigant, and a kind of fertility treatment. The procedure, from an Irish source, but probably an ancient practice, was to "burn the meal of darnel and frankincense on a red stone, and let the smoke pass through a funnel under her [i.e., the woman being treated], and that will prepare her for conception" (T R Forbes).

CONDOR VINE
(*Marsdenia condurango*) see EAGLE VINE

Conium maculatum > HEMLOCK

CONKERS
(see HORSE CHESTNUT, or CHILDREN'S GAMES)

Conopodium majus > EARTH-NUT

Consolida ambigua > LARKSPUR

CONSOUND
COMFREY is the best known consound, a reputation resting in the interpretation of the Greek word which gave the botanical name *Symphytum*; it comes from sumphuo, to grow together. Whether this really is the plant that the Greeks named and described as useful for knitting bones is doubtful, but it was certainly taken as such in medieval times, and has been used for the purpose ever since. It was the glutinous matter of the root that was used, grated, for a plaster that set hard over a fracture. But Comfrey is not the only plant with this reputation. ELM, for instance, was sometimes used (the leaves, boiled with the bark). The generic name for LARKSPUR (*Consolida*) shows that it must have had the same reputation at one time, and DAISY, *consolida minor* to Comfrey's *consolida major*, must have had a similar use at one time. In fact, Turner actually told us so: "the northern men call this herbe a Banwort because it helpeth bones to knyt againe". It was actually called Bone-flower in the north of England (Grigson. 1955).

CONSTIPATION
BUCKBEAN was used in the Outer Hebrides, at least on South Uist. They took the root, cleaned it and boiled it in water all day until the juice was dark and thick. This was strained, and a teaspoonful given to the patient; it was even given to calves for the same complaint (Shaw), though the dose must have been increased. SCAMMONY is a Near Eastern plant, and a drastic purgative. Its gum resin, or the dried, milky juice is collected, and often put with colocynth and calomel, to be used for constipation, worms or dropsy (Lindley). How did it become an ingredient in a Cambridgeshire folk remedy for constipation? (Porter. 1969). BLADDER SENNA is, of course, a laxative, but Sir John Hill's warning has to be heeded – "some

giev an infusion of them [the leaves] as a purge, but they are very rough: they work both upwards and downwards, and are only fit for very robust constitutions …".

CONTRACEPTIVES
New Forest gypsies regarded NETTLES as a contraceptive, although they implemented the belief in a strange way. The man had to put nettle leaves inside his socks before intercourse! But, unlikely as it may sound, nettles have been regarded as aphrodisiacs, perhaps as flagellants (Leland). It was the seeds that "powerfully stimulate the sexual functions", and nettles figured in a Greek remedy for impotence (Simons). PARSLEY beliefs are often connected with conception and childbirth in a very confused way. While acting as an aid to conception, in the Cotswolds area, it is contraceptive (Briggs. 1974). Gerard noted that SPEARMINT acted as a contraceptive: "Dioscorides teacheth, That being applied to the secret part of a woman before the act, it hindreth conception".

In spite of the name, CHASTE TREE (*Vitex agnus-castus*), was believed to be contraceptive. Certainly, Dioscorides said so, and Pomet's comment on the name *agnus-castus* is equally explicit: "and the name of Agnus Castus, because the Athenian ladies who were willing to preserve their Chastity, when there were places consecrated to the Goddess Ceres, made their beds of the Leaves of this Shrub, on which they lay: But it is by way of ridicule that the Name of Agnus Castus is now given to this seed, since it is commonly made use of in the Cure of venereal Cases, or to assist those who have violated, instead of preserv'd, their Chastity". But it was still used by Athenian women as a symbol of chastity in religious rites. And for that reason it became also a symbol of indifference (Pomet).

SAVIN has been known through the centuries as an abortifacient and contraceptive, either by the simple matter of swallowing the berries, or by the decoction of the leaves, or, as in East Anglia, put into the teapot with ordinary tea (M R Taylor. 1929). And its contraceptive properties were used with horses, too. It was said that a stallion would never cover a mare if there was any savin in the stable (G E Evans. 1969). WILLOW is a fruitless tree, and so would be used for contraception. Even in quite modern times, German women believed that drinking willow tea would make them barren (Simons). But Russian folklore has the opposite idea, for it was believed that willow branches put under the marriage bed would ensure a pregnancy (Kourennoff).

Grigson. 1955 reported that the seeds of GROMWELL were being investigated for the contraceptive substance they contain (for they apparently can stop the activities of some hormones (Schauenberg &

Paris). But some of the American Indians had been using native species as an oral contraceptive for a long time (W A R Thomson. 1976). It has apparently been found in Hungary that large amounts of powdered FAT HEN will suppress the oestrous cycle (Watt & Breyer-Brandwijk). In other words, it acts as an oral contraceptive. French women were advised, to avoid a pregnancy, to take, in the first three days after her period, "neuf graines de hiève grimpant", IVY, that is (Loux). American Indians, the Penobscot, for example, used AMERICAN PENNYROYAL for suppressed menstruation (Youngken), still in use, often with sage (Corlett). It is a Hoosier remedy for the condition too, as a form of birth control (Tyler). The Hopi used the powdered dried root of JACK-IN-THE-PULPIT (a teaspoonful in half a glass of cold water) as a contraceptive lasting a week. Two teaspoons of a hot infusion would bring permanent sterility! (Weiner).

Convallaria maialis > LILY-OF-THE-VALLEY

Convolvulus arvensis > FIELD BINDWEED

Convolvulus scammonia > SCAMMONY

Coptis trifolia > GOLDTHREAD

CORIANDER

(*Coriandrum sativum*) When green, the seed has a very disagreeable smell, hence the name Coriander, which is derived from Greek koros, a bug. But when fully ripe, the flavour is aromatic, and the longer they are kept, the more fragrant they become. The seeds are used to flavour curries (Brownlow), pickles, and sauces, and for confectionery. Coriander comfits, for example, used to be made; they were just the seeds coated with sugar (Hutchinson & Melville). They are used in flavouring liqueurs, and were once used in gin and whisky distillation (Fluckiger & Hanbury, G B Foster), "…the Brewers employ it considerably all over Holland, and in some parts of England, to give their strong beer a good 'Relish'" (Pomet).

The offensive smell when handled was the source of the medieval idea that it was poisonous (Fluckiger & Hanbury). But it was one of the many plants supposed to be aphrodisiac (Haining). Albertus Magnus (*De virtutibus herbarum*) includes coriander among the ingredients of a love potion, and its use as an aphrodisiac is mentioned too in the Thousand and one Nights (Clair). From a description quoted by Guazzo, it is clear that coriander seeds were sometimes an ingredient of malevolent witch charms. But in North Africa, it is used to drive away evil spirits, and also as a charm against the evil eye. People in Morocco fumigate themselves with coriander seed as a protection against the evil juun. It is also hung under the roof of a house haunted by juun (Westermarck). The leaves, though, were supposed to cure forgetfulness (Legey). The Chinese have a legend that says it

confers immortality on those who eat it (Sanecki), and it is regarded in Europe as the symbol of hidden worth (Leyel. 1937).

The seeds are still known to herbalists as an efficient indigestion remedy. A few seeds chewed before a meal always help (Conway), and they would be chewed to sweeten the breath, too (G B Foster). The whole plant is used in Chinese herbal medicine in the early stages of measles, to bring the rash out (Geng Junying). The juice, "blown up the nostrils", was used to stop a nosebleed, and when mixed with violets, it was used to sober up a drunk (F J Anderson). The Anglo-Saxon version of Apuleius claimed an extraordinary use for coriander seeds. In translation, it reads "in order that a wife may quickly bring forth, take seed … eleven grains or thirteen, knit them with a thread on a clean linen cloth; let then a person take them who is a person of maidenhood, a boy or a maiden, and hold this at the left thigh, near the natura, and as all parturition be done, remove away the leechdom, lest part of the inwards follow thereafter" (Cockayne). So, according to Apuleius, it helps women in childbirth, but according to Macer Floridus, it virtually has the opposite effect (Gubernatis).

Coriandrum sativum > CORIANDER

Cordyline terminalis > TI PLANT

CORN

see MAIZE, which is more often known as Corn in America.

CORN COCKLE

(*Agrostemma githago*) A cornfield pest at one time, largely eradicated now, and becoming increasingly scarce. Pest, because the seeds contain a poisonous element that has caused the death of domestic animals when feeding stuffs have been contaminated with them (Long. 1924). Animals will usually avoid the growing plant, but dogs and young animals are the most susceptible (Forsyth). It has been claimed that baking or grilling will destroy the toxic principle, but Forsyth denied this strongly. Bread containing it is greyish, tastes bitter and smells bad. No wonder steps were taken to get rid of it: in France, the pulling of cockle from the corn used to form an integral part of the Fête des Brandons, the first Sunday in Lent (Chambers. 1870), and a similar "corn-showing" in England (Vickery. 1995). As Gerard said, "what hurt it doth among corne, the spoile of bread, as well in colour, taste and unwholesomeness, is better knowne than desired"; perhaps these were the "tares" of the parable, that the devil sowed in wheat. Anyway, Shakespeare had a proverb for it: "Sow'd cockle rep'd no corn" (*Love's labour's lost. Act iv; scene 3*). The word cockle is OE coccul, Latin coccum, a berry, the large round seeds being likened to berries. The word used to be applied to any noxious weed growing in the corn, particularly darnel (Ellacombe).

The seeds furnished a divination game in some parts of France. To know if the coming year would be good or bad, the seeds would be pulled off individually, and numbered off one by one in the style of 1. Pain; 2. vin; 3 viande; 4. Foin, and so on (Sebillot).

In spite of its poisonous nature, it had its medicinal uses in times gone by. Incredible as it may sound, the seeds were apparently a quite common laxative in the Middle Ages! Archaeologists have found them, some crushed as if in an apothecary's mortar, in cesspits of the 13th and 14th centuries (Platt). Hill was still recommending them in the 18th century. They "open all obstructions", he said.

CORN MARIGOLD

(*Chrysanthemum segetum*) A cornfield pest, so much of a pest that it became probably the first pernicious weed to claim government attention, for it was during the reign of Henry II that an enactment was published requiring the destruction of the "Guilds Weed". The plant was always plentiful on the light sandy soils of Morayshire, hindering the crops, and fostering the rhyme:

> The guile, the Gordon, an' the hiddie-craw
> Is the three worst things that Moray ever saw.

Again, we find that at Cargill, in Perthshire, the gools, as the plant was known there (gools, guile, guild – they are all names for it, part of a long sequence that starts with the common Scots name, gowan), became such a trouble in the corn that an act of the Baron's Court was passed imposing a quite substantial fine on every tenant for each stock of gool that should be found growing amongst the corn on a particular day. Certain persons called gool-riders were appointed to ride though the fields searching for the plant. Wherever it was found the fine was imposed rigorously (Guthrie).

CORN MINT

(*Mentha arvensis*) Sprigs of this mint were put in corn stacks in Ireland, to keep mice away (E E Evans. 1942). But it prevents milk from curdling, so cheese cannot be made from the milk of cows that have fed on it (Bardswell).

CORN POPPY

(*Papaver rhoeas*) see RED POPPY. This plant always accompanies cereal crops, hence the name. The Greeks always reckoned it to be a companion to the corn, not a weed, and the Romans looked on it as sacred to their corn goddess, Ceres. Often, when cultivation is given up in a certain area, the poppy disappears.

CORN SPURREY

(*Spergula arvensis*) A plant of bare and gravelly places, hence the Shetland observation that a profusion of the plant is a sign that more manure needs to be applied to the soil (Vickery. 1995). The bare ground in which it thrives accounts for the names Poverty-weed (Parish), Farmer's Ruin (Leyel. 1948), and Pickpocket (Grigson. 1955), and their variants. In Shetland, the seeds used to be ground into meal, hence the name Meldi, which means meal (Grigson. 1955), and bread has been made from them in times of scarcity in Scandinavia (C P Johnson).

But the plant has been grown as a fodder crop (Murdoch McNeill), both in Britain and on the Continent, and occasionally in South Africa (Watt & Breyer-Brandwijk). Prior quotes a name, Franke, given to this plant, from "the property it hath to fatten cattle, franke being a stall in which cattle were shut up to be fattened, or, as Halliwell put it, " a small inclosure in which animals (generally boars) were fattened". Any animal so shut up for fattening was said to be franked, and the same term was used after fattening. Parkinson. 1640 used the same term in the names Franck Spurrey, and Franck Spurwort.

CORNFLOWER

(*Centaurea cyanus*) Very scarce in cornfields now, and where it does appear it is more often than not a garden escape. *Cyanus* is named for the Greek youth Kyanos, who worshipped Flora, and was for ever gathering flowers for her altar. When he died, the goddess gave his name to the flower he was always picking (Skinner). For the Russians, the cornflower perpetuates the memory of a young man changed by a jealous nymph into a plant. The young man's name was Vassili, which is why the Russian name for the plant is basilek (Palaiseul). Another legend comes from Belgium, where it is called Blue Baron. Once, a young baron went about constantly wearing a cornflower. He valued the flower so much that he married a girl of very low degree whom he had seen decorating a statue of the Virgin with these flowers (Sebillot). Sienese painters (Sano di Pietro, for example) used cornflowers on the heads of angels and saints, perhaps to show they were heavenly beings (Haig). But in medieval times, the flower was woven into a garland, and worn on the head while dancing (T Wright). It was used, too, to test a lover's fidelity (He loves me, he loves me not …) (Dyer).

Blue cornflowers gathered on Corpus Christi Day would stop nosebleeding, if held in the hand long enough, so the belief ran. Does this mean the flowers could be dried and so used on the other 364 days of the year? Hill recommended the leaves, fresh, bruised, to stop the bleeding of a wound (J Hill). Herbalists use it extensively to remedy disorders of the nervous system, for they say it is at the same time curative and calming (Conway). Its use for an eyewash goes back a very long way – Langham, for instance: for "the pain, redness, inflamamtion, and running of the eies". But it seems that cornflowers were best known for treating scorpion stings, Langham, Parkinson and Culpeper all mentioned this use.

CORNS

MADONNA LILY is a traditional cure for corns. It is suggested that Roman legions planted it round their camps – it certainly grows apparently wild in all the countries that were in the Roman empire (Coats. 1956). IVY leaves are a widespread and favourite remedy for corns, used still from the island of South Uist in the Outer Hebrides to Cornwall. The Hebridean practice is to put a poultice of ivy leaves and vinegar on the corn (Shaw), and exactly the same procedure is used in Somerset (Tongue). They say in Cornwall that corns will drop out in about a week after putting bruised leaves on them (Deane & Shaw); exactly the same is claimed in Ireland – if the corn is still there after a few days, then a handful of ivy leaves is put to steep in a pint of vinegar in a tightly corked bottle for a couple of days. Then the liquid is carefully put on the corn, taking care it does not get on the skin (Logan). They say that John Wesley, who walked enormous distances, used ivy leaves to soothe his tired feet; certainly, in *Primitive Physick*, we find "Corns (to cure) … apply fresh Ivy-leaves daily and in 15 days they will drop out". It is still done in the Lowlands (Rorie); in the Highlands, it was said you had to heat the leaf for corns (Polson). Even a dampened TOBACCO leaf has been used - only for a few days, though (Maloney).

They can be dealt with by a mixture of WILD SOR-REL and lard, according to Illinois practice (Hyatt), and CRANBERRIES are also used there, by applying a poultice of freshly mashed fruit (Hyatt). BITING STONECROP, too, is applied, bruised, not only to corns, but also warts (Flück). Squeeze the juice of RED CAMPION on them, and they will come out (a Somerset remedy) (Tongue. 1965), and the juice of TOUCH-ME-NOT can be used, too (McLeod). Corn-leaves is a name given in Worcestershire to WALL PENNYWORT, commemorating a widespread usage of the leaf as a corn or chilblain plaster (Grigson. 1955). Just applying the leaves of MARIGOLDS is a traditional Scottish way of dealing with the prob-lem (Rorie). PENNYROYAl leaves applied to corns will get rid of them (H M Hyatt). So will the sap of GREATER CELANDINE, though care has to be taken, for this sap is corrosive, and may well raise a blister, or even an ulcer (Flück). They can be dealt with by using TORMENTIL, too. On South Uist it was done by chewing the plant, and then applying it to the corn in a bandage (Shaw).

A French charm for corns is to put a piece of KNOT-GRASS in a pocket on the same side as the corn, and say "Que mon cor s'en aille à l'aide de cette herbe" (Sebillot).

Cornus sanguinea > DOGWOOD

Coronilla varia > CROWN VETCH

Coronopus squamatus > WART-CRESS

Corylus avellana > HAZEL

COSMETICS

A manuscript of 1610, published in the *Gentleman's Magazine*, prescribed BROAD BEAN flowers, distilled, as a lotion with which to wash the face. "and it will be fair". Earlier, Gerard had said that "the meale of Beanes clenseth away the filth of the skin", and Hill, later, repeated the prescription. FUMITORY is even more important as a face wash and skin purifier (Fernie).

> If you wish to be pure and holy,
> Wash your face with fevertory (Dartnell & Goddard).

A leaf infusion is used (C P Johnson), or the whole plant boiled in water, milk and whey (Black). "… its remarkable virtues are those of clearing the skin of many disorders …" (Thornton), especially for babies with scalp trouble (Ireland) (Moloney).

> Them that is fair and fair would be
> May wash them-selves in butter milk and fumitory.
> And them that is black and black would be
> May wash themselves in sut and tea.

The leaves of SWEET BASIL with GREAT BURNET, steeped in boiling water, make a cooling face wash (H N Webster). It seems that MOUNTAIN CLUBMOSS (*Urostachys selago*) was used in the Scottish Highlands as a skin tonic. Women and girls steeped the moss in boiling water. Then the liquid, cooled, was used as a lotion (Beith). ELDER flower in this context is much better known. They are still used as an ingredient in skin ointments (Mabey. 1972). The very first issue of *Illustrated London News* (1842) carried an advertisment for Godfrey's extract of elderflowers for ladies' complexions. The inner bark and the leaves, as well as the flowers, were used in Cambridgeshire for the ointment (Porter), but the lotion was made from the flowers only, for "whitening the skin" (Trevelyan), or for washing off sunburn and freckles (Friend). SILVERWEED, often steeped in buttermilk, was used to remove freckles and brownness of the face (Black). Gerard also advised its use to "take away freckles, spots, pimples in the face, and sun-burning", and it is even said that this herb will remove the marks of smallpox (Billson). Elderflower tea was even drunk in Dorset for the good of the complexion (Dacombe). Washing the face with MELON rind, according to a Kentucky practice, will get rid of freckles (Thomas & Thomas), and so would the sliced roots of BLUE FLAG (Le Strange).

A decoction of OAKAPPLES was recommended for blackening the hair, for sunburn, freckles, pimples, etc., (Thomson. 1978). According to Evelyn, "the leaves (of WAYFARING TREE) decocted to a lie, not only colour the hairs black, but fastens their roots"). A MYRTLE decoction too could be used for dyeing the hair black, and not only that, but "it keepeth them from shedding" (Gerard). Candied myrtle blossoms were supposed to beautify the complexion (Wiltshire). A hair dye can be made from BRAMBLE leaves, too. Boil them in a strong lye; it gives the hair a soft, black colour, reckoned to be permanent (Fernie), something that Culpeper claimed in his herbal. ROSEMARY too provided a rinse for dark hair. It is even claimed that it prevents greying. Fresh rosemary water is reckoned the best (though dried leaves will do at a pinch (Brownlow)), and the way to make it is to simmer a large bunch of the herb, stalks and all, in water, and rain water for preference, for about an hour (Rohde). It is still used to good effect in shampoos to eliminate dandruff (Hemphill). Rather optimistically, it was one of the ingredients of preparations to stop the hair falling out, or to make it grow again (see BALDNESS). Milk baths were once a fashionable cosmetic medication, and rosemary was one of the ingredients in a 17th century example, from Gervase Markham (about 1610). The instructions were to "take Rosemary, Featherfew, Orgaine, Pellitory-of-the wall, Fenell, Mallowes, Violet leaves and Nettles, boil all these together, and when it is well sodden, put to it two or three gallons of milk, then let the party stand or sit in it an hour or two, the bath reaching up to the stomach, and when they come out, they must go to bed and sweat, and beware of taking cold" (Wykes-Joyce). SAFFRON has been used in the Middle East to dye the hair a golden colour (Genders. 1972), and MARIGOLD too has been used. Turner, in 1551, vented his displeasure at people who "make their hayre yelow wyth the flour of this herbe, not beynge content with the natural colour, which God hath gyven them". But they were being used for a popular hair dye for a long time after Turner's day (A W Hatfield, 1973). CAMOMILE is still being used in shampoos for fair hair. Or camomile water can be made at home. Simply steep the dried flowers in boiling water and strain off when cool (Hawke). On the Greek island of Chios, camomile is used to dye the hair a light chestnut colour, almost gold (Argenti & Rosa). The ley of BARBERRY ash and water used as a hair wash will turn it yellow, according to Langham, and after him, Culpeper. From this, it was a small step to believng that it actually made the hair grow. You had to wash your head with the water in which barberry had been boiled – but "take care that the wash does not touch any part where the hair should not grow" (Leyel. 1926).

HONEYSUCKLE as a drug plant does not seem appropriate, but its use as a cosmetic is much more in keeping with the feeling that the plant inspires. Lotions and ointments were made of it. One, from the mid 16th century, orders: "Take a pint of white wine, one handful of woodbine leaves or two or three ounces of the water of woodbine, add a quarter of a pound of the powder of ginger; seethe them all together until

they be somewhat thick, and anoint a red pimpled face therewith, five or six times, and it will make it fair" (Lupton). COWSLIPS enjoyed a similar reputation. An ointment "made with the juice of Cowslips and oile of Linseed" used to be well-known as a good thing to improve the complexion (Dyer. 1889), and is still being recommended – "cowslip leaves in a cold cream base do much to hinder wrinkles and preserve the complexion". Culpeper says "Venus lays claim to this herb as her own, and it is under the sign Aries, and our city dames know well enough the ointment or distilled water of it adds beauty, or at least restores it when it is lost". He agreed with the earlier prescription, for he says the flowers "are held more effectuall than the leaves". Cornish girls believed their complexion could be improved if they rubbed their skin with wild STRAWBERRY leaves. Actually, the belief is widespread – see, for instance, this verse of a version of "Dabbling in the dew":

> Pray whither so trippingly, pretty fair maid,
> With your face rosy-white and your soft yellow hair?
> Sweet sir, to the well in the summer-wood shade
> For strawberry leaves make the young maiden fair (Deane & Shaw).

The leaves are astringent, as are most of the Rosaceae, so there was likely to be sound reason for the practice. An infusion of the whole CHICORY plant has long been used as a cosmetic – applied at bedtime, it will remove blemishes from the skin (Genders. 1976). But Thomas Hill in 1577 wrote that "Ciccorie cureth scabbed places, causeth a faire skin …". Gerard seems to be invoking the doctrine of signatures when he recommended the yellow flowers of TOADFLAX in decoction to take away "the yellownesse and deformitie of the skinne, being washed and bathed therewith". Eating WATER MELON rinds is an Alabama way of achieving a smooth complexion (R B Browne). A young man in Morocco who wants his beard to grow rubs his skin with a piece of water melon, for the juice was thought to produce the desired effect (Westermarck). TANSY soaked in milk had the reputation of "making the complexion very fair" (Dyer); in other words it was used as a cosmetic wash to remove sunburn. DANDELION flowers, boiled for half an hour in water, make a cosmetic wash to get rid of freckles on the face (Palaiseul). Wrinkles were treated with a face-pack of FENNEL, tea and honey (Addison). The water that collected in the cups formed by the fusing together of TEASEL's opposite leaves, "so fastened that they hold dew and raine water in manner of a little bason" (Gerard), was much prized for cosmetic use. Culpeper knew about this, but he described it as the distilled water of the leaves, used by women "to preserve their beauty", hence the names Our Lady's Basin (Macmillan), and Venus's Bath, or Basin (Britten & Holland).

Leicestershire girls washed their faces in this water, in order to make themselves more beautiful (Billson), and the folk use was known in Wales, too – there it was said to be a remedy for freckles (Trevelyan). Gerard recommended WHITE BRYONY for the complexion, for "it scoureth the skin, and taketh away wrinkles, freckles, sun burning, black markes, spots and scars of the face, being tempered with the meale of vetches, or Tares, or of Fenugreeks. It could be used for bruises, too, hence the French name herbe aux femmes battues (Cullum). And the ladies of Salerno were reputed to have steeped bryony roots in honey to put on their faces, "which gives them a marvellous blush" (Withington). Gerard reported that "if the face be washed with juyce (of WATER BETONY), it taketh away the rednesse and deformity of it". BURNET SAXIFRAGE is another plant once used as a cosmetic. In Italy, they used to say that if a woman *eats* it, her beauty will increase (Skinner). More conventionally, Gerard recommended "the juice of the leaves … doth cleanse and take away all spots and freckles of the face, and leaveth a good colour …". The distilled water, too was used as a cosmetic (Leyel. 1937). Gerard wrote "Matthiolus teacheth, that a water is drawne out of the roots (of SOLOMON'S SEAL), wherewith the women of Italy use to scour their faces from Sunne-burning, freckles, morphew or any such deformities of the skinne". What is interesting is that a soap was made until recent times, from this plant, in the Kingsclere district of Hampshire, for just the purpose outlined by Gerard (Read). He was alive, too, to the use to which CUCUMBER could be put in clarifying the skin (a use re-discovered in modern times): "being steeped and outwardly applied in stead of a clenser, it maketh the skin smooth and faire". He was able to indulge his humour on this subject as he went on to say "the fruit cut in pieces or chopped as herbes to the pot, and boiled in a small pipkin with a piece of mutton, … taken … for the space of three weekes together without intermission, doth perfectly cure all manner of sauce flegme and copper faces, red and shining fierie red noses (as red as red Roses) with pimples, pumples, and such like prescious faces". Not only is CHICKWEED a lucky plant for people of the Fen country, it is also said that, if gathered when the dew was still on it, then crushed and applied to the face, it was thought to turn the plainest woman into a beauty (Porter. 1969).

A hair dye, popular in Lancashire in the 17th century, was obtained from BOG ASPHODEL. After the flowers have faded, the stems change to a saffron colour, and the hair dye was extracted from these (Freethy), and an early 17th century auburn hair dye had as its principal, constituents RADISH and PRIVET (Wykes-Joyce). In the 16th century, it was being claimed (by Langham) that the "lee wherein the leaves (of BOX) have been sodden or steeped maketh the haire yelowe being often washed therewith". SAGE tea can ease a sunburnt face, if it is washed with it (Page. 1978), and the tea has been used in Scotland as a hair wash (Gibson, 1959), for cold sage tea has been regarded as a hair tonic, at least in America, and it will also darken the hair (H M Hyatt). Quince-seed lotion, made by stewing the seeds in water, was used as a hair lotion, "for giving ladies' hair a fine wavy appearance" (Savage). MULLEIN flowers were used to dye the hair yellow (Rohde. 1936), at least as early as Parkinson's time, for he tells us that the flowers "boyled in lye dyeth the haires of the head yellow and maketh them faire and smooth" (Parkinson. 1640). One can still buy camomile and mullein shampoo for fair hair. SCENTED MAYWEED is another ingredient in hair rinses, and it seems to have been farmed in one area of Hungary just for this purpose (Clair).

ROSE OF CHINA (*Hibiscus rosa-sinensis*) has beautiful flowers, but the red colour turns black when they are bruised, and then a shoe blacking medium can be made, hence such odd looking names as Shoeflower (Wit), Shoe Black (Campbell-Culver) and Blacking Plant (J Smith. 1882). And, for our purpose here, the blacking can be used for colouring the eyebrows. Dew recovered from the cups of WHITE WATERLILY was used in Hampshire to enhance the appearance of the eyes (Hampshire FWI). DEADLY NIGHTSHADE has also been used for the same purpose. Its alternative name, BELLADONNA, owes its existence to the custom on the Continent for women to use it as a cosmetic to make the eyes sparkle (Brownlow) (atropine is still used by oculists to dilate the pupils).

HENNA leaves, powdered and mixed with whitewash, or water, are laid on the skin as cosmetics all through the orient (Loewenthal), and have been since ancient Egyptian times. Mummies show that women painted their fingernails and finger tips, their palms, and the soles of their feet, as is still done in many areas today. The dye needs renewing every two or three weeks (Moldenke & Moldenke). But it would be wrong to look at the use of henna as a simple cosmetic. It is applied just as often as a protection against supernatural attack, and particularly against the evil eye, red being a good prophylactic colour (see EVIL EYE, and PROTECTIVE PLANTS). PUCCOON roots were used, too, for facial painting. The fresh root would be dipped in grease, and simply rubbed on the face. American Indian groups like the Thompson were reported as using it (Teit). Puccoon's near relative in Britain, GROMWELL, was similarly used. Thornton was able to write that "the root is used by ladies as paint". "The gentlewomen of France do paint their faces with roots [of ALKANET]" (Gerard). It seems to be one of the most ancient of face cosmetics. This use of the roots for making rouge led to the plant becoming known as a symbol of falsehood (Folkard). Oil of SPIKENARD was included with other oils to make spikenard ointment, once used in cosmetics

(Zohary), and Oil of Santal, distilled from the heartwood of SANDALWOOD is another ingredient of Eastern perfumes and cosmetics (Usher). TURMERIC, the typical yellow dye, was used mainly as a cosmetic in the Pacific islands, often with sexual overtones. On Tikopia, it is daubed over mother and child soon after birth, as a mark of attention, or even of honour. It was used to single out individuals who were at the moment of special interest or importance. (Firth).

COSTMARY
see BALSAMINT

Cotinus coggyria > VENETIAN SUMACH

COTTON-GRASS
(*Eriophorum angustifolium*) It appears in a number of Irish and Highland folk tales as a powerful instrument against enchantment, for example in the one called *The three shirts of canach down* quoted by J G McKay. There the sister had to make each of her enchanted brothers a shirt of the moorland canach, which was the Highland name for the plant, ceannabhan mona in Irish. She had to remain completely silent until she herself, after making the shirts, had put them on her brothers to free them from the spell. The Irish version of the tale is quoted by Kevin Danaher in *Folk tales of the Irish countryside*. Nettle is the textile in some versions, a far better proposition for the hard-working sister. But cotton-grass heads were a crop on the Isle of Skye, gathered usually by the children. Dried, they were used to stuff pillows and quilts (Swire. 1961) (St Bride lined the bed of Christ with cotton-grass, even though the birth of Christ occurred in mid-winter (Swire. 1964)). Stuffing pillows must have been a tedious task, for when the heads were used whole they would be lumpy in a pillow, but the trick is to remove the tough base and use only the down. It can be spun like cotton, but the fibres are more brittle than those of cotton, so not so useful. Candle and lamp wicks used to be made of the down (Johnson), while children often use them like powder puffs (Bairacli-Levy).

COUCHGRASS
(*Agropyron repens*) A far-creeping and extremely tenacious weed, the despair of any gardener unlucky to get it in his patch. The answer, so it is claimed, is to sow turnip seed thickly in the infested part of the garden; the couch will disappear (Boland & Boland). But that is a question that has occupied the minds and ingenuity of gardeners for centuries. One answer, from the 10th century collection known as the *Geopontica*, claims that to get rid of "Dog's-tooth grass" one had to plough it up in May, let it dry, and take it away when the moon is 16 days old, in other words, just at the full, "for the antipathy (of the moon for this weed) will contribute towards preventing it growing up again" (Rose).

It did have its uses, for the dried rhizome, in infusion or decoction, has been used medicinally for centuries, for the treatment of kidney and bladder complaints, including cystitis (V G Hatfield), and also for skin eruptions and rheumatic complaints (Fluck). And this is the grass that dogs most often eat when they feel in need of a tonic (Page. 1978).

Most of the many names given to couch stem from OE cwic, alive, an appropriate naming. It comes in many forms – couch itself is a variant of Quitch, or Squitch, and this varies through such forms as Quick, Quack, Twitch, Whick and so on. The Scottish word Kett summarizes the feeling of detestation that any gardener has for the grass. It apparently means filth. Yawl, from the Isles of Scilly, expressesd the same kind of anger, for this is the Old Cornish dyawl, devil. Rack is an old Suffolk name meaning weeds or rubbish. One name, though, is out of this sequence – Grandmother Grass, which is from a children's game that involves cutting the head of a piece of the grass, and sticking it in another head, still on its stem. A flip of the hand holding the stem, and "Grandmother, grandmother, jump out of bed" is recited as the first head springs out (Mabey. 1998).

COUGHS
LEEK juice was often used for whooping cough, or indeed any "old" cough. As Thomas Hill said, "leeke amendeth an old cough and the ulcers of the lungs". It was used either on its own or mixed with something else, as in the Welsh custom of joining it with women's milk for coughs, a recommendation that appears both in the Book of Iago ab Dewi (see Berdoe) and in the Physicians of Myddfai. ONION juice was considered essential to cure a cough or bronchitis centuries before its use in various patent medicines (Camp). Coughs, including whooping cough, have long been treated with TURNIPS, too. The usual country practice was and still is to cut a turnip into thin slices, put them in a dish, and put sugar on them. Leave them for a day or two, and give a teaspoonful of the juice for the cough. That is the Wiltshire remedy (Olivier & Edwards), but it is virtually the same across southern England. NETTLES, whose efficacy in chest complaints was widely believed in, was used for anything from coughs to tuberculosis. Martin, at the beginning of the 18th century, took note of its use in Lewis for coughs. In this case, they used the roots boiled in water and fermented with yeast. Earlier, Gerard had recommended it for "the troublesome cough that children have, called the Chin-cough", whooping cough in modern parlance. A Cambridgeshire cough cure made use of CARAWAY – 2 ounces of the seed boiled in a quart of water down to a pint, half strain off, sweeten with sugar, add a glass and a half of rum. Take a wineglassful every night going to bed (Porter. 1969). COMFREY has been used for coughs for

centuries, from Anglo-Saxon times onwards, either the leaves and root on their own, or with some other ingredient, such as elecampane, or horehound, both renowned cough medicines themselves. ELDER flower tea is good for a cough or for colds and sore throats. Elderflower vinegar used to be saved, too, to use for sore throats (Painter). Coughs can be treated, according to an 18th century prescription from Anglesey, by drinking the result of boiling up a quantity of Liverpool ale with rosemary, honey and salt butter (T G Jones). The nuts, bark and leaves of SWEET CHESTNUT can be used for all kinds of coughs, including whooping cough (Page. 1978), a usage that has been known for a long time. See Gerard, for example, who said that "an electuary of the meale of Chestnuts is very good against the cough and spitting of bloud". MALABAR NUT is an Indian plant, but long cultivated in the tropics, and much used as an expectorant or cough reliever (Thomson. 1976). CAROB molasses is another very popular cough medicine (Bianchini & Corbetta). Native Americans used the inner bark of PIN CHERRY for a cough medicine (Youngken), and the Seneca Indians used SENEGA SNAKEROOT for a cough (Lloyd). BRAMBLE vinegar used to be made in Lincolnshire for coughs (Gutch & Peacock). SLOES were used as a cough cure in North Wales (Friend. 1883); so they were in the Highlands, too, for sloe jelly was reckoned the best cure for relaxed throat (Grant). MARJORAM tea is much used for whooping cough, and can be used as a mouthwash for inflamed mouth or throat (Flück). The Anglo-Saxon version of Apuleius also recommended it for coughs, by the simple expedient of eating the plant (Cockayne). PENNYROYAL tea, too, is good for chills and coughs (Vesey-Fitzgerald), extended to include bronchitis and asthma (Beith). In Morocco, the dried leaves are made into a powder and taken with porridge or milk for coughs and colds (Westermarck. 1926). Wesley recommended it for whooping cough, too.

A root decoction of VIRGINIAN SNAKEROOT was taken for coughs (Corlett), and the Mashona witch-doctor in Zimbabwe made use of the bark of *Ptero-carpus angolensis*, powdered with the inner layer of BLUE LOTUS root. ELECAMPANE's roots have been used to make cough candies for a very long time (Le Strange). The powder was soaked and given to the patient in his porridge (Gelfand. 1956). GORSE also has been used – an infusion of the green tops has featured in many a Scottish Highlands cough cure (Grant), and in Ireland a medicine was made by pack-ing the flowers tightly in a crock, and putting brown sugar on top. The crock would be covered, and put in a saucepan to stew slowly (Lucas). Another Highland remedy consisted of boiling OX-EYE DAISY juice with honey (Fairweather). SAGE tea is famous as a cough cure, and has been since ancient times. It is

still in use, not only for coughs, but also for colds, headaches and fevers (Conway). Cough cures are detailed as early as the 14th century, when we find "Medicina pro tussi. Take sauge and comyne and rewe and peper and seth hym to-gedre in a panne with hony, and ete ther-of a spoone ful a-morwe and at eve a-nuther" (Henslow). WHITE HOREHOUND, too, is just as famous for curing coughs and colds, usually in the form of a leaf infusion, or as a candy. Lozenges made from it are still used (M L Cameron). As long ago as Anglo-Saxon times it was prescribed for colds in the head and coughs. Gerard, too, recommended the infusion, that "prevaileth against the old cough", and the syrup made from the leaves "is a most singular remedy against the cough and wheezing of the lungs …". Syrup of SQUILLS (decocted from the bulb) used to be a common cough cure. See George Crabbe, *The Borough:*

> A potent thing, 'twas said to cure the ills
> Of ailing lungs, the oxymel of Squills.
> Squills he procured, but found the bitter strong
> And most unpleasant, none would take it long.

Gypsies use a CHICKWEED leaf infusion to cure coughs and colds (Vesey-Fitzgerald), and they also used AGRIMONY tea, a very well known cough cure (Conway).

Some Mexican Indians use BLACK MUSTARD seed for a children's cough remedy; the oil from the seed would be heated and rubbed on the chest, which would then be covered with a flannel cloth (Kelly & Palerm). GINGER is a fairly obvious choice for colds and coughs, and in the form of jam it was a favourite medicine (Parihar & Dutt).

An Alabama cough medicine is "shaggy hickory" (SHAGBARK HICKORY) the bark boiled in water, strained off and sweetened. A teaspoonful every few hours is the dose (R B Browne). But the best-known of all the cough cures is that provided by COLTS-FOOT, either in the form of a tea, or a piece of colts-foot rock, to chew, or as coltsfoot wine (M Evans). And it can be smoked, like tobacco. Bechion, the plant in Dioscorides taken to be coltsfoot, was smoked against a dry cough, and it is still smoked in all herbal tobaccos (Grigson. 1955), as it is also in Chinese medicine (F P Smith), for asthma, even lung cancer (Perry & Metzger). Gypsies smoke the dried leaves as beneficial for asthma and bronchitis, and the tea is also taken for coughs (Vesey-Fitzgerald). Cornish miners used to smoke it as a precaution against lung diseases (Deane & Shaw). A sweet paste, called pâté de guimauve, used to be prepared by French drug-gists, from MARSH MALLOW, for coughs and sore throats (M Evans).

COUNTER-POISONS

When a plant is called "holy" or "blessed", i.e., bears the specific name *benedictus*, it means it has power of counteracting poisons, or so it was supposed (herba benedicta, AVENS, that is, is best known for this property). HOLY THISTLE (*Carduus benedicta*) is an another example. Langham could say "the leafe, juice, seede, in water, healeth all kindes of poyson…". Everybody knew it as a heal-all. Langham, indeed, had four pages of recipes under this head, for practically every malady, including the plague. WHITE HOREHOUND is a counter-poison, too, or rather it was thought to be one. There are a number of authorities, though, who were sure of it, the Lacnunga for one – in case a man drink poison: take seed of marrubium; mix with wine. Give to drink (Grattan & Singer), and there were recipes in the 14th century, too (see Henslow).

COVENTRY BELL

Another name for Canterbury Bell (*Campanula medium*). Also recorded are Coventry Rape (Gerard), and Coventry Marian (Britten & Holland)

COWBANE

(*Cicuta virosa*) The *Sium* mentioned in one of Reginald Scot's recipes for a witch ointments is probably Cowbane. Its combination with the others items (sweet flag, cinquefoil) would, so it is claimed, cause great excitement when rubbed on the skin, and might lead to delirium. Bat's blood was another ingredient in this recipe, though it is totally innocuous (Haining).

In the Finnish myth of Lenninkainen, he, in his search for a maiden in the hostile north, drove off all men by the force of his magic songs, except one, a miserable, blind herdsman, who finally shot him with a shaft of cowbane (Turville-Petrie). A comparison with the Scandinavian myth of Baldur and mistletoe is obvious.

COWPEA

(*Vigna unguiculata*) A legume of tropical origin, but they have been grown in America long enough to have just a few superstitions attracted to them. It is said, for instance, that if a married couple throw cowpeas across the road near their home, the woman will become fertile, no matter low long she has been barren (Waring). Another one, specifically from Illinois, says that cooking them on Good Friday will bring luck (H M Hyatt).

The "peas", or seeds, can be of any of six or seven colours. White ones have pigment confined to a narrow "eye", then they are often described as black-eyed beans, which have become identified with the "soul food" of the American deep south (Vaughan & Geissler).

COWSLIP

(*Primula veris*) An unromantic name for such a plant, for cowslip (OE cuslyppe) means cow dung. It must have arisen from observation that a meadow full of cowpats suddenly became full of cowslips as well. Oxlip has a similar derivation. What is clear is that cowslip is not cow's lip, in spite of Ben Jonson's

"Bright dayes-eyes and the lippes of Cowes". It is said that cattle have an aversion to the cowslip, and they will refuse to eat it. It is further said that cowslips would give them the cramp, or colic, and the cattle will become "elfshot" (A R Forbes. 1905).

In Lincolnshire, it was believed that if a cowslip root was set the wrong way up, it would come up a primrose (Gutch & Peacock), and in Cheshire, the result would be that it would come up red (Hole. 1937). Another superstition is that if you dream of them in bloom, it is a sign of a sudden change in your fortunes (Raphael). Unfortunately, the dream book does not say whether for good or ill. Another belief is that you only hear the nightingale's song where there are a lot of cowslips (Swainson. 1886), for they have, according to an old belief, a particular liking for such a place. There is one piece of weather lore – if the cowslip's stalks are short, then we are in for a dry summer (Inwards). Some say, too, that we never get warm, settled weather till the cowslips are finished (Page. 1977), and if they were bloom in winter, it would be an omen of death (Hole. 1937).

There is a kind of divination game that children used to play with cowslips, called Tisty-Tosty, or Tosty-Tosty. Blossoms, picked on Whit Sunday for preference, were tied into a ball. Strictly, the balls were the tisty-tosties, though the growing flowers got the name, too. Lady Gomme mentioned the game as belonging to Somerset, but it had a much wider spread than that. The cowslip ball is tossed about while the names of various boys or girls are called, of the time-honoured Tinker, tailor, soldier, sailor… fashion, till it drops. The name called at that moment is taken to be the "one indicated by the oracle", as Udal puts it, for the rhyme spoken at the beginning is:

> Tisty-tosty tell me true
> Who shall I be married to?

That is quoted as a Dorset rhyme, but the same is recorded in Herefordshire (Leather). The game is also known in Wales, where the purpose is different, for the rhyme there is:

> Pistey, Postey, four and twenty,
> How many years shall I live?
> One, two, three, four … (Trevelyan).

John Clare (*The shepherd's calendar*)called the tisty-tosties cucking balls:

> And cowslip cucking balls to toss
> Above the garlands swinging light …

A different tradition here, obviously. Roy Genders, who used the name cucka-balls, says they were often threaded on twine and hung from one window to another across the street.

Both flowers and leaves have their culinary uses – they have been used in salads since medieval times

(Brownlow), or the leaves can be boiled with other herbs (Jason Hill). Paigle Pudding is mentioned as a Hertfordshire dish (Jones-Baker. 1977), made from the dried petals, flour, etc., Cowslip tea is still made, and has been a country delicacy for a long time. Flora Thompson enjoyed it, and said it is made from the peeps (or pips, a name usually reserved for the dried flowers from which the wine ought to be made (Bloom), by pouring boiling water over them, then letting it stand for a few minutes to infuse. It can then be drunk with or without sugar. Cowslip wine is an excellent sedative, apparently (Grieve. 1931).

Izaak Walton, in *Compleat Angler* (1653) recommended what he called Minnow-tansies. The minnows should be "washed well in salt, and their heads and tails cut off, and their guts taken out and not washed after". They make "excellent Minnow-tansies, that is fried in yolks of eggs, the flowers of cowslips and primroses and a little tansy, thus used they make a dainty dish of meat".

An old name for cowslip is Palsywort, which shows that it must have been used for that complaint. It must have been the trembling or nodding of the flowers that suggested it. Grigson. 1955 pointed out that the medieval *Regimen Sanitatis Salernitanum* had commended the cowslip as a cure for palsy or paralysis (hence another old name, Herb Paralysy). Gerard repeated the prescription – "cowslips are commended against the pain of the joints called the gout, and slacknesse of the sinues, which is the palsie". He goes back to it – "a conserve made with the flours of cowslips and sugar prevaileth wonderfully against the palsie, convulsions, cramps, and all diseases of the sinues". Culpeper, too, mentions it – "because they strengthen the brain and nerves, the Greeks gave them the name paralysis". It still appears in herbal medicine books as a remedy for giddiness, nervous debility or excitement (Wickham), and herbalists still use cowslip leaves as a sedative and pain-killer (Conway).

"An unguent made with the juice of Cowslips and oile of Linseed, cureth all scaldings or burnings with fire, water, or otherwise"(Gerard). This unguent used to be well-known as a good thing to improve the complexion (Dyer. 1889), and is still recommended (Conway). Culpeper says "Venus lays claim to this herb as her own, and it is under the sign Aries, and our city dames know well enough the ointment or distilled water of it adds beauty, or at least restores it when it is lost".

Gypsies use an infusion of the dried flowers to allay convulsions and to lower the temperature (Vesey-Fitzgerald). That is why cowslip wine or tea is taken for measles and other fevers (Hampshire FWI), and herbalists prescribe the root decoction or extract to treat ailments like whooping cough, bronchitis and pneumonia (Schauenberg & Paris). The strangest usage must be this Irish one for deafness: Take the cowslip, roots, blossoms and leaves, clean them well, then bruise and press them in a linen cloth, add honey to the juice thus pressed out, put it in a bottle, and pour a few drops into the nostrils and ears of the patient, who is to lie on his back. Then after some time, turn him on his face till the water pours out, carrying away whatever obstruction lay on the brain (Wilde. 1890).

One series of names for the cowslip starts with Herb Peter. Then follows a whole Bunch of Keys (Macmillan), all from the supposed resemblance to the badge of St Peter – a bunch of keys. The legend is that St Peter once dropped the keys of heaven, and the first cowslip grew up where they fell (Greenoak). So we have St Peter's Keys and St Peter's Herb, or Keywort, and Keys of Heaven, etc., as old names for cowslip.

CRABAPPLE
(*Malus sylvestris*) Roasted and presumably sweetened, and put into ale makes Lamb's Wool, a favourite Christmas drink at one time (Dyer. 1883). Even without the ale, roasted crabs were a favourite fruit in days gone by (Ellacombe), though nobody would take the trouble to cook them these days. However, crabapple tea is still sometimes made.by slicing the apples without peeling them, simmering for an hour and sweetening (Jason Hill). Actually, crabapples are highly astringent, and not fit to be gathered until touched by frost, which cuts the acrid taste considerably (Loewenfeld). A very sour liquid knows as verjuice used to be extracted from crabs, quite popular in the country in the 19th century, but used for veterinary practice long before that. Tusser, for instance, advised the husbandman:

Of vergis be sure
Poore cattle to cure.

It was also used for curdling milk, and for treating sprains (tart, vinegary cider was still being used as a formentation for muscular sprains in the first half of the 20th century (Savage). A drink called wherry in Yorkshire used to be made from the pulp after the verjuice had been expressed (Holloway).

There seems to be very little folklore attached to the tree or its fruit, though it is recorded that Somerset girls would gather them on Michaelmas Day to store till Old Michaelmas (10 October), when they would form them into the initials of their suitors. Those in the best condition would make the best husbands (Tongue) (but when was the inspection made?). There is, too, a suggestion that crabapples were used in some way for sex magic, and that if several were eaten (especially with cucumber and cheese), they would inspire erotic dreams (Haining) (not surprisingly, but one would doubt the eroticism of the dreams). And there is a superstition from Kentucky that must be a virtual impossibility – if you can eat a crabapple, without frowning, you will get the man (or woman) you want (Thomas & Thomas).

There were some medical uses in past times. The verjuice already mentioned was recommended as a remedy "for the falling down of the uvula", and also for sore throats, "and all disorders of the mouth" (Hill. 1754). Wesley's remedy for "eyes blear'd" was to drop in them crabapple juice, while "infantile ague" was dealt with by the Physicians of Myddfai by roasting crabs, and taking "some of the pulp, and half as much honey: let this be the child's only sustenance for a day and a night". Gerard recommended the verjuice to take away "the heat of burnings, scaldings, and all inflammations", and provided it was put on early enough, it stopped, he said, any blistering. He prescribed it too for various skin ailments, for it "taketh away the heat of S. Antonies fire, all inflammations whatsoever", etc.,

CRAMP

American people carry a nut of the YELLOW BUCKEYE around in their pocket to prevent or cure cramp, among other ailments, notably rheumatism (R B Browne), just as Conkers are in Britain. Putting a gall from a DOG ROSE under the pillow was a Norfolk way of curing the cramp (Taylor). Carry a POTATO in your pocket to keep the cramp away (Stout), or, from Somerset, put some YARROW in the shoe (Tongue. 1965). PERIWINKLE was reckoned to be a good remedy for cramp (Grieve. 1931). People used to wear bands of it about the calf of the leg to prevent it (Fernie), and in Lincolnshire a piece of the plant was put under the mattress for the same purpose (Rudkin). A Somerset remedy required the patient to take an infusion of the dried root of STINKING IRIS (Tongue. 1965).

Gerard listed BAY berries as a cure for many complaints, including "cramps and drawing together of the sinues". FIGWORT had a reputation in the Channel Isles as being an efficient remedy for cramp (Garis), probably because it is an anodyne (Mitton). There is a recipe for cramp from Alabama involving BLOOD-ROOT: mix one teaspoonful of crushed bloodroot, half a cup of vinegar and four teaspoons of sugar. Heat to boiling, strain, and give up to one teaspoonful every half hour (R B Browne). GUELDER ROSE is often called Cramp-bark (Youngken), for it has been used for centuries by people who live in marshy country, and suffer from rheumatism and cramp (Wilkinson. 1981).

A strange superstition is recorded from the Pennsylvania Germans. They used to say that a person who is subject to cramps should, immediately on getting up, and without saying a word to anyone, go and plant some SOUTHERNWOOD (Fogel). Just as strange is the claim (Jacob) that CELERY was used by witches to prevent their getting cramp while flying. And why not? It was quite a common medieval remedy for cramp, at least, the earthbound kind.

In parts of France, the plant called SOLOMON'S SEAL was known as l'herbe à la forcure, and it was said that the roots represent all the parts of the body.

It was used for treating muscular cramp, and great care was taken to use that part of the root that doctrine taught represented the part of the body affected (Salle).

CRANBERRY

(*Vacccinium oxycoccus*) A Kansas wart charm is recorded which involves cutting a cranberry into halves. The wart has to be rubbed with each half, and they have then to be buried under a stone (Davenport). Although it is not quoted, surely the wart will disappear as the fruit rots. In a similar way, corns are treated by applying a poultice of freshly mashed cranberries (Hyatt). One interesting American usage is the application of cranberries to shingles. They say such a poultice will cure the condition (Turner & Bell), and cooked cranberries were sworn by in Kansas as effective rheumatic pain relievers (Meade).

The Russian sweet Kissel is made with the juice of stewed cranberries slightly thickend with cornflour. It is served in glasses at the beginning of the meal, or later on with a biscuit (J Hill. 1939).

CRANBERRY SAUCE

Not made with cranberries at all, but with Cowberries (*Vaccinium vitis-idaea*).

Crataegus monogyna > HAWTHORN

Crataegus monogyna ('Praecox') > GLASTONBURY THORN

CREEPING ZINNIA

(*Sanvitalia procumbens*) A well known wound plant in its native central America, where it was often used as a counter-poison, too. Maya medical texts prescribed it, in decoction, for constipation and dysentery, and the leaves were to be boiled and put as a poultice on dislocations (Roys).

Crithmum maritimum > SAMPHIRE

Crocus sativus > SAFFRON

CROSS

Many trees share the odium of supplying the wood of which the Cross was made. It was said to have been made of four woods – cypress, box, cedar and pine. In the Eastern tradition, olive and palm are substituted for box and pine (Child & Colles). The OLIVE supplied the wood for the tablet above Christ's head on which the title was written. ASPEN is another, and its constant shivering is explained in folklore by that belief, making it a cursed tree in some areas, notably in the Scottish Isles, since not only was the Cross made of its wood, but it refused to bow down during the procession to Calvary (Carmichael). Such belief was held too in the west country and Wales, as well as on the Continent. Another tradition has it that the ELDER is the tree of which the Cross was made:

> Bour-tree, bour-tree, crookit rung,
> Never straight, and never strong,
> Ever bush, and never tree,

Since our Lord was nailed t'ye. (Chambers).
(Bour-tree is one of the Scottish names for elder).
"We don't cut elder in a copse, nor do we burn it.
They say the Cross was made of it" (Goddard). That
is the reason for the superstition that the elder is proof
against lightning – it never strikes the tree from which
the Cross was made. In the Greek islands it is the
HOLM OAK that is the unlucky tree. The story went
that a miraculous foreknowledge of the Crucifixion
had spread among the forest trees, which (almost)
unanimously agreed not to allow their wood to serve.
When the foresters came, they either turned the edge
of the axes or bent away from the stroke. Only the Ilex
(the Holm Oak) consented, and passively submitted
to being felled. So now the woodcutters will not soil
their axes with its bark, nor desecrate their hearths by
burning it (Rodd). An Irish belief says that the true
Cross was made of OAK (Ô Súilleabháin).

There is a medieval Jewish tale in which it was told that
every tree had sworn it would not bear Christ's body,
except for a CABBAGE stalk, and it was on this that he
was crucified. There were English traditions according
to which Christ was crucified on a MISTLETOE
(Turville-Petrie). It was christianized in Europe under
the name of Sanctae Crucis Lignum, the wood of the
Holy Cross. As the tree that had borne Christ, it had
dwindled to a lowly shrub, yet had acquired a sacred
force (Grigson. 1955). In Brittany, it was until recently
still looked on as the tree of the Cross (Friend. 1883)
(cf the Scandinavian myth of the killing of the god
Baldur by use of an arrow of mistletoe).

There is an old legend that EARLY PURPLE
ORCHID grew under the Cross, and received spots of
Christ's blood on its leaves, retaining the spots to this
day. The Cheshire name Gethsemane (see Britten &
Holland) is an echo of the legend. Jessamine, recorded
in Warwickshire (Grigson. 1955) looks strangely out
of place for an orchid name, but makes sense if the
word is a variation of Gethsemane. Cross-flower, a
Devonshire name (Britten & Holland), is also a reference to the legend. An old legend tells how BASIL
helped St Helena find the true Cross. She came to a
place where a lot of basil was growing, and the plant's
scent guided her to the right place to find the relic. At
the feast of the Invention of the Cross (originally 14
September, now 3rd May), Greek women bring basil
plants to the church, and the priests distribute twigs to
the congregation. (Argenti & Rose).

CROUP
Buchan prescribed PENNYROYAL for croup:
"Take penny-royal 3 ounces, syrup of althaea and of
poppies, each one ounce, mix them together".

CROWN IMPERIAL
(*Fritillaria imperialis*) This is a native of Persia,
and was brought to Vienna from Constantinople by
Charles de l'Ecluse in 1576, provoking an outbreak

of "Tulipomania", with similarly high prices. Perhaps
the name is an allusion to the Habsburg court in
Vienna (Elliott). Gerard, twenty years later, was still
full of wonder when describing the plant – "in the
bottom of each of these bells there is placed six
drops of most cleare shining sweet water, in taste like
sugar, resembling in shew faire orient pearles; the
which drops if you take away, there do immediately
appeare the like; notwithstanding if they be suffered
to stand still in the floure according to his own
nature, they will never fall away, no, not if you strike
the plant untill it be broken". Such a plant would
engender strange origin legends. One said that it was
once a queen whose beauty, instead of contenting
her husband the king, made him jealous, and in a
moment of anger and suspicion he drove her from
his palace. She, well knowing her own innocence,
wept so constantly at this injustice as she wandered
about the fields, that her very substance shrank to the
measure of a plant, and at last God rooted her feet
where she had paused, and changed her to the crown
imperial, still bearing in its blossoms something of
the dignity and command she had worn as a human
being (Skinner). A Scottish tradition says that the
crown imperial hangs its flowers down for shame at
not having bowed to the Lord, and the spots at the
bottom of the bells are the everlasting tears it sheds in
contrition (Simpson). Yet another legend says it was
white once, but was dyed red by the blood of Christ at
Gethsemane (Bazin).

In at least one area of Buckinghamshire, crown
imperial seems to have ousted native flowers as the
embodiment of the May. The older children each had
one to carry round from door to door in the villages
(Buckinghamshire FWI). There is another side to the
plant, though, for it is toxic, the active principle being
imperialine, described as a heart poison (Whittle &
Cook), also found in *F meleagris*. Even honey made
from the nectar is said to be emetic. The bulbs, so it is
said, have been used for killing dogs (Lindley).

CROWN OF THORNS
HAWTHORN was a sacred tree long before Christian
tradition associated it with the Crown of Thorns, but
that tradition gave rise to the belief in the holy powers
of the wood – rosaries of thorn wood were in great
demand in medieval times, and were treated as if
they were jewels. In parts of France the tree is known
as l'epine noble because of this association with
the Crown of Thorns. It is said that the tree groans
and sighs on Good Friday (Wilks). Like many other
thorny plants, legend has it that REST HARROW
was used to make the Crown of Thorns (Leyel. 1937).
Even BLACKBERRY appeared in the legends and
BLACKTHORN is another candidate for inclusion
(Graves). But the likeliest candidate must be CHRIST-
THORN (*Paliurus spina-christi*), a Mediterranean
plant whose young stems are quite pliant, and

certainly could be woven into a crown (Moldenke & Moldenke). Some legends suggest that BEAR'S BREECH (*Acanthus spp*) was one of the plants used (Perry. 1972).

CROWN VETCH

(*Coronilla varia*) Blacks from the southern states of America used it as a love charm. They would chew it and rub it on the palms. That would give a man power over any woman with whom he later shook hands. More, southern blacks say that no snake will bite you if you carry a piece in the pocket, and moreover, lay a piece in a man's path, and he will never have any more money. It is called Devil's Shoestring there, and mix it with "snail water" (the secretion from a snail when sprinkled with salt), "planted" round the house, and that was reckoned to be infallible in keeping any woman at home. Another "conjure", to bring your wife home, was to get some dried Devil's Shoestring, some dust from her right foot track, and a piece cut from the "hollow" of her right stocking. Mix them together and "plant" them near your house (Puckett).

CRUCIFIXION

There is a legend to the effect that the reddish veining of the flowers of WOOD SORREL arose from the drops of blood shed by Christ on the cross, and the plant was often placed in the foreground of Italian paintings of the Crucifixion, especially by Fra Angelico. This legend is sometimes quoted as the origin of the name Hallelujah or one of its variants (it is called Alleluia in Italy, too (Friend. 1883)). Actually the name comes from the fact that it blooms between Easter and Whitsun. During this time the psalms (113-117) all end with Hallelujah.

CUCKOO

A very common name for LADY'S SMOCK is Cuckoo-flower, because it is in flower when the cuckoos are about. The name can appear quite simply as Cuckoo, but in Wiltshire it is sometimes known as Water Cuckoo, or Wet Cuckoo (Dartnell & Goddard), a comment on its favoured habitat (Dry Cuckoo is *Saxifraga granulata*, the Meadow Saxifrage). A number of other 'cuckoo' names are recorded in English. It is blodau'r gog in Welsh (Hardy), and Glauchsblume or Kukkuksblume in Germany (Grimm), where in some parts they claim that it grows abundantly where the earth is full of minerals. The point is that the cuckoo, by its call, is also thought to perform the same function, for it will show the whereabouts of a mine (Buckland). Cuckoo-spit, another name for Lady's Smock, is the froth enveloping a pale green insect found on the flowers, and the name is sometimes, as in this case, transferred to the plants themselves (see also WOOD ANEMONE, CUCKOO PINT, and RED CAMPION). Few North country children would pick Lady's Smock, for it was unlucky because the cuckoo had spat on them while flying over (Dyer. 1889).

CUCKOO-PINT

(*Arum maculatum*) The scarlet berries are poisonous, and the whole plant has an acrid juice. Tricks used to be played on children and simple people by giving them a small piece of the root to chew. It tastes alright at first, but then the victim experiences a horrible burning sensation that lasts a long time (Carr). That acrid sensation is caused by aroine, an unstable toxin that is produced from the plant and can cause blistering of the skin. But the plant was once cultivated on the Isle of Portland for the tubers, which when cooked yielded Portland Sago, as it was called, used as a substitute for arrowroot. It was quite popular at one time because it was thought to be aphrodisiac, as the very form of the plant proclaimed. Portland Sago was a "pure and white starch", as Gerard said, "but most hurtfull to the hands of the Laundresse that hath the handling of it, for it chappeth, blistereth, and maketh the hands rough and rugged, and withall smarting". This starch was at one time used in the French cosmetic called Cypress powder (Folkard), which, used on the face, made it dazzling white.

The aphrodisiac claim resulted from the form of growth of the plant, the spadix in the spathe, which to the common people stood for copulation. Friend.1883 said the cuckoo-pint is the symbol of zeal or ardour, and this may very well have been his way of giving some respectability to the subject. The very name of the plant, Cuckoo-pint, proclaims its reputation, for 'pint' is short for 'pintel', penis. Another of its names, Wake Robin, seems to say the same. 'Robin', so Prior claimed, is from the French robinet, which he said meant penis (quite possible, as the word in modern French means a tap). Wake Pintel is also recorded; 'wake' probably means alive, so the name actually means erect penis. Shakespeare certainly understood the name perfectly, and so did his audience. He makes the mad Ophelia sing, "For bonny sweet Robin is all my joy". So it is not surprising that cuckoo-pint was reckoned to be aphrodisiac. A quite recent name, Willy Lily (Mabey.1998), is as ribald as any of the many names attached to the plant, a lot of them significantly male + female (Lords-and-ladies is an example), Even some of the superstitions involving the plant have sexual overtones. They used to say in Cambridgeshire that it was very unlucky to bring the plant indoors, for it gave TB to anyone who went near it (Porter. 1969). The real reason may have been forgotten, but this kind of superstition is usually directed against the females of the house, and it would not be TB they would get. A Dorset belief is quite explicit – young girls were told never to touch a cuckoo-pint; if they did, they would become pregnant (Vickery. 1985).

Root preparations have been given for medicinal purposes, in cases of asthma, rheumatism, jaundice and dropsy (Barton & Castle). Such a preparation is still used in homeopathy. Gypsies have always used the plant to cure their ailments. A root decoction, or an infusion of dry powdered flowers, was used to relieve croup or bronchitis (Vesey-Fitzgerald), and a leaf infusion is a gypsy treatment for chills and colds (Boase).

CUCUMBER

(*Cucumis sativus*) Cucumber has been cultivated since prehistoric times, so it is no longer clear where the country of origin lay, though the best guess is somewhere in southern Asia (Moldenke & Moldenke), and certainly it was introduced into Britain from north-west India in 1573 (G M Taylor). So it was quite new in Gerard's time. He was not impressed: "All the cucumbers are of a temperature cold and moist in the second degree. They putrefie soon in the stomacke, and yeeld unto the body a cold and moist nourishment, and that very little, and the same not good ...". He was not alone in his poor opinion. Sir Thomas Browne, in 1646, warned his readers "that they should be so cold, as to be almost poison by that quality ...", which is roughly what Evelyn said, though implying that they were better thought of in his day – "it not being long, since cucumber, however dress'd, was thought fit to be thrown away, being accounted little better than poyson" (Evelyn. 1699).

A cucumber is a phallic emblem, symbolizing fecundity. Of the 60,000 offspring of Sagar's wife in the Buddhist legend, the first was a cucumber, whose descendant climbed to heaven on his own vine (Skinner). In Japanese mythology, the kappa is a spirit associated with the cucumber. It is like a monkey, but has no fur, and is yellow-green in colour. Kappas have a liking for blood, and for cucumbers, and one way of placating them is to throw cucumbers with the names and ages of the family where kappa live. The connection with cucumbers becomes even more significant when considering the belief that these spirits are capable of raping women (J Piggott).

The Pennsylvania Germans say that if you want large cucumbers, a man should plant them – an obvious piece of sexual symbolism. And to get the largest cucumbers, plant them on the longest day (M Baker. 1977), an example of homeopathic magic at work. Similar ideas are quite widespread in America. The seed should be sown before daylight, but also it should be done by a man, preferably naked, and in the prime of life. The size of the cucumbers depends on the "visible virility of the sower". Cucumbers sown by women, children and old men never amount to much. In Maryland, they say that if you plant them with your mouth open, the bugs will not bother them (Whitney & Bullock). Much superstition underlays the actual time of sowing. The longest day, as we have seen, is significant, and in Iowa the 20th or 21st day of June are the favoured days, whereas in Kentucky they prefer a later date, the 4th or 6th of July (Thomas & Thomas). Again, in Alabama, received wisdom

has it that they should be planted on the first of May (R B Browne). There is a Guernsey rhyme that translates:

> Sow your cucumbers in March, you will need neither bag nor sack
> Sow them in April, and you will have few.
> But I will sow mine in May, and I will pick more than you (Garis).

Dreaming of cucumbers is generally a good sign. For a sick person it shows he will recover quickly, and for a single person it is an indication that he or she will make a good marriage (Raphael). It promises success in trade, and to a sailor a pleasant voyage (Gordon. 1985). The Pennsylvania Germans used to say that if a cucumber plant bears only small fruit, it forebodes a funeral (Fogel).

Medicinally speaking, this is a cooling plant, "excellent good for a hot stomach and hot liver", as Culpeper put it. But Thomas Hill, in 1577, really did produce a marvel when he wrote that "if an infant being sick of the Ague, and sucking still of the breast be laid on the bed made of Cucumbers to sleep ... he shall immediately be delivered of the same, for while he sleepeth, all the feverous heat passeth in the Cucumbers"! Even Gerard did not go that far, though he acknowledged that cucumber "helpeth the chest and lungs that are inflamed". He was alive, too, to the use to which cucumber could be put in clarifying the skin (a use-re-discovered in modern times). He was able to indulge his humour, as he went on to say "the fruit cut in pieces or chopped as herbes to the pot, and boiled in a small pipkin with a piece of mutton, being made into potage with Ote-meale ... doth perfectly cure all manner of sauce flegme and copper faces, red and shining fierie red noses (as red as red roses) with pimples, pumples, rubies, and such like precious faces". It is used in Chinese medicine, too, when a sort of cucumber salve is made for skin diseases, and also for scalds and burns (F P Smith). Chinese medicine also makes use of the roots, bruised to apply to a swelling caused by the wound of a hedgehog quill! (F P Smith).

Cucumis melo > MELON

Cucumis prophetarium > GLOBE CUCUMBER

Cucumis sativus > CUCUMBER

Cucurbita foetidissima > BUFFALO GOURD

Cucurbita maxima > GIANT PUMPKIN

Cucurbita pepo > VEGETABLE MARROW

CUDWEED

(*Filago germanica*) Leonard Mascall, *Government of Oxen*, 1587, says that cudwort was given to cattle that had lost their cud (quoted in Grigson. 1974). This plant was the *herba impia*, Wicked Herb, of the old botanists, wicked because of its way of growth. The main stem has its flowers at the top, but underneath this grow two or more flowering shoots, all rising above the main stem. Therein lies the wickedness, for it conveys the idea that children were "undutifully disposed to exalt themselves above the parent flower" (Pratt). Hence, too, Son-before-the-father (Grigson. 1955), and similar names.

Cupressus sempervirens > (COMMON) CYPRESS

Curcuma longa > TURMERIC

Curcuma zedoaria > ZEDOARY

CURTAIN PLANT

(*Kalanchoe pinnata*) A leaf decoction is used in St Kitts to treat hypertension, and a similar decoction is taken in the Dominican Republic to deal with intestinal infections (Laguerre). In Mexico, where the plant is called hoja fresca, they put a green leaf over each temple, for headaches (Kelly & Palerm). In West Africa, it was the custom to squeeze the juice into the mouth of a new-born baby, and an infusion was drunk by both mother and child (Dalziel).

Cuscuta epithymum > DODDER

CUTCH TREE

(*Acacia catechu*) An important dye plant; by boiling chips of the wood, a permanent brown dye can be obtained (Barker). The leaves and shoots can be used as well. The dyestuff is known is known as Cutch, or sometimes Gambir (Le Strange). Much of the catechu exported comes from the Gulf of Cutch, where it has been used as a brown dye for over two thousand years, hence the name. True khaki cloth is made with cutch (Willis).

Cyclamen europaeum > SOWBREAD

Cydonia vulgaris > QUINCE

Cynara scolymus > GLOBE ARTICHOKE

CYPRESS

(*Cupressus sempervirens*) Cypress was important in Greek mythology, being sacred to Hercules, the word, according to some authorities, coming from Cyprus, which was called after Hercules' mother, Cyprian Aphrodite (Graves). Actually, it seems that it may be the other way round, the island being named after the tree (Moldenke & Moldenke), the eventual derivation being from the Hebrew gopher, thence to Greek kyparissos (Grigson. 1974). It is probably this, or one of its varieties, that supplied the gopher wood of Genesis 6; 14, of which Noah's Ark was made. The wood is indeed very durable – the doors of St Peter's in Rome are made of cypress, and after two hundred years still show no signs of decay (Moldenke & Moldenke). Ceres, according to another myth, plugged the crater of Mount Etna with a huge cypress trunk, and thereby imprisoned Vulcan for ever. Etna's

eruptions were looked on as Vulcan's attempts to escape his imprisonment.

Cypress was a sacred tree in Persia, a symbol of the clear light of Ormuszd, and was frequently represented on gravestones with the lion, emblem of the sun-god Mithras (Philpot). But it is better known as a symbol of death and mourning, and for that reason there was a superstition that it should never be cut, for fear of killing it (Evelyn. 1678). The old dream books foretold affliction after dreaming of cypress, in keeping with this symbolism (Gordon. 1985). The classical story of Cyparissos tells how he was stricken with sorrow at having killed his favourite stag, begged the gods to let him mourn for ever, and was transformed by Apollo into a cypress tree (Dyer. 1889). Venus, mourning Adonis, wore a cypress wreath. Both the Greeks and the Romans called it the "mournful tree", because it was sacred to the rulers of the underworld, and to the Fates and Furies. As such, it was the custom to plant it by the grave, and, when a death occurred, to put it either in front of the house or in the vestibule, in order to warn those about to perform a sacred rite against entering a place polluted by a dead body (Philpot). And it is the very durability of the timber (plus the fact that it is insect-proof) that made it desirable from ancient Egyptian times onwards for making the coffins of the rich (modern Greeks still use it for this purpose (Moldenke & Moldenke)). As the wood is incorruptible, and linked with ideas of life after death, it figured a lot in oriental carpet symbolism (Bouisson).

In spite of this death and mourning association, it is still an evergreen tree, and so, inevitably, a symbol of immortality. Gubernatis classes it as an "arbre phallique", and as such is "tout â la fois un symbole de la génération, de la mort et l'âme immortelle …". But on the Greek islands, a cypress is planted when a daughter is born. It was part of her dowry, and when she married it formed the mast for the couple's boat (T B Edwards. 2002/3).

The wood was highly valued by the ancient Greeks, especially for use in temples, more than anything else for monumental doors, as at the Parthenon and the Temple of Asklepios at Epidaurus (Meiggs). It resists decay, and is virtually immune from insect attack, and is also one of the longest-lasting woods. In Addition, it has an attractive scent. It follows from what we have heard of the nature of the timber that "the shavings of the wood laid among garments preserve them from the moths; the rosin killeth Moths, little wormes, and magots" (Gerard). Evelyn described the timber as "a very sonorous wood", enough to account for its use for organ-pipes, harps, and the like.

In medieval times, it was said that the upright of the Cross was made of cypress, the cross-member of palm, "the stock that stood within the earth in which they made a morteys", was of cedar, and the "table above His head … on which the tytle was written" was olive. The cypress was used "so that the smell of His body shal grieve no man that came by …" (Mandeville). Other traditions made the Cross of cypress, cedar, box and pine. It is in the eastern tradition that olive and palm are substituted for box and pine (Child & Colles). Evelyn speaks of a cypress, near the tomb of Cyrus, to which pilgrimages were made. It was believed that the gum turned every Friday into drops of blood. The tree was hollow, and fitted for an oratory (Evelyn. 1678).

The cones have their medicinal uses; they are astringent, and used either internally (finely crushed, infused in water), for menopausal disorders, varicose veins, haemorrhoids, etc., or externally (the decoction applied as a hot compress) on painful haemorrhoids, or used as a footbath, or to combat offensive perspiration (Palaiseul). But apart from all this, the tree itself was seen as a thoroughly healthful being – to quote Evelyn again, "I commend it for the improvement of the air, and a specific for the lungs, as sending forth most sweet and aromatick emissions, whenever it is clipp'd, or handled, and the chips or cones, being burnt, extinguish Moths, and expel the gnats and flies, etc., …".

CYPRESS SPURGE

(*Euphorbia cyparissias*) Lindley mentions its use as a purge, but warned that it was not very safe. It is in fact very violent, but it was still in use in France in recent times (Le Strange). As with other spurges, the juice is used to get rid of warts, but once more a warning has to be given – it can cause blisters on the skin (Schauenberg & Paris). The most spectacular of its uses comes from Russia, where it used to be employed as a rabies cure. It had to be gathered, we are told, in May and September, during the first days of the full moon, and then it was dried and powdered, and anyone bitten by a suspected rabid animal was given a preventive dose of 4 grams in half a glass of some drink or other (Kourennoff).

CYSTITIS

HORSETAIL tea is rich in silicic acid, and is diuretic, so it can be used for urinary problems, including cystitis (M Evans).

D

Dacrydium cupressinum > RIMU

Dactylorchis maculata > HEATH SPOTTED ORCHID

DAFFODIL

(*Narcissus pseudo-narcissus*) "Daffodils that come before the swallow dares, and takes the winds of March with beauty …" (Shakespeare. *The winter's tale* 4, iii). Indeed they do – the church has appropriated it to the Feast of St Perpetua, which falls on 7 March (Geldart). Wild daffodils are out that early in damp woods and meadows. In some parts of Britain, Hampshire, for example, it was generally said that wild daffodils indicated the site of a monastery (Boase). Don't point at a daffodil, though - that would stop it from blooming, so they used to say in Maine (Bergen. 1899).

The longer names given to them, like Daff-a-down-dilly, seem to be reserved for children's rhymes:

> Daff a down dilly has now come to town
> In a yellow petticoat and a green gown (Gutch. 1901).

There is a longer one that has been recorded in Hertfordshire:

> Daff-a-down-dilly
> Has now come to town,
> In a yellow petticoat
> And a green gown.
> Daff-a-down-dilly
> That grows in the dell,
> My father's a tinker,
> My mother can tell.
> My sister's a lady,
> And wears a gold ring.
> My brother's a drummer
> And drums for the King. (Jones-Baker. 1974).

And there is a much shorter version from Gloucestershire:

> Daff-a-down-dilly that grows by the well,
> My mother's a lady, my father can tell (Northall).

Surprisingly, daffodils are not universally the good omen one would expect. Granted, the first daffodil is a lucky one. If you find that, then you will have "more gold than silver" that year (Trevelyan). But one has to be careful about the direction in which the trumpets are pointing. See Herrick's *Hesperides*:

> When a Daffodil I see
> Hanging down her head t'wards me,
> Guess what I may what I must be;
> First, I shall decline my head;
> Secondly, I shall be dead;
> Lastly, safely buried.

In other words, if you see a daffodil with its head bending towards you, it is a sign that you are about to die (Addy. 1895). Dreaming of them may have dire, but different, results - "any maiden who dreams of daffodils is warned not to go into a wood with her lover, or any secluded place, where she might not be heard if she cried out" (Mackay).

As with other spring flowers, notably cowslips and primroses, one has to be careful with daffodils when there is poultry about. There is an old Manx superstition that it is bad luck to a poultry keeper if two or three of the flowers are brought into the house in early spring, before the goslings are hatched (the Manx name for the daffodil shows the connection – it translates to Goose-leek). One finds this superstition in Devonshire, too, while a Cornish belief was that if a goose saw a daffodil before hatching its goslings, it would kill them when they did hatch (Courtney). A Dorset compromise says that you must always take care that the first daffodils brought into the house each season should be a large bunch, for otherwise something is sure to go wrong with the poultry (Udal). Judging from the primrose and cowslip beliefs, what you really should do is to take in quite a bunch – two or three are fatal; the ideal is probably thirteen or more. On the other hand, in parts of Warwickshire, it is a thoroughly unlucky flower, never to be taken indoors.

In Shakespeare's time, daffodils were favourite flowers for chaplets, and there is a quotation from Fletcher's *The two noble kinsmen*:

> " … I'll bring a berry,
> A hundred black eyed maids that love as I do,
> With chaplets on their heads of daffodils".

The narcissus is, of course, the symbol of egotism and self-love. The story of Narcissus, enamoured of his own beauty, becoming spell-bound in front of his own image, is too well-known for comment.

The flowers have not been much used, if we except that of dyeing Easter eggs (Newall). But the bulbs and roots are a different matter. "The roots of Narcissus have such wonderfull qualities in drying, that they consound and glew together very great wounds, yea, and such gashes or cuts as happen about the veins, sinues, and tendons" (Gerard/Langham). It was still being recommended in the mid-18th century (Hill). Gerard went on, "they have a certain clensing facultie … The root … stamped with hony, and applied plaister-wise, helpeth them that are burned with fire, and joineth together sinues that are cut in sunder". This poultice was also good for sprained ankles, "aches and pains of the joints", sunburn, when used with honey and nettle seed, and honey and darnel meal, to "draw forth thorns and stubs out of any part of the body". In the words of Holland's translation of Pliny (1601), " a cataplasm made of the root …, honey and oatmeal, draws forth spills, shivers, arrow-heads, and thorns, and whatever stick within the body" (Seager). But the bulb is actually toxic, and can produce severe poisoning. Indeed, the word narcissus can be traced back to the Greek naike, stupor, an allusion to this narcotic property (Cunningham & Côté). The inference must be that Narcissus himself must have been suffering from similar effects. However, it is used in homeopathy, for bronchitis and whooping cough (Schauenberg & Paris). The Greeks reckoned that narcissus root was one of the cures for impotence (Simons), and there is a Moroccan belief that by smelling a narcissus a person avoids catching syphilis, and, if he has already caught it, can cure himself of it (Westermarck).

A daffodil bulb had other virtues, too – "if men possessed with evil spirits, or mad men, bear it in a clean napkin, they be delivered from their disease", put into verse in a late 14th century manuscript as:

> This herb in a clean cloth, and its root
> Against the falling evil is medicine.
> Affodille in clean cloth kept thus
> Shall suffer no fiend in that house;
> If ye bear it on you day and night,
> The fiend of you shall have no sight (quoted in Rickert).

As Albertus Magnus said, "it suffereth not a devil in the house".

DAGGA

is the southern African version of the drug CANNABIS (HEMP), used even by the San (Bushmen). They usually mix it with tobacco, but occasionally smoke it neat (Schapera). Of course, they cannot grow it, but their Bantu neighbours do.

DAHLIA

(*Dahlia pinnata*) The Navajo Indians use the roots and flowers of dahlia for an orange-yellow dye (Elmore). Broth from the tuber, cooked with pork, is a Chinese remedy for heart trouble (Perry & Metzger).

DAISY

(*Bellis perennis*) Daisy is Day's Eye, OE daeges eage, for it closes its petals at night. It is the same in Welsh, too – Llygad y Dydd. See Chaucer, too:

> That well by reason men call it may,
> The deisie or els the eye of the day,
> The empresse and the floure of flowres all.

It is the symbol of modesty (Hewett), and of fidelity (Friend. 1883). To the Celts, it was the symbol of innocence, and thereby, the newly-born (Bayley. 1919), though it is the woman's flower in Somerset (Tongue. 1965). As the emblem of the newly-born, it represents the infant Christ, and has been used as such in western art since the 15th century. In France, eggs might be painted with a daisy (pâquerette) design, and the egg given to every child attending Easter Mass

(Newall). The legend on the Côte d'Or to explain why the daisies have a golden heart and a pink tinge at the end of the petals, says that when the Three Kings and the Shepherds were bringing their gifts to Christ, a poor herdsman picked a white daisy to give, so that he should not arrive empty-handed. He put it to the lips of the infant, who kissed the flower, which became pink where his lips had been (Sebillot).

Daisies will shut when bad weather is coming (Page. 1977), but there are quite a lot of mere superstitions – for example, if you do not put your foot on the first daisy seen in spring, daisies will grow over you, or someone near you, before the year is out (Folkard). Better known is the saying that spring has not arrived until you can put your foot on twelve daisies (M E S Wright), or nine (Jones-Baker. 1974), or even seven (Sussex - see *Folk-lore. vol 25; 1914 p369*). Note the Scottish proverb:

> Like March gowans,
> Rare but rich (Chambers. 1870).

Daisy roots under the pillow will bring dreams of the beloved and absent ones. To dream of daisies is lucky in spring, but bad in autumn or winter (Folkard). In the Pays de Caux, in Normandy, daisies gathered before sunrise on St John's Day, and then put on the roof, will guard the house against lightning (W B Johnson). Just as extraordinary is a German belief that if a tooth is extracted, the patient must eat three daisies to be free from toothache in the future (J Mason).

Daisy chains, just a children's game usually, are sometimes felt in Somerset to be a protection for children (Tongue. 1965). Certainly, daisies formed one of the three magic posies given to the traveller in the Derbyshire folk tale called Crooker, to protect him from evil. "Take the posy and show it to Crooker". The other posies were of St John's Wort, and primroses (Tongue. 1970). In Lincolnshire, when the underside of a daisy petal in tinged with purple, it is said to be stained with Abel's blood. Everyone knows the divination game played by pulling off daisy petals one by one (he loves me, he loves me not ...) (Vickery. 1995) or "rich man, poor man ...". An American example uses the rhyme:

> One I love, two I love, three I love I say,
> Four I love, with all my heart,
> And five I cast away.
> Six he loves, seven she loves, eight they both love,
> Nine he comes, ten he tarries,
> Eleven he counts, and twelve he marries (Whitney & Bullock).

Another way of playing the game is by naming the alphabet until the last petal is reached. This will give the first letter of the name of the girl's lover (Whitney & Bullock).

The Cumbrian practice of eating two daisies to cure toothache (Newman & Wilson), must be an example of Gerard's dictum that "the daisies do mitigate all kinde of paines". He further advised the "juice of the leaves and roots", to help "the megrim", and "the leaves stamped take away bruises and swellings proceeding of some stroke, if they be stamped and laid thereon; whereupon it was called in old time Bruisewort …".

Naturally, as daisy is "day's eye", use for sore eyes, etc., should be widespread. We certainly find that a decoction of the flowers boiled down was used in Ireland as an eyewash (Wilde. 1890), or the boiled flowers themselves could be dabbed on sore eyes (O'Farrell). Pennant noted that in Scotland "flowers of daisies were thought to be remedies for the oph-thalmia". Gerard had already written that "the juice put into the eies cleareth them, and taketh away the watering of them", but a long time before his day, in the 15th century, there was a leechdom "for the web in the eyes. Take the juice of daisy, avens, southernwood, with water of fennel, and put in the ees" (Dawson. 1934).

Herbalists still use daisies for improving the circulation. They also keep the artery walls soft and flexible (Conway). In Wiltshire, the decoction was used for skin diseases, and the infusion for colds (Olivier & Edwards). As far as the prescription is concerned, it has a long history, for a 15th century leechdom advised that "for the fire of hell (wildfire, St Anthony's fire, i.e., erysipelas that burneth in a man's flesh, take the daisy, crop and root, and bruise it, and lay thereto, and change it often" (Dawson. 1934). A strong decoction of the roots was used for scurvy in the 18th century (Sir J Hill), and another of Gerard's prescriptions was to make a "decoction of the field Daisie (which is the best for physickes use) made in water and drunke", which he said was "good against agues". One old name for a daisy was Banwort (Britten & Holland). "The Northern men call this herbe a Banwort because it helpeth bones to knyt againe" (Turner. 1548). It is a Consound, then (Prior). (Turner has *Consolida minor; Consolida major* would be Comfrey).

Marguerite (Britten & Holland), and Herb Margaret (Dyer. 1889) are other names for the daisy. But which St Margaret is it? There was St Margaret of Antioch, an unlikely choice, or St Margaret of Cortona, or yet another Margaret, St Margaret of Valois (Pratt). Actually, the likeliest candidate is Margaret of Cortona, for her day, 22 February, used to be reckoned as the first day of spring (Jones-Baker. 1974), and that is probably the reason for the name. But there is another explanation: the French 'marguerite' means a pearl. Is that a way of describing the colour of the flower?

Dalbergia melanoxylon > AFRICAN BLACKWOOD

DAMSON

is the cultivated form of BULLACE (*Prunus domestica 'institia'*). Damson is damascena, the pruna damascena of the Romans, so named (by Pliny) from Damascus, in Syria (Grigson. 1974).

DANDELION

(*Taraxacum officinale*) It can be found in bloom virtually all the year round, but the great burst of colour on meadow land occurs in spring, and it is natural that the plant should be an associate of the spring goddess, Brigid, sanctified as St Bridget, or Bride. Milk-yielding plants like dandelion were often allied with the goddess in Scottish folklore (F M MacNeill. 1959). It was a logical connection; as a cowherd in one of her aspects, she would be associated with cattle, and so with such flowers as the dandelion, yielding a milky juice which was believed to nourish the young lambs in spring (E E Evans). It is said in the Highlands of Scotland that "the plant of Bride", i.e., the dandelion, "nourishes with its milk the early lamb" (F M MacNeill. 1959). The flower is one of the three insignia of Bride, the other two being the lamb and the oyster-catcher (Urlin). Its names in Gaelic are "Bearnon Bride", "The little notched flower of Bride", "The little flame of God", and "St Bride's forerunner". St Bride of the shores wears it at her breast, and the sunlight is said to follow.

To dream of dandelions means misfortune (Dyer. 1889), but if a girl wants to dream of her future husband, she should, according to a French belief from the Franche-Comté, put a whole dandelion leaf under her pillow (Sebillot). But you can foretell all sorts of things with dandelions, usually the seed heads – the weather, for example; children use the seed tufts as barometers (Dyer. 1889). When the down is fluffy, then there will be fine weather, but when it is limp and contracted, then there will be rain (Swainson. 1873). (Weather Clock is one of its Somerset names (Macmillan)). The flower heads close directly rain falls, or just before, and always before dew-fall (Rohde. 1936). Or you could tell the time (the multitude of "clock" names, like Time-teller, or Clock-flower, etc., is witness to that belief). Or the dandelion globe will tell children at what age they will get married (Harland & Wilkinson. 1867); it will tell them, too, as in Brittany and Beauce, how many years they have to live, or, as in Côte-du-Nord, how many children they will have (Sebillot), or, as in Kentucky, the number of days until they will get a whipping (Thomas & Thomas). American children will blow a dandelion globe three times. If all the seeds blow away it is sign that their mother does not want them (Bergen. 1899). Another American divination is to award each puff to blow off the seeds a letter of the alphabet; the letter which ends the blowing is the initial of the name of the person they will marry (Bergen. 1896). Yet another is to take note of the direction in which the seeds fly when you blow it, and that is the direction in which to seek your fortune.

Dandelion is a thoroughly useful plant. Not many know that the roots will give a red dye (Dimbleby), but most have at least heard of its use as a food. They were included with the bitter Passover herbs, and the large leaves are certainly bitter (Evelyn. 1699 said they should be "macerated in several Waters, to extract the Bitterness"), but the young ones are not, so it was regarded as "one of the most wholesome spring saladings" (Rohde. 1936). American Indians took full advantage of it as food. The Meskwaki, for instance, cooked them with pork (H H Smith. 1928), and the Menomini cooked them with vinegar made from the last run of sap in the maple tree (H H Smith. 1923). No wonder it was important, for it is credited with having much greater fuel value (calories) than the same bulk of most other standard greens (Sanford).

Dandelion coffee, made by roasting the roots, is well known, and so are other drinks made from the plant, best of all dandelion wine. Fen people liked to gather their dandelions for wine-making on May Day, for these, they said, make the best wine (Porter. 1969). Wiltshire people merely said they should be picked before 24 May (Wiltshire), but the traditional time is St George's Day (23 April). It has always been popular, and quite extraordinary claims have been made for it, as for example this extract from *The Times* of 23 January 1951: "Mr William Weeds, of Caunsall, near Kidderminster, who was 100 last November, died yesterday. For 75 years he had taken a daily glass of dandelion wine, and contended that this beverage enabled his grandmother to live to be 103".

It has been known for a long time that the bitter extract from dandelion roots, preferably dug up in the autumn, has a medicinal effect on the liver (Schauenberg & Paris). Indeed, "if you could ever dig up the whole root … without breaking it or leaving the end in the ground, it would cure anything. But there is a little milky bit at the very end of it you can never get. It breaks off, or the devil bites it away, they say, the way people won't get the cure", Gill. 1963 was told. Dandelion is an ingredient of many popular American "bitters" and "blood purifiers" (Lloyd), naturally used by the Indians as well. There are plenty of examples of similar British use. Dandelion flowers for "the blood", they say in Warwickshire (R Palmer. 1976). In the Scottish Highlands, it used to be the commonest tonic (Grant). With its iron (and copper) content, it must be good for treating anaemia, and skin complaints, too. It was certainly used for anaemia in Ireland (and for consumption), and sometimes nettle leaves were added, as in Dorset, to make a tonic drink for the blood (Dacombe). There are similar records in Welsh popular medicine (Trevelyan), and in published herbals, as in Thornton, in the early 19th century. Dandelion tea was the usual liver medicine, and

dandelion coffee is used, too. Take the roots gathered in autumn, for in the spring they are almost flavourless. Both the tea and the coffee are said to be good for indigestion as well (Browning), and feature in American folk medicine, too – Alabama people used to take the tea for biliousness, for example (R B Browne), just as Suffolk people did for any stomach pain (V G Hatfield. 1994). Dandelion juice put on the tip of the tongue was an Essex indigestion cure (Newman & Wilson).

A glance at the list of local names will confirm dandelion's diuretic qualities, and it is quite often used deliberately for the purpose. The name Pissabed, common enough, although no longer standard English (though pissenlit is still standard French) would shout the fact loudly. German has Pissblume, and Dutch beddepissers, too (Clair). "Children that eat it in the evening experience its diuretic effect, which is the reason that other European nations, as well as ourselves, vulgarly call it Pissabed" (Britten & Holland). Would children have any particular urge to eat dandelions? But it is said that if they even gather the flowers, they will experience the symptoms. Mothers would remove all the dandelions from a child's bunch of wild flowers, and children would tease their fellows by putting the flowers in their pockets (Vickery. 1985). As an indication of the thoroughly contrary nature of superstition, the flowers were given to Fenland children to smell on May Day to inhibit bed-wetting for the next twelve months! (Porter. 1969). Perhaps this is an example of homeopathic magic. But in serious vein, in recent times, Irish country people would brew the leaves, or sometimes the whole plant, roots and all, to increase the flow of urine (P Logan). Hence, of course, its use for gallstones, jaundice, etc., (Schauenberg & Paris); herbalists still prescribe the root tea for stone in the bladder (W A R Thomson. 1978).

The milky juice from the stalks is applied to pimples and spots (Conway), and the flowers, boiled for half an hour in water, give a toilet water to get rid of freckles on the face (Palaiseul). Beware of doing that too often, though. In some parts of France, the result of overuse by young girls would later be that she would have "enfants malingres", i.e., sickly, or ricketty, one of the symptoms of which is just the pale complexion that the mother tried to achieve (Loux). The juice is put on warts in widely separated areas (Rudkin. 1936, Drury. 1991, Polson, Newman & Wilson, Bergen. 1899). Or they rubbed them with a leaf (Gutch & Peacock) (in Somerset they always used to rub dandelion leaves on nettle stings, as well as using dock leaves (Tongue. 1965)). An Irish charm for warts, blending the magical with the practical, was to give one dandelion leaves, three leaves to be eaten on three successive mornings (Dyer. 1889). A stye can be treated, according to another Irish country remedy, by bathing it with the milky juice (P Logan).

Fenland midwives used to give a "pain-killing cake" to women in labour. It was made from wholemeal flour, hemp seed crushed with a rolling pin, crushed rhubarb root, and grated dandelion root. These were mixed to a batter with egg yolk, milk and gin, turned into a tin and baked in a hot oven. At the woman's first groan, a slice of cake would be handed to her (Porter. 1969). Hempseed, rhubarb and gin would have quite an effect, but it is not clear how dandelion fits into the pattern. Some of the receipts seriously offered are mysterious, to say the least. "… Dandelion will staunch blood at the nose, if thou wilt break it, and hold it to the nose that the savour may go into it" (W M Dawson) – that from a 15th century text. Lupton, writing in the middle of the 17th century, recommended an equally odd one – "the herb of dandelion, well sod in water, is counted to be a chief medicine for the joining and knitting of wounds. It is good against ruptures, for them that be broken or bursten". And what should we make of Langham's "bitings venomous, stamp it and apply it"? It sounds as simplistic as the Cumbrian practice of eating dandelion leaves for rheumatism (Newman & Wilson).

Irish country people seem to have made more use of dandelion for medicine than anyone else, or at least there are more records of use. The tea was used for asthma, and a way to "clear" the chest, was to chew the leaves and swallow the juice (P Logan). The recorder went on to say they did the same with ivy leaves. But almost certainly he meant ground ivy, for Lady Wilde. 1890 mentioned a charm for asthma in which the patient had to drink a potion made of dandelion, or of ground ivy, with a prayer said over it before drinking. Dandelion was reckoned to be "good for the heart" in the west of Ireland, and the juice was put on a wound to stop the bleeding (Maloney).

DANDRUFF

PARSLEY makes a good lotion for getting rid of dandruff, and helps to stave off baldness (A W Hatfield). The Wiltshire remedy was to massage the scalp with a NETTLE infusion each day (Wiltshire) (see also Baldness). An American domestic remedy for the condition is to use a lotion made of one part APPLE juice to three parts of water (H M Hyatt). Evelyn favoured a MYRTLE decoction for dandruff, and also for dyeing the hair black. Not only that, but "it keepeth them from shedding". Gerard reported thet the "juyce of the decoction" (of FENUGREEK) "pressed forth doth clense the haire, taketh away dandruffe …", and the "meale", presumably the paste or porridge made from the seeds, he reports as being "good to wash the head …, for it taketh away the scarfe, scales, nits, and all other imperfections".

An American cure for dandruff and falling hair is to make a strong tea from PEACH leaves (H M Hyatt). Native Americans, especially the Pueblo groups, used SOAPWEED (*Yucca glauca*) not only as a soap substitute, but as a ritual hair shampoo in initiation ceremonies, but the Kiowa claimed it was an effective cure for dandruff and baldness (Vestal & Schultes).

DANES

There are a few plants that tradition associates with the Danes, or rather the bodies of Danes killed in battle. PASQUE FLOWER is one, and so is DWARF ELDER. Tradition has it that the latter grows only where blood has been shed. A Welsh name for the plant translates "plant of the blood of man", and there are also relevant English names like Bloodwort and Deathwort. It is associated in England with the Danes – wherever their blood was shed in battle, this plant afterwards sprang up (Dyer). Camden wrote in 1586: "And in those parts of this country which are opposite to Cambridgeshire, lyes Bartlow, famous for four great barrows… And the Wallwort or Dwarf-elder that grows hereabouts in great plenty, and bears red berries, they call by no other name but Dane's-blood, denoting the multitude of Danes that were slain". The opinion of some in Somerset was that the dwarf elder got its noxious properties from growing on the graves of the Danes (Lawrence) (the berries are toxic). The fact that the stems turn red in September presumably gave rise to these traditions (Grigson). Another plant with exactly the same tradition, even to its growing prolifically on the Bartlow barrows, is CLUSTERED BELLFLOWER, again, known as Danes' Blood in Cambridgeshire. FIELD ERYNGO (*Eryngium campestre*) is known as Daneweed in Northamptonshire (A E Baker), for just the same reason. SNAKE'S HEAD LILY is yet another plant said to have grown from a drop of Dane's blood, and that must be the origin of the name Bloody Warrior given in Berkshire to this flower. The dark red juice that exudes from the bruised capsules of TUTSAN was believed in Hampshire to be the sign that the berries originated by germination in the blood of slaughtered Danes (Gomme. 1908). Anyway, this juice was taken as a representation of human blood, and by the doctrine of signatures the plant was apllied to all bleeding wounds (Dyer. 1889).

Daphne laureola > SPURGE LAUREL

Daphne mezereum > LADY LAUREL

DARNEL

(*Lolium temulentum*) A rare casual grass in Britain, but only because generations of farmers have taken good care to eradicate it, for the grain (and only the grain) is toxic, and when ground with the wheat produces a flour that can cause serious illness to those eating bread made from it (Long. 1924). But the plant is notoriously prey to the ergot fungus, *Claviceps*, and it may very well be the ergot infection that gives it its evil reputation. Whatever the cause, ingestion will produce giddiness and confused perception. Hence the German name Schwindel, and the French ivraie, drunk. Gerard said it causes drunkenness in those who eat it in bread hot from the oven, and it is also said to cause blindness; there is a saying, "Darnel for dim sight". And those that were dim-sighted (i.e., dim-witted) were said to have "eaten darnel" (Jacob). Gerard knew about this, for he warns under the heading "The danger", that "Darnell hurteth the eyes and maketh them dim …". See Shakespeare, too:

> Want ye corn for bread?
> I think the Duke of Burgundy will fast
> Before he'll buy again at such a rate;
> 'Twas full of darnel: do you like the taste?
> (*1 Henry VI. Act iii. Sc 2*).

At one time it was used to "fortify" beer (the French name ivraie is a mirror of this practice, for when brewed with barley it acts as a narcotic intoxicant (C P Johnson). But the practice was recognized as highly dangerous, and was prohibited in France as long ago as the reign of St Louis (1226-1270) (Forsyth). There is a Norwegian folk tale that tells how mothers used darnel to put their children to sleep when there was no food to give them (Kvideland & Sehmsdorf). "Sturdy" is a name in use in Ireland and Scotland (Britten & Holland); there is a disease in sheep called sturdy, or staggers. Meal was said to be sturdied when it had a lot of darnel in it. Another name, from Somerset, is Lover's Steps, surely an acknowledgment of the divination game, of the "Love-me, love-me-not" kind (see **RYE-GRASS**).

Datura stramonium > THORN-APPLE

DAY LILY
(*Hemerocallis flava*) In China, it is said that if pregnant women wear Day Lilies at their girdle, the child to come will be a boy (F P Smith). This and other species of *Hemerocallis* are eaten in the east, as gum jum (golden needles) and gum taoy (golden vegetables) (Whittle & Cook).

DEADLY NIGHTSHADE
(*Atropa belladonna*) All parts of the plant, and particularly the berries, are very poisonous, and this makes the naming of the genus after Atropos, the Fate who cut the thread when life was ended, quite understandable. But there is a connection with Hecate, too (Gubernatis), for such a poisonous plant would naturally bring about such an association. There is evidence, if such it can be called, that it was used by the witches themselves as a drug or poison, for Montague Summers was able to quote two formulae for witch ointments involving Deadly Nightshade:

1. De la Bule, de l'Acorum vulgaire, de Quintefeuille, du sang de chauvesouris, de la Morelle endormante, de l'huyle.
2. De graisse d'enfant, de sue d'Ache, d'Aconite, de Quintefeuille, de Morelle, et de suye.

The second formula was taken from Reginald Scot, but they were both written by the 16th century physician Della Porta. Other authorities have said that the salve with which witches anointed their bodies was made of crushed belladonna leaves (De Ropp). With aconite, it would induce excitement and irregular heart action, and could produce the sensation of flying, or the well-known dream symptom of falling through space (discussed, for instance, in Allegro. 1979), for atropine is absorbable even by intact skin. Indeed, it has been noticeable in medicine that belladonna plasters can produce toxic effects (Harnell). Scientists of the 20th century who have made up this witch ointment and tried it, have experienced dreams of flying in spirals, etc., (Harnell). People accidentally poisoned by nightshade substances have imagined themselves to be changed into animals, with a sensation of growing fur (Russell & Russell). In view of the old werewolf beliefs, this is interesting; Rawcliffe considered that continual use of the drug might very well be the reason for such zoomorphic paranoia. Consistent use of these drugs often leads to insanity, with animal delusions and hallucinations, including the sensation of having grown hair or feathers.

Hallucinations could also take on a sexual tone. Large doses of the drug are liable to result in irresponsible sexual behaviour, a fact that gave it a great reputation in medieval times as an aphrodisiac (Rawcliffe). Such beliefs readily gave rise to lesser superstitions. In Normandy, for instance, they used to say that anyone who walks barefoot over the Deadly Nightshade will immediately go mad (W B Johnson), and in the Highlands of Scotland it was said to be used to make people see ghosts (Kennedy). An Irish superstition said that the juice distilled, and given in a drink, would make the person who drank it believe whatever the operator wanted him to (Wilde. 1925). Of course, such a plant is the devil's favourite, he watches over it, but there was a way of getting the plant and putting it to its rightful use, which was, so the legend has it, to rub on a horse, so that the animal gains strength. The way to get it was for a farmer to loose a black hen on Walpurgis Night. The devil will chase the hen, and so be diverted from looking after his plant. Then the farmer can quickly pluck it; this is the only way to make it efficacious (Skinner).

Even well into the 20th century, the bruised leaves were used in Warwickshire to apply for the relief of rheumatism (Savage), and there is a chilblain remedy from Essex – an ointment made from lard and the juice of our plant (V G Hatfield. 1994), probably effective, for a decoction of Deadly Nightshade has been used to keep up the circulation (Brownlow). Herbalists still prescribe it for the treatment of nervous diarrhoea and similar ailments (Flück), and even to treat Parkinson's disease (Schauenberg & Paris). Belladonna plasters are still available; they are occasionally used to relieve the pain of lumbago (V G Hatfield. 1994).

There is one purely magical charm involving the plant. Much as used to be done with elder, deadly nightshade stems were cut up into small pieces, threaded and worn as a necklace by young children, to prevent teething fits (Friend. 1883).

DEAFNESS

LEEK sometimes formed part of quite complicated recipes. The Physicians of Myddfai, for example, conjoined the juice of leeks, goats' gall and honey, mixed in three equal parts, and then put warm in the ears and nostrils. An early leechdom for an ear salve required the doctor to pound sinfull, which is a *Sedum* of some kind, latherwort (probably Soapwort), and leek, "put in a glass with vinegar and wring through a cloth and then drip it into the ear" (Cockayne). A prescription for deafness of the 15th century requires one to "take the juice of leaves of a beech-tree, and good vinegar, even portions, and put thereto powder of quick-lime; and then clear it through a cloth; and of this, when it is cleansed, put hot into the sick ear" (Dawson). CAMOMILE flowers were used in an old recipe for deafness. The patient was to "take camomile and seethe it in a pot, and put it in the ear that is deaf, and wash the ear; and so do for four days or five, and he shall be whole" (Dawson).

BAY leaves, or rather the juice pressed out of them, "is a remedy for for pain in the eares, and deafnesse, if it be dropped in with old wine and oile of Roses …" (Evelyn). Lupton agreed, for "it doth not permit deafness, not other strange sounds to abide in the ear". Squeezing HOUSELEEK juice into the ear is a famous earache remedy, but it is also claimed to be a cure for deafness, and it formed part of a Welsh prescription. The Physicians of Myddfai had a cure that involved "rams' urine, the oil of eels, the houseleek, and the juice of traveller's joy [Old Man's Beard], and a boiled egg. Let him mix and drop into the ear little by little, and it will cure him …".

A strange remedy is this Irish one involving COW-SLIPS. The whole plant, roots, blossoms and leaves is taken, well cleaned, bruised and pressed in a linen cloth. Add honey to the juice pressed out, put it in a bottle, and "pour a few drops into the nostrils and ears of the patient, who is to lie on his back. Then after some time, turn him on his face till the water pours out, carrying away whatever obstruction lay on the brain"! (Wilde. 1890).

DEATH

In spite of its reputation as a protective tree, BIRCH has its more sinister side. Take, for instance, the ballad, The Wife of Usher's Well:

> It fell about the Martinmass,
> When nights are lang and mirk,
> The carlin's wife three sons came hame
> And their hats were of the birk
> It neither grew in dyke nor ditch

> Nor yet in one sleugh;
> But at the gates o' Paradise
> That birk grew fair enough.

In other words, the birch grew at the gates of paradise, and it furnished the ghosts with their "hats o' birk". It is a sacred tree, and traditionally the tree of death. To dream of pulling the "birk sae green" portends death in the Braes o' Yarrow. A "wand o' bonny birk" is laid on the breast of the dead in Sweet William's Ghost (Wimberley).

ROSES too have connections with death, though they are thought of mainly as symbols of love. As far back as ancient Greece, they were used at funer-als, and tombs were decorated with them, under the belief that they protected the remains of the dead. The Romans also believed this, and they even celebrated a festival of the dead under the name of Rosalia. It was a Welsh custom to plant a white rose on the grave of an unmarried woman (Knight), and there is a Scan-dinavian story to the effect that three white roses will grow from the grave of a virgin (Mayhew). It was the custom in parts of England for a young girl to carry a wreath of white roses, and walk before the coffin of a virgin. The wreath would be hung in church after the funeral, above the seat that she had used during her life. It is usually a rose or roses that grow from the graves of two lovers. Evelyn mentions "the custom not yet altogether extinct in my own native county of Sur-rey, and near my dwelling, where the maidens yearly plant and deck the graves of their defunct sweet-hearts with rose-bushes". The custom was so common in Switzerland that cemeteries were often known as rose gardens there (Mayhew). So great is the connection between roses and the dead that apparently spirits at seances are using roses now as symbols of affection and goodwill (Knight). MADONNA LILIES too have an association with death. They grow from the graves of people unjustly executed as a token of the person's innocence (Grimm), and from the grave of a virgin three lilies spring, which may only be gathered by her lover (Dyer). This association with death makes them generally unlucky plants to have in the house (Vickery. 1985), but of course they are often used at funerals. MYRTLE, in spite of its connection with weddings, is also a tree of death, particularly, in Greek mythology, the death of kings (Graves). Perhaps that is why there is such a prejudice against it in America. It is rarely seen there outside cemeteries (H M Hyatt). In modern Egypt, sprigs of myrtle are laid on the graves of relatives, and also apparently placed under-neath the body (Westermarck). In parts of France, too, it was at one time the custom to put the last branch of myrtle that had belonged to the dead man into the coffin with him (Sebillot). Another plant with ambiva-lent virtues is PERIWINKLE. It has its connection with weddings and fertility, but also with death. It is a "Flower of Death" in Wales, and fiore di morte in Italy, where garlands of the plant were put on the coffins of dead children. It was evidently planted on graves in Wales, for there was a belief there that uprooting one from a grave, would very well result in a haunting by the person buried there (Trevelyan). In medieval England, condemned men were forced to wear garlands of periwinkle on their way to the gallows (Emboden. 1974). BLACKTHORN flowers are extremely unlucky things to bring indoors; if it comes to that, virtually any white flowers in the house bring dire results. But more fuss seems to be made about the blackthorn than anything else. Sussex people looked on it as a death token (Latham); in Suffolk, too, they used to say that it would foretell the death of some member of the family (Gurdon). And in Somerset, it would mean that you would hear of a death (Tongue. 1965). SNOWDROP is another of the white flowers that are so unlucky to bring indoors. Taking them into a hospital is even more unlucky. If given to a patient, it was taken to be a death sentence. Nurses would sometimes put a few ivy leaves among them to lessen the fate (Tongue. 1967). It is said that the association of snowdrops with death (they used to be called Death's Flower in Somerset) is because of the flower's resemblance to a shroud (Vickery. 1985). "It looked for all the world like a corpse in its shroud" was how one of Charlotte Latham's informants put it. TUBEROSE is unlucky, too, in American belief, possibly because of its waxy appearance, like death. Others think they emit the odour of death, still others that if you shut yourself in a room with tuberoses, the scent will kill you (H M Hyatt).

BROAD BEANS, besides being a fairy food, are also the food of the dead, or according to another tradition, they contain the souls of the dead (Waring). At any rate, they were sacred to them. They were a favourite offering to the departed in ancient Greece, and for this reason they were forbidden to his followers by Pythagoras. Pliny said that beans were used in sacrificing to the dead because the souls of the dead were in them. Eventually, they were used at funerals in classical times, a use that found its way to northern England, where the tradition used to be (at least in the 1890s) that broad beans should always be buried with the coffin (Pope). Children used to recite:

> God save your soul,
> Beans and all (Tongue).

TANSY, as a word, comes from the Greek for immortality, perhaps because the flowers take a long time to wither, but more likely because the plant was used for preserving dead bodies from corruption (Grieve.1931). Ann Leighton had this quotation: "Samuel Sewall has recorded observing the body of a friend long dead but well preserved in his coffin packed full of tansy". PRIMROSES, too, were used to dress up a corpse in the coffin (Latham), and were strewn on graves.

CYPRESS is the symbol of death and mourning, and for that reason there was a superstition that it should never be cut, for fear of killing it (Evelyn. 1678). Both the Greeks and the Romans called it the "mournful" tree, because it was sacred to the rulers of the underworld, and to the Fates and Furies. As such, it was the custom to plant it by the grave, and, when a death occurred, to put it either in front of the house or in the vestibule, in order to warn those about to perform a sacred rite against entering a place polluted by a dead body (Philpot). And it is the very durability of the timber (plus the fact that it is insect-proof) that made it desirable from ancient Egyptian times onwards for making the coffins of the rich. Modern Greeks still use it for this purpose (Moldenke & Moldenke). Funerary rites are performed by the side of the MAHUA tree (*Bassia latifolia*) by the Gonds of southern India. It is the tree of death, and as such the rites there are the final ones in mortuary ceremonies (Fürer-Haimendorff).

DERMATITIS
In homeopathy, a tincture prepared from the fresh bark of LADY LAUREL is recommended for skin complaints (Schauenberg & Paris). A CHICKWEED poultice was used in Norfolk for quite severe cases of dermatitis or eczema (V G Hatfield. 1994), and in the Balkans a decoction of ivy and ARNICA is used (Kemp).

Deutzia gracilis Apparently an infusion of the whole plant was reckoned to be an aphrodisiac in Scotland (Rorie. 1994).

DEVIL'S PLANTS
GARLIC: when the devil's left foot touched soil outside the Garden of Eden, garlic sprang up, and his right foot gave rise to onions (Emboden). Naturally, with an offensive smell like garlic has, it must be associated with the devil, but taken by and large, it is a protector, holding the devil's works at bay. So too with SOW THISTLE. In Russia it was said that it belonged to the devil, but the Welsh belief was that the devil could do no harm to anyone wearing a leaf from the plant (Trevelyan), and one of the Anglo-Saxon herbaria said (in translation) – "so long as you carry it with you nothing evil will come to meet you" (Meaney). PARSLEY has to be included here, for observation of its germination time has given a number of superstitions. Its seed is one of the longest to live in the ground before starting to come up. Further, the devil is implicated – it is called Devil's Oatmeal sometimes– it goes to the devil nine times and back before it comes up (Northcote) (or some say seven (Clair)). It only comes up partially because the devil takes his tithe of it (M E S Wright). To offset this you can pour boiling water over freshly sown seed to deter the devil (Baker. 1974), or better still, sow it on Good Friday, when plants are temporarily free of the devil's power (Baker. 1980).

NETTLE, not surprisingly, was associated with the devil, as some of the local names imply. They were Devil's Leaf, and Devil's, or Naughty Man's Plaything (Macmillan), Naughty Man being quite a common euphemism for the devil. An Irish name is Devil's Apron, but that may hark back to a textile use. In the Highlands and Islands, nettle's very existence was ascribed to the devil. Tradition had it that they sprang up where Satan and his angels fell to earth on their expulsion from heaven (on 3rd May) (Swire.1964). In spite of bearing names like Archangel and its derivatives, RED DEADNETTLE also is called Bad Man's, or Black Man's Posies, (Grigson. 1955), both euphemisms for the devil. A French tale tells how the devil spun the DODDER at night to destroy the clover; clover was created by God, and dodder is the devil's counter-plant. It is hardly surprising that this parasitic plant was assigned to the devil. Local names in England and Scotland confirm the belief – Devil's Threads, or Devil's Net, for example, and even more expressively, Devil's Guts (Grigson. 1955). DEADLY NIGHTSHADE was the devil's favourite. He watches over it, but there was a way of getting the plant. The farmer was advised to loose a black hen on Walpurgis Night. The devil will chase the hen, and so be diverted from looking after his plant. Then the farmer can quickly pluck it – this is the only way to make it efficacious (Skinner).

ELDER, in spite of other associations, can be said to be one of the devil's plants, especially if one tries burning the wood. Doing that was universally forbidden in England, for "it brings the devil into the house" (Graves). Or, "they dursn't burn 'em if you gave them away – they don't want the devil down their chimbleys" (Heanley), or as they used to say in Lincolnshire, "the devil is in elder wood" (Rudkin). "People say that if you want to keep the devil out of your house, you must never burn elder wood" – that is a Wiltshire belief (Goddard). It would "draw the devil down the chimney" (Boase), or, as in Warwickshire, you would see the devil sitting on the chimney pot (Witcutt). In Needwood Forest (and in Hertfordshire) the expression used is that it would raise the devil (Burne, Jones-Baker). Sometimes the result would be that the person burning it would become bewitched (Vickery. 1985). Around Cricklade, in the north of Wiltshire, it was reckoned safe to burn, provided it was given to you!

BLACKBERRY, without actually figuring as the devil's own plant, has an association with him. A superstition found all over the southern counties of England that blackberries have a sell-by date, after which the devil is supposed to put his foot on them (Graves), or that he "has been on them", or "spat on them" (Widdowson). The devil bears this grudge against blackberries because when he was expelled from heaven (on 10 September, apparently), he landed in a bramble bush on his way to hell. That sell-by date is usually Michaelmas, so that it is unlucky to gather them after this, for in some places they used to say that October blackberries were actually poisonous. But the date is rather variable – it could be interpreted as Old Michaelmas Day (10 October), or in some places the day is given as SS Simon and Jude (28 October) (Folkard). The devil arrives earlier in Devonshire, though – 20 September is the day he spits on the blackberries (Whitlock. 1977). All this is reasonably justifiable, for after the first frosts, blackberries are often tasteless and watery, and so not worth the picking (Widdowson). When they get to this state, the devil is said to have "cast his club over them" in Derbyshire (Addy), and in Scotland "thrown his cloak over them" (Folkard). Another explanation is that the green bug that infests bramble bushes in late autumn is called Pisky in Cornwall, and it is pisky that spoils the fruit (Courtney. 1887).

SWEET BRIAR is also one of the devil's plants, actually planted by him, as is said on some parts of France, the hips being his bread. It is called Rose du diable in those areas, and Rose sorcière sometimes. (Sebillot). But more in keeping with the affection engendered by this rose is the legend that explains why their thorns point downwards. The devil was expelled from heaven, and he tried to regain his lost position by means of a ladder made of the thorns. But when the plant was only allowed to grow as a bush, he put the thorns in their present position out of spite.

STITCHWORT, a perfectly harmless plant, has been given the names Devil's Flower and Devil's Nightcap, in Somerset (Macmillan). They are the Devil's Ears, or his Eyes (Leyel. 1937; C J S Thompson. 1947), and his Nightcap (Grigson, 1955), possibly because there is a connection with snakes in folklore, thus raising doubts as to its harmlessness.

MANDRAKE was the ultimate devil's plant, constantly watched over by him. But the devil himself would appear to the owner of a mandrake to do his (the owner's) bidding, so long as the proper ceremonies were observed when the plant was pulled, and again each time it was consulted (Dyer. 1889).

Dianthus barbatus > SWEET WILLIAM

Dianthus caryophyllus > CARNATION

Dianthus plumarius > PINK

DIABETES

A leaf decoction from BREADFRUIT tree is taken in Trinidad to treat diabetes (Laguerre). The complaint was treated in Russian folk medicine by strict diet, with an infusion of BILBERRY leaves – a tablespoonful to a cup of boiling water, infused for half an hour (Kourennoff). PERIWINKLE used to be an Irish (County Cavan) treatment for the condition (Maloney). In Africa, the seed of the BLUE LOTUS

(*Nymphaea caerulea*) has been used for the condition
(Watt & Breyer-Brandwijk). GINGER has been used
for diabetes, in the form of juice, mixed with sugar
candy! (Parihar & Dutt). In America, an infusion of
SPINY COCKLEBUR plants before they flowered
has been used (Watt & Breyer-Brandwijk). HERB
ROBERT is recorded as being a diabetes remedy
in Ireland (Ó Súilleabháin) – a handful of the herb
to a pint of water, in wineglassful doses, night and
morning (Moloney).

DIARRHOEA

Russian folk medicine recommended a mixture
of dried SAGE, KNAPWEED and CAMOMILE
flowers for all digestive disorders (Kourennoff). Just
chewing the leaves of knapweed was taken as a cure
for diarrhoea in Britain (Page. 1978). BILBERRIES
are very astringent, and are still sometimes prescribed
by herbalists for diarrhoea. A syrup from the blue
berries of MYRTLE was taken in the Western Isles,
a spoonful at a time, for flux (Martin), diarrhoea
in modern terms. Both the leaves and root bark
of BRAMBLE contain a lot of tannin, and so are
astringent enough to be useful in stopping diarrhoea.
QUINCE-seed tea is an American country cure for the
complaint (H M Hyatt), and another American remedy
is the bark tea of CHOKE CHERRY (H H Smith.
1923). A strange use of BEETROOT is recorded in
Corfu, where a small bottle of beetroot juice is corked
and put in the heart of an uncooked loaf of bread.
The bread is baked, and the bottle removed and then
the medicine is drunk in small doses on successieve
days for dysentery or diarrhoea until the patient is
cured (Durrell). A tea made from dried CAT'S FOOT
is also recommended for the condition (Flück).
Africans powder the root bark of WATER MINT, to
be eaten for the relief of diarrhoea (Watt & Breyer-
Brandwijk), and herbalists use it for this all over the
world. Any astringent would do, like ALUM-ROOT
(*Heuchera americana*), still an American country
remedy for the complaint (H M Hyatt). SPEARMINT
tea is a great favourite, too, used for the purpose by
American Indians like the Miwok (Barrett & Gifford),
and still prescribed by herbalists for stomach upsets
and diarrhoea. Gerard described it as "marvellous
wholesome for the stomacke". Country people in
Ireland would tie a sprig round the wrist for the
complaint (O' Farrell), and RHUBARB roots boiled
in a little water is another Irish (County Cavan)
cure (Maloney). TORMENTIL stewed in milk was
another Irish remedy for diarhoea (Foster). Gypsies,
too, use an infusion of the leaves for the complaint
(Vesey-Fitzgerald), and a similar use has been noted
in Fifeshire, too (Rorie). Gerard, too, advised the use
of powdered leaves of this plant, especially if they are
given "in the water of a smith's forge, or rather the
water wherein his steels hath been often quenched
of purpose". CINQUEFOIL will do as a substitute
(Flück), though Tormentil is much more effective,

and the root of AVENS is also effective. Gypsies
would use the crushed root for the purpose (Vesey-
Fitzgerald), but actually it is quite common as a herbal
cure for the condition and for similar ailments
(A W Hatfield).

Dicentra canadensis > TURKEY CORN

Dicentra cucullaria > DUTCHMAN'S BREECHES

Dichrostachys glomerata > CHINESE LANTERN

Digitalis purpurea > FOXGLOVE

Digitaria exilis > HUNGRY RICE

DIME-A-BOTTLE PLANT

INDIAN PHYSIC (*Gillenia trifoliata*) got the name
because of the enormous quantities that went to make
up the "cure-alls" peddled by travelling medicine
salesmen in America.

Dioclea refexa > MARBLES VINE

Diospyros kaki > CHINESE DATE PLUM

Diospyros virginiana > PERSIMMON

Dipsacus fullonum > TEASEL

DISLOCATIONS

Scottish Highlanders had a very strange use for
ROYAL FERN – no less than treating a dislocated
kneecap with it. The roots would be chopped up, put
in a pan and covered with water. After boiling and
cooling, the contents would be applied to the injured
knee. It was claimed that "forthwith the pain stopped,
all swelling disappeared and the kneecap went back to
its original place" (Beith).

DITTANY

(*Origanum dictamnus*) In Crete and Greece, they say
that if the plant thrives on a grave, it is a sign that the
deceased below is both joyful and at peace (Whittle &
Cook). Another belief, from the Balkans, was that sick
people should sleep in the field where dittany grows,
on Ascension Day. The next day, any insect found
under the roots of the plants was given to them as a
medicine (Kemp).

The liqueur Benedictine contained Dittany, which
is still used to flavour vermouth (Baumann). Of the
claims made for the plant, Gerard's description sums
them up : "The juyce taken with wine is a remedy
against the stinging of serpents … The same is thought
to be of so strong operation, that with the very smell it
drives away venomous beasts and doth astonish them
… It is reported likewise that the wild Giats or Deere
in Candy, when they be wounded with arrowes, do
shake them out by eating of this plant, and heale their
wounds …" This is what Marlowe had in mind:

> The forest deer, being struck,
> Runs to an herb that closeth up the wounds …
> (*Edward II. v.1*)

The original story was told of wild goats in Crete by Aristotle, and Pliny repeats it, but applies it to stags (Robin). Gubernatis, too, repeats it, and quotes Plutarch as saying that Cretan women, noticing how it withdraws arrows, used it in easing childbirth. "It is said that a hind … eateth this herb that she may calve easilier and sooner …" (Bartholomew Anglicus). But the plant was dedicated to the goddess Lucine, who watched over women in childbirth (Gubernatis). Theophrastus, too, said that it was used in difficult childbirth (Scarborough).

DIURETICS

DANDELIONS are perhaps the best known of diuretics, and some of the vernacular names hammer it home. Pissabed is common enough, though no longer standard English (though pissenlit is still standard French). "Children that eat it in the evening experience its diuretic effects, which is the reason that other European nations as well as ourselves vulgarly call it Pissabed" (Britten & Holland). Would children have any particular urge to eat dandelions? But it is said that even picking the flowers will bring on the symptoms. Mothers would remove all the dandelions from a child's bunch of wild flowers, and children tease their fellows by putting the flowers in their pockets (Vickery. 1985). But of course, dandelions are quite often exploited deliberately. "The root fresh gathered and boiled makes an excellent decoction to bring away gravel" (Hill. 1754). In recent times, Irish country people would brew the leaves, or sometimes the whole plant, roots and all, to incrase the flow of urine (P Logan). Hence, of course, its use for gallstones, jaundice, etc., (Schauenberg & Paris). Herbalists still prescribe the root tea for stone in the bladder (W A R Thomson. 1978). HORSETAIL tea is rich in silicic acid. and is diuretic, so it can be used for urinary problems, such as bed-wetting, cystitis or inflamed prostrate (M Evans), or anuresis (Schauenberg & Paris).

TURNIPS have long been used for retention of urine, apparently still in use among Irish country people in County Mayo. They pulp a turnip and drink the juice to treat the complaint (Logan). HEATHER too is a diuretic, still in use in homeopathy for the treatment of infections of the kidneys and urinary tract (Schauenberg & Paris); so is PARSLEY, the leaves and roots (Parsley root tea is still prescribed for kidney complaints (Rohde, Hyatt)). Gerard wrote that the leaves are good "to take away stoppings, and to provoke urine; which thing the roots likewise do notably perform". Lupton, in the middle of the 17th century, recommended parsley, stamped in white wine, then strained off, to "cause thee to make water and break the stone". It is reckoned in Russian folk medicine to be a powerful diuretic (Kourennoff). SEA HOLLY, too, was taken to be a diuretic. One of the leechdoms in the Anglo-Saxon version of Apuleius

was "for stirring of the mie …" (Cockayne). BROOM tea is well-known as a diuretic.

JUNIPER berries, and the oil distilled from them, are said to be diuretic, and the juice from the berries is still used in Irish country medicine as such (Logan). It is also recommended for cystitis (Schauenberg & Paris), and a tea can be made from them, or even small pieces of the wood, to be used as a diuretic (Flück). Some of the American Indians used PUMP-KIN seeds as a diuretic; so do people in Alabama, still. They say they are excellent in a tea for kidney troubles, but the odd thing is that Alabama children used to be given the same preparation to stop them bed-wetting (R B Browne). VEGETABLE MAR-ROW seeds, too, make an efficient diuretic, and SWEET FLAG had the same reputation. Already, in the Anglo-Saxon version of Apuleius, we can find (in translation) "If one may not pass water, and the water be at a standstill, let him take roots (of Sweet Flag), and let him seethe (them) in water to a third part; give to drink; then within three days he may send forth the urine; it healeth wondrously the infirmity" (Cockayne). Gerard was still able to claim the same virtue a few hundred years later.

DIVINATIONS. Death divinations:

An odd custom that used to be the tradition in one part of Germany was to make wreaths of MARSH MARIGOLDS and then throw them one at a time on to the house roof. If anybody's wreath actually stayed up, then it would be taken as a sign that he would die before next summer (Hartland. 1909). This took place as part of the St John's Day celebrations, and it is unusual to find death divinations practiced then. Would they still be in bloom at Midsummer? There was a Welsh custom of gathering pieces of ST JOHN'S WORT, and naming each piece for a member of the family, and then hanging them all on the rafters. In the morning, they were examined; those that had withered most represented the person expected to die soonest (Vickery. 1981).

Johann Weyer (*De praestigiis Daemonum et incanta-tionibus ac Veneficiis*,1568) recorded Daphnemancy (using SPURGE LAUREL, (*Daphne laureola*) among his list of divinations ascribed to demonic agency. The divination was by the crackling of the leaves while burning. The leaves were also put under the pillow to induce dreams (Lea). Capnomancy is divination by smoke. One way of using it was by throwing seeds of JASMINE or poppy on the fire, and watching the motion and density of the smoke. If it was thin, and shot up in a straight line, it was a good omen (Adams). There is a gypsy divination to know if an invalid will recover. They put from nine to 21 seeds of THORN-APPLE on a "witch drum", that is, a tambourine covered with an animal skin marked with stripes that have a special meaning. The side of the drum is tapped gently, and according to the position that the seeds take on the stripes, the recovery or death of the patient is predicted (Leland. 1891).

DIVINATIONS. Marriage divinations:

ALMONDS are significant as a fertility symbol at Greek weddings, and a further manifestation of this symbolism lies in the belief fostered by unmarried girls that involved taking some of the wedding almonds and using them for divination. Three of them put under the pillow would ensure dreams of the future husband. The same is occasionally said of PINE needles. One particular tree on the island of Bute served as a "dreaming tree". Some of its needles were put, with some ceremony, under the pillow, for dreams of the future husband or wife (Denham). Divinations were made with LEEKS, too, though details are often lacking. One Welsh charm is properly described, though. The girl had to go out into the garden and uproot a leek with her teeth! Then it had to be put under her pillow, and her lover would appear to her in a dream (Stevens) (or in the flesh, if the proper arrangements had been made). Even TURNIPS have been used to play the divination game. This one comes from west Wales and involves a girl stealing a turnip from a neighbour's field (it must be stolen, not given). She peels it in one continuous strip, in the same way as the much better known apple peel game, taking care not to break the peel, and then buries the peel in the garden. The turnip itself she hangs behind the door. Then she goes and sits by the fire, and the first man who enters after that will bear the same name as her future husband (Winstanley & Rose).

ASH, especially the Even Ash beliefs, were popular once. They are all quite well-known, for the leaf is often used for invoking good luck, as well ("luck and a lover" (Leather)), and there is always a simple rhyme to accompany the charm. One from Cornwall runs:

> Even ash, I do thee pluck
> Hoping thus to meet good luck
> If no good luck I get from thee
> I shall wish thee in that tree.

There are many others recited when the even ash is used for divination purposes:

> If you find an even ash or a four-leaved clover
> Rest assured you'll see your true-love ere the day is over (Dyer).

From Northumberland, there is:

> Even, even ash
> I pull thee off the tree
> The first young man I do meet
> My lover he shall be.

The leaf is then put in the shoe (Denham). A Buckinghamshire charm simply, i e with no accompanying rhyme, to be put in the right shoe – "… the first man you meet you have to marry" (Heather).

> This even ash I carry in hand
> The first I meet shall be my husband
> If he be single then he may draw nigh,
> But if he be married then he may pass by (Friend).

That rhyme is quoted for Somerset (Tongue), with the additional proviso that the leaf had to be thrown in the face of the first man met. Northall recorded:

> Even-ash, even-ash, I pluck thee
> This night my true love for to see;
> Neither in his rick nor in his rear,
> But in the clothes he does everyday wear.

It was important for a girl to know the trade of the future husband, and that could be deduced from the working clothes he was wearing. Sometimes the even ash beliefs did not rely on even florets at all. There is, for instance, an Oxfordshire requirement for "an ash-leaf with nine leaves on" (Thompson). Two leaves were also significant in Wales. Such a leaf had to be found by accident, and should be put under the pillow, so that the finder would dream of the future partner (Howells).

Another tree, HAWTHORN, is used in divinations, but to a much lesser extent. One of them involved a girl hanging a flowering branch at a crossroads on May Eve. She had to go back next morning to see in which direction the wind had blown it. From that direction would come the destined husband. Another was for her to partly break a branch on May Eve, and leave it on the tree. In the morning she must fetch it home, and then she would hope to see an image of the future husband on the way (Eberly). Something more direct from the south of Somerset called for the girl to dance barefoot round a hawthorn on old Midsummer Day before sunrise, to charm her lover into marrying her that year (Tongue). ACORNS, or rather their cups, have been used in Wales: two of them, one named for the lover, and the other for self, were set to float in a bowl of water – watch them. If they sailed together, there would be marriage, but if they drifted apart, then it was obvious what the result would be (Trevelyan).

Cut in two, BRACKEN root was supposed to show the initial letters of a lover's name (Leather, Courtney, Forby, etc.,). A particularly bizarre method of divination involved the NETTLE. It is said that if you beat yourself with one, you will be able to count from the number of blisters you get how many years it will be before you marry (Hald).

ROSES are symbols of love, so it is hardly surprising to find them connected with love divinations. Girls would take a rose leaf for each of their suitors, and name each leaf after one of them. Then she would watch them until one after the other they sank in the bucket of water in which she had put them. The last that sank would represent the future husband (Napier). Or she would carefully pick a rose and lay it under her pillow on Midsummer Eve, when the future husband would appear in a dream (Higgins). Rather more ambitious is the one that requires a girl to pick a rose at Midsummer, fold it in paper, and put it by till Christmas Day. On that day she would wear it to church, and the man who would come and take it from her would be her husband (Opie & Tatem). ROSEMARY, too, is a symbol of fidelity in love, and this too has been the agent in marriage divinations, particularly on St Agnes' Night, when girls were advised to take a sprig of rosemary and one of THYME, and sprinkle them three times with water. In the evening, put one in each shoe, putting a shoe on each side of the bed. When going to bed, they had to say:

> St Agnes, that's to lovers kind,
> Come ease the trouble of my mind.

Then the future husband will appear in a dream. So wrote Halliwell as an example of a charm from the north of England and Scotland. The rhyme, though, has been quoted also as:

> Agnes sweet and Agnes fair,
> Hither, hither, now repair;
> Bonny Agnes, let me see
> The lad who is to marry me (Drury.1986).

Derbyshire girls used to put a sprig of rosemary and a crooked sixpence under their pillows at Hallowe'en, so that they should dream of their future husband (Addy). "If you lay a branch of rosemary under your head, on Easter Eve, you will dream of the party you shall enjoy" – so says an 18th century chapbook (Ashton). Another kind of divination was the fashion in Herefordshire. There they put a plate of flour under a rosemary bush on Midsummer Eve. The initials of the intended lover were expected to appear in the flour next morning (Leather). BAY leaves across the pillow, sprinkled with rose water, were used on St Valentine's Eve, when the girl had to say:

> Good Valentine, be kind to me,
> In dream let me my true love see (Dyer. 1889).

(Substitute St Luke for Valentine in the rhyme, and another divination that uses MARJORAM is the result) (see MARJORAM). Another charm from Devonshire called for five bay leaves, one pinned at each corner and the fifth in the middle of the pillow. The operator of this charm had to say the rhyme seven times, and count seven, seven times over at each interval:

> Sweet guardian angels, let me have
> What I most earnestly do crave –
> A Valentine endued with love,
> Who will both true and constant prove.

The future husband would appear in a dream (Vickery. 1995). The same number of leaves, disposed in the same pattern, was the rule at Gainsborough, in Lincolnshire (Rudkin). Usually, the girl had to put on a clean nightgown for the operation, often inside out (Drury. 1986).

HOLLY is used in various divination games. One from Swansea requires a girl to run seven times round a holly tree in one direction, and seven in the other, when she will see her future husband (Opie & Opie). There are, too, variations on the daisy-petal divinations – in one recorded in Brittany, the prickles on a holly leaf are touched, with "Fille, femme, veuve, religieuse", or "Fils, homme, veuf, religieux" (Sebillot), while a game from Edington, in Warwickshire, is for girls to pick a holly leaf and say:

> I pluck this holly to see
> If my mother does want me.

And then count the prickles – yes, no, yes, no, until the last one (Northall). There is too an American game that consists of naming holly leaves as they are thrown in the fire. The one that pops first has the name of the one that loves you best (Thomas & Thomas). More complicated divinations include an Irish one called "building the house", played at Hallowe'en. Twelve pairs of holly twigs are arranged in a circle, pushed into the ground and tied together at the apex. A live turf representing the fire is put in the centre. Each pair of twigs is named after the boys and girls present, and the pair that first catches fire shows which boy and girl will be the first married (E E Evans). Another divination at Hallowe'en involving holly used to be played (if it can be called as such) in the Border counties. A girl had to take three pails of water up to her bedroom, and to pin three holly leaves to her nightdress before going to bed. She will be roused by three shouts, followed by three hoarse laughs, after which the form of her future husband will appear. If he is deeply attached to her, he will shift the position of the three pails (Banks). The holly leaves, though apparently essential, seem to play no part in the record of the charm. They must have done so at one time, and the way in which the divination has come down to us must be corrupt. IVY too can be used in divinations, but the records are mainly from Scotland, where the plant is held in higher regard than is the case elsewhere. Lanarkshire girls used to wear a piece of ivy against her heart, and repeat the rhyme:

> Ivy, ivy, I love you,
> In my bosom I put you
> The first young man who speaks to me,
> My future husband he shall be (Opie & Opie).

The Argyllshire rhyme was:

> Eevy, ivy,
> I do pluck thee.
> In my bosom I do put thee,
> The first young man I do meet
> Shall my true lover be (MacLagan).

This sort of thing is recorded in the Witney area of Oxfordshire, too. To see her future husband, a girl had to put an ivy leaf in her pocket. The first man she met out of doors whilst carrying it would be the one she would marry. But a most novel form of divination is recorded from Normandy – to be used when unsure of which saint to pray to! A number of ivy leaves are put in holy water, each one bearing the name of a saint; the first one to turn yellow shows the saint you were looking for (W B Johnson). Another example from Witney was to prick her lover's name in a leaf of CHERRY LAUREL. If the prick marks turned pink, it was a sign he would marry her, but if it did not, he would desert her (*Oxfordshire & District Folklore Society. Annual Record; 1952*).

CLOVER can be used, too; there is a well-known rhyme involving a two-leaved clover:

> A clover, a clover of two,
> Put it in your right shoe;
> The first young man you meet,
> In field, street or lane,
> You'll get him, or one of his name (Dyer).

Similarly, there is a Quebec superstition that if you put a four-leaved clover in your shoe, you will marry a man having the first name of the man you meet first after doing so (Bergen. 1899). Another example comes from Michigan: with a four-leaved clover in your shoe, you will meet your lover. Another from the same area is if the finder of a four-leaved clover puts it in her shoe, she will marry the first person with whom she crosses a bridge. From other parts of America, the injunction is to put a four-leaved clover over the door. The first person to pass under it will be your future mate (Bergen. 1899). A two-leaved clover will do as well in Wales, provided it was found by accident. Then it should be put under the pillow, so that the finder should dream of the future partner (Howells). A sprig of MYRTLE, too, put under the pillow will bring on dreams of the future husband (Hawke). Something more complicated was recommended in the Midlands. Take a sprig on Midsummer Eve and lay it on your prayer book with the words "Wilt thou take me (mentioning her name) to be thy wedded wife?". She then had to close the book and sleep with it under her pillow. If the myrtle was gone in the morning, she would marry her present lover (Baker. 1980). SOUTHERNWOOD is another plant to be put under the pillow on going to bed, and then the first man she meets in the morning will be the one she will marry. That is an American belief too, that operates simply by putting a piece down the back, or tucking it in the shoe. Having done that, the first boy she meets will be her future husband (Bergen. 1899). A similar charm is recorded in Guernsey: put two fronds of AGRIMONY, each bearing nine leaflets, crosswise under the pillow, securing them by two new pins, also crossed. The future husband would appear in a dream (MacCulloch).

CORNFLOWERS could be used to test a lover's fidelity, using the he loves me, he loves me not formula (Dyer). If the identification is correct, Northamptonshire children used to pick the leaves of LADY'S SMOCK one by one for a divination of the 'Rich man; poor man' type (Sternberg).

Some divinations using flowers rely on cut pieces re-growing and thus answering the question (or not). Plantain, for instance, or KNAPWEED. A young person would pull the flower from the stalk, cut the top off the stamens with scissors, and put the flower in some secret place where it could not be seen. She would think through the day, and try to dream through the night, of her sweetheart. Then, on looking at the flower next day, would judge of her success in love by whether or not the stamens had shot out to their former length (Henderson). John Clare described a similar charm in *Shepherd's Calendar May*:

> They pull the little blossom threads
> From out the knapweed's button heads
> And put the husk wi' many a smile
> In their white bosoms for a while
> Who if they guess aright the swain
> That loves sweet fancy trys to gain
> 'Tis said that ere its lain an hour
> 'Twill blossom wi a second flower.

That description, though, is of a Buckinghamshire charm. The plant would bloom a second time if they could guess aright the name of the future husband (Leyel. 1926). RIBWORT PLANTAIN is used for a similar type of divination – rid the stalks of the flowers, and if they are blooming again when inspected the following day, the prospect of marriage is good. This is a Midsummer divination in the north of England and Scotland. The procedure differed here and there, but one of them was to take three stalks, strip them of their flowers, and put them in the left shoe, and afterwards under the pillow. In the morning, if the lover was to become the husband, they should again be in bloom. That same game was known in the Faeroe Islands as well, where it was known as a simple wish fulfilment custom (Williamson). The Berwickshire custom was to take two spikes, wrap them in a dock leaf and put them under a stone, or, in Shetland, buried in the ground (Banks). One of the two stalks represented the girl, the other the man. If both were in bloom next morning, it was a good sign for marriage (Denham). A different kind of divination game with Ribwort is an offshoot of the Soldiers game that children play. Soldiers involves trying to knock the head off the opponent's stalk. The divination is of the tinker-tailor-soldier-sailor type (the plant is actually called Tinker-tailor Grass in Somerset). The blow that knocks the head off marks the profession of the future husband (Elworthy. 1888). GREAT PLANTAIN, too, was used for divination in the same way, often connected with the Midsummer festival. SAGE leaves, twelve of them, were used in Yorkshire. They had to be picked at noon on

Midsummer Eve, and put in a saucer, where they would be kept till midnight, then they were dropped out of the bedroom window, one by one with the chiming of the hour. The future husband would be seen, or at least his step heard, in the street below (Morris). A Leicestershire variant of this requires the leaves actually to be picked at each strike of the clock at midnight (Palmer. 1985).

APPLES are important instruments for divination. Peeling the skin off in a continuous piece and throwing it over the left shoulder, whence it was hoped that it would fall in the shape of a letter which will be the initial of the man she will marry, is the best known of these divinations. Or, as in parts of America, the peel had to be put over the door, and the first man to enter through the door will be the husband (Stout). If the peel breaks, she will not marry at all (Waring). Sometimes, as in Lancashire, the peel had to hang on a nail behind the door, and then the initials of the first man to enter the house afterwards would be the same as that of the future husband (Opie & Tatem). The pips too could be used. If, for instance, a girl cannot choose among several of her lovers, she could take a pip, recite the name of one of the men, then drop the pip on the fire. If it pops, well and good, for it shows that the man is "bursting with love for her". Of course, if it is consumed without making any sound, she will know that the man is no good for her (Waring). The rhyme to be spoken is:

> If you love me, pop and fly,
> If you hate me, lay and die. (Waring).

Another divination game involving apple pips is to take one of them between finger and thumb and to flip it in the air, while reciting "North, south, east west, tell me where my love doth rest". You had to watch the direction in which it fell, and then draw your own conclusions (Courtney). Another way of doing it was to stick two pips on the cheek or forehead, one for the girl's choice and the other for another man who was not. The one named for the man she really wanted would stick longest, not all that difficult to manage or to make sure the unwanted man fell first (Opie & Tatem, quoting Gay, *Shepherd's Week* 1714):

> See from the core two kernels brown I take;
> This on my cheek for Lubberkin is worn
> And Boobyclod on t'other side is born.
> But Boobyclod soon drops upon the ground,
> A certain token that his love's unsound,
> While Lubberkin sticks firmly to the last;
> Oh were his lips to mine but join'd so fast.

A Kentucky version requires five seeds on the face, named. Then the first to fall off shows the one that the girl will marry (Thomas & Thomas). Another American children's game merely involves counting the seeds to predict the future:

> One I love
> Two I love
> Three I love I say;
> Four I love with all my heart
> And five I cast away;
> Six he loves
> Seven she loves
> Eight they both love;
> Nine he comes
> And ten he tarries,
> Eleven he courts
> And twelve he marries (Stout).

Similarly, the number of seeds found indicates the number of children you will have (Thomas & Thomas). Even the stalks could be used: the girl has to twist the stalk to find whom she will marry. The game is to twist while going through the alphabet, a letter for each twist. The letter she has reached when the stalk comes off is the initial of the first name of the man she will marry (Opie & Tatem). An Austrian divination involved cutting an apple in two on St Thomas's Eve and counting the number of pips. If it was an even number, then she was soon to marry. But if she had cut one of the pips, she would have a troubled life, and end up a widow (Waring). Even CRABAPPLES were used for the purpose in Somerset. A girl would pick crabs on Michaelmas Day, store them until Old Michaelmas (10 October) and arrange them to form the initials of her various suitors. Later inspection, though quite when is not clear, would reveal the one who would make the best husband, judging by the condition of the apples. An American divination that relied on movement, like apple pips, was played with BOX leaves. Name some leaves and lay them on a hot hearth. The one that swells and whirls towards you will be your future husband or wife. If one turns in the opposite direction, he or she will shun you (Whitney & Bullock). A pod with nine PEAS in it is a lucky find, and can be used in marriage divinations. If a girl finds such a pod, and if she then wrote on a piece of paper:

> Come in my dear
> And do not fear,

to fold and put in the pod, which then had to be laid under the door, the first man who entered the room would become her husband, or in some areas, his Christian name provided the initial of the future husband's name (Opie & Tatem).

Divining by sowing and reaping was probably the commonest form in this country, and the use of Hempseed is the best known, but FLAX could be used, and that used to be quite common in north-east Scotland. The rhyme spoken at the Hallowe'en divining game was:

> Lint-seed I sow ye,
> Lint-seed, I sow ye,

Let him it's to be my lad
Come aifter and pu' me (Gregor).

Dock could be used, too, or at least BUTTER DOCK (*Rumex longifolius*) could. The seeds had to be sown by a young girl half an hour before sunrise on a Friday morning, in some lonely place. She had to strew the seeds gradually on the ground, while saying:

I sow, I sow.
Then my own dear,
Come here, come here,
And mow and mow.

She would see her future husband moving with a scythe (Halliwell. 1869).

MISTLETOE is used at Christmas time, either by putting a sprig (taken from the parish church, in Welsh practice (Trevelyan)) under the pillow, to have a dream of the future husband, or, as with an Irish game, picking ten berries on Christmas Eve. Nine had to be kept, and the tenth thrown away. The nine were put to steep in a liquid composed of equal proportions of wine, beer, vinegar and honey. Then the berries had to be swallowed like pills on going to bed. These, too, would induce dreams about the future (Wood-Martin) (but mistletoe does not grow in Ireland!). In Alabama (R B Browne) and Kentucky (Thomas & Thomas), they used to hang mistletoe over the door to see who would be the first girl to walk in, and she would be the future wife. Another Kentucky divination was to name mistletoe leaves for a boy and a girl, and then put them on a hot stove. If the leaves hopped towards each other, it was taken as a clear sign of marriage between the two. ONIONS were also used at this season. These would be ST THOMAS'S ONIONS, used on the eve of St Thomas's Day (21 December). One game involved peeling a large red onion and sticking nine pins in it, the one in the centre being named for the man the girl really wanted, and the rest radially around this one. As they were put in, the girl recited a rhyme:

Good St Thomas, do me right,
Send me my true love tonight,
In his clothes and his array,
Which he weareth every day,
That I may see him in the face
And in my arms may him embrace.

Normally, it would be an advantage to know in advance in what trade her future husband would be engaged, hence the insistence on "in his clothes and his array, which he weareth every day". Another version of the divination required the girl to cut the St Thomas's onion into quarters, whispering to each quarter the name of the man she was expecting to propose. She waved it over her head while saying the usual rhyme. Pennsylvania German girls used to take four onions, give each a name, and put them under

the bed or the stove in the evening. The one that had sprouted next morning bore the name of the future husband (Fogel).

An American divination game involved MULLEIN. All that had to be done was to twist a stalk nearly off, after naming it for someone. If it lived, then the someone loves you. Afterwards, you had to count the new shoots that sprang up – this will be the number of children you will have (Bergen. 1896). The instructions varied slightly; some say you should break the stalk right off before naming it. Or you had to bend it towards the sun (R B Browne), or in the direction of the boy friend's home, when if the stalk grows straight again, he loves you (Thomas & Thomas).

COWSLIP balls, known as Tisty-tosties, or Tosty-tosties, are the implements in a children's divination game. The blossoms, picked on Whit Sunday for preference, were tied into a ball, and strictly, the balls were the Tisty-tosties, though the growing flower got the name, too. Lady Gomme mentioned the game as belonging to Somerset, but it had a much wider spread than that. The cowslip ball is tossed about while the names of various boys and girls are called, of the time-honoured Tinker, tailor, soldier, sailor, etc., fashion, till it drops. The name called at that moment is taken to be the "one indicated by the oracle", as Udal puts it, for the rhyme spoken at the beginning of the game is:

Tisty-tosty tell me true
Who shall I be married to?

That is quoted as a Dorset rhyme, but the same is recorded in Herefordshire (Leather). The game is also known in Wales, where the purpose is different, for the rhyme there is:

Pistey, Postey, four and twenty,
How many years shall I live?
One, two, three, four … (Trevelyan).

John Clare calls the cowslip balls cucking balls:

And cowslip cucking balls to toss
Above the garlands swinging light ….

A different tradition here, obviously. Roy Genders (Genders. 1976), who used the name cucka-balls, says they were often threaded on twine and hung from one window to another across the street. Games that children play with RYE-GRASS include one, of the Love-me-love-me-not kind, played by pulling off the alternating spikelets, hence names like Yes-or-no. A different version of the game accounts for the name Does-my-mother-want-me, yet another Somerset name (Elworthy). Another game that girls play involves striking the heads together, and at each blow saying Tinker, tailor, soldier, sailor, etc., The blow that knocks the head off marks the profession of the future husband (Grigson. 1959).

YARROW was widely used for love divinations, especially on May Eve and at Hallowe'en. Irish girls would fill a stocking with it, more specifically, the left stocking, tied with the right garter (Cooke), and put it under their pillow, while some recognised rhyme was recited. One Irish example is:

> Yarrow, yarrow. Yarrow,
> I bid thee good morrow,
> And tell me before tomorrow
> Who my true love shall be (Wilde. 1902).

Aberdeen girls went out to the fields on May morning, always in silence, to gather yarrow. They shut their eyes, and pulled what first came to hand, while repeating a local form of the rhyme, such as:

> Good morrow, good morrow,
> To thee, braw yarrow;
> And thrice good morrow to thee;
> I pray thee tell me today or tomorrow
> Who is my true love to be.

Then they would open their eyes, and look around in every direction as far as the eye could see. If a man was visible, the girl who spied him would wed her mate that year. In some districts, they went out on the first night of May (again in silence), carried the yarrow home, and went to bed without speaking a word. During the night, the future husband would appear in a dream (McNeill. 1959), though to be sure of him, he had to be facing the dreamer, if he had his back to her, they would never marry (Beith). Some said that the yarrow had to be picked at the new moon. This applied in Cornwall (Courtney). The yarrow divination travelled to America as well. In Massachusetts, for instance, the formula while walking three times round the yarrow, was:

> Good evening, good evening, Mr Yarrow.
> I hope I see you well tonight,
> And trust I'll see you at a meeting tomorrow.

Then the girls would pluck the head, put it inside their dress, and sleep with it. The first person they met, or spoke to, at church, would be their husband (Bergen. 1896). In Dorset and most of the south of England, the yarrow had to come from a young man's grave (Udal, Watson). Everyone knows the divination game played by pulling off DAISY petals one by one (he loves me, he loves me not ...")(see Vickery. 1995), or (rich man, poor man ...). An American example uses the rhyme:

> One I love, two I love, three I love I say
> Four I love, with all my heart,
> And five I cast away.
> Six he loves, seven she loves, eight they both love,
> Nine he comes, ten he tarries,
> Eleven he counts, and twelve he marries (Whitney & Bullock).

Another way of playing the game is by naming the alphabet until the last petal is reached. This will give the first letter of the name of the girl's lover (Whitney & Bullock).

A gypsy divination using THORN-APPLE seeds, required nine seeds, ploughed-up earth from nine different places, and water from nine more. With these, the girl had to knead a cake, which was laid on a crossroad on Easter or St George's morning. If a woman stepped first on the cake, her husband would be a widower or older man, but if it was a man who trod on it, the husband would be single or young (Leland. 1891).

ORPINE, under another of its names, Midsummer Men, has been used for Midsumer Eve divinations for a very long time. Another name, Love-long, recalls its use by hanging up a piece named after a girl's boy-friends. The piece that lives longest determines the successful suitor. "In Gander Lane we saw in the banks some of the 'Midsummer Men' plants which my Mother remembers the servant maids and cottage girls sticking up in their houses on Midsummer Eve, for the purpose of divining about their sweethearts" (Kilvert). Other divinations are by the bending of the leaves to the right or left, telling whether a lover were true or false (Leather), or, as in America, to tell from what quarter the lover will come (Bergen. 1899). Also, if gathered by two people on Midsummer Eve, and the slips planted, they would know their fortune by the growing or otherwise of the slips. If they leaned towards each other, the couple would marry; if one withered, the person it represented would die (Radford & Radford). Aubrey had noticed the custom in 1686: "Also I remember, the mayds (especially the cooke mayds & Dayrymayds) would stick up in some Chinkes of the joists ... Midsommer-men, which are slips of orpins. They placed them by Paires: one for such a man, the other for such a mayd his sweet-heart, and accordingly as the Orpin did incline to, or recline from the other, that there would be love, or aversion; if either did wither, death". The belief travelled to America, too, albeit in an altered form. A record from New Brunswick advises "take a love-forever leaf, squeeze it to loosen the inner and outer skin. If it makes a balloon as you blow into it, you will be married and live a long time. If it does not, you will be an old maid" (Bergen 1899). ST JOHN'S WORT is another of the Midsummer plants, much used in divinations at that time. In Denmark, girls used to gather it and put it between the beams under the roof, in order to see the future – usually one plant for themselves and one for the boyfriend. If they grew together, it foretold a wedding. Or the plant was set between the beams, and if it grew upwards towards the roof, it was a good sign, but if downwards, sickness and death was forecast (Thorpe). Orkney girls used to gathered the Johnsmas (as Midsummer was known there) flowers, remove the florets, one long, one short, from each flower, wrap what remained in a docken leaf, and bury it overnight. If by morning the florets had re-appeared, it was taken as a happy omen (F M McNeill). Similarly, in parts of Germany, girls fasten sprigs to the walls of their rooms. If they are fresh next morning, a suitor may be expected. If it droops or withers, the girl is destined for an early grave. This custom was also known in Wales, where the plant had to be gathered at midnight, by the light of a glow-worm carried in the palm of the hand (J C Davies). HEMPSEED appears frequently in divinations, sown on Midsummer Eve and at Hallowe'en; often it is necessary for the hemp seed to be harrowed – "steal out unperceived, harrowing with anything you can conveniently draw after you …" (Bell), or else the true love is the one who is supposed to appear to rake it. Sometimes hemp has to be cut, and this figures in the divinations:

> Hempseed I sow, and hempseed I mow,
> And he that is my sweetheart come follow me I trow.

Or the apparition will do the cutting:

> Hempseed I set, hempseed I sow,
> The man that is my true love
> Come after me and mow (Northall).

The significance lies in the fact that freshly-cut hemp is strongly narcotic (hence the bad headaches that the harvesters used to get).

A divination game, which must be modern, is played with a BANANA. To find out whether a boy is being faithful, the question is put, and the lower tip of the fruit is cut off. The answer is found in the centre of the flesh, either a Y, meaning yes, or a dark blob, meaning no (Opie & Opie. 1959). Clearly, the system can also be used to predict the outcome of many other activities, or to solve a problem that requires a simple yes or no answer (Vickery. 1995).

DIVINATIONS. Health divinations

A Yuletide or New Year divination as to future health was by putting an IVY leaf in a bowl of water, and covering it till the 5 January. If it was as green as when it was put into the water, then you would enjoy good health, but if there were black spots on it, then, depending on where the spots were, you would suffer an illness in the part of the body corresponding to the position of the spot on the leaf (Brand/Hazlitt).

DIVINING RODS,

like the magician's wand, are traditionally made of HAZEL. The rod seems to have been introduced into Britain by German miners in the 16th or 17th century. Evelyn's words on the subject, "certainly next to miracle, and requires a strong faith", seem to be an indication of recent introduction. Even today, some three hundred years on, divining is still looked on in some quarters with disbelief. Hazel was the

wood for a wishing-rod, too (Grimm). And the Welsh wishing cap was generally made of its leaves or twigs, though sometimes juniper was used, too. In the Basque country, it was DOGWOOD that was chosen for divining rods (W Webster), rather then hazel. In Yorkshire and Lancashire, divining rods were often made of ROWAN (Besterman), though hazel was usually preferred. In America, or at least Iowa, BOX ELDER seems to be the favourite for the purpose, or the twigs of WITCH HAZEL were used, and that acounts for the common name, because divining was looked on as the result of occult power (Weiner). Guernsey dowsers used TAMARISK, as well as hazel, for their rods (Garis).

Like other magical plants, MISTLETOE was credited once with making the wearer invisible (Dyer. 1889), it will open a lock, and, used as a divining rod, will reveal hidden treasure (Grigson. 1955), just as GOLDEN ROD will, if held in the hand. It has long had this reputation of pointing to hidden springs of water, as well as to treasures of gold and silver (Dyer. 1889).

DOCK

(Includes Butter Dock (*Rumex longifolius*), Curled Dock (*Rumex crispus*) and Broad-leaved Dock (*Rumex obtusifolius*). Every child knows that:

> Nettle out, dock in.
> Dock remove the nettle sting.

One of the Wiltshire variants runs:

> Out 'ettle
> In dock.
> Dock shall have a new smock
> 'Ettle shan't
> Ha' narn (or narrun) (Goddard).

While Irish versions of the charm are:

> Dock, dock, you cure me and I'll cure you

or:

> Dockin, dockin sting nettle

or again:

> Dockin, dockin, in and out,
> Take the sting of the nettle out (Logan).

There is a suggestion that the skin had to be whipped with the dock leaf while the words were recited (Palmer. 1985), but usually placing a dock leaf on the spot stung, or rubbing it was enough. There is nothing extraordinary about the charm, for docks have been used for skin troubles for a very long time. Dock tea, for instance, is an old remedy for boils (Fernie), while Blackfoot Indians made a poultice of the leaves of Curly Dock to apply to the boil. (Johnston). And the condition was treated in East Anglia by drinking an infusion of the seed of Broad-leaved Dock

(V G Hatfield). Gypsies use a decoction of the sliced root taken in elderberry wine to dispel a spring rash (Vesey-Fitzgerald), or you could just rub the dock leaves on (Tongue). Cornish people even treated shingles with a liquid made from dock and bramble leaves (Courtney). We find the leaves being applied to burns and scalds, and for dressing blisters (Fernie). In the Hebrides, dock roots were boiled with a little butter and applied on a bandage to a burn (Shaw). And a dock leaf applied to the forehead will get rid of a headache, too (V G Hatfield).

In the Fen country of East Anglia a root of Broad-leaved Dock tied across the thighs was a cure for the ague (Porter. 1969), though it was probably regarded more of a preventative than a cure. Scrofula, too, was dealt with in Ireland with dock among other plants (Wilde. 1890). Curled Dock, too, has been used for scrofulous conditions (Grieve. 1931). A dock leaf poultice was an Irish treatment for what they called "sore leg". The procedure was to remove the central stem of the leaves, and then to warm the leaves at the fire before applying them. This poultice was also believed to be useful for cleaning out a wound (Logan).

An Irish superstition, obviously homeopathic in origin, was to tie the seeds to a woman's left arm, to prevent her being barren (Wilde. 1902). There was a divination game associated apparently with Butter Dock. Then seeds had to be sown by a young unmarried woman half an hour before sunrise on a Friday morning, in some lonely place. She had to strew the seeds gradually on the ground, while saying:

> I sow, I sow.
> Then, my own dear,
> Come here, come here,
> And mow and mow.

She would see her future husband moving with a scythe (Halliwell. 1869). The name Butter Dock arises from the use of the leaves in which to wrap butter. In fact, any of the large-leaved docks would have been used for that.

Cushy Cows is a name for Broad-leaved Dock from the northern counties of England and the Scottish Border country, given when the dock is in seed. Children "milk" the seeded stems by drawing the stalks through their fingers. The name comes from the call to the cows to get them back at milking time, and is repeated at the start of a milking charm rhyme quoted in the Oxford Dictionary of Nursery Rhymes – "Cushy cow, bonny".

DOCK PUDDING
Made in Yorkshire and Cumbria, like Herb Pudding, but not apparently connected with Easter. It is simply a cheap meal. It contains BISTORT, young nettles, onions and oatmeal. The mixture was simmered till

cooked, strained and allowed to go cold. Slices were then fried, with bacon. The Cumbrian version was more elaborate, with a many different spring leaves (Schofield).

DOCTRINE OF SIGNATURES
A system of medical recognition based on the idea that each plant displayed as its "signature" the ailment for which it formed a cure. Some of them are very simplistic; a plant with a yellow latex would cure the yellow disease, jaundice, as with Gerard's prescription using CORN MARIGOLD for just such a purpose, or GORSE, as another example (Leyel. 1937), or the Irish use of buttercup juice (Ó Súilleabháin), or the sap of GREATER CELANDINE, or the yellow latex of MEXICAN POPPY (*Argemone mexicana)*. BROOM, too, would be used for jaundice, but it is not necessarily doctrine of signatures, for broom tea is a known diuretic, and an infusion not only of flowers, but also stems and roots, has been used in Norfolk. Of course, TURMERIC, with its yellow dye, is the prime example of jaundice medicine, not only in the Pacific islands, but in West Africa (Harley), and even in Britain, for we find Thornton (1810) prescribing it. TOADFLAX, too, could be used for bladder problems, and for jaundice, and Gerard produced another "yellow" remedy – "the decoction of Tode-flax taketh away the yellownesse and deformity of the skinne, being washed and bathed therewith", a cosmetic water, in other words. SAFFRON offers another example of yellow against yellow. "Lay saffron on the navel of them that have the yellow jaundice, and it will help them", was Lupton's advice, but there were earlier examples (see Coulton). The yellow inner bark of BARBERRY is another example. Irish folk medicine recommended the bark in stout, with sulphur, the whole cooked together (Moloney). In Lincolnshire, a tea was made from the twigs and bark for gallstones and jaundice (Gutch & Peacock). BLADDERWORT, of course, could be used for bladder complaints, too (B L Bolton).

Or a red flower (the colour of blood), HERB ROBERT, for instance, showed that it was a wound herb, etc., One red flower, though, RED POPPY, is a jaundice remedy, for it has a yellow juice. Some are entirely esoteric (see Walnut for any brain condition, for example). Others seem to be beyond our comprehension in these days. CYCLAMEN, for example, was used once to aid childbirth. Midwives, in fact, regarded it as invaluable, for old herbals advised women in labour to hang the root around their neck to ensure quick delivery. Coles, the main protagonist for this doctrine, thought that a glance at the leaves was enough to show that the plant belonged to the womb by signature! Much more obvious was EARLY PURPLE ORCHID, with its testicle-suggesting tubers. Salep is a beverage prepared from them, and it enjoyed a reputation for curing impotence (Pliny:

gemina radice, testiculus simili). Equally obvious was BIRTHWORT, whose flower constricts into a tube that opens into a globular swelling at the base. The swelling was interpreted as the womb, the tube as the birth passage. These were the signatures that ensured its use in childbirth, to help delivery, to encourage conception, and to "purge the womb". The generic name *Aristolochia* comes from two Greek words meaning 'best birth', and oddly enough, the plant apparently does have an abortive effect (Grigson). There were anciently a number of strange effects to be encountered by use of this plant. Pliny said it was prescribed for securing a male child (Bonser). Another was described by Guainino (c 1500). He prescribed a kind of medicated pessary made of honey and birthwort, to be introduced before sleeping. The woman was judged to be fertile if on awakening she had a sweet taste in her mouth! (T R Forbes). A plant from the same genus, VIRGINIAN SNAKEROOT, has a different signature. It is called Snakeroot because of its writhed roots, like snakes, and so its early use was as a remedy for snakebite (Lloyd). In general, the Indians simply chewed the root and applied, or spat, it on the bite (Weiner). Similarly, the writhed roots of SENEGA SNAKEROOT provided the signature for the same use. One account says that it grows wherever there are rattlesnakes, so that it was used as a remedy for their bite (Dorman). A number of other drug roots bear the name Snake-root, such as *Rauwolfia serpentina, Asarum canadensis,*etc., The doctrine ensured that ADDER'S TONGUE would be used for snakebite, too. And there is BISTORT, twice twisted, with a root system that looks like a nest of snakes, which is its signature for a snakebite cure.

The most extreme example must surely be that of WALNUTS. Coles, the great presenter of the doctrine, said that "Wall-nuts have the perfect signature of the Head: the outer husk or green covering represent the Peribanium, or outward skin of the skull, whereon the hair groweth, and therefore salt made of these husks … are exceeding good for wounds in the head. The inner woody shell hath the signature of the skull, and the little yellow skin, or Peel, that covereth the Kernell of the hard Meninga and Pia-mater, which are the thin scarfes that envelope the brain. The kernel hath the very figure of the brain, and therefore it is very profitable for the Brain, and resists Poysons. For if the Kernel be bruised, and moystened with the quintessence of wine, and laid upon the Crown of the Head, it comforts the brain and head mightily". But the Walnut tree was involved in so-called cures for madness long before Coles's time. A 15th century leechdom spoke of a sovereign medicine for madness and for men that be troubled with wicked spirits: "upon midsummer night betwixt midnight and the rising of the sun, gather the fairest green leaves of the walnut-tree and upon the same day between sunrise and its going down, distill thereof a water in a still between two basins. And this

water is good if it be drunken for the same malady" (Dawson). Using PEONY as the signature for the head is even more strained. Coles again: "the heads of the Flower … have some signature and proportion with the Head of Man, having sutures and little veins dispersed up and down like unto those which environ the brain. When the flowers blow, they open an outward little skin, representing the skull, and are very available against the Falling-sicknesse…". SKULLCAP, a name awarded as a description for *Scutellaria galericulata,* is also its signature, so that it was to be used for head-related conditions, anything from insomnia to outright madness (Conway). According to Coles, the bulb of SQUILL also has the signature of the head, so it was prescribed for epilepsy, headache, dizziness, etc.,

ASPEN'S use is straightforward enough – the shivering tree to cure the shivering disease – ague. [See also: Cross, Judas, Symbolism, Transference Charms]. Any plant with heart-shaped leaves would be regarded as a legitimate ingredient in cordial preparations (Dyer. 1889). Another ague medicine would be WILLOW, for this tree grows in wet conditions, and ague is caused by damp.

GARLIC relies on the shape of its leaf. The word garlic is OE garleac, where 'gar' means spear, the shape ("taper-leaved" is the accepted way of styling it) recalling that of the spear. So it became useful to cure wounds inflicted by a spear. This use as a wound herb, for which, incidentally, there are sound medical reasons, continued into the twentieth century. It has always been used as an antiseptic for wounds (Brownlow), and during World War 1 the raw juice of garlic was put on sterilized swabs to apply to wounds to prevent them turning septic. TUTSAN has berries that exude a dark red juice, the signature of human blood (they were said to have originated from the blood of slaughtered Danes (Gomme. 1908)). Actually the leaves do have antiseptic properties, and they were certainly used before bandaging became common to cover open flesh wounds (Genders. 1991). Tutsan's close relative, ST JOHN'S WORT, was also a wound herb. The "perforations" in the leaves (not perforations at all, but glandular sacs), constitute the signature. But Coles also interpreted these "perforations" as like the pores of the skin. So they could be used to treat skin problems, too.

The scales of PINE cones were used for toothache because (and this is pure doctrine of signatures) they look like the front teeth (Berdoe). So do the thick scales of TOOTHWORT, so the doctrine says they must be good for the teeth (Gordon. 1977). The pith of ELDER, when pressed with the fingers, "doth pit and receive the impress thereon, as the legs and feet of dropsical persons do", so the juice of the tree was reckoned a cure for dropsy (Dyer). Evelyn mentions it, and so does Gerard, who recommended the seeds for "dropsie, and such as are too fat and would faine be leaner …". But Dawson quoted two leechdoms from a much earlier time, one requiring the patient to drink elderberry juice tempered with wine, and the other to stamp the middle rind and leaves, with flowers or berries, and to drink that after spurge and mastic were added. Earlier still, in the Anglo-Saxon version of Dioscorides, there is a cure for "water sickness" (Cockayne). There are similar reasons for the inclusion of WHITE BRYONY here. The root suggested a swollen foot, so it would be used for such complaints as gout and dropsy.

NETTLE tea could be taken to combat nettle-rash (Berdoe), on the basis of equating that which causes similar burning stings with the ability to cure them. It is still used for this complaint in Cornwall (Deane & Shaw), though one suspects that the value of counter-irritants may be called on in this case, as it would be in an Irish cure from County Down for paralysis of the limbs: "continued excitement" by the nettles on the skin is the form of the treatment (Barbour). A similar usage involves removing a thorn with the aid of HAWTHORN; a very early (14th century) recipe: "For to draw oute a thorne: tak the barke of the hawthorne and stamp him wele in red wyne" (Eberly). To use a thorn to get out a thorn is true homeopathic medicine.

Another group of plants, of which SOW THISTLE is one, have as their signature a milky juice, so it was given to nursing mothers. Coles said it was called Sow Thistle because sows knew that it would increase their flow of milk after farrowing. Perhaps that was why Welsh farmers put some of the leaves in the pig troughs, though it was said that was done to fatten them (Trevelyan). LETTUCE belongs in this company, too. Gerard obviously used the doctrine when he said it "maketh plenty of milke in nurses …" Any plant with heart-shaped leaves would be regarded as a legitimate ingredient in cordial preparations (Dyer. 1889). MILK THISTLE would be an obvious candidate for inclusion in "milk-makers". The white veins of the leaves were said to be due to the Virgin Mary spilling some milk from her breast, hence the common name, and also the specific name, *marianum*. So the doctrine caused it to be taken as a proper diet for wet nurses (Page. 1978).

HORSE CHESTNUT. 'Horse' in a plant name usually denotes largeness or coarseness, and that is probably the case here, though the older writers obviously did not think so. There is the old Latin name *Castanea equina*, and Parkinson wrote "The horse chestnits are given in the East country, and so through all Turkie, unto horses to cure them of the cough, shortnesse of winde, and such other diseases". Skinner suggests the derivation from the likeness to a horse's hoof in the leaf cicatrix, and it may have been from the doctrine of signatures that the nuts, crushed as meal, were given to horses for various diseases.

MAIDENHAIR FERN owes its name ultimately to the myth of Venus arising out of the sea with dry hair. The generic name, *Adiantum*, means unmoistened, for the fern has the property of repelling moisture. So from very early on it has been an ingredient in hair lotions, particularly lotions that claim to keep the hair in curl on damp days. It is hardly surprising that the doctrine of signatures has ensured its use for alopecia – it is the ashes of the fern, mixed with olive oil and vinegar, that are used (Leyel. 1937). DROPWORT (*Filipendula vulgaris*) earns its name from its small, drop-like tubers, the signature of strangury, or gravel, or as Culpeper described the ailment, "such as piss by drops".

The red colour of cosmetic HENNA paste is also the signature of its use against fever, red being a prophylactic colour, just as in Britain, fever patients used to be wrapped in red blankets. In Morocco, henna was applied to the forehead of the sufferer (Westermarck. 1926). In the Balkans, too, in acute fever such as typhoid, henna would be applied to the palms of the hands and the soles of the feet, "in order to draw out the fever" (Kemp). Another "red" signature concerns HOLLYHOCK. It was Gerard who advised "the decoction of the floures, especially those of the red doth stop the overmuch flowing of the monethly courses, if they be boiled in red wine (red, again).

On occasion, the doctrine may be cited in a reverse way, to show the cause of an illness rather than to cure it. In the second edition of Gerard, there is "Bauhine saith that he heard the use of these (POTATO) roots was forbidden in Burgundy (where they call them Indian artichokes) for that they were persuaded the too frequent use of them caused leprosie". Bauhine is Gaspar Bauhin, whose *Prodromos* of 1620 set out the theory. Prejudice against the potato was still apparent a hundred and fifty years later in that area, and its cause was probably to be accounted for by the doctrine of signatures, for the skin of the tuber must have reminded someone of the effects of leprosy.

The shape of SPOTTED MEDICK's leaves give rise to various 'heart' names, Heart Medick, Heart Clover, and so on, much too good for the expounders of the doctrine of signatures to miss. Coles said it is so called "not only because the leaf is triangular like the heart of a man, but also because each leafe doth contain the perfection (or image) of a heart, and that in its proper colour, viz., a flesh colour. It defendeth the heart against the noisome vapour of the spleen".

LESSER CELANDINE, as names like Pilewort and Figwort show, was a country remedy for piles, the reason being the small tubers on the roots, and these are the signs, for pile is Latin pila, a ball. Not only piles were suggested by these small tubers. Growths on the ears and small lumps in the breast were also treated with celandine in Scotland. But like a lot of examples of the doctrine, there does seem to be some curative effect. The specific name, *ficaria*, is another indication, for it derived from the plant's curative efficacy value in the ficus, which means fig, and by extension, piles (hence figwort, a name confined to this plant). So confident were they in Holland as to celandine's efficacy against piles that people there actually wore the roots as amulets against the condition (North), and gypsies say that carrying a sprig or two in the pocket will cure it (Quelch). The same applies to FIGWORT itself, from its knobbly tubers, and its very name, or the signature could point the way to scrofula (*Scrophularia* is the generic name) (Dyer. 1889, who also suggested that the granulated roots of MEADOW SAXIFRAGE provided the signature to use for haemorrhoids, but there is no record that this plant was used in that way). But the real signature of the saxifrage relied on inaccurate observation. The name is derived from Latin saxum, stone, and frangere, to break, "break-stone", in other words. These plants often grow in the clefts of rocks, and it was concluded that the roots had actually broken the rock, and that became its signature – that which breaks stones must also have the power to break stones in the body. The signature of SAMPHIRE is rather similar. The name comes from the French herbe de St Pierre, but 'pierre' also means stone, so if we disregard the attribution to St Peter, we can see that samphire means simply a plant that grows on a rock, and it was this fact that led to its medicinal usage against stone, or general bladder and kidney trouble (Young).

BITING STONECROP "hath the signature of the gums", and so was used for scurvy (Berdoe), and Hill, in the 18th century independently advised that "the juice … is excellent against the scurvy …". Its close relative ORPINE has root tubers that "signal [it] with virtue against the King's Evil" (Grigson. 1955).

It is claimed that SOLOMON'S SEAL gets its name by virtue of a scar to be found from cutting the root transversely, and its similarity to the device known as Solomon's Seal, a 6-pointed star. So this doctrine ensured that the plant was used for "sealing" wounds (Dyer. 1889).

Gerard recommended the decoction of [MELANCHOLY THISTLE] in wine, to "expel superfluous Melancholy out of the body, and make a man as merry as a cricket…" The reference is actually to the hanging flower heads of this thistle, and this became the "signature" of the plant, hence the use for melancholy (Grigson. 1974).

One of the odder examples concerns FIELD BINDWEED. It twines, so the doctrine claimed it to be good for the intestines! (Prest).

GROMWELL (*Lithospermum officinale*) has hard, stony, seed (that is what *Lithospermum* means)

which proclaims its signature perfectly, for it was widely used against the stone. "In case that stones wax in the bladder, and in case that a man may not mie, take of these stones [seeds, that is] … give to drink in wine, it breaketh to pieces the stones, and forth leadeth the mie" (Cockayne). That was translated from an Anglo-Saxon version of Dioscorides, but Gerard was recommending the same in the 16th century, and so was Hill in the 18th. And so it went on. In fact, Pliny wrote "Indeed there is no plant which so instantaneously proclaims at the mere sight of it, the medicinal purposes for which it was originally intended" (see Bonser). It may come as a surprise to find that the plant actually does have diuretic qualities (Schauenberg & Paris).

DRAGON ARUM relies on colours for its signature. Gerard described it as "… with spots of divers colours, like those of the adder or snake …". So this is where the dragon and snake attributions come from. It comes as no surprise, with that kind of signature, to find it being used long ago against snakebite. The Anglo-Saxon version of Apuleius, in the translation by Cockayne, has "for wound of all snakes, take root of this wort with wine and warm it; give to drink". It is easy to combine prevention and cure, and what will cure snakebite will also prevent it. Gerard again – "it is reported that they who have rubbed the leaves or root upon their hands, are not bitten of the viper". He was paraphrasing Pliny, who "saith that serpents will not come neere unto him that beareth dragons about him". The American plant known as CREEPING LADY'S TRESSES (*Goodyera repens*) has white-veined leaves, like snakeskin to the early herbalists, who decided this was the sign of its virtues against snakebite (it is actually called Rattlesnake Plantain sometimes) (Cunningham & Côté). WALL PENNYWORT used to be called Hipwort, reckoned to be given from the resemblance of the leaf to the hip socket, and that, whether it is reasonable or not, is its signature. Coles took it so – "it easeth the pain of the hippes", he said. This same plant got the name Kidneywort, too. If that is in any way descriptive, then its use for kidney ailments and stone was also from the doctrine of signatures. LUNGWORT'S blotched leaves, reminding someone of lung tissue, were the signature for its use as a remedy for various lung diseases; in fact, it actually is of some value (Brownlow).

The "eye" of plants like GERMANDER SPEEDWELL must be responsible for "the leaves stamped with hony and strained, and a drop at sundry times put into the eies, taketh away … dimnesse of sight". Compare this with a Kentish village remedy for cataract; pick the flowers with as little green as possible, boil in rain water that falls in the month of May, pour on to the flowers, stir well, bottle, strain, and drink a cold glassful night and morning. And wash the eyes with it three times a day (Maple. 1962). And the flowers were used in Norfolk to make an eyebath for sore eyes (the plant actually bears the name Sore Eyes there) (V G Hatfield. 1994). There is, too, a Somerset remedy for tired eyes, which was to make a decoction of germander speedwell with eyebright, and then dab the eyes with it (Tongue. 1965).

DODDER

(*Cuscuta epithymum*) Surprisingly, there is virtually no folklore attached to dodder, in spite of the wealth of local names. The word itself is the plural of dodd, which means a bunch of threads, perfectly descriptive of the plant. It is parasitic, of course, and ascribed to the devil. Hence the local name, Devil's Threads, or Devil's Net, very apt. Even more expressive is Devil's Guts, recorded widely throughout England and Scotland (Grigson. 1955). A story from parts of France tells that the devil spun the dodder at night to destroy the clover; clover was created by God, and dodder was the devil's counter-plant.

Scald, or Scaldweed, were names given to it in Cambridgeshire (Grigson. 1955). In this context, scald means scab, or scabies, but it is not clear whether it was thought dodder caused the itch, or whether it relieved it. Perhaps, in the principles of homeopathic magic, both could apply.

DOG ROSE

(*Rosa canina*) It is so-called not, as with many plant names with 'dog' or 'dog's, because it was regarded as inferior in someway to other roses, but because it was supposed to cure the bite of a mad dog (Rowland). As such it is a straight translation from both the Greek and Latin. Aubrey (Aubrey. 1696) quoted Pliny on the belief. He mentions that a woman had a dream that told her to send her son, a soldier, a decoction of the root of a wild rose, "which they call Cynorrhodon". The decoction healed a mad dog's bite. The belief is preserved to this day in the specific name, *canina*. Some, though, say that 'dog' may actually be 'dag', a reference to the dagger-shaped thorns (Freethy). It seems most unlikely, for this theory contradicts the historical record back to the ancient Greeks.

Mid-June was the time for the annual sheep-shearing, after the arrival of warm weather was assured by the blooming of the Dog Rose:

> Must not shear the sheep of its wool
> Before the Dog-rose is at the full (Jones-Baker. 1977).

Another piece of farming lore connected with the Dog Rose comes from Huntingdonshire. They used to predict the date of the harvest as nine weeks from the opening of the first wild rose (Tebbutt).

"… children with great delight eat the berries thereof when they be ripe, make chaines and other prettie gewgawes of the fruit; cokes and gentlewomen make Tarts and such like dishes for pleasure thereof … ; the making whereof I commit to the cunning cookes, and teeth to eat them in the rich man's mouth" (Gerard). The hips contain large amounts of Vitamin C, and they were systematically gathered during World War 2 so that the vitamin content could be exploited. The hips have always been used as a medicine in one way or another. The conserve "is of some efficacy against coughs" (Hill), or a tea made from them was taken for fatigue, and dropsy among other complaints (Fluck), including the common cold (Thomson. 1978). "To prevent a wound going bad: hips of Dog Rose chewed, then let it drop on the wound" (Cockayne). The leaves, too, were used to put on a cut in Essex (V G Hatfield), and a charm from Ireland to cure a stye required the stye to be touched nine times with a rose thorn (Buckley).

The galls made by the gall wasp on the dog-rose enjoyed a great reputation at one time. These "Briar balls", also known by more picturesque names like Robin Redbreast's Cushions in Sussex (Latham), used to be sold by apothecaries to be powdered and taken to cure the stone, as a diuretic, and also for colic. Boiled up with black sugar (the sugar used for curing ham), the result would be drunk for whooping cough (Page. 1978). That is a gypsy remedy, but country people generally used to hang them round their necks as an amulet against whooping cough (Grigson. 1955), or just hanging them in the house (Rolleston), not only for whooping cough, but for rheumatism, too (Bloom), or for piles (Savage). Putting one under the pillow was a Norfolk way of curing cramp (Taylor). In Hereford and Worcester the gall was carried round in the pocket to prevent toothache (Leather), while Yorkshire schoolboys wore them as a charm against flogging (Gutch); that is why they were known as Savelick there. Gerard mentioned the galls, his reference being Pliny, and reckoned that, stamped with honey and ashes, it "causeth haires to grow which are fallen through the disease called Alopecia, or the Foxes Evill".

Stray flowers appearing in autumn were taken as a sign of the plague (Addison. 1985). Another superstition connected with the Dog Rose, from Normandy, was that its smell causes consumption in anyone who does the smelling, or else it sends him mad (Sebillot). It also had the reputation of being able to repel vampires (Valiente). Another odd belief was that scratches inflicted by the prickles were particularly harmful and difficult to heal, causing, as it were, little cancers (Grindon).

Dog Rose seeds constitute the original itching powder with which children amused themselves by putting down each other's necks. They are called Itchy-backs in Leicestershire (Opie. 1959), and in Cheshire Cow-itches (Holland). In Scotland and Ireland, they are Buckie-lice (Grigson. 1955), 'buckie' being the name by which the hips themselves are known.

DOGWOOD

(*Cornus sanguinea*) A Dogwood that was believed to have sprung from the shaft of the javelin that Romulus had thrown from the slopes of the Aventine Hill, stood on the Palatine Hill in Rome, until it was accidently destroyed in the time of Caius Caesar. From what Plutarch says, it appears that this tree was one of the talismans of the city, whose safety was reckoned to be bound up in the tree (Brand). There are signs elsewhere that dogwood is a protective tree. In the Balkans, for instance, women wear it as an amulet against witchcraft (Vukanovic), and in some Serbian villages it was a stick of dogwood that was put in the cradle first, to protect a newborn baby. In America, too, there was recognition of its protective nature, for boats built in Nova Scotia always had dogwood thole-pins (Baker. 1979).

But in England it is something of an unlucky tree, best avoided, especially when it is in flower (Baker. 1977). Quite why is not clear. After all:

When the dogwood flowers appear
Frost will not again be here.

And that in itself shows sympathetic observation. Divining rods were made of it in the Basque country (W. Webster), replacing hazel, which was the favourite dowsing material in Britain. An East Prussian superstition, too, suggests a favourable view of dogwood, for it was said that the sap, absorbed in a handkerchief, would have in some unspecified way the power of fulfilling every wish on Midsummer Day (Dyer. 1889).

Dogwood yields a red dye, and oil from the berries, perfectly edible, was used for lamp-burning at one time. In France, it was used in making soap (C P Johnson). The wood itself, like that of Spindle-tree, makes, in Evelyn's words, "the best skewers for butchers, because it does not taint the flesh", a usage that accounts for a lot of the local names. Skewer-wood is the most obvious (Britten & Holland, Havergal). Skiver is the same word as skewer, and the name appears simply as that in Wiltshire dialect (Dartnell & Goddard). Another name is Pegwood (Miller), indication enough that pegs were another product made from the wood.

DOUBLE NUTS

Double HAZEL nuts have a folklore of their own. Nobody in the eastern counties of England would even think of eating the whole of one of these. One kernel would be passed to a friend, and the two would eat them in silence, wishing a wish that had to be kept secret (Gutch). The practice in Northamptonshire was to eat one and throw the other over the shoulder. In any case, the double nut is a sign of great good fortune to come (Sternberg). They were called St John's nuts in Scotland, and used to throw at witches (Grigson). Their Devonshire

name is loady nuts, and there they were used to cure toothache (Grigson); so they were in Sussex – all you had to do was carry one in your pocket (Latham). In Ireland, that was enough to ensure that you did not get rheumatism or lumbago. In the latter case the nut is acting as a protection against the fairies, for these are elf-shot diseases. A note in the *Gypsy Lore Society Journal; 1957* mentions a mother who gave her son, when he joined the army, a "double hazel nut with three wishes" to ensure his safe return. A triple nut was known as St Mary's nut in Lancashire (R B Peacock).

DOVE'S FOOT CRANESBILL

(*Geranium molle*) Gerard advised the use of the powdered herb and roots to treat hernia, "as my selfe have often proved, whereby I have gotten crownes and credit". He also mentioned its use for "greene and bleeding wounds". Interestingly, the Maori used this same plant for wounds, too, by making a lotion from it, by steeping it in hot water, and using that as an application to open wounds, or it was rubbed on as an embrocation "in case of contusion" (Goldie).

Dracunculus vulgaris > DRAGON ARUM

DRAGON ARUM

(*Dracunculus vulgaris*) Described by Gerard as " …with spots of divers colours, like those of the adder or snake …". So this is where the dragon and snake attributions come from. It comes as no surprise, with that kind of signature, to find it being used long ago against snakebite. The Anglo Saxon version of Apuleius, in the translation by Cockayne, has "for wound of all snakes, take root of this wort, …, with wine and warm it; give to drink". It was still being recommended in the 14th cebtury (Dawson). The fable continued for another two centuries or so, for Gerard repeated the claims: "… it is reported that they who have rubbed the leaves or root upon their hands, are not bitten of the viper", for it is easy to combine prevention and cure, and what will cure snakebite will also prevent it. In fact, as he goes on, "Pliny saith that serpents wil not come neere unto him that beareth dragons about him".

Apuleius Herbarium had other uses for Dragon Arum, one of them requiring you to make a poultice of the roots, with lard, and use it for broken bones – "it draweth from ther body the broken bones"! (Cockayne). Another leechdom recommended a similar poultice for chapped hands. Later on, the roots were used for chilblains – "take the water that dragance rootsd have been seethed in, and bathe well thy kibe [chilblain] therein, and it will heal it (Dawson). Even the plague itself would be held at bay with this plant; "the distilled water hath vertue against the pestilence … being drunke bloud warme with the best treacle or mithridate" (Gerard).

DRAGONS

It was popularly believed in the Middle Ages that dragons, like snakes, could cure their blindness, "apparently a common affliction of dragons" (Hogarth), by rubbing their eyes with FENNEL, or eating it. It was Topsell (1607) who first wrote of the natural history of dragons, and it was his opinion that "their sight many times grows weak and feeble, and they renew and recover it by rubbing their eyes against fennel or else by eating it".

DREAMS

Dream books tell that dreaming of ALMONDS signifies a journey, its success or otherwise depending on whether it was sweet or bitter almonds that were being eaten (Dyer). GARLIC in dreams indicates either the discovery of hidden treasure, or the approach of some domestic quarrel, the one apparently dependent on the other. But to dream of garlic in the home is lucky (Gordon). Similarly with ONIONS; dreaming that you are peeling them foretells domestic strife and impending sickness (Raphael), and if you are eating them it is a sign of finding some valuable treasure, just as dreaming of being in a TURNIP field is a sign of riches to come (Raphael), though the dream books were not unanimous about that. So too with POTATOES – if you dream of digging them, it is a good sign, provided, of course, plenty of them are dug; if there are only few, then there is bad luck coming (Raphael). CABBAGES, too, seem to bring bad luck. To dream of cutting them is a sign that your wife, or husband, or lover as the case may be, is very jealous. If you are actually eating cabbage, then it is a sign of sickness for said wife, husband or lover (Raphael). The American interpretation of dreaming of them is that you will experience a sorrow if the dream is of eating them, or if it is a growing plant, good fortune is coming to you (H M Hyatt). NETTLES have a similar import – they indicate prosperity and good health, but dreaming of being stung by them foretells vexation and disappointment (Raphael). On the other hand, dreaming of gathering them means that someone has formed a favourable opinion of you. If the dreamer is married, then family life will be harmonious (Gordon.1985). Dreams of CELERY are said to be signs of robust good health (Raphael). MADONNA LILIES have similar import – dreaming of them means joy (Mackay), happiness and prosperity (Raphael), just as dreams of RED or WHITE, CLOVER do (Gordon. 1985, Raphael), and OATS (Hewett). BOX is equally lucky, "auguring well for love affairs" (Dyer. 1889), and foretells long life and prosperity, with a happy marriage and large family (Raphael). So, too, with dreams of the grape vine – health, prosperity and fertility (Gordon. 1985). JASMINE, too, is a good omen, especially to lovers (Gordon. 1985). HOPS have the same import; dreaming of a large garden of hops in full leaf is a

sign of wealth. Dried hops, especially when the smell is noted in a dream, shows that the dreamer will soon receive a legacy (Raphael). Dreaming of BANANAS (Dorson. 1964), or RASPBERRIES, is a good sign, for it meant, the latter particularly, success in all things, happiness in marriage, etc., (Gordon. 1985), and the same applies to dreams of APPLES, POMEGRANATES and QUINCES, while VIOLETS mean "advancement in life" (Dyer), but dreaming of their close relative, PANSIES, means heart's pain, quoted by Mackay as one of the popular fallacies of his day, the opposite of Heartsease, presumably. COWSLIPS in bloom signify a sudden change in fortune, but the dream book is silent as to whether the change is good or bad (Raphael). For a man to dream of a BAY tree, it is a sign that he will marry a rich and beautiful wife, but have no success in his business undertakings. It is a good thing for physicians and poets to dream of it (Raphael). After all, wasn't a bay chaplet the proper accolade for a poet? Dreaming of YARROW, too, means good luck in the future (Ireland) (Wood-Martin), and gathering OLIVES denotes peace and happiness. Eating olives means you will rise above your station (Raphael).

To dream of ROSES means happy love, not unmixed with sorrow from other sources (Mackay), but to dream of withered roses means misfortune (Gordon. 1985) – what else could it mean? In spite of its all round virtues, dreaming of HAZEL nuts does not apparently presage any good results. It is a sign of trouble from friends; to the tradesman it is a sign of prison, or at least loss of trade. Dreams of WALNUTS too meant misfortune or unfaithfulness (Dyer), and so does DANDELION (Dyer. 1889) (perhaps the misfortune has something to do with the diuretic effects embraced by names such as Pissabed). Dreaming of RAMPION BELLFOWER also means trouble – it is a sign of an impending quarrel (Folkard), in line with a superstition, in Italy, that the plant among children breeds a quarrelsome disposition that could even lead to murder. Surprisingly, to dream of PRIMROSES means sickness, deceit, sorrow and grief (Raphael), and dreaming of DAFFODILS could have dire results: "any maiden who dreams of daffodils is warned not to go into a wood with her lover, or any secluded place, where she might not be heard if she cried out" (Mackay). It is said that dreaming of ASPARAGUS, gathered and tied in bundles, is an omen of tears. On the other hand, dreaming of it actually growing is a sign of good fortune (Mackay). ELDER berries, though, denote content and riches (Raphael), success in business for the tradesman, and good crops for the farmer. The symbolism here is fairly obvious, for any fruit denotes fulfilment, as with PEARS, provided they are ripe; if unripe, then the future will bring adversity. To a woman, it is a sign that she will marry above her rank (Gordon. 1985). To dream of eating CHESTNUTS was a good sign for an unmarried girl.

It meant that someone would soon be coming to court her. For a married woman, though, it was a sign of sickness (Raphael). SYCAMORE would foretell jealousy to the married, but it promises marriage to the single (Gordon. 1985). A dream of WALLFLOWERS would in some way foretell whether a lover is faithful or not, but to an invalid it is a sign that he or she will soon recover (Gordon. 1985). Dreams of PLUMS have mixed meanings; if they are ripe, it is a good sign. But if the dream involves gathering green fruit, it means a lot of sickness in the family (which result is only reasonable, one must assume). And if one is gathering fallen plums, already rotten, then it is a sign of false friends, unfaithful lovers, and a change in position to poverty and disgrace (Raphael). Dreaming of greengages is a sign of trouble and grief (Raphael), probably a mistake in identification – green, i.e., unripe, plums being a likelier candidate. MEDITER-RANEAN ALOES are a good thing to dream about. If they are not in bloom, it betokens long life; in flower, a legacy (Mackay).

Dreaming of large OAK trees, with good foliage, is always a good sign; but a blasted oak means sudden death. Acorns are also good things to dream about – it shows that health, strength and worldly wealth will be with the dreamer (Raphael). YEW is an unlucky tree to dream about. It foretells the death of an aged person, though he would leave considerable wealth behind him. If you dream you are sitting under one, it is a sign you will not live long, but if you just see and admire the tree, it means a long life (Raphael). At least in Devonshire, dreaming of IVY is a good thing, for it was reckoned a sign that "your friendships are true" (Hewett). Not so in Cornwall, though, for there they say that if you want to dream of the devil, you should pin four ivy leaves to the corners of your pillow (Courtney). PEACHES in one's dreams is a sign of contentment, health and pleasure (Gordon. 1985), and in China, too, such a dream is a very favourable omen, and if you dream of eating beetroot, it is a sign that your troubles will disappear, and that prosperity will follow (Raphael).

To dream of passing through places covered with BRAMBLES means trouble; if they prick you, secret enemies will do you an injury with your friends; if they draw blood, expect heavy losses in trade. Or the latter can mean many difficulties, poverty and privation all your life (Raphael). To dream of passing through brambles unhurt, though, means a triumph over one's enemies (Folkard), or troubles, but only short-lived ones (Raphael). Gathering blackberries is a sign of approaching sickness. If you see others gathering them, you have enemies where you least expect it (Raphael). Dreaming of AURICULAS seems to have had some significance at one time. If they were bedded out, then it was a good luck sign; if they were growing in pots, then it was a promise of marriage.

But if the dreamer was picking the flowers, that was apparently a portent of widowhood (Mackay). It is unlucky to dream of the JUNIPER itself, especially if the dreamer is sickly; but to dream of gathering the berries, if it is winter, is a sign of prosperity to come. To dream of the actual berries means that you will shortly arrive at great honours, and become an important person. To the married, it foretells the birth of a male child (Dyer. 1889). Dreams of NUTMEGS are a sign of impending changes (Gordon. 1985), for better or for worse is not revealed, and to dream of SUNFLOWERS means your pride will be deeply wounded (Mackay). If one has a dream of oneself mourning under a WILLOW over some calamity, it is apparently a happy omen in spite of the context. It forecasts good news! (Gordon. 1985).

Dreaming of GOOSEBERRIES is a sign of many children (Raphael), but if it is a sailor who is dreaming of them, then it is a warning of dangers in his next voyage. To a girl it means an unfaithful husband (Gordon. 1985), and dreams of BROOM means an increase in the family (Mackay). MELONS are lucky. If the dreamer is a young girl, it is a sign of being married to a rich foreigner, and living abroad, presumably because a melon is a foreign fruit. If a sick person dreams of them, then the juicy flesh promises recovery (Gordon. 1985). Dreaming of CUCUMBERS is generally a good sign, too. For a sick person it shows he will recover quickly, and for a single person it is an indication that he or she will make a good marriage (Raphael). It promises success in trade, and to a sailor a pleasant voyage (Gordon. 1985).

A dream of DAISIES is lucky in spring, but bad in autumn and winter (Folkard). Anyway, putting daisy roots under the pillow will bring dreams of the beloved and of absent ones. So will AGRIMONY (see DIVINATIONS). But actually dreaming of the plant foretells sickness in the house (Mackay). MARI-GOLDS were lucky. Dreaming of them means property, and a happy and wealthy marriage (Raphael), and people seem to have had such confidence in marigold-inspired dreams that they tried to induce them, by using the petals in an ointment used on St Luke's Day for the express purpose of bringing prophetic dreams (Wiltshire). GROUND NUTS, though, were unlucky; it means that you will be poor (H M Hyatt). RHU-BARB is lucky; to dream of handling fresh rhubarb is a sign of being taken into favour with those with whom you were not on good terms (Raphael).

BETONY, though, had the reputation of stopping dreams. The Anglo-Saxon Herbal mentions it as a shield against "frightful goblins that go by night and terrible sights and dreams" (Bonser). "For phantasma and delusions: Make a garland of betony and hang it about thy neck when thou goest to bed, that thou mayest have the savour thereof all night, and it will help thee" (Dawson). The first item on Apuleius's list is "for monstrous nocturnal visitors and frightful sights and dreams" (Cockayne). A Welsh charm to prevent dreaming was to "take the leaves of betony, and hang them about your neck or else the juice on going to bed" (Bonser).

DROPSY
see OEDEMA

DROPWORT
(*Filipendula vulgaris*) In the Isle of Man, dropwort was an ingredient of rennet for cheese-making, and for binjeen (Junket) (Killip), while in Sweden the bitter tubers have been dried, ground, and baked into bread (G M Taylor). According to Coles, the name Dropwort was given because of its use in cases of strangury, and Culpeper says it is so called "because it helps such as piss by drops". The real etymology concerns the root, with its small drop-like tubers, so the medical use came from the doctrine of signatures (Grigson. 1959). But Hill was still saying in the 18th century that "the root is good in fits of the gravel".

Dryopteris filix-mas > MALE FERN

Duboisia hopwoodii > PITURI

DUCKWEED
(*Lemna minor*) The weed that covers the surface of stagnant water is Jenny Greenteeth, the nursery bogey who is supposed to drag children down into the pool. Though the role of the weed was well understood, children were told to beware of Jenny Greenteeth (in fact the weed was actually called Jenny Greenteeth in some parts). Jenny, they said, enticed little children into the ponds by making them look like grass, and so safe to walk on. As soon as the child stepped on to the green, it parted, and so of course the child fell through into Jenny's clutches and was drowned. The green weed then closed over, hiding all traces of the child's ever being there. Lancashire children were told that if they did not clean their teeth, they would one day be dragged into one of the pools by Jenny Greenteeth (Vickery. 1983; Vickery. 1995).

A Lincolnshire weather pointer is that duckweed rises in a pond when the weather is going to be fine (Rudkin).

DUTCH RUSH
(*Equisetum hyemale*) "Dutch", because it was imported into Britain in bundles from Holland as a domestic polisher (Grigson. 1974). The American Indians used the native variety in a number of ways – the stalk decoction as a hair wash to get rid of fleas, etc., the heads to cure diarrhoea, etc., (Weiner). The Menomini used the type plant for kidney trouble, by drinking the water in which it had been boiled, while the women used the same medicine to clean up the system after childbirth (Youngken). They were used like other members of the genus (see HORSETAIL) for scouring, cleaning and polishing, hence also the name Gunbright. It is said that the Indians used the plant in polishing their guns (Bergen. 1899).

DUTCHMAN'S BREECHES
(*Dicentra cucullaria*) A North American species, whose common name is descriptive of the shape of the flowers (cf Chinaman's Breeches for Bleeding Heart, *D spectabilis*). It was used in a Menomini Indian love-charm; the young man tries to throw it at the girl he wants, and to hit her with it. Another way was to chew the root, breathing out so that the scent would carry to her. He then circled round the girl, and when she caught the scent, she would follow him wherever he went, even against her will (H H Smith. 1923).

DWALE
is an old word meaning torpor, or trance, and a name accepted since Chaucerian times as synonymous with opiate. See *Reeve's Tale* – when the Miller and his wife went to bed, they had drunk so much ale, "hem needed no dwale". The word was originally used as a general term for sleeping draught, and only later was it restricted to a name for DEADLY NIGHTSHADE, which of course was used for sleeping draughts. So were the other two plants, FOETID HELLEBORE (Voigts & Hudson), and BLACK NIGHTSHADE, bearing the name.

DWARF ELDER
(*Sambucus ebulus*) Tradition had it that dwarf elder grew only where blood had been shed. A Welsh name translates "plant of the blood of man", and there are also relevant names in English, like Bloodwort and Deathwort. It is associated in England with the Danes – wherever their blood was shed in battle, this plant afterwards sprang up. Camden wrote in 1586: "And in those parts in this country which are opposite to Cambridgeshire, lyes Barklow, famous for four great barrows … And the Wallwort or Dwarf-elder that grows hereabouts in great plenty, and bears red berries, they call by no other name but Dane's-blood, denoting the multitude of Danes that were slain". Wallwort, incidentally, has nothing to do with walls. The OE name was wealhwyrt, and it means a foreign plant (the root appears again in walnut, and for various plants described as Welsh. Some Somerset people used to say that dwarf elder gets its noxious properties (the berries are toxic) from growing on the graves of the Danes (Lawrence). The fact that the stems turn red in September presumably gave rise to these traditions (Grigson).

Dwarf elder has been used for medicinal purposes for a very long time. Apuleius knew about it, and in the Anglo-Saxon version of his herbal, it is recommended for three ailments, "in case that stones wax in the bladder, for rent of snake, and for water sickness" (Cockayne), dropsy, that is. In the same collection is mentioned a way of treating piles, which was to heat a quern stone, to lay on top of it and underneath it dwarf elder, mugwort, and brooklime, and then to apply cold water, having the patient poised so that the steam "reeks upon the man, as hot as he can endure it". Gerard repeated the dropsy treatment ("the roots … boiled in wine and drunken…"), and the cure is still recommended by herbalists (Conway), who also prescribe the root tea for kidney ailments. It seems that the Welsh medical text known as the Physicians of Myddfai is referring to dropsy in the prescription: For pain in the feet and swelling in the legs. Take the roots of dwarf elder, and remove the bark, boiling it well, then pound them in a mortar with old lard, and apply as a plaster to the diseased part. "With Buls tallow, or Goats suet this is a remedie for the gout" – "Dr Bullen's remedie for the goute", according to Aubrey. In the Balkans, boils used to be treated with a leaf decoction of dwarf elder, used both externally and internally (Kemp).

DYER'S GREENWEED

(*Genista tinctoria*) A very widespread undershrub, occurring over most of Europe, into Siberia, and naturalised in North America, where they have a saying that when cows have eaten it, their milk is butter (Sanford). This is a dye plant, as both the specific and common names tell. The young shoots and flowering tops are the parts used for a yellow dye. When mixed with woad, it gives the colour known as Kendal Green, used originally by the Flemish weavers who settled at Kendal (J Smith. 1882). The dried flowers are used in Ukraine to colour Easter eggs yellow (Newall).

It has its medicinal uses, too. A tincture made from it was reckoned a laxative, and the seeds were taken to induce vomiting (Sanford), for they are mildly purgative, and a decoction of the plant has been used for oedema, gout and rheumatism (Grieve. 1931). In Russia, it was even used for rabies (Pratt). The gout remedy goes back a long way, for we find a 15th century leechdom quoting the use: "Take flowers of broom and flowers and leaves of woadwaxen [Dyer's Greenweed], equally much, and stamp them with may-butter, and let it stand so together all night: and on the morrow melt it in a pan over the fire, and skim it well. This medicine is good for all cold evils. And for sleeping hand or foot, and for cold gout" (Dawson. 1934).

DYER'S ROCKET

(*Reseda luteola*) This plant has been cultivated for its dye from Neolithic times; it is very often represented in Swiss Neolithic remains (J G D Clark). It is one of the oldest known dyes, and was mentioned by Caesar as being used by the Gauls (Brill), and it was a secondary crop in Cotswold country. Being often biennial, it was often grown with barley as a "catch-crop". "Some … commend it against the punctures and bites of venomous creatures, not only outwardly

applied to the wound, but also taken inwardly in drinke. Also it is commended against infection of the plague …" (Gerard).

DYER'S WOODRUFF

(*Asperula tinctoria*) Important for the red dye that the roots yield, presumable more so than the rest of the genus, for they all provide the dye in more or less satisfactory style.

DYESTUFFS

HOLM OAK is the tree that nourishes the kerm, a scarlet insect not unlike a holly berry (hence Holm Oak and other names meaning holly, including the tree's specific name, *ilex*, which is also the generic name for the hollies). The ancients made their royal scarlet dye from this insect, and the Bible furnishes the oldest evidence of the existence of a scarlet dye; there it is called tola, or tolaat, a word also meaning worm, which is what 'kerm' means. Matthew 27; 28 is translated as "And they stripped him, and put on him a scarlet robe", is assumed by Graves to refer to kerm-scarlet. He must be right, for surely there was no other source of scarlet in biblical times? The roots of TORMENTIL (C P Johnson), and of DYER'S WOODRUFF yield a red dye (Usher), as does the rootstock of BLOODROOT (Speck) and so, in more or less satisfactory style, do all the *Asperulas*. DANDELION roots (Dimbleby) and GOOSE-GRASS roots (Grieve. 1931) also yield a red dye, and so do PUCCOON roots, or perhaps paint would be a better word. The colour could be obtained by simply dipping the fresh root in grease, and rubbing on the object to be painted, which may often be the human face (Teit). But the prime blue dye, at least until 1897, when the German firm of Baeyer produced synthetic aniline dyes, was WOAD, and then INDIGO (see Hurry, Ponting, and Leggett for technicalities). But in Europe, resistance to the introduction of Indigo, has to be seen as a protection for woad. In fact, it was proscribed in Elizabeth I's reign as a dangerous drug, described as "the food for the devil" (Hurry), indeed Devil's Dye is recorded as a name for the plant (C J S Thomson. 1947), and reckoned downright injurious to fabrics, a myth that was sincerely believed in a number of countries (Leggett). In 1609 Henry IV of France issued an edict sentencing to death any person discovered using the deceitful and injurious dye called inde. In fact, it was not until 1737 that French dyers were legally allowed to use it (Hurry). CHICORY leaves will yield a blue dye. too (Hemphill).

TANSY gives a brilliant yellow dye, but with the disadvantage that it is very difficult to fix (Wiltshire), and FENNEL leaves also will dye wool a golden-yellow colour, with chrome mordant (Cullum). GORSE bark will also give a yellow dye (Jenkins. 1966) (green, according to Murdoch McNeill), and the Navajo used BROOMWEED (*Gutierrezia micro-cephala*) for yellow, or a DAHLIA (*D pinnata*) roots for orange-yellow (Elmore). Other Indian groups used the pulp of the stalks, and the roots, of SMOOTH SUMACH to produce yellow (Buhler; Gilmore). Another sumach, VENETIAN SUMACH, will produce a yellow colour from its wood, though it is hardly ever used nowadays, for it is really not permanent at all (Leggett). However, with the proper mordant, it can dye cotton and wool bright yellow through to brown or dark olive. With logwood, it can produce black. POISON IVY roots too will give yellow, and so will ORANGE BALSAM, if the whole plant is used (both H H Smith. 1923). The flowers and leaves of GOLDEN ROD will also give a fine yellow (Barton & Castle). COCKLEBUR leaves are another source of yellow, and are used in China for the purpose (F P Smith). But TURMERIC will be recognised as the best known yellow dye, even if it is certainly not fast to light, and needs a mordant. Curry powder is its best known manifestation. But the Pacific islands and New Guinea are probably the only areas in which turmeric dye is still used in its simplest form, and the main use is in body painting, often with sexual overtones. In Samoa, where yellow is the colour sacred to the gods, the gathering and processing takes the form of a religious rite. SAFFLOWER is one of the oldest dyestuffs (the fruits have been found in Egyptian tombs that are 3500 years old, and since then it has been grown, for centuries past, in India, the Middle East and East Africa, and so for the dye obtained from them (H G Baker). There are in fact two dyes obtainable from them - safflower yellow, similar to true saffron, and often used as a substitute for it (Leggett), and carthamin, a red colour. It has long lost its predominance in Asia, but is still grown in India for local use, such as the dyeing of ceremonial clothes, or in the making of a brush paint for cosmetic rouge, etc., (Buhler). The pigment known as Sap Green is made from BUCKTHORN berries boiled with alum, and, when mixed with lime-water and gum arabic, Bladder Green (Grigson. 1955). As a dye, glovers would use the berries to give a yellow colour to their leather (Aubrey. 1867). The traditional Welsh dyeing practice was to crush the unripe berries in water, and then to evaporate the infusion to a consistency of honey. After boiling the wool in this a green colour was obtained (J G Jenkins. 1966). DYER'S ROCKET, or WELD is one of the oldest dyes known, and has been cultivated for the purpose since Neolithic times (J G D Clark). Caesar mentions it as being used by the Gauls. It produces a golden-yellow dye, or, with alum, a dark yellow. The tradition in the Welsh woollen industry was to cut second year plants in autumn. Boiled for three-quarters of an hour on its own, it would produce pale yellow. If mordanted wool was used, many shades of yellow and light brown could be obtained, and the dye would be fast (J G Jenkins. 1966). Lincoln, or Saxon Green was obtained by dyeing blue with woad, and then yellow from Dyer's Rocket (H G Baker). WHITE HICKORY

will give a yellow dye, if set with alum (R B Browne), and the twigs and leaves give a tan colour, again, if set with alum (S M Robertson). The wood of OSAGE ORANGE (*Maclura pomifera*) will give yellow-orange (Schery). DYER'S GREENWEED young shoots and flowering tops will give a yellow dye. When mixed with woad, it gives the colour known as Kendal Green, used originally by the Flemish weavers who settled at Kendal, in Cumbria, hence the name (J Smith. 1882). AVENS roots, too, will dye wool a permanent dark yellow (Barton & Castle).

BROOM will give wool a moss-green colour (Jenkins. 1966). HOP leaves and flowers will give a brownish, or yellowish brown, dye (C P Johnson), and a light brown colour for wool was sometimes obtained from BIRD CHERRY bark (Fairweather). SLOES will give a slate-blue dye with no mordant. A pound of the berries to a pound of wool must be brought to the boil, and then boiled for a further hour with the wool. The colour comes up when the wool is washed with soap (Coates). This is confusing, for Geraint Jenkins said that if the wool was washed immediately after dyeing in soapy water, it would turn into a slaty green. When alum is used as a mordant, the colour is rose-pink, according to S M Robertson; but the impression Jenkins gives is that rose-pink will result directly from the berries. It should be noted that sloe juice is indelible.

On the Hebridean island of South Uist, WILD SORREL was used on its own to dye wool red; mixed with indigo it could dye blue (Shaw). The roots were used in the Welsh industry too – two parts of sorrel to one part of wool, the whole lot boiled up for three or four hours, to produce a brown colour (Jenkins). ONION skins will dye wool brown (Dimbleby), or yellow, with alum. With a tin mordant the colour will be orange (Coates). [See also EASTER EGGS]. Another unlikely source of a blue dye is BUCKWHEAT straw (Schery), and of course the floweres of MARSH GENTIAN will also give blue (Usher). Even HOLLYHOCK has been used for blue dyes (Northcote, Usher).

An American homemade dye was made with PERSIMMON bark, mixed with that of Red Oak, to give a yellow colour (R B Browne). In the Hebrides, BURNET ROSE was used, with copperas, to make a brown dye (Murdoch McNeill), and elsewhere, using alum as the mordant, the hips were used to dye silk violet (C P Johnson). One of the uses of MARJORAM was as a dyestuff; made from the flowering tops it gives a dark reddish-brown colour, but it fades quickly (Rohde. 1936). PRIVET berries will give a bluish-green dye with alum, and the leaves will give a yellow colour (Coates) with alum, or with chrome a light brown, and a dark brown with copperas (Jenkins), green with copper sulphate, and dark green with iron (S M Robertson).

HENNA can be used as a fabric dyestuff, as well as a protective cosmetic on the person. It is used to dye silk and wool in Morocco (Gallotti), and in India, too, both fabrics and leather are coloured reddish-yellow by its means (*CIBA Review. 63; 1948*). CUTCH (*Acacia catechu*) is an important dye plant; by boiling chips of the wood, a permanent brown dye can be obtained (Barker), and the leaves and shoots can be used as well. The dyestuff is known as Cutch, or sometimes Gambir (Le Strange). Much of the catechu exported comes from the Gulf of Cutch, where it has been used as a brown dye for over two thousand years, hence the name. True khaki cloth is made with cutch (Willis). Navajo Indians used the juice of SOAPWEED (*Yucca glauca*) for a black dye (Kluckhohn), and a black dye, "Sabbath black", it was called, was made by boiling the roots of YELLOW FLAG in water, using iron as a mordant (Macleod). The flowers will give a yellow dye, and the root was used for wool dyeing on the Hebridean island of South Uist. Using alum as the mordant, a shade of blue to steel-grey was obtained (Shaw).

DYSENTERY
An early remedy shows a great mixture of magic and medicine, for it required the herbalist to dig up a BRAMBLE of which both ends were in the earth, and then to take the newer root, cut nine chips on the left hand and sing three times, Miserere mei Deus, and nine times the Our Father. Having completed that part of the preparation, "take then MUGWORT and everlasting and boil these three in several kinds of milk until they become red. Let him then sup a good bowl of it fasting at night, some time before he takes other food. Make him rest in a soft bed, and wrap him up warm …" (Storms). Astringent as they are, bramble leaves and roots, have always been used for diarrhoea and dysentery, and that includes the AMERICAN BLACKBERRY (*Rubus villosus*). A syrup made from the root is an American country medicine for the complaint, and the juice from the fruit, spiced and laced with whisky, is a well-valued carminative drink in Kentucky, the original of the well-known "Blackberry Cordial" (Lloyd). Gerard recommended MYRTLE ("the leaves, fruit, buds and juyce") for dysentery.

A root infusion of the African tree CATCHTHORN (*Zizyphus abyssinica*) is taken for dysentery. There is a lot of tannin in the bark, so that is probably the reason for this treatment (Palgrave & Palgrave). Maoris set great store in HEBE, particularly *Hebe salicifolia*, for curing diarrhoea and dysentery, so much so that the young leaf tips, the astringent part used, were collected and sent out to Maori troops in the Middle East during World War II (C Macdonald). GREAT BURNET root is used in Chinese medicine for the complaint (Geng Junying), as well as for haemorrhages and other conditions.

DYSPEPSIA
CHOKE CHERRY bark has got a reputation in America for dyspepsia; it was usually administered as a cold infusion with syrup (Lloyd), but the important plant at one time was SWEET FLAG. The carminative usage appeared in medieval Latin compilations, and was to be found in New World medicine also. In Alabama, they either chewed the root, or put it in whisky to be used when needed. Or it could be boiled, and the water drunk (R B Browne). The American Indians used it too, for similar purposes. A tisane of WHITE HOREHOUND has often been taken for a weak stomach, lack of appetite, and the like (Flück). Indigestion and dyspepsia were cured with Horehound tea, or, in homeopathy, by a tincture (Schauenberg & Paris). Even Navajo Indians were reported to use this herb for indigestion (Wyman & Harris), and it is certainly an American domestic medicine for dyspepsia still (Henkel). CENTAURY is another plant still used as a popular medicine for dyspepsia, in the form of a leaf infusion (Clair). As long ago as Anglo-Saxon times there were leechdoms for stomach trouble using this plant (Cockayne).

E

EAGLE VINE, or CONDOR VINE

(*Marsdenia condurango*) Well known in the latter half of the 19th century as a "cure" for cancer. According to legend, this poisonous plant, or rather, the bark decoction, was administered by a South American Indian woman to her husband, hoping to kill him, as he was in great agony from a painful growth. Instead of proving fatal the vine rapidly cured him of his cancer. Unfortunately, controlled tests have shown that it is not so valuable as expected, though still quite useful in the early stages of cancer, and it is still used in South America in the treatment of chronic syphilis (Le Strange).

EARACHE

In Ireland, the sap of an ASH sapling was use to cure earache. A sapling would be cut and put into the fire. One end was kept out so that when the stick started to burn, the sap came out and was caught in a spoon. This could be put on cotton wool, and put in the ear. This is actually a very old remedy; take, for instance, this leechdom from the fifteenth century: "Take young branches of ash when they are green. Lay them on a gridiron on the fire, and gather the water that cometh out at the ends of them, an egg-shell full; and of the juice of the blades of leeks, an egg-shell full; and of the drippings of eels. Mix all these together, and seethe them together a little; and cleanse them through a cloth, and put it in a glass vessel. And when thou hast need, put this in the whole ear of the sick man and let him lie on the sore ear. And with [this] juice [used] twice, he shall be whole …" (Dawson). Something similar appears in the Welsh medical text known as the Physicians of Myddfai. Evelyn had heard of it, but misunderstood the usage, for he claimed that "oyl from the ash … is excellent to recover the hearing, some drops of it being distill'd warm into the ears. In America, PERSIMMON sap was used in the same kind of way, merely by letting the sap from a burning branch drip into a spoon, and letting that drop into the ear (R B Browne). Another tree thought to provide relief for earache is WHITE POPLAR, though in this case it was the leaves, or rather "the warm juice" of them "dropped into the eares" (to) "take away the paine thereof" (Gerard), and some American Indians, the Winnebago for example, steeped YARROW, the whole plant, and poured the resulting liquid into the ear (Weiner). West African peoples drip the juice of GREEN PURSLANE into the ear (Harley), and COLTSFOOT juice was used in Ireland (Ó Súilleabháin).

CARAWAY was used, too – the patient was advised to pound up a hot loaf with a handful of bruised seeds, and clap that to his ear (Fernie). A Sussex remedy was to bake a SHALLOT and put that in the ear (Sargent). ONION was used, as early as Apuleius. In an Anglo-Saxon version the prescription required one to boil an onion in oil, and then drip the oil into the ear (Cockayne). The detail may vary, but onions are still being used for the complaint, either, as in America, by putting the heart of a roasted onion in the ear (Bergen. 1899), or, as in Cornwall, by putting a boiled onion in a stocking and holding that to the ear (Hawke). It was treated in Cheshire by warming a small onion and holding it to the ear, or it could be put in flannel and applying that (Cheshire FWI). TOBACCO could be put in the ears to stop the pain (Newman & Wilson), just as a piece could be put on an aching tooth.

A decoction of dried SUMMER SAVORY is used in Russian folk medicine for earache, the method being 2 tablespoonfuls to a glass of water, boiled for a quarter of an hour, and used warm as an ear wash (Kourennoff). Ear drops were made from MULLEIN flowers. The method was to take fresh flowers, steep them in olive oil, leave them for three weeks in a sunny window, then strain off. Two or three drops in the ear will relieve earache quickly (Genders. 1976). Herbalists still use the juice of HOUSELEEK as ear-drops, and squeezing the juice, sometimes mixed with cream, into the ear to cure earache, has been known for a very long time (Briggs. 1974). A name for houseleek used in the north of England is Cyphel (F Grose), or Syphelt, from the Greek juphella, which means the hollows of the ear. In the Middle Ages, the plant was often called Erewort, too (Grigson. 1955), because it was used for deafness. FEVERFEW was apparently used, warm, on the ear, according to a Suffolk record (Kightly. 1984), and a Wiltshire domestic medicine for earache used CAMOMILE. People made a flannel bag and filled it with camomile heads. This was warmed by the fire and held against the ear (Wiltshire). In any case camomile flowers were made into a poultice for the relief of any pain.

An African way of getting relief was to use the juice of a warmed leaf of NEVER-DIE (*Kalanchoe crenata*) (Soforowa).

EARLY PURPLE ORCHID

(*Orchis mascula*) The commonest of the British native orchids. The tubers (and those of most of the *Orchis* species of Europe and Northern Asia), when prepared, yield salep, which had, at least according to the doctrine of signatures (the word orchid means testicle) a reputation for curing impotence (Pliny: gemina radice, testiculus simili), a reputation flourishing in eastern countries particularly, though it is represented too in north-east Scotland, where the root was used as an aphrodisiac, the old tuber being discarded, and the new one used. It would be dried, ground, and secretly administered as a potion (Anson). John Moncrieff of Tippermalloch (in the 18th century) went further:

"Satyrion root holden in the hand", he said, during coitus, would ensure success (Rorie. 1994). There was an old belief, which Gerard ascribed to Dioscorides, that "if men do eat of the great full or fat roots … they cause them to beget male children; and if women eat of the lesser dry or barren root, which is withered or shrivelled, they shall bring forth females". He disclaimed the belief, though – "these are some Doctours opinions only".

The word 'salep', sometimes spelt salop, is Turkish (most of the tubers for its manufacture were imported from Turkey (Coats. 1975)), from an Arabic original, sahlab (Emboden. 1974), apparently meaning fox orchid. The substance was imported at one time from the East. It was used as an article of diet, reportedly very nutritious (Fluckiger & Hanbury), and part of every ship's provisions to prevent famine at sea (Thornton), for the dried material swells to several times its size when water is added, so it was standard starchy food in the days of sailing ships. It forms a jelly with a large proportion of water. A decoction flavoured with sugar and spice, or wine, was reckoned an agreeable drink for invalids though it was acknowledged that it has no medicinal value. Nevertheless, it was used for the treatment for colitis and diarrhoea (Emboden. 1974). It makes a common soft drink, very popular long before the introduction of coffee houses, and mentioned in Victorian books as a common beverage for manual workers (Mabey.1972).

Hebridean children used to call the plants "dappled cows". When found before the flower head had formed, the children would dig round the plant until the tuber was exposed. Then they would "milk" the dappled cow by compressing the tuber, squeezing the juice out. Apparently they had no use for the juice, and did not taste it. Anyway, no harm would be done to the plant.

Ulster fishermen used to put bunches of purple orchis in their windows, on St John's Day, to keep away evil. It is recorded that the plant bloomed in great profusion on some of the bomb sites of Belfast in 1941 (Foster); this habit may be the origin of the belief that it keeps away evil.

EARTH-NUT

(*Conopodium majus*) Its tuberous rootstock is edible, especially when roasted like a chestnut, which it rather resembles. It can be eaten raw, too, scraped and washed (Mabey. 1972). Hardly anybody bothers these days, not even children. But Cuckoo Potatoes is what they are called in Ireland, or else Fairy Potatoes, for this plant belongs to the fairies in general, and the leprechaun in particular. But why should it be ascribed to the devil in Yorkshire? For that is what Bad Man's Bread means, a name made more explicit as Devil's Bread or Devil's Oatmeal (Grigson. 1955).

EASTER

TANSY puddings at Easter were traditional, and were probably originally a Christian adaptation of the bitter herbs of the Passover. In many districts of England they were actually played for on Easter Monday, for a "tansy" can also be a merry-making (Opie & Opie. 1985), but they were actually made to be eaten with the meat course at dinner. One recipe speaks of finely-shredded leaves, beaten up with eggs and fried (Genders. 1977). Pepys provided a "tansy" for a dinner, but this was a sweet dish flavoured with tansy juice. By the 17th century tansies were a kind of scrambled egg made with cream and the juice of wheat blades, violet and strawberry leaves, spinach and walnut buds, plus grated bread, spices and salt, all sprinkled with sugar before serving. Tansy was no longer an ingredient, walnut buds being preferred (Burton), although most recipes that have survived insist on the proper herb ingredient. Sometimes it was the flowers that were used, in custards as well as other sorts of pudding, or the leaves could be steeped in milk to make cheeses and cakes. In Ludlow, Shropshire, there was a customary Easter dish of leg of pork stuffed with Robin-run-in-the-hedge, which is GROUND IVY (Burne. 1883). Devonshire Revel buns were baked on SYCAMORE leaves. These are Easter cakes, each cake being baked on its own individual leaf (Mabey. 1977), the point being that the imprint of the leaf should appear on the cake.

DAFFODILS, by their time of flowering, are associated with Easter. They are called Easter Lilies in Devon (Friend. 1882) and Somerset (Tongue. 1965), where they are also known as Easter Roses (Elworthy. 1888). Then there is Lent Lily, quite common as a name for them, and having many variations.

EASTER EGGS

ONION peel for dyeing Easter Eggs gives various colours from yellow to deep orange and reddish-brown. An 18th century source described how patterns were made by cutting the onion into shapes and sticking them to the shell with egg-white. Yugoslav Macedonians call these onion peels shuski, and use them for dyeing wool as well as for colouring eggs. The peel is soaked for several days in tepid water before the eggs are boiled in it. Afterwards whatever is left is thrown out in the garden, but never where children might walk – it would give them blisters, they say (Newall). SHALLOTS give a particularly rich, deep colour. In Estonia they used to moisten the shell first, roll it in chopped barley and tie it in a cloth before boiling with onion peel (Newall). DAFFODIL flowers have been used for the purpose (Newall), and so has BEETROOT juice, to dye the eggs purple (Newall).

In France, eggs might be painted with a DAISY (pâquerette) design, and the egg given to every child before attending Easter Mass (Newall). For the flower is the symbol of innocence, and thereby, of the newly-born, so of the infant Christ. African Americans in Louisiana boil the leaves of Locust Bean ((*Catalpa bignonioides*) to wrap round eggs at Easter, to colour them (Fontenot).

ECZEMA

BIRCH leaves have always been used for treating skin complaints (Conway), and they can be treated with birch tar oil made up into a soothing ointment (Mitton) or can be used in medicated soaps (Gordon) to treat eczema. The complaint was treated in Dorset with NETTLE tea (Dacombe), which is a well-established East Anglian remedy for any skin complaint (Porter. 1974). The sap of the GRAPE VINE is collected in some country areas when growth starts in spring, to be used for eczema among other complaints (Shauenberg & Paris). Gypsies use an ointment made from the fresh leaves of FOXGLOVE to cure eczema (Vesey-Fitzgerald), and a compress made from MALLOW leaves or flowers is often used. A CHICKWEED poultice is used in Norfolk for quite severe dermatitis and eczema (V G Hatfield. 1994). A tea made from GROUND IVY used to be popular for this complaint in the north of Scotland. It was said that the fairies taught Donald Fraser, of Ross-shire, to use it (R M Robertson). A dozen or so BLACK WALNUT leaves, boiled in a quart of water, with a teaspoonful of sulphur added, is an Alabama eczema cure, and another from the same area is to use PUCCOON root tea for the complaint (R B Browne). The fresh tops of YARROW were used by some American Indian peoples, made into a poultice (Corlett). In Scotland, CABBAGE leaves were made into a cap to put on a child's head to deal with the complaint (Rorie), and the seeds of the LESSER EVENING PRIMROSE have also been used, with some success, it is claimed (T Walker). The infusion of chopped WHITE HOREHOUND has been used in Wales both externally and internally for eczema and shingles (Conway).

EDELWEISS

(*Leontopodium alpinum*) Edel means noble, and weiss, white. It is a name very likely coined for 19th century tourists in the Alps. It used to be a Swiss custom to put wreaths of edelweiss over porches and windows on Ascension Day. It is a lightning plant, and is gathered as a protective charm against lightning (Dyer. 1889). It also acts as a symbol of immortality (Folkard), and is apparently dedicated to St Christopher.

EDIBLE VALERIAN

(*Valeriana edulis*) A species from the north-west coast of North America, and an evil-smelling plant, so much so that the Klamath Indians used to bury it with food in order to discourage animals from digging up stores. Even grizzly bears avoid it (Spier). On the other hand, some groups in that area cooked the roots

in stone-lined pits in the ground, or made them into soup or bread. Not any more, though. For nowadays they are convinced the plant is poisonous (Yanovsky). Some groups used it medicinally. The Menomini, for instance, used it for for cuts and wounds, and it was taken as a tapeworm remedy, obviously successfully, for it is reported that after the worm was expelled it was washed clean, pulverized, and swallowed again, to make the patient fat and healthy once more! (H H Smith. 1923).

EGLANTINE

see SWEET BRIAR (*Rosa rubiginosa*) Eglantine was originally the French designation, probably ultimately from the Latin aculentus, meaning "full of prickles". Or perhaps it was from egla, or egle, which meant a prickle or thorn, but the name is used for the honeysuckle in some parts of the country, and there are no thorns on that (Putnam).

Elaeocarpus dentatus > HINAU

ELDER

This is the familiar shrub with creamy-white flowers in summer, and black berries later on, widespread and very common, and actually planted, Gerard says: "it is planted about cony-boroughs for the shadow of the Conies", though he admits it "groweth every where". So common is it, that there is a tendency to forget how important it was once in various mythologies. It was the tree under which the old Prussian earth-god lived (Farrer), presumably the same as the Latvian Priskaitis, who also lived under an elder, and who ruled over little subterranean beings called Barstukai, or Kantai. If offerings were made to Priskaitis, the little men brought plenty of corn and did the household work (Gimbutas). This idea that a spirit of some kind inhabits the tree has survived in folklore to the present day. Danish peasants believed until very recently that Hyldemoer, elder-mother, lived in the tree, and would avenge all injuries done to it. So her permission had to be asked before an elder was cut (Kvideland & Sehmsdorf). Some accepted form of words was usually employed, as in the ancient German prayer "Frau Ellhorn, give me some of your wood, then I will give you some of mine when it grows in the forest" (Bonser). That kind of prayer, degenerated into a rhyme, has been recorded in Scotland, too. Even if the tradition is quite different, the rhyme "exonerated the gardener from ill intent" (Simpson). There are further similar records from Lincolnshire, where they used to believe that the Old Lady, or the Old Girl, would be offended by cutting elder wood without asking her leave (Burne. 1914). Sometimes it is said in this country that a death in the family would follow the cutting down of an elder (*Notes and Queries; 1941*). You must not make a cradle out of elder wood, in case Frau Helder takes her revenge on the baby (Thompson), and there is a prejudice in Oxfordshire against using the wood for mending anything. In Wales, they say elder bleeds if cut, a belief that was current in Somerset, too. Ruth Tongue cites cases of farmers complaining of hedge-trimmers who refuse to tie up elder into faggots, because of this belief. That is the reason, too, why a lot of hedges in Oxfordshire had their elders left when the rest was being trimmed.

The tree has been associated with witchcraft in this country from early times, for in the 10th century, the canons of King Edgar spoke of "vain practices that are carried on with elders" (A J Evans). It is often said that witches live in an elder, for the change from tree spirit to witch is easily made. In fact, there is a folk tale from north Somerset, called the Elder Tree Witch, where the witch *is* the elder, and moves around the farm in the form of the tree (Tongue). And there is a similar manifestation in the legend of the Rollright Stones. The biggest of the stones is called the King Stone, once a man who would have been king of England if he could have seen Long Compton. A witch said:

> Seven long strides shalt thou take and
> If Long Compton thou canst see,
> King of England thou shalt be.

On his seventh stride a mound rose up before him, hiding Long Compton from sight. The witch then said:

> As Long Compton thou canst not see,
> King of England thou shalt not be,
> Rise up, stick, and still, stone,
> For King of England thou shalt be none.
> Thou and thy men hoar stones shall be,
> And I myself an eldern tree.

The story is at least as old as Gibson's edition of Camden, *Britannia*, 1695. Though where exactly the witch-elder now stands seems to be a matter of conjecture. Nevertheless, the proof that the "elder is a witch is that it bleeds when it is cut" led to the custom, on Midsummer Eve, for people from the area to meet at the stones, and to cut the elder, or at least the tree they thought to be the real one. As it bled, the King Stone was supposed to move its head (Grinsell, 1977).

Christian tradition has it that it is the tree of which the Cross was made (Dyer). Chambers has:

> Bour-tree, bour-tree, crookit rung,
> Never straight, and never strong,
> Ever bush, and never tree,
> Since our Lord was nailed t'ye.

"We don't cut elder in a copse, nor do we burn it. They say the Cross was made of it" (Goddard. 1942–4). That is the reason for the superstition that the elder is proof against lightning – it never strikes the tree from which the Cross was made (Gurdon). Elder wreaths used to be hung up in parts of Germany after sunset, as charms against lightning (Dyer), or

according to another tradition, as a protection against the ravages of caterpillars (Farrer). Perhaps this last accounts for the Sicilan belief that elder sticks kill snakes (and drive away robbers) (Hemphill). Note, too, that witches could be shot only with elder pith (or bullets made of inherited silver).

Elders are plants of St John in German belief (Thonger), and over a much wider area are invested with similar witch-finding properties as the St John's Wort. In the Isle of Man particularly, elder was fixed to doors and windows to protect the house (Moore). According to Train, it was because it was the tree from which Judas hanged himself that such reliance was placed on its powers. Anyway, in his day (1845), practically every cottage on the island had an elder growing by the side of it. It may be a protection, but at the same time they used to say that fairies (the kind equated with the ancestors) actually lived in elder trees (Wentz). They lived there, not like birds in a tree, but somehow in the hollow stems, "as though they were drawn up through its roots from the earth itself" (Gill. 1929), and they would leave if the tree were cut down, and then come back to weep and lament each night (Killip). A German belief, taken to America, was that an elder stick burned on Christmas Eve would somehow or other reveal all the witches in the neighbourhood; another American belief was that if a small piece of elder pith was cut, dipped in oil, lit, and then floated on water, it would point to any witch present. Similarly, to find out if witchcraft was the cause of losses to livestock, the farmer was advised to take six knots of elder wood and to put them "in orderly arrangement" under a new ash bowl or platter. If they were found sometime later "all squandered about", then that was the sign that the cattle were indeed dying from witchcraft (Brockie). In Northumberland, a piece was kept in the chest to protect household linen from evil influences (Denham). An elder hedge secured many a home in Scotland from "undesirable attention" (McPherson); it also protected it from lightning (Tebbutt).

One tradition says that anyone, provided they had been baptised, would be able to see witches in any part of the world if their eyes were anointed with the green juice of the inner bark (Dyer). The berries and pith were sometimes given in the food of anyone thought to be bewitched (Leather). It appears as an animal protector, too, on the Continent, to be fixed in stalls and byres in the same way as is rowan in this country (Sebillot). Also on the Continent, elder trees were cut on Midsummer Eve to "make them bleed", i.e., to let the sap escape. As the elder was particularly the witches' own, this bleeding ceremony helped deprive them of their powers (Urlin); but that is a practice not just confined to the Continent – it took place at the Rollright Stones too, as already mentioned. Hearse drivers were said to favour elder-wood whips in their dangerous association with the dead (Fernie), and for the same reason, the Hildesheim gravediggers used it when measuring the corpse (Bonser). Once, it was buried with a corpse, especially that of a child, to protect it from the attention of witches (Fernie), or to "keep off the fairies" in the hazardous time before the Day of Judgment (Gill. 1929) (why should fairies be afraid of elder leaves when they lived in the tree anyway?). If a twig of elder were planted on a grave, and it grew, it was taken as a sign that the soul of the deceased was happy, "which is the probable reason why the very old Jewish cemetery in Prague was planted full of elders" (Leland).

Elder may have had these protective and anti-witch virtues, but it was still an unlucky tree, with an evil character. The very fact that witches were fond of lurking under it made it dangerous to tamper with after dark (Dyer). And do not sleep under one – the leaves were said to give out a toxic scent which, if inhaled, may send the sleeper into a coma and death (M Baker. 1977). Mending cradles with elder-wood was just as dangerous, as mentioned above, for a Cheshire belief was that it would give witches power to rock from far off so violently that the babies would be injured (Hole. 1937). Again, a child laid in an elder-wood cradle will pine away, or be pinched black and blue by the fairies (Graves); or the fairies may steal it (Grigson); or the Elder-Mother would strangle it (Farrer). In Ireland, elder wood was never used in boat building (Ô Súilleabháin); nor, so it was said in South Wales, should a building of any kind be put up on the spot where an elder had stood (Trevelyan). It is credited with having a harmful influence on any plants growing near it (Rohde. 1936). The flowers were never allowed in the rooms of Fenland houses, because they were supposed to attracted snakes (Porter), and there is a record in the same area of a belief that a wound suffered by contact with the tree, say by driving a sharp stick accidentally into the hand, would inevitably prove fatal.

In the Midlands, and in Sussex, they would never bind elder wood with other faggots (Farrer, Tongue), and it was quite a common belief that beating boys with an elder stick would stunt their growth, though the child might grow up very wise (O Cleirigh). Burning elder, particularly green elder (Forby), was almost universally forbidden in England, for it is a "wicked wood", or "it brings the devil into the house" (Graves, Heanley), or, as they say in Lincolnshire, "the devil is in elder wood" (Rudkin). "People say that if you want to keep the devil out of your house, you must never burn elder wood" – that is a Wiltshire belief (Goddard). It would "draw the devil down the chimney" (Boase), or, as in Warwickshire, you would see the devil sitting on the chimney pot (Witcutt). In Needwood Forest, and in Hertfordshire, the expression used is that it would raise the devil (Burne, Jones-Baker). Sometimes the

result would be that the person burning it would become bewitched (Vickery. 1983), though the Yorkshire wise man would burn it in his efforts to defeat witchcraft (Atkinson. 1886). Derbyshire people would never burn it because of its association, already mentioned, with the Cross (Addy). In lots of other places, it was just thought too unlucky to burn. But around Cricklade, in Wiltshire, it was reckoned safe to burn provided it was given to you (Olivier & Edwards), a piece of Wiltshire pragmatism that embodies the concept of never looking a gift horse in the mouth. But the general appreciation of elder is contained in this rhyme from Warwickshire:

> Hawthorn bloom and elder flowers
> Will fill the house with evil powers (M E S
> Wright).

Perhaps the unlucky associations are a memory of its use in black magic. As early as the 10th century, King Edgar had enjoined that "every priest … extinguish every heathenism; and forbid … vain practices … with elders" (Ewen).

Crowning someone with elder was reckoned in Yorkshire to be about the most insulting thing that could be done (Gutch), for Judas hanged himself on an elder-tree, a belief that Ben Jonson picked up in *Every Man out of his Humour*: "He shall be your Judas, and you shall be his elder-tree to hang on". A fungus that grows on elder is known as Jew's Ear, that is, Judas's Ear (it was looked on as a great medicine for quinsy, sore throat, and the like).

Beekeepers in Yorkshire used to sprinkle the hive with an elder branch, dipped in sugar and water when the bees were ready to swarm (Addy). In Cornwall, too, they say that the inside of hives should be scrubbed with elder flowers to prevent a new swarm from leaving (Courtney). But bees do not seem to like the smell. When they swarm, a sprig of elder is often held about nine inches above them. The idea is that the elder will drive them out of the tree. In any case, elder has insect-repellent qualities, for it is said that people strike fruit trees and vegetables with elder boughs (or better still, drape them over), so that the scent would kill off troublesome insects (Folkard). Indeed, it is described in Wiltshire as a "charm" against flies (Tanner). Hampshire people recommended a wash made from elderberrries steeped in hot water, to deter gnats from biting (Hants FWI). If flies were troubling a horse, Lincolnshire carters would tie a branch of elder in leaf to the harness, and that would keep them away (Rudkin) (though judging by the state of the flowers towards the end of their blooming period, blackfly seem to be immune). And elders used to be grown by a dairy door, or by an outside privy (Vickery. 1993) for exactly the same purpose (Hartley & Ingilby). It is said, too, to be distasteful to snakes. Thomas Hill, in 1577, wrote that "if the Gardener bestoweth the fresh

Elder flowers where the Serpents daily haunt, they will hastily depart the place", as well as destroying the "Mothes, the canker worms and Palmers breeding in the trees".

There is some weather lore connected with the elder. For example, they say in Cheshire that if the weather breaks while the elder flowers are coming out, it will be soaking wet till they fade – or vice versa, for the belief seems to be that the weather never changes while the flowers are in bloom. Anyway, it all seems safe when the flowers are out:

> You may shear your sheep
> When the elder blossoms peep (A C Smith)

or, more obscurely:

> When the elder is white, brew and bake a pack;
> When the elder is black, brew and bake a sack
> (Denham. 1846).

Belgian people once used elder to foretell future weather by putting a branch in a jug of water on 30 December. If buds developed, it would be a sign of a fruitful summer to come; if no buds, then the harvest would be bad (Swainson. 1873).

In spite of all the warnings about it, elder wood was used quite a lot. Shoemakers used to make their pegs of it, for it is soft and easily worked (Nicholson). Skewers too were made of it, but not for poultry, because it was said that its rank smell tainted the flesh (Baker. 1974). It was the material for net-weaving needles, combs, even mathematical instruments (Hemphill). But it is with musical instruments that elder is particularly associated; in fact, *Sambucus* is from Greek sambuca, an instrument played by both Greeks and Romans, and made of elder wood. Halliwell defined a sambuke as a kind of harp. Pliny says that pan-pipes and flutes were shriller when made of elder wood (Hatfield), and young boys in the Hebrides with aspirations to being pipers made their chanters from the young branches (Murdoch McNeill). And Irish people, so it is claimed, used to pour molten lead down an elder stick that had the pith removed – it made a useful protective stick on a journey. A piece of "bored bourtree" was used for blowing up the fire – it was called a pluff (Aitken), or for pop-guns, the pith being so soft and easily removed; they were called Pellet-guns in Northamptonshire (A E Baker), or further north Burtree-guns or Burtree pluffers (Brockett). Rosehips were used as ammunition (Bucks FWI). "In the autumn the pop gun was greatly used and prized. There was always the hunt for a thick piece of elder wood about nine inches long. We scooped all the pith from the centre to make the barrel after which we put half an acorn in each end. Then a firm piece of stick to push (from one's chest) from one end to the other to make a loud pop" (Norfolk FWI).

Various dyes are to be got from the tree – black from bark and root, green (with alum) from the leaves, and various shades of blue and mauve from the berries, with alum, and sometimes with salt (Brownlow), none particularly fast. There is a comment on its durability:

> An eldern stake and black-thorn ether
> Will make a hedge to last for ever.

Another version has:

> An elder stake and a Hazel heather …

Ether, in the first example, and heather in the second, should be "header" – the rods that are put along the top or "head" of the hedge, to fasten the bushes, etc., down (Cobbett. 1825). They say that an elder stake will last in the ground longer than an iron bar of similar size (Akerman) – in fact in Somerset they say that the elder will not be completely destroyed unless it is burned (Tongue).

Folk medicinal uses range from the genuine to the magical, with some usages roughly half way between the two. Such must be a practice apparently stemming from the doctrine of signatures, for the pith, when pressed with the fingers, "doth pit and receive the impress thereon, as the legs and feet of dropsical persons do", so the juice of the tree was reckoned a cure for dropsy (Dyer). Evelyn mentions it, and so does Gerard, who recommended the seeds for "dropsie, and such as are too fat and would faine be leaner". There are records from medieval times onwards for the dropsy remedy, but it comes as a surprise to find a record (from Cambridgeshire) of elderflower tea as a dropsy remedy in the 20th century (Porter). That same tea is good for colds, coughs, (Maloney), and to break a fever (Stout), as well as for sore throats (so is mulled elderberry wine, which is also said to be good for asthma (Hatfield)). A concoction of elderberries was a Highland cold cure (Thornton). Elder flower is still infused in Cornwall and used as a tisane for hay fever and catarrh, while in Cumbria elder flower water is drunk for rheumatism.

The flowers are still used as an ingredient in skin ointments and eye lotions (Mabey. 1972). The very first issue of *Illustrated London News* (1842) carried an advertisement for Godfrey's extract of elderflowers for ladies' complexions. The inner bark and the leaves as well as the flowers were used in Cambridgeshire for the ointment (Porter), but the lotion was made from the flowers only, for "whitening the skin" (Trevelyan), or for washing off sunburn and freckles (Friend). Even drinking the tea was reckoned in Dorset to be good for the complexion (Dacombe). The ointment was used widely for burns, too, which it would cure without leaving a scar, so the belief runs (O'Farrell). All parts seem to be good in this case. The inner bark, for example, "takes out the fire immediately", according to Evelyn. Wesley too prescribed the inner bark for "a deep Burn or Scald".

Rub insect bites with elder leaves, and cover them, rough side for drawing, smooth side for healing (O'Farrell). In Scotland, they used to say that elder leaves gathered on the last day of April, were fine to cure wounds (Banks). Elder ointment was used for jaundice in Ireland (Wilde). In Herefordshire too, the inner rind of the bark, boiled with milk, was taken for the same complaint (Leather), and in Ireland for epilepsy (Barbour).

Elderflower tea is, of course, laxative, though the bark is not, for in Kentucky, tea made from the bark (peeled upward, of course, for if downward it would have the opposite effect) is given for diarrhoea (Thomas & Thomas). There are, too, a number of prescriptions for stomach trouble involving the elder in one way or another. It is of interest to note that the Pennsylvania Germans made a ball of elder bark, and pushed it down a cow's throat when it had indigestion (Dorson). There are a number of other veterinary usages. Gypsies used the leaves to treat a horse's leg – they soak the young shoots from the tips of the leaves in hot water, and bandage them round the lame leg (Boswell). In Ireland, the water in which elder leaves had been boiled was used to dose pigs. Lameness in pigs used to be treated by boring a small hole in the pig's ear and putting in a small plug of elder wood. As the plug withered or fell out, the animal would be cured (Drury. 1985). That sounds very like a transference cure, one of the magical usages in folk medicine, like taking three spoonfuls of the water that has been used to bathe an invalid and pouring it under an elder tree (Dyer). In the same category is the Bavarian belief that a sufferer from fever can cure himself by sticking an elder twig in the ground without speaking. The fever transfers itself to anyone who pulls the stick out (Frazer). And of course virtually all the wart charms are transference cures. The usual practice was like this Welsh one: you take an elder branch, strip off the bark, and split a piece off like a skewer. Hold this near the wart, and rub it either three or nine times with the skewer, while an incantation (of your own composing) is muttered. You then pierce the wart with a thorn and transfix the elder skewer with the thorn, and bury them in a dunghill. The wart would rot away as they decayed (Owen). The Wiltshire version was to strike the warts smartly with the elder twig, saying aloud, "All go away". Then you had to walk backwards towards the midden and throw the twig in without looking (Richardson). Crossing the warts with elder sticks was also a popular charm (Sternberg). There are many others. Both the ague and toothache can be got rid of in the same kind of way. A Welsh charm for the former is simply to stick an elder branch in the ground and bid the ague depart (Trevelyan); the fever sticks to the elder, and fastens on the first person who comes to the spot.

The other kind of magical practice is by the use of amulets, different in that it was usually a prophylactic measure instead of therapeutic. For example, it used to be said in Warwickshire that a child wearing a cross of the white pith of the elder would never have whooping cough (Palmer). Necklaces made of small twigs from a churchyard elder for the same purpose are recorded too (Lewis). That same necklace would have been used to prevent teething fits (Friend), and, in Ireland, for epilepsy (Wilde), and so on. Rheumatism could be cured in a similar way, by carrying the piece of elder, fashioned according to tradition, around with the patient (Hartland). Elder wood in the pocket keeps "the thigh from chafing" (Bergen, Aubrey). Blockwick also mentions an amulet, made from an elder on which the sun never shone, for erysipelas.

After all this, perhaps it is not surprising to find John Evelyn, who was very keen on elder, claiming "an extract, or theriacle may be compos'd of the berries, which is not only efficacious to eradicate the epidemic inconvenience and greatly to assist longevity (so famous is the story of Neander), but is a kind of catholicon against all infirmities whatsoever".

ELDERBERRY WINE
Apparently once called Pop-gun in the south of England (Halliwell), it has always been popular, so much so that whole orchards of the trees were planted in Kent, and the berries have even been used for making ersatz port! (Jordan). Evelyn (in *Sylva*, 1729) was of the opinion that it "greatly assisted longevity", and Cobbett was enthusiastic too – "a cup of mulled elder wine, with nutmeg and sippets of toast, just before going to bed on a cold winter's night, is a thing to be run for". Besides, it is good for sciatica, so it is claimed (Moloney), and according to another Irish belief, drinking it will cure pimples on the face (Maloney). Mulled elderberry wine is certainly good for a sore throat, and also for asthma, so it is claimed (Hatfield).

ELECAMPANE
(*Inula helenium*) The specific name *helenium* seems to suggest some connection with Helen, and one legend says that she was holding a branch of elecampane when she was carried off by Paris. Another holds that the plant was born of the tears she shed (Palaiseul). There is a superstition in Guernsey that it could be used for a love charm. It had to be gathered on St John's Eve, dried and powdered, and mixed with ambergris. Then it had to be worn next to the heart for nine days, after which the person whose love it was wished to obtain had to be got to swallow some of it (MacCulloch). It is a lucky plant, in any case. The Welsh wore it in the hat or cap, because it had the power to frighten robbers. If put in the cap of a deceiver, he or she would immediately get very red in the face. Welsh children had a rhyme:

Elecampane, what is my name?
If you ask me again, I will tell you the same
(Trevelyan).

So, too, in the Balkans, it would be sewn in children's clothes to ward off witchcraft, but it had to be gathered with some ritual, and on special days (Vukanovics).

Elecampane is edible – in fact, the leaves were at one time a popular potherb, and the young roots were cooked, too. The Romans, so it is said, served slivers of the boiled roots as appetisers (A W Hatfield). They were candied, too, like angelica, and eaten as a sweet (Taylor), or sold as elecampane cakes, to sweeten the breath (Genders. 1976). Most London apothecaries stocked them. Another use, apparently, was in the distillation of absinthe in France and Switzerland (Fluckiger & Hanbury).

But it was, and still is, as a medicinal herb that elecampane is famed. The doctor in the mumming plays nearly always resuscitates the dead men with drops from a bottle of elecampane. The word gets transformed into all sorts of nonsense - champagne, Alpine pain, elegant pain, etc., Nowadays, most of the usages are in veterinary medicine. The various 'Horseheal' names are a measure of that. It is used in America for horses' throat ailments (Leighton), and in this country for skin diseases in horses and mules, as well as for scab in sheep (Wiltshire). The leaves, too, are fed to horses to improve their appetite (G E Evans. 1960).

For medicine for humans, the roots have been used for a long time (and apparently still are (Le Strange)) to make cough candies (Brownlow); they were, too, a long-standing remedy for bronchitis (Clair), and even for hay fever (Conway). The infusion of the fresh roots was used for whooping cough (Thornton), or for asthma (Forey). The existence of names like Elf-wort, or Elf-dock, suggested to Grigson that supernatural belief was mixed up with real medicinal value (Grigson. 1952). Be that as it may, the tincture is still sometimes taken for loss of appetite (Flück) (Gerard too said it is "good and wholesome for the stomacke"). From Apuleius as far as Gerard, elecampane was used for worms, even, apparently, by laying the preparation on the stomach. It is not very clear what Gerard wanted his patients to do with it - presumably drink "the juyce ... boyled, for it "driveth forth all kinde of wormes of the belly…".

Gerard recommended the root, "boiled very soft, and mixed in a mortar with fresh butter and the poudre of Ginger", as "an excellent ointment against the itch, scabs, manginesse, and such like". This Unguentum Enulatum was at one time official (Clair) – witness the popular name Scabwort. It has a venerable history, for in Anglo-Saxon times we find this mixture of medicine and magic: "against eruption of the skin: take goose-fat and the lower part of elecampane and viper's

bugloss, bishop's wort and cleavers, pound the four herbs together well, squeeze them out, add a spoonful of old soap to it. If you have a little oil, mix it with it thoroughly, and lather it in well at night. Scarify the neck after sunset, silently, pour the blood into running water, spit three times after it, then say: "Take this disease and depart with it". Go back to the house by an open road and go each way in silence" (Storms). By the middle of the 17th century Lupton was still able to claim that "a sawsfleam or red pimpled face is helped with this medicine …". There is a record from Ireland for its use for burns (Maloney), and, from Sussex, another is the use of the leaves as bandages in wounds (Allen). Like many another prized medicine, even more extravagant claims could be made for it, the best of all being that it is a counter-poison. Gerard quite seriously promoted the idea, and, equally seriously, Thomas Hill wrote that it "is profitable against poyson, against the pestilent aire and plague, etc., …". This is almost capped by the Chinese and Japanese use against cholera, extended more reasonably to malaria and dysentery (L M Perry), and Indiana domestic medicine uses it for rabies. The procedure is to take the roots, powdered, boiled in a pint of milk down to half a pint. Take a third of the result in the morning every other day, and eat no food till 4 pm on those days. It is effective, they say, provided it is started within 24 hours of the accident (Tyler).

There is a record from the Balkans of the use of the roots as a disinfectant. It is not clear exactly what is meant; perhaps it is burned as a fumigant (Kemp), but in Sussex the roots were hung at doors and windows to act as fly catchers (Allen).

Elettaria cardamomum > CARDAMOM

ELIXIRS

In 16th and 17th century Europe, potions for perennial youth were made from TOBACCO, of all things, and it was believed that the leaves had aphrodisiac properties (Brongers). We hear that "essence of balm" (BEE BALM, that is), will preserve youth. Llewellen, prince of Glamorgan, who lived to 108, attributed his long life to it (M Baker. 1980).

ELM

i.e., ENGLISH ELM (*Ulmus procera*) We may mourn the passing of the elm, but it has long been seen as a treacherous tree, hostile to human beings (Ô Súilleabháin):

Elm hateth
Man and waiteth (Wilkinson. 1978).

Kipling knew all about the belief, and wrote in *A tree song*:

Elmen she hateth mankind and waiteth
    Till every gust be laid
To drop a limb on the head of him
    That anyway trusts the shade.

That is probably the nub of this belief. Elms can often, without any warning or sign of decay, shed a limb and cause injury or death. A large elm near Credenhill Court, in Herefordshire, used to be called the Prophet Tree; it was said to foretell each death in the family of the Eckleys, who used to own the place, by flinging off a limb (Leather). It is thought particularly dangerous to shelter under an elm during a thunderstorm (Porter. 1969). "He will wait for me under the elm" is a French proverb, meaning, he will not be there, perhaps because it would be such a stupid place to wait (Wilkinson. 1978).

Distrust goes back a long way – Virgil talks of a great elm [not *U procera*, though] at the entrance to Orcus, and it seems clear that the souls of the dead were thought to inhabit the elm, as dreams or as birds (L B Paton). It should be noted too , in this context, that the elm was the execution tree of the Normans – Tyburn was once known as 'The Elms'.

They say in Somerset that elms pine for their fellows. If two trees out of four are cut down, the remaining two will soon die (Tongue. 1965). Oaks are rather like that, and they will revenge themselves if they can, if any others are cut (Briggs. 1962).

The leafing of the elm has for centuries governed agricultural operations:

> When the elmen leaf is as big as a mouse's ear
> Then to sow barley never fear.
> When the elmen leaf is as big as an ox's eye,
> Then I say, "Hie, boys, hie" (Palmer. 1976).

A variant, from Warwickshire, as is the previous example, runs:

> When elm leaves are as big as a shilling,
> Plant kidney beans, if to plant 'em you're willing.
> When elm leaves are as big as a penny,
> You *must* plant kidney beans, if you mean to have any (Dyer. 1889).

The Cotswold variation has "as large as a farthing" (Briggs. 1974), or "farden" (in Herefordshire), to rhyme with "garden" (Leather). But if the leaves fall before their time it "doth foreshow or betoken a murrain or death of cattle" (Lupton).

Of course, the timber has been used extensively, in spite of the tree's evil reputation. Elm poles made bows (though the best were from Spanish yew) (Rackham. 1976). It endures well under water, and so was the first material for water pipes (Grigson. 1955), and for piles, as in London and Rochester bridges (Rackham. 1976). For the same reason, elm timber was used for coffins, a practice that Vaux wholeheartedly condemned – "the evil practice of making coffins of elm in order to keep the body from the corrupting effects of contact with the earth for as long as possible, the very opposite to which is what all sensible people desire". But elm wood does not burn well:

> Elmwood burns like churchyard mould
> E'en the very flames are cold.

(from *Logs to burn*, trad. English.).

It is quite usual to find purely magical practices masquerading as medical facts, and elm usage is no exception to this rule. From an Irish manuscript of 1509, it appears that an elm wand was used for healing purposes, and a cure for a man rendered impotent by magic was to cut his name in ogham on an elm wand, and then to strike him with it (Wood-Martin).

All parts of the tree, Evelyn said, "asswage the pains of the gout". Rather more optimistic was the assertion in Dioscorides that "the leaves beaten small with vinegar, and soe applied are good for the leprosie ..." (Apuleius Madaurensis). Wesley, too, associated elm with a leprosy cure, but it was the bark he prescribed.

EMBALMING
BASIL is an embalming herb, already used as such in ancient Egypt. This tradition is also met in Keats' poem called *Isabella, or the pot of basil*, a story originally told by Boccaccio. Isabella laid the head of her murdered lover in a pot of basil, which kept it "fairly unspoilt" (Swahn).

EMBLEMS
The LEEK is the national emblem of Wales. But the stories that explain its adoption are mainly from the English point of view, the most scurrilous being that the Welsh were long ago infested with orang-outans. They (the Welsh) asked the English to help them exterminate the apes, but the English killed several Welshmen by mistake. So, in order to distinguish them from the apes, they asked them to stick a leek in their hats (Howells). On the Welsh side, the explanations vary from the very simplistic (St David ate leeks (Hadfield)) to another recognition emblem; in this case the leek commemorates the victory of King Cadwallader over the Saxons in AD 640, when St David made the Britons wear leeks in their caps for purposes of recognition (Friend). It has even been suggested that the vegetable is a visual reminder of the old colours for Wales – white and green, white for the snow on the mountains, green for the pastures of the lowlands (Moldenke). In Greek mythology, the OAK was Zeus's favourite tree, his emblem, and the seat of his divinity (Rhys). It had sheltered him at his birth on Mount Lycaeus (Bayley), on which was a sacred spring to which the priests went in times of great drought to get rain. They did this by touching the water with an oak branch (Rhys). As the tree had sheltered Zeus, it became the symbol of hospitality; to give an oak branch was the equivalent of "you are welcome". The OLIVE was the emblem of Athena, or Minerva, Goddess of medicine and health (Megas).

The ROSE is the emblem of martyrs. The five petals of the red rose typified the five wounds of Christ, the white rose the virginity of Mary. It is said that on opening the tombs of saints, roses were often found in full beauty, but they fade at once when they are touched. St Louis of France was found with a rose in his mouth (Bunyard). But roses appear as part of the emblems of a great number of saints (see Drake & Drake). MADONNA LILIES first and foremost are the emblems of the Virgin Mary, always shown in pictures of the Annunciation. Indeed, the lily of sacred art is always the Madonna Lily, (and see SYMBOLISM also). It is also the emblem of St Catherine of Siena, and is even called St Catherine's Lily occasionally, but another dedication is to St Dominic. St Anthony of Padua, when not with the infant Christ in his arms, invariably has a lily (Haig). And that is not all, for it seems it was the emblem of St John the Baptist and of St Joseph (Woodstock & Stearn). Another emblem of the Virgin Mary was JASMINE (Ferguson), or the CARNATION, used as such in early German painting, and also in Italian art, when it was shown together with the Lily in a vase by the Virgin (Haig). These days, red carnations are the emblems of workers' movements in most European countries (Brouk). The emblem of St Agnes is the CHRISTMAS ROSE, symbol of purity, as St Agnes is the patroness of purity (Hadfield & Hadfield). As with the Madonna Lily, the pure white flowers are the reason for this dedication, as also is its time of blooming. St Agnes' Day is 21 January, at which time the plant should be well in flower. DANDELION is one of the three emblems of St Bridget, or Bride, originally the spring goddess, Brigid. Milk-yielding plants like the dandelion were often allied with the goddess in Scottish folkore (F M McNeill).

THYME was, according to a correspondent of *Notes & Queries; 1873* the emblem of the radical movement in French politics. As such it was also the emblem of Marianne, the figure of the Revolution, known also by the Phrygian cap she wears. VIOLETS, being the symbol of humility, are the emblems of Christ on earth (Haig). So highly was it regarded by the ancient Athenians (it was actually in commercial cultivation for its sweetening properites) that they made it the emblem of their city (Genders. 1971). Closer to our own times, there is another association of the violet – that with the Bonaparte dynasty. When Napoleon left France for Elba, he said that he would return in the violet season; and violet (both the flower and the colour) became a secret emblem of confederates sympathetic to him. When he escaped from Elba, his friends greeted him with violets. BROOM is another symbol of humility, probably because of its use as a domestic implement. Indeed, it is also the emblem of the housewife in a negative sort of way. The display of a broom at the house door showed she was away from home, and her husband would welcome visits from his male friends. Conversely, no respectable housewife would dream of leaving a broom on view if she were at home, for it showed that a man's company would be welcome (M Baker. 1974).

Pope Gregory the Great said the POMEGRANATE should be used as the emblem of the Christian church, because of the inner unity of countless seeds in a single fruit (Haig). It was to be used to symbolize congregations as well. The ecclesiastical emblem for Whitsuntide is, or was, the GUELDER ROSE, and that for Holy Rood Day the BLUE PASSION FLOWER, with its complicated symbolism of Christ's Passion. This is a South American plant, and it is said that the Spaniards, on first seeing it, took it as an omen that the Indians would be converted to Christianity. The difficulty is that virtually every description of the emblem differs. One is that the three styles represent the three nails, the ovary is a sponge soaked in vinegar. The stamens are the wounds of Christ, and the crown, located above the petals, stands for the Crown of Thorns. The petals and sepals represent the Apostles (Bianchini & Corbetta). Another description has it that the ten white petals show the Lord's innocence; the outer circle of purple filaments symbolize his countless disciples; the inner brown circlets the Crown of Thorns; the ovary is either the chalice he used at the Last Supper or the column to which he was tied, or the head of the hammer that drove in the nails; the five anthers are the wounds; the three divisions of the stigma the nails with which he was fastened to the Cross; and the tendrils are the lashes of the scourging, just as the leaves are the hands of those who reached out to crucify him (Whittle & Cook). A REED has been an emblem in Christian art of the Passion, for Christ was offered a sponge soaked in vinegar on the end of a reed (Ferguson). An APPLE is another emblem of Christ, as the new Adam, so when it appears in the hand of the original Adam it means sin, but when it is in the hands of Christ, it symbolizes the fruit of salvation (Ferguson). The GRAPE VINE, too, is an emblem of Christ, especially significant when it is depicted growing from the chalice of the Eucharist (Christ & Colles). Christ said "I am the vine, ye are the branches. He that dwelleth in me, and I in him, the same beareth much fruit, for without me you can do nothing" (John. 15; 5). In Old Testament writings, the vine stands for the Jewish people as a whole. BEAR'S BREECH (*Acanthus spp*) has been used in Christian symbolism as the emblem of Heaven (in the Ravenna Mosaics). DAISY is the symbol of innocence, and, thereby, the newly-born. As such it represents the infant Christ, and has been used thus in western art since the 15th century. The ROSE OF JERICHO (*Anastatica hierochuntica*) is the emblem of the Resurrection, and is actually called the Resurrection Plant, or Flower (Perry. 1972). The reason is that as the seeds ripen during the dry season, the leaves fall off and the branches curve inwards to form a ball which is blown out of the soil to roll about the desert until it reaches a moist spot or until the rainy season begins.

The BAMBOO, Prunus, and PINE together are the emblems of Buddha, Confucius and Lao Tzu – the Three Friends (Savage. 1964). PEACH blossom was the popular emblem of a bride in China. The wood, too has a special quality, for it has the ability to repel evil spirits (so has willow wood) – they are both symbols of immortality, peach being the Taoist emblem (Tun Li-Ch'en). CYPRESS was a sacred tree in Persia, the symbol of the clear light of Ormuszd, and was frequently represented on gravestones with the lion, emblem of the sun god Mithras (Philpot). HINDU LOTUS is the emblem of Lakshmi, the goddess of fortune and beauty. It was worn as a talisman for good luck and fortune (Pavitt), and the SAL TREE (*Shorea robusta*) is the emblem of Indra (Gupta).

The SILVER WATTLE is the Australian national emblem, but that does not stop it from being an unlucky plant to bring indoors, even sometimes unlucky to plant in a garden. There are a few records in England, too, of its bad luck when brought indoors, even being described in one record as "a forewarning of disaster".

Embothreum coccineum > CHILEAN FIREBUSH.

EMETICS
GROUNDSEL – there is a Cornish belief, obviously based on homeopathic magic, that it acts in different ways according to the direction in which the leaves are stripped from the stem. If upwards, that is, beginning from the root, with the knife ascending to the leaf, it makes it good as an emetic; if stripped downwards, it should be used as a cathartic (Hunt). Leighton mentions BLOODROOT as a powerful emetic.

Entada phaseoloides > NICKER BEAN

EPILEPSY
ASH, split, and a child passed through used to be a very widespread and well-known charm for the cure of hernia, and was also apparently used in Suffolk for epilepsy (G E Evans). In Ireland, MELILOT was regarded as a remedy (Moloney), and an Irish manuscript from about 1450 prescribed HONEYSUCKLE leaves for epilepsy – "put salt and white snails into a vessel for three nights, add seven pounds woodbine leaves, and mix them to a paste; a poultice of this applied for nine days will cure" (Wilde). Where was it applied? Along with barley meal and some other herbs, FOXGLOVE was included in an Irish preparation to treat the complaint (Logan). As Thomson said (Thomson. 1976), desperate conditions demand desperate remedies! But it must have been quite a dangerous practice. Quite a spectacular measure was this Irish cure: if the patient fell in the fit, put the juice of absinthe (Wormwood, of course), FENNEL or sage into his mouth, and there would be an immediate recovery! (Hutchings – because they are green, is the implication

in Hutchings's paper). A Cumbrian treatment, very simple, just required the patient to eat SORREL leaves (Newman & Wilson). How many and how often is not divulged, but bearing in mind the oxalic acid content of the leaves, it cannot have been too many at a time.

MISTLETOE has been used as a specific against epilepsy for a very long time. The doctrine of signatures may have played its part, for it has been said that its habit of downward growth recommended it for curing the falling sickness (Browning), but it is certainly not altogether a fancy, since it contains an active principle that is anti-spasmodic, and reduces blood pressure (Grigson. 1955) (always provided, Pliny said, "it has not touched the ground"), and always provided not too much is given, for a large dose would have the opposite effect (Anderson). Culpeper speaks of it as a sure remedy for the condition, and in the 18th century, Hill was also recommending the leaves, "dried and powdered" as a "famous remedy for the falling sickness". A decoction was made in Lincolnshire (Gutch & Peacock), and another folk remedy for the complaint was to take as much powdered mistletoe as would lie on a sixpence early in the morning in black cherry water or beer, and repeat for some days. But then comes the "charm" aspect of the cure – those "some days" had to fall near the full moon (Jacob). There are many more purely magical uses for the same complaint. Finger rings of mistletoe were made in Sweden (Dyer. 1889), and necklaces in Normandy (Sebillot), where also a chaplet was put round a child's head as an antidote for fits and epilepsy (W B Johnson). Another Swedish usage was for victims to get themselves a knife that had a handle of oak mistletoe (Kelly). RUE enjoyed a reputation for treating epilepsy, according to Wesley, who said that all that was necessary was to take a spoonful of the juice, morning and evening, for a month. A recipe from County Cavan for epilepsy requires this plant growing on walls of the old church in Knockbride to be boiled in milk and taken fasting for nine days (Maloney). WALL PENNYWORT had an old reputation for this disease, especially in the west of England (Grieve. 1931).

A volatile oil distilled from BOXWOOD has been prescribed for the condition (Grieve. 1931). A Shona nganga would use HEMP to treat the disease, or at least, one form of it. He would pulverize the leaves and the other plants used, fill a reed with the powder, and when the patient is actually having a fit, he lights the reed and blows the smoke into his face (Gelfand). It is treated in Russian folk medicine by eating as many raw ONIONS as possible, or else by drinking the juice (Kourennoff). Just the scent of THYME was once said to cure the disease (Classen, Howes & Synnott). The distilled water of LIME, according to Evelyn, was regarded as good "against epilepsy, apoplexy, vertigo", and so on. Even sitting under a lime tree is reported as improving the condition of epileptics (M Baker. 1980). Gerard, too, had already noted that "the floures are commended by divers … against the falling sickness, and not only the floures, but the distilled water thereof …". A more barbaric remedy was noted in 18th century Scotland: for the "falling sickness in children"… "take a little black sucking puppy (but for a girl take a bitch whelp), choke it, open it, take out the gall, put it all to the child in the time of the fit, with a little tile-tree flower water, and you shall see him cured as it were by a miracle presently" (Graham). (Tile-tree is Lime, of course, taken directly from the generic name, *Tilia*). JERUSALEM OAK has been used for the purpose in America. An Alabama remedy uses the inner bark of this plant, boiled and mixed with molasses to make a candy (RB Browne).

Equisetum arvense > HORSETAIL

Equisetum hyemale > DUTCH RUSH

Equisetum palustre > MARSH HORSETAIL

Eranthis hyemalis > WINTER ACONITE

Erinus alpinus > FAIRY FOXGLOVE

Eriophorum angustifolium > COTTON-GRASS

Erophila verna > WHITLOW GRASS

Eryngium campestre > FIELD ERYNGO

Eryngium maritimum > SEA HOLLY

Erysimum cheiri > WALLFLOWER

ERYSIPELAS

A Scottish country cure for the condition was to take an infusion of BITING STONECROP (Beith). HOUSELEEK, too, is used, for this is a protector from fire and lightning, and so would be a medicine for the "fiery" diseases, and that includes erysipelas. One of the cures requires pounded houseleek in a little skimmed milk. The rash would be bathed with this several times a day (V G Hatfield. 1994). Gypsies would use the leaves of BLADDER CAMPION externally, as a poultice to cure the condition (Vesey-Fitzgerald). Gerard recommended WALL PENNYWORT for "… all inflammation and hot tumors, as Erysipelas, Saint Anthonies fire and such like". SHEPHERD'S PURSE, made up into an ointment, can be used, too (Vesey-Fitzgerald) (American Indians even used this plant for poison-ivy rashes (H H Smith. 1923)). In Scotland, HERB ROBERT provided the remedy for this condition, also called the "rose". This may be doctrine of signatures, the red plant for the red skin condition (scarlet cloth would be used, too) (Gregor).

Euonymus atropurpureus > WAHOO

Euonymus europaeus > SPINDLE TREE

Euphorbia cyparissias > CYPRESS SPURGE

Euphorbia helioscopia > SUN SPURGE

Euphorbia ingens > CANDELABRA TREE

Euphorbia lathyris > CAPER SPURGE

Euphorbia peplus > PETTY SPURGE

Euphorbia pulcherrima > POINSETTIA

EVERLASTING FLOWERS
see PEARLY IMMORTELLE (*Anaphalis margaretacea*)

EVERLASTING PEA
(*Lathyrus latifolius*) In Northamptonshire, it is called locally Pharaoh's Pea (Vickery. 1995), from a legend that it was brought back from a royal tomb in Egypt. But, unhappily for the tradition, this plant has never grown in Egypt.

EVIL EYE
GARLIC, as a protector against the malign influence of the evil eye, is in widespread use, whether it is stitched in the cap of a new-born baby, hung outside a house, or from the branches of a fruit tree (Abbott). Boats can be protected from envious eyes – long branches of it used to be hung over the stern of Greek and Turkish ships in order to intercept any ill-wishing (Gifford). When a Greek sea-captain first went aboard his new ship, he hung garlic (and laurel) about it, and drank a libation to it. Bunches of garlic are hung about the boats as a charm against storms, as well as the evil eye (Bassett). A Greek mother or nurse walking out with her children would often take a clove of garlic in her pocket, and the formula "garlic before your eyes", or simply the exclamation "Garlic!", was a common expression used by a mother to someone who looks at a baby without using the traditional antidotes (Rodd). It is recognised in Morocco, too, as one of the charms against the evil eye (Westermarck), as was SAFFRON. Evil spirits are said to be afraid of saffron, which is used in the writing of charms against them (Westermarck). Some Hebrew amulets, too, were written with a copper pen, using ink made from lilies and saffron (Budge). It is not an inherent quality in the plant that is exploited. Rather it is the colour that is effective against the evil eye. Another Moroccan belief is that TAMARISK is effective in charms against the evil eye (Westermarck), and so is CORIANDER (Westermarck). In Egypt, if anybody is supposed to have been affected by the evil eye, a mixture of coriander and certain other ingredients is thrown on some live coals, and the smoke made to rise to the sufferer (Westermarck). PINE trees had some means of protection for a child. The way to use them was to sweep its face with a bough from a pine tree (Rolleston). ASH, too, could be used. A twig (from a tree that had a horseshoe buried among its roots) stroked upward over cattle that had been overlooked would soon charm away the evil (Pavitt). Branches

of it were wreathed round a cow's horns, and round a cradle, too (Wilde). English mothers rigged little hammocks to ash trees, where their children might sleep while field work was going on, believing that the wood and leaves were a sure protection against dangerous animals and spirits.

There are records of pieces of GROUNDSEL root being used as amulets against the evil eye (Folkard); it was, too, used as a counter-charm against witchcraft. So was HONEYSUCKLE. In this case it is the way that it grows (spiralling clockwise) rather than any inherent quality it possesses that makes it important in this context. Witches and those with the evil eye are forced to stop whatever they are doing and to follow out every detail of an involved design that they see. So interlacing and complex interwoven braided cords were deliberately made to distract, delay and confuse evil eyes, and were worn specifically for that use. The intricacy of a honeysuckle wreath serves exactly the same purpose (Gifford). MUGWORT, especially if gathered ritually on Midsummer Eve, would protect from the evil eye, but then the belief was that it would protect from evil in general. ROSEMARY is worn in parts of Spain as an antidote to the evil eye (Rowe), and ANISE had a certain reputation in this field (Grieve. 1931), as had PENNYROYAL (Bardswell), and PERIWINKLE (Folkard).

POMEGRANATES too have that effect. The Arabs of Hiaina, when commencing ploughing, always squeezed a fruit on to the horns of one of the oxen, so that the juice would go into any evil eye that looked at the animals, and so render the evil harmless. In the same general area of Morocco, some part of the plough would be made of BAY wood, as an insurance against the evil eye (Westermarck). ROWAN is as efficient against the evil eye as against witchcraft. A sprig of rowan tied to a cow's tail was as often as not to protect it from overlooking as against downright witchcraft (MacLagan). SAGE is a protective plant, at least in Spain and Portugal, where it is thought of as proof against the evil eye (Wimberley). WORMWOOD relies on its strong smell as protection, and gypsies looked on THORN-APPLE seeds as protectors (Leland. 1891). CAYENNE PEPPER, too, is used by Mexican Indians as a remedy for magical malviento and malojo (evil eye) (Kelly & Palerm), while in Amazonia, and Brazil generally, GUINEA-HEN WEED (*Petiveria alliacea*) acts the part. The Ka'apor people of Amazonia make an amulet for infants of the bark, wrapped in cloth. It would ward off the evil divinity (Balée), and they plant it by their doors for protection. In other areas of Brazil amulets are made of the wood, in the shape of the universal figa, usually made with the hand, but wearing a carved one round the neck or waist is much simpler. Brazilian street vendors wear one, or stand one up on their trays so as to protect their goods from the evil eye (P V A Williams).

A different angle to evil eye belief is being able to recognise the victims. According to Brazilian belief, MAIDENHAIR FERN will wilt when looked at by a victim (P V A Williams). PRIMROSES had their dark side (see **UNLUCKY PLANTS**). Giving a child one or two primroses would leave the donor wide open to a charge of ill-wishing (W Jones. 1880).

Palma Christi is an old name for CASTOR OIL PLANT. It means Christ's hand (palma is the palm of the hand), and is a reference to the palmately divided leaves. The open hand, as well as the fist, is a potent instrument for dispelling ill-wishing or the evil eye, and it is interesting to find that castor oil plant leaves have been worn round the neck to ward off devils, because the leaf is like an open hand (C J S Thompson. 1897). The same applied to RUE, famous as a counter to the evil eye. One of the most potent charms in use in Naples was, and still is, the cimaruta, that is cima di ruta, sprig of rue, a representation of the plant these days, but obviously the real thing originally (see **CIMARUTA** for a more detailed description). Rue seems to have been the special protector of women in childbirth; that is why the cimaruta is worn on the breasts of infants in Naples (Elworthy. 1895). Elsewhere in Italy, a newborn baby is washed with a decoction of rue, to make it strong, they say (Canziani), but really to protect it from "overlooking". As far away as Mexico, it is recorded that in the Maya village of Yucatan, a mother may keep a child from ojo (the equivalent of the evil eye) by chewing leaves of rue and rubbing them on the child's eyelids. But it is said that a child who gets this treatment will cause ojo to others when he grows up (Redfield & Villa). In the Ardèche district of France, cart drivers would put rue in their pockets to stop those with evil eye from causing their carts to stop suddenly (Sebillot). All over Morocco, too, rue is carried as a charm against the evil eye, but the real protection there is HENNA, red being a good prophylactic, for fevers as well as more supernatural attacks. Westermarck. 1926 drew together many examples of the use of henna, chiefly by women, but also on special occasions by men, and to new-born babies as well. When afraid of being hurt by the evil eye, they paint spots of henna on the top of their heads. An infant at the age of 40 days had the crown of its head smeared with henna, as a protection against fleas and lice, but also, once again, against the evil eye; the application is repeated frequently till the child gets older. Henna and walnut-root bark are applied to the mother and her newly-born baby.

In the Mayan villages of Yucatán, INDIGO is a protector, for the ordinary amulet to keep off evil spirits and to avert the evil eye is a collection of small objects tied togather with thread. It is this thread that is dyed blue with juice from the plant, and the same dye is used to paint the fingernails of persons who are threatened with death from sickness (Redfield & Villa). It was the practice in Jamaica to bury a

PHYSIC NUT in a field, to keep a neighbour's envious (and therefore evil) eye from a good field crop (Beckwith. 1929). In modern Greece, a method of protecting the household from the evil eye is by fumigation with burning branches of dry OLIVE, blessed during Holy Week (Rodd).

EYESIGHT

An infusion of IVY leaves was still in use in Fifeshire during the 20th century as an eye lotion (Rorie) – interesting, for Gerard recommended the same usage four hundred years ago – "the leaves laid in steepe in water for a day and a nights space helpe sore and smarting waterish eies, if they be washed and bathed with the water wherein they have been infused". Even more interesting is the fact that in homeopathy, a tincture of the young leaves is used to treat cataracts to this day (Schauenberg & Paris). Lady Gregory recorded a belief from the west of Ireland that "a cure can be made for bad eyes from the ivy that grows on a white-thorn bush". A note in Cockayne, acknowledging Pliny, mentioned the use of an amulet of chamacela, which could be either SPURGE LAUREL (*Daphne laureola*) or LADY LAUREL *Daphne mezereum*), to cure pearl (albryo), which is probably cataract, in the eyes, "if the plant is gathered before sunrise, and the purpose outspoken". AGRIMONY tea is another eye lotion for the cure and prevention of cataract, and HEMLOCK was used too. A leechdom for "a Pynn and Webb", which was an earlier term for the condition, was quoted as "take a handful of hemlock and ye white of an egg and a little bay salt altogether very fine and lay it to ye pulce of ye arme on ye contrary side …" (Gutch & Peacock). Red ochre was another ingredient in a similar salve from Suffolk, which was applied to the good eye, not the sore one (Porter. 1974), or, as an earlier leechdom quoted, to the left wrist, and vice versa (Jobson). Buchan congratulated himself on curing a cataract "by giving the patient purges with calomel, keeping a poultuice of fresh hemlock upon the eye, and a perpetual blister on the neck".

GROUND IVY has been famous as an eye medicine. A leechdom for eyestrain from as early as Anglo-Saxon times required it to be boiled in sour beer, and the result used to bathe the eyes (Cockayne), and similar eye recipes are to be found in herbals from that time onwards. The medieval Welsh text known as the Physicians of Myddfai has: "for inflamed eyes. Take the juice of Ground Ivy, and woman's milk, equal parts of each. Strain through fine linen, and put a drop in the painful eye". An example from folk medicine comes from Dorset, and requires an ointment made from the herb (Dacombe). A Warwickshire remedy is to take a large handful of the plant, just cover it with water, and simmer for about 20 minutes, strain it, and use the liquid to bathe the eyes (Vickery. 1995). Wiltshire has a much more localised remedy, in which

water taken from a well in Cley Hill, Warminster, was used to boil ground ivy, as a remedy for weak eyes. The water had a popular reputation as only being valuable for bathing the eyes, and ground ivy had a separate reputation for the same thing (Manley). There is even a story of a fighting cock that got wounded in the eye. Its owner chewed a leaf or two of the herb, and spat the juice in the damaged eye to make it heal quickly! (Palaiseul).

Dragons and snakes were wont to use FENNEL to restore their eyesight; it was Topsell (1607) who first wrote of the natural history of dragons, and it was his opinion that "their sight many times grows weak and feeble, and they renew and recover it by rubbing their eyes against fennel or else by eating it". With a precedent like this, no wonder that the belief arose that "to repair a man's sight that is dim, nothing better than fennel could be found" (Hulme). Already, by the 15th century century, a recipe "for itching and web in the eye" had been recorded (Dawson) – "take the juice of fennel-roots and put it in the sun in a brazen vessel 12 days; and then put it in his eyes in the manner of a collyrium", and in 1542 Boorde could write "the roots of Fenell soden tender, and made in a succade, is good for the lungs and for the syght". A popular rhyme of the time ran:

> Of Fennell, roses, vervain, rue and celandine
> Is made a water good to cleere the sight of eine
> (Gerard).

In the 18th century, Pomet mentioned the distilled water from fresh fennel as "excellent for taking away inflammations of the Eyes". The medicine travelled to America – in Maine (Beck), an infusion of fennel seed is still a domestic remedy for eye trouble. Longfellow remembered the belief, too:

> The fennel, with its yellow flowers,
> In an earlier age than ours
> Was gifted with the wondrous powers
> Lost vision to restore.

RUE is quoted in the rhyme given above, and it has always been supposed that it has a potent effect upon the eyes (Pliny said that painters and sculptors mixed some rue with their food to keep their sight from deteriorating (Baumann). An Arabian writer on eye diseases, Ali ibn Isa, in the 10th century, used a mixture of rue and honey to prevent the development of a cataract (Gifford). The Anglo-Saxon version of Apuleius prescribed rue, "well pounded", laid to the eye, and for "dimness of eyes" it was apparently only a matter of eating rue leaves, or taking them in wine (Cockayne). Gerard continued the recommendation, prescribing rue to be applied with honey and fennel for "dim eies". A mild infusion is still in use as an eyebath and for eye troubles, including cataract (Hatfield). Rue, of course, enjoyed a great reputation against the evil eye, and the claim has been made that it can actually bestow second sight! (MacCulloch. 1911). LESSER CELANDINE is another plant mentioned in the popular rhyme quoted above. It was Pliny who was responsible for the legend that seeks to account for the name celandine, which was Khelidonion in Greek, from khelidon, the swallow, perpetuated in the generic name of the GREATER CELANDINE, *Chelidonium*, which, by the way, belongs to a totally different family, which was no barrier to the use of this "greater" plant's use in all sorts of ways for sore eyes and other eye complaints. The birds used the plant, he says, to restore their sight. An infusion of the flowers was used in Norfolk to treat sore eyes that accompany measles (V G Hatfield. 1994). CLUBMOSS, in Cornwall, was considered good against all diseases of the eyes, if it was gathered properly. It had to be done on the third day of the moon, when the new moon was seen for the first time. Show the moon the knife with which the moss for the charm was to be cut, and repeat:

> As Christ healed the issue of blood,
> So I bid thee begone:
> In the name of … (Courtney. 1890).

An infusion of RIBWORT PLANTAIN leaves has been used for conjunctivitis, as an eyewash (Wickham). So has GREAT PLANTAIN juice (Gerard). Pennant, in Scotland, reported that the flowers were thought to be a remedy for "ophthalmia". SCARLET PIMPERNEL has various "eye" names bestowed upon it. Adder's Eye, for example (Grigson. 1955), or Ox-eye, and Bird's Eye, and a few others as well. These may have some bearing on the medicinal use of pimpernel for eye complaints, in which the usage may well be an example of the doctrine of signatures. The plant was certainly in use for eye diseases. QUINCE, too, has played its part. A decoction of the pips can be applied to inflamed eyes, and it is sometimes added to more usual lotions. From Anglo-Saxon times, there was some belief in YARROW's efficacy against cataract, which Cockayne translates "mistiness of the eyes". The leechdom was for equal quantities of betony, celandine and yarrow juice mixed together, and then applied to the eyes. Yarrow was still being used in the15th century, though the leechdom had become a little more exotic – "for the white that overgroweth the apple of the eyes. Take flowers of yarrow and stamp them with woman's milk, and put it in the eyes, and it shall heal them" (Dawson). Burning MARJORAM and inhaling the smoke was a Moroccan cure for coughs, and the same procedure was reckoned good for eye diseases. There was another way, though. A stalk of marjoram would be lit and the eye regions touched with the glowing tip (Westermarck).

CLARY is one of the more important plants used for eye troubles. Whatever 'Clary' means, it is not Clear-eye, though the latter has been used as a vernacular name and as a book name, with justification, for the seeds swell up when put into water, and become mucilaginous. These can then be put like drops into the eye, to cleanse it (Grigson. 1955). As Gerard said, "the seed of Clarie poudered, finely scarced [sieved, that is] and mixed with hony, taketh away the dimnesse of the eies, and cleareth the sight". Long before his time, though, the seed was being used as an eye salve, and not only the seed, for the leaves are prescribed in a 15th century leechdom, which runs, "for the pearl in the eye [cataract?], and the web: take the leaf of Oculus Christi [a common medieval name for Clary], peeled downward, and hyssop, with a leaf of sage: and drink the juice of these three days, first and last" (Dawson. 1934). How anything taken internally in those days could ever affect the eyesight is not revealed! DAISY must be included here by its very name, which is "day's eye" (OE daeges eage), for it closes its petals at night. But never mind that – with this name, it must be good for sore eyes. We certainly find that a decoction of the flowers boiled down was used in Ireland as an eyewash (Wilde. 1890), or the boiled flowers themselves could be dabbed on sore eyes (O'Farrell). Pennant noted that in Scotland "flowers of daisies were thought to be remedies for the ophthalmia". The Meskwaki Indian name for WAHOO (*Euonymus atropurpureus*) means "weak-eye tree", and they used it for just that. The inner bark was steeped, to make a solution with which to bathe eyes, and a tea was made from the root bark for the same purpose (H H Smith. 1928).

The herbal of Rufinus, dealing with WORMWOOD, has "confortat et caput et visum clarificat" (Thorndike). Both of these usages appear again in English, in the Anglo-Saxon version of Apuleius (Cockayne), and later, in a 15th century leechdom: "… the juice of wormwood oft drunk with honey cleareth a man's sight, and if it be put in his eyes, it doth away the redness and the web that is in his eyes" (Dawson). Gerard's comment is terse: "it is applied … to dim eies". And much later on, we find Wesley recommending: "Eyes inflam'd … Wormwood tips with the Yolk of an Egg. This will hardly fail". Long before Wesley's time, though, egg and wormwood were being prescribed – "wormwood newly stamped, with the white of an egg, and laid over the eyes, takes away the blood and redness thereof, of what humour soever it come" (Lupton). MELILOT is another plant whose juice was used as an eye lotion, or as eye-drops (Schauenberg & Paris). The yellow latex of MEXICAN POPPY has been used for eye problems. It is worth noting that the generic name of this plant, *Argemone,* is derived from argema, the Greek word for cataract (Whittle & Cook).

DEADLY NIGHTSHADE provides the drug atropine, and is used by oculists to dilate the pupils for the examination and treatment of eye diseases. It is said that Deadly Nightshade's other name, Belladona was given because of its use on the Continent as a cosmetic, to make the eyes sparkle (Brownlow).

MARIGOLD water was for a long time a favourite for inflamed eyes (Rohde. 1936). Even just looking at the flowers was thought to help failing eyesight (Page. 1978). "The floures and leaves of Marigolds being distilled, and the water dropped into red and watery eies cureth the inflammation, and taketh away the paine ..." (Gerard). It appears, too, in a manuscript recipe of about 1600 for an "unguent to annoynt under the eyelids, and upon the eyelids, eveninge and morninge; but especially when you ... finde your sight not perfect. Take one pint (of) sallet oyle, and put it into a viall glass, but first wash it with rose-water, and marygold flower water, the flowers to be gathered towards the East. Wash it till the oyle come white; then put it into the glasse, and then put thereto the budds of holyocke, the flowers of marygold, the flowers or toppes of wilde time, the budds of young hazle, and the time must be gathered neare the side of a hill where fayries use to be, and the grasse of a fayrie throne there. All these put into the oyle into the glasse, and sett it to dissolve three dayes in the sunne, and then keepe it for thy use ..." (Halliwell. 1845). An extract from SNOWDROP bulbs has been used to treat glaucoma (Conway).

F

FADDY-TREE

A Cornish name for the SYCAMORE. The Helston Furry was once known as the Faddy, and the tree got the name because the boys would make whistles from its branches at the festival (Deane & Shaw).

Fagopyrum esculentum > BUCKWHEAT

Fagus sylvatica > BEECH

FAIR MAIDS (OF FEBRUARY),

that is, SNOWDROPS, and a reference to the festival of Candlemas (2 February), when it was the custom for the Fair Maids, dressed in white, to walk in the procession at the festival (Prior). A stage from this would mean that the plant was sacred to virgins in general, and at one time the receipt of snowdrops from a lady meant to a man that his attentions were not wanted (Prior).

FAIRY FLAX

(*Linum catharticum*) The fairies use it for their clothes, and its small bells make music that cannot be heard by human ears (Spence. 1949). But this is a medicinal herb, as its specific name implies, and a name in English, Purging Flax, confirms it. It is certainly an effective purge, but like many another herb, thoroughly dangerous to use. But, particularly in the Highlands, it was used regularly for gynaecological and menstrual problems (Beith). Even putting it under the soles of the feet, so it was believed until quite recently in the Hebrides, was an aid to easy childbirth. James Robertson, who toured the Western Highlands and Islands in 1768, noted: "The women are frequently troubled with a suppression of the menses, to remedy which they use an infusion of *Thalictrum minus* [Small Meadow Rue] and *Linum catharticum* (quoted by Beith). We are told, too that "country people boil it in ale, and cure themselves of rheumatic paine" (Hill. 1756), while in Ireland the herb, boiled in beer, was used for jaundice (Moloney).

FAIRY FOXGLOVE

(*Erinus alpinus*) A plant from central and southern Europe, occasionally naturalized in Britain. In the north of England, where it is naturalized, the local tradition is that it only grows where Roman soldiers have trodden (Vickery. 1995). One of its names there is Roman Wall Plant (Mabey. 1998), and it certainly grows in the village of Wall (near Hadrian's Wall).

FAIRY PLANTS

RAGWORT is a fairy plant, dedicated to them in Ireland, and called Fairies' Horse (Friend), for it was believed to be a fairy horse in disguise. If you tread them down after sunset a horse will arise from the root of each injured plant, and will gallop away with you (Skinner). The fairies look to ragwort for shelter on stormy nights, according to Hebridean folklore, as well as riding on it when going from island to island (Carmichael). Yeats quoted a story in which the local constable, when there was a rumour of a little girl's having being abducted by the fairies, advised the villagers to burn all the ragwort in the field from which she had been taken, as it was sacred to the fairies.

FOXGLOVE is just as important as a fairy plant, as many of the local names testify. It is fairy's gloves, fingers, thimbles, hats, dresses, petticoats and so on. When the plant bows its head it is a sure sign that a fairy is passing (Boase). "Fairies have been seen dancing under foxgloves in Cusop Dingle within the memory of some now living there" (Leather). But of course they are seen to a much greater extent in Ireland – see Yeats – "and away every one of the fairies scampered off as hard as they could, concealing themselves under the green leaves of the lusmore [literally "great herb", i.e., foxglove], where if their little red caps should happen to peep out, they would only look like its crimson bells …". Indeed, the Shefro, one of Ireland's more gregarious fairies, is always described as wearing foxglove flowers (Wentz). As so often happens in folklore, the fairies' favourite plant is also the one that offers most protection against them. It can cure any disease the fairies might cause (Logan). It is very useful when dealing with a changeling, a child "in the fairies", as is sometimes said. The child is bathed in the juice, and fairy-struck children had to be given the juice of twelve (or some say ten) leaves of foxglove (Wilde. 1902). Or a piece of the plant could be put under the bed. If it is a changeling, the fairies would be compelled to restore the true child (Mooney). Simpler still, put some leaves on the child itself, and the result would be immediate (Gregory).

Fairy thorns are part of the folklore of HAWTHORN trees. In Ireland, in particular, ancient and solitary thorns are known variously as fairy thorns, gentry or gentle thorns, skeaghs or loine bushes (E E Evans). They are, of course, sacred to the gentle people, and were held in great veneration. It was nothing less than profanation to destroy them or even to remove a bough. Hedgerow thorns are newcomers (though they are said to date from Roman times (Cornish)), so they may be hacked with impunity, but woe betide the man who damages one of the solitary fairy thorns, that is, one not planted by man, but growing on its own. Not even fallen dead branches that would serve as firewood should be taken away. There have actually been examples of branches accidentally broken being carefully tied back in position. The cult of these thorns in Ireland was apparently just as strong in the Protestant north as in other parts of the country (E E Evans), but the belief was not confined to Ireland. For in Somerset the tradition was that you should never cut down hawthorn trees to build your house, for if you did you and yours would never live long (Tongue). In Galloway, too, solitary thorns were left and preserved with scrupulous care (Cromek). Similar beliefs were held in the Isle of Man; it was not advisable to sit too long under one of these trees, and certainly not to sleep under one (Gill). A further result of the fairy thorn belief is the superstition that if thorn bushes are ploughed up, all goodness leaves the land (Tongue). A correspondent of *Notes and Queries; 1941* told a story then current at Berwick St John, in Wiltshire, of the consequences of cutting down a solitary thorn that grew on a prehistoric earthwork nearby. The result was complete loss of fertility over the area, taking in poultry and cows as well as women. Fertility was only restored when the perpetrator planted a new thorn in place of the old one. BLACKTHORN is another fairy tree, under the protection of a special band of them, said by Irish people to guard them especially on 11 November, which is Samhain, old style, and on 11 May, Beltane old style. They would let no-one cut a stick from the tree on those days. If anyone tried to, then he would be bound to suffer misfortune (Wentz). An extension of the fairy belief in Ireland is that they are supposed to blight the sloes at Samhain, just as the devil spits on blackberries at some time usually a little earlier than that.

ELDER is another tree associated with the fairies. According to Danish belief, anyone wandering under an elder at 12 o'clock on Midsummer Eve will see the king of the fairies pass by with all his retinue (Dyer). That belief is also found in Somerset. Fairies become visible under YEW trees, too. One example given (by Sikes) occurs in a wood called Ffridd yr Ywen, which means forest of yews, in Llanwrin. The magical yew tree grows exactly in the middle of the forest. The fairy circle under the tree has the usual legend of a mortal being drawn into the dance and losing all count of time.

ROWAN is as efficient against the fairies as against witchcraft and evil spirits. Craigie quotes a Scandinavian story in which a rowan-tree not only protected a boy from trolls, but did active damage to them. There is too a folk-tale from Ulster that tells of a woman carried off by the fairies. She was able to inform her friends that when she and the fairies were going on a journey, she would be freed if they stroked her with a rowan branch (Andrews). Similarly, it is said that if someone got into a fairy circle, he or she would remain there for a year and a day. After that, the enchanted person could be liberated by someone holding a rowan stick across the circle (T G Jones). One Scottish way of getting rid of a changeling was to build the fire with rowan branches and to hold the suspected changeling in the thick smoke of the fire (Aitken). However, there is another belief that good fairies are kind to children who carry rowan berries in their pockets (Skinner).

BEANS, broad beans, that is, are fairy food. The Green Children captured near Wolfpits, in Suffolk, would only eat beans, but gradually became used to human food. The boy pined and died, but the girl lived, and eventually married a local man. EARTHNUT is another fairy food. The tubers are edible, of course, but nobody seems to bother to collect them, not even children. They are Fairy Potatoes in Ireland, for this plant belongs to the fairies in general, and to the leprechaun in particular. The tops of young HEATHER shrubs were also fairy food according to lowland Scots tradition (Aitken). Perhaps that is why it is rather an unlucky plant to bring indoors. SILVERWEED roots constitute another fairy food (Campbell. 1900). They lived on the roots that were ploughed up in spring (Spence. 1846). HAREBELL is a fairy plant, as the various local names imply. It is called Fairy Bells (Macmillan), and it is their Cap, Thimble and Cup. Stories used to be told of how the fairies drank water from the Virtuous Wells in the Wye Valley, when they danced round them at Hallowe'en. People used to find Harebell flowers (in October?) next morning round the well, withered and thrown away. They used to gather them up and dry them, to use in illness (Eyre). BLUEBELLS are fairy plants, too. In Somerset, they used to say that you should never go into a wood to pick them. If you were a child, you might never come out again, and if you were an adult you would be pixy-led until someone rescued you (Briggs. 1967). ROSEMARY is a fairy plant in the Mediterranean countries, as in Portugal, where it is dedicated to them under the name Alicrum, elfin plant. In Sicily, too, the belief used to be that young fairies are either put to sleep in the flowers (Rohde), or they lie concealed beneath the shrub, under the guise of snakes (Dyer). THYME is another fairy plant, as such dangerous to take indoors (Briggs. 1967). The fairies were particularly fond of it, so it was said. PRIMROSES are fairy plants, but they will protect as well. Manx children used to gather them to lay before the doors of houses on May Eve to prevent the entrance of fairies, who cannot pass them, so it was said (Hull). So they did in Ireland, too (Briggs. 1967), and tied them to the cows' tails (Wilde. 1902). STITCHWORT is another fairy plant, under their protection, and it must not be gathered, or the offender will be "fairy-led" into swamps and thickets at night (Skinner). WOOD ANEMONE, too, is a fairy flower; in wet weather they shelter in them. So it is an unlucky flower to pick (Vickery. 1995). A 17th century recipe for a fairy salve had HOLLYHOCK as one of its ingredients, as a means of conjuring a particular fairy to appear in a glass "meekly and mildly to resolve him truly in all manner of question; and to be obedient to all his (the conjuror's) commands under pain of Damnation" (for the recipe see under **HOLLYHOCK**).

There is a record from Devonshire that CHIVES were looked upon as fairy musical instruments (Whitcombe), like FAIRY FLAX (*Linum catharticum*), which was also used for their clothes (Spence. 1949). LADY'S SMOCK, too, has distinct fairy associations, in spite of belonging to the "Lady", the Virgin Mary. RED CAMPION is another fairy plant, unlucky to pick (Garrad), and actually called Fairy Flower, Blaa ferrish in Manx (Moore, Morrison & Goodwin), as well as bearing a number of "Robin" names, i.e., Robin Goodfellow.

FALSE NUTMEG

(*Pycnanthus kombo*) A West African tree whose seeds are often mistaken for the true nutmeg, but they are less aromatic, and are of more importance as an oil-seed, for they are rich in a vegetable fat, used for lighting and in soap-making. The seed itself, when threaded, will burn like a candle (Dalziel).

FAMINE FOOD

The roots of SILVERWEED were eaten as a marginal or famine food both in the Scottish Highlands (MacGregor), and in Ireland (Drury. 1984). They were roasted or boiled (Fernie), or even eaten raw, or they could be ground into meal to make porridge, and also a kind of bread (Drury. 1984). Perhaps not so marginal, for Carmichael says that it was used a lot before the potato was introduced. Records of cultivation go back to prehistoric times. Particularly remembered for the cultivation of Brisgein (its Gaelic name) is an area of North Uist, in the Outer Hebrides, where a man could sustain himself on a square of ground of his own length. The Gaelic Bliadhna nan Brisdeinan means Year of Silverweed roots. This year was shortly after the Battle of Culloden, and is remembered in Tiree as a year of great scarcity. The land had been neglected in previous years due to the state of the country, and the silverweed sprang up in the furrows, and people made meal of them (Campbell. 1902), the "seventh bread" (MacGregor).

FAT HEN

(*Chenopodium album*) Edible, of course, and actually cultivated as early as the Iron Age in Europe (Brouk). The introduction of spinach from south-west Asia probably put an end to organized food production with Fat Hen. Even in India, it was only eaten as a marginal food (Gammie), though American Indians such as the Hopi made good use of the leaves and seeds (Hough). In Hungary, it seems that a fairly large amount of the powdered herb mixed with food has been shown to suppress the oestrous cycle (Watt & Breyer-Brandwijk). In other words it may act as an oral contraceptive.

FENNEL

(*Foeniculum vulgare*) A Mediterranean plant, in Britain probably a naturalised physic herb, growing mainly by the sea and in some waste places inland.

In Elizabethan times, fennel was used as a symbol of strength (Leyel. 1937), and also flattery, which is what Milton presumably meant, in *Paradise Lost*. Bk xi:

> The savoury odour blown,
> Grateful to appetite, more pleased my sense,
> Than smell of sweetest Fennel.

Shakespeare probably had this in mind when he made Ophelia say, "There's fennel for you, and columbine …" (Dyer. 1883). Grindon, though, had another explanation to offer, linked to the belief that fennel improves the eyesight (see below). She is trying, according to him, "to quicken the royal consciousness", i.e., open their eyes to what is going on, but, to return to the "flattery" theme, the Italian idiom 'dare finocchio' means to flatter (Northcote). In one of the many Italian folk tales involving St Peter, the saint is sent to buy some wine, and allows himself to be persuaded by the wine merchant to eat some fennel seed, so that he cannot distinguish good wine from bad (Crane). It is pointed out that the tale is in all probability the result of folk-etymology – the verb infinocchiare, meaning to impose on one, is very close to finocchio.

'Sow fennel, sow sorrow' is quite a well-known proverb (Wiltshire), pehaps because it tends to inhibit other plant growth (except that of *Eremurus* apparently). And you must not give it away, for that would cause disaster to follow (Wiltshire). Another superstition, from Somerset, is that fennel over the door prevents the house catching fire (Tongue), but that is because it is a protective plant, and powers out of the ordinary have been associated with it. It was used to ward off evil spirits (Emboden. 1979), and to plug keyholes to keep away ghosts (Cullum), and it was hung over the door along with other herbs of St John at Midsummer (C P Johnson). The 'benandanti' of 16th century Friuli, who were the "night-walkers" who fought the witches on a psychic level, carried fennel as their weapon, while the witches carried Sorghum, seemingly some kind of millet, as theirs. It was said that these 'benandanti' ate garlic and fennel "because they are a defence against witches" (Ginzburg). There is even a suggestion that cows' udders were smeared with an ointment or liniment made from fennel to prevent the milk being bewitched (Rowe). Certainly, fennel was used as a personal amulet. The seeds were hung round a child's neck against the evil eye (W Jones), and in Haiti it protects against loupgarous, and also serves to fortify pregnant women (F Huxley). A medieval Jewish protective amulet turns out to be a sprig of fennel over which an incantation had been recounted, which was then wrapped in silk, and then, with some wheat and coins, encased in wax (Trachtenberg). Emboden. 1979 suggested that these beliefs could have arisen from observation of the effect that the distilled oils could produce – epilepsy-like fits of madness, and hallucinations, giving the impression of powers out of the ordinary belonging to the plant.

The Greeks believed that snakes had recourse to fennel to cure blindness (Grieve. 1933), and it was popularly believed in the Middle Ages that dragons could cure their blindness, "apparently a common affliction of dragons", by rubbing their eyes with fennel, or eating it (Hogarth). The other point about snakes and fennel is that " so soone as they taste of it they become young again…" (Hulme). Macer reckoned fennel to be a restorer of youth, quoting the snake belief (Hulme). Perhaps this is relevant to an odd reference in Shakespeare: Falstaff, in Henry IV pt 2, says of Poins, "he plays quoits well, and eats conger and fennel", as if this were a sign of manliness.

Fennel appears in Anglo-Saxon medical receipts as early as the 11th century, probably owing to the active part Charlemagne took in its diffusion through central Europe (Fluckiger & Hanbury). The faith put in it in earlier times is shown by one of the medical maxims from the Book of Iago ab Dewi (Berdoe): "He who sees fennel and gathers it not, is not a man, but a devil". It is said, too, that a certain Comte St Germain became a very rich man by selling a tea that he claimed prolonged life – it was apparently composed of senna and fennel leaves (Thompson. 1897). By the end of the 19th century only the seeds were official in the British Pharmacopeia, and they were used in the form of distilled water, or volatile oil. The chief consumption was then in cattle medicine, and also (the oil) in the manufacture of cordials. But the carminative action had been recognised for a very long time: "Fennel seed drunke asswageth the paine of the stomacke, and wambling of the same, or desire to vomit, and breaketh winde …" (Gerard). In other words, it helps digestion, and relieves flatulence. Fennel tea is the usual carminative these days, made by pouring boiling water on to the crushed seeds. Fennel juice was also used at one time as a sort of anti-fat (Berdoe). For example, "pro stomaco. Who-so have swellyng in his stomake, take ther route of fynel, and the route of arache [orach], and stampe hit with wyn and hit schal helpe and hele hit" (Henslow). It is said that Greek athletes included fennel in their diet for stamina, and as a guard against getting over-weight. In fact, the Greek name for the plant is marathron, from maraino, to grow thin (Sanecki). The Welsh medieval text known as the Physicians of Myddfai has an entry: "To reduce fatness: whosoever is fat, let him drink of the juice of fennel, and it will reduce him". And it is still being prescribed (both the leaves and the seeds) for constipation and obesity (A W Hatfield).

It was mentioned above that dragons were wont to use fennel to restore their eyesight; it was Topsell who first wrote of the natural history of dragons, and it was his opinion that "their sight many times grows weak and feeble, and they renew and recover it by rubbing their eyes against fennel or else by eating it". So did snakes, apparently. With these precedents, no wonder that the belief arose that "to repair a man's sight that is dim", nothing better than fennel could be found (Hulme). Already, by the 15th century, a recipe "for itching and web in the eye" had been recorded – "take the juice of fennel-roots and put it in the sun in a brazen vessel 12 days; and then put it in his eyes in the manner of a collyrium" (Dawson), and in 1542 Boorde could write "the roots of Fenell soden tender, and made in a succade, is good for the lungs and for the syght". Gerard was able to quote a popular rhyme in his exposition of the virtues of the plant:

> Of Fennell, roses, vervain, rue and celandine
> Is made a water good to cleere the sight of eine.

The medicine travelled to America, too – in Maine (Beck), an infusion of fennel seed is still a domestic remedy for eye trouble. Longfellow remembered the belief, too:

> The fennel, with its yellow flowers,
> In an earlier age than ours
> Was gifted with the wondrous powers
> Lost vision to restore (from *Goblet of life*).

The seeds are still used in Chinese medicine for running eyes, and also to treat hernia (R Hyatt).

Another of Gerard's prescriptions was: "the greene leaves of Fenell eaten, or the seed drunke made into a tisan, do fill womens brests with milk". Since it is similar to dill in appearance, fennel has picked up some of dill's attributes, like increasing the flow of milk in nursing mothers, settling babies' stomachs, etc., (G B Foster). Fennel tea is still taken for bronchitis (Flück), and, so it is said, it soothes rheumatic pains (A W Hatfield). It was also part of an ointment to put on the bite of a mad dog, according to the Physicians of Myddfai. Babies with teething difficulties were give fennel tea in America (H M Hyatt). But the most spectacular of its many cures must surely be this one from Ireland: it is a remedy for the falling sickness, and said that if the patient fell in the fit, put the juice of absinthe, fennel or sage in his mouth, and there would be an immediate recovery (Wilde. 1890) (because they are green, is the implication in Hutchings's paper).

FENUGREEK

(*Trigonella ornithopodioides*) The conspicuous horn-like pod containing the seed, which provides the spice, gave the plant its Greek name Keratitis (keras means horn). The seed has been used in a number of different ways. It is one of the chief ingredients of Kuphi, the Egyptian embalming and incense oil (Sanecki), and a Nubian people, the Keruz, used to prepare a ritual drink on the birth of a child. One of the ingredients was fenugreek, which was said to relieve pain, as well as providing nourishment (Callender & El Guindi). The raw seeds smell good but their taste is disagreeable until they are cooked, and then they have many traditional uses, as in the preparation of hot mango pickle and green mango chutney, as well as in curries (Clair).

Fenugreek has been traditionally used to treat catarrh (Schauenberg & Paris), but the fact that the seeds contain a great deal of mucilage in their outer coating make them useful in other ways. They are, for example, cooked in water into a paste or porridge, which is used as a hot compress on boils, abscesses and the like (Flück), or for mastitis (Van Andel). It is also prescribed with anise to treat lactation difficulties, as a tea made with the seeds of both plants (W A R Thomson. 1928). The thick paste was used in Egypt to treat fevers (Clair), and was said to be as good as quinine (Grieve. 1931). The seeds were used in Egypt in mixtures for worms, as well (Dawson. 1929). Gerard had apparently heard that fenugreek seeds, or possibly the herb as a whole, were a good laxative, and he went on to report that "the juyce of the decoction pressed forth doth clense the haire, taketh away dandruffe …", and the "meale", presumably the porridge already mentioned, he reports as being "good to wash the head …, for it taketh away the scarfe, scales, nits, and all other such imperfections".

But the strangest use of fenugreek, even though mixed with a lot of other ingredients, must be the following, that Coulton took from a 14th century manuscript: "For hym that haves the squinansy: tak a fatte katte, and fle [flay] hit well, and clene, and draw out the guttes, and take the gres of an urcheon [hedgehog], and the fatte of a bare, and resynes, and feinygreke, and sauge, and gumme of wodebynd, and virgyn wax: all this nye [crumble] smal, and farse [stuff] the catte within als thu farses a gos, rost hit hale, and gader the grees and anoynt hym therewith".

At one time, fenugreek was commonly prescribed by vets for horses, and it is still used as a vet's medicine for an appetiser (Clair). Suffolk horsemen always used it. G E Evans. 1960 reported that they called the herb Finnigig, and suggested that the name was a deliberate corruption on the part of the horsemen, so that third parties would not be able to recognise the true identity of the ingredient they were buying.

FERTILITY

BIRCH seems to have been a symbol of fertility. Saplings were put in houses and stables, and men and women, as well as cattle, were struck with birch twigs, with the avowed intention of increasing fertility (Elliott). Birch twigs were put over the lover's door on May morning in Cheshire (Wimberley). At one time, when a Welsh girl accepted an offer of marriage, she presented her lover with a wreath of birch leaves; if she refused him, she sent hazel (Trevelyan). But

HAZEL itself is another symbol of fertility. Throwing hazel nuts at a bride and bridegroom had the same significance as rice and confetti have today (Hole. 1957), and until quite recently, Devonshire brides were given little bags of hazel nuts as they left church. The gypsy bridegroom, before the wedding ceremony, had to carry with him hazel wands wreathed in ribbons, "to ward off the influence of water" (Starkie), so it was said, but the real reason was to ensure the fertility of the marriage. A Bohemian saying was that plenty of hazel nuts meant the birth of many bastards (Dyer), but in Somerset it meant fertility in wedlock, too. As the old saying was, "Good nutting year, plenty of boy babies" (Hole. 1957). Ruth Tongue told the story of the Somerset village girl who returned from London in the 1930s to be married. She openly said she didn't intend to be hampered with babies too soon, and would take steps to ensure this. Such talk outraged village morality, and when she got to her new house, she found among the presents a large bag of nuts, to which most of her neighbours had contributed. She had four children very quickly. Anyway, "going a -nutting" is a euphemism for love-making. Another Somerset practice was that of throwing an ONION after the bride, to bring a long family (Tongue). It is certainly unusual to associate onions with fertility.

HINAU (*Elaeocarpus dentatus*) is a New Zealand tree, one particular specimen of which, in North Island, used to be a powerful symbol of fertility. A childless woman embraced this tree while her husband recited the necessary charm. The east side of the tree was the male side, the west the female, and the woman would make her choice of east or west according to whether she wanted a boy or a girl (Andersen). SILVER FIR is another tree with fertility connections. It was sacred in Greece to Artemis, the moon-goddess who presided over childbirth (Graves). As such, it is a symbol of fertility. At Hildeheim, women were struck with a small fir-tree at Shrovetide, and in north Germany, brides and bridegrooms often carried fir-branches with lighted tapers. Elsewhere, firs were planted before a house when a wedding took place. It was also used at weddings in Russia (Hartland. 1909). MISTLETOE, fairly obviously, is a fertility agent. Hartland quoted the maxim ascribed to the Druids that the powder of mistletoe makes women fruitful. In 17th century England, it was certainly regarded as among the most efficacious of medical elements, and in older Celtic lore was thought of as being one and the same as the "Silver Bough". Its berries held the male essence or protoplasm of the god, and so they were regarded especially as conferring powers of fertility. That explains why they were used at Yule as a kind of love-token, nowadays reduced to harmless kissing under the mistletoe. Another point is the way the berries are arranged – they show a likeness to the male parts, further reason for the superstition that it confers fertility (Grigson. 1955). The Welsh medical text known as the Physicians of Myddfai furthered the myth by prescribing a decoction to cause fruitfulness of the body and the getting of children. The Ainos of Japan thought exactly the same (Hartland. 1909). Even wearing mistletoe, without any internal dose, had the same effect, according to Coles: "some women have worn it about their necks or on their arms, thinking it will help them to conceive".

ROSEMARY must have been some kind of fertility agent, for it was at one time much used at weddings, and a symbol of fidelity in love. "Rosemary bound with ribbons" was a token of a bride's love for her husband, but more to the point, as late as 1700, country bridal beds were decked with it (Baker. 1977), and the Welsh Physicians of Myddfai prescribed it as a remedy for barren-ness. WHITE BRYONY, on the strength of its claim to be the "English Mandrake", joins the list of fertility stimulants. In Lincolnshire for instance, it was actually reckoned to be the specific for causing women to conceive (Gutch & Peacock). In East Anglia, a childless woman who wanted a baby would drink "mandrake tea" (Porter. 1969), presumably made from the roots, but not necessarily so. They were even given to mares as an aid to conception (Drury. 1985). (see also **CONCEPTION, aids to, WEDDINGS**, and **SYMBOLISM**).

MANDRAKE had the power, it was said, to put an end to barren-ness, even quite independently of sexual intercourse. See also Genesis 30; 14-16, where Rachel bargained for the mandrakes with her sister Leah. (see Hartland. 1909). Amulets, as figures made from mandrake root, were worn by Palestinians, both men and women, to promote fertility (Emboden. 1979). In the early part of the 20th century, American Jews still believed in the power of the mandrake to induce fertility. They used to import specimens of the root from the Near East just for this purpose (Randolph). CUCUMBERS, being phallic emblems, will naturally symbolize fecundity.

Ferula assa-foetida > ASAFOETIDA

Ferula communis > GIANT FENNEL

FEVERFEW

(*Tanacetum parthenium*) The name is OE feberfuge, from Latin febris, fever, and fugare, to drive away. Yet there is only one specific mention of fever in the list of recorded examples of the plant's medicinal uses, even though that mention seems to pay proper homage to the plant's traditional powers, for according to a Derbyshire belief all you had to do to cure a fever was to put a piece of feverfew in the bed (Addy). But it must have been important, for it was apparently grown commercially for the London markets as a medicinal plant (Grigson. 1955), and gypsies used it extensively, often in place of camomila (Vesey-Fitzgerald). Perhaps it is best known in country

medicine as a painkiller. Evidently, all that had to be done was to boil the plant in water, and drink the resulting liquid (Vickery. 1995), though a Suffolk practice of curing toothache by tying feverfew on to the wrist on the opposite side (V G Hatfield) sounds more like a charm than a remedy. It has been used in cold infusion as a general tonic, and a cold infusion of the flowers as a sedative (Brownlow). Perhaps that was what Gerard had in mind when he recommended it for "such as be melancholicke, sad, pensive, and without speech. It is certainly effective in curing a headache, even migraine, it seems (V G Hatfield), and apparently, it was said, warm, on the ear for earache, according to a Suffolk record (Kightly. 1984). But the dried flowers have been used in home remedies in Europe to induce abortion (Lewis & Elvin-Lewis).

East Anglian horsemen favoured the use of feverfew on their charges. In Cambridgeshire, the way to control unruly horses was to rub the freshly gathered leaves (or those of rue) on their noses (Porter. 1969), while Suffolk horsemen used it for colds, and for giving their horses an appetite (G E Evans. 1960).

FEVERS

NETTLE leaves chopped very small and mixed with whisked egg-white were applied to the temples and forehead in the Highlands as a cure for insomnia, but Martin, in the early 18th century, spoke particularly of fevered patients benefitting from the treatment. A Scottish charm for fever was to pluck a nettle by the root, three successive mornings, before sunrise (Dalyell), saying the name of the patient and his parents. Similarly, on the Aegean island of Chios, quartan fever is subject to St John the Baptist, and the patient invokes the saint frequently. But he can also take the following medicine, fasting: nettle-seed, ears of corn, 8 grains of cinnamon oil, all mixed in a mortar, and water added (Argenti & Rose). MEADOWSWEET is useful, too, for the plant contains some of the chemical constituents of aspirin. Boil the flowers for ten minutes in water, and drink three cupfuls of this a day (Flück).

BIRTHWORT (*Aristolochia clematitis*) was recommended for fevers in the Anglo-Saxon text of Apuleius, and also in the Welsh medical text known as the Physicians of Myddfai ("For intermittent fever. Take the mugwort, the purple dead nettle, and the round birthwort, as much as you like of each, bruising them well in stale goat's milk whey, and boiling them afterwards. Let the patient drink some thereof every morning, and it will cure him"). WORMWOOD was used a lot to treat fevers. The Myddfai text has "for treatment of intermittent fevers. Take dandelion and fumitory, infused in water, the first thing in the morning. Then about noon take wormwood infused in water likewise, drinking it as often as ten times, the draught being rendered tepid". Gerard has "it is often-times a good remedy against long and lingering agues, especially tertians", and there are various complicated leechdoms in Cockayne.

HOLLY leaves do seem to have an effect in relieving fevers and catarrh, and were once stated to be "equal to Peruvian bark" (Dallimore), quinine, in other words. BOX, too, is a febrifuge, still prescribed by herbalists and homeopathic doctors, who treat it as a substitute for quinine in malaria (Palaiseul). Somerset people used to boil the bark of WHITE POPLAR, and drink the infusion for flatulence and fevers (Tongue). A prescription from Alabama is to "take the ashes from burnt HICKORY wood, put them in water, and drink it for fever. Make it very weak, as it will eat the stomach" (R B Browne).

FEVERFEW, by its very name, one would think, would be the best possible medicine for a fever. It is OE feverfugen, from Latin febris, fever, and fugare, to drive away. Yet there is only one specific mention of fever in the list of recorded examples of folk medicine. That came from Derbyshire, and the belief was that all you had to do to cure a fever was to put a piece of feverfew in the bed (Addy). TANSY flower tea was given for fevers (Brownlow), for which the leaf tea was also used in America (Hyatt), where WATER MELON was used, too (Beck). Even the scent alone of PENNYROYAL was thought enough to help patients recover from fevers (Classen, Howes & Synnott).

American Indians used YARROW for fevers, either as a tea, or by putting the flower heads on a bed of live coals, and then inhaling the smoke (H H Snith. 1945). In Britain, there was an odder way of dealing with the problem "For an ague… boil Yarrow in new Milk, 'till it is tender enough to spread as a Plaister. An Hour before the cold Fit, apply this to the Wrists, and let it be on till the hot Fit is over…" (Wesley). COCKLEBUR tea has been used in America to reduce fevers (H M Hyatt). A recipe from the Scottish islands for "burning fevers", prescribed " a tea of WOOD SORREL… to allay the heat" (Pennant). Also in the Scottish Highlands, WATERCRESS tea is taken to reduce a fever (Beith). WALLFLOWERS were once popular for fevers – see Gerard: "The leaves stamped with a little bay salt, and bound about the wrists of the hands, take away the shaking fits of the ague" (cf Yarrow above). MARIGOLDS were used in medieval times for fevers, and as a hot drink to promote sweating (Lloyd); into the 18th century, Hill was still recommending a tea "made of the fresh gathered flowers … as good in fevers; it gently promotes perspiration …". WHITE HOREHOUND has been used in Africa for fevers, especially typhoid (Watt & Breyer-Brandwijk); the Navajo, too, used it for fevers (Wyman & Harris). BEETROOT leaves have been used for fevers since ancient times. One "confection for the fevers" is included in a 15th century collection of medical recipes, and reads "take centaury a handful;

of the root and of the leaves of earthbeet a handful; of the root of clover a handful; of ambrose a handful; and make powder of them, then mix honey therewith, and make thereof balls of the greatness of half a walnut. And give the sick each day one of them fasting, and serve him nine days …" (Dawson). That may have worked, especially as nine days may have been enough to see the fever off naturally, but one would have to question the Balkan practice of treating a fever by laying beet leaves on the skin round the waist, and changing it morning and evening, for three days (Kemp).

Pomet, speaking of CAMPHOR, claimed that "the Oil is very valuable for the Cure of Fevers, being hung about the Neck, in which scarlet Cloth has been dipped ….". HENNA, too is used in the Balkans in acute fevere, like typhoid. It is heated in water, allowed to cool and the juice of some 20 heads of garlic added, the mixture re-heated, and then the henna is applied solid to the palms of the hands and the soles of the feet (Kemp). Herbalists still use the dried rhizome as a febrifuge (it is a good substitute for Peruvian bark). It was well-known in the 18th century. Buchan, for instance, has a rather complicated receipt for intermittent fevers – "an ounce of gentian root, calamus aromaticus, and orange-peel, of each half an ounce, with 3 or 4 handfuls of camomile flowers, and an handful of coriander, all bruised together in a mortar".

A transference charm from the south of France has the fever patient sleeping with his back to a PEACH tree for two or three hours; the tree would gradually get yellow, lose its leaves, and die (Sebillot).

Ficus benghalensis > BANYAN

Ficus carica > FIG

Ficus religiosa > PEAPUL

FIELD COW-WHEAT

(*Melampyrum arvense*) Very rare in Britain now, for its presence in cornfields is not very desirable, as it contains a glucoside something like rhinanthin. If it gets into flour it colours it, and gives a bitter taste and a bad smell. No wonder it is called Poverty-weed in some parts of England, "with reference, no doubt, not only to the way in which it impoverishes the soil, but also to the fact that the seeds becoming mixed with the corn, rendered the latter of small value in the market" (Vaughan).

FIELD BINDWEED

(*Convolvulus arvensis*) There is little folklore attached to this plant. One Scottish belief mentioned by Wentz, in which putting the burnt ends of the stems over a baby's cradle to protect it from the fairies, is almost certainly bundweed, not bindweed, and refers to Ragwort, which is very much a fairy plant. Thunderflower is an interesting name for the bindweed. If you pick them, they say, it will be sure to thunder before the day is out (Vickery. 1985). Red Poppy has this name as well, and it is just possible that the name was given to discourage children from damaging crops in trying to gather them. Would they try to pick bindweed, though? Another "unlucky" name for bindweed is the Scottish Young Man's Death, from Perthshire (Vickery. 1985). In this case the result of picking the flowers, if a girl did it, would be the death of her boy-friend. Just possibly the superstition alluded to the way the flowers fade so quickly.

There are hardly any folk medicinal uses involving this plant. Gypsies, though, use an infusion of the leaves or flowers to expel worms (Vesey-Fitzgerald), and the plant has also been used as a wound dressing (Watt & Breyer-Brandwijk). Because it twines, the doctrine of signatures claimed it to be good for the intestines (Prest). Perhaps the plant has been spurned too much, for in recent times a recommendation as a good tonic has been given; an infusion of the stems, half an ounce to a pint of boiling water, to make a tea (A W Hatfield).

FIELD ERYNGO

(*Eryngium campestre*) Daneweed is a name sometimes given to it (A E Baker), for reasons similar to that other Daneweed (*Sambucus ebulus*) – wherever the blood of Danes was shed in battle, this plant afterwards sprang up, a tradition associated too with a third plant, the Pasque Flower.

FIELD MAPLE

(*Acer campestre*) An old Alsatian legend tells that bats possessed the power of rendering the eggs of storks infertile. So the stork put some branches of maple in its nest, so that every intruding bat was frightened away (Dyer. 1889). Bate's Bush was an old maple at the crossroads at Osebury Rock, in Worcestershire. It was said to be derived from the stake driven through the body of a man named Bate, a suicide buried there (Allies).

Long life would be conferred on children who were passed through the branches of a maple (Friend. 1883). One tree in West Grinstead Park, in Sussex, was in constant use for this purpose, and there was a great outcry when the landlord intended to have the tree felled. Specifically, though, the act of passing children through the branches was to cure them of rickets, or the effect of the evil eye (Thompson. 1897) (cf ASH, for example). Another odd belief, from northern France, was that the leaves became red in the autumn by the action of the fairy that lives in the tree (Sebillot).

It is said that maple leaves, layered with stored apples, carrots and potatoes, have a noticeable preservative effect (M Baker. 1978). Maplin-tree is a name given in Gloucestershire for this tree (Grigson. 1955), and this is a name Alfred Williams also noted as Maypole-ing tree, in some versions of the the wassail song:

Wassail, wassail, all over the town,
Our toast is white and our ale is brown,
Our bowl is made of a maplin tree,
And so is good beer of the best barley.

FIG

(*Ficus carica*) It was probably domesticated in southern Arabia, but its importance and value to the Israelites is illustrated by the fact that the prophets often threatened that the vine and fig crops would be destroyed unless they fell into line. "To sit under one's vine and one's own fig tree" became a proverbial expression among the Jews to denote peace and prosperity. An old tradition says that when Mary sought shelter for the infant Jesus from Herod's soldiers, a fig tree opened its trunk so that they could enter and hide (Moldenke & Moldenke). Figs appear widely in classical mythology, especially in cnnection with Dionysus/Bacchue; in fact one story says that the fig was created by Bacchus. During the bacchanalian feasts, Roman women wore collars of figs as symbols of fecundity, and the men carried statues of Priapus carved from fig-wood (Moldenke & Moldenke). Not only does a fig, like a pomegranate, also used as a fertility symbol, carry a large number of seeds, but it is pointed out that it resembles the womb in shape (Maple). Later, it was used as a symbol of lust (Ferguson). Bearing all this in mind, the choice of a fig-leaf for Adam and Eve was natural, figs being the fruit of the tree of life (Simons). And the traditional apron used by sculptors on their statues of the human figure is a fig-leaf. The fertility aspect of figs has been carried into recent folklore. For example, Bulgarian brides would be presented with dried figs as a promise of many children (M Baker. 1980), and Dutch folk medicine claims that a daily craving for figs during pregnancy ensures that the child will be born quickly and easily (van Andel).

But a certain mistrust of fig trees is evident; to this day, Greeks have a fear of sleeping under one (Kerenyi), and on the island of Chios they say that the shadows of both the fig and the hazel are "heavy", and it is not safe to sleep under either of them. There was a tradition in the south of France that St John the Baptist was beheaded under one of these trees. That is why branches break off so easily, particularly on St John's Day, when anyone who climbs the tree risks a dangerous fall. Similarly, in Sicily, the mistrust lies in the belief that Judas hanged himself on a fig tree (Porteous).

Applying a hot fig (to the tooth or on the cheek?) used to be a Cumbrian remedy for toothache (Newman & Wilson), and the juice of the leaf is sometimes used in East Anglia to put on warts (Hatfield), and boils are treated in Indiana home medicine by splitting a fig and applying it to the boil as a poultice (Tyler). They used to say in Alabama that to remove a birthmark

you should put a fig leaf poultice on the part marked
(R B Browne).

FIG MARIGOLD

(*Carpobrotus equilaterale*) "When any of these
people [the Tasmanians] fall sick, so as to be unable to
accompany the others in their daily removals, they are
furnished with a supply of food as the party happens
to have, and a bundle [*Carpobrotus equilaterale*] of a
plant known in the colony by the name of Pig-faces, ...,
and they are left to perish ..." (Ling Roth, quoting
Backhouse, *Narrative of a visit to the Australian
colonies*, 1843).

FIGWORT

(*Scrophularia nodosa*) Figwort has had its uses,
not least in beer-making, where the bitter principle
was once a valued asset (Wood-Martin). In the
Hebrides, more stress was put on its magical qualities,
especially in its use to protect cows (Carmichael), for
when put in the byre, it has the power of ensuring the
milk supply (J A MacCulloch. 1905). On mainland
Scotland, on the other hand, it was the medicinal
use that was more important. The "fig" of figwort
means piles. "... it is reported to be a remedy against
those diseases whereof it tooke his name, as also the
painefull piles and swelling of the haemorroides...
Some do stampe the root with butter, and set it in a
moist shadowie place fifteeene daies together: then
they do boile it, straine it, and keepe it, wherewith
they anoint the hard kernels, and the haemorrhoid
veines, on the piles which are in the fundament, and
that with good successe" (Gerard). It seems that
the tincture of the fresh plant is still recommended
for piles, as well as for eye complaints and mastitis
(Schauenberg & Paris).

The name of the genus, *Scrophularia*, is the indication of the other medicinal usage in ancient times.
The name comes from the Latin scrofulae, meaning
that swelling of the neck glands we know as scrofule.
It is more than likely that the doctrine of signatures
showed the way to this usage, the signature in this
case being the knobbly tubers (Dyer. 1889). That is
what the name Kernelwort (Gerard) implies, too, and
Scrofula-plant is another name given to it.

On mainland Scotland, the leaf would be applied to
cuts and bruises, and the tuber to sores and tumours
(Carmichael), and in some places to burns (C P Johnson), probably because it is an anodyne, and eases
pain wherever it is applied (Mitton), and that includes
toothache and babies' teething (Gerard). It probably
accounts for its reputation in the Channel Islands for
being an efficient remedy for cramps (Garis).

Filago germanica > CUDWEED

FILBERT

A name given to HAZEL nuts, of either *Corulus
avellana*, or, more accurately, *C maxima*. It appears

in various guises, as Filbeard, widely recorded in the
south and Midlands, sometimes shortened to Beard-
tree, and Filberd, or Filbord, as Evelyn had it. Brouk
suggested that filbert means 'full beard' (from the
fringed husk?), but the usual explanation is that it
comes from a non-existent King Philibert, or from St
Philibert, whose feast day falls on 22 August, when,
it is claimed, the nuts are ripe. But they are certainly
not ripe as early as that. Perhaps we are talking about
St Philibert, old style, giving a date in September,
when there is more likelihood of the claim being true.
Anyway, filbert is a Norman-French word, written
as philbert in the 13th century, and still in use in
Normandy patois at the beginning of the 20th century
(Skeat).

Filipendula ulmaria > MEADOWSWEET

Filipendula vulgaris > DROPWORT

FIRE-MAKING

MAORI FIRE (*Pennantia corymbosa*) is a New
Zealand tree, called kaikomako by the Maori, who
used it for friction fire-making, as the common name
implies. Maui was the deity who taught the people
how to do this (Andersen). See MAORI FIRE for a
version of the myth.

FLANDERS POPPY

The Greeks always reckoned RED POPPIES to be a
companion to the corn, not a weed (Grigson. 1955).
Perhaps, as Grigson suggested, red poppies were
regarded as a life-blood growing along with the
nourishing grain. Certainly, there was no difficulty
in accepting poppies as a natural consequence after
a battle, whether it was after Waterloo, or the battles
of a later war. The Flanders Poppy has become the
symbol of the blood shed there.

FLAX

(*Linum usitatissimum*) Flax seems to have been
regarded as a kind of talisman in central Europe,
against sorcery. In parts of Germany, it was believed
that seven-year old children would become beautiful
by dancing in the flax. It is a lucky plant – when a
German girl gets married, she puts flax in her shoes as
a charm against poverty (Dyer. 1889). Scandinavian
belief also appears to have placed flax in a protective
position. Unbaptised children could be preserved
from harm by "sowing flax seed", though the
authority quoting that does not elucidate (Kvideland
& Sehmsdorf), but sowing flax around the house, on
the road, or by the grave was a common means of
protection against the spirit of the dead in Norway and
Denmark. Flax seed was also put in the coffin, to keep
ghosts away, and round the grave or the house. The
ghost must count every single seed before going any
further, for flax is a magic plant (Rockwell). Popular
explanations assign the power of flax to the belief that
Christ was swaddled in a linen cloth (Kvideland &
Sehmsdorf).

The Hallowe'en divination game involving hemp-seed was also played with flax-seed in north-east Scotland. The rhyme was:

> Lint-seed I sow ye,
> Lint-seed, I sow ye,
> Let him it's to be my lad,
> Come after and pu' me (Gregor).

There was a belief among the Pennsylvania Germans that if you wanted your flax to grow tall you had to show it your buttocks (Fogel). Obviously the same belief occurs in this country, even if more modestly performed, for a Yorkshire farmer sat on the seed bag three times and faced east before sowing, to ensure a good crop. A few stolen seeds in the bag were useful, too (M Baker. 1980). It is best to plant it on Good Friday, at least according to Kentucky belief (Thomas & Thomas). An old tradition tells that flax will only flower at the time of day at which it was originally sown (Dyer. 1889). It was said in Germany that the flax would be sure to prosper if the sun shines on Candlemas Day (2 February). Another popular belief there was that if the sun shone on New Year's Day, the flax would grow straight (J Mason). There are some examples of homeopathic magic connected with the plant. A German practice, for example, involved weakly babies, who would be put naked on the turf on Midsummer Day, and have flax-seed sprinkled over and around them. As the flax grows, so the child will gradually grow stronger. Similarly, in Brandenburg, to cure dizziness, it was recommended that the patient should run naked, after sunset, through a field of flax; the flax will take the dizziness to itself (Dyer. 1889). At a wedding feast, or any other occasion when there was dancing, the Pennsylvania Germans always used to "dance for flax", that is, the higher the feet off the floor, the higher would the host's flax crop grow (Hoffman).

A correspondent of *FLS News. 41; November 2003* draws attention to the account given by Robert Graves, *The Greek myths*, of the flax harvest in the Austrian Alps, where men are not admitted to the harvest. There is a hag-spirit called Harpetsch who attacks men, and the women who beat the flax will chase and surround any stranger who blunders into their midst, and they will attack him with the prickly flax-waste.

The seeds, which are laxative (Schauenberg & Paris), have always been used both as food and as medicine, chiefly in poultices, either by grinding the seed, or by using it as a pulverized cake. Scots travellers used the linseed poultice, with a little mustard on it, to treat even a condition like pneumonia (MacColl & Seeger). In fact, they are often used as soothing poultices, even in times past for sore breasts. Dawson, for instance, quoted a leechdom fom the 15th century: "for women's teats that be swollen: take linseed, and the white of an egg or else the juice of smallage, and lay thereto". They had other uses: a plaster for splints was made in the Balkans from a mixture of flax-seed, egg-white and powdered alum (Kemp). That particular usage is recorded from Ireland, too – "with white and yolk of egg [it is used] by bone setters and makes excellent splinting material when supported with leather" (Moloney).

When the seeds are infused, the result is known as linseed tea, commonly used as a demulcent remedy (Fluckiger & Hanbury). This linseed tea is drunk by gypsy women during pregnancy to ensure an easy birth (Vesey-Fitzgerald). Hill prescribed it as "excellent in coughs and disorders of the breast and lungs…". There is a similar prescription in Alabama folk medicine. It is used for whooping cough there – "take three pounds of flax seed, steep in one quart of water for three hours, mix with two lemons and two cups of sugar or some honey. Give this often as hot as the patient can take it" (R B Browne). This tea was used in Russian folk medicine for kidney complaints, while a decoction was taken for dropsy (Kourennoff). A similar decoction was an Irish country cure for influenza (P Logan). So it was in the Highlands, too, with honey and a little vinegar added (Grant), and bad coughs were treated in the Fen country by boiling some linseed for a few hours, then putting black liquorice in the liquid and drinking that (Marshall).

Linseed oil itself is used for medicinal purposes; for stone in Russian domestic medicine (Kourennoff), and it is taken too for bladder trouble by Afro-Caribbeans in Florida, with cream of tartar (Hurston). It is often used for sprains and bruises (Cullum), while "Carron Oil" (for it was first introduced in the Carron Iron Works) is a mixture of equal parts of linseed oil and lime water, and is an embrocation for small injuries, burns, and for rheumatism and gout (Wickham). Linseed oil has even been used, and not as an embrocation, but taken internally, for tuberculosis in Ireland (P Logan). One might mention that linseed oil is often fed to dogs. It makes their coats silky and bright, and is often given to a dog in the days before a show (Cullum).

As far as curing uses are concerned, we are left with such pieces of wisdom as "to get grit from the eye, put one or two flax seeds in it", from Alabama (R B Browne), and a practice recorded from Dundee, where it was formerly believed that a hank of yarn worn round the loins was a certain cure for lumbago (Fernie).

FLEAS
see **VERMIN**

FLEAWORT
(*Plantago psyllium*) Named "not because it killeth fleas, but because the seeds are like fleas" (Gerard). This is a plant from the Mediterranean area,

sometimes found in Britain as a casual. The seeds have been used in medicine; the mucilage contained in them swells up when they are administered, giving bulk to the intestinal contents, and acting as a laxative (Fernie).

FLOTE GRASS

(*Glyceria fluitans*) In North America, the seeds used to be collected and sold as "manna seeds" (hence the American name Mannagrass (Douglas)), for making puddings and gruel. It was even cultivated there for the purpose (C P Johnson).

Foeniculum vulgare > FENNEL

FOOL'S WATERCRESS

(*Apium nodiflorum*) It seems a little hard to saddle this plant with the name Fool's Watercress, inferring that anyone mistaking the two plants would be a fool indeed. Even if the mistake were made, no great harm would be done, and it is significant that most of the local names given to it belong to the cresses, even though this plant is not a cress. West country people used to collect it at one time, to cook with meat in pies and pasties (Grigson. 1955), more particularly, it seems, in the neighbourhood of Polperro (Quiller-Couch).

FOUR-LEAVED CLOVER

Whoever has a four-leaved clover has luck in all things (even in love potions, according to the Channel Islands (Garis)). He cannot be cheated in a bargain, nor deceived, and whatever he takes in hand will prosper. It brings "enlightenment to the brain, and makes one see and know the truth". But it must never be shown to anyone, or the power would no longer exist (Wilde. 1890), and it must never be taken into a church, for then they would become very *un*lucky (Nelson. 1991). Bretons say it will drive away even the devil himself (it makes by its shape the sign of the Cross), and in the Vosges anyone who has it about him without knowing it can kill a werewolf with a bullet (Sebillot), something that in normal circumstances cannot be done. A four-leaved clover will enable the finder to see the fairies, and to break the powers of enchantment (Vickery. 1995). See, for example, the Irish folk tale in which a travelling magician, or perhaps master of hypnosis would be nearer the mark these days, made simple folk at a fair believe that his cockerel, on the roof of a house, was carrying a long block of timber in his beak. A girl carrying an armful of fresh cut grass came along and saw only normality. The magician was quick enough to realise what was going on, and immediately bought the grass from the girl. At once, she became as enchanted as the others; there was a four-leaved clover in her bundle of grass. It was said that fairy ointment was made of four-leaved clover (see Briggs. 1978).

In a story from Hyde's collection, a widow is told to put a piece of Mary's shamrock, presumably a four-leaved clover, in her sick son's drink. It cured him miraculously. The belief in its powers spread to American folklore – it is lucky to find or keep, but bad to give away. Another American belief is that whereas a four-leaved clover is lucky, a five-leaved one brings nothing but misfortune (Bergen. 1896). But there is a Scots proverb, "he found himself in five-leaved clover", i.e., in very comfortable circumstances (Cheviot). There is a Quebec superstition that if you put a four-leaved clover in your shoe, you will marry a man having the first name of the man you meet first after doing so (Bergen. 1896). Another example comes from Michigan: with a four-leaved clover in your shoe, you will meet your lover. From the same area comes another belief: if the finder of a four-leaved clover puts it in her own shoe, she will marry the first person with whom she crosses a bridge. From other parts of America, the injunction is to put the four-leaved clover over the door. The first person to pass under it will be your future mate (Bergen. 1896).

These days, in America, four-leaved clover is grown commercially. There are clover farms there, and each leaf is enclosed in plastic and sold as "good luck charms"! (Vickery. 1995).

FOXGLOVE

(*Digitalis purpurea*) In Hartland, North Devon, foxgloves are associated with St Nectan. Wherever a drop of his blood fell, a foxglove sprang up. There is a foxglove procession there on the Sunday nearest the patronal festival, 17 June (Vickery. 1995). But everywhere else, the association is with the fairies, as many of the local names for the plant confirm. It is Fairies' glove, fingers, thimbles, hats, caps and dresses, and so on. When the plant bows its head, it is a sure sign that a fairy is passing (Boase). "Fairies have been seen dancing under foxgloves in Cusop Dingle within the memory of some now living there" (Leather). But of course they are seen to a much greater extent in Ireland – see Yeats (*Irish fairy and folk tales*) – "and away every one of the fairies scampered off as hard as they could, concealing themselves under the green leaves of the lusmore, where, if their little red caps should happen to peep out, they would only look like its crimson bells …". Indeed, the Shefro, one of Ireland's more gregarious fairies, is always described as wearing foxglove flowers (Wentz). (Lusmore, the great herb, is the name normally used for foxglove in Ireland).

As so often happens in folklore, the fairies' favourite plant is also the one that offers most protection against them. In Ireland particularly, it can break the fairy spell (but can also cause the individual to be fairy struck (Wilde. 1902)). Nevertheless, it can cure any disease the fairies might cause (Logan). It is very useful when dealing with a changeling, a child "in the fairies", as is sometimes said. The child is bathed in the juice, and fairy-struck children had to be given the juice of twelve (or some say ten) leaves of foxglove (Wilde. 1902). Or a piece of the plant could be put under the bed. If it is a changeling, the fairies would be compelled to restore the true child (Mooney). Simpler still, put some leaves on the child itself, and the result would be immediate (Gregory). And instructions from County Leintrim advised a suspicious parent to "take lusmore and squeeze the juice out. Give the child three drops on the tongue, and three in each ear. Then place it [the suspected changeling] at the door of the house on a shovel (on which it should be held by someone) and swing it out of the door on the shovel three times, saying "If you're a fairy, away with you". If it is indeed a fairy child it will die that night; but if not it will surely begin to mend" (Spence. 1949).

The Irish also used it as an effective charm against witchcraft. The patient was rubbed all over with it – a dangerous practice, and the patient may die of it, especially if tied naked to a stake, as was the custom once (Wilde. 1890).

After all this, it is not surprising to find that it is an unlucky plant to have indoors, and just as unlucky to have on board a ship (M Baker. 1980). It is unlucky to transplant a foxglove, so it was said in Hampshire; but if one grows from a seed it should be nurtured, so that it will set as a lucky mascot (Boase). Gardeners say that foxgloves stimulate growth in plants growing near them, and help to keep them disease-free (Boland & Boland). Foxglove tea, we are told, added to the water makes cut flower last longer (M Baker. 1980).

The dried leaves are the source of a very potent drug that has the effect of reducing the frequency and force of the heart action, so it is given in special cases as a sedative, especially in heart disease. It was Dr William Withering (1741-1799), from Wellington, Shropshire, who first introduced digitalin into general medical practice. He published "an account of the Foxglove and some of its medical uses" in 1788. It is said he got his information from a witch. But this is indeed a dangerous plant, which animals always avoid. All parts are poisonous, but especially the seeds. The leaves are more active before than after flowering (Long. 1924). There was another use of the toxic principle; that was in what the Americans call a "chemical jury". In other words it was used in ordeal trials to test guilt or innocence – if he survived he was innocent! (Thomson. 1976).

Gypsies use an ointment made from the fresh leaves to cure eczema (Vesey- Fitzgerald), and in early times the leaves were used mainly as an external application for wounds and ulcers in the legs (Clair), for the toxic potentialities were recognized very well early on. The gypsy usage for ulcers was certainly known very early, for in the Anglo-Saxon version of Dioscorides (in Cockayne's translation) we have "For inflammatory

sores, take leaves …, work to a poultice, lay to the sore", and also "for a pimply body, take this same wort and fine flour, work to a poultice, lay it to the sore". Much later, there are records from the Highlands for this use of the leaves on boils, and also on bruises (Grant), or to cure erysipelas (Polson. 1926), while on Skye a plaster made from them used to be applied to remove pains that follow fever (Martin).

Along with barley meal and some other herbs, it was included in an Irish preparation to treat epilepsy (Logan), but what herb was not tried at some time or other? As has been said (Thomson), desperate conditions demand desperate remedies! It must have been a quite dangerous practice, but there are other records of foxglove leaf infusions being taken. Gypsies use a very weak infusion of the dried leaves for fevers (Vesey-Fitzgerald), and foxglove tea was apparently a standard domestic remedy for dropsy (Beith). Irish people used to make a tincture for it with gin, and then use a very small quantity on loaf sugar (Egan). But given the known effect on the heart, there should be no surprise at that. But using the leaf infusion as an emetic, as was done in Ireland (Logan) is another matter. Other Irish uses, for lumbago (Ó'Súilleabháin), or for hydrophobia (Wood-Martin), for example, did not need internal consumption. But there is another case of an infusion taken internally; according to the *Times Telescope*, 1822, "the women of the poorer class in Derbyshire used to indulge in copious draughts of foxglove tea, as a cheap means of obtaining the pleasures of intoxication". Actually the practice was far from being confined to Derbyshire.

FOXTAIL
(*Alopecurus pratensis*) i.e., MEADOW FOXTAIL. This is the grass used by children for making "Chinese haircuts". They strip the flowers off the stalk, and twiddle this stalk into the hair of the child sitting in the desk in front. A swift tug would quickly remove all the hair attached – a very painful process (Vickery. 1995).

FRACTURES
In 17th century Skye a mixture of BARLEY meal and white of egg was applied as a first aid measure for broken bones. After that, splints were used (Beith). But it is COMFREY that is the fracture herb par excellence. The glutinous matter of the roots was grated and used (and still is) for a plaster that set hard over a fracture. Sometimes a charm had to be spoken at the same time. In Ireland, it was taken internally, as part of the process of knitting fractures (Moloney). Many of the names for comfrey proclaim the usage; comfrey itself, from Latin confervere, to heal, and its alternative Consound, much applied to this and to plants of a similar reputation, and coming from Latin consolida. Vernacular names include many that advertise its qualities, names like

Knitbone and Boneset, and a few others. Comfrey is *consolida major*, and DAISY is *consolida minor*. "The Northern men call this herbe a Banwort because it helpeth bones to knyt againe" (Turner). Daisies were even called Bone-flowers in the north of England (Grigson. 1955). BUTCHER'S BROOM has been used in a similar way – a decoction of the leaves and berries was made into a poultice applied to help broken bones to knit (Leyel. 1937). Evelyn recommended the use of MYRTLE berries as a consound. In the Balkans, a plaster for splints was made from a mixture of FLAX seed, egg-white and powdered alum (Kemp); that particular usage is recorded from Ireland, too: "with white and yolk of egg [it is used] by bone setters and makes excellent splinting material when supported with leather" (Moloney). SOLOMON'S SEAL is another plant with a reputation for helping broken bones knit, either taken inwardly in ale, or as a poultice (Grigson. 1955), a use mentioned by Gerard, who said "there is not to be found another herbe comparable to it. The root stamped and applied in manner of a pultesse, and laid upon members that have been out of joynt, and newly restored to their places, driveth away the paine, and kintteth the joynt very firmly ..." (perhaps that is why it is known as "seal", conjectured Mrs Leyel) (Leyel. 1937). Culpeper recommended the leaves of BEAR'S BREECH, "bruised or rather boiled and applied like a poultice are excellent good to unite broken bones, and strengthen joints that have been put out".

Fragaria x ananassa > STRAWBERRY

Fragaria vesca > WILD STRAWBERRY

FRANGIPANI
(*Plumeria alba*) According to legend, in the 12th century an Italian called Frangipani, by combining certain volatile oils, created an exquisite perfume. European settlers in the Caribbean 400 years later discovered a plant whose flower had a similar perfume, so it was naturally called Frangipani. In Asia, it is often planted near Buddhist temples (hence the names Pagoda Tree and Temple Flower, often given to it (Leyel. 1937)) in order that the blooms be readily available as temple flowers and as offerings to the gods (*Chinese medicinal herbs of Hong Kong. Vol 3; 1987*). It is also planted in Indian cemeteries, so that the daily fall of white flowers covers the graves (M North).

FRANKE is a word that seems to mean a stall in which cattle were shut up to be fattened. CORN SPURREY (*Spergula arvensis*), which has this name, was certainly grown in Britain to fatten cattle, and probably still is on the Continent (Prior). Halliwell has a slightly different explanation. He described Franke as "a small inclosure in which animals (generally boars) were fattened".

Fraxinus excelsior > ASH

FRECKLES
The water that collects in the cups formed by the fusing together of the TEASEL's opposite leaves was much prized for cosmetic use, and in Wales it was said to be a remedy for freckles (Trevelyan). Another Welsh practice was to use a wart cure from the Physicians of Myddfai to cure freckles incidentally: "take the juice of SHEEP'S SORREL, and bay salt, wash your hands and let them dry spontaneously. Do this again and you will see the warts and freckles disappear. DANDELION flowers, boiled for half an hour in water, give a toilet water to get rid of freckles on the face (Palaiseul), and SILVERWEED was another cosmetic, steeped in buttermilk, to remove freckles and general brownness (Black). Gerard also advised the use of this herb: "the distilled water takes away freckles, spots, pimples in the face, and sun-burning". The sliced roots of BLUE FLAG were once applied to the skin for cosmetic effects, mainly to get rid of freckles (Le Strange).

In America, the roots of WHITE POND LILY (*Nymphaea odorata*) are used for the purpose. They produce a liquid that, mixed with lemon juice, was supposed to remove freckles (Sanford). Another American practice, from Kentucky, is to wash the face with melon rind, to get rid of freckles (Thomas & Thomas).

FRENCH BEAN
(*Phaseolus vulgaris* see KIDNEY BEAN). They were introduced into Britain from France, and the name French Bean was already in use by 1572.

Fritillaria imperialis > CROWN IMPERIAL

Fritillaria meleagris > SNAKE'S HEAD LILY

FROG ORCHID
(*Coeloglossum viride*) It was reported that the Ojibwe Indians in America used this orchid as an aphrodisiac and love charm, details not available (Yarnell).

FROSTBITE
An ointment made from the crushed bulbs of SNOWDROPS has been used as a cure for frostbite and chilblains (Conway).

FUGA DAEMONUM
An old book name for ST JOHN'S WORT, englished into Devil's Flight (chasse diable in French), and given because of the many examples of its power to "cure melancholy" and to drive away all "fantastical spirits". A 13th century writer tells of "the wort of holy John whose virtue is to put demons to flight" (see Summers. 1927). Aubrey. 1696 mentions a case where St John's Wort under the pillow rid a home of the ghost that haunted it. Langham. 1578 was another writer who advised his readers to keep some in the house, for "it suffereth no wicked spirit to come there".

Fumaria officinalis > FUMITORY

FUMITORY

(*Fumaria officinalis*) Best known for its cosmetic uses, particularly as a face wash and skin purifier (Fernie).

> If you wish to be pure and holy,
> Wash your face with fevertory (Dartnell & Goddard).

A leaf infusion is used (C P Johnson), or the whole plant boiled in water, milk and whey (Black). " …its remarkable virtues are those of clearing the skin of many disorders …" (Thornton), especially for babies with scalp trouble (Ireland) (Moloney). Gerard also remarked that it is "good for all them that have either scabs or any filthe growing on the skinne, and for them also that hath the French disease".

> Them that is fair and fair would be
> May wash them-selves in butter milk and fumitory
> And them that is black and black would be
> May wash them-selves in sut and tea.

Herbalists still use it for skin diseases (Schauenberg & Paris).

Earth-smoke, or Fume-of-the earth (Britten & Holland) are old book names for this plant. Fumitory itself is from Old French fumeterre, medieval Latin fumus terrae, i.e., smoke of the earth. One reason is the belief that it did not spring up from seeds, but from the vapours of the earth (Friend. 1883). The root when freshly pulled up gives a strong gaseous smell like nitric acid, and this is probably the origin of the belief in its gaseous origin (Britten & Holland). Another suggestion is that the Greeks and Romans used the juice to clear the sight, and noted that while doing so it would make the eyes water, as smoke would. This use also appears in English herbals, Turner, for instance, claimed that "the juice of thys herbe, which in dede is sharpe, maketh clear eyes". All very well, but perhaps the reason lies simply in the appearance of the plant, for from a distance it does look like smoke.

FUNERALS

ALMONDS, usually associated with weddings, can, under certain circumstances, appear at a funeral in Greece. It would have to be that of a spinster, the symbolism being the same. At such a funeral the almonds are for a parody of a wedding that did not take place in life, and also to mark her wedding to Christ (Edwards). BROAD BEANS, because they were the food of the dead, came to be used at funerals in classical times, and the tradition in the northern counties of England, used to be (at least in the 1890s) that broad beans should always be buried with the coffin (Pope). Children used to recite:

> God save your soul,
> Beans and all.

BOX is evergreen, with many of the associations shared with other evergreens, notably its use at funerals. In the north of England, a basin full of box sprigs was often put at the door of the house before a funeral, and everyone who attended was expected to take one to carry in the procession and then to throw into the grave (Ditchfield). Or a table would be put at the door, with sprigs of rosemary and box, for each mourner to pick up as he came into the house (Vaux). Box grown in Lancashire gardens used to be known as Burying Box (Vaux). It was thrown into the grave in Lincolnshire, too, as a symbol of life everlasting. Small sprigs are sometimes found when old graves are disturbed; though dry and brittle, they are usually quite green (Gutch & Peacock). They have even been found associated with Romano-British burials in Cambridgeshire and Berkshire (Vickery. 1984). Presumably, this practice led to the Dorset superstition that a sprig of box in flower brought indoors meant that death would soon cross the threshold (Udal). Jersey burial customs required the coffin to be covered with BAY and IVY (L'Amy), and there was once a custom in parts of Wales for a woman carrying bay to precede the funeral. She would sprinkle the road with the leaves at intervals (J Mason). ROSEMARY, the symbol of remembrance, must also be the emblem of funerals. Its use by mourners at funerals made it a token to wear in remembrance of the dead, and in the 1930s there was a demand for it for Armistice Day ceremonies (Rohde) along with the more conventional symbolic poppies. In the north of England, mourners at a funeral carry a sprig of the shrub. It was "breach of decorum" for a mourner to attend a funeral without it. The Lincolnshire custom was to put rosemary on the breast of the corpse, and it was buried with him (Gutch & Peacock), just as in France it was once the custom to put a branch of it in the hands of the dead (Thompson. 1897). Around Northwich, in Cheshire, mourners were given funeral biscuits, wrapped in white paper and sealed with black sealing wax. A sprig would be tucked in the folds of the paper, and this was thrown onto the coffin as it was lowered into the grave (Hole. 1937). It is odd to find one herb to have marriage associations, and at the same time to be a funerary herb; there is a story that Hartland has of a widower who wished to be married again on the day of his former wife's funeral, because the rosemary used at the funeral could serve at the wedding too. That reminds one of Herrick, *Hesperides*:

> The Rosemarie branch
> Grown for two ends, it matters not at all,
> Be't for my Bridall, or my Buriall.

CARAWAY seed cake was always given at funerals in Wiltshire (Clark), and in Lincolnshire, seed bread, made with either caraway or tansy seeds, is still traditional at funerals (Widdowson). PARSLEY, in modern Greece, is still a ritual ingredient of funeral food (Edwards). Ancient Greeks put wreaths of it on tombs, for it was said to have sprung from the blood of the hero Archemorus, the forerunner of death. It was used in cemeteries dedicated to Persephone, too (Sanecki). MYRTLE, Aphrodite's plant, was also the tree of death, particularly, in Greek mythology, the death of kings (Graves). Perhaps that is why there is such a prejudice against it in America. It is rarely seen outside cemeteries.

CHRYSANTHEMUMS, favourite flowers in most places, are very unlucky, particularly in Italy, for they are funeral flowers there, and associated with the dead (hence a connection with All Souls' Day). They say that if you give chrysanthemums to anyone it is the equivalent of saying I wish you were dead (Vickery. 1985). So too it is not a flower to take indoors, for it would bring bad luck with it (Vickery. 1995). ARUM LILY is another flower associated with funerals, and so an unlucky plant, not to be taken indoors (Deane & Shaw), and *never* brought into a hospital (Vickery. 1985). RAMPION BELLFLOWER has funeral associations, too. It is not a lucky plant, especially among children. It was said in Italy to give them a quarrelsome disposition, and could even lead to murder (Folkard). There is some mention of strewing MARIGOLDS on a grave at a funeral (Bloom, F G Savage). WILLOWS, besides being symbols of grief and mourning, can be looked on as symbols of resurrection because of their association with water (Curl), and that may be the reason why branches of willow are carried by mourners at a mason's funeral (Puckle).

Solitary HAWTHORNS are often associated with the dead, and they frequently mark graves, or they may mark the spot where a coffin has rested, or where a death had taken place in the open (J J Foster). Some thorns are dedicated to Irish saints, and these figure in burial ceremonies. When funerals pass by, they halt, and stones are placed beside the thorn until they have become cairns. On thorns at crossroads it was the custom in County Wexford to hang small crosses, made of coffin wood, as the funeral procession passed by (E E Evans). BLACKTHORN seems to have been buried with corpses in Ireland (Ó Súilleabháin).

ROSES have connections with death as well as with love (see under **DEATH**). It was the custom in parts of England for a young girl to carry a wreath of white roses, and walk before the coffin of a virgin. The wreath would be hung in church after the funeral, above the seat she had used during life, until the flowers faded (see also **MAIDEN'S GARLANDS**). THYME is an unlucky plant, connected with death, and especially with murder. But it is also used in funeral ritual sometimes. A sprig of it is carried by the Order of Oddfellows (Manchester Unity) at the funerals of one of their brothers, and then cast into the grave. That was a common custom in parts of Lincolnshire, notably at Massingham, when sprigs were dropped on to the coffin (Gutch & Peacock). It is planted on graves in Wales, too (Gordon. 1977). SAGE is a funerary plant in some areas, and graves were planted with it (Drury. 1994), something that Pepys noticed in April, 1662: "To Gosport; and so rode to Southampton. In our way … we observed a little churchyard, where the graves are accustomed to be sowed with Sage". At a gypsy funeral, according to one description, five bulbs of GARLIC were among the articles placed beside the body in its coffin (Sanderson). SWEET SCABIOUS bears such names as Mournful Widow, or Poor Widow in southern England, and it is interesting to find that, under the name Saudade, this plant was much used in Portugal and Brazil for funeral wreaths (Coats).

YEW branches were used at funerals – branches were carried over the coffin by mourners (Ablett), and in Normandy a branch was sometimes put beside a corpse awaiting burial (Johnson). Shakespeare speaks of a "shroud of white, stuck all with yew" (*Twelfth Night*, II. 4). In fact, sprigs of yew were tucked into shrouds at late medieval funerals (Morris). Thomas Stanley, in 1651, wrote:

Yet strew
Upon my grave

Such offerings as you have,

Forsaken cypresses and sad Ewe
For kinder flowers can take no birth
Or growth from such unhappy Earth.

FURZE
see GORSE. Gorse is the standard English name for the shrub, but Furze is locally common as a name, especially in the southern half of the country. It is OE fyrs, and has been rendered in a number of different ways by recorders trying to accomodate them to local pronunciation. Hence the widespread Fuzz, or Fuzzen, which may very well be a plural form in '-en' (Gerard used Furzen, too). Sometimes, furze itself by its sound, is taken as a plural, as in Ireland, when a single bush can be referred to as a Fur, or a Fur-bush (Lucas). Fur is recorded from Lincolnshire, too, or there is Furra, from Norfolk (Grigson. 1955).

FUSTIC

The name comes from an Arabic word, fustuq, which apparently means a small tree. But that Arabic word derives from a Greek one, pistake, which is pistachio (*Pistacia vera*). However, the name is used to define dyestuffs. "Young" Fustic is VENETIAN SUMACH (*Cotinus coggyria*), whose wood, with the proper mordant can produce bright yellow in cotton and wool. There are more fustic names for this small tree, Hungarian Fustic, for instance (Watt & Breyer-Brandwijk), and Zante Fustic (W Miller), Zante being the name of one of the Ionian islands. "Young" fustic, to distinguish it from "Old" fustic, which is African Oak, *Chlorophora excelsa*.

G

Galanthus novalis > SNOWDROP

Galeobdalosn luteus > YELLOW ARCHANGEL

Galium aparine > GOOSE-GRASS

Galium verum > LADY'S BEDSTRAW

GALL WASPS,

responsible for the growths, known as galls, on DOG ROSES. These enjoyed a great reputation at one time. Often known as Briar-balls, they had a number of more picturesque names, such as Robin Redbreasts's Cushions (Latham), or Robin's Pillows, or Robin's Pincushions (Page), as well as Canker-balls (Elworthy. 1888). They used to be sold by apothecaries to be powdered and taken to cure the stone, as a diuretic, and also for colic. Boiled up with black sugar, the result would be drunk for whooping cough. That is a gypsy remedy, but country people generally used to hang them round their necks as amulets against whooping cough (Grigson. 1955) (even merely hanging them about the house (Rolleston) for rheumatism (Bloom), or piles (Savage)). Putting one under the pillow was a Norfolk way of curing cramp (Taylor). In Hereford and Worcester the gall was carried round in the pocket to prevent toothache (Leather), and Yorkshire schoolboys wore them as a charm against flogging (Gutch); that is why they were known as Savelick, or Save-whallop (Robinson), in that area. Gerard mentioned the galls, his reference being Pliny, and reckoned that, stamped with honey and ashes, it "causeth haires to grow which are fallen through the disease called Alopecia, or the Foxes Evill".

GANGRENE

As Culpeper pointed out, "the meal of DARNEL is very good to stay Gangrenes", a usage that was still in vogue in the 20th century in the Balkans, where some darnel is still pushed into wounds (Kemp). One of the names given to MARSH MALLOW was Mortification-root, for the powdered roots make a poultice that would remove obstinate inflammation and prevent "mortification", gangrene in more recent terms, in external or internal injury (AW Hatfield).

GARDEN RADISH

(*Raphanus sativus*) There were odd beliefs about the efficacy of radishes. For example, "if the vintener cutteth a Radish into slices, and bestoweth those pieces into a vessel of corrupt Wine, doth in a short time draw all the evil savour and lothsomeness (if any consisteth in the Wine) and to these the tartness of it like reviveth" (T Hill). Lupton claimed that "if you would kill snakes and adders, strike them with a large Radish, and to handle adders and snakes without harm, wash your hands in the juice of Radishes and

you may do so without harm". It would be interesting to know if anyone actually tried this.

There is an annual radish feast at Levens Hall, about half way between Kendal and Milnthorpe, in Cumbria, held on 12 May (Ellacombe), which is the day on which Milnthorpe used to hold its fair (granted in 1280). Part of the feast seems to have been some kind of initiation ceremony, but the mayor and corporation of Kendal and most of the gentry attended the feast, of radishes and oatbread and butter (Vickery. 1995). Another traditional radish feast was held at the Bull Inn, New Street, in St Ebbe's, Oxford, after the annual meeting for the election of churchwardens (Bloxham).

GARDENERS' WISDOM

GARLIC and roses help each other. A single clove planted beside each rose will keep the greenfly away (Boland & Boland) and will enhance the scent. It is also said to contribute to keeping fruit trees healthy. They say, too, LILIES-OF-THE-VALLEY will only thrive with SOLOMON'S SEAL ("their husbands") growing nearby (M Baker. 1974). FOXGLOVES, too, will stimulate growth on plants growing near them, and help keep them disease free. So does PARSLEY, which often used to be grown as an aid to other cultivation. If it was planted all round the onion bed, for instance, it would keep onion fly away (Rohde). It is grown among roses, too, both to improve the scent and to help repel greenfly. It helps tomatoes and asparagus, too, and it encourages bees into the garden. There are many superstitions about planting or sowing parsley – see **PARSLEY**. NASTURTIUM. too, will keep greenfly at bay. Grow them up the trunks of fruit trees, and plant them in the greenhouse against woolly aphis. And plant them on St Patrick's Day, if you want them to grow fast (Boland & Boland). They say that STRAWBERRIES grow best when planted near nettles (*Notes & Queries. 4th series. vol 19; 1872*).

WALLFLOWERS can be companion plants for an apple tree; they say it encourages the tree's fruiting (M Baker. 1980). A dead tomato plant hung on the boughs of an apple tree through the winter, will preserve it from blight. Or the plant can be burnt under the tree, so that the smoke can ascend among the branches (Quelch). Sow TURNIP seed thickly in a part of the garden infested by couch, and the latter will disappear (Boland & Boland).

> Plant turnips on the 25 July,
> And you'll have turnips, wet or dry

is one of the gardening adages from Kentucky, but they also tell one to plant them on 10 August (and certainly not on the 7 August). To have good luck with them, say as you throw out a handful of seed: "One for the fly, one for the devil, and one for I" (Thomas & Thomas).

Getting rid of NETTLES has always exercised the minds of gardeners. In Herefordshire they used to say that if nettles were well beaten with sticks on the day of the first new moon in May, they would wither and not come up again. The advice given in the old rhyme:

> Cut nettles in June,
> They come up again soon.
> Cut them in July
> They're sure to die (Udal),

seems to be more practical. But gardeners are quite keen on the plant, within reason, for it is said to stimulate the growth of all plants around them while actually growing, as well as being the best thing to hasten the decomposing of a compost heap. Growing controlled clumps of them between currant bushes will help them fruit better (Boland & Boland).
SPEAR PLUME THISTLE:

> Cut thistles in May, they'll grow again some day,
> Cut thistles in June, that will be too soon.
> Cut thistles in July, they'll lay down and die
> (Leather).

COUCHGRASS presents problems, too. The answer, so it is claimed, is to sow turnip seed thickly in the part of the garden that is infested; the couch will disappear (Boland & Boland). Bury a stick of RHUBARB here and there in a bed when planting cabbages, against club-root (Boland & Boland).

If a ROSE is pruned on St John's Eve, it will bloom again in the autumn (Napier), a superstition only in the insistence on St John's Eve. And there is another, common in the Wessex area, that is not strictly folk-lore at all, but an observation of companion planting—people grow onions at the foot of a rose, with the object of making the flowers healthy and improving their perfume (Udal) (cf GARLIC above). MARI-GOLDS are good to have in the garden, for they will keep fly pests away from a vegetable patch (M Baker. 1977). So will AFRICAN MARIGOLDS, which will kill eelworms at three feet distance (M Baker. 1977). Grow them, too, near potatoes and tomatoes (Boland & Boland), as well as carrots and onions (Vickery. 1995), to keep pests away. CAMOMILE is looked on as a "plant physician", restoring to health any sickly plant near which it grows (Bardswell). Of course, one has to be judicious in its use. It should only be planted near the ailing plant for a short time; if the camomile clump grows too big, it has to be moved, otherwise the other plant will weaken again, "as though the patient had become over-dependent on the doctor and tiresomely hypochondriac" (Leyel. 1937).

There were proper times for planting BROAD BEANS, though they seem to have varied rather widely. The favourite seems to have been St Thomas's Day (21 December), but equally well known is the Somerset rule that they should be set in the Candlemas Waddle, that is, the waning of the February moon, or on the February new moon, as some say (Watson),

other wise they would not flourish. Elsewhere a date a little later than this is preferred, according to the rhyme:

> Sow beans or peas on David or Chad
> Be the weather good or bad;
> Then comes Benedict,
> If you ain't sown your beans –
> Keep 'em in the rick (M Baker. 1977).

Warwickshire custom required bean-planting to start on St Valentine's Day, and they agree it must be finished by St Benedict. St Valentine's Day is 14 February, St David's Day 1 March, and Chad the day after. St Benedict is celebrated in 21 March. The leafing of the ELM was a guide to planting:

> When elm leaves are as big as a shilling,
> Plant kidney beans, if to plant 'em you're willing.
> When elm leaves are as big as a penny,
> You *must* plant kidney beans, if you mean to have any (Dyer. 1889).

That was from Warwickshire, as is the following:

> When the elmen leaf is as big as a mouse's ear
> Then to sow barley never fear.
> When the elmen leaf is as big as an ox's eye,
> Then I say, "Hie, boys, hie" (Palmer. 1976).

The Cotswold variation has "as large as a farthing", or farden, to rhyme with 'garden'.

POTATOES have to be planted on a special day, but there is much debate as to the actual day. The Pennsylvania Germans say it should be St Patrick's Day if you wanted a big crop (Dorson), but the general feeling is that the planting should be on Good Friday, an odd choice horticulturally, for there could be as much as a month's variation in the timing. But of course the choice of day has nothing to do with reason – Good Friday is the one day on which the devil has no power to blight the crop. Gardeners have it that SHALLOTS should be sown on the shortest day of the year, and pulled on the longest, with a slight variation in Cheshire, where they say they should be planted on Christmas Day. Opinions differ as to when ONIONS should be set. In Lancashire, they say it should always be on St Gregory's Day (12 March) to ensure a good crop (M Baker. 1980). Sometimes, as in Buckinghamshire, you hear that St Patrick's Day (17 March) is the proper day; in Shropshire, it is Ash Wednesday, though in France it is much later, surprisingly, for Palm Sunday is the proper day there (M Baker. 1977). St Benedict's Day is more reason-able – that is 21 March, and the day the Pennsylvania Germans chose (Fogel). The state of the moon is important, too. Thomas Hill, in the 16th century, gave this advice: "If the Gardner commit seeds to the earth in the wane or decrease of the Moon, he shall pos-sesse small and sowrer ones, if the seeds are sown in the increase of the Moon, then strong and big, and of a moister tast, with the sowrenesss maistred". This is sympathetic magic, of course; as the moon increases so will the size of your plants, but there is no magic about further advice from the Pennsylvania Germans, who say you should bend over the tops of onions on the day of the Seven Sleepers (27 June) to make them grow big (Fogel). Gardeners' advice to set a row of onions between each row of carrots is equally sound – doing so will keep away the carrot pests (M Baker. 1977). CABBAGES should be sown on St Gertrude's Day, which is 17 March, for the best results, say the Germans, and their counterparts in America (Fogel), but in Kentucky, the favoured day is 9 May, and scatter elder leaves over them to keep insects away. Another piece of American wisdom advises sprinkling flour over the plants while the dew is on them, to drive away the worms. Similarly, PENNYROYAL leaves can be used to the same effect (H M Hyatt).

MYRTLE will only take root if planted on Good Friday, they say (Sebillot). Received wisdom in Somerset used to be that when planting it, one should spread out the tail of one's gown, and look proud, or it would not flourish (Tongue). In that same county, it is said to be one of the luckiest plants to have in a window box, although it will not grow there unless planted by a good woman (M Baker. 1980). Again, one has to be proud of it, and water it each morning (*Notes and Queries; 1853*). THYME is different; certainly, when moving house you should take it with you (Whitney & Bullock). But, according to the Pennsylvania Germans, unless you sit on it after planting, it will not grow (Fogel).

There are some pieces of advice about planting WATER MELONS from the southern states of America. They should be set on the 1st May, before sunrise, for good luck (R B Browne), and by poking the seed in the ground with the fingers (Puckett), or, according to Kentucky wisdom, in your night clothes, before sunrise; then the insects will never attack them. And carry the seed out in a wash tub. Then the melons will grow as large as the tub (Thomas & Thomas). CUCUMBERS need special treatment. They are, of course, phallic emblems, and a man should plant them. To get the largest cucumbers, planting them on the longest day (Fogel), is an example of homeopathic magic at work. Similar ideas are quite widespread in America. The seed should be sown before daylight, but also it should be done by a man, perferably naked, and in the prime of life. The size of the cucumbers depends on the "visible virility of the sower". Sown by women or old men they will never amount to much (M Baker. 1977). In Maryland, they say that if you plant them with your mouth open, the bugs will not bother them (Whitney & Bullock). Much superstition unerlays the actual time of sowing. The longest day, as we have seen, is significant, but in Kentucky they prefer a later date, the 4th or 6th of July (Thomas &

Thomas). Again, in Alabama, received wisdom has it that they should be planted on the first of May (R B Browne). There is a Guernsey rhyme that translates:

> Sow your cucumbers in March, you will need neither bag nor sack,
> Sow them in April, and you will have few.
> But I will sow mine in May, and I will pick more than you (Garis).

GARLIC

(*Allium sativum*) When the devil's left foot touched soil outside the Garden of Eden, garlic sprang up, and his right foot produced onions (Emboden). Naturally, with an offensive smell like garlic has (it is said applying it to the soles of the feet will still result in smelling it on the breath), it must be associated with the devil, but taken by and large, garlic is a protector, for all ages and for all purposes. Greek midwives made sure that, at the birth of a child, the whole room smelled of it, and a few cloves of garlic had to be fastened round the baby's neck either at birth or immediately after baptism (Lawson). Palestinian mothers and new born babies had to be protected from Lilith with garlic cloves, for she would otherwise strangle the babies, and frighten the mother into madness (Hanauer).

It is for charms against the evil eye that garlic is most used, whether it is stitched to the cap of a new-born baby, or hung outside a house, or from the branches of a fruit-tree (Abbott). Boats too can be protected from envious eyes – long bunches of it used to be hung over the stern of Greek and Turkish ships in order to intercept any ill-wishing (Gifford). When a Greek sea-captain first went aboard his new ship, he hung garlic and laurel about it, and it could protect boats from storms, as well as from any ill-wishing (Bassett). A Greek mother or nurse walking out with her children would often take a clove in her pocket, and the formula "garlic before your eyes", or simply the exclamation "Garlic!" was a common expression used by a mother to someone who looks at a baby without using the traditional antidotes (Rodd).

Witches are held at bay with it, by putting some under a child's pillow, as was the Polish custom (Leland), while the Bosnian belief was that everyone should taste garlic before going to bed. Every member of a Serbian household would rub garlic on himself, on the chest, the soles of the feet and the armpits at times when it was reckoned that witches would be most active. It was also Serbian practice to put a garlic bulb (or a juniper twig) on the windowsill on the evening of St Thomas's Day (19 October in that calendar). That would keep witches away from that house all the year (Vukanovic). Garlic tied in bundles over a house door will keep out a vampire, and, stuffed into the mouth of a corpse, it would keep the vampire (if there were any suspicions that the deceased might be one) quiet in his grave (Gifford). A Roumanian practice was to anoint door locks and window casements with it, to keep out vampires (Miles). No wonder that throughout eastern Europe anyone who does not eat garlic is liable to come under immediate suspicion! At a gypsy funeral, according to one description, five bulbs of garlic were among the articles placed beside the body in its coffin (Sanderson).

Garlic is recognized in Morocco as one of the charms against the evil eye, but at the same time it is conceded, at least in Tangier, that anyone who eats it will be forsaken by his guardian angels as long as the smell remains in his mouth (Westermarck). On the other hand, according to Thomas Hill, it is the very smell that saves the lives of poultry, for "… neither the Weasel, or Squirrel, will after the tasting of Garlicke, presume to bite any Fowles, by which practice, Pullets and other fowles in the night being sprinkled over with the liquor of the Garlicke, may be defended from harm of either these". Similarly in France, it used to be said that a wolf would not harm the sheep if a clove of garlic had been tied round the neck of the leading ewe (Sebillot). Perhaps in a good many cases of protection mentioned above, it is the smell alone that puts to flight any evil tendency, just as, in Greek mythology, garlic was the herb given by Mercury to Ulysses/Odysseus to protect him from Circe's enchantment (Leland). That smell is useful in the garden, too. Garlic and roses help each other. A single clove planted beside each rose will keep all the greenfly away (Boland & Boland) and at the same time enhance the scent, presumably because of the competition!

Swedish bridegrooms used to sew sprigs of garlic, thyme or some other strongly-scented plants into their clothing to avert the evil eye, and in southern Saudi Arabia the bridegroom wears it in his turban (M Baker). Among gypsy wedding customs was one that required the bride to hang up bundles of garlic in her house – for luck and against evil, for the garlic turns black after attracting all the evil to itself, and so protects her (Starkie). In France (Lorraine) a pregnant woman would be advised to eat plenty of garlic if she wanted a boy (Loux). All these beliefs may help to explain why in some places garlic was reckoned to be the symbol of abundance – material abundance, that is, for it used to be bought at the Midsummer festival in Bologna as a charm against poverty during the coming year. But it used to be regarded as an aphrodisiac at one time. Chaucer's Somnour, who was "lecherous as a sparwe", was particularly fond of it:

Wel loved he garleek, onyons and eek lekes.

It had the same reputation in Jewish folklore (Rappoport). Not unconnected was the central European practice of feeding garlic to dogs, cocks and ganders, in the belief that this would make them fearless and strong (Moldenke & Moldenke). In fact,

it was quite often added to animal feed, but only to protect them from evil (Hohn). But, "if cholerick men eat too much thereof, it is cause of madness and phrensy" (Bartholomew Anglicus).

To dream of garlic was said to indicate the discovery of hidden treasure, or the approach of some domestic quarrel, the one apparently dependent on the other. But to dream of garlic in the home is lucky (Gordon). There are still more odd beliefs concerning it. Lupton, for instance, in the mid-17th century, quite seriously recommended giving a woman "that suspects herself to be with child" a clove of garlic to eat at night when she goes to bed, "and if she feel no savour thereof in the morning then she is with child". Gypsies say that lightning leaves behind it the smell of garlic (Leland), and the smell is important in the belief that if a morsel of a clove is chewed by a man running in a race, it will prevent his competitors from getting ahead of him. Hungarian jockeys, so it is said, sometimes fastened a clove to the bit of their horses; the other horses fall back the moment they come within the range of the offensive smell (Fernie).

The name garlic is OE garleac, where gar means a spear, a recognition of the shape of the leaf ("taper-leaved", in fact, to cite the proper name of the plant). The doctrine of signatures soon got to work on this, and once the name was applied it came to be used against wounds made with spears (Storms). This use as a wound herb, for which there are sound medical reasons, continued into the 20th century. It has always been applied externally as an antiseptic for wounds (Brownlow), and during World War 1 the raw juice was put on sterilized swabs, and applied to wounds to prevent them from turning septic. It has also been included in antiseptic ointments and lotions, just the sort of qualities to be useful in cases of poisoning, or snake- or dog-bites. "The Northern shepherds do drink garlic and stale ale against the bitings of asps" (Topsell), and "garlic is good if a man anoint therewith the biting of a dog or adder or of a snake, it will heal it" (Dawson). Lawrence Durrell reported that in Corfu garlic or onion is applied to insect stings and bites.

But it is for chest complaints that garlic has been most in demand. A Somerset bronchitis remedy was to make an ointment of lard and garlic, and rub it on the soles of the feet at night (Tongue), a recipe that was recorded as recently as 1957. There are similar cures for whooping cough in the North country, either by putting garlic in the stocking (Newman & Wilson), or by making an ointment of it with hog's lard. The soles of the feet (the hands too in a Suffolk country cure (Hatfield)), would then be rubbed with this, two or three times a day, or it could be applied as a plaster (Buchan). Another Cumbrian prescription for whooping cough was to infuse two cloves of garlic in a quarter of a pint of rum for twenty four hours; rub the back and the soles of the feet of the patient for three or four successive nights at bed time, "at the same time abstaining from all animal foods" (Rollinson). In Alabama thay had a way of treating a cold with garlic, either just by eating a spoonful of it, or, if the patient does not like that, by taking a baked potato and putting the garlic inside, or putting gravy on it, and eating the potato without chewing it (R B Browne). Garlic was also used in Ireland for a cold (Moloney). All in all, it seems that garlic cures anything, and that includes "the plague itself" (Thornton). In Alabama domestic medicine, it is taken, cooked, preferably fried, to reduce blood pressure (R B Browne), which is interesting, for it is still prescribed, chopped finely in milk, for arteriosclerosis as well as hypertension (Flück). In Siberia, carrying a whole unpeeled garlic clove on one's person was thought to prevent a heart attack. It must never be changed, even when dried and shrivelled up. Some people wore it on a string round the neck, next to the skin (Kourennoff). Jewish folklore has it that carrying it is efficacious in a epidemic (Rappoport). Even cancer has been treated with it.

"Garlyke ... doth kyll all maner of wormes in a man's body ..." (Boorde), even if its effectiveness operates in strange ways. Louisiana traiteurs give the patient a little ball of garlic to hang round the neck. The worms, they say, are afraid of the smell. It suffocates and kills them (Dorson). In Brittany, a necklace of garlic cloves round children's necks is reckoned to keep them free from worms (Sebillot), and garlic in a cloth round the child's waist would be the African American way of dealing with the problem (Fontenot), while, more reasonably, a tea made from it serves the same purpose in Trinidad (Laguerre). You can even stop a child's bed wetting by feeding him garlic, according to Illinois practice (H M Hyatt).

It has been used for centuries as a cure for sore eyes (Gifford), a practice that may very well owe its origin to the use against the evil eye. In parts of Portugal, onion or garlic is used to cure, if that is the right word, bloodshot eyes. It is in the Algarve area that this purely magical practice is found; you have to pluck onion or garlic without looking at it, and hide them with the eyes shut (Gallop). Deafness, too, was treated with garlic, and so was tinnitus. The Physicians of Myddfai had a leechdom "for noise in the head". There is also a report of veterinary usage of garlic. It seems that an Irish method of treating black leg in cattle is to make an incision in the skin to put in a clove. The wound is then stitched, leaving the garlic inside. Patrick Logan could think of no reason why this should have an effect, so perhaps we are back where we started, and the only reason for the garlic is to drive away the evil influence that caused the disease.

GENTIAN
see JAPANESE GENTIAN (*Gentiana scabra*); MARSH GENTIAN (*Gentiana pneumonanthe*); YELLOW GENTIAN (*Gentiana lutea*)

Gentiana lutea > YELLOW GENTIAN

Gentiana pneumonanthe > MARSH GENTIAN

Gentiana scabra > JAPANESE GENTIAN

Gentianella amarella > AUTUMN GENTIAN

Gentianella campestris > FIELD GENTIAN

Genista canariensis > SCOTCH BROOM

Genista tinctoria > DYER'S GREENWEED

Geranium molle > DOVE'S FOOT CRANESBILL

Geranium pratense > MEADOW CRANESBILL

Geranium robertianum > HERB ROBERT

GERMAN SAUSAGE TREE
(*Kigelia africana*) The common name is given because of the large grey-green fibrous fruits which are often a metre long. The unripe fruits are very poisonous, but are said to be used in cases of syphilis. On the other hand, the ripe fruit, though not edible, is useful. It can be baked and put into beer to help fermentation, and the Tonga use it powdered as a dressing for ulcers and sores (Palgrave & Palgrave; Hobley). It can be put in cattle troughs, too, to guard against leeches (Perry. 1972).

In southern Malawi, the fruit is reckoned a protection against whirlwinds, for it carries a powerful charm, and ensures that no damage is done, if it just hung in a corner of a hut (Palgrave & Palgrave). Similarly, Kikuyu (Kenya) belief had it that if an evil spirit possessed one of the cattle, easily recognised by the animal's behaviour (shaking its head, tears streaming from its eyes), the way to get rid of it would be to get the animal to sniff the smoke of a fire made with the dried fruit of this tree (Hobley). The esteem in which they hold this fruit is shown in the way a piece of it is deposited at the sacred tree (*Ficus capensis* in this case), when a sacrifice is being made (Hobley).

A Shona witch doctor would use the bark of this tree in a medicine prepared for discharging eyes. He needed a spider's web, which he rolled into a ball, and put in water with various other things, including bark of this tree. In the morning, the patient would kneel at the entrance to his hut, open his eyes with the water, and wash his face with it (Gelfand). Of genuine medicinal uses, Yoruba practice is quoted. It is given as a purge to women "to drive the worm from the womb, and the root or bark is one of the ingredients of a medicine for gonorrhea" (Buckley).

GERMANDER SPEEDWELL
(*Veronica chamaedrys*) Whoever could picture this little plant as a dwarf oak? But that is what the specific name means; we are told that it is the shape of the leaves that gives the name, but this remains unconvincing. *Chamaedrys* comes through medieval Latin from the Greek chamaedrua, which

is a corruption of chamai, on the ground, and drus, oak (Grigson. 1955). Germander means the same, for it is simply an englishing of some stage of the development that produced *chamaedrys*. One suspects that shamrock, at least in this one case, is *chamaedrys* in another guise; this plant is one of the many to which the name has been given.

In Irish legend, this is one of seven herbs that nothing natural or supernatural could injure; the others are St John's Wort, vervain, eyebright, mallow, yarrow and self-heal (Wilde. 1902). So Irish people used it to keep evil spirits at bay, and it was sewn on clothes to keep the wearer from accidents (Grigson. 1955). Like pimpernel, Germander Speedwell is a weather forecaster, for it closes its petals before rain, and opens up again when it has stopped (Inwards). People from Cambridgeshire have a saying to the effect that if the flowers close in the morning, it will rain before evening (Porter. 1969).

The leaves were at one time recommended for use as a beverage tea (Curtis), hence Poor Man's Tea, from Cumbria (Grigson. 1955), but the common usage was medicinal. It used to be said that Germander Speedwell was especially good for gout; the Emperor Charles V is supposed to have got benefit from it (Dyer. 1881). It was so sought after for gout in the 18th century that it was, so they said, "made scarce to find through picking for many miles outside London" (Jones-Baker. 1974). Gerard was quite enthusiastic about it, but one of his prescriptions was pure doctrine of signatures. That little white patch in the centre of the flower that produced the varied "eye" names must also be responsible for "the leaves stamped with hony and strained, and a drop at sundry times put into the eies, taketh away ... dimnesse of sight...". Compare this with a Kentish village remedy for cataract; pick the flowers with as little green as possible, boil in rain water that falls in the month of May, pour on to the flowers, stir well, bottle, strain, and drink a cold glassful night and morning. Then wash the eyes with it three times a day (Maple. 1962). The flowers were used in Norfolk to make an eyebath for sore eyes (it actually bears the name Sore Eyes there) (V G Hatfield. 1994). There is, too, a Somerset remedy for tired eyes, which is to make a decoction of our flower with eyebright, and then dab the eyes with it (Tongue. 1965). Bird's Eye is a widespread and common name for Germander Speedwell, obviously descriptive, but there was a belief in some areas that birds would come and peck your eyes out if you picked it, a belief that made Roy Vickery include it in the list of unlucky plants (Vickery. 1985). It is true that there is a sinister side to this little plant. Names like Tear-your-mother's-eyes-out (Macmillan), or Mother-die, etc., owe their being to beliefs like that from Yorkshire, where it is said that if a child gathers the flower, its mother will die during the year

(Dyer. 1889), or there is the Lincolnshire superstition that if anyone picks the flower, his eyes will be eaten (Gutch & Peacock). The name Blind-flower is found in County Durham (Britten & Holland), with the belief that if you look steadily at the flower for an hour, you will become blind. Another result of picking the flowers is a thunderstorm, hence the Cheshire name Thunderbolt (Hole. 1937).

Geum rivale > WATER AVENS

Geum urbanum > AVENS

GIANT FENNEL
(*Ferula communis*) In the Prometheus myth, the fire that he stole was carried in a hollow fennel stalk, and given to men. Giant Fennel stalks, which can be five metres tall, are filled with a dry white pith like a wick, and were still used in the 20th century in parts of Greece to transport fire(Grant). Schoolmasters once used this to whip children (*Ferula* means a cane used to punish slaves and children). Loudon said they were used because "they made more noise than harm") (Coats).

GIANT PUMPKIN
(*Cucurbita maxima*) There is more to pumpkins than Hallowe'en essentials. It is usually associated with North America, but it is more likely that it has South American origins. But it is widely travelled, and pumpkins and gourds generally, though not necessarily this species, appear often as a motif in Chinese art. They are also used as decoration: it was the practice at one time to encase a gourd while small in an intaglio mould, leaving only a hole for the stalk. As the gourd swelled it filled the mould, and the design was impressed into the surface. Gourds were also used in China for cricket cages (Savage. 1964).

Some of the American Indians used pumpkin seed tea as a diuretic; so do people in Alabama still. They say they are excellent in a tea for kidney troubles, but at the same time Alabama children used to be dosed with it to stop them bed-wetting (R B Browne). Another use for the seeds is for a worm powder. They are apparently very efficient when crushed and made into a paste with milk and honey, and then taken three times before breakfast (Page. 1978). As an example of the wrong end of the stick, note the belief in Maine that, though pumpkins may be good for the eyesight, they are apt to breed worms in the stomach which make one itch (Beck).

Gillenia trifoliata > INDIAN PHYSIC

GILLIFLOWER
A name that has served for a number of plants, perhaps modified for convenient identification, as in Clove Gilliflower for carnation, or Stock Gilliflower for stock, and Wall Gilliflower for wallflower, which can be Yellow Gilliflower, too. The word has appeared in a lot of different guises, from Geraflour (Britten & Holland), a Scottish variant, through such close

affines as Gilliver, Gillyfer or Gillyver, and reduced to Gilver in the Isle of Man (Moore, Morrison & Goodwin). To make sure that we understand that the 'g' is soft, a 'j' is often used, hence the Yorkshire Jilliver, and such West country forms as Jilloffer. The variants continue – Jilloffer can become Jilly Offers (Macmillan), and gilliflower easily becomes July-flower, and so on. Grindon notes no less than seventeen variations of the 'gilliflower' name in use in the 15th and 16th century. Actually, the word comes from Caryophyllus (Carnation is *Dianthus caryophyllus*), and so means "clove". So such a form as Clove Gilliflower for carnation is pleonastic.

GINGER

(*Zingiber officinale*) "… great Quantities of it are us'd by the Hawkers and Chandlers in the Country, who mix it with pepper; they reduce it to Powder, and then call it white Spice" (Pomet). Apart from its use as a spice and as a base for alcoholic liquors of one kind or another, ginger has for a very long time enjoyed a reputation for medicinal use, from the prescription of Arabian and Persian doctors for impotence (Dalby), to its still popular reputation as a stomach settler, and this use dates from the earliest records (Lloyd). Ginger tea, even ginger biscuits, help to combat travel sickness, or morning sickness and nausea generally (M Evans). Parihar & Dutt point out that ginger jam is still a favourite for colds and coughs, and it is even used to treat diabetes. In this case, it is ginger juice that is used, mixed with sugar candy.

It was used for asthma in Russian folk medicine (Kourennoff). The recipe given is a pound of ginger grated, put in a quart bottle, which was filled with alcohol. This was kept warm for two weeks, shaken occasionally, until the infusion was the colour of weak tea. This was strained, and the sediment allowed to settle. Then the liquid was poured into another bottle, and the infusion taken twice a day.

Ginger is known as djae, and used in Java as a salve for rheumatism and headaches (Geertz). Similarly in New Guinea, where the usage is more magical: boys at initiation are rubbed all over with ginger, "to give warmth to the body" (La Fontaine). Magic lies, too, behind the Malagasy prohibition on pregnant women eating it. The reason lies in the shape of the root, which is sometimes flat with excrescences like deformed fingers and toes. Nor must she keep the root tied into a corner of her costume, where odds and ends, coins, etc., are usually kept. If she fails to keep these taboos, the foetus will become deformed, with too many fingers or toes; its legs will not grow straight, the deformation making delivery difficult as well (Ruud). Some peoples of the Malay Peninsula have their children wear a piece of ginger tied to a string round their neck to keep harmful spirits away. It is the pungent smell that achieves this (Classen, Howes & Synnott).

GINSENG

(*Panax schinseng,* and *Panax quinquefolium*) The latter is a North American species, a substitute for the real thing when it became rare. It was actually exported to China from the early 19th century (Dalby). This American ginseng was used by native Americans as an ingredient in a love charm. Meskwaki women wanting a mate made up a potion from ginseng roots, mixed with gelatine and snake meat. The Pawnees combined it with wild columbine, cardinal flower and carrot-leaved parsley in their version of the love charm (Weiner).

In domestic medicine, Alabama mothers gave ginseng tea to babies who had colic (R B Browne), and it is, in fact, used in the southern states as a remedy for all kinds of stomach troubles (Thomas & Thomas).

The true ginseng, from northern China, is now rare because of the extensive use of the root in Chinese medicine. The forked root was treated like the human form (like mandrake, in fact; it would seem that the whole of the mandrake legend spread to China, and became attached to ginseng) (G E Smith) (see **MANDRAKE**). It was used as a universal panacea; indeed *Panax*, the name of the genus, has the same derivation as panacea, i.e., "heal-all" (W A R Thompson. 1976). The name All-heal is even recorded in English (Halliwell). Ginseng, the name, is Chinese Jin-chen, which means man-like (W A R Thompson. 1976), and it was because of this supposed resemblance that the doctrine of signatures worked, that is to say that the plant healed all parts of the body. The more closely the root resembled the human form, the more valuable it was considered, and well-formed roots were worth their weight in gold (Schery) – as an aphrodisiac (Simons). It was the Dutch who brought the root to Europe, in 1610, and its reputation as an aphrodisiac came with it. The court of Louis XIV in particular seemed to value this reputation (Hohn). Medicinally, it was recommended for conditions that were characterized by exhaustion and a lack of zest for life. But the Chinese also used it to aid longevity (R Hyatt). The root of *Codonopsis pilosula* or *C tangshen* are often used in Chinese medicine as a substitute for the more expensive ginseng (Hyam & Pankhurst; Perry & Metzger).

GLASSWORT

(*Salicornia europaea*) An annual plant of salt marshes, very rich in minerals (Schauenberg & Paris). Its ashes were used at one time in making glass (and soap). Cattle eat it greedily for its salty taste (Grieve. 1931), and, in any case, steeped in malt vinegar, the young shoots make a good pickle, and were often used as a substitute for samphire (Grieve. 1931), as is obvious when names for it like Marsh Samphire or Rock Samphire (Britten & Holland) are considered, though they say it is inferior to the proper stuff (Hepburn). Nevertheless, it was still collected in the

Eastern counties of England for pickling in recent times (Grigson. 1955), and may even now be gathered still.

GLASTONBURY THORN

(*Crataegus monogyna 'Praecox'*). The legend of the Glastonbury Thorn is well-known. Joseph of Arimathea, on his way to Glastonbury, arrived at Weary-all Hill, to the south of the town, and rested there, having pushed his staff into the ground. The stick took root, and blossomed each year on the anniversary of the birth of Christ (old style, i.e., 6 January, and not 25 December). The difficulty with this is the fact that the legend is of late origin. The first testimony was in a poem called 'The Lyfe of Joseph of Arimathea', apparently written in 1502, though not published until 1520:

> Three hawthornes also that groweth in Wirral
> Do burge and bear greene leaves at Christmas
> As fresh as other in May.

The legend was complete by the early eighteenth century, but it was an innkeeper who first launched the "Weary-all Hill" story (Loomis), and the connection between Joseph's staff and the tree was not established till the early eighteenth century intervention.

Plants grown from the haws of the Glastonbury Thorn do not retain the characteristics of the parents, and the only way of propagating it is by grafting or budding on to other roots, so that must have been achieved as far as the original tree is concerned, and offshoots of the original tree were grown in many gardens in Somerset and also in Herefordshire. Howells, in 1831, wrote of a Christmas blooming hawthorn in a garden at Aberglasney, in Carmarthenshire, modern Dyfed. Apparently the tree was common in Palestine, where it bloomed at the same time as it does here (Williamson), i.e., twice, once in the winter, and again in the spring (hence the alternative specific name, *biflora*). The winter flowers produce no fruit, though the spring ones do.

If a piece of the Holy Thorn were gathered at the Christmas blooming and kept in the house for the rest of the year, it would act as protector from misfortune, but it was also believed that picking the buds or flowers brought very bad luck. Then again, flowers from the thorn were brought in procession in Charles 11's time, on Christmas morning, and presented to the king and queen (Hadfield), and these days Glastonbury parish church is decorated with the flowers at Epiphany (Lawrence). There may be confusion as to the result of picking the flowers, but all sources agree that it is very unlucky to cut down or damage the tree, and there are many stories of the fate of people misguided enough to chop it down.

The Cadnam OAK, a "boundary tree" of the New Forest, according to popular belief, became green on Old Christmas Day, being leafless before and after the day (Bett. 1952). This oak, or rather its descendant, still bears leaves round about the 6 January, though it buds at the normal time as well (Hampshire FWI). One sometimes finds BLACKTHORN cast in the same part, blooming, so it was believed, at midnight on old Christmas Eve (M Baker. 1980).

Other plants had a similar reputation. ROSEMARY, for instance, was thought to bloom exactly at midnight in the eve of Twelfth Day – Old Christmas Eve, that is (Dew). SWEET CICELY (*Myrrhis odorata*) is another, according to belief in the Isle of Man, where it is called Myrrh (Moore), which must be the original for this belief. A watch is still sometimes kept for the flowering (Garrad). According to tradition, the bloom only lasts for an hour.

Glechoma hederacea > GROUND IVY

Gleditsia triacanthos > HONEY LOCUST

GLOBE ARTICHOKE

(*Cynara scolymus*) It was once taken as being aphrodisiac (Campbell-Culver). As Andrew Boorde had it, "they doth increase nature, and doth provoke a man to veneryous actes". Dreaming of them is a sign that you will receive in a short while from the hands of those from whom you would least expect it (Mackay). It was used to treat arteriosclerosis at one time (Schauenberg & Paris), and skin disorders (Wickham).

GLOBE CUCUMBER

(*Cucumis prophetarium*) The prophet of the specific name, given by Linnaeus, is Elisha, for it was once thought that this was the plant of Elisha's miracle, of II Kings; 4 (Moldenke & Moldenke). This is the sacred healing cucumber of the East African Dinka people. Almost every Dinka homestead is likely to have one or two of them displayed. They are intensely bitter, and when ripe turn a greenish-blue colour, suggesting to the Dinka a stormy sky, and hence divinity. Sometimes, if no beast is available for sacrifice, one of these sacred cucumbers may be split, and thrown aside. It is a temporary substitute for an animal victim, and an earnest to provide one when possible (Lienhardt).

Glyceria fluitans > FLOTE GRASS

GOAT WILLOW

(*Salix capraea*) This is the tree that gives "palm" for Palm Sunday, as some of the names given to the tree confirm – Palm itself, and Palm-tree, or Palm Willow (Grigson. 1955; Leather), English Palm (Poole), and, from Dorset, Palmer (Grigson. 1955). Perhaps any catkin-bearing tree could be the palm, for hazel was used, too (Simpson). But the catkins of Goat Willow have always been the English embodiment of 'palm'. In medieval times, a wooden figure representing

Christ riding on an ass was sometimes drawn along in the procession, and the people scattered their branches in front of the figure as it passed (Ditchfield. 1891). Willow sprays used to be put on each seat of Moreton Church, in Dorset, on Palm Sunday (M Baker. 1980). Flora Thompson tells how sprays of sallow catkins were worn in buttonholes for church-going in her day, and how they were brought indoors to decorate the house. They should not be brought indoors before Palm Sunday, for that would be most unlucky. At Whitby, palm crosses were made, and studded with the blossoms at the ends, and then hung from the ceiling (Gutch). Similarly, in County Durham, where the branches were tied together so as to form a St Andrew's cross, with a tuft of catkins at each point, and bound with pink or blue ribbons tied to them (Brockie) – perhaps the ribbons were to hold the cross together. These Durham crosses were kept for the whole of the coming year (M Baker. 1980).

In Subcarpathian Rus' the priest blesses Pussy Willow branches in the church. In some villages they are given to the animals to eat, or they are saved, and if a storm threatens they are thown into the fire. There is a set formula to be said while this is being done, in translation "May the storm vanish in the sky like the smoke from these branches". Another meaning of the rite was recorded elsewhere in the region: "When it thunders, these branches are burnt; they are put in the oven, so that smoke will be produced and the devil not be able to hide in the chimney" (Bogatyrëv).

It is reckoned a lucky tree in Ireland. It is a good thing to take a sallow rod with you on a journey (Grigson. 1955), and they reckoned that the butter would be bound to come if a peeled rod were put round the churn, just as they believed that driving cows with a "sally" rod would ensure a good supply of milk (O'Farrell). When these willows get big, the heartwood turns red, and if kept dry, it is said to last as long as the oak, hence the proverb:

Be the oak ne'er so stout
The sollar red will wear it out (Northall).

And there is another: Sally tree will buy a horse, before an oak will buy a saddle (Leather).

GOATWEED

(*Ageratum conyzoides*) There are records of its use in traditional medicine in widely separate areas. The Chagga, in Africa, drink a decoction of the root for all abdominal upsets. In central Africa (and in the Far East (Perry & Metzger)) the leaf is used to help the healing of wounds and burns (Watt & Breyer-Brandwijk). It is used in traditional medicine in Nigeria, too, for dressing wounds and ulcers, for craw-craw, and as an eyewash, while in East Africa it is used as a styptic (Sofowora). The Mano of Liberia also take it as a wound herb – they squeeze the juice directly into the wound. It is further utilised by these people, by rubbing to a pulp with water, to put on the chest of children with pneumonia. They say the sickness is then transferred to a stick (Harley).

The leaf is used in Yoruba medicine as a worm remedy (there are other ingredients, including snails) (Buckley). A leaf of this plant, and a leaf of *Petiveria alliacea*, with some other, unidentified, leaf, are required in a Yoruba preparation to prevent one from being attacked byanother person. They are all burnt together, and the ashes put into small incisions made on the hand (Verger). Another ritual usage is reported from Brazil. The plant is an ingredient of the ritual baths that form part of Brazilian healing ceremonies (P V A Williams). It is an aid to catching snakes – the leaves are crushed and rubbed on the hands. Presumably the rather unpleasant smell (it is not called Goatweed for nothing) affects the snakes in some way (Harley).

A decoction of the leaves and young stems is used in Chinese medicine for common colds, and eczema (*Chinese medicinal herbs of Hong Kong vol 1*). There is yet another tradition in Africa, to treat a child who cries too often for no known cause, especially at night. Stress is put on the requirement that the plant should be collected at night, especially when witchcraft is suspected. The procedure is described as follows: the plant is found during the day, and in the dead of night the collector approaches the plant and chews 9 or 7 seeds (for male or female respectively) of Melegueta Pepper (*Aframomum melegueta*). The chewed grains are spat on the plant while the appropriate incantations are recited. After that the plant is plucked, taken home and warmed over a fire before the juice is expressed. Palm oil is added to this, and the two mixed together are used to rub the child all over the body (Sofowora).

GOITRE

An American treatment for the condition was to fix a HOP bag round the neck (Hyatt).

GOLDEN ROD

(*Solidago virgaurea*) In England, Golden Rod has long had the reputation, if held in the hand, to find hidden springs of water, as well as to treasures of gold and silver (Dyer. 1889). But sometimes it is seen as an unlucky plant, certainly not to be taken indoors (Vickery. 1995). It is, though, a symbol of precaution, and encouragement (Leyel. 1937). The flowers and leaves give a fine yellow dye (Barton & Castle), but the rest of its uses are purely medicinal, if the Scottish belief that it can heal broken bones (Beith) can be classed as such. Gypsies use an infusion of the leaves for treating gravel and stone (Vesey-Fitzgerald). That usage has a venerable ancestry. Gerard, for instance, reported that it "provoketh urine, wasteth away the stones in the kidneys, and expelleth them …", and Culpeper had virtually the same recommendation.

Another gypsy use is of an ointment made from the fresh leaves, to heal wounds and sores (Vesey-Fitzgerald). Martin also records this use in Lewis, Outer Hebrides. Gerard again: "it is extolled above all other herbes for the stopping of blood in bleeding wounds …". In China, it is the seeds that are used to treat haemorrhages, wounds, etc., (Perry & Metzger). A lotion made from Golden Rod was used in the treatment of ulcers, and it was also said to make loose teeth secure (Addison).

GOLDTHREAD

(*Coptis trifolia*) Both the American Indians and the early settlers used the root as a remedy for sore and ulcerated mouths (Weiner) (one of the names for the plant is Mouthroot (Howes)). The use has continued as a folk remedy into recent times – R B Browne quotes recipes from Alabama, not only for the original use, but also for sore eyes, even burns.

GOOD FRIDAY

The exception to the rule of no work on Good Friday is the very widespread belief that this is the best possible day to do your gardening, for "all things put in the earth on a Good Friday will grow goody, and return to them with great increase" (Bray). Sowing peas and beans should be done on this day, and so should tree grafting; Herefordshire apple trees were always grafted on this day. Anything planted or grafted on Good Friday will grow well. It was even believed that beans sown today will have shoots above ground by Easter morning! (Whitlock. 1977). Somerset people said "they would rise on Easter Day with Christ" (Tongue). Some say that planting anything before Good Friday would be most unlucky (E L Chamberlain).

Observation of PARSLEY'S germination time had given a number of superstitions. Its seed is one of the longest to live in the ground before starting to come up. Further, the devil is implicated – it goes to the devil nine times, or seven, some say (Northcote, Clair). To offset this, various ruses are recommended. But the best is to sow the seed on Good Friday, when plants are temporarily free of the devil's power (Baker. 1980), or do it at the very least on a holy day (Tongue).

Pennsylvania Germans say that this is the proper day to sow cabbage seed (Fogel), and in Kentucky FLAX would be planted now (Thomas & Thomas), and this is certainly the lucky day to plant POTATOES, even though it is an odd choice horticulturally speaking, for there can be as much as a month's variation in the timing. But of course this has nothing to do with reason, and opinions differ anyway, for some say that if potatoes are dibbled on Good Friday, then you could expect nothing but a bad crop (Igglesden), and there was a feeling in Wales that any kind of gardening on this day was most unlucky. Yorkshire gardeners would never disturb the ground with iron today – the soil might bleed, and some of the Pennsylvania Germans

used to say that no gardening should be done between Good Friday and Easter, the period during which Jesus lay buried. Nevertheless, most other people pressed on with it. SPOTTED MEDICK has Calvary Clover as one of its names. A German superstition, taken to America, said that Calvary Clover could only germinate if sown on Good Friday (M Baker. 1977). The red spot on the leaves also help in understanding the name. They are spots of Christ's blood that fell on them at Calvary. In Jamaica, it was said that if you go to a PHYSIC NUT tree at 12 o'clock on Good Friday morning, and stick a penknife in it, the juice that comes out of it will be blood (Beckwith. 1929).

An American (Illinois) belief is that cooking COW-PEAS on Good Friday will bring luck (H M Hyatt).

Churches used sometimes to be hung with "funeral YEW" on Good Friday (Dyer. 1876), and in Hereford-shire, at Whitsuntide, branches used to be fastened to the tops of pews (Leather). The practice continued in Ireland from Palm Sunday to Easter Day, when sprigs of it were worn in the hat or buttonhole (Lowe).

GOOD KING HENRY

(*Chenopodium bonus-henricus*) "King" seems to be intrusive; there is no mention of a king in the specific name, which just means "Good Henry". Perhaps it is meant to distinguish it from *Malus henricus*, Bad Henry, probably Dog's Mercury. Then again, it is the opinion of some that Henry is the 16th century German Guter Heinrich, the name of an elf with a knowledge of healing plants (Grigson. 1974). This plant is perfectly edible, and has been used as a food plant since Neolithic times (Mabey. 1972), and was actually cultivated in gardens right into the 19th century (C P Johnson). Evelyn said the tops could be eaten as "sparagus", but "'tis insipid enough". One of the many names given to it is Blite (Prior), or Bleets, as Tusser had it, from a word meaning 'insipid', justifying Evelyn's comment.

It has its medicinal virtues – in Gloucestershire, an infusion of the fresh leaves was drunk for bladder troubles; and ground up in water, it was a cure for scurvy (Porter. 1974). Gerard advised his readers that the leaves "bruised and layd upon green wounds or foule and old ulcers, doe scoure, mundifie and heale them". In other words, poultices made of the leaves cleansed and healed chronic sores.

GOOD LUCK CHARMS

What is evidently a good luck charm is recorded in Cheshire, where when children first see the flower heads of RIBWORT PLANTAIN in spring, they repeat the rhyme:

> Chimney sweeper all in black,
> Go to the brook and wash your back,
> Wash it clean or wash it none;
> Chimney sweeper, have you done? (Dyer. 1889)

Goodyera repens > LADY'S TRESSES, more
properly CREEPING LADY'S TRESSES

GOOSE-GRASS

(*Galium aparine*) "This weed is considered
excellent food for goslings, who are very fond of it"
(Akerman). The roots will dye red (Grieve. 1931),
and the seeds, roasted, have been used as a coffee
substitute (Barton & Castle), and still are, sometimes,
in Ireland (Usher). Another use of the seeds, in the
green state, was to adorn the tops of lacemakers' pins;
the young seeds were pushed on to the pins to make a
sort of padded head (Mabey. 1977).

This is a plant traditionally used to soothe wounds
and ulcers (Schauenberg & Paris). In Ireland a whole
mass of the herb would be applied to ulcers, while the
juice was given internally at the same time (Moloney).
In country medicine, it has long been given for skin
diseases, acne in particular, and also scurvy, etc.,
(Grigson. 1955), because it is said to rejuvenate
the tissues (A W Hatfield). There is an early charm
"against eruption of the skin" that involved elecampane, viper's bugloss, bishop's wort and goose-grass.
The patient was told to "pound the four herbs together
well, squeeze them out, add a spoonful of old soap to
it". Then he was told to "scarify the neck after sunset,
silently pour the blood into running water, spit three
times after it, then say: "Take this disease, and depart
with it". Go back to the house by an open road, and go
each way in silence" (Storms).

A poultice of goose-grass is still advised for boils in
Somerset (Tongue. 1965), and they use it in France
for sores and blisters (C P Johnson). Gypsies say
that an infusion drunk very hot last thing at night is
a remedy for a cold in the head (Vesey-Fitzgerald).
The tops have been used as a country "spring drink"
(Grieve. 1931). That is probably what Evelyn meant
when he said the tender shoots, with young nettle
tops, were used in "Lenten pottages" (Evelyn. 1699).
Another use for the plant is as a solvent of stone or
gravel in the bladder (Quelch). It does actually act on
the kidneys and bladder, and is mildly laxative (A W
Hatfield). American Indians also used it for kidney
and bladder trouble (H H Smith. 1945). Gypsies claim
that goose-grass is a very ancient remedy for cancer
(Vesey-Fitzgerald), and Thornton does record its use
for tumours in the breast. It appears to be one of the
ingredients of a recipe for abortion used among the
gypsies of the former Yugoslavia (Clébert). Lastly, a
slimming remedy from Gerard: "Women do actually
make pottage of Cleavers with a little mutton and
Ote meale to cause lanknesse, and keepe them from
fatness".

GOOSEBERRY

(*Ribes uva-crispa*) Simply known as Berries in
Yorkshire, for there, gooseberries are berries, par
excellence (Hunter), that grow on a Berry-tree.

Gooseberry-pies are berry-pies. Perhaps it is not
surprising that this is so, for gooseberries in the north
have long been the subject of esteem and competition.
Goosegog is another very common name for them.
The fairy that guards the unripe fruit is known as Awd
Goggie in Yorkshire (and, more sedately, in the Isle of
Wight as the Gooseberry Wife).

According to the Victorian dream books, to dream of
them is a sign of many children (Raphael). Of course,
English babies are found under gooseberry bushes. In
parts of Germany, though, children (especially girls)
grow on a tree (O'Neill). To revert to dream prognostications – if it is a sailor who dreams of gooseberries,
then it is a warning of dangers on his next voyage
(Raphael). To a girl it means an unfaithful husband
(Gordon. 1985). But gooseberries apparently served
as symbols of anticipation (Leyel. 1937).

Green gooseberry pie used to be a traditional Whit
Sunday dish (Savage), and besides being used as a
dessert fruit, or in tarts, etc., they have traditionally
been used to mix into a sauce with sorrel and sugar,
to be eaten with a young goose (Grigson. 1955),
hence the derivation of the name, according to some.
It would be a green goose, cooked when the gooseberries were ripe, and eaten with gooseberry sauce.
Another similar use was to have stewed gooseberries,
puréed, with mackerel (Mabey. 1972). That there is
nothing new in this cook's practice, see Gerard: The
fruit is used in divers sauces for meat, as those that
are skilfull in cookerie can better tell than my selfe
…". Not that he thought gooseberries were entirely a
good thing, for "they nourish nothing or very little"
"not worth a gooseberry berry" was a phrase common enough in Shakespeare's time), and Gerard saw
but little of their "vertues". Certainly, the juice "of
the green gooseberries cureth all inflammations,
erysipelas, and S Anthonie's fire", and, too, the young
leaves eaten raw as a salad "provoke urine, and drive
forth the stone and gravell", a claim that was to be
repeated often. Culpeper judged the berries to be
"excellent good to stay Longings of Women with
Child". But any remaining medicinal usages belong
firmly in the world of charms. An Irish cure for warts,
for instance, is to prick them with a gooseberry thorn
passed through a wedding ring (Fernie). Similarly, and
also from Ireland, a charm for a stye in the eye or a
whitlow is to point a gooseberry thorn at it nine times
(three x three), in the name of the Trinity, or rather get
a twig with nine thorns on it, and point each thorn in
turn at the stye, saying "away, away, away". The thorns
had then to be thrown over the left shoulder, and the
stye will presently vanish (Wilde. 1902; O'Farrell;
Cooke). The Scottish practice was to lay a wedding
ring over the wart, which was then pricked (through
the ring) with a gooseberry thorn. Ten thorns would
be picked, the other nine simply being pointed at the
wart, then thrown over the shoulder (Beith).

GOPHER WOOD

Probably CYPRESS. The word cypress itself, at least according to Grigson. 1974, derives eventually from Hebrew gopher, though its more immediate origins are from Greek kyparissos and Latin cupressus. It is probably our cypress that provided the gopher wood of Genesis 6; 14, of which Noah's Ark was made. Cypress wood is indeed very durable.

GORSE

(*Ulex europaeus*) Gorse is the standard English name for this shrub, Furze being a regional name (Jones & Dillon), while the other common name, Whin, is confined to those parts of Britain that were under Scandinavian occupation in times past. There is an old rhyme highlighting the difference in habitat of the gorse and broom:

> Under the gorse is hunger and cold,
> Under the broom is silver and gold (Northall).

There seems to be no great unanimity about that, though. One of Roy Vickery's informants from Cumbria mentioned a local saying: "where there's bracken there's gold; where there's gorse there's silver; where there's heather there's poverty" (Vickery. 1995). It is upgraded even more in County Kerry; the Gaelic saying is translated "gold under furze, silver under rushes and famine under heath" (Lucas).

There are equally mixed views about the "luck" of the flowers. A typical west country belief (though not confined to that area, apparently) is that it is very unlucky to take gorse into the house, just as unlucky as hawthorn, say, or lilac, for "to carry furze flowers into the house – carrying death for one of the family" (Opie & Tatem), and there are other similar sayings of the "gorse in, coffin out" variety. Giving the flowers to someone is also unlucky, but without such dire results. But the act would be bound to provoke a quarrel between the two people involved in a short time (Vickery. 1995). Or, in Scotland, it is said to be a sign of anger if gorse is given (Simpkins). Indeed, in the language of flowers, gorse is the symbol of anger; but at the same time it symbolizes love for all seasons, and enduring affection (Leyel. 1937), for is it not said that "kissing is out of fashion when the whin is out of bloom" (Nicholson), or its Scottish equivalent, "when the gorse is oot o' bloom, kissin's oot o' fashion" (Simpkins), for isn't gorse in bloom virtually the whole year through? And there is another Scottish proverb to emphasise the point:

> When the whin gangs out of bloom
> Will mean the end of Edinburgh town(Addison).

A story from Guernsey amplifies the view; a man on his death-bed asked his wife not to marry again while the furze was in bloom (MacCulloch). In Poitou, France, they ask the question "En quelle saison l'ajonc n'est-il pas en fleur?" The reply is "A l'époque ou les femmes ne sont pas amoureuses". In Brittany, to promise to love while the furze is in bloom is to promise to love forever (Sebillot). All this underlines the old country custom of putting a spray of gorse in the bridal bouquet (Grieve. 1931), and a Somerset version of the wedding dress rhyme runs:

> Something old,
> Something new,
> Something borrowed,
> Something blue
> And a sprig of vuz.

For the belief is that it brings gold to the house (Raymond).

There are other superstitions that hint at a deeper level of belief, such as the Irish saying that carrying a sprig of furze on the person will save the traveller from getting lost (Wood-Martin), and the plant forms part of the "summer" brought in on May Day in County Cork. At May Day, and also at Midsummer, furze bushes were burned to protect cows and crops (Ó Súilleabháin), and it was also used to hang over the door on May Day "to keep luck in the house" (Wood-Martin), or it was stuck in the roof, for the same reason (Lucas). It is significant too that according to Welsh belief, the fairies cannot penetrate a hedge of furze (Sikes), implying that it is more than the prickles that keep them out. The plant had some significance in Devonshire Christmas festivities, too, for they had a version of the "kissing bush" that involved a small furze bush, dipped in water, powdered in flour, and studded all over with holly berries (Whitlock. 1977).

In the Hebrides, it is said that the result of cows eating the young shoots is a rich yellow colour in the butter (Murdoch McNeill), and a yellow dye can be got from the bark (Jenkins. 1966). The flowers were widely used for colouring Easter eggs yellow (A R Wright, Gill. 1963), but the best known and most widespread use of the plant was as fuel, especially for firing baking ovens (Dacombe, Carew). It was used in Ireland, too, for domestic fuel, and it was actually brought into the city for sale there at one time; indeed, in parts of County Carlow, faggoting (a faggot being furze tied in bundles by briar bands) was a regular trade (Lucas). It was actually cultivated, at least into the 19th century, for fuel and fodder. It was cut with some ceremony in certain areas. The *West Briton* for 30 July 1858 reported that gorse harvesting (for fuel) was still carried out on the Lizard peninsula then. The fuel was carried in trusses on horseback, the truss being a large bundle bound round with ropes, raised on end by putting a pole through it, and raised to the horse's back, which was completely covered with coarse cloth, animal skins being once used. The custom was called "leading furze" (Barton. 1972). When cut and dried it changed its name to 'bavin' (Grindon). One of the oddest late harvests was that of gorse and fern on Berkhamstead Common, in Hertfordshire.

No cutting was allowed before 1st September, but by tradition people gathered on the common late on the night before. They listened for the chimes of St Peter's Church, over a mile away, and on the stroke of midnight the cutters pegged out their claims, and returned at daybreak to do the actual cutting (Jones-Baker. 1977). It was just as important in Yorkshire, and especially important for the poor, who were often allowed special access to land to use the gorse, or sometimes furzy land was set aside for the poor, even after enclosure (Harris). There are examples, too, of more industrial use of gorse as a fuel, for brick making, for instance, or for lime burning.

It used to be equally important as feed for horses and cows, the former in particular, after the prickles were dealt with, of course. There were special gorse-cutters in Wales (gordd eithin), heavy mallets with cruciform blades (Davies & Edwards), and later, gorse mills were installed, some of which are still in existence, if only as museum specimens. But it is the young shoots that are particularly succulent for stock, after the old foliage is burned off, and the cows put into the gorse land when the consequent new shoots appear. This was particularly the case in Ireland (Ulster farmers said that an animal fed on whins would never be content with grass again (St Clair)), but it was common Welsh practice as well. The gorse was grown especially for the purpose, and called either *eithin ffrengig*, French gorse, or *eithin bras* coarse gorse. On some farms, a field of gorse was regarded as being as valuable as a field of hay. Pembrokeshire children were often given the job of collecting gorse seed, and gorse sales were commonplace in the 19th century – one year old gorse commanded quite a high price (Jenkins. 1976).

Its other uses include that of stopping a gap in a fence or hedge just by putting a bush in the opening, or by using it as a dead hedge to protect the base of a haystack from damp (Harris). In the south of Ireland it was employed as a quick hedge too. At the beginning of the 19th century fences there were either stone walls topped with earth, and furze growing on it, or banks of earth only, similarly planted (Lucas). It was used in cottage building sometimes – Flora Thompson described the way furze and daub was employed for the walls. A further use was for bedding for animals. A layer was well firmed by trampling, and straw put on top of that (Lucas). The commoners of Ashdown Forest, in Sussex, had the right to cut sticks out of gorse bushes. They would bundle them up and sell them to umbrella makers to make handles (Sargent).

There is a very ancient regard for gorse, particularly the seed, as a means of getting rid of fleas. An Anglo-Saxon version of Dioscorides is translated (by Cockayne): "Against fleas, take this wort with its seed, sprinkle it into the house; it killeth the fleas". Something similar was being advised in the 15th century:

"For fleas and lice to slay them. Take gorse and seethe it in water, and sprinkle that water about the house, and they will die" (Dawson). In more recent times, it has been used in medicine for scarlet fever, and also jaundice, the latter surely being the result of the doctrine of signatures (yellow flowers to cure the yellow disease). The flower infusion is an old Wiltshire remedy for dropsy (Wiltshire). The green tops featured in many a Highlands cough medicine (Grant), as it was in Ireland too (Maloney), whooping cough as well (St Clair). It was also used there for asthma, by steeping it overnight and drinking the water, and furze used to be the agent in an Irish worm cure, both for children and horses. In the first case, it was enough to boil a handful of the flowers in milk, and give that to the child to drink (Vickery. 1995). For horses the tops were cut and pounded on a block, and this would be given to the horse, often with a pint of linseed oil (Logan).

GOSPEL OAK

So-called because passages from the Gospels for Rogation Day were recited by the priest under them during parish perambulations, or "beating the bounds", always carried out at Rogationtide. The best known of the many Gospel Oaks is in the Suffolk village of Polstead, where an annual service is still held, an event that has been going on for a thousand years (Wilks). By their nature, Gospel Oaks usually served as Boundary Oaks too. Hertfordshire people used to keep acorns from Gospel Oaks in their pockets as some kind of health-giving charm (Jones-Baker).

GOUT

The primary remedy for gout has to be GOUTWEED, whose very name proclaims its virtue, and not only the common name, for the specific name *podagraria* means 'good for gout', from *podagra*, gout in the feet. It was even cultivated once specifically for the treatment (Beith). Nowadays a tea might be prescribed, but Culpeper even believed that "the very bearing of it about one easeth the Pains of the Gout, and defends him that bears it from the Disease". Colchicine, the drug obtained from MEADOW SAFFRON (*Colchicum autumnale*), has been used (in small doses, for it is extremely toxic) to treat gout and rheumatism. Gypsies use a very weak infusion of the sliced roots for the condition (Vesey-Fitzgerald). This is an ancient usage going back at least to the Arab physicians, but it probably still stands as the best alleviation for gout (Thomson. 1976). CANDYTUFT seeds have long been a traditional remedy for the condition (O P Brown), and in Indiana eating half a cupful of CHOKE CHERRIES each day was reckoned to be a cure (Tyler). In Wales, HERB ROBERT was used (Dyer. 1889), and, so it is said, GERMANDER SPEEDWELL, is especially good; the Emperor Charles V is supposed to have got benefit from it. It was so sought after for gout in the 18th century that it was, so they said, "made scarce to find through picking for many miles outside London" (Jones-Baker. 1974).

BIRCH tea (boiling water on a couple of tablespoon-fuls of the chopped leaves) has been used for gout, as well as urinary complaints in general (Fluck). Gerard reckoned that CUCKOO-PINT "hath a peculiar vertue against the gout, being laied on with Cowes dung". "With Buls tallow, or Goats suet this [DWARF ELDER] is a remedie for the gout" – "Dr Bullen's remedie", according to Aubrey. So is MUGWORT; take a handful of it, "and seeth it in sweet oil olive, until the third part of the oil be consumed; then anoint therewith any part that is tormented with the gout, and the pain thereof will be quickly gone or put away" (Lupton). A tincture of BIRTHWORT is used in Russian folk medicine for gout (Kourennoff). TANSY used to be a favourite remedy for the condition. Gypsies used a hot fomentation or an infusion for it (Vesey-Fitzgerald), and in Scotland it was the dried flowers that were used (Fernie). Gerard confirms its use in his day: "The root preserved with hony and sugar, is an especiall thing against the gout, if every day for certaine space, a reasonable quantity thereof be eaten fasting". And two hundred years before his time tansy was already being used for gout, accord-ing to leechdoms of the 14th century (Henslow). A decoction of DYER'S GREENWEED has been used for gout, as well as other ailments (Grieve. 1931), a remedy that goes back a long way, for we find a 15th century leechdom quoting the use: "Take flowers of broom and flowers and leaves of woadwaxen [i.e., dyer's greenweed], equally much, and stamp them with may-butter, and let it stand so together all night; and on the morrow melt it in a pan over the fire, and skim it well. This medicine is good for all cold evils, and for sleeping hand or foot, and for cold gout" (Dawson. 1934).

Duke of Portland's powder was well known at one time for treating gout. It was made up of BIRTHWORT, GENTIAN, GERMANDER, GROUND PINE and CENTAURY (Paris). It used to be said that drinking water from a well beside a PEAR tree was helpful for gout (Cullum). TEASEL seems to have been linked with gout in popular imagination, so it is not surprising to find the plant used in the treatment of that complaint (Blunt), and an early prescription for the complaint required the sufferer to lay pounded MULLEIN to the sore place. "… within a few hours it will heal the sore so effectively that [the gouty man] can even dare and be able to walk" (Cockayne). Dawson. 1934 also quoted a medieval prescription for the same ailment – "seethe mullein in wine, and it helpeth the hot gout; and seethe it in water, and it helpeth the cold gout". Aubrey 1686/7 quoted an instance of the use of WHITE BRYONY'S leaves for gout: "Take the leaves of the wild vine; bruise them and boyle them, and apply it to the place grieved, laid

in a colewort leaf". As it was used for dropsy as well, one suspects that this was a case of doctrine of signatures, for the root could be taken as resembling a swollen foot. BLACK BRYONY'S roots used to be applied as a plaster for both gout (Whitlock. 1992) and rheumatism, rather dangerous, one would have thought, for these roots are irritant and acid. Evelyn recommended ELM; all parts of the tree, he said, "asswage the pains of the gout". LILY-OF-THE-VALLEY has been used for gout, and Gerard, though doing little more than repeat earlier prescriptions, did print a novel way of tackling the disease. He enjoined the practitioner to put the flowers in a glass, "and set it in a hill of ants, close stopped for the space of a month, and then taken out, therein you shall find a liquor that appeaseth the paine and griefe of the gout, being outwardly applied …".

Russian folk medicine prescribed the stalks and leaves of SUNFLOWER, infused in vodka, and given three times a day, for gout (Kourennoff). An ointment made from BROOM flowers has been used for the condi-tion (C P Johnson). Folk medicine in Indiana advised eating lots of ASPARAGUS, which, so it is claimed, brought relief in just a few days (Tyler).

GOUTWEED

(*Aegopodium podagraria*) A common white umbellifer, not a true native to Britain, but introduced in the Middle Ages or even earlier, by the Romans, according to some authorities (Huxley, for instance) as a potherb and medicinal plant. By now it is widespread and common, a pernicious weed in the garden (Ground Elder to the exasperated gardener). The young leaves are perfectly edible if boiled like spinach (see Jordan).

Primarily, though, and as the common name shows, this is a medicinal herb, and a gout cure in particular, or at least a treatment for gout, over the centuries. It was even cultivated once, specifically for that treat-ment (Beith). The specific name *podagraria* means good for gout, from *podagra*, gout in the feet. Gerard knew the cure, of course, for it was known long before his time: Herb Gerard with his roots stamped, and laid upon members that are troubled or vexed with the gout, swageth the paine …". Nowadays a tea might be prescribed, but Culpeper even believed that "the very bearing of it about one easeth the Pains of the Gout, and defends him that bears it from the Disease". The name Gerard used in the quote above, Herb Gerard, has nothing to do with the 16th century herbalist, but with a Saint Gerard, probably not one of the two or three saints of that name recognized as such, but per-haps merely apocryphal, and the patron of gout suf-ferers, once invoked to cure the disease. The point is that the Dutch for the ailment and the plant is geraert, and the German Giersch for the plant and Gicht for gout. Bishop's Weed (Britten & Holland), or Bishop's Elder, from the Isle of Wight (Grigson) are two of the many names for the plant. Were bishops particularly prone to gout? Probably so, but the real reason for the names is likely to be the fact that this plant is so often found near ecclesiastical ruins; it was said that monks introduced it (Grieve).

Herbalists still prescribe it as a diuretic and sedative, hence as a painkiller (Le Strange). Drinking the infusion can help aching joints, and sciatica can be treated with it; the practice in the Highlands (Beith) and in Ireland (Moloney) was to make a poultice of the crushed herb. Eczema can be cured by drinking daily a half pint of the tea (A W Hatfield), and in East Anglia, the juice was squeezed on warts (V G Hatfield).

GRAPE VINE

(*Vitis vinifera*) The vine is sometimes used as an emblem of Christ. As such it has had the highest honour in the decoration of churches (Haig), especially when it is growing from the chalice of the Eucharist (Child & Colles). Christ said "I am the vine, ye are the branches. He that dwelleth in me, and I in him, the same beareth much fruit, for without me you can do nothing" (John. 15; 5). In Old Testament writings, the vine stands for the Jewish people as a whole. Dreaming of vines denotes health, prosperity and fertility (Gordon. 1985).

It is said in Iowa that a vine leaf in the hat will prevent sunstroke (Stout), and in Kentucky, they say that rubbing the sap from a grape vine on the hair will make it grow (Thomas & Thomas). Sap collected when growth starts in the spring is used for eczema in some country areas, and drops of it are also used for eye infections (Shauenberg & Paris).

GREAT PLANTAIN

(*Plantago major*) There is a well-known legend describing the persistent way that Great Plantain follows the tracks of man. More specifically, one superstition says that it follows Englishmen, and springs up in whatever part of the world he makes his home (Leyel. 1926). In this case, "White Man's Foot", which is what the native Americans called the plant (see Longfellow's Hiawatha), becomes Englishman's Foot. The German story is that it was once a woman, who waited by the wayside for her lover (Grimm). But once in seven years it becomes a bird, either the cuckoo, or the cuckoo's servant, the "dinnick" (Devonshire) (wryneck, according to Swainson), or, in German, Wiedhopf (which must be the hoopoe), which follows its master everywhere (Dyer. 1889). Plantago, and so 'plantain', is derived from Latin planta, the sole of the foot.

The belief that a coal was to be found under mugwort on St John's Day was extended to the plantain. Aubrey mentions that he saw young women looking for this coal on St John's Day, 1694, so that they could put it under their pillows to dream of their future husbands.

On St John's Day in Shetland, two stems were picked, one for the boy, and one for the girl, to foretell if they would love and marry (Grigson. 1955). The procedure was to pick the florets and then lay the heads under a flat stone. If the florets re-appeared before the heads withered, they would be sure to marry (Marwick). The Shetland names Johnsmas Pairs and Johnsmas Flowers refer to this practice. In Berwickshire, the divination was performed by removing all the visible anthers, wrapping the two scapes in a dock-leaf, and putting them under a stone till the next day. Then if more anthers have appeared, love is certain (Grigson. 1955). These are too close to the Ribwort Plantain divinations to be sure this particular plantain is meant.

In France, it is one of the more important herbs of St John. In some places, it has the power of disordering the wits, like fern-seed. It was believed in some parts that toads cured themselves of their ills, specifically, spider's stings (Berdoe), by eating the plantain. Another odd superstition was recorded in Iowa – pull a plantain leaf and the number of ribs shows the number of lies told during the day (Stout).

The leaves contain a mucilage that affords rapid relief after the stings of wasps and mosquitoes (Clair), and they can be applied to burns and scalds, too (A W Hatfield). In Somerset, they even say you could treat rheumatism by getting bees to sting you, and using the plantain leaves to ease the stings (Tongue. 1965). Some American Indian groups made poultices of the leaves to reduce swellings, to bring boils to a head, or to draw out thorns and splinters (Sanford), the kind of usage that is almost universal, for that virtue of the leaves has always been well-known.

The Anglo-Saxon version of Apuleius has cures ranging from headaches to snake-bites (Cockayne). In the latter case, the patient was advised to eat the plant! But native American peoples in general agreed with the European verdict; the Ojibwe, for instance, actually carried plantain about with them for immediate emergency protection (Densmore). Mad-dog bite was also dealt with by "rubbing (plantain) fine and lay it on" (Cockayne). In the 15th century the prescription was to make a plaster of this plant and the white of an egg, and laying that on the bite (Dawson. 1934). The Welsh text known as the Physicians of Myddfai also had a prescription for mad-dog bite, in this case using a handful of sheep's sorrel (*Rumex acetosella*) as well as the plantain, well pounded in a mortar with the white of an egg, honey and lard, to make an ointment to put on the bite. It is a wound herb, too, effective simply by laying the leaves on the wound (Leask, Egan, V G Hatfield), or sometimes they are mashed up and put on, to stop the bleeding. Hence names like Healing Blade, or Healing Leaf (Britten & Holland).

Another of the Anglo-Saxon prescriptions was that "if a man's feet in a journey swell, take then waybread the wort (Great Plantain, that is), pound in vinegar, bathe

the feet therewith, and smear them" (Cockayne). More recent practice is simpler. In Scotland, the plantain was used just by putting the leaves under the foot, or inside the stocking (Browning). There are some very strange ailments in the Anglo-Saxon Apuleius, as "in case that a man be ill-grown in wamb", or "in case one wishes to make a man's wamb dwindle" (and who does not?). What does "in case a man's body be hardened" mean? Or, "in case a spreading wart wax on a man's nose or cheek", or "of all strange bladders that sit on a man's face"?

Gerard listed many ailments to be treated with Great Plantain ("of all the plantains the greatest is the best …"). Among them, "… fluxes, issues, rheumes, and rottennesse, and for the bloudy flux", which is dysentery, and it is still used in Chinese herbal medicine for that complaint (*Chinese medicinal herbs of Hong Kong*), and plantain tea is still being recommended for diarrhoea (A W Hatfield). Jaundice is another ailment to be cured with plantain in modern times, but which had already appeared in a much earlier age. The treatment was known in folk medicine, in this case, in Cambridgeshire (Porter. 1969), but undoubtedly over a much wider area as well. The tea is still used for complaints as different as piles and asthma (A W Hatfield), and bronchitis can also be treated in this way. A leaf poultice was used for corns and ulcers (Vickery. 1995), and boils too (Stout), but that is a very old recipe – *Reliquae Antiquae* has "take the rotes of red nettilles and playntayne, and stamp them wele in ale, and do thereto cray [chalk] that hir parchemeners [paper-makers] wirkes withall, and ger hym drynk hit" (see I B Jones). But the list of ailments for which this remarkable plant has been recommended seems endless. Only a few have been mentioned here.

GREAT BURNET

(*Sanguisorba officinalis*) By tradition, a stauncher of blood, perhaps from the colour of its flowers, which are of a dark crimson (the generic name, *Sanguisorba*, comes from Latin sanguis, blood). "Burnet is a singular good herb for wounds … it stauncheth bleeding and therefore it was named *Sanguisorba*, as well inwardly taken as outwardly applied" (Gerard), in other words, surface wounds as well as internal haemorrhages. The plant was actually called Bloodwort (Clair), or Burnet Bloodwort (Prior). An interesting fact, whether this use is from flower colour or not, is that it is taken in Chinese medicine for haemorrhages, too (Geng Junying), as well as for dysentery and other ailments. The leaves of Sweet Basil and Burnet steeped in boiling water make a cooling face wash (H N Webster).

GREAT PRICKLY LETTUCE

(*Lactuca virosa*) Medieval naturalists reckoned that eagles kept their eyes keen by eating hawkweed and wild lettuce. Another piece of medieval belief

was that dragons used the juice as a spring tonic (Hulme. 1895). This plant has the milky juice common to all members of the genus, but it has been actually cultivated both in Britain and in Europe for it (Usher). For this juice hardens when exposed to the air, and produces a gum called lactucaria, or lettuce opium (Brouk), which has distinct narcotic and soporific qualities (Brownlow). All lettuces are slightly narcotic, but none more so than this one. Gerard reported that it "procures sleepe", and also that "the seed taken in drinke, like as the garden lettuce, hindreth the generation of seed and venerous imaginations". The last is interesting, for Ibykos, the Pythagoran poet, called the lettuce by the name eunuch; in other words, it puts to sleep, that is, renders stupid and impotent (Gubernatis. 1872). They say that if vinegar is added to lettuce, its soporific virtue is destroyed (Leyel. 1926). The leaf extract used to be recommended in small doses for dropsy (Thornton).

GREATER CELANDINE

(*Chelidonium majus*) The name Celandine (see also Lesser Celandine, *Ranunculus ficaria*) derives from Pliny, for he was responsible for the legend that seeks to account for the name, which was Khelidonion in Greek, from Khelidon, the swallow. The birds used the plant, he says, to restore their sight. But he also said the flowering of the plant coincided with the swallow's stay in Europe, but that would not agree with conditions as we know them. Theophrastus says it bloomed when the swallow-wind blew (Browning).

The doctrine of signatures, in this case the plant's yellow sap, ensured that it would be used to cure jaundice. It is the leaves that are used in Suffolk, boiled in water (V G Hatfield. 1994), and earlier usage, as in Gerard's advice, prescribed the roots: "the root cureth the yellow jaundice which comes of the stopping of the gall…". Even putting some leaves in the shoe was supposed to cure the disease (Tynan & Maitland).

The legends connected with this plant and with lesser celandine explain its use, from the earliest times (in ancient Chinese medicine, for instance), as a specific for sore or weak eyes (Grigson. 1974). It continued into Anglo-Saxon times, when it was prescribed for "dimness of eyes and soreness and obstruction …" (Cockayne). The orange latex of celandine is very irritant, but after drying or heating the acrid property is much reduced or destroyed. In this state it has been used successfully "since time immemorial to remove films or spots from the cornea (M L Cameron). A medieval Jewish work recommended the sap for spots in the eye, and for cataracts (Trachtenberg). The pre-scriptions carried on into Gerard's time. He said that "the juice of the herbe is good to sharpen the sight, for it clenseth and consumeth away slimie things that cleave about the back of the eye, and hinder the sight".

As well as to "eat away … opacities in the cornea", it was used to "eat away warts…" (Thornton). Some of the names given to the plant reflect this usage – see, for instance, Wart-plant, Wartweed, Tetterwort, Fellonwort, etc., and it is called in French herbe aux verrues (Schauenberg & Paris). Gypsies use the juice as an outward application both for corns and warts (Vesey-Fitzgerald). Somerset advice, too, was to rub the juice on the corn, which would eventually come out with the blister (Tongue. 1965) (that was also from a gypsy). The juice on the skin will raise a blister; and it may cause ulcers, too (Flück), but the Pennsylvania Germans used the juice on bee stings, or on poison ivy rashes (Fogel). But then it was the bruised leaf that was employed (Radford & Radford). The plant is used in East Anglia for toothache (Fernie), an ancient usage, for Gerard repeated it from much earlier herbals: "the root being chewed is reported to be good against the toothache". It is even reported that a decoction of the plant has been used in East Anglia to treat liver cancer (V G Hatfield. 1994).

GRECIAN HAY
See FENUGREEK, which is a corruption of *foenum-graecum,* an alternative specific name for the herb, i.e., *Trigonella ornithopodioides. Foenum graecum,* of course, means Grecian Hay. It has been suggested that it got this name because it was used to scent inferior hay (Grieve. 1931).

GREEN HELLEBORE
(*Helleborus viridis*) Just as poisonous as the other hellebores, and that makes John Josselyn's report in 1638 interesting. He said that the root was used by young Indian braves in an ordeal to choose the chief – "he whose stomach withstood its action the longest was decided to be the strongest of the party, and entitled to command the rest" (Weiner).

Gerard reported that it "… is thought to destroy and kill lice, and not onely lice but sheepe and other cattell …" But not quails, though, for there was an ancient belief that quails grew fat on poisonous plants, particularly hellebore, apparently springing from Aristotle, *de Plautis,* v – "henbane and hellebore are harmful to men, but food for quails". Lucretius, *de Natura Rerum*, also says "hellebore is a violent poison to us, but it fattens goats and quails" (see Robin. 1932). Beware of eating quails, then – it would be liable to give you epilepsy! (Hare).

Like Stinking Hellebore, this was used for expelling worms in children (Grigson. 1955), and it is equally dangerous. But the dried rhizome and roots were used, perfectly seriously, to slow the heart's action and to soothe the nerves. It was official in America, where it very rapidly naturalized, up to 1960, and was also used to lower blood pressure (Weiner).

GREEN PURSLANE
(*Portulaca oleracea*) A food plant, valued highly by peoples as wide apart as the Navajo Indians, who eat the seeds (Elmore), and the Mano people of Liberia, who recognize it as an accessory green food, specially prescribed for malnutrition (Harley). There are a number of other medicinal uses throughout the world. The Navajo use the green plant for stomach ache (Elmore), and the Mano too recognize it as an indigestion remedy (Harley). In Central America, Maya medical texts prescribed the crushed plant, rubbed on the body, for tuberculosis. The juice is given for giddiness, and an infusion is used as a bath for convulsions (Roys). In West Africa it is prescribed for local application to swellings and bruises, or as a poultice for abscesses or boils. The juice is sometimes dropped in the ear for earache, and is also used for toothache. Skin diseases are treated in West Africa, as well as in China, with purslane, but in Ghana they eat the leaves along with tiger nuts as the remedy (Dalziel). The Mano look on it as a sore throat remedy, too. They take a large handful, beaten up with root ginger. It has to be mixed with water from a "talking stream", and meat and salt are added to make a soup (Harley).

Purslane is a children's good luck charm in West Africa, and a symbol of goodwill (Dalziel).

GREENGAGE
(*Prunus domestica 'italica'*) To dream of eating greengages is a sign of trouble and grief (Raphael), probably a mistake in identification, for it is more likely that the trouble and grief would be caused by green, i.e., unripe, plums.

GROMWELL
(*Lithospermum officinale*) "The root is used by ladies as paint" (Thornton) (cf PUCCOON, an American species). The leaves are used to make Croatian, or as it is sometimes called, Bohemian, tea (Perry. 1972 ; W A R Thomson. 1976). Medicinal uses lean strongly on the doctrine of signatures. Its stony seeds (that is what *Lithospermum* means) proclaim its signature perfectly, for it was widely used against the stone. "In case that stones wax in the bladder, and in case that a man may not mie, take of these stones [seeds, that is] … give to drink in wine, and forth leadeth the mie" (Cockayne). That was translated from an Anglo-Saxon version of Dioscorides, but Gerard was recommending the same in the 16th century, and so was Hill in the 18th. And so it went on. In fact, Pliny said "Indeed there is no plant which so instantaneously proclaims at the mere sight of it, the medicinal purposes for which it was originally intended" (see Bonser). After all this, it comes as a surprise to find that the plant actually does have diuretic qualities (Schauenberg & Paris). It can also, apparently, stop the activities of some hormones (Grigson. 1955), which is presumably why Grigson reported that the seeds were being investigated for the contraceptive substance they contain. But some of the American Indians had been using native species as an oral contraceptive for a long time (F P Smith).

GROUND CYPRESS

(*Santolina chamaecyparissus*) The whole plant is aromatic after rain, and the leaves are used to keep moths from clothes (Brownlow) (it is Garde-robe in French). It is known as a stimulant, similar to wormwood in its properties, and it is also used for ringworm and as a general vermifuge (Macleod). Lavender Cotton is another general name for it.

GROUND ELDER

(*Aegopodium podagraria*) see GOUTWEED

GROUND IVY

(*Glechoma hederacea*) In the past, Welsh milkmaids wore ground ivy when first milking the cows in the pastures (Trevelyan), and according to a story quoted by Lady Wilde, ground ivy carried in the hand gave protection against attacks by fairies (Wilde. 1902) (cf PEARLWORT). In the Tyrol, rue, worn with agrimony, maidenhair, broom straw and ground ivy, was said to confer fine vision, and to point out witches (Dyer. 1889). So, too, in Germany on May Day, it was said that by putting a bunch of ground ivy on the breast, or a chaplet of it on the head of a virgin going to church, one would be enabled to recognise and name witches (Lea). Probably, the Welsh custom of making a poultice of the leaves, and applying it to sore eyes (Trevelyan), should be included in this context, though see below for eye medicines.

It is a herb of St John in France. In Poitou, they would say that an old woman who makes a waistband of ground ivy while her friends and relations are away dancing round the Midsummer fires, and who preserves the girdle to the end, will escape the pains and sorrows of old age (Sebillot). More recently, it seems that chaplets of ground ivy were worn on the head, as the people danced round the Midsummer fires (Palaiseul).

In Ludlow, Shropshire, there was a customary Easter dish of leg of pork stuffed with Robin-run-in-the-hedge, which is ground ivy (Burne. 1883). According to Genders. 1972, the principal ingredients of Elizabethan snuffs were ground ivy, camomile and pellitory-of-the-wall. But it is in the domain of medicine that ground ivy is most important, for it is a real cure-all (it was even given to the insane (Leyel. 1937)). A tea made from it used to be popular for eczema in the north of Scotland. It was said that the fairies taught Donald Fraser, of Ross-shire, to use it (R M Robertson). It was one of the cries of London, and, drunk as a tea, sold as a "blood purifier" (Thornton), and was always used in this way in Dorset (Dacombe) and Hampshire (Hampshire FWI). And it was used for asthma – an Irish recipe advised the patient to drink of a potion made of ground ivy (or dandelion), with a prayer said over it before drinking (Wilde. 1890). In Scotland, snuff from the dried leaves was used for the complaint, and for headaches (Beith). The infusion was given in Ireland for bronchitis, and

"boil ground ivy and drink the water" is an Irish cold cure (Moloney). Actually, "Gill-tea", as it was called, mixed with honey or sugar to take away the bitterness, has always been a favorite remedy for coughs and colds (Clair). It could be combined with wood sage in a tea to treat a cold – that is a New Forest gypsy remedy (Boase).

Ground Ivy has been just as famous as an eye medicine. A leechdom for eyestrain from as early as Anglo-Saxon times required it to be boiled in sour beer, and the result used to bathe the eyes (Cockayne), and similar eye recipes are to be found in herbals from that time onwards. Folk medicine took it up, too. For example a Dorset remedy for sore eyes is to make an ointment with it (Dacombe). A Warwickshire remedy is to take a large handful of this herb, just cover it with water, and simmer for about 20 minutes, strain it, and use the liquid to bathe the eyes (Vickery. 1995). Wiltshire has a much more localised remedy, in which water from a well on Cley Hill, Warminster, was used to boil ground ivy, as a remedy for weak eyes. The water had a popular reputation as only being valuable for bathing the eyes, and the ground ivy had a separate reputation for the same thing (Manley). There is even a story of a fighting cock that got wounded in the eye. "Its owner chewed a leaf or two of ground ivy, and spat the juice in the damaged eye to make it heal quickly" (Palaiseul).   AU:1

From its former use in brewing, for its tonic bitterness, such names as the very widespread Alehoof (Grigson. 1955, Bloom), which means literally that which causes ale to heave, or work (Britten & Holland). Tunhoof is another, related, name, tun being a cask (of ale). The place where such medicated beer was sold was known as a gill-house (Barton & Castle), and Gill, or Jill, so we know that the 'g' of Gill is soft, are two of the many names using this word (J D Robertson, Britten & Holland). Gill-ale was used as a name for this plant in Devonshire (Friend. 1883), and there are many further, picturesque, names, of which Gill-go-by-the-ground (Prior) will serve as an example.

GROUND NUT

A name given to EARTH-NUT (*Conopodium majus*) (Grigson. 1955), which has edible tubers, but there are more important plants, notably *Arachis hypogaea*, with the same name. This latter plant seems to be unlucky in many ways. If you dream of it, you will be poor, so it was believed in Illinois (H M Hyatt). And African-Americans in the southern states of America say it is unlucky to eat peanuts when you are going to play a game of any sort, and peanut hulls scattered about the door mean that you will go to jail (Puckett). In Madagascar, they are roasted like chestnuts in the home, but in many places there, peanuts are taboo. Nuts lying on the ground remind people of souls that lay their eggs on the ground. So they are taboo to pregnant women, as they will cause a miscarriage (Ruud). They are equally well known as Peanuts, of course, and Monkey Nuts, nearly as ubiquitous, though Grigson. 1974 suggested that the only reason for the name was that they are bought to feed monkeys in the zoo.

GROUNDSEL

(*Senecio vulgaris*) An ubiquitous weed, apparently blooming twelve months a year. But it has an association with witchcraft, both good and bad; in the Western Isles it was, in Martin's day, used as a counter-charm, in particular when milk was being stolen by witchcraft (Polson), and there are records of the use of pieces of the root as amulets against the evil eye (Folkard). On the other hand, it seems that in the Fen country, the belief was that the witches were actually responsible for the weed itself. A small patch growing beside an old trackway showed that a witch had stopped there to urinate; large patches meant that a number of them had met to plot. Groundsel growing in the thatch was a sign that a witch had landed there during a broomstick flight. It was also believed that witches could never die in winter, but only when the groundsel was in flower (even though it seems to be in flower all the year round). The point was that the witch could then take with her a posy of the flowers, by which the devil would recognise her as his follower (Porter). Burning groundsel was a way of driving evil spirits from the house, and incidentally would get rid of vermin in clothes and bedding.

Uses of groundsel in domestic medicine have been many and varied, either as charms or as genuine, if misguided, medicines. An example of the charm is its use for ague. A woman suffering from an ague was recommended to tell her husband to tie a handful of groundsel to her bare bosom while the charmer spoke the necessary incantation (Black). It had to remain there, and as the herb withered, the ague would go away. Wesley prescribed another charm for the same sickness – "For the Ague … take a Handful of Groundsel, shred it small, put it in a Paper Bag, four Inches square, pricking that Side which is to [go] next the skin full of Holes. Cover this with thin Linnen, and wear it on the Pit of the Stomach, renewing it two Hours before the Fit: Tried". That charm was being used in Cornwall long after Wesley's time (Deane & Shaw).

There is a typical wart charm from Devonshire: rub a wart with groundsel to make it go. The leaves should then be thrown over your head, and afterwards they should be buried by someone else. As the leaves rot, so will the wart (Crossing). On the other hand, groundsel poultices were quite common and widespread for boils (Dacombe; Randell; Grant; Foster) and abcesses (Hampshire FWI), and one finds it being recommended in Apuleius for "sore of loins" (lumbago, that is), in Cockayne's translation. The same source recommended groundsel for wounds, pounded "with old lard, lay it to the wounds", and "if any one

be struck with iron", when the plant had to be taken at early morning, or at mid-day, pounded and mixed with "old lard". There are many examples of its use as a wound herb, up to and including Gerard's time, and another of his prescriptions stated that "the leaves of Groundsel boiled in wine or water and drunke, heale the pain and ach of the stomacke that proceeds of Choler …". Gypsies were using it for the same purposes until very recently, i.e., the bruised leaf and stems to relieve colic and inflammations (sprains too) (Vesey-Fitzgerald). There is a Cornish belief, obviously based on homeopathic magic, that groundsel acts in different ways according to the direction in which the leaves are stripped from the stem. If upwards, that is, beginning from the root, with the knife ascending to the leaf, it makes it good as an emetic; if striped downwards, it should be used as a cathartic (Hunt).

A Norfolk remedy used groundsel to bring relief to rheumatic sufferers. All they had to do was soak their feet in water in which the plant had been boiled for ten minutes (Randell). The plant enjoyed the reputation of being able to soften water, if it is poured, boiling, on the plant. Such water would be used as a skin wash (Pratt). There are very many more ailments for which groundsel has been used in earlier times, and there is even a record from Germany of using it as a children's vermifuge, and an infusion is still used in Cornwall for jaundice, and to relieve obstructions of the bladder and kidneys (Deane & Shaw).

GUELDER ROSE

(*Viburnum opulus*) Well established in Britain, but it is not a native, the introduction being from Gueldres, hence the common name, though some say it is a corruption of 'elder rose'. Gerard, though, knew it as Gelders Rose. It has a connection with May Day and Whitsuntide, as its names reveal. It is King's Crown in Gloucestershire, the King of the May being crowned with it (Britten & Holland). Maypole, May Tassels, or May Tossels are all from Devonshire (Macmillan). 'Tossels' takes us further, for Guelder Rose is the May Tosty, or Snow Toss, in Somerset, and they all refer to Tisty-tosty, used across the south-west of England (Tennant; Elworthy. 1888; Friend. 1882). Tisty-tosty is usually a ball of cowslips or primroses for the May garland, but Guelder Rose is evidently used, too. As for Whitsun, for which Guelder Rose is the ecclesiastical symbol), it is known as Whitsun Boss (J D Robertson; Leather), or Whitsun Balls, Whitsun Flower, Whitsun Tassels, and Whitsun Rose (Macmillan). That leaves us with Club Bunches, from Berkshire, and particularly Hagbourne in that county. Guelder Rose flowers were used to decorate the president's chair at the club dinner on Feast Day, the second Tuesday after Whitsun (Berkshire FWI).

Another name given to the tree is Cramp-bark (Youngken); it has been used to treat the complaint for a long time by people living in marshy country (Wilkinson. 1981).

GUINEA-HEN WEED

(*Petiveria alliacea*) Originally from Venezuela and Amazonia, but it grows in the Caribbean region, and also in West Africa. It will give a garlic flavour to milk if cows eat it, and it has been used as a fish poison (Perry. 1972), and it functions as an insecticide (Dalziel). A Yoruba preparation to prevent being attacked by someone, required a leaf of this plant, a leaf of *Ageratum conyzoides*, and some other, unidentified, leaf, all burnt together, and rubbed into small incisions on the hand (Verger).

Ka'apor people of Amazonia made an amulet of the bark for infants, wrapped in cloth. It would ward off the evil divinity (Balée), and they plant it by their doors as an apotropaic protector. It is used as an ingredient in the ritual baths that are part of Brazilian healing ceremonies (P V A Williams), and amulets are made of the wood, in the shape of the universal figa. The figa gesture is usually made with the hand, but wearing a carved figa round the neck or waist is much simpler. Brazilian street vendors wear one, or stand one up on their trays so as to protect their goods from the evil eye. Petiveria plants are widely used in Brazil for repelling the evil eye and for curing in general, and the leaves are popular in the ritual that accompanies the recitation of a curing prayer (reza). Three leaves can be worn behind the ear as an amulet (P V A Williams).

In West Africa, this provides a whooping cough remedy (Dalziel), and it was used for toothache by slaves in Jamaica (Laguerre). The leaf decoction is used for abortion in Guyana, where it is called Gully-root, and that same decoction is taken for arthritis in Barbados, and for headaches in Jamaica (Laguerre).

GUMBOIL

Sucking a SLOE is said to cure gumboils (Addison & Hillhouse). A poultice made from CAMOMILE flowers or leaves was a Scottish treatment for them (Gibson. 1959).

GUNBRIGHT

is one of the names given to DUTCH RUSH (*Equisetum hyemale*), which is one of the Horsetails. They were used for scouring and polishing, and it is said that the American Indians used it in polishng their guns (Bergen. 1899).

Gutierrezia microcephala > BROOMWEED (1)

H

HAEMORRHAGES

GREAT BURNET (*Sanguisorba officinalis*), besides being a wound herb in the usual sense, is able to deal with internal bleeding. As Gerard said, "Burnet is a singular good herb for wounds ... it stauncheth bleeding, and therefore it was named Sanguisorba [Latin sanguis, blood], as well inwardly taken as outwardly applied ...". But SHEPHERD'S PURSE is "the great specific for haemorrhages of all kinds", in Mrs Leyel's words (Leyel. 1937). A 17th century physician, Symcott by name, was treating a pregnant woman for blood loss. Then "a baggar woman told me that she would recover if she took shepherd's purse in her broth". She was cured (Beier). It is still being recommended for similar conditions, and even if we did not know that, old names like Stanche and Sanguinary would quickly point out the use.

HALITOSIS

Chew ORRIS-ROOT to neutralize the smell of liquor, garlic or tobacco on the breath (Moldenke & Moldenke).

HALLUCINOGENS

Some uses of SWEET FLAG stem directly from the fact that the oil expressed from the roots can produce an LSD-like experience. The Cree Indians, for instance, have long used it as a hallucinogen (Emboden 1979). In Europe it seems to have been connected with witchcraft (one formula for a witch ointment was "De la Bule, de l'Acorum vulgaire,, de la Morelle endormante, et de l'Huyle" (Summers. 1927)).

MORNING GLORY

(*Rivea corymbosa*), the revered ololiuqui of the Aztecs, was one of the most important hallucinogens of ancient Mexico at the time of the conquest. The seeds, which contain alkaloids closely related to LSD (Wasson, Hofmann & Ruck), were used to induce a ritually divinatory trance (Norbeck). Another New World hallucinogen is COHOBA, growing in the northern part of South America, particularly in the Orinoco basin. A snuff, known as yopy, or parica, is made from the powdered seeds, and is one of the most famous New World hallucinogens. The original inhabitants of Haiti made this narcotic snuff, which they took through a bifurcated tube (Youngken), and in fact it was much used in religious ceremonies over most of the area (Hostos). Lewin described the use among one of the Brazilian tribes: "... begin to take Parica snuff ... they assemble in pairs, everyone with a (bamboo) tube containing Parica in his hand, and ... everyone blows the contents of his tube with all his strength into the nostrils of his partner. The effects produced in these generally dull and silent people

are extraordinary. They become very garrulous and sing, scream and jump about in wild excitement..." *Banisteriopsis caapi* is another South American narcotic, inseparably submerged in the total culture of the people who take it. Ayahuasca, a Quechua word meaning 'vine of the dead', or; 'vine of the souls', is its Peruvian name. Partakers often experience a kind of "death", and the separation of body and soul. Those who "die" are reborn in a state of greater wisdom. It serves, too, for prophecy, divination, etc., But it may be taken at funeral ceremonies, and., in other contexts, by a shaman (or ayahuasquero) to diagnose an illness or divine its cure, especially for those who believe themseleves bewitched, or to establish the identity of an enemy (Reichel-Dolmatoff), when evil magic would be "returned to its perpetrator" (Dobkin de Rios. 1970). The effects may be violent and with unpleasant after-effects, especially when the bark is boiled, and certainly when other toxic plants are mixed in. Nausea and vomiting are almost always early characteristics; this is followed by pleasant euphoria and visual hallucinations, but few have ever admitted that they find it a pleasant experience, for they drink it to learn about things, persons or events which could affect the society as a whole, or its individuals (Kensinger).

Hamamelis virginica > WITCH HAZEL

HAND OF GLORY

is a magical torch made from a dead man's hand, usually cut from a criminal on the gibbet, to cast people into deep sleep, a charm much used by thieves. It is mentioned in this context because VERVAIN played a part in its preparation, for one set of instructions (Radford & Radford) required the hand to be wrapped in a piece of winding sheet, drawing it tight so as to squeeze out the little blood that might remain. Then it had to be placed in an earthenware vessel with saltpetre, salt and pepper, all well dried and carefully powdered. It should remain a fortnight in the pickle, and then it had to be exposed to the sun in the dog days, till completely parched, or it could be dried in an oven heated with VERVAIN and fern.

HANGOVER

A sailor's cure for a hangover, from South Uist, was to pull a bunch of THRIFT, roots and all, and boil it for an hour or more. It had to be left to cool, then it was drunk slowly (Shaw). Sufferers in Norfolk would cure a hangover by the simple expedient of chewing CELERY. (V G Hatfield. 1994). RED POPPY, too, was a hangover treatment in Norfolk, (V G Hatfield. 1994), an interesting choice, for the underlying folklore has it that these plants will actually cause a headache. As John Clare said:

Corn poppys that in crimson dwell
Call'd 'head achs' from their sickly smell.

HAREBELL

(*Campanula rotundifolia*) It is said that the plant is dedicated to St George; in former times, blue was worn on St George's Day, 23 April (Friend):

> On St George's Day, when blue is worn,
> The blue harebells the field adorn.

That is all very well, but harebell is not normally in flower in April – it is a flower of high summer. Perhaps another plant is meant – the common bluebell, perhaps?

As some of the local names suggest (Witch Bells, Witches' Thimble, etc.,), this is a witch plant, not to be picked (Coats. 1975). Perhaps the belief arose by a misunderstanding of the common name; hares have always been witch animals. Harebell is OE hara, hairy, a reference to the thread-like stalks.

HARTSTONGUE

(*Phyllitis scolopendrium*) Both in Wales and in Scotland, hartstongue leaves have been used for a wound application (C P Johnson), and, made into an ointment, it was a Highland cure for burns (Beith).

HASHISH

is an Arabic word, meaning hay or dried herb (Grigson. 1974), but it is a term with many meanings, though it seemed to be applied by Burton to a form of CANNABIS (HEMP) taken or used voluntarily. "Tis composed of hemp leaflets whereunto are added aromatic roots and somewhat of sugar; then they cook it and prepare a kind of confection which they eat, but whoso eateth it, (especially if he eat more than enough), talketh of matters which reason may in no wise represent" (quoted by Lloyd). It is the resin obtained from the glandular leaves and floral parts of the female plant. The name appears, too, in the name of a Persian form Hashishin Rus (some would say it actually derived from that name). Al-Hasan ibn-al-Sabah (the "old man of the mountains"), a 12th century charismatic dissenter from orthodox Moslem thought, founded a new sect called Hashishin, a name that also produced the word assassin (Emboden. 1969). In Egypt and the Middle East, hashish is smoked in special pipes called josies (De Ropp). See also HEMP.

HAWTHORN

(*Crataegus monogyna*) A sacred tree, treated as such long before Christian tradition associated it with the Crown of Thorns. It was said that the tree groans and sighs on Good Friday (Wilks). In some parts of France it was not unknown for mothers to kneel before the tree, and pray to it for a child's good health. If they lived a long way from the church, they would go and say their prayers to the tree (Devlin). In medieval times rosaries made of thorn wood were in great demand, and were treated as if they were jewels. But in Ireland particularly, ancient and solitary thorns, known variously as fairy thorns, gentry or gentle (gentry or gentle being fairies – the gentle people) thorns, skeaghs, or lone bushes (E E Evans), were always held in great veneration. It was nothing less than profanation to destroy them or even to remove a bough. A lady dressed in a long white robe (the banshee, perhaps) was often supposed to come from them, and elves and fairies were seen among the branches. Hedgerow thorns are newcomers (even though they are said to date from Roman times (Cornish), so they may be hacked with impunity, but woe betide the man who damages one of the solitary fairy thorns, that is, one not planted by man, but growing on its own. Not even fallen dead branches that would serve as firewood should be removed. There have even been examples of branches accidentally broken being carefully tied back in position (E E Evans). The cult of these thorns in Ireland was apparently just as strong in the Protestant north as in other parts of the country, but the belief was not confined to Ireland, for in Somerset too the tradition was that you should never cut down hawthorn trees to build your house, for if you did you and yours would never live long (Tongue). In Galloway too, solitary thorns were left and preserved with scrupulous care. There is a story of two lads who were ploughing a field that had one of these thorns in it, and they carefully ploughed in a circle round the tree. They found a green table placed at the end of the furrow, heaped with cheese, bread and wine (Cromek). Similar beliefs were held in the Isle of Man; it was not advisable to sit too long under one of these trees, and certainly not to sleep under one (Gill). The fairies danced round these thorns at night (Beard), and:

> By the craggy hillside,
> Through the mosses bare,
> They have planted thorn-trees
> For pleasure here and there.
>
> Is any man so daring
> As dig them up in spite,
> He shall find their sharpest thorns
> In his bed at night.

(William Allingham, a stanza from *The fairies* 1850).

A further result of the fairy thorn belief is the superstition that if thorn bushes are ploughed up, all goodness leaves the land (Tongue). A correspondent of *Notes and Queries; 1941* told a story then current at Berwick St John, in Wiltshire, of the consequences of cutting down a solitary thorn that grew on a prehistoric earthwork nearby. The result was complete loss of fertility over the area, taking in poultry and cows as well as women. Fertility was only restored when the perpetrator planted a new thorn in place of the old one. Another Irish belief is that hawthorn trees grow over graves or hidden treasures (Ô Súilleabháin); similarly there is a Cornish tradition that those who buried treasure always planted a hawthorn over it (Wilks). Solitary thorns are often associated with the dead, and they often mark graves, or they mark a spot where a coffin was rested, or where a death had taken place in the open (J J Foster). They are often associated with death cairns, too, for the thorn was sometimes planted where a death had taken place. These are often called Monument Trees (M Mac Neill. 1946). In Galway, Ireland, these thorns were said to have sprung from dead men's dust scattered through the world (Fitzgerald). Hence the idea that the soul becomes a tree. Some thorns are dedicated to Irish saints, and these figure in burial ceremonies. When funerals pass by, they halt, and stones are placed beside the thorn until they become cairns. On thorns at crossroads it was the custom in County Wexford to hang small crosses, made of coffin wood, as the funeral procession passed by (E E Evans). One of these dedications to the saints is St Leonard's bush (Cran san Lionairt) at Dunnamaggan, County Kilkenny, held in such veneration in the nineteenth century that no native would emigrate without carrying a chip of the tree with him as a protection against shipwreck (Lucas). Another function of these thorns was to mark a well, and they were very common. One near Tinshally, County Wicklow, was known as Patrick's Bush. Devotees attended on 4 May, rounds were made about the well, and offerings were made to the thorn (Wood-Martin), if that is the right way to describe the ritual of hanging rags or trinkets on the tree that is companion to the holy well. The same inviolability applies to these thorns. One growing beside St Laghteen's Well, Knockyrourke, County Cork, is said to be impervious to fire, and the men who tried to cut it down were seized with violent pains. There is a decorated tree at Appleton Thorn, in Cheshire, where there is the well-known ceremony of "bauming the thorn", bauming being a dialect word meaning adorning. The ceremony takes place in July, when an old thorn tree in the village is decorated with red and white flowers and ribbons, and the children dance round it (Baker).

Another aspect of the hawthorn is as an abode, not just of fairies, but of witches, too. It was an accepted belief in the Channel Islands that witches used to meet under solitary hawthorns (MacCulloch), and there used to be quite a widespread superstition that it was dangerous to sit under a hawthorn on Walpurgis Night, May Eve in our terms, because it was then that a witch was most likely to turn herself into a thorn tree (Jacob). On the other hand, and quite in accordance with accepted belief, hawthorn would also protect against witchcraft. In Monmouthshire (Gwent) tradition, one of the commonest ways of breaking a witch's spell was reckoned to be putting a cross of whitethorn (or birch) over the house door (Roderick); far from there, the Serbs believed that a cradle made

from hawthorn wood would be a most powerful protective device (Vukanovic), and in the same area, a small hawthorn peg may be driven into a grave, to prevent the corpse from turning into a vampire, or a stake of the same wood was used in Serbia to "kill" the vampire (P Barber). To "drive witches out of milk" by beating it with hawthorn used to be a Pennsylvania German saying (Fogel).

The "authority" of hawthorns is illustrated in another aspect, that of justice. In the Lake District, they were apparently associated with places of trials and courts of justice. Two at least, so it is claimed, still survive where courts were held (Rowling). They were markers of meeting places, too, and the existence of "thorn" in a place name, especially an old hundred name, is often taken as an indication of such a hawthorn tree there. The old Hundred of Spelthorne, in Middlesex is an example. Some Irish thorns were known as "Mass bushes", the marks of assembly for Catholic congregations during the persecutions of the 17th and 18th centuries (Cornish). Until the early 19th century, a thorn stood as the "Luck" of Earlstoun, or Erceldoune, in Scotland. Thomas of Erceldoune prophesied that:

> This thorn tree, as lang as it stands,
> Earlstoun shall possess a' her lands.

A further range of hawthorn folklore arises from its inclusion in the band of lightning plants, presumably because of its red berries. In many parts of England, hawthorn gathered on Holy Thursday (whether that means Maundy Thursday or Ascension Day, both of which bear the name, is not clear), was used as a protection against lightning (Burne). It is peculiarly a Shropshire tradition to gather your hawthorn on Holy Thursday to protect the house. In Touraine it was cut, fasting, on May morning, for the same purpose (Sebillot), and the Greeks too hang hawthorn blossom from their doors on May morning (Wilks). "The white thorn is never stricken with lightning", said Langham in 1578, and Mandeville had "White thorn hath many virtues: for he that beareth a branch thereof upon him, ni thunder nor tempest may hurt him; and no evil spirit may enter in the house in which it is, or come to the place that it is in". There are many folk rhymes to remind one of this, most of them very similar. This particular version was recorded in Fittleworth, in Sussex:

> Beware of an oak,
> It draws the stroke;
> Avoid an ash,
> It courts the flash.
> Creep under the thorn
> It can save you from harm (Opie & Tatem).

In Normandy, they still say that a twig of hawthorn will protect him who carries it (Johnson). On the other hand, and as is often the case, the direct opposite is sometimes found, as with Cornish people, who thought it dangerous to stand under a whitethorn during a thunderstorm (Deane & Shaw). And there is a Welsh belief that the tree itself, or at least one particular tree, will cause the storm. The tree is the old thorn at Ffynan Digwg, Caernarvon, and thunder and lightning would be the result if it were cut down (F Jones). This would be a guardian tree of a holy well, so perhaps it was not the tree, but a higher entity, that would answer the sacrilege.

Mandeville's remark that "no evil spirit may enter the house in which it is" has already been noticed; it is an ancient belief, and even in Roman times a sprig of hawthorn was attached to the cradle of a newborn baby (Palaiseul). Burgundian mothers would carry their sick child to a flowering hawthorn, for they believed their prayers would ascend better to heaven in company with the fragrance of the flowers. At the other end of life's span, there is a recorded superstition from Portslade, in Sussex, that a dying perrson can recover if carried three times around, and be bumped three times against, a particular ancient thorn (*Sussex Notes and Queries. Vol 7; 1938–9*). But in spite of all this, it is an unlucky tree, particularly unlucky to bring indoors. It "brought illness, etc.,' with it, according to Devonshire belief (*Devonshire Association. Report and Transactions; 1971*), and in Somerset, may well cause death in the house into which the blossom is brought (Elworthy). Cheshire children are forbidden to bring it indoors, the belief being that their mother will die if this is done (Hole. 1937):

May in,
Coffin out

in fact (Igglesden). It has been suggested that the ill-luck may be something to do with the May-goddess. The hawthorn was sacred to her, and the may that can be taken inside on May morning perhaps represents a ritual breaking of the taboo on the May goddess's festival (Graves). In any case, a tradition in some parts of Ireland is that you should never bring hawthorn flowers into the house in May, for it would bring bad luck with it. You must wait till June (O Cleirigh). A superstition recorded in Suffolk says that sleeping in a room with the whitethorn bloom in it during the month of May will be followed by some great misfortune (Gurdon). Here there is confusion between the month and the name of the tree (May); the month of May is always an unlucky one. There is a Devonshire belief that it is unlucky for hawthorn to be in bloom before 1st May (W Jones), not that that is likely to happen. And 'Ne'er cast a clout till May is out' means do not put on any new clothes till the unlucky month of May is over, the month being represented by the tree:

Hawthorn tree and elder flowers
Will fill the house with evil powers.

No wonder that the haws were called poires du diable in Brittany (Sebillot).

There was a tradition in some places in England that hawthorn flowers preserved the stench of London during the plague. They contain trimethylamine, and this is an ingredient of the smell of putrefaction (Grigson). It is often said that the hawthorn has "a deathly smell". The scent has another interpreta-tion – that of sex. It was said to arouse sexual desire (Anderson), and the tree itself was used constantly in medieval love allegory. It is the *arbor cupidatitis*, the symbol of carnal love as opposed to spiritual love, and was used as such throughout the literature of the Middle Ages (Eberly). Like the hazel (another lightning plant), hawthorn has from early times been connected with marriage rites, either, as in Greece and Rome, as an ingredient of the bridal wreath, or as decoration for the altar. Even in this country we find traces of hawthorn propitiation at the time of a marriage. At Polwarth, in Berwickshire, newly-weds, with their friends, had to dance round the two ancient thorns in the parish (Spence. 1947). But at May-tide it was associated with young girls generally. In France, for instance, it was set outside the windows of every young girl (Grigson). And note Herrick's verse from *Corinna going a-Maying*:

and coming, mark
How each field turns a street, each street a park
Made green, and trimm'd with trees; see how
Devotion gives each house a bough,
Or branch; each porch, each door, ere this,
An ark, or tabernacle is
Made up of whitethorn neatly enterwove …

"Dew from the hawthorn tree" has special properties at this time (in fact, any May dew has):

The fair maid who, the first of May
Goes to the fields at break of day,
And washes in dew from the hawthorn tree,
Will ever after handsome be.

In Suffolk, any maidservant who could bring in a branch of hawthorn in full bloom on May morning (it would surely have to be old style), was entitled to a dish of cream for breakfast:

This is the day,
And here is our May,
The finest ever seen,
It is fit for a queen,
So pray, ma'am, give me a cup of your cream.

There are love divinations involving hawthorn, but first, something more direct – there was a belief in the south of Somerset that a girl should dance barefooted round a whitethorn on old Midsummer Day before sunrise to charm her lover into marrying her that year

(Tongue). The divinations themselves include one that involved a girl hanging a flowering branch at a cross-roads on May Eve. She should go there next morning, to see in which direction the wind had blown it. From that direction would come the destined husband. Another was for her to partly break a branch on May Eve, and leave it on the tree. In the morning she must fetch it home, and then would hope to see an image of the future husband on the way (Eberly).

One piece of weather lore is very well known – if there are a lot of haws, there is a hard winter to come. The belief is expressed in succint rhymes, such as the Scottish:

> Mony haws
> Mony snaws

or,

> A haw year
> A snaw year

(Swainson), and so on. The haws, though, have a reputation in folklore for being useless, being "unprofitable, and sour to eat, and fit for nothing". In fact, by Chaucer's time they had given rise to a common expression, "not worth a haw". Chaucer had the Wife of Bath, commenting on her husband's moralizing, say:

> But al for noght, I setts noght an hawe
> Of his proverbes n' of his old saws …

But it is another matter when medicinal usages are studied. Russian folk practitioners always treat angina pectoris with an infusion of haws (a glassful three times a day, at meals) (Kourennoff). In Germany, the infusion in alcohol is considered to be the only effective cure for angina. A tea made from haws was also said in Wiltshire to be good for heart disorders, and so are the flowers (Leete). A decoction, taken instead of tea or coffee, is used for high blood pressure (Kourennoff), for it helps to prevent arteriosclerosis – this is a traditional medicine in Scotland (Beith). Herbalists warn, though, that the effect is noticeable only after a prolonged course of treatment (Fluck). Haws were at one time prescribed for stone. Of course, this may very well have been from the doctrine of signatures – the stone fruits to destroy the stone. One usage that was almost certainly doctrine of signatures was this very early (14th century) recipe: "For to draw oute a thorne: tak the barks of the hawthorne and stamp him wele in red wyne" (Eberly). To use a thorn to get rid of a thorn is true homeopathic medicine.

In County Clare, people used to pick and chew the bark of an ancient hawthorn at a holy well as a cure for toothache (Westropp), a practice that could be classed either in the same category as the willow bark as a primitive aspirin, or else simply as a charm, because of the connection with the holy well. One that is certainly a charm, and a transference charm at that, is this French prescription to get rid of a fever. The patient is advised to take bread and salt to a hawthorn, and say:

> Adieu, buisson blanc;
> Je te porte du pain et du sel
> Et la fièvre pour demain.

The bread has to be fixed in a forked branch, and the salt thrown over the tree. Then he has to return home by a different road to that from which he set out. If there was only one door, then the patient had to get back in through a window (Sebillot).

HAY FEVER
Surprisingly, dried SWEET VERNAL GRASS has been used to cure hay fever! Particularly surprisingly, because this grass, dried, gives the typical coumarin scent of new-mown hay (Leyel. 1937). Gypsy lore has it that to cure hay fever permanently, one should pick some fresh SPEARMINT, and put it in a muslin bag in one's pillow, so that the scent can be inhaled during sleep. Also one should wear some each day (Vickery. 1995).

HAZEL
(*Corylus avellana*) A tree of countless virtues, a fairy tree, in fact. In the Grimms' *Aschenputtel*, a hazel sapling grows up on a mother's grave, and her bones transform it into a powerful wishing tree to work her daughter's revenge, for the tree shakes down the gold dresses and silk slippers that this Cinderella wears to the ball, and it shelters the doves, who act as her protectors (see Warner. 1994). Coll is the Celtic name for it, and a hazel grove is Calltuin in Gaelic, and there are Caltons in Edinburgh, Glasgow, and other places. The Edinburgh Calton was a fairy mound (Mackenzie). Nine hazels grew over a well in the Celtic land of promise, and it was a hazel that grew over the source of the river Shannon (F Jones). This was a life-giving tree in the Irish Elysium (Mackenzie), so it should be no surprise to find that sticks of hazel were laid on or under burials in Sweden, and deposits including "hazel nuts and the twigs of fruit-bearing trees" were noted in the Kennet avenue at Avebury (R Morris). It is the tree of wisdom (Graves), even a god in its own right if we accept the remark in Keating, *History of Ireland*, vol 1, that "Coll was god to MacCuill". In a roundabout way, it was the hazel that was responsible for Finn McCool's all-seeing wisdom, for that was the tree that grew over Connla's well in County Tipperary, and it was the nuts that fed the sacred salmon (F Jones). It was from one of these salmon that Finn gained his wisdom. Scottish children born in autumn were lucky, because they could be given "milk of the nut" as their first food (Mackenzie), just as *Ficus sycomorus* was given in Greece. Massingham suggested that the hazel was

sacred to the Celts because of its "oily exudations", which probably means the same as the "milk" of the nuts; in Scotland, this is the milk-yielding tree, the "milk" being in the green nut, and an elixir given to weakly children is "comb of honey and milk of the nut" (Mackenzie).

Another aspect of the mythology of the hazel lies in its association with Thor, for it is a lightning tree, an actual embodiment of the lightning. Hence the common belief that hazels are never struck (Kelly), and so offer the greatest protection, in all circumstances. Christianity adopted the myth in the story of the Holy Family taking refuge under a hazel during the flight into Egypt (Dyer). But on the Aegean island of Chios they say the shadows of both the fig and hazel are "heavy", and it is not good to sleep under either (Argenti & Rose). This seems an aberrant view, though, but Bartholomew Anglicus agreed that "the shadow of the Nut-tree grieveth them that sleep thereunder". In Lincolnshire, hazel was often used as "palm" on Palm Sunday, and kept green the year round by putting it in water. In the south of the county, these "palms" were preserved for the express purpose of protection from thunder and lightning (Gutch & Peacock). In Somerset, they say you should make a cross in the ashes with a hazel twig on May Day, and put a branch outside the house (Tongue). The connection between the lightning tree and the robin, itself associated with fire, is expressed in the west of France by the custom, long since dead, of killing a cock robin on Candlemas Day, and running a hazelwood stick through the body, which was then put by the fire. It would turn by itself, so the belief was (Swainson. 1886). German farm labourers would cut a hazel twig in spring, and make the sign of the cross with it over every heap of grain as soon as the first thunderstorm broke. The idea was to keep the corn good for many years. Hazel twigs were sometimes put in window frames during a heavy shower, and, in the Tyrol, it was reckoned to be an excellent lightning conductor (Dyer).

In Sweden, they said that snakes would lose their venom by a touch of a hazel wand (Fiske), and elsewhere it was believed that snakes cannot approach the tree (Kelly). It was by use of a hazel stick that St Patrick drove the snakes out of Ireland (Wilde. 1890). To this day, Dartmoor people say that if a dog is bitten by an adder, a hazel wand should be twisted into a ring and put round the animal's neck (St Leger-Gordon). A very long way from Dartmoor, in the Balkans in fact, they say that a young hazel twig, cut after sundown on St George's Day (a significant saint in view of this belief) should be used to rub a snakebite wound, and to draw a magic circle round it (Kemp). But in spite of all this, there was a German belief in something like a hazel serpent – a crowned white snake that lived beneath the hazel tree (Rowling). It is

pointed out that the snake is traditionally the symbol of wisdom, while the crown of course represents sovereignty. The hazel too is the tree of wisdom, so it looks as if this snake belief actually belongs to the older Celtic myth. There was too a Welsh belief that a snake found under or near a hazel on which mistletoe grew, would have a precious stone in its head (Trevelyan), presumably another way of describing the traditional wisdom of the snake.

Hazel was the medieval symbol of fertility. Throwing hazel nuts at the bride and groom is sometimes the practice at Greek weddings, and sugar coated nuts are known too to take the place of the better known sugared almonds (the word used for sugared almonds is the same as for the nuts in these circumstances) (T B Edwards). Until quite recently, Devonshire brides were given little bags of hazel nuts as they left church. These had the same significance as rice and confetti have today (Rowling). The gypsy bridegroom, before the ceremony, had to carry with him hazel wands wreathed in ribbons, "to ward off the influence of water" (Starkie), so it was said, but the real reason was to ensure the fertility of the marriage. A Bohemian saying was that plenty of hazel nuts meant the birth of many bastards (Dyer), but in Somerset it meant fertility in wedlock too (Tongue). Double nuts, of course, presaged a number of twins (Leather). As the old saying was, "Good nutting year, plenty of boy babies" (Hole. 1957). But in France it is "année de noisettes, années de filles" (Loux). Ruth Tongue told the story of the Somerset village girl who returned from London in the 1930s to be married. She openly said that she didn't intend to be hampered with babies too soon, and would take steps to ensure this. Such talk outraged village morality, and when she got to her new house, she found among the presents a large bag of nuts, to which most of her neighbours had contributed. She had four children very quickly. The Somerset girl who goes nutting on a Sunday will meet the devil, and almost certainly the baby will come before the wedding (Tongue); in Oxfordshire too they say you should never go nutting on a Sunday, for the devil will go with you to bend the branches down to your hand (Hole. 1957). "Going a-nutting" is a euphemism for love-making, anyway. The symbolism contained in the Breton saying that when you break a hazel wand with the little finger, you will be married within the year (Sebillot), is not very difficult to decipher.

In Ireland, hazel was included in the "summer" brought into the house on May morning (Ó Súilleabháin), and there is recognition in France of a connection with the Midsummer Fire (Kelly), both occasions when the fertility of the livestock and the land are in the mind. A few hazel nuts used to be mixed sometimes with the seed corn on German farms, to ensure its being prolific (Dyer). Yorkshire people used to stick hazel twigs with the catkins on them into various objects round the fireplace, the object apparently being in some way to help the sheep at lambing time (Gutch).

There seems no end to the magical properties of the tree. A hazel stick is the most effective protection that Irish folklore remembers against fairies or spirits, or the very generalization of evil (Ô Súilleabháin) ("If you cut a hazel and bring it with you, and turn it round about now and then, no bad thing can hurt you") (Gregory). Presumably this is why burials with hazel wands and leaves used to be so widespread. These are "measuring rods". Pennant. *Tours in Wales,* quotes an account of burials at Tal-y-llyn, Merioneth. Along the graves and coffins were laid hazel rods, with the bark on, and a hazel rod with the bark on was found in graves during the restoration of Chester cathedral. This rod would be cut to the exact length of the dead man's body, and put beside him in the grave. It would then actually represent the person buried. In the case of the "holy length", the measure is that of Christ, however that could be computed, and not of the deceased. These rods in some cases were to benefit the dead, in others to protect the living. But there are examples of its use in a healing rite, and in general it is a means of gaining power over that which is measured (Rees).

The Somerset practice of putting a hazel branch outside the door, and making a cross in the ashes with a hazel twig on May morning, seems to be purely protective in intent (Tongue). Among the many German beliefs about hazel is one that shows that a twig cut on Good Friday gives you power to strike a person who is absent (Dyer). Good Friday is the day, in the Tyrol, on which a hazel divining rod must be cut (Dyer). So it is in western Scotland, where St John's Day is added (Banks). Swedish folklore says that the nuts have the power of making you invisible. Obviously related to a legend of invisibility is a Scottish story concerning a certain "cave of gold" (Uamh an oir). The belief was that if any fugitive ran to the cave and struck the rock with a hazel stick it would open and let him in, and shut again before the pursuers could get there (Polson. 1891). In the Book of St Albans (c 1496), a recipe is given for making oneself invisible by carrying a hazel rod one and a half fathoms long, with a green hazel twig inserted in it (Graves). And of course the magician's wand was often made of this wood. Anyone, according to Irish belief, can draw a protective circle round himself with the hazel wand, provided it was cut on May morning, before sunrise. That would make it powerful enough to ensure that no evil thing could enter the circle (Wilde). The divining rod seems to have been introduced into Britain by German miners in the 16th or 17th century (Grigson). "Divinatory rods for the detecting and finding out of minerals; (at least, if that tradition be no imposture) is very wonderful;

by whatsoever occult virtue, the forked stick, so cut, and skilfully held, becomes impregnated with those invisible steams and exhalations; as by its spontaneous bending from an horizontal posture, to discover not only mines, and subterranean treasures, and springs of water, but criminals, guilty of murder, etc., … made out so solemnly, and the efforts thereof, by the attestations of magistrates, and divers other learned and credible persons (who have critically examined matters of fact) is certainly next to miracle, and requires strong faith". Evelyn's words seem to be an indication of recent introduction; it could be argued that a practice of some antiquity would not be regarded as "next to miracle" by someone with his critical faculties. Hazel used to be the wood for a wishing rod, too (Grimm). And the Welsh wishing cap was generally made of the leaves or twigs, although sometimes juniper was used. They had to be gathered at midnight, and at new or full moon, and made up as quickly as possible (Trevelyan). This cap was worn for good luck, too, particularly by sailors, or anyone connected with the sea and ships (R L Brown).

Irish horse handlers always used to have hazel as the breast band on the harness, to keep the horse from harm (Ô Súilleabháin); in much the same way, Somerset drovers always used a hazel stick to drive cattle and horses, though in moist places rowan was often preferred. For a horse that had over-eaten, the remedy was to bind its legs and feet with hazel twigs to relieve the discomfort (Drury. 1985). Note also the purely magical use in this Welsh charm: "if calves were scouted overmuch, and in danger of dying, a hazel twig the length of the calf was twisted round its neck like a collar, and it was supposed to cure them" (Owen).

Welsh people used to look on the nuts as an emblem of good fortune. In the south they were always kept in the house until brown with age, and when quite rotten, they were burnt on the fire to ensure prosperity (Trevelyan) – to burn them in the house rather than to throw them out would seem the logical way to keep the good fortune inside. But to dream of the nuts is a sign of trouble from friends; to the tradesman it is a sign of prison, and decay of trade (Raphael).

Double nuts have a special folklore of their own. Nobody in the east of England would even think of eating the whole of one like this. One kernel would be passed to a friend, and the two would eat it in silence, wishing a wish that had to be kept secret (Gutch). The practice in Northamptonshire was to eat one and throw the other over the shoulder. In any case, the double nut is a sign of great good luck to come (Sternberg). They were called St John's nuts in Scotland, and used for throwing at witches (Grigson). Their name in Devonshire is loady nut, and there they were used to cure toothache (Grigson); so they were in Sussex – all you had to do was to carry one in your pocket (Latham). In Ireland, that was enough to ensure you did not get rheumatism or lumbago. In this case, the nut is acting as a protection against the fairies, for these are elf-shot diseases. A note in the *Journal of the Gypsy lore Society. 1957* mentions a mother who gave her son, when he joined the army, a "double hazel nut with three wishes" to ensure his safe return.

One piece of weather lore connected with hazel nuts has to do with the thickness of the shells – the thicker they are, the harder the winter to come; conversely of course, thin shells, mild winter. (Conway). An American version expects a large crop of nuts to be followed by a hard winter (Hyatt).

There used to be a saying in Boston, Lincolnshire, that the devil goes a-nutting on Holy Rood Day, which is 14 September. Better to keep well clear of hazel trees then; at Ormsby, in the same county, they reckoned that nutters on that day were certain to come to grief (Gutch). Elsewhere, on the other hand, Holy Rood Day was reckoned to be the proper day to go nutting:

> Tomorrow is Holy Rood Day
> When all a-nutting take their way.

That is from *Grim the Collier, act ii, sc 1* (1662) (Britten).

With hazel's supernatural background, one would expect the medicinal use to be hinged firmly on to magical practices. So they are, but there one or two that seem to be pragmatically genuine. Herbalists still maintain that hazel nuts improve the condition of the heart, and prevent hardening of the arteries (Conway), while they say in Wiltshire that they are good for curing coughs (Wiltshire). Ignoring certain early prescriptions that seem to be half way between the real and the magical, we are firmly dealing with charms in such a wart cure that involved cutting notches in a hazel twig, one for each wart, which would disappear as the notches grew out of the twig (Newman & Wilson). The very soil from under a hazel bush was valuable. In Yorkshire, it was given to cows that lost their cud (Hartley & Ingilby).

Some of the names given to the nuts need a little explanation. Filbert, with its variations, is the name under which the nuts of *Coryluys maxima* are known. Brouk suggested that it meant 'full beard' (from the fringed husk?), but the usual explanation is that it comes from a non-existent King Philibert, or from St Philibert, whose feast-day falls on 22 August. The nuts would certainly not be ripe then, so it is probable that the saint's day is old style, bringing the day into September, when there is more likelihood of their being ripe. Anyway, filbert is a Norman-French word, written as philbert in the 13th century, and still in use in Normandy patois at the beginning of the 20th century. Another name to explain is Cob-nut,

with its variations. Cob-nut is actually what a game played with the nuts is called. One of them is very like conkers, and the winner is called the cob-nut (Hunter). Hence the Cornish Victor-nut (Jago) and the Devonshire Crack-nut (Britten & Holland). Halliwell's description of the game shows that there was more than one version. The one like conkers he considered the most recent, but he says the older consisted of pitching at a row of nuts piled up in heaps of four, three at the bottom and one at the top of each heap. All the nuts knocked down became the property of the pitcher, and the one used for pitching was the Cob. Finally, one must comment on the fact that hazel is the only British tree bearing edible nuts (walnut gives its origins away by its very name, for the first element means 'foreign'). No, this is the Nut-tree par excellence, and whenever that name was used, there could only be one recipient, the hazel.

HEADACHE
Taking a drink made from the infused flowers of HONEYSUCKLE seems a pleasant way of getting rid of a headache (V G Hatfield), as is ROSEMARY tea, in a simple infusion of the leaves and flowers (Hill), or you could just rub the forehead with a handful of the herb (Newman & Wilson), and one can do the same with PEPPERMINT leaves, or drink the tea (Vesey-Fitzgerald). BISTORT tea is used in Cumbria (Newman & Wilson). Probably the best of all the headache remedies is LIME-BLOSSOM tea. It is used a lot in France, where "tilleul" is taken, a slightly sedative drink (F G Savage). It is a very pleasant drink, and is taken a lot for insomnia, too (Tongue. 1965). MARIGOLD water was for a long time a favourite for a headache (see Rollinson for an example from Cumbria). PEACH leaves bound round the head will bring relief (Puckett). A freshly cut slice of raw POTATO held to the temples is a headache cure (R B Browne), just as a HORSERADISH leaf, bruised and wetted, could be tied to the head (Thomas & Thomas). That is the American way, but the cure is simpler in Britain. All you need do is smell it, which is the Norfolk claim (V G Hatfield), or in Sussex, just holding the scrapings tight would do the trick (J Simpson). Country people in Essex used to rub the forehead with a handful of SEA WORMWOOD to cure a headache (Newman & Wilson). Its near relative, SAGEBRUSH, need only to be smelt to cure a headache, at least according to the Navajo (Elmore). Using CONKERS, powdered, as a snuff was a way of curing catarrh or headache. The Pennsylvania Germans used it that way (Fogel), but this was quite an early habit (Thornton). The idea was to make one sneeze. Apparently it was also recommended as an infusion or decoction to take up the nostrils.

FEVERFEW is an effective headache remedy. A cold infusion of the flowers will do the trick, possibly because it is a painkiller, and also because it is mildly sedative. Even migraine will succumb, so it is claimed (V G Hatfield). RED CLOVER, being mildly sedative, has been traditionally used for curing a headache (Conway), and RED POPPY too, being sedative, is also a headache and migraine remedy. It is actually called Headache (Grigson. 1955 etc.,), and the belief was that picking it could cause the headache, so it is interesting to find that the cause can also be the cure. The leaves of the CASTOR OIL PLANT have been used for headaches. African peoples like the Mano of Liberia rub the leaves in water, and bathe the head with the result (Harley), while in the southern states of America, a similar practice merely involves wrapping the forehead in the leaves, which will treat a fever as well (Puckett). A RHUBARB leaf held to the forehead will relieve the headache (V G Hatfield. 1994), though any large leaf will probably do as well. A prescription from the Physicians of Myddfai is more of a charm than a genuine remedy. We are told that we should take an apronful of sheep's sorrel, and boil it in the milk of a one-coloured cow till it is nearly dry, and apply it as a plaster to the head, "the patient keeping his bed, being covered with clothes, so as to cause him to perspire". The Anglo-Saxon Apuleius prescribed RUE for many ills, including simple headache, which is to be cured by drinking it in wine, or dabbing the head with rue and vinegar. Modern folk medicine still recognises chewing a leaf of rue as being a cure for nervous headaches (Brownlow), and in the Middle Ages we find that a plaster of rue with ground ivy and laurel was prescribed for headache, especially for " an ache that endureth long" (Dawson). Even HEMLOCK has been used. See Wesley: for "a chronical Head-ache … wear tender Hemlock-leaves, under the feet, changing them daily" ! Another from the Lacnunga has a prescription involving BEET-ROOT. Roots of beet, pound with honey, wring out, apply the juice over the nose. Let him (the patient) be face upward toward the hot sun and lay the head downward until the brain be reached. Before that, he should have butter or oil in the mouth, the mucus to run from the nose. Let him do that often until it be clean (see Grattan & Singer). BASIL has been used for centuries to counter headaches and colds, either by an infusion, taken hot at night, or by taking it as snuff. Dried basil leaves have been used in this form for a very long time (Quelch, Hamphill).

African cures for a headache included the use of leaves of NEVER-DIE (*Kalanchoe crenata*) in the Yoruba Ewé ritual, just as in Mexico the leaf of another species, CURTAIN PLANT (*Kalanchoe pinnata*), known as hoja fresca there, would be put over each temple (Kelly & Palerm).

HEART DISEASE
FOXGLOVE must be mentioned first in this connection. The drug from the dried leaves is very potent, and has the effect of reducing the frequency

and force of the heart action, so it is given in special cases as a sedative (especially in heart disease). It was Dr William Withering (1741–1799), from Wellington, in Shropshire, who first introduced digitalin into general medical practice. He published "an account of the Foxglove and some of its medical uses" in 1788. It is said he got his information from a witch! LILY-OF-THE-VALLEY provides a drug (convallotoxin) that has been used as a substitute for digitalin (Lloyd), which can act as a heart stimulant, thoiugh it is less powerful then digitalin. The root was used in Ireland for heart disease (Maloney), and it was widely prescribed in Russian folk medicine (Kourennoff). OLEANDER (*Nerium oleander*) has an active principle that operates in much the same way as digitalis, and is used in the treatment of heart conditions, particularly in Russia, where it is included in the official pharmacopeia (Thomson. 1976). Another plant with a digitalin-like action is the American shrub WAHOO (*Euonymus atropurpureus*), and it became a popular heart treatment in domestic medicine (Weiner). Another American plant, SENEGA SNAKEROOT, was used by the Meskwaki Indians for the condition (H H Smith. 1928). MISTLETOE may be a magical plant, with magical medicinal uses, but the strange thing is that some of them are perfectly genuine. It has been known since ancient times that it has a beneficial effect on the heart (Thomson. 1978), and it also reduces blood pressure. One of the DAHLIAS (*D pinnata*) is used in China for heart disease. They make a broth from the tubers, and cook it with pork - that is the medicine (Perry & Metzger).

Herbalists maintain that HAZEL nuts improve the condition of the heart, and prevent hardening of the arteries (Conway). ROSEMARY tea, made as an infusion of leaves and flowers, is recommended for weak hearts (Hole. 1937). Mrs Wiltshire called it a "supreme heart tonic" (Wiltshire), and BROOM tops are a very old popular remedy for the problem, perfectly justified, too, apparently. WATERCRESS, eaten raw, is said in Ireland to be good for the condition (Vickery. 1995). WOODRUFF, with a high coumarin content, and so an anti-coagulant, has been useful for drugs used in heart disease (Mabey. 1972). CAMPHOR has been used in the Far East for heart disease for centuries (Tosco). Herbalists are still using MARIGOLD flowers for heart disease; they benefit the arteries and veins (A W Hatfield. 1973). Gerard was recommending them for heart trouble four hundred years ago – "conserve made of the floures and sugar taken in the morning fasting, cureth the trembling of the heart".

HEARTSEASE
The alternative name for WILD PANSY, though apparently given to the wallflower originally (Ellacombe). There is another form from Devonshire,

Heart-pansy (Friend. 1883). Heartsease itself became Hearts-at-ease, or even Heartseed. But the word itself means that which can make a cordial, and must have somehow been given by mistake, for this reputation as a medicine for heart trouble is not borne out by domestic practice, and must properly belong to another plant, presumably the wallflower.

HEATH SPOTTED ORCHID

(*Dactylorchis maculata*) A gypsy love potion was made from it, if the identification is correct. They were dried and crushed, and then the girl mixed this with her menses. This was introduced somehow into the food of the person whose love she was trying to get (Ireland. 1891). There is a Swedish belief that has some bearing on this. It was known there as Maria's Keys, and it used to be put in the pregnant woman's bed as an amulet for an easy delivery. A prayer to be said at the same time refers to the Virgin's keys, and asks for the loan of them during childbirth - hence the name (Kvideland & Sehmsdorf).

A poultice used to be made in the Highlands from this plant, for drawing thorns and splinters out of the flesh (Beith). A belief from Norway suggests that, mixed with woody nightshade and "tree sap", it was a remedy for protecting people and animals against the demonic (Kvideland & Sehmsdorf).

HEATHER

(*Calluma vulgaris*) A Welsh comment on heather's habitat translates: Gold under the bracken, silver under the gorse, famine under the heather (Condry), though, like silverweed, the tops of young heather shrubs were fairy food, according to Lowland Scots tradition (Aitken). Perhaps that is why it is an unlucky plant in Welsh belief. Bringing it into the house was a token of misfortune, even death (Trevelyan). All over Scotland, it is said that burning the heather in spring would bring down the rain (Banks).

"In the Highlands of Scotland, the poor inhabitants make walls for their cottages with alternate layers of heath, and a kind of mortar, made of black earth and straw. They also make beds of it, and their houses are thatched with it …" (Taylor). George Buchanan, in 1582, approved highly of beds made of heather: "… they form a bed so pleasant, that it may vie in softness with the finest down, while in salubrity it far exceeds it …" (quoted in Beith) (see also Hartley & Ingilby, for similar usages in the Yorkshire Dales). Besoms too were made of it, and it has been a dye plant for a very long time, giving yellow to orange, and, with indigo, green (Pennant, Shaw). It was wound into ropes called Gadd on the Isle of Man, strong enough to be used for mooring boats (Mabey. 1998).

There is a widespread belief that the Danes made heather beer, and that the secret of how to make it died with them. The story is that there remained alive only a father and his son. When pressed to tell the secret, the father said, "Kill my son, and I will tell you our secret"; but when the son was killed, he (the father) cried, "Kill me also, but our secret you will never know". This is an Irish story, but very similar ones occur all over Scotland, there relating to the Picts. Nevertheless, Pennant, after visiting Islay in 1772, said that "ale is frequently made in this island from the tops of heath, mixing two-thirds of that plant with one of malt", and he repeats Boethius's story of the loss of the secret of making ale, with the extinction of the Picts, the inference being that the ale being made in the 18th century was not the famous heather beer.

Heather used to be quite important in medicine; it is a diuretic, and is still used in homeopathy for the treatment of infections of the kidneys and urinary tract (Schauenberg & Paris). An infusion of four or five flower sprays in a pint of boiling water is drunk as a tea for insomnia; it was even just applied to the head for insomnia, and a heather pillow is still used to give refreshing sleep (Beith). It is also good for calming the nerves and the heart. A stronger brew sweetened with heather honey is an old Highland remedy for coughs and colds (A W Hatfield). (see also WHITE HEATHER)

HEBE

A New Zealand genus, featuring in Maori mythology, and much used by them (especially *Hebe salicifolia*), for medicinal purposes. In particular, the young leaf tips were well regarded as a cure for diarrhoea and dysentery, so much so that they were collected and sent out to Maori troops in the Middle East during World War II (C Macdonald). They also took a weak infusion of the leaves as a tonic, and used them also in the steam bath (Goldie).

Hedeoma pulegioides > AMERICAN PENNYROYAL

Hedera helix > IVY

Helenium autumnale > SNEEZEWEED

Helianthus annuus > SUNFLOWER

Helianthus tuberosus > JERUSALEM ARTICHOKE

Helleborus foetida > STINKING HELLEBORE

Helleborus niger > CHRISTMAS ROSE

Helleborus viridis > GREEN HELLEBORE

Hemerocallis flava > DAY LILY

HEMLOCK

(*Conium maculatum*) Tradition has it that it was a decoction of this drug that was drunk by Socrates, but how could he have died so quickly? Hemlock poisoning is a relatively slow progress towards death, so Socrates must have had some narcotic like opium (Rambosson) mixed with it. The purple spots on the stems were said to reproduce the marks on Cain's brow after he had killed Abel (M Evans).

Herbivorous animals do not seem to be poisoned by eating it (it is said that a Cambridgeshire way of controlling unruly horses was to pound hemlock in a mortar until it was finely powdered, and then to rub it on their noses (Porter. 1969)), but it seems that carnivores are more susceptible (Sanford). It is also said to be most poisonous in the southern part of its range (Salisbury. 1964). Martin. 1703 described the effects of eating hemlock: "Fergus Caird, an empiric, living in the village Taliste (Talisker?), having by mistake eaten hemlock-root, instead of the white wild-carrot, his eyes did presently roll about, his countenance became very pale, his sight had almost failed him, the frame of his body was all in a strange convulsion, and his pudenda retired so inwardly, that there was no discerning whether he had been male or female. All the remedy given him in this state was a draught of hot milk, and a little aqua-vitae added to it, which he no sooner drank, but he vomited presently after, yet the root still remained in his stomach. They continued to administer the same remedy for the space of four or five hours together, but in vain; and in about an hour after they ceased to give him anything, he voided the root by stool, and then was restored to his former state of health…".

A plant as poisonous as this would naturally have an association with witches. "Root of hemlock digged in dark" was one of the ingredients in the witches' cauldron in Macbeth, and Summers claimed that it was used by them either as a poison or as a drug, favoured mainly because of its soporific effects. The soporific effect is uppermost in a story told by Coles that "if asses do chaunce to feed upon Hemlock they will fall so fast asleepe that they will seeme to be dead. In so much that some thinking them to be dead have flayed off their skins, yet after the Hemlock had done operating they have stirred and wakened out of their sleep, to the griefe and amazement of the owners, and to the laughter of others". On the fringes of the association with witchcraft is an Irish love charm that consisted of taking ten leaves of hemlock, dried and powdered, and mixing this powder in food or drink (Wilde. 1902). Some say that it is the purple blotches on its stem that gives it a bad name, for these streaks are copies of the brand put on Cain's brow when he had committed murder (Skinner).

The fruits are the only convenient source of the alkaloid coniine (or Conia), which was introduced into British medicine in 1864. The pure drug has been used sometimes in soothing cancer pain (Schauenberg & Paris). The plant itself was used in Anglo-Saxon times, and is mentioned as early as the 10th century, in the vocabulary of Aelfric, Archbishop of Canterbury, as Cicuta, hemlic (Fluckiger & Hanbury). It was, in fact, an ingredient in the narcotic drink called dwale (Voigts & Hudson). It has been used in the past in dealing with a cataract. Buchan, in the 18th century,

congratulated himself on curing a cataract "by giving the patient purges with calomel, keeping a poultice of fresh hemlock constantly upon the eye, and a perpetual blister on the neck".

In parts of Ireland, it was used for giddiness (Barbour), and an Irish recipe of about 1450 recommended it for the falling sickness, epilepsy (Wilde. 1890). Boiled hemlock was widely used in Ulster (and in the Isle of Man (Moore, Morrison & Goodwin)) to reduce swellings in men and animals. It could not be used if there was a cut or scratch near the swelling (Foster). On the other hand, the leaves, dried and powdered, were used in Essex to be put on cuts (V G Hatfield. 1994), and in Ireland a pain-killing poultice used to be made by mixing hemlock leaves with linseed meal (Moloney), a preparation also used on boils (Maloney).

In 1790, a Cornish blacksmith, Ralph Barnes, was supposed to have cured himself of a cancer by taking immense quantities of hemlock juice (Deane & Shaw) (primitive chemotherapy?). Equally unlikely is Buchan's recommendation of it for the King's Evil, but there is another usage that is perhaps not so unlikely: Granny Gray, of Littleport, in Cambridgeshire, used to make up pills from hemlock, pennyroyal and rue. They were famous in the Fen country for abortions (Porter. 1969) in the mid-19th century.

HEMP
(*Cannabis sativa*) Long grown for its fibre, and one of the oldest areas of cultivation in Britain is in the Fenland borders of Cambridgeshire. By custom, women were not allowed to work in these fields; it was thought that merely touching the plant made a young woman barren, while older ones would get a severe rash on their arms. It was the devil's plant – certainly, cutting the crop gave the workers a nasty headache (Porter. 1969).

But the plant is also cultivated for the drug obtained from the flower heads. This drug is treated in different ways in different parts of the world, and is known by different names – hashish in Arabia; beng in Persia; kif in Morocco, dagga in South Africa; charas, gangha, or bhang in India; in Liberia diamba, and in Mexico and America marijuana, which is Portuguese maran guango, meaning intoxication (Emboden. 1969). Cigarettes containing marijuana are known as reefers, mooters, muggles, greeters, or gates (Farnsworth). As a drug plant it is extremely ancient, for it was well known to the Chinese emperor Shen Neng, whose work on pharmacy dates from 2737 BC (De Ropp). There is mention of it under the name Ma in the Chinese book known as Rh-Ya, of the 15th century BC. Herodotus mentions it, too. In describing the funeral rites of the Scythians, he talked of the way they purified themselves after the burial by putting hempseed on hot stones inside a small felt-covered structure, and inhaling the smoke it produced. It is pointed out that this purification rite must have been a form of shamanism, for the smoke of the hemp seeds produced a trance state in those inhaling it (Balazs).

Ancient Chinese medicine used cannabis as a treatment for gout, rheumatism, malaria, and so on (the list included absent-mindedness!) (Emboden. 1972), and it is still used in Chinese herbal medicine as laxative or sedative, and to treat asthma. It was in AD 220 that the Chinese physician and surgeon Hua-To performed surgery using cannabis resins mixed with wine, as an anaesthetic, rather like the early Greek use of mandrake. Both of them are pain-relievers (Emboden. 1969). Later, it was done by putting resin in an incense burner in which myrrh, balsam and frankincense had been mixed. The action resembles that of opium in many respects, but there are fewer after-effects. The use as an anaesthetic was known in India, too, in ancient times. Hemp is described in the Athaveda as a protector, and the gods are said to have three times created the herb. Indra has given it a thousand eyes, and conferred on it the power of driving away all diseases, and of killing all monsters (Lloyd). A Mashona nganga in Africa will use it to treat one form of epilepsy. He will pulverize the leaves of hemp and the other plants used, fill a reed with the powder, and when the patient is actually having a fit, he will light the reed and blow the smoke into his face (Gelfand. 1964).

The most potent of all the hemp drugs is said to be charas, produced in Yarkand, Central Asia, and imported into India via Tibet. In Tibet itself, they made momea, consisting of charas incorprated into human fat. In Bombay, charas was often made into a sweet called aajun, popular with women (De Ropp). Bhang, according to Burton "the Arab Banj and the Hindu bhang", is the word used most frequently in reference to the drug cannabis. Hashish is Arabic, meaning hay or dried herb (Grigson. 1974), but it is a term with many meanings. Al-Hasan ibn-al-Sabah (the "old man of the mountains") was a 12th century charismatic dissenter from orthodox Moslem thought. He founded a new sect called Hashishins, a name that also produced the word assassin (Emboden. 1969). In Egypt and the Middle East, hashish is smoked in special pipes called josies (de Ropp), and in India, it was apparently never smoked without tobacco; the two were kneaded together with the thumb on the palm of the hand. If there was no pipe, it was enough to make a small hole in the ground, put the mixture in that and lie down to inhale it. Dagga is the Southern African version of the drug, used even by the San (Bushmen). They too usually mix it with tobacco, but occasionally use it neat (Schapera).

In English tradition.hemp was one of the plants that provided the witch's broomstick (Emboden. 1972). It was, too, probably the herb that the witches smoked

in their clay pipes (Eckenstein). Connected with the witch tradition is the love divination in which hemp seed is sown, on Midsummer Eve and at Hallowe'en; often it is necessary for the hemp seed to be harrowed – "steal out unperceived, harrowing with anything you can conveniently draw after you ..." (Bell), or else the true love is the one who is supposed to appear to rake it. Sometimes hemp has to be cut, and this figures in the divinations:

> Hempseed I sow, and hempseed I mow.
> And he that is my sweetheart come follow me and mow.

Or the apparition will do the cutting:

> Hempseed I set, hempseed I sow
> The man that is my true love
> Come after me and mow (Northall).

The significance lies in the fact that freshly-cut hemp is strongly narcotic (hence the bad headaches, already mentioned, that the harvesters used to get).

Hemp has got its uses, though. The seeds are fed to cage birds, and the oilcake makes cattle-feed (Schauenberg & Paris), but the plant is best known for its fibres. Huckaback, for instance, is made of coarse hemp fibre (F G Savage). Male hemp was used for ropes, sacking and the like; female hemp for sheets and other domestic uses (Peacock). Perhaps the best known application was for gallows rope; indeed, it seems to have stood as a kind of symbol for the gallows. For, so we are told, to show a man how unpopular he was because he had broken the Fen code of never betraying a fellow Fenman, they drew a sign on his door, a stem of hemp and a willow stake, with the words "Both grown for you". The hemp to hang him with, or rather, to hang himself with, and the willow stake to drive through his heart on burial (Porter. 1969). Given this use for hemp, an American remedy for epilepsy assumes macabre significance, for the way to do it, according to Kentucky belief, was to wear a hemp string round the throat (Thomas & Thomas). Cf the Somerset name Neckweed (Elworthy. 1888), and the more explicit name Gallow-grass.

Old "handywomen" in the Fen country would recommend hemp leaves to get an abortion (Porter. 1969), So they did in Wiltshire, too, and there they said that if no hemp was available, green horse-radish leaves would do (Whitlock. 1988). The aim in either event would be to cause severe vomiting, often enough to result in a miscarriage. In addition, the mother would be given a "groaning cake", one of the ingredients of which was hemp seed, obviously to ease labour pains (Whitlock. 1992). French mothers were said to use hemp in some way to stop their flow of milk (Loux).

HENNA

(*Lawsonia inermis*) The root yields a dark red dye, known in East Africa, Arabia, India, Indonesia, and also north Australia, since the earliest times. Egyptian mummies show that women painted their fingernails and finger tips, their palms, and the soles of their feet, with henna. All through the orient, henna leaves, powdered and mixed with whitewash, or at least water, are laid on the skin, or rather bandaged on the skin overnight, as cosmetics (Loewenthal). The colour ranges from yellow to red, and the cosmetic is given still, just as it was in the time of the ancient Egyptians, to put on finger nails, toenails, finger tips, the palms of the hands, and the soles of the feet. The dye needs renewing every two or three weeks (Moldenke & Moldenke). It is also used for dyeing the hair, and men's beards, even horses' manes (*CIBA Review. 63; 1943*). In more recent times, henna paste is made by mixing the powdered leaves with catechu (an extract of the wood of the eastern Indian *Acacia catechu*) (Moldenke & Moldenke).

Henna can be used as a fabric dyestuff - it is used to dye silk and wool in Morocco (Gallotti), and in India, too, both fabrics and leather are coloured reddish-yellow by its means (*CIBA Review. 63; 1948*). Henna flowers, with a scent something like that of roses, were sold in the streets of Cairo as bouquets, and in India they are used as offerings in Buddhist temples (Moldenke & Moldenke). It is said to be Mahomet's flower, and the cosmetic has extra meaning as a protection from evil influences, and even as a medicine (Westermarck. 1926). For instance, in Morocco, where, mixed with water, it is applied to the forehead of a person suffering from fever, to the head of a boy troubled with ringworm, to the hair of women to promote its growth, or to chapped feet and hands. In the Balkans, too, in acute fever, like typhoid, henna is heated in water, allowed to cool, and the juice of some twenty heads of garlic added, the mixture re-heated, and then the henna is applied solid to the palms and soles of the feet, in exactly the same way as for staining, "in order to draw out the fever" (Kemp). This use of a red dye to allay fever is in all probability an example of the doctrine of signatures. Just as in Britain at one time, fever patients were wrapped in red blankets to allay the symptoms. Palestinian men setting out on a long journey will get their wives to spread some on the soles of their feet to prevent them getting sore, and riders use it to prevent saddle soreness (Crowfoot & Baldensperger).

In any case, red is almost as good a prophylactic colour as the saffron to be got from *Crocus sativus*. This is one of the reasons, if not the main reason, why henna is used, chiefly by women, but also on special occasions by men, as a preventive against the evil eye, and it is applied to new-born babies as well. Westermarck. 1926 recorded many examples of this use of henna in Morocco. Some painted spots of henna on the tops of their heads. An infant of 40 days had the crown of its head smeared with it, as a protection against fleas and lice, but also against the evil eye, an application that is repeated frequently till the child gets older. Greyhounds also had their heads and feet smeared with it. Henna and walnut-root bark are applied to the mother and her newly-born baby as protection. Some say that the reason is "so that she may enter Paradise as a bride if she dies in childbirth"; such a woman is said to become a Houri after death. There was a belief, perhaps still is, in Pakistan and Iran that dogs hate henna. Actually it is the henna-ed palm of the outstretched hand that keeps dogs at bay. The reason is that the dog represents the devil (Loewenthal).

Similar beliefs are held among the Nubian people called the Kanuz. Almost all life-crises, whatever their nature, involve the use of henna, invariably used in sacrifice to the Nile spirits. It is applied to the hands, feet and forehead of a groom on his wedding night, while the bride's entire body was henna-ed. Girls snatched any of the henna paste left over, to ensure their own marriage in the near future. A newly-born baby was bathed and rubbed with henna, while onion juice was dropped into his eyes, and sweetened water put into his mouth. And it was to be used at the final life-crisis; it was applied to a corpse as soon as death took place, and also to the deceased's clothes (Brain).

Heracleum sphondyllium > HOGWEED

HERB PUDDING

was made from BISTORT leaves on Easter Day (or more properly at Passion-tide) boiled in broth with barley, chives, etc., and served to accompany veal and bacon. Easter Giants, or Easter Mangiants, both from the French manger, to eat, are other names for the pudding, and there is Ledger Pudding, as well. There a number of recipes. One given by Grigson, and said to come from Cumbria, runs: "pick young Easter Ledger leaves, and drop them with leaves of Dandelion, Lady's Mantle, or Nettle into boiling water and cook for 20 minutes. Strain and chop. Add a little boiled barley, a chopped egg (hard-boiled), butter, pepper, and salt. Heat in a saucepan and press into a pudding basin. Serve with veal and bacon". From the same general area, there is "equal quantities (about 1lb) of young bistorts and young nettles, a large onion, a teacup of barley, ½ teaspoon salt. Chop greens and onions and sprinkle washed barley among them, add salt. Boil in a muslin bag for about two hours. Before serving beat mixture in a dish with one egg and butter and flavour with salt and pepper …" The final mixture was sometimes fried and eaten with bacon and eggs (Rowling). Another recipe is to wash and chop the leaves, boil them with onions and add oatmeal to thicken it up. Many people make enough to freeze, to have for breakfast on Christmas Day (J Smith. 1989). (see also **DOCK PUDDING**)

HERB ROBERT

(*Geranium robertianum*) Who was Robert? Many
suggestions have been made in the past. It is said to
be dedicated to St Robert, whose feast day falls on
29 April. Perhaps it is Robin, and so Robin Hood? Or
Robin Goodfellow? The seed vessels, with their sharp
needles, are known as Pook Needles. In Germany, it
was used to cure a disease known as Ruprechtsplage,
said to be named after a Robert, Duke of Normandy.
Another suggestion is that Herb Robert is derived
from Robert, an 11th century Abbot of Molesne.
Another opinion, more plausible than the rest, is that
the name is from Latin ruber, red, i.e., a herb of a red
colour.

This is an unlucky flower. Snakes would come from
the stems if you went to pick it, witness the names
Snake Flower, from Somerset, and Snake's Food,
from Dorset (Macmillan; Vickery. 1995). Even more
telling is Death-come-quickly, from Cumbria, for
this is one of the flowers that if picked by children,
would result in the death of one of their parents. That
superstition is quite common. See, for instance, the
name MOTHER-DIE, applied to a dozen plants, and
always with an injunction not to pick the flowers (see
Watts. 2000).

In Wales, Herb Robert was used as a remedy for gout,
and it is recorded as a diabetes remedy in Ireland
(Ô Súilleabháin) – a handful of the herb to a pint
of water, in wineglassful doses, night and morning
(Moloney). Culpeper says that it will heal wounds and
stay blood. This sounds like doctrine of signatures, for
the whole plant has a red look about it, particularly
the stems and the fading leaves. In the same way, it
was used in Scotland for erysipelas, or "rose". Scarlet
cloth was also used (Gregor). But the medicinal use
went beyond this, and still is used, by herbalists, to
treat any skin eruption, herpes, etc., (Schauenberg &
Paris), even skin cancer (Beith).

HERNIA

One of the best known and widespread of charms is
that of passing children through holes in an ASH tree
as a remedy for hernia. In Cornwall, the ceremony
had to be performed before sunrise, and a further
Cornish belief was that the child would recover only
if he were washed in dew collected from the branches
on three successive mornings (Deane and Shaw).
The Wiltshire practice was to pass the child through
a maiden ash, i.e., one that had never been pruned,
at sunrise on 1 May, head towards the sun (Eddrup).
Gilbert White reported that it was customary to split
an ash, and to pass ruptured children through. If any
injury should happen to the split tree, the child would
suffer accordingly. The practice of planting a tree to
commemorate the birth of a child may well be a relic
of this belief that the life of an individual is bound up
in that of a tree. Perhaps this is the reason why it is

always so unlucky to break a branch off an ash. The
rules given for the split ash ceremony in Suffolk are:

1. Must be early in the spring before the leaves
   come
2. Split the ash as near east and west as possible
3. Split exactly at sunrise
4. The child must be naked
5. The child must be put through the tree feet first
6. The child must be turned with the sun
7. The child must be put through the tree three
   times (Gurdon).

The Somerset rules include 2,4 and 6, but go on
further to say that the child must be handed in by a
maiden, and received by a boy (Mathews), though the
"maiden" requirement may be an error, and refer to
the tree rather than a participant in the ceremony.
A maiden ash is one grown from seed, and never
topped. Evelyn mentions the split ash rites, which
continued in England at least to 1830 (Graves) –
indeed, it was still being done in Essex in 1925
(Mason). Rarely, there are records of HOLLY being
used in the same way. One such cure is from Surrey
(Clinch & Kershaw): the usual slit was made in the
tree, the two sections being held apart by two people,
while two women, one at the child's head and one at
the feet, passed him naked several times backwards
and forwards through the slit. And in Portugal the
same was done with a REED. The child's injury will
heal while the injury to the plant heals. There is the
usual ritual to be observed: in this case it had to be
performed at midnight on St John's Eve (23 June), by
three men of the name of John, while three women,
each called Mary, spun, each with her own spindle, on
one and the same distaff (Gallop).

Another spectacular way of dealing with the problem
can be quoted from the Cree Indians, who used the
powdered root of the poisonous AMERICAN WHITE
HELLEBORE (*Veratrum viride*) for the purpose. The
procedure was to raise the patient up on a platform,
in a horizontal position, when he would take a good
pinch of the hellebore snuff, and during the violent
sneezing that followed someone would be standing
ready to push the hernia back with his fist! (Corlett).

Comfrey is a fracture herb, *the* fracture herb in fact,
but the Pennsylvania Germans extended the virtue to
include hernia also. It is really a magical cure – you
have to hold it on the hernia until it is warm. Then the
comfrey has to be planted. If it grows, the hernia will
be cured (Fogel). Gerard advised the use of the powdered herb and roots of DOVE'S FOOT CRANES-
BILL to treat hernia, "as my selfe have often proved,
whereby I have gotten crownes and credit".

Heuchera americana > ALUM-ROOT

Heuchera bracteata > NAVAJO TEA

Hibiscus rosa-sinensis > ROSE OF CHINA

HICCUPS

An Alabama cure for the hiccups is to drink a tablespoonful of QUINCE juice (R B Browne). Wickop, or Wicopy, are American names for ROSEBAY WILLOWHERB (*Chamaenerion angustifolium*). They mean hiccup, for which the root was used as a cure (Sanecki). ANISE, according to Gerard, "… helpeth the yeoxing or hicket, both when it is drunken or eaten dry …". An infusion of WHITE HELLEBORE has been used in Russian folk medicine. It would "stop hiccuping immediately" (Kourennoff). Naturally, with a plant as poisonous as this, doses would have to be very small. A few drops of CAJUPUT oil on sugar will quickly end hiccups (Mitton). Maya medical texts prescribed KIDNEY BEANS for the hiccups (Roys).

HICKORY

See SHAGBARK HICKORY (*Carya ovata*), and WHITE HICKORY (*Carya tomentosa*).

HINAU

(*Elaeocarpus dentatus*) Hinau is the Maori name for this New Zealand tree. One particular specimen, in North Island, was famous with the Maori as a fertility smbol. A childless woman embraced the tree while her husband recited the necessary charm. The east side of the tree was the male side, the west the female, and the woman would make her choice of east or west according to whether she wanted a boy or girl child (Andersen).

HINDU LOTUS

(*Nelumbo nucifera*) This is the Sacred Lotus (Brouk), introduced to Egypt about 500 BC, and believed to contain the secrets of the gods. In particular, it was consecrated to the sun. The Egyptians both worshipped it, and worshipped with it, offering the flowers on their altars. One Egyptian creation myth relates how a lotus flower rises out of the waters; when its petals open the calyx of the flower is seen to bear a divine child, who is Ra. Another version is that the lotus opens to reveal a scarab (symbol of the sun); the scarab then transforms itself into a boy, who weeps; and finally, his tears become mankind. As the lotus is a flower that opens and closes every day, so it could easily be associated with the cult of the sun god (Larousse). It no longer grows beside the Nile, but it is still venerated in all countries to the east, and representations of it appear in the structure of temple buildings, in decorations, prayers and songs from Syria to Tibet, through Indonesia, China and Japan (R Hyatt).

It is one of the emblems of Lakshmi, the goddess of fortune and beauty, and it is worn as a talisman for good luck and fortune, and to avert all childish diseases and accidents (Pavitt). As in China, it is the Japanese symbol of purity. Buddha is often depicted as seated on a lotus (Savage. 1954).

Hippocrepis comosus > HORSESHOE VETCH

HIPS

is the usual name for the fruit of roses. It is OE heope, with a long first vowel that has become short in modern English. The word varies in local use to forms like Heps, widely used, or in northern Scotland, Haps (Britten & Holland), pronounced hawps and written as such sometimes. Hips are called Choops in northern England and Scotland, spelt sometimes Choups or Shoups (Nodal & Milner, Carr). In addition, they are often known as Cat-choops or Dog-choops, with many variations (Watts. 2000).

As far as Dog Rose hips are concerned, they are edible, and " … children with great delight eat the berries thereof when they be ripe, make chaines and other prettie gewgawes of the fruit; cooks and gentlewomen make Tarts and such like dishes for pleasure thereof …; the making whereof I commit to the cunning cooke, and teeth to eat them in the rich man's mouth" (Gerard). Dog Rose hips contain large amounts of Vitamin C, and they were systematically gathered in England during World War 2. They have always been used as a medicine in some way or another. The conserve "is of some efficacie against coughs" (Hill), or a tea made from them was taken for fatigue and dropsy among other complaints (Fluck), including the common cold (Thomson. 1978). They were used, too, for their pulp as an ingredient of pill-making, etc., (Fluckiger & Hanbury, and "to prevent a wound going bad: hips of Dog Rose chewed, then let it drop on the wound" (Cockayne).

HIPWORT

is WALL PENNYWORT (*Umbilicus rupestris*), whose leaves were reckoned to resemble the hip socket. So by the doctrine of signatures, Coles was able to claim that "it easeth the pain of the hippes" (see also DOCTRINE OF SIGNATURES).

HOGWEED

(*Heracleum sphondyllium*) The young shoots are edible, boiled as in cooking asparagus (they say it actually tastes not unlike asparagus). Later in the season the stalks can be used as a green vegetable, but then the bitter outer parts have to be peeled off. Pigs like it – why else should it be called Hogweed? All over the country in days gone by, the plant has been gathered as free food for pigs.

There is one superstition recorded that seems on the face of it dubious. Scrammy-handed, we are told (Waters) is a Wye Valley dialect expression meaning left-handed, or awkward. Children called the hogweed scrammy-handed, because they believe that they will become left-handed if they pick the plant with the left hand. Elsewhere, this plant, along with other umbellifers like hemlock, have the name Scabby Hands. The superstition sounds very like a rationalization, though "scram" certainly means awkward. Some of the names given to the plant are interesting, Kex, for instance, and its many variants. All the umbellifers have this name, which really belongs to the dried, hollow stalks. A number of proverbial phrases in English refer to kex, or kecksies. 'As dry as a kex' is the best known, but there are others, like 'as light as a kex', and 'as hollow as a kex', the last used in Yorkshire to describe a deceitful man. (Easther). Hogweed's hollow stems provide another source of local names. You can, for instance, drink your cider through them at a pinch, hence Wippul-squip. Surely the same applies to the picturesque Somerset name Lumper-scrump, or Limper-scrimp (Britten & Holland), and so on, to the Devonshire Humpy-scrumple (Macmillan). Another use for these hollow stems is made manifest in the Scottish name Bear-skeiters. 'Bear' is not 'bear', but 'bere', so the whole name means barley-shooters, for children use the stems as pea-shooters. Instead of peas, they shoot out barley, or oats (then it is Ait-skeiters).

It has its medicinal uses: it has been shown to affect blood pressure, and the juice was used in East Anglia to cure warts (V G Hatfield. 1994). It has also, so it is claimed, been shown to have a distinct aphrodisiac effect (Schauenberg & Paris). Gerard reckoned the seed "scoureth out flegmaticke matter through the guts, it healeth the jaundice, the falling sickness, the strangling of the mother, and them that are short winded …".

HOLLY

(*Ilex aquifolium*) In Lady Gregory's translation of the *Tain bo Cuailgne*, Nachtrantal is sent against Cuchullain, and he "would bring no arms with him but three times nine holly rods, and they having hardened points". Holly appears again in mythology in the early Irish romance of Sir Gawain and the Green Knight, the latter an immortal giant whose club is a holly bush. He and Gawain (Cuchullain) make a compact to behead each other at alternate new years (i.e., midsummer and midwinter), but when it comes to the point, the Green Knight, the holly knight, spares Gawain, the oak knight. In Christian terms, St John the Baptist is the midsummer/oak representative, while Jesus, "as John's merciful successor", took over the holly king's role (Graves). That is why the holly was glorified above the oak:

> Of all the trees that are in the wood,
> The holly bears the crown.

To revert to holly as a weapon, there is an example, again from Irish mythology, of its use as a "misdi-rected weapon", the best known instance of which, in Scandinavian mythology, is the mistletoe in the Baldur sequence. In the Irish case, the weapon is a hardened holly spear with which Fergus is killed by Ailell.

The scarlet berries ensure that holly is looked on as a lightning plant, with all the protective power that such a plant always has. In East Anglia, for instance, a holly tree growing near a house is regarded as a protection against evil (Graves). Holly hedges surounding many Fenland cottages were probably planted originally with the same idea in mind (Porter. 1969). And being a lightning plant, it must protect from lightning. "Lightening never struck anyone if you were under a holly tree. Lightening never struck a holly tree", a statement from Devonshire. In Germany, it is a piece of "church-holly", i.e., one that has been used for church decoration, that is the lightning charm (Crippen). Pliny says that holly flowers cause water to freeze, evidently an association of the tree with winter. Another of holly's attributes is to bring back runaway cattle. All one had to do was throw a holly stick after them; it did not even have to touch them (Dyer). Drovers' sticks were often made of holly wood (Burne. 1914).

Holly is equally efficacious against witchcraft and fairies. Fenland belief had it that a holly stick in the hand would scare any witch, and coachmen never used to like to drive at night unless their whip handle was of holly wood (Porter. 1969). But you should never cut a holly branch for a whipstock – it was safer to pull up the long shoots around the trunk (Baker. 1974). Builders liked to use holly wood to make external door sills, for no witch could cross it (Porter. 1969). Collars of holly (and bittersweet) saved horses from witchcraft (Aubrey: "Take Bittersweet, and Holly, and twist them together, and hang it about the Horses neck like a garland: it will certainly cure him [of being Hag-ridden]. Probat"). Needfire in the Highlands was lighted by "spinning an oaken augur in a holly beam" (C M Robertson). Perhaps belief in its antagonism to witchcraft, which was very wide-spread, stemmed from its name, which was seen as another form of 'holy'. The Wiltshire tradition is that the Christmas holly and bay wreath hung on the front door is to keep the witches outside. They would stay there counting the berries indefinitely, because witches could only count to four, and then have to start again (Wiltshire). Compare the gypsy name God's Tree, and their liking in past times for pitching their tent by a holly tree – it provided divine protec-tion (Groome). Sometimes pieces of holly were put up with rowan (they are both red-berried) over the stable to prevent the entrance of the nightmare (McPherson), and in the Highlands, decorating the house with holly on New Year's Eve, or putting a sprig over the dairy door, was a recognised means of keeping the fairies out (McGregor). (see also CHRISTMAS DECORA-TIONS). In the West of England it was said that a maiden should adorn her bed with a sprig of berried holly on Christmas Eve, otherwise she might receive an unwelcome visit "from some mischievous goblin" (Crippen). In Lancashire, a sprig of holly used to be hung in a new building, to bring luck. A branch also used to be put in a house in which someone had died (Crosby), to "purify" the house, so it was said, but the reason must have been more fundamental than that.

Holly is, under certain circumstances, an unlucky plant, as in the Welsh superstition that bringing it into the house during summer was supremely unlucky, holly flowers in particular (Baker. 1980). If you pick a sprig of holly in flower, there will be a death in the family (Trevelyan). There is a French tradition that says that when Christ invented the bay, the devil wanted to imitate it, but he could only manage the holly – that is why it has prickles. In Brittany, it is the devil's counterpart of the oak (Sebillot). Sterile holly is unlucky at any time, and dangerous to man and beast. In a year when there are no berries, it is wise to put a sprig, perhaps with the berries reddened with raddle (Drury. 1987), or box, into the holly wreath, to break the bad luck (Tongue).

In Normandy, holly wood used to be put in milk, to stop it turning sour (W B Johnson). But this is really an anti-witch example, for they were the ones held responsible for souring it. Also, the Wiltshire belief that it is lucky to have a holly tree growing in the garden (Wiltshire) is another example of its protective virtues; so it becomes obvious that it is very unlucky to fell such a holly, and you should never cut down a holly that has seeded itself. It is unlucky too to burn holly branches while they are green (Wiltshire, D Lewis). To stamp on a holly berry and crush it is certain to bring bad luck (Igglesden). In Northern Ireland, holly is a fairy plant. There is a story of a farmer's house being damaged by flying stones because he had swept his chimney with a holly bough (L'Amy). Holly is lucky to men, as the ivy, being the female plant, is to women; or at least berried holly is the traditional male symbol, while the smooth varie-gated kind, and the ivy, were the female luck symbols. That accounts for a custom known in some parts of Kent for the village girls to make guys of holly branches to burn on Shrove Tuesday, the boys retaliat-ing with guys of ivy (Dallimore). Holly and ivy seem to be interchangeable in other respects, though, for a holly bush hung outside a house meant exactly the same as ivy – that wine was to be had there (Briggs. 1974). The fact that it is evergreen is celebrated in the Scottish saying, about a habitual liar, that he "never lies but when the holyn's green". It is the symbol of foresight, in the language of flowers (Dallimore).

Holly is used in various divination games. In Swansea, it is said that if a girl runs round a holly tree seven times in one direction and then seven in the other, she will see her future husband (Opie & Opie). Is that why holly is the symbol of foresight? Then, too, there are variants on the daisy-petal divinations – in Brittany, for instance, the prickles on the leaf are touched, with "Fille, femme, veuve, religieuse", or "Fils, homme, veuf, religieux" (Sebillot), while a game from Eding-ton, in Warwickshire, is for girls to pluck a holly leaf, saying:

I pluck this holy leaf to see
If my mother does want me,

and then count the prickles – yes, no, yes, no, until the last one (Northall). There is too an American game that consists of naming holly leaves as they are thrown into the fire. The one that pops first has the name of the one that loves you best (Thomas & Thomas). More complicated divinations include an Irish one, called "building the house", played at Hallowe'en. Twelve pairs of holly twigs are arranged in a circle, pushed into the ground, and tied together at the apex. A live turf representing the fire is put in the centre. Each pair of holly twigs is named after the boys and girls present, and the pair that first catches fire shows which boy and girl will be the first to be married (E E Evans). Another Hallowe'en divination involving holly was played (if that is the right word) in the Border counties. A girl has to take three pails of water up to her bedroom, and to pin three holly leaves to her nightdress before going to bed. She will be roused by three yells, succeeded by three hoarse laughs, after which the form of her future husband will appear. If he is deeply attached to her, he will shift the position of the three pails (Banks. 1937). The holly leaves, though apparently essential, seem to play no part in the record of the charm. They must have done so at one time, and the way in which the divination has come down to us must be corrupt.

There are still a few more superstitions connected with holly. It is a common belief that a good crop of holly berries is a sign of a hard winter to come, put succinctly in the Somerset saying: Many berries, much snow (Tongue). But this, of course, applies equally well to hawthorn. A boy whipped at Hog-manay in the Highlands with a branch of holly can be reassured by the belief that he will live a year for every drop of blood he loses (Banks. 1937). But whipping with holly approaches closely some medi-cal, or at least pseudo-medical, usages, for the cure of chilblains over quite a wide area is to thrash them with holly until they bleed. This is a logical enough cure, for chilblains are caused by poor circulation, and thrashing them deals with this. But there is a variant to the belief – the feet should be crossed while this is done (Igglesden), and this must surely be an illogical superstition. Beating with holly as a cure for rheu-matism is also recorded in Somerset (Mathews). On the other hand, though, and this is a Cornish supersti-tion, "rheumatism will atack the man who carries a walking stick made of holly" (Courtney), a belief that seems quite out of character, given the many virtues of holly. A more conventional chilblain cure, from Wiltshire this time, is to make an ointment of holly berries mixed with goose fat or lard. For toes and fingers, the best thing is to apply the ointment liberally, wrap the part tightly in an old stocking, and then toast it in front of the fire till the heat becomes unbearable (Whitlock. 1988).

Holly wood is hard, tough enough to be used in the past for flail swingels (G E Evans. 1956), and the whitest of hard woods, polishing well. It has even been dyed black and used as a substitute for ebony, and it was used too "by the inlayer, especially under thin plates of ivory, to make it more conspicuous" (Evelyn). It was used in building work, too, especially for beams that were close to chimneys for this is a wood that does not catch fire easily (Dallimore). Aparently holly has even been used for tying round bacon and salt beef put away for the winter, the idea being that the prickles would keep rats and mice away (Palaiseul). Sheep were browsed with ash and holly in the Lake district, well into the 19th century, and undoubtedly still are (Mabey. 1998). In 1774, there is an account that says "at the shepherd's call the flock surround the holly-bush, and receive the croppings at his hand, which they greedily nibble up and bleat for more. The mutton thus dressed has a fine flavour" (Rollinson). Cattle too were fed with dried and bruised holly during the winter in France (Dallimore), and goats love it.

So-called medical practices vary considerably, from the purely superstitious, like taking medicine, in this case just new milk, for whooping cough, from a cup made from a variegated holly (*Notes & Queries; 1851*), to the magical. Take for instance the practice of passing a child through a holly as a charm for rupture. It must be rare, for this is almost always associated with the ash rather than holly. Nevertheless, there is a record, from Surrey, of such an incident, if we are to believe the informant (Clinch & Kershaw). A slit was made in the tree and the two sections held apart while two people, one at the child's head and the other at the feet, passed him naked several times backwards and forwards through the opening. The berries and leaves do have medicinal effects, of course – the berries are violently emetic, and were used once for colic (Dallimore). When dried and powdered, herbalists still sometimes use them for dropsy (A W Hatfield), which may explain why in times gone by they were used for the stone. Lupton claimed to treat it by "seeth(ing) an handful of holly berries in a pint of good ale, till half the ale be consumed; then strain it, putting then a little butter to it, and let the party drink thereof ...". Evelyn too reported the value against the stone, this time of the leaves, which certainly do relieve fevers and catarrh, and were once stated to be "equal to Peruvian bark" (Dallimore). But using them for toothache seems to bring us back to the magical, especially

when the cure involved getting rid of worms that were believed to be responsible some hundreds of years ago. "For toothwark, if a worm eat the tooth, take an old holly leaf ... boil two doles [i.e., two of worts to one of water] in water, pour into a bowl and yawn over it, then the worms shall fall into the bowl" (Black). The Welsh medical text known as the Physicians of Myddfai took up the same theme, but expounded it more rationally: "For the toothache, take holly leaves and boil in spring water till they are tough, then remove the pot from the fire, and put a kerchief about your head, holding your mouth over the pot in order to inhale the vapour. It will cure you". There are other ancient remedies using holly, from Anglo-Saxon times onwards (see Grattan & Singer). There was even a theory that a decoction of the root bark would help dislocations and fractures (Dallimore).

HOLLYHOCK

(*Althaea rosea*) A 17th century recipe for the manufacture of a fairy salve reads "an unguent to anoint under the eyelids, and upon the eyelids, evening and morning: take a pint of sallet-oyle, and put it into a vialle-glass; but first wash it with rose-water and marygold water; the flower [to] be gathered towards the east. Wash it till the oyle come white; then put thereto the budds of holyhocke, the flowers of marygoldm the flowers or tops of wild thime, the budds of young hazel: and the thime must be gathered near the side of a hill where Fayries use to be; and "take" the grasse of a fayrie throne [i.e., ring] there. All these put into the oyle, into the glasse: and set it to dissolve three dayes in the sunne, and then keep it for thy use, it supra". After this there follows a form of incantation, conjuring a fairy named "Elaby Gathon" to appear in the glass, "meekly and mildly to resolve him truly in all manner of question; and to be obedient to all his commands under pain of Damnation" (L Spence. 1946 and 1949).

Fibre from the stalks has been used for cloth manufacture. In the 1830s, about 280 acres near Flint were planted with hollyhocks. It also yields a blue dye (Northcote), and the darker varieties green (Usher). In its native China it is grown as an economic plant, the stem being like hemp, the dried petals as the source of a blue-black dye, the leaves as a potherb, the flowers as vegetables, and the boiled roots as a popular remedy for chest troubles (Whittle & Cook).

Dried hollyhock flower tea is good for a cold (H M Hyatt), and it is still prescribed for bronchitis, and sometimes as a laxative (Flück), and the plant is one of the ingredients of an early leechdom for lung diseases (Cockayne). The root decoction, Hill said, "is good in the gravel", and the same decoction, Lupton wrote, "with honey and butter, being drank, doth marvellously ease the pain of the colic and of the back". Gerard only observed that "the decoction of the floures, especially those of the red "(of course, for this is a doctrine of signatures remedy) "doth stop the overmuch flowing of the monethly courses, if they be boiled in red wine" (red, again).

HOLM OAK

(*Quercus ilex*) Holm, or Holly, because this is the tree that nourishes the kerm, a scarlet insect not unlike a holly berry. The ancients made their royal scarlet dye (and an aphrodisiac elixir) from this (Graves). The Bible furnishes the oldest evidence of the existence of a scarlet dye; there it is called tola, or tolaat, a word also meaning worm, which is what kerm means. So the eastern peoples were well aware of the animal origin of kermes, but the Greeks and Romans believed it to be vegetable, so the Greek name for kermes was kokkos, which means a berry. The fallacy has persisted into modern times; the specific name *ilex* is also the generic name for the hollies, and of course the common name means 'holly oak'

This is an evergreen tree, so, like all evergreens, it is a funerary symbol and at the same time a symbol of immortality. But it is also an unlucky tree, at least in the Greek islands, because, so the story goes, it was from its wood that the Cross was made. A miraculous foreknowledge of the Crucifixion had spread among the forest trees, which unanimously agreed not to allow their wood to serve, and when the foresters came, they either turned the edge of the axe, or bent away from the stroke. Only the ilex consented, and passively submitted to be cut down. So now mountain woodcutters will not soil their axes with its bark, nor desecrate their hearths by burning it (Rodd).

HOLY BASIL

(*Ocimum tenuifloru*) see **TULSI**

HOLY HERB

A name for VERVAIN (*Verbena officinalis*); in folklore it is *the* Holy Herb. The Romans gave the name verbena, or more frequently, the plural form verbenae to the foliage or branches of shrubs and herbs, which, for their religious association, had acquired a sacred character. These included laurel, olive and myrtle, but Pliny makes us think that the herb now known as verbena was regarded as the most sacred of them all. The Greeks also looked upon it as particularly sacred (Friend), and burned it during invocations and predictions (Summers). The same sacredness was part of Persian belief (Clair), while it was called "the tears of Isis" by the priest-physicians of Egypt (Maddox). Vervain is herbe sacree in French, too. Holy Vervain is sometimes used (Northall) instead of Holy Herb, and Devil's Bane or Devil's Hate follow necessarily.

HOLY ROOD DAY

(14 September) (HAZEL) It was the custom at one time to go nutting on Holy Rood Day; in fact in some places it was looked on as unlucky *not* to go (Hull):

Tomorrow is Holy Rood Day,
When all a-nutting take their way.

In Glamorgan, large parties of men and boys used to go out nutting, and brought the nuts back to inns well known for keeping up the Mabsant, the Welsh name for a patronal festival. The nuts were portioned out equally among those present, and various games were played involving forfeits, which were paid with the nuts (Trevelyan). But, as is quite normal in folklore, quite the opposite view is held in other places. In Lincolnshire, for instance (Gutch & Peacock), and in Suffolk (Gurdon), it was said that it was the devil who went nutting on Holy Rood Day. At Owmsby, in Lincolnshire, it was believed that nutters going out today would be certain to come to grief in some way or other. And the Michaelmas taboo on BLACKBERRY eating was, in some areas, applied to Holy Rood Day (Banks).

The ecclesiastical symbol of Holy Rood Day, whether the autumn festival, or the spring celebration (3 May), is the BLUE PASSION FLOWER (*Passiflora caerulea*), with its wider symbolism of Christ's Passion (see EMBLEMS).

HOLY THISTLE

(*Carduus benedictus*) When a plant is called "holy" or "blessed" (Blessed Thistle is recorded for this (Ellacombe)), it means it has the power of counteracting poison, or so it was supposed; (herba benedicta, Avens, that is, is better known for this property). Langham could say "the leafe, juice, seede, in water, healeth all kindes of poyson…". Everybody knew it as a heal-all. Langham, indeed had four pages of recipes under this head, for practically every malady, including plague, for which it was regarded as a specific. Thomas Hill had a very similar list. Culpeper, too, was enthusiastic. Wesley was more restrained, but could still say, "Coldness of the Stomach. Take a spoonful of the Syrup of the Juice of Carduus Benedictus, fasting, for three or four Mornings. A warm infusion is used for bad colds or intermittent fevers". That same infusion is used in America for dyspepsia and as an appetite restorer (Henkel), which probably is the equivalent of John Wesley's advice above. But Shakespeare knew another use, for he has Margaret, in *Much ado about nothing*, say: "Get you some of this distilled Carduus Benedictus and lay it to your heart: it is the only thing for a qualm". It is still prescribed by herbalists as a tea for liver disorders (W A R Thomson. 1978), and it is used in Ireland for asthma, by boiling a few leaves in milk, and then drinking the milk, and it is also used for a cold (Maloney). Bartholomew Anglicus found yet another use for it. The juice, he said, "cureth the falling of the hair" (Seager).

HONEY LOCUST

(*Gleditsia triacanthos*) A tea made from the roots of this tree is used in Indiana folk medicine for kidney trouble (Brewster).

HONEYSUCKLE

(*Lonicera periclymenum*) The very familiar plant of woods, hedges and bushy places, anywhere that gives it a base from which to carry out its clockwise climbing. So familiar, and well-loved, that it is difficult to realise that it has a distinct association with witches, either as a means to thwart, or as the exact opposite, a plant for witches to use for their own ends. Certainly, its evil-averting powers outweigh its evil-working claims. On May Day, when, so it seems, there was always a lot of ill-wishing about, honeysuckle took care of the butter and milk, and the cows (Grigson). It was a favourite in Scotland (pregnant women were advised to wear it (Dempster)), along with rowan, and was looked on as a mighty barrier to the ingress of witches, to the extent of being the subject of a popular rhyme. To be mentioned in the same breath as rowan is praise indeed:

> The ran-tree an' the widd-bin
> Haud the witches on come in (McPherson).

The widd-bin twig would be wound round a ran-tree wand, and then put over the byre door (Milne). D A Mackenzie reckoned it was its spiral-growing habit that gave it such a reputation. Maclagan quotes a correspondent from north Argyllshire who said that a cure for the evil eye lay in getting a good long bit of the iadh-shlait [honeysuckle, that is, though some dictionaries give ivy as well], "take it and twist it this way round the whole body", while reciting some form of words not actually given. The point about this is that the protection lies not in any inherent quality itself, but in the way it grows. For there was a firm belief that witches or those that have the evil eye are forced to stop whatever they are doing and to follow out every detail of an involved design that may be presented to their eyes. So interlacing and complex interwoven braided cords were deliberately made to distract, delay and confuse evil eyes, and were worn specifically for that purpose. The intricacy of a honeysuckle wreath serves exactly the same purpose (Gifford). Even when the wreath was used by a witch to cure an ailment, such as was certainly done in 18th century Scotland to remedy children with "hectic fever", as well as consumptive patients, it did not stop a witch from being charged. There is a record of Janet Stewart, in 1597, standing her trial for just this kind of healing (Rorie. 1994). She was charged with healing sundry women "by taking ane garland of green wood-bynd, and causing the patient pas thryis throw it, quhilk thereafter scho cut in nyne pieces and cast in the fire". The method of use was to let the wreath down over the body from head to feet.

But of course, as in all magical patterns, exactly the opposite can be quoted. The witches themselves will use honeysuckle against victims who will use the same plant for protection. In the ballad of Willie's Lady, for instance, the witch tries various means of preventing the birth of the Lady's child, including a "bush o' woodbine" planted between her bower and the girl's. Once this restricting, constricting, plant has been removed, the birth proceeds normally (Grigson). It is worth noting that both honeysuckle and plants that have a similar habit – like, woody nightshade – are named after the elves in Germany, e g Alprauke, Alp-kraut, etc., (Grimm). Very often, it was the witch who had to use the plant in order that its curative powers could succeed – see the case of Janet Stewart above.

This might explain why woodbine is an unlucky plant. From Scotland to Dorset there are records of a general belief that to bring it indoors is very unlucky; in Dorset they say it brings sickness into the house with it, and in west Wales it was believed it would give you a sore throat (Vickery. 1985). Honeysuckle was never brought into a Fenland house where there were young girls; it was thought to give them erotic dreams, especially if it were put in their bedrooms. If any of it *was* brought in, then it was said that a wedding would shortly follow (Porter) – not surprisingly, if the girls' minds were concentrated in that direction. Sussex boys bound a honeysuckle bine round a hazel stick, and when after several months the wood was twisted like barleysugar, its possession gave instant, and presumably magical, success in courtship (F R Williams). Presumably it is this twining habit that makes honeysuckle a symbol of constancy (Tynan & Maitland).

Of course, the berries are poisonous, but really only if excessive amounts of them are eaten. Such excess would produce severe vomiting and diarrhoea (Jordan).

The plant is still in use as a "heart tonic", and for chest colds, coughs and asthma, rheumatism, liver trouble (the leaves are still used for jaundice in Ireland (Maloney)), sore throat, etc., (Conway) this last known in Gerard's time – "… the water … is good against soreness of the throat". Gypsies use the juice from the berries to cure the condition, and also canker in the mouth (Vesey-Fitzgerald). Coulton quoted a 14th century manuscript, prescribing for "hym that haves the squynancy" a remarkable amount of disgusting rubbish, but containing as an ingredient "gumme of wodebynde".

A recipe for asthma is to mke a conserve of the flowers, and beat it up with three times their weight of honey; a tablespoonful dose is to be taken night and morning, to relieve the condition (Hatfield). An ointment is used for the treatment of ulcers (Vesey-Fitzgerald), while the bark, used in some unspecified way, is useful for gout (Barton & Castle). Another use of the bark is for dropsy; a heaped tablespoonful of thin flaked bark to a pint of cold water, brought just to the boil, is taken in wineglassful doses three times a day to cope with dropsical and glandular complaints (Hatfield). Another decoction of half an ounce of the leaves to a pint of water, is to be taken to relieve constipation and diseases of the liver and spleen, the latter figuring in Gerard's quite enthusiastic catalogue of the virtues.

The Welsh medical text known as the Physicians of Myddfai records an extraordinary leechdom for "pain in the eye" – "seek the gall of a hare, of a hen, of an eel, and of a stag, with fresh wine and honeysuckle leaves, then inflict a wound upon an ivy-tree, and mix the gum that exudes from the wound therewith; boiling it swiftly, and straining it through a fine linen cloth; when cold, insert a little thereof in the corner of the eyes, and it will be a wonder if he who makes use of it does not see stars in midday…". A 15th century leechdom was still recommending honeysuckle for eye trouble: For a web in the eyes or spots in the eyes: take the juice of wild teazle and the juice of woodbine and put them in the eyes when thou goest to bed. And it shall break it well" (Dawson).

This is still not the end of the claims made for the curative powers of honeysuckle. Lupton confided a sure cure for warts by using woodbine leaves, "stamped and laid on …, using them six times …". You can put the juice from the leaves on stings, too (Page. 1978). There are some Irish manuscripts that prescribe it in one form or another for much more serious complaints, one, from about 1450, for epilepsy, no less (Wilde). All these seem inappropriate to the popular conception of the honeysuckle; we cannot really perceive it as a drug. Much more in keeping with the vague feeling that the plant inspires is its use as cosmetic, either a lotion, or an ointment. One example was written in the mid 16th century: "Take a pint of white wine, one handful of woodbine leaves or two or three ounces of the water of woodbine, add a quarter of a pound of the powder of ginger; seethe them all together until they be somewhat thick, and anoint a red pimpled face therewith, five or six times, and it will make it fair" (Lupton). And taking a drink made from the infused flowers is a pleasant way of getting rid of a headache (V G Hatfield).

HOPS

(*Humulus lupulus*) A British native, but Long reckoned that most of the so-called Wild Hops here are probably escapees from cultivation. They have been cultivated in Europe for flavouring malt liquors since the 9th century (Lehner), but the date of its introduction into England for the same purpose is quite problematical. It is agreed that it arrived in the 16th century (Faulkner), but the precise date is elusive. 1542 is often quoted. But that is apparently not true. In any case, after its introduction there was strong popular prejudice against their use, reinforced by by an injunction of Henry VIII to the growers not to put any hops or brimstone into ale (Barton & Castle). He was a great lover of spiced ale, that is the old English unhopped ale (Jones-Baker. 1974). But, before that, in 1519, the corporation of Shrewsbury had forbidden the brewers to use "that wicked and pernicious weed", under a penalty of 6s. 8d (Bett. 1950). A popular rhyme is supposed to show the date of introduction:

> Hops and Turkey, Carp and Beer
> Came into England all in one year.

That is not helpful. The truth is that no-one is actually sure when it made its first commercial appearance here (Haydon). Another of these introduction sayings was "Heresy and hops came in together".

As hops became a more and more important crop, some weather lore attached itself to the plant:

> Till St James's Day be come and gone,
> There may be hops, or there may be none (Dyer).

St James's Day is 25 July, which seems rather a late day to judge the well-being of the crop. Perhaps we are talking about St James the less (1 May). Another rhyme seems more realistic:

> Rain on Good Friday and Easter Day,
> A good crop of hops, but a bad one of hay
> (Leather).

This is from Herefordshire. Another saying might apply to any crop:

> Plenty of ladybirds, plenty of hops (Dyer).

Old beliefs connected with hops include a dream interpretation: a large garden full of hops in full leaf is a sign of wealth. Dried hops, especially when the smell is noted in a dream, shows that the dreamer will soon receive a legacy (Raphael), and there was a general belief in the good fortune brought by hops. Wreaths made of them were commonly seen over the mantelpiece in country areas, put there for luck, and for ensuring the household's prosperity (Opie & Tatem). At one time, it was the custom for anyone visiting a Kentish hopfield to contribute "foot money" to stop the luck leaving the field. In the language of flowers, a garland of hops was a symbol of hope, often worn in Elizabethan times by suitors (Quennell), and we learn from Gubernatis that there used to be a Russian custom of putting hop leaves on a bride's head, "en signe de joie, d'ivresse et d'abondance". There was once a belief that the nightingale's song is only heard where hops grow (Dyer).

Hop pillows have been used as a soporific for a long time. The secret of success against insomnia lies in not packing them too tightly, and the dried hops should be renewed every four to six weeks (Thomson. 1976). The sedative action of a hop pillow was used to combat other conditions, too. Lindley tells us that they were prescribed for "mania". George III is said to have slept on one (Genders. 1971). One was used for rheumatism in Worcestershire, according to a correspondent of the *Gentleman's Magazine* in 1855 (oddly enough, it was to be laid *under* a patient's bed). There are records, too, for its use for toothache, earache and neuralgia (Lehner), and it is even recommended for asthma sufferers (Hyatt). Being slightly narcotic, hops have been smoked as a tobacco substitute, something that was tried as recently as World War II (Swahn).

Hordeum sativum > BARLEY

HORSE CHESTNUT

(*Aesculus hippocastanum*) 'Horse' in a plant name usually denotes largeness and coarseness, and that is probably the case here, though the older writers obviously did not think so. Note, for example, the old name *Castanea equina*. Parkinson says "The horse chestnits are given in the East country, and so through all Turkie, unto horses to cure them of the cough, shortnesse of winde, and such other diseases". Gerard also accounts for the name in the same way. Skinner suggests the derivation from the likeness to a horse's hoof in the leaf cicatrix, and it may have been from the doctrine of signatures that the nuts, crushed as meal, were given to horses for various diseases. Actually, horses do not seem to like them, though deer and cattle do (Barber & Phillips), and the nuts are a good food for sheep (Lindley). But they are definitely not for human consumption; there have been cases of poisoning due to children eating the green outer cases of the nuts, and there have even been reports of fatalities in America. But the nuts are not pleasant to the taste, so children do not usually eat enough of them to produce toxic symptoms (Kingsbury. 1967).

There are cases of horse chestnuts being taken as the May Bush (see **MAY GARLAND**). In Carrick-on-Suir, in Ireland, it was branches of this tree that were set up on the morning of the first of May, and hung over doors of the byres, where the cattle were kept (*Béaloideas. Vol 15; 1945 p 283-4*).

There has been some weather lore recorded in Wales – if the leaves spread like a fan, then, so it was believed, warm weather would come; but long before rain, the leaves begin to droop and point downward (Trevelyan). The other instance of country belief involving carrying a nut around in the pocket, is recorded not only in Britain but also in America. For instance, people in Kansas used to carry it (and Yellow Buckeye) for rheumatism, either to prevent or to cure it. In New Hampshire, it was said that after a few years, the chestnut would become hard and dark from "absorbing the rheumatic germs" (Meade). Some said the chestnut had to be either begged or stolen (Sackett & Koch). Norfolk people believed in chestnuts, made

into a necklace, as a charm to ward off rheumatism. They had to be gathered by children who had never suffered from the ailment (Porter. 1974). Some kind of horse chestnut decoction was drunk in Essex for lumbago, spoken of there as rheumatism (Newman & Wilson). Elsewhere, including Spain (H W Howes), it is piles that is reckoned to be cured or prevented by carrying them around (W B Johnson; Tongue; Fogel), and that is interesting, because it is known that extracts of horse chestnut are rich in Vitamin K, and so is useful in treating circulatory disorders like piles, varicose veins and chilblains (Conway). A fluid extract made from the nuts is also used to protect the skin from the harmful effects of the sun (Schauenberg & Paris).

Another usage of conkers was as a snuff to cure catarrh and headache. The Pennsylvania Germans used it that way (Fogel), but this was quite an early habit (see Thornton), and the idea was to grate them up and use the powder to make one sneeze. Apparently it was recommended not only as a powder but also as an infusion or decoction to take up the nostrils. The leaves and flowers have occasionally been used, too (and so has the bark). The leaves are narcotic; an infusion of them has been used for insomnia (Conway), and a tincture of the flowers is sometimes given for rheumatism (Perry. 1972). The bark has been used for fevers, and externally, for ulcers (Wickham).

The nuts are Conkers, or Conquerors, sometimes shortened to Conks, and in parts of England lengthened to Oblionkers, or Hoblionkers (Salisbury). The reference, of course, is to the children's game with the nut on a string, to be struck by the opponent's similarly strung conker in an effort to establish a conquering, or champion, nut. So popular has it been that there is actually an organised conker competition, at Barnstaple, held on 20 October (M Baker. 1980). But since 1965, a World Conker Championship has been held at Ashton, Northamptonshire, on the second Sunday in October (Mabey. 1998). A variation on the game was described by George Bourne. It was called Mounters, and consisted in whirling a conker on its string round and round, and then letting it go mounting up into the sky. Not much of a game, perhaps, but he said the tree was actually called Mounter-tree, at about the time of year the game took place (Bourne. 1927). The petioles are known as Knuckle-bleeders in Norfolk, evidence of another boys' game (Britten & Holland).

The flowers are Candles in the west country, or Christmas Candles (Macmillan) (the tree itself is a Christmas Tree). The name one would expect to find, but do not, is Fairy Candles, for the story is that it was the horse chestnut that kept its candles burning to light the fairies home after their dance (Barber & Phillips).

HORSE MINT

(*Mentha longifolia*) "For fleas and lice to slay them: take horsement and strew in thy house, and it will slay them" (Dawson. 1934). In Nepal, the leaf juice is used as an antiseptic for cuts and wounds, and a leaf decoction is taken for a sore throat. It is said that the Basuto treat colds by stuffing the fresh leaves in the nostrils (Ashton).

HORSERADISH

(*Armoracia rusticana*) Fenland couples who wanted to know the sex of an unborn child put a piece of horseradish under each of their pillows. If the husband's piece turned black before the wife's, it would be a boy, and vice versa (Porter. 1958). Slightly more comprehensible was the belief that a piece of horseradish carried in the pocket would ensure that the owner would never to be without money (Swahn).

Horseradish roots are high in Vitamin C content, almost double per 100 grams as orange juice (G B Foster). Older writers stressed the sulphur content. This was apparently why it was used for chronic rheumatism, as a plaster instead of mustard (Rohde. 1926) – perhaps as a counter-irritant? But a report in *Notes and Queries; 1935* shows a Welsh rheumatism cure that is very different – shredded horseradish in a bottle of whisky, which was then buried in the ground for nine days. Then the dose was three spoonfuls daily. Another method for the same complaint comes from Russian folk medicine. Equal amounts of horse radish and paraffin were mixed, to be used as a quick rub-down before going to bed (Kourennoff). Chilblains could be treated by wrapping grated horseradish round the finger/toe, and kept in place with a piece of lint (Rohde. 1926).

Headaches can be cured with it, by bruising a leaf, wetting it and tying it to the head (Thomas & Thomas). That is the American way, but the cure is simpler in Britain. All you need do is smell it, so they claim in Norfolk (V G Hatfield), or in Sussex, just holding the scrapings tight would do the trick (J Simpson). The cure in Gloucestershire was also to smell it, better put as inhaling the vapour from the grated root (Vickery. 1995). It is even said that sniffing the juice will cure baldness! (Page. 1978). It will relieve toothache, too, if bound on (Newman & Wilson). That was in Essex, but in Norfolk the grated root had to be put on the opposite wrist for twenty minutes (V G Hatfield). Horseradish figures quite a lot in Fenland medicine. Wearing a bag filled with the grated root round the neck was a Cambridge ague pre-ventive, and Fen people claimed that a slice applied to a cut stopped the bleeding and drew the edges of the skin together quickly so that the minimum of scarring resulted. It is also said that the root, shredded and infused in hot water provided a powerful emetic to sober up a drunk (Newman & Wilson).

Grated horseradish is used in Russian folk medicine as a compress on the calves of both legs, for insom-nia. Dry mustard was sometimes added (Kourennoff). Another use in Russia is for asthma – half a pound of fresh horseradish, finely grated, mixed with the juice of two or three lemons. The dose would be half a tea-spoonful, twice a day. Horseradish in malt whisky was an Irish cure for pleurisy (Buckley), a pleasant way to be cured. It is good for kidney or bladder trouble and is used as such in Alabama (R B Browne), where a tea made from the root is taken for "a weak back". Does that mean lumbago, perhaps? It was certainly used in Britain for that complaint. Fens people grated and mixed it with boiling water, and this would be immediately applied to the patient's back on going to bed. The resultant blister was treated the next day by removing the plaster, baking it in the oven until it was powdery, then mixing it with flour, the mixture being dusted over the blister (Porter. 1958). Gerard recom-mended it for sciatica, asserting that "it mitigateth and asswageth the paine of hip or haunch, commonly called Sciatica".

The water in which horseradish had been boiled was used as a cough medicine in East Anglia (V G Hatfield). Lastly, it should be noted that eating the green leaves three times a day was a valued means in the Fen country of causing an abortion (Porter. 1958), knowledge evidently not confined to East Anglia, for Whitlock mentions it as a Wiltshire remedy, if that is the right word to use.

HORSESHOE VETCH

(*Hippocrepis comosus*) "Horseshoe" in this case is given from the shape of the pods. That is the reason why the plant had the reputation of being able to unshoe any horse that trod on it (Dyer. 1889) (hence the name Unshoe-the-horse, used once) (Prior). Sebillot recorded the same superstition from Savoy, and, using plain commonsense, adds that this plant grows in the sort of stony ground that leads to accidents.

HORSETAIL

i.e., COMMON HORSETAIL (*Equisetum arvense*) Mature horsetails have a large amount of silica in them, and that makes them extremely hard when dried (Forsyth). That is why the stems were used by cabinet-makers as a fine grade "sandpaper" (Salisbury. 1964), and they served as pot-scourers, too (North). They were "…employ'd by artificers for polishing of vessels, handles of tools, and other utensils: it is so hard that it will touch iron itself" (Camden). Many of the names given to it mirror this usage. Gerard, for instance, had Shave-grass, which is "not without cause named Asprelle, if his ruggedness, which is not unknown to women, who scoure their pewter and wooden things of the kitchen therewith". Pewterwort is another of these names, and so is Scouring Rush, used in America (Youngken).

Coles advised his readers that "the young buds are dressed by some like Asparagus, or being boyled are often bestrewed with flouer and fryed to be eaten". Re Asparagus, W Miller gave the name Fox-tailed Asparagus to an "*Equisetum maximum*". It was used as fodder sometimes, too, and the Blackfoot Indians were reported as using it as autumn and winter forage for horses (Johnston).

Horsetail tea is rich in silicic acid, and is diuretic, so it can be used for urinary problems, including bed-wetting, cystitis, or anuresis (M Evans; Schauenberg & Paris). The tea was used in Russia for all menstrual disorders (Kourennoff), and this same tea can be taken for oedema (Flück).

HOTTENTOT FIG

(*Carpobrotus edulis*) In South Africa, it was said that if babies were washed with the juice of this plant, it would make them "nimble and fleet of foot" (Schapera).

HOUSELEEK

(*Sempervivum tectorum*) or Welcome-home-husband-though-never-so-drunk, which must be the most picturesque of plant names. "Preserves what it grows upon from Fire and Lightning", Culpeper said; "as good as a fire insurance", they say in Wiltshire (Wiltshire). John Clare noted that in his native Northamptonshire "no cottage ridge about us is without these as Superstition holds it out as a charm against lightning". In Somerset the cover is extended to include witches (Tongue. 1965). So it is in the Isle of Man, where it is encouraged to grow as near the door as possible (Gill. 1932). It is luibh an toetean (fire herb) in Irish (Grigson. 1955), and if it grows on the thatch it will preserve the occupants of the house not only from the dangers of fire, but from related mishaps such as burns and scalds, so long as it remains untouched (Wilde. 1890) – a belief specifically recorded in County Clare by Westropp. 1911. Exactly the same idea turns up in East Anglia (G E Evans. 1966); in fact, during the last half of the 20th century there have been instances recorded of families moving from old or condemned houses to new council houses, and carefully taking some with them to put on the roofs of their new homes (Porter. 1969). Even the outside toilet had to be similarly protected (Lancashire FWI). Further back in time, we find it even said that Charlemagne himself ordered that every dwelling in his empire should have one growing on its roof, to protect it from fire and lightning (Conway). The belief was around even further back, for to the Romans, this plant was Diopetes, "fallen from Zeus", to protect the house from the lightning he wielded (Grigson. 1955).

There is an extension to this superstition – what protects from fire will protect from heat. Albertus Magnus described it by saying that he who rubs his hands with the juice of houseleek will be insensible to pain when required to take red hot iron in his hands (Folkard). This is mentioned in the *Boke of Mervayles of the World* – "spell to prevent the hands burning with a Red-hot iron. Wne redde Arsenicum is taken, broken and confected, or made with the juice of a herb called Housleeke, and the gall of a bull, and a man anointeth his hands with it, and after taketh hot irone it burneth them not". One test for an accused witch was to make her take a bar of red-hot metal in her hands – she would be innocent if it did not burn her. Actually, the "redde Artsenicum", red lead, that is, might just offer some protection against heat, but certainly not to the extent of handling with no effect metal at red heat.

It follows that these superstitions are mirrored in a number of houseleek's medicinal uses. "They take away the fire of burnings and scaldings ..." (Gerard), as one would be led to suspect. The leaves were used in Scotland to put on burns, like a plaster (Jamieson), and so they were in America (O P Brown). A Yorkshire remedy for burns and scalds was to use the leaves bruised with cream (Morris); and in fact the freshly gathered leaves are used to this day like this (and for corns, too). Similar ideas would account for its use against the fiery diseases – "they are good against S. Anthonie's fire, the shingles, and other creeping ulcers and inflammations ..." (Gerard). A gypsy remedy for ringworm is to boil houseleek, and then to dab the affected part with the water, and there is an erysipelas cure that requires pounded houseleek in a little skimmed milk. The rash should be bathed with this several times each day (V G Hatfield. 1994). Rather more exotic is a shingles remedy from Hertfordshire, which was to mix hair from a black cat's tail with the jucie of houseleek and cream, warm it all, and apply it three times a day (Jones-Baker. 1974). It sounds like a fusion of the two traditions, for styes on the eye used to be treated in old wives' lore by rubbing them with the tail of a black cat, as Parson Woodforde well knew (Woodforde). Probably, the Balkans use of the juice for a scorpion sting (Kemp) has its origin in the same idea.

Presumably because of its virtues against lightning, it is looked on as a good luck plant in a lot of places, whether, as in Dalmatia, the luck is integral with its medicinal uses (Kemp), or whether, as in parts of France, it brings the luck to the house on whose roof it is planted (Sebillot) (perhaps that is why it was used as a symbol of domestic industry, and also of vivacity (Leyel. 1937)), and long life to the people living in it (houseleek is called meure-jamais in Berry, and there are lots of "evergreen" names). In other places, it was used as a prophylactic in some way or other, by cutting the flowering stalks, for instance, and making them into crosses to hang over stable doors (Sebillot). In England, though, it was once said that the flowers were unlucky, and they were cut off before they could bloom (Conway). Nevertheless, there is negative evidence from this country of the luck of the houseleek, for they used to say in Sussex that you would bring trouble to the house if you pulled up the houseleek from the roof (Latham); in other words, the trouble would come with the destruction of the luck.

There is one very odd Breton superstition that said if a man put houseleek in his pocket, and made a girl smell it, it would have her running after him (Sebillot). It sounds like an error, and was probably meant for something like southernwood, which, with a name like Lad's Love, has been famous for the courting belief. Nevertheless, there is an example of love divination from America. You stick two pieces of the plant in a wall, naming them after a man and a girl. If the pieces grow towards each other, the couple will love each other (Whitney & Bullock).

Besides the medicinal uses that stem from its magical, protective qualities, houseleek has been used for more rational purposes. Herbalists still use it, for example, for eye- and ear-drops (Conway), the latter an old Cotswold remedy – squeezing houseleek juice, sometimes mixed with cream, into the ears to cure earache has been known for a very long time (Briggs. 1974; Helias). The medieval Welsh medical text known as the Physicians of Myddfai prescribed it "for deafness. Take ram's urine, the oil of eels, the house leek, and the juice of traveller's joy [*Clematis vitalba*], and a boiled egg. Let him mix and drop into the ear little by little, and it will cure him …". Bathing sore eyes with the juice is well known in Ireland (P Logan. 1972), and in Wessex (Rogers). Gerard knew all about this, for he recommended the juice to cool the "inflammation of the eyes, beng dropped therein, and the herb bruised and layd upon them". Sore lips could be treated with it, too; there is a Lincolnshire cure which involved holding a leaf between the lips, and bruising it so that the "cream" comes out. In fact, an application of houseleek, especially made into an ointment, is good for any sore place, it was said (Rudkin), including bed sores. It was used for the last named well into the 20th century, and is good for shingles and burns, too (Beith). The juice is still used to put on insect bites (M Baker. 1974), while there is a report from Norfolk that it is used in some unspecified way to cure cramp (Taylor. 1929). In Ireland, the juice is used to stop bleeding from a cut (Maloney), and from East Anglia to the Scottish Highlands the crushed leaves have been applied to bruises (Randell; Grant), and as a poultice for headaches (Gibson. 1959). In Cumbria, warts are cured by rubbing them with houseleek (Newman & Wilson), and they formed part of an odd cure for wens, too – "anoynt them with oil of snails, oil of swallows, and houseleek, e g parts, but do this not on a Friday or the first quarter of the moon" (Jeffrey). Houseleek poultices were used in East Anglia for treating boils and abscesses (V G Hatfield. 1994), and there is a report of a leg ulcer being treated (on gypsy advice) with houseleek, but it would be no use unless the dried herb was made up

into an ointment using fresh dairy-cream as the base (G E Evans. 1960). It was used in veterinary medicine, too. Irish people used to treat their horses' blistered feet (as long as the blisters were not too severe) with melted goose grease to which turpentine and the juice of houseleek had been added (P Logan. 1972).

A surprising report was given by Roy Vickery in *Folk-lore.vol 96; 1985 p 253*, in which he said that an Irish use of houseleek was to cause abortion. Some of the plants would be boiled, and the water given to the girl to drink. Later on, she was to climb a high wall and jump down, and that would do the trick. One assumes that the second part of the treatment wouild have been the only operative one, yet the plant appears in a list of abortifacients in Dutch folk medicine (Van Andel), so perhaps there is some virtue in that direction.

HUCKABACK
A textile made of coarse HEMP fibre. Male hemp was used for ropes, sacking, and the like; female hemp for sheets and other domestic uses (Peacock).

HUMAN ORIGINS
In Scandinavian mythology, when Ragnarok draws near, it is said that the ASH tree, Yggdrasil, will tremble, and a man and woman, Lif and Lifthrasir, will survive the ensuing holocaust and flood. They stand alone at the end of one cycle and the beginning of another in the world of time and men. From these two, the earth will be re-peopled, and Yggdrasil itself will survive Ragnarok. In other words Yggdrasil is the source of all new life (Crossley-Holland). The Eddas describe the stars as the fruit of Yggdrasil, and also say that all mankind descended from the ash and the elm (Dyer). According to Hesiod, the men of the third age of the world (the Bronze Age race) grew from the ash tree (Rydberg); Hesychius too said that the Greeks believed that the human race was the fruit of the ash (Philpot). Teutonic mythology also recognized that the first men came from the ash (Rydberg). In Greece, the OAK too was reckoned the tree from which men first sprang; they called it the "first mother", which fed man, mother-like, with her own acorns (Ovid). Piedmontese children used to be told that it was from the trunk of an oak tree that their mothers had taken them when they were born (Gubernatis). In Irish belief, the first woman sprang from a ROWAN tree, and the first man from an ALDER (Wood-Martin).

Some of the earlier peoples of India regarded the CASSIA as the origin of human life (Porteous. 1928), Some Negrito groups of Malaysia used the symbol of the banana to explain man's mortality. When the deity gave to one of the superhumans some "water life-soul" to give to the humans they had made, it was inadvertently lost. So the superhuman borrowed some from a banana plant. This was "wind life-soul" that he then gave to the inert bodies. "Wind life-soul" is a "short" life-soul, whereas what had been lost (the "water life-soul") was a "long" life-soul. And that would have made man immortal, whereas the one that the human beings eventually received was merely borrowed and thus provides only temporary life. Some say that not only the life-soul but also the heart and blood were borrowed from a banana plant, and this is supposed to account for the resemblance, in colour and viscosity, between coagulated banana plant sap, which dries to a dark brown colour, and dried human blood (Endicott). In West Africa, men and women, so it is said, descended to earth from the branches of a huge mythical IROKO (*Chlorophora excelsa*) tree (Parrinder).

An Egyptian creation myth relates how a LOTUS flower rises out of the waters; when its petals open the calyx of the flower is seen to bear a divine child, who is Ra. Another version is that the lotus opens to reveal a scarab (symbol of the sun); the scarab then transforms itself into a boy, who weeps; and finally, his tears become mankind (Larousse).

Humulus lupulus > HOPS

HUNGRY GRASS

(*Alopecurus myosuroides*) A crop of Hungry Grass is said to spring up if people who have eaten in the fields fail to throw some of the food away for the fairies (Patterson).

HUNGRY RICE

(*Digitaria exilis*) The staple food of many African people, especially in northern Nigeria, where it is grown in open fields or in terraces. The grain is very small, hence the common name, presumably, and is used as a porridge mainly. But it has never been a popular grain in Africa, judging from Dogon mythology, for it was said that eight different grains were given to the first ancestor, who rejected the eighth, which was Hungry Rice, on the grounds that it was too small, and too difficult to prepare. Indeed, the grain is hard and very small, yielding very little nourishment for a great deal of labour. It was said that Dogon women would ask for a divorce rather than undergoing all the labour needed to prepare it. Harvesting it has to be done in a great hurry, the moment it is ripe, for the stalks will bend with the slightest breeze and let the grain fall off. So it has to be stacked and threshed very quickly, if possible on the same day. In practice, this is done in the early hours of the night, and both men and women take part. This has led to extraordinary ritual behaviour, with men and women exchanging joking sexual songs. The grain is forbidden to a great number of men, and only the "impure" could eat it with impunity. Priests could be seriously defiled by merely touching it; possibly there is some Dogon symbolism connecting it with women's menstruation. Respectable women could only be persuaded to pound it with difficulty, and winnowing could only be done outside the walls, in places set aside for the purpose (Griaule). For Europeans, Hungry Rice has been a substitute for semolina.

Hyacinthoides nonscripta > BLUEBELL

Hydrocotyle vulgaris > PENNYWORT

HYLDEMOER

In Danish folklore, Hyldemoer, elder-mother, lives in the ELDER tree, and will avenge all injuries done to it. So her permission has to be asked before an elder is cut. The idea of some spiritual being, fairy or witch, inhabiting elders is very widespread. (Cf the legend of the Rollright Stones, for example). One of Hans Andersen's fairy tales tells that those who drink elderflower tea will see Hyldemoer in their dreams.

Hypericum androsaemum > TUTSAN

Hypericum elodes > BOG ST JOHN'S WORT

Hypericum perforatum > ST JOHN'S WORT

Hypericum pulchrum > UPRIGHT ST JOHN'S WORT

HYPERTENSION

GARLIC In Alabama domestic medicine, garlic is taken, cooked, preferably fried, to reduce blood pressure (Browne). Actually it is still prescribed, chopped finely in milk, for arteriosclerosis, as well as for hypertension (Fluck). Like true garlic, wild garlic, or RAMSONS, as it is generally known, has been used for hypertension, either by eating the fresh leaves, or by drinking a tea made from the dried leaves (Flück). Actually, all the *Alliums* can be used to reduce blood pressure, including SHALLOTS, and particularly ONIONS (Schauenberg & Paris). MADAGASCAR PERIWINKLE (*Catheranthus roseus*) is used in Haiti to bring down blood pressure (F Huxley), and that is one of its uses in Chinese medicine, too.

A decoction of HAWS, was traditionally taken in Scotland instead of tea or coffee, as a medicine for high blood pressure (Kourennoff), for it helps to prevent arteriosclerosis (Beith). Haws, in various preparations, have been used for angina pectoris both in Russia (Kourennoff) and in Germany, where it was said that the infusion in alcohol was the only effective cure for angina. LIME-FLOWER tea is another favourite, for it is reckoned to be a cure for arteriosclerosis, so improving the circulation, and being helpful for hypertension as well (M Evans). A cupful a day of dried RED CLOVER flowers, with alfalfa hay added, is taken for high blood pressure in parts of America, and raw CABBAGE, chopped fine and mixed with salt, pepper, vinegar and sugar, eaten once a day, was also said to bring blood pressure down (H M Hyatt). SWEET CICELY is also used in herbal medicine to lower blood pressure (Schauenberg & Paris), and MISTLETOE has long been used for the purpose (Grigson. 1955), always provided, Pliny said, "it has not touched the ground". OLIVE leaf tea is prescribed by herbalists for the condition (Thomson. 1978).

The use of the yellow leaves of BREADFRUIT, in decoction, to treat high blood pressure, is reported from Guyana (Laguerre), and the juice squeezed from the unripe fruit of CHINESE DATE PLUM is taken in China for the condition (L M Perry). GREEN HELLEBORE was official in America, where it very rapidly naturalized, up to 1960, and was used to slow the heart's action and to lower blood pressure (Weiner). WHITE HELLEBORE (*Veratrum album*) is a highly poisonous plant, causing among other symptoms a fall in blood pressure. That property has been appropriated to medicinal use – a tincture is prescribed to treat hypertension (Schauenberg & Paris). AMERICAN WHITE HELLEBORE (*Veratrum viride*) supplies a similar extract for the same purpose

(W A R Thomson. 1976). ARNICA, in minute doses, has also been prescribed (Wickham). WATER-MELON is said in Americe to be good for the condition, the treatment being to drink a tea made from the seeds (R B Browne). SHEPHERD'S PURSE has been adopted in Chinese medicine to treat hypertension, as well as colds and fevers, and enteritis (*Chinese medicinal herbs of Hong Kong* vol 2 1981).

HYSTERIC ALE

BOXWOOD turners' chips were used by herbalists as the basis of "hysteric ale", which also contained iron filings, and a variety of herbs. It was recommended to be "taken constantly by vaporous women" (Wilkinson. 1973).

I

Iberis amara see CANDYTUFT

IBOGA

(*Tabernanthe iboga*) A West African shrub, socially important as a hallucinogen. It is said that the discovery of the plant's properties was due to wild pigs. The natives saw that the pigs would dig up the roots, only to go into a wild frenzy (Pope). It is the root bark that contains the hallucinogenic principle, and it is in the Bwiti cult of the Fang people that the shrub becomes important. Bwiti is a cult of the dead, the object being principally to create a satisfactory relationship with the dead; it offers its members the experience (through iboga) of a passage over to the afterlife, and by doing that to come to terms with death itself (Fernandez). The root bark of Iboga, powdered and taken as an infusion, causes an altered state of consciousness (Emboden. 1979). In small doses it acts as a stimulant in much the same way as coca, enabling the user to double the length of a day's march, for example, or, one report claimed, allowing a hunter to sit awake and motionless for as long as two days while waiting for game (Pope). Larger doses, several times larger, produce euphoria, so that the user must "endure intense and unpleasant central stimulation in order to experience the hallucinogen effects" (Pope). Initiates into the Bwiti cult are given a massive dose, said to be 40 to 60 times the normal amount, the object being to effect contact with the ancestors through collapse and hallucination. This can, and sometimes does, cause death. (For a full account of the use of iboga in the Bwiti cult, see FERNANDEZ, J W *Tabernanthe iboga*: narcotic ecstasis and the work of the ancestors *in* FURST, P C editor *Flesh of the gods* New York, Praeger 1972).

Ilex aquifolium > HOLLY

Illicium anisatum > JAPANESE ANISE

Illicium verum > TRUE ANISE

Impatiens capensis > ORANGE BALSAM

Impatiens noli-me-tangere > TOUCH-ME-NOT

Impatiens walleriana > BUSY LIZZIE

IMPOTENCE

The testicle-suggesting tubers of EARLY PURPLE ORCHID would ensure that Salep, a preparation made from them, enjoyed a reputation for curing impotence. Salep, which has medicinal value, was extremely popular, particularly in eastern countries, for the purpose. It was treated as an aphrodisiac. ANISE was used as a Greek cure for the condition; ointments were made of the root of narcissus mixed with the seeds of anise or nettles (Simons). The early Persian and Arabian doctors prescribed GINGER for impotence

(Dalby), 'hot' making the reason fairly obvious, and SESAME seed, blended with crow's gall, made an embrocation for impotence (Lehner & Lehner).

INCENSE

CINNAMON has been used as incense – a little burned on charcoal, it is claimed, is good for "aiding meditation and clairvoyance" (Valiente). Egyptians used FENUGREEK seeds to make an incense oil (Sanecki), and powdered dried PATCHOULI leaves are sometimes introduced into incense (Schery). It is said that the Chinese used the sawdust of SANDALWOOD to make incense, mixed with swine's dung(!) (Moldenke & Moldenke). CARDAMOM seeds are sometimes burned to produce an incense-like atmosphere (Valiente).

INDELIBLE INK

SLOE juice is indelible, as careless handling during the making of sloe gin will prove. Juice squeezed out of unripe sloes was sold at one time under the name of German Acacia, and used to mark linen – an ideal laundry marking, in fact. American Indian groups used POISON IVY to make an indelible ink (Sanford).

INDIAN PHYSIC

(*Gillenia trifoliata*) This was the principal ingredient of "cure-all" nostrums sold by travelling medicine salesmen in America. So much of it was sold that the plant got the name Dime-a-bottle Plant (Mitton).

INDIAN SHOT

(*Canna indica*) The seeds, round and bullet-like (that is why the plant is called Indian Shot), have been used as coffee substitutes, and to make rosaries (Coats. 1975). According to Burmese legend the canna sprang from sacred blood. Dewadat tried to kill the Buddha by pushing a great boulder on him as he passed by below. The boulder fell at the Buddha's feet, bursting into a thousand fragments. A single fragment striking the Buddha's toe, drew blood, from which the canna arose (Skinner). On the other hand, it is said in Ghana that a witch keeps the soul and blood of the people she has killed in a pot. It is not real blood, but the seed of *Canna indica*, one for each person killed (Debrunner). The analogy is with the red flower, red = blood.

INDIAN TOBACCO

(*Lobelia inflata*) In large doses, this is a poison (Weiner), but in small ones, it is useful for asthma (one of its names in America is Asthma-weed). It was actually introduced into Britain for this purpose around 1830 (W A R Thomson. 1976). The leaves and tops of Thorn-apple are mixed with this herb to make the "asthma powders" commonly sold for the relief of the complaint. A little nitre is included to make it burn, and the smoke is inhaled. The mixture is often made up into cigarettes, for convenience (Hutchinson & Melville). The plant has also been used to treat falling hair. The Indiana practice is to fill a bottle with the pulverized herb, then pour in equal parts of brandy or whisky and olive oil. Let it stand for a few days, then bathe the head once a day with the liquid (Tyler).

INDIGESTION

Gypsies use a root and herb infusion of FIELD GENTIAN to relieve indigestion (Vesey-Fitzgerald), and TANSY leaves, chopped up and added to bread dough and cake mixtures, were a popular indigestion remedy in Cambridgeshire (Porter. 1969). In Ireland flatulence used to be cured by taking a tansy leaf decoction with salt added (Egan), and another Irish remedy is the simple expedient of eating CELERY (Maloney). LEMON VERBENA leaves, fresh or dried, are widely used as a tea for indigestion (Macleod), and FENNEL is used in the same way, and has been for a very long time. "Fennel seed drunke asswageth the paine of the stomacke, and wambling of the same, or desire to vomit. And breaketh winde …"(Gerard). A Middle English rimed medical treatise prescribed wither betony or fennel for the digestion, fennel to be taken "in droge after meat". "Drogges" were a kind of digestive powder for weak stomachs, and were used by Chaucer's 'Doctour':

> Ful redy hadde he his apothecaries
> To serve him drogges and his letuaries

(Prologue to *Canterbury Tales*). CORIANDER seeds are still known to herbalists as an efficient indigestion remedy. A few seeds chewed before a meal always help (Conway).

Oil of JUNIPER used to be sold (and is probably still available) as a carminative, for it has a beneficial effect on the appetite and digestion (Schauenberg & Paris). The juice of a couple of berries in hot water is a gypsy remedy for indigestion (Tongue. 1965). Buchan reckoned them to be "the most celebrated among the class of Carminatives". NUTMEG has always enjoyed a reputation as a carminative, and has been used as such until quite recently. It was in the first half of the 19th century that they were used most, to relieve flatulence and dyspepsia. Nearly every middle-aged woman carried one to grate over her food and drink. They were carried in the pocket in small wooden or metal cases with a grater at one end (Newman & Wilson). Evening drinks, known as possets, were commonly taken, and these nearly always contained large amounts of nutmeg and other spices. The drinks themselves were regarded as carminatives. A recipe for indigestion that comes from Alabama requires one to steep a pinch of YARROW blossom in a cup of water, and to drink a little of this several times a day for three days (R B Browne). ANISE was much used for indigestion. Gerard recommends it as "good against belchings and upbraidings of the stomacke". The Romans offered an anise-flavoured cake at the end of rich meals to ease indigestion. MARJORAM has often been included

in indigestion remedies, and the decoction used to be an Irish remedy for indigestion and acidity (Egan). In the 18th century, Hill was prescribing an "infusion of the fresh tops [to] strengthen the stomach, and [it] is good against habitual colic". DANDELION tea or coffee are said to be good for indigestion (Browning), or BETONY tea (Newman & Wilson). GREEN PURSLANE is used around the world for indigestion. The Navajo Indians, besides taking the seeds as food, recognise the green plant as a cure for stomach ache (Elmore), and in West Africa they take it for indigestion by beating up the leaves with water and adding a little salt (Harley).

WORMWOOD tea is good for indigestion, though the dose should not be continued for more than a day or two. It is made with a handful of fresh leaves, over which a pint of boiling water is poured. After infusing for a few minutes, a wineglassful is to be drunk (Brownlow). On the island of Chios, wormwood is still the great standby for stomach upsets; there is a saying there that translates "bitter on the lips, sweet to the heart" (Argenti & Rose). As far off as Morocco, we find it being used for heartburn and stomach-ache: it is boiled in water, sugar added, and the mixture drunk warm. Mixed with tea, it is also supposed to promote proper digestion after a meal; but if persons have such tea on two or three consecutive days, they will quarrel and separate (Westermarck), surely a recognition of the deleterious effect of the continued dosage. A tisane of WHITE HOREHOUND is often taken for indigestion, and in homeopathy a tincture of the plant is used (Schauenberg & Paris). Even the Navajo Indians were reported to use this herb for the condition (Wyman & Harris), and it is certainly an American domestic remedy for dyspepsia still (Henkel).

INDIGO

(*Indigofera anil*) the name Indigo nicely pinpoints the plant's area of origin, for this is Latin indicum (H G Baker), and India was the oldest centre of indigo dyeing. 'Indicum' originally was used to define all imports from India, and only later was applied to the blue dye; it replaced the Arab word al-nil, blue, the ancestor of the word aniline (Leggett). It has been known as a dyeplant since antiquity, though it was rare in Europe throughout the Middle Ages. The earliest evidence of tropical indigo as a dye comes from Italy – it was described in Genoa in 1140, and again at Bologna in 1194. It reached England in 1274, and France in 1288 (Hurry). It was not until Vasco da Gama discovered the sea route to the East Indies in 1498, with the resultant setting up of trading settlements, that indigo began to challenge and gradually replace woad as a blue dye (Ponting).

But, as a protection for woad, indigo was proscribed in Elizabeth 1's reign as a dangerous drug, and was described as "the food for the devil" (Hurry) – indeed,

Devil's Dye is recorded as a name for the plant (C J S Thomson. 1947), and reckoned downright injurious to fabrics, a notion sincerely believed in a number of countries (Leggett). In 1609, Henry IV of France issued an edict sentencing to death any person discovered using the deceitful and injurious dye called inde. In fact, it was not until 1737 that French dyers were legally allowed to use it (Hurry). For the technical processes of indigo dyeing, see Hurry, and Leggett.

Indigo must have been introduced into (or is it indigenous to?) tropical America, for there are records of its use, not always as a dyeplant, in the Mayan villages of Yucatán. It is a protector there, for the ordinary amulet to keep off evil spirits and to avert the evil eye is a collection of small objects tied together with thread. It is this thread that is dyed blue from the plant, and the same dye is used to paint the nails of persons who are threatened with death from sickness (Redfield & Villa). Another record affirms that these Maya crush the leaves and put them in a bath to cure convulsions (Roys). But in ancient times it was classed as an astringent medicine, as it cleaned wounds and was used for ulcers and inflammations. As late as the 17th century it was used internally in some way, though in Elizabethan times it had been denounced as a dangerous drug (Leggett). That denunciation was probably not made with public health in view, but rather the protection of the woad industry.

Indigofera anil > INDIGO

INFERTILITY
SKULLCAP is one of the herbs traditionally held to cure infertility (Conway). STINKING ORACH (*Chenopodium vulvaria*) is another plant that was supposed at one time to cure barrenness – just by the smell of it! (Grieve. 1931).

INFLUENZA
Boil MEADOWSWEET flowers in water for ten minutes, then take three cupfuls a day (Fluck). It will have the same effect as aspirin, for some of the chemical constituents of the drug are also present in the plant. A LINSEED decoction was an Irish country cure for influenza (Logan). So it was in the Highlands of Scotland, too, with honey and a little vinegar added (Grant). One way to deal with 'flu, according to Alabama belief, is to put an onion in a pan under the bed (R B Browne), and the complaint was tackled in Scotland in a similar way – one half at an open window, and the other above the door. The onion would attract the disease, and so turn black (Beith).

INSECT REPELLENTS
ELDER has insect-repellent properties, for it is said that people strike or whip fruit trees and vegetables with elder boughs (or better still, drape them over), so that the scent would kill off troublesome insects (Folkard). Elder is described in Wiltshire as a "charm" against flies. Hampshire people recommended a wash made from elderberries steeped in hot water, to deter gnats from biting (Savage). If flies were troubling a horse, Lincolnshire carters would tie a branch of elder in leaf to the harness, and that would keep them away (Rudkin). And elders used to be grown by a dairy door, or by an outside privy (Vickery. 1993), for exactly the same purpose (Hartley & Ingilby). Elder twigs stuck in along the rows of broad beans will keep blackfly away, though judging from the state of elder blossom towards the end of their flowering period, blackfly are immune there. But they say that elder blossoms picked at full moon and stuck into fruit, drive away weevils (Baker). You can get rid of ants, too, by pouring a strong decoction of elder leaves over the nests. And you can actually keep fly off the turnip by making an elder-bush harrow to draw over young turnip crops (Savage). Of course, insect bites should be rubbed with an elder leaf (Tongue). BLACK WALNUT leaves, too, will keep house-flies away (Bergen. 1899), and a solution of WALNUT leaves and bark was used to keep greenfly at bay (Allen). The powdered rhizome of WHITE HELLEBORE (*Veratrum album*) can be sprinkled on currant and gooseberry bushes to protect them against insect pests. As the powder loses its toxic property after three or four days, it is safe to apply to ripening fruit (M Baker. 1977). That powder is highly toxic to fleas and lice, too (Flück). AMERICAN WHITE HELLEBORE (*Veratrum viride*) is equally efficacious as an insecticide (Lloyd), and so is RED RATTLE (*Pedicularis palustris*).

MUGWORT keeps the flies away. You can either wear a sprig (Genders), or keep an infusion to sponge over the face and arms (Cullum). The very name of the plant confirms it. Mugwort is OE muogwyrt, from a Germanic base meaning a fly or gnat. Midge is the same word. Mugwort's relative, SOUTHERNWOOD relies on its smell to keep insects at bay. The French name for the plant is Garderobe; moths will not attack clothes in which Garderobe has been laid (Grieve. 1933). GROUND CYPRESS has the same name in French, and is just as efficient at keeping moths from clothes (Macleod). WORMWOOD and SEA WORMWOOD have the same effect. TANSY is another plant that keeps flies away. Rubbing the surface of raw meat with tansy leaves will protect it from flies (Hemphill), and bunches of it used to be hung in the windows of farm kitchens for the same purposes. Sprigs were put in bedding at one time, to discourage vermin (Drury. 1992). Or use PEPPERMINT to keep flies and midges away – rub the face and hands with the leaves. Mint (or parsley) grown in a window sill is also said to keep insects out of a kitchen. Bruise the leaves every now and then to release more odour (Boland. 1977). A sprig of SPEARMINT in the kitchen will keep the bluebottles out (Vickery. 1993). An infusion of FEVERFEW, allowed to dry on the skin, is a good gnat or mosquito repellent (Quelch). PARSLEY serves the same purpose; grow it on the kitchen window-sill, or make parsley baskets to hang up in a porch or window, to keep flies and other insects away (Boland. 1977). It does the same job in the garden, too – old gardeners used to plant it round the onion bed, to keep onion fly away (Rohde), and it is grown among roses, too, not only to improve their scent, but also to keep greenfly at bay (Boland & Boland). In Mediterranean countries a pot of BASIL is kept on windowsills to keep flies out of the room (G B Foster), and a sprig in the wardrobe will keep moths and other insects away (Conway). People working in the Cambridgeshire harvest fields used to garland themselves with sprays of WHITE BRYONY to keep off flies, and leaves of it were put in the privy pits of Fen cottages as a deodorant in hot summer weather (Porter. 1969). A Devonshire way of keeping flies out of the house was to burn a handful of PLOUGHMAN'S SPIKENARD each day during the summer (Hewett).

The leaves of Chinaberry Tree (*Melia azedarach*) can be used to drive off flies and fleas, and a traditional American usage is to put the seeds in with dry fruit to keep insects away (R B Browne). MALABAR NUT (*Adhatoda vasica*), an Indian plant long grown throughout the tropics, has the reputation of being insecticidal (L M Perry). The oil distilled from the leaves of CAJUPUT (*Melaleuca leucadendron*) is a good mosquito repellent (Chopra, Badhwar & Ghosh). The juice of MEDITERRANEAN ALOE (*Aloe vera*) rubbed on the skin will protect from insect bites (G B Foster).

INSOMNIA

A HOP pillow is the best known soporific and has been for a long time. The secret of success against insomnia lies in not packing it too tightly, and renewing the dried hops every four to six weeks (Thomson. 1976). The pillow's sedative action was used to combat other conditions, too. Lindley tells us that they were prescribed for "mania". George III is said to have slept always on a hop pillow (Genders. 1971). Pillows are stuffed with the dried leaves of CATMINT, too, for the smell is supposed to help the sleepless (Sanford). NETTLE seems an unlikely plant to be associated with cures for insomnia, but it certainly is in both the medical sense and also as a charm. In the Highlands, we are told that nettle leaves chopped very small and mixed with whisked egg-white used to be applied to the temples and forehead. This remedy was mentioned as early as the beginning of the 18th century by Martin, who particularly speaks of fevered patients benefitting from the treatment. In Wales, they used to put a bunch of nettles in broth, both to induce appetite and to promote sleep (Trevelyan). Grated HORSERADISH root is used in Rusian folk medicine as compresses on the calves of both legs, for insomnia. Dry mustard was sometimes added. It is said that this causes a flow of blood away from the brain, and so induces sleep (Kourennoff).

An infusion of four or five HEATHER flower sprays to a pint of boiling water is drunk as a tea for insomnia; it was even just applied to the head for that condition. And a heather pillow is still used to give refreshing sleep (Beith). Another tea that can be taken is from the dried flowers of SCENTED MAYWEED (Flück). The root infusion of PRIMROSE, if taken last thing at night, has a distinct narcotic tendency (Leyel. 1926), and so is good for insomnia. THYME, too, has been used for insomnia, as well as melancholy and nightmare (Hill, J *The family herbal*, claimed that "the night mare is a very troublesome disease, but it will be perfectly cured by a tea made of this plant"). VIOLETS were associated with sleep, but by a misunderstanding. It is well known that the smell of violets is fleeting. When first coming on them, the fragrance is obvious, but it soon seems to go. "To smell the smell out of violets" is a proverbial saying, and there is factual basis for it, for the fragrance contains ionine, which has a soporific effect on the sense of smell. This effect was recognised, but misunderstood, as the gift of sleep. One 16th century herbalist (Ascham) said, "for them that may not sleep for a sickness seethe violets in water and at even let him soke well hys temples, and he shall sleepe well by the grace of God". Even TOMATOES have been recommended for the condition (Ackermann).

In some parts of Europe, leaves of BLACK NIGHT-SHADE used to be put in babies' cradles, the idea being that they (the leaves) would soothe the babies to sleep (Grieve. 1931). There was some justification for this – the generic name, *Solanum*, comes from a word neaning 'to soothe'. Some South American Indian peoples use Black Nightshade for insomnia, by steeping a small quantity of the leaves in a lot of water (Weiner). Ancient writers refer to MANDRAKE as a remedy for insomnia (Randolph). Even the smell of the plant was said to have the same effect. Dodoens says that "the smell of the mandragora apples causeth sleepe, but the juyce of the same taken into the bodie doth better" (Ellis). But of course mandrake was used as an anaesthetic; indeed it was probably the very first anaesthetic in Italy. Amulets for the prevention of insomnia were made by binding OAK twigs into the form of a cross (Leland. 1898).

A very strange leechdom from the 15th century offered treatment "for those who speak in sleep. Take SOUTHERNWOOD and stamp it, and mingle the juice with white wine or with vinegar, and give the sick to drink when he goeth to his bed, and it shall let him for speaking in his sleep" (Dawson). Just as odd was a very similar prescription by the Physicians of Myddfai. They make more sense when it is realised that it is insomnia they are trying to cure, not just talking in the sleep. Even WHITE HOREHOUND has been used in a sleeping draught in the Fen country. Rue was added too, followed by a good dose of gin mixed with laudanum. It is quoted as being a last resort means of stopping a mother giving birth on 1 May (an unlucky day). It just put her to sleep for twenty-four hours (Porter. 1969).

LIME-BLOSSOM tea is the best known cure for insomnia (Tongue. 1965). A hot bath with lime flowers in it, is another remedy (Quelch). This infusion is valued for headaches, too, and seems to be a sort of cure-all.

Inula conyza > PLOUGHMAN'S SPIKENARD

Inula helenium > ELECAMPANE

INVISIBLE INK
One of the "notable things" reported by Lupton was its use for "writing that cannot be read without putting the paper in water. Take the juice of SPURGE LAUREL, put it into a little water wherein alum has been dissolved, and if you write with it, it will appear as nothing on the paper, but being put into water, the letters will appear plain and legible".

Ipomaea batatas > SWEET POTATO

Iris foetidissima > STINKING IRIS

Iris germanica > BLUE FLAG

Iris germanica "Florentina" > ORRIS

Iris pseudo-acarus > YELLOW IRIS

IROKO
(*Chlorophora excelsa*) A large deciduous tree of African forests and savannas, and a sacred tree throughout West Africa (Parrinder), often marked with a piece of white cloth tied round the trunk. Sacrifices were often made to it in Yoruba practice, for they not only held it to be sacred, but to be inhabited also by some powerful spirit. Men fear having the tree near their dwelling, for the spirit that inhabits it makes terrible noises. Furniture made of its wood can also make disturbing noises in the house, and doors made of its wood can fling open of their own accord (Awolalu). The surroundings of such a tree were often meeting places for the religious guilds (Tampion).

It is called Loko in Dahomey, and is the centre of one of the most ancient cults, Men and women, it is said, descended to earth from the branches of a huge mythical Loko. If a shoot grows in a compound or street, it is taken as a sure sign that the god wishes a cult to be founded there. But it is believed that an iroko will not grow if planted deliberately. Anyway, it is unlucky to plant a tree, for the planter will not live to see it grow (Parrinder). And in Nigeria, as Talbot said, nearly every Ibo compound has its sacred tree, more often than not an iroko, called there Ojji (the other sacred tree is the Oil Bean, *Pentaclethra macrophylla*).

The souls of the dead are thought to live in the tree while awaiting reincarnation. When in the course of time it falls, the family to whom it belongs carefully marks the place where it stood; no garden may be made on that spot again. These are "ghost trees"

then, but it is only the souls of good men who await reincarnation in these trees. The ghost of an evil man would be driven away from the tree by the souls of the good men already there. It is believed that no storm can ever damage them, so that the souls continue to live in peace (Talbot).

The small masks called 'ma' made by the Mano of Liberia were the especially sacred ones, and some of their sanctity seems to derive from association with iroko trees. It is not said from what wood these masks are made, but before being worn they had to be washed in water containing iroko bark. Whether to be worn or not, the washing had to be done every new moon (Harley).

IRONWEED
(*Vernonia noveboracensis*) A North American plant. Chew it for toothache (R B Browne), and chew it for diarrhoea, too (H M Hyatt).

Isatis tinctoria > WOAD

ITCHING POWDER
The seeds of DOG ROSE constitute the original itching powder with which children amused themselves by putting down each others' necks. They are called Itchy-backs (Opie & Opie. 1959) and Cow-itches (Holland). In Devonshire they are Ticklers, or Tickling Tommies (Macmillan), and in Scotland and Ireland, Buckie-lice (Grigson. 1955), buckie being the name under which the hips themselves are known. Another source of children's fun lies in using the fruits of LONDON PLANE for itching powder. The spicules fill the air round the trees, and can be the cause of health problems.

IVY
(*Hedera helix*) An important plant in mythology, dedicated to Bacchus/Dionysos, an association retained into modern times. Ivy, after all, is the bush that good wine does not need. Dionysos had the name Kissos, or ivy, in Attica. He was, in fact, the "ivy-crowned", and so the ivy wreath was worn in his cult, and it is said that initiates even had themselves tattooed with the mark of the ivy leaf (Otto). An ivy bush (a tod of ivy (Ellacombe)) was the sign of a wine-tavern. Aubrey said the custom died out around 1645 ("the dressing the tavern bush with Ivy-leaves fresh from the plant was the custome 40 yeares since, now generally left off for carved work"). It was still hung outside taverns in Normandy and Brittany at the end of the 19th century, and indeed was reported (by W B Johnson) to have still been seen in Normandy in 1929. It was still in use at Ashburton, Devonshire, though as a conscious anachronism, in 1950, at the annual ale-tasting ceremony, and the ivy was hung to show that the ale was of good quality. Nevertheless, there is a Somerset saying, "put a trail of ivy across a drunkard's path, and he will become a sober citizen" (Tongue). For ivy is a "cool" plant, in contrast to the

heat of the vine. Its coolness has the power, so it was said, of extinguishing the heat of the vine, and that is the reason for Dionysos telling his fellow celebrants to wreathe themselves with it (Otto). And if the labourer who pruned the vine wore an ivy wreath while doing it, the vine would bear well, according to a 10th century collection known as the *Geopontica* (Rose).

Ivy ale was a highly intoxicating medieval drink, still made, according to Graves, at Trinity College, Oxford, apparently in memory of a Trinity student murdered by Balliol men. He goes into no details, but it is to be assumed that this is a conventional brew laced with ivy leaves, in which case it certainly would be highly intoxicating, for ivy leaves are poisonous, producing a narcotic effect not unlike that of atropine (Jordan).

Tacitus said the Jews decorated themselves with ivy at festival times (Bayley). It was accepted by the earliest Christian church as the emblem of life and immortality; for instance, they laid it in coffins as a symbol of new life in Christ (Leland). All evergreens by their nature are emblems of immortality, but ivy's use extended into an association with Christmas, and, as the companion of holly, into Christmas decorations (see CHRISTMAS DECORATIONS). Ivy let Christ hide in it, so an English gypsy tale goes, and the reward was that it should be green all the year. Elder, though, betrayed him, and the punishment was "You shall always stink" (Groome). The rivalry between ivy and holly in medieval carols has suggested, at least to Robert Graves, a domestic war of the sexes. Holly is the male plant and ivy the female. The explanation seems to lie in the fact that the last sheaf in harvest was bound with ivy, and was eventually called the Ivy Girl, a name that came to be synonymous with a shrewish wife, but the sentiments of this medieval song bring no emphatic confirmation with that idea , for "ivy is soft and mek of speech" (Gubernatis). The dark man who had to be the first-foot on New Year's morning was called the Holly Boy, and elaborate precautions were taken to keep the women out of his way, a recipe for opposition (Graves). They say you should be careful not to let ivy predominate in the Christmas evergreen decorations, as it is an unlucky plant (Sternberg) (unlucky for men, perhaps? though it is a genuinely unlucky plant) (see below). Ivy saved from the Christmas decorations in church was fed to the ewes to encourage the conception of twin lambs (Drury. 1985). Goats, too, are said to be very fond of ivy – one Gaelic saying translates "sickness alone would keep goats from eating ivy" (A R Forbes). As early as ancient Greek times, we find that priests presented an ivy wreath to newly-married people – it is said, too, that ivy is a symbol of fidelity (Grieve) (because it clings?). Perhaps not unrelated to this is the Lincolnshire practice of hanging some ivy over the bed on New Year's Eve while the occupant is asleep, so that the ivy would be the first thing he or she sees

on waking on New Year's morning (Rudkin). On the other hand, to avoid pregnancy, the woman should take, in the first three days after her period, "neuf graines de hièvre grimpant" (Loux).

Though in the Highlands, ivy has some protective powers (a piece would be nailed over the byre door to prevent witches harming the cattle, and also to protect them from disease (MacGregor)), superstition generally looks on it as an unlucky plant. In County Armagh there is a saying, "The house where ivy grows will surely fall" (Bergen). In Wales, too, if ivy growing on an old house begins to fall away from the walls, and becomes shrivelled, people start talking of the financial disaster or death soon to overtake the owner (Trevelyan). It should never be grown indoors either (Hyatt). Nevertheless, in other places, they said that ivy growing on a house would protect it from witchcraft or misfortune (R L Brown). It seems that this protective belief may be confined to Scotland as far as the UK is concerned, but it has certainly been used as a witch-finder in the Balkans. In Dalmatia they used to make ivy-wood cups from which to serve drinks at Christmas. If a woman refuses to drink from it, it is a sign that she is a witch. Ivy-wood appears in other Balkans charms: encircle the church when it is full of worshippers on Christmas night, with white thread, then bury the thread under its threshold, together with a little spoon that the operator had made from ivy-wood, so that no witch could leave the church (Vukanovic).

In the Fen country, ivy is one of the unluckiest plants to bring into a house (Porter), and so they say in America, too, for it is looked on in the same way in Alabama (R B Browne). Massachusetts people say it is unlucky as a gift, and nearby, in Maine, they say that a person who keeps ivy will always be poor (Bergen). A 1610 manuscript says it is "unwholesome to sleepe under the ivie, or in a ivie-bush" (*Gentleman's Magazine*). At least, dreaming of it is a good thing, for in Devonshire that is a sign that "your friendships are true" (Hewett). Not so in Cornwall, for there they say that if you want to dream of the devil, you should pin four ivy leaves to the corners of your pillow (Courtney).

The Scottish regard for ivy continues in a couple of marriage divinations. Lanarkshire girls used to wear a piece of ivy against the heart, and repeat the rhyme:

> Ivy, ivy, I love you,
> In my bosom I put you,
> The first young man who speaks to me,
> My future husband he shall be (Opie & Opie).

The Argyllshire rhyme was:

> Eevy, ivy,
> I do pluck thee.
> In my bosom I do put thee.
> The first young man I do meet
> Shall my true lover be (MacLagan).

This sort of thing is recorded in the Witney area of Oxfordshire, too. To see her future husband, a girl had to put an ivy leaf in her pocket. The first man she met out of doors whilst carrying it would be the one she would marry. A Yuletide or New Year divination as to future health, was by putting an ivy leaf in a bowl of water and covering it till the 5 January. If it is as green as when it was put in the water, then you would enjoy good health, but if there were black spots on it, then, depending on where the spots were, an illness in the part of the body corresponding to the position of the spot on the leaf would occur (Brand/Hazlitt). But a most novel form of divination is recorded from Normandy – to be used when in doubt as to which saint to pray to! A number of ivy leaves are put in holy water, each one bearing the name of a saint; the first one to turn yellow shows the saint you were looking for (W B Johnson).

The berries and leaves are poisonous, and they can cause vomiting and diarrhoea, though serious only in small children (North). Of course, birds eat them, and it is doubtful whether animals have been poisoned with ivy. In some areas it used to be strewn on the fields for cattle and sheep to eat, and it is often given to sick animals (see below). Perhaps they would have to eat a very large amount before feeling any effects (Long. 1910). The berries have always been used medicinally. The 1610 manuscript already quoted claims that ivy berries would cure the pestilence, as well as boils, carbuncles and other skin diseases. The berries have to be taken from high up the climber, "not fro that which is foudne lowe by the grounde". They had to be kept carefully until needed, when they would be powdered in a mortar and given in white wine.

A widespread, and favourite, use of ivy leaves is for corns. From the island of South Uist in the Outer Hebrides, to Cornwall, the remedy is still in use. The Hebridean practice is to put a poultice of ivy leaves and vinegar on the corn (Shaw), and Ruth Tongue reported exactly the same procedure in Somerset. They say in Cornwall that corns will drop out in about a week after putting bruised ivy leaves on them (Deane & Shaw); exactly the same as claimed in Ireland – if the corn is still there after a few days, then a handful of ivy leaves is put to steep in a pint of vinegar in a tightly corked bottle for a couple of days. Then the liquid is carefully put on the corn, taking care that it does not get on the skin (Logan). It is said that John Wesley, who walked enormous distances, used ivy leaves to soothe his tired feet (A W Hatfield); certainly, in his *Primitive Physick*, we find "Corns (to cure) … apply fresh ivy leaves daily and in 15 days they will drop out". It is still done in the Lowlands (Rorie. 1914); in the Highlands, it was said you had to heat the leaf (Polson). Warts, too, are occasionally treated by applying an ivy leaf to them (Newman & Wilson), and a wart charm was still in

use in Essex in the 1950s. A hole was pricked in an ivy leaf for each wart, and the leaf was then impaled on a thorn in the hedge (Mabey. 1998).

A cap of ivy leaves worn on the head was supposed to stop the hair falling out (Leather), or to make it grow again when illness had caused it to fall. In Fifeshire, it is put on a child's head for eczema (Rorie), and on the island of Colonsay, the leaves were sewn into children's caps to stop sores (Murdoch McNeill). Gerard had been told something rather different – "the gum that is found upon the trunke or body of the old stock of ivie, killeth nits and lice, and taketh away haire" as against the ivy-leaf cap which stopped the hair from falling out.

It is said that the best way to treat mumps in adults is to take ivy leaves and berries internally, and to apply them externally (Page). Ivy vinegar from the berries was very popular at one time. It was used in London, for instance, during the plague (Putnam), like the strong-smelling herbs people carried round with them, and which are still in existence in the ceremonial nosegays carried by the monarch, and by judges. There are still more conditions that ivy was supposed to cure. A leaf infusion was still in use in the 20th century as an eye lotion (Rorie), interesting, for Gerard recommended the same treatment four hundred years ago. Even more interesting is the fact that in homeopathy, a tincture of the young leaves is used to treat cataract to this day (Schauenberg & Paris). Lady Gregory recorded a belief from the west of Ireland that "a cure can be made for bad eyes from the ivy that grows on a white-thorn bush". Even in the 15th century the berries (with the vinegar panacea) were being recommended for toothache – "for aching of the hollow tooth. Take ivy-berries and seethe them well in vinegar and [take] the soup of the liquor all hot and hold it in the mouth till it be cold and then cast it out and take more. And do so three or four times and it shall heal without doubt, for it is a principal medicine therefor" (Dawson).

But some of the so-called medical practices are purely magical. From the Gironde, in France, there is a record of ivy-root necklaces (they had to be green, and an odd number of pieces) put round a baby's neck to help teething (Sebillot). Little ivy-wood cups used to be made, and children with whooping cough were made to drink their milk from them (Grigson) (cf the similar practice, using holly), These ivy cups were very extraordinary – Browne mentions a belief that they would separate wine from water, the wine soaking through, but the water not! It seems that there was an Irish belief that a cow's sore eyes could be cured simply by hanging an ivy leaf over the chimney. As it dried, so would the soreness go (O'Farrell).

Ixia viridiflora > AFRICAN CORN LILY

J

JACK-IN-THE-PULPIT

(*Arisaema triphyllum*) The spadix is the 'Jack', and the spathe is the 'pulpit' (Parson-in-the-pulpit is another name for it). Keep a piece in the pocket as a preventive against rheumatism (H M Hyatt). African-Americans in the southern states of the USA look on it as a protective plant. They would take the leaves and rub them on the hands, and that would blind an enemy. But they use it to make charms to bring security and peace, and to protect them from enemies (Puckett). The leaves are luck-bringers, if you carry them on the person. For centuries it was regarded as an aphrodisiac (Whittle & Cook) (the spadix in the spathe is expressive enough). But this is a poison, inasmuch as the rhizomes are extremely acrid. They are certainly edible after boiling, for that reduces the poison, though the acrid principle is never entirely eliminated.

It has been used as a medicine by both native American Indians and Caucasian immigrants, and also by African Americans, who take the root tea for kidney and liver problems (Fontenot). The Hopi used the powdered dried root (a teaspoonful in half a glass of cold water) as a contraceptive, lasting a week. Two teaspoons of a hot infusion would bring permanent sterility, so they said (Weiner). Another use of the root, pounded, is for a poultice to put on sore eyes. Small doses of the partially dried rhizome are used to treat chronic bronchitis, asthma and rheumatism, the latter also treated by the Pawnees with the powdered rhizome (Corlett).

JACKFRUIT

(*Artocarpus heterophyllus*) A near relative of Breadfruit, but the fruits are much larger, and, astonishingly, may weigh as much as 26 pounds or more (Tosco). The tree is now grown in all tropical countries, but it is economically important only in tropical Asia. The enormous size of these fruits convinced the Negritos of Malaysia that this was the original, or archetypal, fruit, from which all others are descended (Endicott).

Southern Indian symbolism equates the Jackfruit tree with the life-span of the family and its ancestral house, so it is a sacred tree there (Rival). One will be planted at the south-west corner of a house, to be associated, not with the ghost of an original family member, but with all the ancestors of the house and the past and present family group that it shelters. The tree therefore is a "metaphor" for the "longevity of the house-family group" (Uchimayada). But it is feared in Jamaica, where no driver will agree to carry them – "they think it attracts duppies, and will cause an accident" (Newall. 1981).

JAPANESE ANISE

(*Illicium anisatum*) The branches are used to decorate Buddhist graves (Hyam & Pankhurst).

JAPANESE GENTIAN

(*Gentiana scabra*) A plant that is much used in Chinese medicine. The root is prescribed for fevers and rheumatism, and is of benefit to the liver. It is also given as an analgesic, and to treat eye disorders (R Hyatt), even to strengthen the memory (F P Smith).

JAPANESE MINT

(*Mentha arvensis "Piperascens"*) In the 18th and early 19th centuries, the dried powdered leaves of this mint were specially imported, to be carried in small silver boxes fastened to the belts of gentlemen, who would inhale a pinch whenever they felt like it (Genders. 1972).

Jasione montana > SHEEP'S BIT

JASMINE

(*Jasminum officinale*) A symbol of grace, elegance and amiability, and an emblem of the Virgin Mary (Ferguson). A south Indian folktale tells of a king whose laugh would spontaneously spread the fragrance of jasmine for miles around (Classen, Howes & Synott). It is one of the many plants once supposed, mistakenly, to be aphrodisiac (Haining). In Italy, it is woven into bridal wreaths. There is a proverb that says that a girl who is worthy of being decorated with jasmine is rich enough for any husband (McDonald). Around Menton, however, they believed the bride whose husband offered her a bouquet of jasmine would die within the year (Sebillot). To dream of jasmine is generally a good omen, especially to lovers (Gordon. 1985), and in ancient times, one of the ways of practising capnomancy (divination by smoke) was by throwing seeds of jasmine or poppy in the fire, and watching the motion and density of the smoke. If it was thin, and shot up in a straight line, it was a good omen (Adams).

Jasminum nudiflorum > WINTER JASMINE

Jasminum officinale > JASMINE

Jasminum sambac > ARABIAN JASMINE

Jatropha curcas > PHYSIC NUT

JAUNDICE

An infusion of GROUNDSEL is still in use in Cornwall for jaundice (Deane & Shaw). But most of the older prescriptions are nothing more than aspects of the doctrine of signatures – yellow flowers to cure the yellow disease. Gerard's recommendation of CORN MARIGOLD for jaundice is a case in point, and the use of GORSE is another example (Leyel. 1937), as is the Irish use of buttercup juice (Ô Súilleabháin), or the yellow sap of GREATER CELANDINE. TOADFLAX would be used, too, and so would SAFFRON. "Lay saffron on the navel of them that have the yellow jaundice, and it will help them", was Lupton's advice. The yellow juice of RED POPPY is the reason for its use in Chinese medicine for the complaint (F P Smith), and the same applies to the yellow latex of MEXICAN POPPY (*Argemone mexicana*) (See also DOCTRINE OF SIGNATURES). WORMWOOD was prescribed in the old herbals for jaundice. It is possible that this is also doctrine of signatures (it has yellow flowers), but it does not seem all that likely. But the most obvious example is that of TURMERIC, with its yellow dye, and not only in the Pacific islands, but in West Africa (Harley), and even in Britain (Thornton, *New family herbal*, 1810: "Turmeric, when taken internally, tinges the urine a deep yellow colour, and acts as a gentle stimulant. It has been celebrated in diseases of the liver, jaundice, etc.,"). SAFFLOWER, the yellow dyestuff, is used in the Philippines to treat it (Perry & Metzger). A root infusion of TREE CELANDINE (*Bocconia frutescens*) has been used in Colombia for the complaint (Usher), and BARBERRY bark would be used in Europe.

The Welsh medical text known as the Physicians of Myddfai has a cure for jaundice involving HAW-THORN: "take the leaves which grow on the branches of the hawthorn and the mistletoe, boiling them in white wine or good old ale, till reduced to the half, then take it off the fire and strain. Drink this three times a day". ELDER ointment was used for jaundice in Ireland (Wilde). In Herefordshire, too, the inner rind of the bark, boiled with milk, was taken for this complaint (Leather). HONEYSUCKLE leaves are still used in Ireland for jaundice (Maloney), and Grieve said that a root decoction of BUTCHER'S BROOM was a favourite medicine for jaundice, still being used in Ireland in her day . The complaint was also treated with MISTLETOE, either the leaves, as a receipt from the Welsh text known as the Physicians of Myddfai prescribes, or the berries, which, according to the old usage in the bocage country of Normandy, had to be soaked in a male baby's urine, and then put on the patient's head, while an unspecified, and doubtless secret, charm was spoken (Sebillot). A "breakstone" like PARSLEY PIERT would naturally be used for this complaint, as it is a powerful diuretic, and much in demand for bladder and kidney troubles (Brownlow).

The condition has been treated by herbalists with GREAT PLANTAIN in modern times (A W Hat-field), but Gerard had already recommended it, as "the rootes … with the seed boiled in white wine and drunke, openeth the conduits or passages of the liver and kidneys, cures the jaundice, and ulceration of the kidneys and bladder". And the treatment was known in folk medicine too; in Cambridgeshire, "the decoctions of the leaves … is a sure remedy for the diseases of the bladder, being drunk night and morning" (Porter. 1969). WOOD SAGE used to be known as Gulseck-girse in Orkney (Leask), and as jaundice was called

gulsa in the far north of Britain, one can assume that the plant was used for the complaint there. AGRI-MONY tea is another medicine with a reputation for curing the condition (Brownlow). CHICORY enjoyed a similar reputation; gypsies used a root decoction (Vesey-Fitzgerald).

Jeffersonia diphylla > TWIN-LEAF

JENNY GREENTEETH
The name of a nursery bogey that is actually the same as DUCKWEED. Little children were warned to stay away from ponds that were covered with the green weed, otherwise Jenny Greenteeth would drag them down and drown them. Duckweed was actually called Jenny Greenteeth in some parts (see DUCKWEED).

JERUSALEM ARTICHOKE
(*Helianthus tuberosus*) A North American plant, and the name Jerusalem artichoke has no geographical significance. It is merely a corruption of the Italian girasole articiocco. Girasole means turn-sun, a reference to the quite imaginary practice of the sunflowers in turning their heads to follow the sun.

Cutting an artichoke in half and rubbing the cut side on the roots of the hair was an old country remedy against baldness (Quelch).

JERUSALEM OAK
(*Chenopodium botrys*) It gets its common name because the young leaves look like miniature versions of those of the oak. This is a vermifuge (Watt & Breyer-Brandwijk) – see, for example, the prescription from Alabama: for worms, one teaspoonful of the seed or the stalk tea mixed with syrup, three times a day (R B Browne). There is, too, a remedy, using the inner bark of this plant, boiled and mixed with molasses to make a candy. It also seems to have been used in some way for tuberculosis (R B Browne).

JIMSONWEED
That is, Jamestown Weed, one of the names given to THORN-APPLE (*Datura stramonium*). The oft-quoted story is that English soldiers, sent to Jamestown to put down the uprising known as Bacon's Rebellion, in 1676, gathered young plants of this species and cooked them as potherbs, "the effect of which was a very pleasant Comedy; for they turn'd natural Fools upon it for several days …" (Safford).

JOHN THE CONQUEROR ROOT
Said to have been a love charm much used by men. It was a piece of dried root with a prong or pike growing out of it, an obvious piece of phallic symbolism. It was carried in a little chamois leather bag, or one made of red cloth. The origin of the root was kept secret by the people who sold the charm, but it is said to have been BOG ST JOHN'S WORT (Valiente).

JUDAS
There are a number of examples of trees on which Judas was said to have hanged himself, beliefs usually

connected with some characteristic of the tree itself. ASPEN is one example; Russian folklore explained the constant trembling of the leaves by taking it to be the tree of Judas. The JUDAS TREE (see below) got its purple-rose flowers, when it burned with shame when Judas hanged himself on it. FIG TREES are viewed with mistrust in the Mediterranean area; in Sicily it was because, so it was said, Judas hanged himself on one. Crowning someone with ELDER was reckoned in Yorkshire to be about the most insulting thing that could be done (Gutch), for Judas hanged himself on an elder tree, a belief that Ben Jonson picked up in *Every man out of his humour*: "He shall be your Judas, and you shall be his elder-tree to hang on". A fungus that grows on elder is known as Jew's Ear, that is Judas's Ear (it was looked on as a great medicine for quinsy, sore throat and the like (Leask)). TAMARISK "was of old counted infelix, and under malediction, and therefore used to wreath, and be out on the heads of malefactors" (Evelyn. 1678), for this is one of the trees said to be the one on which Judas hanged himself. Once a tall and beautiful tree, it is now reduced to a shrub by divine malediction (A Porteous). They say that the ghost of Judas contiually flits around the tamarisk.

JUDAS TREE
(*Cercis siliquastrum*) The purple-rose flowers tell how the tree burned with shame when Judas hanged himself on it (Skinner), originally a Greek tradition, so it has been said, but the name Judas Tree first appeared in Gerard. (Hutchinson & Melville). It is naturally, given the association with Judas, the symbol of betrayal (Leyel. 1937). Another result of the association is that we are advised to avoid it, especially when it is flower (M Baker. 1977). A lot of people are reluctant to cut it, especially after dark. They used to say in Friesland that the Judas Tree was a favourite haunt of witches (Dyer. 1889). Any tree with a Judas association would have a corresponding witch connection – compare, for instance, the elder, which is also a very unlucky tree to cut. There is, too, a superstition that it would be death to fall into a Judas tree (Folkard), an obvious re-enactment of the hanging of Judas.

Wilkinson. 1973 suggested that Judas Tree should actually be Judaea Tree. So did Barber & Phillips. It was called Tree of Judaea in France, and it was from France that the tree came to England, but it had become Judas Tree by Shakespeare's time.

Juglans nigra > BLACK WALNUT

Juglans regia > WALNUT

JUJUBE
(*Zizyphus jubajuba*) A Chinese species, but long cultivated there and also in the Mediterranean area, and in southern USA. They have an edible, olive-sized fruit, known as French jujubes (Willis). These berries have been famous since ancient times for cold cures and for bronchitis. They used to be made up into lozenges, and were widely exported; such lozenges are still called jujubes (Mitton). "This plum is an excellent Pectoral, and opens the Body … It expectorates tough Flegme, and is good against Coughs, Colds, Hoarseness, Shortness of Breath, Wheezings, Roughness in the Throat and Wind-Pipe, Pleurisies … (Pomet). The seeds are used in Chinese medicine to give sleep, and also to benefit the nervous system (R Hyatt).

Juncus conglomeratus (Compact Rush) > RUSH

Juncus effusus (Soft Rush) > RUSH

JUNIPER
(*Juniperus communis*) This is a protective tree, indeed the very symbol of protection (Leyel. 1937). The wood and berries have been used all over Europe as a protection against evil influences and in containing witchcraft (Westermarck). Juniper canopied Elijah in his flight from Jezebel, and there is a legend that it saved the lives of the Virgin and Jesus when they fled into Egypt. In order to screen her son from Herod's men, the Virgin hid him under certain plants and trees, which naturally received her blessing in return for the shelter given. Among these plants, the juniper was believed to have been particularly invested with the power of putting to flight the spirits of evil, and of destroying charms (Friend. 1883). Italian stables are protected from demons and thunderbolts by a sprig of juniper (Fernie). So they are in north Germany, where "Frau Wachholder", the juniper spirit, is invoked to discover thieves, apparently by means of the bending down of certain of its branches (Elworthy. 1895). Like box, juniper growing by the door protected the house from witchcraft, for the witch had to stop and count every leaf before proceeding (M Baker. 1977). One could hang some in the beehive, too, to protect the bees from adverse magic (Boland. 1977), and in the far north of Scotland, a baby's teething ring would be made from juniper wood, more for its protective qualities than anything else (Rorie. 1994). In France, branches of juniper act as a Christmas tree; they are put round the chimney breast, and presents for the children are hung from them (Salle). In Italy, too, juniper is hung up at Christmas time (Elworthy. 1895).

In the Highlands of Scotland, juniper was specially used for "saining", a form of blessing, on New Year's Day. Branches were set alight, and carried through the house, the smoke spreading into a thick, suffocating cloud (McNeill, 1961), produced by closing up every window, crevice and keyhole in the house, for the smoke, besides protecting the house and its occupants from evil influenes, was also supposed to have the ability to dispel infection (Camp). Juniper was burned before the cattle, too (Davidson. 1955), or it was boiled in water, to be sprinkled over them. Some ceremony was necessary before the juniper could be used. It had to be pulled by the roots, with its branches made into four arms, and taken between the five fingers, while an incantation was spoken, given in translation as:

> I will pull the bounteous yew
> Through the five bent ribs of Christ,
> In the name of the Father, Son, and Holy Ghost,
> Against drowning, danger and confusion
> (Campbell. 1902).

As the charm makes clear, it was regarded as a special protection both by sea and by land, and no house in which it was taken would take fire. It was used for saining, not only at New Year, but also at Shrovetide, and as a protective against the evil eye (Banks. 1946).

After all this, it almost goes without saying that one should not cut down a juniper. According to Welsh belief, he who does so will die within the year, and aged junipers are preserved (Trevelyan). Grimm recognized it as a wishing tree in German belief, and pointed out the fact that it is known as the Tree of Wishes in India. Another German belief was that a kind of spirit called Hollen was connected with juniper trees, obviously of the helpful fairy type. When small children get ill the parents carry wool and bread to a juniper growing on a neighbour's ground, and they say:

> Ihr Hollen und Hollinen,
> Hier bring ich euch was zu spinnen,
> Und was zu essen.
> Ihr sollt spinnen und essen
> Und meines Kindes vergessen (Runeberg).

It is unlucky to dream of the tree itself, especially if the dreamer is sickly; but to dream of gathering the berries, if it is winter, is a sign of prosperity to come. To dream of the actual berries means that you will shortly arrive at great honours, and become an important person. To the married, it foretells the birth of a male child (Dyer. 1889). Another superstition, from Somerset, is that you should never tell a secret by a juniper tree. Everyone will know it within a week (Tongue. 1965).

The wood was sometimes burned indoors (without any idea of saining) to give rooms a sweet smell (Grigson. 1955), and in Scotland the young twigs used to be burnt for smoking hams, giving them a slightly turpentine-like flavour (C P Johnson). Burton, *Anatomy of Melancholy*, tells that "the smoke of juniper is in great request with us to sweeten our chambers", and Ben Jonson has: "He doth sacrifice twopence in juniper to her every morning before she rises, to sweeten the room by burning it" (Rimmel). Sprays were often strewn over floors so as to give out, when trodden on, the aroma which was supposed to promote sleep; Queen Elizabeth's bedchamber was sweetened with them (Fernie).

Juniperus gave the French name genièvre, which became geneva, or gin for short, since proper Hollands gin has its flavour from juniper berries.

They are used for Steinhäger, too (Brouk), and there are records of their use in the Hebrides for flavouring whisky (Murdoch McNeill). There is a French beer called Genièvre (Leyel. 1937), surely flavoured with the berries, while Elizabeth Raper wrote down a recipe for porter that required two pints of the berries to give it the proper flavour.

It has been quite important for medical uses since ancient times, for the berries were commonly used by the Greeks and Romans, as well as by Arab Physicians. The berries and the essential oil from them are said to be diuretic, and the juice from the berries is still used in Irish country medicine as such (Logan). It is also recommended for cystitis (Schauenberg & Paris), and a tea can be made from them, or even small pieces of wood, to be used as a diuretic (Flück). In much the same way, the Cree Indians stewed the berries and ate them for the same purpose (Corlett), and infused the root for cases of gravel. Oil of juniper, a carminative, has also been used in domestic medicine for rheumatism (two or three drops on a lump of sugar every morning) (V G Hatfield. 1994). Juniper is an American domestic medicine for colds and colic (Bergen. 1899), and Evelyn praised it, "… the berries swallow'd only instantly appease the wind-collic, and in decoction most sovereign against an inveterate cough. They are of rare effect, being stamp'd in beer; and in some northern countries, they use a decoction of the berries as we do coffee and tea". But this use for coughs is ancient indeed, for "Dioscorides reporteth that this being drunke is a remedie against infirmities of the chest, coughs, windinesse, gripings, and poisons …; the decoction of these berries is singular good against an old cough, and against that with which children are now and again extremely troubled, called the Chin-cough" (Gerard).

But the claims made for juniper have been much more ambitious; it was even reckoned at one time to be rejuvenating (Gerard), and there are a number of notices of its power as a counter-poison. Pomet noted their ability "to prevent infectious Airs", in connection with the French habit of his time to "make comfits of [the berries] which they call St Roch's Comfits, and carry them in their Pockets, that they may chew two or three of them in a Morning, to prevent infection, and make the Breath sweet". In the Highlands of Scotland, a child cutting its first teeth was given a piece of juniper wood to chew, to prevent toothache, it was said (Polson. 1926). Scottish travellers used it for exactly the same purpose, but they thought there was something in juniper twigs that cut the teeth quicker (MacColl & Seeger).

Juniperus californica > CALIFORNIAN JUNIPER

Juniperus communis > JUNIPER

Juniperus monosperma > ONE-SEEDED JUNIPER

Juniperus sabina > SAVIN

Juniperus virginiana > VIRGINIAN JUNIPER

K

Kalanchoe crenata > NEVER-DIE

Kalanchoe pinnata > CURTAIN PLANT

KAPOK

(from a Malay word, kapoq) is the very light fibre that covers the seeds of silkcotton trees, used for stuffing pillows, lifebelts, etc., The Red Silkcotton Tree (Bombax ceiba) is grown in the West Indies (though it is not indigenous to the area), and they say there that it is the haunt of ghosts and other spirits, hence the names Jumbie Tree or Devil's Tree. Doubtless it is the kapok itself that suggests it. Anyone trying to cut one down could expect harm (Bell), unless he had propitiated the duppy first with rum and rice put round the root. But the usual tree producing kapok is *Ceiba pentandra*, from West Africa, Malaysia, etc., The floss is harvested chiefly from cultivated trees in Java, Sri Lanka and the Philippines (Everett). This is a sacred tree in Yoruba belief (J O Lucas), and in Maya cosmology, a sacred silkcotton tree (another name for the Kapok Tree) stood at each of the four cardinal points, fertilising and feeding life in the four directions. Each tree had its colour (red – east; white – north; black – west; yellow – south), and in each tree a bird nested. There was also a central tree, green in colour to represent the fountain of all life (I Nicholson). In Ghana, it is often used in sorcery, and sorcerers themselves are believed to have the power to change themselves into a silkcotton tree, which is a place of assembly for witches. Any big forest tree like this is also said to be a place where tutelary spirits reveal themselves (Debrunner).

KAPOORIE TEA

A Russian beverage, made by adulterating ordinary tea with the leaves of Rosebay Willowherb (*Chamaenerion angustifolium*) (Leyel. 1937). The spelling varies somewhat, as Kapa, or Kappair, etc., (Usher).

KAPPA

A spirit in Japanese mythology, much associated with the CUCUMBER. It is like a monkey, but has no fur, and is yellow-green in colour. Kappas have a liking for blood, and for cucumbers, and one way of placating them is to throw cucumbers with the names and ages of the family where kappa live. The connection with cucumbers, phallic emblems, becomes even more significant when considering the belief that these spirits are capable of raping women (J Piggott).

Khaya ivorensis > AFRICAN MAHOGANY

Khaya nyassica > RED MAHOGANY

KIDNEY BEAN

(*Phaseolus vulgaris*) In Italy, beans are distributed among the poor on the anniversary of a death (Grieve. 1931). In the west of England they say that kidney beans will not grow unless they are planted on a special day, the 3 May in some places (Waring), but there is no great agreement on the actual day. It could be as late as Stow-on-the-wold Fair Day, which is 12 May, or as early as St George's Day, (23 April) (Pope). Kidney Beans are good for the kidneys, so it is often said, but this is a well-known remedy. They have been used in France till very recently for kidney and bladder disorders (Palaiseul). Maya medical texts prescribe them for, among other complaints, hiccups (Roys).

KIDNEY COMPLAINTS

PARSLEY root tea is still prescribed for kidney trouble (Rohde, Hyatt). Gypsies used the leaves for the same purpose. Gerard wrote that the leaves "… take away stoppings, and … provoke urine; which thing the roots likewise do notably perform …". The herb is, indeed, a well known diuretic. WHORTLEBERRIES are used sometimes; both the berries and the leaves are used in an Irish cure for kidney troubles (Ô Súilleabháin). In the north of Scotland, they were used for dissolving kidney stones (Beith), as the name Kidneywort, or Kidneyweed, proclaim. WALL PENNYWORT was used particularly against kidney trouble and stone. Indiana folk medicine advised a tea made from the roots of HONEY LOCUST (*Gleditsia triacanthos*) for kidney trouble (Brewster).

KIDNEY VETCH

(*Anthyllis vulneraria*) The common name states that it was used for kidney troubles, and the specific name pinpoints its use as a wound herb. In the Highlands of Scotland an ointment was made from it (Fairweather) for cuts and bruises, and in the Channel Islands, the leaves were used to stop the bleeding from wounds (Vickery. 1995).

Kigelia africana > GERMAN SAUSAGE TREE

KING'S EVIL

(see Scrofula) Scrofula was so called because the kings of England and France claimed to have the ability to heal it by their touch, a gift conferred by God through the oil used at their coronation. Edward the Confessor was the first English monarch to touch for scrophula, and the last was Queen Anne, but the ritual remained in the Book of Common Prayer till 1744 (Simpson & Roud).

KINNIKINNICK

An Algonquin word, apparently meaning "that which is mixed", usually referring to tobacco. BEARBERRY, for example, actually has this as a name, for American Indians smoked the dried leaves, either with ordinary tobacco, making the mix that warrants the name, or without any tobacco (Sanford). The Chippewa smoked the mixture, in the ordinary way, and also, they claimed, "to attract game" (Densmore). Bearberry was not the only plant to be called kinnikinnick – SILKY CORNEL (*Cornus amomum*) was another, and so was RED-OSIER DOGWOOD (*Cornus stolonifera*). The inner bark of the former was used, by the Menomini Indians, for example, as a tobacco substitute (H H Smith. 1923), and so was the bark of the latter (see Bergen. 1899, Chamberlin. 1911, etc.,).

KISSING COMFITS

SEA HOLLY roots were preserved in sugar and sold as Candied Eringo, or Kissing Comfits, Colchester being the centre of the trade. They were sold until as late as the 1860s (Grigson. 1955), and were said to be good for those that "have no delight or appetite to venery", and were "nourishing and restoring the aged, and amending the defects of nature in the younger" (Gerard). Likewise René Rapin, in a Latin poem on gardens, 1706 (in translation):

> Grecian Eringoes now commence their Fame
> Which worn by Brides will fix their Husband's Flame
> And check the conquests of a rival Dame.

There is, too, a reference in Dryden's translation of Juvenal's Satires; he is talking about libertines:

> Who lewdly dancing at a midnight ball
> For hot eryngoes and fat oysters call.

The best known quote is from Shakespeare. Falstaff, in *Merry wives of Windsor*, says: "Let the sky rain potatoes; let it thunder to the tune of 'Green Sleeves'; hail kissing comfits, and snow eringoes".

KNAPWEED

(*Centaurea nigra*) There are some love divination games played with knapweed. A young girl or man would pull the flower from the stalk, cut the top off the stamens with scissors, and lay the flower somewhere secret, where it could not be seen. She would think through the day, and try to dream through the night, of her lover, and then, on looking at the flowers next day, would judge of her success in love whether or not the stamens had shot out to their former length (Henderson). John Clare also described a similar charm in *Shepherd's Calendar. May:*

> They pull the little blossom threads
> From out the knapweed's button heads
> And out the husk wi' many a smile
> In their white bosoms for a while
> Who if they guess aright the swain
> That loves sweet fancy trys to gain
> 'Tis said that ere its lain an hour
> 'Twill blossom wi a second flower …

That description, though, is of a Buckinghamshire charm. The plant would bloom a second time if they could guess correctly the name of her future husband (Leyel. 1926).

Knapweed has a varied list of ailments for which it was used as a cure, from a general tonic to a cure for a carbuncle. The tonic, a decoction, was used in Sussex until recent times (Allen), but the carbuncle cure is ancient, and comes from the Welsh medical text known as the Physicians of Myddfai: "For a carbuncle … take the flowers of the Knapweed or the leaves, pounding with the yolk of an egg and fine salt, thene applying thereto, and this will disperse it …".

Russian folk medicine recommended a mixture of dried sage, knapweed and camomile flowers for all digestive disorders – one teaspoonful of the mixed herbs to a glass of water, boiled for 15 minutes and strained. The dose was a tablespoonful every two hours during the day (Kourennoff). Just chewing the flowers was taken as a cure for diarrhoea in Britain (Page. 1978). It has been used for rheumatism in Cumbria, the prescription just being described as "boiled horse-knops" (Newman & Wilson), horse-knop being one of the names for this plant. Lastly, there is an Irish gem: "this is dwareen (knapweed) and what you have to do with this is to put it down, with other herbs, and with a bit of three-penny sugar, and to boil it and to drink it for pains in the bones, and don't be afraid but it will cure you …" (Gregory).

KNOTGRASS

(*Polygonum aviculare*) Dyer. 1889 says that knotgrass is probably so called from some unrecorded character by the doctrine of signatures, that it stops growth in children (presumably if they eat it). Cf Beaumont & Fletcher, *Burning Pestle*: "… and say they would put him into a strait pair of gaskins, 'twere worse than knotgrass, he would never grow after it". They used the concept in *The Coxcomb*, too:

> We want a boy extremely for this function,
> Kept under for a year with milk and knot-grass.

Presumably they wanted an undersized lad, for the 'function' must have been nefarious. Shakespeare knew about it, too:

> "Get you gone, you dwarf,
> You minimus, of hindering knot-grass made,
> You bead, you acorn …"

(*Midsummer Night's Dream. Act 3. Sc 2*). But, more probably, the name is from the knots of the intricately jointed stem (*Polygonum* is from Greek meaning "many joints", or more accurately, "many knees", for Greek gonu meant knee (Potter & Sargeent)).

This is an astringent, useful in diarrhoea and other such ailments (Grieve. 1931), which would include haemorrhages. A Somerset remedy for nosebleed is to rub the plant into the nostrils (Tongue. 1965). It has been used for many other ailments, as recommended by the early herbalists. There is a leechdom from the 15th century for earache, for example, using the juice in the ear, "and it shall take away the aching wondrously well" (Dawson). In Chinese medicine, the juice is used in skin diseases, and for piles (F P Smith), and also for bladder complaints (Geng Junying). A French charm for corns is to put a piece of knotgrass in a pocket on the same side as the corn, and say "Que mon cors s'en aille à l'aide de cette herbe" (Sebillot).

KOLA

(*Cola nitida*) Tropical West Africa is its real habitat, but it has been introduced into many parts of the world, because of the commercial value of the kola nut, which is a source of flavour and caffeine in many "cola" drinks (Schery). They can be eaten, too, and are recognised as valuable fatigue- or hunger-inhibitors (Harley), either chewed or boiled to make a beverage (Emboden. 1979). Students take them before an examination, and they are given to horses before a race. Kola is also a sexual stimulant, so it is said, and promotes conception in women (Lewin), but excessive doses act as a depressant (Schauenberg & Paris). A Kola leaf, with leaves from three other shrubs, all ground up with black soap, are ingredients in a Yoruba Ewé cure for madness (Verger).

Kola nuts, or kula, to give them their African name, are also included as magical elements in many medicines, as in Yoruba practice, which requires sweet or bitter elements. Kola is taken to be bitter, while used as complete foods they are sweet (Buckley). They are used, too, in simple Yoruba divination procedures. The split segments, usually four, are thrown, and the different configurations are interpreted as positive or negative answers to a question (Buckley). In addition, two of the four lobes are known as male, and two are female (Awolalu).

But eating kola is more important as a recognised social institution, like chewing betel elsewhere, or drinking tea. They have their role, for instance, in Igbo culture, not only socially, but ritually as well. They are symbols of luck, social distinction and prosperity, and it is a great honour to be offered the best nuts. What Uchendu calls a "four-cotyledon kola nut" is most important for ritual purposes, for four is a sacred number among the Igbo (they have a four-day week, for instance, and in divination, the number four "count" is auspicious). The kola nut is the greatest symbol of Igbo hospitality – to be presented with a kola nut is to be made welcome, and the presentation itself is an important ceremony. It is the host's privilege to give the nut, which is passed through a chain of men who represent different segments of the lineage, until it reaches the principal guest, who then starts a relay through his own party, until the nut finally gets back to the host who then ritually breaks it and eats the first share, to demonstrate that it is wholesome and free from poison. Then each member of the guest's party, and finally each of the host's party, take their shares (Uchendu).

The nuts are classified according to colour, red, white or pink. White are preferred, and command a much higher price, besides having special significance in the ritual (Dalziel).

L

LABRADOR TEA

(*Ledum groendlandicum*) A North American evergreen shrub, whose leaves contain some narcotic substance (Turner & Bell), though it appears that the Indians were unaware of any such property, although the Ojibwa did use it as a substitute tobacco (Jenness. 1935). Certainly, strewn among clothes, the leaves will keep moths away, and in Lapland, branches are put among grain to discourage mice (Grieve. 1931). Bergen. 1899 mentions the leaf tea as an American domestic medicine for stomach disorders, and it has also been drunk for rheumatism (E Gunther).

LABURNUM

(*Laburnum anagyroides*) Poisonous, of course, particularly the bark and seeds. So poisonous, that merely carrying the twigs has been known to induce the symptoms, for, like nicotine, the alkaloid cytisine is readily absorbed through the skin (Jordan). If anyone builds a fishpond under a laburnum tree, the fish will be poisoned by seeds dropping in it (Chaplin). But a tincture made from the leaves and fresh flowers is used in homeopathic medicine to treat depression, cramps, dizziness, and some digestive disorders (Schauenberg & Paris).

In Herefordshire, they say that lilac and laburnum mourn if any tree of the same kind is cut down near them, and it was believed that they would not bloom the following year (Leather). A laburnum winter is so called when a spell of cold weather coincides with the flowering of the tree (Vickery. 1995).

Laburnum anagyroides > LABURNUM

LACTATION

One of Gerard's prescriptions for FENNEL was: "the greene leaves of Fenell eaten, or the seed drunke made into a Ptisan, do fill womens brests with milke". Since it is similar to dill in appearance, fennel has picked up some of dill's attributes, like increasing the flow of milk in nursing mothers, settling babies' stomachs, etc., (G B Foster).

In Morocco, when a newly delivered mother has no milk, she is given a kind of very liquid porridge made of pounded LUCERNE seeds (given to cows, too) (Legey). FENUGREEK seeds are used, with ANISE seeds, as a tea, to treat lactation difficulties (W A R Thomson. 1978).

Lactuca sativa > LETTUCE

Lactuca virosa > GREAT PRICKLY LETTUCE

LAD'S LOVE

A well-known name given to SOUTHERNWOOD (*Artemisia abrotanum*). There have been a number of attempts to explain the name, one being the use of an ointment that young men used to promote the growth of a beard (Leyel. 1937). That was certainly done, for we have Gerard's prescription: "the ashes of burnt Southernwood, with some kind of oyle that is of thin parts … cure the pilling of the hairs of the head, and make the beard to grow quickly". The likeliest explanation seems to lie in the courting customs of Fenland youths, described by Porter. 1969. A few sprigs of the plant were generally added to the nosegay that courting youths used to give the girls, and the plant was quite prominent in other customs in that area. A youth would cut a few sprigs to put in his buttonhole. He used to walk, sniffing ostentatiously at his buttonhole, through groups of girls. If the girls went by and took no notice, he knew he would have to try again, but if they turned and walked slowly back towards him, then he knew they had noticed his Lad's Love. He would then take his buttonhole and give it to the girl of his choice. If she was willing, she would also smell the southernwood, and the two would set out together on their first country walk.

LADY LAUREL

(*Daphne mezereum*) The specific name, *mezereum*, comes from a Persian word, Madzaryon, which means destroyer of life, for the red berries are poisonous (although they were once used as a pepper substitute (Hyam & Pankhurst). And Russian peasants used to take up to thirty of the berries as a purge, while the French regarded fifteen as a fatal dose (Le Strange). The bark is poisonous, too. It is said that Hampshire beggars used to produce artificial sores by infecting a wound with this plant (Read).

Like other red-berried plants, it was dedicated to Thor in Scandinavia, and thus is a lightning plant. The plant was an ingredient in a Norwegian remedy against the supernatural, the others being Woody Nightshade and *Dactylorchis maculata*, with "tree sap" (Kvideland & Sehmsdorf). Apart from its use as a dyeplant (Pratt), its fame rested on widespread and dangerous medicinal employment, particularly the bark, which could be collected in March and Aprtil, and dried in the shade. It is violent in action, and always dangerous. Even external application (to warts) has to be carried out with great care (Flück). Presumably the violence of its action is the reason why it appears in a list of plants used as abortifacients in Dutch folk medicine (van Andel). According to Dodonaeus, it is so strong that it had only to be applied on the belly to kill the child. On the Continent, the bark, soaked in vinegar and water, used to be applied with a bandage as a blistering agent (Fluckiger & Hanbury). It has even been used for snakebite (Grieve. 1931).

In homeopathy, a tincture prepared from the fresh bark is recommended for dermatitis (Schauenberg & Paris), and a tincture of the berries was occasionally given in German practice for relieving neuralgia (Le Strange), while in Lincolnshire, particularly in a village called Willoughton, the berries were swallowed like pills as a remedy for piles! (Rudkin). Gerard noted a quite original use for them – "… if a drunkard doe eat one graine or berry of this plant, he cannot be allured to drinke any drinke at that time, such will be the heate in his mouth and choking in the throat". There is one other medicinal use to notice, and this time it is the root that is used – for toothache (Pratt).

LADY'S BEDSTRAW

(*Galium verum*) The medieval legend was that the Virgin Mary lay on a bed of bracken and bedstraw. Bracken refused to acknowledge the child, and lost its flowers. This bedstraw, on the other hand, welcomed the child, and, blossoming at that moment, found its flowers changed from white to gold. Other plants with the same claim are Woodruff and Groundsel. Another version of the legend says that it was the only plant in the stable that the donkey did not eat (Grigson. 1955). Sudeten women would put it in their beds to make childbirth easier and safer, a belief that follows the early legend naturally. And since women who have just had a child are susceptible to attack from demons, they would not go out unless they had some Lady's Bedstraw with them in their shoes (Grigson. 1955). There is some evidence that it was used in love philtres (Dyer. 1889). See Gerard: "the root … drunke in wine stirreth up bodily lust, and the flowers smelled unto works the same effect".

It is a sign of rain if Lady's Bedstraw becomes inflated and gives out a strong smell (Inwards). There are a number of Cheese-rennet, or curdling names given to this plant, reminders that it served this purpose in the past. Gerard, for instance: "the people of Cheshire, especially about Namptwich, where the best cheese is made, doe use it in their Rennet, esteeming greatly for that Cheese above other made without it". Double Gloucester cheese has nettle mixed with the Lady's Bedstraw (M Baker. 1980). Dried plants could be used for lining wardrobes, to deter moths (Vickery. 1995). The root was used on South Uist, and on Jura, for dyeing wool orange red (Shaw; Pennant. 1772). The tops, with alum, give a yellow dye (SM Robertson), and the same preparation will serve as a hair dye (Leyel. 1937).

Lady's Bedstraw was commonly used as a styptic (Grigson. 1955). Culpeper recommended it (as bruised flowers) to put up the nostrils to stop a nosebleed. Gerard, too mentioned the property: "the floures … is used in ointments against burnings, and it stauncheth blood". Externally, the infusion is used as an application to wounds and skin eruptions (Flück). This infusion is still a popular remedy in gravel, stone and urinary diseases (Grieve. 1931).

LADY'S SMOCK

(*Cardamine pratensis*) Any plant that has "Lady" in its vernacular name must be associated in some way with the Virgin Mary; indeed, this is dedicated to her, for it appears in bloom round about Lady Day (25 March) (Inwards). Nevertheless, it has distinct connections with the fairies. For this reason, it was never included in the May garland, nor probably in any other bouquet. In some districts it used to be said that if a few sprigs of it got into the May garland by mistake, it was not enough to pull them out. The whole lot had to be taken to pieces and remade (Hole. 1976). The exception is Oxford, the one place that allowed them in the garland (Hole. 1975).

Is this the reason for the French superstition recorded by Sebillot, that it was the favourite flower of snakes? Mothers warned children not to touch them, for fear of being bitten by a snake some time in the coming year. They are certainly unlucky flowers. Oxford children may put them in their May garlands, but they would not be allowed to bring them indoors (*Oxfordshire and District Folkore Society. Annual Record; 1951*). If the identification is correct, it seems that Northamptonshire children used to pick the leaves one by one for a divination of the 'Rich man poor man' type. The last leaf would indicate the condition of their future partners (Sternberg).

The young leaves have proved a useful anti-scorbutic in their time. Hill. 1754 thought the juice of the fresh leaves "an excellent diuretic, and … good for the gravel". They have been used for hysteria, and epilepsy, too (Hulme). Thornton rather ambitiously reported that "St Vitus's dance … has yielded to these flowers…" In Russian folk medicine, it is sometimes combined with an infusion of haws for angina pectoris remedies (Kourennoff), but it is the haws that is the important element in this case. In the Highlands, it was reckoned good for reducing fevers (Beith).

A smock, from the Middle Ages to the 18th century, was a woman's undergarment, usually linen, worn next to the body. The word was replaced in the 18th century by 'shift', because of the innuendo 'smock' had attracted to itself. By the next century, 'shift' had lost its refinement, and 'chemise' took its place (Buck). 'Smock' had come to have coarse associations quite early on, probably in the 17th century, and 'smick', which was the same as smock, gave the word to smicker, which means "to have amorous looks and purpose" (Grigson. 1955). Smick-smock as a name for this plant, is recorded quite widely in the south of England, and there is even Smell-smock, too (Grigson. 1955, Morley). Nevertheless, this is still the Virgin's smock that is being celebrated. Smock becomes 'flock', or 'cloak', as Lady's Flock and Lady's Cloak. 'Flock', though, is very descriptive. The word means tufts of wool, and Lady's Smock can give this appearance when seen at a distance.

The other very common name for this plant is Cuckoo-flower, very widespread, because it is in flower when the cuckoos are about. It can appear quite simply as Cuckoo (Grigson. 1955), but in Wiltshire it is sometimes known as Water Cuckoo, or Wet Cuckoo (Dartnell & Goddard), a comment on its favoured habitat (Dry Cuckoo is Meadow Saxifrage, *Saxifraga granulata*). There are more possibilities – Cuckoo-bread, for instance, from Devon and Somerset (Friend. 1882), or Cuckoo-bud (Miller), which Rydén says may have been Shakespeare's invention 'Cuckoo' to 'cuck-old' is easily reached, and, it is suggested, 'bud' has the sense of 'horn', which was supposed to grow on a cuckold's head. It is blodau'r gog in Welsh (Hardy), and it is Glauchsblume, or Kukkuksblume in Germany (Grimm), where in some parts they claim that it grows abundantly where the earth is full of minerals. The point is that the cuckoo, by its call, is also thought to perform the same function, for it will show the where-abouts of a mine (Buckland). Cuckoo-spit is another name for Lady's Smock. Cuckoo-spit is the froth enveloping a pale green insect found on the flowers, and the name is sometimes transferred to the plants themselves. Few North country children would pick these flowers, for it was unlucky because the cuckoo had spat on them while flying over (Dyer. 1889).

LADY'S TRESSES

(*Goodyera repens*) More properly, Creeping Lady's Tresses, an American species. The Mohegan Indians applied the mashed leaves to prevent thrush in infants (Weiner). It is also known as Rattlesnake Plantain in America, for the white-veined leaves looked like snakeskin to the early herbalists, who took it as the sign for virtues against snakebite (Cunningham & Côté).

LAMB'S TONGUE PLANTAIN

(*Plantago media*) It will cure blight on fruit trees. Rub a few green leaves on the affected part of the tree, and the cure will be immediate. It is often found growing underneath trees in orchards (Grieve. 1931).

LAMB'S WOOL

A favourite Christmas drink at one time, made by roasting CRABAPPLES and putting them (sweetened) into ale, often mulled. (Dyer. 1883).

Lamium album > WHITE DEADNETTLE

Lamium maculatum > SPOTTED DEADNETTLE

Lamium purpureum > RED DEADNETTLE

LAMMAS

(1 August) The festival of the first fruits of harvest. The word is 'Loaf' mass Day, OE hlaf-maesse. Bread made from the early grain, the first fruits, was offered, as the name of the festival implies. In Welsh, it is Gwyl Awst (August Feast), but in Ireland it was the feast of the sun-god Lugh, or Lleu, best known as Lughnassad, the celebration of which had a definite bearing on the yield of corn, milk, fruit, etc., throughout the land, and the results would be bad if any of the rites were neglected (Spence. 1949).

The first Sunday in August is Lammas Sunday, the first Sunday of Harvest, often called Garland Sunday, or even Garlic Sunday (Hull). Other names for the day in Ireland are more concerned with our main purpose, for this is BILBERRY Sunday, or Blaeberry, Fraughan, Whort, Hurt or Heatherberry, or even Mulberry, Sunday, all names for bilberries, and that includes mulberry, which is what bilberries are called in parts of County Donegal (Mac Neill). This is the day for the ritual picking of bilberries on the mountain tops in Ireland, which would be put into little rush baskets made on the spot. The day was recognised as a legitimate time for courting, as the comment that "many a lad met his wife on Blaeberry Sunday" shows (E E Evans. 1957). It was the custom for the boys to make bracelets of bilberries for the girls; they were worn while they were on top of the hill, but left behind when it was time to go home.

Laportea stimulans > MALAY NETTLE TREE

LARCH

i.e., COMMON, or EUROPEAN, LARCH (*Larix decidua*) After a mare had foaled, it was the custom in parts of Yorkshre to hang the afterbirth on a larch tree, to bring good luck and good health to the foal (Drury. 1985). According to Beard, a sacred larch stood at Nauders, in the Tyrol, until 1859. Whenever it was cut, it was thought to bleed, and it was held that the woodman's axe inflicted an invisible wound on his own body at the same depth as that he had given to the tree, the effects of which would not cease until the bark had closed over the gash in the trunk. It was said that children, especially boys, were brought forth from this tree. Crying or screaming near it was taken to be a serious misdemeanour, and quarrelling or cursing was looked on as an offence that called to heaven for instant punishment. It is said that God created two trees when he created the earth and man: a male, larch, and a female, a fir (Joseph Campbell).

Larix decidua > LARCH

Larix sibirica > SIBERIAN LARCH

LARKSPUR

(*Consolida ambigua*) A poisonous plant, with effects similar to those of Monkshood. English folklore has shown the connection between Mugwort and the Midsummer fires, but according to Grimm (or his translator), Larkspur and Monkshood were used in addition in Germany at the rites: "… whoso looketh into the fire thro' the same, hath never a sore eye all that year; he that would depart home unto his house, casteth this his plant into the fire, saying, 'So depart all mine ill-fortune and be burnt up with this herb'".

It provides two pigments, delphinine for blue, and kaempferol for yellow (Schauenberg & Paris). The tincture of the seeds was used in cosmetic prepara-tions, and also to treat asthma (Lindley). The seeds have also been used to destroy lice and nits in the hair (Grieve. 1931), a usage reported also in Alabama (R B Browne), hence a name given to the plant, Lousewort. The generic name, *Consolida*, shows that it must have been used as a consound at some time (Prior).

Lathraea squamaria > TOOTHWORT

Lathyrus latifolius > EVERLASTING PEA

Lathyrus montanus > BITTERVETCH

LAUREL
(*Laurus nobilis*) see BAY; if *Prunus laurocerasus,* see CHERRY LAUREL

Laurus nobilis > BAY

Lavandula angustifolia > LAVENDER

LAVENDER
i.e., English Lavender (*Lavandula angustifolia*) In Welsh folklore, it is lucky to wear a sprig of lavender blossoms, which had the capability of bewildering witches and evil spirits (Trevelyan). It was scattered over rushes, with rosemary, on Cheshire farmhouse floors on May Day (Hole. 1937), just as, in Spain and Portugal, it was strewn on the floors of churches and houses on feast days, and to make bonfires on St John's Day. In Tuscany, it was used to counter the effect of the evil eye on little cbildren (Dyer. 1889). Carry lavender, and you will have the ability to see ghosts (Boland. 1977). The Welsh said it quickened the wits of dull-minded people, and cleared the brains of poets and preachers (Trevelyan). As with sage and rosemary, lavender is a plant that shows that "the mistress is master" if it flourishes (Briggs. 1974).

There was once a belief that adders lived in lavender, (but see Cogan, *Haven of Health*). "The settting of lavender within the house in floure pots must needes be very wholsome, for it driveth away venomouis wormes, both by strewing and by the savour of it …". He also said that "being drunke in wine it is remedie against poyson" (see Hulme. 1895).

Lavender water "purifies the face" (Trevelyan). Gerard recommended this water "smelt unto, or the temples and forehead bathed therewith, is refreshing to them that have the Catalepsy, a light megrim, and to them that have the falling sicknesse, and that use to swoune much …". A sprig of lavender worn under the hat would drive away a headache (Leyel. 1937).

LAVENDER COTTON
see GROUND CYPRESS (*Santolina chamaecyparissus*)

Lawsonia inermis see HENNA

LAXATIVES
RASPBERRY leaves, in one form or another, constituted a Dorset remedy for constipation (Dacombe). There was an ancient use of VIOLETS as a laxative. Gerard prescribed a syrup intended to

"soften the belly …"; he also claimed that "the leaves … inwardly taken do … make the belly soluble…". The use predates Gerard by a very long time, for we find, in the Anglo-Saxon version of Dioscorides, "for hardness of maw, take blossoms mingled with honey, and soaked in very good wine" (Cockayne). Again, we find this syrup being recommended a hundred years before Gerard's time: "a laxative. Take the juice of Violet or of the flowers, a good quantity of sugar; and mingle them together, and put them in a glass, and stop it; and set it in the sun, and take the (sediment) thereof, and keep it well in a box, and use it first and last" (Dawson).

FLAX seeds, better known as LINSEED, are laxative (Schauenberg & Paris), and have always been used, both as food and as medicine. CASTOR OIL is too well known for further comment. But one of the most extraordinary laxatives in the Middle Ages were the very poisonous seeds of CORN COCKLE. Archae-ologists have found them, some crushed as if in an apothecary's mortar, in cesspits of the 13th and 14th centuries (Platt). Hill was still recommending them in the 18th century! CAMOMILE tea, that great stand-by for almost any ailment, is used as a laxative, too (V G Hatfield. 1994).

A very odd way to tackle the problem comes from Alabama – you had to boil YARROW and thicken it with meal, and then apply it to the stomach (R B Browne). But the best known of all laxatives is the Cascara Sagrada, the bark of the CALIFORNIAN BUCKTHORN (*Rhamnus purshiana*). This sacred bark was so named by the Spanish pioneers, who took notice that the Californian Indians (see Schenk & Gifford; Spier) used the bark infusion as a physic.

Ledum groendlandicum > LABRADOR TEA

Ledum palustre > MARSH ROSEMARY

LEEK
(*Allium porrum*) An important vegetable in ancient times. The OE leac-tun meant a kitchen garden, implying that leeks were in the majority. Similarly, a leac-ward, literally leek keeper, meant a gardener. The Hebrew word for leek literally means "herb", implying it was the herb par excellence (Moldenke). But leeks have always been looked down upon, in spite of an Irish legend that they were created by Saint Patrick (Swahn). Certainly in the Orient it has always been seen as the food of the poor (that is why it is a symbol of humility (Moldenke). Nero, though, was apparently fond of them, and so they were raised to respectability, if only temporarily, for after his death he was referred to as derisively as Porrophagus, leek-eater (Moldenke). The same contempt is shown in English relations with the Welsh on the subject. Everyone knows that the leek is the national emblem of Wales, but the stories that explain its adoption are mainly from the English point of view, and they are

usually scurrilous (see Howells for some examples), mostly to do with a recognition emblem. On the Welsh side, the explanations vary from the very simplistic (St David ate leeks (Hadfield)) to other recognition signals; as for instance the victory of King Cadwallader over the Saxons in AD 640, when St David made the Britons wear leeks in their caps for purposes of identification (Friend).

The leek is a lucky plant to grow in the garden, and when worn it was said to scare away both evil spirits and human enemies. If a fighter wore a leek, it would make him victorious without a wound (Trevelyan), a belief that may well have something to do with the origin myths quoted, and which links this plant with its relative, garlic, for garlic is the greatest protector of them all. It is difficult to imagine the result of throwing leeks into the loving cup, but apparently that was done at Courts Leet in Glamorgan as late as 1850 (Trevelyan). But the Welsh were not completely in love with them, for it is said in old Welsh medical texts that "they produce fearful dreams" (Gerard).

Divinations were, so it is claimed, made with leeks, though the details are usually lacking, though one charm is recorded: the girl had to go out into the garden and uproot a leek with her teeth! Then it had to be put under her pillow, and her lover would appear to her in a dream (Stevens) (or in the flesh if the proper arrangements had been made). It is said that at Plouer, on the Cotes-du Nord, it was the custom to hang a leek from one of the joists of the kitchen ceiling before the fishing fleet set out. If the plant kept alive, this was regarded as a good omen, but if it dried up and died, then it could be taken as certain that a member of the family had died at sea (Anson).

Women who wanted children were told to eat leeks, and not only in Wales, though Evelyn noted that "the Welch, who eat them much, are observed to be very fruitful", and the medieval Welsh medical text known as the Physicians of Myddfai lists among the virtues of leeks that "it is good for women who desire children to eat [leeks]" Perhaps the reason for such claims lies in the Germanic peoples' belief that leeks contribute to "manly vigour" (Wimberley), clearly derived from their upright growth. A proverb from Normandy runs: Femme stérile/ Mangeant poireau/ Son ventre gros/ Devient fertile (Loux).

Folk medicine claimed many uses for them. In some places, the juice was mixed with cream, and used to cure chilblains (Dyer), and in Fifeshire poultices of chopped leeks were often used for whitlows (Rorie). Ulcers were also said to benefit by such a poultice (Physicians of Myddfai). Boils too could be "matured" in the same way. Piles were treated with leeks, too ("take the roots of leek and stamp and fry them with sheeps tallow, and as hot as he may suffer it, bind to the fundament oft …") (T Hill).

Leek juice was often used for whooping cough in past times, or indeed any "old" cough. As Thomas Hill said, "leeke amendeth an old cough and the ulcers of the lungs". It was also used for deafness, sometimes forming part of quite complicated recipes. The Physicians of Myddfai, for example, conjoined the juice of leeks, goats' gall, and honey, mixed in three equal parts, and then put, warm, in the ears and nostrils. An early leechdom for an ear salve required the doctor to pound sinfull, which is a *Sedum* of some kind, latherwort (probably Soapwort), and leek. Put into a glass with vinegar, wring through a cloth, and drip into the ear (Cockayne). The juice was also recommended for headaches (T Hill), and a surprising early usage was as a wound herb. A Middle English medical treatise claimed that leeks with salt "helpes a wounde to close some" (I B Jones), and the Physicians of Myddfai included a prescription "to restrain bleeding from recent wounds". Nosebleeds were also dealt with by using leeks, but not in a way easily forecast. Lupton, in the mid-seventeenth century, ordered the patient to take nine or ten fresh leeks, and to put a thread through the midst of them, "but cut off the tops of the leaves, then hang them round the party's neck that bleeds, so that the leaves be upward to the nose, and the heads of them downwards …".

There are still more ailments that were treated with leeks, one way or another. One is toothache (Henslow), and another for "the deliverance of a dead child" (Dawson). Andrew Boorde, in 1542, recommended them to "provoke a man to make water", but he went on "they make and increase evyll blode". The most astonishing caution comes from the Physicians of Myddfai, who warned that eating raw leeks occasions intoxication!

Lemna minor > DUCKWEED

LEMON VERBENA
(*Lippia citriodora*) Best known for the oil, obtained by distillation of the leaves, for which it is cultivated in the south of France and in Algeria (Whittle & Cook). This oil is collected commercially for the perfume industry. But this plant can be used medicinally, too. The leaves, either fresh or dried, may be used in tisanes for reducing fevers, or as a sedative, and also for indigestion, for which it is widely employed (Macleod).

Lens esculenta > LENTIL

LENT LILY
A common name for DAFFODIL, obviously related to its time of flowering. It has a number of variations – Lenty Lily, Lent Cups, Lents, even Lentils. One related name, Lentcocks, is interesting. One of the more barbarous of Shrovetide games was cock-throwing, or cock-squailing, as it was called in Dorset (Udal). A bird was tied to a stake, and sticks were

thrown at him until he was killed. Lentcocks for daffodils would seem to be an allusion to this "game". Were the flowers decapitated when the real cock-throwing died out?

LENTIL
(*Lens esculenta*) This is one of the oldest legumes, known to have been cultivated in the Bronze Age (Bianchini & Corbetta), and may be a lot older than the Bronze Age, for carbonized seeds, which are the edible part of the plant, have been found in the Near East dating back six to seven millenia BC (Zohary). The seeds are made into a porridge or soup, and are the best source of protein apart from meat (Schery). There are references in the Bible to the Hebrew adashim, and to Arabic adas; both mean lentil (Zohary). Was this the mess of pottage that Esau sold to Jacob?

Application of lentil paste was taken to be a remedy for measles, chickenpox, even to smallpox, or any other skin eruption (Farooqi).

Leonotis leonurus > LION'S TAIL

Leontopodium alpinum > EDELWEISS

LEPROSY
In the second edition of Gerard, there is "Bauhine saith that he heard the use of these (POTATO) roots was forbidden in Burgundy (where they call them Indian artichokes) for that they were persuaded the too frequent use of them caused leprosie". Bauhine is Gaspar Bauhin, whose Prodromos of 1620 set out the theory. As late as 1761 this prejudice against the potato was still apparent in that area, and its cause was probably to be accounted for by the doctrine of signatures, the skin of a potato reminding someone of the effects of leprosy.

Dioscorides asserted that the leaves of ELM, "beaten small with vinegar, and soe applied are good for the leprosie ..." (Apuleius Madaurensis). Wesley, too, associated elm with a leprosy cure, but it was the bark he prescribed. The leaves of PHYSIC NUT are used, externally, in Chinese medicine, to make an ointment to treat skin diseases, even leprosy (Chinese medicinal herbs of Hong Kong. Vol 3; 1987). In southern India, the dried root of SCARLET LEADWORT (*Plumbago indica*) used to be highly regarded as a leprosy (and syphilis) cure (P A Simpson).

SNAKE'S HEAD LILY (*Fritillaria meleagris*) has the name in some areas of Leopard Lily, which looks very strange, until another name, Lazarus Bell, is considered. It is named after the small bells that lepers were made to carry about with them, so that they could warn the healthy of their approach. Leopard Lily is just a corruption of leper's lily.

LESSER CELANDINE
(*Ranunculus ficaria*) Pliny was responsible for the legend that seeks to account for the name celandine, which was Khelidonion in Greek, from khelidon,

a swallow. The birds used the plant, he says, to restore their sight. The name appears again in Theophrastus, who says that the flower blooms when the swallow wind blows. It is extraordinary how such a fancy as Pliny's hung on. The rhyme that Mrs Hewett quoted from Devonshire retains the idea, though in an extended form:

> Fennel, rose, vervain, celandine and rue,
> Do water make which will the sight renew.

An infusion of the flowers was once used in Norfolk to treat the sore eyes that accompany measles (V G Hatfield. 1994).

Celandine roots have some symbolic significance; they resemble cows' teats, or so it is claimed, so farmers used to hang them in the cowshed in a magical effort to make the cows produce more milk (North).

As names like Pilewort and Figwort show, this plant was a country remedy for piles. Gerard mentioned the use, and it was certainly still being practiced in the Hebrides in Murdoch McNeill's time (1910), and it was a common remedy in Dorset, too (Dacombe). One wonders if it had any effect, for the reason for its use was almost certainly doctrine of signatures – there are small tubers on the roots, and these are the signs, for pile is Latin pila, ball. Not only piles were suggested by these small tubers. Growths on the ears and small lumps in the breast were also treated with it in Scotland. The juice was applied externally, but carefully, of course, for the juice is acrid. In the case of swellings in the breast, the Highland practice was to put celandine roots under the arms (Beith). But like a lot of examples of the doctrine, there does seem to be some curative effect. Fernie certainly thought so, and Conway confirmed its value, especially in the form of an ointment made from the crushed roots. The specific name, *ficaria*, is another indication, for it derived from the plant's curative value in the ficus, which means fig, and by extension, piles (hence Figwort, a name not confined to this plant). So confident were they in Holland as to celandine's efficacy against piles that people there actually wore the roots as amulets against the condition (North), and gypsies say that carrying a sprig or two in the pocket will cure it (Quelch). Logan talks about the Irish use of wild buttercup (presumably he means this) for piles, either by grinding up the roots, boiling them with lard and then making an ointment, or by boiling the leaves and drinking the water.

A few other medicinal uses obviously rely on the same virtues. A manuscript of somewhere around 1680, from Lincolnshire, advised that "for a teter or ringe worme, stampe celandine and apply it to the (griefe?) and it will quickly cure you". A leechdom in use in Anglo-Saxon times for toothache prescribed, in Cockayne's translation, "nether part of raven's foot boiled in wine or vinegar. Drink as hot as possible".

If we really are talking of lesser celandine, for there seems no reason why raven's foot should be applied just to this (after all, Crowfoot is a name given to all the buttercups), then the efficacy must have relied on the counter-irritant principle. (see BUTTERCUP)

LESSER EVENING PRIMROSE

(*Oenothera biennis*) The tap roots have been used as a vegetable, boiled, which makes them quite nutritious, but they were little used after the introduction of the potato (C P Johnson). The taste is not unlike parsnips (Loewenfeld), or even salsify, so it is claimed (Kearney). There are a few medicinal uses. The American Indians, or at any rate the Ojibwe, used to soak the whole plant in warm water to make a poultice that would heal bruises (H H Smith. 1945). But there are recognized herbal remedies involving the bark and leaves, which are known to be sedative and astringent, so they have been used for gastro-intestinal disorders in particular, and also for asthma and whooping cough (Grieve. 1931). More recently, the seeds have been successfully used to treat eczema (T Walker). Oil of Evening Primrose helps menopausal changes and pre-menstrual problems, and it has been recommended to help arthritis, and even to slow down changes in multiple sclerosis (M Evans).

LESSER YELLOW TREFOIL

(*Trifolium dubium*) Is this the true shamrock? Very likely, it seems (Britten). A recent (1988) survey showed that more people (46% of those replying) were convinced that this was the true shamrock rather than other clovers (Nelson). One difficulty about this is the belief that the shamrock is peculiar to Ireland and will not grow out of that country. But this plant is common all over Europe. Another Irish belief is that shamrock does not flower. This is understandable, given that the demand for it is in March, and no clover will be in flower then. (see also SHAMROCK)

LETTUCE

Lactuca sativa describes the cultivated forms, of which *var. capitata* is Cabbage Lettuce, or Head Lettuce, because it forms a head, like cabbage, hence the name *capitata*. The other well-known lettuce is *var. longifolia,* Cos lettuce, called after the Greek island of that name. Another name for this kind is Romaine Lettuce. The word lettuce itself is through Old French laiture, from Latin lactuca, itself from the Greek. In that language milk is gala, and the Latin equivalent is lac – hence lactuca, which simply means milk plant (Potter & Sargent). All species contain a milky latex, which in the case of *L. virosa*, Great Prickly Lettuce, is actually gathered for medicinal purposes.

Its name implies that it was known to the ancient Greeks, and it figures in their mythology, too. It was much used as a funeral food, especially in memory of Adonis (or, according to another legend, Phaeon of Lesbos). Adonis had been hidden by Aphrodite beneath a lettuce (!), a boar ate the leaves, and by so doing managed to wound Adonis mortally (Gubernatis. 1878). Another version says that Adonis was struck by the boar after having eaten a lettuce (Gubernatis. 1872), something that probably means that he lost his sexual powers after eating it. (Gerard was saying the same thing a very long time afterwards, and so was Thomas Hill). This belief is an inversion of the original one, for lettuce was thought to have the power of arousing love and of promoting childbearing if eaten by a wife. But by the 19th century we have a saying "O'er much lettuce in the garden will prevent a young wife's bearing" (*Notes and Queries. vol 7; 1853 p152*). Women were wary of lettuce, for it would cause barrenness. The belief probably arose because it was thought that the plant itself was sterile, and was regarded as antaphrodisiac, according to the 10th century Geopontica (Rose). It is recorded that women in Richmond, Surrey, would carefully count the lettuce in the garden, for too many would make them sterile (R L Brown). What the maximum acceptable number was is not revealed. As against this, lettuce is included in the long list of plants quite mistakenly supposed to be aphrodisiacs (Haining). The Romans certainly thought of it as promoting sexual potency (R L Brown), though much later, Gerard said the juice "cooleth and quencheth the naturall seed if it be too much used, but procureth sleepe".

All lettuces have narcotic properties, though cultivation has lessened the effects considerably, but it is still used for making skin lotions useful in sunburn, and roughness of the skin (Grieve. 1931). These narcotic properties have been recognised for a long time: for example, from Bartholomew Anglicus – "For the frenzy – ... the temples and forehead shall be anointed with the juice of lettuce". Gerard, later on, pointed out that "it causeth sleepe" (it was called Sleepwort in Anglo-Saxon times (Ellacombe)). It has been used for liver complaints since early times, but what should we make of an American belief that eating lettuce will prevent smallpox? (H M Hyatt).

Leucanthemum vulgare > OX-EYE DAISY

LEUCORRHOEA
Gerard advised that "the white flowers (of WHITE DEADNETTLE) ... stay the whites" (which sounds suspiciously like doctrine of signatures).

LEUKEMIA
Colchicine, the drug obtained from MEADOW SAFFRON (*Colchicum autumnale*) turned out to be one of the first to bring acute leukemia under control (Thomson. 1976). Other authorities, though, say it is far too toxic to be of use (e.g., Schauenberg & Paris).

LEVANT WORMSEED
(*Artemisia cina*) This is the Biblical wormwood (Clair). As its name proclaims, this provides a medicine for intestinal worms, obtained from the

flower heads (Hutchinon & Melville), minute as they are, and often called seeds. They are made up into tablets for worms, but large doses have been known to be lethal (Le Strange) – to the patient, that is.

LICE
Gerard reported that GREEN HELLEBORE "… is named in the German tongue Lowszkraut, that is Pedicularis, or Lowsie grasse: for it is thought to destroy and kill lice …". LARKSPUR seeds have been used to destroy lice and nits in the hair (Grieve. 1931). Hence the name Lousewort for the plant. MARSH ROSEMARY (*Ledum palustre*) was once used as an anti-parasitic, to treat lice, scabies, etc., (Schauenberg & Paris). See also VERMIN CONTROL

LIGHTNING PLANTS
Yggdrasil, the ASH tree, was sacred to Odin (Graves), and that would be enough to make ash a lightning tree:

> Avoid an ash,
> It courts a flash.

BRACKEN must be included here – it will protect the house from lightning if hung up inside (but if you cut or burn it, it will bring on rain) (Waring). HAWTHORN too is a lightning tree, thought to avert lightning. In many parts of England, hawthorn gathered on Holy Thursday (whether that means Maundy Thursday or Ascension Day is not clear), was used as a protection against lightning (Burne). "The white thorn is never stricken with lightning", Langham said with confidence. There are many folk rhymes to remind one of this, most of them very similar. This particular version was recorded in Fittleworth, Sussex:

> Beware of an oak,
> It draws the stroke;
> Avoid an ash,
> It courts the flash;
> Creep under the thorn,
> It can save you from harm (Opie & Tatem).

In Normandy, they still say that a twig of hawthorn will protect him who carries it (Johnson). On the other hand, and as is often the case, the direct opposite is sometimes found, as with Cornish people, who thought it dangerous to stand under a whitethorn during a thunderstorm (Deane & Shaw). And there is a Welsh belief that the tree itself, or at least one particular tree, will actually cause the storm. The tree is the old thorn at Ffynon Digwg, Caernarvon, and thunder and lightning would result if the tree were cut down (F Jones). This would be a guardian tree of a holy well, so perhaps it was not the tree, but a higher entity, that would answer the sacrilege.

An aspect of the mythology of the HAZEL lies in its association with Thor, for it is a lightning tree, an actual embodiment of the lightning. Hence the common belief that hazels are never struck (Kelly),

and so offer the greatest protection at all times. Christianity adopted the myth in the story that the Holy Family took refuge under a hazel during the flight into Egypt (Dyer). The connection between the lightning tree and the robin, itself associated with fire, is expressed in the west of France by the custom, long since dead, of killing a cock robin on Candlemas Day, and running a hazel wood stick through the body, which was then put by the fire. It would turn by itself, so the belief was (Swainson. 1886). In Lincolnshire, hazel was often used as "palm" on Palm Sunday, and kept green the year round by putting it in water. In the south of the county, these "palms" were preserved for the express purpose of protection from thunder and lightning (Gutch & Peacock).

ELDER, so the story goes, is proof against lightning – it never strikes the tree from which the Cross was made (G E Evans). Elder wreaths used to be hung up in parts of Germany after sunset on Good Friday, as charms against lightning (Dyer). MUGWORT too, ritually gathered on Midsummer Eve, serves as a preventive against evil in general, and that includes lightning (Le Strange). Another protector is HOUSE-LEEK, which "preserves what it grows upon from Fire and Lightning", Culpeper said. John Clare noted that in his native Northamptonshire "no cottage ridge about us is without these as Superstition holds it out as a charm against lightning". It is actually called (in translation) Fire Herb in Irish (Grigson. 1955), and if it grows on the thatch it will preserve the occupants of the house not only from the danger of fire, but from related mishaps such as burns and scalds, so long as it remains untouched (Wilde. 1890). Charlemagne, so it was said, ordered that every dwelling in his empire should have one growing on its roof, to protect it from fire and lightning (Conway). The belief is even older, for to the Romans, this plant was Diopetes, "fallen from Zeus", to protect the house from the lightning he wielded (Grigson. 1955).

OAK was dedicated in ancient times to Zeus/Jupiter, and to Thor in Scandinavian mythology. Earlier, it was consecrated to the thunder god, Perun, in Teutonic mythology. Kelly suggested that the lightning connection occurred because of the red colour of the fresh-cut bark, and that oak was the wood most usually used for kindling need-fires (C M Robertson). Whatever the reason, the superstition was that the oak was more frequently struck by lightning than most trees. It "draws thunder", as was said in Hampshire (Read). In Wales, it was considered dangerous to take shelter under an oak during a thunderstorm, for the lightning penetrated fifty times deeper into them than any other tree (Trevelyan). Hence King Lear's allusion to "oak-cleaving thunderbolts", and Prospero's:

> "To the dread rattling thunder
> Have I given fire, and rifled Jove's stout oak
> With his own bolt".

But the connection with Thor meant that it was believed to give protection to shelterers (except on Thursdays, Thor's own day (Tongue)), even though the tree itself was struck. Indeed, oaks known to have been struck were often visited, so that pieces could be taken away to be attached to buildings for proetction (Wilks). Acorns too were a charm against lightning, and ornamental designs used to be made with them and put in cottage windows (Lovett). According to one school of thought, it was a very bad sign if an oak was struck (Rambosson), and a belief recorded in Hampshire said that the oak actually drove away thunderstorms; it was even thought that the iron in the oak drew the lightning away from the town (Boase). That rarity, oak MISTLETOE, is doubly significant. Mistletoe, in its own right, is an embodiment of the lightning (a Swiss name for it is Donnerbesen, thunder besom); there it is supposed to protect from fire, and in Britain it was often used in the same way as rowan, to protect not only from lightning and fire, but also from witchcraft and the like. BAY is another lightning tree, and a protector from lightning, which was believed powerless to hurt a man standing by this tree (Dyer. 1889), one of the "vulgar errors" listed by Aubrey. 1686. But people have been known to carry branches of bay over their heads in a storm (Waring). "He who carrieth a bay leaf shall never take harm from thunder" (Browne. 1646), and Culpeper added to the belief – "… neither Witch nor Devil, Thunder nor Lightning, will hurt a Man in the Place where a Bay-tree is". As garlic protected the boats from storms and the evil eye, so bay protected them from lightning (Bassett). It was said (by Pliny) that the emperor Tiberius wore a laurel chaplet during thunderstorms for this reason.

HOLLY, by virtue of its scarlet berries, is looked on as a lightning plant, with all the protective power that such a plant always has. In East Anglia, for example, a holly tree growing near a house is regarded as a protection against evil (G E Evans. 1966). Holly hedges surrounding many Fenland cottages were probably planted originally with the same idea in mind (Porter). And being a lightning plant, it must protect from lightning. "Lightening never struck anyone if you were under a holly tree. Lightening never struck a holly tree", was a Devonshire maxim. As far back as Pliny's time there are records of holly being planted near the house for that reason. In Germany, it is a piece of "church holly", i.e., one that has been used for church decoration, that is the lightning charm (Crippen). Another red-berried plant, LADY LAUREL, is a lightning plant in Scandinavian belief, for it was also dedicated to Thor. BEECH must be mentioned too. It was supposed to be proof against lightning (Dyer), or put another way, lightning never strikes it (Sebillot). ROWAN's red berries make it a lightning plant, and rowan is perhaps more than any other the embodiment of the lightning, from which the tree was sprung

(Dyer. 1889). That makes it a prophylactic against lightning (Graves). In Ireland, a twig of rowan would be woven into the thatch, so protecting the house from fire for a year at least (Wilde. 1902), and in Northumberland the house would be secured with a rowan-wood pin (Denham). EDELWEISS is a lightning plant too, gathered as a protective charm against the lightning (Dyer. 1889), and so was MARSH MARIGOLD, which used to act as a charm against lightning during May storms (Baker. 1980). PEONY, too, protected from lightning (Tynan & Maitland), as did RED POPPY; if you had poppies growing on the roof of a building, they would protect it from lightning (Wiltshire). On the other hand, picking them would be likely to cause a thunderstorm. Names like Thunderflower, Thunderbolt, and Lightning Flower testify to the belief. ST JOHN'S WORT is a lightning plant, too. If gathered on St John's Eve, it was kept as a charm against thunder; that is from both France and Germany; also in Holland, where it should be gathered before sunrise (Dyer. 1889), and in Scotland, where it should be burnt in the Midsummer Fires (Banks. 1937).

Ligusticum scoticum > LOVAGE

Ligustrum vulgare > PRIVET

Lilium candidum > MADONNA LILY

Lilium martagon > TURK'S CAP LILY

LILY-OF-THE-VALLEY

(*Convallaria maialis*) Associated in Britain with Whitsuntide, when the churches would be decorated with them (J Addison). But on the Continent, they are earlier blooming apparently, and are the emblems of May Day, muguet de mai in France, and Maiblume, or Maiglöckchen in Germany. It is only fair to add, though, that Gerard knew it as May Lily, and Mayflower Lily is another British name for it (Tynan & Maitland). They are, too, traditionally worn by participants in the Helston Furry Dance (Vickery. 1995), usually held on 8 May. Lilies-of-the-valley are a customary May Day gift in Paris. Large quantitites of them are sold for the purpose (Hole. 1976). Both the flower and the scent ("muguet") are widely advertised as May Day approaches (M Baker. 1977). Gardeners' wisdom has it that lilies-of-the-valley only thrive when Solomon's Seal ("their husbands") are growing nearby (M Baker. 1974). But in Devonshire superstition, it is unlucky to plant a bed of these plants, as the person to do so will be sure to die within the year (*Notes and Queries; 1850*). Is this a relic of the German belief, quoted by Lea, that said that lilies-of-the-valley buried under the threshold of a stall will bewitch the cattle and milk? It does, though, belong to the group of white flowers, like snowdrops and white lilac, that will cause death if brought into the house – and lily-of-the-valley is always unlucky for girls. It is the girl child who will die, so they say in Somerset, if they are brought indoors (Tongue. 1965).

The Sussex folk tale called the Basket of Lilies starts: "There was a woman who loved Lilies-of-the-valley. She'd be always looking for them or sending her little daughter to find a bunch to bring home, so of course the little girl sickened and died, as everyone knew she would …" (Tongue. 1970). This plant was sometimes used as an ingredient in love potions (Coats), for it was said to encourage virtue and faithfulness. As such, it was sometimes included in a bride's bouquet (A W Hatfield).

They can be toxic, though, and are dangereously poisonous if eaten by animals or poultry (Forsyth). The drug is actually convallotoxin,which has been used for treating patients recovering from a stroke (Conway), and it has often been been used as a substitute for digitalin (Lloyd), and so as a heart stimulant, though it is less powerful than digitalin. The dose in Russian folk medicine (and not only in folk medicine, for it was widely prescribed throughout the Soviet Union (Thomson. 1976)), is one tablespoonful of the flowers and leaves to a pint of boiling water, infused for an hour, and a tablespoonful of this taken once or twice a day (Kourennoff). The root was used in Ireland for heart disease (Maloney).

A Wiltshire nerve tonic was a leaf infusion, or a root decoction, taken in small doses night and morning. It was also reckoned to strengthen the memory, and to be a good tonic remedy after a stroke (see above), and to help restore lost speech, and to reduce high blood pressure (Wiltshire), which is interesting, for American domestic medicine recognises that a tea made from the plant is good for heart trouble (H M Hyatt). In Russia, it was used for dropsy (Lloyd), and it was taken with some effect for treating men who had been gassed in World War I (Grieve). It is a wound plant, too – there are records of treating cuts and abrasions by binding a leaf on them (Vickery. 1995).

LIMA BEAN
(*Phaseolus lunatus*) Probably better known as Butter Bean. African-Americans in the southern states of America had a superstition about these bean hulls – they had to be thrown into the road. They could not be burned, for that would inhibit the next crop, neither could they be fed to the cows or pigs, for that would mean the animals would get into the garden and eat the growing plants (Puckett). Split a Lima Bean in half and rub it over a wart. Toss the bean into a well, and the wart will disappear (H M Hyatt) (cf BROAD BEAN).

LIME
i.e., COMMON LIME (*Tilia x vulgaris*) In German mythology, the Elf King lived in the Linden tree (Bayley. 1919), and dwarfs loved to haunt the tree. Heroes fell into enchanted sleep beneath them. In both Hungary and Germany, magical properties were ascribed to the tree. In some villages it was usual to plant one in front of a house to stop witches entering.

It was sacred to Venus among the Greeks, due, it was said, to the heart-shaped leaves. This was transmitted to Christian symbolism; the penance laid on Mary Magdalene by Christ was that "she should have no other food, and sleep on no other bed, save one made of its leaves". "For Magdalene had loved much, and therefore her penance was by means of that which is a symbol of love" (Leland. 1898).

What was evidently a sacred Lime tree, known as the Wonderful Tree, once grew in Ditmarschen, near the bridge at Suderheistede. An old prophecy said that as soon as the Ditmarscheners lost their freedom, the tree would wither. This happened, but it was said that a magpie would one day build its nest in the branches, and hatch five white birds, and then the tree would begin to sprout again, and the country recover its ancient freedom (Thorpe). There was a "family" Lime tree at Cuckfield Hall, in Sussex. It regularly dropped a branch when there was going to be a death in the family (M Baker. 1980).

In Germany, lime flowers were never brought indoors – it gave the girls in the house erotic dreams, so it was said (M Baker. 1977). Lime is a feminine tree in Lithuanian folk belief (oak is the male tree par excellence). The souls of women moved into lindens or firs, the other feminine tree, at death, and women's graves were marked by a linden cross (Gimbutas. 1958).

The bast used by gardeners for tying up plants, and for packing goods, is obtained from the inner bark. In Europe, the use of bark fabrics was established in prehistoric times, particularly the bast of the Lime tree, which at one time grew in large forests. Shoes of plaited bast were still worn in very recent times in eastern Europe, particualrly in the Volga district (Buhler. 1940). The wood, to quote J Taylor. 1812, is "soft, light and smooth; close-grained, and not subject to the worm, and of a spongy texture. It is used for making lasts and tables for shoemakers. It also makes good charcoal for gunpowder". All the Grinling Gibbons carvings are in lime, which is ideal for the task (Ablett). "The flowers afford the best honey for bees, and the gummy sap or juice, when repeatedly boiled and clarified, produces a substance like sugar" (J Taylor. 1812). Lime-flower honey is indeed very good; in fact at one time it cost three or four times as much as ordinary honey (Ablett).

Lime-blossom tea is valued for headaches. It is used a lot in France as "tilleul", a slightly sedative drink (F G Savage). Besides being very pleasant, it is given in Somerset for insomnia (Tongue. 1965). A hot bath with lime-flowers in it, is another insomnia remedy, and it is also good for nervous irritability (Quelch). The tea is "good against giddiness of the head, tremblings of the limbs and all other lighter nervous disorders" (Hill. 1754). The distilled water, according to Evelyn, was regarded as good "against epilepsy, apoplexy, vertigo, trembling of the heart …". Even sitting under a lime tree is reported as improving the condition of epileptics (M Baker. 1980). Gerard had already noted that "the floures are commended by divers … against … the falling sickness…". A more barbaric remedy was noted in 18th century Scotland: for the "falling sickness in children": "take a little black sucking puppy (but for a girl take a bitch whelp), choke it, open it, take out the gall, put it all to the child in the time of the fit, with a little tile-tree flower water, and you shall see him cured as it were by a miracle presently" (Graham) (Tile-tree is Lime, of course, taken directly from the generic name, *Tilia*, which was the Latin name for the tree).

It is also said that the infusion of the flowers is good in the treatment of arteriosclerosis, for it thins the blood, and so improves the circulation (Palaiseul), and so it is useful for hypertension as well (M Evans). That same infusion, used as a lotion, will act as a hair and scalp conditioner (Conway).

Linaria vulgaris > TOADFLAX

LINDEN
A common, though poetical, synonym for LIME. The OE linde, sometimes used for a tree in general (Halliwell. 1881), is the source. Other versions of linden are Lin (F K Robinson), Lind, or Lynd, both Scottish variations (Jamieson). Place names in England include Lindfield, in Sussex, which means lime-tree land, Lindsell, in Essex – huts built round lime-trees (Wilks). Lyndhurst, in the New Forest, is another example (Rackham. 1986).

LINSEED
see FLAX. Lin, or something like it, is a widespread alternative name for flax, used mainly in Scotland and the north of England (see Carr for example). Lint, used at least as early as Turner's time, is the version most often found. The original owes its existence to the generic name *Linum*. In Dorset, a man in the flax trade was called a lin-man, and it is genertally assumed that flax is the name reserved for the plant, and lin for any product of the plant. Nevertheless, the plant is still called Lint, and other versions are Line (Gerard), or Lyne (Fernie). The bird known as the linnet got its name from its partiality for flax seeds.

Linum catharticum > FAIRY FLAX

Linum usitatissimum > FLAX

LION'S TAIL
(*Leonotis leonurus*) Dagga is the South African term for cannabis, but it applies equally to this plant, which has come to many subtropical areas as an ornamental shrub. There is a dark green resinous exudate that can be got from the leaves that is smoked with tobacco. Or the young shoots that are about to flower are pinched out and smoked as a tobacco substitute under the name dagga-dagga. It is a narcotic, producing mild euphoria. It is also used for diseases such as

leprosy, cardiac asthma, epilepsy, and snakebites (Emboden. 1979).

Lippia citriodora > LEMON VERBENA

Lithospermum angustifolium > PUCCOON

Lithospermum officinale > GROMWELL

LIVER DISORDERS
AGRIMONY is the plant most associated with a medicine for liver complaints. Culpeper assured us that "it openeth and cleanseth the liver …", and in fact that is what it was used for in Gaelic folk tradition (Beith). Gerard had already recommended the leaf decoction as "good for them that have naughty livers…", and herbalists still prescribe it as a liver tonic (it is sometimes known as Liverwort).

LOBED CUDWEED
(*Artemisia ludoviciana*) A North American species, useful to the Indians; the seeds are edible, and used to form part of their normal diet (Yanovsky). And it formed the basis of a number of medicines – the leaf tea would be taken for tonsilitis and sore throat. The Meskwaki used the leaves as a poultice to put on old sores (H H Smith. 1928). In the same way, the Comanche used the leaves for insect and spider bites (D E Jones). The Paiute stuffed wads of the fresh plant into the nostrils to stop a nosebleed, and Paiute women took a strong tea to help delivery (Youngken).

Lobelia cardinalis > CARDINAL FLOWER

Lobelia inflata > INDIAN TOBACCO

LOCUST
CAROB pods may have been the "locusts" eaten by St John the Baptist – "now John wore a garment of camel's hair, and a leather girdle round his waist; and his food was locusts and wild honey" (Matthew 3; 4). The pods are known as Locust Beans, and this is the name under which they are imported into Britain (Wit). Locust is Latin locusta, meaning a lobster, hence by inference a locust, and a description of the shape of the pods. Carob is not the only tree bearing the name. *Robinia pseudacacia*, the FALSE ACACIA, pod is Black Locust, while members of the genus *Gleditsia*, and *Hymenaea* too, bear the name (Watts. 2000). *Catalpa bignonoides* is also known as the Locust Bean.

LOCUST BEAN
(*Catalpa bignonioides*) A tree from the south-eastern states of America, whose wood has been prized for, among other things, chess sets, musical instruments, and coffins. African Americans in Louisiana boil the leaves to wrap around eggs to colour them at Easter (Fontenot). A decoction of the fruits has been used as a cough cure (Lindley). Other names for it include Cigar Tree (Harper), Indian Cigar and Smoking Bean (Hyam & Pankhurst) (it is said that the Indians actually smoked the capsules (Perry. 1972).

Lolium perenne > RYE-GRASS

Lolium temulentum > DARNEL

LONDON PLANE

(*Platanus x acerifolia*) It used to be said that it thrived in London because it shed its bark, the argument being that in doing so the tree could "breathe" (Ackermann), a fallacy, of course, for trees do not "breathe" through their bark. Many people believed that the London Plane was an unhealthy tree to live near. It may perhaps be true, for there is a theory that the small spicules which form the fruit swarm in the air surrounding the trees and cause bronchial catarrh and pneumonia, as well as certain throat affections (Brimble). Certainly, children make itching powder from the fruits.

LONDON PRIDE

(*Saxifraga x urbinum*) A hybrid, *S spathularis x S umbrosa*, introduced in the 18th century by Mr London, the royal gardener of the time. So, to make any sense, it ought to be London's Pride. But nobody ever worried about it, and almost from the moment it was introduced it has been a favourite garden plant, especially in cottage gardens. One of the many Devonshire names given to it is Bird's Eye (Friend. 1882), descriptive, of course, as is more obviously the case when the name is given to the speedwells. There was a belief that if you gather Bird's Eye, the birds will come and peck your eyes out, or it will be your mother who suffers the fate. That was certainly the case with the speedwells, but it does not necessarily follow that London Pride protects itself in the same way.

Lonicera periclymenum > HONEYSUCKLE

LOUSEWORT

(*Pedicularis sylvatica*) Not because it can deal with lice, but because, so it was thought, causes them. "It filleth sheep and other cattell that feed in medowes where this groweth, full of lice", according to Gerard, but it is a belief still current.

LOTUS TREE

(*Zizyphus spina-christi*) A small tree from the Near east and eastwards into India. This is a sacred plant, often used to mark boundaries between lands of different villages. Some believe that the hedge surrounding Paradise is made of this. When a Lotus-tree reaches the age of 40 years, it often becomes the abode of some dead saint, so it is therefore a dangerous thing to cut down an old tree (Hanauer).

LOVAGE

(*Ligusticum scoticum*) A reputed aphrodisiac, but only in all probability as a result of misunderstanding the name Lovage, which was Loveache in Middle English, levesche in Old French (modern French is livèche), levisticum in late Latin, and thence to Ligusticum. It appears in Roumanian folklore as a protective plant that could be used as well as wormwood or hedge hyssop to repel evil forces (Beza).

It was used quite extensively in earlier times for medicinal purposes, particularly (in Scotland) to combat scurvy (Grigson. 1955), for the leaves and stalks are edible, "a plant much in use in the western parts [of Scotland and the Isles] as a food" (Pennant). The leaf stalks were blanched once like celery (the leaves taste rather like celery, too), and the young stems were also candied, like Angelica (Rohde). The roots can be candied, too, and the seeds furnish an oil that is used to flavour candy (Sanford).

Lovage is used for a sore throat remedy in Indiana. Cut up the root and fry it in lard, and apply that to the throat as a poultice (Tyler).

LOVE CHARMS

Fenland girls used YARROW as a love charm, by pinning it on the dress, and then taking every opportunity to get as near as possible to young men, in order to declare their love by means of the flowers. If a girl found that the man she was interested in ignored the hint, then she was likely to wait for a full moon, go to a patch of yarrow and walk barefoot among them. She would then shut her eyes, bend down and pick a bunch. If she found next morning that the dew was still on the yarrow, then all was not yet lost – it was a sign that he would soon come courting in earnest. If the flowers were quite dry, on the other hand, she could wait till the next full moon and try again (Porter. 1969), or look elsewhere, of course. VALERIAN, once supposed to be aphrodisiac, was used by Welsh girls. They used to hide a piece of it in their girdle, or inside their bodice, to hold a man's attention (Trevelyan). A Breton charm was for a man to put HOUSELEEK in his pocket, and to make a girl smell it, and it would have her running after him (Sebillot).

There was a belief in ELECAMPANE as a love charm in Guernsey. It had to be gathered on St John's Eve, dried and powdered, and mixed with ambergris. Then it had to be worn next to the heart for nine days, after which the person whose love it was wished to obtain had to be got to swallow some of it (MacCulloch). Quite similar was a native American charm, using RUSSIAN TARRAGON. Winnebago men chewed the root and put it on their clothes. The effect was supposed to be secured by getting to windward of the object of desire and letting the wind waft the odour (Youngken). Such a procedure doubled with a hunting medicine. The same procedure was adopted by young Menomini men, using DUTCHMAN'S BREECHES They would chew the root, breathing out so that the scent would carry to the girl. He would then circle round her, and when she caught the scent, she would follow him wherever he went, even against her will. Another charm was to try and throw the plant at the girl, and to hit her with it (H H Smith. 1923). HEMLOCK was used in Ireland; ten leaves, dried and powdered, mixed in food or drink (Wilde.

1902). THORN-APPLE (*Datura stramonium*, or virtually any of the genus) seeds, in both the old and new worlds, were administered in various ways as love potions (Safford), and the roots were used, too, according to Haining; he says they were burned at the Sabbats in order to excite (and also to overcome) women for sexual motives. CINQUEFOIL appears in a witch philtre for love or hate, composed of adders, spiders, cinquefoil, the brains of an unbaptised baby, and so on (Summers). That charm, with only cinquefoil, makes an unlikely appearance with the Pennsylvania Germans. "To gain the admiration of girls, carry cinquefoil in your pocket" (Fogel).

At one time, CARAWAY was an essential ingredient in love philtres (so was CHICORY seed), because it was supposed to induce constancy (Clair). It was thought, too, to confer the gift of retention (including retention of husbands – even a few seeds in his pocket would prevent the theft of a husband (Baker)). CORIANDER is another spice that had this reputation. Albertus Magnus (*de virtutibus herbarum*) includes it among the ingredients of a love potion, but then coriander was one of the many plants that were supposed to be aphrodisiac (Haining). If the identification is correct, HEATH SPOTTED ORCHID was used in a gypsy love philtre. The roots were dried and crushed, and then the girl mixed this with her menses, and the result introduced somehow into the man's food (Leland. 1891). Another doubtful identification is WATER GERMANDER, involved in a French belief. It was said that if a woman wanted to make a man love her, she had to put a piece of this plant in his pocket without his knowing it (Sebillot). One of the odder examples of a love charm concerns BLADDER CAMPION. The inflated (bladder) calyx snaps when suddenly compressed. That fact was used as a charm, according to Coles. The degree of success depended on the loudness of the pop.

The American GINSENG was used by some of the Indian peoples as a love charm. For instance, Meskwaki women wanting a mate made their potion from the roots, mixed with gelatine and snake meat. The Pawnees combined it with other plants, such as wild columbine, cardinal flower and carrot-leaved parsley in their version of the love charm (Weiner). The Ojibwe apparently used FROG ORCHID in a love charm, but no details were given (Yarnell). The American plant BLOODROOT formed another Indian love charm. A man would rub some of the root on the palm of his hand, and then contrive to shake hands with the girl he wanted, in the belief that after five or six days she would be willing to marry him (Corlett). African-Americans in the southern states of America would chew CROWN VETCH and rub it on their palms. That would give a man power over any woman with whom he later

shook hands (Puckett). MANDRAKE roots were the most famous of charms. Amulets made of them were worn by Palestinian men and women (Budge) as fertility charms (see MANDRAKE). AMBOYNA WOOD (*Pterocarpus indicus*) had some reputation as an aphrodisiac at one time, and was certainly used as a man-attracting charm (C J S Thompson. 1897). In Haiti, WHITE BROOMWEED (*Parthenium hysterophorus*) was used. You take seven small bushes, tie them together and throw them into a river, while saying the correct prayer, and this would give you the love of the girl you want (F Huxley).

John the Conqueror Root was a love charm favoured by men. It was a dried root with a sprong or spike growing out of it, an obvious piece of phallic symbolism. It was carried in a little chamois leather bag, or one made of red cloth. What the root was, was kept secret by those who sold the charm, but it is said to have been BOG ST JOHN'S WORT (Valiente). PEARLWORT was said to have the power to attract lovers, for girls in the Scottish Highlands drank the juice, or at least wetted their lips with it, and if they had a piece in their mouth when they were kissed, the man was bound for ever (Grigson. 1955). Pearlwort may seem an insignificant little plant to be used in this way, but this was, according to most observers, the mystical plant Mothan, very important in Gaelic speaking communities.

LUCERNE
(*Medicago sativa*) A fodder plant, but large quantities may be poisonous to livestock; it is known that it may cause jaundice in horses (North), but owing to its vitamin and mineral content, it has a reputation for increasing the speed and stamina of racehorses (it has long been used by Arabs to feed their purebred horses (Schauenberg & Paris), and human athletes. It is a body-builder, and reduces acidity (A W Hatfield). In Morocco, when a newly delivered mother has no milk, she is given a kind of very liquid porridge made of pounded lucerne seeds, for this is supposed to give milk to cows (Legey).

LUCKY BEANS
The seeds of PRECATORY BEAN (*Abrus precatorius*) are the lucky beans in Africa. But *Erythrina abyssinica* is the Lucky Bean Tree, too.

LUCKY HAND,
or St John's Hand, so called because it had to be prepared on St John's Eve. It is made from the roots of MALE FERN, to protect a house from fire. When it was dug up, all but five of the unrolled fronds were cut away, so that what remained looked like a gnarled hand with hooked fingers. It was then smoked and hardened in one of the Midsummer bonfires, and then hidden away in some corner of the house. As long as it stayed there, the house would be safe from fire and a good many other perils (Hole. 1977). The young fronds,

too, were reckoned to be a protection against sorcery (Gordon. 1985).

LUCKY PLANTS

LEEKS were lucky plants to grow in the garden, protectors from both evil spirits and human enemies, especially when worn (Trevelyan). WHITE HEATHER is lucky, too – "happy is the married life of her who wears the white heather at her wedding" (Cheviot). It is a lucky plant in France, and in Belgium, at least in the Liege district, where it is said that the girl who finds it will have good luck in her household management (Sebillot). There used to be a general belef in the good fortune brought by HOPS. Wreaths made of them were once commonly seen over the mantelpiece in country areas, put there for luck, and for ensuring the household's prosperity (Opie & Tatem). RAMSONS cloves used to be planted in Ireland on thatch over the door, for good luck (Opie & Tatem) in general, but in particular to ward off fairy influences (Mooney). HOUSELEEK, too, in the thatch, is a lucky plant, for it not only protects from fire and lightning, but brings good luck and long life to the occupants of the house. The negative aspect is shown in a Sussex saying that bad luck would arrive if the houseleek was removed from the roof (Latham). SAGE is a lucky plant, and FLAX is too. In parts of Germany, when a young woman gets married, she puts flax in her shoes as a charm against poverty (Dyer. 1889). In Scotland, CLUBMOSS is something that brings luck to a house. As long as a piece of it is in the house, bad luck cannot enter (Gregor. 1888). MYRTLE is a plant to be proud of. It was said in Somerset that when planting it, one should "spread out the tail of one's gown", and look proud, or it would not flourish (Tongue). In the same area, it is said to be one of the luckiest plants to have in a window box, although it will not grow there unless planted by a good woman (Baker. 1980). Again, one has to be proud of it, and water it each morning. The luck in having a flourishing bush of myrtle is often expressed, and not only in this country, for a Greek rhyme translates:

> Who passeth by a myrtle bush, and pluscketh not a twig,
> O may he not enjoy his youth, although he's tall and big (Argenti & Rose).

CARNATIONS are lucky plants. They often figured as such in oriental carpet symbolism (Bouisson), and the good luck extended to protection from witchcraft in Italian folklore. On St John's Eve, when witches were specially active, all you had to do was to give them a flower. For any witch had to stop and count the petals, and long before she had done that, you were well out of reach. Finding a green-topped RUSH is just as lucky as finding a four-leaved clover:

> With a four-leaved clover, a double-leaved ash, and a green-topped Seave,
> You may go before the queen's daughter without asking leave (Burton).

"Seave" is a Yorkshire name for a rush (Hartley & Ingilby). DAFFODILS are difficult to categorize. Granted, the first daffodil is lucky. If you find that, then you will have more gold than silver that year (Trevelyan). But many aspects of daffodil lore place it in the distinctly unlucky category. (see UNLUCKY PLANTS)

In Ireland, GOAT WILLOW, or SALLOW is taken to be a lucky tree. It is a good thing to take a sallow rod with you on a journey, and in times past it was believed that the butter would be bound to come if a peeled rod were put round the churn (Grigson. 1955), just as they believed that driving cows with a "sally" rod would ensure a good supply of milk (O'Farrell).

CHICKWEED is a lucky plant, at least to the people of the Fen country of England, where it used to be grown in pots, to bring good luck to the house (Porter. 1969).

The UPRIGHT ST JOHN'S WORT is known as Luck-herb in the Isle of Man (Moore, Morrison & Goodwin), and recognized as a bringer of good luck. It was especially prized when found in the flocks' fold, for this would augur peace and prosperity to the herds throughout the year (Carmichael).

LUMBAGO

NETTLE stings are still believed in some places to cure lumbago (Rollinson), just as they are used for sciatica, and, much more widespread, rheumatism. GROUNDSEL for lumbago is an ancient usage, for it is prescribed in the Anglo-Saxon version of Apuleius "for sore of loins" (lumbago, that is), in Cockayne's translation. FOXGLOVE leaves, applied externally, were used for lumbago in Irish tradition (Ó Súilleabháin), and a decoction of CELERY seed is still being taken for lumbago and rheumatism (Newman & Wilson). In Alabama, a tea made from HORSERADISH root is taken for "a weak back" (R B Browne). Does this mean lumbago? The root was certainly used in Britain for the complaint. Fens people grated and mixed it with boiling water, and this would be immediately applied to the patient's back on going to bed. The resultant blister was treated the next day by removing the plaster, baking it in the oven until it was powdery, then mixing it with flour, so that the whole thing was dusted over the blister (Porter. 1958).

A practice recorded in Dundee claims that FLAX, in the form of a hank of linen yarn worn round the loins was a certain cure for lumbago (Fernie).

Carrying a NUTMEG around in the pocket, mainly as a charm for rheumatism, but also to cure a backache was a Lincolnshire belief (Rudkin) (or more specifi-cally to cure lumbago (Opie & Tatem)).

The Zezuru, a central African people, used the roots of MIMOSA THORN (*Acacia karroo*) to cure lum-bago. They burned them to charcoal, which was then finely powdered. Small cuts would be made in the flesh over the painful area and this powder rubbed in. (Palgrave & Palgrave).

LUNGWORT

(*Pulmonaria officinalis*) The name comes from the spotted leaves, the signature of the lungs, it was felt, so by that doctrine, they were used for diseased lungs, further boosted by the generic name *Pulmonaria*. Actually, it has been claimed that they are of some value (Brownlow), and they are still used in infusion for lung infections and respiratory disorders (Schauenberg & Paris). A leechdom for lung disease was actually used in Anglo-Saxon times (Cockayne), and in the mid-eighteenth century, Hill was still prescribing the leaf decoction for "coughs, shortness of breath, and all disorders of the lungs".

The legend that accounts for these spotted, or blotched, leaves, is that during the flight into Egypt, some of the Virgin's milk fell on the leaves while she was nursing the infant Jesus, causing the white blotches on them, hence names like Virgin Mary's Milkdrops (Macmillan), and Spotted Mary (Grigson. 1955), among others. Another version of the legend tells that it was her tears that spotted the leaves, for the plant was growing on Calvary, at the foot of the Cross. That, incidentally, is why it is unlucky to dig it up from the garden (Britten & Holland). The relevant names are Lady Mary's Tears, from Dorset, Virgin Mary's Tears (Macmillan). Double, or even treble, names for flowers are often references to two-coloured, or changing coloured, flowers, and Lungwort is one of them, so there are names like Adam-and-Eve, or Joseph-and-Mary for this plant, even Faith, Hope and Charity, from Dorset (Udal). Twelve Apostles is an old name from Somerset (Tongue. 1965), and there is a folk song with this title:

> The Twelve Apostles in the garden plot do grow,
> Some be blue, some be red and others white as snow.
> They cure the ill name of every man, whatever ill it be,
> But Judas he was hanged on an elder tree.

LUSMORE

is the Irish name for the FOXGLOVE, literally the 'great herb'

Lychnis flos-cuculi > RAGGED ROBIN

Lycopersicon esculentum > TOMATO

Lycopodium inundatum > CLUBMOSS

M

MABINOGION

Early medieval Welsh tales from the Red Book of Hergest, translated into English by Lady Charlotte Guest in the 19th century. Any aspirant to bardic rank was called a Mabinog, and the traditional lore he had to acquire came to be known as Mabinogi, of which Mabinogion is the plural, used by Lady Charlotte as the general title of the twelve tales she published. In the story of Llew Llaw Gyffes, Gwydion and Math made Blodeuwedd from "the flowers of the OAK, the flowers of the BROOM, and the flowers of the MEADOWSWEET". Blodeuwedd means "face flower".

Maclura pomifera > OSAGE ORANGE

MADAGASCAR PERIWINKLE

(*Catheranthus roseus*) This is an important plant, which has been used in cancer research, particularly with regard to leukemia in children. But apart from that, it brings good luck to a house in Haiti, where it is used for hypertension (F Huxley), as it also is in Chinese medicine (*Chinese medicinal herbs of Hong Kong. Vol 3*). It was noted during medical research that a side effect of its use was euphoria and hallucinations. When this became generally known, there was an outbreak of Catheranthus smoking in Miami, where it grows like a weed. But the side effects of smoking it are pretty severe (Emboden. 1979).

MADONNA LILY

(*Lilium candidum*) Probably the oldest domesticated flower, known to have been grown in Crete from about 3000BC (Woodcock & Stearn). An old name for this lily is Juno's Rose. The legend is that Jupiter, to make his infant son Hercules immortal, put him to the breast of Juno. The drops of milk that fell to the ground became white lilies, while those that went into the sky became the Milky Way. But this lily is better known, as the common implies, for its association with the Virgin Mary. The Buckinghamshire name Lady Lily is relevant; in fact any plant name referring to 'Lady' or 'Lady's' must always show there was a connection with the Virgin. But it is this lily that is the Virgin's plant par excellence, and her emblem, always shown in pictures of the Annunciation. Indeed, the lily of sacred art is always the Madonna Lily, and after the twelfth century it is always used as the symbol of purity, associated with the Virgin. Later, after the fourteenth century, it is very occasionally shown in the hand of the infant Christ, again as the symbol of purity (Haig). It is also the symbol of chastity (Haig) (it will only grow for "a good woman" (M Baker. 1977)), beauty (Zohary), and celestial bliss (Woodcock & Stearn), for to early medieval artists and theologians this was the flower of heaven. It is

used also as an emblem of St Catherine of Siena, and is even called St Catherine's Lily occasionally, but another dedication is to St Dominic. St Anthony of Padua, when not with the infant Christ in his arms, invariably has a lily (Haig). And that is not all, for it seems it was the emblem of St John the Baptist and of St Joseph (Woodcock & Stearn). Long before this, it seems to have been sacred to the Minoan goddess, for a lily appears at her feet when she is enthroned (Willetts).

With such an impressive list of dedications, it comes as no surprise that this lily has been used against evil influences (and as an antidote to love philtres (Napier)), and these beliefs are carried on to ordinary superstitions. Madonna lilies are often seen in old gardens – in Wiltshire it was said they were there to keep ghosts away (Wiltshire). If a man treads on a lily, he will crush the purity of the womenfolk of the house (Radford). From the graves of people unjustly executed, white lilies are said to spring as a token of the person's innocence (Grimm), and from the grave of a virgin, three lilies grow, which no-one but her lover may gather (Dyer). Because white lilies have this association with the commemoration of the dead, they are generally unlucky plants to have in the house (Vickery. 1985), but of course they are often used at funerals (Drury. 1994). Nevertheless, to dream of lilies means joy (Mackay), happiness and prosperity (Raphael).

There are one or two strange beliefs connected with this lily. One of them, recorded in Dorset, claimed to foretell the price of wheat by studying the lily. People counted the number of blossoms on the majority of spikes, each blossom representing a shilling a bushel (Udal). The other interpretation suggests that the more flowers, the cheaper the corn (M Baker. 1977). The other belief, even stranger, is old and concerns the investigation of the sex of an unborn child. Take a lily and a rose to the pregnant woman. If she chooses the lily, it will be a boy; if the rose, a girl (Cockayne).

The bulbs are edible, though very bitter, and are cooked and eaten in parts of Asia, Japan, for instance (Le Strange). They are used medicinally, too; pulped, they are used as a poultice, for boils, carbuncles, etc., (Gutch & Peacock; Maloney). There is a long list of historical references to this type of medicament. Hill, for example, declared that such a poultice is "excellent to apply to swellings". Boiled with milk and water, it was used as "an emollient cataplasm to broken breasts" (Thornton) (cracked nipples, did he mean?). Gerard, of course, had a number of examples in which lily roots were declared to be efficacious, and recorded too far more ambitious claims for it, such as it "expelleth the poison of the pestilence and causeth it to break forth in blisters in the outward part of the skin", and took note of "a learned gentle-man … who … hath cured many of the dropsie with the juice thereof tempered with Barley meale, and baked in cakes, and so eaten ordinarily for some month or six weeks together with meat, but no other bread during that time".

The petals may be used, too. Macerated in alcohol, they make a good antiseptic, good for sores and burns (Palaiseul). To make what used to be known as "Brandy leaves" in Sussex, petals, not leaves, were used (Parish). They were steeped in brandy, and bound to abscesses, sores and ulcers. Exactly the same process was in use in Dutch folk medicine, applied there to sore or inflamed breasts (Van Andel). In Dorset, they used the petals for gatherings and whitlows, by scalding the petal and simply applying it. There were more complicated methods, like taking forty petals and steeping them in a quarter pint of brandy before using (Dacombe). The simpler method was used in Hampshire for cuts and bruises, but the practice occurred over a much larger area than this might suggest (Vickery. 1995), and it was known in Ireland, too (Maloney). In fact it is still used by herbalists for similar purposes (Palaeseul), and if using it for burns and scaldings, it was said to leave no scar (Wiltshire). A thirteenth century monk prescribed the lily as a sovereign remedy for burns, for "it is a figure of the Madonna, who also cures burns, that is, the vices or burns of the soul" (Haig). Traditionally, this lily is a cure for corns, and it is suggested that Roman legions planted it round their camps – it certainly grows apparently wild in all the countries that were in the Roman empire (Coats. 1956). It has travelled to South America, too, for it was used as a fumigant and as an ingredient in ritual baths, in Brazilian healing ceremonies (P V A Williams).

MAHOGANY
see AFRICAN MAHOGANY (*Khaya ivorensis*), or RED MAHOGANY (*Khaya nyassica*)

MAHUA
(*Bassia latifolia*) An Indian tree, known also as Butter Tree (Coon). When the flowers fall to the ground in April and May, they are eaten by the indigenous people. In anticipation, they may burn the ground under the trees, to make it easier to pick them up. Some they eat fresh, others they dry, boil and ferment, and, so it seems, now distil in a simple still consisting of two pots and a bamboo tube. The Gonds, or Konds, also distil a very strong liquor, "something resembling Irish whisky" (Chopra, Badhwar & Ghosh) from the flowers, a drink important enough to figure in their mythology (Fürer-Haimendorff). But the tree has an importance for them unrelated to the drink, for this is the Kor tree, by the side of which funerary rites are performed. It is the tree of the dead, and as such the rites there are the final ones in mortuary ceremonies (Fürer-Haimendorff). The Gonds hung the dead bodies of their relatives on a branch of this tree before burying them (Upadhyaya).

MAIDENHAIR FERN
(*Adiantum capillus-veneris*) *Adiantum* is from a Greek word meaning unmoistened, because the fern has the property of repelling moisture, a peculiarity that was attributed to the hair of Venus (*capillus-veneris*), who when she rose from the sea came out with dry hair. So, ever since these legends arose, it has been used in hair lotions, and particularly in lotions to prevent the hair going out of curl on damp days. The doctrine of signatures ensured that it should be used for alopecia; it is the ashes of the fern, mixed with olive oil and vinegar, that are used (Leyel. 1937). It was used too for lung complaints, like coughs and breathing difficulties, and it was also recommended for jaundice and swollen joints (Addison. 1985).

In Brazilian belief, maidenhair fern will wilt if looked at by by the victim of the evil eye (P V A Williams).

MAIDENS' GARLANDS
It was the custom in parts of England for a young girl to carry a wreath of white roses before the coffin of a virgin. The wreath would be hung in church after the funeral, above the seat that she had used during her life, till the blooms faded. But if the wreath was made with artificial flowers, when it is known as a "maiden's garland", it could be kept in church for a long time. The church at Abbot's Ann, in Hampshire, has its walls hung with these "maidens' garlands" of paper or linen roses; the earliest of them dates from 1716 (Mayhew).

MAIZE
(*Zea mays*) The name Corn is used more frequently than Maize in America (A W Smith), often varied to Indian Corn, or even to Indian Wheat (Britten & Holland) in England. It has also been dubbed Welsh Corn, Asiatic Corn or Turkish Corn (Turner, in 1548, also called it Turkish Millet). "Asiatic", and "Turkish", because the early herbalists of the 16th century believed the plant had been brought by the Turks from Asia. The Turks invaded Europe about this time, and brought many new plants into the west. Anything unusual was labelled "Turkish", or perhaps it was confused with buckwheat, which was at one time specified as turcicum for some reason (Bianchini & Corbetta). "Welsh", of course, must simply mean "foreign" (OE walch, or something like it).

A supremely important plant like maize was bound to attract many origin myths, often, as with the other cereal grasses, with an incident of human or divine sacrifice involved. The idea was that one death, that of the god itself, would provide life to countless thousands by the gift of agriculture. Sometimes the death would be not of a divine personage, but of an often unnamed hero or precursor of the human race. Such a myth is this Menomini Indian one: an old man had corn, which he kept hidden from mankind. It took a young boy, his nephew, to kill the old man and by so doing release the corn for the benefit of all (A Skinner). An example of a different type of origin myth is this one from Mexico: a childless woman, who went to fetch water, saw the reflection of an egg on the cliff above. Her husband fetched it, and in seven days it hatched a small child with golden hair, which was soft and silky, like maize. The child grew quickly and was well developed after seven days. But people teased him because he was "only a little egg taken out of the water". Venetia Newall described this as "an inversion of the idea that the life-giving water causes the grain to grow, so that there must be its origin". Naturally, there would be some ritual, leading to superstition, in the planting of the grain. You should plant maize with a full stomach, Malayan peasants say, and use a thick dibber, as this will swell the maize ear (Skeat) (and see Hatt for further examples of the origin myth).

Maize is apparently toxic to cattle (Kingsbury. 1964), and a narcotic effect is produced when Peruvian Indians inhale the smoke made by burning the female styles. The effect is intense mental excitement, close to delirium (Schauenberg & Paris). The Tewa Indians used maize in a remedy for glandular enlargements in the neck. An ear of corn was laid on the warm hearth near the fire, and the patient put his foot on it and rubbed it to and fro. In the course of two or three days, so it was said, the swellings would subside (Youngken), surely an example of searching for an exterior cause of a natural cure. In Trinidad folk medicine, a tea made from the husks is taken for amenorrhea (Laguerre), and African-Americans made "corn shuck tea" to treat their influenza and colds (Fontenot).

Derbyshire well-dressing displays (see under WELLS) used the grain, called Hen Peas there, often dyed, in the pictures produced for the displays (Porteous).

MALABAR NUT
(*Adhatoda vasica*) An Indian plant, but long cultivated in the tropics, and much used as a cough reliever and dilator of the bronchial tubes. A synthetic derivative of the active principles was put on the market under the name of bromhexine (Thomson. 1976). The plant is said, too, to be insecticidal, and that it has antiseptic properties (L M Perry).

MALARIA
(see also AGUE) In Africa, it is common to see hedges of NEEM TREE (*Melia indica*) grown close to houses, because of its reputation as a cure for malaria (Sofowora). In the Balkans, it was dealt with by steeping SAGE leaves and stems in brandy, and then straining it off (Kemp). OPIUM POPPY, and opium itself, used to be the standard medicine for malaria, or ague, as it was called, in the Fen country of England. Doctors said that it had more effect than quinine (V G Hatfield. 1994). Every Fenland garden had a patch of these poppies growing, and "Poppy

tea", made from the seeds, was a general fever remedy there. BUCKBEAN has been used for the complaint, perhaps doctrine of signatures, for this plant prefers wet, marshy ground. Hill, in the mid-18th century, mentions this use for the dried leaves, and it also crops up in Russian domestic medicine. Four or five tablespoonfuls of the dried herb in a gallon of vodka, kept for two weeks, and one small wineglassful to be taken daily (Kourennoff). Presumably, the fact that Buckbean is a sedative would help.

In Sierra Leone, the leaf of a BAOBAB is used as a prophylactic against the disease (Emboden. 1974), and in central Africa, a decoction of BARWOOD (*Pterocarpus angolensis*) root is used to cure, not only malaria, but also blackwater fever (Palgrave).

MALAY NETTLE TREE
(*Laportea stimulans*) A vicious tree – walking with bare bodies through these nettle-trees in wet weather can be fatal. In spite of this (or perhaps because of it?), the leaves were sometimes strung on a cord and tied to the portal of a Malay home to scare evil spirits away (Gimlette).

MALE FERN
(*Dryopteris filix-mas*) After bracken, this is the best-known fern in Britain, widespread and common in woods and hedgerows. The Lucky Hand, or St John's Hand (so called because it had to be prepared on St John's, or Midsummer, Eve), is made from the root of Male Fern, to protect a house from fire. When it was dug up, all but five of the unrolled fronds were cut away, so that what remained looked like a gnarled hand with hooked fingers. It was then smoked and hardened in one of the Midsummer bonfires, and then hidden away in some corner of the house. As long as it stayed there, the house would be safe from fire and a good many other perils (Hole. 1977). The young fronds, too, were reckoned to be a protection against sorcery (Gordon. 1985).

The root had other, more genuine, uses, for it served as a vermifuge. In the 19th century, oil of fern, made from this plant, could be bought to do the job (C P Johnson). The root was apparently marketed in the 18th century by a Madame Noufleen "as a secret nostrum", for the cure of tapeworm. After he had paid a lot of money to buy it, Louis XV and his physicians discovered that it had been used ever since Galen's time (Paris). But, though used quite a lot in folk medicine, the roots are poisonous, and can even be fatal (Tampion). Perhaps that is why the dried leaves are used in Ireland for the purpose (Maloney). Although the root is occasionally used in tincture in homeopathic medicine, to treat septic wounds, ulcers and varicose veins, the chief use these days is in veterinary practice, for expelling tapeworms (Wickham).

MALLOW
i.e., COMMON MALLOW (*Malva sylvestris*) This has been a plant used to decorate graves (Drury. 1994).

In spite of the extraordinary belief that "in the month of February, eat no pottage made of hocks, for they are venomous" (Dawson. 1934), mallow leaves are perfectly edible, either boiled like spinach, or better, for they are very glutinous, made into soup. In Arab countries, the leaves of an almost identical species are the basis of the famous soup, melokhia (see Mabey. 1972 for receipt).

Mallow's medicinal uses are many and varied, and some are quite ancient. The Anglo-Saxon version of Apuleius lists a number of leechdoms, some of which were still in use hundreds of years afterwards. They include one "for sore of bladder" (Cockayne), for which about a pound of the leaves would be needed. Directions follow, but it is really only mallow tea that is to be taken. One can take note of an "Account of a large Stone voided through the Urinary Passage by a Woman", communicated by D Richard Beard, FRS, Physician at Worcester, and published in the *Philosophical Transactions* of the Society in Jan/Mar 1727. "A poor woman in the parish of Fladbury in this County …was afflicted with the normal symptoms of a Stone in the Kidnerys, and afterwards in the Bladder … Finding relief … by a plentiful use of Mallow tea …".

The Anglo-Saxon Apuleius also prescribed mallow for "sore of side", and also for "sore of sinews". Sprains, perhaps? Certainly sprains in both humans and animals used to be treated in Dorset by steeping the leaves in boiling water and using the result as the foundation for a poultice (Dacombe). Ulcers were treated in Somerset either with an infusion made by pouring boiling water on the leaves, or simply using them as a poultice (Tongue. 1965). Boils, abscesses and the like were dealt with by a compress made from the decoction (Flück), perhaps as a matter of common knowledge – they were certainly treated that way in the Highlands of Scotland (Grant), and in East Anglia (V G Hatfield. 1994), and also as far away as the Greek island of Chios (Argenti & Rose), where mallow was also used for throat, chest and stomach ailments. Eczema can be treated with a compress made from the leaves or flowers.

Gerard was quite enthusiastic about the virtues of mallow leaves. They are good, he says, against bee or wasp stings, even "the stingings of Scorpions", and "if a man be first anointed with the leaves stamped with a little oile, he shal not be stung at all …". One prescription from the Welsh text known as the Physicians of Myddfai, is for piles. It required the mallow to be boiled in wheat ale or in spring water. It is reported from Norfolk that children used to chew the fruits as a laxative (V G Hatfield. 1994). According to Somerset practice, varicose veins could be treated with an ointment made from mallow roots and unsalted lard, a treatment equally useful for sore feet (Tongue. 1965). The mucilage of the plant was, and probably still

is, used in Cornwall to relieve diseases of the chest, intestines and kidneys (Deane & Shaw), while the kidney ailment known as Bright's Disease was treated in Ireland with mallow juice (Maloney).

In both Somerset and Wiltshire, mallow is called Dock, or Round-dock. According to both Akerman (for Wiltshire), and Jennings (for Somerset), it is mallow that is the anti-nettle sting leaf in this area, and not the convenient species of *Rumex* that the rest of England uses.

Malus domestica > APPLE

Malus sylvestris > CRABAPPLE

Malva sylvestris > MALLOW

Malvastrum coromandelianum > BROOMWEED (2)

MAMEY

(*Mammea americana*) This is an evergreen tree with edible fruit from the West Indies and central America. The flowers are used in the making of a liqueur known as eau de Créole (Willis). In Jamaica it was reckoned to be extremely unlucky (fatal, in fact) to plant the seed (*Folk-lore. vol 15; 1904 p 94*, in a series of papers called Folk-lore of the negroes of Jamaica).

Mammea americana > MAMEY

Mandragora officinalis > MANDRAKE

MANDRAKE

(*Mandragora officinalis*) Elliot Smith derived the word from Greek mandros, sleep, and agpora, object or substance, the whole meaning 'sleep-producing substance'. The English word mandrake comes through the Latin from Greek mandragora, though folk etymology tends to associate the 'drake' part of the word with 'dragon' (Latin draco). The association man + dragon then is probably an allusion to the man-like form of the plant's root. Sorcerers of the Middle Ages looked on it as something half-way between the vegetable and human kind (Valiente). The human analogy is repeated in the medieval belief that when pulled, the plant gives a dreadful shriek, which brought madness or even death to anyone who heard it. Shakespeare knew the belief, and quoted it in *Romeo and Juliet, iv. Iii*:

> And shrieks like mandrake torn out of the earth,
> That living mortals, hearing them, run mad.

It actually does make a small sound when pulled from the ground, as irises and most tuberous plants do (H F Clark).

Pliny and Theophrastus said that people pulling up mandrakes first trace circles on the ground with a sword, and cut it while looking towards the west (Borges). This is probably the earlier ritual, but popular tradition involved the use of a dog. The herbalist who wanted the plant would tie the dog to it, and then retire to a safe distance and stop his ears with wax. Then he should keep to windward of the plant, in case the smell overwhelmed him (Frazer. 1917). Then he would throw a piece of meat just out of the dog's reach; the dog tugged, and the plant was pulled up – but it was the dog that died (Clair). According to Grimm, the dog used had to be a black one without a single white hair on its body, and the plant had to be pulled on a Friday, which is the day of Venus, a fact that will have significance later on. The first recorded legend of the mandrake's scream occurs in a biblical story. Reuben, tending his father's ass during the harvest, tethered the animal to the root of a mandrake. When he returned, he found that in struggling to get loose, the ass had managed to uproot the plant, with fatal consequences (H F Clark).

The 1st century AD Jewish historian Flavius Josephus said that in the dungeons of the castle of Machaeras at Baras, there grew a root that was flame coloured and shone like lightning on anyone who attempted to approach. When the intruder drew near, the root retreated and could only be brought to a standstill by the exercise of some rather unpleasant rites. However, he says that if the hunter was skillful enough he could lasso the root and attach the ends of the rope to a dog. Aelian, in the 2nd century AD, described a plant which he called Aglaophotis, because it shone like a star at night. The assumption is that both these plants were mandrakes, which have always been said to have the power of shining in the night. A 13th century Arab herbalist called Ebn Beitan reckoned that there may be some basis in fact for the belief – for some reason, its leaves are attractive to glow-worms, something that may very well explain the Arab name that means 'devil's candles' (H F Clark).

Another set of traditions links mandrakes with the gallows. It grew beneath them (Clair), where a man had been unjustly hanged for theft (Randolph). The plant springs from the urine voided just before death, a belief mentioned in a sentence in Pio Baroja (*La leyenda de Juan de Alzate*): "En Errotazar, hacia el aldo de tierra, hay, colgando, tres ahorcados, y debajo de ellos crecen esplendidamente unas mandragoras con las deyecciones de los cadaveres …". The word 'deyecciones', though, carries the sense of any bodily detritus, rather than just urine, it may be blood (Elliot Smith), or the sex of the root. Dioscorides said that the male mandrake was white and the female black (in English folklore they are known as Mandrake and Womandrake (Frazer. 1917)), for the sex of the root was determined by the sex of the malefactor hanging on the gallows (H F Clark). If not under the gallows, mandrake could be found at cross-roads, where suicides had been buried.

People went to great lengths to get this gallows mandrake, for it was said to become a sort of familiar spirit, speaking in oracles when properly consulted,

and bringing good luck to the household (Elton). Or it was said that being Satan's plant, and constantly watched over by him, the devil himself would appear to do one's bidding, so long as the proper ceremonies were observed when the plant was pulled, and again each time it was consulted. (Dyer. 1889). In 1579 a prosperous Leipzig burgher wrote to his brother in Riga; the latter had complained at the sudden death of his cattle, and the souring of his wine in the cellar. The Leipzig brother enclosed a mandrake which he had got from the town hangman for 65 thalers. He advised his brother to receive the mandrake into his house, to bathe it in warm water, and to sprinkle this water on the cattle and on the thresholds of his house. He had to do this every day for four years, keeping the mandrake wrapped in a silk cloth, and put in with his best clothes (H F Clark). Grimm confirmed the general belief, and said that a mandrake was so much of a valued family possession in Germany that it was customary for it to pass on the death of the father to the youngest son, on condition that he buried a morsel of bread with his father's body in the coffin. Such beliefs would provoke deep suspicion among those in authority – for instance, one of the accusations levelled at Joan of Arc at her trial was that of having a mandragora in her possession (Randolph). The plant was associated in medieval German belief with the Alraun, a devilish spirit in human form, and also the early name of the mandrake there. The Alraun revealed all secret things touching the welfare of the family; it made them rich, removed their enemies, doubled every piece of coin laid under it, and seems to have had an influence on the fertility of marriages (H F Clark), though you must not overwork it, otherwise it would go stale, or even die (Frazer. 1917). This gift of doubling money has an echo in Scandinavian and Icelandic folklore; there it was said that if the owner of a mandrake root were to steal a coin from a widow during a celebration of mass at either Christmas, Easter or Whitsun, and put it under the root, this coin would draw to itself from the pockets of the congregation all those of a similar denomination (H F Clark). There is a somewhat similar French belief. Mandrake, it was said, was to be found at the foot of oaks that bore mistletoe, as deep in the earth as the mistletoe was high on the tree. The man who found it was under an obligation to give it meat or bread every day, a service which must never cease, otherwise the mandrake would kill him. To make up for this, the plant would give back twofold next day what had been given it the day before (H F Clark).

If the Alraun looked after the fertility of German married couples in its care, in Greece the mandrake became a symbol of the golden apples of Aphrodite, or of Aphrodite herself. Elliot Smith discussed this connection at some length. As its name Love-apple (this is what the Hebrew name dudaim means) suggests, it was reckoned to have aphrodisiac as well as narcotic properties. This was the reason it was dedicated to the goddess. So also, it should be gathered on a Friday, the day of the goddess (Bonser). Theophrastus, in the 4th century BC, recommended the root, scraped and soaked in vinegar, as an aphrodisiac (Dyer. 1889). One of the Egyptian names means phallus of the field, and an Arabic name was devil's testicles (Ellis). It was the fruit that was to be used; it had the power, so it was said, to put an end to barrenness, quite independently of sexual intercourse. See Genesis 30; 14-16. Rachel bargained for the mandrake with her sister Leah (by giving up her husband to her). She afterwards bore, though she had previously been barren, her first-born, Joseph (Hartland. 1909). The plant had been found in the fields by Reuben, and given by him to his mother Leah on the night of the conception of Issachar (Bonser). Palestinian women quite often used to bind a piece of the root to their arm (for it could only exert its magical influence if worn in contact with the skin) (G E Smith), to promote their fertility (Emboden. 1979), and figures cut from the roots were also worn as amulets by both men and women (Budge). Henry Mandrell, travelling in Palestine in 1697, was told that it was then customary for women who wanted children, particularly male children, to put mandrake under the bed. The Persians used it for the same purpose – they called it man's root, or love-root, and used it as an amulet (Hartland.1909). Even in the early part of the 20th century, American Jews still believed in the power of the mandrake to induce fertility. They used to import specimens of the root from the Near East just for this purpose (Randolph).

Of course, mandrake had been known from early times in Greece as a medicinal plant, but probably all the complicated ritual for pulling it up was to ensure its efficacy as a love charm. By the 17th century the root had become so sought after that itinerant hawkers began to make mandrake figures, which they sold very profitably to childless women. The carvers became so skillful that these images were preferred to natural roots. They are still sold in the Near East – they reveal hidden treasures underground, and cure their owner of chronic illnesses, by absorbing it into themselves. But a new owner was apt to contract the malady that the previous owner had transferred to the plant (Frazer. 1917). In modern Cairo, drug sellers give pills made from the roots to young couples about to be married, and wanting a large family of boys. The ultimate testimonial to mandrake's aphrodisiac effects occurs in a Middle English bestiary of about the middle of the 13th century. Speaking of elephants, it says "they are so cold by nature that they think of no lasciviousness till they partake of a plant whose name is mandragora" (Bonser). It was the female that brought the mandrake to the male. Then, legend has it, the two journey eastward to Paradise, where they eat the root, copulate, and the female immediately conceives, remains pregnant

for two years, and finally gives birth – but only once. Medieval bestiaries would depict the elephants standing near the mandrake root (see Hassig).

In countries where the mandrake was not a native, it quickly became identified with plants that had some similarity, usually in the way the root grew. In Britain, Black Bryony, Cuckoo-pint, Enchanter's Nightshade, and White Bryony, particularly the last named, were known in local dialects as mandrake (Watts. 2000), and drew many mandrake superstitions to themselves (see below). True mandrake is not common as a wild plant, nor is it easy to grow, facts that added to its assumed value. In all probability it was either in order to protect it from indiscriminate use and consequent possible extermination that the myth of danger in digging the plant was allowed to grow up (H G Baker), or to keep the common man from getting hold of the drug by his own efforts, and so destroying the market (Emboden. 1979). That drug was important. The root juices were extracted by boiling; it was probably the first anaesthetic, and could be prepared simply by steeping the root, or boiling it in wine, then a draught could be given before surgery, and the pain possibly dulled (H F Clark). A 5th century work has the observation "If anyone is to have a member amputated, cauterized, or sawed, let him drink an ounce and a half in wine; he will sleep until the member is taken off, without either pain or sensation", and so on through the early herbals (see, too, Blunt & Raphael). Mixing it with lettuce seed and mulberry leaves would make it more potent. It is said that the first volatile anaesthetic was a sponge boiled in such a mixture and held under the face of the patient. Roman women, out of compassion, used to offer a soporific sponge of this "gall" to the victims of crucifixion, and it has been said that the sponge of vinegar offered to Christ on the Cross was in fact this anaesthetic (H F Clark) – it has even been said that it put Christ into trance state for three days, and he never died at all (Emboden. 1979). Mixed with morphine, mandrake has been used to produce "twilight sleep" (H G Baker). But of course, it is a powerful poison, and too much of it could bring on madness, paralysis or death- what the myths in fact said about the dangers of pulling it up.

For a very long time, it has been known that mandrake has a smell which could have some effect on the nervous system (H F Clark), like depriving one of the power of speech (Borges). Pliny warned his readers about it. In the Song of Solomon there is written, "the mandrakes give a smell, and at our gates are all manner of pleasant fruits, new and old, which I have laid up for thee, O my beloved". No suggestion there of unpleasant effects produced by the smell; it is probably a reference to a belief that the smell itself is a sleeping draught. Throughout the Middle Ages, it was the custom to put a mandrake at the head of a labour bed, more from the belief in its magic than with any idea of dulling the pain. It was used also for mental disorders of various kinds, the idea being generally to produce sleep. Many ancient writers alluded to it as a remedy for insomnia (Randolph). The leaves have been used in various ways; they have been read as tea leaves, applied to ulcers, taken as emetics (so has the root). Even today, they are sometimes used in ointments (Brownlow); they seem to have a soothing and cooling effect, good for erysipelas and other similar complaints (Randolph). But the modern mandrake of the pharmacists is that of the North American *Podophyllum peltatum*, the May-apple.

Anthropomorphism is carried to its extreme in one of the early medical texts. St Hildegard, inspired by the shape in which the root habitually grew, says in her *Physics*, "if a man suffers from any infirmity in the head, let him eat of the head of this plant; or, if he suffers in the neck, let him eat of the neck", etc., (Bonser).

But in England it was BRYONY, both WHITE BRYONY (*Bryonia dioica*) and BLACK BRYONY (*Tamus communis*), two entirely unrelated plants, that was taken to be mandrake, and credited with the same powers and attributes. Imitation mandrake puppets used to be made out of the roots, often used by witches in malevolent charms: "they take likewise the roots of mandrake, according to some, or, as I rather suppose, the roots of briony, which simple folk take for the true mandrake, and make thereof an ugly image, by which they represent the person on whom they intend to exercise their witchcraft" (Coles). People could open the earth round a young bryony, taking care not to disturb the lower fibres, and then put a mould round the root, after which it could all be covered up again, and then left. The mould, of course, would have to bear some resemblance to a human figure. Bryonies grow very quickly, and the object was generally accomplished in one summer. The leaves were also sold for those of mandrake, though there is no resemblance. The chief use of the root in England was as a fertility stimulant, and in some parts, Lincolnshire, for instance, it was looked on as a specific for ensuring that women conceived (Gutch & Peacock). In East Anglia, a childless woman would drink "mandrake tea" (Porter. 1969), presumably made from the roots, but not necessarily so. They were even given to mares as an aid to conception (Drury. 1985). All this had by 1646 attracted the attention of Sir Thomas Browne, who dismissed it like any other superstition: "… for the roots [of mandrake] which are carried about by impostors to deceive unfruitful women, are made from the roots of canes, briony, and other plants; for in these, yet fresh and virent, they carve out the figures of men and women, first sticking therein the grains of barley or millet where they intend the hair to grow; then bury them in sand until the grains shoot forth their roots which, at the longest, will happen in twenty days; they afterwards clip and trim these

tender strings in the fashion of beards and other hairy teguments…".

English Mandrake is a widespread name for white bryony (Brownlow), though Mandrake itself is even more widespread. There is a Lincolnshire form, Woman Drake, that shows that the word has been misunderstood. Granted, bryony is dioecious, so the male and female are different plants, but is this sufficient explanation of the name? Especially as in that county it is the Black Bryony (*Tamus communis*) that is the mandrake (Rudkin).

The forked root of GINSENG was treated like the human form, just like Mandrake; it would seem that the whole of the mandrake legend spread to China, and became attached to ginseng (G E Smith).

-*Manihot esculenta* > CASSAVA

MANIOC > CASSAVA

MANNA SEEDS
In some parts of Europe, and in America, the seeds of FLOTE GRASS (*Glyceria fluitans*) used to be collected and sold as "manna seeds" (hence the American name for the plant, Mannagrass (Douglas)), for making puddings and gruel. It was even cultivated here and there for the purpose (C P Johnson).

MAORI FIRE
(*Pennantia corymbosa*) A New Zealand tree, called kaikomako by the Maori, who used it for friction fire-making, as the common name implies. Maui' was the deity who taught the people how to do this. Andersen's version of the myth runs as follows: "… from a kaikomako he broke dry branches, and from them he fashioned fire-sticks. While at his request a man held one stick firmly on the ground with his foot, Maui' rubbed the second, sharpened to a point, briskly to and fro on the one so held. First it heated, and formed a little ball of black powder; then the powder smoked; then it glowed. Maui' took dry moss, wrapped the powder in it, waved it in the air, when lo! A flame! The people did likewise …" (Andersen).

MAPLE
see FIELD MAPLE (*Acer campestre*); SUGAR MAPLE (*Acer saccharum*)

MARBLES VINE
(*Dioclea reflexa*) Perhaps of American origin, but it grows in West Africa on sandbanks or seashore, and the seeds are used by children all over the area in a game played like marbles (Dalziel), hence the common name.

MARGUERITE,
sometimes Herb Margaret (Dyer. 1889), common names for the DAISY, obviously a St Margaret, but which one? There was a St Margaret of Antioch, an unlikely choice, or St Margaret of Cortona, or yet another Margaret, St Margaret of Valois. Actually,

St Margaret of Cortona is the likeliest candidate, for her day, 22 February, used to be reckoned as the first day of spring (Jones-Baker. 1974), and that probably is the reason for the name. But there is another possibility: the French word marguerite means a pearl (the colour of the flower? (Skinner)).

MARIGOLD

(*Calendula officinalis*) The name of the genus derives from calends, for the Romans believed that marigolds flowered all the year round. In other words, they were to be found blooming on the first day of each month, the calends. Cf the French fleur de tous les mois, and the Italian fiore di ogni mese. Presumably its very ubiquitousness accounts for the Wiltshire name Nobody's-flower (Macmillan).

What did Bloom mean when he said marigold was "a favourite at funerals"? F G Savage does mention strewing graves with marigolds, but the only other record of a graveyard use comes from Bavaria, where, according to Frazer, it was the tradition to decorate tombs with it on All Souls' Day. Some confirmation of this seems to be implicit in the acceptance of marigolds as symbols of grief (Leyel. 1937). But apart from this, beliefs in marigolds seem to be as sunny as the plant's own nature. To dream of them means property, and a happy and wealthy marriage (Raphael), and people seem to have had such confidence in marigold-inspired dreams that they tried to induce them, by using the petals as an ingredient in an ointment used on St Luke's Day for the express purpose of bringing prophetic dreams (Wiltshire). Marigolds are good to have the gardens, too, for it is said that they keep fly pests away from a vegetable patch (M Baker. 1977).

Marigold is a sun follower, something that is implicit in names like Husbandman's Dial, and Summer's Bride, and explicit in the Guernsey name Soucique, which comes directly from the Latin solsequium. One would expect some weather lore from such a plant, and indeed, they say that if it does not open its petals by seven in the morning, the signs are that it will rain or thunder that day. Marigold also closes up before a storm (Swainson. 1873). Measles-flower is a Wiltshire name. Children were warned against picking them in the garden. That would give them the measles. The reason for the name is likely to be just the opposite.

The petals have been used as a colouring agent, for cheese and butter (Grieve. 1931), and as a substitute for saffron, Maud Grieve says. Edith Brill gave an account of the way they were used for butter: the flowers were put in an earthenware pot and covered with a layer of salt, alternate layers of flowers and salt being added till the pot was full. Then it was covered to keep it airtight. When the flowers were needed, they were pounded with a pestle in a wooden mortar, the juice strained after being mixed with a little skim milk, and then squeezed through muslin into the cream

before it was put in the churn. But that was not the only occasion when marigolds were used for colouring – Turner, in 1551, weighed in against people who "make their hayre yelow wyth the flour of this herbe, not beynge content with the natural colour, which God hath gyven them". But they were being used for a popular hair dye for a long time after Turner's day (A W Hatfield. 1973).

The flowers used to be candied, too, and preserved, even made into wine (Clair). Marigold puddings had the finely chopped petals as an ingredient (Clair), and marigold buns were made, too (Grieve. 1931 has a recipe). The petals were added to cordials, too, and given in possets to treat a cold. And they were at one time a common feature in salads (Rohde. 1936). In fact, the flowers were so useful that as a matter of course they were dried and stored away for winter use; they were particularly popular for boiling in soups, stews and broths, especially mutton broth (*Oxfordshire & District Folklore Society. Annual Record. 7; 1955*). The petals give quite a distinctive taste to stews, and indeed they do to salads, and even to porridge (A W Hatfield. 1973). Pot Marigold is a name sometimes given to the plant (E Hayes) – nothing to do with growing it in a pot, rather it is because it is a potherb. Mrs Leyel gave a recipe for marigold "cheese", too (Leyel. 1973). Gerard mentions this use of what he called the "yellow leaves", particularly popular in Holland, and sums it up with "… no broths are well made without dried Marigolds".

Marigold water was for a long time a favourite for a headache (Rollinson), and for inflamed eyes (Rohde. 1936). Even just looking at the flowers was thought to help failing eyesight (Page. 1978). "The floures and leaves of Marigolds being distilled, and the water dropped into red and watery eies, cureth the inflammation, and taketh away the paine …" (Gerard). It appears, too, in a manuscript recipe of about 1600 for an "unguent to annoynt under the eyelids, and upon the eyelids, eveninge and morninge; but especially when you … finde your sight not perfect…" (Halliwell, 1845). That ointment was used too for wounds and peristent ulcers, but marigold is still used as a tincture in the same way as arnica, as it has the same properties (Schauenberg & Paris). Indeed, if marigold tea is taken after an accident, it brings out the bruises and prevents internal complications. A lotion would be applied to sprains and bruises as well (Moloney). It is said, too, that they should be used for burns. Not only does it cure, and help to relieve the pain, but it will also prevent the formation of scars. And it is even claimed that it will take away existing scars (Leyel. 1937).

Marigolds were used in medieval times for fevers (Lloyd), and an infusion of the flowers has long been a country remedy for whooping cough (V G Hatfield. 1994). Marigold tea and cider was given in Dorset, and in Scotland, though without the cider (Simpkins), to those who had measles (Dacombe). The marigold, it was said, helped to bring out the rash. Herbalists are still using the flowers, especially for heart disease; they benefit the arteries and veins (A W Hatfield. 1973), and it is prescribed for chickenpox and shingles, too (Warren-Davis). Gerard was recommending them for heart trouble four hundred years ago.

The tincture is used as a wound application (Flück), though the leaf itself would do just as well; it stops bleeding quickly, and just wrapping a leaf round a cut finger is quite effective for a surface cut, but never for a deep one (Painter & Power). The petals are still rubbed on bee or wasp stings to bring relief, and were a remedy for warts (Rohde, 1936). A Victorian cure is given as "the bruised leaves … mixed with a few drops of reduced vinegar". The Romans were using the juice for just this purpose in their day, hence the old Latin name, Verrucaria. Chewed marigold leaves bandaged on like a poultice, helps sores (Page. 1978), and eczema (W A R Thomson. 1978), and just applying the leaves is a traditional Scottish way of dealing with corns (Rorie). Bathing the feet in warm water in which the flowers had been infused was a Somerset remedy for varicose veins and sore feet (Tongue. 1965). The Russian use for jaundice smacks rather of doctrine of signatures, but is given as a folk remedy, particularly for children. The idea was to add marigold flowers to warm baths. Half a pound of dried flowers would be boiled for half an hour in a gallon of water, then this is strained into the tub (Kourennoff). Herbalists are still prescribing the infusion for this condition (Flück).

MARIJUANA
The name used in Mexico and America for the drug cannabis (see **HEMP**)

MARJORAM
(*Origanum vulgare*) A divination game played with marjoram, with other herbs, was a St Luke's Day (18 October) charm: "Take marigold flowers, a sprig of marjoram, thyme, a little wormwood; dry them before a fire, rub them to powder. Then sift it through a piece of fine linen; simmer these with a small quantity of virgin honey, in white vinegar, over a slow fire; with this anoint your stomach, breasts, and lips, lying down, and repeat these words thrice:

> St Luke, St Luke, be kind to me,
> In dream let me my true love see.

This said, "hasten to sleep, and in the soft slumber of night's repose, the very man you shall marry shall appear before you." (Dyer. 1889). In Greece and Rome, young married couples were crowned with marjoram. It was said, too, that the sheets of lovers should be perfumed with marjoram (Boland. 1977). Virgil says that when Venus carried off Ascanius

to the groves of Idalin she laid him on a bed of marjoram.

There was a saying that marjoram would only grow on a grave if the dead person was happy (Wiltshire) (Cf DITTANY). People were wary of it in Portugal, for there was a saying there that if you use your nose to smell it, your nose would drop off. Always stroke it with your hand, and smell that (Jacob). A very strange belief of older times was that a tortoise fortifies itself with marjoram before starting a fight (Jacob). This probably came from another, even older, belief that the tortoise immediately made for marjoram when it had eaten a viper, to purge itself (Albertus Magnus). In Morocco, they apparently used to burn marjoram on the Midsummer fires, presumably as a censing agent. Midsummer marjoram was also kept as a medicine. Coughs are cured by burning the herb and inhaling the smoke, which will also cure eye diseases. The stalk is lit and the eye regions touched with the glowing tip. Jaundice was also treated like this (Westermarck).

Marjoram is much used for flavouring. For example, it is put into home-brewed ale to give it a flavour; freshly gathered flowering tops would be used for this (Rohde. 1936), though they actually become sweeter as they dry (Mabey. 1972). Another use was as a dye plant; made from the flowering tops it gives a dark reddish-brown colour, but it fades quickly (Rohde. 1936).

In medicine, the infusion is given for whooping cough, and is used as a mouthwash for inflammation of the mouth and throat (Flück). The Anglo-Saxon version of Apuleius also recommended it for coughs, by the simple treatment of eating the plant (Cockayne). Marjoram tea was a Wiltshire remedy against infection of the lungs and chest (Wiltshire); it was actually a general tonic, right up to the 19th century (Rohde. 1936). A tisane, or tincture, of marjoram will act as a sedative to prevent sea sickness (Leyel. 1937).

It has often been included in indigestion remedies, and widely used in that way (for example, see Argenti & Rose). Certainly, the decoction used to be an Irish remedy for indigestion and acidity (Egan). In the 18th century, Hill was prescribing an "infusion of the fresh tops [to] strengthen the stomach, and [it] is good againsy habitual colic". A long time before Hill, the herbal of Rufinus advised: "Vinum decoctionis eius digestionem confortat; dolorem stomaci et intestinorum excludit" (Thorndike), which is virtually exactly the same.

The Pennsylvania Germans cured scofula by wearing marjoram roots as a necklace. They would dig the roots, cut them crosswise, and thread an odd number of pieces.The necklace had to be removed on the ninth day, and other pieces of root threaded on. This was repeated twice, and each time they had to be buried under the eaves (Fogel).

Marrubium vulgare > WHITE HOREHOUND

Marsdenia condurango > EAGLE VINE

MARSH GENTIAN

(*Gentiana pneumonanthe*) The flowers are used to make a blue dye (Usher), and it has the usual gentian medicinal uses, though the early ones are a bit unusual. Gerard, for instance, reported that "the later Physicians hold it to be effectual against pestilential diseases, and the bitings and stingings of venomous beasts". One of the Saxon leechdoms, translated by Cockayne, advised the use of this plant (under the name 'marsh maregall') if "a worm eat the hand". The patient was required to "boil marsh maregall, red nettle, dock, ... in cow's butter. Then shake three parts of salt on. Shake up, and smear therewith. Lather with soap at night".

MARSH HORSETAIL

(*Equisetum palustre*) The croaking of frogs is said to be the playing, by the frogs, of the hollow pipes of this horsetail, like musical instruments (Gibbings).

MARSH MALLOW

(*Alcea officinalis*) The plant gave the sweet its name. Today it (the sweet) is made from starch, gelatine and sugar, but once it was produced from the roots of Marsh Mallow, for they contain starch, albumen and a crystallisable sugar, a fixed oil, and gelatine matter. It was said in Lincolnshire that eating the sweet would cure ague and rheumatism! (V G Hatfield. 1994), and fishermen's wives were engaged in gathering it along the east coast (Mabey. 1972). The root can be eaten raw in salads, and made into a tea (Usher), and in France the young tops and leaves were added to spring salads (Fernie).

The roots are used in the cosmetics trade, but in France there is a more immediate use. Dried, and known as Hochets de Guimauve, they are sold in chemists' shops as teethers. They are hard and fibrous enough for a baby to chew on, but slowly soften on the outside as their mucilage is released (Mabey. 1977). There is one interesting piece of folklore. In the Isle of Man, marsh mallow was used (both internally and externally) to remove the result of walking or lying on "bad ground", and "bad ground" meant ground affected by the fairies (Gill. 1963).

Marsh Mallow is still one of the best known cough cures (Conway), taken in various ways. In France, druggists used to prepare a sweet paste, called pâté de Guimauve, for coughs and sore throats (M Evans). The practice in Ireland was to boil the seeds in milk, and then drink the liquid (Maloney). It can be prepared as a gargle for sore throats, made from shredded root or leaves, in water (Thomson. 1978). The root can be taken as a demulcent (it is about 30% mucilage); it is also sometimes applied as an emollient poultice (Fluckiger & Hanbury) ("avec un onguent tiré des feuilles ..., on frotte en Esthonie les membres du corps attenits par quelque magie" (Gubernatis)). East Anglian people used to treat rashes, grazes and pimples with ointment made by boiling marsh mallow (or elder flowers or periwinkle leaves), then mixing them with goose grease or lard (Randell). An Irish cure was very similar (O'Farrell), and a root decoction (2 cups a day, hot) is often prescribed in Russian folk medicine for pneumonia (Kourennoff).

Marsh Mallow tea was taken in Cumbria for rheumatism, and probably is of wider significance than this suggests (see, for instance, C P Johnson). In the Cotswolds country, a fomentation of the leaves, soaked in boiling water, is used for mumps and swollen glands (Briggs. 1974). The same is used in Somerset for an abscess (Tongue. 1965), and in Hampshire for boils (Hampshire FWI). The Physicians of Myddfai prescribed the same treatment, and this hot poultice is a gypsy remedy for toothache (Vesey-Fitzgerald). Marsh Mallow ointment, made from the crushed roots, is good for chapped hands (Page. 1978), sore feet and varicose veins (Vesey-Fitzgerald), and to put on a burn (V G Hatfield. 1994). The same ointment was used in horse doctoring, for sores, and sprains (Boase). East Anglian horsemen, too, used marsh mallow to cure a horse "with a pricked foot" (G E Evans. 1969).

The older herbalists were just as keen on marsh mallow as their more recent counterparts. The Anglo-Saxon version of Apuleius recommended the juice for slimming, "in case that a man be overwaxen in wamb", or "if one hreak up blood much" (Cockayne). Gerard listed among the "vertues" the fomentation "against pain of the sides or the stone" (a decoction of the roots in water was an Irish remedy for gravel (Egan)), as a painkiller, the root decoction "Helpeth the bloody flix", and so on.

There are one or two charms to mention. Dyer. 1899 notes a German antidote against the "hurtful effects of any malicious influence". It was an ointment made from the leaves. In parts of France, a necklace of marsh mallow (roots?) is said to keep children free from toothache (Sebillot). One should take note of the name Mortification Root. The point is that the powdered roots make a poultice that would remove obstinate inflammation and prevent "mortification", gangrene in more recent terms, in external or internal injuries (A W Hatfield).

MARSH MARIGOLD

(*Caltha palustris*) The name Marsh Marigold probably came about by confusion. The OE was marsc meargalle. 'Mearh' is horse, and 'gealla' is a blister (Grigson. 1955). Like most of the members of the Ranunculaceae its juice is acrid, hence the use of the term for blister. All parts of the plant contain this acrid, irritant toxin; the juice can produce large areas of swelling and inflammation on the skin, and blistering, too, if one is allergic enough (Jordan). Cattle have died from eating it, and it will certainly cause loss of milk production in cows (Long. 1924).

One festival at which Marsh Marigolds were used (and still are, in Ireland and the Isle of Man) is May Day. They were hung on doorposts in Shropshire, stalk upwards, it seems (Baker. 1980), perhaps on the analogy of the horseshoe, but it was unlucky to bring them indoors before the first of May (Burne. 1883), though it was alright to put bunches of them outside on the window sill (Vickery. 1985). George Bourne dscribes the use of the "Broad Buttercup", as he called Marsh Marigold, in the garlands in Farnham, Surrey, where he lived. "The chief value, … of broad-buttercups and milkmaids [*Cardamine pratensis*] was for making garlands for May Day. Not that any of us Sturts ever carried garlands. That was a sign of poverty …" (Bourne. 1927). The role of the flowers was prophylactic, and they are still used in Ireland rather like rowan elsewhere, to protect the cattle (Grigson. 1955) and the home. There it is the Beltane flower par excellence, used more than any other in the May Garlands, and also for strewing before doors and on the threshold (W R Wilde). Children in County Antrim used to gather marsh marigolds (which they called May-flowers), and push one of them through the letter box of every house in the village, so that the flower could protect the house from evil. The children were rewarded, of course. Paton records the use on the Isle of Man on May Day; its Manx name is Lus y Voaldyn, the herb of Beltane, and there it was a protection against witches (Killip). Primroses and marsh marigolds still appear on the island in a jar of water on that special day (Garrad), and it was used as a charm against lightning during May storms (Baker. 1980). This protective influence stretches up as far as Iceland; there is a strange story that marsh marigolds will prevent whoever is carrying it from having an angry word spoken to him (Dyer. 1889) – always provided the right words and ceremony were used when it was picked.

Equally odd is the report that in one part of Germany (Saxony, perhaps), they used to make wreaths of marsh marigolds and then throw them one at a time on to the house roof. If anybody's wreath actually stayed up, then it would be a sign he would die before next summer (Hartland. 1909). It should be said that this happens as part of the St John's Day celebrations, and it is unusual to find death divinations practiced then. In any case, would marsh marigolds still be in flower at Midsummer? Perhaps it is a recognition that the plant is actually poisonous, though parts of it can be eaten. The unopened flower buds are sometimes pickled and used like capers or a spicy condiment (Sanford). Hulme pointed out that to be safe they would have to be soaked first in vinegar for a long time. Some native American peoples, the Menomini, for instance (H H Smith. 1923), boiled the leaves to eat them as spinach. Boiling would get rid of the sharp and biting taste.

Very occasionally, the plant has been used medicinally. An infusion of the flowers has been recommended for treating fits, and a tincture from the whole plant has been used for anaemia (North). Turner actually recommended chewing the leaf to relieve toothache! It sounds extremely hazardous.

MARSH ROSEMARY

(*Ledum palustre*) It used to be employed as an anti-parasitic to treat lice, scabies, etc. It is also a traditional abortifacient (Schauenberg & Paris). It is certainly narcotic enough to cause trance, or at least deep sleep. There seems to have been a custom among Finns, on the eve of a wedding, of plying the groom with beer containing Marsh Rosemary, enough of it to cause him to fall into a deep sleep, during which the bride took the opportunity to crawl between his legs. Apparently, this was done as a means of ensuring easy childbirth when the time came (J B Smith). The groom was naturally unaware of what was happening, and the custom sounds very like a transference ritual, an attempt to foist the pains of childbirth on to the father. Siberian shamans used the plant to induce a trance state. It was done by inhaling the smoke of this shrub, or some other resinous plant. Some dried stalks or leaves would be put on some glowing embers on an iron plate, and the shaman would bend over the smoke produced, which would be dense and strong-smelling (Balazs).

MASTERWORT

(*Peucedanum ostruthium*) A native of southern Europe, introduced into British physic gardens in the Middle Ages, and now naturalized beside streams and in damp meadows. It enjoyed the reputation, as spurious as all the rest of them, of being an antidote to poison (Grigson. 1955), repeated in Gerard: "… good against all poison, …, singular against all corrupt and naughty aire and infection of the pestilence …". A Scottish cure for ague was "a little bit of ox-dung drunk with half a scruple of masterwort" (Graham), if such a mixture can be imagined. Gerard too recommended it for the ague among other ills, including a prescription that "helpeth such as have taken great squats, bruses, or falls from some high place…". Hill used the root, telling us that it is "good in fevers, disorders of the head, and of the stomach and bowels …". In fact, the root is still used in homeopathic medicine, in tincture, for stomach ailments and for dermatitis (Grigson. 1955).

MASTITIS

Dutch folk medicine recognised FENUGREEK seeds cooked in water until the whole consistency is like a paste or porridge, to use as a hot compress to treat this condition (Van Andel).

Matricaria matricarioides > RAYLESS MAYWEED

Matricaria recutita > SCENTED MAYWEED

MAY

HAWTHORN is known as May, and thereby lies the confusion between tree and month. "Ne'er cast a clout till May is out" means do not put on any new clothes till the unlucky month of May is over, the month being represented by the tree. May, the tree in flower, is just as unlucky, particularly unlucky to bring indoors. "It brought illness, etc.," with it, according to Devonshire belief (Devonshire Association. Report & Transactions; 1971), and in Somerset may well cause death in the house into which the blossom is brought (Elworthy). Cheshire children are forbidden to bring it indoors, the belief being that their mother will die if it were done (Hole. 1937):

> May in
> Coffin out,

in fact (Igglesden). The hawthorn that can be taken in the house on May morning might represent a ritual breaking of the taboo on the May festival, and he, or more usually she, who breaks it might be entitled to a reward. In Suffolk, for example, any maidservant who could bring in a branch of hawthorn in full bloom on May morning (it would surely have to be old style), was entitled to a dish of cream for breakfast:

> This is the day,
> And here is our May,
> The finest ever seen,
> It is fit for a queen,
> So pray, ma'am, give me a cup of your cream
> (Gurdon).

"Dew from the hawthorn tree" has special properties at this time (in fact, any May dew has).

In some parts of England, Devon and Cornwall for instance, May is the regular word for a sprig of ELM, not hawthorn, gathered early in the morning of May Day (Friend. 1882).

MAY DAY

In Ireland, HAZEL, a powerful symbol of fertility, was included in the "summer" brought into the house on May morning (Ô Súilleabháin) MARSH MARIGOLDS were used at the May Day festivities, and still are in Ireland and the Isle of Man. The Manx name for it is Lus y Voaldyn, the herb of Beltane (Paton), the May flower par excellence. They were hung on door posts in Shropshire, stalk upwards, it seems (Baker. 1980), perhaps on the analogy of the horseshoe, but it was unlucky to bring them indoors before the first of May, but it was alright to put bunches of them outside in the window sill (Vickery. 1985). Children in County Antrim used to gather marsh marigolds (which they called Mayflowers),

and push one of them through the letter box of every house in the village, so that the flower could protect the house from evil. The children were rewarded, of course. J C Foster reported that GREATER CELANDINE was the Mayflower in parts of Ulster, almost certainly an error. Greater Celandine would not be in flower at the beginning of May.

In Europe, LILIES-OF-THE-VALLEY are the emblems of May Day, muguet de mai in French, Maiblume or Maiglöckchen in German. They are a customary May Day gift in Paris, large quantities of them being sold for the purpose (Hole. 1976). They are, too, traditionally worn by participants in the Helston Furry Dance (Vickery. 1995), usually held on 8 May.

MAY GARLAND

A branch of May, dressed as a garland as part of the May-time ceremonies, used to be seen as a memory of some ancient tree or agricultural cult, whether it was carried about, hung outside a house, or put at the top of the maypole. It was the symbol of the death of the spirit of vegetation in winter, and its resuscitation in spring, and was important enough for the day itself to be known as Garland Day in some parts, Sussex, for example (Sawyer. 1883). In more northern parts, spring comes later, so the same celebration does too, e.g., on St John's Eve in Sweden, where the equivalent of the maypole is called Maj Stanger (Elworthy. 1895).

It is clear, though, that the real function of the May Garland is to protect the people and their stock. Scot wrote that "the popish church … to be delivered from witches … hang in their entries … haythorne, otherwise white thorne gathered on Maie daie …". The three primary May plants in England were HAWTHORN, MARSH MARIGOLD and ROWAN, all protectors and averters of evil (Grigson. 1959). (See Bourne. 1927 for his memory of the use of "broad-buttercup", as he called Marsh Marigold, in the May Garland). Rowan was more important later in the year, when its berries were formed, and these three were certainly not the only plants used. In much of County Cork, SYCAMORE was the favourite May bough, and the tree was actually called the Summer Tree (Danaher). It was used in Cornwall, too, as well as hawthorn (Borlase). In fact, any tree in blossom or young leaf could be regarded as the "May"; the modern use of the word as describing just the hawthorn did not apply at all. In some parts of England, May was the regular word for a sprig of ELM, not hawthorn, gathered early in the morning of May Day, as in Worcestershire, where it was an elm bough that was put up (Opie & Tatem). HAZEL was included in Ireland in the garland called the "summer", brought into the house on May Day (Ô Súilleabháin), or GORSE in County Cork, and HORSE CHESTNUT was used in Ireland, too. There is a note in *Béaloideas. Vol 15; 1945 p 283-4*, describing the proceedings at Carrick on Suir, Ireland, when branches of this tree were set up on the morning of May Day, and hung over byre doors, to protect the cattle. BIRCH was used in Wales and Herefordshire. Crosses made of birch, rowan, hawthorn and COWSLIPS were put over the cottage doors, "for luck", and also to repel witches (Baker. 1980). Sycamore, and even LILAC have been called May, and Mayflower is a name given to at least eight plants, including Marsh Marigold, cowslip, primrose, stitchwort and even Lady's Smock. The refrain of a May song from south Lancashire runs "the baziers are sweet in the morning of May", so it looks as if they were once part of the garland (Britten & Holland). Baziers is bear's ears, in other words, AURICULAS.

The green boughs were not necessarily set up just at the house door (Brighton fishermen used to hoist them up to the masts of their boats, for instance (Sawyer. 1883)) though that was the usual place to find them. The garland seems to serve as a general luck-bringer, or, negatively, as an averter of evil, though in Staffordshire bringing the May in uninvited was said to be unlucky (Hackwood), or rather it was reckoned to bring bad luck to the household. Occasionally, the garland's virtues are thought to be more specific. In some parts of Ireland, for example, they say that the green bough of a tree fastened against the wall on May Day will ensure plenty of milk in the summer (Camden). All over Germany, the may-bush was brought into the house, too, rather than being fixed outside; the householder could never get the garland himself – someone else must bring it in (Grimm), so creating an obligation that, in similar circumstances in Cornwall, is met by the present of a dish of cream. The maidservant who brought in a branch of haw-thorn in bloom on May morning received her dish of cream as something that was her unquestioned due (Courtney. 1890). So important was this that around St Ives, May Day was called Cream Day (Barton. 1974). FURZE was sometimes used, and around Landrake and Liskeard in east Cornwall, anyone who picked a piece of BRACKEN was given as much cream as would cover it (Courtney. 1890); an account that makes a little more sense tells that if the frond of bracken was long enough to cover the cram bowl, the youngsters were given slices of bread and cream, after they recited:

Here's a fern
To measure your shern.
Please give me some milk and cream.

(a shern is a cream-bowl (Deane & Shaw)). Children from Hatfield in Hertfordshire, dressed in white, and holding branches of hawthorn or BLACKTHORN if the season was late, used to go from door to door singing a local version of the May song which began:

A bunch of May I bring unto you
And at your door I stand.

Come pull out your purse,
You'll be none the worse,
And give the poor Mayers some money (Jones-Baker. 1979).

Sometimes flowers replaced the greenery as the embodiment of May. Marsh marigolds were used in Ireland more than any other flower in the garland, and were strewn plentifully before doors and on the threshold (W R Wilde). They could not be taken indoors before May Day, though – that would be very unlucky, according to Shropshire belief (Burne. 1883). COWSLIPS too were sometimes featured in the May garland, and CROWN IMPERIAL seems to have been the May in Buckinghamshire, where in one area of the county, the older children each had one to carry round from door to door (BUCKING-HAMSHIRE FWI), and in Leicestershire they were attached to the top of the garland, as a finishing touch (Ruddock). Many other flowers were used around the country, GUELDER ROSE, for instance, but on no account was the CUCKOO-FLOWER (or LADY'S SMOCK) put in the May wreath, presumably because of its fairy associations (Hull) – except in Oxford, where the children nearly always included it, in spite of the fact that they knew it was a vaguely unlucky flower, especially indoors.

The garland progressed from a single bough, or bunch of greenery, to the idea of fixing the bunches on a stick, and carrying that around, as at Leckhampstead, in Buckinghamshire, where the boys carried these around, while the girls carried the two wooden hoops, or a child's small chair, decorated in some way with greenery, or, as at Combe, in Oxfordshire, where the girls carried their bunches of flowers on sticks round the village, singing their May song:

Gentlemen and ladies,
We wish you a happy day,
We've come to show you our garlands,
Because it is May Day (Baker. 1980).

From the simple tying of a bunch of flowers on to a stick, the garland got ever more elaborate, start-ing with two hoops joined together and decorated, much in the style of the Christmas kissing bough. The King's Lynn garland was described in 1894 as being made of two wooden hoops, fastened together at right angles, and supported on the end of a pole, and that was how the Saxby (Leicestershire) garland was made, too (Ruddock). Flowers and green boughs were arranged over the hoops, and a strand of bird's eggs hung from the whole thing (Newall. 1971). In some places, the garland had three hoops in its make-up, so that the end product was in the shape of a ball. Sometimes, in Lincolnshire for example, the traditional shape of the garland was oval; it was made up with cowslips, wood anemones, crab blossoms, wallflowers, primroses and daisies (Gutch. 1908), and there is an isolated example, from a Lincolnshire village, of MISTLETOE being used in the garland. Simpler than these are the garlands made in Oxford, called cross garlands, a plain wooden cross that is then covered with flowers. There was an example, at Charlton-on-Otmoor, of a cross kept above the rood screen in the church, actually called the garland. It was taken down and dressed as such each May Day (Hole. 1975).

The introduction of a doll into the garland marked the next development. Sometimes the doll was called the "May Queen", or they were "May-babies" (Vaux). In parts of Cambridgeshire the dolls were known as May-ladies, and the custom itself as May-dolling. Children of the village of Hanby, in Lincolnshire, contented themselves with taking their dolls around in a basket, but custom usually demanded something more elaborate than that, either putting the doll in the centre of crossed hoops, or making some kind of hooped structure, a doll put inside, and then the whole thing covered with a white cloth. A rope would be stretched across the road from one tree to another, and the hoops hung on it. On May morning, the girls gathered at the spot, and asked any passer-by for money, the consideration being a sight of the doll, done by lowering the rope. Inevitably, the garland deteriorated in course of time. Brighton children, for instance, were still going round collecting in the 1930s. But their garland had by that time degenerated into a few paper flowers and some ribbons stitched to their clothes. But for the traditional custom see the description of the preparation of the garland by Flora Thompson in *Lark Rise,* in her area of Oxfordshire.

But it is clear that "garland" did not necessarily mean some kind of structure of plants and greenery, but often meant a "pyramidal pile of decorative valuables usually carried on the head" (Judge). Perhaps such a concept arose from the traditional Huntingdonshire garland, which was a pyramid contrived from parallel hoops supported and kept at the right distance from each other by upright poles. It could be five or six feet high, smothered in flowers, greenery and ribbons, and having a May Doll fixed on its front (Hole. 1976). But the pyramids carried on the head grew larger and larger. The Tatler for 2 May, 1710 had a letter that said: "May 1 – I was looking out of the parlour window this morning, and receiving the honours which Margery, the milkmaid to our lane, was doing me, by dancing before my door with the plate of half her customers on her head". Not surprisingly, later on, a porter would carry the heavy garland, and process, even dance, with the milkmaids (Phillips. 1952). Eventually, the word became associated with the Jack-in-the-Green, the green being the garland. But, quite early on, May Day in London had become the sweeps' festival, when they went round the streets, and always had their Jack-in-the-Green with them, a custom

known in some areas as sooty-bobbing. So the garland became the Jack-in-the-Green, called also Green Jack, or Green George, who was a man carrying, or rather inside, a framework covered with a thick mass of leaves, with a hole left to see through, " a locomotive mass of foliage with his black face shining through an aperture in the leaves" (quoted in Jones-Baker. 1974).

Sometimes the May Garland took on a more personal significance, according to the plant or tree used. BIRCH, in Wales, was meaningful if you happened to be courting. It was a love emblem, or an indication that you had been accepted. The custom was to have sprigs of birch and rowan, decorated with flowers and ribbons, and to leave them where they were most likely to be found by the person intended, on May morning (Davies. 1911). Another Welsh custom, kept up until about 1870, was for the young men of the village to decorate a large bunch of rosemary with white ribbons on May morning, and to put it at the bedroom window of the girls they admired (Trevelyan). In Europe, lilies-of-the-valley are presented to ladies on May morning, as a compliment, presumably. This is a typical Parisian custom, and vast quantities are sold for the purpose. (Hole. 1976). But May garlands could be insulting, too, Usually, the distribution of these garlands represented the honest opinion of the villagers, and was intended as a warning. But of course it could serve for spite, or revenge, and then could do great harm. It all depended on the plants used. For example, in Northamptonshire, while hawthorn was left to show the greatest esteem, elder, crab, nettles, thistles, sloes, etc., marked the different degrees of disrespect in which some were held. BLACKTHORN, according to one authority (Tynan & Maitland), was reserved for a shrew. In Cheshire, the plants chosen were supposed to rhyme with the word that best described the recipient – pear, fair; plum, glum; owler (alder), growler, or scowler – in other words, bad-tempered; thorn, scorn; lime, prime, and so on. Gorse in bloom over a woman's door was the worst of insults (Hole. 1941). At Hitchin, Hertfordshire, elder and nettles were used to show contempt, and blackthorn was put over a shrew's door in some places (Tynan & Maitland). The custom was known as May-booing in Lancashire, and there a thorn (other than hawthorn, that is, for that is universally a highly complimentary gift) showed scorn; rowan, affection (via dialectal forms – wicken, chicken, a sign of affection (Hole. 1976)); holly, folly; briar, liar, etc. Salt sprinkled before a door suggested a great insult (A R Wright). Potato peelings carried the same message in France (Salaman).

MAYPOLE
The Welsh maypole was always made of BIRCH (Trevelyan) – in fact the Welsh word bedwen serves both for birch and maypole. There are other uses connected with May Day, too. In Herefordshire a birch tree was brought into the farmyard on May Day,

decorated with red and white rags, and then propped against the stable door to protect the horses from being hag-ridden. In some parts of Sweden, it was the May custom for boys to go round with the village fiddler, each with a bunch of freshly gathered birch twigs. They sang songs that had the general theme of asking for fine weather, good harvests, and the like. At every cottage they got the proper reward, and then they would decorate the door with one of their birch sprays (Philpot).

MAYWEED

(*Anthemis cotula*) It is still used in the Hebrides for swellings and inflammations, and a domestic medicine from Alabama uses Chigger-weed, which is our plant, with Pleurisy-weed (*Asclepias tuberosa*). They are boiled together and applied as a poultice for pleurisy (R B Browne). The tea is taken for stomach trouble in Kentucky (Thomas & Thomas), and herbalists prescribe it for migraine (Mitton & Mitton).

MEADOW CRANESBILL

(*Geranium pratense*) "… Fuchsius saith that Cranesbill with the blew floure is an excellent thing to heale wounds" (Gerard). That was still the opinion well into the 19th century, when one or two geranium leaves bruised would be applied to the cut (Dodson & Davies). It was used in the Highlands of Scotland to stop bleeding after a tooth had been pulled out (Fairweather).

MEADOW SAFFRON

(*Colchicum autumnale*) If Meadow Saffron bloomed on a grave, it was taken as a good sign for the deceased (Friend. 1883). In Lorraine, apparently, the flowers were crushed and put on the heads of children who had a lot of hair, in the belief that they would destroy the vermin that could not be reached by normal combing (Sebillot). There was an idea that these plants were fatal to dogs (cf the French names Mort-aux-chiens (Pratt) and Tue-chien (Bardswell). Certainly, all parts of the plant are poisonous, both in the green state and when dried, as it might possibly be in hay. Horses, cattle and pigs are vulnerable. Gerard's advice to anyone eating it by mistake was to "drinke the milke of a cow, or else death presently ensueth".

It is, however, used in small doses in the treatment of gout and rheumatism. Gypsies use a very weak infusion of the sliced roots for gout (Vesey-Fitzgerald), and it still stands as the best alleviation in herbal medicine of the pain from the condition (Thomson. 1976). Rheumatism, too, has been treated with the drug, and this is probably the ailment rendered as "sore of joints" by Cockayne, in his translation of Apuleius. The drug colchicine has been used for acute arthritis, too (Schauenberg & Paris). Another of the Anglo-Saxon Apuleius leechdoms was for "granulations", by which was meant pimples. If they grow on a woman's face, she was advised to "take roots …

and mingle with oil; then wash afterwards therewith" (Cockayne). But colchicine has another application. It turns out to be one of the first drugs in bringing leukemia under control, particularly the acute form. It has an action that arrests the growth of young white blood cells, hence its effect in this disease (Thomson. 1976). Other authorities, though, say it is far too toxic to be of use (e.g., Schauenberg & Paris).

The flowers rise naked from the earth, hence a series of well-known names, including Naked Ladies (Prior), Naked Nannies (Dartnell & Goddard), Naked Boys (Aubrey. 1847; Nall), and several others. The same phenomenon accounts for names like Upstart (Prior), or Son-before-the-father (Britten & Holland). It should be borne in mind that this plant is not a crocus, nor does it have anything to do with saffron. However, the name Autumn Crocus seems to have stuck (Grigson, 1959).

MEADOW SAXIFRAGE

(*Saxifraga granulata*) The name Saxifrage is derived from the Latin, rock, and frangere, to break – "breakstone" in other words, for these plants often grow in clefts of rocks. Inaccurate observation led to the conclusion that the roots had actually broken the rock, and that became the signature of the plant – that which breaks stones must also have the power to break stones in the body. See, for instance, the Anglo-Saxon version of Apuleius: "in case that stones wax in the bladder, pound it in wine; give … to drink … It is said by those that have tried it that it breaketh to pieces the calculi the same day, and tuggeth them out, and leadeth the man to his health" (Cockayne). Gerard, a long time later, repeated the prescription, and Hill, in the mid-18th century, was saying exactly the same a long time after Gerard.

Dyer. 1889 suggested that the granulated roots of this species provided another signature. Presumably he had in mind the similar roots of plants like Lesser Celandine, though in the latter case it was haemorrhoids it was supposed to cure, and there seems to be no record that Meadow Saxifrage was used in a similar way.

MEADOWSWEET

(*Filipendula ulmaria*) "Meadowsweet" as the common name is firmly entrenched, but it was originally "Meadwort", a name that surfaces occasionally today, in Somerset, for instance, and there is another form, "Meadsweet" that was used in Cornwall and elsewhere (Britten & Holland). The dried leaves were once used to give an aromatic bouquet not only to mead, but also to port and claret (Mabey. 1972).

In the Mabinogion story of Lleu Llaw Gyffes, Gwydion and Math made Blodeuwedd from "the flowers of the oak, the flowers of the broom, and the flowers of the meadowsweet", and it was with meadowsweet that Aine was traditionally crowned in Irish mythology (Bayley. 1919). It was her plant, and it owed its fragrance to her (Fitzgerald). Christian usage has appropriated the plant to St Christopher, for not very clear reasons.

It is an unlucky, even fatal, plant in Welsh superstition. If someone fell asleep in a room where many of these flowers were put, death was inevitable. It was even dangerous for anyone to fall asleep in a field where there was a lot of meadowsweet growing (Trevelyan). This sounds as if it were an extension of the fear of bringing any white flower – hawthorn, lilac, etc., – indoors. According to Icelandic belief, it will, if put under water on St John's Eve, reveal a thief; floating, the thief is a woman, and sinking, a man (Dyer. 1889).

The flowers have a hawthorn-like scent, and the leaves, having oil of wintergreen in them (Genders. 1971), are very aromatic. It was Queen Elizabeth's favourite strewing herb. "Queen Elizabeth of famous memorie did more desire meadowsweet than any other sweet herbe to strewe her chambers withal" (Parkinson. 1629). Gerard was enthusiastic about that usage: "the leaves and floures farre excell other strewing herbes for to decke up houses, to strew in chambers, halls, and banqueting houses in the summer time; for the smell thereof makes the heart merrie, delighteth the senses; neither does it cause head-ache, or lothsomeness to meat, as some other sweet smelling herbes do". These aromatic leaves can be added to soups (Genders. 1971). So why should it be stigmatized as a symbol of uselessness? (Leyel. 1937).

Most of the medicinal uses derive from the fact that the essential oil obtained from the plant contains methyl salicylate and salicylic aldehyde, which are part of the compound now known as aspirin (Palaiseul). So it is a good pain reliever, prepared by boiling the flowers in water for ten minutes, the dose being three cupfuls a day (Page. 1978). For the same reason, it is useful in treating rheumatism and arthritis (Schauenberg & Paris), and also influenza (Fluck), and even a common cold (Thomson. 1978). "E W[yld], Esq, tells me of a woman in Bedfordshire who doth great cures for agues and fevers with meadsweet, to which she adds some green wheat" (app to Aubrey. 1686/7). In the Highlands of Scotland, it is traditionally used for treating fevers and headaches. The Gaelic name means Cuchullain's belt, for the saga tells that Cuchullain, ill with fierce fevers, was cured by being bathed with meadowsweet (Beith).

An infusion of the flowers or leaves will settle an upset stomach (Conway), and a root decoction can be given for diarrhoea (Fluck). There does seem to be a great variety in the ills that this plant is reckoned to cure, and the list continues with a Somerset cough cure in which dried meadowsweet is added to a parsley infusion (Tongue). The eyes were also treated

with it, by distilled water of the leaves, which was supposed to strengthen them, and to prevent itching (Genders). Even sunburn can be treated with an infusion of the flowers (V G Hatfield).

MEASLES
NETTLE tea has been much used for skin complaints (Porter. 1974), including eczema, boils, even measles (in Ireland (Ó Súilleabháinn)). COWSLIP wine or tea can be taken for measles for either has the ability to lower the temperature, so they can be taken for any fever (Hampshire FWI). YARROW tea, made from the flowers, is another folk remedy for the complaint (V G Hatfield. 1994). SAFFRON tea is the medicine used in American domestic medicine to cure the condition in young children (R B Browne), or, as it was put in Ireland, "to bring out the rash" (Moloney). Lemon and sugar are added in Alabama.

A Wiltshire name for MARIGOLDS is Measlesflowers; it is said that children were warned that picking garden marigolds would give them measles (Dartnell & Goddard). The reason for the name, though, is likely to be just the opposite, for nearby, in Dorset, marigold tea and cider was given as a medicine (Dacombe), and in Scotland, too, though without the cider (Simpkins), there is a record from Suffolk, too (V G Hatfield. 1994). The marigold, it was said, helped to bring out the rash, which was what RED DEADNETTLE roots were reckoned to do, when boiled in milk for the children to drink (Vickery. 1995) (that was an Irish remedy), or CORIANDER, which is used in Chinese herbal medicine for that purpose (Geng Junying).

A decoction of the fruit of the TAMARIND tree is reported to be taken in Guyana as a measles remedy (Laguerre).

MEASURING RODS
Usually of HAZEL, with the bark left on, and found in graves. This rod would be cut to the exact length of the dead man's body, and put beside him in the grave, then would actually represent the person measured. In the case of the "holy length", the measure is that of Christ, on some agreed computation, and not of the deceased. These rods in some cases were there to benefit the dead, in others to protect the living. But there are examples of their use in a healing rite, and in general terms the ritual is a means of gaining power over that which is measured (Rees).

Medicago arabica > SPOTTED MEDICK

Medicago sativa > LUCERNE

MEDITERRANEAN ALOE
(*Aloe vera*) Originating probably in the Middle East, this grows about a metre tall, and has yellow or reddish flowers, unlucky to look at, according to French herbalists (Boland. 1977). Dreaming of them, though, not in flower, betokens a long life; in flower,

a legacy (Mackay). How often do they bloom? The whole leaf steeped in water had been used to make a soothing drink in tropical countries (G B Foster), but the main use of the plant is medicinal. The drug alone, better known as bitter aloes, is extracted from plants introduced into the West Indies, and the jelly-like juice from the leaves is used to treat cuts and burns, like its New World relative, *Aloe chinensis* (see **BURN PLANT**). When rubbed on the skin, it protects one from insect bites, and the juice relieves sunburn, too, although the plant itself does not flourish in direct sunlight (G B Foster).

MEDLAR

(*Mespilus germanica*) In parts of France, a branch of medlar was reckoned to keep sorcery at bay; it was put in the cattle byre, and in a baby's cradle (Sebillot). The dream books suggested that dreaming of medlars was a very good omen, and denoted riches, power and success in law (Raphael). The fruit is eaten "bletted", that is partially decayed, as in medieval times. So Shakespeare, because the fruit is only fit to be eaten when rotten, applies it to a loose woman. See *Measure for Measure. Act iv, sc 3*: "they would else have married me to the rotten medlar".

MEETING SEEDS

Seed heads of both FENNEL and DILL were taken to church at one time, so people could nibble them during the over-long sermons, so they got the name of Meeting Seeds. The name apparently travelled to America, and was used by the Puritan settlers in New England (Wiltshire; Rowe).

Melaleuca leucadendron > CAJUPUT

Melampyrum arvense > FIELD COW-WHEAT

MELANCHOLY THISTLE

(*Cirsium heterophyllum*) Gerard was able to prescribe the "decoction of the Thistle in Wine [to] expel superfluous Melancholy out of the Body, and make a man as merry as a Cricket ...", but the reference is actually to the hanging flower heads and this became the "signature" of the plant, hence the use for melancholy (Grigson. 1974) (and see **DOCTRINE OF SIGNATURES**).

MELEGUETA PEPPER

(*Aframomum melegueta*) Melegueta Pepper figures quite prominently in some traditional Yoruba sacrifical rites. Sometimes it accompanies Kola-nuts, and is eaten with them. It is also believed to facilitate the efficacy of a prayer or a curse. The officiating priest chews the pepper as he prays for a supplicant, or curses an offender (Awolalu). Various African peoples recommend that seven seeds of Melegueta Pepper, with a piece of paw-paw root, should be chewed during labour. It is supposed to cause an immediate delivery (Sofowora). The Mano people of Liberia had many uses for the seeds. Soaked in gin or

palm wine, they were taken as a cure for colic. The seeds and buds are used for pneumonia, and for pain in the back a handful of buds is beaten up with white clay. Four grains of *Piper guineense* are added, also powdered. The patient sits on the floor, and the doctor puts the medicine on the sole of the foot, and makes four (three if the patient is a woman) strokes down the patient's back. Then he gives the rest of the medicine to the patient to rub himself as he likes. They tame a snake with this, too. The bud leaves are chewed, and spat on to the snake's head, to quiet it (Harley).

Melia azedarach > CHINABERRY TREE

Melia indica > NEEM TREE

Melilotus officinalis > MELILOT

Melissa officinalis > BEE BALM

MELON

(*Cucumis melo*) Dreaming of melons is reckoned quite lucky. If the dreamer is a young girl, it is a sign of being married to a rich foreigner, and living abroad (Raphael). presumably because the melon is a foreign fruit. If a sick person dreams of them, then the juicy flesh promises recovery (Gordon. 1985). There is one other superstition, from China, in this case. There it is said that if you eat the pulp too much it will cause pimples, bring on fever, and cause general debility (F P Smith). In Chinese medicine, they are prescribed for cancer of the stomach. An oddity is the Kentucky practice of washing the face with melon rind to get rid of freckles (Thomas & Thomas).

MEMORY

ROSEMARY, the symbol of remembrance, was reckoned to stimulate the memory, and was prized as such. Wear rosemary on the person, and memory will be strengthened, wrote a Derbyshire correspondent of Notes and Queries in 1871, and in the 1930s Mrs Leyel was still able to say that rosemary tea, besides curing a nervous headache, would by constant use improve a bad memory (Leyel. 1937).

MENTAL ILLNESS

Given that WALNUT, by the strange doctrine once current bears the signature of the head and brain, it follows that it must be used for mental cases, from depression and mental fatigue to outright insanity. But the tree was involved in so-called cures for madness long before the doctrine of signatures was fashionable. A 15th century leechdom spoke of a sovereign medicine for madness and for "men that be troubled with wicked spirits: upon midsummer night betwixt midnight and the rising of the sun, gather the fairest green leaves of the walnut-tree, and upon the same day between sunrise and its going down, distill thereof a water in a still between two basins. And this water is good if it be drunken for the same malady" (Dawson). Another nut featured in cures for insanity is KOLA. But it is the leaf of the Kola, together with leaves of three other shrubs, all ground up and mixed with black soap, that makes a Yoruba Ewé cure for mental illness (Verger). The roots of CHRISTMAS ROSE were used as a purge, a dangerous practice, for all the hellebores are poisonous, but this treatment was also prescribed for the insane. As Gerard had it, the treatment was to be given to "mad and furious men ... and all those that are troubled with blacke choler, and molested with melancholy", i.e., for nervous disorders and hysteria. A DAFFODIL bulb had virtues in this field – "if neb possessed with evil spirits, or mad men, bear it in a clean napkin, they be delivered from their disease" (Albertus Magnus), put into verse in a late 14th century manuscript as:

> This herb in a clean cloth, and its root
> Against the falling evil is medicine.
> Affodille in clean cloth kept thus
> Shall suffer no fiend in that house;
> If ye bear it on you day and night,
> The fiend of you shall have no sight (quoted in Rickert).

As Albertus Magnus said, "it suffereth not a devil in the house".

It was believed once, that CLOVER, gathered with a gloved hand, and brought into the house in which there was a lunatic, without anyone knowing, would cure madness by its very presence (Wood-Martin). PEONY, when worn on the person, was long considered an effective remedy for insanity. The Anglo-Saxon version of Apuleius has: "For lunacy, if a man layeth this wort over the lunatic, as he lies, soon he upheaveth himself whole; and if he hath (this wort) with him, the disease never again approaches him" (Cockayne). MANDRAKE was used for mental disorders of various kinds, the idea being generally to produce sleep (Randolph) (see **MANDRAKE**). Wesley advised his readers to take the decoction of AGRIMONY for "Lunacy", for which it was to be given four times a day.

Brazilian curanderos, or traditional healers, put a ZINNIA leaf (*Z elegans*) on top of a patient's head to cure madness. It was also an ingredient in the ritual bath that accompanies Brazilian healing ceremonies (P V A Williams).

Mentha aquatica > WATER MINT

Mentha arvensis > CORN MINT

Mentha arvensis "Piperascens" > JAPANESE MINT

Mentha longifolia > HORSE MINT

Mentha x piperata > PEPPERMINT

Mentha pulegium > PENNYROYAL

Mentha spicata > SPEARMINT

Menyanthes trifoliata > BUCKBEAN

Mespilus germanica > MEDLAR

Metrosideros excelsa > POHUTUKAWA

MEXICAN POPPY

(*Argemone mexicana*) The seeds are used as a narcotic in several areas of northern Mexico (Emboden. 1979), and so they are in East Africa as well. They are described as producing a degree of intoxication at least as great as cannabis (Raymond). The yellow latex is sometimes used for removing warts (Gooding, Loveless & Proctor), and the juice is also used in the treatment of jaundice (doctrine of signatures – yellow juice), dropsy and as a cure for eye diseases (Chopra, Badhwar & Ghosh). It is worth noting that the generic name, *Argemone*, is derived from argema, the Greek word for cataract, and the plant's yellow latex was reported to soothe the condition (Whittle & Cook).

MIDSUMMER MEN

A name given to ORPINE (*Sedum telephium*), especially when associated with the Midsummer Eve *divinations* that involved this plant (for which see ORPINE).

MIGNONETTE

i.e., WILD MIGNONETTE (*Reseda lutea*) The name is the diminutive of French mignon, darling. The name of endearment was given to the plant by Lord Bateman in 1742. In the Oise district of France, mignonette put over a girl's door on May Day "annonce une rupture. Reseda, je te laisse là" (Sebillot). On the other hand, French brides believed that mignonette in their bouquet will hold a husband's affection (M Baker. 1979).

MIGRAINE

It is claimed that the condition can be allayed by holding a freshly cut slice of raw POTATO to the temples (R B Browne). BAY berries, too, at least according to Gerard, "stamped with a little Scammonie and saffron, and labored in a mortar with vinegar and oile of Roses to the form of a liniment, and applied to the temples and fore part of the head, do greatly ease the pain of the megrim", and he also advised "the juice of the leaves and roots" of DAISY to help "the megrim". CAMOMILE tea will help, both for migraine and any sort of headache (Schauenberg & Paris), and PELLITORY-OF-SPAIN was also used once. A leechdom from a 15th century collection advises sufferers to "take pellitory of Spain, and stone-scar [lichen] and hold long between thy teeth on the sore side; and chew it and it will run to water" (Dawson. 1934). The root of STINKING IRIS has the reputation of being a painkiller, and a migraine remedy (Conway).

MILK THISTLE

(*Silybum marianum*) The white veins of the leaves were believed to be due to the Virgin Mary spilling milk from her breast, hence the common name, and also the specific name, *marianum*. So the doctrine of signatures caused it to be taken as a proper diet for wet nurses (Page. 1978). Another remedy derived from the doctrine is its use to cure stitch in the side, for this plant has numerous prickles (Dyer. 1889). But herbalists prescribe the plant for many other disorders, particularly as a tincture made from the seeds, for liver disorders, jaundice, gallstones, etc. It can be chewed, too, as an anti-depressant (Boland. 1977), a fact known by Gerard's sources. An Irish country remedy for whooping cough was to boil it in water, and then drink the liquid (Maloney) and the juice was put on warts in East Anglia (V G Hatfield).

MILKWORT

(*Polygala vulgaris*) In Wales, it was once thought that milkwort in a pasture would increase the milk yield of the cattle grazing on it (Gibbings. 1941). Another belief from Wales was that it cures slight dog- or snake-bites (Trevelyan), and it enjoyed the same reputation in Leicestershire (Billson), probably taking the cue from the roots (cf SENEGA SNAKEROOT). The roots secrete a milky fluid, hence the name Milkwort, that was, and probably still is, used for rubbing on warts.

Gerard knew the names Cross-flower, Procession-flower, Gang-flower, and Rogation-flower for this plant, for they were the "floures the maidens which use in the countries to walke the Procession", and they made of them "garlands and nosegaies". All these names, then, refer to the use of the flower in the Roga-tion-tide processions, probably a continental tradition, Grigson thought. "Gang" is OE gangen, to go - the 'going' being about the parish, beating the bounds, etc.

MIMOSA THORN

(*Acacia karroo*) A central African species, where the Zezuru used the roots to cure lumbago. They burned them to charcoal, which was then finely powdered. Small cuts would be made over the painful area and this powder rubbed in. The same people chewed the roots as an aphrodisiac (Palgrave & Palgrave).

MINT

see WATER MINT (*Mentha aquatica*); CORN MINT (*Mentha arvensis*); JAPANESE MINT (*Mentha arvensis "Piperascens"*); HORSE MINT (*Mentha longifolia*); PEPPERMINT (*Mentha x piperita*); SPEARMINT (*Mentha spicata*)

MINT JULEP

Fresh PEPPERMINT leaves are used to flavour the Kentucky Whisky drink Mint Julep, made by inverting (tops down) a small branch of young mint sprouts in the sweetened, diluted whisky, so imparting the aroma of the leaves, but not the bitterness of the broken stalk (Lloyd). Another recipe is to crush together a teaspoonful of sugar and the same of crushed mint, add a sherry glass of whisky, fill up with shaved ice, and decorate with whole strawberries and mint leaves (Leyel. 1926).

MISTLETOE

(*Viscum album*) Most frequently parasitic on apples and poplars, but very rarely, at least these days, on oak. It is chiefly confined to the south and west of England though not to Devonshire, according to local folklore. They used to say that their county had been cursed by the Druids, and that mistletoe would never grow there. It was even claimed that a particular orchard straddled the Devon/Somerset border. The apple trees on the Devon side were completely devoid of mistletoe, while those on the Somerset side were full of it (Hewett). Neither does it grow in Ireland, nor has it ever been known to grow there, at least according to a report in the mid-19th century (*Notes & Queries; 1853*).

The word itself is OE mistiltan, where mistil is the name of the plant, and tan means a twig, or perhaps a shoot of a plant is better. Bearing in mind the fact that the German word Mist means dung, it is not surprising that there seems to be a connection between the name of the plant and the time-honoured belief that mistletoe seed cannot germinate until it has passed through a bird, a belief that started with Pliny. That bird, at least in popular belief, can only be the mistle-thrush, which is known as Grive du Gui in France, Misteldrossel in Germany, and Viscado in the Italian speaking parts of Switzerland (Swainson. 1886). The berries, "of a clammy or viscous moisture" (hence the generic name, (*Viscum*)), are such "whereof the best Bird-lime is made, far exceeding that which is made of the Holm or Holly bark" (Gerard). Bird Lime was used up to medieval times for catching small birds in the branches of trees, and also for catching hawks, which were decoyed by a bird tethered between the arches of a stick coated with the stuff (J G D Clark. 1948).

Perhaps its connection with birds, the messengers of the gods, together with its abnormal method of growth, conspired to make this a sacred plant in early times. It is claimed that it was closely associated with Druidical ritual, and its gathering was a definite function of the druidic religion (Kendrick), the cutting being made with a golden sickle (Stukeley described the implements as "upright hatchets of brass, called Celts"). Pliny said that the Druids "held nothing more sacred than the mistletoe", and that they believed it an antidote to all poisons (Aubrey 1686/7); furthermore, that it imparted fecundity to all animals. Hartland. 1909 quoted the maxim ascribed to the Druids that the powder of mistletoe makes women fruitful. In 17th century England it was certainly regarded as among the most efficacious of medical elements. Its berries held the male essence or protoplasm of the god, and so they were regarded especially as conferring powers of fertility. That explains why they were used at Yule as a kind of love-token, nowadays reduced to harm-less kissing under the mistletoe. The Welsh medical text known as the Physicians of Myddfai furthered the belief by prescribing a decoction to cause fruitfulness of the body and the getting of children. The Ainos of Japan thought exactly the same thing (Hartland. 1909). Even wearing mistletoe, without any internal dose, had the same effect, according to Coles: "some women have worn it about their necks or on their arms, thinking it will help them to conceive". On the other hand, African-Americans in the southern states of America used to wear mistletoe round their necks and dangling between the breasts to dry up the milk after weaning (Puckett). In parts of France, it was used in cases of difficult labour (Salle), while Welsh farmers linked mistletoe with luck in the dairy. In a scarce season they would say "No mistletoe, no luck", and they put a branch of it beside the first cow that calved after the first hour of the New Year (Trevelyan). Surely the extreme of the superstition is the belief sometimes quoted that if a girl is not kissed under the mistletoe before her marriage, she will be barren all her days (Waring).

It is difficult to decide whether fertility, or just luck, or even protection, is involved in some of the beliefs. In Germany, for instance, it was certainly used as a protective device for the cattle, just as rowan was (Macdonald. 1893). No mistletoe, no luck – but the opposite was believed, quite naturally, when it was unusually abundant, for then an exceptionally prosperous year could be looked forward to (Trevelyan). It follows, then, that to cut down a mistletoe-bearing tree could mean a very unpleasant end to the perpetrator (M Baker. 1980).

All evergreens are connected with the idea of immortality, and in European folklore mistletoe stands for life in the midst of death; it wards off death, being, as they say, indestructible by fire or water. Virgil's Golden Bough is the mistletoe, also indentified with that plucked in the grove at Aricia, by Lake Nemi – Frazer's Golden Bough. When it was found growing on an oak, a rare occasion, its presence there was attributed specially to the gods, and as such was treated with the deepest reverence (Dyer. 1889). Oak mistletoe was thought to be a panacea, or "heal-all" (Elton). Any mistletoe was, too, an embodiment of the lightning (the Swiss name is Donnerbesen, thunder besom, which protects from fire). In Sweden it is hung up in farmhouses, like rowan in Scotland (Dyer. 1889). In Austria, a branch fastened over the bedroom door kept away nightmares (Palaiseul). In Scandinavian mythology, Baldur was killed with a shoot of mistletoe, at least according to Snorri's version. Saxo thought of the weapon as a sword (and there was a famous sword called mistletoe in Norse tradition) (H R E Davidson. 1964). In any case, the plant must have been sacred to Teutonic peoples as well as to Celts, and in Greek mythology, Persephone had to use a mistletoe wand to open the gates of the Underworld (Baumann). There is a medieval Jewish tale in which

it was told that every tree had sworn it would not bear Christ's body, except for a cabbage stalk, and it was on this that he was crucified. There were English traditions according to which Christ was crucified on a mistletoe (Turville-Petrie). It was christianized in Europe under the name Sanctae Crucis Lignum, the wood of the Holy Cross. As the tree that had borne Christ, it had dwindled to a lowly shrub, yet acquired a sacred force (Grigson. 1955, and *Notes and Queries; 1853*). In Brittany, it was until quite recently still looked on as the tree of the Cross, capable of curing fevers (Friend. 1883).

The oak, too, was a lightning tree, so the rare oak mistletoe would be doubly significant. Equally important is the idea that when the oak lost its leaves, its spirit would have gone into the mistletoe (G Henderson. 1911). Any mistletoe was quite often used in Sweden against epilepsy and as a poison antidote (Fiske). It was referred to as "all-heal", not only for its curative properties, but also because it was the plant of peace, under which enemies were reconciled in ancient Scandinavia (Hole, *Christmas and its customs*). It was worshipped as All-heal in Persia, too, where it was sacred, and termed the Ethereal plant, because alone among the vegetable creation it springs ethereally in mid-air, and not from the earth" (Bayley. 1919).

But mistletoe is unfailingly associated with Christmas, perhaps because its berries appear close to the winter solstice, a long time after the flowers, which bloom in March to May usually. Every house had to have its mistletoe (it is unlucky *not* to have some at Christmas), but that was the only time it could be cut, for it was said in some parts that it would be very unlucky to gather it any other time (Drury. 1987). But, with a few exceptions, it was forbidden in churches, because of its pagan associations. Most of the exceptions occurred in the Black Country, where churchwardens' accounts show the practice to be quite widespread. In the seventeenth century, the incumbent at Bilston blessed and distributed it (A R Wright); so did the priest at St Peter's Wolverhampton (Hackwood). A bunch used to be hung inside the tower of Ashton-under-hill, in Worcestershire, and left there till replaced the following Christmas (Vaux). It was evidently used in Welsh churches, for a sprig used in the Christmas decoration of churches would bring good luck for the coming year. Another reference to it lies in the belief that if an unmarried woman put a sprig of mistletoe from the parish church under her pillow, it would cause her to dream of her future husband (Trevelyan). But the best known exception was at York, where it was carried ceremonially to the Minster on Christmas Eve, and laid on the high altar. After that there was a general pardon, and liberty for all was proclaimed for as long as the branch lay there (Hole). Mistletoe was the harbinger of peace, anyway (see above), a bunch hanging over a door was a sign of the friendly intentions of the inhabitants, and not just at Yule, for such a branch would hang the year round as a sign that guests greeted under it were safe in the house (Boland).

Mistletoe grows well in Worcestershire apple orchards, and there it is said to be unlucky to cut it at any other time than at midnight on Christmas Eve. A bunch was often kept to burn under next year's Christmas pudding, or to be hung round the neck as a witch repellent. Never take the whole bush, for that would ensure the worst of bad luck (Waring). Often the mistletoe was ornamented, traditionally with apples, nuts and ribbons, and, in Worcestershire, cut by the last male domestic to enter the family. When decorated it was hung in the centre of the room, sometimes with a cord attached to a pulley, to let the lady lower it and pluck a berry after she had been kissed. The berry was then thrown over the left shoulder (Grigson. 1955). In Herefordshire the mistletoe bough was cut on New Year's Eve, and hung up at twelve o'clock; the old one that had hung there through the year was at the same time taken down and burned. In Devonshire too they say that mistletoe should never be removed with the other Christmas decorations. But should be left until next Christmas, so that the house shall not be struck by lightning (Brown. 1964), or so that the house will never be without bread (*Devon Association. Report; 1955*), a reason also given in Huntingdonshire for leaving it up as long as possible:

> Mistletoe dead,
> Never want bread (Tebbutt).

But that could not have been the case everywhere, for the fact that Worcestershire girls ought to be given one of the berries after being kissed surely shows that a strictly limited life was accorded to its Yule fertility-giving powers. But to give some idea of the importance of the Christmas mistletoe, in those parts where it was unobtainable, the Lincolnshire marshes for example, some other evergreen was conscripted in its place, having exactly the same functions and privileges as the real thing (Gutch & Peacock).

Like other magical plants, it was credited once with making the wearer invisible (Dyer. 1889), it will open a lock, and, used as a divining rod, will reveal hidden treasure (Grigson. 1955). In Sweden, the divining rod had to be cut on Midsummer Eve (Philpot). Most of the other beliefs about mistletoe circle round its protective function, like the Black Country custom of hanging a small bag of it round the neck as a safeguard against witchcraft and evil spirits (Hackwood). In America, the seeds alone were adequate, apparently, for there is a record from Alabama of the practice of putting these mistletoe seeds above the door to keep off evil spirits (R B Browne), and mistletoe is guaranteed in some parts of Normandy to keep the bed free from fleas (W B Johnson). An extreme

example of faith in its powers is the French belief that a train could not be derailed if a piece of mistletoe was put in one of the wagons (Sebillot). Welsh people, though, as late as the 19th century, had their doubts about it. For it to have any power, they said, it had to be shot or struck down with stones, from the tree on which it grew (Trevelyan). It comes as a surprise, though, to find that in Devonshire it was a thoroughly unlucky plant. There was a tradition that farmers were forbidden to allow it to grow in their orchards (*Notes and Queries; 1872*), a Druidic ban, according to earlier belief (but did it actually grow there? – see above). From the same county the belief is recorded that if you plant mistletoe and it grows, your daughters will never marry (*Devonshire Association. Report; 1959*).

A Christmas marriage divination game has already been noted, and another Welsh one involving mistletoe was to gather a sprig on St John's Eve, or at any time before the berries appeared, and to put it under the pillow, when it would induce dreams of the future, good or bad (Trevelyan). In both Alabama and Kentucky (R B Browne, and Thomas & Thomas), they used to hang mistletoe over the door, to see who would be the first girl to walk in, and she would be the future wife. Another Kentucky divination was to name mistletoe leaves for a boy and a girl, and then put them on a hot stove. If the leaves hopped towards each other, it was taken as a clear sign of marriage between the two.

One would expect such a magical plant as mistletoe to have a lot of medicinal uses, but the surprising thing is that many of them are genuine. In spite of the fact that the berries are toxic (if enough are eaten they will cause gastro-enteritis), mistletoe has been known since ancient times to have a beneficial effect on the heart (W A R Thomson. 1978), and also on tumours. Application of the juice on tumours is a well-known form of treatment (W A R Thomson. 1976), and it has been used as a specific against epilepsy for a very long time. The doctrine of signatures may have played its part, for it has been said that its habit of downward growth recommended it for curing the falling sickness (Browning), but it is certainly not altogether fancy since it contains an active principle that is anti-spasmodic, and reduces blood pressure (Grigson. 1955) (always provided, Pliny said, "it has not touched the ground"; and always provided not too much is given, for a large dose would have the opposite effect (Anderson)). Culpeper speaks of it as a sure remedy for apoplexy, palsy and the falling sickness, and in the mid 18th century, Hill was also recommending the leaves, "dried, and powdered" as a "famous remedy for the falling sickness". A decoction was made in Lincolnshire (Gutch & Peacock), and another folk remedy for the complaint was to take as much powdered mistletoe as would lie on a sixpence early in the morning in black cherry water or beer, and repeat for some days. But then comes the "charm" aspect of the cure – those "some days" had to fall near the full moon (Jacob). There are many more purely magical uses for the same complaint. Finger rings were made of mistletoe in Sweden (Dyer. 1889), and necklaces in Normandy (Sebillot), where also a mistletoe chaplet was put round a child's head as an antidote for fits and epilepsy (Johnson), and it seems to have been used in Sussex, along with peony root, as an amulet to be hung round an infant's neck as a cure for fits (Rolleston). Hawthorn mistletoe was used for dropsy (Sebillot), and also in Lincolnshire for St Vitus's Dance (Gutch & Peacock).

Jaundice was another ailment treated with mistletoe, either the leaves, as a receipt from the Welsh text known as the Physicians of Myddfai prescribes, or the berries, which, according to the old usage in the bocage country of Normandy, had to be soaked in a male baby's urine, and then put on the patient's head, while an unspecified, and doubtless secret, charm was spoken (Sebillot). In the same region, hawthorn mistletoe was used for children's coughs, after it had been soaked in mare's milk. Something described as a broth of mistletoe was said to remove adenoids (W B Johnson). Mistletoe tea has been regarded in America as a good remedy for menstruation that has been stopped by a cold (H M Hyatt).

MOLES

If you plant CAPER SPURGE (*Euphorbia lathyris*) in the garden, it will keep moles away. That is a widespread belief, and may very well be true. Anyway, the practice spread to America, and there they call the plant Mole-tree, or Mole-plant (G B Foster; Bergen. 1899).

MONKEY NUT

Nearly as common as peanut as a name for GROUND NUT. Grigson. 1974 suggested that the only reason for the name was that they are bought to feed monkeys at the zoo.

MONKEY PUZZLE TREE

(*Araucaria araucana*) Equally well-known is the name CHILE PINE. Introduced into Britain at the end of the 18th century, long enough ago to have a little folklore attached to it. "Never speak while passing a monkey puzzle tree" (Opie & Opie) is a widely known children's proscription, from Scotland to London, where the result of breaking the commandment is said to be bad luck for three years (Opie & Tatem). It was an old Fenland belief that if one of these trees was planted on the edge of a graveyard, it would prove an obstacle to the devil when he tried to hide in the branches to watch a burial. In spite of that, though, it is a generally unlucky tree throughout the area (Porter. 1969).

The name Monkey Puzzle is said to have arisen from a remark by Charles Austin during the ceremonial planting of one of these trees in the gardens at Pencarrow, in Cornwall, in 1834 – "it would be a puzzle for a monkey to climb that tree" (Grigson. 1974). Another derivation claims that the name came about because of the way the branches curl into the shape of a monkey's tail (Campbell-Culver).

MONKSHOOD

(*Aconitum napellus*) An extremely poisonous plant, "the force and facultie" of whose poison "is deadly to men, and all kindes of beasts" (Gerard). Shakespeare (in 2 Henry IV) compared the effects of the poison to gunpowder:

> "...though it do work as strong
> As Aconitum or rash gunpowder".

In the language of flowers, it is the symbol, not surprisingly, of crime (Rambosson). It has been used for medicinal purposes. Pliny called it "plant arsenic", while Rambosson referred to it as "le poison suprême", atavisha, as it was known in India. In classical tradition, monkshood sprang from the foam of Cerberus, when dragged to light by Hercules. It is mentioned by Ovid, *Metamorphoses*, as the principal ingredient in the poisonous draught prepared by Medea for the destruction of Theseus. It was apparently dedicated to Hecate (Grigson. 1959), and was used by witches for drugs and poisons (Summers. 1927). According to Scot. 1584, it was an ingredient in witch flying ointment, details of which he cites, and had taken from Della Porta, a 16th century physician. The effects of such ointments have been analysed frequently in modern times. If such an ointment ever existed, it could give the impression of flying through the air. Some witches who claimed they flew to sabbats were known to have been in their beds the whole time (Donovan). It was a common belief at one time that poisons were the antidote to poison, so, for example, one finds it was used against scorpion stings (Ellacombe). Another strange belief was that they were the cause of fevers; but it has been used as a febrifuge for a very long time. Rheumatism has been treated in Russian folk medicine by a liniment made from the dried roots. Care was taken in its application; if the patient had any heart trouble, the amount used was decreased significantly. This liniment was rubbed in until the skin was dry, and then the area was covered with flannel (Kourennoff).

MOON PLANT

(*Sarcostemma acidum*) A plant from the tropical parts of India, one of the plants used as a substitute for "soma", in Vedic sacrifices. In fact, some believe it to be the true "soma" from which the intoxicating drink was prepared. It contains a large amount of milky sap, acid, and often used by Indian travellers too allay thirst (Chopra, Badhwar & Ghosh).

MOONWORT

(*Botrychium lunaria*) Moonwort, so called because of the crescent shape of its leaflets, so it has always been associated with moonlight and magic. It was believed to wax and wane like the moon, and to shine at night (Coats. 1975). Like other "moon" plants, it was supposed to have the power of repelling snakes, and of opening locks (R T Gunther. 1905). Any horse that trod on one would cast a shoe, and burglars would make an incision in the palm of their hand and put a piece of moonwort under it, for this would give them a hand that would open bolts and bars at a touch (Coats. 1975).

MORNING GLORY

(*Rivea corymbosa*) This plant provided one of the most important sacred hallucinogens of the Mexican Indians at the time of the conquest, the revered ololiuqui of the Aztecs. The seeds, which contain alkaloids closely related to LSD (Wasson, Hofmann & Ruck), were used to induce a ritually divinatory trance, and are still so used in modern Mexico (Norbeck). Aztec priests also ground up the seeds as a potion which they used in ointments to make themselves fearless, or to appease pain (Dobkin de Rios).

MOTHAN

A mystical herb called Mothan or Moan in Scotland and Ireland. Lewis Spence (Spence. 1945) reckoned it was the THYME-LEAVED SANDWORT (*Arenaria serpyllifolia*), though a likelier candidate is PEARLWORT (*Sagina procumbens*); but the Bog Violet, or Butterwort (*Pinguicula vulgaris*) has also been put forward. Spence described the Mothan as being given to cattle as a protective charm, and people who ate cheese made from the milk of a cow that had eaten the plant were secure from witchcraft. The Mothan was to be picked on a Sunday as follows: three small tufts to be chosen, and one to be called by the name of the Father, one by that of the Son, and one by the Holy Ghost. The finder would then pull the tufts, saying (in translation):

> I will pull the Mothan
> The herb blessed by the Domnach;
> So long as I preserve the Mothan
> There lives not on earth
> One who will take my cow's milk from me.

The three tufts were then pulled, taken home, rolled up in a piece of cloth, and hidden in a corner of the dairy. Gaelic has a number of versions of the incantation, some very lengthy. Such a plant had to be pulled with the suitable verses if only out of respect (see, for examples J G Campbell. 1902; Carmichael).

Mothan was put in the milkpail as a means of restoring virtue to the milk, and at the time of mating, a piece was put in the bull's hoof, so that no witch could touch the calf's milk. It protects from fire, and from the attacks of fairy women (J G Campbell. 1900). Anyone carrying this plant, or having drunk the milk

of a cow that had eaten it is "immune from harm" (McGregor). There is an old saying in the Hebrides that, when a man has a miraculous escape from death, he must have drunk of the milk of a cow that ate the mothan (McGregor).

MOTHER-DIE

When a plant bears this name, then it is taboo for children to pick it. Breaking that taboo would bring about the death of the child's mother. No less than twelve plants bear the name, though only FOOL'S PARSLEY (*Aethusa cynapium*) is poisonous. That makes the ban more understandable, but the others are generally unlucky plants in one way or another. The fact that COW PARSLEY is on this list makes it possible that HEMLOCK was the original plant proscribed. CHERVIL is another example, as is SWEET CICELY. Any white-flowered umbellifer is unlucky, simply because there is a superficial resemblance to hemlock. Cow Parsley is unlucky to bring indoors; so are HAWTHORN and YARROW, which are both white flowered. SHEPHERD'S PURSE is another Mother-die plant. There is a children's game played with the seed pod. Children hold it out to their companions, inviting them to "take a hand o' that". It immediately cracks, and the first child shouts "You've broken your mother's heart" (Cf the names Mother's Heart and Pick-your-mother's-heart-out) (Vickery. 1985, J D Robertson). The CAMPIONS are unlucky too- if you pick them, one of your parents will die – father if it is Red Campion, mother if white. The other plants with the name include GERMANDER SPEEDWELL, very unlucky to bring indoors, and ALEXANDERS. One is alerted by a Cumbrian name for HERB ROBERT, Death-come-quickly, to expect a similar superstition to that marked by Mother-die. Sure enough, there is the warning that if picked by children, the result would be the death of one of the parents.

MOTHER-OF-THOUSANDS

(*Saxifraga sarmentosa*) This plant propagates itself by sending out trailing runners, and that is the significance of the common name. Even more graphic is the name Hen-and-chickens, giving the impression of a mother plant with children around her. That name is given to a number of other plants (see Watts. 2000), and the STRAWBERRY is a good example of this kind of reproduction. That is why the saxifrage is called Strawberry Geranium, or Strawberry Begonia in some areas. A rather more mystical name given to this plant, still referring to the runners, is Thread-of-life, from Northamptonshire (Britten & Holland), and these threads also account for a number of "wandering" names, like Creeping, Wandering, or Roving, Sailor, the source of a belief that this ia an unlucky plant, for an accident to the plant would ensure a mishap to any relative who is a sailor (*Folk-lore. vol 37; 1926 365–6*).

MOTHERING SUNDAY

or Mid-Lent Sunday, the fourth in Lent, when traditionally children away from home returned to visit their parents. Servants were given a holiday for the same purpose. A bunch of VIOLETS is the traditional gift, and carrying violets gave rise to the saying: "Go a-mothering, and find Violets in the lane" (Friend).

MOUNTAIN ASH

An alternative name for *Sorbus aucuparia* – see ROWAN

MOUNTAIN CLUBMOSS

(*Urostachys selago*) In the Scottish Highlands, this clubmoss was, we are told, used as an abortifacient. James Robertson, who toured the West Highlands and Islands in 1768, noted that "the Lycopodium selago is said to be such a strong purge that it will bring on an abortion" (quoted in Beith). Another, quite separate, Highland usage was as a skin tonic. Women and girls steeped the moss in boiling water. The cooled liquid was used as a lotion(Beith).

MOURNING

WEEPING WILLOWS are fairly obvious symbols of death and mourning, often used in some stylised form embossed on Victorian mourning cards, though they first appeared at the end of the previous century as a tomb decoration, usually with the figure of Hope, or the widow weeping and clinging to the urn beneath its boughs (Burgess). CYPRESS too symbolises death and mourning.

MUGWORT

(*Artemisia vulgaris*) A common plant of waste places and roadsides in this country. Common it may be, but this is one of the most important plants in the folklore of Britain; its ritual importance emphasised by its particular association with Midsummer. It is actually known as John's Feast-day Wort in the Isle of Man (bollan feaill Eoin (Moore)), and in Europe, too, it is known as St John's Herb, and also as St John's Girdle – this is the medieval cingulum Sancti Johannis (St Johannesgurtel), and Sonnewendgurtel (Storms). It was believed that John the Baptist actually wore a girdle of it in the wilderness. It does keep flies away if worn, and grows in the right area (Genders), but this girdle was actually worn on St John's Eve to serve as a protection against ghosts or magic. Then the girdle would be thrown into the Midsummer Fire, and the ill-fortune of the wearer is burnt along with it (Bonser). Mugwort would be ritually gathered on Midsummer Eve to serve as a preventive against evil in general and witchcraft in particular, and to be hung over doors to purify the house of evil spirits (Jones), and to protect it from lightning (Le Strange). Cows were protected from fairy interference by having mugwort put in their byres at St John's (Spence. 1948), and chaplets were made of it in the Isle of Man, to be worn by both man

and beast for the same purpose (Moore). Sprigs would be worn in button-holes, too. Soldiers attending the Tynwald on Old Midsumer Day (4 July) used to wear them; the custom was revived in 1925, and sprays are distributed among the people coming up the hill (Paton), and they have become a conspicuous feature of what is a National Day on the island (Mabey.1998), its ritual importance emphasised by its particular association with Midsummer. "Muggwith twigs" were used in the south of Ireland as late as 1897, after they had been singed in a St John's fire, to guarantee protection against disease (Davidson. 1955). It is not surprising that mugwort served as a symbol of happiness, and of tranquillity (Leyel. 1937).

People in Germany made hats of the plant, and, as with the girdles, threw them into the fires. "… Whoso looketh into the fire thro' the same, hath never a sore eye all that yeare: he that would depart home unto his house, casteth this his plant into the fire, saying, so depart all mine ill-fortune and be burnt up with this herb" (Grimm). In Britain, too, it was said that if the flames of the Midsummer fires were viewed through a branch of mugwort, it would ensure good vision for the year.

Still connected with Midsummer, there was a widespread delusion about the roots of mugwort – see Lupton: "It is certainly and constantly affirmed that on Midsummer Eve there is found, under the root of Mugwort, a coal which saves or keeps them safe from plague, carbuncle, lightning, the quartan-ague, and from burning, that bear the same about them …". From the *Practice of Paul Barbette*, 1676, a more rational view is taken – "for the falling sicknesse some ascribe much to coals pulled out (on St John Baptist's Eve) from under the roots of mugwort; but those authors are deceived, for they are not coals, but old acid roots, consisting of much volatile salts, and are almost always to be found under mugwort : so that it is only a certain superstition that those old dead roots ought to be pulled upon the eve of St John Baptist, about twelve at night".

This plant was noted for its magical ability to open locks, like a "springwort". Another of its attributes was to enable the traveller who carried it never to feel weary, a belief that was made much of in earlier times. See Gerard: "Pliny saith, That the traveller or wayfaring man that hath the herbe about him feeleth no wearisomeness at all; and that he who hath it about him can be hurt by no poysonesome medicines, nor by any wilde beast, neither yet by the Sun itselfe". Coles wrote that if a footman take mugwort and put it in his shoes in the morning he may go forty miles, and not be weary" (quoted in J Mason). The same had already been written in the Anglo-Saxon version of Apuleius, and Neckem, *De naturis Rerum*, of late 12th century date, mentions the superstition, which is still current in Somerset – "Put mugwort in your shoes and you can run all day" (Tongue).

Mugwort was in demand, too, to "put to flight devil sickness, and in the house in which he has it within, it forbiddeth evil leechcrafts, and also it turneth away the (evil eyes of evil men) (Cockayne).That use of mugwort is found as far away as China. At the time of the dragon festival, people hang it up to ward off evil influences (F P Smith). In Wiltshire, people used to say that the leaves always turned to the north, and north was supposed to be the devil's quarter (Wiltshire). Crystal gazers found, so they say, that they got better results by drinking some mugwort tea during the operation (Kunz), and the tea is quite seriously believed to be an aid in the development of clairvoyance – in Japan, for instance, if a house were robbed in the night, and the burglar's footmarks were visible next morning, the householder would burn mugwort in them, so hurting the robber's feet and making his capture easy (Radford). Perhaps it is rather a question of protection; mugwort will hold evil at bay, or make it easier to overcome. Hence the Ainu make mugwort images which they put upside down into holes in order to bring misfortune upon their enemies (E A Armstrong). This concept makes it easier to understand the Chinese belief that mugwort is one of the weeds that had to be entirely destroyed before people could settle in a new place (E A Armstrong), for there would always be the suspicion that the mugwort was there to look after someone else's interests.

Miwok Indians of California protected themselves against the attention of ghosts by wearing a necklace of mugwort. Corpse handlers rubbed themselves with it, otherwise they would be haunted by the ghost of the deceased (Barrett & Gifford). It was recommended in the *Grete Herball* (1526) to produce merriment, though it becomes clear as we read on that this is not exactly what is meant – it should be laid under the door of the house, for if this is done "man or woman can not annoy in that house" (homme ne femme ne pourra nuire en cette maison) (Arber).

After all these wonders, it almost sounds too prosaic to mention that mugwort will keep the flies away. You can either wear a sprig (Genders. 1971), or keep an infusion to sponge over the face and arms (Cullum). The very name of the plant confirms it. Mugwort, O E muogwyrt, from a German base meaning a fly or gnat. Midge is the same word. It is a vermifuge too; the dried flower heads used to be sold by herbalists as "wormseed" (Earle). One of the French names, mort des vers, sums it up. And not only flies and worms – it will keep snakes at bay, too, if we are to believe Thomas Hill's quote – "… Serpents in the Garden ground or elsewhere will not lodge or abide, if the owner sow or plant in borders about, or in apt corners of the Garden, either the Woirm-wood, Mugwort, or Southern-wood …" (T Hill).

One would expect a benevolent herb like mugwort to be well-favoured for medicinal purposes; in fact only betony seems to have a longer list of ailments for which it is prescribed. Only a few of the important uses can be discussed here. Chinese medicine uses it extensively, and recognizes several forms in commercial use, including a cautery use that is additional to that which forms part of normal acupuncture treatment. One of the best-known pieces of folklore in this country involving mugwort is the Clyde legend that as the funeral procession of a young woman who had died of consumption was passing along the high road, a mermaid surfaced, and said:

> If they wad drink nettles in March,
> And eat Muggons in May,
> Sae mony braw maidens
> Wadna gang to the clay (Chambers).

Similarly, from Galloway, there is a story of a young girl close to death with consumption, and a mermaid who sang to her lover:

> Wad ye let the bonnie May die i' your hand,
> An' the mugwort flowering i' the land?

The lad cropped and pressed the flower tops, and gave the juice to the girl, who recovered (Cromek). A Welsh rhyme takes up the theme:

> Drink nettle-tea in March, mugwort tea in May,
> And cowslip wine in June, to send decline away.

But why a mermaid in Scotland? Benwell & Waugh came up with an interesting answer – Artemis (The generic name of mugwort is *Artemisia*) was also a fish goddess, sometimes depicted with a fish tail. So it was the goddess herself, and by extension the plant itself, that was advertising its own benevolence.

Mugwort tea is still often taken for rheumatism (Deane & Shaw), and the Physicians of Myddfai were recommending something similar, though considerably more complicated, for intermittent fevers. They also prescribed the leaves boiled in wine to "destroy worms", for mugwort is a known vermifuge. The use of the plant for palsy crops up several times in the old herbals, and even epilepsy, which herbalists still use for the disease in the form of root tincture (Schauenberg & Paris); there is an Irish record of a similar use (Moloney).

The Physicians of Myddfai made the extraordinary claim, "if a woman be unable to give birth to her child, let the mugwort be bound to her left thigh. Let it be instantly removed when she has been delivered, lest there should be haemorrhage. Clearly this is a garbled version of an older, probably genuine, usage of the plant. There is a 17th century recommendation of its use, steeped in wine, as an aid to conception (K Knight). Lastly, there are a group of leechdoms to relieve sore feet. Cockayne's translation of the Anglo-Saxon version of Apuleius has "take the wort, and pound it with lard, lay it to the feet". He also has, for the same complaint, "eat root of mugwort. Mixed with oil, and also bind on elder, waybread and mugwort leaves pounded". To round it off, the Physicians of Myddfai had "For weariness in walking. Drink an eggshell of the juice of mugwort, and it will remove your weariness". Just carrying mugwort would do, for there is a Scottish belief that travellers who carried it would not tire (Beith).

MULLEIN,

or **GREAT MULLEIN** (*Verbascum thapsus*) Mullein seeds are narcotic, and have been used to stupefy fish (H C Long. 1924), as have those of other mulleins around the world. The leaves, too, are similarly useful, at least to get rid of lice, for the Pennsylvania Germans always used to to put them in the pigsty for just that purpose (Fogel). The flowers were used to dye hair yellow (Rohde. 1936), at least as early as Parkinson's time, for he tells us that the flowers "boyled in lye dyeth the haires of the head yellow and maketh them faire and smooth" (Parkinson. 1640). One can still buy camomile and mullein shampoo for fair hair.

There was once some idea of protection against evil connected with this plant. See, for instance, the Anglo-Saxon version of Apuleius, as translated by Cockayne: "if one beareth with him one twig of this wort, he will not be terrified with any awe, nor will a wild beast hurt him, or any evil coming near". It is also pointed out by Apuleius that what came to be regarded as mullein was the plant given by Mercury to Ulysses to neutralize the evil magic of Circe. In much the same vein, there is recorded an Irish charm to get back butter that had been witched away – all that is required is to put mullein leaves in the churn (Gregory. 1970). It was used in a divination game in America. All that had to be done was to twist a stalk nearly off, after naming it for someone. If it lived, then the someone loves you. Afterwards, you had to count the new shoots that sprang up – this will be the number of children you would have (Bergen. 1896). The instructions varied slightly; some say you should break the stalk right off before naming it. Or, you had to bend it towards the sun (R B Browne) or in the direction of the boy friend's home, when if the stalk grows straight again, he loves you (Thomas & Thomas).

The name Hag Taper does not mean witch's candle, but hedge taper; in fact, the name Hedge Taper is actually recorded (Macmillan). Taper is straightforward enough, for the stems and leaves used to be dipped in tallow or suet and burnt to give light at outdoor country gatherings, or even in the home (Genders. 1971). They were used at funerals, too, and in the French Fête de Brandons, held on the first Sunday in Lent.

The flannelly leaves had their own uses, for they could actually be used like flannel, and wrapped round the throat to help relieve coughs and colds (Sanford), or, more simply, cut to be used as shoe-liners to keep the feet warm in winter (Forey). Smoking the dried leaves has long been an Irish remedy for asthma or bronchitis (Ô Súilleabháin), and it is used like that in America, too, for catarrh as well as asthma (Henkel). Several American Indian groups gathered the leaves to smoke as tobacco (Corlett), though not, apparently, for any remedial purpose, though they certainly used the roots medicinally, for lung disease, presumably as an expectorant (H H Smith. 1923; Youngken. 1925), and consumption was treated with mullein, in cattle as well as in humans (Grigson. 1955; Ô Súilleabháin; Wood-Martin). An Irish remedy for tuberculosis was to boil an ounce of the dried leaves in a pint of milk, and give the result to the patient several times a day (Moloney). Alabama folk remedies include hot mullein tea for a cold, and a root tea, or eating the root with sugar, for the croup. Another one is for bronchial infections: heat mullein leaves in warm vinegar, and put them on the back and chest. Then drink a cup or two of peppermint tea (R B Browne). Kentucky practice was to take mullein tea for colds, bronchitis and croup (Thomas & Thomas). Eating the root appears again in the Balkans – it was taken there for a swelling in the throat (Kemp).

The Pennsylvania Germans used to say that to ease piles, what you had to do was to sit on mullein leaves (Fogel). That particular usage has a venerable history. A medieval record states that "for piles, take a pan with coals and heat a little stone glowing and put thereon the leaves of … mullein; and put in under a chair or under a stool with a siege, that the smoke thereof may ascend to thy fundament as hot as thou mayest suffer" (W M Dawson. 1934). Gerard, later on, was still able to recommend not only mullein leaves, but also the flowers – "the later Physitions commend the yellow floures, being steeped in oile and consumed away, to be a remedie against the piles".

Culpeper recommended "the juyce of the Leaves and Flowers" for warts (Drury. 1991), and the powder of the dried root was also prescribed for the same condition. It has also been used down the centuries for various skin conditions, for mullein leaves rubbed on the skin will produce a fine complexion, or so it was believed in America (Bergen. 1896), or, as in Alabama, the leaf tea could be used, with some glycerine added, as a wash three or four times a day (R B Browne). They used to make a salve, too, by steeping mullein blossoms with lard (Bergen. 1896). Ear drops have been made from the flowers. The method is to take fresh flowers, steep them in olive oil, leave them for three weeks in a sunny window, then strain off. Two or three drops in the ear will relieve earache quickly (but one has waited three weeks already!).

A very odd thing about this preparation is that a few drops in warm water before bedtime actually, or so it is claimed, cures children of bed-wetting (Genders. 1976). Boils were treated by Irish country people by applying a mullein leaf roasted between dock leaves, and moistened with spittle, as long as the spittle be that of an Irishman (Egan).

It is used for rheumatism in America, either by boiling the root and mixing it with whisky, or wild cherry bark, and drinking this as needed, or by dipping a cloth in mullein leaf tea, and binding it on the affected part (R B Browne), which reminds one of a very early cure for gout, which required the sufferer merely to lay the pounded herb to the sore place, "… within a few hours it will heal the sore so effectively that [the gouty man] can even dare and be able to walk" (W M Dawson. 1934). There are still some fantastic claims made for mullein. Even carrying it about with one helps epileptics, so Gerard reported, though he took care to say he did not believe it, but he does not disclaim the belief that wearing the leaves under the feet day and night "in manner of a shoe sole or sock, brings down in young maidens their desired sicknesse, being so kept under their feet that they fall not away". Certainly, the leaves often used to be put into children's shoes, but for quite a different reason – it was done when the soles were wearing thin and so would delay the time when they had to be replaced (Genders. 1976). Finally, an American usage – a mullein leaf is reckoned to be good for a parrot's bite ! (Bergen. 1896).

MULTIPLE SCLEROSIS
One of the claims made for oil of EVENING PRIMROSE is that it can slow down changes in multiple sclerosis (M Evans).

MUMPS
Strong CINNAMON tea taken at the beginning of mumps will, it is claimed, reduce the violence of the complaint and prevent complications. CAT'S FOOT tea is another treatment for the disease (Grieve. 1931). In the Cotswolds area, a fomentation of the leaves of MARSH MALLOW, soaked in boiling water, is used for mumps and swollen glands (Briggs. 1974). PRIVET offers a cure. The berries have to be boiled until the juice is out of them, and the juice poured into a small bottle with cream from the top of the milk. When it has been cool for at least three hours, take a teaspoonful of the juice and one berry, once daily, and after food (Vickery. 1995).

Musa x paradisiaca > BANANA

MUSICAL INSTRUMENTS
The simplest, and probably the most ancient, musical instrument, is made from REEDS, and called variously Pan-pipes, Shepherd's Pipes (though these were usually made from oat straw), or Syrinx. The reeds are of different sizes, placed side by side, each

stopped at the bottom end (F G Savage). *Sambucus,* the generic name for the elders, is from the Greek sambyke, the name given to an ancient musical instrument, rather like a harp, played by both Greeks and Romans, and made of ELDER wood (Halliwell), and supposed to be the Biblical sackbut (N D G James), though whatever it was, a sackbut was not a harp. Pliny said that pan-pipes and flutes were shriller when made of elder wood. Young boys in the Hebrides who had aspirations to be pipers, made their chanters from the young branches (Murdoch McNeill).

There is a record from Devonshire that CHIVES were looked upon as fairy musical instruments (Whitcombe).

LABURNUM wood, a hardwood, was used for all purposes where strength and elasticity were needed, for parts of musical instruments, for example (Brimble). Flutes were once made from it (Ablett). Irish harps were usually made of WILLOW wood, for these trees have a soul in them which speaks in music (Wilde. 1890). Evelyn. 1678 described CYPRESS timber as "a very sonorous wood", enough to account for its use for organ-pipes, harps, etc. And the wood of LOCUST BEAN (*Catalpa bignonioides*) was prized for making musical instruments.

MUSK THISTLE
(*Carduus nutans*) The thistledown used to be collected by country people for stuffing into cushions and pillows, usually with the dried aromatic leaves of wormwood and camomile (Genders. 1971).

MUSTARD TREE
see TOOTHBRUSH TREE (*Salvadora persica*)

Myristica fragrans > NUTMEG

MYROBALAN
(*Terminalia catappa*) A useful tree from India and the Andamans. It is highly esteemed by Hindus, and it is said that when Indra was drinking nectar in heaven, a drop of it fell on earth, and became this tree (Pandey). The flowers are used in Indian dyeing practice for a yellow colour when mixed with mango tree bark and alum (Gillow & Barnard). The fruit is also used for dyeing, and, in eastern India, to colour the teeth black (Perry. 1972).

Myrrhis odorata > SWEET CICELY

MYRTLE
(*Myrtus communis*) An evergreen, used by the Greeks as a symbol both of love and immortality. Because they regarded their great men as immortal, both Greeks and Romans put myrtle wreaths on the heads of their successful poets and playwrights (Moldenke). In the Scriptures, myrtle is referred to chiefly as a symbol of divine generosity, and throughout the Bible it is emblematic of peace and joy, and also of justice (Moldenke). But it is the love symbolism that is the most important. It was held to have the power of

creating and perpetuating love (Philpot), and as such was a symbol also of married bliss (Baker. 1980). It was often used in Europe in the bridal wreath, the forerunner of orange blossom as the bridal emblem (Baker. 1977). Since the mid-nineteenth century, British royal brides have carried in their bouquets a sprig of myrtle from a bush said to have grown from a piece carried by Queen Victoria on her wedding day, growing in the gardens of Windsor Castle (Higgins). The habit spread, until it was quite common for a sprig of myrtle from the bridal wreath to be planted in the bride's garden, but always by a bridesmaid, never by the bride herself (Baker. 1980). If the sprig did not strike, then the destiny of the planter was to stay an old maid, an unlikely fate in this context, because myrtle roots very easily. The future blooming of the bush was the portent of another wedding in the family (Leather). But Elworthy says the connection with Venus is the reason for its *exclusion* from a maiden's wreath. Perhaps the real reason for his condemnation is the fact that myrtle used to be one of the constituents, with roses and olive, in black magic for sexual motives (Haining).

During Lent, lovers in Tuscany used to break a small myrtle branch in two, each taking a piece and keeping it. Whenever they happened to meet, they greeted each other with the challenge Fuori il verde - out with your green branch. If either failed to respond, the affair was broken off, for it was a sign of misfortune. The custom usually ended in marriage, though, on Easter Sunday, if possible (Friend. 1883). Welsh brides used to wear sprigs of myrtle, and in some parts of Wales, young girls wore it when going to their first communion. Sprigs were also put in cradles, to make babies happy (Trevelyan), though the real reason was probably to protect them from witchcraft, just as the hen blackbird was once said to spread myrtle twigs on her nest to preserve her young from evil influences (Hare). Another Welsh belief was that if myrtle grew on both sides of the door, blessings of love and peace would never leave the house. Conversely, of course, to destroy a myrtle would destroy both love and peace (Trevelyan). The bush also acts as a talisman to protect the house from lightning, according to Staffordshire folklore (Burne. 1897). Pliny recommended a myrtle stick for a long journey, for then "he shall never be weary (Holland's translation of 1601 – see Seager).

There was a Cornish superstition that only old maids could rear a myrtle, and that they would not bloom when trained against houses where there were none (Courtney. 1890), an odd belief when the plant's connection with the bridal wreath is considered. There are some other examples of advice as to its cultivation. In parts of France, for example, they said that a myrtle would only take root if planted on Good Friday (Sebillot). Received wisdom in Somerset used to be

that when planting it, one should spread out the tail of one's gown, and look proud or it would not flourish (Tongue. 1965). In that same county, it is said to be one of the luckiest plants to have in a window-box, although it will not grow there unless planted by a good woman (Baker. 1980). Again, one has to be proud of it, and water it each morning (*Notes and Queries. vol 7; 1853*). The luck in having a flourishing bush of myrtle is often expressed, and not only in Britain, for a Greek rhyme translates:

> Who passeth by a myrtle bush, and plucketh not a twig,
> O may he not enjoy his youth, although he's tall and big (Argenti & Rose).

Closer to home, it is said that if the leaves crackle in the hands, that shows your future partner will be faithful (Bardswell), which brings us to direct divination using myrtle. Simply to dream of it promises not only many lovers but also a legacy (Gordon. 1985). Myrtle put under the pillow will bring dreams of the future husband (Hawke). In the Midlands and North Country, a girl would take a sprig on Midsummer Eve and lay it on her prayer book with the words 'Wilt thou take me (mentioning her name) to be thy wedded wife?' She then had to close the book and sleep with it under her pillow. If the myrtle was gone in the morning, she would marry her present lover (Baker. 1980).

Although it was Aphrodite's plant, it was also the tree of death, particularly, in Greek mythology, the death of kings (Graves). Perhaps that is why there is such a prejudice against it in America. It is rarely seen outside cemeteries. Never let it grow around the house, they say in Illinois, or there will be sickness and trouble there as long as it is growing (H M Hyatt). In modern Egypt, sprigs of myrtle are laid on the graves of relatives, and also apparently put underneath the body, too (Westermarck). In parts of France it was also once the custom to put the last branch of myrtle that had belonged to the dead man into the coffin with him (Sebillot).

Candied myrtle blossoms are supposed to beautify the complexion (Wiltshire), and Evelyn recommended the use of the berries for inflammations of the eyes, and as a consound, which is an older version of the word to consolidate, and refers to the knitting of broken bones. A syrup from the blue berries was taken in the Western Isles, a spoonful at a time, for flux (Martin), diarrhoea in modern terms. Bloody flux is dysentery, and Gerard mentions that myrtle ("the leaves, fruit, buds and juyce") could be taken for that ailment. Among other recommendations, he said dandruff could be treated with the decoction, and he agreed with Evelyn that it could dye the hair black, and not only that, but "it keepeth them from shedding".

Myrtus communis > MYRTLE

N

Narcissus pseudo-narcissus > DAFFODIL

NARD
see SPIKENARD

Nardostachys jatamansi > SPIKENARD

Narthecium ossifragum > BOG ASPHODEL

NASTURTIUM
(*Tropaeolum majus*) A South American plant, brought from Peru to France in 1684, and called La Grande Capuchine. Surprisingly, it is listed as the symbol of patriotism (Leyel. 1937). *Tropaeolum*, the botanical name, means a trophy; it was the shield-like leaves and helmet shaped flowers that suggested the name. Old gardeners claim that growing nasturtiums up the trunks of fruit trees will get rid of aphis on the latter, and planting them in the greenhouse also guards against woolly aphis (Boland & Boland). If you want them to grow fast, plant them on St Patrick's Day (Thomas & Thomas).

Nasturtium leaves are antiseptic, and have been recommended in cases of scurvy (Thornton). Eating the flower petals is an oriental custom, but in Europe they are often added to salads and sandwiches these days (and so are the leaves). Evelyn reckoned that that mustard made from the finely ground nasturtium seeds had the best flavour of all. The flower buds and fruits, or "nuts", as they are sometimes called, are often pickled in vinegar and used like capers (Hemphill). Nasturtium is sometimes used in folk medicine as a stimulant in cases of scrofula, and it is also given for influenza (Schauenberg & Paris). The name nasturtium comes from Latin nasus torsus, twisted nose, a reference to the peppery taste of the edible parts (Painter & Power), hence local names like Nose-smart, and Nose-tickler, both from Somerset, and Nose-twitcher, from Dorset (Macmillan).

Nasturtium officinale > WATERCRESS

NAUSEA
GINGER will help to reduce nausea generally, and in particular travel sickness and morning sickness. Ginger tea, even ginger biscuits, are useful in these cases (M Evans).

NAVAJO TEA
(*Heuchera bracteata*) A Navajo toothache medicine consists of the crushed leaves of this plant and of *Aplopappus lanuginosus*, mixed with water. The resulting mass is held to the aching tooth with a heated stone. It is sometimes chewed to relieve sore gums (Elmore).

NAVELWORT
A name once given to WALL PENNYWORT, (*Umbilicus rupestris)*, for in medieval times it was umbilicus-veneris. It is Venus-Nabel in German, too, and Venus's Navelwort was used in England (Parkinson. 1640).

NEEM TREE
(*Melia indica*) The tree is pervaded throughout with a strongly antiseptic resin, used a lot in Indian domestic medicine and across Africa, where hedges of it are grown close to houses because of its reputation as a cure for malaria. The leaves, which are very bitter, are also used as an antiseptic (Lewis & Elvin-Lewis). Some Hindu castes put the bitter neem leaves in their mouth when returning from a funeral, as an emblem of grief (Pandey).

Touching the tongue with a smouldering Neem twig used to be part of the ritual of re-admittance into a caste in India, as well as the giving of a feast at which initiates served the company. Although it is in no way a sacred tree, it is acknowledged that a Bhut (elemental) may take up residence in one. Shitala, the goddess of smallpox, lives in the tree, and the leaves of the tree are used to relieve the condition. The leaves are used to drive away evil spirits. If a man is possessed by any spirit, he is made to experience the smell of smoke rising from the burnt leaves (Upadhyaya). But its therapeutic properties are well-known (Codrington).

Nelumbo nucifera > HINDU LOTUS

Nepeta cataria > CATMINT

Nerium oleander > OLEANDER

NERVOUS DISORDERS
Herbalists use CORNFLOWERS extensively to remedy disorders of the nervous system, saying that they are at the same time curative and calming (Conway). BEE BALM "comforts the heart, and driveth away all melancholy and sadnesse ..." (Gerard). (It was still in use in the 20th century for nervous complaints and depression (Boland. 1979).

NETTLE
(*Urtica dioica*) Nettle is a most useful, though a much maligned, plant. True, minds are exercised in trying to get rid of them. Country wisdom is of no great help, though in Herefordshire they used to say that if nettles are well beaten with sticks on the day of the first new moon in May, they will wither and not come up again (Leather). The advice given in the old rhyme:

> Cut nettles in June,
> They come up again soon.
> Cut them in July,
> They're sure to die (Udal),

seems rather more practical. More advice is given in a Scottish nursery tune:

> Gin ye be for lang kail,
> Cou' the nettle, stoo the nerrle,
> Gin ye be for lang kail,
> Cou' the nettle early

and so on (Chambers. 1870). Local names like Devil's Leaf (Macmillan), Devil's Apron, or Naughty Man's Plaything (Parish) show how disliked the plant is. Naughty Man is a common euphemism for the devil. The gardener will perhaps ask himself where they came from in the first place. The answer from the Highlands and Islands is either that they sprang up where Satan and his angels fell to earth on their expulsion from heaven (on 3rd May) (Swire. 1964), or that they grow from the bodies of dead men, or more specifically, from "dead men's bones" (Beith). Similarly in Denmark, for there they used to claim that nettle clumps grow from the shedding of innocent blood. Nearer the mark is the Lincolnshire belief that they grow spontaneously where human urine has been deposited (Gutch & Peacock), nearer, because nettles thrive in nitrogen-rich soil, particularly around deserted human habitations. In Scandinavian mythology the nettle stands for the god Thor, and that may account for the old custom of throwing nettles on the fire during a thunderstorm to protect the home from lightning (Sanecki), and not only in Scandinavia, for exactly the same is recorded in the Tyrol, and further south into Italy (Folkard). Nettles gathered before sunrise were used in Germany to drive evil spirits from the cattle (Folkard), possibly because it is an evil plant in itself. In India, it is the symbol for a demon, for the great serpent poured out its poison on it. But evil is an antidote to evil, so a nettle held in the hand is a safeguard against ghosts (Leland).

When cows eat nettles, they are apt to give bloody milk, and then, at least in Northumberland, they used to say the cow was bewitched (Denham). This belief, though, could not have been confined to that area, for there was a counter-charm involving nettle that was used with considerable ritual (J Spence). In the Carpathians, nettles were put in the cask into which ewe's milk was to be poured, "so that the sheep will not fall ill". Another explanation was given as "the flock will be like the plant, which no-one touches", i.e., the power of the nettle is transmitted by contact to the milk, and from there to the sheep themselves (Bogatyrëv). Dreaming of nettles is a sign of good health and prosperity, but to dream of being stung by them foretells vexation and disappointment (Raphael) (after all, nettle has been used as the symbol of envy (Haig), and of spitefulness (Leyel. 1937)). On the other hand to dream of gathering them means that someone has formed a favourable opinion of you. If the dreamer is married, then family life will be harmonious (Gordon.1985).

One of the most unlikely beliefs connected with the nettle was that it is an aphrodisiac, to be used also as a flagellant (Leland). The seeds, so it was claimed, powerfully stimulate the sexual functions, and they figured too in a Greek remedy for impotence, when an ointment was made from the roots of narcissus with the seeds of nettle or anise (Simons). On the other

hand, "to avoid lechery, take nettle-seed and bray it in a mortar with pepper and temper it with honey or with wine, and it shall destroy it …" (Dawson). In other words, exactly the opposite of the aphrodisiac claim is reported here, and apparently the New Forest gypsies also had their doubts. They regarded nettle as a contraceptive, although they implemented it in a strange way. The man had to put nettle leaves inside his socks before intercourse! Not unrelated to all this are the marriage divinations constantly to be met in folklore. A particularly bizarre one involved the nettle. It is said that if you beat yourself with a nettle, you will be able to count from the number of blisters you get how many years it will be before you marry (Hald). Another one from the same source requires a girl to plant a nettle in wet sand. If it bends in the course of the night, it will show from which quarter the suitor will come, though if the tips bend downwards it is a sign of an early death.

In the Fen country, a patient suffering from smallpox, right to the end of the 19th century, would be rubbed all over with freshly gathered nettle leaves, and then with dock leaves. This was supposed to prevent scarring of the skin (Porter.1963). But dock, of course, has always been looked on as an antidote to nettle stings:

Nettle out, dock in –
Dock remove the nettle sting (Dyer. 1889),

and there are very many of these rhymes or chants, usually admonishing in some way. But in Somerset and the neighbouring parts of Wiltshire, it is the mallow, and not any member of the genus *Rumex* that is the dock (see *Malva sylvestris*).

There is a lot more to nettles than all this, though. Is it not said that the plant was blessed by St Patrick as useful both to man and beast? It is said, too, that St Columcille saw a poor woman picking nettles for food, and he resolved to live on nettle broth from then on (A T Lucas. 1964). Gardeners are quite keen on it, within reason, for it is said to stimulate the growth of all plants around them, as well as being the best thing to hasten the decomposing of a compost heap. Grow controlled clumps of them between currant bushes, which will then fruit better (Boland & Boland). But it is as a textile plant that nettles are best known, The fibres were used for cloth certainly up to the 18th century in Scotland (Grigson), and in the Tyrol it was in use for nettle linen as late as 1917 (Hald). The earliest known literary reference to its use as a textile is in Albertus Magnus, in the 13th century, and the old German name for muslin was Nesseltuch, i.e., nettle cloth (C P Johnson). Fabrics made from the fibres were famous in many parts of northern and central Europe at the end of the 18th century for their fine, gauze-like qualities (J G D Clark). Alvard Jivanyant, in *FLS News. 31; June 2000* said that Russian girls did not ask their mothers for old rags with which to dress their dolls, but they spun nettle fibres to weave into tiny frocks and coats. Cf the Hans Andersen fairy tale in which the sister wove twelve nettle shirts to save her bewitched brothers (see also COTTON-GRASS, also used for the same purpose in another version of the folk tale).

Nettle is a perfectly good vegetable, provided that only the young shoots are used (the older plant has a marked diuretic effect (Hald)). A belief from the New Forest carries the same warning. It is that the devil gathers the nettles to make his shirts on May Day, so they are unfit for eating after that date (Boase). Young nettle leaves chopped and served as salad made a common Cambridgeshire "spring tonic", while Thornton advised his readers that "nettle broth is good against scurvy". But apart from that they were popular enough as a vegetable, particularly in Yorkshire (Nicholson). In Scotland, nettle broth was made in hard times (Chambers), by boiling the leaves with a little oatmeal and adding any other plant that might be to hand (Drury), and there is the Scottish nettle pudding (Jordan), as well as nettle haggis (Mabey. 1972).

Nettle's medicinal virtues are many and varied, ranging from consumption (eating nettle seeds was an Irish way of combatting the disease (Maloney)) to the doctrine of signatures usage that causes nettle tea to be taken for nettle-rash (Berdoe), on the basis of equating that which causes similar burning stings with its ability to cure them. It is still used in Cornwall (Deane & Shaw), though one suspects that the value of counter-irritants may be called on in this case, as it would be in an Irish cure from County Down for paralysis of the limbs: "continued excitement" by them on the skin is the form of treatment (Barbour). Young nettles have the effect of increasing the haemoglobin of the blood, and have a high content of salicilic acid (Bircher), hence the various "spring tonic" uses one finds in folk medicine, whether it is called that or a "blood-purifier". The medicine usually takes the form of a tea, and sometimes dandelion leaves were included, as in a Dorset "blood tonic" recipe (Dacombe). An Irish cure for dropsy was to boil down nettles from a churchyard and drink the result (Wilde), while the root infusion was a common dropsy remedy in Russian folk medicine (Kourennoff).

Nettle seems a most unlikely plant to be associated with cures for insomnia, but it certainly is in both the medical sense and as a charm. In the Highlands, we are told that nettle leaves chopped very small and mixed with whisked egg-white were applied to the temples and forehead (Beith). This remedy was mentioned as early as the beginning of the 18th century by Martin, who particularly spoke of fevered patients benefitting from the treatment. In Wales, they used to put a bunch in broth, both to induce appetite and to promote sleep (Trevelyan). A Scottish charm for fever was to pluck a nettle by the root, three successive mornings, before sunrise (Dalyell), saying the name of the patient and his parents. The seeds have also been

prescribed to "reduce excessive corpulency" (a few daily is apparently enough) (Dawson). Soaked in gin, the seeds used to be taken for the relief of rheumatism (Savage), but there were a number of different ways that nettle could be used for that complaint, by eating the leaves, for instance, which was a Cumbrian practice (Newman & Wilson). But more spectacularly, the way to deal with it was to thrash oneself so that the nettle-stings took their toll, on the basis, presumably, of the counter-irritant principle. Urtication, it was called, and it continued to be a recognised treatment for rheumatism right into the 20th century (Coats. 1975). You can lie down on them, too, and the more you are stung, the better the cure, for the evil is said to come out with the blisters (Hald). There is still a belief in some places that nettle stings will cure lumbago (Rollinson). Sciatica, too, can be dealt with, according to the Cornish practice, by a nettle poultice (Deane & Shaw), and a Kentish village remedy for arthritis required nettles to be cooked and eaten, and then the water in which they had been boiled had to be drunk (Hatfield. 1994).

Homeopathic doctors still prescribe nettles, or at least an extract of them, for eczema and nose-bleeding, and it is also used in lotions to make the hair grow (see above) (Schauenberg & Paris); but all of these are ailments anciently treated with nettles in folk medicine. Eczema. for example, was treated in Dorset with nettle tea (Dacombe), which is a well-established East Anglian remedy for skin complaints (Porter.1974), while Irish country people used boiled nettles to cure boils (O' Farrell), even measles (Ô Súilleabháin). Nosebleed, the other ailment quoted as cured by the homeopathic use of nettles, is also quite traditional. Martin spoke of the cure on Gigha in 1703, the roots being chewed and held to the nostrils, and earlier still Gerard claimed that "being stamped, and the juice put up into the nosthrills, it stoppeth the bleeding of the nose".

Domestic medicine agrees on nettle's efficacy in chest complaints, whether for a cough or for something more serious. Martin took note of its use in Lewis for coughs. In this case they used the roots boiled in water and fermented with yeast. Earlier, Gerard had recommended it for "the troublesome cough that children have, called the Chin-cough", whooping cough as it is now known. Asthma was treated by "a spoonful of Nettle-juice mixt with clarified Honey, every night and morning" (Wesley), and in Russian folk medicine they use a tea made from the dried leaves (Kourennoff). The roots were used in the Highlands for consumption, even (Polson), while in Somerset an infusion of the seeds served to treat the same condition (Tongue). Nettle beer was the specific in East Anglia (Hatfield). Even the smoke from burning dried nettles is said to be good for the chest (Page).

Patrick Logan records the Irish use of boiled nettle roots for worms. There is even a Welsh record of the belief that eating nettles in spring would cultivate a good memory (Trevelyan). There are many more ailments that have been treated with this most useful plant, but the list would be wearisome. But a Devonshire charm should be mentioned – it is a remedy for sore eyes. A woman who had never seen her father had to blow into the patient's eyes through a hole made in a nettle leaf, before she had put her hand to anything for the day! (Friend).

NETTLE-LEAVED BELLFLOWER

(*Campanula trachelium*) Once much in demand for treating sore throat and tonsilitis (Grigson). The specific name, *trachelium*, shows it was a remedy for inflammation of the windpipe, or trachea (Fisher). Gerard called it Throatwort, or Uvula-wort, and other names like Haskwort or Halswort (Cockayne), which is straight from the German Halskraut, where Hals means the throat, or neck, are evidence enough of the usage. Hence, too, the early name Neckwort (Storms). Earlier still, it was taken to be a wound herb, quoted as such in the Anglo-Saxon Apuleius.

NETTLE RASH

BARWOOD (*Pterocarpus angolensis*) bark is used in central Africa to cure nettle rash (Palgrave).

NEURALGIA

RED CLOVER flowers have been traditionally used for neuralgia or headache, for they are mildly sedative (Conway). There used to be something called a spice plaster to put on parts affected with rheumatism or neuralgia. The way to make it is to crush an ounce or so of whole allspice, and boil it down to a thick liquor, which is then spread on linen ready to be applied (A W Hatfield).

NEVER-DIE

(*Kalanchoe crenata*) A West African species, much used in medicine in the region. In Ghana, for example, the leaves are boiled or macerated in water, to be used as a sedative for asthma (Dalziel). It also formed part of a Yoruba headache remedy in the Ewé ritual (Verger), and the leaves, either on their own or combined with other ingredients formed a medicine to treat not only headache but also more serious illnesses, including smallpox. They are also viewed there as a poison antidote.

NICKER BEAN

(*Entada phaseoloides*) This is a large tropical climber, whose pods can be a metre or more in length. The bean is sometimes found washed up on the Atlantic shore of the Hebrides. Hence the name Barra Nut (Duncan), Barra being one of the islands of the Outer Hebrides. It brought luck to the finder, who would carry it about with him (Shaw).

Nicotiana rustica > TURKISH TOBACCO

Nicotiana tabacum > TOBACCO

NOAH'S ARK

Genesis 6; 14: "Make thee an ark of gopher wood: rooms shalt thou make in the ark, and shalt pitch it within and without with pitch". But what is gopher wood? Probably CYPRESS. Grigson. 1974 is of the opinion that the very word cypress stemmed eventually from the Hebrew gopher. Cypress wood is indeed very durable.

NOSEBLEED

LEEKS were used, but not in a way easily foreseen. Lupton, in the mid-seventeenth century, ordered the patient to take nine or ten fresh leeks, and to put a thread through the midst of them, "but cut off the tops of the leaves, then hang them round the party's neck that bleeds, so that the leaves be upward to the nose, and the heads of them downwards …" The homeopathic use of NETTLES for nosebleed is quite traditional. Martin noted the use on Gigha in 1703, the roots being chewed and held to the nostrils, and earlier still, it was claimed that "being stamped, and the juice put up into the nosthrills, it stoppeth the bleeding of the nose" (Gerard). The Physicians of Myddfai also recommended it, and Wesley prescribed the same cure. Lupton, in the mid-17th century, too, said "let the party that bleedeth chew the root of a nettle in his mouth, but swallow it not down, and without doubt the blood will staunch; for if one keep it in his mouth, he can lose no blood". A leechbook of the 14th century includes "for bledyng of the nose. Take the bark of (HAZEL), and branse it and blow the powder in thi nose" (Henslow), a remedy that would probably work quite well, but would be far too long-winded, unless, of course, one had a stock of the powdered bark.

CORNFLOWERS, gathered on Corpus Christi Day, would stop nose bleeding, if held in the hand long enough. What about the other 364 days in the year? Perhaps the flowers could be dried and could be preserved for future use. A bunch of BROOM flowers round the neck will stem a nosebleed, at least according to Scottish belief (Beith). Rubbing KNOTGRASS into the nostrils was the Somerset way of stopping a nosebleed (Tongue. 1965).

YARROW has a special place in any account of nosebleeding. "The leaves being out in the nose do cause it to bleed, and easeth the pain of the megrim" (Gerard). The plant is actually called Nosebleed over a wide area of England. The French, too, have saigne-nez. Prior claims that it got this application by mis-translation, the plant actually referred to being the horsetail. Perhaps so, but it is firmly fixed in yarrow's folklore. The propensity was used to test a lover's fidelity. In East Anglia, for instance, a girl would tickle the inside of a nostril with a leaf of yarrow, saying at the same time:

Yarroway, yarroway, bear a white bloe;
If my love love me, my nose will bleed now.

Bergen also quotes this use in America, where the girl says:

Yarrow, yarrow, if he loves me and I loves he,
A drop of blood I'd wish to see.

Another Suffolk rhyme is:

Green arrow, green arrow, you bears a white blow,
If my love love me my nose will bleed now;
If my love don't love me, it 'ont bleed a drop.
If my love do love me, 'twill bleed every drop
(Northall).

SHEPHERD'S PURSE (in tincture form) is the "great specific for haemorrhages of all kinds" (Leyel. 1937), and that includes nosebleed, as well as all other internal haemorrhages. It has been so used through the centuries, and is still prescribed.

A very odd cure is recorded from Japan. There, an infusion of the leafy shoot of CHICKWEED, with sugar, is given internally, to stop a nosebleed (Perry & Metzger). CORIANDER juice, "blown up the nostrils", was used, too (F J Anderson).

Nuphar lutea > YELLOW WATERLILY

NURSERY RHYMES

One of the many names for LADY'S SMOCK is Lucy Locket, used mostly in the Derbyshire area of England. When little girls pick it, they sing:

Lucy Locket lost her pocket
In a shower of rain,
Milner fun' it, milner grun' it
In a peck of meal (grain?) (Northall).

NUTMEG

(*Myristica fragrans*) The fruits are rather like apricots in appearance, and inside the orange-yellow pulp is the large brown seed (the nutmeg), which is surrounded in turn by crimson arillus fingering. This is mace, which serves as a distinct spice, and contains quite different essential oils from those of the nutmeg seed.

Nutmeg in large doses (more than one teaspoonful according to Farnsworth) is a hallucinogen. A few of the kernals will relieve weariness and bring euphoria eventually, for it may take up to five hours to take effect (Farnsworth), though nausea and dizziness often follow.. The earliest clear mention of the narcotic effect of nutmeg occurs in Lobel's *Plantarum seu stirpium historia*, 1576. Later on, there were reports of its causing respiratory distress, dry mouth, swollen lips and serious constipation. There were stories of people sleeping under nutmeg trees (in the Banda Islands), and waking with such hangovers that they were taken to be drunk. In 1657, a woman was found dead in her bed with a basket full of nutmegs

in front of her. It was concluded that she had eaten so many of them that she sank into a fatal stupor (Weil). Apparently, by far the greatest number of people poisoned by nutmegs have been women. One to three nutmegs will cause restlessness, dizziness, coldness of the extremities, nausea and abdominal pain, and sometimes unconsciousness. The only fatality, apart from the 1657 record above, ever attributed to it occurred when an eight year old boy ate two whole nutmegs, became comatose, and died within twenty-four hours. One to two ounces of ground nutmegs seems to produce prolonged delirium and disorientation, all similar to alcohol inebriety, although the response may be quite diverse. Freshly grated nutmeg produces the deepest intoxication (Emboden. 1979), but the hangover is likely to be extreme nausea, headache, dizziness, and dry nose and throat (Furst). On the other hand, the view of Bartholomew Anglicus was that "the Nutmeg holden to the nose comforteth the brain…" (Seager).

There are a few superstitions concerning nutmeg: to dream of them is a sign of impending changes (Gordon. 1985), but to carry one about on the person may be just a good luck charm (Vickery. 1995), or at least a good health charm. According to St Hildegard, whoever received a nutmeg on New Year's Day and carried it in the pocket could fall as hard as he wished during the coming year without breaking the smallest bone. Nor would he suffer a stroke. Nor be afflicted by haemorrhoids, scarlet fever or "boils in the spleen" (Swahn). In the majority of cases they are carried as a preventive, particularly in America, though not exclusively so, for Irish country people carried one in the pocket to cure rheumatism (O'Farrell). In Missouri, it would prevent the complaint, so it was believed (Bergen. 1899), while in Alabama the nutmeg would be worn round the neck, and that would prevent rheumatism from the waist up (R B Browne). Worn that way in Maine, it would prevent croup and neuralgia, and elsewhere in America, cold sores, earache and headache (Bergen. 1899), even asthma (Whitney & Bullock) and quinsy (Thomas & Thomas), or, in Alabama, styes (Bergen. 1899). It is a rheumatism curer or preventer in Britain, too (Vickery. 1995 has listed several examples), and more – it will cure backache, too, according to Lincolnshire belief (Rudkin), or lumbago, to be more specific (Opie & Tatem). Boils too can be cured by wearing one round the neck, and nibbling it nine mornings fasting (Hawke). In Devonshire, a little more ritual was needed. The patient must be given a nutmeg by a member of the opposite sex. He carries it in his pocket and nibbles it from time to time. Only when the nutmeg has quite disappeared will the boils have gone (see *Devonshire Association. Report. vol 91; 1959 p199*). Indeed, a nutmeg in the purse is a safeguard against almost anything; in Brazil, they are worn round the neck against what is thought of as an illness brought about by an evil wind (P V A Williams). Even representations of them were carried about for luck, and a gilt nutmeg was at one time a common Christmas gift, and, encased in silver, they were worn at night as an inducement to sleep. Nutmeg tea was an old insomnia remedy. Victorian women often carried a silver grater and nutmeg box on their châtelaines. One nutmeg crushed was enough for a pint of the tea, and the dose was a small cupful on going to bed (Leyel. 1926).

They were reckoned to be aphrodisiac at one time, standard ingredients in love potions, and widely used. They still are, apparently, for Yemeni men take them even now to enhance their potency (Furst). They were used to cause abortion – the women in London who were the practitioners were actually known as "nutmeg ladies". This usage must surely have arisen from folk medicine, for no physician had claimed it as such, just the opposite, in fact. The earliest exact reference to nutmeg's medicinal virtues, rather than to its use as a charm, is in the work of an Arab physician at the end of the 9th century. He recommended it as a carminative (a reputation it has retained ever since), and for freckles and skin blotches. By the 11th century Arab physicians were using it for kidney disorders, and as an analgesic (Weil). As far as western medicine is concerned, nutmeg was regarded as something of a "cure-all". It was in the first half of the 19th century that they were used most, particularly to relieve

flatulence and dyspepsia. Nearly every middle-aged woman carried one to grate over her food and drink. They were carried in the pocket in small wooden or metal cases with a grater at one end. The better-off used silver boxes, but the poorer classes used a wooden one, an inch or so in length, so large enough to contain a full-sized nutmeg (Newman & Wilson). Evening drinks, known as possets, were commonly taken, and these nearly always contained large amounts of ground nutmeg and other spices – apart from anything else, these drinks were supposed to correct the effect of alcohol. A nutmeg poultice on the chest is used in Indiana for the croup or a bad cold (Tyler).

False Nutmeg is *Pycnanthus kombo,* a West African tree, whose seeds are often mistaken for the real thing, but they are less aromatic, and are of more importance as an oil-seed, for they are rich in a vegetable fat, used for lighting and soap-making. The seed itself, when threaded, will burn like a candle (Dalziel).

NUTMEG LADIES
Nutmegs were used to cause abortion, and the women in London who were the practitioners were actually known as "Nutmeg Ladies" (Emboden. 1979).

Nymphaea alba > WHITE WATERLILY

Nymphaea caerulea > BLUE LOTUS

Nymphaea odorata > WHITE POND LILY

O

OAK

(*Quercus robur*) In Greek mythology, the oak was Zeus's favourite tree, his emblem, and the seat of his divinity (Rhys). It had sheltered him at his birth on Mount Lycaeus (Bayley), so making the oak the symbol of hospitality. To give an oak branch was the equivalent of "you are welcome"; one of the earliest references to the tree is the story of Abraham's hospitable entertainment of the angels under the oak of Mamre (possibly *Q pseudo-coccifera*) (see Genesis. xviii). Compare this with Taylor's description of the Fairlop Oak, "in a glade of Hainault Forest, in Essex, about a mile from Barkingside ... Beneath its shade, which overspreads an area of three hundred feet in circuit, an annual fair had long been held, on the second of July, and no booth is suffered to be erected beyond the extent of its boughs ... ".

The famous oracle at Dodona was the sacred oak there – Zeus lived in the tree, and the rustling of the leaves was his voice, though exactly what constituted his voice is debatable. Homer knew the oracle, which probably was the most ancient in Greece. The attendants at the oracle were the priestesses of Dione, identified with the earth goddess, and the priests of Zeus (Flaceliere). From the context in the Odyssey, the implication is that the will of Zeus was audible from the oak itself, and that it was the tree itself that spoke (not the rustling of the leaves nor the birds in the tree). See the Homeric story of the Argo. When it was being built, Athena took a timber from the Dodona Oak and fitted it into the keel, with the result that the Argo itself could speak, and so guide or warn the Argonauts at critical moments (Parke).

After Greece, Rome, and there too the oak was sacred to Zeus, but under the name Jupiter. The dedication to Zeus/Jupiter was mirrored in Scandinavian mythology, where the oak was sacred to Thor, and under his immediate protection. So it was sacrilege to mutilate it even in the smallest degree (Dyer). Gimbutas speaks of a sacred oak near Vilnius, in Lithuania, at which people congregated and to which they gave offerings. Both the church authorities and the secular in the 18th century saw its very sacredness as a seat of dissent, and saw to it that it was cut down. But it took a foreigner to do it, for no local would. The connection with Thor accounts for various lightning superstitions. They say the oak is more frequently struck by lightning than most trees. It "draws thunder", as they say in Hampshire (Read). In Wales, it was thought dangerous to take shelter under an oak during a thunderstorm, for the lightning penetrates fifty times deeper into them than any other tree (Trevelyan). Hence King Lear's allusion to "oak-cleaving thunderbolts", and Prospero's:

To the dread rattling thunder
Have I given fire, and rifted Jove's stout oak
With his own bolt.

But the association with Thor meant that it was believed to give protection to shelterers, even though the tree itself was struck. Not on Thor's day, though – Somerset children used to say you should never picnic under an oak on Thursday (Tongue). Oaks known to have been struck were often visited, so that pieces could be taken away to be attached to buildings for protection (Wilks). Acorns too were a charm against lightning, and ornamental designs used to be made with them and put in cottage windows (Lovett). Anyway, one school thought it a very bad sign if an oak were struck (Rambosson), and a belief recorded in Hampshire said that the oak actually drove away thunderstorms; it was even thought that the iron in the oak drew the lightning away from the town (Boase).

One of the three revered trees in early Irish tradition was Omna, oak. The Welsh Lleu took refuge after his "death" in a great oak growing on a plain, a common site for trees of cult importance. The tree cannot be soaked by rain nor destroyed by fire (Ross). There is a connection, too, with the Herne the Hunter legend if, as has been claimed (by Murray) that Herne is a pre-Christian deity, and the Hunter legend a memory of this, it would be natural that he should be associated with an oak. Herne's oak was, according to one theory, blown down in 1863; another claims it was destroyed accidentally in 1796. Shakespeare, in the *Merry Wives of Windsor iv, 4,* gives:

There is an old tale goes that Herne the Hunter,
Sometime a keeper in Windsor Forest,
Doth all the winter time, at still midnight,
Walk round about an oak, with great ragg'd horns ...
... There want not many that do fear
In deep of night to walk by this Herne's oak.

It is well-known that the oak was a sacred tree to the Celts,and that their priests, the Druids, were particularly associated, both by name and by ritual, with the tree. John Evelyn knew the belief well: "For in truth the very tree itself was sometimes deified, and that Celtic statue of Jupiter no better than a prodigious tall oak". According to Pliny, "... the Druids ... have nothing which they hold more sacred than the mistletoe and the tree on which it grows, provided only that it be an oak. But apart from that, they select groves of oak. And they perform no sacred rite without leaves from that tree ... For they believe whatever grows on these trees to be actually sent from heaven, and to form a mark in each instance of a tree selected by the god himself ..." (actually, oak mistletoe is very rare). There is no direct evidence that the Irish Druids performed their rites in oak groves, but it has been argued (by Frazer) that this may be inferred from the fact that on the introduction of Christianity, churches and monasteries were sometimes built in oak groves or near solitary oaks, as though the choice of the site was determined by the sanctity of the tree (see also Kendrick). Kildare is relevant here. It is cill-dara, church of the oak tree (Hyde), founded by St Brigit. Many shrines to this saint were built under oak trees (J A MacCulloch). St Boniface cut down an oak tree sacred to Thor at a place called Geismar; the wood was then used to build a chapel dedicated to St Peter (R Morris). At Brinso, in Yorkshire, there is a very old and large oak on a tumulus, and it is always known as Brinso Church (Gutch. 1901). Marriages were once celebrated under so-called "marriage oaks", until the Church forbade the practice. Even so, dancing three times round an oak tree after church weddings was the practice for a long time (Wilks).

Oak veneration was resuscitated in England after Charles II's incident of the Boscabel Oak, giving a holiday, appropriately named Royal Oak Day (or Oakapple Day), 29 May. The Mile Oak, near Oswestry, another sacred tree, was popularly associated with St Oswald, King of Northumbria, killed in battle AD 642. When it was cut down by the agent of the lord of the manor in 1824, a ballad-lament was made:

> To break a branch was deemed a sin,
> A bad-luck job for neighbours,
> For fire, sickness or the like
> Would mar their honest labours.

Note Aubrey's remarks on the tree: "When an oake is falling, before it falles it gives a kind of shriekes or groanes, that may be heard a mile off, as if it were the genius of the oake lamenting …". The idea of the oak as a reward is still extant (the oak leaf for *Mentioned in Despatches*' for example), following its ancient use.

Another aspect of the mythology of the oak concerns its life-giving properties. The ash is usually reckoned the tree from which men first sprang. But in Greece, the oak too was venerated for that reason; they called it the "first mother", which fed man, mother-like, with its own acorns (Porteous). Piedmontese children used to be told that it was from the trunk of an oak tree that their mothers had taken them when they were born (Gubernatis). Oak boughs were carried during Roman wedding ceremonies as symbols of fecundity, the point of reference being the acorn in its cupule, a phallic emblem. Marriage oaks have already been mentioned, but there was one at Bampton, in Cumbria, until the 1860s. It was good luck for the bride and groom to embrace and dance under this tree (Brown); for good luck, read fertility in marriage.

Boundary oaks are famous. Grimm said that boundaries were defined by oak trees from very ancient times, even from those of mythology. The Cadnam Oak, a few miles from Lyndhurst, in the New Forest, was a "boundary tree" of the forest. According to popular belief, it became green on Old Christmas Day, being leafless before and after the day (Bett. 1952). This oak, or rather its descendant, still bears leaves round about 6 January, though it buds at the normal times as well (Hampshire FWI). Cross oaks were planted at crossroads so that people suffering from ague could peg a lock of their hair in the trunk, and by wrenching themselves away might leave the hair in the tree, together with the illness (Fernie).

Gospel Oaks were so-called because passages from the Gospels for Rogation Day were recited by the priest under them during parish perambulations, or "beating the bounds", always carried out at Rogationtide. Best known of these Gospel Oaks is in the Suffolk village of Polstead, where an annual service is still held, an event that has been going on for a thousand years (Wilks). Hertfordshire people used to keep acorns from Gospel Oaks (or Boundary Oaks) in their pockets as some kind of health-giving charm (Jones-Baker. 1977). Selly Oak, in Birmingham, is said to be a corruption of Sarah's Oak, Sarah being the name of a witch hanged from the tree there. A variation is that the tree sprang from the stake driven through her heart (Palmer). Many other individual oaks throughout the country have legends attached to them, and some ancient trees were quoted as being old in William the Conqueror's time, an apparently overblown age for an oak, but it is nothing compared to the implied age of the trees in a Gaelic saying, translated by A Forbes as:

> Three ages of a dog the age of a horse;
> Three ages of a horse the age of man;
> Three ages of a man the age of a deer;
> Three ages of a deer the age of an eagle;
> Three ages of an eagle the age of an oak tree.

He calculates the result as 2800 years!

There is a legend that all the oaks at Newburgh, Yorkshire, were decapitated by Cromwell's order, as a punishment for the loyalty of the owner, the punishment being transferred from the lord to his trees. Only by this propitiation, so it is said, did Cromwell consent to give his daughter in marriage to Lord Fauconberg (Gutch). According to a prophecy of Thomas the Rhymer, the fortunes of the Hays of Errol were bound up with the fate of a particular oak:

> While the mistletoe bats on Errol's aik
> And the aik stands fast,
> The Hays shall flourish, and their good gray hawk
> Shall nocht flinch before the blast.
> But when the root of the aik decays
> And the mistletoe dwines on its withered breast,
> The grass shall grow on Errol's hearth-stane
> And the corbie romp in the falcon's nest
> (Wimberley).

There are other examples of this apparent binding of family fortunes with that of oaks, and more particularly their leaves, for it is written that "some are of opinion that divers families of England are preadmonished by oaks strange leaves" (Heath, *Description of Cornwall*, 1750), for instance, an oak in Lanhadron park, in Cornwall, is supposed to bear speckled leaves before a death in the family (Puckle).

The legend of the Wandering Jew tells that he can only rest where he shall happen to find two oaks growing in the form of a cross (Dyer). An Irish belief holds that the true Cross was made of oak (Ô Súilleabháin), while in Italy, amulets for the prevention of insomnia were made by binding oak twigs into the form of a cross (Leland. 1895). Other superstitions include that, from Ireland, that says it is unlucky to use oak in a house roof (Ô Súilleabháin), a belief that can only have local distribution; perhaps it stems from the deification of the oak, or at least from the belief, as in Devon and Cornwall, that the elves lived in them (Philpot). A superstition from Yorkshire says that when you see a large hole in an oak, be sure the tree has been haunted (Gutch). We have already seen, as Aubrey said long ago, that if you cut down an oak, you will hear it scream. But Somerset people go on to say that the sound will cause death within a year, or at least that you will be taken seriously ill. Oaks always resent cutting, and a coppice that has sprung up from the stumps of cut-down oaks is generally avoided as hostile to man. See the Somerset folk song:

> Ellum do grieve
> Oak do hate
> Willow do walk
> If you travels late (Briggs. 1978).

But the oak will be your guardian if it likes you (Tongue). To dream of a large oak, with beautiful foliage, is always a good sign; but a blasted oak means sudden death (Raphael).

Just like "many haws, many snows", an acorn year was everywhere considered "a bad year for everything" – Many acorns, a long, hard winter:

> Année de glande,
> Année du cher temps

And:

> Anno ghiandoso
> Anno cancheroso,

which is, in French:

> Année glanduleuse,
> Année chancreuse

From Germany, in similar vein: Viel Eicheln lassen strenger Winter erwarten (Swainson. 1873).

There are one or two more pieces of weather lore connected with the oak. From Wales, there is a saying that when oak leaves curl up, it will get very hot (Trevelyan). There are a series of rhymes about the result of ash coming into leaf before oak, and vice

versa. A rhyme from Surrey is a little different from most:

> If the oak before the ash come out,
> There has been, or there will be, a drought.

The most succinct of them is from Kent:

> Oak, smoke,
> Ash, squash (Northall).

Prognostication used to be taken from oakapples, too. The insects found in them provided the sign, each apparently having its special meaning (T Browne). For instance, from Wales, a fly found in one was a sure sign of a quarrel, a worm was a token of poverty, while a spider was a sign of illness (Trevelyan).

Acorns, too, have their own folklore. In some parts of the Continent, they are put in the hands of the dead (Friend). Their cups and stems are the pipes smoked by leprechauns (Ô Súilleabháin), and the cups are fairies' shelter. See Shakespeare: "All their elves for fear, creep into acorn-cups, and hide them there" (*Midsummer Night's Dream*). Carrying one around in one's pocket or purse is a way to keep oneself youthful, and to preserve health and vitality (Waring), or to prevent rheumatism (Thomas & Thomas). Dreaming of them is a good sign – it shows that health, strength and worldy wealth will be yours (Raphael), for acorns were in ancient times the symbol of fecundity – the acorn in its cup was one of the earliest phallic emblems (the acorn is the masculine, and the cup the feminine (Wellcome). But in some parts of America, Maryland for instance, a plentiful crop of acorns presages a poor corn crop next year (Whitney & Bullock). There was a form of marriage divination connected with them, or rather, their cups – two of them were taken, one named for the lover, and the other for one's self. Then they were set to float in a bowl of water; watch them – if they sailed together, there would be marriage, but if they drifted apart then it was obvious what the result would be (Trevelyan).

Oak bark's high tannin content accounts for a number of medicinal uses, such as making an infusion to be used as a gargle for sore throat (Beith). Argenti & Rose noted that on the island of Chios, diarrheoa was cured with a potion of oak-galls in wine. Acorns too are astringent, and so used for diarrhoea – they even used to be worn round the neck for the purpose (Lovett). The tannic acid would serve to restrain bleeding if taken in decoction internally. This decoction was also recommended for blackening the hair, for sunburn, freckles, pimples, etc., or for skin diseases like eczema, when a compress made from bark, boiled, can be used (Thomson. 1978). A Suffolk cure for ague was a mixture of beer, gin and acorns, and another Suffolk remedy was to use the powdered acorns to cure diarrhoea (V G Hatfield). Welsh practice mixed a little magic with the known astringency of the bark, a piece of which had to be rubbed on the left hand, in silence, on Midsummer Day, when, so it was believed, it would heal all open sores (Trevelyan).

People were not so enthusiastic about acorns in Ireland, it seems – they were believed to cause black-quarter in cattle (Ó Súilleabháin). They had every reason to be distrustful of them, it seems, for though cows are very fond of acorns, they can easily fall victim to acorn poisoning. Wiltshire children were given the job of collecting buckets-full of acorns, ostensibly to feed to the pigs, but in reality to get rid of them before the cows got at them (Whitlock, 1988). The leaves too can be poisonous to cattle. There is a condition known in France as Mal de Bron, or Maladie des Bois, which can be fatal, and which has been recognised as such for centuries (Long).

OAKAPPLE
Prognostications used to be taken from oakapples. The insects found in them provided the sign, each apparently having its special meaning (Sir T Browne). For instance, from Wales, a fly found in one was a sure sign of a quarrel, a worm was a token of poverty, while a spider was a sign of illness (Trevelyan).

OAKAPPLE DAY
(29 May) see ROYAL OAK DAY

OATS
(*Avena sativa*) Dreaming of oats is an omen of success (Hewett). It used to be said that a girl who wants children should wear a straw garter. She can even determine the sex of the child – wheat straw will bring a boy, but oat a girl (Waring). A drop of rain or dew hanging on an oat head is a sign of a good crop to come (Addison. 1985).

Oats are much used in domestic medicine, especially in Scotland, where oatmeal poultices serve for a number of complaints. For example, minor boils and suppurations are treated with an oatmeal poultice made with water and a slight dressing of salt butter. For "difficult" boils urine would take the place of butter (Beith). Oatmeal used to be reckoned good for colic, too, and there is a recipe for sore eyes from Scotland. A handful of oats would be put into a bowl of water, stirred, and left for a while. Then the mixture would be strained through a cloth, and the meal water put to the eyes (Beith, Parman). A bath prepared with about a pound of oat straw is prescribed for neuralgia (Thomson. 1978). There is even a wart charm using oats. You could cure your own warts by taking 81 (9 × 9, that is, a magical number) stems of oats (which are bumpy, like warts), binding them in 9 bundles of 9 each, and hiding them under a stone. As the stems rotted, so the warts would disappear (Beith, Parman). Freckles and spots can be washed away with a preparation of oatmeal boiled in vinegar (Addison. 1985).

Ocimum basilicum > BASIL

Ocimum tenuiflorum > HOLY BASIL – but see TULSI, rather.

OEDEMA
Putting boiling water on a couple of tablespoonfuls of chopped BIRCH leaves makes a tea for urinary complaints, especially dropsy (Fluck). An Irish cure for it was to boil down NETTLES from a churchyard and drink the result (Wilde), while the infusion was enough in Scottish practice (Rorie), and a root infusion was a common dropsy remedy in Russian folk medicine (Kourennoff). DWARF ELDER has been used for dropsy for a very long time. The Anglo-Saxon version of Apuleius recommended it for "water sickness" (Cockayne), a prescription copied by Gerard – "The roots of Wall-wort [Dwarf Elder] boiled in wine and drunken are good against the dropsie", a remedy still recommended today by herbalists (Conway). It seems the Welsh medical text known as the Physicians of Myddfai is referring to dropsy in the prescription: For pain in the feet and swelling in the legs – take the roots of dwarf elder, and remove the bark, boiling it well, then pound them in a mortar with old lard, and apply as a plaster to the diseased part. Gerard also recommended the seeds for this complaint. HONEYSUCKLE bark was also used – a heaped tablespoonful of thin flaked bark to a pint of cold water, brought just to the boil, and taken in wineglassful doses three times a day (A W Hatfield).

The doctrine of signatures comes into play when ELDER itself is taken for dropsy, for the pith, when pressed with the fingers, "doth pit and receive the impress thereon, as the legs and feet of dropsical persons do". Evelyn mentions it, and so does Gerard, who recommended the seeds for "dropsie, and such as are too fat and would faine be leaner". Dawson quoted two leechdoms from a much earlier time, one requiring the patient to drink the juice of the berries tempered with wine, and the other needing the middle rind, leaves and blossoms or berries, all stamped together, with the additon of "esula [spurge] and mastic". Earlier still, in the Anglo-Saxon version of Dioscorides, there is a leechdom "for water sickness" using elder. Lupton, in the 17th century, and Wesley, in the 18th, were still recommending it. But it comes as a surprise to find a record (from Cambridgeshire) of elderflower tea as a dropsy remedy (Porter) in the 20th century. WHITE BRYONY'S roots have been taken as resembling a swollen foot, so the doctrine of signatures ensured that it would be a cure for gout and dropsy. Lupton recommended it: "If you seeth briony in water and use to drink the same, it helps and cureth them that have dropsy". BUCKBEAN was used for the complaint in the Hebrides; there is a recipe from South Uist that involved cleaning and boiling the whole plant, putting the juice in a bottle, to be drunk daily (Shaw). This may be doctrine of signatures, given the plant's preference for wet, marshy ground.

A dropsy remedy using IVY was known in Anglo-Saxon times. "For water sickness or dropsy, take 20 grains, rub them in a sextarius of wine, and of the wine administer to drink 2 draughts for 7 days …" is a translation of the Anglo-Saxon version of Apuleius (Cockayne). ALEXANDERS was also used for the complaint, a use dating from Dioscorides' time. A complicated 15th century leechdom has Alexanders seed as one of its ingredients: "for all manner of dropsies: take sage and betony, crop and root, even portions, and seed of alexanders, and seed of sow thistle, and make them into powder, of each equally much; and powder half an ounce of spikenard of Spain, put it thereto, and then put all these together in a cake of white dough and put it in a stewpan full of good ale, and stop it well; and give it the sick to drink all day …" (Dawson). FOXGLOVE tea used to be a standard domestic remedy for dropsy (Baker. 1980), a use also noted in Gaelic medical tradition (Beith). Irish people used to make a tincture for it with gin, and then use a very small quantity on loaf sugar (Egan). RED CLOVER tea, made from the dried flowers, was taken as a domestic cure in parts of America (H M Hyatt). A GORSE flower infusion is an old Wiltshire remedy for the complaint (Wiltshire). Herbalists still use PANSY tea to cure dropsy and the like (Flück), and REST HARROW is used, too (Flück), just as the early herbalists recommended. SQUILL enjoyed a reputation as a remedy for the condition. The Anglo-Saxon version of Apuleius had a leechdom for "water sickness" (Cockayne). But long before that, it seems it was used by the Egyptians for the complaint, under the name of "Eye of Typhon" (Thompson. 1947); the Delta people found the squill so useful for oedema that they are said to have built a temple in its honour. It continued to be prescribed through the ages of the herbals. Thomas Hill mentioned that it "amendeth the dropsie", and Gerard too recommended the bulb, roasted or baked, and mixed with other medicines, not only for dropsy but also for jaundice, and for "such as are tormented with the gripings of the belly". He also recommended the leaves and roots of WALL PENNYWORT, which "prevaile much against the dropsie".

BROOM tea is a well-known diuretic, so it follows that it would be used to cure oedema, and there are widespread reports of its use in domestic medicine in Britain not necessarily just by drinking it, for one treatment required the patient to soak the feet in a bath of the hot tea. HORSETAIL tea is another known diuretic, so useful for this complaint (Flück). A decoction of DYER'S GREENWEED has been used for the condition, as well as for gout and rheumatism (Grieve. 1931). WHITE HOREHOUND is another herb used for the complaint. A 16th century recipe from France reads: "pisser, neuf matins sur le marrube avant que le soleil l'ait touché; et à mesure que la plante mourra, le ventre se desenflera" (Sebillot) – but that is a simple transference charm.

A root infusion of TREE CELANDINE (*Bocconia frutescens*) has been used in Colombia for the complaint (Usher).

Oenothera biennis > LESSER EVENING PRIMROSE

OIL BEAN

(*Pentaclethra macrophylla*) One of the sacred trees of West Africa. As Talbot points out, in Nigeria, nearly every compound has its sacred tree, usually Chlorophora excelsa, but often a *Pentaclethra*. The souls of the dead live in these trees while awaiting re-incarnation.

The long pods explode loudly when they open, and the beans, known as oil-seeds, produce up to about 45% oil from the kernels, rich in protein, but poor in starch. They are eaten after roasting, more as a condiment than a staple food.

Olea europaea > OLIVE

OLEANDER, or ROSE BAY

(*Nerium oleander*) Very fragrant, a quality much appreciatd in Victorian times, when it was often the fashion to have an oleander kept at the foot of the stairs in the front hall, so that its perfume could pervade the whole house (Kingsbury. 1967). But this is a highly poisonous plant, the leaves and stems of which can be dangerous to children and animals. Eating a single leaf, or even eating meat skewered with oleander during cooking has been recorded as deadly (Tampiom). It was even thought at one time that the very perfume was toxic, and having the plant in a closed room might poison a person (Kingsbury. 1967). It is used in India for homicidal and suicidal purposes, and also for abortions (P A Simpson); it is known there as 'horse-killer' (Folkard), from the Sanskrit name for the shrub, something of which they have been aware since ancient times, for we are told that during the Persian campaign, Alexander's army lost horses that had fed on the shrub, and some soldiers, too – those that had grilled their meat on skewers made from the wood. Gerard described it in the usual way: "the flowers and leaves kill dogs, asses, and very many other foure footed beasts", though it is said on the Greek island of Chios that the leaves are so bitter that even the goats cannot eat them (Argenti & Rose).

Gerard finished his sentence with "but if men drinke them in wine they are a remedy against the bitings of Serpents", better still if rue is added. Wishful thinking, but there have been medicinal uses. The active principle, oleandrin, has an effect like digitalis, and is used in the treatment of heart conditions (Thomson. 1976). There is, too, a wart cure involving the plant on the Greek islands, but this is a charm cure; on Chios, the practice is to put a leaf on each wart. Then, in the wane of the moon, the leaves are put under a stone in a river-bed. The patient has to go away without looking back (Argenti & Rose).

There are examples from North Africa of its use as a protective plant. The box in which a Moroccan bride was transported, on the back of a mule, to the bridegroom's house, was made of oleander twigs, to avert the evil eye. For the same purpose, twigs would be put between the horns of ploughing oxen, and at the bottom of stacks of reaped corn. At Midsummer, such twigs would be hung on their fig trees. The so-called "sultan of the oleander", a stalk with a cluster of four pairs of leaves round the stem, is always endowed with "baraka" (protective power), but the power is greatest when it has been cut immediately before midsummer. When brought into the house, the branches must not touch the ground, for that would make them lose their "baraka" (Westermarck. 1905). In cases of sickness thought to have been caused by an evil eye, the leaves are burned, and the patient lets the smoke pass underneath his clothes, inhaling it as it comes through. Written upon, the leaves serve as charms, and pens are made of the wood. Anyone who is ill from an attack by "juun" has to be rubbed for two months and twenty days with oil of oleander, mixed with various medicines, and to be beaten with oleander twigs on either side of the body (Westermarck. 1926). The origin myth recorded in Morocco states that the shrub is the child of Fatima, the daughter of the prophet. Her husband took a second wife. When she saw her coming into the house she spat on the ground in contempt, and the oleander sprang up at once. Since then it is said "a rival is bitter as the oleander" (Legey).

OLIVE

(*Olea europaea*) Olive oil is mentioned so frequently in the Bible that it must have been important to the ancient Hebrews. We know that it was used in holy ointments for kings and priests, and for anointing the sick (Zohary), as well as for sacrificial purposes, and as fuel for lamps, as a tonic for hair and skin, and medicinally in surgical operations. It formed the base of the perfumed ointments sold in classical Greece and Rome (Moldenke & Moldenke), for it was equally prominent in classical times. The ancient Greeks used the oil for cooking, washing, and for lighting. Even today, in Greece, few people eat butter; bread and olives is the usual. Oil was used in practically every dish. The Greeks used no soap, but rubbed themselves with oil.

Among the Greeks, the olive was the symbol of wisdom, abundance and peace (Dyer. 1889). It is still the symbol of peace and friendship, possibly because, in the Noah's Ark story, the dove is said to have brought back an olive leaf as an indication that God's wrath, in the form of the Flood, was abating (the dove, too, carries this symbolism) (Moldenke & Moldenke), and the olive branch is still a symbolic demand for peace. A

garland of olive was given to Judith when she restored peace to the Israelites by the death of Holofernes (Judith. xv. 13, in the Apocrypha). Conversely, in ancient Rome, if an olive was struck by lightning, it was taken to be an augury of breaking the peace (Rambosson). It was, too, the emblem of Athena, or Minerva, goddess of medicine and health. In some parts of Greece to this day, an olive branch is always put on the New Year table as a symbol of health, together with coins, for happiness, etc., (Megas). To dream of gathering olives denotes peace and happiness, and dreams of eating olives means you will rise above your station (Raphael).

It was sacred to Athena, and was the gift of the goddess (Haig). As such it was judged greater than Poseidon's gift of the horse (Moldenke & Moldenke). Every sanctuary and temple to Athena had its olive tree (Philpot). In modern Greece, a method of protecting the household from the evil eye is by fumigation with burning branches of dry olive blessed during Holy Week (Rodd). In modern Italy, an olive branch hung over a door is supposed to keep out witches and evil spirits (Moldenke & Moldenke). Similarly, olive leaves blessed on Palm Sunday are still hung in a corner of a Maltese house, or occasionally burned to purify the house and ward off evil (Boissevan). On the Greek island of Chios, it is said to be a holy tree, for Christ blessed it (Argenti & Rose). After the birth of a boy, a prophylactic wreath of olive branches was hung outside the house as an amulet (a girl got a fillet of wool) (Halliday). An olive stick is believed in Morocco to be a charm against the juun (Westermarck). An olive crown was the prize at the Olympic Games, and in Rome, newly-married couples carried olive garlands; the dead were also crowned with olives (Rambosson).

It is a sacred tree among the Palestinians. Legend says that at the death of Mohamed most of the trees went into mourning by shedding their leaves as they do in winter. When the others were asked why they did not do the same, the olive, as their elder and spokesman, replied: "you show your sorrow by external signs, but our grief … is no less sincere, though inward. Should you cleave my trunk open, for instance, you will find that at its core it has become black with grief" (Hanauer). It is sacred in Morocco, too, because the name of God is supposed to be written on its leaves (Legey).

In medieval times, it was said that the Cross was made of four sorts of wood – cedar, palm, cypress, and "the tablet above his head … on which the title was written, of olive". The oil is reckoned effective in treating skin diseases (F J Anderson), and the leaf tea is prescribed by herbalists for hypertension (Thomson. 1978).

OMENS

Omens of prosperity for the year could be drawn from the number and size of the new crop of POTATOES, more especially those of the first digging, when a stem each was taken for each member of the family (north-east Scotland). It is said that at Plouer, on the Cote-du-Nord, it was at one time the custom before the fishing fleet set out, to hang a LEEK from one of the joists of the kitchen ceiling. If the plant kept alive, this was regarded as a good omen, but if it dried up and died, then it could be taken as certain that a member of the family had died at sea (Anson). CABBAGES, too, provided omens. If one ran to seed the first year, or one grew with two heads on one stalk, then that was a sign of death (Whitney & Bullock; Fogel). If one of them has white leaves, it meant a funeral (Stout).

When plants bloom out of season, it often foreboded some disastrous event. Thus, when BURNET ROSE acted in this way, the belief around the Bristol Channel was that it was an omen of shipwreck (Radford & Radford). COWSLIPS or PRIMROSES blooming in winter were an omen of death (Hole. 1937), as are PLUM trees blooming in December (M Baker. 1980). When BROAD BEANS grow upside down in the pod, it is an omen of some kind. They did so in the summer of 1918 apparently, and then it was remembered that the last time they did that was the year the Crimean War ended. Unfortunately the correspondent of Notes and Queries who reported it dated his letter in 1941, when once more they were growing upside down, but that war took a few more years yet to end.

If a RUSH-light curled over, it was a sign of death, and if a "bright star" appeared in the flame, a letter could be expected (Burton).

There are a number of death omens connected with APPLES, particularly with out of season blooms. If it happens when there is fruit on the tree, it is a sign of death in the family, put into rhyme as:

> A bloom upon the apple tree when apples are ripe,
> Is a sure termination of somebody's life (Baker. 1980).

The "somebody" being a member of the owner's family, it must be understood. Never leave a last apple on the tree, for that too would mean a death in the family. Not in Yorkshire, though, for there one *must* be left, as a gift for the fairies. But if one stayed on the tree until the spring, that too was a sign of death in the family (Gutch. 1911). A better-known, and ancient, death omen is connected with BAY trees. Shakespeare voiced the superstition:

> 'Tis thought the king is dead; we will not stay,
> The bay-trees in our country all are wither'd
> (*Richard 11, ii. 4, 7*).

See also Holinshed (*Chronicles*): "In this year 1399 in a maner throughout all the realme of England, old baie-trees withered, and contrary to all men's thinking grew greene againe, a strange sight, and supposed to import some unknowne event". This belief is traceable at least to Roman times. It was the custom for a successful general to plant a bay at his triumph in a shrubbery originally set by Livia. This tree, too, was believed to fall after his death.

The dropping of the leaves of a PEACH TREE is a bad sign (apart from being a bad sign for the tree, that is), for it was said to forecast a murrain (Dyer. 1889).

ONE-SEEDED JUNIPER

(*Juniperus monosperma*) A North American species, growing roughly from Kansas to Mexico. The seeds are edible, and were eaten or used for seasoning meat by the Indians of New Mexico (Yanovsky). The Zuñi Indians used the bark as tinder to ignite the sparks from the fire sticks employed for making the New Year fire. They roasted thin twigs, roo, and then steeped them in hot water to make a tea to be drunk before childbirth, the idea being to promote muscular relaxation. The tea could not be taken long before labour, otherwise, they said, the child would be dark in colour (Stevenson). A Navajo legend tells how in ancient days a woman was seated under a juniper tree finishing a basket as she always did, when she thought how much better it would be if she could make a more beautiful margin. As she sat there, a juniper branch was thrown into her basket by one of the gods, so she imitated the peculiar fold of the leaves, and that is why Navajo baskets still have this characteristic rim (Elmore).

ONION

(*Allium cepa*) It is very likely the oldest cultivated vegetable, certainly known in ancient Egypt, where the people swore by it, and perhaps, so it has been suggested even worshipped it. They saw it as symbolizing the universe, since in their cosmogony the various spheres of heaven, earth and hell were concentric, like the onion's bulb layers. They also saw it as an emblem of the moon, whose different phases could be seen in the bulb when it was cut through. As such, it was dedicated to Isis, the moon goddess (Leland. 1899).

Presumably the onion is especially connected in some way with St Thomas, for it is on St Thomas's Eve that a well-known divination is practiced. Derbyshire girls, for instance, peeled a large red onion and stuck nine pins in it, one in the centre being named for the man she really wanted, and the rest radially. As they were being put in, the girl would recite:

> Good St Thomas, do me right
> Send me my true love tonight,
> In his clothes and his array,
> Which he weareth every day,
> That I may see him in the face,
> And in my arms may him embrace (*Notes and Queries; 1871*).

Normally, it would be an advantage to know in advance in what trade her future husband would be engaged, hence the insistence on seeing him "in his clothes and his array, which he weareth every day". Another version of the divination required the girls to cut the St Thomas Onion into quarters, whispering to each the name of the young man she was expecting to propose. She waved it over her face, while saying the above rhyme, or another very like it. Pennsylvania German girls used to take four onions, give each a name, and put them under the bed or the stove in the evening. The one that had sprouted next morning bore the name of the future husband (Fogel). There was a sanction in London about St Thomas onions, for it was said that when buying them you should always go in at one door, and come out at another (*Notes and Queries; 1853*). There is another, far more sinister, association of onions in this connection. To get an erring lover back, it was recommended that the girl should roast an onion which had been stuck full of pins (pins that had never been through paper), which were to pierce his heart (Henderson).

Among the many beliefs connected with onions, one can include weather forecasting. When the skin is thin and delicate, we can expect a mild winter; if it is thick, it foretells a hard season (Inwards, Stout). Put into verse, we have:

Onion's skin very thin,
Mild winter's coming in;
Onion's skin thick and tough,
Coming winter cold and rough (Krappe).

The Pennsylvania Germans used to work out what the year's weather would be by taking twelve onions, naming them one for each month, hollowing them a little and filling them with salt. The wet months would be indicated by those onions in which the salt was dissolved (Fogel).

Onion peelings should be destroyed immediately, for it is unlucky to keep part of an onion in the house (Leather). Similarly in America, where it is said that you should never leave a cut or peeled onion in the house, for it will bring bad luck (Bergen. 1899), or will become impregnated with all the bacteria floating about (Igglesden), nor, in some parts, should you leave whole onions around (Whitney & Bullock). Actually, though, onions (and garlic) are good antibiotics (Cameron). Carried or worn, it will keep off disease. In Cheshire, they say that a peeled onion set on the mantelpiece during an epidemic will cause the infection to fly to the onion, and so spare the inhabitants of the house (Hole. 1937). The onion was supposed to turn black if there was any infection present (Palmer. 1976). Yoruba belief is somewhat similar, but put another way. They say that onions use their smell to kill disease (Buckley). When there was an outbreak of scarlet fever in Warwickshire, onion peelings were

buried where they could not be disturbed, in order to carry the fever from the house (Bloom). There are still a lot more examples of the belief. Clearly the cure did not rely on the onion's being peeled, for we are told simply that a row of onions should be hung over the door, so that they will absorb all diseases from anyone who comes in (Bergen. 1899). That was from Massachusetts, but a Somerset belief is exactly the same, and indeed goes farther, for the bunch of onions over the door will keep away not only illness, but also witches (Tongue. 1965). There is even a story from India of people who always hung an onion by their house-doors. When plague visited the area, they were the ones to survive (Igglesden), It was suggested, perhaps facetiously, that in the time of plague in London a ship freighted with peeled onions should be sent up the Thames to purify the air and carry the infection out to sea (Coats. 1975). During plague years in London, three or four onions left on the ground for ten days were believed to gather all infection in the neighbourhood (F P Wilson). One way to deal with 'flu, according to Alabama belief, is to put an onion in a pan under the bed (R B Browne), and the disease was dealt with in Scotland in a similar way – one half at an open window, and other above the door. The onion would attract the disease, and so turn black (Beith). Animals could be safeguarded as well as humans. During the disastrous foot and mouth disease outbreak in Britain in 1968, on one Cheshire farm which escaped, although in the midst of the infection, the farmer's wife had laid rows of onions along all the windowsills and doorways of the cowsheds, and attributed the farm's escape to this precaution (M Baker. 1980); apparently it was once standard practice in Yorkshire to hang four or five onions round a distempered cow's neck. A few days of this, and the cow's nose would run, and so the disease would be cured. The onions had to be buried deep after removal (Gutch. 1911).

There are other examples of onion's protective role, even if they involve wishful thinking. If, for example, you rub the schoolmaster's cane with one, it will split when he strikes you (Addy. 1895), a superstition that onions share with green walnut shells; or you will not feel the cane if you rub an onion across the palm of the hand (Gutch. 1911). In similar vein, in parts of Morocco, a horseman who went to war put into his saddlebag a couple of onions, to prevent bullets from hurting him (Westermarck. 1926). An onion carried in the pocket will relieve the bad luck of meeting a single magpie (Tongue. 1965; Whitlock. 1992). Or, in Kentucky, carrying one would bring good luck. Similarly, there is a Pennsylvania German belief that rattlesnakes will not bite you if you have onions in your pocket, or if you have rubbed your legs with one (Fogel). It will ward off smallpox, too, so Kentucky practice would have us believe (Thomas & Thomas). Dogon people of Mali regard onion smell as the loveliest of all aromas. Young people will fry onions in butter, and rub it all over their bodies as a perfume (Classen, Howes & Synnott). Was this originally a protective measure?

Dreaming that you are peeling onions foretells domestic strife and impending sickness (Raphael); dreaming that you are eating them is a sign of finding vauable treasure. The American experience tends to confirm this, or at least it is regarded as a good omen (Bergen. 1896), and there they say that eating an onion before going to bed, or simply keeping one over the bed head, will make you sleep well (H M Hyatt). Fertility is not usually associated with onions, but there is a Somerset practice of throwing an onion after the bride, to bring a long family (Tongue. 1965). There are still some more obscure examples of onion superstition – it is said, for instance, that when the devil's left foot touched soil outside the Garden of Eden, garlic sprang up, and his right gave rise to onions (Emboden. 1974). A very odd onion belief is quoted from an Italian source; it is a prescription for divining the health of a person far distant: gather onions on Christmas Eve and put them on an altar, and under every one write the name of one of the persons about whom information is sought. When planted, the onion that sprouts first will clearly announce that the person whose name it bears is well (Hartland). An equally odd magical practice involving onions or garlic is recorded in Portugal. In the Algarve area, bloodshot eyes are cured, if that is the word to use, by plucking onions or garlic without looking at them, and then hiding them with the eyes shut (Gallop). An American belief was that you should always burn onion peelings. If you just throw them away, you will die poor (Whitney & Bullock).

Opinions differ as to when onions should be set. In Lancashire, they say it should be on St Gregory's Day (12 March) to ensure a good crop (M Baker. 1980). Sometimes you hear that St Patrick's Day (17 March) is the proper day; in Shropshire, it is Ash Wednesday, though in France it is much later, surprisingly, for Palm Sunday is the proper day there (M Baker. 1977). St Benedict's Day is more reasonable – that is 21 March, and the day the Pennsylvania Germans chose (Fogel). The state of the moon is important, too. Thomas Hill, in the 16th century, gave this advice: "If the Gardner commit seeds to the earth in the wane of decrease of the Moon, he shall possesse small and sowrer ones, if the seeds in the increase of the Moon then strong and big, and of a moister tast, with the sowreness maistred". This is sympathetic magic, of course; as the moon increases so will the size of your plants, but there is no magic about further advice from the Pennsylvania Germans, who say you should bend over the tops of onions on the day of the Seven Sleepers (27 June) to make them grow big (Fogel). Gardeners' advice to set a row of onions between each row of carrots is equally sound – doing so will

keep away the carrot pests. There is one more piece of received wisdom from Alabama. To stop your eyes watering when peeling onions, start peeling from the root and carry on upward. Or put a match in your mouth while paring them (R B Browne).

Onions are useful in more ways than the obvious. The skins will dye wool brown (Dimbleby), or yellow, with alum. With a tin mordant, the colour will be orange (Coates). Onion peel for dyeing Easter eggs gives various colours from yellow to deep orange and reddish brown. In Estonia, the practice used to be to moisten the shell first, rill it in chopped barley and tie it in a cloth before boiling with onion peel. An 18th century source described how patterns were made by cutting the onion into shapes and sticking them to the shell with egg-white (Newall. 1971).

The humble onion has been one of the greatest standbyes in domestic medicine for a very long time, for anything between dealing with a boil to treating epilepsy. Earache was cured, as early as Apuleius, with an onion. In the Anglo-Saxon version the prescription required one to boil an onion in oil, and then drip the oil into the ear (Cockayne). The detail may vary, but onions are still being used for the complaint, either, as in America, putting the heart of a roasted one into the ear (Bergen. 1899), or, as in Cornwall, putting a boiled onion in a stick and holding that to the ear (Hawke). An onion poultice (with or without black treacle (Thomson. 1976)) is a widespread treatment for a boil in Europe and America. The Scottish Highland practice was to use an onion poultice for toothache, in which case it was applied to the cheek. For pneumonia, two onions were boiled, and one put in each armpit, and for diphtheria, when the poultice would be put on the neck (Beith). On Corfu, water from boiled onions is applied as hot fomentations for piles (Durrell. 1945), and in Britain warts were drawn and broken with an onion poultice (Drury. 1991), being treated as if they were boils. Corns, too were dealt with using equal portions of soft soap and roasted onions, as a poultice (Page. 1978). Onion juice rubbed on a bee or wasp sting was an old Witlshire bee-keeper's remedy (Wiltshire), but there is nothing either local or esoteric about that. Chilblains used to be treated in Ireland simply by rubbing onion juice on them (Maloney), or by putting a roasted onion on them as hot as could be borne. Bruises, too, can be treated with them. In Norfolk, the papery outer skin put in a cracked lip was claimed to heal it (V G Hatfield. 1994).

So much for superficial complaints. But internal ailments are treated with onions, too; the juice was considered essential to cure a cough or bronchitis centuries before its use in various patent medicines (Camp). In the Fen country colds, sore throats and coughs were always cured with "onion sorrel" ("with plenty o' pepper in it") (Marshall). In Alabama, they say that eating onions, cooked or raw, would cure the 'flu, but you had to drop the skin on the floor and leave it there (R B Browne). Indiana practice was to hang an onion in every room, and that, they say, will prevent the family from having colds (Brewster). An Irish country remedy for tonsilitis was simply to eat roasted onions (Maloney), and chopped onions boiled in milk was another Irish remedy, but for asthma (Logan). A cold in the chest used to be dealt with in America either by taking "onion syrup", or with a hot onion poultice, put not necessarily where you would suppose, for in Maine the practice was to put them on the wrists and ankles (Bergen. 1899) (in Iowa, fried onions would be put either on the chest or on the soles of the feet (Stout)), and there is too a cure for pneumonia that involves putting an onion in each armpit (Beith). Earache was treated in Cheshire by warming a small onion and holding it to the ear, or put in flannel and applying that (Cheshire FWI).

Onion cures go a lot further than this. The juice will lower blood pressure (in fact all the Alliums have this property), and improve the blood in cases of anaemia (Schauenberg & Paris). Epilepsy is treated in Russian folk medicine by eating as many raw onions as possible, or else by drinking the juice (Kourennoff), and children's convulsions were cured in Yorkshire by rubbing a raw onion on the palms of the hands (Gutch. 1901). An apparently equally inconsequential placing of the curative onion occurs in a folk cure for whooping cough, where the juice was rubbed on the soles of the feet (Camp). Hollands gin in which onions have been macerated was given as a cure for gravel and dropsy (Grieve. 1931), and a baby's colic was dealt with in Alabama by roasting an onion in hot ashes, and squeezing the juice out so that the baby receives a few drops at a time. The juice was also given there to make the baby sleep. While on the subject of babies, it is worth noting that Hungarian gypsies usesd to say that a mother who has no milk should eat plenty of onions (Erdõs).

Finally, something that is really strange – onions were always regarded as a drunkard's cure (Leyel. 1926), perhaps with some justification. But the oddest usage of all is to get the juice "anoynted upon a pild bald head in the Sun", which will bring "the hair again very speedily" (Gerard), a belief that dates back to the Middle Ages (Thomson. 1976), and which was still in use as an Irish country cure in the 20th century (Logan).

Onobrychis sativa > SAINFOIN

Ononis repens > REST HARROW

Ophioglossum vulgatum > ADDER'S TONGUE

OPIUM POPPY

(*Papaver somniferum*) The medical properties of opium were discovered a very long time ago, for the poppy was already domesticated for its seeds during Neolithic times in Switzerland; it was also grown in Spain at a time when metal was first coming into use (Clark. 1952). Opium was almost certainly the active principle of the drug nepenthes, described by Homer as the "potent destroyer of grief". Homer attributed its discovery to the ancient Egytians; the Ebers papyrus actually has a chapter devoted to a "remedy to prevent the excessive crying of children", by giving them what can only be opium, a remedy still in use, both in Egypt and Europe, in the 20th century. Children are still soothed with the aid of the drug, quite often fatally (Lewin). Virgil calls it the "Lethean poppy", and the flower heads were the emblems of Morpheus, the god of sleep. That is why poppy seeds, along with others, were put into coffins, to make a corpse stay in his grave – they would encourage the deceased to "sleep", rather than walk (Barber). In addition, the great number of its seeds made it a type of fertility and wealth symbol. Hence the gilt poppy heads, once common in Italian apothecaries' shops, which were originally amulets to attract money (Leland. 1898). In Christian art it also symbolises ignorance and extravagance (Ferguson), again presumably from the great number of its seeds.

Syrup of poppies was being recommended in the 10th century as a sedative for catarrh and coughs, and is still commonly used (Fluckiger & Hanbury). Opium was extensively used in the Fen country of England for malaria or ague, as it was called. Doctors said that it had more effect than quinine (V G Hatfield. 1994). It was taken as "poppy tea". Every garden had a patch of these poppies, the seeds of which were boiled, and the resulting liquor given to drink for fever, and general aches and pains. Even teething children were given the tea, or a few poppy seeds to suck, tied in a piece of linen (Porter. 1974), or a dummy dipped in poppy seeds (V G Hatfield. 1994).

ORACLES

The famous oracle at Dodona was the sacred OAK there – Zeus lived in the tree, and the rustling of the leaves, according to one source, was his voice. Homer knew the oracle, which was probably the most ancient in Greece. The attendants of the oracles were priestesses of Dione, the earth goddess, with Selli, priests of Zeus (Flacelière). From the context in the Odyssey, the implication is that the will of Zeus was audible from the Dodona oak itself, and that it was the tree itself that spoke (not the rustling of the leaves, nor birds in the tree). See the Homeric story of the Argo: when it was being built, Athena took a timber from the Dodona oak and fitted it into the keel, with the result that the Argo itself could speak, and so guide or warn the Argonauts at critical moments (Parke). On Greek coins and other works of art, Zeus is frequently shown crowned with oak leaves, or sitting or standing beside an oak tree.

BAY was the important ingredient at the Delphi oracle, Apollo's own sanctuary. The bay staff of a reciting poet was assumed to assist his inspiration, just as the bay rod in the hand of the prophet and diviner was assumed to help him to see hidden things. That is why the use of bay played an essential part in the oracular ceremony at Delphi (Philpot). The priests carefully observed the smoke of incense, and the crackling of bay wood and barley in the fire (Flacelière).

There is a legend of a speaking tree in India. When Alexander the Great reached "the furthest forests of India" the inhabitants led him to an oracular tree which could answer questions in the language of anyone who addressed it. The trunk was made of snakes, animal heads sprouted from the boughs, and it bore fruit like beautiful women. Apparently, the tree warned Alexander of the futility of invading India in order to dominate it. It is known as the Waq-waq Tree in Islamic tradition, and was often shown on four thousand year old Harappan seals. The association of India with oracular trees derives from tales of the tree-worship which has flourished there since very early times. European maps of India Ultima from the 12th century and later show the Speaking Tree (Lannoy).

ORANGE BALSAM
(*Impatiens capensis*) A North American species, used by some Indian groups to make an orange-yellow dye (H H Smith. 1923). It was used medicinally, too. To cure a headache, the fresh juice could be rubbed on the head (H H Smith. 1945); other groups used the fresh plant for a poultice to cure sores (H H Smith. 1928; Johnston). The juice, or a poultice, was applied to poison ivy rash (Brewster), and the juice is still used as a wart cure (Bergen. 1899). Some of the names used in America for this balsam, like Snapweed, or Kicking Colt (Grigson. 1955), are comments on the plant's explosive method of seed ejection.

Orchis mascula > EARLY PURPLE ORCHID

ORDEAL TRIALS
FOXGLOVE, a powerful poison, has been used in what the Americans call a "chemical jury", in other words ordeal trials to test guilt or innocence – if he survived he was innocent! (Thomson. 1976). The Jagga people in Africa used *Datura* (*stramonium* perhaps, but the exact species is not revealed in the account). Two handfuls of THORN-APPLE herb would be put into rather less than a pint of water, along with banana blossoms. (Why banana?). The litigant would address a solemn magical formula to the herb while putting his right hand into the vessel, and then the mixture was boiled, and eight snail-shells full handed to the person to be tested. The plaintiff described the offence, and urged the decoction to make the defendant fall down if he was

guilty, but otherwise to spare him. The accused, with the container at his mouth, would assert his innocence and utter a corresponding wish. If any of the liquid dripped, it was taken as a preliminary sign of guilt. After the potions had all been drained, the defendant was ceremonially taken for a walk. A few minor rites would be celebrated, and finally, the decoction would produce the desired effect of putting the drinker into a trance-like state in which he soliloquized, confessed his guilt or denied it, or vehemently resented the indignity of the test. Only if he made a clean breast of his guilt was he convicted and condemned to pay all the requisite fees. On the following day he would be given an emetic to purge him of the poison. But even so, the effects would probably not wear off for over a month (Lowie).

Not a trial, but certainly an ordeal, was the use of poison to test powers of leadership. John Josselyn reported in 1638 that the root was used by young Indian braves in an ordeal to choose a chief – "he whose stomach withstood its action the longest was decided to be the strongest of the party, and entitled to command the rest" (Weiner).

ORIENTAL PLANE
(*Platanus orientalis*) Some authorities say that this is the "chestnut" of the Bible (for example, Hutchinson & Melville); whether true or not, the tree was held in the greatest esteem in ancient times. Philosophers taught beneath the tree, and so it acquired a reputation as one of the seats of learning (Dyer. 1889). It was too the symbol of genius, and magnificence, and also of charity, in Christian art (Ferguson). It must have been a universe tree, for Durrell records the Corfu legend that on the ten days preceding Good Friday, all the Kallikanzaroi in the underworld are engaged simultaneously upon the task of sawing through the giant plane tree whose trunk is supposed to hold up the world. Every year they almost succeed, except that the cry "Christ has arisen" saves us all by restoring the tree, and driving them up into the real world. It is a protector, too, as Evelyn points out: "Whether for any virtue extraordinary in the shade, or other propitious influence issuing from them, a worthy knight, who stay'd at Isphahan in Persia, when that famous city was infested with a raging pestilence, told me, that since they have planted a greater number of these noble trees about it, the plague has not come nigh their dwellings".

When Greek lovers had to separate for a while, we are told, they used to exchange the halves of a plane leaf. Each half was carefully kept until they met again, and the whole leaf had to be made up (Gubernatis).

Origanum dictamnus > DITTANY

Origanum vulgare > MARJORAM

ORIGIN LEGENDS

Cereals in particular have attracted many origin myths, because of their supreme importance in the maintenance of life. There is often an incident of human or divine sacrifice involved. The idea was that one death, that of the god itself, would provide life to countless thousands by the gift of agriculture. Sometimes the death would not be of a divine personage, but of an often unnamed hero or precursor of the human race. Such a myth is this Menomini Indian one. An old man had CORN (MAIZE) which he kept hidden from mankind. It took a young boy, his nephew, to kill the old man and by so doing release the corn for the benefit of all (A Skinner).
An example of a different type of myth is this one from Mexico: a childless woman, who went to fetch water, saw the reflection of an egg on the cliff above. Her husband fetched it, and in seven days it hatched a small child with golden hair, which was soft and silky, like maize, which is personified by the child. (Newall, and see also Hatt for other corn origin myths).

One origin myth for the POMEGRANATE says that when Agdos, the hermaphroditic son of Zeus, was emasculated, the plant sprang from him (Freund).

Ion is the Greek for VIOLETS, and the legend of its origin is, in Lyte's words, "after the name of that sweete girle or pleasant damoselle Io, which Iupiter turned into a trim Heyfer or gallant Cow, because that his wife Iuno (being both an angry or jealous Goddesse) should not suspect that he loved Io, as also for her more delicate and wholesome feeding, the earth at the commandment of Iupiter brought forth violets, the which, after the name of the well-beloved Io, he called in Greeke Ion". Gerard had this story too, but another legend says that the Greeks adopted the name Ion after certain nymphs in Ionia had made an offering of the flowers to Jupiter (Browning). A quite different origin myth is that it sprang from the blood of Attis when Cybele changed him into a pine tree.

Cyanus, the specific name for CORNFLOWERS, is named for the Greek muse Kyanos, who worshipped Flora, and was for ever gathering flowers for her altar. When he died, the goddess gave his name to the flowers he was always picking (Skinner). The Russian name basilek for the cornflower is explained by invoking the memory of a young man, Vassili, who was changed by a jealous nymph into the plant (Palaiseul). One of the Greek origin legends for SAFFRON tells of a youth named Crocus (the word crocus is always taken to be saffron in early accounts) who was changed into the flower after being accidentally killed by Mercury. Another tradition has it that the crocus sprang from the spot on which Zeus once rested. The anemone, more specifically the POPPY ANEMONE (*Anemone coronaria*) owes its origin to the death of Adonis, one legend saying that he was changed into the flower, and another that it sprang from the mixture of the blood of Adonis and the tears of Venus (Rambosson).

In Morocco, the origin of OLEANDER is said to be the child of the spittle of Fatima, daughter of the prophet. Her husband took a second wife, and when Fatima saw her entering the house she spat on the floor in contempt. The oleander sprang up at once. Since then it is said "a rival is bitter as the oleander" (Legey). BREADFRUIT, cultivated widely throughout Asia, originated in Polynesia, and several of the islands have stories of the origin of the tree, all quite similar. The version from Hawaii tells how a man called Ulu died during a famine, and his body was buried near a spring. During the night, his family, who remained indoors, could hear the the sound of dropping leaves and flowers, and then heavy fruit. In the morning, they found a breadfruit tree growing from the grave, and the famine was over (Poignant). Ulu in fact signifies an upright, i e male, breadfruit, which is called ulu-ku. The low, spreading tree whose branches lean over, is ulu-ha-papa, and is regarded as female (Beckwith. 1940). There are, of course, perfectly rational stories of the introduction of breadfruit from other islands, though some of them suggest that it was taken from an island inhabited only by gods, and preserved for human use. The origin of TARO, too, is celebrated in Hawaiian mythology. One legend is of a daughter who had a child by her father, the god Wakea. It was born not in the form of a human being, but of a root, and it was thrown away at the east corner of the house. Not long after, a taro plant grew from the spot, and afterwards, when a real child was born to them, Wakea named it from the stalk (ha) and the length (loa). Another version says that the child of Papa was born deformed, without arms or legs, and was buried at night at the end of the long house. In the morning there appeared the stalk and leaves of a taro plant, which Wakea named Ha-loa (long root stalk), and Papa's next child was named after that plant (Beckwith. 1940). COCONUT, too, has many origin legends ascribed to it in the Pacific islands, involving the head and "eyes", and an eel shape that has a strong sexual content (for details, see **COCONUT**).

CROWN IMPERIAL (*Fritillaria imperialis*) was once a queen, so a legend has it, whose beauty, instead of contenting her husband the king, made him jealous, and in moment of anger and suspicion, he drove her from his palace. She, well knowing her innocence, wept so constantly at this injustice as she wandered about the fields, that her very substance shrank to the measure of a plant, and at last God rooted her feet where she had paused, and changed her to the crown imperial, still bearing in its blossoms something of the dignity and command she had worn as a human being (Skinner). A Scottish tradition says that the crown imperial hangs its blossoms down for shame at not having bowed to the Lord, and the spots at the bottom of the bells are the everlasting tears it shed in contrition (Simpson). Yet another legend says it was white once, but was dyed red by the blood of Christ at Gethsemane (Bazin).

There is a Burmese legend accounting for the origin of INDIAN SHOT (*Canna indica*). Dewadat tried to kill the Buddha by pushing a great boulder on him as he passed by below. The boulder fell at the Buddha's feet, bursting into a thousand fragments. A single fragment striking the Buddha's toe, drew blood, from which the Canna arose (Skinner).

One of the legends of the origin of SNOWDROPS is that an angel was comforting Eve after the Fall. No flower had bloomed since the expulsion from Paradise, and it was snowing. The angel caught a snowflake in his hand, breathed on it, and it fell to earth as the first snowdrop (Gordon. 1977).

Ornithogalum umbellatum > STAR OF BETHLE-HEM

ORPINE

(*Sedum telephium*) Orpine is a strange name, deriving from Latin auripigmentum, pigment of gold, hardly appropriate for this plant. Perhaps it was originally applied to Yellow Stonecrop. In the Gironde region of France, it was said that orpine was sensitive to the presence of sorcery – if there was a witch around it would wither as soon as it was brought near her (Sebillot). In Massachusetts, they said that it brings prosperity if it grows around the house. Another superstition from Ohio says that it only blooms once in seven years (Bergen. 1899).

But Orpine, under the name Midsummer Men, is chiefly associated with divinations. Another name, Love-long, recalls its use by hanging up a piece after a girl's boy-friends. The piece that lives longest determines the successful suitor. "In Gander Lane we saw in the banks some of the 'Midsummer Men' plants which my Mother remembers the servant maids and cottage girls sticking up in their houses on Midsummer Eve, for the purpose of divining about their sweethearts" (Kilvert). Other divinations are by the bending of the leaves to the right or left, telling whether a lover were true or false (Leather), or, as in America, to tell from what quarter the lover will come (Bergen. 1899). Also, if gathered by two people on Midsummer Eve, and the slips planted, they would know their fortune by the growing or otherwise of the slips. If they leaned towards each other, the couple would marry; if one withered, the person it represented would die (Radford & Radford). Aubrey had noted the custom in 1686: "Also I remember, the mayds (especially the cooke mayds & Dayrymayds) would stick up in some Chinkes of the joists … Midsummer-men, which are slips of orpins. They placed them by Paires: one for such a man, the other for such a mayd his sweet-heart, and accordingly as the Orpin did incline to, or recline from the other, that there would be love, or aversion; if either did wither, death". From a tract called *Tawny Rachel* (about 1800): "… she would never go to bed on Midsummer Eve without sticking up in her room the well-known plant called Midsummer men, as the bending of the leaves to the right or to the left, would never fail to tell her whether her lover was true or false". The belief travelled to America, too, albeit in an altered form. A record from New Brunswick advises "take a love-forever leaf, squeeze it to loosen the inner and outer skin. If it makes a balloon as you blow into it, you will be married and live a long time. If it does not, you will be an old maid" (Bergen. 1899).

In Iowa, they say that "a live-forever plant kept in the room will prevent cancer" (Stout). The shape of the root tubers "signal orpine with virtue against the King's Evil" (Grigson. 1955), or scrofula, and the plant was recommended for fevers, sterility in women and too profuse menstruation (Flück). Coles advised that the leaves, "bruised and applied., to the throat cureth the Quinsy …".

Oryza sativa > RICE

OSAGE ORANGE

A native of the south-western states of USA, but much planted as a hedge (it is actually called just Hedge (Sackett & Koch), or Hedge Balls (H M Hyatt)). The "oranges" are not edible, but warts can be treated by anointing them with the "milk" from the fruit. Sometimes, a charm is developed by anointing the wart three times and then burying the "orange". When it rots, the wart will leave (H M Hyatt).

OSIER

(*Salix viminalis*) Osiers seem to have enjoyed some prophylactic reputation in the English Fen country, for there is a record that at Hallowe'en, peeled osier twigs used to be put "at all the ways into the houses because no witch dared cross over them …" (Barrett. 1964). There is a traditional dance known as "strip the willow", the name apparently coming from the "peeling off" movement of the dancers (Addison & Hillhouse).

Osmunda regalis > ROYAL FERN

OX-EYE DAISY

(*Leucanthemum vulgare*) There is very little folklore attached to the Ox-eye, which is surprising in view of its frequency. We do find that it was used in a spell to bring back an unfaithful lover in Somerset (Tongue. 1965), and there is a record of its use in north-east Scotland in a charm to increase one's own milk supply without injuring that of one's neighbours. The way to do it was to boil white gowans, the local name for the plant, and to wash all the milk utensils with the decoction (Gregor), which seems merely homeopathic in intention, using colour as the link. Ox-eye had its medicinal uses, though they too are few. In the Highlands, the juice boiled with honey was used for coughs, and the same preparation was applied to wounds, while the plant was sometimes made into a tea to treat asthma. Elsewhere, in Russia, it was used as a household remedy for external haemorrhages (Watt & Breyer-Brandwijk), while in America, the

Menomini Indians used it for fevers (H H Smith. 1923). There is one veterinary usage, and a very odd one it is. Coughs in cattle were treated at one time by putting a piece of the root in a hole made in the cow's ear or dewlap (Drury. 1985).

ORRIS

(*Iris germanica "Florentina"*) Mixed with anise, orris was used in England as a perfume for linen as early as 1480 (see Wardrobe accounts of Edward IV). It smells like violets, and in fact is sometimes called violet powder (Hemphill). The root was also crushed and used as a substitute for dried violet in sachets and powder. It was once used for scenting tooth powders, too (Rimmel). It was used for this purpose in Alabama until quite recently – a tablespoonful of orris with seven tablespoonfuls of chalk; mix thoroughly, and dampen before use (R B Browne).

Orris-root would be thrown into the fire to produce a pleasant smell, and it was chewed to neutralize the smell of liquor, garlic or tobacco on the breath (Moldenke & Moldenke). Another use for the root was as an aid to babies' teething, as a "coral" (Bloom), and in Illinois a root worn round the neck served the same purpose (H M Hyatt). In London, too, Whitechapel Jews used it, but only if the piece of root had some fancied resemblance to the human figure, more or less suggestive of male or female forms – the he-root was used for girl babies, and the she-root for boys (letter published in *Folk-lore. vol 24; 1913 p 120*).

It was once popular for bronchitis and oedema (Camp), and Hill recommended it as "good against disorders of the lungs, coughs, hoarseness, and all that train of ills …". Gerard also prescribed it for oedema, and for cramp, convulsions, and snakebite, as well as for gonorrhea. Earlier still, there was advice "to cause hair to grow. Take water of flower-de-luce, and washe thy head therewith, and it shall cause hayre to grow" (see Gentleman's Magazine Library: popular superstitions 1884).

Oxalis acetosella > WOOD SORREL

P

Paeonia mascula > PEONY

PAIGLE

A common alternative name for COWSLIP. Hazlott recorded "as blake as a paigle" in his collection of proverbs (blake is yellow). Chambers's dictionary is honest about this word –"derivation unknown" is fair, but there have been a number of attempts at explaining it. A verb 'to paggle', unknown to Halliwell, is sometimes quoted. It apparently meant 'to bulge', or 'swell', according to one informant. Grigson. 1955 saw a different meaning – to paggle, he said, when applying it to a cow's neck, meant to hang and shake, and he saw the analogy with the loosely hanging flowers. Yet another attempt at the derivation saw the original as French 'paillette', a spangle. Whatever it was, the word itself went through a number of changes, from Peagle, to Piggle, Peggle (Hazlitt; Macmillan) and Paggle (Tusser), even Beagle (Tynan & Maitland). The ultimate variant must be Pea Gull (J Smith. 1882).

PAIN RELIEVERS

An essential oil obtained from MEADOWSWEET contains methyl salicylate and salicylic aldehyde, which are part of the compound now known as aspirin (Palaiseul). So it is a good pain reliever, prepared by boiling the flowers in water for ten minutes, the dose being three cupfuls a day (Page. 1978*)*. FEVERFEW is best known in country medicine as a painkiller, in spite of its name pronouncing the banishing of fevers. All that had to be done was to boil the plant in water, and drink the resultant liquid (Vickery. 1995). CAMOMILE has been used virtually for any ailment, understandably, for it is a good pain-reliever, and the root of STINKING IRIS has the reputation, too (Conway). Irish practitioners used WALLFLOWER blossoms steeped in oil as an anodyne (Moloney), and another Irish practice was to use HEMLOCK leaves mixed with linseed meal as a painkilling poultice (Moloney). FIGWORT has a certain anodyne value, and eases pain wherever it is applied (Mitton), and that includes toothache and babies' teething (Gerard). It probably accounts for its reputation in the Channel Isles for being an efficient remedy for cramp (Garis). THORN-APPLE has long been valued in American domestic medicine as a pain-reliever, in the form of a poultice or ointment made from the pulp of the bruised green leaves. In Essex, the method of use was to cut the top of the fruit off, and to pulp the inside, adding vinegar. Inhalation of the fumes brought relief (V G Hatfield, 1994).

Some of the hallucinogens have been used as anodynes, MORNING GLORY (*Rivea corymbosa*) being one example. Aztec priests would grind up the seeds as a potion which they used in ointments to appease pain (Dobkin de Rios). Like MANDRAKE, HEMP has been used as an anaesthetic, and both these plants are pain-killers (Emboden. 1972).

Paliurus spina-christi > CHRIST-THORN

PALM (PALM SUNDAY)

In Lincolnshire, HAZEL was often used as "palm" on Palm Sunday, and kept green the year round by putting it in water. In the south of the county, these "palms" were preserved for the express purpose of protection from thunder and lightning (Gutch & Peacock). See also LIGHTNING PLANTS

YEW acted as "palm" in many parts of Britain, and was actually called Palm in a number of areas. In 1709 a "palm-tree" was planted in the churchyard of St Dunstan's, Canterbury, and the accounts of Woodbury, in Devonshire, for 1775 refer to " a yew or palm tree planted ye south side of the Church" (Tyack). But it was the Goat Willow (*Salix capraea*) whose catkins were most often used as 'palm', and was the English embodiment of the tradition. In medieval times, a wooden figure representing Christ riding on an ass was sometimes drawn in procession, and the people scattered their branches in front of the figure as it passed (Ditchfield. 1891). Flora Thompson tells how sprays of sallow catkins were worn in buttonholes for church-going in her day, and how they were brought indoors to decorate the house. They should not be brought in before Palm Sunday, though – at least, that was the belief in Hampshire, for that would be most unlucky (Boase). At Whitby, palm crosses were made, and studded with the blossoms at the ends, and then hung from the ceiling (Gutch). Similarly in County Durham, where the branches were tied together so as to form a St Andrew's cross, with a tuft of catkins at each point (Brockie). These Durham crosses were kept for the whole of the coming year (M Baker. 1980).

PALSYWORT

An old name for COWSLIP, which shows that it must have been used for that complaint. It must have been the trembling or nodding of the flowers that suggested it (Grigson. 1955). The *Regimen Sanitatus Salernitanum* had commended the cowslip as a cure for palsy or paralysis (hence another old name, Herb Paralysy). Gerard repeated the prescription – "cowslips are commended against the pain of the joints called the gout, and slackness of the sinues, which is the palsie".

PANACEA is defined as a universal medicine, a "heal-all", like GINSENG. Indeed the generic name for ginseng, Panax, has the same derivation as panacea. Other plants, besides ginseng, have been given the name of All-heal, or Heal-all, MUGWORT, for example, perhaps as a tribute to the esteem with which it was regarded. SELF-HEAL (*Prunella vulgaris*) is another, as is VALERIAN. MISTLETOE also bears the name, which is only to be expected, given that it is a magical plant in folklore, endowed with extraordinary powers. PENNYROYAL may as well be included here, for it has been used for so many illnesses and conditions. In fact, they used to say in Iowa that a lotion made from it is good for any ailment (Stout).

TUTSAN (*Hypericum androsaemum*) is French toutesaine, wholly sound, or perhaps healing all, a concept that was well known in Gerard's time, when it was regarded as a panacea. HOLY THISTLE (*Carduus benedictus*) has a particular place here, for any plant called "holy" or "blessed" was taken to be a counterpoison. Langham could say "the leafe, juice, seede, in water, healeth all kindes of poyson …". In addition, everybody knew it as a heal-all. Langham, indeed, had four pages of recipes under this head, for practically every malady, including the plague, for which it was regarded as a specific. CAMOMILE tea is a virtual panacea, used for a remarkable number of unrelated ailments. It is made simply by infusing ½ or 1 ounce of fresh flowers, less of the dried, in a pint of boiling water. Galen prescribed APPLE wine as a cure-all (Krymow).

Panax quinquefolium/Panax schinseng > GINSENG

PANSY

(*Viola tricolor*) There are detailed beliefs about planting pansies in America. Put them on the north side of the house, they say, or they will not flower. They must be planted exactly two inches deep, or they will not grow. Plant them at 6 a.m, and always water them at 6 a.m. (Hyatt).

Like a good many other plants, pansies were once thought to be aphrodisiac. Shakespeare, of course, knew this, for didn't Oberon tell Puck to put a pansy on the eyes of Titania? And it was dedicated to St Valentine; all this accounts for the numerous "love" names, including the one given by Shakespeare – "Cupid's Flower", and a lot of examples of the Kiss-me-love-at-the-garden-gate type. On the principle of homeopathic magic, that which causes love will also cure it. That was why it was prescribed for venereal diseases. Gerard noted the belief, and prescribed "the distilled water of the herbe or floures given to drinke for ten or more daies together, three ounces in the morning, and the like quantitie at night, doth wonderfully ease the paines of the French disease, and cureth the same, if the patient be caused to sweat sundry times". Culpeper too regarded it as "an excellent cure for the French disease, the herb being a gallant Antivenerean", the latter remark being contrary to the accepted belief of his time.

Pansy leaves in the shoe were said to cure the ague. "It is good, as the late Physitions write", said Gerard, "for such as are sicke of an ague, especially children and infants, whose convulsions and fits of the falling

sickness it is thought to cure". Langham (*The garden of health … 1578*) has this too. Nursing mothers used it, – a handful of the fresh herb, boiled two hours in milk, is to be strained, and taken night and morning (Thornton). American domestic medicine recommended "a tea made of pansy (to) relieve gravel misery" (R B Browne). Herbalists in England still use pansy tea to cure dropsy and the like (Fluck), as well as children's skin eruptions and diarrhoea (Schauenberg & Paris). It has even been used for asthma and epilepsy (Leyel. 1937).

Dreams of pansies means heart's pain, quoted Mackay as one of the popular fallacies of his day, the opposite of heartsease, presumably; and some of the names given to the flower, as Love-in-idleness, which can only mean love-in-vain, a name actually in use (Grigson. 1955). One more odd belief, this time from Wales – if you pick pansies on a fine day (Trevelyan), or while they still have the dew on them (Baker. 1974), then you would cause it to rain very soon, or, much worse, in the latter case, you would cause the death of a loved one (Addison. 1985).

Pansies were used as symbols of remembrance and meditation in Christian art (Ferguson).

Papaver rhoeas > RED POPPY

Papaver somniferum > OPIUM POPPY

PAPAW

(*Asimina triloba*) The seeds have one strange use in Alabama. For teething, nine seeds are strung and the child then wears it round the neck (R B Browne). Why nine? Presumably it is the memory of a mystical number – 3 × 3.

Parietaria judaica > PELLITORY-OF-THE-WALL

PARK LEAVES

is a very common name for *Hypericum androsaemum* – TUTSAN (Macmillan). It is, of course, a corruption of *Hypericum*.

PARKINSON'S DISEASE

Apparently, THORN-APPLE can be used in the treatment of the disease (Scarborough). Herbalists also prescribe DEADLY NIGHTSHADE as a form of treatment (Schauenberg & Paris).

Paronychia sp > WHITLOW-WORT

PARSLEY

(*Petroselinum crispum*) Old gardeners always planted some parsley, either for use, or as an aid to other cultivation, for it was said that if it were planted all round the onion bed, it would keep the onion fly away (Rohde). It was grown among roses, too, both to improve their scent and to help repel greenfly, and tomatoes and asparagus are helped by its presence, which also encourages bees into the garden, so they say (Boland & Boland). Another tip is that it keeps insects and flies out of a kitchen if it is grown on a window sill (Boland. 1977).

Parsley is a symbol of festivity (Leyel. 1937), though quite why is not very clear. The Greeks used it as a symbol of great strength, and crowned the winners of the Isthmian Games with chaplets made from it (Hemphill). They made wreaths of parsley to put on tombs, for it was said to have sprung from blood, that of the hero Archemorus, the forerunner of death. It was also used in cemeteries dedicated to Persephone (Sanecki). It is still used in modern Greece as an ingredient of funeral food (Edwards). It is strange that such a useful plant should have this association, but it lasted until quite modern times. Cf the proverb 'Welsh parsley is a good physic'. By 'Welsh parsley' was meant the gallows rope (Young). It is, in fact, a very unlucky plan, not to be transplanted. If it were, bad luck would follow, or even death, to yourself or your relations, within the year (Farrer). One correspondent of *Notes and queries* in 1853 quoted a saying he had collected: "Where parsley's grown in the garden, there'll be a death before the year's out". He must have got it slightly wrong – "grown" must mean "transplanted". Twenty years later, in the same publication, it was confirmed that it was "most unlucky to transplant parsley" (except on Good Friday, though – see below). Ruth Tongue mentioned a Lancashire man who used to pay any passing tramp to plant his parsley for him (Tongue. 1967). This belief was carried to America: the Pennsylvania Germans say that someone in your family will die if you transplant parsley into pots (Dorson), and African-Americans in the southern states think so, too (Puckett), while the Kentucky belief is succinct – "plant parsley, plant sorrow" (Thomas & Thomas). Maryland belief is that you should never take parsley with you if you move house; get it from someone else (Whitney & Bullock). Of course, all this means that you should grow parsley from seed, and never move it about, but conventional sowing of the seed is frowned upon in parts of America. If you have to, blow it from a Bible, or from a gatepost (Whitney & Bullock). Country people still give it away only very reluctantly, though anyone wanting some can safely help themselves; it is bad luck to receive it, too, according to Maryland belief. In Canada, the taboo was extended to forbidding thanking the owner (Baker. 1974). The source of these beliefs (apart from the Greek myth) is probably a medieval idea that you could condemn your enemy to sudden death if you pronounced his name while in the act of pulling up a root of parsley (Palaiseul). It seems, too, that the Greek expression that translates as "to be in need of parsley" meant that a patient was *in extremis* (Baker. 1972). There is another medieval magical use, for Reginald Scot recorded a recipe for a witch ointment that consisted of aconite, boiled with leaves of poplar and parsley, and mixed into an ointment with soot and fat. The monkshood is the important element. It is a poison, and it is said that a fifteenth of a grain of alkaloid from the root is lethal. Rubbed on, the ointment would produce a tingling sensation, followed by numbness on the part of the body on which it had been applied, and thence to light-headedness and visions. But would parsley have more than a symbolic effect in such a mixture? There is no suspicion of a poisonous quality about it (except to birds, apparently (A W Hatfield)), and most animals relish it. It has even been claimed that if parsley is thrown into a fishpond, it will heal sick fish, but that is an ancient belief, quoted by Thomas Hill from classical sources.

There are many other parsley superstitions, a good many of them connected with conception and childbirth, summed up in the saying "Sow parsley, sow babies" (Waring). If a young woman sows parsley seed, she will have a baby, so it used to be said in Lincolnshire (Rudkin) among other areas. In some parts of Wiltshire it was thought safest if only the mistress of the house, i.e., a married woman, should sow it (Wiltshire), though in Guernsey folklore it is the man who should wear parsley under the arm, though both should drink large quantities of parsley tea (Garis). There is negative confirmation from Essex – if parsley will not grow in your garden you will never have children (Chisendale-Marsh). The Pennsylvania Germans had a similar belief, though not quite so pointed. If you sow parsley and it comes up it means an addition to the family; if it does not sprout, you can expect a death (Fogel). The parsley bed, as well as the gooseberry bush, were once "the euphemistic breeding grounds of babies" (Gordon. 1977), or at least girl babies will be found there (Baker. 1977). Obviously, then, there was a connection with conception, but Cambridgeshire girls would eat it three times a day to get an abortion (Baker. 1980). It is actually quite a widespread belief that eating lots of parsley acts as an abortifacient (Waring). French mothers used an application of parsley to stop their milk (Loux). Some ambivalence is shown in the Cotswolds, where some consider it a fertility plant, and some sterilising, or at least contraceptive (Briggs. 1974). Perhaps the reason lies in the well-known idea that where parsley thrives, the missus is master, put into rhyme as:

Where the mistress is the master
The parsley grows the faster.

That happens to be from Gwent (Wherry), but similar sayings are very widespread. The sight of it must have had an inhibiting effect on men, and so mothers with daughters did not like to see parsley flourishing, as it was apt to condemn the girls to spinsterhood (Rohde). The belief undoubtedly accounts for the Berkshire saying, "never pick parsley when you are in love; it will kill the love". On the other hand, parsley wine is an aphrodisiac, or so they used to say in Gloucestershire (Baker. 1977).

Thomas Hill interpreted various classical sources as postulating a male and female parsley, the latter with "crispeder" leaves, the male with blacker leaves and shorter roots. Eating parsley would have deleterious effects on the like sex ("… the Female eaten, doth procure the woman barren, as the Male the men"). So it would be wise for pregnant women not to eat parsley, advice given also to epileptics of both sexes. Perhaps this is the basis of parsley's power of predicting the sex of an unborn child, a belief at least as old as Galen, who advised putting a piece of parsley on a woman's head, without her knowledge. If after that the first person she spoke to was a male, it meant she would have a son (T R Forbes). To make sure the coming child would be a boy, Fenmen used to keep an eye on the parsley in the garden, and see that it did not grow too thick or too tall. If it did, it showed the wife was "master" of the house, and would bear a girl (Porter. 1969).

Observation of parsley's germination time has given a number of superstitions. Its seed is one of the longest to live in the ground before starting to come up. Further, the devil is implicated – it goes to the devil nine times before it comes up (Northcote) (or some say seven (Clair)). So it takes an honest man to grow parsley (or only a wicked one, depending on the point of view). It only comes up partially because the devil takes his tithe of it (M E S Wright); in fact, they say in Wiltshire that you must sow four times the amount you need (Wiltshire). To offset this you can pour boiling water over freshly sown seed to deter the devil (Baker. 1974), or better still, sow it on Good Friday, when plants are temporarily free of the devil's power (Baker. 1980), or do it at the very least on a holy day, which Somerset people say (Tongue. 1965), giving themselves a little leeway. In Ireland, naturally, they say it should be sown on St Patrick's Day, which in most years would not be far removed from Good Friday. It may even be safe to move parsley on Good Friday (Baker. 1974), and parsley sown then bears a heavier crop than that sown on any other day; or some say in Sussex it will come up curly (Simpson), or double, in Suffolk (Bardswell). Better still, according to Wiltshire wisdom, when there is a rising moon on Good Friday (Wiltshire). All this is explained in Somerset by saying that it had to be sown on a holy day or the fairies would get it (Tongue. 1965), while in Essex people reckoned there were extra benefits that would accrue if you sowed it on Good Friday – it would ensure luck and happiness in the coming year (Rohde).

The Pennsylvania Germans say that the way to make parsley grow is to piss in the hole in which you are going to plant it (Fogel). In any case, and this is from the Cotswolds, it will always grow better if sown with curses, and better for a bad man than a good one (Briggs. 1974), an obvious reference to the devil's

influence. There seems no end to sowing advice where parsley is concerned. From America, more particularly from Maryland, comes the advice that parsley seed should never be hand sown, but rather blown from a Bible or a gatepost onto the seed bed. In Texas it is scattered on the kitchen floor and tossed casually on to the garden with the dust. That, they say, is the only way to make it grow (Baker. 1977). A gardening adage from Normandy tells that to sow parsley in the shade is to run the risk of reaping hemlock (W B Johnson), and an interesting ecological superstition is the Fenland belief that it should be sown in drills running due north and south; they get the right direction by sowing at night and using the Pole Star (Porter. 1969). Having got it to grow, it becomes a valuable commodity, if we are to believe the dictum never to take a house with an established garden in which parsley is not growing, or you will never see the year out (Boland. 1977).

Other odd beliefs concerning the herb include one from 16th century France that even touching parsley, let alone eating too much of it, would be harmful to the sight (Sebillot), and another French superstition is that it would shatter any glass with which it came into contact. In Poitou it was said that a woman only had to touch a spray for this power to be transmitted to her (Sebillot) (temporarily, presumably).

Eating parsley seed helps to avoid drunkenness, so it is claimed (Page. 1978; Camp), and another piece of country wisdom tells that eating parsley will take away any garlic smell and taste (Browne). That is probably true, for parsley has been described as a "good natural deodorant" (Page. 1978). There is, or was, a limited medicinal demand for parsley seed, and a few acres were raised in Suffolk each year for the distillation of an essential oil, with apiol as its main constituent. It is the apiol that is used for kidney complaints (Clair), either in the form of this oil, or as parsley tablets that herbalists sell. The seed is also useful in the treatment of malaria (Hatfield), and they also "…, open, provoke urine, dissolve the stone, break and wast away winde, are good for such as have the dropsie, draw down the menses, bring away the birth, and after birth …"(Gerard) (Berkshire wise women used to prescribe plenty of parsley to recuperate quickly after childbirth). But Gerard's recommendation of parsley seed to "draw down the menses" is still apparently current (Opie & Tatem). In fact, herbalists issue a warning that large doses must be avoided by pregnant women (Fluck), and it was certainly taken as an abortifacient in the Fen country (Porter. 1958).

Parsley-root tea is still prescribed for kidney complaints (Rohde, and in America, Hyatt). It is reckoned in Russian folk medicine to be a powerful diuretic (Kourennoff), while gypsies certainly recognised its value. They used the leaves for treating kidney and liver trouble, and for dropsy and jaundice (Vesey-Fitzgerald). Parsley tea used to be a rheumatism remedy (Rohde). Even chewing the leaves is still thought of as a means of warding off rheumatism (Camp). Actually, parsley for rheumatism is a very ancient medicine; there is a leechdom in the Anglo-Saxon version of Apuleius "for sore of sinews…" (Cockayne). Chapped hands were cured in the Fens by rubbing on a salve made from finely-chopped parsley mixed with the fat of a roasted hen (Porter. 1969). It seems, too, that parsley was used for snakebite in the past (Cockayne). Indeed, it once enjoyed the reputation of being able to destroy poison, probably, as one suggestion has it (C P Johnson), because it can overcome strong smells. The crushed leaves make an antiseptic dressing for insect bites, scratches and bruises, or boils (V G Hatfield). It was even recommended for baldness as far back as Pliny's time (Bazin), repeated a long time afterwards as "powder your head with powdered parsley seed three nights every year, and the hair will never fall off" (Leyel. 1926). Actually, it really does make a good lotion for getting rid of dandruff, and would help to stave off baldness (Hatfield).

Parsley had its veterinary uses, too, for country people used to feed it in large amounts to sheep to cure foot rot (Drury. 1985), and a way of quieting a vicious horse, according to old Irish horse-dealers, was to give it as much parsley as it could be persuaded to eat (Logan).

PARSLEY PIERT
(*Aphanes arvensis*) Parsley (Piert) refers to the form of the leaves, not any relationship to parsley. The common name is from French perce-pierre, meaning breakstone (Prior) and it is actually called Parsley Breakstone (Grigson. 1955) (cf SAXIFRAGE). By sympathy, it was much used against stone in the bladder. Gypsies use an infusion of the dried herb for gravel and other bladder troubles (Vesey-Fitzgerald). It was well-known as a powerful diuretic in Camden's time, and it was in great demand during World War 11, being used for bladder and kidney troubles, and it is also valuable for jaundice (Brownlow). A decoction with sanicle was used for stomach complaints, but it was especially recommended, powdered and with a little cochineal, for bowel complaints, especially bowel-hive, an inflammation of the bowel, occurring in children. It was even called Bowel-hive, or Bowel-hive Grass (Britten & Holland), once. Colicwort is another relevant name, from Herefordshire (Grigson. 1955).

PARSNIP
(*Pastinaca sativa*)

If you want a parsnip good and sweet,
Sow it when you sow the wheat. (Leather).

The leaves were regarded as poisonous once (Graves), and the opinion in America was that it was poisonous when growing wild (Bergen. 1899). It seems to have been used in an English formula for witches' flying ointment. It is given as the fat of a newly born infant, eleoseline (which is wild celery), skiwet (identified as wild parsnip), and soot (Graves).

Wild parsnip was used in Anglo-Saxon medicine for a difficult labour. It was, too, recommended as an emmenagogue (M L Cameron). Interestingly, native Americans, for example, the Pillager Ojibwe, used a minute quantity of the root mixed with four other kinds of root to make a medicinal tea for female troubles (H H Smith. 1945). In the 18th century, Wesley recommended it for "a cancer … stamp the Flowers, Leaves and Stalkes of wild Parsnips, and apply them as a plaister, changing it every twelve hours. It usually cures on a few days". Cancer here is probably canker. At about the same time, Sir John Hill was declaring that a strong decoction "opens all obstructions. It is good against the gravel and the jaundice, and will bring down the menses"; the last part harks straight back to Anglo-Saxon medicine.

Parthenium hysterophorus > WHITE BROOMWEED

Passiflora caerulea > BLUE PASSION FLOWER

Pastinaca sativa > PARSNIP

PATCHOULI
(*Pogostemon patchouli*) The famous eastern perfume, from a plant whose scent is very powerful, and in an unadulterated form very unpleasant, smelling strongly of goats. But when the attar is diluted with attar of roses, the unpleasant quality goes completely (Genders. 1972). Patchouli oil is obtained by distillation of the leaves. Patchouli perfume (not the same as the oil) is also made from the leaves – it has the reputation of being aphrodisiac. Powdered dried patchouli leaves are sometimes introduced into incense (Schery). The perfume first became known in Britain about 1820, when it was used to impregnate Indian shawls which became so fashionable that the designs were copied by Paisley weavers for export to many other parts of the world. But they could not sell them if they did not smell of patchouli (Genders. 1972).

Patchouli leaves are regarded in India as a prophylactic against disease. Malays put the dried leaves among their clothes to protect them from insects, and they used them medicinally in an oil prescribed for smallpox (Leyel. 1937).

PEA, or GARDEN PEA
(*Pisum sativum 'hortense'*). Originally, the name was PEASE, the reduction of which to Pea is an example of a "false singular", arising from the mistaken belief that Pease was a plural. Cf the word 'cherry', which arose from a similar belief about the French word 'cerise' (Putnam).

Sow peas and beans on David and Chad,
Be the weather good or bad.

That is, the first and second of March. But study the weather a month earlier than that, and:

On Candlemas Day if the thorns hang a drop,
Then you are sure of a good pea crop.

Cambridgeshire farmers used to say that it would be a good year for peas if the hedges dripped on St Valentine's morning (Vickery. 1995). Tusser gave different advice, autumn instead of spring for sowing, but he also said they must be sown in the wane of the moon:

Sow peas and beans in the wane of the moon.
Who soweth them sooner he sowed too soon.

It is lucky to find nine peas in a pod, not only lucky but useful in marriage divinations. If a girl finds such a pod, and if she then wrote on a piece of paper:

Come in my dear
And do not fear,

and folded it and put in the pod, which then had to be laid under the door, the first man who entered the room would become her husband. In some areas the pod had to be hung up, and in the Midlands and the west of England, the Christian name of the first man to enter provided the initial of the future husband's name (Opie & Tatem). In America, the good luck of a nine-pea pod is recognised, so it is put over the door (Stout). Another way to celebrate the find is to throw one over the right shoulder and make a wish. It will come true (Waring). Against all logic, a pod with just one pea in it is also lucky (Waring).

Warts can be charmed by touching each one with a green pea, which would then be wrapped in paper and buried. As the pea decays, the warts will go (Allen).

PEACH

(*Prunus persica*) Peach originated from China, and was held in great esteem there. Peach blossom was at one time the popular emblem of a bride, and a sprig of blossom was put over the front door to keep evil spirits away (Waring). Dreaming of peaches was a very favourable omen, a superstition that has found its way into European dream books. It is a sign of contentment, health and pleasure (Gordon. 1985). Peach wood had a special quality in China, for it was said that it has the ability to repel evil spirits (so has willow wood) – they are both symbols of immortality, peach being the Taoist emblem (Tun Li-Ch'en). The Chia, a spirit writing instrument akin to the western planchette, was usually made of one of these two woods, to prevent the instrument being used by the wrong sort of ghost (L B Paton).

In some places, the dropping of the leaves of the peach tree is a bad sign (apart from being a bad sign for the tree, that is), for it was said to forecast a murrain (Dyer. 1889). Pennsylvania Dutch settlers horse-whipped their peach trees before breakfast on Good Friday, in order to encourage them to bear more fruit (M Baker. 1974), just the treatment that walnuts have always received in Britain.

Peach leaves were once recommended for children with worms (Black). "The leaves of the Peach Tree … being applied plaisterwise unto the navel of young children, they kill the worms, and drive them forth. The same leaves boiled in milke do kill the worms in children very speedily" (Gerard), who also said that the leaves "being dried and cast upon green wounds cure them", something that seems to have been known on the Greek islands, too, for people on Chios recommend the leaves, chopped small, for wounds and bruises (Argenti & Rose). In the southern states of America, they reckon that peach leaves bound round the head will stop a headache (Puckett), and as a poultice they were reckoned to cure rheumatism and neuralgia (Thomas & Thomas). It was even claimed there that a strong tea made from the leaves would cure dandruff and falling hair (H M Hyatt).

One recipe from Alabama sounds very doubtful. It was for morning sickness, and involved cracking a peach stone and extracting the kernel, which had to be beaten and made into a tea to give to the patient (R B Browne). The seeds are used in Chinese medicine, too. They are given for coughs, and as a demulcent for sore throats (R Hyatt), while they are quoted as being described as a "blood invigorator", and are also given for constipation (Geng Junying). Of course, the drug is toxic in large doses. The rest of the medicinal uses are simply charms, or transference superstitions. A Sicilian one for the cure of scrofula was to chew the bark, either on Ascension Eve or St John's Eve. If it dried up and withered, it was a sign that it had taken the disease to itself (Sebillot). Fernie quotes the same belief, except that the disease was goitre, and the peach was eaten. A peach sprout was used in Maryland, put either over the door or on the sill. Passing under or over it would cure the ague (Whitney & Bullock), and there is a charm from Indiana for a chill: the patient had to tie a string to a young peach, walk away, and not look back (Brewster) – this sounds like an incomplete example of a transference charm. In Marseilles, they used to get rid of a fever by sleeping with the back against a peach tree for two or three hours; the peach would gradually go yellow, lose its leaves and die (Sebillot). The same idea is inherent in an Italian way of dealing with warts – the leaves had to be applied to them, and then buried, so that they and the wart would perish at the same time (Fernie). Exactly the same thing is reported from America, or you could cut as many notches in a young peach tree as you had warts. In seven days the wart would go. Some say that three notches have to fill up before the wart will be gone (Thomas & Thomas).

PEANUT

See GROUND NUT

PEAPUL, PIPAL

(*Ficus religiosa*) Better known, perhaps as the Bo-tree, the tree of enlightenment, or perfect knowledge, for this is the most sacred tree to both Hindus and Buddhists, and using it as fuel is strictly prohibited (Upadhyaya). A good Hindu who sees a peapul tree will take off his shoes, and walk five times anticlockwise round it, while repeating a verse in praise of the tree (Pandey). In southern India it is worshipped as an embodiment of Vishnu, who was born under a pipal tree, and he is said to have taken the form of that tree (Simoons). It is said, too, to symbolize male creative power. The Peapul tree is often associated with the mangosa (*Melia azedarach*), as the symbolical male to the latter's female. The two trees are often planted together in "marriage", so that the peapul's vines can twine round the limbs of its mate (Lincoln). But it is the Banyan that is the male to this tree's female according to some authorities (Pandey), and in some mythologies, the marriage of the Peapul with the Tulsi (*Ocimum sanctum*) is "ordered" (Pandey). The tali, a gold ornament tied round a girl's neck at puberty, is shaped like a leaf of the peapul tree (Lincoln). Pipal leaves are used in the marriage ceremony to ward off the evil eye. The wedding robe of the bride is used to decorate a pipal tree in Gujarat. Appeals by barren wives are made to the tree, and it is believed that the spirits who live in the tree have power to make a woman fertile (Simoons).

This, of course, is the tree under which Buddha achieved perfect knowledge, and it is by the Gautama's special injunction that it is venerated above all others. Its leaves quiver ceaselessly, like those of an aspen. O'Neill wondered whether this "perpetual life-motion and whispering of the leaves" is a reason for its holiness, hence the concept of the "talking tree", though every tree inhabited by a spirit or deity is a talking tree (Simoons). In every Indian village, a bo-tree grows near a Hindu or Buddhist temple, surrounded by a mud platform on which meetings or meditation are held (Upadhyaya). It does have its secular uses – the bark, for instance, is a common medicine for diarrhoea in India. But because it is a sacred tree, care has to be taken to rub butter on the trunk's scar, to prevent the accumulation of sin for having harmed it (E B Harper).

PEAR

(*Pyrus communis*, which is Choke Pear, one of the parents of the cultivated pear). It is the symbol of affection, and of comfort (Leyel. 1937). Dreaming of the ripe fruit is a sign of riches and happiness; if unripe, adversity; if baked, success in business. To a woman, a dream of pears means that she will marry above her rank (Gordon. 1985). Bringing pear blossom into the house is unlucky. It signifies a death in the family (Vickery. 1995). Bartholomew Anglicus seems to have looked on pears with even greater mistrust. "Always", he said, "after eating of Pears, wine shall be drunk, for without wine Pears be venom"! It is said that water drunk from a well beside a pear tree is helpful for gout, and that pears are the natural antidote to mushroom poisoning (Cullum). (see also WARDEN PEAR)

PEARLWORT

(*Sagina procumbens*) A tiresome weed in gardens, but far from tiresome in the Scottish Highlands and Islands, for it was said to have been blessed by Christ, St Bride and St Columba, and it was also reckoned that this was the first plant Christ stepped on when he came to earth, or perhaps when he arose from the dead (J G Campbell. 1902). Most authorities agree that this is the mystical plant known as Mothan, which protects from fairy changing, etc., (Wentz; Carmichael). It was put in the milkpail as means of restoring virtue to the milk, and at the time of mating, a piece was put in the bull's hoof, so that no witch could touch the calf's milk (J G Campbell. 1902). It protects from fire, and from the attacks of fairy women (J G Campbell. 1900). Anyone carrying this plant, or having drunk the milk of a cow that had eaten it is "immune from harm" (MacGregor). There is an old saying in the Hebrides that, when a man has a miraculous escape from death, that he must have drunk of the milk of a cow that ate the mothan (MacGregor), and cows were given a little mothan to eat around Beltane Eve (31 April) and Samhain (31 October), to give them protection against the fairies at these times when they would be most active (MacGregor). Put on the door lintel, it prevents the sprits of the dead from entering the house, or for more generalised good luck (Murdoch McNeill).

Of course, such a plant had to be pulled with the proper incantations, of which the Gaelic has a number of examples (for examples, see J G Campbell. 1902; W Mackenzie; Carmichael). W G Stewart gave further instructions: "Go to the summit of some stupendous cliff or mountain where any species of quadruped never fed or trod, and gather of that herb in the Gaelic language called mothan … The herb you will give to a cow, and of the milk of the cow you are to make a cheese, and whosoever eats of that cheese is for ever after, as well as his gear, perfectly secure from every species of fairy agency".

Pearlwort also gives the power of fluent speech. The fairies would bestow this gift on a child who "has drunk of the milk of the cow that ate the mothan" (McGregor). Girls drank the juice of this plant, or at least wet their lips with it, to attract lovers. If they had a piece in their mouth when they were kissed, the man was bound for ever (Grigson. 1955), and, under the right knee of a woman in childbed, it soothes her mind and protects her child and herself from the fairies (J G Campbell. 1902).

PEARLY IMMORTELLE

(Anaphalis margaretacea) An American plant, introduced to Britain very early, and one of the first New World plants to escape. It is now well established in some areas, notably the South Wales valleys, and also in Cambridgeshire, where, particularly around Newmarket, they used to say it was lucky to wear "everlastings", or even artificial flowers (Burn).

Cheyenne Indians made use of it in a magical way; the dried powdered flowers were put on the sole of their horses' hooves, and also between their ears, to make the animal long-winded and spirited (Johnston). Similarly, they used it to provide the "strong medicine" for themselves. The flowers were dropped on hot coals and the smoke was used to purify gifts which were left on a hill for the sun or the spirits. Before going into battle each man chewed a little of it, and rubbed it over his body, arms and legs, for it was supposed to give strength and energy, and so protection from danger (Youngken). Some of the Ojibwe people used it in a similar way, that is, they sprinkled the flowers on live coals, to be inhaled by anyone who had had a paralytic stroke (H H Smith. 1945).

Pedicularus palustris > RED RATTLE

Pedicularis sylvatica > LOUSEWORT

PELLITORY-OF-SPAIN

(*Anacyclus pyrethrum*) A medicinal plant, a favourite in the East, and long exported from North Africa and southern Spain to India via Egypt. It was a popular remedy for ague once, but is hardly ever used now (Lloyd). Its chief use in Europe was for toothache relief (Fluckiger & Hanbury). One example from the 15th century in which the plant plays a distinctly subordinate role runs: "for aching of the hollowtooth. Take raven's dung (!) and put it in the hollow teeth and colour it with the juice of pellitory of Spain that the sick recognise it not nor know what it be; and then put it in the tooth and it shall break the tooth and take away the aching, and as some men say, it will make the tooth fall out" (Dawson. 1934). Another from the same collection runs: "For toothache, a fine medicine: take long pepper, pellitory of Spain, nutgalls, lichen, and seethe them in vinegar from a quart to a pint: and put therein a pot of treacle of a pennyworth, and then take dregs of ale in a vial. And then take [it] once a week, and refresh thy gums therewith". There are more recipes in other sources (see Lupton, for example). It was still being prescribed in the middle of the 18th century by Hill: "the root … is good against the tooth-ache. It is also good to be out into the mouth in palsies …". Another ailment catered

for in the 15th century collection already quoted is migraine, for which sufferers were advised to "take pellitory of Spain, and stone-scar [lichen] and hold long between thy teeth on the sore side…".

PELLITORY-OF-THE-WALL

(*Parietia judaica*) Pellitory is a strange-sounding name, but it comes eventually from latin paries, a wall. Irish schoolboys used to grasp pellitory hard when the need arose, saying:

> Peniterry, peniterry, that grows by the wall,
> Save me from a whipping, or I'll pull you, roots and all (Wood-Martin).

According to Pliny, when the Acropolis of Athens was being built, a slate fell from the top on to Pericles. Athena made known to Pericles in a dream, that this pellitory would effectively heal his wounds. Since then, the Greeks call it parthenion in remembrance of the goddess Athena Parthenios. Greek country people still use it in compresses for bruises and swellings (Baumann). Gubernatis quoted a Tuscan custom of gathering the plant on Ascension Day, and keeping it hanging in the bedroom until Lady Day (8 September). The plant often blooms after it has been picked. "Cette floraison d'une herbe coupé est, pour le peuple, in miracle, unce bénédiction spéciale de la Madonne; si au lieu de fleurir, le plante se dessèche, c'est in présage de malheur. Une malédiction divine".

The plant had its cosmetic uses, witness a 17th century milk bath, recorded by Gervase Markham, about 1610: "Take rosemary, Featherfew, Orgaine, Pellitory of the Wall, Fennel, Mallowes, Violet leaves and Nettles, boil all these together, and when it is well sodden, put to it two or three gallons of milk, then let the party stand or sit in it an hour or two, the bath reaching up to the stomach, and when they come out, they must go to bed and sweat, and beware of taking cold" (quoted by Wykes-Joyce). The plants that grew on Oswestry church tower were used locally as fomentations for pains in the back (Burne. 1883). Four handfuls of the tops of self-heal and pellitory-of-the-wall, boiled for three hours in three pints of water, and the liquid drunk, is an old Irish treatment for gravel (Logan). A similar treatment was used for stone, and dropsy (Quelch), in fact, this herb has been used for the purpose through the ages. Gerard, Wesley and Hill all had similar recipes for the condition.

It was used in the Balkans for scrofula (the leaves, cooked with salt, vinegar and honey) (Kemp). A poultice made from it was put on wounds and bruises in Guernsey, where a tea made from it was used for diabetes (Vickery. 1995). In Italy, it is used for eye diseases. The bruised leaf is applied to the eyelids, which are rubbed until the blood flows, and an invocation to St Lucia being said meanwhile (Gubernatis). Lupton had already offered a recipe: "the white of an egg, and the juice of pellitory of the wall, well beaten together and skimmed, and then one drop of that liquor put into the eye, doth heal the web in the eye".

Pennantia corymbosa > MAORI FIRE

PENNYROYAL

(*Mentha pulegium*) Pennyroyal is a corruption of Puliol Royal (Latin pulices – fleas), for this is a good plant to use against them. The "royal" part of the name, so it is said, came about because royal palaces were not immune (Genders. 1971). It was supposed to purify stagnant water, too, and that is why sailors took it to sea with them (Bardswell).

It is said that this plant was used in witchcraft to make people see double (Folkard), though why it should, and why that should be the aim, is not clear. But on the whole this is a protective plant. In Italy, it counteracts the evil eye, and in Sicily, it was hung on fig trees to prevent the figs falling before they were ripe (Bardswell). Also in Sicily, it was given to husbands and wives who were always quarrelling (Folkard). Children there put sprigs of it in their caps (and in the cribs) (Gubernatis) on Christmas Day, believing that at the exact moment that Christ was born, they would come into bloom (Folkard). In Wales, pennyroyal had to be gathered on Whit Sunday or St John's Eve, for the benefit of a "person who has lost consciousness in consequence of illness" (Physicians of Myddfai). In parts of Morocco, it is picked and put in the rafters, as a protection against evil, but it had to be gathered before Midsummer Day (Westermarck. 1926). It used to be burnt in the Moroccan fires, too, lit between the animals, the smoke presumably acting as a protection (Westermarck. 1905).

Pennyroyal tea is good for chills and coughs (Vesey-Fitzgerald), extended to include bronchitis and asthma in Scotland (Beith). In Morocco, the dried leaves are powdered and taken with porridge or milk for coughs and colds (Westermarck. 1926). In Wiltshire, it was used as an infusion for all chest and lung complaints (Wiltshire). Wesley recommended it for whooping cough: "Chin-cough or Hooping-cough … give a spoonful of Juice of Pennyroyal mixt with Sugar-candy, twice a day", and Buchan prescribed it for croup. It is used in folk medicine "with much confidence in obstruction of the courses, or when these are attended with pain of hysteria" (Thornton). In other words, it is a known emmenagogue; it is an abortifacient, too (V G Hatfield. 1994). Gypsies used to peddle it in remote country districts for just this purpose (Wiltshire). But this is a very old concept. The Anglo-Saxon version of Apuleius has, for instance, a leechdom for use "if a dead-born child be in a wife's inwards, take three sprouts [of pennyroyal], and let them be new, so of the strongest scent, pound in old wine; give her to drink" (Cockayne). That scent, and nothing more, was supposed at one time to help

patients recover from fever, while the aroma of the seeds was recommended in cases of speech loss (Classen, Howes & Synnott).

The juice rubbed on the skin prevents insect bites (Vesey-Fitzgerald). It was used for itch, too (Cockayne), and a plaster is good for a burn (Moore, Morrison, and Goodwin). The leaves applied to corns will get rid of them (H M Hyatt). It was applied externally to wounds, in Morocco (Westermarck. 1905), and they say in parts of America that pennyroyal tea will bring out the measles (R B Browne; Beck), and it is taken in Alabama as a cure for diarrhoea (R B Browne).

The older herbalists used pennyroyal for many more conditions. No wonder they used to say in Opwa that a lotion made from it is good for any ailment (Stout). The Anglo-Saxon Apuleius had remedies for "sore of bladder, and in case that stones therein wax", cramp, sea sickness, "for sore of inwards", tertian fever, coughs, and loss of voice. Gerard repeated a lot of these, and added a few of his own, such as "a garland of Pennie Royal made and worne about the head is of great force against the swimming of the head, and the paines … thereto". No wonder that Brazilian curanderos used it in their healing ceremonies, as a fumigant, and as an ingredient in the ritual baths that always form part of their healing ritual (P V A Williams).

Turner called this plant Pudding-grass, and Pudding-herb is recorded from Yorkshire (F K Robinson). They refer to the practice of using pennyroyal to make stuffings for meat, formerly called puddings (Prior).

PENNYWORT

(*Hydrocotyle vulgare*) A plant with a very bad reputation, regarded with suspicion, quite rightly, for it is extremely damaging to livestock. As Gerard said, "husbandmen know well, that it is noisome unto Sheepe, and other Cattell that feed thereon, and for the most part bringeth death unto them …". Many of the local names reflect this reputation, for it is Penny Rot, or Farthing Rot, even Shilling Rot in Scotland, an inflated value. Many of the diseases caused by this plant are collectively called 'rot'. There are Sheepbane, and Flookwort too, the plant that causes the fluke worms that result in liver rot in sheep, and so on. The only good recorded of the plant is a sunburn remedy, prescribed by the Physicians of Myddfai: "for a sunburn … take Marsh Pennywort and cream, half and half, and boil on a slow fire till it becomes a thick ointment, keep it in a box covered".

Pentaclethra macrophylla > OIL BEAN

PEONY

(*Paeonia mascula*) *Paeonia* as the name was established by Theophrastus, who chose it in honour of Paeon, who first used it medicinally, and was said to have cured with it the wounds that the gods received during the Trojan war (Blunt. 1957). Peon first received the flower from the mother of Apollo on Mount Olympus, and with it cured Pluto of a wound he had received in a fight with Hercules. A plant of divine origin, then, so why should it be the symbol of bashfulness? (Leyel. 1937). It was looked on, too, as an emanation of the moon, endowed with the property of shining, or at least glowing, in the night (like mandrake), of chasing away evil spirits, and of protecting the houses near which it grew (Henderson). It was an antidote to sickness caused by demonic possession (Dalyell), such as the falling sickness. Culpeper speaks of its virtues for this complaint, and the powdered root was often given for it (see Thornton, for instance). In Sussex, a necklace turned from the roots was worn by children to prevent convulsions, and to help teething (Latham) (probably elsewhere, too, for Berdoe mentions it), and a necklace of peony seeds was used in parts of France to keep children free from fits (Sebillot). The Pennsylvania Germans used to say that to prevent convulsions, you should wash the child with a rag that had been tied over a peony flower (Fogel). Gerard also mentioned a necklace made from the roots "tied about the neckes of children" as an effective "remedy against the falling sicknesse…". The use went further: it "heals such as are thought to be bewicht …". Langham, too, went beyond the falling sickness; he claimed that it protected against "the haunting of the fairies and goblins". These necklaces were known as 'Anodyne necklaces' in the 17th and 18th centuries. According to Pliny, such necklaces also protected against nightmares. Even someone as rational as Bacon subscribed to the view that it protected from "the incubus we call the mare", a belief repeated many times, by Coles, Gerard, Lyte, etc.,

Peony protected from lightning, too (Tynan & Maitland), and when worn on the person, it was long considered as an effective remedy for insanity. The Anglo-Saxon version of Apuleius has: "For lunacy, if a man layeth this wort over the lunatic, as he lies, soon he upheaveth himself whole; and if he hath (this wort) with him, the disease never again approaches him" (Cockayne). Other superstitions centred round the peony include an example that enjoins one to count the flowers on the plant. If there is an odd number on each plant, it is a sign that there will be a death in the house before the year is out (*Notes and Queries; 1873*). Another from the Pennsylvania Germans said that it is very bad luck to give a peony plant to someone as a present – somebody in your family will die within the year (Fogel).

To revert to so-called medicinal uses, and the doctrine of signatures, Coles opined that "the heads of the Flower … have some signature and proportion with the Head of Man, having sutures and little veins dispersed up and down like unto those which environ the brain. When the flowers blow, they open an outward little skin, representing the skull, and are very available against the Falling sicknesse …". Wesley, nearly a hundred years later, was advising his people to take a "Tea-spoonfull of Piony root dried and grated fine, Morning and Evening for three Months" for falling sickness. Hill also recommended it, and so did Burton (*Anatomy of Melancholy*), and it is also recorded in Aerivan domestic medicine (H M Hyatt).

Peony roots were prescribed for sciatica (Cockayne), and jaundice too was treated with it, at least according to Gerard, who also quoted Dioscorides when claiming that the root "is given to women that be not well clensed after their deliverie, being drunke in Mead or honied water to the quantitie of a beane …". That is interesting, for there is a record that Hungarian gypsies took the leaves in wine, with Rose Bay leaves and ergot, for abortive purposes (Erdǒs).

PEPPERMINT

(*Mentha x piperita*) Fresh mint is used to flavour the Kentucky whisky drink Mint Julep, made in inverting (tops down) a small branch of young peppermint sprouts in the sweetened, diluted, whisky, so imparting the aroma of the leaves, but not the bitterness of the broken stems (Lloyd). This is the mint in Crême-de-menthe (Forsell), and it is also used for scenting soap (a sprig of mint used to be put in a bath at one time (Gordon. 1977)).

Use peppermint to keep flies and midges away – rub the face and hands with the leaves. Mint (or parsley) grown on a window sill is also said to keep flies and insects out of a kitchen. Bruise the leaves occasionally to release more odour (Boland. 1977). Applied to the temples, it will relieve headaches, and it can also be used for a queasy stomach and indigestion. It is an antiseptic (Genders. 1971). Gypsies use the tea for headaches (as well as laying the leaves on). A drop of the juice on an aching tooth will relieve the pain (Vesey-Fitzgerald). Peppermint tea is also a sedative (Bircher), and is used in Russian folk medicine as a "heart strengthener" (Kourennoff). In Alabama, they used to give peppermint tea to babies who had a cold (R B Browne).

PERFUMES

The essence of ROSEMARY was used for scenting; it smelt, according to Rimmel, very like camphor, but one perfume became extremely well known. In 1709, an Italian chemist named Farina concocted a mixture of orange, alcohol, bergamot, lemon oil and rosemary. This proved very popular with the Germans, and Farina marketed it under the name Kölnisches Wasser – later known as eau de Cologne (Wykes-Joyce).

The distilled oil from the leaves of LEMON VERBENA are used commercially in the scent industry. It is blended with citrus oils, orris, rose and heliotrope (Whittle & Cook). BEE BALM is another plant with a lemon fragrance, used as a strewing herb, and even as an ingredient in furniture polish, to give the wood a sweet perfume. The leaves of SWEET FLAG can produce by distillation a volatile oil, used as an ingredient in perfumery (Genders. 1972). CLARY is cultivated these days for its oil, which is used in the cosmetics industry as a perfume fixer. It is a highly aromatic oil, with a scent resembling ambergris (Clair). CASSIA oil was one of the precious perfumes, an ingredient of the holy oil of the Old Testament. It was also used as part of the incense burnt in the Temple (Zohary).

However disagreeable to modern taste, the smell of VALERIAN used to be held in quite high esteem, for the root was put among clothes as a perfume in the 16th century (Fluckiger & Hanbury), hence perhaps the name English Orris, and it is still used as a perfume in the East (Lloyd). According to legend, an Italian called Frangipani in the 12th century created an exquisite perfume. European settlers in the Caribbean 400 years later discovered a plant whose flower had a similar perfume, so it was naturally called FRANGIPANI. PATCHOULI is a famous Eastern perfume, obtained from the leaves of the Indian plant of the same name. Unadulterated, the plant smells strongly and unpleasantly of goats. But when the oil is diluted with attar of roses, all traces of the unpleasant quality disappear completely. Patchouli perfume, not the same as the oil, is made from the leaves – it has a reputation of being aphrodisiac. Powdered leaves are sometimes put into incense (Schery). The perfume first became known in Britain about 1820, when it was used to impregnate Indian shawls which became so fashionable that the designs were copied by Paisley weavers for export to many other parts of the world. But they could not sell them if they did not smell of patchouli (Genders. 1972). SANDALWOOD is another Eastern perfume, A fragrant oil, known as Oil of Santal, is distilled from the heartwood for use in perfumery and cosmetics. It is a good fixative for other perfumes. The Chinese make joss sticks from the wood (Usher), and incense from the sawdust, mixed with swine's dung(!) (Moldenke & Moldenke).

ORRIS-ROOT was largely used in ancient Greece and Rome in perfumery, and Macedonia, Elis and Corinth were famous for their unguents of iris. Mixed with anise, it was used in England as a perfume for linen as early as 1480. It smells like violets (and in fact is sometimes called violet powder (Hemphill)). It was also crushed and used as a substitute for dried violet in sachets and powder. It was once used for scenting tooth powders (Rimmel).

PERIWINKLE
i.e., GREATER PERIWINKLE (*Vinca major*) and LESSER PERIWINKLE (*Vinca minor*) Periwinkle is the symbol of sincere friendship (Friend. 1883), and

it is also said to typify excellence, as in an old ballad, a noble lady is called "the parwenke of prowesse". In Germany, it is a symbol of immortality, as befits an evergreen plant (Fernie).

There is a superstition that if the leaves are eaten by a man and wife, it will cause them to love each other. This is in Albertus Magnus, where it was said that houseleek had to be taken as well. A 14th century manuscript says that 'Pervinca' powdered with earthworms induces love between husband and wife, if they take it first in their food. Very similar to this is the Fenland belief that if a young married couple plant a patch of periwinkle in the garden of their first home, they would have a happy life together (Porter. 1969). In Gloucestershire the "something blue" that a bride wears is periwinkle. Some say that it must be worn in the garter for fertility (Vickery. 1995). It is one of the flowers believed by people in Cambridgeshire to wither quickly if worn as a buttonhole by a young flirt or an unchaste wife (Porter. 1969).

In Wales, it was said to be very unlucky to uproot one of these plants from a grave – you ran the risk of being haunted by the person buried there (M Baker. 1980), and to have terrible dreams for a year (Trevelyan). There is a connection with death in Italy, too, for garlands of the plant were put on the coffins of dead children. These are "flowers of death" (Fiore di morte), and in medieval England, condemned men were forced to wear garlands of periwinkle on their way to the gallows (Emboden. 1974).

Periwinkle soothes nettle rash, they say in the Fen country (Porter. 1969), and an ointment made with it was used for bruises and persistent skin irritation in Scotland (Beith). The roots were a popular colic cure in the Fen country (Porter. 1969), and periwinkle used to be an Irish (County Cavan) treatment for diabetes (Maloney). The leaves laid on gatherings and boils is an Oxfordshire remedy (*Oxfordshire and District Folklore Society. Annual Record. 1951*). It was reckoned to be good for sore breasts in Lincolnshire, the leaves being crushed and applied to the part (Gutch & Peacock); a poultice of the roots applied to a cow's udder was said in Cambridgeshire to cure milk fever (Porter. 1969). It is said to be a good remedy for cramp, too (Grieve. 1931). People used to wear bands of green periwinkle about the calf of the leg to prevent it (Fernie), and in Lincolnshire a piece was put between the bed and mattress for the same purpose (Rudkin).

Lesser Periwinkle has been used as a vulnerary, that is, as a treatment for bleeding wounds (Grigson.1955), or for that matter, any issues of blood. Gerard, for instance, advised its use: "a handfull of the leaves stamped, and the juice given to drinke, in red wine, stoppeth the laske and bloudy flix, spitting of bloud, which never faileth; it likewise stops the inordinate course of the monethly sicknesse". Add nosebleed to that list (Coats. 1975). There is some suggestion that it is good for the eyes. It is called Old Woman's Eye in Dorset (Macmillan) (one of its Italian names is Centocchio), and, more interesting, another Dorset name is St Candida's Eyes. St Candida's well is at Morecombelake, in Dorset, and the water is said to be a certain cure for sore eyes; it is on Stonebarrow Hill, where the wild periwinkles are called St Candida's Eyes (Dacombe).

PERSICARIA

(*Polygonum persicaria*) In Gaelic, it is: lus chrann censaidh – the herb of the Crucifixion; it grew under the Cross, and the leaves were spotted by Christ's blood (Grigson. 1955), a belief also known in Lincolnshire (Gutch & Peacock). Another version of the legend is that the Virgin Mary rejected the plant, leaving her mark on the leaf. Throwing it aside, she said, "This is useless", and useless it has been ever since. It actually bears the name 'Useless' in Scotland, but other relevent names are Virgin's Pinch and Pinchweed (Grigson. 1955), even Devil's Pinch in Dorset (Macmillan).

PERSIMMON

(*Diospyros virginiana*) Its uses are mainly medicinal, but there is a note that the bark, mixed with that of Red Oak, gives a yellow dye (R B Browne). Like sassafras, persimmon wood is unlucky to burn in the house (they both pop and crackle a lot when burning). If you throw it in a man's fireplace, he will soon move away. So runs a belief that was current in all the southern states of America (Puckett). Another belief, or rather charm, recorded there is to string the seeds and wear them as a necklace, to keep off disease (R B Browne).

But the fruit has got genuine medicinal value. It is astringent, and was listed as such in the US Pharma-copeia for a while (Weiner), though the Cherokee Indians had been using a boiled fruit decoction for diarrhoea a long time before that. Southern states domestic medicine used the sap quite a lot. For teeth-ing, the juice from a burned branch was put on the gums, and the same procedure was used for earache; just let the sap drop in a spoon, and then drop this sap into the ear. It was used for thrush, too – stew persim-mon bark, mix with honey, and wash the mouth with the juice. Sometimes a small piece of alum would be added in the cooking (R B Browne). Then there is a real oddity; if you have a tootheche, walk round a persimmon tree, and don't think about an opossum, and the tooth will get well. It is a humorous cure, for opossums are nearly always associated with this tree (R B Browne).

PERUVIAN MASTIC

(*Schinus molle*) An evergreen tree that yields a resin, known as American Mastic, as well as the more common name Peruvian Mastic. It features in the widely used technique in Mexico of spiritualist healing, borrowed from traditional folk medicine known as the 'limpia', or cleansing. Flowers, and especially whisks made from twigs of this tree (called Pirul in Mexico) serve in the ceremony. It is a semi-sacred tree anyway, probably because of the pungency of its foliage (M Kearney). It is used, too, as an ingredient of the ritual baths for Brazilian healing ceremonies. In fact, it seems to be the essential ingredient, as it is the only herb that appears in all the recipes quoted (P V A Williams).

PEST CONTROL

People used to powder the roots of STINKING HELLEBORE and mix them with meal to lay down and poison the mice (Drury. 1992). All the Hellebores, of course, are poisonous.

Petiveria alliacea > GUINEA-HEN WEED

Petroselinum crispum > PARSLEY

PETTY SPURGE

(*Euphorbia peplus*) This is a poisonous plant, with effects similar to those of Caper Spurge (Long. 1924), and it is even said to have been fatal to man in some cases (P North). Certainly it is dangerous to animals, and there have been a number of records from Australia and New Zealand of fatalities among horses, cattle and sheep (Forsyth). Yet it is known in Gaelic as lus leighis, "healing herb". It is true that it has been clinically investigated for use in cancer treatment. Perhaps the Highland physicians recognised its usefulness in treating skin cancer centuries ago (Beith). Names like Wartwort, or Wart Grass (Drury. 1991) confirm the use of the juice as a wart cure.

Peucedanum ostruthium > MASTERWORT

Phaseolus lunatus > LIMA BEAN

Phaseolus vulgaris > KIDNEY BEAN

PHLEBITIS

MELILOT has been given by herbalists for this disorder (Thomson. 1978).

Phleum nodosum > CATSTAIL

Phleum pratense > TIMOTHY

Phragmites communis > REED

Phyllitis scolopendrium > HARTSTONGUE

PHYSIC NUT

(*Jatropha curcas*) or BARBADOS NUT, as it is often called, is a drastic, and dangerous, purgative, often planted round houses in Haiti to keep out evil spirits. It is the Voodoo god Legba's plant (F Huxley). In Jamaica, a physic nut used often to be buried in a field, to keep a neighbour's envious (and therefore evil) eye away from a good field crop. Horse-eye Bean (*Dolichos ensiformis*) would be planted at the top and bottom of the field, too. (Beckwith. 1929). Also from Jamaica, if you go to a Physic Nut tree at 12 o'clock on Good Friday morning, and stick a penknife in it, the juice that comes out will be blood (Beckwith. 1929).

Phytolacca decandra > POKE-ROOT

PICKPOCKET, or Pickpurse and their like, are names given either because the plant so named flourishes in unproductive land (in the profit sense), e.g., plants like CORN SPURREY (*Spergula arvensis*),or because it is such a rampant weed as to damage the farmer's chance of profit. Cf similar names, like Beggarweed, Poverty, etc., SHEPHERD'S PURSE, another plant that bears this name (Ô Súilleabháin), grows in similar habitats. There is a rhyme from Northamptonshire:

Pickpocket, penny nail.
Put the rogue in jail (A E Baker).

Pickpocket-to-London is an elaboration of the name, from Yorkshire:

Pick-pocket to London,
You'll never go to London (Grigson. 1955).

A Scottish equivalent is Rifle-the-ladies'-purse.

PICNIC THISTLE
(*Cirsium acaulon*) The too familiar and ubiquitous stemless thistle is a weather forecaster; in the Alps they say that when the flower is open, there is good weather to come, and the opposite when it is closed (Gubernatis). It does have some medicinal use, for herbalists prescribe the root infusion to treat dropsy, bronchitis and also prostate problems (Schauenberg & Paris).

Picris echioides > BRISTLY OX-TONGUE

PIGEON PEA
(*Cajanus indicus*) Probably a native of Africa, but it grows in tropical and sub-tropical countries, particularly in India, Central Africa, and the East and West Indies. In all the areas it now grows, it is an important part of the diet, eaten either as the green, or often maroon, pods or as ripe seeds, used as a pulse (Brouk).

"Here is how to make a zombi out of a dying man. Take a white pot, fill it with twenty-one seeds of *pois congo* [i.e., Pigeon Pea] and a length of string knotted twenty-one times, and slip it under his pillow. After his death, leave the pot in a small dark room. The string will turn into a spider, and the spider is the zombi. You must treat it carefully. Feeding it with just enough food and water, but no salt ..." (Haiti) (F Huxley). On the other hand, there is this belief from Jamaica: "To "keep the ghost down", plant pigeon peas on the grave – as the roots grow downward, so the ghost will be prevented from going upward. In

the west of the island they boil the peas, for, as the peas cannot shoot out of the ground, so the ghost must remain in the ground" (Beckwith. 1929).

PIGMENTS
BUCKTHORN berries, boiled with alum, provide Sap Green, and when mixed with lime-water and gum arabic, Bladder Green (Grigson. 1955). As Gerard said, "there is pressed forth of the ripe berries a juyce, which being boyled with a little Alum is used of painters for a deepe greene, which they call Sao-greene ...". WOAD supplied a blue pigment used by artists, especially in the illumination of missals. It was got from the scum that floated on the surface of the vat, known as the "flower" of woad, or "flory". It was used by Italian artists from the 13th to the 16th centuries. Evidently, there were two shades of blue that could be got, for there was a distinction made between "indigo" and "azure" (Hurry). Lupton gave advice on how "to make a green that will not fade away. Take the flowers of FLEUR-DE-LUCE, stamp them and strain them, then put the juice thereof intogum water, and dry it in the sun". This green colour was known as Verditer, "us'd by the Painters in Miniature" (Pomet).

PILES
Piles could be treated with LEEKS – "take the roots of leek and stamp and fry them with sheeps tallow, and as hot as he may suffer it, bind to the fundament oft ..." (Dawson). ONIONS too were used, at least in the Greek islands. On Corfu, for example, water from boiled onions is applied as hot fomentations (Durrell). Culpeper had this to say about CUCKOO-PINT: "for the emeroids and figge in the fundament, boyle it with Mulline in wine or water ...". The fresh juice from ROWANS is used in Russian folk medicine, taken sweetened with honey three times a day, and followed by a glass of hot water, as a remedy for piles (Kourennoff), and CAROB was used too in Russian folk medicine (half a pound of the pods, chopped, in a quart of vodka, taken twice a day). The patient was warned that no other alcohol should be taken (Kourennoff). The best known medicine for piles was supplied by LESSER CELANDINE, as names for it like Pilewort and Figwort show. But this is a case of doctrine of signatures – there are small tubers on the roots, and these are the signs, for pile is latin pila, a ball. Not only piles was suggested by these small tubers. Growths on the ear and small lumps in the breast were also treated with it in Scotland. The juice was applied externally, but carefully, of course, for it is acrid. But like a number of examples of the doctrine, there does seem to be some curative effect. Fernie certainly thought so, and Conway confirmed its value. The specific name, *ficaria*, is anothet indication, for it derived from the plant's curative value in the ficus, which means fig, and by extension, piles (hence Figwort, a name not confined to the

celandine). So confident were they in Holland as to its efficacy against piles that people there actually wore the roots as amulets against the condition, and gypsies say that carrying a sprig or two in the pocket will cure it (Quelch). FIGWORT itself relies on the same physical conditions (knobbly tubers) for the doctrine of signatures to ensure its use against piles. Its very name proclaims it. The same set of circumstances brings its other use to the fore. The generic name is *Scrophularia*, hence its use for scrofula, the same 'knobbly tubers' sufficing for scrofula's symptoms.

In the Anglo-Saxon version of Apuleius, there is mentioned a way of treating piles, which was to heat a quern-stone, to lay on top of it and underneath it DWARF ELDER, MUGWORT and BROOKLIME, and then to apply cold water, having the patient poised so that the steam "reeks upon the man, as hot as he can endure it" (Cockayne). The American Dispensatory of 1852, recommended a strong decoction of POKE-ROOT leaves, for such treatment "is of much benefit in haemorrhoids" (Lloyd). In Lincolnshire, particularly in a village called Willoughton, the highly toxic berries of LADY LAUREL were swallowed like pills as a remedy for this complaint (Rudkin). There is, too, an oil distilled from BOX wood that has been used for piles (and epilepsy, as well as toothache!) (Grieve. 1931). The Pennsylvania Germans used to say that to ease the condition, what you had to do was to sit on MULLEIN leaves (Fogel). That particular usage has a venerable history. A medieval record states that "for piles, take a pan with coals and heat a little stone glowing and put thereon the leaves of mullein; and put it under a chair or under a stool with a siege, that the smoke thereof may ascend to thy fundament as hot as thou mayest suffer" (Dawson. 1934). Gerard, later on, was able to recomment not only the leaves "a little fine Treacle spread upon a leafe of Mullein, and layd to the piles … cureth the same"), but also the flowers – "the later Physitions commend the yellow floures, being steeped in oile and set in warme dung untill they be washed into the oile and consumed away, to be a remedie against the piles". The Physicians of Myddfai prescribed MALLOW for the condition. The leaves had to be boiled in wheat ale or spring water.

Carrying CONKERS around is well known as a cure or prevention of rheumatism, but in some places (including Spain), it was piles that were reckoned to be cured or prevented by carrying them around (W B Johnson, Tongue, Fogel), and that is interesting, because it is known that extracts of horse chestnut are rich in Vitamin K, and so is useful in treating circulatory dieases like piles, varicose veins or chilblains (Conway). The American equivalent of Conkers is the nut of YELLOW BUCKEYE, and that too is carried around in the pocket for, among other ailments, piles. Gypsies say the powdered nuts of SWEET

CHESTNUT are good for piles. That may be true, but they even wear a nut in a little bag round their neck as a preventive charm, but the bag, they say, must never be made of silk (Vesey-Fitzgerald). CYPRESS cones were used in infusion, to be taken internally, or externally, especially if the condition was painful (Palaiseul). A compress made of SILVERWEED is said to be good for piles (Thomson. 1978), and a strong infusion is used to stop the bleeding (Wickham). The ailment was treated in East Anglia by infusing RED DEADNETTLE in white wine for an hour. A winegassful would be taken two or three times a day (V G Hatfield. 1994). AVENS is an astringent, and will clot blood to stop bleeding (A W Hatfield), so in some French country districts, the root, gathered before sunrise, is put in a linen bag, to be worn round the neck as an amulet to stop all bleeding, particularly haemorrhoids (Palaiseul). A tea made from dried PERSICARIA is used in Russian folk medicine for the condition (Kourennoff).

Pimento dioica > ALLSPICE

Pimpinella anisum > ANISE

Pimpinella saxifraga > BURNET SAXIFRAGE

PIN CHERRY

(*Prunus pennsylvanicum*) A North American species, the inner bark of which was used by native Americans for a cough medicine, and they also took advantage of the astringency of the roots to make a decoction to cure their stomach upsets (Youngken).

PINE

(i.e., Scots Pine, *Pinus sylvestris*) In ancient Greece and Rome, pine cones had a phallic significance, so they were symbols of fertility. It is interesting to record a superstition from the Highlands of Scotland to the effect that a lot of illegitimate births could be blamed on the large numbers of pine trees growing in the district (Begg). But they were also emblems of Cybele, and used to be fixed on a pole in Italian vineyards, to protect them from blight and witchcraft. Perhaps this was the origin of decorative pine cones on gateways (Pavitt).

There is mention of protecting a child against the evil eye by sweeping its face with a bough from a pine tree (Rolleston), and one occasionally finds evidence of the use of pine needles for divination. There was, for instance, a pine on the island of Bute that served as a "dreaming tree". Some of its needles were put, with some ceremony, under the pillow, for dreams of the future husband or wife (Denham). But they were unlucky trees in the Channel Isles. Guernsey belief had it that whoever planted a row of them ran the risk of losing the property, or letting it pass from the rightful heir to a younger branch of the family. There was also a belief there that if you fell asleep under a pine tree you would never wake up (Garis).

There is one piece of weather lore attached to the cones. They stay open when the weather is to turn fine, but as soon as they close, then it is believed that rain is on its way (Waring). But it was in the field of folk medicine that pines really came into their own. They are a source of turpentine, and as such had been known from ancient times, though by the nineteenth century turpentine had for a long time come from America, from the Shortleaf Pine, (*Pinus taeda*). But it is still in use, as are the oil and the resin, in homeopathic remedies for a number of ailments (see Schauenberg & Paris). Stockholm Tar, obtained by the distillation of stem and roots, was used for chest complaints such as bronchial coughs (Putnam). Similarly, the dried young shoots, taken as a tea, were used for bronchitis, etc., while in Russian folk medicine an inhalation of the needles and cones was recommended for asthma sufferers. (Kourennoff). Even the smell of pine trees, so it was said, could help in chest complaints (Logan). That is why so many were planted round chest hospitals. The cones were used for toothache because (and this is pure doctrine of signatures) the scales resemble the front teeth!

PINEAPPLE WEED

A name given to the RAYLESS MAYWEED (McClintock & Fitter), for when the leaves are crushed it gives off a pleasant scent, so making it valuable as a strewing herbs in times past. Indeed, its abundance by waysides is in part due to its tolerance of being trampled underfoot (Mabey. 1977).

PINK

(*Dianthus plumarius*) In medieval art, pinks were usually the symbol of divine love, and also signified that a lady was engaged to be married – see Memling's *Lady with a pink*, and a painting with the same name by Rembrandt (Genders. 1971). It was also a symbol of courtesy (Reeves. 1958). In Italy, it was said that pinks sown on the morning of St John's Day will flower every month (Canziani).

PINKROOT

(*Spigelia marilandica*) It is said that the Cherokee Indians prepared a worm remedy by boiling a lot of the root of this plant in water. In the early 1700s, it was taken up by the white settlers, and was used for worm medicine till the early 1900s (Weiner), and it was known as Worm Grass (Hyam & Pankhurst), as well as the more usual Pinkroot.

Pinus sylvestris (Scots Pine) > PINE

Pisum sativum 'hortense' > PEA

PITURI

(*Duboisia hopwoodii*) An Australian hallucinogen plant that had an important place in the culture of the desert aborigines. Small doses of the plant give rise to hallucinations and time-space detachment. Like coca, it has the ability to quell hunger and thirst and let the aborigines travel the long distances needed to find the bare essentials of life. There were so-called "pituri roads", with trade networks across the desert, for such things as spears, boomerangs, nets, etc., were exchanged by tribal groups lacking the plant in their own habitat. The leaves were packed tightly into woven bags and traded over hundreds of miles. Nevertheless, they were aware of its extreme toxicity, and they even used it to poison emu (Lewin). Pituri would be given to strangers as a token of friendship, or chewed as part of social interaction behaviour. It could be used as payment for certain ceremonials, like puberty rites, or used by old men who acted as seers, as well as simply as a pick-me-up and comforter. The only form of written communication known to the aborigines was linked to pituri trading, for incised message sticks were used to indicate to neighbouring tribes that they wanted to trade the plant. With the advent of Europeans and the introduction of processed tobacco, cigarettes slowly eliminated its use, and today the pituri culture is only a memory (Dobkin de Rios).

PLAGUE

RUE's use in medicine in times past has been widespread and general, with claims for almost anything from warts to the plague. In the latter case, Alexis of Piedmont gave the following recipe, anglicised a long time ago as: "Take the toppe of Rue, a garlicke head and half a quarter of a walnutte and a corne of salt. Eat this every morning, contynuing so a munneth together and be mery and jocund". Gerard too gives a recipe for the plague with the leaves of rue. Thornton repeated the belief: "it is supposed to be antipestilential, and hence our benches of judges have their noses regaled with this most foetid plant". Bunches of it used to be hung in windows to protect the house against entry of the plague (especially east facing windows, for it was thought that was the direction from which plague came). So powerful was rue considered that thieves looting plague-contaminated houses would risk entry if they carried it, even if corpses still lay there (Boland. 1977). Thomas Dekker's, *Wonderful Yeare*, 1603, spoke of persons apprehensive of catching the plague, when "they went (most bitterly) miching and muffled up and downe, with RUE and WORMWOOD stuft into their eares and nosthrils, looking like so many bore's heads stuck with branches of rosemary, to be served in for brawne at Christmas". Wormwood steeped in vinegar and kept "in a close-stopped pewter piece" was commonly carried in plague years, to be sniffed in dangerous places (Painter). This, of course, is a case of strong smells drowning infection; southernwood was used for the same purpose in Orkney.

GARLIC cures anything, including "fevers of the typhoid type, and ... the plague itself" (Thornton).

A protection against plague, Galen said, was to eat garlic with butter and salt at breakfast (Wilson). TOBACCO was another plague protector, either by smelling it, or by taking it fasting in the morning, "provided, that presently after taking thereof, you drinke a deepe draught of six shilling Beere, and walke after it" (Wilson). ARCHANGEL root, the root of the Holy Ghost (radix Sancti Spiriti), was chewed during the Great Plague in an attempt to avoid the infection. It was a case of the name governing the use, for the plant was credited with wonders regulated with its "angelic virtues", as Culpeper had it. Archangel's wild relative, WILD ANGELICA, was also recommended as a protector from plagues, as well as poison, but only as an inheritance from its august cousin. BOX hedges used to be planted as a plague preventive, particularly in Dorset. It is said that traces of these borders planted in the 16th century can still be seen in Netherbury (Dacombe). VALERIAN, too, was reckoned in the Middle Ages to be a medicine against the plague (Lloyd), and Gerard was still recommending it in his time: "the dry root … is put into counterpoysons and medicines preservative against the pestilence …". "The distilled water (of DRAGON ARUM) hath vertue against the pestilence … being drunke bloud warm with the best treacle or mithridate" (Gerard).

ZEDOARY (*Curcuma zedoaria*), a close relative of Turmeric, was mentioned in Anglo-Saxon medicine, but being rather unusual, was recommended for magical medicine, and as late as the mid-17th century, was marvel enough to be prescribed for the plague. Lupton, for instance, has "The root of [Zedoary] (but be sure it be perfect and good) mixed with raisins, and a little liquorice, champed with the teeth and swallowed, preserves them that do so, unhurt, or without danger of the plague …". The name actually appears in this text as Zeodary. Is it a misprint, or did he really call it that? SAFFRON could keep the plague at bay, according to Gerard, in a mixture of walnuts, figs, sage leaves, a mithridate and pimpernel water, such mixture "given in the morning fasting, preserveth from the pestilence, and expelleth it from those that are infected".

Plantago coronopus > BUCK'S HORN PLANTAIN

Plantago lanceolata > RIBWORT PLANTAIN

Plantago major > GREAT PLANTAIN

Plantago maritima > SEA PLANTAIN

Plantago media > LAMB'S TONGUE PLANTAIN

Plantago psyllium > FLEAWORT

PLANTAIN-LEAVED EVERLASTING
(*Antennaria plantaginifolia*) Another American species that is supposed to cure rattlesnake bite (Sanford). But the Indians used it for genuine medicinal purposes. They made a tea from the leaves for women to drink for two weeks after giving birth (H H Smith. 1928), and also a tea from LESSER CAT'S FOOT (*Antennaria neodioica*) (H H Smith. 1945). African-Americans in the southern states used to steep the leaves with some corn meal, and use the mixture to ease menstrual pains (Puckett). 'Ladies' Tobacco' is a name given to this species (Sanford) (and also to Cat's Foot); 'Indian Tobacco' is another (Grieve. 1931), these names imply that the dried plant must have been smoked as a tobacco substitute, and found to be too mild, if "Ladies" is to be believed. Love's Test is another name (Bergen. 1899), for this was a plant used in love divinations. A leaf has to be taken by the ends, while the operator thinks of someone of the opposite sex. The leaf then has to be pulled apart. If the down on its underside is drawn out long, much love is indicated. Or both ends may be named, and the one whose end has the longer down on it shows who is the more ardent lover.

Platanus x acerifolia > LONDON PLANE

Platanus orientalis > ORIENTAL PLANE

PLEURISY
Domestic medicine from Alabama uses Chigger-weed, which is *Anthemis cotula,* with Pleurisy-weed (*Asclepias tuberosa*). They are boiled together and applied as a poultice for pleurisy (R B Browne).

PLOUGHMAN'S SPIKENARD
(*Inula conyza*) A handful of Ploughman's Spikenard burnt each day during summer will keep houseflies out of the house (Hewett). The root has a spicy scent, and used to be hung up in cottages to scent the musty air, and was until recent times burnt on ale-house fires to counteract the often unpleasant atmosphere (Genders. 1971). Or it could be used as a strewing herb. Turner claimed that "either strewed upon the ground, or in a perfume with the smoke of it [Coniza] driveth away serpents and gnats amd kills fleas"(quoted by Fletcher). This practice of burning it to release the aromatic perfume accounts for the name Spikenard (the true Spikenard is *Nardostachys jatamansi*, well known for its scented roots). The common name of our plant really means poor man's spikenard). Gerard claimed that "the smell thereof provoketh sleep".

PLUM
(*Prunus domestica*) Like a lot of the Prunuses, it flowers early, in fact in Chinese symbolism, it is the emblem for January, and thus for spring (Savage. 1954). But a Welsh superstition says that a plum blooming in December foretells a death in the family (M Baker. 1980). Dreaming of plums has mixed meanings; if they are ripe, it is a good sign. But if the dream involves gathering green fruit, it means a lot of sickness in the family (which result is only reasonable, one must assume). And if one is gathering fallen plums, already rotten, then it is a sign of false

friends, unfaithful lovers, and a change of position to poverty and disgrace (Raphael). Nevertheless, it is a symbol of fidelity, and independence, in Christian art (Ferguson).

Plumbago indica > SCARLET LEADWORT

Plumbago zeylanica > CEYLON LEADWORT

Plumeria alba > FRANGIPANI

PNEUMONIA

There is a remedy for pneumonia that involves putting an ONION in each armpit (Beith). The Mano of Liberia use GOATWEED (*Ageratum conyzoides*), reduced to a pulp with a little water, to rub on the chests of children with pneumonia. They also use the seeds and buds of MELEGUETA PEPPER for the complaint (Harley). A powder made by pounding the leaves and bark of the African tree CATCHTHORN (*Zisyphus abyssinica*), is used to rub into incisions made on the chest in cases of pneumonia (Palgrave & Palgrave). A root decoction of MARSH MALLOW (2 cups a day, hot) is often prescribed in Russian folk medicine for the illness (Kourennoff). ELDERFLOWER wine was used in Wales for the illness ("drink it as fast as you can; 'tis better then any doctor" (R Palmer. 2001)).

Pogostemon patchouli > PATCHOULI

POHUTUKAWA

(*Metrosideros excelsa*) The Maori name for the tree, which is also known as the New Zealand Christmas Tree (Leathart). In Maori myth, the roots of this tree lead down into the underworld (Andersen) (cf Yggdrasil).

POINSETTIA

(*Euphorbia pulcherrima*) Named after its discoverer, Dr J R Poinsett, in Mexico, 1828. This is the species with scarlet bracts that is often grown as a house plant. In recent years it has become an accepted Christmas plant in America. Evidently it was regarded as such in Mexico before Dr Poinsett discovered it, for its Mexican name translates as: Flower of the Holy Night. The Indians use it as an altar decoration on Christmas Eve (Kelly & Palern), the Holy Night of the name, or noche buena.

The latex is highly irritant, and is used as a depilatory in Mexico and Brazil. In South China, where it also grows, it has been used as a fish poison (Watt & Breyer-Brandwijk).

POISON IVY

(*Rhus radicans*) It is so poisonous that even smoke wafted from the burning trees can affect a sensitive skin. The plant known as ORANGE BALSAM, or Jewel-weed rubbed on the skin is said to be a cure for its effects (Sanford). In spite of this vicious toxicity, it has been a useful plant in its day, particularly useful to the American Indians. The Navajo, for instance, used it as one of the ingredients in their arrow poison (Wyman & Harris). It was used as a dyeplant, too, by some groups by boiling the roots to give yellow. They used to dry and store the berries. When they were wanted, an infusion in water gave a drink like lemonade (H H Smith. 1923). They recognised its worth as a medicine, albeit a dangerous one, only to be used by the most skilled of the medicine men. Some groups used a mash of the leaves to treat ringworm (Weiner), and boils were treated by the Kiowa simply by rubbing the leaves over them. The resultant dermatitis lasts about as long as the boil. So the disappearance of the two afflictions together may explain this usage (Vestal & Schultes). GREATER CELANDINE juice was another way of treating a poison ivy rash, which the Pennsylvania Germans chose (Fogel), and some North American Indians used SHEPHERD'S PURSE (H H Smith. 1923).

POKE-ROOT

(*Phytolacca decandra*) is a common American plant of waste places and roadsides. Children make a red ink from the purple berries, a time-honoured use in the United States, for the Indians used it to give a red stain with which to colour their ornaments (Sanford), and it has also been used (in Portugal) to deepen the colour of port! (Whittle & Cook). The berries will also give, with alum as the mordant, red or tan colours, depending on the length of time of immersion, and the state of the dyebath.

The root, seeds and leaves are poisonous, but perfectly edible when boiled (Allan). Poke greens used to be a common potherb in the early spring in the eastern states of the United States (Emboden. 1979). And the young shoots (but only the young shoots! – older plants develop a poison), are still sometimes eaten as asparagus. They could be bought in markets under the name of "sprouts" (Le Strange).

In the autumn, the root was dug up for use in many home remedies, despite the knowledge of its toxic nature; an overdose leads to paralysis of the respiratory systen, and subsequent death (Emboden. 1979). In Alabama they used to say that pokeberry-root tea was good for tuberculosis, and taking a poke-root bath was helpful for the itch (R B Browne). They even used the root for thrush by, of all things, putting it in whisky and giving it to the child to drink! The berries mixed with whisky form a cure for rheumatism in Kentucky, where another practice was to use the dried berries made into a tea, or just eat the raw berries, for the same complaint (Thomas & Thomas). The Indiana remedy for arthritis is to eat fresh or dried pokeberries daily (Tyler). It still continues in use as a Cancerroot, and for many of the same problems for which belladonna was prescribed (Emboden. 1979), and in Indiana, the patient with tuberculosis was advised to drink Poke-root juice (Brewster).

According to the *American Dispensatory* of 1852, a strong decoction of the leaves "is of much benefit in haemorrhoids (Lloyd). The earlier *Every man his own doctor*, of 1816, advised caution, "being a plant of great activity, operating both as an emetic and cathartic", but went on to recommend the leaf juice as being more effective than the berries in application to skin diseases. In the Caribbean area, the leaves are used as a blood purifier (whatever that means) for adults (Laguerre).

POLIOMYELITIS

SNOWDROP has been prescribed for polio, even though it has actually very little therapeutic value.

Polyanthes tuberosa > TUBEROSE

Polygala senega > SENEGA SNAKEROOT

Polygala vulgaris > MILKWORT

Polygonatum multiflorum > SOLOMON'S SEAL

Polygonum aviculare > KNOTGRASS

Polygonum bistorta > BISTORT

Polygonum hydropiper > WATER PEPPER

Polygonum persicaria > PERSICARIA

Polypodium vulgare > POLYPODY

POLYPODY

(*Polypodium vulgare*) A fern, widespread in Britain in woods, especially in the west, where it often grows on trees. It is known in America as Licorice Fern, for the rhizomes have a strong licorice taste, and were once used as a sugar substitute (Turner & Bell). In Scotland, it was made into a medicine for catarrh (Beith), but it was also used for chest complaints, including tuberculosis (Quelch). The Indians of the Pacific north-west of America chewed the rhizomes for stomach troubles, sore throats and colds, as well as eating them as food, either fresh or sun-dried and stored for winter use (Turner & Bell). If this is what is meant by "pollypodden", it was used in Ireland for burns. The procedure was to boil the stems with butter. The green juice sets to a jelly, and this is put on the burn (Maloney).

POMANDERS

Usually a perforated globe or box in which perfumes were carried, often as a preservative against infection. They usually contained CINNAMON amongst other aromatic substances.

POMEGRANATE

(*Punica granatum*) References to pomegranates have been found as far back as the Mosaic writings, and on the sculptures at Persepolis and in Assyria, as well as on the monuments of ancient Egypt; the actual fruits have been found in Egyptian tombs (W M Dawson. 1929). The Eber Papyrus recommends the rind beaten up and taken in water as a vermifuge.

In fact, the value of the root-bark as a tapeworm remedy is well-known in Africa. The Hausa also use the flowers in infusion as a vermifuge (Dalziel). The peel was also in common use among the Romans for tanning leather, and it still is so used. It is red Morocco leather that benefits most from tanning with the unripe fruit rind, which yields a red dye – so do the flowers (Moldenke & Moldenke). The bark and rind were used as ingredients of ink once (Zohary). The astringency that makes the rind useful in tanning ensures that it would be recommended for diarrhoea and dysentery (Thornton). It was often combined with opium and an aromatic like cloves in India, for the same complaints (Fluckiger & Hanbury), and Huxley records the same use in Haiti, where it is also used for chills and asthma.

In a Persian story, Khodaded was one of fifty children begotten by a childless nomad upon his fifty wives, after eating as many pomegranate seeds. He had incessantly prayed for offspring, and was commanded in a dream to rise at dawn, and to go, after saying certain prayers, to his chief gardener, who was required to get him a pomegranate. He had to take from it as many seeds as seemed best to him (Hartland. 1909). The fruit itself symbolized the womb in a state of pregnancy, and the immense number of seeds made it suitable as an emblem for a prolific mother goddess (Simons). It is the symbol of the feminine, and of Aphrodite in particular (Grigson. 1976), and of fertility. The mother of Attis was a virgin, Nana (which means Earth (Freund)), who conceived by putting a pomegranate (or a ripe almond, according to another version) in her bosom (Frazer iv). It is ironic that, given the creative symbolism inspired by that mass of seeds, they should also have supplied the name of a deadly weapon. When the explosive shell that strewed metal particles over a wide area was invented, the French called it grenade, from the seed-scattering pomegranate (Lehner). Not so ironic perhaps, after all, when one considers the Persephone myth. Hades agreed to let Persephone go, but secretly gave the girl a pomegranate seed to eat, so that she would have to come back to him, since the fruit was sacred to the underworld (Grant), or, to put it a better way, was the food of the dead. In all myth and later folklore, partaking of the food of the dead condemned the indiscreet to live perpetually with the dead, just as taking fairy food made it impossible ever to rejoin the world of human beings. But the seed is also the symbol of reproductive power, so Persephone has in a sense undergone a puberty initiation (Lindoln).

One origin myth for the pomegranate says that when Agdos, the hermaphroditic son of Zeus, was emasculated, the plant sprang from him (Freund). At wedding ceremonies in Rhodes, a pomegranate was put on the threshold, to be crushed by the bridegroom's foot as he entered (Rodd), though T B Edwards. 2003 says that it is the bride who crushes the fruit, with her right foot, as she enters her new home. Gregory the Great said the pomegranate was to be used to symbolize congregations, because of its many seeds, and also to be the emblem of the Christian church, because of the inner unity of countless seeds in a single fruit (Haig), the same imagery as that used to symbolize fertility. In much the same way, that multitude of seeds stood for peace and prosperity. Jewish New Year dinners always include foods that are in some way suggestive of prosperity and happiness. One of them is pomegranates, "that our merits may be as numerous as its seeds" (Trachtenberg), and Greek New Year tables too always have the symbolic pomegranate on them (Megas). Breaking a pomegranate over the threshold is a very ancient New Year custom in Greece, just as a Moroccan practice, cutting a pomegranate open and throwing it on the ploughshare, will magically affect the fertility of the soil, and the ears of corn will be as loaded as the pomegranate (Legey). "Full as a pomegranate" is a saying that explains the custom and its object (Megas). Dreaming of them is supposed to be a sign of good fortune and success (Gordon. 1985). A New Year's custom from the Aegean island of Lesbos has a similar import. It is that of breaking a pomegranate on a stone that had been gathered from the seashore, literally a "woolly stone", i.e., a stone with seaweed on it. The pairing of symbolized faithfulness with the stone's symbolic steadfastness (for it must have lain a long time in one place for seaweed to grow on it) was just what was needed to ensure a good year to come (Rouse).

There was some belief in the protective power of the juice. The Arabs of Hiaina, when commencing ploughing, always squeezed a fruit on to the horns of one of the oxen, so that the juice would go into any evil eye that looked at the animals, and so render the evil harmless (Haining).

It was said in ancient times (also in the 10th century *Geoponica*), that pomegranate and myrtle have such affection for each other that their roots will entwine even if they are not very near, and both will flourish (Rose).

POPPY

see RED POPPY (*Papaver rhoeas*), OPIUM POPPY (*Papaver somniferum*)

POPPY ANEMONE

(*Anemone coronaria*) Moldenke & Moldenke felt that this is the "lily of the fields" which surpassed "Solomon in all his glory". There is a legend that the flower originated at the death of Adonis, who was changed into the flower, or that it sprang from the mixture of the blood of Adonis and the tears of Venus (Rambosson). The name *Anemone* itself is, Frazer said, derived probably from Naaman, or "darling", an epithet of Adonis. The Arabs still call it "wounds

of Naaman". More directly, it is the Greek "daughter of the wind" (Blunt. 1957). The death-of-Adonis myth makes this anemone the symbol of sorrow or death (Ferguson), and so it is found in Christian art in paintings of the Crucifixion.

Populus alba > WHITE POPLAR

Populus tremula > ASPEN (*Populus tremuloides* is the American Aspen)

Portulaca oleracea > GREEN PURSLANE

Portulaca sativa > PURSLANE

POTATO

(*Solanum tuberosum*) The fruit is poisonous, as are the vines, sprouts or peelings (Kingsbury. 1964), and so is the potato tuber itself if left to turn green on the surface (Young). That poison is solanine, which can be fatal to children (Duncalf). Potato originally came from the Peruvian Andes, and was held in such regard as to be almost sacred. Dorman describes a ceremony in which a lamb was sacrificed and its blood poured over potatoes. At the higher Andean levels, where the cold climate made preservation difficult, potatoes were converted into a flour called chuya, by an elaborate process of alternating freezing and warming which breaks down starch-containing cells. This "flour" could be preserved for long periods (Forde).

By the time potatoes had reached Europe, some very odd beliefs had become attached to them. Whoever would have thought of them as aphrodisiac? But Shakespeare was only echoing popular belief when he had Falstaff say: "Let the sky rain Potatoes... and hail Kissing-comfits, and snow Eringoes". Almost certainly he was talking about sweet potatoes, but no matter, the idea lingered after the introduction of our potato, and all because of a fundamental error. Being a tuber, it was mistaken by the Spanish who first came across both the potato (papa) and sweet potato (batata), for a truffle, and the truffle was the trufa, eventually meaning testicle, and so an aphrodisiac (Wasson). The other Spanish term for the truffle was turma de tierra, even more explicitly 'earth testicle'.

Putting a potato (or the peelings) outside a girl's door as an expression of contempt for her on May Day (Salaman) is probably a relic of the aphrodisiac belief, though a whole range was used then to express the villagers' opinion, ranging from hawthorn, a great compliment, through to nettles, which certainly were not.

Eat potatoes for good luck, they say in Alabama (R B Browne), probably a simplification of the undoubted dependence on potatoes in many communities, and not just in Ireland. When they were dug for the first time in the north-east of Scotland, a stem was put for each member of the family, the father first, next the mother, and the rest in age order. Then omens of prosperity for the year could be drawn from the number and size of the potatoes under each stem (Hartland). It was important for each member of the family to have a taste of the new crop, otherwise it would rot (Baker. 1977). If you dream of digging potatoes, and finding plenty of them, then that is obviously a good sign, with "gain and successes"; if there are only a few of them, then there is bad luck coming (Raphael). There is debate about the luckiest day for planting them. The Pennsylvania Germans say it should be St Patrick's Day if you wanted a good crop with large tubers (Dorson), but the general feeling is that the planting should be done on Good Friday, an odd choice horticulturally, as there can be as much as a month's variation in the timing. But of course the choice of day has nothing to do with reason – Good Friday is the one day on which the devil has no power to blight the growth of plants. Anyway, it seems to be agreed, from the Hebrides (Banks. 1937-41) to southern England (Goddard. 1942), that Good Friday is the great day for planting them. The Irish in County Galway say when you should *not* plant them – on a Cross Day, apparently, for that is unlucky. Cross days are any fourth day following Christmas, counting Christmas Day as the first of the four (Salaman). In a number of areas in Ireland church ritual is invoked to help the crop, on Ascension Day, for instance, when holy water is sprinkled on the growing crop. A burning faggot from the Midsummer Fires is thrown into the potato plot; moulding up is always done on St John's Day itself (Salaman). According to Iowa folklore, you should not plant them "in the light of the moon", for that would make them "go to vines" (Thomas & Thomas); but the best time to get them in was at the new moon (Hoffman).

Pregnant women should avoid potatoes, especially at night, if they want their child to have a small head (Salaman). Such a superstition is understandable once it is accepted that some ritual for getting a good, big crop of potatoes could have a similar effect on the head of the child in the womb.

Irish boxty bread was made from potato flour, and cooked on the griddle, and tattie bannocks used to be a Shetland delicacy, especially if they were dusted with fine oatmeal before being cut into rounds and baked on the griddle (Nicolson). Actually, if we are to believe Cobbett, potatoes for "flour" was widespread: "… think a little of the materials of which a baker's loaf is composed. The alum, the ground potatoes, and other materials; it being a notorious fact, that the bakers in London at least, have mills wherein to grind their potatoes, so large is the scale upon which they use that material" (Cobbett. 1822).

To keep the cramp away, or to be free of rheumatism, or even to prevent train sickness (Stout), carry a potato in your pocket, a well-known belief that has travelled to America (see Davenport; Thomas &

Thomas, etc.,). It is reported in Spain, too. There it was believed that the disease would actually go into the raw vegetable (H W Howes). Two potatoes were needed in Maryland, one for each pocket (Whitney & Bullock). In Ireland, it is said that as the potato dried up the rheumatism will go away (Mooney). It has to be a new potato, kept until it has turned black and is as hard as wood (Waring). "It will draw the iron out of the blood", as a Somerset belief had it (Whistler). In parts of France, it was carried around as a general charm against pain (Sebillot), or, in Kentucky, to prevent a chill (Thomas & Thomas), or kidney trouble, as Illinois belief had it (Hyatt). Andrew Lang said that the potato had to be stolen, or the cure would not work (Lang). Devonshire superstition also required some ritual. Here, a member of the opposite sex had to be asked to put the potato, unseen, in one of your pockets. You could change the pocket at will after this had been done, but the potato had to be worn continuously, or the charm would not work (Hewett). Similarly, a peeled potato, if carried in the pocket on the same side as an aching tooth, will cure it as soon as the potato itself was reduced to crumbs (Salaman) – a long time to endure the toothache, surely! Another potato charm is practiced in South Africa by Europeans for, of all things, delirium. Slices of raw potato would be tied behind the ears with a red cloth till they turned black (Watt & Breyer-Brandwijk), and in the same way a potato bound to the knee for rheumatism had to be kept there without change until the thing had become offensive through decay (Mooney). Sometimes the potato was changed annually, when the new crop was being dug (Heather). A potato was used for warts, too. Just rub it on and then bury it, without looking, some say. As the potato rots away, so will the warts disappear. Or you can throw it over your left shoulder. Another piece of wisdom tells that the wart should be rubbed with the potato each night, letting the juice dry in the warts. Another method is to tie two pieces of potato together and put them under the eaves. When the potato rots, the warts will disappear (Stout). An Irish charm was to cut a potato into ten slices, count out nine and throw the tenth away. Rub the warts with the nine and then bury them (Haddon). Even the water in which potatoes had been boiled was used. A notice that appeared in the *Surrey Gazette* for 5 January 1864, and quoted in *FLS News.15; July 1992*, publicized a rheumatism cure that involved bathing the affected part with water in which potatoes had been boiled, as hot as could be borne, just before going to bed (see also V G Hatfield).

These are all charms, of course, but mistaken observation may account for some so-called medicinal effects. In the 2nd edition of Gerard, there is "Bauhine saith that he heard the use of these roots was forbidden in Burgundy (where they call them Indian artichokes) for that they were persuaded the too frequent use of them caused leprosie". Bauhine is Gaspar Bauhin, whose

Prodromos of 1620 set out the theory. As late as 1761 this prejudice against the potato was still apparent in that area, and its cause is probably to be accounted for by the doctrine of signatures, for the skin of a potato reminded someone of the effects of leprosy. As the disease disappeared in western Europe, the potato acquired the signature of scrofula.

Are there any genuine medical uses of potatoes? Lady Wilde spoke of a plaster of the scraped tubers, constantly applied for a burn, and giving, she said, great relief (Wilde. 1890), while a record from County Cavan requires the victim simply to rub a slice of raw potato on the burn (Maloney), provided the burn was a small one, when Sussex practice would agree. But if the burn were larger, then the potato had its inside scraped out. This would be mashed, and the result would go on the burn (Sargent). The opposite is said to work, too, for an Alabama treatment for frostbite is to put a warm roast potato on the spot (R B Browne). It is still being claimed that slices of raw potato, freshly cut and applied to the temples, soothe headaches and migraines (Palaiseul), and from the same source, there is a note enjoining us, to make the hands soft and white, to rub them for several consecutive evenings with a paste made by boiling and mashing very white floury potatoes, adding a little milk, and, if you like, a few drops of glycerine and rose water. There is another use for the starchy juice of the potato – ferment it and distil it for pure alcohol, the cheapest kind of alcohol in fact, at least, in Europe. It can be diluted with water and drink as vodka, or it can be used as a cheap method of producing fortified wines and liqueurs. Most of the cheaper alcoholic drinks are now prepared from potato alcohol (Brouk).

Some of the names for potato are interesting. The word itself appeared in 1565. It is from Spanish patata, and that came from Haitian batata (Barber). Pratie, used in Ireland, is claimed to be Gaelic, but almost certainly is derived via prata from patata (Salaman). So is Frata, another Irish name (Salaman). Potatoes are Taters in Wales, Tateys in Yorkshire (a Tatey garth is where you grow them (Robinson)), and Tetties in Devonshire. Crokers is a well-known Irish name, in use as early as 1640. The reason, it is said, is that they had been first planted in Croker's field at Youghal (Salaman). A nod towards Ireland, or at least to the Irish surname Murphy, is implicit in the use of that name for potatoes, rendered as Murfeys sometimes (Salisbury). But the best known of all potato names is Spud. The word originally meant some kind of spade or digging-fork, more particularly the three-pronged fork used to raise the potato crop, which are thus "spudded-up". Spuddy is a slang term for a man who sells bad potatoes (Salaman).

Potentilla anserina > SILVERWEED

Potentilla erecta > TORMENTIL

Potentilla reptans > CINQUEFOIL

POVERTY, or **POVERTY-WEED**, etc., Such names usually mean that the plants grow in the poorest soil, and hence are a sign to the farmer that they may well keep him in poverty, plants like CORN SPURREY (*Spergula arvensis*), or PEARLWORT (*Sagina procumbens*), and their like. Cf Beggarweed, Pickpocket, etc., all of which carry the same import. FIELD COW-WHEAT is another plant bearing the name. Very rare now, but at one time it grew in cornfields, and if it got into flour it coloured it, made it taste bitter and also gave it a bad smell, owing to the glucoside in it. The name is given "with reference, no doubt, not only to the way in which it impoverishes the soil, but also to the fact that the seeds becoming mixed with the corn, rendered the latter of small value in the market" (Vaughan).

PRECATORY BEAN
(*Abrus precatorius*) A tropical climber, with red to purple flowers, followed by bright red and black seeds, which are the African LUCKY BEANS. But they are dangerous if crushed and then swallowed, for they contain a toxic agent, abrin, which is very poisonous. The beans have a notorious history in India as an agent of criminal poisoning (Grieve. 1931), chiefly against livestock, but also frequently against human beings, and abrin is an ingredient of some arrow poisons (Reynolds).

The beans are always associated with dangerous magic in most parts of southern Africa. When they are found decorating an object, it may safely be identified as being used in sorcery, witchcraft, etc., (Reynolds), though in the West Indies, it seems, they are used simply for ornamental beadwork (Gooding, Loveless & Proctor). But the best known uses of the seeds are as prayer beads (precatory, in the common name, is straight from the Latin precari, to pray) – the names Prayer Beads (Grieve. 1931), Rosary Pea (Kingsbury. 1964), and Paternoster Bean (Howes) emphasise the concept.

The root was used in India and Java as a substitute for liquorice (F P Smith); the leaves taste of it, too (the plant is sometimes known as Indian Liquorice, or Wild Liquorice; the word is reduced to Lick in a Jamaican name, Lickweed (Beckwith., 1969)). The seeds weigh about one carat, and have been used in India for centuries for weighing gold, under the name Rati (Grieve. 1931). They were also the basic unit in the Ashanti system of weights. The smallest brass weight was called ntoka, and weighed approximately the same as ten abrus seeds (Plass).

PREGNANCY
FIG features in Dutch folk medicine. A craving for figs during pregnancy ensures that the child will be born quickly and easily (Van Andel). QUINCES, too, had some significance in pregnancy. In 17th

century England, it was reckoned that "the woman with child that eateth many during the time of her breeding shall bring forth wise children and of good understanding" (Boland. 1977). HORSERADISH features in a very strange piece of folklore. Fenland couples who wanted to know the sex of an unborn child, put a piece of horseradish under each of their pillows. If the husband's piece turned black before the wife's, it would be a boy, and vice versa (Porter. 1958). But pregnant women should avoid potatoes, especially at night, if they want their child to have a small head (Salaman). Such a superstition is understandable once it is accepted that some ritual for getting a good, big crop of potato could have a similar effect on the head of the child in the womb. Beware of eating too many STRAWBERRIES during preganancy, for East Anglian superstition held that the birthmarks known as "strawberry" marks were caused by the mother eating too many of them (Porter. 1974).

Magic lies behind the Malagasy prohibition on pregnant women eating GINGER. The reason lies in the shape of the root, which is sometimes flat with excrescences like deformed fingers and toes. Nor must she keep the root tied into a corner of her costume, where odds and ends are usually kept. If she fails to keep these taboos, the foetus will become deformed, with too many fingers or toes; its legs will not grow straight, the deformation making delivery difficult as well (Ruud). So, too, with GROUND NUTS, which are taboo there to pregnant women. The thinking seems to be that peanuts lying on the ground remind people of souls that lay their eggs on the ground. So they will cause a miscarriage (Ruud).

It is said in China that if pregnant women wear DAY LILIES (Hemerocallis spp) at their girdle, the child will be male (F P Smith).

PRICKLY ASH

(*Zanthoxylum americanum*) Many Indian tribes used the root and bark to cure their toothache (Weiner), and the shrub was actually called Toothache Tree. D E Jones says his Comanche medicine woman uses a root infusion for fevers. The powdered root is also used in the treatment of toothache and burns. For toothache the powder is wrapped in a piece of cloth and this tiny bundle is put next to the bad tooth, and this causes a numb feeling in the mouth. For burns, the powdered root is sprinkled dry on the burnt area. The plant is used in domestic medicine, too. Alabama people looked on the bark tea as a good tonic (R B Browne), and African-Americans used this as a blood purifier (Berdoe).

PRICKLY COMFREY

(*Symphytum asperum*) A species from Iran and the Caucasus, introduced into Britain 1779 as a fodder plant (Sanecki). Rare escapes are still found in drier places. The claim made for it was that it prevented, as well as cured, foot and mouth disease (Quelch), before wholesale slaughter was required by law.

PRIMROSE

(*Primula vulgaris*) Primroses are fittingly fairy flowers, at least in Welsh and Irish folk tradition. But Milton must have been aware of the belief, too. His "yellow-skirted fayes" wore primroses. But fairy flowers can give protection from the fairies, too. Manx children used to gather them to lay before the doors of houses on May Eve to prevent the entrance of fairies, who cannot pass them, so it was said (Hull). So they did in Ireland, too (Briggs. 1967), and tied them to the cows' tails (Wilde. 1902), for no evil spirits can touch anything protected by these flowers (Buchanan. 1962). A primrose ball over the threshold served the same purpose in Somerset (Tongue. 1965). Those powers of protection went further – they could be used against the evil eye, for example (Wood-Martin). In the Derbyshire folk tale called Crooker, primroses formed one of the three magic posies given to the traveller to protect him from the evil Crooker. The others were St John's Wort and daisies (Tongue. 1970). Another of the Somerset beliefs was that you should keep some primroses under a baby's cot, or in its room, but always more than thirteen flowers (Tongue. 1965), but that proviso comes by confusion with another belief that will be mentioned shortly. The Welsh for primrose is Briallu. Perhaps Davies was right when he gave as its derivation bru, which means dignity, and gallu, power. He suggested too that the Druids used it in their mystical apparatus, so it is interesting to find that offerings of milk and primroses used to be made at a prehistoric burial chamber called the Water Stone, at Wrington, Avon (Grinsell. 1976).

Primroses were used as love charms in many places. Browne is talking about them when he says "maidens as a true-love in their bosoms place" (quoted by Dyer. 1889). Indeed, they symbolise wantonness in folk tradition, as Shakespeare well knew when he has Hamlet say "himself the primrose path of dalliance treads". But in the language of flowers, it was associated with melancholy (Webster).

Primroses were not always entirely welcome, for they had their dark side. To dream of them, for instance, means sickness, deceit, sorrow and grief (Raphael). A primrose blooming in June is a sign of trouble and bad luck, according to Welsh belief (Trevelyan), and if it blooms in winter, then it is a death omen. Bringing them indoors – well, it all depended on how many were gathered. Two or three brought into a poultry-keeper's house in early spring, before the chicks were hatched, meant bad luck to the sittings, but it would be alright if there were thirteen or more flowers, or "no less than a handful". In Devonshire, they said that the number of primroses brought in would agree with the number of chickens reared (Friend. 1883; W Jones. 1880; Gill. 1963), for thirteen is the number traditional to a clutch of eggs placed under a hen during the spring (G E Evans. 1966). There was a similar belief in France – if you threw the first primroses you found before the goslings, it would kill them, and if you took the flowers indoors, the goslings would die before being hatched (Sebillot). It was even unlucky to include primroses (and hazel catkins) in the posy carried to church on Easter Sunday. Violets had to be put in too, to compensate for the primroses (Tongue. 1965). But it was probably a lot more serious than it seems, at least in some areas, those in which primroses were looked on as a death token, just as snowdrops are. One explanation from Sussex is that it was used to strew on graves, and to dress up corpses in the coffin (Latham). Certainly, quarrels have been recorded as arising from this belief, and it could lead to charges of ill-wishing. Anyone giving a child, say, one or two primroses, would leave himself open to such a charge (W Jones. 1880).

In Lincolnshire, it was believed that if primroses were planted the wrong way up, the flowers would come red (Gutch & Peacock). They say exactly the same thing about cowslips, too. Northamptonshire people would claim that a common primrose fed with bullock's blood will become deep red (Baker. 1977). Christina Hole had a note that the brown marks in the middle of primroses were supposed to be the rust marks left by the keys of heaven when St Peter carelessly lost them, and they were left out all night on the primroses – something else that really belongs to cowslips.

Primrose leaves and flowers were used in salads (when, so it is claimed, they will help to keep off arthritis (Page. 1978)), and as pot-herbs. The leaves were often used, too, in herbal medicines. "Primrose leaves stamped and laid on the place that bleedeth stauncheth the blood", said Lupton, and Culpeper agreed – "of the leaves of primroses is made as fine a salve to heal wounds as any that I know". The flowers and young leaves boiled in lard make an ointment for healing cuts and chapped hands, and they say in Dorset that an ointment made with bramble tips and primroses is excellent for getting rid of spots and pimples on the face (Dacombe), and something known in Scotland as "spring rashes" was treated with the juice of primroses used as a lotion (Gibson. 1959). Burns and cuts would be treated with a salve made from the leaves (Beith), while in Suffolk, the leaves were dried, soaked in linseed oil, and put on the burn, which would heal in two or three hours after that treatment (Hatfield. 1994). The leaves themselves are often rubbed on a cut by men working in the fields (Hampshire FWI).

In modern herbal medicine, it is the root infusion that is used, in tablespoonful doses, as a good remedy against nervous headaches (Grieve. 1931). If taken last thing at night it has a decided narcotic tendency (Leyel. 1926), and so is good for insomnia. A 15th century recipe recommended boiling lavender and primrose in ale, and drinking the result "for trembling hands, and hands asleep" (Dawson. 1934). Gerard included among the virtues of the flowers "sodden in vinegar", and applied, the ability to cure the King's Evil [scrofula, that is], "and the almonds of the throat and uvula, if you gargarise the part with the decoction thereof". Even more remarkable is a prescription included in the Welsh medical text known as the Physicians of Myddfai: "whosoever shall have lost his reason or his speech, let him drink of the juice of the primrose, within two months afterwards, and he will indeed recover".

Primula auricula > AURICULA

Primula veris > COWSLIP

Primula vulgaris > PRIMROSE

PRIVET

(*Ligustrum vulgare*) Privet flowers belong to the large group that bring misfortune if brought indoors (Vickery. 1995), and another superstition once current was that diphtheria could be caught from privet leaves (Vickery. 1985), probably invented as a warning to children not to touch, or eat the poisonous berries. These berries give a bluish-green dye with alum, and the leaves will give a yellow colour (Coates) with alum, or with chrome a light brown, and a dark brown with copperas (Jenkins), green with copper sulphate, and dark green with iron (S M Robertson). An early 17th century auburn hair-dye had as its principal constituents radish and hedge-privet (Wykes-Joyce).

To cure sore lips one is advised to chew privet leaves and let the juice flow over the sore lip (Vickery. 1995). A cure for mumps is to boil privet berries till the juice is out of them, and tip the juice into a small bottle with cream from the top of the milk. When it has been cool for at least three hours, take a teasponful of the juice and one berry, once daily and after food (Vickery. 1995). A French charm for thrush in infants was to put a piece of privet in flower over the chimney piece. When this dries up, the child's thrush will also have dried (Loux).

PROTECTIVE PLANTS

SOW THISTLE – a Welsh belief was that the devil could do no harm to anyone wearing a leaf from this plant (Trevelyan), or as one of the Anglo-Saxon herbaria said (in translation) – "so long as you carry it with you nothing evil will come to meet you" (Meaney). CLOVER, too, is a protective plant, able to drive witches away (Dyer). Anyone carrying it about his person will be able to detect the presence of evil spirits (Wood-Martin). If a farmer carries one, all will be well with his cattle at that most difficult time, May Day.

HAWTHORN is ambivalent in this regard. While providing shelter and abode for both fairies and witches, it will also protect from witchcraft. In Gwent, one of the commonest ways of breaking a witch's spell was reckoned to be putting a cross of whitethorn (or birch) over the house door (Roderick); far from there, the Serbs believed that a cradle made from hawthorn wood would be a powerful protective device (Vukanovic). "Drive witches out of milk" by beating it with hawthorn used to be a Pennsylvania German saying (Fogel). A BLACKTHORN stick is a protection, and not just in the physical sense, for in Irish folklore it was used to overcome evil spirits (Ô Súilleabháin); so it does in Slav folklore, too, where in addition bits of the plant would be caried sewn into the clothing (Lea). HAZEL too, in spite of being a fairy tree, provides the most effective protection that Irish folklore remembers against fairies and spirits (Ô Súilleabháin) ("If you cut a hazel rod and bring it with you, and turn it round about now and again, no bad thing can hurt you") (Gregory). The Somerset practice of putting a hazel branch outside the door, and making a cross in the ashes with a hazel twig, seems to be purely protective in intent (Tongue). DOGWOOD, from the time of ancient Rome, has similar powers. The safety of the city was bound up in one particular tree on the Palatine Hill. There are signs elsewhere that it is a protective tree. In the Balkans, for instance, women wear it as an amulet against witchcraft (Vukanovic), and in some Serbian villages it was a stick of dogwood that was put in the cradle first, to protect a newborn baby, just as FLAX did in Scandinavian belief. Unbaptised children could be preserved from harm by "sowing flax seed", though the authority quoting that does not elucidate (Kvideland & Sehmsdorf), but sowing flax seed around the house, on the road, or by the grave, was a common means of protection against the spirit of the dead in Norway and Denmark. Flax seed was also put in the coffin, and, to keep ghosts away, round the grave or house. The ghost must count every single seed before going any further, for flax is a magic plant (Rockwell). Popular explanations assign the power of flax to the belief that Christ was swaddled in a linen cloth. PEARLWORT, in its role as the mystical plant Mothan, is one of the prime protectors. It will protect from fairy changing, and from all fairy activity. Put on the door lintel, it prevents the spirits of the dead from entering the house (J G Campbell. 1902) (see MOTHAN).

HOLLY's scarlet berries ensure its inclusion among lightning plants, with all the protective power such a plant always has. In East Anglia, for example, a holly tree growing near a house is regarded as a protection against evil (G E Evans. 1966). Holly hedges surrounding many Fenland cottages were probably planted originally with the same idea in mind (Porter). And being a lightning plant it must protect from lightning. "Lightening never struck anyone if you were under a holly tree. Lightening never struck a holly tree", as a Devonshire informant said. As far back as Pliny's time, there are records of holly being planted near the house for that reason. In Germany, it is a piece of "church-holly", i.e., one that has been used for church decoration, that is the lightning charm (Crippen). Holly is equally efficacious against witchcraft and the fairies. Fenland belief had it that a holly stick in the hand would scare any witch. Builders used to like to make external door sills of holly wood, for no witch could cross it (Porter). Wiltshire tradition has it that the Christmas holly and bay wreath hung on the door is to keep witches out. In the west of England it was said that a young girl should put a sprig of holly on her bed on Christmas Eve, otherwise she might receive an unwelcome visit "from some mischievous goblin" (Crippen). YEW also protects, and if some branches are kept in the house, it will preserve it against fire and lightning (Elliott). WALNUT, too, is a protective tree, in spite of being looked on as unlucky in another aspect. In Bavaria, where the Easter Sunday fire used to be lit in the churchyard with flint and steel, every household would bring a walnut branch, which after being partially burned, would be carried home to be laid on the hearth as a protection against lightning (Dyer). Walnut leaves, gathered before sunrise on St John's Day, were believed in parts of France to protect from lightning too (Sebillot). People from Poitou used to jump three times round the Midsummer Fires with a walnut branch in their hands. That branch would be used to nail over the cowhouse door, as a protection for the beasts (Grimm). Moslem belief also recognised its protective influence, particuary the root and bark, with which Moroccan women used to paint their lips and teeth a brownish colour (Westermarck). Another tree that is a protector is the ORIENTAL PLANE (Platanus orientalis), as Evelyn pointed out: "Whether for any virtue extraordinary in the shade, or other propitious influence issuing from them, a worthy knight, who stay'd at Ispahan in Persia, when that famous city was infested with a raging pestilence, told me, that since they have planted a greater number of these noble trees about it, the plague has not come nigh their dwellings".

IVY seems to be welcome as protection in Scotland, whereas elsewhere it is an unlucky plant. In the Highlands, a piece in the form of a circlet, often with honeysuckle and rowan, would be nailed over the byre door to prevent witches harming the cattle, and also to protect them from disease (MacGregor). Ivy kept evil away from the milk, butter, and the animals, and was put under the milk vessels, at such times, like May Eve (Grigson), when such protection was deemed necessary.

ROSEMARY was always taken to be a powerful "disperser of evil" (Baker. 1978). That is why they are planted near the house, so that "no witch could harm you" (Opie & Tatem), and to carry a piece with you was to keep every evil spirit at a distance (Hartland. 1909). In Spain, it is worn as an antidote to the evil eye (Rowe). A Jamaican belief was that if a house was haunted, burn rosemary, cow dung and horn, "and the duppies will leave" (*Folk-lore. vol 15; 1904*). It seems to have been looked upon as a general protector in all diseases, even if it was only treated as a charm rather than as a drug. A Sarajevo doctor told Kemp that women would throw down a sprig of rosemary as a protection against the doctor, and therefore all the illnesses with which he was associated, not least the plague. SWEET BRIAR had a similar reputation. In Normandy, hanging it over the door used to be regarded as a certain protection against witchcraft, but it also protected from fevers (W B Johnson), which must have been thought of as a result of malevolence. ELECAMPANE also protects. It is, of course, a famous medicinal plant, but in addition it would be sewn in children's clothes to ward off witchcraft, though it had to be gathered with some ritual, and on special days. That belief is centred on the Balkans (Vukanovics). AUTUMN GENTIAN protects, too. It is a herb of St John, and, according to Gubernatis, they used to say that whoever carried it about with him would never incur the wrath of the Czar. PERIWINKLE was used in charms against the evil eye (Folkard). Worn in the buttonhole, or carried dried in a sachet, it was a great protection against any witch not carrying it herself (Boland. 1977).

The root of MALE FERN, prepared as the Lucky Hand, guards the house from fire and many other perils (Hole. 1977) (see LUCKY HAND), and a Somerset belief is that FENNEL over the door will also prevent the house catching fire (Tongue). That is because fennel is a protective plant, and powers out of the ordinary have been associated with it. It was used to ward off evil spirits (Emboden), and to plug keyholes to keep away ghosts (Cullum), and it was hung over the door together with other herbs of St John at Midsummer. The 'benandanti' of 16th century Friuli, who were the "night-walkers" who fought witches on the psychic level, carried fennel as their weapon. It was said that these benandanti ate garlic and fennel "because they are a defence against witches" (Ginzburg). It was used as a personal amulet, too. The seeds were hung round a child's neck against the evil eye (W Jones), and in Haiti it protects against loupgarous, and also serves to fortify pregnant women (F Huxley). A medieval Jewish amulet turns out to be a sprig of fennel over which an incantation had been recited, and which was then wrapped in silk, with some wheat and coins, and then encased in wax (Trachtenberg). SAGE, too, is a protective plant, at least in Spain and Portugal, where it is thought of as proof against the evil eye (Wimberley), and it appears that MULLEIN too offered some protection against evil. See, for instance, the Anglo-Saxon version of Apuleius,

as translated by Cockayne: "if one beareth with him one twig of this wort, he will not be terrified with any awe, nor will a wild beast hurt him, or any evil coming near". It is also pointed out by Apuleius that what came to be regarded as mullein was the plant given by Mercury to Ulysses to neutralize the evil magic of Circe. In much the same vein, there is recorded an Irish charm to get back butter that had been witched away – all that is required is to put mullein leaves in the churn (Gregory. 1970).

VERVAIN was taken to be a great protector, either of the home (plant it on the roof and it will guard the house against lightning (Sebillot)), or of the person. Even in ancient times, it served in the purification of houses, and it was a Welsh custom to cut it, in the dark, to bring into a church, there to be used as a sprinkler of holy water (Clair). In Sussex, vervain leaves were dried and put in a black silk bag, to be worn round the neck of sickly children (Latham), probably to avert witchcraft rather than to effect a cure. Adults could be protected from the fairies and their spells by brewing a tea made from it, and drinking that (Spence. 1949). This belief in the power of vervain goes back a long way. The Romans hung it in their houses to ward off evil spirits (Thompson. 1897). Gerard tells us that "the Devil did reveal it as a sacred and divine medicine", and there are several versions of the couplet that Aubrey quotes:

> Vervain and dill
> Hinders witches from their will.

YARROW is another of these protective plants, doubly effective because it is also a herb of St John, and it is on the eve of the saint's festival that it would be hung up in Ireland to turn away illness (Grigson. 1955). It was also believed to have the power of dispersing evil spirits (Dyer. 1889), and in the Fen country it protected against evil spells, too; if it were strewn on the doorstep, no witch would dare enter the house (Porter. 1969). It would be hung up in the toolshed, "for safety" (it is a wound herb), but also to stop entry by thieves (Boland). A bunch of it was tied to the cradle to protect both baby and mother (R L Brown), or to make babies grow up happy and even-tempered (Porter. 1969). When going on a journey, pull ten stalks of yarrow, keep nine, and throw the tenth away (as the spirit's tithe, of course). Put the nine under the right heel, and evil spirits will have no power over you (Wilde. 1902). ST JOHN'S WORT is another protector against witches and fairies. When hung up on St John's Day together with a cross over the door of houses, it kept out evil spirits (Napier), and the Pennsylvania Germans fasten a sprig to the door to keep out witches (and flies) (Fogel). In Essex, they said that if it was hung in the window it would prevent witches looking in (C C Mason), while in the Western Isles the emphasis was on preventing ordinary folk seeing the witches (or "grisly visions", as it was described); it had to be sewn into the neck of a coat (Bonser), and left there, for if it were interfered with in any way, it would lose its power (Spence. 1959). But to be effective as an amulet it had to be found accidentally (J A MacCulloch). It was given to Irish children on St John's Eve to avert sickness (Ó Súilleabháin. 1942). These are all passive amulets, but it is said that St John's Wort had an active role, too. A white witch's "unwitching medicine" consisted of, among other things and rituals, three leaves of sage and three of Herb John, steeped in ale, to be taken night and morning (Seth).UPRIGHT ST JOHN'S WORT (*Hypericum pulchrum*) is just as active, provided it was found accidentally, "when neither sought for nor wanted", and then it should be put secretly in the bodice, if the finder was a woman, or if a man, in his waistcoat under the armpit (Banks. 1937), when it would ward off fever, and keep its owner from being taken in his sleep by the fairies (J G Campbell).

PRIMROSES are fairy flowers, but fairy flowers can give protection from the fairies as well. Manx children used to gather them to lay before the doors of houses on May Eve to prevent the entrance of the fairies, who cannot pass them, so it was said (Hull). In Ireland they tied primroses to the cows' tails (Wilde. 1902), for no evil spirits could touch anything protected by them (Buchanan). A primrose ball over the threshold served the same purpose in Somerset (Tongue. 1965). Those powers of protection went further – they could be used against the evil eye, for example (Wood-Martin). DAISY chains, just a children's game usually, are sometimes felt in Somerset to be a protection for children (Tongue. 1965), and daisies formed one of the three magic posies given to the traveller in the Derbyshire folk tale called the Crooker, to protect him from evil. "Take the posy and show it to Crooker". The other posies were of St John's Wort, and primroses (Tongue. 1970). PENNYROYAL will protect against the evil eye in Italy, and in Sicily it is hung on fig trees to prevent figs falling before they are ripe (Bardswell). Also in Sicily, it was given to married couples who were always quarrelling (Folkard). In parts of Morocco, it is picked and put in the rafters as a protection against evil, but it had to be gathered before Midsummer (Westermarck. 1926). It used to be burnt in the Moroccan Midsummer fires, too, lit between the animals, the smoke presumably acting as a protection (Westermarck. 1905). POMEGRANATES were protectors against the evil eye under certain circumstances. The Arabs of Hiaina, for example, always squeezed one of the fruit over the horns of the oxen when commencing ploughing. The juice would run into any evil eye, and so render the evil harmless. HENNA paste is a cosmetic, of course, but that use is overlaid by its protective role against evil eye. Red is a good prophylactic colour, anyway. Westermarck. 1926 outlines many cases in Morocco in which henna plays its protective role. It is used chiefly by women, but also on special occasions by men, and

it is applied to new-born babies as well. At the age of 40 days the infant had the crown of the head smeared with henna, as a protection against fleas and lice, but also against the evil eye; the application is repeated frequently till it gets older. The mother would have her hands and feet painted with it. Similar beliefs are held among the Nubian people known as the Kanuz. Almost all life-crises, whatever their nature, involve the use of henna. It was applied to the hands, feet and forehead of a groom on his wedding night, while the bride's entire body would be hennaed. SAFFRON had its protective side. In Morocco, for instance, it was one of the plants used to drive away the juun, and it was also used as a charm against the evil eye. Evil spirits are said to be afraid of Saffron, which is used in the writing of charms against them (Westermarck). Some Hebrew amulets, too, were written with a copper pen, using ink made from lilies and saffron (Budge). But it is not an inherent quality in the plant that is exploited. Rather it is the colour that is effective. Another example of protective colour is the blue of INDIGO in the Mayan villages of Yucatán. The ordinary amulet to keep off evil spirits and to avert the evil eye is a collection of small objects tied together with thread. It is this thread that is dyed blue with juice from the plant, and the same dye is used to paint the fingernails of persons who are threatened with death from sickness (Redfield & Villa).

In Polynesia, it is the leaves of the TI PLANT (*Cordyline terminalis*) that are the safeguard against attacks from the angry dead. Carrying food, especially at night, is regarded as very dangerous in Hawaii, and so they tie a green ti leaf to the container as a protective charm, which commands the ghost to fly away. Similarly, women wear a ti leaf as a protection when they approach particularly dangerous places (Beckwith. 1940).

ROWAN is the most important of the protective trees, whether against lightning, witchcraft, the dead or the fairies, or any evil influence whatever. It was the most powerful antidote to witchcraft known in the British Isles, particularly with regard to livestock and the general fertility of the farm. The list of usages is too long to be included here, but see ROWAN, where the customs are described fully. BIRD CHERRY assumes rowan's role in Wester Ross, Scotland. A walking stick made from the wood, for example, prevented the bearer from getting lost in mist (C M Robertson). One rhyme from the area shows bird cherry and rowan working together:

Hagberry, hagberry, hang the de'il,
Rowan-tree, rowan-tree, help it weel (Denham).

WYCH ELM is used in Scotland in very much the same way as Rowan. In the Cromarty legend of Willie Miller, who went to explore the Dropping Cave – "he sewed sprigs of rowan and wych-elm in the hem of his waistcoat, thrust a Bible into ine pocket and a bottle of gin into the other …"

(H Miller). Smollett knew of Wych Elm's prophylactic powers, for he has a character in *Humphrey Clinker* say "As for me, I put my trust in the Lord, and I have got a slice of witch elm sewed in the gathers of my under petticoat". In some villages of central Europe it was usual to plant a LIME tree in front of a house to stop witches entering.

MARSH MARIGOLD acts as a protector at a particular time of year, i.e., May Day. In Ireland, they act like the rowan does elsewhere, to protect the cattle (Grigson. 1955) and the house from evil influences. In County Antrim, children used to gather the flowers and push one through the letterbox of every house in the village, so that the house would be protected. The children, of course, were suitably rewarded. The pure white flowers of the CHRISTMAS ROSE make it the symbol of purity, and so it would be a protector from evil spirits. No evil could enter a house near which this plant was growing. JUNIPER, both the wood and the berries, is a protective tree, indeed the very symbol of protection (Leyel. 1937). Like many another plant, juniper offered the Virgin and the infant Jesus protection when hiding from Herod's men, and received her blessing in return for the shelter given. Among these plants, juniper was believed to have been particularly invested with the power of putting to flight the spirits of evil, and of destroying charms. Like box, juniper growing near the door protected the house from witchcraft, for the witch had to stop and count every leaf before proceeding (M Baker. 1977). In the Highlands of Scotland, juniper was specially used for "saining" on New Year's Day, and at Shrovetide. Branches were set alight, and carried through the house, the smoke spreading into a thick, suffocating cloud (McNeill. 1961), produced by closing up every window, crevice and keyhole in the house, for the smoke, besides protecting the house and its occupants from evil influences, was also supposed to have the ability to dispel infection (Camp). Juniper was burned before the cattle, too (Davidson. 1955), or it was boiled in water, to be sprinkled over them. ARCHANGEL (*Angelica archangelica*) by its very name, ensured its value as a protection, not only against epidemic diseases like plague, but also against other evil spirits and witchcraft, and the cattle disease elf-shot (Prior). Cornish folklore still regards it as a strong witch repellent (Deane & Shaw). Wearing a piece of WALL PENNYWORT keeps you from harm, so it is said in Dorset (Dacombe). Even AGRIMONY is a protector, according to a complicated Guernsey charm, to be worn round the neck: take nine bits of green broom, and two sprigs of the same, which you must tie together in the form of a cross; nine morsels of elder, nine leaves of betony, nine of agrimony [the number nine is significant; it is three times three, a magical number] and a little bay salt, salammoniac, new wax, barley, leaven, camphor and quicksilver. The quicksilver must be enclosed in cobbler's wax. Put the whole into a new linen cloth that has never been used, and sew it well up so that

nothing will fall out. Hang this round your neck. It is a sure preservative against the power of witches (MacCulloch). BETONY, too, is a protector. Sown round the house it protects it from witchcraft. "The house where Herba Betonica is sowne, is free from all mischeefes" (Scot). The Anglo-Saxon Herbal mentions it as a shield against "frightful goblins that go by night and terrible sights and dreams" (Bonser). "For phantasma and delusions: Make a garland of betony and hang it about thy neck when thou goest to bed, that thou mayest have the savour thereof all night, and it will help thee" (Dawson). The first item on Apuleius' list is "for monstrous nocturnal visitors and frightful sights and dreams" (Cockayne). A Welsh charm to prevent dreaming was to "take the leaves of betony, and hang them about your neck, or else the juice on going to bed" (Bonser).

ONION protects by attracting to itself epidemic diseases. In any case, they (and garlic) are good antibiotics. In Cheshire, they used to say that a peeled onion set on the mantelpiece during an epidemic will cause the infection to fly to the onion, and so spare the inhabitants of the house (Hole. 1937). The onion was supposed to turn black if there was any infection about (Palmer. 1976). Yoruba belief is somewhat similar, but put another way. They say that onions use their smell to kill disease (Buckley). Clearly the cure did not rely on the onion's being peeled, for we are told simply that a row of onions over the door would absorb all diseases from anyone who came in (Whitney & Bullock). That came from Massachusetts, but a Somerset belief is exactly the same, and indeed goes further, for the bunch of onions over the door will keep away not only illness, but also witches (Tongue. 1965).There is a story from India of people who always hung an onion by their house-doors. When plague visited the area, they were the ones to survive (Igglesden), and in plague years in London, three or four onions left on the ground for ten days were believed to gather all the infection in the neighbourhood (Wilson). Animals could be safeguarded as well. During the disastrous foot and mouth disease outbreak in Britain in 1968, on one Cheshire farm that escaped, although in the midst of the infection, the farmer's wife had laid rows of onions along all the windowsills and doorways of the cowsheds, and attributed the farm's escape to this precaution (M Baker. 1980). There are other examples of the protective role of onions, even if they involve wishful thinking, If, for example, you rub the schoolmaster's cane with an onion, it will split when he strikes you (Addy. 1895), a superstition that onions share with green walnut shells; or you will not feel the cane if you rub an onion across the palm of the hand (Gutch. 1911). An onion carried in the pocket will relieve the bad luck of meeting a single magpie (Tongue. 1965, Whitlock. 1992), and so on. WORMWOOD relies on its very strong smell in serving as a protection against evil spirits. Roumanian folklore recognises malignant spirits, generally figured as three female divinities, who haunt fountains and crossroads, and can raise whirlwinds. They sing to lure people to their doom. The protection against them is wormwood, bunches of it to hang on doors and windows, or at the belt (Beza). Arnaldo de Vilanova, writing in the early 14th century, said that wormwood put at the door will act as a protection against sorcery (Lea). So, too, in Somerset, where it was used against the evil eye. Another protector by smell is GINGER, some Malaysian people have their children wear a piece of ginger round their neck, so that the pungent smell will keep harmful spirits away (Classen, Howes & Synnott).

SWEET FLAG seems to have had a protective reputation in Japanese folklore. In the various stories of the serpent-bridegroom, the plant is used to thwart the evil supernatural (Seki). It was a protector in China, too; it was put up at the side of outdoor gates (so was MUGWORT) to avert the unpropitious (Tun Li Ch'en). BASIL is kept in every Hindu home, to protect the family from evil (A W Hatfield). GROUND IVY, too, has some protective powers. According to a story quoted by Lady Wilde, ground ivy carried in the hand gave protection against attacks by the fairies (Wilde. 1902), and it was one of the plants used on the Continent to enable the wearer to see and name witches (Lea). African-Americans in the southern states of USA look on JACK-IN-THE-PULPIT as a protective plant. They would take the leaves and rub them on the hands, and that would blind an enemy. But they use it to make charms to bring security and peace, and to protect them from enemies (Puckett). The leaves are luck-bringers, if carried about on the person (Puckett).

PRUNES
Stewed prunes were once apparently kept in brothels, and were supposed "to be not only a cure but a preventative of the diseases contracted there" (Toone).

Prunus amygdalus > ALMOND

Prunus avium > WILD CHERRY

Prunus domestica > PLUM

Prunus domestica 'institia' > BULLACE

Prunus domestica 'italica' > GREENGAGE

Prunus laurocerasus > CHERRY-LAUREL

Prunus padus > BIRD CHERRY

Prunus pennsylvanicus > PIN CHERRY

Prunus persica > PEACH

Prunus spinosa > BLACKTHORN

Prunus virginiana > CHOKE CHERRY

PRY
One of the names of SMALL-LEAVED LIME (*Tilia cordata*), and preferred by Rackham. 1986 as the common name for this tree.

Pteridium aquilinum > BRACKEN

Pterocarpus angolensis > BARWOOD

Pterocarpus indicus > AMBOYNA WOOD

PUCCOON
(*Lithospermum angustifolium*) One of the North American species whose roots yield a red dye, or paint, rather, often used by the Indians for painting dressed skins. The fresh root would be dipped into deer's grease and rubbed on the object to be painted, which may often be the human face (Teit). So the name Indian Paint (Elmore) is easy to understand. People like the Cheyenne used the root by grinding it fine, and rubbing the powder on an affected part as a remedy for "paralysis", or they made it into a tea to rub on the head or face of a patient who was irrational as a result of illness (Youngken). The Zuñi used it with some ceremony, by grinding it in the morning in the patient's room, on a special grinding stone, used only for this ritual. The remainder of the plant would be made into a tea. This medicine was for sore throat, and for swelling of any part of the body (Stevenson). The plant was also used in Alabama domestic medicine, as a lotion after ther roots had been boiled in vinegar. This was to treat eczema, and the root tea would be used for rheumatism (R B Browne).

Pulmonaria offiicinalis > LUNGWORT

Punica granatum > POMEGRANATE

PURL
At one time, WORMWOOD was used in the preparation of all sorts of medicated wines and ales. One of them was called purl, "which hard drinkers are in the habit of taking in the morning to go through their hard day's labours" (Thornton) – this was wormwood mixed with ale. (see also ABSINTHE)

PURSLANE
(*Portulaca sativa*). See also GREEN PURSLANE (*Portulaca oleracea*). One of the many supposed to be aphrodisiac (Haining). It has certainly been used as a medicine for a long time, as a cure for erysipelas, while Thomas Hill was of opinion that it "helped the shingles". He also recommended it as helping "the burning Fever…", for worms and for toothache. It also "helpeth swolne eyes, and spitting of bloud … it stayeth the bleeding at the Nose, and the head-ache …" (T Hill. 1577).

PUSSY WILLOW
A very common name for GOAT WILLOW, or SALLOW (*Salix capraea*), and the reference is to the fluffy catkins (they are actually called Fluffy Buttons in Somerset (Macmillan)).

Pycnanthus kombo > FALSE NUTMEG

Pyrus communis > PEAR

Q

QUAKING GRASS

(*Briza media*) A common grass with dozens of local names, all of the "quaking" or "trembling" type, such as Shaky Grass, Wiggle-waggle Grass, Dothering Dillies, and Shickle-Shacklers, to name only a few. Wiltshire children used to be told that if the spikelets ever stop trembling they would change into silver sixpences or shillings. In France, it was reckoned to be a St John's Herb, picked on St John's Eve and purified in the smoke of the Midsummer Fires (Grigson. 1959). For what purpose, though? But there are some traditions of ill luck connected with it – it is unlucky to the man who picks or wears it, for example, and it is also bad luck to bring it into the house. If it is laid in a baby's cradle, the child will be rickety (Tynan & Maitland). But it has its mundane uses. In Yorkshire, for example, it used to be tied in bundles, and hung on the mantelpiece, to ward off mice (Drury. 1992).

Quercus ilex > HOLM OAK

Quercus robur > OAK

QUINCE

(*Cydonia vulgaris*) As a symbol of happiness and fertility it was dedicated to Venus (Fluckiger & Hanbury), who is often represented holding one in her right hand (Ellacombe). Sending quinces as presents, or eating them together, were recognized love tokens; so was throwing them at each other. Dreaming of them was reckoned to be a sign of successful love, or it could be interpreted as speedy release from troubles and sickness (Gordon. 1985). In 17th century England, it was reckoned that "the woman with child that eateth many during the time of her breeding shall bring forth wise children and of good understanding" (quoted by Boland. 1977). Perhaps all this is why they were claimed to be the Golden Apples of the Hesperides.

Quince, in one form or another, was a favourite medieval stomachic, the "confection of quinces" being recommended against sea-sickness, for example (Withington). An Alabama folk remedy for stopping the hiccups was simply to take a tablespoonful of quince juice (R B Browne). A decoction of the pips is still sometimes used as an application in skin complaints, like chilblains, chapped skin, and burns (Schauenberg & Paris), and for eye inflammations. Indeed, it is sometimes added to more usual eye lotions. Quince-seed lotion, made by stewing the seeds in water, was used as a hair lotion, "for giving ladies' hair a fine wavy appearance (Savage), and quince-seed tea is an American country cure for diarrhoea (H M Hyatt). There was a medieval notion that quinces prevented drunkenness. The *Hortus Sanitatis* noted that it could be achieved by taking "syrup of quinces at the second course after wine" (Seager).

QUINSY

SQUINANCYWORT, which is actually Quinsy-wort, provides an astringent gargle to treat the complaint. In much the same way, BLACKCURRANTS, also good for the condition, were known as Quinsy-berries (Newman & Wilson). HONEYSUCKLE – Coulton quoted a 14th century manuscript, prescribing for "hym that haves the squynancy" a remarkable amount of disgusting rubbish, but containing as an ingredient "gumme of wodebynd". SANICLE leaves, which are astringent, can be used in infusion as a gargle for sore throat and quinsy (Wickham), and Coles advised that the leaves of ORPINE "bruised and applied to the throat cureth the Quinsy …". Inhaling an infusion of WOOD SAGE was a Yorkshire remedy for the complaint (Hartley & Ingilby), and it is still used in homeopathy for the same illness (Grieve. 1931). American Indians would make a decoction of the fruit of SMOOTH SUMACH to use as a gargle for the complaint (Lloyd). CAT'S FOOT (*Antennaria dioica*) is another plant used for the condition (Grieve. 1931).

R

RABIES

In Glamorgan, the roots and leaves of BUCK'S HORN PLANTAIN used to be made into a decoction, sweetened with honey, and given as a cure for hydrophobia (Trevelyan). Sir John Hill had heard of this, but gave it no credit: "it is said also to be a remedy against the bite of a mad dog, but this is idle and groundless". RIBWORT PLANTAIN was given for hydrophobia in Ireland (Denham) (it was being prescribed for snakebite in the Anglo-Saxon version of Apuleius). In Ireland, BOX leaves were used as a remedy (Wood-Martin); compare this with the 14th century recipe: "For bytyngge of a wood hound. Take the seed of box, and stampe it with holy watyr, and gif it hym to drynke" (Henslow). Wood-Martin records the use in Ireland of WILD ANGELICA as a cure for hydrophobia, probably only as an inheritance from its august relative, ARCHANGEL. BLACKCURRANTS were used in Ireland for the disease (Wood-Martin). A Russian cure uses CYPRESS SPURGE. It had to be gathered in May and September, during the first days of the full moon, and then it was dried and powdered. Anyone bitten by a suspected rabid animal was given a preventive dose of 5 grams in half a glass of some drink or other (Kourennoff). Another Russian folk remedy used DYER'S GREENWEED for the task, so it is said (Pratt).

VIRGINIAN SKULLCAP has been called Mad-dog, or Mad-dog Herb (House, Lloyd), for it was used to treat the condition, after a Dr van der Veer experimented with it in 1772 (Weiner). Hoosier home medicine uses ELECAMPANE. The roots have to be boiled in a pint of milk down to half a pint, and the patient has to take a third of the result every other morning, and eat no food until 4 pm on those days. It is effective, they claim, provided it is started withing 24 hours of the accident (Tyler).

RAGGED ROBIN

(*Lychnis flos-cuculi*) Surprisingly, in view of the relatively large number of local names, Ragged Robin has virtually no associations in folklore or folk medicine. "Ragged" in the common name refers to the typically tattered appearance of the flowers. The other part of the name, Robin, is the diminutive of Robert, and much used in earlier times in wild flower and bird names. So for Ragged Robin we can find Cock Robin and Red Robin, both from Somerset (Macmillan, Grigson); from Cumbria there is Rough Robin, and, more widespread, Robin Hood. Bobbin Joan, from Devonshire, (Tynan & Maitland) is probably connected with this series of names, though the name itself has other connotations, notably with the bobbin-shape of the spadix of Cuckoo-pint. "Ragged" names include Ragged Jacks (Elworthy. 1888), Ragged

Urchin, from Devon, or Ragged Willie, from Shetland (Grigson), and so on. Thunder Flower is recorded from Yorkshire, and is reminiscent of the Red Poppy, with all the superstitions attached to that plant, and none for this.

RAGWORT

(*Senecio jacobaea*) Severely toxic to animals (Forsyth –"Ragwort alone probably causes more annual loss to the livestock industry, than all the other poisonous plants put together …"). It causes cirrhosis of the liver, from which no recovery is possible. The condition is known in Canada as the Pictou cattle disease, and it is known scientifically as seneciosis. The trouble is that the animals will not normally eat it, so it flowers and seeds undisturbed, and the effect is eventually to produce more ragwort than grass in a pasture (Long).

Nevertheless, this is a fairy plant, dedicated to them in Ireland, and called Fairies' Horse (Friend), for it is believed to be a fairy horse in disguise. If you tread them down after sunset a horse will arise from the root of each injured plant, and will gallop away with you (Skinner). The fairies look to ragwort for shelter on stormy nights, according to Hebridean folklore, as well as riding on it when going from island to island (Carmichael). Folklore equates fairies with witches in many particulars, the belief in their using ragwort to ride on at midnight being very widespread (Henderson, Hunt, Wentz). "As rank a witch as ever rode on ragwort" was a common saying in those areas where the belief held (Cromek) (perhaps any Kecksies when dry would do (Briggs. 1978)). Yeats quoted a story in which the local constable, when there was a rumour of a little girl's being taken by the fairies, advised the villagers to burn all the ragwort in the field from which she had been taken, as it was sacred to the fairies. Lady Gregory, too, has a story about burning ragwort, and thereby bringing a protest from those who used the plant as horses.

In the Isle of Man, it was used as a protection against infectious diseases (Friend). When visiting a sick person, you were advised to smell a piece of ragwort before actually going into the sick room (Gill). Was that a genuine medicinal usage, or a charm against witchcraft? The latter, probably, for there was a Scottish belief that if a mother takes bindweed and puts it burnt at the ends over her baby's cradle, the fairies would have no power over her child (Wentz). Although this is given as bindweed, it is almost certainly ragwort (bundweed) that is meant.

Gerard passed on the recommendations of others for using ragwort in a number of ways for healing "greene wounds", ulcers and the like, and also, in the form of an ointment, "to helpe old aches and paines in the armes, hips and legs", including the "old ache in the huckle bones called Sciatica". That is recorded as an Irish treatment, too. Hot fomentations with decoctions of the whole plant in water would have been used. Bruises were treated with the bruised leaves in lard as an ointment (Egan).

Jacobaea is the specific name, and so we find book names like Jacoby, St James's Wort, or St James's Ragwort. In the vernacular, James's Weed is used in Shropshire (Grigson). Perhaps it is because the plant is in full flower on St James's Day, which is 25 July, but it is pointed out that St James is the patron saint of horses, and the use of the plant in veterinary practice is confirmed by other names, such as Staggerwort, that is, the herb that cures the staggers in horses. This disease is an acute form of selenium poisoning (Drury. 1985), and it also occurs in cattle and sheep, one of the principal symptoms being a giddiness in the head (Sternberg). All this is very confusing – ragwort's toxic properties have already been mentioned, and Sir Edward Salisbury, for one, stated clearly that staggers was actually caused by ragwort. If both views are correct, here must be an example of homeopathic magic, of like curing like. Prior had different views, regarding it more probable, as the name Seggrum suggests, that ragwort was applied to stop the bleeding of newly castrated bulls, called seggs or staggs (see Watts for other similar names).

RAMPION BELLFLOWER

(*Campanula rapunculus*) Not a British native, but established here and there, usually as an escape. The specific name, *rapunculus*, means a little turnip, and the roots are quite edible, either raw, or sliced in salads, or cooked, when they taste rather like parsnips.

The heroine of one of Grimm's tales is named Rapunzel, called after the herb, and the tale is woven round the theft of Rampion roots, and there is a Calabrian legend of a village girl who gathered a root in a field and found that the hole left led down to a place in the depths of the earth (Rohde). But this is not a lucky plant, for it is a funeral root, and in Italy there was a supersition that rampion among children gives them a quarrelsome disposition, and may even lead to murder. So, to dream of it is a sign of an impending quarrel (Folkard).

RAMSONS

(Allium ursinum) Ramsons is an interesting word. Its ultimate origin was the Greek kromon, but more relevant is the OE hrama, whose plural was hramaton. That plural noun was later on, when its derivation was forgotten, taken as a singular. So a new plural was formed with 's', so giving us a rarity of a double plural. The word occurs in several place-names, notably Ramsey in Essex, and in Huntingdonshire – ramsons valley.

It is a British native, and our only broad-leaved garlic, widespread and locally frequent in damp woods, and in the north sometimes in open situations, where their presence in pastures may lead to difficulties, for they spoil the taste of butter if cows eat them. Children in Yorkshire used to be paid to "knock down ramps", to save the butter (Hartley & Ingilby). The white flowers show from April to June. In the ecclesiastical calendar, it is appropriated to St Alphege, whose feast day is 19 April (Geldart). The church using ramsons as decoration must have overwhelmed its congregation with the smell.

Cloves of ramsons, or wild garlic as it is often known, used to be planted in Ireland on thatch over the door, for good luck (Opie & Tatem) in general, but in particular to ward off fairy influences (Mooney), which is one of true garlic's functions. Of the plant's medicinal virtues, an old west of England proverb gives a summary:

> Eate leekes in Lide, and ramsins in May,
> And all the yeare after Physitians may play"
> (Aubrey. 1686).

"Lide" is March. Of the rhyme Gerard says "the leaves may very well be eaten in Aprill and May with butter, of such as are of a strong constitution, and laboring men", but the only real good he had to say of it was that the distilled water "breaketh the stone, and driveth it forth", a usage that was well-known in the Highlands, too. But there is more to ramsons than that, for, like real garlic, it is prescribed by herbalists for arteriosclerosis (the fresh leaves, or a tea made from the dried leaves), hypertension, diarrhoea and distension (Flück), the last-mentioned being well-known in medieval times, for the Welsh medical text known as the Physicians of Myddfai has "for a swelling of the stomach. Take goats' whey, and pound the herb called ramsons, mixing together and straining. Let it be your only drink for three days". Pennant found on the Isle of Arran that "an infusion of Ramsons … in brandy is esteemed a good remedy for the gravel".

Ranunculus acris (Meadow Buttercup) > BUTTERCUP

Ranunculus bulbosus (Bulbous Buttercup) > BUTTERCUP

Ranunculus ficaria > LESSER CELANDINE

Ranunculus flammula (Lesser Spearwort) > SPEARWORT

Ranunculus lingua (Greater Spearwort) > SPEARWORT

Ranunculus repens (Creeping Buttercup) > BUTTERCUP

Ranunculus sceleratus > CELERY-LEAVED BUTTERCUP

Raphanus raphanistrum > WILD RADISH

Raphanus sativus > GARDEN RADISH

RASPBERRY

(*Rubus idaea*) An ever popular and healthy fruit. Even to dream of them was reckoned a good sign, for it meant success in all things, happiness in marriage, and the like (Gordon. 1985). Raspberry leaves were used in the same way as those of bramble, for sore throats and stomach upsets. "The leaves of Raspis may be used for want of Bramble leaves in gargles …"(Parkinson. 1629). The leaves, boiled with glycerine and the juice drunk, is an Irish remedy for thrush (Maloney), and raspberry leaf tea was an old remedy for relieving morning sickness; it was also said to help labour, in fact it is a general country drink taken to ensure easy childbirth. It should be started, so it is said, three months before the birth is due, and taken 2 or 3 times a week (Page. 1978; Beith). Powdered leaves, in tablet form, can be bought – they help relaxation in childbirth, so they say, and the fruit will have the same effect. Gerard wrote that "the fruit is good to be given to those that have weake or queasie stomackes", something that had already appeared in Langham. Distilled raspberry water was given in Scotland as a cooling drink to feverish patients (Beith). Then there is raspberry vinegar, made by pouring vinegar repeatedly over successive quantities of the fresh fruit – this was at one time a favourite sore throat medicine (Fernie).

RATI

The seeds of Precatory Bean (*Abrus precatorius*), which weight about one carat, have been used for centuries for weighing gold, under the name Rati (Grieve. 1931).

RAYLESS MAYWEED

(*Matricaria matricarioides*) This is sometimes referred to as Pineapple Weed (McClintock & Fitter), for the scent of the crushed leaves made it valuable as a strewing herb. Indeed, its abundance by waysides in is part due to its tolerance of being trampled underfoot (Mabey. 1977).

RED CAMPION

(*Silene dioica*) An unlucky plant to pick (Tongue. 1965), one of the plants that bear the name Mother-die (Grigson. 1955), always an indication that children are advised never to pick them. To reinforce the injunction, it also has the name Devil's Flower (Britten & Holland). It is a Fairy Flower, too (Moore, Morrison & Goodwin), in the Isle of Man, another reason why it can never be picked (Garrad). But it is called Robin (another indication that it is a fairy flower) or some variant over most of England. Another reason for the ill-luck is an association with snakes – it is Blodyn neidi, snake flower, in Welsh – if you pick it you will be attacked by snakes (Vickery. 1985). Another Welsh name is Blodyn Taranan, thunder flower. Thunder and lightning will be the result if you gather it.

Squeeze the juice on corns, and they will come out; that is a Somerset remedy. And also rub it on a wart (Tongue. 1965). But that is the only record of medicinal uses. We can discount Gerard's claim that "the weight of two drams of the seed … purgeth choler by the stoole, and it is good for them that are stung or bitten of any venomous beast".

RED CLOVER

(*Trifolium pratense*) The commonest of the red or purple clovers. The seed was largely imported from Holland when it was being grown as a fodder plant, not only as cattle fodder, it seems, for Henry Mundy (1680), speaking of the Irish, said they "nourish themselves with their shamrock (which is the purple clover), are swift of foot and of nimble strength" (Britten). To dream of clover is a happy sign, indicating health and prosperity (Gordon. 1985). Pliny said that the leaves stand upright at the approach of a storm (Gerard), but actually the leaves close and droop when rain is coming on (Pratt).

Herbalists use red clover to keep the blood flowing freely; this is due to the presence of an anticoagulant drug in the plant (V G Hatfield). It is mildly sedative, too, and the flowers have been traditionally used for headaches, neuralgia and gastric trouble (Conway), and a syrup made from the flowers once enjoyed a fine reputation in the treatment of whooping cough (Fernie). Actually, the plant is well-known as a cough cure, and a remedy for bronchial affections (V G Hatfield). Dried, it has been mixed with coltsfoot leaves to be smoked as a herbal tobacco (Savage). It is even believed in parts of America that red clover tea will cure tuberculosis (H M Hyatt). American domestic medicine claims that a tea made from the dried flowers are good for dropsy, while the same tea with alfalfa hay added is taken, a cupful a day, for hypertension (H M Hyatt).

Herbalists claim that the plant relieves gout and rheumatism, and it has even been recorded as giving relief in cases of cancer (V G Hatfield), a use that is known in American domestic medicine from Alabama – "for cancer of the breast three or more quarts of red-clover-blossom tea a day" (R B Browne).

RED DEADNETTLE

(*Lamium purpureum*) There is an American saying that if deadnettles are in abundance late in the year, it is a sign of a mild winter to come (Inwards). A very odd belief is quoted by Lupton: "Cast the water of any sick person, newly made at night, on red nettles, and if the nettles be withered and dead in the morning after, then the sick party is like to die of that disease; if they be green still, then he is like to live".

When children had measles, red deadnettle roots were boiled in milk for them to drink. This was reckoned to bring out the rash (Vickery. 1995). That was an

Irish remedy, and there is one from East Anglia for piles, when the plant was infused in white wine for an hour. A wineglassful would be taken two or three times a day (V G Hatfield. 1994). The rest of the medicinal uses are taken from the earlier herbalists. For example, from the 15th century, "to heal wounds full of blood. Stamp red nettle in a mortar with red vinegar (?), and lay on the wound: and it shall do away the blood and cleanse the wound" (Dawson. 1934). This is in fact an old remedy for "stopping the effusion of blood" (Pratt), and Hill, in the 18th century, was recommending such a cure. There was a recipe for boils in *Reliquae Antiquae* (14th century), and Gerard recommended it for the King's Evil, which is scrofula, and also as a poultice for wens and hard swellings. Another of Lupton's suggestions was for constipation, using red deadnettles (and mallows). The plants had to be boiled in water, and then the party was advised to "…sit close over the same, and receive the fume thereof up his fundament, and it will help him certainly and speedily (God willing)".

RED MAHOGANY

(*Khaya nyassica*) It has an astringent bark, resembling quinine in its nature. The bark infusion is taken to relieve a cold. The seeds are used, too, crushed and boiled to extract the oil, which is then rubbed into the hair to kill vermin (Palgrave & Palgrave). As it is one of the biggest and most imposing of trees, it is used in medicines for strength. It features in an Ambo chief's medicine, the inclusion of this strengthens the medicine to such an extent that it cannot be overcome by another, stronger medicine (Stefaniszyn).

RED POPPY

(*Papaver rhoeas*) It almost always accompanies cereal crops, and often disappears when cultivation is given up. The Greeks always reckoned it to be a companion to the corn, not a weed (Grigson. 1955), the flower of Aphrodite as goddess of vegetation, just as the Romans looked on it as sacred to their corn goddess, Ceres. Perhaps, as Grigson suggested, red poppies were regarded as a life blood growing along with the nourishing grain. Certainly, there was no difficulty in accepting poppies as a natural consequence after a battle, whether it was after Waterloo, or the battles of a later war. The Flanders poppy has become the symbol of the blood shed there.

Perhaps that is why there is something uncanny about poppies. Irish women had a dread of touching them (Grigson. 1955), and they are certainly unlucky flowers to bring indoors, for they can cause illness (Waring). Across northern Europe, and in Belgium, it was said that picking them would bring a thunderstorm (Sebillot); so they said in Wiltshire, too, but if you had poppies growing on the roof of a building, they protected against lightning (Wiltshire). There is no great difficulty here - a flower as red as a poppy would naturally be taken as a lightning plant.

Another superstition recorded around Liège shows a way of ensuring that children did not pick the flowers. If they did, they would wet the bed. This was still current at the beginning of the 20th century, and was recorded as early as the 16th (Sebillot). Around Dinan, they say that God punished the poppy for being too proud of its form by allowing the devil to touch it; the black patches are the marks of the devil's hand. Hence the Warwickshire proscription – children touching them would get warts on their hands (Savage). If corn poppies were put to the eyes they would cause blindness, hence such names as Blind Eyes, Blind Man, or Blindy-buff. Another group of names like Headache, or Headwork, show that they were used to cure headache, and from Norfolk, a hangover, but the underlying folklore shows that they were believed to cause it. As John Clare said:

> Corn poppys that in crimson dwell
> Call'd 'head ache' from their sickly smell.

In Wiltshire, they said that if you picked poppies from the corn, you would either have a bad headache or there would be thunder and lightning (Wiltshire) (cf names like Thunderflower, (Dartnell & Goddard), Thunderbolt, etc.,). Ear-ache is another of these names; if picked and put to the ear, ear-ache would be the result, but in Somerset a poultice made of poppies laid against the ear was used to cure the condition (Tongue. 1965).

One rather odd divination practice has been recorded from Switzerland. A poppy petal used to be put in the palm of the hand, and smartly slapped. If it burst, making a noise, then all was well – he was sincere, but if there was no sound, then she would have little to hope for (Friend. 1883). To find the sex of the first child, take a poppy bud, "sépare les deux sépales, et on regarde la forme des pétales qui dépassent; si elles sont divisées comme un pantalon, cela sera un garçon, si les sépales sont unis, cela sera une fille" (Loux). Poppy wine used to be made in Wiltshire from the petals. "Very heady", Jefferies said it was (Jefferies. 1880). The petals had medicinal uses, too, if for nothing else, they have served to colour medicines (Brownlow). But one keeps finding direct references to the flowers in prescriptions. In Chinese medicine, for instance, it is said that both flower and root are used for jaundice (F P Smith). But that sounds like doctrine of signatures, the plant's yellow juice to cure the yellow disease. The other doctrine of signatures usage is based, so it is said, on the shape of the capsules, vaguely like a skull; so they were regarded as appropriate for "diseases of the head" (Coles), by which was probably meant headache or migraine. A 15th century herbal says "also yff a man have the mygreme or hed-ache payn [pick] thys herbe and temper hit with aysell [vinegar] and make a plaster and ley to the fore-hede and to the templys *et sessabit*" (Grigson. 1974).

A syrup made from the flowers is even yet sometimes used as a sedative to soothe coughs (Schauenberg & Paris), and in France, young babies with whooping cough (coqueluche) were given infusions made from red poppy (Loux). Not all that long ago mothers on South Uist made a liquid from the flowers to help babies in their teething (Beith), and an infusion used to be given to relieve the pain of earache (V G Hatfield. 1994). Corn Poppy certainly has some sedative effect, though not as much as *P somniferum*. An old Welsh sleeping draught recipe reads: Boil poppy heads in ale, and let the patient drink it, and he will sleep (Ellis). In the Highlands the juice used to be put in children's food to make them sleep (Beith). Hill recommended syrup of red poppies as a sleep-procurer (Hill. 1754). This presumably is why red poppies were once used to treat mental illness. One of Lady Mildmay's courses for the treatment of "frenzy and madness" relied heavily on them - take one spoonful of the distilled water of red field poppy …, give this for one potion. Also every night apply wool to the temples and forehead, wet in oil of poppy. Also put up a feather into each nostril wet with the same oil … (Pollock).

RED RATTLE

(*Pedicularis palustris*) A poisonous plant, especially for sheep (Long. 1924). It is a powerful insecticide, too (Schauenberg & Paris).

RED SAGE

(*Salvia coccinea*) Mexican Indians like the Totonac use this as one of the seven herbs that are prepared in a bath to cure people of "fright" occasioned by the dead (Kelly & Palerm). On a more mundane level, the Maya crush the stalk and leaf and press it into the cavity of a decayed tooth (Roys).

RED SILKCOTTON TREE

(*Bombax ceiba*) A kapok tree, originally from India and Burma south-eastwards to Australia, but now grown in the West Indies, where they say it is the haunt of ghosts and other spirits, probably the kapok suggesting it. Hence its names Jumbie Tree and Devil's Tree. Anyone trying to cut one down could expect harm (Bell), unless, that is, he had propitiated the duppy with rum and rice put round the root.

In China, the flowers have been used to apply to boils, sores and the itch, while in Indonesia juice from the roots (gathered before sunrise) is taken to lower fever, and serves as a mouthwash and gargle to treat thrush in infants (L. M. Perry).

REED

(*Phragmites communis*) Useful for bedding of animals, and also for thatching, for this is the most durable of all for roofing. It may be the most expensive, but it has a life of up to a hundred years, provided the roof pitch is forty-five degrees or steeper (Jenkins. 1976). For centuries the reed thatcher (med

Latin arundinator) has been a different profession from the straw thatcher (cooperator). In Somerset, they used to call it Pole-reed (Elworthy.1888), possibly a corruption of pool-reed, but it is usually known to thatchers as Norfolk Reed (Jenkins. 1976). Around Abbotsbury, in Dorset, where the thatching tradition has been strong until recent times, it is known as Spear (Nash & Nash). Coopers at one time used specially grown reeds to put between the staves to make a barrel watertight. Delivery men apparently carried supplies of reeds in case barrels sprang a leak during transit (Brill).

The simplest, and probably the most ancient, musical instrument is made from reeds, and called variously Pan-pipes, Shepherd's Pipes (though these were usually made from oat straw), or Syrinx. The reeds are of different sizes, placed side by side, each stopped at the bottom end (F G Savage). Children in the Fen country used to put the leaf between their palms and blow, to make a piercing whistle (though that is a universal occupation, using virtually any leaf). Another pastime there was to make boats out of the leaves. "You took the leaf with its hard little stalk still on it and folded each end back. Then you split the folded ends into three and tucked one of the outside ones through the other outside ones, leaving the middle one flat for the little boat to sail on, and the stalk would stick up in the middle like a real little mast" (Marshall).

In Shakespeare's time, reeds were symbols of weakness, for they are tossed about by the wind, bending to a superior force (Ellacombe), hence also its use as a symbol for imbecility (Grindon). It is also, in Christian art, an emblem of the Passion, for Christ was offered a sponge soaked in vinegar on the end of a reed (Ferguson).

The rootstock is edible, and has a sweet flavour. Dry them first, then grind them coarsely and make porridge of them (Loewenfeld). They are used in China for fevers and coughs (Geng Junying). The practice of passing a ruptured child through a split sapling, well known in Britain, is actually done with a reed in Portugal. The child's injury will heal while the injury to the plant heals. There is the usual ritual to be observed; in this case it had to be performed at midnight on St John's Eve (23 June), by three men of the name of John, while three women, each called Mary, spun, each with her own spindle, on one and the same distaff (Gallop).

RENNET

In the Isle of Man DROPWORT (*Filipendula vulgaris*) was an ingredient in the preparation of rennet for cheese-making, and for binjeen (junket). Best known, though, is LADY'S BEDSTRAW, and it bears a number of names to prove it. Rennet itself (Runnet in Kent (Parish & Shaw)) is one of them, and Cheese-rennet, Cheese-renning, which Gerard used, and Cheese-running (Prior) are further examples. It is Keeslip in Scotland (Grigson. 1955), with the same meaning, and Curdwort is also recorded (Britten & Holland). Gerard wroite that "the people of Cheshire, especially about Namptwich, where the best cheese is made, doe use it in their Rennet, esteeming greatly of that Cheese above other made without it".Double Gloucester cheese traditionally has Lady's Bedstraw, and nettles, in its composition (M Baker. 1980).

Reseda lutea > WILD MIGNONETTE

Reseda luteola > DYER'S ROCKET

REST HARROW

(*Ononis repens*) Cammock is the proper English name for this plant. Rest Harrow is actually 'arrest harrow', brought in from 'remore aratri', literally plough hindrance. There is also 'resta bovis', "because it maketh the Oxen whilst they be plowing to rest or stand still" (Gerard), and of course, it is only pulled up with some effort.Thus, in the Carpathians, it was deemed to possess great power. So if a new-born baby is bathed in water in which this plant had been put, the child will get a portion of the power of the plant (Bogatyrёv). Another reason why farmers disliked the plant was because it tainted dairy produce (Grigson. 1955). There is a Wiltshire name for that – cammocky, or gammotty (Dartnell & Goddard). But it is a favourite food of donkeys (hence the generic name *Ononis*, from Greek onos, ass) (Grieve. 1931), and Gerard recommended it for human consumption: "The tender sprigs or crops of this shrub before the thornes come forth, are preserved in pickle, and be very pleasant sauce to be eaten with meat as a sallad". The thorns he spoke of make this one of the many plants from which legend says the Crown of Thorns was made (Leyel. 1937).

Wild Liquorice, Liquory-stick and Spanish-root are names given to this plant in the north of England and Scotland, for children used to dig up the root and eat it, and not just children either, for labourers would suck the juice from the roots to quench their thirst (Vickery. 1995). Sometimes the children would put the root in a bottle of water, shake it up, and then drink the result – Spanish water.

Rest Harrow is still used medicinally; herbalists prescribe it for dropsy and related conditions (Flück), just as the early herbalists did. It had been used for dysentery, too, for which one extraordinary leechdom has survived: "take cammocks, roots and all, and seethe it in water, and wash thy feet therewith to the ankle, and take the seed of cress a penny-weight, and drink it with red wine and be whole" (Dawson. 1934). The leaf juice was used in Ireland to apply to chapped and rough hands (Moloney).

RESURRECTION PLANT

A name given to the ROSE OF JERICHO (*Anastatica hierochuntica*), for the leaves fall off as the seeds ripen during the dry season, and the branches curve inwards to make a lattice-like ball which is blown out of the soil to roll about the desert until it finds a moist spot, or until the rainy season begins. The Greek word anastasis, after which the generic name is taken, means resurrection.

Rhamnus cathartica > BUCKTHORN

Rhamnus purshana > CALIFORNIAN BUCKTHORN

Rheum rhaponticum > RHUBARB

RHEUMATISM

Cornish people used to carry a piece of ASH wood in their pockets as a rheumatism cure (Deane & Shaw), or in Essex a HAZEL nut (Newman & Wilson). Carrying conkers, the nuts of HORSE CHESTNUT, is very well known, not only in England, but also in America (Sackett & Koch), where YELLOW BUCKEYE or BLACK WALNUT nuts would be used as well (Bergen. 1899), or keep a piece of JACK-IN-THE-PULPIT in the pocket (H M Hyatt). The idea is either to prevent, or cure. Some said that the conker had to be begged or stolen. Norfolk people favoured having the nuts in the form of a necklace, and they had to be gathered by children who had never had the ailment (Porter. 1974). A POTATO, too, or half an APPLE (Foster), can be carried in the pocket (two potatoes in Maryland, one for each pocket) (Whitney & Bullock). In Ireland, it is said that as the potato dries up the rheumatism will go away (Mooney) – it will "draw the iron out of the blood", as a Somerset belief had it (Whistler). The same belief was found in France and Belgium, where the potato was carried about as a general charm against pain (Sebillot). Andrew Lang said the potato had to be stolen, or the cure would not work, and Devonshire superstition also required some ritual. Here, a member of the opposite sex had to be asked to put the potato, unseen, in one of your pockets.You could change the pocket at will after this had been done, but the potato had to be worn continuously, or the charm would not work (Hewett). Some say that the potato had to be bound on to the affected part, but it had to be kept there without change until the thing had become offensive through decay (Mooney). Sometimes the potato was changed annually, when the crop was being dug (Heather). A piece of WHITE BRYONY root replaced a potato, at least in Norfolk (M R Taylor. 1929). NUTMEGS, too were carried about as a charm either to prevent or to cure rheumatism, particularly in America. In Alabama, it would be worn round the neck, and that, so it was thought, would prevent rheumatism from the waist up (R B Browne). Another charm is to put slices of green PEPPER under the fingernails. It is the smell, they say, that drives the "spirit" of rheumatism away (Waring). Apparently, ASH leaves are still used to treat the complaint. So are BIRCH leaves, in the form of birch tea (boiling water on a couple of tablespoonfuls of the chopped leaves) (Grieve). NETTLE seeds soaked in gin used to be taken for the relief of rheumatism (Savage), but there were a number of different ways that nettles could be used, by eating the leaves, for instance, which was a Cumbrian practice, as was eating DANDELION leaves (Newman & Wilson). But, more spectacularly, the way to deal with the complaint was to thrash oneself so that the stings took their toll (Vickery. 1992), on the basis, presumably, of the counter-irritant principle. Urtication, it was called, and it continued to be a recognised treatment for rheumatism right into the 20th century (Coats. 1975). You can lie down on them, too, and the more you are stung, the better the cure, for the evil is said to come out with the blisters (Hald). Similarly, in Somerset, they said that you could treat it by getting bees to sting you, and then use GREAT PLANTAIN leaves to ease the stings (Tongue. 1965). HOLLY was used in a similar way. Better known as a chilblain cure, but in some places, Somerset for one, rheumatism was dealt with by beating the part with a holly branch (Mathews). On the other hand, and this is a Cornish superstition, "rheumatism will attack the man who carries a walking stick of holly" (Courtney), a belief that seems quite out of character, given the many virtues of the tree. Another Cornish practice was to wrap a CABBAGE leaf round the affected part (Hawke), which was what some native Americans did with a leaf of the poisonous AMERICAN WHITE HELLEBORE (*Veratrum viride*) to relieve the pain; others, including the Quinault, boiled the whole plant, and drank it in necessarily small doses (E Gunther).

GROUNDSEL was used in the Fen country to bring relief. All they had to do was to soak their feet in water in which the plant had been boiled for ten minutes (Randell). CHICKWEED is another plant that can be used. Homeopathic doctors prescribe what is described as an essence of the fresh plant, to be taken to relieve the pain, and it can be used externally as a rub (Schauenberg & Paris), or, as in Sussex, it can be crushed and laid on as a poultice (Allen), and in Scotland an ointment made from it is used to like effect (Vickery. 1995). WILD STRAWBERRY leaf or root tea (V G Hatfield. 1994), and also COMFREY root tea can be drunk for rheumatism (Painter), and GOUTWEED has also been used, either drunk as an infusion (A W Hatfield), or by applying hot fomentations of the leaves to to rheumatic joints (Le Strange). MUGWORT tea is still often taken, and a Cornish cure is a draught of the herb, made by pouring boiling water over a handful of it, taken

hot, and sweetened with treacle (Deane & Shaw). A similar decoction of BIRTHWORT has also been advised (Schauenberg & Paris), and so has FENNEL tea (A W Hatfield), or AGRIMONY tea (Conway), or, in Cumbria, MARSH MALLOW tea (Newman & Wilson). and, in American domestic medicine, a tea from the leaves of LABRADOR TEA (E Gunther). PARSLEY is another herb well recommended for the complaint, either in the form of a tea (Rohde), or by simply chewing the leaves, which will ward it off (Camp). Actually, parsley for this complaint is of ancient origin; there is a leechdom in the Anglo-Saxon version of Apuleius for "sore of sinews, take parsley, pounded, lay it to the sore" (Cockayne). People in the Dursley district of Gloucestershire used to pick the leaves of WOOD SAGE in spring, and dry them, for a tea against the complaint (Grigson. 1955). The tops, according to Hill. 1754 "drank for a continuance, is excellent against rheumatic pains". The decoction of the bast of SMALL-LEAVED LIME has also been used (Schauenberg & Paris). CELERY, too, is still recommended, either as a decoction of the seeds (Newman & Wilson), or as a tea (Browning). An American cure was to use celery seed worn in a bag round the neck. A wedding ring in the bag would clinch the cure (Whitney & Bullock). Russian folk medicine agrees that celery is good for rheumatism, but there it is the stem that is used (Kourennoff). MEADOW SAFFRON (*Colchicum autumnale*) has been used in the form of colchicine, for rheumatism, and this is probably the ailment rendered as "sore of joints" by Cockayne, in his translation of the Anglo-Saxon version of Apuleius.

POKE-ROOT berries mixed with whisky form a Kentucky rheumatism cure. In the same area the practice was to use the dried berries made into a tea, or just eating the raw berries (Thomas & Thomas), while people in Kansas used to swear by cooked CRANBERRIES as an effective rheumatic pain reliever (Meade). Another American remedy was to boil MULLEIN root and mix it with whisky, to be drunk as needed, or to dip a cloth in mullein leaf tea, and bind it on the affected part (R B Browne). A tea from PUCCOON root used to be another treatment in the same area (R B Browne). The bark of CHINABERRY TREE was used in Indian domestic medicine for the complaint (Codrington). Older writers stressed the sulphur content of HORSERADISH, apparently the reason why it was used for chronic rheumatism, as a plaster instead of MUSTARD (Rohde. 1926), perhaps as a counter-irritant? But a report in *Notes and Queries; 1935* shows a Welsh treatment of a very different kind. Like Poke-root berries in Kentucky, shredded horseradish was put in a bottle of whisky, which was then buried in the ground for nine days. The dose was three spoonfuls daily. Another method for the same complaint comes from Russian folk medicine. Equal amounts of horseradish juice and paraffin were mixed, to be used as a quick rub-down before going to bed (Kourennoff). CANDYTUFT seeds have long been a traditional remedy for the condition (O P Brown).

Like any poisonous plant, WOODY NIGHTSHADE was said to have great powers of evil. Such a witch plant would actually give people rheumatism! But at the same time it was used in medicine to cure the condition, as Lindley confirmed. The stalks had been introduced into medical practice by the German physicians and botanists in the 16th century, and were still being used in rheumatic or skin affections in the 19th century (Fluckiger & Hanbury). It was the year-old shoots, gathered in May or June, that were used, boiled in water, and drunk between meals over two or three days (Palaiseul), presumably as a sort of homeopathic dose. It appears in Thornton's herbal, too, as a recommendation for rheumatic swellings and a number of other complaints. Even a virulent poison like DEADLY NIGHTSHADE or MONKSHOOD has been used to combat rheumatism. In the former, a Warwickshire cure was to apply the bruised leaves (Savage), and in the latter case Russian folk medicine prescribed a liniment made from the dried roots: 4 ounces of the ground root was put in a quart of very strong vodka, and kept in a warm place for 3 or 4 days. After that it was strained, and ready for use. A large spoonful was considered the maximum for a single application, and if the patient had any heart trouble, no more than a teaspoonful. The liniment was rubbed in until the skin was dry, and then the area was covered with flannel (Kourennoff). Two or three drops of oil of JUNIPER on a sugar lump every morning is another recommendation for treating rheumatism (V G Hatfield. 1994). A decoction of the seeds of BUCKBEAN was once taken to treat, or even prevent rheumatism (Sargent). Even TOBACCO has been tried (Brongers); plasters could be made by damping the leaves, or even using cut-up pipe tobacco. They relieved the pain, and reduced the swelling, so it is said. The poultice had to be kept damp, usually by means of a wet bandage. Another plaster – a spice plaster, it was called, was made with ALLSPICE. The way to make it is to crush an ounce or so of whole allspice, and boil it down to a thick liquor, which is then spread on linen ready to be applied (A W Hatfield).

A poultice of SWEET VERNAL GRASS can be used, applied as hot as can be borne, or a bath with the grass added can be used, too (W A R Thomson. 1978). The American plant TWIN-LEAF (*Jeffersonia diphylla*) is also known as Rheumatism-root, for it provides a popular American remedy for the complaint (Berdoe), and it is said that Pawnee Indians used the powdered rhizome of JACK-IN-THE-PULPIT to cure their rheumatism and muscular pains (Corlett).

Rhodiola rosea gt; ROSEROOT

RHUBARB

(*Rheum rhaponticum*) To dream of handling fresh rhubarb is a sign of being taken into favour with those with whom you were not on good terms (Raphael). There is a very odd belief from Kentucky that if a woman wears a bag of rhubarb round her neck, her children will not have club-feet (Thomas & Thomas). Bury a stick of rhubarb here and there in the bed when planting cabbages, against club-root (Boland & Boland). In Oxfordshire, they say that the first rhubarb of the year should be eaten on Easter Sunday (*Oxfordshire and District Folklore Society, Annual Record; 1955*).

Fenland midwives used to give a "pain-killing cake" to women in labour. It was made from wholemeal flour, hemp-seed crushed with a rolling pin, crushed rhubarb root, and grated dandelion root. These were mixed to a batter with egg-yolk, milk and gin, turned into a tin, and baked in a hot oven. At the woman's first groan, a slice of the cake would be handed to her (Porter. 1969). Rhubarb roots boiled in a little water is an Irish (County Cavan) diarrhoea remedy (Maloney), and a root preparation is used to strengthen nails. Distilled water of rhubarb was recommended to remove scabs, to relieve earache, and as a gargle for sore throats. The seeds are supposed to ease stomach pains (Addison. 1985). You can cure a headache by applying a rhubarb leaf to the forehead (V G Hatfield. 1994), but one suspects that any large leaf would do (a cabbage leaf is certainly used so). It merely provides a cooling application. Rhubarb juice on a wart will cure it (Stout).

Rhus glabra > SMOOTH SUMACH

Rhus radicans > POISON IVY

Rhus trilobata > THREE-LEAF SUMACH

Ribes nigrum > BLACKCURRANT

Ribes uva-crispa > GOOSEBERRY

RIBWORT PLANTAIN

(*Plantago lamceolata*) Divination games were played at Midsummer (hence the Shetland names Johnsmas Flowers and Johnsmas Pairs, Midsummer Day being also St John's Day) in the north of England and in Scotland. The procedure differed a little here and there, but they were all basically the same; one of them involved taking three stalks of Ribwort, stripping them of their flowers, and putting them in the left shoe, and afterwards under the pillow. In the morning, if the lover was to become the husband, they should again be in bloom; if he were untrue, they should remain flowerless. That same game was known in the Faeroe Islands, where it was looked on as a simple wish fulfilment custom (Williamson). In Berwickshire, two spikes were taken in full bloom, wrapped in a dock leaf and put under a stone, or, in Shetland, buried in the ground (Banks). One spike represented the girl, the other the man. Next morning, if both were in bloom, it was a sign of true love between the two (Denham).

Cheshire children, when they first see the flower heads in spring, repeat the rhyme:

> Chimney sweeper all in black,
> Go to the brook and wash your back,
> Wash it clean or wash it none;
> Chimney sweeper, have you done?

It is probably a good luck charm (Dyer. 1889). Chimney Sweeper, or Sweeper, are widely distributed as local names for the plant (Grigson. 1955; Dyer. 1889; A E Baker). So is Sweep's Brushes (Macmillan, Britten & Holland; A E Baker).

The flower heads are used as "soldiers" or "fighting cocks" in a children's game. One description of the "Soldiers" game from Kintyre has one child holding out a duine dubh (black man), and his opponent tries to decapitate it with another. If one soldier takes the head off another, it is called a Bully of one, Bully of two, and so on (MacLagan). Hebridean children knew the ribwort stalks as "giants", but the game was played in exactly the same way. Girls find the "giants" useful for making daisy chains – they pick the daisy tops and string them on the tallest giants they can find (Duncan). The "soldiers" game is also known as Black Man (the duine dubh already mentioned), Cocks-and-hens, Hard Heads, or Knights (Opie & Opie. 1969). "Cocks", for the plant as well as the game, is the name in northern England (Brockett), and in Kent it is apparently known as "Dongers" (Mabey. 1998). The game had another significance in Somerset, where the plant is given the name (among many others) of Tinker Tailor Grass. The blow that knocks the head off marks the profession of the future husband, in the tinker, tailor, soldier, sailor tradition (Elworthy. 1888), a divination game also played with Rye-grass. The "soldiers" game is also known as Kemps in the northern counties and in Scotland. To kemp is to fight (OE campe, soldier, with similar words in Scandinavian languages). In some parts of Scotland, the game is called Carldoddie, probably from the names of Charles Stuart and King George (MacLagan) – carl is Charles (the Prince) and doddies were the supporters of King George, doddie being the local name for George.

Ribwort is a wound herb: "Plantain ribbed, that heals the reaper's wounds". The leaves are simply applied to the cut, and are used that way in Ireland, Scotland, and there is a record from India, too (Watt & Breyer-Brandwijk). A refinement to the technique is the traditional Irish remedy for stopping bleeding from a cut – by chewing the plant before applying it (O'Toole). An ointment for wounds using ribwort was in use in the 15th century: "take a pint of juice of ribwort, and a pint of vinegar, and a pint of honey, and

boil them together to the thickness of the honey, and keep it, for it is full precious" (Dawson. 1934). It must have been, for a pint of ribwort juice must have taken some collecting!

Ribwort is mentioned as a Highland remedy for boils and bruises (Grant), and in the west of Ireland, for a "lump" (Gregory. 1925). A leaf tea is used for bronchitis or asthma (Conway), and as a gargle it soothes sore throats (Schauenberg & Paris). A record from South Uist shows that the leaves were applied to relieve sore feet (Shaw). The seeds, left in water for two hours to swell, are a mild purgative (Flück), and a cold decoction of the plant was a Russian folk remedy for constipation (Kourennoff), but a leaf infusion was used in Norfolk for just the opposite effect – to cure diarrhoea (V G Hatfield). A similar preparation has been used for conjunctivitis, as an eyewash (Wickham). In earlier times, e.g., the Anglo-Saxon version of Apuleius, this plant was prescribed "for bite of snake", for a "quartan agus", and "for uselessness of the ears". As far as the snakebite remedy is concerned, it should be noted that Ribwort was given for hydrophobia in Ireland (Denham). Perhaps the most ambitious prescription comes from Russian folk medicine, for an infusion of ribwort seeds was taken for sterility (Kourennoff).

RICE
(*Oryza sativa*) It seems that sailors once thought that the regular use of rice as food would lead to blindness, and they called it Strike-me-blind (Brand/Hazlitt). To Hindus, Lakshmi is the goddess of wealth, and so rice is identified with the goddess. In rural areas, a basketful of unhusked rice is represented as the goddess herself (Gupta). Brahmans, when performing marriage ceremonies, after the prayers, consecrate the union by throwing rice flour coloured with saffron over the newly-married couple (Pandey).

Ricinus communis > CASTOR OIL PLANT

RICKETS
The split ASH (see entry under Hernia) ceremony has been performed for rickets in places as far apart as Norfolk and Jersey (Glyde; Le Bas). In Bavaria, a decoction of acorns, referred to either as acorn coffee or acorn cocoa, was taken as a cure for rickets (Rolleston).

An old name for WALL RUE was Tentwort or Taintwort, 'tent' being the name rickets was known as at one time (Coats. 1975). It follows, then, that this fern must have been used to treat the condition, in some way that has been lost now. Ô Súilleabháin quotes an Irish use of SHEPHERD'S PURSE for rickets.

RIMU
(*Dacrydium cupressinum*) A tall New Zealand conifer, with red timber, stained, in Maori mythology, by the

blood of the water monster, Tuna-roa, killed by Mairu (Andersen). Rimu, the name by which it is usually known, is what the Maori call it. They pulped the bark to apply to burns, and the gum, which, incidentally, is the reason for naming the genus *Dacrydium* (from Greek Dakrydion, a tear), which is a reference to the drops of resin (Leathart), was used to stop the flow of blood from a wound. A lotion for bathing wounds was made by cutting the bark into pieces and boiling them in water. Early settlers in New Zealand rubbed the juice from cut stems over bald heads; they found it an excellent hair restorer (C Macdonald).

RINGWORM

A manuscript of somewhere around 1680, from Lincolnshire, advised that "for a teter or ringeworme, stampe (LESSER CELANDINE) and apply it to the (griefe (?)) and it will quickly cure you (Gutch & Peacock). In Iowa, a tea made from TOBACCO juice was drunk to cure the condition (Stout). American Indians would use the decoction of the fruit of SMOOTH SUMACH as a wash for the complaint (Lloyd). Mexican Indians have used TREE CELANDINE leaves soaked in alcohol to bathe ringworm of the scalp (Kelly & Plerm). HOUSELEEK has been used, too, for this is a protector from fire and lightning, and so would be used against the "fiery" diseases, too. A gypsy remedy for ringworm is to boil houseleek, and then to dab the affected part with the water. Even POISON IVY has been used to treat the condition; various Californian Indian groups used a mash of the leaves for the purpose (Weiner). SUN SPURGE is another plant used in decoction for this condition (Dyer. 1889; Trevelyan), and so is GROUND CYPRESS (*Santolina chamaecyparissus*). In Portugal, the root of WHITE ASPHODEL is a specific against ringworm (Gallop), a treatment that seems genuinely beneficial.

Rivea corymbosa > MORNING GLORY

ROAST BEEF PLANT

A well-known name for STINKING IRIS. (Prior), given, so it is said, from the smell of the bruised leaf. On the other hand, the leaves "being rubbed, are "of a sticking smell very lothsome" (Gerard).

ROB

A fruit syrup, made from such fruits as WHORTLEBERRY. Gerard's recipe for this runs "the juyce of the blacke Wortleberries is boyled till it become thicke, and is prepared or kept by adding honey or sugar to it; the Apothecaries call it Rob". ELDERBERRY Rob was very popular, and taken as a cold cure, the dose being one or two spoonfuls in a glass of hot water (Sanecki).

ROGATIONTIDE

Celebrated on the Sunday, Monday and Tuesday before Ascension Day, the time for processions intended to establish annually the boundaries of the parish ('beating the bounds'). They were actually the last of the religious processions to survive, but the practices died out as the need for them ceased, i.e., as the lands were enclosed. MILKWORT (*Polygala vulgaris*) was carried in the processions; "the maidens which use in the countries to walke the procession do make themselves garlands and nosegaies of the milkwort" (Gerard). Cf the names Cross-flower, Procession-flower, Gang-flower, and Rogation-flower, all of them known to Gerard. But BIRCH leaves and branches were also prominent – Gerard again: "It serveth well to the decking up of houses and banqueting rooms, for places of pleasure, and for beautifying the streets in the Gosse and Gang weeks, and such like". 'Gang', OE gangen, to go, is another of the names given to the Rogation processions, others being Cross, Procession or Perambulation week, corrupted sometimes to Rammalation week. Birch served another purpose here – boys were whipped with them (and of course the boundaries were beaten) by tradition, to fix the boundaries in their minds for future years.

ROLLRIGHT STONES

(Oxfordshire) The legend tells that the biggest of the stones, called the King Stone, was once a man who would have been King of England if he could have seen Long Compton. A witch said:

> Seven long strides shalt thou take, and
> If Long Compton thou canst see,
> King of England thou shalt be.

On his seventh stride a mound rose up before him, hiding Long Compton from him. The witch then said:

> As Long Compton thou canst not see,
> King of England thou shalt not be,
> Rise up, stick, and still, stone,
> For King of England thou shalt be none.
> Thou and thy men hoar stones shall be,
> And I myself an eldern tree (A J Evans).

Though where exactly the witch-elder now stands seems to be a matter of conjecture. Nevertheless, the proof that the "elder is a witch is that it bleeds when it is cut" led to the custom, on Midsummer Eve, for people from the area to meet at the stones, and to cut the elder, or at least the tree they thought to be the real one. As it bled, the King Stone was supposed to move its head (Grinsell).

ROMAN WORMWOOD

(*Artemisia pontica*) It has a similar effect when used medicinally, though milder, as *A absinthium*, when used for colds, or as a tonic, or to expel worms (Sanford). "If a root of this wort be hung over the door in any house, then may not any man damage the house" (Cockayne).

Rorippa nasturtium-aquaticum > WATERCRESS

Rosa canina > DOG ROSE

Rosa pimpinellifolia > BURNET ROSE

Rosa rubiginosa > SWEET BRIAR

ROSARY PEA

One of the names given to PRECATORY BEAN (*Abrus precatorius*) in its capacity of prayer beads. Precatory, in the common name, is straight from Latin precatio, prayer. They are actually known as Prayer Beads (Grieve. 1931), and another name is Paternoster Bean (Howes).

ROSE

(*Rosa*) There are a number of different legends of the origin of the rose, a mirror of the popularity of the plant. In the East, they still say that the first rose was generated either by a tear, or, according to another legend, the sweat, of Mahomet. For that reason Moslems never allow rose petals to lie on the ground. There are two or three Greek myths about the plant. One of them says that the goddess Cybele, to take revenge on Venus, found no better way to do it than by creating the rose, whose beauty she likened to that of the goddess. Another story is again connected with Venus, who was born of the sea. At the same moment, the rose sprang for the first time on earth (Bunyard). Another Greek legend says that prior to the death of Adonis, all roses were white. Venus, while hurrying to Adonis, trod on a rose thorn, and some drops of her blood (or, as yet another version tells, that of Adonis) fell on the petals, and so that the plant became the parent of all the red roses. A Moslem version says the red colour was from the blood of the prophet, and so, of course, it is a sacred flower. Another Greek myth says it was Bacchus who dropped the wine on the rose, while an English legend has it that it was the blood of Christ that made white roses red. A German story tells of its becoming red after Adam and Eve had eaten the forbidden fruit. Another German legend says that all roses were originally red, but then Mary Magdalen wept for Christ, her tears fell on a rose and bleached the petals.

A Moslem version of the legend of how the red rose came into being incorporates the famous story of the rose and the nightingale. The flowers complained to Allah that the lotus, then queen of the flowers, slept at night, and they demanded a new queen. Allah answered by creating the white rose, and to protect it gave it thorns. The nightingale fell deeply in love with the new flower, and flying towards it, was pierced by its thorns, and so the white flower was changed to red.

Father Rapin, the seventeenth century Jesuit, quoted another origin legend connected with Greek mythology. It concerns a queen of Corinth named Rhodanthe, who was so beautiful that no-one could look on her without falling madly in love. She had countless suitors, and, wishing to escape them, she took refuge in a temple consecrated to Diana. But three of the suitors, bolder than the rest, followed her there and tried to abduct her. Rhodanthe defended herself and appealed to the people for help. They came, but were so overcome by her beauty that they cried that Diana was no longer goddess of the temple; from then on they would pay homage to the beautiful Rhodanthe. They were on the point of overturning Diana's statue when Apollo, furious at this outrage against his sister, appeared. He turned Rhodanthe into a rose tree, and changed the three suitors into a worm, a fly and a butterfly.

There must be no end to these tales of the origin of the rose. There is yet another that tells of the rose's thorns. When first created it had no thorns, but as man's wickedness increased, so the plant grew them. Others say that the thorns appeared only after the appearance of Adam and Eve. Finally, there is a story in Mandeville's *Travels* about a Bethlehem girl falsely accused of unchastity, and sentenced to the stake. She called on God to help her and to demonstrate her innocence in the eyes of all. Then she stepped into the fire, and the flames at once died down. Those brands that were already burning changed into red roses, and the others into white.

A familiar piece of Greek mythology is that Eros presented a rose to the god of silence, for the rose has been a symbol of silence since very early times, a belief still with us in the expression *sub rosa*. It is said that many of the great houses with plaster ceilings had a rose as the central ornament, a reminder that matters talked of at table "under the rose" must not be repeated outside that room. A rose is often out over confessionals in Catholic churches, for the same reason (Ingram). But to be born "under the rose" is said to mean being illegitimate, a rose being also a symbol of secrecy, so the wild rose is sometimes used to signify illicit love (Briggs. 1974). To dream of red roses was said to foretell success in love (Higgins); and florists say that more red roses are sold for St Valentine's Day than for any other day of the year (Baker. 1977).

With all this love symbolism, it is hardly surprising to find roses connected with divinations. Girls would take a rose leaf for each of their suitors, and name each leaf after one of them. Then she would watch them until one after another they sank in the bucket of water into which she had put them. The last that sank would represent the future husband (Napier). Or she could carefully pick a rose and lay it under her pillow on Midsummer Eve, when the future husband would appear in a dream (Higgins). Rather more ambitious is the one that requires the girl to pick a rose at Midsummer, fold it in paper, and put it by till Christmas Day, when she would wear it to church, and the man who would come and take it from her would be her husband (Opie & Tatem).

Roses have connections with death as well as with love. As far back as ancient Greece, they were used at funerals, and tombs were decorated with them, under the belief that they protected the remains of the dead. The Romans also believed this, and they even celebrated a festival of the dead under the name of Rosalia, forerunner of the Roumanian Rousalia (Knight). It was a Welsh custom to plant a white rose on the grave of an unmarried woman (Trevelyan), and it was the practice in parts of England for a young girl to carry a wreath of white roses, and walk before the coffin of a virgin. The wreath would be hung in church after the funeral, above the seat that she had used during her life, till the blooms faded. The church at Abbot's Ann, in Hampshire, has its walls hung with "maidens' garlands" of paper or linen roses; the earliest of them dates from 1716 (Mayhew). It is usually a rose or roses that grow from the graves of two lovers, and they will intertwine. Evelyn mentions "the custom not yet altogether extinct in my own county of Surrey, and near my dwelling, where the maidens yearly plant and deck the graves of their defunct sweet-hearts with rose bushes". The custom was so common in Switzerland that cemeteries were often known as rose gardens there (Mayhew). So great is the connection between roses and the dead that apparently spirits at seances are using roses now as symbols of affection and goodwill (Knight).

If a white rose blooms in autumn, it is regarded in the west of Scotland as a token of an early death for someone. But if a red rose blooms then, it is a sign of an early marriage (Napier). On the other hand a red rose whose petals fall while it is being worn or carried, is death omen (Waring). A Welsh belief was that a summer rose blooming in November or December was a sign of trouble and bad luck (Trevelyan). Throwing rose leaves on a fire brings good luck, according to some (Napier), but the Italian belief was that to scatter them on the ground, more especially those of a red rose, was very unlucky, and would result in an early death (W Jones). For dreaming of roses, see above; dreaming of withered roses means misfortune – what else could it mean? Kentucky superstition insists that dreaming of flowers of any kind means bad luck, but especially so of white roses (Thomas & Thomas). A Dorset belief was that smelling a white rose would be injurious to health, but smelling a red one beneficial (Udal).

The provision of a rose as a condition of tenure was quite widespread. Rent day was often Midsummer Day, as with the Crown & Thistle in Loseby Lane, Leicester. Under a deed of 1626, an annual rent of two pennies and a damask rose is still paid. But the time of payment varied. Some land at Wickham, County Durham, for example, was held by the service of one rose at Pentecost "si opetatur" (if required),

and, rather more difficult, the manor of Crendon, Buckinghamshire, was held by the service of one chaplet of roses at Christmas (Blount). Hungerford has a condition of presenting a red rose to the reigning sovereign whenever he or she passes through the town (Berkshire FWI).

The rose was used in Germany as a charm against haemorrhage of every kind. It is not clear what the colour of the rose had to be; presumably it had to be red, in which case the charm must have sprung originally from the doctrine of signatures. The virtue of a charm, though, lay in the words spoken, and there are several versions in this case. One was "Abek, Wabek, Fabek; in Christ's garden stand three red roses – one for God, the other for God's blood, the third for the angel Gabriel; blood, I pray you, cease to flow". Another version ran "On the head of our Lord God there bloom three roses; the first is his virtue, the second is his youth, the third is his will. Blood, stand thou in the wound still, so that thou neither sore nor abscess giveth" (Dyer). One very strange charm was used in order to find the sex of an unborn child. All one had to do was take a lily and a rose to a pregnant woman. If she chose the lily, it would be a boy; if the rose, a girl (Cockayne).

ROSE BAY
(*Nerium oleander*) see OLEANDER

ROSE-NOBLE
A name given to plants like WATER BETONY and FIGWORT, as well as to Hound's Tongue and Lesser Celandine. The first two belong to the genus Scrophularia, from Latin scrofulae, which means that swelling of the neck glands we know as scrofula, once known in England as the King's Evil, for the touch of the anointed monarch was taken to be the only cure for the condition. Rose-noble was the name of an English gold coin, with the figure of a rose on it. More importantly in this context it would have the king's head too, and so the implication is that the touch of the coin was a substitute for that of the king himself.

ROSE OF CHINA
(*Hibiscus rosa-sinensis*) All over the tropics, women use the flowers for adornment, and Hawaiian welcome necklaces are made from them (Perry. 1972). But among the names given to the flower, some are distinctly odd-looking, like Shoeflower (Wit), used in Jamaica, or Shoe Black (Campbell-Culver), or Blacking Plant (J Smith. 1882). The truth is that the red of the flowers becomes black when they are bruised, and can then be used for blacking shoes (or for colouring the eyebrows). In India, though, it was thought to be a medium for witches' evil-doing. It was quite common to see the flowers put at crossroads, etc., so that people who touched them would become possessed by evil spirits (Upadhyaya).

ROSE OF JERICHO

(*Anastatica hierochuntica*) A white-flowered plant from North Africa to the Middle East, brought to Europe as an emblem of the Resurrection, or as a symbol of new life (Crowfoot & Baldensperger). The Greek word anastasis, from which the generic name is taken, means resurrection, and the plant is known as Resurrection Plant (Perry. 1972). The reason is that as the seeds ripen during the dry season, the leaves fall off and the branches curve inwards to make a round lattice-type ball which is blown out of the soil to roll about the desert until it reaches a moist spot, or until the rainy season begins. (Cf the American TUMBLEWEEDS).There are records of its use in sympathetic magic when there is a difficult birth. They put it in water, so that it will slowly open, thus sympathetically bring about the same result for the woman in labour.

The name Rose of Jericho is confusing, for it is not a rose, and neither does it grow in Jericho, in spite of the specific name, which affirms it is from that city. But it is also known as Rose of the Virgin (Ackermann), or Mary's Flower (Moldenke & Moldenke), because of a legend that tells that all the plants of this species expanded, became green, and blossomed again at the birth of Jesus, and still do so in commemoration. More, they were supposed to close at the Crucifixion, and to blossom again at Easter (Perry. 1972). Actually, were not all plants supposed to have opened up and blossomed when Christ was born? (see A Huxley).

ROSEBAY WILLOWHERB

(*Chamaenerion angustifolium*). The leaves are edible, and can be eaten as a vegetable, Linnaeus actually recommending that the young shoots should be served like asparagus (Grigson. 1955). Some North American Indian groups boiled the young leaves to make a beverage tea, and the roots, also edible, would be split open and eaten raw (Turner & Bell). The leaves are used in Russia to adulterate ordinary tea, the result being called Kapoorie Tea (Leyel, 1937, Usher). They are also used to flavour beer, and a sweet tisane can be made from the dried flowers (Schauenberg & Paris). In Greenland, apparently, the leaves were used copiously to spice seal blubber (Swahn).

A root decoction has been taken as a cure for whooping cough, and asthma, too. In America, there are records of aboriginal use for various ailments; the Menomini used the roots to make a wash for swellings (H H Smith. 1923). And the Ojibwe used this for boils and carbuncles (H H Smith. 1945).

Rosebay has been called Fireweed in America, for its seeds are almost the first to colonise an area cleared of vegetation by fire, etc., (House). An interesting development on this is the name given over south-east England, Bombweed (Mabey. 1998), from the way it so quickly colonised bomb sites during World War 2. Perhaps this characteristic is the reason for its assignment to 'pretension' in plant symbolism (Leyel. 1937).

Mother-die is one of the names given to Rosebay (Vickery. 1995; Opie & Tatem), and shared with many other plants (see MOTHER-DIE). Any plant bearing this name is taboo to children. They must never pick them, for breaking that taboo would be to bring about the death of the child's mother.

ROSEMARY

(*Rosmarinus officinalis*) Tradition has it that rosemary was introduced into Britain by Philippa of Hainault, Edward II's queen, though it probably took place much earlier than that. One belief is that it never grows higher than the height of Christ during his earthly life; after 33 years the plant may increase in girth but never in height (Rohde). Another legend tells how, during the flight into Egypt, the Virgin threw her cloak over a rosemary bush while she rested beside it. For ever afterwards the flowers, which had been white, turned the blue colour of her cloak (Rohde). There was certainly a definite association between rosemary and the Virgin Mary (Baker. 1980), if only by a misunderstanding of the common name. But that was a good reason why it was always taken to be such a powerful "disperser of evil" (Baker. 1978), and that is why it is planted near the house, so that "no witch could harm you". A remedy for illness caused by witchcraft "used and prescribed by the cunning man was to put rosemary, balm and many gold flowers in a bagg to the patients brest as a charm and to give them inwardly a decoction made of the same in a quarrt of ale and their own blood …" (taken from a deposition to the Assizes, Leicester, 1717 (Ewen). Like the Glastonbury Thorn (in fact) rosemary was thought to bloom exactly at midnight on the eve of Twelfth Day – Old Christmas Eve, that is (Dew).

Rosemary is the symbol of remembrance. Shakespeare has Ophelia say "there's rosemary – that's for remembrance", and earlier, Sir Thomas More wrote: "As for rosemarine, I lett it runne all over my garden walls, not onlie because my bees love it, but because 'tis the herb sacred to remembrance, and therefore to friendship; whence a sprig of it hath a dumb language that maketh it the chosen emblem of our funeral wakes and in our burial grounds". Its use by mourners at funerals made it a token to wear in remembrance of the dead, and in the 1930s there was a demand for rosemary for Armistice Day ceremonies (Rohde), along with the more conventional symbolic poppies. In the north of England, it was a "breach of decorum" for a mourner to attend a funeral without it. It was also the custom for a table to be put by the door, on which would be put the sprigs of rosemary and box (another funerary token), for each mourner to pick up as he came into the house (Vaux). The Lincolnshire custom was sometimes to put the rosemary on the breast of the corpse, and it was buried with him (Gutch & Peacock), just as in France it was once customary to put a branch of it in the hands of the dead (Thompson. 1897). Around Northwich, in Cheshire, mourners were given funeral biscuits, wrapped in white paper and sealed with black sealing wax. A sprig would be tucked in the folds of the paper, and this was thrown on to the coffin as it was lowered into the grave (Hole. 1937). Not surprisingly, given the symbolism, rosemary was also reckoned to stimulate the memory, and was prized as such (Macleod). Wear rosemary on the person, and memory will be strengthened, wrote a Derbyshire correspondent of *Notes and Queries* in 1871, and in the 1930s Mrs Leyel was still saying that rosemary tea, besides curing a nervous headache, would, by constant use improve a bad memory (Leyel. 1937).

Rejuvenation is also symbolised by rosemary, at least in the French language of flowers, where it stood for the power of re-kindling lost energy (Rohde). Bancke's Herbal of 1528 carried the advice "make a box of the wood and smell it and it shall preserve thy youth" (Arber). It was with rosemary that they tried to wake the Sleeping Beauty (Rohde). Yet another of rosemary's symbolic virtues is that of fidelity in love. As such, it was appropriate as a gift from the bride to the groom. "Rosemary bound with ribbons" was a token of a bride's love for her husband (Higgins), and a favourite valentine used to be a sprig of rosemary painted on a heart (Webster). Rosemary was worn at weddings, and appears to have been considered as the insignia of a wedding guest; on these occasions the sprigs were often gilded, and dipped in scented water (Andrews). Polish brides and bridegrooms wore a tiny wreath of rosemary on their head. It was blessed by the priest and carefully kept, to be boiled in the water to be used as the first baby's first bath (Kennedy). Sprigs of rosemary in the bride's bouquet signified happiness (Webster), and it was also woven into the bridal coronet. Anne of Cleves, when she arived at Greenwich as a bride, wore "on her head a coronet of gold, and precious stones, set full of branches of rosemary". Gerard described the plant under the name *Rosmarinus coronarium*, "because women have been accustomed to make crowns and garlands thereof". As well as the coronet, the flowers, tied with coloured ribbons, were put in the bride cup and carried ceremonially before her (Rohde). It was also strewn before her, and a sprig dipped in the wine before the health of the bride and groom was drunk (Leyel. 1937). In addition, young men in Wales used to put a large bunch of rosemary, tied with white ribbons, at the bedroom window of a girl they admired, as a May morning tribute (Roheim). Rosemary, then, was obviously a fertility agent. As late as 1700, country

bridal beds were decked with it (Baker. 1977), and the Welsh medical text of the Physicians of Myddfai prescribed it as a remedy for barren-ness. Hartland reported that it still had that reputation in Belgium in his day. At Hildesheim, women were struck with branches of rosemary at Shrovetide, much as in the Roman festival of the Lupercalia. The plant also appears in marriage divinations. For example, on St Agnes' Night, take a sprig of rosemary and one of thyme, and sprinkle them three times with water. In the evening put one in each shoe, putting a shoe on each side of the bed. When ready to go to bed, say:

St Agnes, that's to lovers kind,
Come ease the trouble of my mind.

The future husband will appear in a dream. So wrote Halliwell. 1869, as an example of a charm from the north of England and Scotland. The rhyme, though, has been quoted also as:

Agnes sweet and Agnes fair,
Hither, hither, now repair;
Bonny Agnes, let me see
The lad who is to marry me (Drury. 1986),

if we can believe that a country girl could use such language. Derbyshire girls used to put a sprig of rosemary and a crooked sixpence under their pillows at Hallowe'en, so that they should dream of their future husbands (Addy). "If you lay a branch of rosemary under your head, on Easter Eve, you will dream of the party you shall enjoy", so says an 18th century chapbook (Ashton). Another kind of divination was in fashion in Herefordshire. There they put a plate of flour under a rosemary bush on Midsummer Eve. The initials of the intended lover were expected to appear in the flour next morning (Leather). The bad atmosphere created by marital quarrels could be exorcized by burning rosemary in the house, as was the custom in parts of Portugal (Gallop).

It is odd to find one herb to have marriage associations, and at the same time to be a funerary plant; there is a story that Hartland has of a widower who wished to be married again on the day of his former wife's funeral, because the rosemary used at the funeral could serve at the wedding too. That reminds one of Herrick, *Hesperides*:

The Rosemarie branch
Grown for two ends, it matters not at all,
Be't for my Bridall, or my Buriall.

There are many other superstitions concerning rosemary. In Guernsey, it was thought unlucky not to have one in the garden, but it should never be bought, or stolen, but grown for you and presented by a friend (MacCulloch), preferably with a gracious speech, something on the lines of "I give you this rosemary gladly, and I hope it will grow with you" (Baker. 1978). In America, it is good luck to dream of it (Whitney & Bullock). Another common superstition is that where rosemary grows, the woman rules, or "the grey mare is the better horse" (Hone). Nevertheless, Dr Hackett, in 1607, wrote: "Rosmarinus is for married men; the which, by name, nature and continued use, man challengeth as properly belonging to himself. It overtoppeth all the flowers in the garden, boasting man's rule ..." (quoted in Tyack). That, though, is very much a minority view, the consensus being that "it only grows where the missis is master".

North country people in England say that a sprig of rosemary in the buttonhole will give the wearer success in all his tasks, and of course will particularly help his memory (Waring). There are plenty of beliefs about it in Wales. It was used there for chronic drunkenness, for example, and for this reason an infusion of it was often put in a cask of beer. It also kept the beer from souring (Trevelyan). Welsh people also thought that the smell of the burning bark could release a person from prison, that the leaves pressed and applied as a poultice kept wounds from running, that a spoon made from the wood would make whatever was eaten from it nutritious, and that when put under a doorpost it would keep snakes away (Trevelyan).

There was an old belief that rosemary made people merry. Banckes' Herbal of 1528 has " ...take the flowers and make powder thereof and bynde it to the ryfgt arm in a lynen clothe, and it shall thee bryght and mery". Langham, in 1579, repeated the belief – "carry powder of the flower about thee, to make thee merry, glad, gracious, and well beloved of all men". Banckes went on with more practices, the strange mixed with the practical: "take the floures and put them in a chest among your clothes or among bokes, and moughtes [moths] shall not hurt them ... Also boylle the leves in whyte wyne and wasshe thy face therewith – thou shall have a fayre face. Also put the leves under thy beddes head, and thou shall be delivered of all evyll dremes ...", and so on.

Essence of rosemary was used for scenting; it smelt, accrding to Rimmel, very like camphor, but one perfume became extremely well known. In 1709, an Italian chemist named Farina concocted a mixture of orange, alcohol, bergamot, lemon oil and rosemary. This proved very popular with the Germans, and Farina marketed it under the name Kölnisches Wasser – later known as eau de Cologne (Wykes-Joyce). Rosemary was the chief ingredient of Hungary water, too. The name came from Elizabeth, Queen of Hungary, whose use of it is said to have had such a rejuvenating effect that the King of Poland wanted to marry her (she was 72). She refused him (Leyel). A hair rinse for dark hair is another toiletry for which the herb is used. It is even claimed that it prevents greying. It is

still used to good effect as a shampoo to eliminate dandruff (Hemphill), and it is an ingredient in many current hair restorers (Porter. 1969). A manuscript of 1610 advised rosemary water for baldness – "if thou wash thy head with the ... water, and let it drye on agayne by itselfe, it causeth hayre to growe if thou be balde" (quoted in *Gentleman's Magazine Library: popular superstitions*). Equally optimistic was "a wash to prevent the hair from falling off", noted in the *Housekeeper's and Butler's Assistant* of 1862; it required "a quarter of an ounce of unprepared tobacco leaves, two ounces of rosemary, two ounces of box leaves, boiled in a quart of water in an earthen pipkin with a lid, for twenty minutes ...". A pomade was made from rosemary, too. In the Fen country, it was picked in April and May, pounded with the kidney fat of ewes that had died lambing, and then strained into pots, to be used on the hair when required (Porter. 1969).

Rosemary tea, made simply by pouring boiling water on to a handful of the leaves and flowers, is recommended for a number of ills, particularly for a cold, to clear the head (Rohde), but also for headaches, "tremblings of the limbs, and all other nervous disorders" (J Hill), or for weak hearts (Mrs Wiltshire called it a "supreme heart tonic"). It is also claimed to be good for the liver. Coughs can be treated, according to an 18th century prescription from Anglesey, by drinking the result of boiling up the herb with a quantitiy of Liverpool ale with honey and salt butter (T G Jones). Headaches are cured by rubbing the forehead with a handful of the herb (Newman & Wilson), as well as by drinking rosemary tea.

There seems to be some belief in rosemary as a general protector in all diseases, even if it is only treated as a charm rather than a drug. In the Balkans, for example, a doctor in Sarajevo told Kemp that women would throw down a sprig of rosemary as a protection against the doctor, and therefore all the diseases with which he was associated. In Brazil, too. its virtues are well appreciated in healing ceremonies. It is believed that anyone passing near a bush should stop and savour its scent:

> Whoever passed by the rosemary
> And did not smell it
> If he was ill
> Became worse (translation by P V A Williams).

Gerard, of course, was enthusiastic about its powers. It is given, he said, "against all fluxes of bloud; it is also good, especially the floures thereof, for all infirmities of the head and braine", and he cites Dioscorides as his authority for using it for jaundice. He recommended it for a lot more ills, from "stuffing of the head, that commeth from coldness of the braine", to halitosis, and to comfort the "weake and feeble braine in most wonderfull manner". Not least of the marvels assigned to rosemary was its supposed powers against the plague. It was an old custom to burn it in sick chambers, for this reason (Grieve. 1933), and there is the oft-quoted passage from Thomas Dekker, *Wonderful Yeare*, 1603, speaking of persons apprehensive of catching the plague: "They went (most bitter) miching and muffled up and downe, with rue and rosemary and wormwood stuft into their eares and nosthrills, looking like so many bores' heads with branches of rosemary to be served for brawne at Christmas".

ROSEROOT

(*Rhodiola rosea*) When the root is cut, a rose-like smell is given out, and rose-scented water can be made from it (Genders. 1976). It was used in the Highlands of Scotland to treat malignant ulcers, and also scurvy (Fairweather).

Rosmarinus officinalis > ROSEMARY

ROUGH CHERVIL

(*Chaerophyllum temulentum*) It was believed in Wales at one time that if children ate chervil (identified as this plant, but it may very well be a mistake for true chervil), they would become epileptic, and see everything double, or blurred (Trevelyan).

ROWAN

(*Sorbus aucuparia*) A very familiar small tree, with creamy flowers, and, more important in folklore, scarlet berries, which were the food of the Tuatha de Danaan, mythical early inhabitants of Ireland; they had the sustaining power of nine meals. They healed the wounded, and added a year to a man's life (MacCulloch. 1911). In the romance of Diarmuid and Grainne, the berries are described as the food of the gods. Possibly, the berries were once used in the brewing of an intoxicating drink – that would explain the ancient idea that they were the food of the gods, who jealously guarded the berries until outwitted by a human (Rhys. 1892).

In Irish belief, the first woman sprang from a rowan tree, the first man from an alder (Wood-Martin). It was a holy tree in Scandinavian mythology, too, sacred to Thor, and called Thor's Helper, because it bent to his grasp while he crossed a flooded river in the Vimur (Farrer); hence the proverb which translates "the rowan is the salvation of Thor". Possibly, the wife of Thor was once conceived in the form of a rowan, to which the god clung (Turville-Petrie). It was long the custom for shipbuilders to put at least one plank of rowan into the hull of every ship, in the belief that Thor would look after his own (Skinner). In Scotland, it was put up as a protective charm, because it "cannot be removed by unholy figures" (Banks). Everywhere, the rowan is a sacred tree, and nowhere more so than in Iceland (Turville-Petrie).

Indo-European tradition ensures that the rowan's red berries make it a lightning plant – any plant with red berries qualifies, but rowan perhaps more than any other is the embodiment of the lightning, from which the tree is sprung (Dyer.1889). Thus it is also a prophylactic against lightning (Graves). In Ireland, a twig of rowan would be woven into the thatch, so protecting the house for a year at least (Wilde. 1902). Much better known is rowan's protective role against witchcraft – it was the most powerful antidote known in the British Isles. There is a saying recorded in Herefordshire: "the witty [rowan] is the tree on which the devil hanged himself" (Leather). The need for it was felt most around the farm, where the stock in particular was judged to be at risk. The peg of the cow shackle would be made of rowan, so would the handle and cross of the churn-staff. Collars of it were made for securing the cattle at night, and this collar would be passed three times round the chimney crook each time the cow visited the bull (Campbell). Strathspey people made a rowan hoop, and they made all their sheep pass through it every Samhain and Beltane (1 November and 1 May) evening and morning (Ramsay). In Germany, rowan twigs were put over the stables (Dyer. 1889), and milkmaids in the Highlands used to carry a rowan cross, usually fastened with red thread (see below), and put between the lining and cloth of their dress, "against any unforeseen danger" (Stewart). No less an authority than King James VI and I recorded in his *Daemonologie* the charming of cattle "from evill eyes by knitting rountree to the haires and tailes of the goodes" [i.e., cattle]. Pigs in Lincolnshire used to have a rowan garland hung round their neck when they were to be fattened, to prevent their being bewitched (Gutch & Peacock). To restore a cow's milk, a rowan stick would be used to stir the boiling of whatever milk the cow still had, while pins, and, at boiling, rusty nails, were thrown into the pail (Cromek). On May Eve on Islay, tar was put on the horns and ears of the cattle, and rowan berries on their tails, in order to protect them from the evil eye (Banks). Rings were made of it, for amulets (Owen), and rowan switches were used for driving cattle and horses (Burne. 1914), for they could never injure any of the animals, no matter how hard they were struck. Before the cows were put out to pasture for the first time after calving, a red thread, and sometimes a rowan cross, were tied round their tails (McPherson). That rowan cross (in the shape of a St Andrew's cross) was cut each summer to put over the cowhouse door (Macdonald), and one was put over house doors at Lammas old style, 12 August, that is, in Aberdeenshire. Here, it had to be done in secret and in silence (Banks). 2 May was another special day, called in Yorkshrie Witchwood Day, when pieces of rowan were ceremonially cut, taken home and fixed over the door. One of J C Atkinson's

Yorkshire parishioners told him that several pieces of "witchwood" were needed for proper protection – one for the upper sill of the house door, one for the door of the stable, and cow byre, etc., one for personal use, one for the head of the bed, etc., They must be cut on St Helen's Day (18 August or 8 October?), and must be cut with a household knife, and from a tree the cutter had never seen or knew of before. They must be taken home by a completely different route from that used to start. Kilvert noted in his diary that "this evening being May Eve [1870] I ought to have put some birch and wittan [rowan] over the door to keep out th' old witch'. But I was too lazy to go out and get it …".

Gypsies and showmen always had a piece of it incorporated in the caravans they had built for them, "to keep the devil away" (Blount), and people would put a piece on the pillows to keep evil spirits and witches away (Porteous). In Ross and Sutherland, branches of rowan decked with heather used to be carried by the young men round the May festival fires (Banks), while in other parts of Scotland branches were ceremonially burnt in front of houses to keep witches away (Simpkins). One of the Irish beliefs was that a witch could use the first smoke from a chimney on May morning to bring bad luck to a house. The safeguard was to put a bunch of rowan leaves up the chimney to dry, and set fire to them first thing on May morning. So it would be the smoke from the rowan that emerged from the chimney, and "the witches can do nothing with it" (Duncan). In Ireland, too, a rowan branch was mixed with the timber of a boat, to ensure that no storm would upset it and that no-one on board would be drowned for at least a year (Wilde; Ramsay). Scots fishermen would tie twigs of rowan to their lines, and to the thole-pin of their oars (Anson). Bassett has a statement that on the face of it looks contradictory: "in Iceland, it is deemed unlucky to use the wood of a cetain tree, called sorb, in building ships. When used for such purposes, the ship will sink, unless willow or juniper wood is used to counter its ill infuences, as they are inimical woods". Highlanders agree with the last bit, and say that no boat should go to sea if there is anything made of rowan on it, or juniper – the hate the two woods have for each other would be enough to split the boat and sink it (Swire. 1963).

Getting back to more familiar ideas, there is a Yorkshire story of a witch who stopped a lad's plough-team in the middle of a field. But the boy had a wicken-tree whip-stock, with which he touched each horse in turn, and so broke the spell (Gutch). There is a proverb from the same area, "woe to the lad without a rowan-tree gad", and a rhyme:

> If your whipstock's made of rowan
> You may ride your nag through any town
> (Denham).

When a cart was stopped by magic on a highway, it could be freed by thrashing the wheels with a whip or branch of rowan (Hole. 1947). Similarly, a Derbyshire story tells of a man who came to a toll bar one night. The old woman who kept the gate would not open up for him, so he struck the gate with the rowan stick he was carrying, and the gate immediately flew open (Addy. 1895).

No evil spirit could ever pass a rowan tree or even a piece of the tree. So according to the tales, anyone on a dangerous errand would fill his pockets with pieces of the bark (Grice). If the tree were what Runeberg called a "flying rowan", by which he meant a rowan growing parasitically on another tree, so much the better, for the power is especially concentrated in such a plant. You only had to shake a branch at witch-cats to make them disappear. One of the many stories of a man trying to shoot a witch hare, in which he is usually advised to load his gun with a silver sixpence, in a version from Bedd Gelert has the advice to put a small branch of mountain ash and a bit of vervain under the stock of the gun (D E Jenkins).

The red (anti-lightning) association is carried further in the anti-witchcraft properties:

> Rowan-tree and red thread
> Put the witches to their speed

is well-known, but there are variations. In Teesdale, there used to be a line before the couplet:

> Black luggie, lammer bead.

The black luggie was a small wooden dish with only one handle, out of which children took their porridge. "Black", because it was made of bog oak. Lammer beads are amber beads, the initial preservative against a variety of diseases, particularly asthma, dropsy and toothache (Brockie). There are other variants – Chambers, for instance, has:

> Rowan-tree and red thread
> Make the witches tyne [i.e., lose] their speed.

From Northumberland there is recorded:

> Rowan-tree and red thread
> Haud the witches a' in dread (Denham).

From the Northumberland ballad "Laidley Worm":

> The spells were in vain, the hag returned
>   To the queen in sorrowful mood,
> Crying that witches have no power
>   Where there is a row'n tree wood.

Honeysuckle was linked with the rowan in the Highlands in the same way. The honeysuckle twig would be wound round the ran-tree (Milne), and they would be put over the byre-door inside:

The ran-tree and the widd-bin
Haud the witches on come in (McPherson).

Holly, too, was associated with rowan in Scotland. Branches of either were put over the door to prevent nightmares (Sebillot). Both of these have red berries, too, but in Germany, mistletoe was used with rowan to protect the cattle (Macdonald). Red thread was wound round the rowan on Rowan-tree Day (2 May) in the northern counties of England – in fact, dozens of yards of it would be used round a branch, which was then hung in the window (Denham). In Scotland, quarter-days, being particularly unlucky, meant that the house had to have the special protection that hanging some rowan on it could give (Banks).

In some parts of Ireland, a sprig of rowan is stuck upright in the midden, sometimes decorated with May flowers (Buchanan), (or in the four corners of the cornfield (Wilde. 1852)) on May Eve to protect the farm – the muck symbolizes the fertility of the farm (E E Evans. 1957). The dunghill's fertilising powers can be increased by making a thick porridge and stirring it with a rowan stick, then burying the stick with the porridge clinging to it, in the top of the midden (on Maundy Thursday, apparently (Banks)). Coincidentally, it used to be claimed in Germany that in a year when there were many rowan berries or hazel nuts (much more widespread), there would be many children born (T R Forbes. 1966).

Yorkshire people used to distinguish between heder-wicken and sheder-sicken. Heder and sheder, usually applied to lambs, mean simply male and female. The heder variety would be used for ailments caused by male witching, the sheder for those attributed to female witches (Davidson). "There's heder wicken, and there's sheder wicken, one has berries, and the tother has none; when you thought you were overlooked, if the person was he, you get a piece of sheder wicken; if it was she, you got heder wicken, and made a T with it on the hob, and then they could do nowt at you" (Cole). "There's a special twig on a Witchin Tree, which, if you can get it an' keep it allers on you, then you can *witch* as well as bein' safe from *bein' witched*" (Rudkin). The reason why witches shunned any contact with the rowan is given in the Wilkie *MS* as "a witch who is touched by a branch of rowan by a christened man, will be the victim carried off by the devil when he comes next to claim his right" (McPherson). It is as efficient against the evil eye as against witchcraft. A sprig of rowan tied to a cow's tail was as often as not to protect it from overlooking as against downright witchcraft (MacLagan). And it was just as efficacious against the dead – for instance, a stake of this wood hammered through a corpse would immobilize its ghost (Graves). Ô Súilleabháin too notes its use to keep the dead from rising. Coffins were made of it in some places (Greenoak), and a rowan tree used to stand in most

Welsh churchyards, like the yew in England. There is a tale from Glamorgan that tells us that a fiddler could exorcize a ghost, if his fiddle had a back made from rowan wood (Roderick). Another connected use at one time was to ensure sound sleep – a piece of the wood was put over the bed (Harland & Wilkinson). Of course, nightmares were caused by witchcraft, or at least by evil spirits (Gutch).

When cock-fighting was popular in Wales, it was the custom to put a few crossed twigs, or a ring, of mountain ash in the cockpit, to prevent any evil ill-wishing to affect the courage of the birds (Trevelyan). It makes the best rod for a fisherman. If he takes with him:

Ragged tackle,
A stolen hook
And a crooked wicken-rod,

he is most likely to be in luck (Campbell). In Ireland, it was used to make a collar for a hound, to give it greater speed (Ô Súilleabháin). Divining rods were made of it in Yorkshire and Lancashire (Grigson. 1955), though hazel was usually preferred. But it is still its protective nature that is all-important, against evil, witchcraft, the dead, and the fairies, too. Craigie quotes a Scandinavian story in which a rowan tree not only protected a boy from the trolls, but did active damage to them. There is, too a folk-tale from Ulster which tells of a woman carried off by the fairies. She was able to inform her friends that when she and the fairies were going on a journey, she would be freed if they stroked her with a rowan branch (Andrews). Similarly, it is said that if someone got into a fairy circle, he or she would remain there for a year and a day. After that, the enchanted person could be liberated by someone holding a rowan stick across the circle (T G Jones). One Scottish way of getting rid of a changeling was to build the fire with rowan branches, and to hold the suspected changeling in the thick smoke of the fire. The brat would disappear up the chimney, and the true child would be returned (Aitken). However, there is another belief that good fairies are kind to children who carry rowan berries in their pockets (Skinner).

Other folk tales and ballads carry on the theme. There is a Yorkshire legend known as The Maid of the Golden Shoon, a rambling, complex affair about the fight against three witches who were stealing infants in mysterious ways. The wise man gave the villagers instructions that included several references to the protective anti-witchcraft nature of rowan. They were, for example, told to build a bower for the Maid of the title, who was to come to their rescue. It had to be "woven together in the thickest part of the wood, and when finished, not a glimmer must enter". There must be wicken-wood worked into the entering place, obviously as a protection. At the same time, they were told that a magic sign had to be written, "with charcoal from burned wicken-wood", by the house of the witch the Maid had to combat. There are more examples in this story of the part that rowan was to play, but this tale is full of magical symbols. Rowan's protective ability is again made plain in the Cromarty legend of Willie Millar, who went to explore the Stopping-cave. "He sewed sprigs of rowan and wych-elm in the hem of his waistcoat, thrust a Bible in one pocket, and a bottle of gin in the other, and a staff of buckthorn (surely blackthorn?) which had been cut at the full of the moon ... he set out for the cave" (Miller). Again, in the ballad of Willie and Douglas Dale, a woman, about to give birth in the forest, asks her lover to get her a "bunch of red roddins", because of the belief that protective measures must, at childbirth, be taken against evil influences:

> He's pu'd her a branch o' yon red roddins,
> That grew beside yon thorn,
> Put an a drink o' water clear,
> Intill his hunting horn" (Wimberley).

The beneficent attitude of rowan continues to be seen with other superstitions, like the Manx one that a stick of this wood could be used to stir the fire on Good Friday – of course, no iron could come in contact with the fire on this day (Dyer. 1876). Why, then, should there be a belief, Irish, it seems, that it prevents childbirth? (see Ô Súilleabháin).

Don't plant a rowan near an apple, they say on Exmoor, as one will kill the other (Tongue).

ROYAL FERN

(*Osmunda regalis*) Scottish Highlanders had an odd use for this fern – no less than treating a dislocated kneecap with it. The roots were chopped up, then put in a pan, and covered with water. After boiling and cooling, the contents would be applied to the injured knee. It is claimed that "forthwith the pain stopped, all swelling disappeared and the kneecap went back to its original place" (Beith).

ROYAL OAK DAY

(29 May) A holiday in Britain once, popularly supposed to be in commemoration of Charles II's famous escape from the Boscobel Oak, when the custom used to be to wear sprigs of oak in the hat, and to decorate houses with oak boughs, which were often gilded (Jones-Baker. 1974) for greater effect. The Charles II theme is particularly apparent at the Royal Hospital, Chelsea, for 29 May is Founder's Day there. Every pensioner wears a spray of oak leaves as a memorial to the royal founder (Hole. 1941). But another legend connects the oak with Charles I rather than with his son. Borlase has the story – at the time of the battle of Braddock Down, Charles set up his standard in an oak tree in Bocconoc Park. After the execution of the king the leaves, so it is said, suddenly changed colour to white. Braddock

means 'broad oak', so there may have been some earlier tradition of a venerated tree there (Grigson. 1955). It seems, in fact, that some kind of celebration was taking place long before the time of either of the royal Charleses. Aston-on-Clun, in Shropshire, has a 29 May festivity that may well be the oldest of them. There is a BLACK POPLAR there, called the Arbor Tree, growing at the crossroads at the village centre, and this tree is decorated (Hole. 1976). A recent piece of folklore claims that the twigs of the Arbor Tree have powers of bestowing fertility. They were sent to brides all over the country, when asked for (Hole. 1976). This superstition eventually forced the parish council to ban the ceremony, because of the publicity it had received. The ban did not seem to have lasted very long, for the parish council still hangs flags from eight large branches, or rather from eight long poles attached to the branches (Rix).

The flags and decorations are left on the tree until they are replaced on next Arbor Tree Day, and this has its parallel in the decoration of church towers with some kind of garland, as at Castleton, in Derbyshire; again, the garland would be left there until the day came round again (Vaux), or at least until all the flowers had withered away, when the framework would be taken down and kept for another year (A Burton). There are quite a lot of records of OAK branches being hoisted to the top of church towers, in widely separated areas. Bell-ringing was a feature of many of these festivities, and the ringers were active in Dorset in particular. At Whitchurch Canonicorum in that county, the men and boys used to go out at three bo'clock in the morning to cut oak boughs. One was put on the church tower, and another on a post driven into a plot of land near where the war memorial now stands. The church bells were rung, and then the ringers and boys went round the village putting oak boughs over the doors (Dacombe). The ringers at Upton Grey, in Hampshire, did the same (Partridge. 1912), and started ringing at 6 a m (Bushaway). Oak leaves were used to decorate Hampshire door knockers, and people wore them in their hats (Boase).

The schoolboys' celebration of Royal Oak Day is the best known of the customs. The sprig of oak was worn for self-protection. In some places, the sprig was enough, in others the sprig had to bear an oakapple (Opie & Opie. 1959). In Dorset, the boys gilded their oak leaves (Udal) – how is not very clear, but oak-apples were gilded in Gloucestershire, at Minchinhampton, by covering them with gold leaf, specially bought for the purpose (Partridge. 1912). Newcastle boys used to taunt those not wearing the oak in their hats, with:

> Royal Oak,
> The Whigs to provoke.

But there the rival faction answered back:

Plant tree leaves
The Church-folk are thieves (Tongue. 1965).

So wearing the oak can be regarded as a badge of loyalty, not only to the king, but also to the establishment (A Smith. 1968). Even railway engines were decorated with oak leaves in the 19th century (Opie. 1954). In Sussex not wearing the oak meant one would be pinched, preferably on the bum, hence the local name Pinch-bum Day, or just Pinching Day (Simpson. 1973). Similarly, the day was Bumping Day in Essex (Chisendale-Marsh), and Rump Day in Yorkshire, for oak leaves, even when still on the tree, were called Rump there (Easther).

Northamptonshire children carry bunches of NETTLES, with which they attack anyone foolhardy enough not to be wearing the oak. In Hertfordshire, the day was known as Half-day Stinging Day (half-day because, like April Fools' Day, there was a noon closure on the proceedings). "Roundheads", that is, those not wearing the oak, were greeted with the chant:

Poor King Charles lies hidden in a tree (this line
sung thrice)
Show your oak-apple
Or I'll sting thee (Jones-Baker. 1977).

The punishment varies in different parts, hair-pulling being substituted for stinging in parts of Cumbria, while the song:

Nob him once,
Nob him twice,
Nob him till he whistles thrice

is sung, with actions to suit (Rowling).

There were colourful names for the day in Wessex, where words like Shick-sack, Shick-shack, Shit-sack, and so on, were used. As far as Royal Oak Day is concerned, the shick-shack was a piece of oak with an oakapple on it. An extension was to use the Shick-shack name in any of its many variants as a name for anyone not sporting the oak. In the afternoon, the oak could be replaced with ASH leaves ("even ash", according to Norman Rogers).

Rubus fruticosus > BLACKBERRY

RUE (*Ruta graveolens*)

Plant your rue and sage together,
The sage will grow in any weather.

That is a piece of gardeners' wisdom from Warwickshire (Northall) that is not however held to be true by other old wives' tales, some of which say that it can make sage planted near it "positively poisonous", or will simply kill it, as it may basil (Boland & Boland). Others, though, say that it will be the rue that suffers near basil (Bardswell). But most

will agree that rue is bad for cabbages (Boland & Boland).

Rue is the symbol of regret and repentance, and to rue means to be sorry for. The trouble is that the two words have no relation with each other and have quite different derivations. Rue, the flower, is from Latin ruta, and before that Greek, while rue, to be sorry for, has a Germanic origin. The symbolism, then, must have been at worst mistaken, and at best punning. Thompson (C J S Thompson. 1897) confused the matter still further by suggesting that rue, the herb, could be derived from the same root as Ruth, meaning sorrow or remorse, for ruth (with no capital letter) is another word from Germanic origins, and has no connection with rue, the herb. Nevertheless, the association is a lasting one. Rue is still carried by judges at the assizes, traditionally to ward off fever, but probably originally as a symbol (Brownlow). The series of names based on Herb Grace (Herb of Grace, Herbgrass, Herbygrass etc.,) may very well have this symbolism as their origin (Rohde.1936), but another explanation may be that holy water was once sprinkled from brushes made of rue at the ceremony usually preceding the Sunday celebrations of High Mass (Grieve. 1931). In any case, rue (repentance) may evoke the grace of God (gratia dei).

In Lithuanian folklore, rue is the symbol of chastity (Gimbutas. 1963). As the rue thrives, so does the girl:

If you will wish me, rue
Good fortune,
Branch out, rue,
Up to the tenth branchlet!
If you will wish me, rue
Ill fortune,
Dry up, rue,
From the white roots (Gimbutas. 1958).

Something similar has been recorded in Italy. If the plant flourished, all went well with the girl's love affair, but if it withered it was a sign that the lover she desired had failed her (Hartland. 1909). That may be why in this country it was a funeral herb, when posies of rosemary and rue were put on the coffin of a *chaste* wife (Tongue. 1967). When Lithuanian girls marry, they wear rue as a wreath. Note too the Herefordshire custom for a jilted girl to curse the man involved by waiting in the church porch during the wedding, and throwing rue at him as he came out. The curse would be something like "May you rue this day as long as you live" (Leather). (see also WALL-RUE)

It used to be said that rue prevented anything evil from growing anywhere near it in a garden (Leighton). It was also said to be twice as valuable if stolen from someone else's garden (Wiltshire), and it would flourish better, too. "Rue stolen thriveth the best". If floors are regularly rubbed with it, no witch can enter, and you may be preserved from witchcraft by green rue leaves consecrated on Palm Sunday (Gordon. 1985). It is particularly as a counter to the evil eye that rue is famous. One of the most potent charms in use in Naples was, and still is, the cimaruta, that is cima di ruta, sprig of rue, a representation of the plant these days, but obviously the real thing originally (see CIMARUTA, where the charm is described in more detail). Rue seems to have been the special protection of women in childbirth, and the cimaruta is worn on the breast of infants in Naples (Elworthy. 1895). Elsewhere in Italy, a new-born baby is washed with a decoction of rue to make it strong (Canziani. 1928). As far away as Mexico, it is recorded that in the Mayan village of Yucatan, a mother may keep a child from ojo (the equivalent of the evil eye) by chewing leaves of rue and rubbing them on the child's eyelids. In the Ardèche district of France, cart drivers would put rue in their pockets to stop those wih the evil eye from causing their carts to stop suddenly (Sebillot). All over Morocco, too, rue is carried as a charm against the evil eye, and also to drive away the evil spirits they know of as juun (Westermarck). Sore eyes caused by this "fascination" were to be cured by a woman bending a branch of rue into a wreath, tied with a red ribbon. But it had to be done in another room, for neither the patient nor any child or animals should watch the process. As the wreath was being made, a rhyme was said:

"I prepare this wreath
To place it on the eyes
Of that sufferer,
That his sight I may restore,
And he may never suffer more" (Gifford).

The patient would have to look through the wreath three times, and say "Santa Lucia, Santa Lucia, Santa Lucia, from the evil eye guard me". St Lucia is of course the defender against the evil eye (Lucia – lux, lucis, light). Cesalpino (1519-1603) recommended this wreath of rue to defend a child against "fascination", and the usage was still popular in Tuscany as late as 1890 (Gifford).

No plant had more virtues ascribed to it in ancient times. In addition to those well-known powers against the evil eye, there were some that were lesser known. Soldiers, for example, at one time used to heat the point of their swords in the fire and then smear them with the juice of rue. That was supposed to make them invulnerable (Clair). The idea survived into the firearms age, for we hear that gun flints were boiled with rue and vervain. Doing that would ensure that the shot would reach the intended victim, no matter how poor the aim (Moldenke). It was said in Jersey that rue could give second sight, in some unspecified way (L'Amy). It was one of the most active of medicinal plants; no less an authority than Pliny claimed "every species of rue employed by himself has the effect of an antidote if the leaves are bruised and taken in

wine". It is good for snakebite – "so much so, in fact, that weasels when about to attack them, take the precaution first of protecting themselves by eating rue" (Elworthy. 1895). Only a weasel could face the basilisk unharmed, provided it ate the rue first. There is a misericord in Westminster Abbey showing a basilisk flanked by two small animals with foliage in their mouth (Rowland). So efficacious against all manner of venomous beasts was it thought to be that Estienne advised that great numbers of the plant should be planted near "sheep coates, and houses for your foule and other cattle … for adders, lizards and other venomous beastes will not come neere unto rue by the length of the shadow of it" (Clair). Not only poisonous creatures, either, if we are to believe Topsell – "it is reported that if wild rue be secretly hung under a hen's wing, no fox will meddle with her". Antonius Mizaldus tells a story of a man who died after rubbing his teeth and gums with sage. On investigation it was found that a "great toad, … which infected the same sage with his venomous breath" was living under the sage bed. The judge advised people to plant rue round their sage, "for toads by no means will come nigh unto rue" (Lupton). But as we have seen, the sage might be deadly for the rue. It is supposed to heal bee and wasp stings, and particularly mad-dog bites. Such a use is mentioned in a 12th century manuscript of Apuleius. It was said that at Cathorpe, in Lincolnshire, rue dramatically cured the whole town of the bites of a mad dog that had run through the houses attacking everyone (Baker. 1980), date not divulged.

Rue was once taken to cure people of talking in their sleep (Gordon. 1985), and according to the Physicians of Myddfai, "if you would preserve yourself from unchaste desires, eat rue in the morning". Veterinary use includes a Yorkshire record of boiling rue in ale to give to horses to cure farcy, or glanders, to give the disease another name. A little rue juice would be put in the horse's ears, too (Gutch. 1911). It was used by horsemen in the Fen country. For example, the way to control an unruly horse is to rub freshly gathered rue leaves on its nose (Porter. 1969), and a sprig or two given to horses would make them well, and make their coats shine (Randell) (this is an old Fenland custom). Norfolk turkey breeders also used rue to make the birds eat and put on weight (Randell), while a leaf is given to poultry to help in curing croup (Brownlow). This is actually a 17th century usage from Lincolnshire and other areas, when the owner was told to chop the herb very finely and form it into piles with butter, and so feed it to the sick hens (Drury. 1985).

Rue's use in medicine in times past has been widespread and general, with claims for almost anything from warts to the plague. In the latter case, Alexis of Piedmont gave the following recipe, anglicised a long time ago as: "Take the toppe of Rue, a garlicke head and half a quarter of a walnutte and a corne of salt. Eat this every morning, contynuing so a munneth together and be mery and jocunde". Gerard too gives a recipe for the plague with the leaves of rue. Thornton repeated the belief: "it is supposed to be antipestilential, and hence our benches of judges have their noses regaled with this most foetid plant". Bunches of it used to be hung at the sides of windows to protect the house against entry of the plague (especially east facing windows, for it was thought that was the direction from which the plague came). So powerful was rue considered that thieves looting plague-stricken houses would risk entry if they carried it, even if corpses still lay there (Boland.1977). It was a strewing herb, too, for it kept away fleas. In 1750, it was used to strew the dock of the Old Bailey as a protection against jail fever then raging at Newgate Prison, and the custom continued right into the 20th century (Genders. 1972). We find that in the Balkans, rue is one of the plants burnt to provide fumigation in epidemics (Kemp), and it is used as a fumigation herb in Brazilian healing ceremonies too (P V A Williams).

It has always been supposed that rue has a potent effect upon the eyes (Pliny said that painters and sculptors mixed some rue with their food to keep their sight from deteriorating (Baumann)). A 10th century Arabian writer on eye diseases used a mixture of rue and honey to prevent the development of a cataract (Gifford). The Anglo-Saxon version of Apuleius prescribed rue: "well pounded, laid to the eye", and for "dimness of eyes" it was apparently only a matter of eating rue leaves, or taking them in wine (Cockayne). A mild infusion is still in use as an eyebath and for eye troubles, including cataract (Hatfield). We have already seen its great virtue against the evil eye, and the claim has been made that rue can actually bestow second sight (MacCulloch. 1911).

There is usually a warning given that rue should never be taken in large amounts, and not at all if the patient is pregnant (Gordon. 1977) – with good reason, for it has often been used for abortions (Clair). French folklore insists that there was a law forbidding its cultivation in ordinary gardens. It was said that the specimen in the Paris Botanical Garden had to be enclosed to prevent pregnant girls from stealing it. In the Deux-Savres district of France it was believed that it caused any woman who merely touched it with the hem of her dress to miscarry (Sebillot). Granny Gray, of Littlepool, in Cambridgeshire, so Enid Porter reported, used to make up pills from rue, pennyroyal and hemlock. They were famous in the Fen country for causing abortion. A strong infusion of rue and horehound, followed by a good dose laced with poppy juice and laudanum was a noted Fenland sleeping draught. It is quoted as being a last resort means of stopping a mother giving birth on the first of May, an unlucky date. It just put her to sleep for twenty-four hours (Porter. 1969).

Rue tea for indigestion is a well-known medicine with ancient origins. American domestic medicine recognized rue's worth for this sort of complaint. From Alabama, for instance, one is prescribed rue and sugar for kidney colic (R B Browne), whatever that is. Rue tea is much esteemed on the Greek island of Chios, too, for stomach upsets, and also for checking excessive periods (Argenti & Rose). With vinegar, it cured the stitch (Gerard), and it was used for worms, too. Modern folk medicine still recognizes chewing a leaf of rue as being a cure for nervous headaches (Brownlow). Fever is another condition dealt with by using the plant, usually a leaf decoction. Coughs and colds are also treated with rue tea (Vickery. 1995), and bronchitis too, though in this case it is by the agency of hot compresses, cotton cloths saturated in a strong infusion of rue, applied to the chest (Hatfield). Even epilepsy could be treated with it, according to Wesley, simply by taking a spoonful of the juice, morning and evening, for a month, and trying to control children's fits with it was quite common, either with syrup (Thornton), or leaf juice (Watt & Breyer-Brandwijk), or by more esoteric methods. In Herefordshire, for instance, rue was tied to children's hands, wrists and ankles, to prevent the convulsions (Leather), while a Bulgarian practice was to rub the back of an infant's neck, and behind the ears, with rue, daily, for the complaint (Rolleston). It was even used to "invigorate the memory". Lyte recommended it for this – "the juice, with vinegar, doth revive and quicken such as have the forgetful sickness" (Grindon).

RUSH

i.e., COMPACT RUSH (*Juncus conglomeratus*) or SOFT RUSH (*Juncus effusus*) "Not worth a rush" is a common enough saying, but it is far from the truth, for rushes used to supply, besides rush-lights, floor covering, including, in England, the stage, (J Mason), mats, chair seats, ropes and toys (Burton). They were strewn, along with herbs and flowers, at weddings, from the house of the bride to the church (J Mason). Rush-lights (there is an American name for rushes, Lampwick Grass) were made from second-year plants (those in flower, that is), and they were trimmed to about 12 – 15 inches. They were then peeled, while still leaving on a thin strip of rind to support the fragile pith. After drying, they were passed through hot fat. Of course, the resulting rush-lights were very fragile, so they were stored in a cylindrical tin called a "rush-bark" (or "candle bark"), originally a piece of bark, as the name implies, which could be hung on the wall. There was a custom in County Antrim, Ulster, when someone lay dying, for twelve rush-lights to be lit and stuck in a bowl of meal. These were left burning until death occurred. Then the rush-lights would be extinguished, and the bowl of meal given to the first poor person who came to the door (St Clair). If a rush-light curled over, it was a sign of death, and if a "bright star" appeared in the flame, a letter could be expected (Burton).

Twisted into ropes, rushes were used for securing thatched roofs, and for trussing hay and straw. It was claimed that they were stronger and more durable than ropes of hemp. They were used for bedding, too. An 18th century ballad has:

> "Fair lady, rest till morning blushes,
> I'll strew for thee a bed of rushes".

Rushes were once made into bridal rings, originally probably for bethrothal rings, but later for mock marriages. As early as 1217, Richard, Bishop of Salisbury, had to issue his edict against the use of "Annulum de junco" (Savage). Hassock is a Lancashire name for a rush, reed, or any coarse grass. The name was then applied to mats and cushions made from the stems, on which people kneel at church.

Finding a green-topped rush is as lucky as finding a four-leaved clover:

> With a four-leaved clover, a double-leaved Ash, and a green-topped Seave,
> You may go before the queen's daughter without asking leave (Burton).

"Seave" is the Yorkshire name for a rush (Hartley & Ingilny). Children make "arrows" of the stems, described as half-peeling a strip of the outer skin away from the pith, balancing the stem across the top of the hand, then pulling sharply on the half-peeled strip, which propels the arrow at a target. Girls make "Lady's Hand-mirrors", by bending the stem sharply at the middle at two points about a quarter of an inch apart. The stem is then plaited by bending each side in turn sharply over the other at right angles, and so on (Mabey. 1998).

A Devonshire charm for the thrush was to take three rushes from any running stream, and pass them separately through the mouth of the infant, then plunge the rushes again into the stream. As the current bears them away, so will the thrush go from the child. There is also a Cheshire charm for warts involving rushes. Take a long, straight one, tie three knots in it,

and make it into a circle, Draw it over the wart nine times, say the required formula (not divulged), and the wart will disappear within three months (Burton).

RUSSIAN TARRAGON

(*Artemisia dracunculoides*) Native Americans, the Ojibwa in particular used this plant for a "deer medicine". The hunter had to find the plant on his right hand, for if it lay on his left it had no virtue. He buried its stem in the ground with a little tobacco, chewed the root and rubbed the mingled juice and saliva over his eyes. Then he could approach a deer close enough to kill it with a tomahawk (Jenness), or so it was traditionally stated. Winnebago men also used it as a hunting charm, but in addition, they could use it as a love charm. They chewed the root and put it on their clothes. The effect was supposed to be secured (in both cases) by getting to windward of the object of desire and letting the wind waft the odour (Youngken).

But it was a food plant for some of the Indians, too, especially the Hopi, who apparently used to gather the leaves in early spring, to bake between hot stones. They would then eat them as greens after dipping them in salt water (Hough). There were genuine medicinal uses, too. The Mescalero and Lipan, for example, made a liniment by pounding the root and soaking it in cold water, and using that for bruises (Youngken).

Ruta graveolens > RUE

RYE-GRASS

(*Lolium perenne*) There is a series of names bestowed on this grass arising from children's games. One kind, the Love-me, Love-me-not variety is played by pulling off the alternating spikelets, hence Yes-or-no, from Somerset, or Aye-no Bent, from Gloucestershire. A different version of the game accounts for Does-my-mother-want-me, another Somerset name (Grigson. 1959). Another game that girls used to play involved striking the heads together, and at each blow saying Tinker, tailor, soldier, sailor, and so on. The blow that knocks the head off marks the profession of the future husband (Elworthy. 1888). Hence Tinker-tailor Grass, from a wide area of the west of England (Grigson. 1959), and, from Wiltshire, Soldiers-sailors, tinkers-tailors (Dartnell & Goddard). Presumably the Sussex name What's-your-sweetheart (Grigson. 1959) fits in here.

S

Saccharum offficinarum > SUGAR CANE

SAFFLOWER

(*Carthamus tinctorius*) One of the oldest dyestuffs known. The fruits have been found in Egyptian tombs that are 3500 years old, and since then it has been grown, for centuries past, in India, the Middle East and East Africa, for the flower heads, and so for the dye obtained from them (H G Baker). There are in fact two dyes obtainable from them – safflower yellow, similar to true saffron, and often used as a substitute for it (Leggett), and carthamin, a red colour. Both are simple to extract, and direct, but not fast. China and Japan made early use of the dyes, at least from the end of the second century BC. It has long lost its predominance in Asia (Buhler), though it is still cultivated in India for local use (Leggett), such as in the dyeing of ceremonial clothes, or in the making of a brush paint for cosmetic rouge, and so on (Buhler) (one of the names given to the plant is Rouge Plant (Howes)). It was apparently introduced into the south-western states of North America by the Mormons, in about 1870. They cultivated it in gardens that were irrigated by narrow trenches, but people like the Hopi Indians had access to the plant, for they used the flowers to colour their waferbread yellow (Weiner). Pomet, in the 18th century, noted that "this Saffron is in great vogue among the feather-sellers, and for making Spanish-red …".

The dried flowers are used in Chinese medicine as a "blood invigorator" (Geng Junying), whatever that means, though one meaning is certainly to promote menstruation, which is presumably why the flowers have the reputation in China of causing abortion (F P Smith). In the Philippines, the flowers are used to treat jaundice, and the seeds serve as a remedy for apoplexy and dropsy (Perry & Metzger).

SAFFRON

(*Crocus sativus*) The original home is doubtful, but it has been cultivated from very early times; it was an article of trade on the Red Sea in the first century AD, being exported from Egypt to southern Arabia. The word Crocus, always taken to be saffron in early accounts, is from the Greek Krokos, the adjective from which, krokotos, means yellow, or dyed with saffron (Potter & Sargent). Earlier than the Greek word, though, is the Hebrew Carcom, or Karkom (Genders. 1972), mentioned in the Song of Solomon. The Arabic is Kurkum. One of the Greek origin legends tells that a youth named Crocus was changed into the flower after being accidentally killed by Mercury. Another tradition says that the crocus sprang from the spot on which Zeus once rested (Rimmel).

Saffron was being cultivated in Spain as early as AD 961, and according to legend was introduced in the reign of Edward III into England (Fluckiger & Hanbury), where it continued to be grown extensively in the eastern counties (Crafton, in Berkshire, commemorates saffron in its first syllable), though Saffron Walden is the more famous place name. The growers there were known as "crokers" (Swahn), and were cultivating it from the 14th to the end of the 18th century, and sold it as a choice drug. It was once used as a dyestuff, especially in Britain as an alternative to gold thread for church vestments. Henry VIII's knights at the Field of the Cloth of Gold had their garments coloured with saffron, and monks used it instead of gold leaf for illuminating manuscripts. "The chives steeped in water serve to illumine or (as we say) limne pictures and imagerie …" (Gerard). Cennini gave instructions for making this pigment (see Il libro dell-Arte). Saffron was even used to dye the hair a golden colour in the Middle East (Genders. 1972).

But it was more popular as a spice than as a dye, and as such was one of the chief trade commodities – it is still part of the arms of Saffron Walden. It was used a lot in cookery – warden pies were coloured with saffron in Shakespeare's time, in the same way as pears are coloured with cochineal now. In Cornwall particularly, it was used for colouring cakes, but Cornish fishermen viewed it with suspicion, for it would bring bad luck, they said; saffron cake in a boat spoiled the chance of a catch (Vickery. 1995). In 17th century England it was fashionable to use starch stained yellow with saffron, and in a cookery book of that time, it is directed that "saffron must be put into all Lenten soups, sauces, and dishes ; also that without saffron we cannot have well-cooked pies" (Fernie). It is still used for soups on the Continent, especially for the Marseilles soup called bouillebaise, in which it is an essential ingredient, as it is for the Spanish paella. In the East, it is used extensively in the more expensive curries such as chicken biriana (Brouk). And saffron itself is expensive – in Devon it is used as a figure of speech for anything costly, and a common Lincolnshire saying used to be "as dear as saffron" (Gutch & Peacock). Nevertheless, sheets in Ireland used to be dyed with it, apparently to preserve them from vermin (and to strengthen the limbs of those who lay in them) (Fernie). It is recorded that Wolsey, in order to purify the foetid air, spread the floors of his rooms at Hampton Court with rushes strongly impregnated with saffron (Dutton). But then he could afford it. It is difficult to understand why it should be a symbol of charity, which apparently it was, according to the system of Raban Maur (Haig).

Like coca, saffron enjoyed in the Aztec court the reputation of being an aphrodisiac (claimed by de Ropp, but was saffron grown in Mexico at that time?). However unlikely that may sound, there are comparable beliefs in the Old World (see Leland. 1891: "Eos, the goddess of the Aurora, was called … the one with the saffron garment. Therefore the public women wore a yellow robe). Not unconnected was the practice of drinking saffron tea (or was it a drink merely known by that name?). It was supposed to make one vivacious and optimistic (Brouk). The Physicians of Myddfai also noted this belief: "If you would be at all times merry, eat saffron in meat or drink, and you will never be sad; but beware of eating over much, lest you die of excessive joy". See too another quote from Christopher Catton (in C P Johnson) – "Saffron hath power to quicken the spirits, and the virtue thereof pierceth by and by to the heart, provoking laughter and merriment …". (Perhaps that was why saffron was particularly used in Lent). A habitually cheerful person was said to have slept on a sack of saffron (Coats. 1975). But it is said that saffron tea was drunk to cause an abortion. Lowestoft is mentioned as a place where this was done (M R Taylor).

It had its protective side, too. In Morocco, for instance, it was one of the plants used to drive away the juun, and it was also used as a charm against the evil eye. All evil spirits are said to be afraid of saffron, which is used in the writing of charms against them (Westermarck). Some Hebrew amulets, too, were written with a copper pen, using ink made from lilies and saffron (Budge). But it is not an inherent quality in the plant that is exploited. Rather it is the colour that is effective, especially against the evil eye. Yellow, the colour of gold, is a most effective prophylactic. So a person who wears yellow slippers, to take one example, had little to fear from other people's glances (Westermarck). The same really applies to saffron's medical uses. There is no active principle at all in saffron, and if it appears in the Pharmacopeia, it is there solely for its use as a colouring agent (Fluckiger & Hanbury). Nevertheless, it has been used quite extensively in folk medicine, as, for instance, for a leechdom for swollen eyes, or what Thomas Hill called "the distilling of eyes". The 15th century recipe required the practitioner to "stamp violet with myrrh and saffron, and make a plaster and lay thereto". Gypsies used to cure sore eyes with a wash made of spring or well water and saffron (Leland. 1891).

The most obvious example of the use of colour in saffron is in its use as a remedy for jaundice – doctrine of signatures in other words, in which the colour yellow must be regarded as the cure for the yellow disease. "Lay saffron on the navel of them that have the yellow jaundice, and it will help them" was Lupton's advice, and there are earlier jaundice cures, too. Coulton quoted two from the 14th century. Saffron also appears to have been used as a prophylactic against jaundice, for in Westphalia,

an apple mixed with saffron was given on Easter Monday, against the disease (Fernie).

Saffron has been mentioned as a local remedy for tuberculosis (Fernie), but this is actually an old usage. Gerard, for instance, quotes it as a special remedy for those suffering from the disease, "and are, as wee terme it, at deaths doore, …". Even plague could be kept at bay, according to Gerard, with saffron as one of the ingredients of a concoction that "preserveth from the pestilence, and expelleth it from those that are infected". Lesser ailments were also treated with it, boils, for instance (Wesley), or asthma (Wesley, again). The early writers were also quite sure that it "strengtheneth the heart". That was Gerard, and Parkinson. 1629 agreed. It was apparently useful to sober people up. The Physicians of Myddfai had a remedy "if you would remove a man's drunkenness, let him eat bruised saffron with spring water".

In more recent times it was widely used as an abortifacient (Schauenberg & Paris). Surely there could not have been any physiological action to make women believe that? But saffron tea is still used in American domestic medicine as a mouth wash in cases of thrush, and as a drink to cure measles in young children, or, as it was put in Ireland, "to bring out the rash" (Moloney). Lemon and sugar are added in Alabama (R B Browne). It is still used in East Anglia for fevers in children. It acts as a diaphoretic, that is, it induces sweating and so cools down the patient (V G Hatfield. 1994).

It should be borne in mind that the plant called MEADOW SAFFRON (*Colchicum autumnale*) has nothing to do with saffron, nor for that matter is it a Crocus, though the name Autumn Crocus is commonly used for it.

SAGE
(*Salvia officinalis*) A funerary plant in some areas, and graves were planted with it (Drury. 1994), something that Pepys noticed in April, 1662: "To Gosport; and so rode to Southampton. On our way … we observed a little churchyard, where the graves are accustomed to be sowed with Sage".

Gardeners' wisdom in America holds that it is unlucky to grow it from seed (Whitney & Bullock), and in Britain we are told to:

Plant your sage and rue together,
The sage will grow in any weather (Northall).

In Wiltshire, they say that the rue will prevent the sage being poisoned by toads (Wiltshire). But the sage will stimulate the growth of the rue, and will perform the same office for rosemary. It likes to grow with marjoram, too. If grown among cabbages, it will protect them against the cabbage white butterfly (Baker. 1980). The season of planting is important. According to a Wiltshire rhyme:

Sage set in May
Will never decay (Whitlock. 1992).

Although, in the Appledore area of Kent, there was
a saying that if sage blooms, misfortune would
quickly follow (Parish & Shaw) (it doesn't usually
flower, for household requirements would keep it
well cut), it is generally looked on as a lucky plant,
able to strengthen the memory, and to impart wisdom
(Waring), another way of saying that it grows best for
the wise (Gordon. 1977). It is the symbol of domestic
virtue (Leyel. 1937), for:

If the sage tree thrives and grows,
The master's not master, and that he knows
(Northall).

As with parsley (and rosemary) sage was said to
flourish where "the missis is master". In East Anglia
they used to say they could foretell the sex of a baby
by studying these plants in the parent's garden. They
all flourished more profusely where the mother
was the dominant partner. Such a woman was more
likely to have girl babies, so if a boy was wanted, the
growth of the plants was restricted (Porter. 1974). An
aberrant view is that sage will flourish only so long as
the master of the house is in good health (Whitlock.
1992).

As well as being a lucky plant, it is a protective one,
at least in Spain and Portugal, where it is thought
of as proof against the evil eye (Wimberley). There
is a Yorkshire divination with sage leaves. Twelve
of them had to be gathered at noon on Midsummer
Eve, and put in a saucer, where they would be kept
till midnight, then they would be dropped out of the
bedroom window one by one with the chiming of the
hour. The future husband would be seen, or at least his
step heard, in the street below. A Leicestershire variant
of this requires the leaves to be picked at each strike
of the clock at midnight (Palmer. 1985).

Medicinally speaking, sage had an extraordinary
reputation:

He that would live for aye
Must eat sage in May.

There is a medieval Latin version of that couplet:

Cur moriatur homo cui
Salvia crescit in horto? (Clair).

In other words, how can a man die who has sage
growing in his garden? After all, salvia means good
health, or health-giving. It is said that a white witch's
unwitching medicine consisted of, among other things
and rituals, three leaves of sage and three of Herb
John (*Hypericum perforatum*), steeped in ale, to be
drunk night and morning (Seth). A more recent claim
is that sage will reduce memory loss in Alzheimer's
patients (R Palmer. 2001).

Its most remarkable (and genuine) property is its
ability to stop perspiration. The action apparently
starts two hours after the dose is taken, and can be
prolonged for several days (Schauenberg & Paris).
Perhaps that is why it has been used so much for
fevers:

Sage helps the nerves, and by its powerfull might,
Palsies and Feavers, sharp it puts to flight.

The doggerel is by Coles. 1657. In Sussex, it used
to be said that to eat nine sage leaves on nine
consecutive mornings (Thompson. 1947) is a cure
for the Ague, a fever akin to malaria. In the Balkans,
a fever is dealt with by steeping the leaves and stems
in brandy, and then straining it off (Kemp), and in
America a remedy for a common cold is to drink hot,
sweetened sage tea in bed, after soaking the feet in hot
water (Stout).

Sage has also got the reputation of increasing
women's fertility, quite deservedly, and it has been
used for that purpose since ancient Egyptian times
(Schauenberg & Paris). "Sauge is good to helpe a
woman to conceive", as Andrew Boorde said (Boorde.
1542), and it seems to do so by arresting lactation. So
a strong infusion has been used to dry up the breast
milk for weaning (Fernie). Given all this, it is difficult
to understand the old wives' tale that drinking salvia
cooked in wine would ensure that a woman would
never conceive (Boland. 1977).

It seems to have some antiseptic properties. Certainly,
on the Greek island of Chios, it was used to cleanse
a wound. Sage boiled with a little water is used, and
the wound washed with the liquid, and the leaves were
soaked and put on the cut (Argenti & Rose). But this
use as a lotion has never been confined to that area,
for the remedy is noted in recent herbal lists (Flück).
The leaf infusion is used, too, as a mouthwash and
gargle, and the teeth can be whitened and the gums
strengthened by rubbing them with sage leaves (Page.
1978), which are still used for the purpose (Vickery.
1995). Some gypsy groups make their own toothpaste
with chopped sage and salt.

Sprains and swellings are helped by a poultice of the
bruised fresh herb. The *Housekeeper and Butler's
Assistant, 1862*, had the advice "bruise a handful of
sage leaves and boil them in a gill of vinegar for five
minutes; apply this in a folded napkin as hot as it can
be borne to the part affected". Bathing the forehead
with hot water in which sage had been boiled is a
Cambridgeshire headache remedy (Porter. 1969).
Other external uses include easing a sunburnt face
by washing it with sage tea (Page. 1978). This was
also used in Scotland as a hair wash (Gibson. 1959),
for cold sage tea is regarded as a hair tonic (at least
in America), and will also darken it (H M Hyatt).
Skin disorders were treated with it, too. Wesley, for

instance, prescribed it for erysipelas, or St Anthony's Fire, as it was known in his day: "… boil a handful of Sage, two Handfuls of Elder-leaves (or bark), and an ounce of Alum in two Quarts of Forge Water, to a Pint. Anoint with this every Night". Extravagant claims were made for its healing properties, not least in an Irish remedy for epilepsy, when the juice of absinthe, fennel or sage put in the patient's mouth when he fell in the fit would effect an immediate cure (Wilde. 1890). It was even suggested as a cure for paralysis in a medieval Jewish work (Trachtenberg).

A mixture of sage, knapweed and camomile flowers was a Russian folk remedy for all digestive disorders – a teaspoonful of the mixed herbs to a glass of water, boiled for 15 minutes, and strained (Kourennoff); a similar use is noted in Scotland (Gibson. 1959). Sage is good for dyspepsia, anyway – that is why it has always been been used in a stuffing for rich meat (Knight), and worms have been dealt with in America by taking sage tea (H M Hyatt). But the best known medical use is that for coughs, and has been since ancient times. It is still in use, not only for coughs, but also for colds, headaches and fevers (Conway). Cough cures are detailed as early as the 14th century, when we find "Medicina pro tussi. Take sauge and comyne and rewe and peper and seth hym to-gedre in a panne with hony, and ete ther-of a spoone ful a-morwe and at eve a-nuther" (Henslow). Another, taken from a manuscript of similar date advises an extraordinary procedure for "hym that haves the squinsy: tak a fatte katte, and fla [flay] hit well. and clene, and draw out the guttes, and take the fles of an urcheon, and the fatte of a hare, and resynes, and feinegreke, and sauge, and gumme of wodebynd, and virgyn wax; als this nye [crumble] smal, and farse [stuff] the catte within als ther a farses a goss, rost hit hale, and gader the gres and enoynt hym therewith" (Cockayne).

SAGEBRUSH

(*Artemisia tridentata*) A North American species, much used medicinally by the Indians. The Klamath, for instance, took the decoction for diarrhoea (Spier), as did the Coahuilla and the Tewa, who also chewed and swallowed the leaves for a cough (Youngken). The Gosiute (Chamberlin) and the Navajo used it for colds and fevers. The same people used it for headaches, which they still say can be cured simply by smelling the plant (Elmore).

The Klamath had a sort of ceremonial use for the plant. At the birth of a first child, both parents wore sagebrush bark belts for five days. This was the belt worn also in mourning for a child, and by a girl at puberty (Spier).

Sagina procumbens > PEARLWORT

SAINFOIN

(*Onobrychis sativa*) Sainfoin means 'healthy hay', in other words a crop, dried, that is good for cattle

(Grigson. 1974), and not only for cattle, for horsemen were at one time fond of giving their horses sainfoin seed to make them fat and their coats sleek. But the opening syllable of the name has been misunderstood enough for the plant to be called Saintfoin (A S Palmer), and Holy Clover has also been used (Wit). A spurious saint has been invented, with a legend to go with it, in Hertfordshire. It was said there that sainfoin grew only in places once owned by the Church. This was recorded by Nathaniel Salmon in his *History of Hertfordshire* of 1728, and he it must have been who invented St Foyne as the plant name (Jones-Baker. 1977).

ST BARNABAS'S DAY

(11 June) Ox-eye Daisy has some connection with this feast day, presumably only because it is the Midsummer Daisy, and as such a herb of St John. St Barnaby's Thistle is *Centaurea solstitialis*, and must have got the name because it blooms about now, or is said to bloom about now, for surely this is a July/August flower? Rose, lavender, woodruff and box were used for church decoration on St Barnabas's Day, and the officiating clergy wore wreaths of roses; so, indeed, did the choristers (Tyack). See, for instance, the churchwardens' accounts of St Mary-at-Hill, in London, during Edward IV and Henry VI: "For rose garlandis and woodroff garlandis on St Narnabe Daye, xj d". "Item, for two doss di boise garlands for opresetes and claekes on Sr Barnabe Day, js. Vd".

One of the versions of the Glastonbury Thorn legend is that Joseph of Arimathea had a walnut staff, which he planted. It rooted, and always flowered on St Barnabas's Day (Tongue. 1965). The earlier description, by Camden, has the tree only putting forth its leaves on this day.

ST BRIDGET, or BRIDE

Milk-yielding plants like the DANDELION, which, though in bloom virtually throughout the year, has its great burst of colour in spring, so it is naturally an associate of the spring goddess, Brigid, sanctified as St Bridget, or Bride. It was a logical connection; as a cowherd in one of her aspects, she would be associated with cattle, and so with flowers like the dandelion, yielding a milky juice which was believed to nourish the young lambs in spring (E E Evans). It is said in the Scottish Highlands that "the plant", i.e., the dandelion, "nourishes with its milk the early lamb" (F M McNeill). The flower is one of the three insignia of Bride, the other two being the lamb and the bird called the oyster catcher (Urlin). Its names in Gaelic are: "Bearnon Bride", the little notched flower of Bride (F M McNeill), "The little flame of God", and "St Bride's forerunner". St Bride of the shores wears it at her breast, and the sunlight is said to follow.

ST GEORGE'S DAY

(23 April) was the traditional day for picking DANDELION flowers for making Dandelion

wine, and blue was the traditional colour to wear on this day. Hence the claim that BLUEBELLS (or HAREBELLS) should be worn, and used to decorate churches. Both seem highly unlikely.

ST JOHN'S EVE & DAY

(23/24 June) The most widely held view was that all herbs, even poisonous ones, lose their evil on St John's Eve – they are all purified by St John's dew (Gubernatis). In fact, they actually gain power. See the Portuguese proverb:

> Todas ces erbas têm prestimo
> Na manhã de São João.

That is, all herbs have power on St John's Morning (Gallop), for in Spain, Portugal and Brazil, herbs for curing should be gathered on the eve of Midsummer, for the season is at its most powerful on that day (P V A Williams). Even NETTLES, gathered tonight from the premises of a suspected witch, and put under milkpails, would undo any mischief that the witch had done (Marwick). There were a few, a very few, beliefs that tended to belie the generally held opinion; for example, it is said that German witches chose St John's Eve to gather Hexenkraut, ENCHANTER'S NIGHTSHADE, that is, for their nefarious purposes (Runeberg). In some parts of the Continent, it was believed that all poisonous plants came up through the ground on Midsummer Eve (Grimm), the inference here being that they would not be purified by St John – just the opposite. Lastly, it was said that gathering herbs on St John's Day could be dangerous. In Altmark, they say it would give you cancer (Gubernatis). These, though, are very much minority opinions. Far more widespread is the belief that the plants themselves acquire protective qualities because they are in bloom at this season. Midsummer is the time when the power of the fairies is at its highest, and only flowers and herbs can then protect the people (Moore); they were often worn on the person; Penzance children, for instance, used to wear garlands of flowers on the afternoon of Midsummer Eve (Courtney). Among gypsy marriage customs was one that required the bride to burn flowers gathered on St John's Eve as a protection against sickness (Starkie).

There are many charms and superstitions connected with the name of St John, and his day was often spent in collecting herbs for some secret purpose or other (W G Black), but usually a medicinal one. According to the Anglo-Saxon Herbal, many plants used for a medicine had to be gathered at Midsummer (Bonser), and see the Lacnunga (an 11th century collection of medical lore):

> "Gather all those herbs together three nights before summer comes to town (i.e., before midsummer) … And then in the night when summer comes to town in the morning, then the man who wishes to take the drink must keep awake all night; and when the

cock crows for the first time, then he must drink once, a second time when day and night divide (i.e., at the first streak of dawn), a third time when the sun rises, and let him rest afterwards" (Storms).

> "a sovereign medicine for madness and for men that be troubled with wicked spirits. Upon midsummer night betwixt midnight and the rising of the sun, gather the fairest green leaves on the walnut tree, and upon the same day between sunrise and its going down, distil thereof a water in a still between two basins. And this water is good if it be drunken for the same malady" (Dawson. 1934).

The herbalists recognised that herbs gathered on Midsummer's Day had special powers: "Against liver complaint gather this same herb (he was talking about VERVAIN) on midsummer day" (Storms). A complicated charm cure:

> Iarum
> Origanum
> Herba benedicta
> Allium
> Nigella
> Excrementa diaboli
> Succisa.

(The initial letters spell the name Iohannes). These were plants to be collected on St John's Day (Bergen. 1899). Similarly, young people of the Abruzzi, Italy, used to gather the galls on ELM trees on this day, for then they contained the oil of St John, to be put on the hair to make it grow strongly and quickly. The insects of the gall were used with oil as an ointment for wounds (Canziani. 1928).

St JOHN'S WORT, as its name implies, is one of the plants associated particularly with this festival. In Germany, it was worn as an amulet on this day, the idea being that it would expose the work of witches (Dyer. 1899); the inference must be that anyone with the plant could then see them. It was hung up in houses and barns to keep the people and cattle free from enchantment (Runeberg). Stow, *Survey of London*, 1598 says it was hung over the house doors in his day, together with green birch, long fennel, orpine, white lilies, etc. Some also had wrought iron baskets, containing lights (Brand). The herb was gathered before sunrise, with the dew, itself a magical substance, still on it. They would be smoked in the Midsummer fires, and so became more potent for medicine and protection from evil (Grigson. 1955). In Wales, there used to be a custom for young girls to go out at midnight to gather the plant by the light of a glow-worm held in the palm of the hand. The bunch was taken home and hung in the bedroom. Next morning, if the leaves were still green, it was a good sign – the girl would marry that year; but if the leaves were dead, she would not be married (some said she would die) (J C Davies; Owen).

WORMWOOD was a St John's herb in France – the old French saying, "Herbe Saint Jean, tu portes bonne encontre", refers to this plant (Beza). Its close relative, MUGWORT, is just as important as St John's Wort at this time of year. It is known as Johannesgürtel, or Sonnenwendgürtel, in Germany (Storms). "Muggwith twigs" were being used in the south of Ireland as late as 1897 as preventatives against disease, after being singed in a St John's fire (T D Davidson. 1960), and it was widely used as a protection against witchcraft. In the Isle of Man, for instance, chaplets were made of it, and worn on the heads of man and beast, to protect them from evil influences (Moore; Train). Chaplets made of GROUND IVY were worn in France while the people danced round the fires (Palaiseul). MONKSHOOD and LARKSPUR were used in much the same way in Germany (Grimm). Another belief shared with mugwort by PLANTAIN mentions that a rare coal was to be found under the roots of mugwort at noon or midnight today. This coal, if worn on the person, would protect the wearer against the plague, carbuncles, fever and ague (Radford & Radford). Aubrey mentions that this was a current belief when he described it in 1694.

Chaplets of VERVAIN were made for the young women in parts of Spain. As with St John's Wort, this was done early in the morning, while the dew was still on the plants. Then the women's fortunes were told, according to the length of time the dew stayed on the plants (Beza). Piedmontese belief was that if young men gathered vervain today, any girl with whom he shook hands while it was in his possession would fall in love with him (Canziani. 1913).

FENNEL was a herb of St John, too, and was put over doors this night (Dyer. 1889). See Stow: "On the vigil of St John the Baptist and St Paul the Apostle, every man's door being shadowed with green birch, long fennel, St John's Wort, orpin, white lilies, and such like …". ORPINE was hung up inside houses, too. It stays green a long time, if watered a little. A common name for it was Midsummer Men (Brand), and it was used in divinations tonight (see **ORPINE**). There is a record, too, of a Swiss charm in the form of a cross made from dried pieces of GOAT'S BEARD (*Spiraea aruncus*) and MASTERWORT (*Astrantia major*). They were made on St John's Eve, and taken to church on St John's Day to be blessed by the priest, and were reckoned efficacious against lightning, fire and storms. During severe storms, a small piece of the cross would be taken and burned for added protection (Broadwood).

FERNS (according to Bock's Herbal, the Royal Fern, *Osmunda regalis*) was supposed to bloom at midnight tonight, and to seed shortly afterwards. The seeker must neither touch it with his hand, nor let it touch the ground. A white cloth would be put under the plant (more specifically, in Russia, a white towel that had

been used at Easter) (Gubernatis). This fern seed was the agent that conferred invisibility. Ben Jonson refers to this in *New Inn*:

"… I had
No medicine, Sir, to go invisible,
No fern seed in my pocket."

And see, too, Shakespeare, *1 Henry IV:*

Gadshill: … We have the receipt of fernseed, and we walk invisible.
Chamberlain: Nay, by my faith, I think you are more beholding to the night time than to fern seed for your walking invisible.

The Lucky Hand was made from the root of Male Fern on St John's Eve (see **LUCKY HAND**). Herodias, the daughter of Herod, who danced before her father so that she might be rewarded with the head of St John on a charger, was later confused with Herodias, the witch or fairy queen, and it was this confusion that brought the association of fern seed with St John's Day (Spence. 1948), for the queen of the fairies was the custodian of the seed (Wright & Lones), and whoever tried to gather it ran enormous risks from the attacks of the spirits. One way of getting the seed was to shoot at the sun at mid-day today. Three drops of blood would fall, and these had to be gathered up and preserved, because this was the fern-seed (Dyer. 1889) though it was said in Germany that fern-seed shone like gold on this night (J Mason). It was believed in Ireland until quite recently that the roots of BRACKEN and LILIES gathered on St John's Eve, provided they were cut after the proper incantations, would show a young women her true lover's name (Wood-Martin, though he does not say how they were going to do it). But the idea of fern as a love charm is found in England, too. It is said that young people went to Clough, near Moston, in Lancashire, to gather the "seed of St John's Fern", as a love-charm (Wright & Lones). The OAK-TREE had a similar connection – it was believed to bloom on Midsummer Eve, the blossoms withering before daybreak. In Shropshire, a girl was advised to spread a white cloth under the tree, and in the morning she would find a little dust, which was all that remained of the fallen blossoms. A pinch of this dust put under her pillow would make her dream of her future husband (Radford & Radford).

There was a custom in Northumberland and Durham of dressing up stools with a cushion of flowers at Midsummer. A layer of clay was put on the stool, or else it was first covered with calico or silk, and the flowers stuck in to form a cushion. They were put on show at house doors, and the attendants begged money from passers-by (a set rhyme was used in the begging (Denham & Hardy)), so that they could have an evening feast, and dancing (Brand). A similar type of custom existed in the Channel Isles, where the "jonquière", a sort of divan, made usually of dried ferns, was decorated with flowers, preferably in a formal pattern. It seems to have been the custom once to elect a girl to sit in state on the jonquière, where, as La Môme, she received the homage of all the assembled guests (MacCulloch).

YARROW is a herb of St John, and in Ireland it would be hung up at this time to turn away illness (Grigson. 1955). In one part of Germany (Saxony, perhaps), they used to make wreaths of MARSH MARIGOLDS, to throw one at a time, on to the house roof. If anybody's wreath actually stayed up, then it would be a sign he would die before next summer (Hartland. 1909). It should be said that this happens as part of the St John's Day celebrations, and it is unusual to find death divinations practiced then. In any case, would marsh marigolds still be in flower at Midsummer? Midnight on St John's Eve was the ritual time for Portuguese children to be passed through a split REED to cure their hernia (see under HERNIA, and cf ASH)

It was said that the woodpecker got its strength by rubbing its beak against a plant that flowered only at midnight on Midsummer Eve. CHICORY, against all observation, was one of the candidates for this identification (Hare).

ST JOHN'S WORT

(*Hypericum perforatum*) '*Perforatum*' means pierced with holes, but the holes are spurious in this instance. Hold a leaf up to the light and one sees dots that look exactly like holes. But they are oil sacs, and give the plant its aromatic smell when bruised (Salisbury. 1954). The "holes" were said to have been made by the devil, in anger at the power of the plant to thwart him (Browning). Another view is that they are drops of the saint's blood, appearing every year on St John's Day (Hole. 1976), or, with greater logic, on 29 August, the day of the beheading of the saint (Folkard). The sap is reddish, too, and that has been described as a representation of the saint's blood (Genders. 1971).

This is a plant with almost supernatural qualities, for in pagan mythology, the summer solstice was a day dedicated to the sun, and was believed to be the day upon which witches held their festivities. St John's Wort was their symbolical plant. In Scandinavian mythology, it was the property of Baldur, the sun-god (Browning). Christians dedicated this day to St John the Baptist, and the sacred plant was named after him, becoming a talisman against evil. In one ballad a young lady falls in love with a demon, who tells her:

> Gin you wish to be leman mine
> Lay aside the St John's Wort and the Vervain (Napier).

In Ireland, it was one of the seven herbs that nothing natural or supernatural could injure; the others were

vervain, speedwell, eyebright, mallow, yarrow and self-heal (Wilde. 1902). In the Derbyshire folk tale called Crooker, St John's Wort made up one of the three magic posies that protected the traveller from the evil Crooker. "Take the posy and show it to Crooker". The other two were primrose and daisy (Tongue. 1970).

It was often carried in Scotland as a charm against witches and fairies. When hung up on St John's Day, together with a cross over the doors of houses, it kept out evil spirits (Napier), a custom known in Wales, too (T G Jones), as well as in Ireland (O' Farrell). It was taken to America – the Pennsylvania Germans fasten a sprig of the plant to the door to keep out witches (and flies) (Fogel). In Essex, they said that if it was hung in the window it would prevent witches looking in (C C Mason), while in the Western Isles the emphasis was on preventing ordinary folk from seeing witches (or "grisly visions", as it was described); it had to be sewn into the neck of a coat (Bonser), and left there, for it it were interfered with in any way, it would lose its power (Spence. 1959). But to be effective as an amulet it had to be found accidentally (J A MacCulloch. 1905). It was given to Irish children on St John's Eve to avert sickness (Ô Súilleabháin. 1942), and Burton (*Anatomy of Melancholy*) mentions it as an amulet, worn as a remedy for "head-melancholy … gathered on a Friday in the horn of Jupiter". A white witch's unwitching medicine consisted of, among other things and rituals, three leaves of sage and three of Herb John, steeped in ale, to be taken night and morning (Seth).

A 13th century writer tells of "the wort of holy John whose virtue is to put demons to flight" (see Summers. 1927), and Aubrey. 1696 mentions a case where St John's Wort under the pillow rid a home of a ghost that haunted it. Langham was another writer, some hundred years before Aubrey, who advised his readers to keep some in the house, for "it suffereth no wicked spirit to come there". Fuga daemonum is an old book name, anglicised as Devil's Flight (chasse-diable in French). In Aberdeenshire, it was quite common to gather the plant on St John's Day, and put it under the pillow. The saint would appear in a dream and give his blessing, which would act as insurance against death for the year (Banks. 1937–41); indeed, gathering it on Midsummer Day and keeping it in the house would give luck to the family in all their undertakings, especially those begun on that day (Napier). It was quite a common practice to gather the plant before sunrise on St John's Eve, with the dew still on it, and then smoke it later in the Midsummer fires (Grigson. 1955).

St John's Wort was much used, too, in Midsummer divinations. In Denmark, girls used to gather it and put it in between the beams under the roof, in order to see the future – usually one plant for themselves and one for their boyfriend. If they grew together, it foretold a wedding. Or the plant was set between the beams, and if it grew upwards towards the roof, it was a good sign, but if downwards, sickness and death was the forecast (Thorpe). Orkney girls used to gather the Johnsmas flowers, as they were known there, remove the florets, one long, one short, from each flower, wrap what remained in a docken leaf, and bury it overnight. If by morning the florets had re-appeared, it was taken as a happy omen (F M McNeill. Vl). Similarly, in parts of Germany, girls fasten sprigs to the walls of their rooms. If they are fresh next morning, a suitor may be expected, If it droops or withers, the girl is destined for an early grave. This custom was also known in Wales, where the plant had to be gathered at midnight, by the light of a glow-worm carried in the palm of the hand (J C Davies). The plant was also credited with fertility-inducing powers, and was used in love charms and in remedies for barrenness. If a childless wife walked naked and in silence in the garden at midnight (on St John's Eve) to gather the flowers, she would have a child before the next twelve months were past (Hole. 1976). But they could apparently be used for death divinations, too. There was a Welsh custom of naming each piece of the plant gathered for a member of the family, before being hung on a rafter. In the morning, the pieces were examined; those whose plant had withered the most were expected to die soonest (Vickery. 1981).

There are some really outlandish beliefs connected with this plant. Grimoires suggest that sex "will greatly improve if you do give a maiden to wear a girdle which has been anointed with the oil of the St John's wort plant" (Haining). Another is from Aubrey. 1696, who said that "against an evil tongue, take unguent" containing this plant "and a put a red-hot iron into it. You must anoint the backbone, or wear it on your breast". It does not seem clear who has the evil tongue that must be stopped. Is this self-service? And St John's Wort was used in the ritual to conjure back a suicide. Apparently, a consecrated torch was used, and that had to be bound with this plant (Summers. 1927).

Almost as wonderful are the medicinal virtues ascribed to it. Inevitably, there is a remedy that owes its origin to the doctrine of signatures. It is those "perforations" that constitute the signature; they look like holes, and that was enough to let the plant be taken as a wound herb (one of the Gaelic names for the plant translates as bloodwort (Beith)). Nevertheless, it is still prescribed in homeopathy for painful cuts and wounds, as well as for some insect bites and for piles (Homeopathic Development Foundation). One often comes across a "balm" made from the plant (note the name Balm of the Warrior's Wound (Macmillan)). In Sicily, for example, it

was gathered on St John's Eve and put in oil. The consequent balm was considered infallible for the cure of wounds (Gubernatis). Gerard also speaks of an oil made of the plant, "a most precious remedie for deep wounds". Another interpretation by Coles of the "perforations" was that they resembled the pores of the skin. So they could be used to treat skin problems (Vickery. 1981).

Gypsies use the St John's Wort as a hair dressing, to make it grow (Vesey-Fitzgerald), a usage that is known too in Somerset, where the leaf infusion was used not only to make the hair grow, but to heal cuts, and to make a poultice for sprains (Tongue. 1965), and that same ointment is also good for burns, and for throat and lung complaints (Wickham). In Scotland, the plant is used in a herb mixture for coughs (Simpkins). The infusion had further uses, one being for children's bed-wetting (Fernie); in Russian folk medicine, centaury and St John's Wort were mixed in equal amounts for this (Kourennoff). The juice is sometimes used for warts, and carbuncles were also treated by direct application of the plant (Physicians of Myddfai). It is still used for bed sores, too, in an ointment prepared from the flowers and leaves, and mixed with olive oil (Genders. 1971), and that same ointment is also good for burns (Tongue. 1965). There is even an ointment for fractures involving this plant, though it was only called into use after the splint had been removed.

ST THOMAS'S ONION

Apples and onions seem to have been the chosen agents for St Thomas's Eve (20 December) love divinations. It was even said that the onion was sacred to St Thomas (Moldenke & Moldenke), and one of the old cries of London was, "Buy my rope of onions – white St Thomas's onions" (Dyer. 1876). St Thomas's Onion, then, became the recognized term for the one used tonight in one way or another. You could perhaps peel it, wrap it in a clean handkerchief, and put it under your pillow, all the while saying:

> Good St Thomas, do me right,
> And see my true love come tonight,
> That I may see him in the face,
> And him in my kind arms embrace.

Or the girls could quarter the onion, and then whisper to her quarter the name of the man she expected, or hoped, would propose to her. She would then wave her bit of onion over her head, and recite a rhyme very similar to the above. She had to be in bed by the stroke of twelve, and then would expect satisfactory dreams. Another way to conduct the ritual was to stick nine pins in the onion, with eight pins round one in the centre. The name of the lover was given to this central pin. Then the whole lot was put under the pillow (Northall). While the pins were going in, the usual words were recited:

Good St Thomas do me right,
Send me my true love tonight,
In his clothes and his array,
Which he weareth every day,
That I may see him in the face,
And in my arms may him embrace.

The significance of lines three and four in the rhyme is fairly obvious. The girl would want to know what trade her future husband was to follow, and this would be recognizable from his working clothes, so she would want to see him "not in gay apparel, fine array …", as a Suffolk rhyme has it (Porter. 1974) but in "the clothes he walks in every day". To get the expected dream in this Suffolk case, she had to go to bed backwards, and to maintain strict silence till morning (Gurdon).

ST VITUS'S DANCE
(Chorea) Hawthorn MISTLETOE was used in Lincolnshire for the complaint (Gutch & Peacock). Thornton claimed that "St Vitus's Dance, and spasmodic asthma, have yielded to these flowers" (LADY'S SMOCK). WOOD SAGE has been used for the condition (Vickery. 1995). There is a recipe from Alabama for St Vitus's Dance : "one ounce skullcap (VIRGINIAN SKULLCAP, *Scutellaria laterifolia*), one ounce feverweed (*Verbena syriaca*, perhaps?), one ounce Lady's Slipper (*Cypripodium Calceolus*). Take half of each one, put in a quart jar filled with boiling water, and seal. Let it stand for two hours, then take a wineglassful three times a day" (R B Browne). A Warwickshire prescription was to take the root of STINKING IRIS, and boil it in ale (porter, to be specific), strain it off, and drink the liquor (F G Savage).

SAL, or SAUL TREE
(*Shorea robusta*) A valued hardwood in India, taken in Bengal as the seat of the gods, and an emblem of Indra (Gupta). The mother of Buddha is represented as holding a branch of this tree in her hand, and it was under this tree that Buddha passed the last night of his life on earth (Pandey). In Indian villages, witches were discovered with the aid of this tree. The names of all the women in the village over twelve years of age were written on the bark of the branches. These branches were then steeped in water for a given length of time (four and a half hours, apparently). If after that time one of them withered, the woman whose name was written on it was deemed to be a witch (Porteous).

SALEP
EARLY PURPLE ORCHID (*Orchis mascula*), and indeed most of the *Orchis* species of Europe and northern Asia, have tubers that when prepared yield Salep, which had, at least according to the doctrine of signatures (the word orchid means testicle) a reputation for curing impotence. (see also APHRODISIAC). The

word itself, sometimes spelt 'salop', is from Turkish, originally Arabic sahlab (Emboden. 1974). The substance was imported at one time from the East, and used as an article of diet, reported to be very nutritious (Fluckiger & Hanbury), part of every ship's provisions, to prevent famine at sea, for the dried material swells to several times its size when water is added, so it was standard starchy food in the days of sailing ships. It forms a jelly with a large proportion of water. A decoction flavoured with sugar and spices, or wine, was reckoned an agreeable drink for invalids, though it has no medicinal value. Nevertheless, it was used in the treatment of colitis and diarrhoea (Emboden. 1974). It makes a common soft drink, very popular long before the introduction of coffee houses, and mentioned in Victorian books as a common beverage for manual workers (Mabey. 1972).

Salicornia europaea > GLASSWORT

Salix spp > WILLOW

Salix capraea > GOAT WILLOW

Salix viminalis > OSIER

SALLOW, SALLY
Common names for Goat Willow (*Salix capraea*). The word is OE sealh, willow, still preserved in the Wiltshire place name Zeals (Jones & Dillon). Both the tree and the name were familiar enough for the first Australian settlers to name some of their eucalyptus trees Sallee.

Salvadora persica > TOOTHBRUSH TREE

Salvia coccinea > RED SAGE

Salvia horminoides > WILD SAGE

Salvia officinalis > SAGE

Salvia sclarea > CLARY

SAMPHIRE
(*Crithmum maritimum*) The thick, fleshy leaves are edible, and quite sought after. This is the plant whose cliff-face gathering Shakespeare called a "dreadful trade" in King Lear (iv. Sc 5 line 12 …). From Dover and the Isle of Wight, samphire was despatched in casks of brine to London, where in the 19th century, wholesalers would pay up to 4 shillings a bushel for it (Mabey. 1972). There are various ways of preparing it – on the Yorkshire coast it is cooked and eaten cold with bread (Grigson. 1955), or it can be gathered, boiled for ten minutes, then served with lemon and butter. But the usually favoured way of dealing with it is as a pickle. Gerard knew about this: "the leaves kept in pickle, and eaten in sallads with oile and vinegar, is a pleasant sauce for meat". So, indeed, are the young shoots of GLASSWORT (*Salicornia europaea*), actually called Samphire, or Marsh Samphire, etc., (Britten & Holland). Experts say, though, that it is inferior to the proper stuff (Hepburn). Nevertheless,

it was still being collected in the Eastern counties into recent times (Grigson. 1955), and may still be gathered even now.

The name Samphire comes from the French 'herbe de St Pierre'; which is 'herba di San Pietro' in Italian, too (Ellacombe). 'Pierre' also means stone, so if we disregard the attribution to St Peter, we can see that samphire means simply a plant that grows on a rock (Young), and it was this fact that led to its medical use against stone, or general bladder and kidney trouble. The doctrine of siugnatures is presumably responsible, too, for Gerard's "The leaves, seeds, and roots, as Disocorides saith, boiled in wine and drunk, provoke urine and womens sicknesse, and prevaile against the jaundice".

SANDALWOOD

(*Santalum album*) A fragrant oil, called Oil of Santal, is distilled from the heartwood for use in perfumery and cosmetics. The paste that can be got by rubbing the wood on a stone with a little water is used for painting the body after bathing, and is also used for making caste marks, especially in south India (Pandey). The Chinese make joss sticks from the wood (Usher), and incense from the sawdust, mixed with swine's dung(!) (Moldenke & Moldenke). The oil is used for the treatment of urinary complaints and sexually transmitted diseases, and heartwood is sliced and used in Chinese medicine for abdominal and chest pains, i.e., angina (Geng Junying). In Indian mythology, it is described as surrounded by snakes, but it is a sacred tree, and the devotees of Vishnu apply a paste made from it on their foreheads. Hindu funeral pyres are made from it if the family is rich enough to afford such a luxury (Upadhyaya).

Sanguinaria canadensis > BLOODROOT

Sanguisorba officinalis > GREAT BURNET

SANICLE

(*Sanicula europaea*) Sanicle means a healing herb (from Latin sanare, to heal), in spite of attempts to see in it a corruption of St Nicholas (Fernie). It has been widely used as a medicinal herb for a very long time, chiefly as a wound herb, though numerous other complaints have been treated with it. These days it is usually prescribed in the form of a compress made from the roots, for wounds (Thomson. 1978), but in earlier times it was taken internally, as a "wound drink". Gerard for example said that "the juice being inwardly taken is good to heale wounds". A 15th century wound drink was made from sanicle, yarrow and bugle, pounded, and given with wine. "This is the vertu of this drynke: bugle holdith the wound open, mylfoyle clenseth the wound, sanycle helith it". But it was emphasised that sanicle must not be given for wounds in the head, or a broken skull, for fear of killing the patient (Grigson. 1955). There was a veterinary use, too – after a cow had calved, country

people used to feed it with sanicle leaves, to help the expulsion of the afterbirth, and to stop any bleeding (Drury. 1985).

The astringent leaves can be used in infusion to treat catarrh (Conway), and as a gargle for sore throats and quinsy (Wickham), as well as stomach or lung complaints (Usher). It formed an old treatment for dropsy, and was once used as a lotion for chilblains (Mitton). All in all, the French rhyme is understandable:

Qui a la Bugle et la Sanicle
Fait aux chirurgiens la nicle.

(faire la nicle means to thumb one's nose at).

Sanicula europaea > SANICLE

Santalum album > SANDALWOOD

Santolina chamaecyparissus > GROUND CYPRESS

Sanvitalia procumbens > CREEPING ZINNIA

Sapindus saponaria > SOAPBERRY

Saponaria officinalis > SOAPWORT

Sarcostemma acidum > MOON PLANT

Sarothamnus scoparius > BROOM

SAVIN

(*Juniperus sabina*) It has been known through the ages as an abortifacient and contraceptive, either by the simple matter of swallowing the berries, or by the decoction of the leaves; or, as in East Anglia, put into the teapot with ordinary tea (M R Taylor. 1929); all common knowledge, of course, as is shown by the fact that it is spoken of in plays of the Elizabethan period. See Middleton, for instance, in *A game of chess*, act 1, sc 2:

To gather fruit, find nothing but the savin-tree,
Too frequent in nuns' orchards, and there planted
By all conjecture, to destroy fruit rather.

Some of the common names for the tree bear witness to this usage, names like Cover-shame and Bastard Killer. And its contraceptive properties were used with horses, too. It was said that a stallion would never cover a mare if there was any savin in the stable (G E Evans. 1966). East Anglian horsemen also used it to smarten a horse's coat, though they knew quite well that it was poisonous, and so took great care with it. It was mixed with the horse's food, either boiled, strained, and then sprinkled on, or it was baked and powdered (G E Evans. 1969). Savin was mentioned as a veterinary drug by Marcus Portius Cato, as early as 200 BC (Lloyd). It is a powerful uterine stimulant, producing in overdoses very serious effects.

The fresh tips, gathered in spring, can be extremely poisonous, but they can be used externally as an ointment for blisters and skin troubles (Brownlow), and it is still being recommended, either fresh or dried, by herbalists, for warts (Flück). Internal use of it is extremely rare these days, but in past times children were regularly wormed with it; such a remedy survived well into the 19th century, dangerous as it must have been. Earlier, Gerard published a receipt that required the physician to "anoint their bellies therewith", safer, though hardly efficient.

Saxifraga granulata > MEADOW SAXIFRAGE

Saxifraga sarmentosa > MOTHER-OF-THOUSANDS

Saxifraga x urbinum > LONDON PRIDE

SCABIES
MARSH ROSEMARY (*Ledum palustre*) has the reputation of being a good anti-parasitic, for treating lice, scabies, etc., (Schauenberg & Paris).

Scabiosa atropurpurea > SWEET SCABIOUS

SCAMMONY
(*Convolvulus scammonia*) An Asian species, whose gum resin, or the dried milky juice, is a drastic purgative (Pomet), and has been known as such since ancient times. The drug is collected in Asia Minor, chiefly around Smyrna, and often put with colocynth and calomel, to be used for constipation, worm cases and dropsy (Lindley), even for "rheumatic pains" (Porter. 1969). But however did it become an ingredient in a Cambridgeshire folk remedy for constipation, unless it reached the folk level via Academe?

SCARLET LEADWORT
(*Plumbago indica*) *Plumbago* derives from the Latin plumbum, lead, so the common name of the species is Leadwort. There was some feeling at one time that the plant was a cure for lead poisoning (Hyam & Pankhurst). Scarlet Leadwort is an Indian and Malaysian species. The root is an abortifacient, well-known as such, apparently by introduction into the vagina. But this is dangerous, sometimes lethal (P A Simpson), for that use would cause violent local inflammation (Gimlette). This is the source of the poison known as Lal Chitra in India. The root bark, rubbed into a paste with flour and water, and applied to the skin, will blister it. Some Indian peoples use the root to cure toothache, and in southern India, the dried root used to be highly regarded as a syphilis and leprosy cure (P A Simpson).

SCARLET PIMPERNEL
(*Anagallis arvensis*) It is a magical plant, known in Ireland as the "blessed herb" (seamair mhuire), and having the power, so it was seriously thought, to move against a stream. There were more wonders ascribed to it. If you hold it, it gives you the second sight, and you can understand the speech of birds and animals (Grigson. 1955). It has the power of drawing out thorns or splinters, and can protect against witchcraft,

if hung over the door or porch of a house (C J S Thompson. 1897). As usual, there was a formula that had to be recited when pulling it, and it had to be repeated for fifteen days together, twice a day, morning early fasting, and in the evening full:

> Herb pimpernel I have thee found
> Growing upon Christ Jesus' ground;
> The same gift the Lord Jesus gave unto thee,
> When he shed his blood upon the tree.
> Rise up, pimpernel, and go with me,
> And God bless me,
> And all that shall wear thee. Amen. (Harland & Wilkinson).

A glance through the local names of the pimpernel will confirm its status as a weather forecaster. As Gerard said, "... the husbandmen having occasion to go unto their harvest worke, will first behold floures of Pimpernel, whereby they know the weather that shall follow the next day after; as, for example, if the floures be shut close up, it betokeneth raine and foule weather; contrariwise, if they be spread abroad, faire weather".

> Pimpernel, pimpernel, tell me true,
> Whether the weather be fine or no;
> No heart can think, no tongue can tell,
> The virtues of the pimpernel (M E S Wright).

This is quite true; the flowers open when it is going to be sunny, and close when it is going to rain (Page. 1977). It will forecast twenty four hours ahead, so it is claimed (Trevelyan). No wonder it has attracted such a lot of names like Weather-teller, Weather-flower, and Farmer's, Ploughman's, Countryman's, Shepherd's, and Poor Man's, Weatherglass, and many more besides.

Country people used it medicinally, too, notably for complaints of the eyes. Perhaps the pimpernel's habit of closing its petals at dusk suggested a connection with the eyes (Conway). Anyway, the use is quoted from ancient Greece onwards. 15th century leechdoms have examples for "the web in the eyes". One prescription required the patient to "take pimpernel a good quantity and stamp it, and wring the juice through a cloth; and take swine's grease, and as much of hen's grease; and melt together and put the juice thereto, and keep it in boxes and anoint the eyes therewith when thou goest to bed" (Dawson. 1934). Something similar to this preparation, i.e., a lotion made from the plant with hog's lard, has been used as a cure for baldness (Page. 1978).

The bruised leaves were used for dog- and snake-bites. They were prepared by powdering, and, with the root, were made into an infusion. A teaspoonful of this, or about 20 grains of the powder, were put into a cup, with 15 drops of spirits of hartshorn, and a dose given every six hours. This would be continued for 15 days. Gerard agreed that "it is good against the stingings of Vipers, and other venomous beasts". After that, it comes as no surprise that it was "especially applied for poyson … (T Hill. 1577), or that it was looked on as a plague remedy (C J S Thompson. 1897).

Gerard recommended it for toothache, "being snift up into the nosthrils". In Somerset, warts are rubbed with the juice (Tongue. 1965). Gout and dropsy have also been treated with pimpernel in India (Dawson. 1934). There is even a leechdom "to know the life of a wounded man, whether he shall live or die". Some pimpernel had to be stamped in a mortar and mixed with water or wine. This was to be given to the wounded man to drink, "and if it come out at the wound he shall die; if it come not out of the wound he shall live"! (Dawson. 1934).

SCENTED MAYWEED
(*Matricaria recutita*) An annual of waste places, with white flowers whose petals hang down soon after opening, a fact that makes sense of the specific name recutita, circumcised. A glance at the profile of the flower will soon tell one why. The main use of the plant is medicinal, though it is harvested in the Debrecsan region of Hungary, where it grows in great quantities, for inclusion in hair rinses (Clair). Mayweed, or Maydweed, is from OE maythe, from the word meaning maiden, for this and other similar plants bearing the name have been used in complaints like painful menstruation (Flück). But the plant has many other uses, as a tea made from the dried flower heads is used for all stomach upsets (Flück), as well as for such varied complaints as insomnia, rheumatism, sciatica, gout, and so on. Externally, the infusion is used in compresses applied to slow-healing wounds, and for skin eruptions like shingles and boils (Flück), or eczema (W A R Thomson. 1978).

Schinus molle > PERUVIAN MASTIC

SCIATICA
PEONY roots were prescribed for sciatica: "for Hip bone ache [sciatica], take some portion of a root, and with a linen cloth, bind it to the sore" (Cockayne). LESSER SPEARWORT's blistering propensities were used at one time in the Hebrides for sciatica. Martin (1703) described the procedure: "Flammula Jovis, of Spirewort, being cut small, and a limpet shell filled with it, and applied to the thighbone, causes a blister to arise about the bigness of an egg: which being cut, a quantity of watery matter issues from it: the blister rises three times, and being emptied as often, the cure is performed …". Of course it will raise a blister – all the buttercups will. Hill, in the 18th century, wrote that "the leaves [of CANDYTUFT] are recommended greatly in the sciatica; they are to be applied externally, and repeated as they grow dry". A MUSTARD plaster can be used for rheumatism,

sciatica, etc. Homeopathic doctors prescribe POISON IVY for this complaint (Homeopathic Development Foundation).

Scilla maritima > SQUILL.

Scirpus acutus > TULE

Scirpus lacustris > BULRUSH

Scorzonera humilis > VIPER'S GRASS

SCOTCH BROOM
(*Genista canariensis*) Why "Scotch"? The specific name is accurate enough, for it is a native of the Canary Isles, though it was taken to central America in post-Conquest times, which explains the shaman's use of the plant. It contains cytisine, which in high doses can cause nausea, convulsions, hallucinations, and even death from respiratory failure (Furst. 1976). But Yaqui shamans used to smoke the yellow flowers. Several cigarettes produced the desired "long relaxation, intellectual clarity and physical ease, coupled with psychological alertness" (Emboden. 1979).

SCROFULA
Irish people used to apply the bruised leaves of WATERCRESS as a poultice to scrofulous swellings (Egan), and in the southern states of America, CHINABERRY TREE (*Melia azedarach*) roots and POKE-ROOT (*Phytolacca decandra*) made a cure for scrofula. They were boiled together, with a piece of bluestone in the water, and then strained off. The sores had to be salved with this mixture, and then anointed with a feather dipped in pure hog lard. This would bring the sore to a head, so that the core could be pressed out (Puckett). In Orkney, the crushed leaves of BUCKBEAN were once applied (Leask). NASTURTIUM, too, has been used in folk medicine against the disease (Schauenberg & Paris). Lady Wilde recorded a much more complicated scrofula cure from Ireland. It required burdock roots, common dock, buckbean and rose-noble [probably figwort] boiled in water, of which the patient had to drink three times a day (Wilde. 1890). The very name of the genus (*Scrophularia*) to which WATER BETONY and FIGWORT belong, proclaim their efficacy against this disease, once known here as the King's Evil, for the touch of the anointed monarch was taken as the only cure for the condition. The shape of the root tubers "signal ORPINE [to be] with virtue against the King's Evil" (Grigson. 1955). In the Balkans, WATER MINT is used, cooked in water, and the hot plant applied externally, while the liquid is drunk (Kemp).

A Pennsylvania charm for scrofula was to wear MARJORAM roots as a necklace. They would dig the roots, cut them crosswise, and thread an odd number of pieces. The necklace had to be removed on the ninth day, and other pieces of root threaded on. This was repeated twice, and each time they had to be buried under the eaves (Fogel).

Scrophularia auriculata > WATER BETONY

Scrophularia nodosa > FIGWORT

SCURVY

LOVAGE was eaten at one time in Scotland to combat scurvy (Grigson 1955), and in the Highlands ROSEROOT was also used (Fairweather), but it is BUCKBEAN that is most often associated with the disease, and it seems to have been of prime importance in combatting that once-dreaded condition. BITING STONECROP "hath the signature of the gums" (!), and so was used for scurvy (Berdoe). Hill, in the 18th century independently advised that "the juice … is excellent against the scurvy and all other diseases arising from what is called foulness of the blood", and a strong decoction of DAISY roots was used for this disease in the 18th century (Hill. 1754), as were the leaves of GOOD KING HENRY, ground up in water (Porter. 1974). The combination of BROOKLIME and WATERCRESS has been praised in the past as being very good for scurvy (Gerard; Hill, etc.,)

Scutellaria galericulata > SKULLCAP

Scutellaria lateriflora > VIRGINIAN SKULLCAP

SEA ASTER

(*Aster tripolium*) Ancient Greek writers reported that it changed its colour three times a day (hence *tripolium*), white in the morning, purple at noon and crimson in the afternoon (Pratt). One of the old names for this plant was Toadwort (Hulme. 1895). The *Ortus Sanitatis* has "when a spider stings a toad and the toad is becoming vanquished, and the spider stings it thickly and frequently, and the toad cannot avenge itself, it bursts asunder. If such a burst toad be near the toad-wort, it chews it and becomes sound again: but if it happens that the wounded toad cannot get to the plant, another toad fetches it and gives it to the wounded one". Topsell vouches for this having been actually witnessed! The error seems to have been the confusion between bubo and bufo. The latter is the toad, but bubo is a swelling in the armpit or groin (Prior), presumably to be treated with this herb.

SEA HOLLY

(*Eryngium maritimum*) "They report of the herb Sea Holly, if one goat take it into her mouth, it causeth her first to stand still, and afterwards the whole flocke, untill such time as the shepherd take it from her mouth" (Gerard).

Sea Holly roots were preserved in sugar and sold as Candied Eringo, or Kissing Comfits, Colchester being the centre of the trade, and they were sold until as late as the 1860s (Grigson. 1955). They were said to be good for those "that have no delight or appetite to venery", and were "nourishing and restoring the aged, and amending the defects of nature in the younger" (Gerard). Likewise Rapin, in a Latin poem on gardens, 1706 (in translation):

> Grecian Eringoes now commence their Fame
> Which worn by Brides will fix their Husband's Flame
> And check the conquests of a rival Dame.

There is, too, a reference in Dryden's translation of Juvenal's Satires, he is talking about libertines:

> Who lewdly dancing at a midnight ball
> For hot eryngoes and fat oysters call (quoted in Leyel. 1926).

But best known is a quote from Shakespeare. Falstaff, in *Merry Wives of Windsor*, says:

> Let the sky rain potatoes; let it thunder to the tune of 'Green Sleeves';
> hail kissing comfits, and snow eringoes.

It had a wide variety of medicinal uses in the older herbals, as a diuretic, to treat stone, gout or snakebite, etc. One, from the Physicians of Myddfai, is worth quoting, it is for toothache. "Take a candle of mutton-fat, mingled with seed of sea-holly; burn this candle as close as possible to the tooth, holding a basin of cold water beneath it. The worms … will fall into the water to escape the heat of the candle". But the only prescription with any claim to more recent folk use comes from Ireland, where it was used for asthma (Ô Súilleabháin).

SEA PLANTAIN

(*Plantago maritima*) A wound herb, according to Gerard, "such an excellent wound herbe, that it presently closeth or shutteth up a wound, though it be very great and large". But then he went on, "if it be put into a pot where many pieces of flesh are boyling, it will soder them together" !

SEA SICKNESS

QUINCE, in one form or another was a favourite medieval stomachic, the "confection of quinces" being recommended against sea sickness, for example (Withington). A tisane, or tincture, of MARJORAM will act as a sedative to prevent sea sickness (Leyel. 1937).

SEA WORMWOOD

(*Artemisia maritima*) "It is reported by such as dwell neere the sea side, that the cattell which doe feed where it groweth become fat and lusty very quickly" (Gerard), who also noted its value, along with most other members of the genus, in keeping insects at bay: "the herbe with his stalkes laid in chests, presses and wardrobes, keepeth clothes from moths, and other vermin". Country people in Essex used to rub the forehead with a handful of sea wormwood to cure a headache.

SECOND SIGHT

RUE has been a famous protection against the evil eye, but some say that it also has the ability to bestow the second sight (MacCulloch. 1911; L'Amy). PIMPERNEL, too, had the reputation, among other wonders, of giving the second sight, if it were held in the hand (Grigson. 1955). In the Hebrides, it was said that putting a leaf of YARROW against the eyes would give you the second sight (M Baker. 1980). It is said, too, that THORN-APPLE played some part in the development of second sight (Trevelyan).

SEDATIVES

Herbalists still use COWSLIP leaves as a sedative and pain-killer (Conway). The plant has always had this reputation of inducing sleep. Hill said "the flowers have a tendency to procure sleep, and may be given in tea or preserved in form of a conserve".

> If you need rest,
> Lettuce and cowslip wine probatum est (Pope wrote).

A root infusion of PRIMROSES, too, has a decided narcotic tendency if taken last thing at night (Leyel. 1926). RED POPPY also has some sedative effect, though not as much as OPIUM POPPY. A syrup made from the flowers of the former is still sometimes used as a sedative to soothe coughs (Schauenberg & Paris), and not all that time ago mothers on South Uist made a liquid from the flowers to help babies in their teething (Beith). An old Welsh sleeping draught recipe reads: Boil poppy heads in ale, and let the patient drink it, and he will sleep (Ellis). In the Highlands the juice was put in children's food to make them sleep (Beith), and Hill. 1754 recommended syrup of red poppies as a sleep-procurer. But of course, it is the OPIUM POPPY (emblem of Morpheus, the god of sleep, and the symbol of sleep itself) that provides the sedative par excellence, whether it is by opium or some preparation of the seeds. A syrup of poppies was being recommended in the 10th century as a sedative in catarrh or cough, and for insomnia (the Anglo-Saxon version of Apuleius has a recipe "for sleeplessness, take ooze, smear the man with it" (Cockayne)). Poppy tea used to be a very popular sedative in the Fen country of England. Even teething children were given the tea, or a few poppy seeds to suck, tied in a piece of linen! (Porter. 1974). THORN-APPLE seeds have been prescribed as a sedative (Fluckiger & Hanbury), and of course VALERIAN is used extensively (Emboden. 1979). It is usually taken as a root tea, or used in a bath to be taken before bed time (W A R Thomson. 1978). PEPPERMINT tea is a sedative (Bircher), and LIME-FLOWER tea is also taken for insomnia, for it is slightly sedative. A hot bath with lime-flowers in it, is another insomnia remedy (Quelch).

In Chinese medicine, the seeds of the JUJUBE (Zizyphus jubajuba) are used to give sleep, and also to benefit the nervous system (R Hyatt).

Sedum acre > BITING STONECROP

Sedum telephium > ORPINE

Sempervivum tectorum > HOUSELEEK

SENEGA SNAKEROOT

(*Polygala senega*) The doctrine of signatures established its root as a remedy for snake-bite. One account says that it grows wherever there are rattlesnakes, so that it was used as a remedy for their bite (Dorman), and, of course, the Seneca Indians always used it for the purpose (Lloyd). A number of other drug roots bear the name Snakeroot; cf *Rauwolfia serpentina*, *Eryngium aquaticum*, *Asarum canadensis*, for example.

The Seneca Indians used to give it for coughs (Lloyd), and among the Meskwaki, it was the chief remedy for heart trouble (H H Smith. 1928). The Chippewa just carried it about them for general health and safety on a journey (Densmore).

SESAME

(*Sesamum indicum*) In spite of the specific name, this is an African plant, though it was taken to India at an early date. From there it spread to China (Brouk), and, in times of slavery, to Brazil. Sesame oil is known as benne in West Africa, teel (til) or gingili in India, and sim-sim in East Africa. A confection of sesame and honey has always been popular in China, and in America "bene" candy is traditional (G B Foster). The seed, soaked in sparrow's eggs, and cooked in milk, was used for centuries as an aphrodisiac. Sesame oil was a cosmetic, it was mixed with vinegar as an ointment for the forehead, "to strengthen the brain"!, and, blended with crow's gall, as an embrocation for impotence (Lehner & Lehner).

A correspondent of *Folk-lore. vol 65; 1954 p 51-2* says that in the Bahamas, if someone has to travel along dark roads at night, Benne (sesame) seeds should be carried. Sooner or later a spirit will try to catch the traveller, and then he should drop a few (actually more than ten) of the seeds in order to be quite safe, for the spirit has to stop and count them. As no spirit can say ten, it would have to start all over again when it had reached nine … and so on.

Sesamum indicum > SESAME

SEXUALLY TRANSMITTED DISEASES

The Thompson Indians from British Columbia made a decoction of SMOOTH SUMACH that was claimed to be a powerful remedy for syphilis (Teit). The bark of CHOKE CHERRY (*Prunus virginiana*) was used by other groups (Lloyd). EAGLE VINE (*Marsdenia condurango*), once thought to be a perfect cancer cure, is still used by the South American Indians in the treatment of chronic syphilis (Le Strange). There is a Moroccan belief that by smelling a NARCISSUS a person avoids catching syphilis, and, if he has already caught it, can cure himself of it (Westermarck). In Africa, the so-called GERMAN SAUSAGE TREE (*Kigelia africana*) whose fruits, often a metre or more long (hence the common name), are said to be used in cases of syphilis. In Yoruba practice, the root or bark of this tree is one of the ingredients of a medicine for gonorrhea (Buckley).

SHAGBARK HICKORY

(*Carya ovata*) "Shagbark", or "Shellbark" (Everett), because the bark curls away in strips (Barber & Phillips). An Alabama cough medicine is "shaggy hickory" bark boiled in water, strained off, and sweetened. A teaspoonful every few hours is the dose (R B Browne). The tea is good for an ordinary cold, too (H M Hyatt).

SHALLOT

(*Allium ascalonium*) Gardeners have it that to do well, shallots should be sown on the shortest day of the year, and pulled on the longest, with a slight variation from Cheshire, where they say they should be planted on Christmas Day. Shallot is claimed as a heart tonic, and it is used as such in Chinese medicine. Its action is to lower blood pressure, but in fact any Allium will do that. An earache was treated in Sussex by baking a shallot, and putting it in the ear (Sargent).

SHAMANISM

SWEET FLAG is used by the Dusun (Borneo) shamaness, who exploits the narcotic nature of the rhizome to contact certain beings. For a further description of the ritual, see under **SWEET FLAG**. Siberian shamans would use MARSH ROSEMARY to achieve trance state. It was done by inhaling the smoke of this shrub (or some other resinous plant). Some dried stalks or leaves would be put on some glowing embers on an iron plate, and the shaman would bend over the smoke produced, which would be dense and strong-smelling (Balazs). BULRUSH pollen was a sacred substance to the San Carlos Apache, where the medicine man would rub it on the affected parts of his patients, and then sing, after which he would extract the disease in the usual sucking way of a shaman. Sometimes he would put a little pollen in the patient's mouth (Youngken).

Some of the South American hallucinogens, *Banisteriopsis caapi*, for instance, are used by shamans to diagnose an illness or divine its cure, or to establish the identity of an enemy (Reichel-Dolmatoff). Herodotus mentions HEMP in describing the funeral rites of the Scythians. He talked of the way they purified themselves after the burial by putting hemp seed on hot stones inside a small felt-covered structure, and inhaling the smoke it produced. This procedure must have been a form of shamanism, for the smoke of the hemp seeds would produce a trance-state in those inhaling it (Balazs).

Yaqui shamans used SCOTCH BROOM flowers, smoked as cigarettes. After several cigarettes, the desired "long relaxation, intellectual clarity and physical ease, coupled with psychological alertness" would be achieved (Emboden. 1979).

SHAMROCK

Is WHITE CLOVER (*Trifolium repens*) the shamrock, the emblem of Ireland? That belief is not as old as may be supposed. In fact, it is first mentioned in Tudor times. Campion's *Historie of Ireland* (1571) refers to "shamrokes, water cresses and other herbs they feed on". A hundred years later, shamrock is clearly identified as an Irish emblem. Thomas Dinely, who made a tour of Ireland in 1681, wrote in his journal of "shamroges" (Sheehy). All this in spite of the legend that had St Patrick himself using the trefoil leaf of a clover as an illustration of the Holy Trinity (it is still used as the emblem of the Trinity (Haig)). But, as Nelson says "the shamrock had not been invented in 700 AD". Shamrock is actually Irish seamrog, the diminutive of seamar, clover (Lockwood). But the identity of the plant that truly bears the name has been a vexed question of a great many years. The white clover is, or used to be, worn as such on St Patrick's Day, though *Medicago lupulina* was also sold as the shamrock. Threlkeld, the earliest writer on the wild flowers of Ireland, gives Seamaroge as the Gaelic name for white clover, and expressly says this is the plant worn by the people in their hats on St Patrick's Day. Colgan said that *T repens* and *T dubium* (Lesser Yellow Trefoil) were equal candidates for the true shamrock. A recent survey carried out by Nelson showed that *T dubium* was recognised by most people as "shamrock". 46% claimed this while 35% claimed White Clover. Sometimes it was thought to have been *Oxalis acetosella*, WOOD SORREL, probably because the Gaelic name, seamroge, does for both (Sheehy). BLACK MEDICK, *Medicago lupulina*, has also been claimed by a few. GERMANDER SPEEDWELL (*Veronica chamaedrys*) also bears the name, and one suspects, in this one case only, it is *chamaedrys* in another guise.

SHEEP'S BIT

(*Jasione montana*) Picking the flowers is supposed to give one warts, so a Cornish superstition holds (Vickery. 1995).

SHEPHERD'S PURSE

(*Capsella bursa-pastoris*) Yorkshire children would open a seed vessel. If the seed inside is yellow, you will be rich, but if it is green, you will be poor (Opie & Tatem). Another children's game, if it can be called a game, was played with the seed pod. They hold it out to their companions, inviting them to "take a haud o' that". It immediately cracks, and there follows a triumophant shout, "You've broken your mother's heart" (Dyer. 1889), or, in Middlesex, "You've picked your mother's heart out" (Vickery. 1995). The plant actually bears the name Pick-your-mother's-heart-out (Grigson. 1955), while names like Mother-die and Mother's Heart (Vickery. 1985; J D Robertson) are relatively common. Pickpurse is a name used by Gerard, and Pickpocket is quite common (Jones & Dillon; Tynan & Maitland, etc.,). These and others like them are usually the result of the barren soil in which the plant thrives; in a sense, it robs the farmer by stealing the goodness of his land. There is a rhyme fom Northamptonshire:

> Pickpocket, penny nail,
> Put the rogue in jail (A E Baker).

Then there is Pickpocket-to-London, from Yorkshire:

> Pick-pocket to London,
> You'll never go to London (Grigson, 1955).

From Scotland, Rifle-the-ladies'-purse (Grigson. 1955).

Clappedepouch (Prior), or Rattlepouch (Grieve. 1931) are both allusions to the licenced begging of lepers, who stood at crossroads, with a bell and clapper. They could receive their alms in a cup or basin at the end of a long pole, likened in these names to the purses of Shepherd's Purse in this context (Prior). St James's Wort is another name for this plant (see Grigson. 1955; Vesey-Fitzgerald), deriving, according to Tynan & Maitland, from the leather pouch carried by the poorer pilgrims en route to the shrine of St James at Compostella. All these purses, pouches, pounces, bags, scrips, belonging to shepherds, gentlemen, poor men, or ladies are descriptive names, the shape of the seed vessels suggesting the names.

Shepherd's Purse belongs to the cress family, and it is perfectly edible, and was at one time eaten as a potherb. It tastes something like cress, and when dried, the leaves make a peppery flavouring for soups and sauces (Loewenfeld). Apparently, it was specially grown (in good soil, so that it was very much bigger) in America, as a green vegetable (C P Johnson). It is, too, a great medicinal herb, still used to stop bleeding. A 17th century physician, Symcott by name, was treating a pregnant woman for blood. Then "a beggar woman told me that she would recover if she took shepherd's purse in her broth". She was cured (Beier). It is still being recommended for similar purposes. The powdered plant mixed with the normal diet has been used to inhibit the oestrous cycle (Watt & Breyer-Brandwijk). Langham had a different method of achieving the same end: "nothing is better to stop the flowers, than to make a fomentation or moyst-bath thereof, and to sit over it close, and to drinke of the same classified in red wine". Mrs Leyel called it "the great specific for haemorrhages of all kinds". Such old names as Stanche and Sanguinary tend to confirm her claim.

All this is far from the total of its virtues. Gypsies, for instance, use a leaf infusion as an ingredient of medicines that are employed in the treatment of oedema, jaundice, as well as kidney and liver complaints. As an ointment, it is good for erysipelas (Vesey-Fitzgerald), and North American Indians used it as a cure for another skin problem, poison-ivy rash (H H Smith. 1923). The Cheyenne also took a cold infusion of the powdered leaves for a headache (Youngken). It has even been used for malaria in Europe, while Ô Súilleabháin quotes its use as an Irish remedy for rickets.

SHINGLES

HOUSELEEK is a protector against fire and lightning, so similar ideas would account for its use against the fiery diseases – "they are good against S Anthonie's fire [erysipelas], the shingles ... (Gerard), who made the same claim for OLIVE "branches, leaves and tender buds". A Hertfordshire remedy was to mix the blood from a black cat's tail with the juice of houseleek and cream, warm it all, and apply it three times a day (Jones-Baker. 1974). It sounds like the fusion of two traditions, for styes on the eye used to be treated in old wives' lore by rubbing them with the tail of a black cat. (see Woodforde for example). Homeopathic doctors sometimes prescribe POISON IVY for the complaint (Homeopathic Development Foundation). A compress made of the infusion of SCENTED MAYWEED is another possibility (Flück). Thomas Hill, in 1577 was of opinion that PURSLANE "helpeth the shingles", so, he claimed, does CHICORY. WHITE HOREHOUND was made up in Wales as an infusion, to be used both externally and internally, for eczema and shingles (Conway), and MARIGOLD is still used by herbalists for both chickenpox and shingles (Warren-Davis).

SHITTAH

It is generally agreed that the hebrew Shittah must be either *Acacia seyal* or the Umbrella Thorn, *Acacia tortilis*, for they are the only timber trees of any size that grow in the Arabian desert. The references in the Bible are always in connection with the ark of the tabernacle, which, along with its altar and table, was ordered to be made of this wood. The Greek for shittah tree means wood not liable to rot (Moldenke & Moldenke). In some parts of California in the early days, the term Shittim Bark was applied to the bark of CALIFORNIAN BUCKTHORN. It was said locally to be the Shittim wood of which the Hebrew ark was made (Maddox).

Shorea robusta > SAL, or SAUL TREE

SHRUBBY WORMWOOD

(*Artemisia arborescens*) Gubernatis said that it was this plant that the women of Avola, in Sicily, used to make Ascension Eve crosses. They put them on house roofs, believing that Jesus, on ascending to heaven, would bless them. The crosses were kept for a year, and put in the stables to be used to "calmer les bêtes indomptables".

SIBERIAN LARCH

(*Larix sibirica*) The larch was a sacred tree among the Siberian Tungus. It is the tree Tuurn, in which the soul of the shaman is reared. The souls lie in nests in the boughs, and the higher the nest in the tree, the stronger will be the shaman who nests in it. The rim of the shaman's drum is cut from a living larch. At each séance the shaman plants a "tree" with one or more cross-sticks in the tent where the ceremony takes place, and this tree too is called Tuurn. According to their belief, the soul of the shaman climbs up this tree to God when he shamanizes, for the "tree" grows during the rite, and invisibly reaches the summit of heaven (Joseph Campbell).

Silene dioica > RED CAMPION

Silene latifolia > WHITE CAMPION

Silene vulgaris > BLADDER CAMPION

Silphium laciniatum > COMPASS PLANT

Silybum marianum > MILK THISTLE

SILVER FIR

(*Abies alba*) A symbol of fertility, for it was sacred in ancient Greece to Artemis, the moon-goddess who presided over childbirth (Graves). Hildesheim women were struck with a small fir-tree at Shrovetide, and in north Germany, brides and bridegrooms aften carried fir-branches with lighted tapers, while at Weimar, firs were planted before a house when a wedding took place. It was also used at weddings in Russia (Hartland). In Orkney, mother and child were "sained" soon after delivery with a burning fir-candle whirled three times round the bed (Graves).

On Midsummer Night in Germany, fir-branches are decorated with flowers and coloured eggs, and the young people dance round them, singing rhymes (Dyer. 1889). South Germans use this as their Christmas Tree. In fact, Christmas Tree is an accepted name for this species. One piece of weather lore is recorded. The cones will close for wet, and open for fair, weather (Wright). Dreaming of being in a forest of firs is a sign of suffering (Gordon. 1985).

SILVER WATTLE

(*Acacia dealbata*) In spite of the fact that this is the Australian national emblem, it is a generally unlucky plant to bring indoors, even sometimes unlucky to plant in gardens. There are a few records in England, too, of its being unlucky to bring indoors, even being described in one record as "a forewarning of disaster".

SILVERWEED

(*Potentilla anserina*) The roots of silverweed were a marginal or famine food in the Scottish Highlands

(Grigson. 1955), and in Ireland (Drury. 1984), and they were known as a food of the fairies, too (MacGregor). The roots were roasted or boiled (Ferneie), or even eaten raw, or they could be ground into meal to make porridge, and also a kind of bread (Drury. 1984). Perhaps not so marginal, for Carmichael says that it was used a lot before the potato was introduced; it was cultivated, so it grew quite large. Records of cultivation go back to prehistoric times, and the Anglo-Saxons are known to have grown silverweed as a root crop (Jordan). Particularly remembered for the cultivation of Brisgein (its Gaelic name) was an area in North Uist, Outer Hebrides, where a man could sustain himself on a square of ground of his own length (Carmichael). The Gaelic Bliadhna nan Brisdeinan means Year of the Silverweed roots. This year was shortly after Culloden, and is remembered in Tiree as a year of great scarcity. The land had been neglected in previous years because of the state of the country, and the silverweed sprang up in the furrows, and people made meal of them (Campbell. 1902), the "seventh bread" (MacGregor). Martin records the use on Tiree, as does Duncan. Children were still digging it up in recent times. They know that putting the roots for a moment on red hot cinders makes them swell a little, and makes them taste sweeter, rather like parsnips (C P Johnson).

Campbell. 1900 mentions silverweed roots as fairy food. They lived on the roots that were ploughed up in spring (Spence. 1946). The tops of young heather shoots were fairy food, too, but this was a Lowland Scots tradition (Aitken). Another superstition connected with silverweed is that of putting a sprig inside each shoe to prevent blisters when walking long distances (Freethy), hence the names Traveller's Ease and Traveller's Leaf.

It is known that silverweed has been used to try to remove the marks of smallpox (Billson), and it was certainly used, steeped in buttermilk, as a cosmetic, to remove freckles and brownness (Black). Gerard also advised its use: "the distilled water takes away freckles, spots, pimples in the face, and sun-burning: but the herb laid to infuse or steep in white wine is far better: but the best of all is to steepe it in strong white wine vinegar, the face being often bathed or washed therewith". The cosmetic use was recorded even earlier, in Anglo-Saxon times (Cockayne).

A compress made from the chopped herb is said to be good for piles (Thomson. 1978), and a strong infusion will stop their bleeding (Wickham). It has been used for stomach cramp (Fernie), and a decoction is claimed to be a cure for mouth ulcers (Wickham). Boiled in salted water, it "dissolves clotted and congealed bloud in such as are hurt or bruised by falling from some high place" (Gerard).

Sison amomum > STONE PARSLEY

SKULLCAP

(*Scutellaria galericulata*) The common name is descriptive, and indicates its signature. According to this doctrine, the plant would be used to treat head-related ailments, anything from insomnia to madness. Other names given to it are equally descriptive, as for instance, Helmet-flower, or Hoodwort (Grieve. 1931). It is one of the herbs traditionally held to cure barrenness (Conway).

SLIMMING

Using GOOSE-GRASS (or CLEAVERS, which is as commonly used for *Galium aparine*.) see Gerard: "Women do actually make pottage of Cleavers with a little mutton and Ote-meale to cause lanknesse, and keepe them from fatness".

SLIPPERY ELM

(*Ulmus fulva*) An American tree, best known as the medicinal slippery elm, the inner bark or cambium, the new tissues of which are formed each spring, and taken for among other complaints, stomach ulcers. It could be chewed for a cough (Corlett), or worms could be dealt with by taking it – "makes the intestines so slippery the worms can't hold on" (H M Hyatt). The American Indians made a soothing drink for fevers from it, and they also used it as an external application in some skin diseases (Lloyd), a practice that lasted in domestic medicine. For example, mixed with lard, slippery elm bark is applied to a boil in Alabama (R B Browne), or a paste could be made by pouring boiling water over the bark, and that could be put on the boil (H M Hyatt). One oddity from Alabama is a recipe for blood poisoning resulting from spiders, insects or snake injuries. The bark has to be powdered with that of sassafras, in equal parts, and this should be put into a pan with enough water to be absorbed. Then the mixture is to be put into a gauze bag and applied to the affected part (R B Browne). But in Illinois they claim that blood poisoning can be cured simply by drinking slippery elm tea (H M Hyatt).

SLOES

are the fruit of the BLACKTHORN (*Prunus spinosa*). The name is OE sla(h). Robert Graves pointed out that the words sloe and slay are closely related in early English, something that may account for the unlucky label given to the tree. The name appears in various guises as dialectal forms. Slue is a Wiltshire variant (Dartnell& Goddard), and the word has appeared also as Slow (Drury. 1992), or Slea (A E Baker). Slon, or Slan (Grigson. 1955), are modern singulars of the ME plural slon and OE plural slan, the well-known plural ending in '-n', is still in use in words like children. Then we have a double plural, for in Wiltshire the plurals of Slon and Slan, are Slons and Slans.

SMALL-LEAVED LIME

(*Tilia cordata*). The flowers are used in infusion to treat catarrh, and the decoction of the bast for rheumatism (Schauenberg & Paris). Rackham. 1986 uses PRY as if it were the common name.

SMALLAGE

An older name for CELERY. Ach, or Ache, is a comon archaic word for the umbelliferous plants, with parsley perhaps the main recipient of the name. It is probably an O French word, directly from *Apium*. Smallage is a name derived from Ach, and celery is very commonly known by this name. Its variants are Smalach or Smallache (Britten & Holland), along with Smalladge and Smalledge (Parkinson).

SMOKE TREE

see **VENETIAN SUMACH** The tree probably is better known in Britain as Smoke Tree rather than as Venetian Sumach. The name, of course, is descriptive of the tree when it is in bloom.

SMOOTH SUMACH

(*Rhus glabra*) Much used by native Americans, mainly as a dye plant. The Ojibwa, for example used the pulp of the stalk to produce yellow (Buhler), while the Omaha and Winnebago used the roots for the same purpose (Gilmore). The Plains Indians used to dry the autumn leaves for smoking (Gilmore). They used the shrub widely, too, for medicinal purposes. One was to make a styptic wash from the boiled fruit to check bleeding (Sanford), especially to stop bleeding after childbirth (Corlett). The powdered seeds would also have been applied to wounds, and to treat piles. The juice of the fresh fruit was used for warts and for skin diseases like tetter, while the fruit decoction was taken as a gargle for quinsy, mouth and throat ulcers, and as wash for ringworm (Lloyd). It was even said that the Thompson Indians of British Columbia made a decoction that was claimed to be a powerful remedy for syphilis (Teit).

Kiowa Indians claimed that it is not a medicine in itself. But they were deeply involved in the peyote cult, and sumach, they said, was used to "purify" the body and mind so that peyote, the real medicine, could effect a cure more easily (Vestal & Schultes).

Smyrnium olusatrum > ALEXANDERS

SNAKE'S HEAD LILY

(*Fritillaria meleagris*) Snake's-head, from the shape of the flower buds, while Fritillary comes from the Latin fritillus, a dice-box; it is a reference to the chequered flowers, and to the chequer-board on which the dice were thrown. One of the names for this plant is Bloody Warrior, which looks odd; it seems that it was one of the flowers which were supposed to grow from a drop of Dane's blood. Another name is Leopard Lily, and that looks just as strange, until Lazarus Bell is considered. It is named after the small bells that lepers were made to carry about with them, so that they could warn the healthy of their approach. Hence, too, Dead Man's Bell (Leyel. 1948). As Friend. 1883 pointed out, Leopard's Lily is just a corruption of leper's lily.

SNAKES

In Sweden, they said that snakes would lose their venom by a touch of a HAZEL wand (Fiske). Elsewhere it was said that snakes cannot approach the tree (Kelly). It was by use of a hazel stick that St Patrick drove the snakes out of Ireland (Wilde. 1890). To this day, Dartmoor people say that if a dog is bitten by an adder, a hazel wand should be twisted into a ring and put round the animal's neck (St Leger-Gordon), or, according to another authority, hold the dog's head over a cauldron in which ELDER flowers were being boiled (Whitlock. 1977). A very long way from Dartmoor, in the Balkans in fact, it is said that a young hazel twig, cut after sundown on St George's Day (a significant saint in view of this belief) should be used to rub a snakebite wound, and to draw a magic circle round it (Kemp). But in spite of all this, there was a German belief in something like a hazel serpent – a crowned white snake that lived beneath the hazel tree. The snake is a symbol of wisdom, and the hazel too bears that symbolism. There was also a Welsh belief that a snake found under or near a hazel on which mistletoe grew, would have a precious stone in its head (Trevelyan), presumably another way of describing the traditional wisdom of the snake. GREAT PLANTAIN, too, was thought to provide protection against snakes. The Anglo-Saxon version of Apuleius advised the patient who had been bitten to eat the plant! (Cockayne). But American Indian peoples in general agreed with the European verdict; the Ojibwe, for instance, actually carried plantain about with them, for immediate emergency protection (Densmore). CENTAURY had a great reputation in early times for healing snake-bites. From the Anglo-Saxon version of Apuleius, there is "for bite of snake, take dust of this wort, or itself pounded; administer this to the patient in old wine" (Cockayne), and in the 15th century, there is a leechdom for "biting of an adder. Grind centaury and butter and give the sick to drink and it will help both man and beast". Topsell added his weight. For snake-bite, he wrote "… after cupping glasses and scarifications, there is nothing that can be more profitably applied than centaury …". AGRIMONY was another plant with a great reputation. One Anglo-Saxon leechdom advised the victim to make a ring with this plant "about the bite, it (the poison) will not pass any further …" (Storms). Culpeper recommended it, and Lupton noted that agrimony by itself was enoug:. "… with a wonderful facility [it] healeth the bites of serpents and other venomous beasts".

ASH has the power to repel serpents, such was the belief, from Pliny's time onwards. He said that snakes would rather creep into a fire than come in contact with it. Bartholomew Anglicus repeated the belief: "And if a serpent be set between a fire and Ash-leaves, he will flee into the fire sooner than into the leaves" (Seager). Evelyn knew about the "old imposture of Pliny's, who either took it upon trust, or we mistake the tree", and Gerard also repeated the belief, less critically: "The leaves of this tree are of so great virtue against serpents that they do not so much as touch the morning and evening shadows of the trees, but shun them afar off". Trees were actually planted round houses, to keep adders away. In Devonshire, they said it only needed a circle drawn round an adder with an ash stick to kill it, and a circle of ash twigs round the neck cured adder bite.

ELDER, so it is said, is distasteful to snakes. Thomas Hill, in 1577, wrote that "if the Gardener bestoweth the fresh Elder where the Serpents daily haunt, they will hastily depart the place". The Physicians of Myddfai prescribed "for the bite of a snake … let the juice of the elder be drunk, and it will disperse all the poison". PARSLEY also had a reputation of being able to destroy poison, so would cure snakebite (see Anglo-Saxon Apuleius again – "for bite of adder, take some very small dust of … parsley, by weight of a shilling, give it to drink in wine; then take and lay to the wound the wort pounded" (Cockayne). Topsell suggested that "apium seed", CELERY, that is, was used for snakebites. PIMPERNEL leaves were used for dog- and snake-bites. They were prepared by powdering, and, with the root, were made into an infusion. A teaspoonful of this, or about 20 grains of the powder, were put into a cup, with 15 drops of spirits of hartshorn, and a dose given every six hours. This would be continued for 15 days. Rags steeped in the decoction were applied to the wound, and the underlinen was soaked, too (Trevelyan). Lupton, in the 17th century claimed that "if you would kill snakes and adders, strike them with a large RADISH, and to handle adders and snakes without harm, wash your hands in the juice of Radishes and you may do so without harm", Did anyone actually try that?

Sometimes it is the physical characteristics of a plant that makes it an antidote to snake-bite. ADDER'S TONGUE, for example, was used, for the doctrine of signatures ensured that this was appropriate. Another example involves DRAGON ARUM. Gerard described the plant as having "spots of divers colours, like those of the adder or snake …". It comes as no surprise, with that kind of signature, to find it being used long ago against snakebite, and, as it is easy to combine prevention and cure, what will cure snakebite will also prevent it. Gerard again: "it is reported that they who have rubbed the leaves or root upon their hands, are not bitten of the viper". He ascribed this to Pliny, who "saith that serpents ill not come neere unto him that beareth dragons about him". The white-veined leaves of the American plant known as CREEPING LADY'S TRESSES, looked like snakeskin to the early herbalists, who took it is as a sign for the virtues against snakebite (Cunningham & Côté). That is why it got the name Rattlesnake Plantain. The rhizomes of WATER ARUM (*Calla palustris*) also contain an acrid compound used as an antidote to snake venom (Schauenberg & Paris). Given the common name of *Scorzonera* humilis, VIPER'S GRASS, the herb was bound to cure snakebite. See Gerard: "It is reported by those of great judgement, that Viper's-grasse is most excellent against the infections of the plague, and all poysons of venomous beasts, and especially to cure the buitings of vipers…". A plant like VIRGINIAN SNAKEROOT, whose roots are writhed, and look like snakes, would similarly provide the remedy, though it must be said that native Americans were using it first. All they did was chew the root and apply it, or spit it, on to the bite (Weiner), and chewing the leaves is still an Indiana way of treating it (Tyler). The same applies to SENEGA SNAKEROOT. One account says it grows wherever there are rattlesnakes, so naturally it would be used to cure their bite (Dorman). Similarly, SUNFLOWER seeds were used by the American Indians – the seeds were chewed and made into a poultice, and this was taken to be an antidote to rattlesnake-bite (Stevenson). GOATWEED (*Ageratum conyzoides*) is an aid to catching snakes – the leaves are crushed and rubbed on the hands. Presumably the rather unpleasant smell (the plant is not called Goatweed for nothing) affects the snakes in some way, and the Mano tame a snake by using the bud leaves of MELEGUETA PEPPER. They are chewed, and spat on the snake's head to calm it (Harley).

The Greeks believed that snakes had recourse to FENNEL to cure blindness (Grieve. 1933), and it was popularly believed in the Middle Ages that dragons could cure their blindness, "apparently a common affliction of dragons", by rubbing their eyes with fennel, or eating it (Hogarth). The other point about fennel and snakes is that " so soone as they taste of it they become young again …" (Hulme. 1895). There was a French superstition that LADY'S SMOCK was the favourite flower of snakes. Mothers warned children not to touch them, for fear of being bitten by a snake some time in the coming year (Sebillot). They are certainly unlucky flowers, not to be brought indoors. STITCHWORT, too, must not be gathered, for it is a fairy plant. In addition, Cornish children say they will be bitten by an adder if they pick it (Macmillan). There are a variety of "snake" names given to this plant, Snake-grass, Adder's Meat, Hagworm-flower, and so on. LAVENDER, so it was thought at one time, was the home of snakes, but other early writers assured their readers that the herb actively discouraged "venomous wormes" from coming near it (Hulme. 1895). In central Africa, the roots of CHINESE LANTERN (*Dichrostachys glomerata*) are chewed and macerated, to put

on snake or scorpion bites to remove the poison (Palgrave).

SNAPDRAGON

(*Antirrhinum majus*) Wear the seeds round your neck and you will never be bewitched (B L Bolton). It is said that snapdragons are grown on the roof in parts of Ireland to bring luck, and to guard against fires (Vickery. 1995).

SNEEZEWEED

(*Helenium autumnale*) It was the custom with some American Indian people to dry the almost mature flower heads, and keep them in a loose bunch, which is hung from the rafters of the house. When wanted for use, it was pulverized and used as snuff. It makes one sneeze violently several times, and so would be used to loosen up a cold. The name of the genus, *Helenium*, has been explained away by claiming that these are the flowers that sprung from the tears of Helen of Troy, and the genus has been known sometimes as Helen Flower (W Miller). Actually, it is from the Greek name, helenion, for another plant, perhaps *Inula helenium*.

SNEEZEWORT

(*Achillea ptarmica*) "The smel of this plant procureth sneezing" was Gerard's comment. Indeed, the powdered leaves were once used for that purpose, eiher medicinally, or as a cheap snuff substitute (Coats. 1968). The French equivalent is 'herbe à éternuer'; the specific name tells the same story, for *ptarmica* is from Greek stamos, sneezing. The chief medicinal use is against toothache, as a native substitute for Pellitory-of-Spain (Grigson. 1955).

SNOWBERRY

(*Symphoricarpos albus*) A North American shrub, once much used by native Americans. A root decoction could be taken for colds and stomach ache (Barrett & Gifford), and bark was scraped into warm water to be used as an eyewash. They were also in the habit of rubbing the berries on rashes, sores and burns (Turner & Bell) – something obviously never heard of in Kent, where a modern belief is that the juice of the berries would actually cause warts (Vickery. 1995).

SNOWDROP

(*Galanthus nivalis*) One of the origin legends is that an angel was comforting Eve after the Fall. No flower had bloomed since the expulsion from Paradise, and it was snowing. The angel caught a snowflake in his hand, breathed on it, and it fell to earth as the first snowdrop (Gordon. 1977). But it is an unlucky plant; more specifically, it is unlucky to bring indoors. Some say the bad luck applies only to cut snowdrops, and not to those grown in pots indoors (Vickery. 1995). Others, in Wales, say the sanction applies only to snowdrops taken indoors on St Valentine's Day (L Davies). The result of such rash actions were equally variable, ranging from the death of someone living in the house, as in Wiltshire and North Wales (Vickery. 1995) to the cows' milk being watery and affecting the colour of the butter (Burne. 1883). In Somerset the belief was that the girl child in the house would die within the year (Tongue. 1968). There was a similar superstition there about Lily of the Valley and white lilac, but this is so widespread that there is no point in trying to localise it. Taking snowdrops into a hospital is even more unlucky. If they were given to a patient, it was often taken to be a death sentence. Nurses would sometimes put a few ivy leaves with them to lessen the omen (Tongue. 1967). It is said that the association of snowdrops with death (they used to be called Death's Flower in Somerset) results from the flower's resemblance to a shroud (Vickery. 1985). "It looked for all the world like a corpse in its shroud" was how one of Charlotte Latham's informants put it (Latham), though one's imagination would have to be heightened to recognise it. Another reason given to her was that "it always kept itself close to the earth, seeming to belong more to the dead than to the living". Again, we hear that the reason is that they are so often found growing in old graveyards (Vickery. 1985).

But the plant is not always associated with ill luck. Along the Welsh border, when the Christmas decorations were removed at Candlemas, a bowl of snowdrops was sometimes brought in to drive out evil, and to mark the beginning of the spring season, giving the house what was known as "the white purification" (Drury. 1987). It was said, too, in lowland Scotland, that the finder of a snowdrop before the first of January would be in luck all the coming year (Tynan & Maitland). The mention of "purification" recalls the plant's connection with Candlemas (it is called Candlemas Bells in a number of areas (Macmillan; Britten & Holland; Banks. 1937)), for Candlemas is the festival that falls on 2 February, when we can expect snowdrops to be in bloom. The feast is also called the Purification of the Virgin Mary, hence the flower is known as Purification Flower (Prior). Fair Maids of February, or February Fair-maids (Macmillan), reduced in Norfolk to Fair Maids (Britten & Holland), are all references to the February festival of Candlemas, when it was the custom for the Fair Maids, dressed in white, to walk in procession to the feast (Prior). A stage on from this would mean that the plant was sacred to virgins in general, and at one time the receipt of snowdrops from a lady indicated to a man that his attentions were not wanted (Prior). Near Hereford Beacon (the one place on Britain where it may be native), a bowl of snowdrops brought into the house on Candlemas Day (evidently the only day on which such a thing could be done with safety), was thought to purify the house (Coats).

Snowdrop has been known to be prescribed for polio, though it has actually very little, if any, therapeutic

value, but a decoction of the bulb has been used as an emetic in eastern Europe (Schauenberg & Paris). An ointment made from the crushed bulbs is used as a cure for frostbite and chilblains, and an extract from the bulbs has been used to treat glaucoma (Conway).

SNUFF
Tobacco was not the only herb taken as snuff, and there were many cheap substitutes at one time, the powdered leaves of SNEEZEWORT (*Achillea ptarmica*) being one of them (Coats. 1968). The American plant SNEEZEWEED (*Helenium autumnale*) was used as snuff by some of the American Indian peoples. The almost mature flower heads would be dried, and hung up somewhere to be used as necessary. When wanted for use they were pulverized and taken as snuff. It makes one sneeze violently several times, and so is used to loosen up a cold (H H Smith. 1923/1928). The dried and powdered rhizome of SWEET FLAG has been used in snuff – the French 'snuff à la violette' apparently contains some of it (Barton & Castle). CAMOMILE was popular in Elizabethan times as the principal ingredient (along with ground ivy and pellitory-of-the-wall) in snuffs (Genders. 1972). Dried BASIL leaves can be taken as snuff, but this is to get rid of a headache, or a head-cold (Hemphill). Many early snuffs contained GROUND IVY, especially during Elizabethan times (Genders. 1972).

SOAP
The roots of the American WHITE POND LILY were used by native American groups as a soap substitute (Sanford). Another American plant that gives a soapy lather is *Chlorogalum pomeridianum*, called SOAP PLANT, or SOAPROOT (Schery). See also SOAPWEED below. SOAPBERRY (*Sapindus saponaria*) has fruits that lather well with water, and are used for washing clothes, etc., while the fruits of Indian members of the genus (*S mukorossi*, for example), are reserved for washing valuable fabrics (see SOAPBERRY below). At one time SPEARMINT oil was used to perfume soap (Fluckiger & Hanbury), and OLIVE oil was always used instead of soap in ancient Greece.

MILKWORT is known as Fairy Soap in Donegal, for it was believed there that the fairies make a lather from the roots and leaves (Grigson. 1955). But the best known soap plant is SOAPWORT (*Saponaria officinalis*), from which a lather can be got by rubbing the leaves in water. The species was used by the Greeks and Romans for washing clothes, and probably it was the Romans who introduced it into Britain. They certainly knew about its water softening properties (Thomson. 1976). It was used in the East as far back as the 10th century for cleaning clothes and carpets. It is still regarded as valuable for cleansing and restoring old tapestry without damaging the fabric

(Brownlow). In the Swiss Alps, sheep were washed with it before they were shorn (Grigson. 1955).

SOAPBERRY

(*Sapindus saponaria*) A New World tree, with white flowers followed by brown fruits, whose pulp lathers with water, and can be used as a soap substitute. They are much used for the purpose in Mexico (Roys), but they are poisonous, like those of the rest of the genus, for they contain a high percentage of saponin, and cause dermatitis (Kearney) (Indian members of the genus, like *S mukorossi*, are used as fish poison). These berries have some ritual significance, for Mexicans use them for necklaces and rosaries, and in ancient times the temple court was swept with the leaves of this tree in connection with the Maya baptismal ceremony (Roys).

SOAPWEED

(*Yucca glauca*) As with other members of the genus, the root was used like soap by the native Americans. Once the root bark is stripped off, the root can be pounded in cold water to make a lather. Blankets were washed in this way (Stevenson); in fact, the Navajo, in washing wool, prefer to use Yucca roots, because there is no grease or fatty substance in it, and they also say that they have a greater cleansing power than soap (Elmore). As with Datil (*Yucca baccata*) the special, ceremonial use is for hair shampoo – the Pueblo Indians used it as part of the ritual in initiation ceremonies (La Fontaine), though people like the Kiowa claimed it was an effective cure for dandruff and baldness (Vestal & Schultes).

The Navajo have other uses for Soapweed; the juice will produce a black dye, and the fibres provide them with basketry materials (Kluckhohn). Occasionally, it was used as a medicine. The Pomo, for instance, used the root to make a lotion to put on a poison ivy rash (Weiner), or sometimes, as with the Neeshenam, it was used to heal and cleanse old sores, being heated and laid on hot (Powers). Apparently, the Hopi used it as a strong laxative (Whiting). But it was also used as a fish poison.

SOAPWORT

(*Saponaria officinalis*) One can get a lather by rubbing the leaves in water. It was used in the East as far back as the 10th century for cleaning clothes and carpets, and it was being cultivated in Syria in recent times for the purpose (Grigson. 1955). It is still a valuable means of cleansing and restoring old tapestries without damaging the fabric (Brownlow). In the Swiss Alps, sheep were washed with it before they were shorn (Grigson. 1955). Gypsies use a decoction of the root to apply to a bruise or a black eye (Vesey-Fitzgerald), or a freshly dug root put on the black eye would do (Campbell-Culver, though that might take longer to achieve the desired effect). The root is useful for skin diseases, too (Mitton & Mitton),

particularly for the itch (C P Johnson), though boils are also treated with it (Schauenberg & Paris). Wesley specially mentioned "Ring-wormes", for which he advised his readers to "wash them with a Decoction of Soap-wort".

Solanum carolinense > CAROLINA NIGHTSHADE

Solanum dulcamara > WOODY NIGHTSHADE

Solanum nigrum >BLACK NIGHTSHADE

Solanum tuberosum > POTATO

Solidago virgaurea > GOLDEN ROD

SOLOMON'S SEAL

(*Polygonatum multiflorum*) On cutting the roots transversely, scars resembling the device known as Solomon's Seal, a 6-pointed star, can be seen – so, from the doctrine of signatures, it was used for "sealing" wounds (Dyer. 1889). It is said that Lilies-of-the-valley only thrive with Solomon's Seal, "their husbands", nearby (M Baker. 1977).

Gypsies use an ointment made from the leaves to apply to a bruise or black eye (Vesey-Fitzgerald). So did Fenland peoples (Porter. 1974), and it is said that a Sussex pub landlord grew beds of this plant specially to put on the black eyes received after a pub fight. The ointment was made by bruising the roots or leaves and mixing them with lard (Sargent). Gerard mentioned their use for bruises "gotten by falls or womens wilfulness, in stumbling upon their hasty husbands fists, or such like". In Scotland, the treatment was described as "the root of Solomon's Seal, grated, and sprinkled on a bread poultice" to remove "bruise discolourations" (Gibson).

It was used for broken bones, too, either taken inwardly in ale, or as a poultice (Grigson. 1955), a use mentioned by Gerard, who said "there is not to be found another herbe comparable to it". See also: "The root stamped and applied in manner of a pultesse, and laid upon members that have been out of joynt, and newly restored to their places, driveth away the paine, and knitteth the joynt very firmely …", perhaps that is why it is known as "seal", conjectured Mrs Leyel (Leyel. 1937).

It is used in the Balkans for a sore throat, either as a decoction to be drunk, or as a poultice round the neck and chest (Kemp). In parts of France the plant is known as 'l'herbe à la forcure', and it was said that the roots represent all the parts of the human body. It was used for muscular cramp, and great care was taken to use that part of the root which doctrine taught represented the part of the body affected (Salle). The ointment had a veterinary use, too, for ulcers and wounds in horses and cattle (V G Hatfield. 1994). There is one more usage to record. Gerard wrote "Matthiolus teacheth, that a water is drawne out of the roots, wherewith the women of Italy use to scoure their faces from Sunne-burning, freckles, mirphew or any such deformities of the skinne". What is interesting is that a soap was made until recent times, from Solomon's Seal, in the Kingsclere district of Hampshire, for just the purpose outlined by Gerard (Read).

SOMA

is defined either as a plant or as its intoxicating juice, used in ancient Indian religious ceremonies, and regarded as a god in its own right, or at least as divine power. Then it is as nectar, the drink of the gods, a symbol of immortality and victory over death. At first, the juice was confined to the gods and the priests, but gradually it was extended to the three chief castes, giving to the gods and men strength, to the poet inspiration, to the old renewal of virility and ultimately "a blissful immortality, together with all the blessings it bestowed" (James. 1966). Some say that the MOON PLANT (*Sarcostemma acidum*) is the true soma, or at least one of the plants used as a substitute for it. It contains a large amount of milky sap, acid, and often used by Indian travellers to allay thirst (Chopra, Badhwar & Ghosh). There is certainly a connection between the moon and soma. The plant was gathered by Vedic priests by moonlight, and was washed in water and milk in order to identify it with its counterpart, the moon. "We have drunk soma; we have become immortal; we have gone to the light; we have found the gods" (Rig-veda), and both soma and the moon are associated with rain and fertility (James. 1966)

SON-BEFORE-THE-FATHER

is a name given to all plants that flower before the leaves appear, COLTSFOOT, for instance (Grigson. 1955). It is Filius ante patrem, and the Dutch have Zoon-voot–de-vader (Clair). MEADOW SAFFRON (*Colchicum autumnale*) is another, with Daughter-before-the-mother (Grigson. 1959), as an alternative. CUDWEED, too, and as a result it was called *herba impia*, Wicked Herb, by the old botanists, wicked because of its way of growth. The main stem has its flowers at the top, but underneath this grow two or more flowering shoots, all rising above the main stem. Therein lies the wickedness, for it conveys the idea that children were "undutifully disposed to exalt themselves above the parent flower" (Pratt).

Sonchus oleraceus > SOW THISTLE

SORE EYES

A charm from Devonshire: a woman who had never seen her father had to blow into the patient's eyes through a hole made in a NETTLE leaf, before she had put her hand to anything for the day (Friend).

Gerard recommended BUCK'S HORN PLANTAIN (*Plantago coronopus*) for sore eyes: "the leaves of buckes-horne boyled in drink, and given morning and evening for certaine daies together, helpe most wonderfully those that have sore eies, watering or blasted, and most of the griefes that happen unto the eies …". ROWAN leaves have also been used, made up into a salve (Cromek). CAMOMILE flowers were used for sore eyes. They had to be gathered before dawn, and the gatherer had to say why he was taking it; "let him next take the ooze, and smear the eyes therewith" (Cockayne). Gerard noted that "the juice (of CENTAURY) is good in medicines for the eies …", and long before his time the Anglo-Saxon Apuleius had "for sore of eyes, take this wort its juice; smear the eyes therewith; it heals the thinness of sight. Mingle also honey thereto; it benefits similarly dim eyes, so that the brightness (of vision) is restored". GROUND IVY leaves could be used as a poultice to be applied to sore eyes (Trevelyan). Actually, Ground Ivy was famous as an eye medicine, and its use was widespread in Britain. A Dorset remedy for sore eyes was to make an ointment with it (Dacombe), and a Warwickshire cure was to take a large handful of the herb, just cover it with water, and simmer for about 20 minutes, strain it, and use the liquid to bathe the eyes (Vickery. 1995).

In central Africa, an extract of the leaves of CHINESE LANTERN (*Dichrostachys glomerata*), mixed with salt, is applied to sore eyes (Palgrave).

SORE FEET

There are a group of leechdoms involving MUGWORT to relieve sore feet, some dating from Anglo-Saxon times, in the version of Apuleius. One, for example, simply required the patient to "take the wort, and pound it with lard, lay it to the feet" (Cockayne). The Physicians of Myddfai claimed that "for weariness of walking. Drink an eggshell of the juice of mugwort, and it will remove your weariness". Just carrying mugwort would do, for there is a Scottish belief that travellers who carried it would not tire (Beith). In Somerset, an ointment used to be made from MALLOW roots and unsalted lard, equally efficacious for varicose veins, they say (Tongue. 1965).

SORE THROAT

HONEYSUCKLE has been used to treat a sore throat (Conway), a use going back at least to Gerard's time: "[it] is good against soreness of the throat". Gypsies use the berries to cure the condition, and also canker in the mouth (Vesey-Fitzgerald). Gargles can be made from SANICLE's astringent leaves (Wickham), or those of SUMMER SAVORY and RIBWORT PLANTAIN (Schauenberg & Paris). Verjuice, the very sour liquid made from CRABAPPLE, was once used for sore throats "and all disorders of the mouth" (Hill). BRAMBLE vinegar used to be made in Lincolnshire for coughs (Gutch & Peacock), and the decoction of the tips with honey was an old sore throat remedy (Hill) (so is blackberry jam (Page. 1978)). Langham's *The garden of health* was written in 1578, and we can find something very similar there: "the new sprigs … doe cure the hote and evill ulcers of the mouth and throat and the swellings of the gums, uvula and almonds of the throat, being often chewed…". In North Wales, SLOES were used for a cough cure (Friend. 1883); so they were in the Highlands, too, for sloe jelly was reckoned the best cure for a relaxed throat (Grant), while the juice of boiled sloes was an East Anglian gargle for a sore throat (V G Hatfield. 1994). Perhaps the best known cure is provided by BLACKCURRANTS. They were used in folk medicine long before cultivation, and are still so used. A wine or jelly used to be made in Yorkshire from the fruit, and set aside for the complaint (Nicholson). In France, druggists used to prepare a sweet paste, called pâté de guimauve, from MARSH MALLOW, for coughs and sore throats (M Evans). The practice in Ireland was to boil the seeds in milk, and then drink the liquid (Maloney). It can be prepared as a gargle for sore throat, made from the shredded root or leaves, in water (Thomson. 1978).

SOLOMON'S SEAL is used in the Balkans for a sore throat, either as a decoction to be drunk, or as a poultice round the neck and chest (Kemp). The pulp of WATER MELONS will relieve sore throat or sore mouth (Perry & Metzger). American Indian groups like the Zuñi used PUCCOON roots, with some ceremony, by grinding it in the morning in the patient's quarters, on a special grinding stone, used only for this ritual. The remainder of the plant would be made into a tea. The medicine would be for sore throat, and for swelling on any part of the body (Stevenson). The Mano people of Liberia look on GREEN PURSLANE as a sore throat remedy. They take a large handful, beaten up with root ginger, which then had to be mixed with water from a "talking stream", and meat and salt are added to make a soup (Harley).

A sore throat remedy from Indiana is made from the powdered leaves and root of COCKLEBUR. Mixed with a little flour and water, the result would be put on the back of the tongue, so that it would drip into the throat (Tyler). Another Hoosier remedy is to use LOVAGE. Cut up the root and fry it in lard, and apply that to the throat as a poultice (Tyler).

SORREL

(*Rumex acetosa*) (Garden Sorrel is *Rumex scutatus*). There is not much folklore recorded for sorrel, none but the fact that Yorkshire children believed that it grew only where dead men had been (Nicholson). But it did have one or two folk uses, one of which was to use the juice to take rust marks out of linen (M McNeill). And of course it was used for dyeing wool. On South Uist, it was used on its own to colour red; mixed with indigo, it could dye blue (Shaw). The roots were used in the Welsh industry, too – two parts of the sorrel to one part of wool, the whole lot boiled

up for three or four hours, to produce a brown colour (Jenkins).

The main use of wild sorrel in the past has been culinary, and it is still quite often eaten, always remembering the oxalate in the leaves, which can make it poisonous if a lot of it is eaten. But it was employed as we now do lemons, especially for a green sauce with fish. The usual way of making it was by pulping the leaves and mixing them with sugar and vinegar. The boiled leaves are said to go well with pork or goose in lieu of apple sauce (Grigson), a use well enough known to Gerard, who advised his readers that "the juice hereof in Somer time is a profitable sauce in many meats, and pleasant to the taste: it cooleth an hot stomacke, moveth appetite to meat, tempereth the heat of the liver, and openeth the stoppings thereof". Sour-sab pie (sour-sab being a west country name for wild sorrel) is a Cornish dish made of the tender leaves and stalks. It is eaten with sugar and cream (Jago), and sorrel pie is a favourite in America, as is stewed sorrel (Sanford). In Lancashire, green sauce dumplings were made by chopping Wild Sorrel, mixing it with suet and flour, and boiling, as a dumplings, in broth or water (Crosby).

It is still used in folk medicine externally for skin disorders, or internally as a purge (Schauenberg & Paris). "Externally" must presumably mean the use of a leaf poultice. Corns can be dealt with by a mixture of bruised sorrel and lard, according to Illinois practice (H M Hyatt), but there have been some remarkable claims for its efficacy in most unlikely cases. An Irish cure for cancer, for instance, was to drink a decoction of the dried flowers (Egan). Equally optimistic was a Cumbrian epilepsy treatment, in which all that was needed was to eat sorrel leaves (Newman & Wilson). How many and how often is not divulged, but bearing in mind the oxalic acid content of the leaves, it cannot have been too many at a time.

SOT-WEED

An old, and expressive, name for TOBACCO (Halliwell), presumably deriving from its narcotic properties.

SOUTHERNWOOD

(*Artemisia abrotanum*) The smell of Southernwood is well enough known; like that of other members of the genus it is insect-repellent. Hence the French name Garderobe, for moths will not attack clothes in which Garderobe has been laid (Grieve. 1933). Bees do not like it either, and it is even claimed that southernwood, or wormwood for that matter, will keep adders from garden borders (M Baker. 1977). Roses do not mind the strong scent, apparently, for Cornish people claim that they grow best near southernwood (M Baker. 1977). The smell of the burning leaves will give a pleasant scent throughout the whole house, and it is enough to throw a pinch of the dried and

powdered leaves on the stove or fire (Webster). The
leaves used to be put between the pages of a Bible
(Banks. 1937), like Tutsan; in fact, in Scotland, it was
the tradition to carry it to church in a Bible (Aitken)
("in their hands they carried their Bibles along with
sprigs of appleringie and peppermint") (Kennedy).
It is also said that appleringie kept one awake in
church, "however dreich the sermon" (Gibson. 1959).
[Appleringie is a Scottish name for southernwood;
apple is from the old word aplen, church, and ringie
is Saint Rin's, of St Ninian's, wood (E Simpson).
Appelez Ringan, pray to Ringan, became first
Appleringan, and then appleringie (Aitken). Jamieson,
though, gave another explanation. It is, he says, from
French apile, strong, and auronne, southernwood,
from *abrotanum*].

There is some symbolism attached to it. St Francis
de Sales refers to sprigs of the herb being included
in bouquets given to each other by lovers, the
message being fidelity even in bitter circumstances
(Brownlow) – this is presumably the same as the
red rosebud surrounded by forget-me-nots and
southernwood referred to by Webster. The general
idea is conveyed by the very common name, Lad's
Love. Bergen. 1899 reckoned, almost certainly
incorrectly, that the origin of the name Lad's
Love lay in some love divinations she recorded in
Massachusetts. One said very simply that if a girl puts
a piece of southernwood down her back, or tucks it
in her shoe, she will marry the first boy she meets.
Another requires a girl to put a bit under her pillow on
going to bed, and then the first man she meets in the
morning is the one she is to marry. The name probably
come about because a few branches of southernwood
were generally added to the nosegay that courting
youths used to give the girls. But the girls used to
carry it about, too – see the note in Elworthy. 1888 –
"a very great favourite with the village belles. In the
summer, nearly all carry a spray of it half wrapped in
a white handkerchief, in the lane to church. In fact, a
village church on a hot Sunday afternoon quite reeks
with it". It was quite prominent in Fenland courting
customs. A youth would cut some sprigs and put
them in his buttonhole. He used to walk, sniffing
ostentatiously at his buttonhole, through groups
of girls. If the girls went by and took no notice,
he knew he had no chance, but if they turned and
walked slowly back towards him, then he knew they
had noticed his Lad's Love. He would then take his
buttonhole and give it to the girl of his choice. If she
was willing, she would also smell the southernwood
and the two would set together on their first country
walk (Porter. 1969). Denham recorded the following
southernwood rhymes in the north of England:

> Lad's love, lasses's delight,
> If t'lads doesn't come
> The lasses'll flite.

and

> Lad's Love is lasses' delight
> And if the lads don't love
> Lasses will flite [flite = scold].

It was also said in Gloucestershire that if Lad's Love
were put in the shoe, you would come across our lover
by chance (J Lewis). Another explanation of the origin
of the name Lad's Love, is that the name comes from
its use in an ointment that young men used to promote
the growth of a beard (Leyel. 1937). Gerard among
others, had something to say about southernwood's
anti-baldness properties: "the ashes of burnt Southernwood, with some of oyle that is of thin parts …
cure the pilling of the hairs of the head, and make the
beard to grow quickly". That is to be found too in the
Gentleman's Magazine, where the correspondent is
quoting a 1610 manuscript. "To remedye baldnes of
the heade". After burning and powdering the herb,
"mix it with oile of radishes and anoynte the balde
place, and you shall see great experiences".

It seems to be an unlucky plant to have in the garden,
at least in Devonshire. It was recounted that when
a woman was ill, a neighbour came to see her and
found a lot of Boy's Love in her garden. She pulled
it all up, and the patient recovered (*Devonshire
Association. Report and Transactions. vol 103*). A
strange superstition is recorded from the Pennsylvania
Germans. They used to say that a person who is
subject to cramps should, immediately on getting up,
and without saying a word to anyone, go and plant
some southernwood (Fogel). This must have been an
almost forgotten portion of a genuine, if one can call
it that, remedy, for Gerard recommends it in his list
of the virtues of the plant: "The tops, floures, or seed
boyled, and stamped raw with water and drunke, helpe
them that cannot take their breaths without holding
their neckes straight up, and is a remedie for the
cramp …".

Most of the medical uses of southernwood derive
entirely from ancient sources, very few being from
folk tradition, and then usually an echo from these
early herbals. However, carrying a piece of this herb
(or tansy), stuck between the nose and upper lip,
was an Orkney way of resisting infection, obviously
because of the powerful smell (Leask). Rubbing the
forehead with a handful of southernwood leaves
used to be an Essex cure for a headache (Newman
& Wilson). A 15th century leechdom has a remedy
for "tingling in a man's ear", tinnitis in our terms.
The remedy was to "take the juice of southernwood
and put it in the ear: and stop the ear with leaves
of the same herb, and drink the juice at even and at
noon, and it shall stay the tingling. And the juice
of wormwood doth the same" (Dawson). There are
remedies, quite fantastic at first sight, "for those who
speak in sleep", using southernwood (Dawson), and

another from the Physicians of Myddfai "to cure one who talks in his sleep". It makes more sense when it is realised that it is insomnia they are trying to cure, not just talking in the sleep. A sort of candy was once made for sleeplessness, by pounding four ounces of very finely chopped fresh southernwood leaves with six ounces of white sugar, until the whole was a paste. A piece the size of a nut was to be taken three times a day" (Rohde).

SOW THISTLE

(*Sonchus oleraceus*) Despised these days as a food plant, but it was not always so – "whilst they are yet young and tender, they are eaten as other pot-herbes are" (Gerard). And so it has been all over the world. All the sow thistles are edible (though not very interesting). They are probably best in soups and casseroles. Or cooked as a vegetable with something else with a stronger flavour. Some people actually eat them raw, as an ingredient in salads (Jordan).

Unlikely as it may seem, this is a plant with a distinguished background. In Wales, it used to be said that anyone who carried it in his hat would be able to run and never get tired. In like manner, it was tied to the tails of horses before a ploughing match. But it had its dark side, for if one man used it, it would take the strength out of his companion, and if by accident a man gave some of the leaves to his wife, one of them would waste away and die (Trevelyan). In Russia, the plant was said to belong to the devil (Dyer), but the Welsh belief was that the devil could do no harm to anyone wearing a leaf from the plant (Trevelyan), or as one of the Anglo-Saxon herbaria said (in translation) – "so long as you carry it with you nothing evil will come to meet you" (Meaney).

Sow Thistle would disclose hidden treasure, too, if properly treated, and like other plants that could do that, it opened locked doors as well. These sesame qualities are well to the fore in Italian folklore concerning the plant (Gubernatis). There are magical uses recorded from Macedonia – see Abbott for some examples. It has got a milky juice, so by the doctrine of signatures, it was given to nursing mothers. Coles said it was called Sow Thistle because sows knew that it would increase the flow of their milk after farrowing; perhaps that was the reason why Welsh farmers put some of the leaves in the pig trough, though it was said the reason was to fatten them (Trevelyan). A separate tradition associates the plant with hares rather than with pigs. Various old names appear (Hare's Palace, Hare's Thistle (Dyer; Henslow), and Hare's Colewort, the last a translation of *Brassica leporina*, one of the old Latin names (Britten & Holland)). An old legend tells how this plant gave strength to hares when they were overcome with the heat. This is from the Anglo-Saxon Herbal – "of this wort, it is said that the hare, when in summer by the vehement heat he is tired, doctors himself with this

wort". It is more likely, given the Somerset names Rabbit's Meat and Rabbit's Victuals (Macmillan, Grigson), that hares and rabbits actually eat them. Cowper wrote that his pet hares were very fond of Sow Thistle.

According to Culpeper, "its juice is wonderfully good for women to wash their faces with, to clear the skin, and give it a lustre". Similarly, "the herb bruised or the Juice is profitably applied to all hot inflammations in the Eyes …" (Coles). Actually, the juice has been quite seriously used as eye drops, and the use for the complexion might very well be the result of its having been taken for liver troubles, or as a blood purifier (Watt). Another use for the juice was for "to kill a wen", at least according to a 17th century manuscript from Jersey, and of course the juice has been used to put on warts (Barbour).

SOWBREAD

(*Cyclamen europaeum*) Most people will call this plant Cyclamen, the English name being but rarely used now. Sowbread, however, was given quite simply because it was used as a food for swine. "Cyclamen, id est, panis porcinus" (Rufinus, in Thorndike). Such a usage strains belief these days, and it is quite possible that truffles, which pigs certainly like, were meant.

Friend claimed that cyclamen was used as a charm against bad weather, though in some parts of America it was not a favourite to keep indoors; anyone who did so would get chills, so it was said (Hyatt). But it certainly enjoyed a reputation as an aphrodisiac, unlikely as that sounds, from ancient times. In fact, it became the very symbol of voluptuousness (Haig). According to Gerard, the root should be "beaten and made up into trochisches, or little flat cakes", when "it is reported to be a good amorous medicine to make one in love, if it be inwardly taken". Another opinion was that it should be burned and the ashes marinated in wine and formed into little balls which could then be concealed in soups or stews (Haining). Expectant mothers should avoid it, especially stepping over it (Friend). Gerard fenced his cyclamens in for this very reason. Midwives, by the same extended logic, regarded it as invaluable, for old herbals advised women in labour to hang the root around their neck to ensure quick delivery (Vickery. 1995). Parkinson, in turn, reported that "it is used … for women in long and hard travels, where there is danger, to accelerate the birth, either the root or the leaf being applied" (Parkinson. 1629). This is actually doctrine of signatures. A glance at the leaves, Coles thought, was enough to show that the plant belonged to the womb by signature, logic that is not easy to follow.

Cyclamen, like many another plant, used to be taken as a counterpoison. Gerard again – ".. very profitable against all poisons, and the bitings of venomous beasts, … outwardly applied to the hurt place". The plant was actually grown in the house as a protection against poison (Withington). If there is any virtue in such a claim, perhaps it is as a homeopathic remedy, for the tubers are certainly toxic, even in small quantities.

SPEAR PLUME THISTLE

(*Cirsium vulgare*) It is said that these thistles are graze-resistant, in other words, they are not eaten by cattle, and they often flourish where competitors are removed by browsing cattle. Large numbers of these thistles in an old field are taken as firm evidence of overgrazing (Southwick). One piece of received wisdom is worth quoting:

> Cut thistles in May, they'll grow again one day,
> Cut thistles in June, that will be too soon.
> Cut thistles in July, they'll lay down and die
> (Leather).

Gerard reported that "… being stamped before the floure appeareth, …, and the juice pressed forth, causeth the haire to grow where it is pilled off, if the place be bathed with the juyce". Whoever would have thought of thistles as a hair restorer? But that is not all, for he also passed on the information that this thistle chewed, "is good against a stinking breath". The roots, boiled in wine and drunk "… take away the ranke smell of the body and arme-hole". You could even apply the root like a poultice for the same purpose.

SPEARMINT

(*Mentha spicata*) Spire Mint (Prior) is the correct name (*spicata*), not Spearmint, but the latter has persisted while the former has fallen into disuse. Mint is shy; one should never look at it for a month after planting. It needs time to settle down, or so they say in Somerset (Tongue. 1965). A sprig of mint in the kitchen will keep bluebottles out (Vickery. 1993). Put some in the barn before putting any grain there, to keep rats out of it (Fogel). A sprig in the milk keeps it from turning, and mint rubbed over a new hive will keep the bees from deserting it (Tongue. 1965).

If the women of the Abruzzi region in Italy came across a mint plant, they would bruise a leaf between their fingers to ensure that on the day of their death, Jesus would help them (Clair).

In the form of an essential oil, like peppermint, it is used by confectioners, chiefly in America, and at one time perfumed soap contained this oil (Fluckiger & Hanbury). One of the commonest uses now is for flavouring chewing gum. In the form of an aromatic tea, it has been a great favourite in domestic medicines. American Indians like the Miwok used this tea to relieve stomach trouble and diarrhoea (Barrett & Gifford), and herbalists still prescribe it for these complaints (Flück). Gerard described it as

"marvellous wholesome for the stomacke", and in the 18th century Sir John Hill declared it as "excellent against disorders of the stomach". Country people in Ireland tie a sprig round the wrist for a stomach upset (O'Farrell). It is used in Russian folk medicine for heart complaints (Kourennoff).

Gerard noted that "it is applied with salt to the bitings of mad dogs", and he also quoted a contraceptive use: "Dioscorides teacheth, that being applied to the secret part of a woman before the act. It hindreth conception". Gypsy lore has it that to cure hay fever permanently, one should pick some fresh mint each day, and put it in a muslin bag in one's pillow, so that the scent can be inhaled during sleep. Also one should wear some each day (Vickery. 1995).

SPEARWORT

i.e., LESSER SPEARWORT (*Ranunculus flammula*), and GREATER SPEARWORT (*Ranunculus lingua*) The plant was used medicinally for blistering, especially in the Highlands and Hebrides (Grigson. 1955). Martin (1703) has this description of its uise for sciatica: "Flammula Jovis, or Spire-wort, being cut small, and a limpet shell filled with it, and applied to the thighbone, causes a blister to arise about the bigness of an egg; which being cut, a quantity of watery matter issues from it: the blister rises three times, and being emptied as often, the cure is performed …". Of course it will raise a blister – all the buttercups will. But a Somerset entry records the use of the juice to raise a blister on the hands deliberately, and apparently ponies were treated this way for some unspecified ailment (Tongue. 1965). Lindley knew the distilled water as an emetic, but apparently it did not need this sophistication – a little of the infusion was drunk in Skye in melted fresh butter for an efficient purge (Martin).

"The common cure for farcy in horses was sparrow-weed (i.e., GREATER SPEARWORT) stewed with garlic" (Ireland) (Foster). It was sometimes used for swine fever there, too (Moloney).

SPEEDWELL

(*Veronica spp*) The name means goodbye. We are asked to believe that the plants got this widely used name because the blossoms when full fall off and fly away. Forget-me-not, now applied to quite a different plant, was apparently once given to the speedwell, for the same reason. But Andrew Young suggested that if "speed" means the same as in "speed the plough", then perhaps the name means good luck. (see also **GERMANDER SPEEDWELL**).

Spergula arvensis > CORN SPURREY

SPICE PLASTER

Once used to put on parts affected by rheumatism or neuralgia. The way to make it is to crush an ounce or so of ALLSPICE berries, and boil it down to a thick liquor, which is then spread on linen ready to be applied (A W Hatfield).

SPIDERWORT

(*Tradescantia x andersoniana*) Ignoring the obvious descriptive value of the name Spiderwort, it was claimed that it was given as a tribute to its supposed efficacy against the bite of a certain kind of spider (Nash); hence too Spider Lily, and Parkinson's Day Spiderwort, the first element of which accurately describes the duration of each flower, or at the very least, it described flowers that will only open during daylight – so with Flower-of-a-day, or One-day Flower (Howes), and with the ironic reference Life-of Man, recorded in Dorset (Macmillan).

Spigelia marilandica > PINKROOT

SPIKENARD

(*Nardostachys jatamansi*) This is the Spikenard of the Bible (Moldenke & Moldenke), the "spike" being the ear that grows from the rhizome (Dalby). It has scented roots, which are used in India as a perfume for the hair (Moldenke & Moldenke), and the oil is used to improve hair growth (Lewis & Elvin-Lewis). Distillation of this root produces oil of spikenard, although all parts contain the aromatic oil, especially the rootstock, whose fragrant oil is mixed with other oils to make spikenard ointment, once used in cosmetics and in medicine for the treatment of nervous disorders (Zohary). The 'ointment of spikenard' Mary used to anoint Jesus's feet before the last supper is thought to have been derived from this plant (Lewington), but there are other candidates, the desert Camel grass (*Cymbopogon schoenanthus*) being one.

SPINDLE TREE

(*Euonymus europaeus*) There is a tradition that the best drawing charcoal is made from spindle twigs. The French obviously thought so, for their name for the shrub is Fusain, which also means charcoal (Young). Evelyn (*Sylva*) recalls its use for violin bows, and for inlaying. But it is best known for making skewers (see such names as Prickwort, or Skiverwood). Aubrey. 1847 said. "the butchers doe make skewers of it, because it doth not tainte the meate as other wood will doe …". English gypsies also used the wood for knitting needles and clothes pegs (Jordan), and, of course, there is the common name for the shrub – Spindle Tree. It seems that spindles were indeed made from this wood, but not necessarily in England, where there was no great tradition of spinning. The name is actually imported, for Turner. 1548 mentioned the Dutch name Spilboome, without implying an English equivalent. However, they say the wood is ideal for spindles, being tough, heavy, and very hard (Mabey. 1977).

"…three or foure of the berries purge both by vomit and siege …", observed Evelyn. That is not surprising, for they are poisonous, even if only relatively mildly so, for it is the leaves and bark that could be dangerous to children and animals. Gerard had no hesitation in warning that "this shrub is gurtfull to all things, and namely to Goats …". Why goats in particular? The answer may, just possibly, be found in another common name for spindle – Gatter-tree, and its derivatives. 'Gatter' is OE gat, goat.

Spindle is called Death Alder in Buckinghamshire (Grigson. 1955), and as such is reckoned unlucky to bring in the house. And, of course, the berries are mildly poisonous. There are two other miscellaneous connections with death: Evelyn produced the odd information that parricides were scourged with rods of spindle, and apparently Pliny said that too many spindles flowers foretell a plague.

SPINY COCKLEBUR

(*Xanthium spinosum*) An infusion of the plant before flowering has been used in the treatment of diabetes (Watt & Breyer-Brandwijk), and in Kentucky they say "cockle-burr tea is good for the phthisick" (Thomas & Thomas), tuberculosis, presumably?

SPLEENWORT

(*Asplenium trichomanes*). Obviously, it must have been used to treat spleen disorders in some way, but in Scotland it was taken for tuberculosis as well (Beith).

SPOTTED DEADNETTLE

(*Lamium maculatum*) The leaves of this deadnettle usually have a large pale blotch, hence another name given to it in Lincolnshire, Jerusalem Nettle, because of the legend that, like virtually all of the plants with spotted leaves, it got that way from a drop of the Virgin's milk falling on it (Gutch & Peacock).

SPOTTED MEDICK

(*Medicago arabica*) Calvary Clover is a name given to this plant (McClintock & Fitter). A German superstition, taken to America, said that Calvary Clover could only germinate if sown on Good Friday (M Baker. 1977). The red spots on the leaves also help in understanding the name. They are spots of Christ's blood that fell on them at Calvary. The shape of the leaves give rise to various 'heart' names – Heart Medick (Curtis), or Heart Clover (Britten & Holland), and so on, all too good for the expounders of the doctrine of signatures to miss. Coles says it is so called "not only because the leaf is triangular like the heart of a man, but also because each leafe doth contain the perfection (or image) of a heart, and that in its proper colour, viz., a flesh colour, it defendeth the heart against the noisome vapour of the spleen".

SPRAINS

Sprains in both humans and animals used to be treated in Dorset by steeping MALLOW leaves in boiling water and using the result as the foundation of a poultice (Dacombe). But that is an ancient practice, if we interpret "sore of sinews" as a sprain in the Anglo-Saxon version of Apuleius (Cockayne).

SPRING TONIC

Young NETTLE leaves have the effect of increasing the haemoglobin in the blood (Bircher), hence the various "spring tonic" uses one finds in folk medicine, whether it is called that or a "blood purifier". The medicine usually takes the form of a tea, sometimes with dandelion leaves added, as in a Dorset "blood tonic" recipe (Dacombe), or in the East Anglian use to "clear pimples" (Hatfield. 1994). WOODRUFF has a high coumarin content, so it is a blood-thinner, so a tea made from it was taken for a spring tonic (G B Foster) in the days when that was thought to be necessary. CAMOMILE is listed as a bitter stomachic and tonic, and as such is recorded in Hampshire as good for "clearing the blood" (Hampshire FWI), by which a spring tonic is probably meant. Wiltshire people used the powdered roots of AVENS in boiling water as a spring pick-me-up. It was said that these roots should be dug on 25 March, from dry ground (Wiltshire). A tonic known as "Spring Juice" was fresh BROOKLIME and SCURVY-GRASS, cut and beaten in a mortar, and left to steep for twelve hours. Then it was strained and an equal amount of the juice of Seville oranges was added. A wineglassful taken fasting each morning for a week was the spring tonic (Quelch).

SPRINGWORT

A magical, but mythical, plant, though it is sometimes identified as Caper Spurge (*Euphorbia lathyris*) (Cockayne; Grimm). Springwort has the power of opening doors and locks, even medicinal ones – "if anyone is bound by herbs, give him springwort to eat and let him sip holy water ..." (Meaney). Grimm quotes Springwurzel as an explosive root. The Italian name ferrocavallo is given because its power over metals is so great. This name implies that a horse stepping on the plant will leave its shoe behind (Swainson. 1886). It is a lightning plant, best got hold of by using a woodpecker, itself a lightning bearer, as an intermediary. One should plug the hole in the tree where the young are. The parent bird flies off to find the plant, and on its return holds it against the plug, which immediately falls out. A red cloth should already have been spread to catch the plant when the woodpecker drops it, for it is said the bird will mistake it for fire, and its object is to burn the plant before it falls into anyone's hands. Presumably the original idea was that the bird must return the plant to the element from which it was obtained (Kelly).

SPURGE LAUREL

(*Daphne laureola*) Johann Weyer (*De praestigiis Daemonum et incantationibus ac Veneficiis*, 1568) recorded Daphnemancy among his list of divinations ascribed to demonic agency. The divination was by the crackling of the leaves while burning. The leaves were also put under the pillow to induce dreams (Lea). Presumably the association with demons is a result of the poisonous nature of both the bark and berries, poisonous enough to be dangerous to children (North).

One of the "notable things" reported by Lupton was its use for "writing that cannot be read without putting the paper in water. Take the juice of spurge laurel, put it into a little water wherein alum has been dissolved, and if you write with it, it will appear as nothing on the paper, but being put into water, the letters will appear plain and legible".

Of the medicinal, the most notable was the use of the bark in the treatment of cancer (Grigson. 1955), though this was only a cottage remedy, and there seems no record of how it was done, nor whether it was really cancer (and not canker) that was being treated. The bark (and the berries) was used also in some parts – a dangerous practice for humans, but a useful one for horses (Garrad). The very acrid roots were used in some parts to relieve toothache (Vickery. 1995), but surely this must have been a dangerous remedy. Hill, in 1754, noted that it was "a powerful remedy against the dropsy", although he was careful to point out that "it is not every constitution that can bear such a medicine". An Anglo-Saxon leechdom, acknowledging Pliny, quoted the use of an amulet of chamacela, which could be either this shrub or *D. mezereum*, to cure pearl (albryo), which is probably cataract, in the eyes, "if the plant is gathered before sunrise, and the purpose outspoken".

SQUILL

(*Scilla maritima*) According to Arnaldo de Vilanova (early 14th century), squills hung up are a cure for the results of sorcery (Lea), and Dalyell also said that they "precluded the accesss of sorcery. The 10th century *Geopontica* said that squills serve as a means of foretelling the harvest, for if the flower is long, "like a stick", and does not soon fall, the crops will be good (Rose. 1933).

The bulb is effective as a rat poison because the vermin like it and will readily eat it (Kingsbury. 1967). As Red Squill Powder, it has long been used in America as a rat poison. According to Coles, the bulb has the signature of the head (see DOCTRINE OF SIGNATURES), so it was prescribed for epilepsy, headache, dizziness, etc. But it was as a remedy for dropsy that squill was used with greatest continuity. The Anglo-Saxon version of Apuleius had a leechdom for "water sickness" (Cockayne), and it seems it was used by the Egyptians for the complaint, under the name of the "Eye of Typhon" (Thompson. 1947); the Delta people found the squill so useful for dropsy that they are said to have built a temple in its honour (Withington). It continued to be prescribed for dropsy through the centuries and the ages of the herbals. Thomas Hill mentioned that it "amendeth the dropsie", as well as "the defaults of the Liver, Ague, etc.,", and Gerard too recommended the bulb, roasted or baked, and mixed with other medicines, not only for dropsy, but also for jaundice, and for "such as are tormented with the gripings of the belly".

SQUINANCYWORT

(*Asperula cynanchica*) Squinancywort is actually Quinsy-wort, for this small plant provides an astringent gargle which at one time was used to treat the complaint. In much the same way, blackcurrants, also good for the condition, were known as Quinsy-berries. Like most of the genus, a red dye can be got from the root (Grieve. 1931).

Stachys officinalis > BETONY

STAR ANISE

Another name for TRUE ANISE (*Illicium verum*), whose fruit grows in the form of a star.

STAR OF BETHLEHEM

(*Ornithogalum umbellatum*) It is said that the flowers resemble the pictures of the star that indicated the birth of Jesus, hence the common name (Geldart). Some say, too, that these are the "lilies of the field, that toil not, neither do they spin…" (Vickery. 1995).

Stellaria holostea > STITCHWORT

Stellaria media > CHICKWEED

STERILITY

An American belief is that if a married couple throw COWPEAS across the road near their home, the wife will become fertile, no matter how long she has been barren (Waring). Weiner reported that Cherokee women chewed and swallowed the highly toxic root of AMERICAN COWBANE for four consecutive days to induce permanent sterility!

STINKING CAMOMILE

See MAYDWEED (*Anthemis cotula*)

STINKING HELLEBORE

(*Helleborus foetidus*) Poisonous, of course, like all the hellebores. Even so, it was a protective plant for the homestead in early times, when it was also believed to be an antidote against madness (Genders.1971). Considering the toxic nature of the plant, one wonders how it was used, and how many people were killed while being treated. Still, it did have its domestic uses, in pest control, for people used to powder the roots and mix them with meal to lay down and poison the mice (Drury. 1992). There were veterinary uses too, some of them lasting until quite recently, like using it to condition horses, and sheep too, apparently, in the Chipping Camden area (Brill). It was used for the same purpose in East Anglia, where the horsemen were proud to have their horses' coats shine (G E Evans. 1969). More seriously, cattle were treated with it when they coughed. It was done by making a

hole in the dewlap with a setter, or thread (hence the name Setter-grass, or Setterwort (Grigson. 1955)), and a length of hellebore root inserted to irritate the flesh and keep it running. Or sometimes a few rolled-up leaves were used (Hartley & Ingilby). Green Hellebore could be employed equally well for this operation. It was the way, too, to treat a pig that had been bewitched, by making a hole in its ear and putting a piece of the stem there (Lovett).

Used in human medicine, the results are extremely uncertain. Hellebrin and helleborin, both contained in this plant, have been used as a cardiac stimulant (Cullum), but the really notorious usage was for worms in children. Thornton wrote that a decoction of about a drachm of the green leaves, or 15 grains of the dried, is given to children and repeated three mornings, when it seldom fails expelling the round worms, or "a tea-spoonfull of the juice, mixed with syrup, may be given for that purpose". Gilbert White mentions this use, and recognised it as "a violent remedy … to be administered with caution". In the following century it was condemned as "far too uncertain in the degree of its action to be safely administered" (C P Johnson).

STINKING IRIS

(*Iris foetidissima*) The leaves, being rubbed, are "of a stinking smell very lothsome" (Gerard), although the same smell accounts for the name Roast Beef Plant, surely not all that loathsome. The root is a painkiller, and a migraine remedy (Conway). One Somerset remedy is for cramp, requiring the patient to take an infusion of the dried root (Tongue. 1965). Another cure involving this plant comes from Warwickshire, and was for, of all diseases, St Vitus's Dance. Again, the part used was the root, boiled in ale (porter, to be specific), strained off, and the liquor drunk (F G Savage). Earlier recommendations were for scrofula (the King's Evil) (Gerard), while Langham had a recipe for freckles – "seethe the root of stinking segs in cowes milke, and use it". It even featured as an ingredient in a poison cure (Dawson).

STINKING ORACH

(*Chenopodium vulvaria*) "Stinking" is right – when handled it smells like rotting fish. That unpleasant smell is trimethylamine (like hawthorn), and there is also ammonia in the smell (Genders. 1971). Nevertheless, they can be cooked and eaten, like spinach. Hill described it as "an excellent medicine in all hysteric complaints", and there were similar recommendations from all the older herbalists. Indeed, it was even supposed at one time to cure infertility. Just by the smell of it! (Grieve. 1931).

STITCH

Holding a CABBAGE leaf to the affected side, will cure a stitch, they say (Page. 1978). But the more important plant must be STITCHWORT itself, called thus "for its property in helping stitches and pains in the side" (Coles). Gerard also made the point: "they are wont to drinke it in wine with the powder of Acornes, against the paine in the side, stitches, and such like". An ointment made from CAMOMILE, so Martin reported in 1703, was used on the Isle of Skye for stitch!

STITCHWORT

i.e., GREATER STITCHWORT (*Stellaria holostea*) A fairy flower, under the protection of the fairies, and it must not be gathered, or the offender will be "fairy-led" into swamps and thickets at night (Skinner). Cornish children say they will be bitten by an adder if they pick it (Macmillan), and that may explain the variety of "snake" names given to the plant (Snake-flower, Snake Grass, Adder's Meat, etc.,), from areas much more widespread than Cornwall, which suggests that the belief was once commoner than it is now.

The plant is called Stitchwort "for its property on helping stitches and pains in the side" (Coles). Gerard also made the point: "they are wont to drinke it in wine with the powder of Acornes, against the paine in the side, stitches, and such like". He also reported, on the authority of Dioscorides, "that the seed of Stitchwort being drunke, causeth a woman to bring forth a man childe".

STOMACH ULCERS

The inner bark of SLIPPERY ELM is well known as a treatment for stomach ulcers, and needs no further comment. Not so well known is the fact that herbalists use CHICKWEED in the treatment of the ulcers, and as an aid to digestion (Conway). Gypsies use an infusion of GROUND IVY leaves to cure stomach ulcers (Vesey-Fitzgerald).

STOMACH UPSETS

A leaf tea from LABRADOR TEA used to be an American domestic medicine for stomach disorders (Bergen. 1899). Hopi Indians used BROOMWEED (*Gutierrezia microcephala*) tea to cure the complaint (Weiner). CAMOMILE tea is a great standby for an upset stomach, not only in Britain, but also in France, Italy and Greece as well, and probably more. SPEARMINT tea is another well tried remedy. The Miwok Indians used it to relieve stomach trouble and diarrhoea (Barrett & Gifford), and herbalists still prescribe it for these complaints (Flück). Gerard described it as "marvellous wholesome for the stomacke". Country people in Ireland simply tie a sprig round the wrist for a stomach upset (O' Farrell).

STONE

The name SAXIFRAGE is derived from Latin saxum, stone, and frangere, to break – "break-stone", in other words, for these plants often grow in clefts of rocks. Inaccurate observation led to the conclusion that the roots had actually broken the rock, and that became

the signature of the plant – that which breaks rocks must also have the power to break stones in the body. So the plant was prescribed for the purpose, from Anglo-Saxon times right up to the mid-18th century, and probably later. Hill, for example, in 1754, was still repeating Gerard's prescription: "The root of White Saxifrage boiled in wine and drunken, provoketh urine, clenseth the kidneys and bladder, breaketh the stone and driveth it forth …". Similarly, another "Breakstone", PARSLEY PIERT, was much used, by sympathy, against stone in the bladder. Gypsies use an infusion of the dried herb for gravel and other bladder troubles (Vesey-Fitzgerald). It was well-known as a powerful diuretic in Camden's time, and it was in great demand during World War11, being used for bladder and kidney troubles (Brownlow). Another example of the doctrine of signatures concerns GROMWELL (*Lithospermum officinale*), whose stony seeds (that is what *Lithospermum* means), proclaim its signature perfectly, and it was used for the stone from Pliny onwards, at least to the mid-18th century, for Hill repeated the prescription that had been in vogue for cenuries before his time.

STONE PARSLEY

(*Sison amomum*) It was once believed that this plant could break a goblet or tumbler if rubbed against one (Fernie).

STRAWBERRY

(*Fragaria x ananassa*), includes **WILD STRAWBERRY**

Fragaria vesca, which is often confused with the Barren Strawberry, which is no relation, and belongs to an entirely different genus). Nobody seems to agree about how strawberry got its name. Of course, we put straw around the plants to protect the fruit, but the name was streawberige in OE times. The *Oxford Dictionary of English Etymology* admits it does not know the derivation. Anyway, the fruit has always been much appreciated; to quote Aubrey – "strawberries have a most delicious taste, and are so innocent that a woman in childbed, or one in a feaver, may safely eate them". But to show that nothing can unequivocally be good, he goes on, "but I have heard Sir Christopher Wren affirm, that if one that has a wound in his head eates them, they are mortall".

The commendations are echoed in the symbolism that has attached itself to strawberries. They were used a great deal in pictures representing the garden of Heaven, and in the "enclosed garden" of the Virgin Mary. They are shown with fruit and flowers together, and in this way represent the good works of the righteous. It is the emblem of "the righteous men whose fruits are good works", or the symbol of "perfect righteousness" (Haig), and, apparently, of foresight (Leyel. 1937). But "righteousness" seems the more important; as such it was used chiefly in early Italian paintings of the Adoration, where the infant Christ is laid upon the grass. In these Italian paintings, the symbolical strawberry is usally accompanied by the violet, as if to point out that the truly fruitful soul is always humble (Haig).

It is said that strawberries grow best when planted near nettles (*Notes and Queries. 4th series. vol 19; 1872*), and gardeners say also that BORAGE is a good companion plant for strawberries (Boland & Boland).

Cornish girls believed that their complexion could be improved if they rubbed their skin with wild strawberry leaves. Actually, the belief is widespread – see, for instance, this verse of a version of "Dabbling in the Dew":

> Pray whither so trippingly, pretty fair maid,
> With your face rosy-white and your soft yellow hair?
> Sweet sir, to the well in the summer-wood shade
> For strawberry leaves make the young maiden fair
> (Deane & Shaw).

There may have been some reasonably sound reason for this; the leaves are obviously astringent, as indeed are many of the *Rosaceae*, of which the strawberry is a member. Hill described the infusion of the fresh leaves as a "good liquor to wash a sore mouth or throat". It was also said to keep tartar off the teeth (Fairweather). But we find the fruits themselves connected with receipts for the complexion. "The ripe strawberries quench thirst, and take away, if they be often used, the rednesse and heate of the face" (Gerard). Other prescriptions probably refer to the leaves, though it is by no means clear, as there is no separate word for plant and fruit, as there is for blackberry, for instance, when bramble is reserved for the stems and leaves, while the berry name can only mean the fruit. What are we to make of this receipt, ascribed by Lupton to Conradus Gesnerus? "Many have been helped that have had foul and leprous faces, only by washing the same with distilled water of strawberries; the strawberries first put into a close glass, and so purified in horse dung"!

But strawberries were not always viewed with unmitigated delight; the colour of the fruit seems to be the reason. Some birthmarks are known as "strawberry" marks, and East Anglian superstition tells that they were caused by the mother's eating too much of the fruit during her pregnancy (Porter. 1974). The dislike of the colour is found in Italy, too. No child would be given red fruits like strawberries to eat, or red shoes to wear. The effect of the colour on a small child would be to cause in later life some antisocial behaviour like bad temper or even uncontrollable rage (Camp).

STREWING HERBS

RUE was important as such, for not only did it keep fleas away, but it was a protection too against plague (see **PLAGUE**). In 1750 it was used to strew the dock of the Old Bailey as a protection against the jail fever then raging at Newgate Prison, and the custom continued right into the 20th century (Genders. 1972). MEADOWSWEET, with its very aromatic leaves, was Queen Elizabeth 1's favourite strewing herb. "Queen Elizabeth of famous memorie did more desire meadowsweet than any other sweet herbe to strewe her chambers withal" (Parkinson. 1629). Gerard was enthusiastic about that usage: "the leaves and floures farre excell other strowing herbes for to decke up houses, to strew in chambers, halls, and banqueting houses in the summer time; for the smell thereof makes the heart merrie, delighteth the senses; neither does it cause head-ache, or lothsomeness to meat, as some other sweet smelling herbes do". Sprays of JUNIPER were often strewn over floors so as to give out, when trodden on, the aroma which was supposed to promote sleep. Queen Elizabeth's bedchamber was sweetened with them (Fernie). It was also burned in rooms to sweeten the air, just as PLOUGHMAN'S SPIKENARD was. But that, too, could be used as a strewing herb to counteract musty atmospheres (Genders. 1971).

SWEET FLAG roots are aromatic, with a violet-like fragrance that is brought out as they are trodden on. Hence its former use for strewing on the floors of churches and the houses of the rich. Ely and Norwich cathedrals had their floors covered with it at festival times, as the plant grows in the Fens (Genders. 1971; Fletcher. 1997), where it was actually harvested, and not only for floor covering, for it seems that they were used for thatching too, especially for churches (A W Hatfield). One of the charges of extravagance brought against Cardinal Wolsey was that he ordered the floors of his palace at Hampton Court to be covered mch too often with rushes and flags, since they were expensive and difficult to get (Genders. 1972). Even more expensive was SAFFRON. He used rushes strongly impregnated with saffron to strew at Hampton Court (Dutton). But then he could afford it. BASIL was used, too (Brownlow). RAYLESS MAYWEED (*Matricaria matricarioides*) is a plant that gives off a pleasant scent when crushed (hence one its names, Pineapple Weed), so was valuable as a strewing herb. Indeed, its abundance by waysides is in part due to its tolerance of being trampled underfoot (Mabey. 1977). BEE BALM leaves, with their lemon fragrance, were used, too (Clair), with a "quasi-medicinal effect", as one writer put it (Fletcher). Mint, too, was used – WATER MINT, for example, "the savor or smell (of which)…rejoyceth the heart of man, for which cause they use to strew it in chambers" (Gerard).

STYE

A stye can be treated, according to an Irish country remedy, by bathing it with the milky juice of the DANDELION (P Logan). A charm, also from Ireland, is to point a GOOSEBERRY thorn at it nine times in the name of the Trinity, or rather get a twig with nine thorns on it, and point each thorn in turn at the stye, saying "away, away, away". The thorns had then to be thrown over the left shoulder, and the stye will presently vanish (Wilde. 1902). A leechdom from Anglo-Saxon times advised the patient to toast FIELD GENTIAN, and apply it as hot as possible (Cockayne).

SUCCORY

Another version of CHICORY, though some say it comes from the Latin succerere, to run under, because the roots go so deep (Hemphill).

SUGAR BEET

(*Beta vulgaris 'Altissima'*) Sugar beet leaves have found a use in the treatment of erysipelas in pigs. When the leaves are fed to them, the blisters disappear and the pigs improve quickly (V G Hatfield. 1994).

SUGAR CANE

(*Saccharum officinarum*) Malaysian planters would say the time to plant sugar cane is at noon. This will make it sweeter, by drying up the juice and leaving the saccharine matter. If you plant it in the morning its joints will be too long, and if later in the day they will be short (Skeat. 1900).

SUGAR MAPLE

(*Acer saccharum*) Many maples yield a sugary sap, but only this one does so in sufficient quantity to justify exploitation. It was known to the Indians of north-east America, and sap extraction was practiced there long before the arrival of the Europeans (H G Baker). "The old trees produce the most and the richest juice ... It is a common remark (in Kentucky) that whenever they meet with a black tree ... they are sure it will prove a rich one. The blackness proceeds from the incisions made in the bark by the pecking of the parraquet, and other birds, in the season of the juice rising, which, oozing out dribbles down its sides, and stains the bark, which, in the course of time, becomes black" (J Taylor). Effective production of maple sugar is limited to the time of the spring thaw, around March. If maples are wounded at that time their sap flows freely with pressures generated in their root system. The sap must be collected and concentrated the same day, or it will ferment and turn sour (Edlin).

Maple sugar was one of the most important Menomini foods, and was used in almost every combination of cooking. With them, it took the place of salt (H H Smith. 1923). So important was it that they have an origin myth about this tree. Mä'näbus was travelling about over the earth, when one day he noticed the maple tree, which was not one of his creations. He saw that its sap was pure syrup, and it ran very slowly. He realised that it was too slow and tedious to be of much use to humans, so he urinated into the tree, thus thinning the sap and making it flow more freely (A Skinner). The Menomini would establish their sugar camps in late March or early April, and start tapping the trees at the proper time. None could be spilled or wasted, for that would be an offence to Mä'näbus; the sugar would shrink as a punishment.

SUMACH

Rhus glabra see SMOOTH SUMACH; *Rhus radicans* see POISON IVY; *Rhus trilobata* see THREE LEAF SUMACH

SUMMER SAVORY

(*Satureia hortensis*) Is *Satureia* from satyr? Some have thought so, and it was certainly cultivated in ancient times (Hemphill), and, like many another plant, was given aphrodisiac virtues (Palaiseul), encouraging virility, to put it another way (Hatfield. 1973).

Both summer and winter savory (*Satureia montana*) are used to rub on bee or wasp stings, to get quick relief (Clair), and the infusion is an effective gargle for sore throats (Schauenberg & Paris). Even the Welsh medieval text known as the Physicians of Myddfai noticed the herb, and recommended it "to hasten a tedious labour, take the juice of savory, and administer it in water". The decoction of the dried herb was used in Russian folk medicine for earache (Kourennoff). The juice was used an eye lotion at one time, while, mixed with oil of roses, it was claimed to relieve tinnitus (Addison. 1985).

SUN SPURGE

(*Euphorbia helioscopia*). The common garden weed, which generations of children have gathered, to put the milky juice on the warts on their fingers. If they suck their fingers afterwards, there will be an acute burning sensation in the mouth (Forsyth). This use on warts is very widespread. In East Anglia, sun spurge, dandelion and greater celandine were the three plants used (G E Evans. 1966). But not in that order, for sun spurge is the least popular. It is used in Brittany, too, but it has to be picked from the path along which a funeral has passed, and it has to be found by chance, (Sebillot). There was ritual in the use in Ireland, too – in County Clare the juice had to be applied seven times, with appropriate prayers being said, and there was a further proscription, for it had to be picked at a particular unspecified period of the sun and moon in August (Westropp. 1911). There is recorded a domestic use of a decoction of sun spurge for ringworm (Dyer. 1889, Trevelyan). Gerard recommended it "to stop hollow teeth, being put into them warily, so that you touch neither the gums, nor any of the other teeth in the mouth ...".

It is known as Whitlow-grass in Lincolnshire (Britten & Holland), but it is difficult to imagine anyone willingly putting spurge juice on a whitlow.

SUNBURN

Use the juice of the MEDITERRANEAN ALOE (*Aloe vera*) on the skin to bring relief (G B Foster). The treatment in Indiana is to crush the leaves of BLACK NIGHTSHADE and stir them in a cup of cream. When ready, put the cream on the sunburned area (Tyler). Gerard recommended the berries of BLACK BRYONY "to take away the Sunn-burne and other blemishes of the skin".

SUNFLOWER

(*Helianthus annuus*) To dream of sunflowers means your pride will be deeply wounded (Mackay), but to have them growing in the garden will bring good luck (H M Hyatt). The American Indians used the plant medicinally to a large extent. It was considered an antidote to rattlesnake-bite, the method being to chew the seeds in order to make a poultice of them (Stevenson). They also boiled the seeds to get oil to dress their hair, and the whole flower heads were boiled by, among others, the Dakota Indians, for lung trouble (Gilmore), which is interesting, for the plant is used in Russian domestic medicine for the same complaint, listed as bronchitis, laryngitis and pulmonary disorders. The same medicine is used there for nervous disorders (Kourennoff). Russian folk medicine also prescribed the stalks and leaves, infused in vodka, and taken three times a day, for gout. To cure a wart, the Navajo Indians burned some powdered sunflower pith on the wart (Wyman & Harris). Herbalists still use the tincture for bruises, in the same way as arnica (Schauenberg & Paris).

The seeds are a specific for whooping cough, and Sunflower Gin is often made for that complaint. The recipe is to boil 2 ounces of the seeds in a quart of water till it is reduced to a pint. Strain and add 6 ounces of Hollands gin and sugar to taste. The dose is a teaspoonful at a time (Leyel. 1937). Much simpler is a syrup made in Iowa, of stewed seeds and brown sugar, and regarded there as a good remedy for the complaint (Stout). They are smoked, too, (dried of course) as a hay fever remedy in America (H M Hyatt). Having them growing in the garden will prevent typhoid, so it was believed in Indiana (Brewster).

SWEET BRIAR

(*Rosa rubiginosa*) Just as well known under the name Eglantine. "Sweet", because of its inherent fragrance, which, so it is said, is fresher before rain (Trevelyan). This rose has attracted legends, just as the others have. One of them explains why their thorns point downwards. The devil was expelled from heaven, and he tried to regain his lost position by means of a ladder made of the thorns. But when the plant was only allowed to grow as a bush, he put the thorns in their present position out of spite. In some parts of France, it is said that the sweet briar was actually planted by the devil; and the hips are his bread. It is called 'Rose du diable' in those areas, and 'Rose sorcière' in Anjou and La Mayenne. French mothers use an injunction to stop their children picking the flowers. If they touch them, they say, a thunderbolt will fall on them (Sebillot).

There is a Derbyshire superstition that where the sweet briar flourishes in a garden, the mistress is ruler (Addy. 1895), a belief that is associated with quite a few other plants, notably rosemary, and even sage. But this rose is also a symbol of chastity, sometimes being painted in portraits of ladies of the Elizabethan period. There is one of Elizabeth Brydges, for instance, in which a spray of eglantine appears in the top left hand corner, and also in her hair, and, as an embroidery, on her cuffs (Hearn).

"Gather Sweetbriar in June, for it promoteth cheerfulness", is an old Scottish saying. Twigs of it were often used for sprains. But the protection Sweet Briar gives is more often found to be spiritual. In Normandy, for instance, hanging it over the door used to be regarded as a certain protection against witchcraft, and also against fevers (W B Johnson), which must have been thought of as a result of malevolence.

SWEET CHESTNUT

(*Castanea sativa*) The English tree, introduced by the Romans, is not one of the special nut varieties developed by Italian growers (see Rackham. 1976). The nuts sold in England are nearly always imported from Spain (hence the name Spanish Chestnut, for it was said that the best nuts came from there). In Provence, chestnuts are regarded as female fertility symbols (Alford. 1978), and such a belief was probably wider spread than that. The dream books, for instance, told us that to dream of eating chestnuts was a good sign for an unmarried girl. It meant that someone would soon be coming to court her. For a married woman, it was a sign of sickness (Raphael). A Portuguese method of divining whether a baby will be a boy or girl is to spit on a chestnut and throw it into the fire. If it bursts, the child will be a boy, while if it only fizzles feebly it will be a girl (Gallop). As the husk is surrounded by thorns, it is the symbol of chastity (Ferguson).

Theophrastus reported that the wood was used for roof timbers in the ancient world, and cited a case where chestnut beams in a public bath gave warning before they broke, so that all the people in the building were able to get out before the roof collapsed (Meiggs). Taylor. 1812 reckoned there was a "deceitful brittleness in it which renders it unsafe to be used as beams". He also said that bread had been made with the nuts, and flour is still made in Italy, where chestnuts assume great importance as a foodstuff (Rackham. 1986).

The nuts, bark and leaves can all be used for coughs of any kind, including whooping cough (Page. 1978). The tannin content of both the bark and leaves make them (and the nuts) useful for diarrhoea, and an infusion can be used as an expectorant for bronchitis. Gypsies say the powdered nuts are good for piles. That may be true, but they even wear the nut in a little bag round the neck as a charm to prevent piles, but the bag, they say, must never be made of silk (Vesey-Fitzgerald).

SWEET CICELY

(*Myrrhis odorata*) It is one of the plants that are supposed to bloom on Old Christmas Eve, so it was said in the Isle of Man, where it is called "myrrh" (Moore). A watch is still sometimes kept for the flowering. According to tradition, the bloom only lasts for an hour (Garrad). Another superstition about Sweet Cicely was that it increased "the lust and strength of the old" (Camp). It is a symbol of sincerity (Leyel. 1937). But it really does have its uses – most parts are edible, the leaves and roots being used in salads, the leaves raw, and the roots boiled and sliced (Rohde. 1936). They have a mild aniseed flavour, ideal for flavouring stewed fruits like gooseberries and plums (Mabey. 1972). It is apparently one of the main ingredients of Chartreuse (Clair). The seeds too are edible, at least in the green state, before they become black and hard; children like to chew them, for they taste like liquorice (Foster).

The whole plant is very attractive to bees, and was often rubbed over the inside of hives to induce swarms to enter (Northcote). In popular medicine it is taken for flatulence and any digestive ailment (Clair), and is a remedy for chest troubles or bronchial colds (Gibson), and is still used to lower blood pressure (Schauenberg & Paris).

SWEET FLAG

(*Acorus calamus*) Some uses of this plant stem directly from the fact that oil expressed from the roots can produce an LSD-like experience. The Cree Indians of North America have long used it as a hallucinogen (Emboden. 1979). In Europe, it seems to have been connected with witchcraft (one formula for witch ointment was "De la Bule, de l'Acorum vulgaire, de la Morelle endormante, et de l'huyle") (Summers. 1927), but in Asia the influence of Sweet Flag was on the whole good. It seems to have had a protective reputation in Japanese folklore. In the various stories of the serpent-bridegroom type, the plant is used to thwart the evil supernatural. The bride becomes pregnant, but a bath in water that has Sweet Flag taken at the proper festival time in it will kill the unwanted child (Seki). It was a protector, too, in China; it was put up at the side of outdoor gates (so was mugwort) to avert the unpropitious (Tun Li Ch'en). But the use made of it by the Dusun of the northern parts of Borneo seems closer to North American Indian practice. They use the hallucinogen of the rootstock to contact certain spirit beings, in usual shamanic style. Women shamans, or ritual specialists, as anthropologists term them, call upon these spirits for divination, and also during restoring and curing rituals, using the power of the root to do so. The spirit can be acquired for life, and she is believed to descend and ascend with the rootstock to the lower and upper halves of the universe in her search for disease givers or souls of the dead, responsible for crises. The root is specially cultivated by these women, and is "brought to life" through a ritual, then hung to dry in the house eaves. Before the root can be used in any major ritual, it must be "given the power" through a long formal ceremony, when several shamanesses participate in the recitation of verses and magical acts known to ensure power from the plant. The shamaness calls up her familiar spirit from the root by running the tip of her right forefinger along the blade of a magically endowed knife. As she moves her finger she chants (in translation):

> Wake up, sacred plant, wake up.
> Sacred plant, I wake you to tell me,
> I wake you to tell me
> If I may serve this person.

When the spirit awakes in the root, her finger stops at a point along the blade; this point is an indication of the nature of the disease giver or soul of the dead causing the crisis. From this, she will choose the types of ritual to be recited. Small bits of the root are worn on clothing, or on a necklace, to frighten away disease givers by its smell (T R Williams. 1965). Rather similar is a North American practice – the Chippewa once used the root of this and *Aralia nudicaulis* to make a decoction to put on fish nets, or to "rattle snakes away" (Densmore).

The rootstock may be pungent enough to frighten off disease, but the leaves can best be described as aromatic, with a violet-like fragrance that is brought out as they are trodden on. Hence its former use for strewing on the floors of churches and the houses of the rich. Ely and Norwich cathedrals had their floors covered with it at festival times, as Sweet Flag grows in the Fens (Genders. 1971), where it was actually harvested, and not only for floor covering, for it seems that they were used for thatching too, especially for churches (A W Hatfield). One of the charges of extravagance brought against Cardinal Wolsey was that he ordered the floors of his palace at Hampton Court to be covered much too often with rushes and flags, since they were expensive and difficult to obtain (Genders. 1972).

Not only does the dried and powdered root make a refreshing talcum powder, which the ancient Greeks sprinkled over beds and clothes (Genders. 1972), but it has been used in snuff – the French 'snuff à la violette' apparently contains some of it (Barton & Castle). It is an early example of a substance used to kick the smoking habit (Leyel. 1937). Cut up and boiled in syrup, as one would do with ginger, the rootstock made a confection, as well as a mild stimulant to aid digestion (Sanford). It is quite a favourite on the Continent, and is sold in Indian bazaars as a sweetmeat to comfort dyspepsia (A W Hatfield).

Sweet Flag was one of the constituents that Moses was commanded to make for use in the Tabernacle (Exodus xxx 23-5: "Take thou also unto the principal spices, of pure myrrh five hundred shekels, and of sweet calamus two hundred and fifty shekels/ and of cassia five hundred shekels .../ And thou shalt make of it an oil of holy ointment ...: it shall be an holy anointing oil"), while the prophet Ezekiel says of the commerce of Tyre: "Bright iron, cassia, and calamus were in thy market" (Ezekiel xxvii. 19). Calamus is still official in most pharmacopeias, though it is not much used now. It can be given as a mild aromatic stimulant, and herbalists still use the dried rhizomes as a febrifuge (it is a good substitute for Peruvian bark, quinine in other words); up to about 1940 it was well used by vets (Sanecki). Calamine lotion is still familiar enough – that, too, comes from the root of Sweet Flag (Genders. 1976).

It has been a favourite medicine in India from the earliest times, valuable for children's bowel complaints (Sanford), but its best known use in former times was for dyspepsia, loss of appetite, etc., (W A R Thomson. 1978). The carminative usage appeared in medieval Latin compilations, and was to be found in New World medicine also. In Alabama they either chewed the root, or put it in whisky to be used when needed. Or it could be boiled, and the water drunk (R B Browne). In Ohio, this was drunk as a cure for diarrhoea (Bergen. 1899). The American Indians used it, too, for similar purposes. People like the Menomini cured cramps in the stomach with it, though it was reckoned very powerful, to be used only in minute quantities. The measure of a dose was the length of a finger-joint (H H Smith. 1923). It was even claimed that the Blackfoot Indians used it to cause abortions (Johnston); in fact, it was a general cure-all: indigestion, toothache, fever (it was even said to be beneficial in typhoid as a stimulant), and for religious purposes. A warrior rubbed a paste made from it on his face to make him immune to fear (Sanford), and they took it to alleviate fatigue on a long journey (Emboden. 1979), just as plants more narcotic than this, coca for instance, were used. In Indiana, the roots mashed to a consistency of mashed potatoes and spread on a bandage, were bound on a carbuncle as a poultice (Tyler).

SWEET POTATO

(*Ipomaea batatas*) Maoris used the plant medicinally, as well as for food. The whole plant was boiled and used internally for a low fever, and externally for various skin diseases (Goldie). There are some strange beliefs recorded from America, too, one of which claims that you can have rosy cheeks if you eat the skins of sweet potatoes (H M Hyatt) (but that claim is also made for ordinary potatoes). Malaysian peasants always planted sweet potatoes on a starry night; it is a magical charm, to ensure that they fill out properly, by getting plenty of eyes, like stars in the sky (Skeat).

SWEET SCABIOUS

(*Scabiosa atropurpurea*) Names like Widow's Flower (Folkard), Poor Widoe, or Mournful Widow (E M Wright) are fairly common in the south of England, and it is interesting to find that, under the name Saudade, this plant was much used in Portugal and Brazil for funeral wreaths (Coats). The name *Scabiosa* has an interesting derivation. It is from Latin scabere, to scratch. Scabies got its name from the plant, which was used as a remedy. Or is it the other way round? Young suggested that it was named from the scab. The OED seems to confirm this (med Latin scabiosa herba, named as specific against itch). Anyway, it was actually grown in gardens to use as a cure for the itch (Folkard).

SWEET VERNAL GRASS

(*Anthoxanthum odoratum*) "Sweet" because it is aromatic when dried, and it is this that gives the characteristic smell of new-mown hay. Surprisingly, provided it has been dried first, it is given to cure hay fever! (Leyel. 1937). Rheumatism is treated either with a poultice made from this grass, applied as hot as can be borne, or by taking a bath with the grass added (W A R Thomson. 1978).

SWEET WILLIAM

(*Dianthus barbatus*) The name has exercised the imagination of commentators in the past. The French name is oeillet (oeillet de poète, Rambosson calls it), and this may have been corrupted into Willy (Friend. 1883). Archdeacon Hare said that it is dedicated to St William, whose feast day is 25 June, and that "sweet" is a substitute for "saint". Others thought that the Childing Pink (*Dianthus prolifer*, or *Kohlrasuchia prolifera,* as it is called these days), is the original of the Saint Sweet William, for it was said that the "saint" was only dropped since the demolition of St William's shrine in Rochester Cathedral (Dyer. 1889).

SWOLLEN GLANDS

In the Cotswolds area, a fomentation of MARSH MALLOW leaves, soaked in boiling water, was used for mumps and swollen glands (Briggs. 1974).

SYCAMORE

(*Acer pseudo-platanus*) Sycamore was introduced into England towards the end of the 16th century, recent enough for Gerard to call it "a stranger to England". It withstands sea and mountain winds better than most timber trees, and so was widely planted in upland and exposed parts as a windbreak for farmhouses (Hoskins). Some say, though, that sycamore will tend to kill other trees in its vicinity (Wookey).

There was a belief in mid-Wales that sycamore trees keep the fairies away and stop them spoiling the milk, probably because it was a fairly late introduction, and also because of its association with the farmyard (E E Evans). Devonshire Revel buns were baked on sycamore leaves. These are Easter cakes, each cake baked on its own individual leaf (Mabey. 1977), the point being that the imprint of the leaf should appear on the cake. To dream of sycamores foretells jealousy to the married, but it promises marriage to the single (Gordon. 1985).

Solitary sycamores in Ireland were treated with the same respect as solitary hawthorns. There was one at Killadoon, known as "Honey-tree", for example, which was certainly venerated in the same way (Wood-Martin). A lone sycamore once grew a mile or two from Dover, and was known as the Lone Tree. It was said to be a witness to murder. Legend has it that a soldier from the Dover garrison killed a comrade with a staff he was carrying, and which he stuck in the ground, saying that the crime would never be discovered until the day the dry wood took root. He served abroad for many years unsuspected, but when he came back, and was stationed again at Dover, he found that the staff had become a flourishing tree. So the crime was discovered by confession, and he was hanged (Bett. 1952).

The name sycamore (and the French sycomore) are derived by a mistake from the name of the Fig, *Ficus sycamorus* which means fig-mulberry. The shape of the leaves is vaguely similar (Barber & Phillips). In Cornwall, this tree used to be called Faddy-tree. The Helston Furry was at one time known as the Faddy, and the sycamore Faddy-tree, for the boys would make whistles from its branches at the festival (Deane & Shaw).

SYMBOLISM
CYCLAMEN was taken as the symbol of voluptuousness (Haig), for it had the reputation since ancient times of being an aphrodisiac, however unlikely that sounds. CUCKOO-PINT is another matter. Friend said that it is the symbol of zeal and ardour. It was probably his way of giving some respectability to the subject, for to the common people the very form of this plant, the spadix in the spathe, stood for copulation. That is the reason for the many male + female names given to it, and the sexual overtones of many of the rest. ELDER too is a symbol of zeal, according to the ideas of Raban Maur (Haig), though doubtless not with the same

undertones. HAWTHORN too became a symbol of carnal love as opposed to spiritual love. It is the *arbor cupiditatis*, used as such throughout the literature of the Middle Ages. The symbolism probably arose from the scent, the trimethylamine of which causes the smell of putrefaction. But the scent has this other interpretation as well – that of sex. The smell was said to arouse sexual desire (Anderson), and the tree itself was used constantly in medieval love allegory. The Chaste Tree (*Vitex agnus-castus*) was used by Athenian women as a symbol of chastity in religious rites. For that reason it also became the symbol of indifference. There is a suggestion that the leaves may be contraceptive; certainly Dioscorides said so, and Pomet's comment on the name *agnus-castus* is equally explicit: "and the name of Agnus Castus, because the Athenian ladies who were willing to preserve their chastity … made their beds of the Leaves of this Shrub, on which they lay: But it is by way of ridicule that the Name of Agnus Castus is now given to this seed, since it is commonly made use of in the care of venereal cases, or to assist those who have violated, instead of preserv'd their Chastity". Another symbol of chastity is provided by SWEET CHESTNUT – the husk is surrounded by thorns, hence the imagery (Ferguson). But in some parts, chestnuts are regarded as female fertility symbols (Alford. 1978).

FIGS, like POMEGRANATE, carry a large number of seeds, and so they are both symbols of fertility. Doubly so, in the case of figs, for it is pointed out that it resembles the womb in shape (Maple). The association stems from classical times, but it has been carried over into more recent folklore. Bulgarian brides, for example, used to be presented with dried figs as a promise of many children (M Baker), and Dutch folk medicine claims that a craving for figs during pregnancy ensures that the child will be born quickly and easily (van Andel). At wedding ceremonies in Rhodes, a pomegranate was put on the threshold, to be crushed by the bridegroom's foot as he entered (Rodd), an obvious piece of symbolism. In a Persian story, Khodaded was one of fifty children begotten by a childless nomad upon his fifty wives, after eating as many pomegranate seeds. ALMONDS, too, carry a similar symbolism of fertility. Pink almond blossom would be heaped at a gypsy bride's wedding in Spain. An almond tree, so it is said in Greece, will always bear fruit well near the home of a bride, and almond paste usually appears on wedding cakes (M Baker. 1977). Almonds are equally prominent at Indian weddings, where they may indicate both prosperity and children, i.e., money fertility and sexual fertility. (A secondary view of almond symbolism is that of hope, for this is an extremely early flowering tree, and hope is needed if the tree blooms in January, as it often does, for a subsequent frost will ensure that there will be no crop). SILVER FIR, sacred in Greece to Artemis, the moon-goddess who presided over childbirth (Graves), would also be a symbol of fertility. In some parts of Germany, women were struck with a small fir-tree at Shrovetide, and brides and bridegrooms often carried fir branches with lighted tapers. Firs were often planted before a house when a wedding took place. BIRCH, too, was a symbol of fertility. Saplings were put in houses and stables, and men and women, as well as the cattle, were struck with birch twigs, with the avowed intention of increasing fertility (Elliott). HAZEL was yet another tree with this symbolism, even more powerfully than the others. Throwing nuts at the bride and groom is sometimes the practice at Greek weddings, and until quite recently, Devonshire brides were given little bags of hazel nuts as they left the church (see also **WEDDINGS**). PINE cones had some phallic significance in ancient times, so they were also symbols of fertility. It is interesting to record a superstition from the Highlands of Scotland to the effect that a lot of illegitimate births could be blamed on the large number of pine trees growing in the district (Begg). WALNUTS were the symbol of <u>marriage</u>, according to Pliny. Sexual magic was performed with them. Arnold de Villeneuve gave a receipt for "tying the knot". One takes a walnut, separates the two halves, and puts them in the marriage bed. The counter charm is to stick the two halves together, crack the nut, and then the couple eat it (Bouisson). It is because they are of two halves that the marriage symbolism exists. The tree remained a bridegroom's symbol in Germany until Evelyn's time.

ASPEN symbolizes <u>fear</u>, from the constant shivering of its leaves, but it is also the symbol of <u>scandal</u>, by a different interpretation of the same characteristic, for an old saying tells that its leaves were made of women's tongues. Gerard knew the saying – " … it is the matter whereof womens tongues were made … which seldome cease wagging". OAK is the symbol of <u>hospitality</u>, Zeus's tree and his emblem, the tree that had sheltered him at his birth on Mount Lycaeus. One of the earliest references to the tree (possibly *Quercus pseudo-coccifera*) is the story of Abraham's hospitable entertainment of the angels, given under the oak of Mamre (Bayley).

GARLIC, in Bologna, was the symbol of <u>abundance</u>. The abundance was reckoned material, for it used to be bought at the Midsummer festival as a charm against poverty during the coming year. But the closely related LEEKS were taken as the symbol of <u>humility</u>, for they have always been rather looked down upon, in spite of an Irish legend that they were created by St Patrick (Swahn). Certainly in the Orient the vegetable has always been seen as the food of the poor, hence this symbolism, which is shared by VIOLETS (as such it would be the emblem of Christ on earth). BIRCH, too, was a symbol of <u>meekness</u> (Leyel), "as bare as the birch at Yule Eve"

is a proverb used once of anyone in extreme poverty (Denham).

NETTLE is a symbol of <u>envy</u> (Haig), and of <u>spitefulness</u> (Leyel. 1937), while MONKSHOOD, not surprisingly, given its extremely poisonous nature, is the symbol of <u>crime</u> (Rambosson). ROSES have been used as the symbol of many things, <u>love</u>, mainly, but also of <u>silence</u>, and also of <u>secrecy</u>. Silence, is an idea still with us in the expression *sub rosa*. It is said that many of the great houses with plaster ceilings had a rose as the central ornament, a reminder that matters talked of at table "under the rose" must not be repeated outside that room. But "under the rose" had another meaning. To be born "under the rose" is said to mean being illegitimate, a rose also being a symbol of secrecy, so the wild rose is sometimes used to signify illicit love (Briggs. 1974).

HONEYSUCKLE is a symbol of <u>constancy</u>, presumably because of its twining habit (Tynan & Maitland), and MUGWORT symbolises <u>happiness</u> and <u>tranquillity</u>, as does the PEAR for <u>affection</u> and <u>comfort</u> (Leyel. 1937). The MADONNA LILY is the symbol of <u>purity</u> (Haig), <u>chastity</u> (also Haig) (it will only grow for "a good woman" (M Baker. 1977), of <u>beauty</u> (Zohary), and <u>celestial bliss</u> (Woodcock & Stearn), for to early medieval artists and theologians this was the flower of heaven. But of course, this lily is the emblem of the Virgin Mary, so the symbolism stems from that. PARSLEY seems to have been a symbol of <u>festivity</u>, though quite why is not clear. To the Greeks it also symbolised <u>strength</u>, and they crowned the winners of the Isthmian Games with chaplets made of it. FENNEL, too, is a symbol of strength, and also of <u>flattery</u> (Dyer). The Italian idiom 'dare finocchio' means to flatter (Northcote). SWEET CICELY represents the opposite, <u>sincerity</u>, while LETTUCE is the emblem of <u>temperance</u>, according to the ideas of Raban Maur (Haig). Why should MEADOWSWEET be stigmatized as the symbol of <u>uselessness</u>? It is far from true. Equally odd is Shakespeare's use of BLACKBERRIES as symbols of <u>worthlessness</u>, though the way it grows makes it an emblem of <u>lowliness</u>. Yet another of its symbolic attributions is that of <u>remorse</u>, from the "fierceness with which it grips the passer-by" (Dyer. 1889). In Christian symbolism it stands for the purity of the Virgin Mary, who "bore the flames of divine love without being consumed by lust" (Ferguson).

BALSAMINT symbolises <u>impatience</u> (Leyel. 1937), while VERVAIN was the very symbol of <u>enchantment</u>, not surprisingly when the reputation of that plant is recalled. WHITE POPLAR, with its leaves that are white on the undersides, and blackish-green above, symbolises <u>time</u>, for those leaves show the alternation between night and day. That and the fact that they always seem to be in motion, were enough to excite somebody's imagination into inaugurating this symbolism. <u>Hope</u> is symbolised by HOPS (a garland of them was worn by suitors in Elizabethan times), and also by WILD THYME (Quennell), as well as <u>activity</u>, and <u>courage</u>. Those who eat it are said to become braver, and in the Middle Ages, a sprig of thyme was given to knights by their ladies, to keep up their courage. Even a scarf embroidered with a bee alighting on the plant was supposed to have the same effect (Leyel. 1937). Watercress is the symbol of <u>stability</u>, and also of <u>power</u> (Leyel. 1937). RUE is the symbol of <u>regret</u> and <u>repentance</u>; to rue is to be sorry for, even though the two words have different derivations (see **RUE**).

Evergreen trees are at the same time funerary emblems and symbols of <u>immortality</u> (see for instance HOLM OAK, or YEW). EDELWEISS also has this symbolism (Folkard).

MYRTLE'S symbolism is complicated. The Greeks used it for <u>love</u> and for <u>immortality</u> (it is evergreen). Throughout the Scriptures, it is the symbol of <u>divine generosity</u>, and also of <u>peace</u> and <u>joy</u>, and of <u>justice</u>. It was held to have the power of creating and perpetuating love (Philpot), and as such was a symbol also of <u>married bliss</u> (Baker. 1980). But on the other hand, myrtle was a symbol of <u>death</u>, and of <u>war</u> (Gordon. 1977).

Abundance	OLIVE, in ancient Greece
Affection	BURNET SAXIFRAGE (Leyel. 1937)
Amiability	JASMINE (Ferguson)
Anticipation	GOOSEBERRY (Leyel. 1937)
Bashfulness	PEONY. Why? It was claimed to be of divine origin
Betrayal	JUDAS TREE, naturally, if Judas was supposed to have hanged himself on this tree
Charity	1. WHITE WATERLILY, according to the ideas of Raban Maur (Haig) 2. SAFFRON, according to the system of Raban Maur (Haig) 3. ORIENTAL PLANE (*Platanus orientalis*), in Christian art (Ferguson)
Chastity	SWEET BRIAR, which was sometimes painted in portraits of ladies of the Elizabethan period
Compassion	ALLSPICE (Leyel. 1937)
Coquetry	LADY LAUREL (Ingram)
Courage	BAMBOO, in Japanese symbolism (J Piggott)
Courtesy	PINK
Death	1. CYPRESS. Both Greeks and Romans called it the "mournful tree", because it was sacred to the rulers of the underworld.

It was planted by a grave, and, when a death occurred, it was the custom to put it in front of the house so that those about to perform a sacred rite would be warned against entering a house polluted by a dead body. And of course, coffins were made of cypress wood (if you were rich enough), the wood being incorruptible (and insect proof)

2. POPPY ANEMONE, for the myth has it that the plant originated at the death of Adonis, who was changed into the plant according to one version, or that it sprang from the mixture of his blood and the tears of Venus (Rambosson). It is also the symbol of sorrow

Divine love	PINK
Domesticity	HOUSELEEK, probably because a growing plant in the thatch brought luck and order to the house
Egotism	NARCISSUS. The story of Narcissus, enamoured of his own beauty, becoming spell-bound in front of his own image, is too well-known for comment
Elegance	JASMINE
Encouragement	GOLDEN ROD
Energy in adversity	CAMOMILE, perhaps from the truism that "the more it is trodden, the faster it grows"
Excellence	PERIWINKLE, as in an old ballad, in which a noble lady is called "the parwenke of proesse"
Extravagance	OPIUM POPPY, in Christian art (Ferguson). Presumably from the great number of its seeds
Falsehood	ALKANET, for its roots, for making rouge, seem to be one of the most ancient of face cosmetics
Fascination	CARNATION
Fertility	QUINCE, which was dedicated to Venus, who is often represented holding one in her right hand (Ellacombe). Sending quinces as presents, or eating them together, were recognized love tokens; so was throwing them at each other. In 17th century England, it was reckoned that "the woman with child that eateth many during the time of her breeding shall bring forth wise children and of good understanding" (Boland. 1977)

Fidelity	1. BROOKLIME (Leyel. 1937) 2. DAISY (Friend. 1883) 3. PLUM, in Christian art (Ferguson)		

Fidelity in misfortune — WALLFLOWER (Friend. 1883), perhaps suggested by the legend of the Scottish girl who was in love with the heir to a rival clan. She was planning to escape with him, by climbing from a high window to join him, "but the silken cord untied; she fell, and bruised, and there she died", and was turned into "the scented flower upon the wall"

Forsaken love — WILLOW. See Shakespeare: *Much ado about nothing*, when Benedick says "I offered him my company to a willow-tree, either to make him a garland, as being forsaken, or to bind him up a rod, as worthy to be whipped"

Foresight — STRAWBERRY (Leyel. 1937), though the usual symbolism is Righteousness

Friendship — PERIWINKLE

Frugality — CHICORY (Leyel. 1937)

Genius — ORIENTAL PLANE (*Platanus orientalis*), for it is said that philosophers taught beneath this tree, and so it acquired a reputation as a seat of learning (Dyer. 1889)

Goodwill — GREEN PURSLANE, in West Africa (Dalziel)

Grace — JASMINE

Grief — 1. WILLOW, especially that of disappointed love
2. MARIGOLD, though it seems odd for such a popular flower. Perhaps there is some connection with a rather vague concept of its use at funerals

Happiness — QUINCE

Hatred — BASIL (Leyel. 1937). The Romans used to sow the seeds with curses through the belief that the more it was abused the better it might grow. The Greeks, too, supposed basil to thrive best when sown with cursing – this explains the French saying "semer la basilic", as signifying slander (Fernie). In Italian folklore, basil always stands for hatred

Health — WORMWOOD, so famous a medicine for an enormous list of ills over the centuries, for plague and cholera down to lesser ailments (Painter)

Hidden worth — CORIANDER (Leyel. 1937)

Humility — BROOM (Friend. 1883), possibly only because of its use as a humble domestic implement. See also EMBLEMS (Housewife)

Ignorance — OPIUM POPPY, in Christian art (Ferguson)

Imbecility — REED, probably for the same reason as led to the symbol of weakness (they are tossed about by the wind, and have to bow before a superior force (Grindon)

Immortality — 1. PEACH, in China, for it is Taoist emblem (Tun Li-Ch'en)
2. CYPRESS, for it is evergreen, a symbolism shared with most evergreens
3. PERIWINKLE, in Germany, for it is evergreen (Fernie)

Innocence — DAISY, and thereby the emblem of the newly-born (Bayley. 1919), and so of the infant Christ

Independence — PLUM, in Christian art (Ferguson)

Indifference — BLACK MUSTARD

Longevity — BAMBOO, in Chinese and Japanese systems. In the latter it also represents tenacity and courage (J Piggott)

Magnificence — ORIENTAL PLANE (*Platanus orientalis*)

Marriage — CARNATION

Melancholy — PRIMROSE

Modesty — DAISY

Mourning — 1. BASIL, which in the eastern Mediterranean countries is a herb of grief, to be put on graves
2. WILLOW, used in some stylised form embossed on Victorian mourning cards, though they first appeared as such at the end of the 18th century as a tomb decoration, usually with the figure of Hope, or the widow weeping and clinging to an urn beneath its boughs

Patriotism — NASTURTIUM (Leyel. 1937) Why?

Peace — OLIVE, as it still is. Possibly because in the Noah's Ark story, the dove is said to have brought back an olive leaf as an indication that God's wrath, in the form of the Flood, was abating (the dove, too, carries this symbolism)

Pleasantry	BEE BALM, because of its aromatic fragrance
Precaution	GOLDEN ROD
Pretension	ROSEBAY WILLOWHERB, possibly because of the way it so quickly colonises areas cleared of vegetation by fire, etc. Cf the name Bombweed
Protection	JUNIPER, for this was one of the plants that offered shelter to the Virgin and Child during the flight into Egypt. It naturally received the Virgin's blessing, and the power of putting to flight the spirits of evil. (see also PROTECTIVE PLANTS)
Purity	1. CHRISTMAS ROSE – the result of its pure white flowers. It is also the emblem of St Agnes, the patroness of purity. Her feast day is 21 January, when the plant should be well in bloom (Hadfield & Hadfield)
	2. LILY-OF-THE-VALLEY, partcularly in early Netherlandish and German paintings (Haig)
	3. MADONNA LILY (Haig). This lily is the emblem of the Virgin Mary, so the symbolism stems from that
	4. HINDU LOTUS, in Japanese thought. Buddha is often depicted as seated on a lotus
Rendezvous	CHICKWEED (Leyel. 1937). Why?
Riches	BUTTERCUP, obviously, from the golden coloured flowers
Righteousness	STRAWBERRY. The plant was used a great deal in paintings representing the garden of heaven, and the "enclosed garden" of the Virgin Mary. They are shown with fruit and flowers together, and in this way represent the good works of the righteous. It is the emblem of "the righteous men whose fruits are good works" (Haig). As such

	it was used chiefly in early Italian paintings of the Adoration. In these paintings, the symbolical strawberry is usually accompanied by the violet, as if to point out that the truly fruitful soul is always humble (Haig)
Sadness	WILLOW
Sleep	OPIUM POPPY, for it is sedative, and the very emblem of Morpheus
Sorrow	1. WILLOW
	2. POPPY ANEMONE, from the myth of the death of Adonis, from which the plant sprang, and the sorrow of Venus, whose tears when mixed with the blood of Adonis produced the flower according to another version of the myth
Surprise	BETONY
Tenacity	BAMBOO, in Japanese symbolism (J Piggott)
Vengeance	AFRICAN MAHOGANY, in Yoruba belief. The tree may not be cut down unless the tree Spirit has first been propitiated by the offering of a fowl, or some palm wine (J P Lucas)
Vivacity	HOUSELEEK
Wantonness	PRIMROSE, which was used as a love-charm in many places. Browne is talking about them when he says "maidens as a true-love in their bosoms place". Shakespeare has Hamlet say "himself the primrose path of dalliance treads"
Weakness	REED, for they are tossed about by the wind bending to a superior force (Ellacombe)
Wisdom	OLIVE, in ancient Greece

Symphoricarpus albus > SNOWBERRY

Symphytum asperum > PRICKLY COMFREY

Symphytum officinale > COMFREY

Syzygium aromaticum > CLOVE

T

TABASCO PEPPER

(*Capsicum frutescens*) The fruit produces Cayenne Pepper, though this is "an indiscriminate mixture of the powder of the dried pods of many species of capsicum, but especially of the *capsicum frutescens* …, which is the hottest of all" (Thornton). Cayenne pepper is standard seasoning among Mexican Indians, the Totonac among them, who also use it as a remedy for magical malviento and malojo [evil eye] in children (Kelly & Palerm). In Africa, a way of detecting witches is by Guinea Pepper, which may be *C annuum*, or perhaps this *C frutescens*. "You light a fire under the tree where the witches habitually meet, and you put pepper into the fire. If there is a witch on the tree and the smell entangles her, she can't fly away and will be found on the tree in the morning" (Debrunner).

Tabernanthe iboga > IBOGA

Tagetes erecta > AFRICAN MARIGOLD

TAMARIND

(*Tamarindus indicus*) This tree folds its leaflets at night and in overcast weather, an ideal abode of the rain god in Burmese thought (Menninger). Indians have a prejudice against sleeping under a tamarind, probably because of the dampness in the tree, which certainly affects the canvas of tents pitched near it (Leyel. 1937). When an Oraian (India) woman is in difficult labour, a man goes to find a tamarind tree that has been singed by lightning, stands by it, and strips off a piece of bark where it touches his waist. He takes the bark back to the door of the delivery room, and pushes the piece of bark through a hole in the door, and stands holding his end of the bark. The woman has to fix her gaze on this bark to facilitate delivery. As soon as it takes place the man takes out the bark, otherwise inversion of the uterus is believed to occur. Another belief of the same people is that a child born after three sisters or three brothers will bring misfortune to its siblings or parents. To neutralize the ill effect the father of the newborn child has to go to a tamarind tree and make three strokes on it with some weapon (Gupta). A medicinal use is reported from Guyana, where a decoction of the fruit is taken as a measles remedy (Laguerre).

Tamarindus indicus > TAMARIND

TAMARISK

(*Tamarix gallica*) "It was of old counted infelix, and under malediction, and therefore used to wreath, and be put on the heads of malefactors" (Evelyn. 1678), because it was one of the trees reckoned to be the one on which Judas hanged himself. Once a tall and beautiful tree, it is now reduced to a shrub by divine malediction (A Porteous). They say that the ghost

of Judas always flits around the tamarisk. But in Morocco, it was used as a charm against the evil eye (Westermarck), and in Egypt it was a holy tree, for it was chosen to overshadow the tomb of Osiris; the chest containing his body was found by Isis lodged in its branches (Elworthy. 1895). On one tomb, a bird is depicted perched among the branches of the tamarisk, with the legend "the soul of Osiris". It shows that the spirit of the dead god was believed to haunt his sacred tree (Frazer).

Tamarix gallica > TAMARISK

Tamus communis > BLACK BRYONY

Tanacetum balsamita > BALSAMINT

Tanacetum parthenium > FEVERFEW

Tanacetum vulgare > TANSY

TANSY

(*Tanacetum vulgare*) The clue to the meaning of 'tansy' lies in the claim that it was dedicated to St Athanasius (Dyer). Actually, it was the plant itself that was called, in medieval Latin, athanasia, from the Greek word for immortality (athanatos means undying). Athanasia became in Old French tanesil, and thus to tansy in English. The immortality name might come from the fact that the flowers take a long time to wither (Brownlow), but more likely because it was used to preserve dead bodies from corruption (Grieve. 1933). Ann Leighton had this quotation: "Samuel Sewall has recorded observing the body of a friend long dead but well preserved in his coffin packed full of tansy". Not unrelated to this is the claim that rubbing the surface of raw meat with tansy leaves will protect it from flies (Genders. 1977); that, too, is why bunches of it were put on the windows of farm kitchens. Sprigs were put in bedding, to discourage vermin (Drury. 1992), and in Elizabethan times it was a favourite strewing herb (Macleod). Another belief, from Maryland, is that you should never sow tansy seed. If you do, there will be a death in the family (Whitney & Bullock).

Tansy leaves are perfectly edible, though very bitter (a piece about a quarter of an inch square is enough to flavour a salad) (Rohde. 1936). So eating a tansy salad as a means to procure a baby, as was certainly done by childless women wanting a child, must have been no mean undertaking. No less an authority than Culpeper advised it: "Let those Women that desire Children love this herb, 'tis their best Companion, their Husband excepted". He recommended it either bruised and laid on the navel, or boiled in beer, and drunk to stay miscarriages. But the real authority as far as Fenland couples were concerned was rabbits. They used to say that where there were wild rabbits, there was sure to be tansy. And everybody knows what large families they produce, so the plant must have the same effect on humans. On the other hand, many unmarried pregnant girls would chew tansy leaves to procure an abortion (Porter. 1969). Indeed, tansy's poisonous oil has long been taken to induce abortions (Grigson). Nevertheless, it was still being used in the mid-20th century for women's illnesses, presumably because of its stimulant and tonic effects (Henkel, Fernie).

Tansy puddings at Easter were traditional, and were probably a Christian adaptation of the bitter herbs eaten at Passover (Newall. 1973). In many districts they were actually played for on Easter Monday (Opie & Opie. 1985), but they were actually made to be eaten with the meat course at dinner. One recipe speaks of finely shredded leaves, beaten up with eggs and fried (Genders. 1977). Pepys provided a "tansy" for a dinner, but this was a sweet dish flavoured with tansy juice. By the 16th century, tansies were a kind of scrambled egg made with cream and the juice of wheat blades, violet and strawberry leaves, spinach and walnut tree buds, plus grated bread, cinnamon, nutmeg and salt, all sprinkled with sugar before serving. Tansy was no longer an ingredient, walnut buds being preferred (Burton), although most recipes that have survived insist on the proper herb ingredient. Tansies seem to be some kind of milk pudding, with local variations on whether or not tansy is included, though it usually is. Sometimes, as in Oxfordshire, it was the flowers that were used, in custards as well as other sorts of puddings, or the leaves could be steeped in milk to make cheese and cakes.

When laid to soak in buttermilk, tansy had the reputation of "making the complexion very fair" (Sanford); in other words, it was used as a cosmetic wash to remove sunburn. One other non-medicinal use – it apparently gives a brilliant yellow dye, but with the disadvantage that it is very difficult to fix (Wiltshire).

In spite of the fact that the plant is used as a poison (Flück), it has been used through the centuries for a large number of medicinal purposes. It is an aromatic strong bitter, long esteemed, for instance, as an anthelmintic. It is the young tops and the seed that are used. Gypsies have always used the infusion to expel worms (Vesey-Fitzgerald). Martin mentions this use in the Isle of Skye, and so does Leask in Orkney, while it is certainly used that way in Ireland (Moloney); it was common enough in America, too (Bergen. 1899) (they even wore a tansy bag round the neck in New England for children's worms (Beck)). Perhaps this is the place in which to mention that rubbing a dog's coat with tansy helps to get rid of fleas (Hemphill).

Gout was another ailment for which tansy was a favourite medicine. Gypsies use a hot fomentation or an infusion for it (Vesey-Fitzgerald). In Scotland it was the dried flowers that were used (Fernie). Gerard confirms its use in his day, and two hundred years before his time tansy was already being used for gout (Henslow). Tansy flower tea was given for colds (Palmer. 1976) and fevers, for which the leaf tea was also used in America (Hyatt), and for nervous afflictions (Brownlow). In Wiltshire it was taken as a general tonic in all heart weaknesses, and for coughs and chest complaints (Wiltshire).

If you wear a sprig of tansy inside your boots, you will never get the ague, so runs a gypsy belief (Vesey-Fitzgerald), probably New Forest gypsies, for Hampshire farm labourers believed this too (Boase). There may be something in it, for the oil produced from the flower heads is still in use for application to the skin to treat rheumatism (Schauenberg & Paris), and in Newfoundland tansy flowers are used to make poultices and to bathe sprains. In Ohio, they say that a poultice made by moistening bruised tansy with vinegar will take the soreness out of a dog-bite (Bergen), and it is used in Ireland, boiled in unsalted butter, strained and kept for later use, to put on wounds (Maloney).

There have been a few veterinary uses of tansy. On South Uist, for example, they treat red water in cows by boiling the entire plant, putting the juice in a bottle, then pouring it down the cow's throat (Shaw). 17th century Yorkshire shepherds used tansy, finely chopped and mixed with fresh butter, to heal the wound on castrated lambs. As the butter healed the wound, the tansy would keep the flies away (Drury. 1985). East Anglian horsemen knew its worth for making horses' coats shine. They would sprinkle a little of the dried, powdered leaves now and then into the horses' feed (G E Evans. 1960).

TAPIOCA
A preparation made from CASSAVA tubers, by peeling, expressing the juice, grating and soaking the pulp, then heating it. This causes the starch to form the small lumps characteristic of tapioca (Kingsbury. 1964).

Taraxacum officinale > DANDELION

TARO
(*Colocasia antiquorum*) Important as a food plant, particularly for its turnip-like root. The very large leaves (celebrated with the name Elephant's Ears) are eaten as vegetables, there are some restrictions on their use in Madagascar, where the plant is known as saonjo. For example, local custom there forbids the use while the rice crop is ripening (Ruud). Taro is important enough there to be used in ritual, and also to be subject to taboo. Malagasy bridal couples had to eat taro (and also bananas). Both are symbols of fertility and multiplication. The traditional blessing is "Plant saonjo and bananas" (Ruud). But walking across a field planted with saonjo is taboo for a pregnant woman. Sometimes, the large leaves are torn and ragged along the edges, and this will cause the child to be born with a hare-lip.

A Hawaiian origin myth tells how the plant grew from the embryo child of Papa and Wakea, two of the

Hawaiian pantheon (Beckwith. 1940). Similarly, there is a legend of Hoohuku-a-ke-lami, who had a child by her father, Wakea. It was born not in the form of a human being but of a root, and it was thrown away at the east corner of the house. Not long after, a taro plant grew from the spot, and afterwards, when a real child was born to them, Wakea named it from the stalk (ha) and the length (loa). Another version says that the child of Papa was born deformed, without arms or legs, and was buried at night at the end of the long house. In the morning there appeared the stalk and leaves of a taro plant, which Wakes named Ha-loa (long root-stalk), and Papa's next child was named after that plant (Beckwith. 1940).

TARRAGON

(*Artemisia dracunculus*) Holding a piece of tarragon root between the teeth will cure toothache (Grieve. 1931).

TATTOOING

On Easter Island, the burned leaves of the Ti Plant (*Cordyline terminalis*) serve as the pigment for tattooing (Englert). The acrid juice of CEYLON LEADWORT (*Plumbago zeylanica*) is used in parts of Africa. After several hours contact with the flesh, it is said to leave an indelible mark (Scudder).

Taxus baccata > YEW

TEA SUBSTITUTES

The leaves of GERMANDER SPEEDWELL were at one time recommended for use as a beverage tea (Curtis), as the Cumbrian name Poor Man's Tea (Grigson. 1955) will testify.

TEAK

(*Tectona grandis*) A large, and valuable timber tree from tropical Asia, whose leaves are large enough to serve as rainhats in southern India (Fürer-Haimendorff).

TEASEL

(*Dipsacus fullonum*) A plant of prime importance in the cloth-making industry, the uses of which are mirrored in the names Fuller's Teasel and Burler's Teasel. Fulling is the process of raising the nap on woollen cloth, to give a softer feel to the fabric, and a burl is a knot in wool, so to burl is to remove these knots. Both processes were achieved by use of the dried flower heads of teasel, for "no substitute for their gentle action on the finest cloths has been found" (Ryder). The teasels were set in a wooden frame, usually known as a handle, but sometimes called a card. Hence the name Card Thistle, but 'card' comes from the Latin carduus, which means a thistle – so the pleonasm is revealed. For more information on the process, see RYDER, N L Teasel growing for cloth raising *Folk Life. Vol 7; 1969*, and ROGERS, K H *Warp and weft* Buckingham, 1986.

The dried stalks were used in Ireland as a sort of thatch, or at least they would be laid at right angles on the purlins, before the grass scaws and then the sods were put on the roof (E E Evans. 1942). Then there is one odd record of the use of a teasel as a weather forecaster. It was said that, when hung up in the house, "upon the alteration of cold and windy weather [it] will grow smoother, and against rain will close up its prickles" (Dyer. 1889). In Wales, it was used in some unspecified way as a protection against witches (Trevelyan), the idea probably being that the prickles would discourage them, much as brambles do. In Somerset people used to cut open the heads, where they would often find a worm. Any odd number of these worms, carefully put away, was believed to charm away sickness (Brill).

Rather more widespread was the belief in the efficiency of water which collected in the cups formed by the fusing together of the plant's opposite leaves, "so fastened that they hold dew and raine water in manner of a little bason" (Gerard), water that was much prized for cosmetic use; Culpeper knew about this, though he described it as the distilled water of the leaves, used by women "to preserve their beauty". Leicestershire girls washed their faces in this water, in order to make themselves more beautiful (Billson), and the folk use was known in Wales, too – there it was said to be a remedy for freckles (Trevelyan). It was said also to cure warts on the hands (Curtis). There was yet another use for this water, for sore eyes (Dyer. 1881); in fact, teasel has been recommended for diseases of the eyes since Anglo-Saxon times.

Teasel was used in some parts as a remedy for ague, or malaria (Black). The *Gentleman's Magazine* for 1867 mentions a remedy where the patient had to gather some teasels and carry them about with him. But the remedy lay apparently in what was found inside the teasel – the small worms that we have met already. In Lyte's translation of Dodoens, 1586, he says "the small wormes that are founde within the knops of teasels do cure and heale the quartaine ague, to be worne or tied about the necke or arme" (see Hulme. 1895). Gerard, though, was very scathing about the efficacy of this. What are we to make of a 15th century receipt "for the frenzy, a medicine. Take "'Shepherd's yard [i.e., teasel], and stamp it and lay it on his head when it is shaven"? (Dawson).

Tectona grandis > TEAK

TEETHING

GROUNDSEL, "the leaves stamped and strained into milke and drunke, helpe the red gum and frets in children" (Gerard), or for teething babies (Hatfield). Babies with teething difficulties were given FENNEL tea in America (H M Hyatt). IVY has been used, too, but as a charm, for in the Gironde, in France, ivy-root necklaces (they had to be green, and an odd number of pieces) used to be put round a baby's neck, to help teething (Sebillot). So were the stalks of

CAROLINA NIGHTSHADE (*Solanum carolinense*) in America (Puckett. 1926), or, sometimes, a necklace of ALLSPICE (H M Hyatt), and WOODY NIGHTSHADE, or DEADLY NIGHTSHADE stems (FRIEND. 1883) and PEONY root (Latham), in England. And the juice of the latter, mixed with oil of roses, was a 16th century medicine for teething pains (Phaer, *The boke of chyldren* 1545, quoted in *FLS News. 35; Nov 2001*). A piece of the root of BITTER GOURD (*Colocynthus vulgaris*) set in a gold or silver case, was at one time hung round a baby's neck as a teething amulet, a charm certainly known as early as the 6th century AD. In Alabama, the necklace was made from nine PAWPAW seeds (R B Browne). In Iowa a bag of ASAFOETIDA tied round the baby's neck will help it to cut teeth without pain (Stout). Domestic medicine in the southern states of America used PERSIMMON sap for teething, the juice from a burned branch was dropped on the gums (R B Browne). Mothers from South Uist once made a liquid from RED POPPY flowers to help babies in their teething (Beith), and OPIUM POPPY was used, too, probably with more effect. Fenland mothers gave them "Poppy tea", or a few poppy seeds to suck, tied in a piece of linen (Porter. 1974), or even a dummy dipped into poppy seeds (V G Hatfield. 1974). FIGWORT has a certain anodyne value, and will ease pain wherever it is applied, and that will include babies' teething (Gerard).

Hochets de Guimauve, that is, pieces of dried Marsh Mallow roots, used to be sold in French chemists' shops as teethers. They are hard and fibrous enough for a baby to chew on, but slowly soften on the outside as their mucilage is released (Mabey. 1977). Bits of the dried rhizome of BLUE FLAG were once given to babies to help their teething (Schauenberg & Paris). ORRIS-ROOT was used in Warwickshire (Bloom), as a "coral".

TENURES
The provision of a ROSE as a condition of tenure was quite widespread (see Blount for examples not quoted here). Rent day was often Midsummer, as with the Crown & Thistle in Loseby Lane, Leicester. Under a deed of 1626, an annual rent of two pennies and a damask rose is still paid. In other cases, the time of payment varied. Some land at Wickham, County Durham, for example, was held by service of one rose at Pentecost "si petatur" (if required), and, rather more difficult, the manor of Crendon, Buckinghamshire, was held by service of one chaplet of roses at Christmas.

CARNATIONS also were used as conditions of tenure, though less frequently. Lands and tenements in Ham, Surrey, were once held by John of Handloo of the men of Kingston on condition of rendering them three clove gillyflowers at the king's coronation (Friend. 1883). The manor of Mardley, in Welwyn parish (Hertfordshire) was held for the annual "rent of a clove gilliflower", while two of them plus 3s. 6d. paid the yearly rent of 100 acres and a 40-acre wood in Stevenage. And in Berkhamstead a tenant of the royal manor provided one clove gilliflower "at such times as anie King or Queen shall be crowned in the Castle" (Jones-Baker. 1974).

CLOVES must have been extremely expensive when introduced into Europe in the Middle Ages, valuable enough to pay for a year's rent on a manor, for the manor of Pokerley in County Durham, was held by the provision of one clove on St Cuthbert's Day, annually (Blount).

Terminalia bellirica > BASTARD MYROBALAN

Terminalia catappa > MYROBALAN

TETANUS
The Pennsylvania Germans claim that crushed BEET leaves put in a rag and bound on a wound will cure lockjaw (Fogel).

Teucrium scordium > WATER GERMANDER

Teucrium scorodonia > WOOD SAGE

TEXTILES
COTTON-GRASS can be spun like cotton, but the fibres are more brittle than those of cotton, so not so useful, and extremely tedious. But it features in a number of Highland folk tales as a powerful instrument against enchantment, for example, in one tale called *The three shirts of canach down*, quoted by J G McKay. There the sister had to make each of her enchanted brothers a shirt of moorland canach, which was the Highland name for the plant, ceannabhan in Irish. She had to remain completely silent until she herself, after making the shirts, had put them on her brothers to free them from the spell. In other versions, NETTLE is the chosen textile, a far better proposition for the sister, for nettle is best known as a textile plant. The fibres were used for cloth, certainly up to the 18th century in Scotland (Grigson), and in the Tyrol it was in use for linen cloth as late as 1917 (Hald). The earliest known literary reference is in Albertus Magnus, in the 13th century. The old German name for muslin was Nesseltuch, i.e., nettle cloth (Johnson). There is plenty of evidence for the use of nettle fibre among more primitive groups of northern Europe and Eurasia. The threads were taken from the outer skin of the stalks, which, after being moistened, were peeled off by means of bone or wooden chisels. The fibres were then broken down by beating or pounding in a trough, and after being rubbed in the palms of the hands or swingled by a wooden knife were spun on a wooden spinning stick (J G D Clark). American Indians of the North-west Pacific coast made fishing nets of nettle fibre, strands being twisted on a spindle to form the desired weight of twine for the net (Inverarity). The people of Kamchatka also used it for

nets. But nettle fibre was also capable of being spun and woven into fabrics, which were famous in many parts of northern and central Europe at the end of the 18th century for their fine, gauze-like qualities.

The fibres of HOPS can be made into cloth, as used to be done in Sweden (C P Johnson). HEMP, too, is a textile plant, its coarse fibres being used in the making of Huckaback, for instance. Male hemp was used for ropes, sacking and the like, female hemp for sheets and other domestic textiles (Peacock). The use of TEASELS in the finishing of woollen textiles is well known, and need not be expounded here.

THATCH

Common REED provides the most durable of all materials. It may be the most expensive, but it is said that it has a life of up to a hundred years, provided the roof pitch is forty-five degrees or steeper (Jenkins. 1976). For centuries the reed thatcher (medieval Latin arundinator) has been a different profession from the straw thatcher (coopertor). This reed is usually known to thatchers as Norfolk Reed. Around Abbotsbury, in Dorset, where the thatching tradition was strongest, it is known as Spear (Nash & Nash). When reeds were in short supply in Cornwall, the leaves of YELLOW FLAG were used as a substitute (Barton. 1972).

THISTLE

see, rather, **HOLY THISTLE** (*Carduus benedictus*), or **MUSK THISTLE** (*Carduus nutans*), etc.,

THORN-APPLE

(*Datura stramonium*) Found throughout most of the temperate world, but the fact that native Americans called it White Man's Plant (Sanecki) suggests that it is not native to North America. The capsules were, and, so it is claimed, still are, used in black magic (Summers), and it was certainly considered to be clearly associated with magic, witches, and also in the development of second sight (Trevelyan). In Puritan times, those who grew it in their gardens were in danger of being burned as witches. Significantly, in Maryland folklore, it was said it would unlock any door, if dipped in honey, and used at the proper times and seasons (Whitney & Bullock), a property assigned to any "magical" plant.

In both the old and new worlds, datura seeds (of most of the species) were administered in various ways as love potions (Safford), and the roots were used, too, according to Haining; he says they were burned at the Sabbats in order to excite, and also to overcome, women for sexual motives. Similarly, a 17th century medical report claims that the seeds given to anyone will cause that victim to be at the complete mercy of the practitioner for 24 hours, "and you can do what you like with him; he notices nothing, understands nothing", and will remember nothing (Haining). That sounds very like Voodoo magic, and what was

certainly Voodoo practice was to pound the seeds up with the dried head of a snake. The mixture was used "to produce a mysterious and baffling blindness" (Puckett. 1926). The oft-quoted report that English soldiers, sent to Jamestown (hence the name Jamestown-weed, better known as Jimson-weed) to put down the uprising known as Bacon's Rebellion in 1676, gathered young plants of this species and cooked them as a potherb, "the effect of which was a very pleasant Comedy; for they turn'd natural Fools upon it for several days…" (Safford).

The seeds were used in divinations, too. For example, there is a gypsy love divination that required nine thorn-apple seeds, ploughed-up earth of nine different places, and water from nine more. With these, the girl had to knead a cake, which was laid on a crossroad on Easter or St George's morning. If a woman stepped first on the cake, her husband would be a widower or older man, but if it was a man who trod on it, the husband would be single or young (Leland. 1891). There is also a record of gypsy divination for a different purpose – to know if an invalid would recover. They put from 9 to 21 seeds on a "witch drum", that is, a tambourine covered with an animal skin marked with stripes that have a special meaning. The side of the drum is tapped gently, and according to the position that the seeds take on the stripes, the recovery or death of the patient is predicted (Leland. 1891). But the gypsies also looked on the seeds as protectors against the evil eye. After a wedding ceremony, it was the custom to throw water over the couple, and to rub them with a bag made of weasel skin. Inside this bag there had to be thorn-apple seeds to give the necessary protection (Starkie). There is even a belief (from Kentucky) that putting the juice in the eyes will make light eyes turn dark (Thomas & Thomas).

Thorn-apple has always been valued for medicinal as well as narcotic purposes. The active principle seems to be identical with atropine, and has been used commercially as a substitute for it (Lloyd). It has long been valued as a pain-reliever in American domestic medicine, in the form of a poultice or ointment made from the pulp of the bruised green leaves, and it was used for the same purpose in Essex at one time. The method there was to cut the top of the fruit off, and pulp the inside, adding vinegar. Inhalation of the fumes brought relief, so they claimed (V G Hatfield. 1994). It was used for rheumatism and headache, bee stings and bruises, and for carbuncle, or any minor skin irritation (R B Browne); warts too can be treated by rubbing the leaves over them, whether the leaves are buried afterwards as a charm, or not (H M Hyatt). Sores used to be treated in Kentucky with "jimsonweed and fat meat on a penny" applied. It would stop the pain (Thomas & Thomas). Gerard talks about the "juice of Thorn-apples boiled with hogs grease to the form of an unguent or salve", to cure "all inflammations whatsoever, all manner of burning or scalding … as my selfe have found by my daily practice, to my great credit and profit". That was in use in East Anglia until very recently. The green fruit was boiled in pork fat to make an ointment for inflammations, burns and scalds (V G Hatfield. 1994). Alabama women treated bruised or caked breasts by boiling jimsonweed with lard, about half as much weed as lard, and rubbing on the result as hot as possible (R B Browne). In fact, the treatment was quite well known, for King's *American Dispensatory* of 1852 mentions that a poultice of the fresh leaves bruised, or the dried leaves in hot water, is beneficial as a local medication to "all species of haemorrhoidal tumours" (Lloyd). That ointment was used in veterinary medicine, too. In Ohio and Illinois it is rubbed on the fetlocks of horses for "scratches" (Bergen. 1899). Pieces of thorn-apple, rubbed or bound on the sores, used to be an English cure for galled horses, particularly in Littleport, Cambridgeshire (Porter. 1969).

Minor operations have been performed with the patient under the influence of this drug, as the pain is reduced a little (Sanford), and the seeds have been prescribed as a sedative (Fluckiger & Hanbury). It is used, too in the treatment of Parkinson's disease (Scarborough).

Thorn-apple was put to quite a different use in Africa, where its narcotic properties were employed by the Jagga in trial by ordeal. The potion was prepared by putting two handfuls of a Datura herb (*stramonium* perhaps, but the exact species is not revealed in the account) into rather less than a pint of water, along with banana blossoms. The litigant would address a solemn magical formula to the herb while putting his right hand into the vessel, and then the mixture was boiled, and eight snail shells full handed to the person to be tested. The plaintiff described the offence, and urged the decoction to make the defendant fall down if he was guilty, but otherwise to spare him. The accused, with the container at his mouth, would assert his innocence and utter a corresponding wish. If any of the liquid dripped, it was taken as a preliminary sign of guilt. After all the potions had been dripped, the defendant was ceremonially taken for a walk, after which various minor rites were celebrated. Finally, the decoction would produce the desired effect of putting the drinker into a trance-like state in which he soliloquised, confessed his guilt or denied it, or vehemently resented the indignity of the test. Only if he made a clear breast of his guilt was he convicted and condemned to pay all the requisite fees. On the following day he would be given an emetic to purge him of the poison, but even so, the effects would probably not wear off for over a month (Lowie).

The American *Daturas*, and in particular *D meteloides*, the Downy Thorn-apple, probably better known under its Mexican name, Toloache, assumed great importance in the life of native Americans. Toloache had its medicinal uses, the leaves, for instance, being used to make an anodyne. The seeds, too, were ground and mixed with pitch to help in setting broken bones (Safford). The Californian Indians regarded it as the prime medicament against rattlesnake poison (Emboden. 1979). But the real use of *Datura* among the aboriginal Californians was as an intoxicant and hypnotic. Yokuts shamans employed it that way, for good purposes, though it was firmly believed that malicious shamans would use it as a poison, mixed with some other, unidentified, herb (Gayton). Similarly, though a piece of the root was often carried about by the shaman, the ordinary Yokuts believed that an evil shaman could make himself invisible by holding this amulet in his hand. Then he could eavesdrop, or poison people without anyone seeing him (Gayton).

A decoction of the plant was given to the young women to stimulate them in dancing, though in some localities only men were allowed to drink it (Driver). It is also used in the ceremonial initiation of young men. Some Apache mixed the root with their corn beer to make it more intoxicating (La Barre), but the plain decoction is usually enough for shamans to reach the state of exhilaration that enables them to prophesy the future or to make supernatural beings visible. The Zuñis, too, ascribed the power of second sight and prophecy to this *Datura* (Safford), as in discovering the whereabouts of stolen objects (Lewin). The Navajo also chewed the root or drank infusions of it to produce narcosis, either as an anodyne during minor operations, or in order to prophecy, or even indulge in witchcraft. They fully recognised the poisonous qualities, and handled it with caution (Wyman & Harris). For *Datura* can kill, and it can apparently also be administered by an experienced person in such amounts and in such ways to bring about temporary derangement and even permanent insanity – no wonder it has entered the practice of witchcraft (Furst).

Kroeber has given the best account of the cult, known variously as the Jimsonweed, or Toloache, cult, as applied to *D. meteloides*. In California, it appears to have originated in the coast or island area of the southern region, and to have spread as far north as the Yokuts of San Joaquin Valley. The cult is similar to a secret society in that initiation is a fundamental ceremony. So in some measure, toloache ceremonies are clearly puberty rites, applied in some instances to girls as well as boys. The avowed intent and purpose is to render each candidate hardy, strong, lucky, wealthy and successful (and to have the ability to dodge arrows) (C Grant). The Yokuts boys' initiation centred round the narcotic, and so supernatural, effect of the drug. Initiation was into manhood rather than to membership of any organisation. So the boys' puberty ceremony was given its distinctive character by intoxication. The drug is not only a narcotic but also a hallucinogen, procuring visions, so explaining the tremendous supernatural power ascribed to it. The vision-producing effect would have been enhanced by the preceding ceremony of fasting, when the participants were withdrawn from the public, usually remaining in a separate hut for a six day fast, during which the old men in charge of the ceremony gave instruction to the boys. In the next stage, when the boys had drunk the decoction they were taken away from the village to places where they would not be disturbed, and where older men could watch over them. If a boy vomited his dose of the drug, it was taken as a sign that he would die, and his relatives paid the old master of ceremonies for praying to avert this fate.

The children expected to see an animal in the visions, and they were instructed to place their entire confidence in this animal, when it would defend them from all future dangers. At times they learn from their animal a song which they keep as their own. Also they will not kill any individual of the species they have seen in their vision, a trait reminiscent of totemism. All in all, Emboden. 1979 sums up the uses as providing a trance state for the passage of youth into manhood, or to sustain a person during grief, or to simulate the death and resurrection of the shaman. Only in the trance state can there be a communion between man and god.

The Navajo used jimsonweed for finding thieves or recovering missing property. They would address the jimsonweed and explain why they wanted a piece of its root. Then an offering of turquoise would be made, after which a hole would be dug by the side of the plant, and a piece of root removed. The whole plant was never picked, for to be ritually effective it must continue to live. The owner of the missing property either chewed it or drank some of it in a solution, and then would either achieve trance state, or have a visionary experience or an audible hallucination. In trance state he would wander about and meet the thief or locate the lost articles. In the visionary type the man saw the location of the property or the identity of the thief would be revealed, and in the auditory type voices directed the man to the hiding place (W W Hill).

THORNS AND SPLINTERS

A poultice used to be made in the Highlands from HEATH SPOTTED ORCHID (*Dactylorchis maculata*) for drawing out thorns or splinters from the flesh (Beith). One of the wonders ascribed to SCARLET PIMPERNEL was that it had the power of drawing out thorns and splinters (C J S Thompson. 1897), and PRIMROSES too had the power – it "draweth forth of the flesh any thorne or splinter, or bone fixed therein" (Gerard). A 15th century leechdom reckoned that "a plaster of southernwood" with fresh grease will "draw out stub or thorn, that sticketh in the

flesh" (Dawson). DAFFODIL bulbs or roots, used as a poultice with honey and darnel meal, would "draw forth thorns and stubs out of any part of the body" (Gerard).

THREE-LEAF SUMACH

(*Rhus trilobata*) American Indian groups like the Navajo and Apache used the branches of this sumach more extensively than those of any other plant, except willow, for baskets. The Zuñi reserved them for the very best baskets (J Taylor), while the Navajo made their sacred baskets from them, and they used the twigs, with the leaves and berries, boiled, in dyeing processes (Yarnell). It provides black for basketry and leather dyes. Tradition required that basketry dyes should be stirred with a sumach stick (H H Smith. 1923). The berries found their way into American domestic medicine. They are given in Alabama to stop bed-wetting (R L Browne).

THRIFT

(*Armeria maritima*) In spite of being so common, there is very little folklore attached to the plant. "For sympathy, give thrift" (Freethy), is one doubtful piece of symbolism. Its use in folk medicine is equally limited, though it has long been used for epilepsy and obesity (Schauenberg & Paris), and the leaves are still used in slimming foods (Usher). Even Gerard could produce no "vertues". The only other record comes from South Uist, where a sailor's cure for a hangover was to pull a bunch, with its root, and boil it for an hour or more. It had to be left to cool, and then was drunk slowly (Shaw). The name Thrift apparently means that which thrives, or is evergreen (Grigson. 1974), and has nothing to do with economy, though, as a pun, the plant figures on the back of the old English threepenny bit.

THROMBOSIS

An infusion of MELILOT has been prescribed by herbalists as a preventive (Flück).

THRUSH

A Devonshire charm for the thrush was to take three RUSHES from any running stream, and pass them separately through the mouth of the infant, then plunge the rushes back in the stream. As the current bears them away, so will the thrush go from the child (Burton). A French charm for thrush in infants was to put a piece of PRIVET in flower over the chimney piece. When this dries up, the child's thrush will also have dried (Loux).

In Indonesia, root juice of the Red Silkcotton Tree (*Bombax ceiba*) serves as a mouthwash and gargle to treat thrush in infants (L M Perry), and in Alabama, the root of POKE-ROOT was used for the complaint, by, of all things, putting it in whisky and giving it to the child to drink (R B Browne). In the same area, PERSIMMON bark was stewed and mixed with honey for use as a mouthwash (R B Browne).

SAFFRON tea is used in American domestic medicine as a mouth wash in cases of thrush. Mohegan Indians used the mashed leaves of CREEPING LADY'S TRESSES (*Goodyera repens*) to prevent thrush in infants (Weiner).

THUNDER-FLOWER

A name given to a number of plants, viz FIELD BINDWEED (*Convolvulus arvensis*), OX-EYE DAISY (*Leucanthemum vulgare*), RAGGED ROBIN (*Lychnis flos-cuculi*), WHITE CAMPION (*Silene latifolia*), RED POPPY (*Papaver rhoeas*), and STITCHWORT (*Stellaria holostea*). Usually, such a name implies that if a child picks any of the flowers, thunder will ensue before the day is out. Perhaps the prohibition acted as a way to stop children damaging a crop by trying to gather the cornfield flowers. Would they want to pick bindweed, though?

THYME

(*Thymus vulgaris*) A bee plant, always grown near hives, and the leaves would be rubbed on their hives (Gordon. 1977). It is a symbol of hope (Reeves), and of activity (Haig), and also of courage (Leyel. 1937). Those who eat it are said to become more courageous, and in the Middle Ages, a sprig of thyme was given to knights by their ladies, to keep up their courage. Even a scarf embroidered with a bee alighting on the plant, was supposed to have the same effect (Leyel. 1937). It was also, according to a correspondent of *Notes and Queries; 1873*, the emblem of the radical movement in French politics, particularly of the Marianne section (Marianne was the statuette of the Republic, wearing the red Phrygian cap).

But it is an unlucky plant, connected with death; the souls of murdered men inhabit it (Gordon. 1977), and the smell is that of a murdered man's ghost. Or the scent is a sign that a murder has been committed at that spot at some time. There is a place in Coate, near Bishop's Cannings, in Wiltshire, known as the "Thyme Tree" (where there is neither thyme nor tree), where the passer-by gets just two whiffs (and no more) of the scent of thyme (Wiltshire). That same scent was said to be present on a footpath leading from Dranfield to Stubley, in Derbyshire. The tradition was that a man murdered his sweetheart there as she was carrying a branch of thyme (Addy. 1895). Staining Hill, in Lancashire, has its ghost, with the smell of thyme (Addy. 1895). A sprig of thyme is carried by the Order of Oddfellows (Manchester Unity) at the funeral of one of their brothers, and then cast into the grave, and it was a common custom at funerals at Massingham, in Lincolnshire, to drop thyme on the coffins of the dead (Gutch & Peacock). It is planted on graves in Wales, too (Gordon. 1970). As a corollary to all this, thyme has got itself a bad name, too unlucky to be taken indoors, for it will bring death or illness to some member of the family (Tongue. 1967). Gypsies believe this, too, but it

can be used outdoors, for medicinal purposes, for example (Vesey-Fitzgerald). Again, in Somerset, it is dangerous to keep indoors, because it smells of death (Tongue. 1965). Another reason why it is dangerous to take indoors is that it is a fairy plant (Briggs. 1967), and it was said that the fairies were particularly fond of it. On the other hand, it was once said that the smell of thyme would cure epilepsy (Classen, Howes & Synnott).

There was an old belief that thyme could strengthen the brain and preserve the aged by somehow reversing the aging process. A report in *FLS News 15; July 1992 p 8* says that work has been done at the Scottish Agricultural College on the effect of plant extracts on reversing the aging process, and that both thyme and sage produce volatile oils containing antioxidants that have been tested for such effects. Whether this is a matter of belief or not, one has always been advised to drink thyme tea when suffering from exhaustion (Tynan & Maitland), just as, in the Highlands, it has always been considered the most potent of tonics (Beith). It was used to cure depression, too (Dyer. 1889), and in the Highlands again, it was said to prevent bad dreams, either as tea taken last thing (Grigson. 1955), or even by putting a sprig under the pillow (Beith).

"Take thyme with you when you move house" is an American dictum (Whitney & Bullock), and another piece of wisdom, from the Pennsylvania Germans, is that unless you sit on thyme after planting it, it will not grow (Fogel). There is a divination charm that includes the use of thyme. On St Agnes' night, take a sprig of rosemary and one of thyme, and sprinkle them three times with water. In the evening, put one in each shoe, putting a shoe on each side of the bed. On going to bed, St Agnes has to be invoked:

St Agnes that's to lovers kind,
Come ease the troubles of my mind,

and the future will be revealed in a dream (Halliwell. 1869). The divination seems to be from northern England and Scotland, and a different version of the rhyme is:

Agnes sweet and Agnes fair,
Hither, hither now repair;
Bonny Agnes let me see
The lad who is to marry me (Drury. 1986).

Thyme's medicinal reputation rests largely on the highly aromatic essential oil, described as camphor of thyme by the Berlin apothecary Neumann in 1725, and called thymol by Lallemand in 1853 (Lloyd). It has been a popular ingredient of domestic liniment, and is often used in veterinary medicine. It is antiseptic (more than 12 times as powerful as carbolic acid, so one folklorist claimed (Wiltshire)), and vermifuge (Flück). Due to the thymol, it has been used for whooping cough (Camp), and is much used in gargles and mouthwashes (Sanecki). Gypsies used the plant for whooping cough too – the decoction would be prepared outside (see above), a little sugar added, and drunk cold (Vesey-Fitzgerald).

Gerard had a long list of ailments to be cured by thyme. They include ague, strangury, hiccough ("it stayeth the hicket"), stone in the bladder, as well as "lethargie, frensie and madnesse", and so on. A regular heal-all in his eyes, particularly as he went on to claim that "it helpeth against the bitings of any venomous beast, either taken in drinke, or outwardly applied". And Andrew Boorde wrote that "Tyme brekyth the stone; it doth desolve wyndes, and causeth a man to make water".

THYME-LEAVED SANDWORT

(*Arenaria serpyllifolia*) Lewis Spence (Spence. 1945) suggested that this was the mystical herb known as Mothan or Moan in Scotland and Ireland. He described the Mothan as being given to cattle as a protective charm, and people who ate the cheese made from the milk of a cow that had eaten the plant were secure from witchcraft. It is said to be found where no quadruped had ever trodden, on the summit of a cliff, or mountain. The Mothan was to be picked on a Sunday as follows: three small tufts to be chosen, and one to be called by the name of the Father, one by that of the Son, and one by the Holy Ghost. The finder would then pull the tufts, saying (in translation):

> I will pull the Mothan
> The herb blessed by the Domnach;
> So long as I preserve the Mothan
> There lives not on earth
> One who will take my cow's milk from me.

The three tufts were then pulled, taken home, rolled up in a piece of cloth, and hidden in a corner of the dairy.

Thymus vulgaris > THYME

TI PLANT

(*Cordyline terminalis*) This is the source material of so-called grass skirts. They are made from its leaves, which also serve as plates and food wrappers at feasts in Polynesia (H G Baker). The burned leaves made the pigment for tattooing on Easter Island (Englert). But Ti Plant is at least as important for ritual reasons, for this is a protective shrub (hence presumably the English name Good Luck Plant). It is planted in graveyards in Malaysia, and occasionally at the four corners of a house, to drive away ghosts and demons (in keeping with its specific name, which must imply a celebration of boundaries – hence the West Indian name Boundary Mark (Howes)). Its leaves are also used for the ceremonial brush used to sprinkle a paste made of rice-flour and water on objects like house posts, that need special protection (Skeat). In Hawaii, ti plants were grown round family altars, once set up

to the god Kane (Beckwith. 1940). Hawaiians have always been afraid to disturb human bones, for the dead may enter any object, and especially bones. Similarly, they fear to talk of sacred things lest they anger these spirits of the dead, who will then work them mischief. In each case, ti leaves are a safeguard. Carrying food, especially at night, is very dangerous, and so they tie a green ti leaf to the container as a protective charm, which commands the ghost to fly away. This is called placing a law (kanawai) upon the food. But unless the leaves are fresh the law will not work. Similarly, women wear a ti leaf as protection when they approach particular places (Beckwith. 1940).

Tilia cordata > SMALL-LEAVED LIME

Tilia x vulgaris > LIME

TIMOTHY (GRASS)

(*Phleum pratense*) The name is given after one Timothy Hanson, who did much to promote its cultivation in America, round about 1720, and who brought it over to Britain, under the quite erroneous idea that it only grew in America. It is reported that an American domestic practice is to give Timothy seed tea to stop vomiting (H M Hyatt). This is one of the grasses that London children use to give "Chinese haircuts" (see **MEADOW FOXTAIL**).

TINNITUS

Mixed with oil of roses, the juice of SUMMER SAVORY (*Satureia hortensis*) has been claimed to relieve tinnitus (Addison). BAY leaves, too, according to Lupton, were used for ear drops, for it "doth not permit deafness, nor other strange sounds to abide in the ear". A 15th century leechdom offered a cure for "tingling in a man's ear. Take the juice of SOUTHERNWOOD and put it in the ear: and stop the ear with leaves of the same herb, and drink the juice at even and at noon, and it shall stay the tingling. And the juice of wormwood doth the same" (Dawson). Gerard recommended the juice of GROUND IVY dripped into the ear against the "humming noyse and ringing sound of the eares…".

TISTY-TOSTY

There was a kind of divination game that children used to play with cowslips, called Tisty-Tosty, or Tosty-Tosty, Blossoms, picked on Whit Sunday for preference, were tied into a ball (hence another name, quite simply Cowslip-Ball). Strictly, the balls were the Tisty-Tosties, though the growing flowers got the name, too, as did GUELDER ROSE. Lady Gomme mentioned the game as belonging to Somerset, but it had a much wider spread than that. The cowslip ball is tossed about while the names of various girls and boys are called, till it drops. The name called at that moment is taken to be the "one indicated by the oracle", as Udal puts it, for the rhyme spoken at the beginning is:

Tisty-tosty tell me true
Who shall I be married to?

Instead of actual names, some used the time-honoured
Tinker, tailor, soldier, sailor sequence (Opie & Tatem).
The game is also known in Wales, where the purpose
is different, for the rhyme there is:

Pistey, Postey, four and twenty,
How many years shall I live?
One, two, three, four … (Trevelyan).

A Gloucestershire rhyme has:

Tisty Tosty, cowslip ball,
Tell me where you're going to fall?
Dursley, Uley, Coaley, Cam,
Frampton, Fretherne, Arlingham?

John Clare called them cucking balls:

And cowslip cucking balls to toss
Above the garlands swinging light.

A different tradition here, obviously. Roy Genders,
who used the name cucka balls, says they were often
threaded on twine and hung from one window to
another across the street.

TOADFLAX

(*Linaria vulgaris*) The name has been rather a puzzle.
Gerard's explanation is that it has a "mouth like unto
a frog's mouth …". Skinner's offering is that it was
originally bubonina, because it cured buboes, then
becoming corrupted into bufoanium, from bufo, toad,
and so to the meaningless toadflax. Neither of these
sounds very convincing. We would expect to find
Toad's Mouth if Gerard's elucidation were correct,
but there is nothing – plenty of other "mouth" names,
but not a toad's. So let us pass on to more enlightened
expositions. One is that "toad" means "dead", i.e., it
is the German word tot, conveying the idea of flax
that is dead, that is, useless for the purpose to which
proper flax is applied (A S Palmer). Grigson comes
to the same kind of conclusion, but says the name is
a simple translation of Krŏtenflachs, a wild, useless,
flax, a flax for toads. It was, of course, at one time a
destructive weed of the flax fields, as is implicit in
Gerard's name of Flaxweed.

It has had its medicinal uses in the past, even though
strongly derivative of the doctrine of signatures. As
Grigson put it, yellow suggests yellow, so one should
not be surprised to find the early herbalists prescribing it for bladder problems. Parkinson. 1640, for
instance, said, "the Tode Flaxe is accounted to be
good, to cause one to make water". Earlier, Gerard
had claimed that the decoction would "provoke urine,
in those that pisse drop by drop", and it would unstop
the kidneys and bladder. The same decoction was
used for a second ailment, jaundice, also obviously
from the same doctrine. Gerard produced yet another
"yellow" remedy – "the decoction of Tode-flax taketh

away the yellownesse and deformitie of the skinne,
being washed and bathed therewith". It can be used
for warts, too – just rub it on (Tongue. 1965).

TOBACCO

(*Nicotiana tabacum*) The name 'tobacco' comes from
the Spanish tabaco, which in turn is derived directly
from the Arawak term for the cigar. More accurately,
it comes from an implement used by the Carib
Indians, called a tabaco. They strewed dry tobacco
leaves on the embers of a fire, and inhaled the smoke
through a hollow forked reed, the two ends of which
were put in the nostrils. This reed was the tabaco. By a
misunderstanding, the name became transferred to the
herb, and so gave tobacco (Hutchinson & Melville).

Like many another narcotic plant, it is used in a
religious way, as well as secular. Most shamans use
it, both to establish rapport with their spirit-helpers,
and to drive away disease from a patient's body.
Tobacco was a part of almost every American Indian
public religious ceremony (Driver). At meetings of
ambassadors, councils of nations, etc., the calumet,
or pipe of peace, was always circulated (Safford). Not
all tobacco used for religious purposes was smoked,
or chewed or snuffed. A considerably amount was
burned as incense, thrown into the air or on the
ground, or buried (Driver). It was an important part
of Plains culture, medicine bundles nearly always
containing a pipe and tobacco. The pipe was smoked
as a part of the ritual whenever the bundle was
unwrapped and put to its religious use. It is a sacred
plant, esteemed by the Iroquois as one of the blessings
bestowed upon them by the Creator, and would be
burned in practically all rites, individual or social.

It is a sacred plant in Peru, too, and in fact is known
as *the* sacred herb, venerated for its invigorating
effect. It is called the holy herb in Brazil because it
induces visions in which spirits are seen (Dorman).
The ancient Mexicans regarded it in the same light,
and they used it as incense in religious rites. In
Europe, many French fishermen believed in its magical properties. Spitting the juice into the sea was
believed to attract fish. Others believed the smoke
lured fish towards the boat, and they often lit their
pipes expressly for this purpose (Anson).

In 16th and 17th century Europe, potions for perennial
youth were made from it, and it was believed that the
leaves had aphrodisiac properties (Bringers), while
the Creek Indians linked it with sex to account for its
power of giving peace.

An old name for tobacco was Sot-weed (Halliwell),
presumably deriving from its narcotic purposes, and
Parkinson called it Indian Henbane; it is of course
a poison, used in India to kill fish (Heizer), and
long esteemed by gardeners as an insecticide. It did,
though, have its medicinal uses. In 1560 Nicot, after
whom the genus is named, was French ambassador to

Portugal, and claimed that tobacco healed boils and running sores. Even quite recently, in Iowa, a tea was made from tobacco juice and drunk to cure ringworm (Stout), and carbuncles have been dealt with by using a tobacco leaf as a poultice (Thomas & Thomas), while in Scotland, a chewed leaf has been used to cure a whitlow (Rorie. 1914). Ointments and syrups were made from it by infusing the leaves in water, milk or urine. Wounds and tumours were treated with the fresh leaves (cuts were until very recently treated in Ireland and Scotland by binding on a tobacco leaf to stop the bleeding and to heal it (Egan), or a dampened leaf could be put on a corn for a few days (Maloney)). Plasters for rheumatism could be made by damping the leaves, or even using cut-up pipe tobacco (Hutchinson & Melville). The seeds, taken with molasses, was an Indiana remedy for worms (Brewster).

The tobacco-smoke enema-syringe was a favourite instrument, apparently adopted in Europe from the Central American Indians. Used at first to combat a wide variety of diseases, it was, during the 18th century, even used to resuscitate the apparently drowned, and was still known up to about 1850 (Brongers). Putting tobacco in the ears, or on a tooth, was quite a common earache or toothache remedy (Newman & Wilson). ("Jane Josselin treated herself for toothache with tobacco" (Beier)). It was used as a plague protector, too, either by smelling, or by taking it fasting in the morning, "provided, that presently after the taking thereof, you drinke a deepe draught of six shilling Beere, and walke after it" (F P Wilson).

TOBACCO SUBSTITUTES

BUCKBEAN leaves used to be shredded and smoked in the Faeroes in times of tobacco scarcity (Williamson). The Lapps believed that ARCHANGEL roots prolonged life, and they chew it and smoke it in the same way as tobacco (Leyel. 1937), just as gypsies smoke the stems of WILD ANGELICA. BLACKTHORN leaves were an Irish substitute (Ó Súilleabháin), as were LABRADOR TEA leaves for the Ojibwa Indians (Jenness. 1935), while Plains Indians used to dry the autumn leaves of SMOOTH SUMACH for smoking (Gilmore). BITTERVETCH roots were chewed in the Scottish Highlands as a tobacco substitute (G M Taylor). Devil's Tobacco is a name given to HOGWEED, but there is also Boy's Bacca, from Devonshire, where the stems were actually smoked as a tobacco substitute, and not only by boys, for apparently gypsies smoked them, too. One of the names for CAT's FOOT in America is Ladies' Tobacco, implying that experiments have been made in smoking the dried herb, and that it was found to be very mild in character – hence "Ladies" (Leighton). PLANTAIN-LEAVED EVERLASTING is a close relative, and that, too, has the name Ladies' Tobacco as well as Indian Tobacco. The real INDIAN TOBACCO is *Lobelia inflata,* and its chief use as a substitute is as an asthma remedy, in small doses,

for this can be toxic if enough is consumed. The leaves and tops of Thorn-apple would be mixed with this plant to make "asthma powders", commonly sold for the relief of the complaint. A little nitre is included to make it burn, and the smoke is inhaled. The mixture is often made up into cigarettes, for convenience (Hutchinson & Melville). The existence of names like Cigar Tree (Harper), Indian Cigar and Smoking Bean (Hyam & Pankhurst) for the LOCUST BEAN (*Catalpa bignonioides*) is enough to show that American Indians actually smoked the capsules (Perry. 1972). YELLOW GENTIAN has been used as an antidote to or substitute for, tobacco (Lloyd), and MELILOT was included with ordinary tobacco as an aromatic (Flück). BEARBERRY leaves were used as a tobacco substitute by American Indians (Sanford). Keres Indians mixed them with tobacco in the ordinary way (L A White), and so did the Chippewa, who also claimed that they smoked them "to attract game" (Densmore). The North-west Coast Indians used the leaves to make the smoking substance kinnikinnick, which is also used as a name for Bearberry. It is an Algonquin word meaning "that which is mixed", usually tobacco (Emboden. 1979).

ARNICA leaves (or indeed all parts) are used to make a tobacco substitute, known in France as 'tabac des savoyards', 'tabac des Vosges', or 'herbe aux prêcheurs' (Sanecki). The young shoots of LION's TAIL that are about to flower are pinched out and smoked as a tobacco substitute in Africa, under the name dagga-dagga. Dagga is cannabis, and whether this is smoked, or the dark green resin that the leaves exude is smoked with ordinary tobacco, the effect is the same, for this is a narcotic, producing mild euphoria (Emboden. 1979).

COLTSFOOT should be included here. But this is smoking with a purpose, as a cough cure, or for asthma. Bechion, the plant in Dioscorides taken to be coltsfoot, was smoked against a dry cough, and it is still smoked in all herbal tobaccos (Grigson. 1955), as it is also in Chinese medicine (F P Smith), for asthma, and even lung cancer. Gypsies smoke the dried leaves for asthma and bronchitis, and it was smoked by Cornish miners as a precaution against lung disaeses (Deane & Shaw).

TOMATO

(*Lycopersicon esculentum*) There was a belief in the 19th century that eating tomatoes caused cancer. That strange belief must have been widespread enough for Ackermenn to take pains to refute it. On the contrary, his informant says, they are "probably the most health-giving fruit in the kingdom. They sweeten the blood", and he even recommended them in dealing with insomnia. A similar belief has been recorded in more recent times. Children may be forbidden to eat them, and the reason given is that birds never peck them, and the worms would never eat them (Vickery. 1995).

The Totonac Indians of Mexico used tomatoes as a febrifuge; the raw tomato is put on a castor leaf, which is then applied to the abdomen (Kelly & Palerm), and the Maya used the crushed leaf for skin complaints, and for an inflamed throat (Roys). In Alabama, they say that you should eat tomatoes to cure a "torpid liver" (R B Browne).

Gardeners say that a dead tomato plant hung on the boughs of an apple tree through the winter, will preserve it from blight. Or the plant can be burnt under the tree, so that the smoke can ascend among the branches (Quelch).

Love-apples is an old name for tomatoes, arising from a mis-reading. The original Italian name was 'pomo dei moro' (apple of the Moors), and this later became 'pomo d'ore' (hence Gerard's Gold-apples). It was introduced to France as an aphrodisiac, and the French mis-spelled its name as 'pomme d'amour'. So the tomato eventually reached England under the name 'pome amoris' – love-apple, which name went back to America with the colonists (Lehner & Lehner).

TONICS
CENTAURY earned itself a reputation as a tonic (Vickery. 1995) and gypsies used it as such (Vesey-Fitzgerald). The juice (in whisky?) is described on South Uist as excellent for the one in need of a nerve tonic or for weakness following an illness (Shaw). Better known is one made with FIELD GENTIAN, as "bitters". In Switzerland and Germany, the fresh root is used for the production by fermentation of a "gentian brandy", very popular as a pick-me-up (Clair).

TONSILITIS
Eating roasted ONIONS was taken in Ireland to be a cure for tonsilitis (Maloney).

TOOTHACHE
Treating toothache by picking at the decayed tooth with a sharp twig of WILLOW, until it bled was recorded in Wales. After that the twig had to be thrown into a running stream. Simply chewing some willow bark would have been useful, for it contains salicin, from which salicylic acid was obtained. Later, this was compounded into acetyl-salicylic acid – aspirin, in a word.

Applying a hot FIG (to the tooth or the cheek?) used to be a Cumbrian remedy for toothache (Newman & Wilson), but the strangest remedy must be the use of pine cones. The scales were the part needed, because (and this is pure doctrine of signatures) they resemble the front teeth! (Berdoe). A transference charm for the toothache involved BIRCH. It was recorded in Suffolk, and the sufferer was instructed to clasp the tree in his arms, and then cut a slit in it. A piece of his hair had to be cut from behind the ear, with the left hand, and this had to be buried in the slit. When the hair had disappeared, so would the toothache (Burne). ELDER was used in various charms for the condition.

One from Denmark and Germany involved putting an elder stick in the ground (or a twig held in the mouth) while saying something like "Depart, evil spirit" (Dyer). In Ireland, clay from under an elder tree is applied to an aching tooth, and toothpicks were made in Warwickshire from elder wood, to ensure protection from toothache (Ô Súilleabháin; Palmer). A peeled POTATO, if carried in the pocket on the same side as an aching tooth, will cure it as soon as the potato itself is reduced to crumbs (Salaman).

In County Clare, people used to pick and chew the bark of an ancient HAWTHORN at a holy well as a cure for toothache (Westropp), a practice that could be classed in the same category as the WILLOW bark as a primitive aspirin, or else simply as a charm, because of the connection with a holy well. HOLLY leaves were once used, the cure involving getting rid of worms that were believed to be responsible some hundreds of years ago. "For toothwark, if a worm eat the tooth, take an old holly leaf ... boil two doles [i.e., two of worts to one of water] in water, pour into a bowl and yawn over it, then the worms shall fall into the bowl" (Black). Again, "if a worm eat the teeth, take holly rind over a year old and root of CARLINE THISTLE, boil in water, hold in the mouth as hot as thou hottest may" (Black). SEA HOLLY seeds, we are told, could be used to treat toothache. A remedy from the Physicians of Myddfai enjoins the sufferer to "take a candle of mutton-fat, mingled with the seed of sea-holly; burn this candle as close as possible to the tooth, holding a basin of cold water beneath it. The worms ... will fall into the water to escape the heat of the candle"! WALNUT leaves, strongly astringent, used to be bound on to the cheek for toothache (a practice that could certainly harm one's face (Helias).

The Physicians of Myddfai record the use of HONEYSUCKLE leaves for the complaint: "Take the inner bark of the IVY and the leaves of honeysuckle, bruising them well together in a mortar, expressing them through a linen cloth into both nostrils, the patient lying on his back, and it will relieve him". Turner actually recommended chewing MARSH MARIGOLD leaves to relieve toothache. It sounds extremely hazardous, as does the use of the very acrid roots of SPURGE LAUREL (Vickery. 1995). CAJUPUT oil can be used as a counter-irritant by just rubbing it on the gums (Mitton). BURNET SAXIFRAGE root, too, was chewed to relieve the pain of toothache (Grieve. 1931), or the roots of GREATER CELANDINE, as Gerard advised. HORSERADISH leaves, bound on, will relieve the condition (Newman & Wilson). That was in Essex, but in Norfolk the cure was to grate the root and put it on the opposite wrist for twenty minutes (Hatfield). A hot CABBAGE leaf, sprinkled with pepper, put to the cheek was another recommendation from Essex. Early prescriptions include the use of BUCK's HORN PLANTAIN for toothache ("Shave hartshorn and seethe it well in water, and with the water wash the teeth, and hold it hot in thy mouth a good while. And thou shalt never have the toothache again") (Dawson). Another charm making the same claim, from France, was to make a necklace of MARSH MALLOW (roots?) for children to wear (Sebillot). Native Americans would put a piece of the root of VIRGINIAN SNAKEROOT into their hollow teeth to try and stop the toothache (Coffey), while the California Indians used the root of CALIFORNIAN BUCKTHORN (*Rhamnus purshiana*), heated as hot as could be borne, to put in the mouth against the aching tooth, and gripping it tightly between the teeth (Powers). Holding a piece of TARRAGON root between the teeth will cure toothache, too (Grieve. 1931), while chewing MUSTARD seeds used to be a common way of relieving the condition, or, in Norway, WATER PEPPER could be chewed (Barton & Castle) Gerard recommended SCARLET PIMPERNEL for toothache, "being snift up into the nosthrils". An oil distilled from BOXWOOD was once recommended for toothache (Hill. 1754), as well as for ailments as diverse as epilepsy and piles. Earlier herbalists seem to be agreed that CINQUEFOIL offers a cure for toothache. Albertus Magnus says so, among sundry other doubtful receipts. Gerard also recommended it, and before his time, there was a 15th century leechdom "for aching of the teeth. Take the root of cinquefoil and seethe it well in vinegar or in wine, and hold it as hot as he may suffer it a good while, in his mouth. And it shall take away the ache" (Dawson).

"Most men say that the leaves (of YARROW) chewed, and especially greene, are a remedy for the Toothache" (Gerard), something that was well known in Saxon times, for Cockayne has, from Apuleius Herbarium, "for toothache, take a root, give to eat, fasting". An old Irish remedy advised the patient to chew the leaves (Moloney). The Salish Indians of Vancouver Island agreed. They just hold a leaf in the mouth to stop the pain (Turner & Bell), and in southern Malawi, the roots and leaves of CHINESE LANTERN (*Dichrostachys glomerata*) are used as a toothache cure (Palgrave). Yarrow's close relative, SNEEZEWORT, was also used, as a native substitute for Pellitory-of-Spain, just chewing a leaf and holding it in the mouth, or by mixing the juice with vinegar and holding that in the mouth (Gerard), who also suggested a mouthwash made from the decoction of BEE BALM. Putting TOBACCO on the tooth was quite a common remedy, just as putting it in the ear would stop earache (Newman & Wilson), while an Irish remedy is to put a piece of CAMOMILE root on the aching tooth (Vickery. 1995). FIGWORT has some anodyne value. It will ease pain wherever it is applied, and that includes toothache and babies' teething.

By the principle of the counter-irritant, BUTTERCUPS were used at one time. Gerard, in a burst of humour, wrote that "Many do use to tie a little

of the herbe stamped with salt unto any of the fingers, against the pain in the teeth; which medicine seldome faileth; for it causeth greater paine in the finger than was in the tooth …". PELLITORY-OF-SPAIN was a favourite for the relief of toothache at one time, and a number of prescriptions from the 15th century onwards are recorded. Lupton, for example, advised the sufferer that "the root of pellitory of Spain, chewed between the teeth a good while, will purge the head and gums very well, and fasten the teeth; it helps the head-ach and tooth-ach, if it be used four or five times a day, two or three days together". There are records of the use of this plant from a collection in a 15th century leechbook, some reasonably straightforward, but what are we to make of this one?: "for aching of the hollow teeth. Take raven's dung and put it into the hollow teeth and colour it with the juice of pellitory of Spain that the sick recognise it not nor know not what it be; and then put it into the tooth and it shall break the tooth and take away the aching, and some men say, it will make the tooth fall out" (Dawson. 1934).In some parts of Scotland, the island of Mull being one (Beith), the root of YELLOW IRIS was chopped up and used for the relief of toothache. A 14th century recipe prescribed the leaves, stamped with honey, and applied to the cheek (Henslow). Wesley, too: "to cure the Tooth-ach, … chew the root of the yellow Flower-de-Luce". In one case, the instruction was to put the juice in the ear on the same side (Langham).

There was a belief in Germany that if a tooth was extracted, the patient must eat three DAISIES to be free from toothache in the future (J Mason). There was, too, a practice in Cumbria of eating two daisies to cure toothache (Newman & Wilson), possibly an example of Gerard's dictum that "the daisies do mitigate all kinde of paines". Gypsies would cure toothache by dropping the juice of PEPPERMINT on to the aching tooth to relieve the pain (Vesey-Fitzgerald)., and apparently slaves in Jamaica used the GUINEA-HEN WEED (*Petiveria alliacea*) (Laguerre).

TOOTHBRUSH TREE

(*Salvadora persica*) Toothbrush Tree, so called because the twigs are used as such in Africa (Dale & Greenway), apparently with good reason, for when chewed, the stem releases juices that seem to have a protective anti-bacterial effect. An extract from this tree is now incorporated in some commercial toothpastes (Lewington). When a Kikuyu (Kenya) smith forged a sword or spear, he would rub it with a piece of kianduri wood, kianduri being the local name for this tree, and at the same time speak an incantation over it to the effect of: "If the owner of this meets with an enemy, may you go straight and kill your adversary; but if you are launched at one who has no evil in his heart, may you miss him and pass on either side without entering his body" (Hobley). This is also known as Mustard Tree, because the berries are slightly aromatic, and pungent, like cress. It is said to be commonly used in the Near East just like mustard, and in the opinion of some, this is the mustard tree of the Bible (F G Savage).

TOOTHPASTE

Teeth can be whitened and the gums strengthened by rubbing them with SAGE leaves (Page. 1978), which are still used for the purpose. Some gypsy groups make their own toothpaste with chopped sage and salt (Vickery. 1995).

TOOTHWORT

(*Lathraea squamaria*) The common name is given because of the tooth-like thick scales on the rhizomes. So, by the doctrine of signatures, it was reckoned good for the teeth (Gordon. 1977). This leafless parasite was often called Corpse-flower (Grigson. 1955), grimly descriptive.

TORMENTIL

(*Potentilla erecta*) A bunch of tormentil root, burnt at midnight on a Friday, would force an errant lover to return. It featured, too, in an exhibition of charms and amulets in London, according to a *Times* report of 5 March 1917. This has been a useful plant, the roots giving a red dye to leather as well as to wool (C P Johnson), and also being used in tanning as substitute for tanner's bark, usually oak, hence one of its names in Shetland, Earth-bark. It is said that the tannin in a pound of tormentil is equal to seven pounds of oak bark. The use is recorded for Ireland (E E Evans), and the Western Isles, where fishermen tanned their nets with the roots (Murdoch McNeill). It was used, too, in Orkney (Martin), and Shetland (Hibbert), particularly in connection with the sheepskins once worn by Shetland fishermen over their ordinary clothes.

It is this tannin that has made it important in folk medicine. The name Tormentil is from French tormentille, Lain tormentum, the rack, or tormine, colic – the plant was used to reduce the pain. It survived as a medicine against the colic until quite recently in isolated parts, notably Northumberland and the Hebrides (Grigson. 1955). Pennant. 1772 records its use on the Isle of Rum – "if they are attacked they make use of a decoction of the roots … in milk". Gerard notes that the powdered roots cure diarrhoea, and the "bloody flux" (dysentery) especially if they are given "in the water of a smith's forge, or rather the water wherein his steele hath been often quenched of purpose. Tormentil stewed in milk was an Irish remedy for diarrhoea (Foster). Gypsies, too, use an infusion of the leaves for the purpose (Vesey-Firzgerald). And yet it has also been recommended as a laxative – "for costiveness. Take tormentil a good quantity, and mouse-ear, and five leaves of dittany and scabious and bruisewort …" (Dawson). With Yellow Pimpernel, this plant was used in Ireland as a treatment for insomnia (Moloney), while on South Uist corns were treated by chewing the plant, and then applying on a bandage (Shaw), and they

would chew it for a sore lip, and it is made into a paste and applied to any suppurating sore.

TOUCH-ME-NOT

(*Impatiens noli-me-tangere*). The common name and the specific are borrowed from the words Christ spoke to Mary Magdalene after the Resurrection (see John. xx. 17). The same caution applies in a French superstition recorded by Sebillot. 1906; It seems that girls were made to touch this plant. If she was not a virgin the flower would recoil and fade at once.

It has the usual balsam method of explosive seed distribution, hence names like Jumping Betty, or Jumping Jack (Parish), as well as Quick-in-the-hand, that is, alive in the hand (Prior). It has been used as an external application to get rid of corns and warts (McLeod), and a decoction is prescribed in herbal medicine to treat piles (Schauenberg & Paris).

Tradescantia x andersoniana > SPIDERWORT

TRAMMAN

The Manx name for ELDER

TRANSFERENCE CHARMS

A method of passing an ailment to a tree or another person by some ritual, simple or complicated. Take ASPEN – its constant shivering, by the doctrine of signatures, was taken as a sign that it could cure the shivering disease, ague, or malaria. But in some areas in France, the fever could be transferred by the simple rite of the patient tying a ribbon to the tree (Sebillot). Cross OAKS were planted at crossroads so the people suffering from ague could peg a lock of their hair in the trunk, and by wrenching themselves away might leave the hair in the tree, together with the illness (Jones-Baker. 1977). A feature of a lot of the charms connected with ASH is that illnesses would be handed over to the tree. So too with warts:

> Ashen tree, ashen tree,
> Pray buy these warts of me.

That is a Leicestershire rhyme to accompany the charm, which was to take the patient to an ash tree, and to stick a pin into the bark. Then that pin would be pulled out, and a wart transfixed with it till pain was felt. After that the pin would be pushed back into the ash, and the charm spoken. Each wart was treated, a separate pin being used for each (Billson) (see **WART CHARMS** for further examples).

Getting rid of warts by rubbing a snail on them and then impaling it on a BLACKTHORN used to be common practice; or, from East Anglia, you could rub the wart with a green sloe, and then throw the sloe over your left shoulder (Glyde). They are both transference charms; cattle doctors in Worcestershire used to cure foot rot by cutting a sod from the spot on which the animal was seen to tread with its bad foot, and then to hang the turf on a blackthorn. As the sod dried out, so would the foot heal (Drury. 1985).

Toothache could be treated in a similar way, according to a charm recorded in Suffolk. The sufferer was instructed to clasp the tree (BIRCH, in this case) in his arms and then cut a slit in it. He then had to cut a piece of hair from behind the ear with his left hand, and bury it in the slit. When the hair had disappeared so would the toothache (Burne). Another tree used in a similar way is HAWTHORN, according to a French prescription to get rid of a fever. The patient is advised to take bread and salt to the tree, and say:

> Adieu, buisson blanc;
> Je te porte du pain et du sel
> Et la fièvre pour demain.

The bread has to be fixed in a forked branch, and the salt thrown over the tree. Then he has to return home by a different road to that from which he set out. If there was only one door, then the patient had to get back in through a window (Sebillot). In much the same way, an offering of bread and butter was put under the BLACKBERRY arch, usually for whooping cough (see **BLACKBERRY**), after the child had been passed through. The patient had to eat some of the bread and butter while the adults present recited the Lord's Prayer. The rest of the food was given to an animal or bird on the way home – the animal would die, the disease dying with it (Baker. 1980).

ELDER is used in this same kind of magical way. Taking three spoonfuls of the water that has been used to bathe an invalid, and pouring it under an elder tree (Dyer) is an obvious kind of transference charm. In the same category is the Bavarian belief that a sufferer from fever can cure himself by sticking an elder stick in the ground in silence. The fever transfers itself to anyone who pulls the stick out (Frazer).

A Sicilian charm for the cure of scrofula was to chew PEACH bark, either on Ascension Eve or St John's Eve. If it dried up and withered, it was a sign that the tree had taken the disease to itself (Sebillot). Fernie quoted the same belief, except that the disease was goitre. In Marseilles, they used to get rid of a fever by sleeping with the back against a peach tree for two or three hours; the peach would gradually grow yellow, lose its leaves and die (Sebillot).

TRAVEL SICKNESS
GINGER will help to reduce the complaint, as it will deal with nausea generally. Ginger tea, or even ginger biscuits, are useful in these problems (M Evans).

TRAVELLER'S EASE
Names like this usually mean that there was a belief that putting a leaf in the shoe would help a traveller in some way. SILVERWEED is an example with this name. Putting a sprig of the herb inside each shoe will prevent blisters when walking long distances (Freethy). Then there is TANSY, which is called Traveller's Rest in Wiltshire, and there must have

been a belief at one time that putting the leaves inside a boot would stop tiredness on a journey, but the only belief that has come down is from New Forest gypsies, who put a sprig of tansy inside the boot to prevent the onset of ague. GOOSE-GRASS is called Traveller's Comfort sometimes, perhaps because it was traditionally used to soothe wounds and ulcers.

TREE CELANDINE
(*Bocconia frutescens*) A small tree from central America south to Peru. The Totonac use it for ringworm of the scalp, the leaves being soaked in alcohol and the affected parts bathed; it is also used for tuberculosis, when the leaves are boiled and the liquid drunk, or as a bath (Kelly & Palerm). A root infusion is used to treat jaundice and oedema in Colombia, while in Mexico the same infusion is used to cure warts and ophthalmia (Usher). The bark contains an alkaloid that has been used as an anaesthetic by Mexican surgeons (Perry. 1972).

TREE OF LIFE
See also Yggdrasil, for the Tree of Life also grew on the surface of the earth, with roots below and branches above, uniting the earth and heaven, and the underworld. All over the Near East trees were planted in the temples, particularly the fig, cypress, apple, sycamore and olive, as well as the COCONUT palm which was at one time the prime candidate for identification as the tree of life. As Raleigh said, "the Earth yeeldeth no plant comparable to this", and the tree "giveth unto man whatsoever his life beggeth at Nature's hand"(Prest). This one tree yielded wine, oil, vinegar, butter and sugar, while threads from the bark could be spun and woven into cloth, the nuts provided coir, as well as, dried, copra, and the fronds too have numerous uses. The SYCAMORE FIG was another tree identified as the Tree of Life. Adam and Eve, it will be remembered, sewed fig leaves together to hide their nakedness. It was only in the Middle Ages that the tradition grew up that the tree was an apple (Baring & Cashford).Whatever the tree, it was essentially the source and giver of life in all its different forms and aspects, and it tended to have its roots in paradise, whether this was located in some earthly or celestial abode, on isles of the blest, or far away in the west. This is one of the features that distinguishes it from other sacred trees that were venerated for some specific reason (James. 1966).

Trifolium dubium > LESSER YELLOW TREFOIL

Trifolium pratense > RED CLOVER

Trifolium repens > WHITE CLOVER

Trigonella ornithopodioides > FENUGREEK

Tropaeolum majus > NASTURTIUM

TRUE ANISE

(*Illicium verum*) The fruit of this Asian tree grows in the form of a star (hence another name, Star Anise). Dried, it is sold in Asia as a spice, and as a remedy against a number of diseases. In the West, the (unripe (Hyam & Pankhurst)) fruit is used to aromatize cordials and liqueurs like anisette, and it is also used in the perfume industry (Wit).

TUBERCULOSIS

Parts of a PINE tree were used for a number of chest complaints. Even the smell of them was said to be helpful. That is why so many were planted around chest hospitals. But it is MUGWORT that has pride of place in folklore. There is a very well-known legend from the Clyde area of Scotland, in which the funeral procession of a young woman who had died of consumption was passing along the high road when a mermaid surfaced, and said:

> If they wad drink nettles in March
> And eat Muggons in May,
> Sae mony braw maidens
> Wadna gang to the clay (Chambers).

Similarly, from Galloway, there is a story of a young girl close to death with tuberculosis, and a mermaid who sang to her lover:

> Wad ye let the bonnie May die i' your hand
> An' the mugwort flowering i' the land?

The lad cropped and pressed the flower tops, and gave the juice to the girl, who recovered (Cromek). A Welsh rhyme takes up the theme:

> Drink nettle-tea in March, mugwort tea in May,
> And cowslip wine in June, to send decline away.

But why a mermaid in Scotland? Benwell & Waugh came up with an interesting answer – Artemis (the generic name for mugwort is *Artemisia*) was also a fish goddess, and is sometimes depicted with a fish tail. So it was the goddess herself, and by extension the plant itself, that was advertising its own benevolence.

An Irish remedy was to boil an ounce of the dried leaves of mugwort in a pint of milk, and give the result to the patient several times a day (Moloney). In the Highlands of Scotland the liquid in which WOOD-RUFF had been boiled was given to consumptives (Grant) – indeed, the Gaelic name for the plant means "wasting plant" (Beith). SPLEENWORT, by its very name, must have been used to treat spleen disorders. But in Scotland it was taken for tuberculosis as well (Beith). CAMOMILE has been used – even the dew shaken from the flowers was taken (in Wales) for consumption (Trevelyan).

POLYPODY has been used for chest complaints, including tuberculosis (Quelch), and, in America, the bark tea from CHOKE CHERRY was taken (Lloyd).

Kentucky home remedies insisted that "COCK-LEBUR tea is good for the phthisick" (Thomas & Thomas), tuberculosis, presumably(?). SAFFRON has been mentioned as a local remedy for the disease (Fernie), but this is actually an old usage. Gerard, for instance, quotes it: "it is also such a speciall remedy for those that have consumption of the lungs, and are, as wee terme it, at deaths doore, and almost past breathing …". Central American Mayan medical texts prescribed crushed GREEN PURSLANE, rubbed on the body, for the disease (Roys). Mexican Indians like the Totonac use TREE CELANDINE (*Bocconia frutescens*) leaves, boiled, and the liquid drunk, or they could be used in a bath (Kely & Palerm). And patients in Indiana were advised to drink POKE-ROOT juice (Brewster).

TUBEROSE

(*Polyanthes tuberosa*) An unlucky flower, in American belief, possibly because of the waxy appearance, like death. Others think they emit the odour of death, still others that if you close yourself in a room with tuberoses, the scent will kill you (H M Hyatt).

TULE

(*Scirpus acutus*) The yellow pollen of this plant was called by the Apache "hoddentin". The medicine men applied this powder to the foreheads of the sick, and in the form of a cross, to the breast. No Apache would go on the warpath without a bag of this powder on him. When very tired while out on the warpath, they would put a pinch of hoddentin on the tongue and one on the crown of the head. Babies had a bag of it round their necks. When an Apache girl attained the age of puberty, they threw hoddentin to the sun, and scattered it about her. The very first thing an Apache did in the morning was to blow a pinch of it to the dawn (Bourke).

TULSI

(*Ocimum tenuiflorum*) This is the tulasi, or tulsi, the sacred basil of India, the most sacred in Hindu religion (Pandey), a protector from all misfortunes and disease (Folkard), sacred to Vishnu (Brouk), and Krishna (Pandey), and a goddess in its own right, as Vishnu's beloved, hence Hindus call the plant tulsi-mata, "mother tulsi". It is grown in Hindu houses as a symbol of what O'Neill called "the divine Universe-tree". It is perceived as the place where heaven and earth meet (Simoons). Flowers and rice are offered to it, and a pot of it is also put at the foot of the village Pipul-tree. The care and worship of tulsi plants have long been regarded as religious duties. Salvation is assured to the person who waters and cares for the plant daily. The responsibility rests with women, for often their ritual activities as Hindus relate entirely to the tulsi plant. A string of beads or a necklace made from the stems or roots of this plant is an important symbol of initiation into the cult of Vishnu, and it places the wearer under the god's protection. It is a sacrilege to boil its leaves in hot water, because it torments the soul of the plant (Upadhyaya). But tulsi is one of the fuels used in a Hindu funeral pyre. The Puranas say that even persons guilty of many sins are absolved if they are cremated with tulsi twigs (Simoons). According to Gubernatis, one of the Sanskrit names for it was Apetorakshasi, "la plante qui éloigne les monstres". It is repellent of dangerous organisms. It can drive away mosquitoes, and other insects as well as poisonous reptiles, and is an antidote to snake bites and scorpion stings. It is believed to purify the air, and so has been used against respiratory illnesses, from colds to tuberculosis (Simoons).

It "grants children to the childless, wealth to the needy, and opens the gates of heaven to the devout worshipper" (Simoons). Girls may worship the tulsi in order to get a good husband , and a plant may be put on their marriage altar. It may be used, too, to make certain that a pregnant woman does not suffer a mis-carriage. For example, the Kol, a central Indian tribe, put tulsi leaves on a pregnant woman's abdomen, to pray that she remain safe during that time of danger when evil forces can cause a miscarriage. But tulsi seed is thought to quell sexual desire.

TURK'S CAP LILY

(*Lilium martagon*) *Martagon* is derived from a Turkish word for a kind of turban. Hence the common name. But *'martagon'* was used for all lilies with strongly recurved petals, and it was this characteristic that produced such vernacular names as Turncap and the odd-looking Turn-again-gentlemen, widely used in the southern counties of England. The roots of this lily have provided a traditional Irish remedy for boils and swellings (Maloney).

TURKEY CORN

(*Dicentra canadensis*) A North American species, with greenish-white flowers, a colour that may account for the Indians' name for it of "ghost-corn". They used to say that the souls of the dead raised it in the abandoned fields of the living (Hewitt). That never stopped them from raising the plant themselves, for the tubers were useful as food (Yanovsky).

TURKISH TOBACCO

(*Nicotiana rustica*) A yellow-flowered hybrid, derived from two wild species growing on the western side of the Andes near the borders of Ecuador and Peru. It spread by diffusion and migration over a much larger area than did *N tabacum*, though the latter is the kind raised commercially today (Driver) (see TOBACCO). The fumes of Turkish tobacco are strong and almost intoxicating, for the plant contains four times as much nicotine as in modern cigarette tobacco. Some southern Indians raise it still, always in remote plots and away from women. Four solemn puffs to the four directions was the custom, as the pipe was passed ceremonially from hand to hand.

Sometimes two puffs were added for the zenith and the nadir (Underhill). The curing shamans of Aztec-speaking communities used this tobacco to put themselves into a state in which they could call on the spirits to help in a patient's cure (Furst). Being so much more powerful, this was the tobacco that was more widely used in ritual. The Huichola refer to this as the "proper tobacco of the shaman", while the Seneca of New York called it "real tobacco" (Furst).

TURMERIC

(*Curcuma longa*) Grown for its bright yellow dye, which is not fast to light, let alone washing, and so has to have some mordant with it. It was used in surprising ways, if we are to believe Pomet – "the Founders employ it to tinge their Metals, and the Button-makers to rub their wood with, when they would make an Imitation of Gold".

Where the simplest preparation is still in use, in the Pacific islands and New Guinea, it serves mainly for cosmetics or for painting wood, etc., (Buhler). Body painting of one form or another was the main use of the yellow powder in Polynesia, and there are distinct sexual overtones to its use. On the Marquesas it was used in quantity by adolescents, particularly during orgiastic ceremonies and other situations involving sexual activity. The smell was supposed to have a sexually stimulating effect (Suggs). To the Muria, of India, the yellow colour makes it both a ghost scarer and a sexual symbol. The oil with which it is mixed is another sex symbol, recalling the oil traditionally put on a lover's mat "to make it slippery" (Elwin). The rubbing of turmeric and oil on the bride and groom is an essential part of a marriage festival in India (Pandey).

Elsewhere, though, the yellow colour has a different meaning. On Tikopia, turmeric is daubed over mother and child soon after birth, as a mark of attention, or even of honour. It is used to single out individuals who are at the moment of special interest and impor-tance (Firth). Yellow is a colour sacred to the gods in Samoa, so the gathering and processing of the roots became a religious ceremony, with its prescribed rites. Turmeric powder is used as a medicine, as well as a dye. Mixed with oil, it is rubbed on inflamed parts, especially over recent tattooing, to soothe the pain. It was used in Samoa as a dusting powder for babies (Buck). In the Marquesas it is used as an insect repel-lent, and in Java it is the commonest laxative in use (Geertz). But the best known use in medicine is pure doctrine of signatures, for it is very comonly pre-scribed for jaundice, and a long way from Polynesia, too. The Mano, of Liberia, for example, use it in this way. The patient has to drink daily a cupful of the root infusion (Harley). Thornton, in *New family herbal*, 1810, noticed the use for jaundice, too. In Chinese herbal medicine, the root is used as a stomachic and diuretic (R Hyatt), but it is also classed there as a "blood invigorator", and can be used for a form of mental derangement (Geng Junying).

Perhaps the most recognisable use of turmeric (in Britain) is as a colouring agent for food. Curry powders contain it, as well as ginger and fenugreek, but it is the turmeric that accounts for the distinctive colour (H G Baker), so much so that the plant is often called Curry (Schauenberg & Paris). Pickles used to be coloured that way (Clair), and probably still are. Because the preparation of the yellow dye is lengthy, the powder was quite valuable, and in some Pacific areas it was used as a unit of currency right up to the 1940's. The value of anything was expressed as so many taik cakes, taik being the name of the powder (Einzig).

TURNIP

Turnips seem to have been first grown in the London area in the 16th century, but Norfolk was the first county in which they were extensively cultivated for cattle feed (G M Taylor). Gerard, at the end of that century, was rather disparaging about them: "the … root …is many times eaten raw, especially of the poore people in Wales, but most commonly boiled. The raw root is windy, and engendreth grosse and cold bloud; the boiled doth coole lesse … yet it is moist and windy". The Regimen Sanitatis Salernii was equally scathing: "Turnips cause flatulence and spoil the teeth, stimulate the kidneys, and when ill cooked cause indigestion" (Hickey). There seems to have been some doubt early on as to what one should do with them – English travellers in Scotland in the 17th century complained that they got turnips (neeps) as dessert. ("The Scots had no fruit but turnips" (Graham)).

Nevertheless, this is a thoroughly useful vegetable, and its solid worth is recognized in superstitions, too, for to dream of being in a turnip field is a sign of riches to come, so it is said (Raphael), though the dream books were not unanimous about that. It is recorded that African Americans in the deep south of the USA used to scatter turnip seed round the house to keep witches away (R B Browne); they must have been growing right up to the doorstep! There is even a divination game played with turnips. It comes from west Wales, and involves a girl stealing a turnip from a neighbour's field (it must be stolen, not given). She peels it in one continuous strip, in the same way as the better known apple peel game, taking care not to break the peel, and then buries the peel in the garden. The turnip itself she hangs behind the door. Then she goes and sits beside the fire, and the first man who enters after that will bear the same name as her future husband (Winstanley & Rose).

Sow turnip seed thickly in a part of the garden infested with couch, and the latter will disappear, so it is claimed (Boland). Another piece of gardeners' wisdom comes from America:

Plant turnips on the 25 July,
And you'll have turnips, wet or dry,

is one of the adages from Kentucky, but they also tell one to plant them on the 10 August (and certainly not on the 7 August). To have good luck with them, say as you throw out a handful of seed: "One for the fly, one for the devil, and one for I" (Thomas & Thomas).

Turnips are diuretic, and were recognized as being so some centuries ago, and are still used for retention of urine among Irish people in County Mayo. They pulp a turnip and drink the juice (Logan). But in the southern states of America, and among black women in particular, there was a belief that if the mother eats turnip green while the baby is young, then the baby will die (Puckett), possibly because of the connection between greens and mother's milk, although Gerard was of the opinion that "they do increase milke in womens breasts". "Turnepes boyled and eaten with flessche, augmentyth the seeds of man" (G M Taylor) was a very early (1542) fiction.

Gerard again: "They of the Low countries do give the oile which is pressed out of the seed … to young children against the worms, which it both killeth and driven forth". The early Welsh text known as the Physicians of Myddfai echoes this, though the procedure is different – "to destroy worms in the stomach or bowels. Take the juice of turnips, foment therewith, and they will come out". They provide a very well known cough cure, the usual practice being to cut one into slices, put them in a dish, and put sugar on them. Leave them for a day or two, and give a teaspoonful of the juice for the cough. That is the Wiltshire remedy (Olivier & Edwards), but it is virtually the same across southern England from Cornwall to Buckinghamshire (Hawke; Heather, for instance). Whooping cough was treated in the same way, and dried turnip grated and mixed with honey is an American cold cure (Stout).

Tussilago farfara > COLTSFOOT

TUTSAN

(*Hypericum androsaemum*) The black berries are often viewed with suspicion by country people. In the Hebrides, for example, they say that if you eat them, you will go mad (Murdoch McNeill). The specific name, *androsaemum*, comes from two Greek names meaning man and blood, the reference being to the dark red juice that exudes from the bruised capsules. There is a belief in Hampshire that tutsan berries originated by germination in the blood of slaughtered Danes (Gomme. 1908). Anyway, this juice was taken as a representation of human blood, and by the doctrine of signatures the plant was applied to all bleeding wounds (Dyer. 1889). Actually, the leaves do have antiseptic properties, and they were certainly used to cover open flesh wounds before bandaging became common (Genders. 1971). Gerard said that "it stauncheth the blood and healeth them" (the wounds). He also mentioned "broken shins and scabbed legs" as conditions that tutsan leaves could heal, "and many other hurts and griefes, whereof it took his name Toute-saine or Tutsane, of healing all things", a panacea, in other words. He also recommended it for burns and there are other ailments that have been treated externally by either the leaves or roots in some kind of ointment. They range from chilblains to carbuncles, both Welsh usages, the former in the medieval text known as the Physicians of Myddfai ("… boil the roots, and pour upon curds. Pound the same with old lard, and apply as a plaster"). The carbuncle usage seems to be confirmed by one of the Welsh names for the plant, Dail fyddigad, carbuncle leaves (Awbery). Perhaps the strangest of the conditions to be treated with tutsan is, not surprisingly, in Gerard, and supposed to be a remedie for sciatica: "the seed … beaten to pouder, and drunke to the weight of two drams, doth purge cholericke excrements, … and is a singular remedie for the Sciatica, provided that the patient drinke water for a day or two after purging".

When dried, the leaves have a very sweet smell, likened to ambergris. Picking the leaves and pressing them in books used to be a favourite pastime, resulting in the names Bible Flower or Bible Leaf (Grigson. 1955). Book Leaf, too, is known in Dorset (Macmillan). Incidentally, you can put them amongst clothes, too, to keep moths away (Genders. 1976).

The name Park-leaves, listed as a Somerset name (Macmillan), is by now used as the common name almost as much as tutsan. It is, of course, a corruption of *Hypericum*.

TWIN-LEAF

(*Jeffersonia diphylla*). This plant is also known as Rheumatism-root, for it provides a popular American remedy for the complaint (Berdoe).

TYPHOID

HENNA, in addition to its cosmetic and protective roles, is also used as a medicine (see Westermarck. 1926), as for instance in Morocco, where, mixed with water, it is applied to the forehead of a person suffering from fever. In the Balkans, too, in cases of acute fever, like typhoid, henna is heated in water, allowed to cool and the juice of some twenty heads of garlic added, the mixture re-heated, and then the henna is applied solid to the palms of the hands and the soles of the feet, in exactly the same way as for cosmetic staining, "in order to draw out the fever" (Kemp). This use of a red dye to allay fever is in all probability an example of the doctrine of signatures, just as in Britain at one time, fever patients were wrapped in red blankets to allay the symptoms. WHITE HOREHOUND (*Marrubium vulgare*) is used for fevers, especially typhoid in Africa (Watt & Breyer-Brandwijk). Having SUNFLOWERS growing in the garden would prevent typhoid striking, so it was once believed in Indiana (Brewster).

U

ULCERS

LEEK poultices were put on them (Physicians of Myddfai). A Somerset treatment used MALLOW leaves, either by an infusion, or simply by using the leaves as a poultice (Tongue. 1965). Green CABBAGE leaves were a favourite Irish country way to treat an ulcer, by poultice, perhaps? But not necessarily, for the leaves could simply be applied (Logan). Herbalists prescribe cabbage juice for stomach ulcers (Thomson. 1978), for which BUCKBEAN infusions were given in Scotland, successfully, apparently (Beith). Another Scottish remedy, from the Highlands, was to use ROSEROOT on ulcers (Fairweather). An ointment has been made from WOAD to heal ulcers (Brownlow), and gypsies used the juice of fresh COLTSFOOT leaves in making such an ointment (Vesey-Fitzgerald). GOOSE-GRASS is traditionally used to soothe wounds and ulcers (Schauenberg & Paris). In Ireland a whole mass of the herb would be applied, while the juice was given internally at the same time (Moloney).

Ulex europaeus > FURZE

Ulmus fulva > SLIPPERY ELM

Ulmus glabra > WYCH ELM

Ulmus procera > ELM (i.e., ENGLISH ELM)

Umbellularia californica > CALIFORNIAN LAU-REL

Umbilicus rupestris > WALL PENNYWORT

UNIVERSE TREE

See YGGDRASIL, but there are others, of which ORIENTAL PLANE (*Platanus orientalis*) is an example. A legend from Corfu tells that on the ten days preceding Good Friday, all the Kallikanzaroi in the underworld are engaged simultaneously upon the task of sawing through the giant plane tree whose trunk is supposed to hold up the world. Every year they almost succeed, except that the cry "Christ has arisen" saves us all by restoring the tree, and driving them up into the real world (Durrell).

UNLUCKY PLANTS AND TREES

In Greek folklore, people have a fear of sleeping under a FIG tree. On the island of Chios they say that the shadows of both fig and hazel are "heavy", so it is not good to sleep under either of them (Argenti & Rose). Another aspect of this mistrust of a fig tree comes from the belief in the south of France that John the Baptist was beheaded under one. That is why the branches break off so easily, particularly on St John's Day, when anyone who climbs the tree risks a dangerous fall. Similarly in Sicily, the mistrust lies in the belief that Judas hanged himself on one.

WALNUT trees are unlucky in some areas, the Abruzzi, in Italy, for instance. Anyone who plants one will have a short life (Canziani), or, as in Portugal, he would die when the tree attains his own girth (Gallop). Perhaps it is one of the cases of the man's life going into a tree which is very long-living.

PINE trees were looked on as unlucky in the Channel Isles. Guernsey belief had it that whoever planted a row of them ran the risk of losing the property, or letting it pass from the rightful heir to a younger branch of the family. There was also a belief there that if you fell asleep under one you would never wake up (Garis). HAWTHORN, too, was unlucky, in spite of its being a sacred tree and of offering its protection against witches. It is particularly unlucky to bring indoors, perhaps because of the belief that Jesus was crowned with these thorns. It "brought illness, etc.," with it, according to Devonshire belief, and in Somerset it may cause death in the house into which the blossom is brought (Elworthy). Cheshire children are forbidden to bring it in, the belief being that their mother will die if they did (Hole. 1937).

> May in.
> Coffin out,
>   in fact (Igglesden).

ELM, too, was long regarded as a thoroughly treacherous tree, hostile to human beings (Ô Súilleabháin):

> Elm hateth
> Man and waiteth (Wilkinson. 1978).

Kipling knew all about the belief, and wrote in *A tree song*:

> Elmen she hateth mankind and waiteth
>   Till every gust be laid
> To drop a limb on the head of him
>   That anyway trusts the shade.

That is probably the nub of this belief. Elms can often, without any warning or signs of decay, shed a limb and cause injury or death. "He will wait for me under the elm" is a French proverb, meaning he will not be there, perhaps because it would be such a stupid place to wait (Wilkinson. 1978).

BLACKTHORN is a thoroughly unlucky plant. Even its early blooming brings talk of a "blackthorn winter" (see **WEATHER LORE**), and the more sloes there are in autumn, the worse the winter to come. The flowers are extremely unlucky to bring indoors, as are most white flowers, but more fuss seems to be made about blackthorn than anything else. It is just as bad to wear it as a buttonhole. Sussex people looked on it as a death token (Latham); in Suffolk, too, they used to say that it would foretell the death of some member of the family (Gurdon). And in Somerset, it would mean you would hear of a death (Tongue. 1965). Of course, all this might possibly amount to preventive superstition – Vickery pointed out that a scratch from the fierce thorns could very well cause blood poisoning. Against this is a Lincolnshire belief that it is not quite a death token – at Alford, in that county, the belief was that a blackthorn flower indoors would result in the relatively lesser misfortune of a broken arm or leg (Gutch & Peacock).

ELDER, with many protective and anti-witch virtues, is still an unlucky tree, with an evil character. The very fact that witches were fond of lurking under it made it dangerous to tamper with after dark (Dyer). And do not sleep under one – the leaves were said to give out a toxic scent which if inhaled may send the sleeper into a coma and even death (Baker. 1977). Mending cradles with elder wood was just as dangerous, for a Cheshire belief was that it would give the witches power to rock it from afar so violently that the baby would be injured (Hole. 1937). Again, a child laid in an elder-wood cradle would be pinched black and blue by the fairies (Graves); or the fairies may steal the child (Grigson); or the Elder-mother may strangle it (Farrer). In Ireland, elder wood was never used in boat-building (Ô Súilleabháin) nor, so it was said in South Wales, should a building of any kind be built on the spot where an elder had stood (Trevelyan). It is credited with having a harmful influence on plants growing near it (Rohde). The flowers were never allowed in the rooms of Fenland houses, because they were supposed to attract snakes (Porter), and from the same general area there is a record of the belief that a wound suffered by contact with the tree, say by driving a sharp stick accidentally into the hand, would inevitably prove fatal. It was quite a common belief that beating boys with an elder stick would stunt their growth (Ô Súilleabháin). Burning elder wood, particularly green elder (Forby), was almost universally forbidden in England, for "it brings the devil into the house" (Graves). "They dursn't burn 'em if you gave them away – they don't want the devil down their chimbley" (Heanley). Or, as in Lincolnshire, "the devil is in elder wood" (Rudkin). There are many other examples. JUDAS TREE (*Cercis siliquastrum*) is a similarly unlucky tree. It is the symbol of betrayal, for Judas hanged himself on one of them. It should be avoided, especially when in flower (M Baker. 1977), many people are reluctant to cut it, particularly after dark, and it is a favourite haunt of witches (Dyer. 1889). In the southern states of America, PERSIMMON is also an unlucky wood to burn in the house. Like SASSAFRAS, it pops and crackles a lot while burning, and perhaps that is the reason for the belief. Throw a piece of it in a man's fireplace, and he will soon move away. So runs a belief that was current in all the southern states of America (Puckett). HONEYSUCKLE is another plant with an ambivalent reputation. It is a witch plant, and at the same time an anti-witch protection, and that might explain its

reputation as an unlucky plant. From Scotland to Dorset there are records of a general belief that to bring it indoors is very unlucky; in Dorset they say it brings sickness into the house with it, and in west Wales it was believed that it would give you a sore throat (Vickery. 1985). It was never brought into a Fenland house where there were young girls; it was thought to give them erotic dreams, especially if it were put into their bedrooms. If any of it *was* brought in, then it was said that a wedding would shortly follow (Porter) – hardly surprising, if the girls' minds were concentrated in that direction. MYRTLE is another of these ambivalent plants. It is lucky to have one, so long as one is visibly proud of it. But on the other hand, the shrub is connected with death, which makes it unlucky, particularly in America, where it is rarely seen outside cemeteries. Never let it grow around the house, or there will be sickness and trouble there as long as it is growing (H M Hyatt). TUBEROSE is another unlucky flower in American belief, possibly because of its waxy appearance, like death. Others think they emit the odour of death, still others that if you shut yourself in a room with tuberoses, the scent will kill you (H M Hyatt). CUCKOO-PINT is equally unlucky to have indoors, for it gave TB to anyone who went near it (Porter. 1969). The real reason may have been forgotten, but this kind of superstition is usually directed against the females of the house, and it would not be TB that they got. A Dorset belief is quite explicit – young girls were told never to touch a cuckoo-pint; if they did, they would become pregnant (Vickery. 1985), a belief quite in keeping with the plant's sexual display.

MEADOWSWEET is equally unlucky in Welsh superstition. If someone fell asleep in a room where many of these flowers were put, death was inevitable. It was even dangerous for anyone to fall asleep in a field where there was a lot of it growing. This sounds as if it were an extension of the fear of bringing any white flower – hawthorn, lilac, etc., indoors, to which PEAR blossom must be added. That too would cause a death in the family (Vickery. 1995). PRIMROSES were not always entirely welcome when brought indoors – it all depended on how many were gathered. Two or three brought into a poultry keeper's house in early spring, before the chicks were hatched, meant bad luck to the sittings, but it would be alright if there were thirteen or more flowers, or "no less than a handful". In Devonshire they said that the number of primroses brought in would agree with the number of chickens reared, and the same was said in Norfolk (Friend. 1883), for thirteen is the number traditional to a clutch of eggs placed under a hen. There was a similar belief in France – if you threw the first primroses you found before the goslings, it would kill them, and if you took them indoors, the goslings would die before being hatched (Sebillot). It was even unlucky to include primroses (and hazel catkins) in the posy carried to church on Easter Sunday. Violets had to be put in too, to compensate for the primroses (Tongue. 1965). But it was probably a lot more serious than it seems, at least in some areas, those in which primroses were looked on as a death token, just as snowdrops are. One explanation from Sussex is that it was used to strew on graves, and to dress up corpses in the coffin (Latham). Certainly, quarrels have been recorded as arising from this belief, and it could lead to charges of ill-wishing. Anyone giving a child, say, one or two primroses, would leave himself wide open to such a charge (W Jones. 1880). Another spring flower that is unlucky for poultry keepers is the DAF-FODIL, though there is a certain ambiguity about its luck. Granted, the first daffodil is a lucky one. Welsh belief had it that if you find the first daffodil you will have more gold than silver that year (Trevelyan). But one has to be careful about the direction in which the trumpets are pointing. See Herrick, *Hesperides*:

> When a Daffodil I see
> Hanging down her head t'wards me,
> Guess what I may what I must be;
> First, I shall decline my head;
> Secondly, I shall be dead;
> Lastly, safely buried.

In other words, if you see a daffodil with its head bending down towards you, it is a sign that you are about to die (Addy. 1895). As with primroses, one has to be careful with daffodils when there is poultry about. There is an old Manx superstition that it is bad luck to a poultry keeper if two or three of the flowers are brought into the house in early spring, before the goslings are hatched (the Manx name for the daffodil shows the connection – it translates to Goose-leek). One finds this superstition in Devonshire, too, while a Cornish belief was that if a goose saw a daffodil before hatching its goslings, it would kill them when they did hatch (Courtney). A Dorset compromise suggests that you must always take care that the first daffodils brought indoors each season should be a large bunch, for otherwise something would be sure to go wrong with the poultry (Udal). Judging from the primrose belief, you should always take in quite a large bunch – two or three are fatal; the ideal is probably thirteen or more. On the other hand, in parts of Warwickshire, daffodils are thoroughly unlucky flowers, never to be taken indoors.

In some parts, particularly Italy, CHRYSANTHE-MUMS are seriously unlucky flowers. They are funeral flowers there, and so are associated with the dead (hence a connection with All Souls Day, too). They say that if you give chrysanthemums to anyone, it is the equivalent of saying I wish you were dead (Vickery. 1985). Obviously, then, it is not a flower to have indoors, for it would bring very bad luck with it (Vickery. 1995). ARUM LILY is another funeral

flower, and so thoroughly unlucky. That too should not be taken indoors (Deane & Shaw), and *never* be brought into a hospital (Vickery. 1985). WATER ARUM, too, is a thoroughly unlucky plant to have in the house (Bergen. 1899). BOX is an unlucky tree in one sense – its association with death and funerals. A sprig of box in flower brought indoors meant that death would soon cross the threshold (Dorset) (Udal). But in spite of this connection, it is generally a plant that brings good luck. Sawn WILLOW was unlucky in the house; if the timber was admitted at all, it would have to be shaped with an adze (Whitlock. 1982).

HEATHER, at least according to tradition in Wales, is unlucky to bring indoors. It brings misfortune with it, even death (Trevelyan). Perhaps that was because the young tops were a fairy food, but that is a Lowland Scots tradition (Aitken). It is very unlucky to take GORSE into the house, just as unlucky as hawthorn or lilac, for "to carry furze flowers in the house – carrying death for one of the family" (Opie & Tatem), and there are other similar sayings of the "gorse in, coffin out" variety. Giving the flowers to someone is also unlucky, but without such dire results. But the act would be bound to provoke a quarrel between the two people involved, in a short time (Vickery. 1995). BROOM, too, especially in its role as a domestic implement, can sometimes be very unlucky:

> If you sweep the house with blossomed broom in May,
> You're sure to sweep the head of the house away.

It is still believed to be an unlucky flower to bring indoors (Widdowson), as, for instance, in the Isle of Man (Gill. 1932), or in Sussex, where they say "it sweeps someone out of the house" (Vickery. 1985). RED POPPIES are equally unlucky to bring indoors, or even touch. Irish women had a dread of touching them (Grigson. 1955), but it is likely that proscriptions were designed to stop children picking them, i.e., getting into the growing corn and causing possible damage. They were told that if they picked poppies they would wet the bed, or it would provoke a thunderstorm, or give themselves a headache, etc., Even GOLDEN ROD has been seen as unlucky, certainly not to be taken indoors (Vickery. 1995). The same applies to WOOD ANEMONES; picking them would provoke a thunderstorm (it was actually called Thunderbolt in Staffordshire (Vickery. 1995)). BLUEBELLS are equally unlucky to bring indoors (*Devonshire Association. Transactions. vol 65; 1933*). Devonshire superstition has it that it is unlucky to plant out a bed of LILY-OF-THE-VALLEY, as the person to do so will be sure to die within the year (*Notes and Queries; 1850*). It belongs to the group of white flowers, like snowdrop and white lilac, that will cause death if brought into the house – and lily-of-the-valley is always unlucky for girls. It is the girl child who will die if they are brought inside (Tongue. 1965). The Sussex folk tale called the Basket of Lilies starts: "There was a woman who loved Lilies-of-the-valley. She'd be always looking for them ore sending her little daughter to find a bunch to bring home, so of course the little girl sickened and died, as everyone knew she would ..." (Tongue. 1970).

SOUTHERNWOOD, at least according to a Devonshire belief, is an unlucky plant to have in the garden. It was said that when a woman was ill, a neighbour came to see her and found a lot of this plant in her garden. She pulled it all up, and the patient recovered (*Devonshire Association. Report and Transactions. vol 103*). Eating CELERY will bring bad luck, at least according to Kentucky belief (Thomas & Thomas).

Most evergreen trees are funerary emblems, and at the same time symbols of immortality. That is so with HOLM OAK, but in the Greek islands it is an unlucky tree, because it was of its wood, so the story goes, that the Cross was made. A miraculous foreknowledge of the Crucifixion had spread among the forest trees, which agreed (almost) unanimously not to allow their wood to serve. When the foresters came, they either turned the edge of their axes, or bent away from the stroke. Only the Ilex consented, and passively submitted to being felled. So now the woodcutters will not soil their axes with its bark, and not desecrate their hearths by burning it (Rodd). YEW, of course, is the ultimate unlucky tree. It is the "funeral yew". In Shakespeare's words, it is the "dismal yew" – slips of it "slivered in the moon's eclipse" were among the ingredients in the witches' cauldron. Derbyshire farmers will not cut yew trees down, and they reckon it is unlucky to burn, too (Addy), and of course it is poisonous.

SILVER WATTLE (*Acacia dealbata*) may be the Australian national emblem, but that does not stop it from being thoroughly unlucky to bring indoors, even sometimes to plant in a garden. There are a few instances in England, too, of its being unlucky to bring indoors, even being described in one record as " a forewarning of disaster".

MOTHER-OF-THOUSANDS (*Saxifraga sarmentosa*) is an unlucky plant, apparently on the basis of its attributed names, such as: Creeping, Wandering, or Roving Sailor. It was said that an accident to the plant would ensure a mishap to any relative who was a sailor (*Folklore. vol 37; 1926 365–6*). But, of course, these names are actually given as a description of the plant's method of reproduction, by sending out runners from the parent plant. French herbalists used to say that looking at the flowers of the MEDITERRANEAN ALOE (*Aloe vera*) is unlucky (Boland. 1977). RED CAMPION (or WHITE CAMPION) are other examples of a flower unlucky to pick. It has the name Mother-die in Cumbria (Grigson. 1955), always an indication of an injunction against picking it. It is a fairy flower, too, another

reason why it should never be picked, and in Wales it is Blodyn Neidi, snake flower, another sanction, for if you pick it you will be attacked by snakes (Vickery. 1985). Another Welsh name is Blodyn Taranan, thunder flower. Thunder and lightning will be the result if you gather them. The reason for these sanctions is that campions often grew in the corn, and if children were allowed to search for them, they could very well damage the crops in doing so. HERB ROBERT is another of these flowers that are unlucky to pick. It is called Snake Flower in Somerset, and there are other "snake" names for it. If you pick it, snakes would come from the stems (Vickery. 1985). But more significantly, the name 'Death-come-quickly' is recorded from Cumbria, for this is one of the flowers that, if picked by children, would result in the death of one of the parents. So with GERMANDER SPEEDWELL, a normally cheerful little plant. But it has its sinister side, as names like 'Tear-your-mother's-eyes-out' will testify. If you pick it, your mother will die during the year (Dyer. 1889), or it will result in a thunderstorm, etc., Even QUAKING GRASS has to be included here, for it is unlucky to the man who picks it or wears it, and it is also bad luck to bring it into the house. If it is laid in a baby's cradle, the child will be rickety (Tynan & Maitland). SPINDLE TREE was called Death-alder in Buckinghamshire (Grigson. 1955), and as such is reckoned unlucky to bring in the house, and, of course, the berries are mildly poisonous.

SNOWDROPS are another example of white flowers unlucky to bring indoors (like PRIVET in flower). Some say the bad luck applies only to cut snowdrops, and not to those grown in pots indoors (Vickery. 1985). Others, in Wales, say the sanction applies only to snowdrops taken indoors on St Valentine's Day (L Davies). The result of such rash actions were equally variable, ranging from the death of someone living in the house to the cows' milk being watery and affecting the colour of the butter (Burne. 1883). In Somerset, the belief was that the girl child in the house would die within the year (Tongue. 1968). There was a similar superstition regarding lilies of the valley and white lilac, beliefs that are very widespread. Taking snowdrops into a hospital is even more unlucky. If they were given to a patient it was often taken to be a sentence of death. Nurses would sometimes put a few ivy leaves with them to lessen the omen (Tongue. 1967). It is said that the association of snowdrops with death (they used to be called Death's Flower in Somerset) results from the flower's resemblance to a shroud (Vickery. 1985). "It looked for all the world like a corpse in its shroud" was how one of Charlotte Latham's informants put it. Another reason given to her was that "it always kept itself close to the earth, seeming to belong more to the dead than to the living". Again, we hear that the reason is that they are so often found growing in old graveyards (Vickery. 1985).

The West Indian tree called MAMEY (*Mammea americana*) must be mentioned here, not that the tree is itself unlucky, but in Jamaica it was reckoned unlucky (even fatal) to plant the seed (*Folk-lore. vol 15; 1904 p 94,* in a series of papers called *Folklore of the negroes of Jamaica*). GROUND NUTS are unlucky, too. If you dream of them, it is a sign that you will be poor (H M Hyatt). And African Americans in the southern states of the USA say it is unlucky to eat peanuts when you are going to play a game of any sort, and the hulls scattered about the door mean that you will go to jail (Puckett). In Madagascar, they are taboo to pregnant women, as they will cause a miscarriage. They say that nuts lying on the ground remind people of souls that lay their eggs on the ground, and that is the reason for the taboo (Ruud).

UNSHOE-THE-HORSE

A name given both to HORSESHOE VETCH (*Hippocrepis comosus*) and to HONESTY (*Lunaria annua),* both of which were supposed at one time to have the power of unshoeing horses. Horseshoe Vetch was so called because of the shape of the pods, and any horse that trod on it would be unshod. Very possibly, the real reason may be that the plant grows in the sort of stony ground that could lead to accidents. Honesty earns the name because of its ancient magical fame as a pick-lock. So great was its power that it could draw the nails from horses's hoofs (Watts. 2000).

UPRIGHT ST JOHN'S WORT

(*Hypericum pulchrum*) The Gaelic name for the plant is Achlasan Chalum Chille, which seems to be untranslatable. This is St Columba's plant, and the point is that the plant was worn under the left armpit as a protective charm, and that was the way Columba carried it (Banks. 1937). The traditional story is told of a young herd boy whose nerves had been upset by the long nights out on the hills with the cattle, and was brought to St Columba for a cure. The saint is said to have put this herb in the boy's armpit, and the boy very soon began to recover the balance of his mind (Beith). Putting the healing agent under the armpit is not unknown in other diseases. Cf, for instance, a later cure for pneumonia that involved putting an onion in each armpit. This St John's Wort was only reckoned to be effective when it was found accidentally, "when neither sought for nor wanted", and then it should be put secretly in her bodice, if it was a woman who found it, or, if a man, in his waistcoat under the armpit (Banks. 1937), when it would ward off fever, and keep its owner from being taken in his sleep by the fairies (J G Campbell). Or, according to Mackenzie. (1895), it could be very lucky, for prosperity and success would follow in its train (so it was thought in the Isle of Man, too, for it is known there as Luckherb (Moore, Morrison & Goodwin), and recognised as a bringer of good luck). It was especially prized when found in the flocks' fold, for this would augur peace and prosperity to the herds throughout the year (Carmichael).

Urostachys selago > MOUNTAIN CLUBMOSS

Urtica dioica > NETTLE

Utricularia vulgaris > BLADDERWORT

V

Vaccinium myrtillus > WHORTLEBERRY.

VALERIAN

(*Valeriana officinalis*) This is one of the many plants once thought to be an aphrodisiac (Haining), and it is said that Welsh girls used to hide a piece of it in their girdles, or inside their bodices, in order to hold a man's attention (Trevelyan). It is sewn into children's clothes in the Balkans, though for quite a different reason. There it is an amulet to ward off witches (Vukanovic).

Cats are said to be very fond of it. Topsell reported in 1607 that "the root of the herb Valerian is very like to the eye of a cat and wheresoever it groweth, if cats come thereto, they instantly dig it up for the love thereof, as I myself have seen in mine own garden, for it smelleth moreover like a cat". They certainly love it, the root seeming to act on them as an intoxicant as they roll over and over on the plant. There is no evidence that there is any medicinal reason for their behaviour (M Baker. 1980), unlike Catmint (*Nepeta cataria*). However disagreeable to modern taste, the smell used to be held in quite high esteem, for the root was put among clothes as a perfume in the 16th century (Fluckiger & Hanbury) (hence perhaps the name English Orris), and it is still used as a perfume in the East (Lloyd). Rats like it too, and it was much used for baiting traps (C P Johnson). In fact, it has been suggested that the secret of the Pied Piper's success in rat-charming lay more in the valerian in his pockets than in his music (M Baker. 1980).

Normally the root is the only part used medicinally, but there is one record from Ireland when a decoction of the flowers was taken for consumption (Maloney). The root should be gathered in September or October, to be dried in the shade (Palaiseul). When boiled, it makes a strong nervine that was famous for promoting sleep – it is still one of Europe's principal sedatives (Emboden. 1979). Oil of valerian has a depressant effect on the central nervous system, but it is usually taken as a tea, as a calming medicine, to prevent hysteria (Grigson. 1955), or to treat nervous diseases "peculiar to females" (Lloyd).

In such high esteem was valerian held that in the Middle Ages it was even reckoned to be efficacious against the plague itself (Trevelyan). As Gerard said, "the dry root … is put into counterpoysons and medicines preservative against the pestilence …". It was used for many slighter ailments, too. Thomas Hill, in 1577, listed some of them: "valerian provoketh sweat, and urine, amendeth stitches, killeth mice, moveth the termes, prevaileth against the plague, helpeth the straightnesse of breath, the headache, fluxes, and Shingles, procureth clearnesse of sight and healeth the piles". Gerard was able to add more, such as cuts

and wounds, jaundice, cramp, and so on. No wonder it was called 'All-heal' in the west of England, and 'Guérit-tout' in France. It was called Herb Bennett, too (Macmillan), Bennett being 'benedicta', herba benedicta, 'blessed herb' being the the name given to any plant with a reputation as a counter-poison.

Valeriana edulis > EDIBLE VALERIAN

Valeriana officinalis > VALERIAN

VAMPIRES

GARLIC tied in bundles over a house door will keep out a vampire, and stuffed in the mouth of a corpse will keep the vampire (if there were suspicions that the deceased might be one) quiet in his grave (Gifford). These are Balkan superstitions, and there were Roumanian beliefs that vampires left their graves on St Andrew's Eve, and walked about the houses in which they used to live. So before nightfall every woman took some garlic and anointed the door locks and window casements with it (Miles). In Hungarian folklore, a vampire's grave could be recognised by two holes on the tombstone. Stop these holes with garlic and you could make sure the vampire would stay there.

The correct wood to be driven through a vampire's heart is ASH, though hawthorn and rowan are also mentioned. In the Balkans, it was a BLACKTHORN stake that impaled a vampire (Kemp).

VARICOSE VEINS

According to Somerset practice, an ointment made from MALLOW or MARSH MALLOW roots and unsalted lard was used to treat the condition, equally useful for sore feet, too (Tongue. 1965; Page 1978), and bathing the feet in which the flowers of MARIGOLD had been infused, was another Somerset way of dealing with the problem (Tongue. 1965). At one time the roots of BRISTLY OX-TONGUE were used, bound to the affected part (see Tusser). CYPRESS cones were also used (Palaiseul).

VEGETABLE MARROW

(*Cucurbita pepo*) Marrow seeds, in the form of an infusion as well as in a pulp, have for long been a domestic remedy for internal parasites (Lloyd). As with pumpkin seeds, they are an efficient diuretic. The American Indians knew this perfectly well, too – the Menomini, for instance, used the pulverized seeds for any kind of urinary complaint (Corlett). A decoction of the flowers is given in Trinidad as a measles remedy (L M Perry).

VENETIAN SUMACH

(*Cotinus coggyria*) Probably better known in Britain as Smoke Tree, a good descriptive name. This plant provides a dyestuff, so it is sometimes referred to as Young Fustic, "young", to distinguish it from Old Fustic, which is *Chlorophora tinctoria*. It is the wood, reddish-orange in colour, that is the source of a yellow dyestuff, hardly ever used now, as the dye is hardly permanent at all (Leggett). However, with the proper mordant, it can dye cotton and wool bright yellow through to brown or dark olive. With logwood, it can produce black.

Veratrum album > WHITE HELLEBORE

Veratrum viride > AMERICAN WHITE HELLEBORE

Verbascum thapsus > MULLEIN

Verbena officinalis > VERVAIN

VERJUICE

A very sour liquid extracted from crabapples, quite popular in the English countryside in the 19th century, but used for veterinary practice long before that. Tusser, for instance, advised the husbandman:

> Of vergis be sure
> Poore cattel to cure.

It is described in Tusser Redivivus as being "for Strength and Flavour … little short, if not exceeds Lime juice". It was also used for curdling milk, and for treating sprains (tart, vinegary cider was still being used as a fomentation for muscular sprains in the first half of the 20th century (Savage).

VERMIN CONTROL

Crushed roots of WHITE BRYONY used to be stuffed into rat holes in barns, to drive the vermin away (Porter. 1969). The bulb of SQUILL is effective as a rat poison because the vermin like it and will readily eat it (Kingsbury. 1967). As Red Squill Powder, it has long been used in America as a rat poison. They like VALERIAN, too, and it was much used for baiting traps (C P Johnson). In fact, it has been suggested that the secret of the Pied Piper's success in rat-charming lay more to the valerian in his pockets than to his music (M Baker. 1980). Cats may love CATMINT, but rats hate it. Plant it thickly around the walls of a rat-infested house, and they will soon be cleared (Quelch). ANISE is said to be a good mice bait, if smeared in traps; it will destroy lice, too. In Lapland, branches of LABRADOR TEA are put among grain to discourage mice (Grieve. 1931), just as in Ireland, sprigs of CORN MINT used to be put in corn stacks to keep mice away (E E Evans. 1942) (so was SPEARMINT in America (Fogel)). It was said in Yorkshire that a bunch of QUAKING GRASS hung on the mantelpiece will ward off mice (Drury. 1992). Loudon said that a decoction of SPINDLE TREE leaves was used to wash dogs free from vermin. This has always been the accepted reason for the name DOGWOOD (*Cornus sanguinea*), a name, incidentally, given to Spindle as well, but Hart says the name derives from the traditional use of both the shrubs for skewers – a "dog" means a sharp spike. But the pesticide derivation persists, and, it seems, small boys' nits were treated, as well as dogs' coats

(Grigson. 1955). As Evelyn said, "… the powder made of the berry, being bak'd, kills nits, and cures scurfy heads". Just to prove the point, there is the name louseberry, and the tree is the Louseberry-tree (Fernie).

BROOM juice mixed with oil of RADISHES is an old recipe for killing lice and other pests (Drury. 1992), and the powdered rhizome of WHITE HELLEBORE is highly toxic to fleas and lice (Flück). In parts of France, it is said that the flowers of MEADOW SAFFRON crushed, and put on the heads of children who had a lot of hair, would destroy the vermin that could not be reached by normal combing (Sebillot). PENNYROYAL is a corruption of Puliol Royal (Latin pulices, fleas), for this is a good plant to use against them. The "royal" part of the name, so it said, shows that royal palaces were not immune from the vermin (Genders. 1971). It is said that fleas will not come into a room where the herb WATER PEPPER is kept (Fernie).

In Normandy, it was once said that fleas could be acquired by the ill-wishing of a witch. The only way to break the spell was to go down to the river before sunrise, and to beat one's shirt for an hour with a branch of BLACKTHORN (W B Johnson).

Strewn among clothes, the leaves of LABRADOR TEA will keep moths away (Grieve. 1931); so will the dried leaves of TUTSAN (Genders. 1976). CYPRESS wood is virtually insect-proof, and "the shavings of the wood laid among garments preserve them from the moths: the rosin killeth Moths, little Wormes, and magots" (Gerard).

Vernonia noveboracensis > IRONWEED

Veronica beccabunga > BROOKLIME

Veronica chamaedrys > GERMANDER SPEEDWELL

VERTIGO

In some parts of Germany, to cure dizziness, it was recommended that the patient should run, naked, after sunset, through a field of FLAX; the flax will take the dizziness to itself. (Dyer. 1889). The seeds of MELEGUETA PEPPER, with a leaf of Sweet basil, ground together and cooked in palm oil, form a Yoruba (Nigeria) medicine to cure giddiness (Verger). The juice of GREEN PURSLANE is recommended in Mayan medical texts for giddiness (Roys).

VERVAIN

(*Verbena officinalis*) This is a holy herb, or to be more accurate, *the* Holy Herb. The Romans gave the name verbena, or more frequently, the plural form, verbenae, to the foliage or branches of shrubs and herbs which, for their religious association, had acquired a sacred character. These included laurel, olive and myrtle, but Pliny makes us think that the herb now known as verbena was regarded as the most sacred of all of them (Browning). He said it was gathered at sunrise after a sacrifice to the earth as an expiation. When it was rubbed on the body, all wishes would be gratified. It dispelled fevers and other maladies. And it was an antidote against serpents (MacCulloch. 1911). The Greeks also looked on it as particularly sacred (Friend. 1883), as it was in Persian belief (Clair), while it was called "the tears of Isis" by the priest physicians of Egypt (Maddox).

In Rome, it was carried as a symbol of inviolability by the state envoys when dealing with an enemy, but that did not stop both the Greeks and Romans dedicating it to the god of war (C J S Thompson. 1897), and it was sacred to Thor in Scandinavia too. Leland. 1898 says it was a plant of Venus. In other words, it was used as an aphrodisiac, or some kind of love philtre (Folkard). As such it was planted at the door step in the southern states of America to attract lovers (M Baker. 1977). Pillows stuffed with verbena were recommended for their strong aphrodisiac scent (Boland. 1977); perhaps just a sprig in a pillow would do. Piedmontese belief had it that if young men gathered it on Midsummer Eve, any girls they shook hands with would fall in love with them (Canziani. 1913). In the Fen country, courting couples used to exchange vervain leaves to keep carefully in their Bibles. If the leaves kept green, then the love of both was true, but if they turned brown, it was a sign that one of them was false (Porter. 1969. The old rhyme was:

A verbena leaf sent to a lover
Carries a message; you need no other.

Vervain was one of the ingredients, in Celtic mythology, of Ceridwen's cauldron. It was usually gathered, we are told, at the rise of the Dog-star, "without being looked upon either by the sun or the moon" (Spence. 1945), and with the usual expiatory sacrifices of fruit and honey made to the earth (Wilde. 1890). According to old Irish belief, vervain was one of the seven herbs that nothing natural nor supernatural could injure; the others were yarrow, St John's Wort, eyebright, speedwell, mallow and self-heal (Wilde. 1902). Naturally, with such a background, vervain was taken to be a great protector, either of the home (plant it on the roof and it will guard the house against lightning (Sebillot)), or of the person. Even in ancient times, it served in the purification of houses (Browning), and it was a Welsh custom to cut it, in the dark, to bring into a church, there to be used as a sprinkler of holy water (Clair). At one time in the Isle of Man, neither the mother nor a newborn baby were let out of the house before christening day, and then both had a piece of vervain sewn into their underclothes for protection (Gill. 1963). In Sussex the practice was to dry the leaves and put them in a black silk bag, to be worn round the neck of sickly children (Latham), probably rather to avert witchcraft than to effect a cure, and it was sewn

into children's clothing to keep fairies away. Adults could be protected from fairies and their spells by brewing a tea made from it, and drinking that (Spence. 1949). Welsh tradition, too, recognised its value as an amulet. There they dried and powdered the roots, to be worn in a sachet round the neck (Trevelyan).

This belief in the extraordinary powers of vervain goes back a long way. The Romans, for instance, hung it in their houses to ward off evil spirits (C J S Thompson. 1897). Gerard tells us that "the Devil did reveal it as a sacred and divine medicine", and there are various versions of the couplet that Aubrey quotes:

Vervain and dill
Hinders witches from their will (Aubrey. 1696).

The expanded version runs:

Trefoil, vervain, John's wort, dill,
Hinders witches of their will (Gutch. 1901.

This coupling of vervain and St John's Wort occurs in a charm that Aubrey quoted, against "an evil tongue" – "take unguent … and vervain and hypericon, and put a red-hot iron into it. You must anoint the backbone, or wear it on your breast". Another recipe to see spirits, or, put another way, to bestow second sight, was to anoint the eyes for three days with the combined juices of dill, vervain and St John's Wort (Hewett). There is also the well-known and oft quoted rhyme:

Fennel, rose, vervain, celandine and rue
Do water make which will the sight renew.

Like many another healing plant, there was a special time for gathering it, the "Sun being in the Sign of the Ram", according to Albertus Magnus, when, "put with grain or corm of Peony of one year old", it would have the power of healing "them that be sick of the falling sickness". Another example for vervain is:

Between mydde Marche and mydde Aprille
And yet awysyd muste ye be
That the sonne be in arrete (I B Jones).

It was well known that a witch hare could not be shot, unless a bent silver coin or something like that were used as a bullet. One piece of advice from Bedd Gelert tells the man with the gun to put a small piece of rowan and one of vervain under the stock (D E Jenkins). Vervain's protective powers can be seen on a lower level, too. Manxmen, for example, would never start a journey, or any other enterprise, without a sprig of vervain (Killip), and there is another illuminating record: "… a couple of years ago a young singer at a Manx musical-guild competition held a leaf of it in her hand while singing, and won first prize (Gill. 1932).

Besides being used as a protector from supernatural practices, vervain, in true homeopathic style, was used by the witches themselves for their own ends, if only as a protection for the practitioner. The Great Grimoire makes this quite clear in describing a protective device to be used in the working of black magic – "… two vervain crowns and left sides of the triangle within the circle …". It was used, too, in a tiara (with cypress) to be worn on such an occasion (Haining). According to D B Wyndham Lewis, in his *Gilles de Rais*, vervain was bound with garlands round the special sword used by sorcerers for drawing on the ground the magic circle within which they must stand for their own safety. The Physicians of Myddfai took up this theme. "If one goes to battle let him seek the vervain, and keep it in his clothes, and he will escape from his enemies". It was used in the preparation of the Hand of Glory: "wrap the hand in a piece of a winding sheet, drawing it tight so as to squeeze out the little blood that might remain. Then place it in an earthenware vessel with saltpetre, salt and pepper, all well dried and carefully powdered. Let it remain a fortnight in the pickle and then expose it in the sun in the dog days, till completely parched, or dry it in an oven heated with vervain and fern" (Radford & Radford). One of the charges against the witches was that they went invisible by night, and vervain was thought to confer that invisibility. A belief from Pliny's time tells that by smearing his body all over with the juice of this plant, the operator could have whatever he wished, be able to reconcile his greatest enemies, cure diseases, and perform any other magical feat (C J S Thompson. 1897). After all this it comes as no surprise that it was believed to have the power of opening locks. Indeed, it was the very symbol of enchantment (Ingram).

It was used in sex magic, too, and appeared in most witch philtres. "Place vervain in thy mouth, and kiss any maid saying these words, "Pax tibi sum sensum conterit in amore me" and she shall love thee (Haining). Katharine Briggs (Briggs. 1962) quoted from a 1662 ms: "ffavour to have. Gather vervain on midsummer even ffastinge and out of deadlye sime with 3 paternosters 1 Aves and 2 Credo and beare it about thee". Again, "Rubbe vervain in the ball of thy hand and rubbe thy mouth with it and immediately kysse her and it is done"

There are still a few minor superstitions connected with the plant. One is that it should be bought or stolen. If it is offered as a gift, it can be accepted after refusing it twice (*Notes and Queries. vol 67; 1941*). Another involves the use of oil of vervain; Fenland belief said that if the oil was put in mid-stream and allowed to float down river, it would attract large numbers of eels, and by so doing mark the spot where a drowned body lay. Fenmen were fond of using this oil as bait, often by steeping the worms in it; wildfowl hunters often baited their snares and traps with the crushed leaves (Porter. 1969).

Medicinal uses are many and varied, though most of them should rather be described as magico-medicinal. Take Gerard, for example: "It is reported to be of sin-

gular force against the Tertian and Quartane fevers; but you must observe mother Bombie's rules, to take just so many knots or sprigs, and no more, lest it fall out so that it do you no good, if you catche no harm by it". Then there are the various recipes for dealing with scrofula, or the King's Evil, as it was termed. "To cure the King's Evil, bake a toad, and when dried sufficiently to roll into a powder, beat it up in a stone mortar, and mix with powdered vervain. Sew in a black silk bag and wear round the neck" (Moloney). This idea of putting vervain (with or without the toad) in a bag and wearing it was in favour for a long time. It even appeared in a 19th century supplement to the London Pharmacopeia as a scrofula cure – "necklaces of vervain roots , tied with a yard of white satin ribbon" (Leyel. 1926). It also appeared in a 17th century manuscript from Jersey, where the magic was continued by burning the rest of the root and hanging the leaf up the chimney. As the leaf dried, so would the disease dry up (Le-Bas). Brand's editor published similar practices he had been told about – "Squire Morley of Essex used to say a prayer which he hoped would do no harm when he hung a bit of Vervain-root from a scrophulous person's neck. My Aunt Freeman had a very high opinion of a baked Toad in a silk bag, hung round the neck". Similar practices were advised for many other complaints, ranging from snakebite to headaches, and including wounds, stone, dropsy, "bleared eyes", childbirth problems, suppressed lactation, and so on. There is one early prescription: "to prevent dreams, take the vervain, and hang it about a man's neck, or give him the juice on going to bed, and it will prevent his dreaming" (Physicians of Myddfai). Before dismissing it as fantasy, one should bear in mind that vervain tea is a sedative.

VETERINARY USES OF PLANTS
GARLIC – an Irish method of treating black leg in cattle is to make an incision in the skin and put in a clove of garlic. The wound is then stitched, leaving the garlic inside. Patrick Logan could think of no reason why this should have any effect, so perhaps the only reason for the garlic is to drive away the evil spirits that caused the disease, for in folklore garlic is the prime agent for combatting evil influences. ASH-Devon farmers believed that feeding infected cattle with ash leaves was a cure for foot and mouth disease, and BLUE COMFREY (*Symphytum caucasicum*), or PRICKLY COMFREY (*Symphytum asperum*) was also believed to help prevent the disease, as well as acting as a cure for it before wholesale slaughter was enforced by law (Macleod; Quelch). Foot rot in cattle was cured in Worcestershire by cutting a sod of turf from the spot on which the animal was seen to tread with its bad foot, and then hanging the turf on a blackthorn. As the sod dried out, so would the hoof heal (Drury. 1985). BRACKEN has its uses in this connection; powder made from it apparently cured the galled necks of oxen (Tynan & Maitland), and

it was used on South Uist for something called the dry disease in cows. The roots would be boiled, and the juice given to the cow to drink. They often used TANSY there to treat red water in cows, by boiling the entire plant, putting the juice in a bottle, then pouring it down the cow's throat (Shaw). Bracken has always been said to make the best litter for horses and cattle, and it was cut in huge quantities for the purpose. After a cow had calved, country people used to feed it with SANICLE leaves, to promote the expulsion of the afterbirth, and to stop any bleeding (Drury. 1985). Sanicle is a great wound herb, used for centuries as such. When cattle coughed, they used to be treated by the use of STINKING HELLEBORE. It was done by making a hole in the dewlap with a setter, or thread (hence the name Setter-grass, or Setterwort, given to the plant), and a length of hellebore root inserted to irritate the flesh and keep it running (Grigson. 1955). Or sometimes a few rolled-up leaves were used (Hartley & Ingilby). It was the way, too, to treat a pig that had been bewitched, by making a hole in its ear and putting a piece of stem there (Lovett). A poultice of the roots of LESSER PERIWINKLE would be applied to a cow's udder to cure milk fever, at least in Cambridgeshire (Porter. 1969).

RAGWORT is a problem. It is well-known as being severely toxic to animals (Forsyth), causing cirrhosis of the liver, from which the animal cannot recover. The trouble is that animals will not usually eat it, so it flowers and seeds undisturbed, and the effect is eventually to produce more ragwort than grass in a pasture. The condition is known in Canada as the Pictou cattle disease, and the scientific name is seneciosis. The problem is that the plant seems to have been used as a curing agent; this is St James's Wort, and St James is the patron saint of horses. The use of Ragwort in veterinary practice seems to be confirmed by other names, such as Staggerwort, that is, the herb that cures staggers in horses. Sir Edward Salisbury, for one, stated clearly that staggers was actually caused by Ragwort. If both views are correct, here is an example of homeopathic magic at work, of like curing like. East Anglian horsemen favoured the use of FEVERFEW on their charges. A way to control unruly horses was to rub freshly gathered leaves (or rue) on their noses (Porter. 1969), and they used it for curing colds, and for giving their horses an appetite (G E Evans. 1960), just as TANSY would make their coats shine, by sprinkling a little of the dried, powdered leaves now and then into their feed (G E Evans. 1960). The same source reported that Suffolk horsemen used the herb FENUGREEK as a horse medicine, to give them an appetite. They called it Finnigig, which Evans suggested was a deliberate corruption on the part of the horsemen, so that third parties would not be able to recognise the true identity of what they were buying. The HELLEBORES were used for the same purpose. ROWAN berries were fed to pregnant

mares to ensure an easy birth, and, one may be quite sure, to protect the foal. 17th century Yorkshire shepherds used tansy, finely chopped and mixed with fresh butter, to heal the wound on castrated lambs. As the butter healed the wound, the tansy would keep flies away (Drury. 1985). BOX leaves were at one time fed to horses to cure them of bots, and oil of JUNIPER was put into a drench for the same purpose (Drury. 1985). MARSH MALLOW ointment, made from the crushed roots, besides being used for a number of human conditions, was used in horse doctoring, too, for sores and sprains (Boase). East Anglian horsemen used marsh mallow to cure a horse "with a pricked foot" (G E Evans. 1969).

In some parts of Ireland, GREATER SPEARWORT (Foster), or better, COMFREY is used to treat swine fever, by boiling the roots in milk and adding everything, roots and all, to the pig's feed. This has to be kept up for some weeks (Logan). But Norfolk pigkeepers added comfrey leaves to the pig's feed to keep them in good health. That was not the whole story, though – the comfrey feed had another function, that of ensuring that the pig could not be bewitched (Randell). SUGAR BEET leaves have found a use in treating erysipelas in pigs. When they are fed to them, the blisters disappear and the pigs improve quickly (V G Hatfield. 1994).

Irish horse handlers used HAZEL in the breastband of the harness, to keep the horse from harm (Ô Súilleabháin); in much the same way, Somerset drovers always used a hazel stick to drive cattle and horses, though in most places ROWAN was preferred. For a horse that had over-eaten, the remedy was to bind its legs and feet with hazel twigs to relieve the discomfort (Drury. 1985). Another purely magical use was recorded in Wales as a charm: "if calves were scoured over much, and in danger of dying, a hazel twig the length of the calf was twisted round the neck like a collar, and it was supposed to cure them" (Owen). The very soil from under a hazel bush was valuable. In Yorkshire it was given to cows that had lost their cud (Hartley & Ingilby, and recorded as late as 1966). GREATER SPEARWORT was used in Ireland to cure farcy. It was stewed with garlic (Foster), and there is a Yorkshire record of boiling RUE in ale to give to horses to cure farcy, or glanders. A little rue juice would be put in the horse's ears, too (Gutch. 1911). Horsemen in the Fen country would use it to control unruly horses, by rubbing it, freshly gathered, on its nose (Porter. 1969), and a sprig or two given to horses would make them well, and their coats shine (Randell). Norfolk turkey breeders used rue to make the birds eat, and put on weight, while a leaf is given to poultry to help in curing croup (Brownlow). This is actually a 17th century usage from Lincolnshire and other areas, when the owner was told to chop the herb very finely and form it into piles with butter, and so feed it to the sick hens (Drury. 1985). The Pennsylvania Germans made a ball of ELDER bark, and pushed it down a cow's throat when it had indigestion (Dorson). Gypsies use the leaves to treat a horse's leg – they soak the young shoots from the tips of the leaves in hot water, and bandage them round the lame leg (Boswell). In Ireland, too, the water in which elder leaves had been boiled was used to dose pigs. One way to treat a horse that cannot urinate is to strike it gently with an elder stick, and to bind some leaves to its belly. Lameness in pigs used to be treated by boring a small hole in its ear and putting in a plug of elder wood. As the plug withered or fell out, the animal would be cured (Drury. 1985). A similar usage was to cure coughs in cattle by putting a piece of OX-EYE DAISY root in a hole made in the cow's ear or dewlap (Drury. 1975). Manx vets still use ALEXANDERS as a treatment for animals with sore mouths (Garrad). Martin, in his account of the Western Isles, reported that horses were wormed with WILD SAGE, and "a quantity chewed between one's teeth, and put into the ears of cows and sheep that become blind, cures them, and perfectly restores their sight, of which there are many fresh instances both in Skye and Harris, by persons of great integrity". ELECAMPANE has been a famous medicinal plant in its day. Nowadays, though, most of the usages are for veterinary medicine. The various 'Horseheal' names given to this plant are witness to that. It is used in America for horses' throat ailments (Leighton), and in Britain for skin diseases in horses and mules, as well as for scab in sheep (Wiltshire). The leaves, too, are fed to horses to improve their appetite, and BLACK BRYONY root was put in feed to bring up the gloss on their coats. But they believed it had supernatural powers as well - the association with Mandrake (see **WHITE BRYONY**) was evident here, for they said it had aphrodisiac qualities for both man and horse (G E Evans. 1960). An ointment made from THORN-APPLE is used to rub on the fetlocks of horses for "scratches" (Ohio and Illinois – see Bergen. 1899). Pieces of it, rubbed or bound on the sores, used to be an English cure for galled horses, particularly in Littleport, Cambridgeshire (Porter. 1969). Irish people used to treat horses' blistered feet (as long as the blisters were not too severe) with melted goose grease to which turpentine and the juice of HOUSELEEK had been added (P Logan).

BRAMBLE leaves were used in Ireland to cure scour in cattle, just as they have been used, because of their high tannin content, to combat diarrhoea and dysentery in humans. WHITE BRYONY root was given to mares as an aid to conception, but only because this was the English Mandrake (see **MANDRAKE**). They were also given to horses in their feed, to make them look sleek. It was certainly used in East Anglia for that purpose, and the chopped leaves, gathered before

the flowers appeared, were also used (Porter. 1969). But it was always known to be a dangerous practice. See the rhyme:

> Bryony if served too dry,
> Blinded horses when they blew (G E Evans. 1966).

During the disastrous foot and mouth disease outbreak in Britain in 1968, on one Cheshire farm which escaped, although in the midst of the infection, the farmer's wife had laid rows of ONIONS along all the windowsills and doorways of the cowsheds. The farm's escape was attributed to this precaution (M Baker. 1980). Apparently it was standard practice in Yorkshire to hang four or five onions round a distempered cow's neck. A few days of this, and the cow's nose would run, and so the disease would be cured. The onions had to be buried deep after removal (Gutch. 1911). Is foot-and-mouth disease the modern term for murrain? BLACK BRYONY bears the name Murrain-berry (Britten. 1880), or Murren-berry (W H Long), surely indicating that some part of the plant was used to contain the disease.

Aubrey noted the use of BROOM in Hampshire and Wiltshire to prevent rot in sheep (Aubrey. 1847). He knew of "carefull husbandmen" who cleared their land of broom, "and afterwards their sheep died of the rott, from which they were free before the broom was cutt down". So then they made sure of leaving some plants of it round the edges of their land just for the sheep to browse on, "to keep them sound". An ointment made from SOLOMON'S SEAL was used for treating ulcers and wounds in horses and cattle (V G Hatfield. 1994), and East Anglian horse handlers used the juice of BISTORT to rub round horses' teeth to prevent decay (A W Hatfield).

Viburnum lantata > WAYFARING TREE

Viburnum opulus > GUELDER ROSE

Viburnum prunifolium > BLACK HAW

Vicia faba > BROAD BEAN

Vigna unguiculata > COWPEA

Vinca major/minor > PERIWINKLE

Viola odorata > VIOLET

Viola tricolor > PANSY

VIOLET,
or SWEET VIOLET (*Viola odorata*) Posies of violets was a fashion actually set by Queen Victoria – over 4000 plants, it is said, were grown under frame at Windsor, to provide the Court with posies for evening wear (Genders. 1971). A bunch of violets is the traditional Mothering Sunday gift (Opie & Opie. 1959). Their scent is fleeting. When first coming upon them, the fragrance is obvious, but it soom seems to go. "To smell the smell out of violets" is a proverbial saying, and there is a factual basis for it, for the

fragrance contains ionine, which has a soporific effect upon the sense of smell. This effect was recognised, but misunderstood, when the gift of sleep was ascribed to the violet. One 16th century herbalist (Ascham) said, "for them that may not sleep for a sickness seethe violets in water and at even let him soke well hys temples, and he shall sleepe well by the grace of God". Violets were once used as strewing flowers (Genders. 1971), but the scent is not wholly approved. Towards the end of the 19th century a French scientist named de Parville said that their scent had a harmful effect on the voice, and there is a record that the singer Marie Sass could take no part in a concert after smelling a bunch of Parma violets given to her. Singing teachers still forbid the use of perfume made from them, or flowers that have a similar aroma, like mimosa (Genders. 1972). It is even said that a lot of violets in a room can cause convulsions (Pratt. 1913), and some say that a bunch of violets will attract fleas or other vermin into a house (Vickery. 1993).

The ionine contained in the smell reminds us that ion is the Greek for violet, and that the legend of its origin is, in Lyte's words, "after the name of that sweete girle or pleasant damoselle Io, which Jupiter turned into a trim Heyfer or gallant Cow, because that his wife Iuno (being both an angry or jealous Goddesse) should not suspect that he loved Io. In the honour of which his Io, as also for her more delicate and wholesome feeding, the earth at the commandment of Iupiter brought forth Violets, the which, after the name of well-beloved Io, he called in Greeke Ion". Gerard had this story, too, but another legend says that the Greeks adopted the name Ion after certain nymphs in Ionia had made an offering of the flowers to Jupiter (Browning). A quite different origin myth is that it sprang from the blood of Attis when Cybele changed him into a pine tree.

In the language of flowers, violet's association was with death (Webster), possibly because it was the colour of mourning (Haig), or even perhaps because of its use as a soporific. But it is really thought of as the symbol of humility. As such it would be the emblem of Christ on earth. For the same reason, it was given also as the emblem of confessors (Haig). In addition, it has been taken as a symbol of constancy (Dyer. 1889), and in France, "de la modestie, de la pudeur et de l'innocence" (Rambosson). In Britain, it is the white violet that is the symbol of innocence (Friend. 1883). So highly was it regarded by the ancient Athenians (it was actually in commercial cultivation there for its sweetening properties) that they made it the emblem of their city (Genders. 1971). Closer to our own times, there is yet another association of the violet – that with the Bonaparte dynasty. When Napoleon left France for Elba, he said he would return in the violet season, and violet (both the flower and the colour) became a secret emblem of confederates sympathetic to him. When he escaped from Elba,

his friends greeted him with violets. He is referred to as "le père de la violette" in a French soldiers' song:

> Chantons le père de la violette,
> Au bruit de sons et de canons.

Byron also uses the imagery of the violet in his *Napoleon's farewell to France*:

> The violet grows in the depths of thy valleys,
> Though wither'd, thy tears will unfold it again.

Dreaming of violets means "advancement in life" (Dyer. 1889). Another superstition connected with them is that when they and roses bloom in autumn, there will be an epidemic the following year (Dyer. 1889).

In medieval times, violets were grown as a salad herb. It was the flowers that were eaten, raw, with onions and lettuce. Byrne says that the buds were still eaten as a salad in Elizabethan times. They could be cooked, too, with meat and game (Genders. 1971). Wine could be made with them – very popular with the Romans, it seems (Hemphill), and apparently vinegar was also made with them: "Vinegar acquires a very agreeable colour and taste by infusing in it some petals of this odoriferous flower" (Thornton). But it was probably as a confection that it achieved its greatest popularity. Gerard says "there is ... made of Violets and Sugar certaine plates called Sugar violet, Violet tables, or Plate, which is most pleasant and wholesome, especially it comforteth the heart and the inward parts", sold especially in the 17th century as a remedy for weak lungs (Pratt. 1913).

Violets were used quite extensively in medicine, usually the leaves, but the roots (and the seeds) were said to be purgative, and country people use them still as such. Gypsies make a poultice of the leaves steeped in boiling water, for cancerous growths. An infusion of the leaves, they say, will help internal cancers (Vesey-Fitzgerald). This is also found in Welsh folklore (Trevelyan), and as a Dorset herbal cure (Dacombe). It is often used, too, in Russian folk medicine (Kourennoff). Violet buds eaten in salads were said to be taken for the same purpose (Tongue. 1965). The leaf plaster was used for ulcers too, or boils – there is a recipe from the 15th century for "hot botches", which are described as inflamed boils: "Take violet, and stamp it with honey and vinegar, and make thereof a plaster; and anoint the head [of the botch] in the beginning of its growing with the juice of violet, and then lay on the plaster" (Dawson. 1934). Herbalists are still recommending this leaf plaster to help heal any wound (Flück). People in Dorset use the leaves to put on stings (Dacombe), and the leaves are often applied to bruises (Pratt. 1913). One of the oddest medical uses of violets must be this one, taken from the Book of Iago ab Dewi – "to ascertain the fate of a sick person, bruise violets and apply them to the eyebrows; if he sleep, he will live, but if not he will die" (Berdoe).

VIPER'S GRASS

(*Scorzonera humilis*) The common name shows that this would be a herb with which to treat snakebite. See Gerard, for example: "It is reported by those of great judgement, that Viper's-grasse is most excellent against the infections of the plague, and all poysons of venomous beasts, and especially to cure the bitings of vipers, if the juyce or herbe be drunke…".

VIRGINIAN JUNIPER

(*Juniperus virginiana*) Sometimes known as Pencil Juniper or Pencil Cedar – no other wood has been found that has just the right physical properties for the casing of lead pencils (Harper). But by the end of World War II, it had become extremely scarce, so it had to be replaced for pencil wood by Red Cedar (*Calocedrus decurrens*). (Lewington). Clothes chests are made of it, too, for the smell of the wood repels moths. Smoking crushed juniper berries is an American domestic medicine for catarrh (H M Hyatt), and earlier, Indian peoples had used it for a variety of ailments. Both leaves and berries boiled together were taken for coughs. Twigs were burned and the smoke inhaled for a cold in the head (Gilmore). The Kiowa chewed the berries as a remedy for canker sores in the mouth (Vestal & Schultes), while the Natchez used it in some way for mumps (Weiner).

VIRGINIAN SKULLCAP

(*Scutellaria laterifolia*) Cherokee women used to drink the herb infusion to cure suppressed menstruation, and there is a recipe for St Vitus's Dance from Alabama: "one ounce skullcap, one ounce feverweed (*Verbena syriaca*, perhaps?), one ounce Lady's Slipper (*Cypripedium*). Take half of each one, put in a quart jar filled with boiling water, and seal. Let it stand for two hours, then take a wineglassful three times a day" (R B Browne). This plant has been called Mad-dog, or Mad-dog Herb (House; Lloyd), because it was used to treat rabies, after a Dr van der Veer experimented with it in 1772 (Weiner).

VIRGINIAN SNAKEROOT

(*Aristolochia serpentaria*) An American plant of the same genus as Birthwort. The early use was as a remedy for snakebite (LLoyd), which is pure doctrine of signatures, though the native American had never heard of such a theory. The point is that the roots are writhed, like snakes, and that accounts for both the common and specific names. In general, the Indians simply chewed the root and applied it, or spat it, on the bite (Weiner), and chewing the leaves is still an Indiana way of treating it (Tyler). Some groups also blew the root decoction on to fever patients (Coffey). That same preparation was taken for coughs (Corlett), and the root was also used to put in a hollow tooth to cure toothache (Coffey). It was the root, powdered, that was used by slaves in America to combat pneumonia (Laguerre). It is a well-known emmenagogue, especially in the form of a tincture known as Hiera Pina (Hikey Pikey in East Anglia), of aloes, snakeroot and ginger (V G Hatfield).

Viscum album > MISTLETOE

Vitex agnus-castus > CHASTE TREE

Vitis vinifera > GRAPE VINE

W

WAHOO

(*Euonymus atropurpureus*) An American species whose name comes from the Dakota word wan-hu. It has a digitalin-like action on the heart, and it became a popular heart medicine in American domestic medicine (Weiner). The Indians had already used it for other medicinal purposes – Winnebago women, for instance, used to drink a decoction made from the inner bark for uterine troubles (Gilmore), and the Meskwaki, whose name for the shrub means "weak-eye tree", used it for just that. The inner bark is steeped, to make a solution with which to bathe the eyes, and a tea was made from the root bark for the same purpose (H H Smith. 1928).

WALL PENNYWORT

(*Umbilicus rupestris*) Locally common in the west of England, but not nearly so elsewhere, and certainly not where Gerard found it: "upon Westminster Abbey, over the doors that leadeth from Chaucers tombe to the old palace". Put a piece of Wall Pennywort under your pillow, they used to say, and you will dream of your love (Tynan & Maitland), probably because this plant used to be known as Venus's Navelwort (*umbilicus-veneris*), the reference being to the navel-like dimples in the middle of the leaves, above the stalk. On that subject, there is an odd legend about Anne Boleyn from County Clare: she was supposed to have enjoyed influence over Henry VIII by means of the Pennywort. When she "was sent to jail she couldn't get the plant, and they hanged her" (Westropp. 1911). The only other piece of folklore attached to the plant comes from Dorset, where they say that wearing a piece of it keeps you from harm (Dacombe). There are, though, a number of examples of folk medical usage (but it will only cure Protestants, they say in County Clare (Westropp. 1911)), at least one of which relies on the doctrine of signatures, for the name Hipwort, reckoned to be given from the resemblance of the leaf to the hip socket, advertises the use Coles ascribed to it: "… it easeth the pain of the hippes".

It is used in Cornwall to treat spots and pimples, and there is a more widespread use in treating chilblains (Tynan & Maitland), a practice that is as "good for kibed heels, being bathed therewith, and one or more of the leaves laid upon the heele" (Gerard). Gerard also recommended the juice as a "singular remedy against all inflammation and hot tumors, as Erysipelas, Saint Anthonies fire and such like". The juice and extract had an old reputation for epilepsy, especially in the west of England. In Wales it is applied to the eyes (Grieve. 1931). But, as the names Kidneywort and Kidneyweed imply, it is against kidney trouble and stone that Wall Pennywort is best known.

WALL RUE

(*Asplenium ruta-muraria*) One sometimes finds cases of using rue-fern instead of rue itself in ritual. It could be used, for instance, by a jilted girl, who could wait in the church porch while the man was being married to someone else. She could then throw a handful of wall-rue at him when they came out, with "May you rue this day as long as you live" (Leather) (see also RUE).

WALLFLOWER

(*Erysimum cheiri*) The dream books mentioned wallflowers – dreaming of them is said to foretell a lover's faithfulness; to an invalid such a dream is a sign that he will soon recover (Gordon. 1985). One piece of modern folklore concerns its use as a companion plant – they say that planting wallflowers near an apple tree encourages the latter's fruiting (Baker. 1980).

There are some medicinal uses: traditionally, they were used as a purgative, and for liver disorders (Schauenberg & Paris). In Somerset, they used to say you should eat plenty of wallflower buds in salads and jams, for apoplexy (Tongue. 1965), and that harks straight back to Parkinson, who recommended "… a conserve made of the flowers … for the Apoplexie and Palsie". They were popular for fevers, too – see Gerard: "The leaves stamped with a little bay salt, and bound about the wrists of the hands, take away the shaking fits of the ague". Later, Wesley's prescription was substantially the same, except that he wanted the medicine to be applied "to the Suture of the Head". Irish practitioners used the flowers steeped in oil as an anodyne, and in infusion (one ounce to one pint of water) for nervous troubles (Moloney).

Gilliflower was a name applied to a number of plants, perhaps modified for convenient identification, as in Clove Gilliflower for carnations, Stock Gilliflower for stocks, and Wall Gilliflower for wallflower, etc. Bloody Warrior is another old name from the west country. But "warrior" is not warrior at all, but "wallyer"; in fact Bloody Wallyer exists in its own right (Halliwell). A popular legend is said to account for the 'wall' names. It is the story of a Scottish girl who had given her heart to the heir of a hostile clan. She arranged to climb out of a high window and escape with him, but:

> Up she got upon a wall,
> Attempted down to slide withal.
> But the silken twist untied;
> She fell, and bruised and there she died.
> Love in pity to the dead,
> And her loving luckless speed,
> Turn'd her to this plant we call,
> The scented flower upon the wall.

It is even claimed that the building is identifiable, and Neidpath Castle, near Peebles, is suggested. Perhaps this legend accounts for the symbolism connected with the wallflower – it is a symbol, they say, of fidelity in misfortune (Friend. 1883).

WALNUT

(*Juglans regia*) Not a native British tree, as the common name shows. The first syllable derives from OE wealh, foreign. Nevertheless, it has been growing in Britain since ancient times; possibly it was introduced by the Romans, from its native regions in Asia Minor or the Balkans. *Juglans*, the generic name, is joviglans, Jovis glans, the fruit of Jove, which was the Latin name for acorn, or rather walnut was the acorn of Zeus in ancient Greek times. That has led to confusion in translation between walnuts and chestnuts, for exactly the same name is given to the latter.

Like many another tree, it has its magically protective associations. In Bavaria, where the Easter Sunday fire used to be lit in the churchyard with flint and steel, every household would bring a walnut branch, which, after being partially burned, would be carried home to be laid on the hearth as a protection against lightning (Dyer. 1889, Kelly). Walnut leaves, gathered before sunrise on St John's Day, were believed in parts of France to protect from lightning, too (Sebillot). People from the French region of Poitou used to jump three times round the Midsummer fires with a walnut branch in their hands. The branch would be used to nail over the cowhouse door, as a protection for the beasts (Grimm). Moslem belief also recognised its protective influence, particularly the root and bark, with which Moroccan women used to paint their lips and teeth a brownish colour (Westermarck). Henna and walnut root and bark are protection against supernatural dangers. Some groups believed that the henna and walnut root were applied "so that she may enter Paradise as a bride if she dies in childbirth" (Westermarck).

Sexual magic was performed with walnuts. Arnold de Villeneuve, the Catalan physician and alchemist, who lived from 1235 to 1311, gave a receipt for "tying the knot". One takes a walnut, separates the two halves, and puts them in the marriage bed. The counter charm is to stick the two halves together, crack the nut, and then the couple eat it (Bouisson). It is because the nuts are of two halves that they were symbols of marriage. Walnuts were scattered at Roman weddings, and Pliny described them as symbols of marriage and protectors of resultant offspring. The tree remained a bridegroom's symbol in Germany until at least Evelyn's time: "in several places … in Germany, no young farmer whatsoever is permitted to marry a wife, till he bring proof that he hath planted, and is a father of such a stated number of walnut trees, as the law is inviolably observed to this day…". It is mentioned as an aphrodisiac in Piers Plowman (I B Jones). There is a particular walnut tree in the region of Creuzay,

France, that is kissed by brides on their wedding day, to "les faire devenir bonnes nourrices" (Loux).

The Benevento tree, already mentioned, had many legends connected with it, one of which said the nuts were triangular in shape. The original tree, or at least a very ancient one, was destroyed by St Barbatus in 663. It was re-planted in the 8th century and was standing in the 16th (Summers. 1927). Guazzo mentions the "wizard walnut tree" of Benevento. Pope Paschal II ordered a large walnut in the Piazza del Popolo to be cut down and burnt owing to the superstitions that had arisen around it, one of which was apparently that the evil soul of Nero was living in its branches (Skinner). Even in England, where it is not native, there are mentions of ancient individual trees. Camden speaks of one in Glastonbury Abbey churchyard "which never buds before the feast of St Barnabas, and on that very feast-day shoots out leaves" (Wilks). And there was a particular walnut tree at St Germans, Cornwall, that formed the central part of the May Fair (held on 28 May) (Barton. 1972). Walnut trees seem to have had special treatment in East Anglia. Evans. 1966 quotes a Suffolk farmer as saying that men who travelled round felling walnut trees that had been sold to them, often found a gold coin buried near the roots.

There are other superstitions connected with walnuts. A heavy crop means a fine corn harvest next year (Waring). On the other hand, dreaming of a walnut tree means misfortune, or unfaithfulness (Dyer.1889). A belief from the Abruzzi, in Italy, says that he who plants a walnut tree will have a short life (Canziani. 1928), or, as in Portugal, he would die when the tree attains his own girth (Gallop). Perhaps it is one of the cases of a man's life going into a very long-lived tree, (hence a belief that if the tree dies, or is blown down, it is a most unlucky event (Campbell-Culver)). One of the most engaging of walnut superstitions, for it is nothing more than that, is the belief that Yorkshire schoolboys once had. They said that if their hands were rubbed with a green walnut shell, they would not feel the schoolmaster's cane – indeed, the cane would split (Halliwell. 1869). Another one, also connected with pain killing, is that wearing a walnut in a bag round the neck would stop one from getting toothache (Waring).

According to Culpeper, burnt walnut ash, or green walnut husks mixed with oil and wine, and applied to the hair, would make it fair. From hair on the head to the head itself, and the most extreme of the medical usages taken from the doctrine of signatures: its use for mental cases, from depression and mental fatigue to outright insanity. Coles was the great presenter of the doctrine, and in his words "Wall-nuts have the perfect signature of the Head: the outer husk or green covering represent the Peribanium, or outward skin of the skull, whereon the hair groweth, and therefore salt made of these husks or basks, are exceeding good for wounds in the head. The inner woody shell hath the signature of the skull, and the little yellow skin, or Peel, that covereth the kernell of the hard meninga and Pia-mater, which are the thin scarfes that envelope the brain. The Kernal hath the very figure of the brain, and therefore it is very profitable for the Brain, and resists Poysons. For if the Kernell be bruised, and moystened with the quintessence of wine, and laid upon the Crown of the Head, it comforts the brain and head mightily". But the walnut tree was involved in so-called cures for madness long before Coles's time. A 15th century leechdom spoke of a sovereign medicine for madness and for men that be troubled with wicked spirits: Upon midsummer night betwixt midnight and the rising of the sun, gather the fairest green leaves of the walnut tree, and upon the same day between sunrise and its going down, distill thereof a water in a still between two basins. And this water is good if it be drunken for the same malady" (Dawson. 1934). Perhaps this accounts for a Sussex belief that it was not safe to sit under a walnut tree, for it might damage the mind, and sleeping under it might very well result in madness or even death:

> He that would eat the fruit must climb the tree.
> He that would eat the kernel must crack the nut.
> He that sleepeth under a walnut doth get fits in the head (Allen).

But Andrew Boorde wrote that "the walnut and the Banocke [i.e., bannut] … do comfort the brayne if the pyth or skyn be pylled of …".

Even in Evelyn's time, the distillation mentioned "with honey and wine", was being used hopefully to "make hair spring on bald-heads", still harping on the connection with the head. A propos of Coles's mention of counter-poisons, it should be recorded that it is one of the chief ingredients in the antidote to poison attributed to King Mithridates (C J S Thompson. 1897). The idea occurs, too, in Neckham's late 12th century *De Natura Rerum* – all poison in herbs could be nullified by the walnut: it had merely to be placed among the most deadly plants for all poison to be expelled.

Skin diseases have long been treated with walnut leaves in one form or another. In parts of America they say that ringworm can be cured by rubbing it with green walnuts (Sackett & Koch; Stout). The Pennsylvania Germans do the same to get rid of a wart, and then the nut has to be buried under the eaves (Fogel). But the leaves are strongly astringent anyway, and have been used to treat a wide variety of ailments, including earache (Dyer. 1889), and even toothache, by binding on to the cheek (a practice that could certainly harm one's face).

WARDEN PEAR

Believed to have originated at Woburn Abbey (at one time called Warden Abbey) in Bedfordshire, first grown by Cistercian monks in the 12th century. Three

warden pies appear in the arms of Woburn Abbey. Perhaps, though, as in the opinion of some, the word is the Anglo Saxon 'wearden', meaning 'to keep' (Genders. 1971). The pies were a local delicacy, sold on the feast of SS Simon and Jude (28 October), when the cry was:

> Who knows what I have got?
> In a pot hot?
> Baked wardens … all are hot.
> Who knows what I have got?

Warden pies were coloured with SAFFRON in Shakespeare's time, in the same way as pears are coloured with cochineal now.

WART-CRESS

(*Coronopus squamatus*) Called wart-cress not because it was used to deal with warts, but because of the characteristic warty appearance of the seed vessels.

WARTS

Many and varied are the remedies in medical folklore for getting rid of warts, some by direct application, and some by a ritual that can only be described as a charm, the best known being those connected with the ASH:

> Ashen tree, Ashen tree,
> Pray buy these warts of me.

That is a Leicestershire rhyme to accompany the charm, which was to take the patient to an ash tree, and to stick a pin into the bark. Then that pin would be pulled out, and a wart transfixed with it till pain was felt. After that the pin would be pushed back into the tree, and the charm spoken. Each wart was treated, a separate pin being used for each (Billson). An East Anglian cure was to cut the initial letters of both one's Christian and surnames on the bark of an ash that has its keys. Count the exact number of the warts, and cut the same number of notches in the bark. Then, as the bark grows, so will the warts go away (Gurdon; Glyde). Another method is to cross the wart with a pin three times, and then stick the pin into the tree (Northall). The Cheshire cure was to steal a piece of bacon, and to rub the warts with it, then to cut a slot in the bark and slip the bacon underneath. The warts would disappear from the hand, but would make their appearance as rough excrescences on the bark of the tree (Black). They could be treated by cutting notches in a HAZEL twig, one for each wart, which would disappear as the notches grew out of the twig (Newman & Wilson). They could be charmed by touching each one with a green PEA, which would then be wrapped in paper and buried.. As the pea decays, the warts will go (Allen). Or you could rub the wart with WATER PEPPER (*Polygonum, hydropiper*), and throw the plant away (Stout).

The use of ELDER in wart charms is widespread, most of them being typical transference cures, like this Welsh one – you take an elder branch, strip off the bark, and split a piece off like a skewer. Hold this near the wart, and rub it either three or nine times with the skewer, while an incantation (of your own composing) is muttered. You pierce the wart with the thorn, and then transfix the elder skewer with the thorn, and bury them in a dunghill. The wart would rot away as they decayed (Owen). The Cricklade (Wiltshire) version was to strike the warts smartly with the elder twig, saying aloud, "All go away". Then you had to walk backwards towards the midden, and throw the twig in without looking (Richardson). Similarly, from Derbyshire – take a green spit of elder and a penknife. Touch one of the warts with the tip of the penknife, and then cut a notch in the elder stick. Then bury the stick after all the warts have been touched, and as the stick decays, the warts will decay also (Addy). Another Welsh cure was to rub the warts with elder leaves picked at night, and then burnt, after which the warts would disappear (Owen). Crossing the warts with elder sticks was also a popular charm (Sternberg). HONEYSUCKLE leaves were used, too, without the charm, for Lupton confided a sure cure by using woodbine leaves, "stamped and laid on … using them six times …". PEACH leaves are used, too. An Italian charm requires the leaves to be applied to the warts, and then buried, so that the leaves and the warts would perish together. Exactly the same idea is recorded in America, or you could cut as many notches in a young peach tree as you have warts. In seven days the warts would go. Some say that three notches have to fill up before the wart will be gone (Thomas & Thomas). The water that collected in the cups formed by the fusing together of TEASEL's opposite leaves, "so fastened that they hold dew and raine water in manner of a little bason" (Gerard), would cure warts (Curtis).

There is a typical charm from Devonshire, using GROUNDSEL. Rub the wart with it to make it go. The leaves should then be thrown over your head, and afterwards they should be buried by someone else. As the leaves rot, so will the wart (Crossing). A POTATO could be used in the same way (Stout). Similarly, warts can be rubbed with the furry inside of a BEAN pod. Sometimes one finds the relics of a charm attached to this. For instance, in Essex, there is the injunction to throw the pod down a drain after rubbing the warts (Newman & Wilson).

> As this bean shell rots away,
> So my warts shall soon decay (Hardy. 1878).

So, also, with LIMA BEANS. Split one in half and rub it over the wart, and then toss the bean into a well, and the wart will disappear (H M Hyatt). HOUSELEEK can be rubbed on a wart (Newman & Wilson), and so can a leaf of THORN-APPLE, or the fruit of OSAGE ORANGE, whether it would be buried afterwards as a charm, or not (both H M Hyatt).

An Irish charm was to get ten knots of BARLEY straw (or more usually ten slices of POTATO), count out nine and throw away the tenth. Rub the wart with the nine, then roll them up in a piece of paper, and throw them before a funeral. Then the wart would gradually disappear (Haddon). Or cut an APPLE in two, rub one half on the wart, give it to the pig to eat, and eat the other half yourself (Choape). Similarly, rub the warts nine times with an apple cut in two. Re-unite the sections, and bury them where no human foot was likely to tread. In Northumberland, the warts were opened to the quick, or until they bled, and then they were rubbed well with the juice of a sour apple, which was then buried (Drury. 1991). An American charm of the same nature involved cutting a CRANBERRY in half, rubbing the warts, and burying the fruit under a stone (Davenport). OATS are used, too. You had to take 81 stems (9×9, that is, a magical number), which are bumpy, like warts, binding them in 9 bundles of 9 each, and hiding them under a stone to await the decaying. A Cheshire charm using RUSHES involved taking a long, straight rush, tying three knots in it, and making it into a circle. It had to be drawn over the wart three times, while the required formula was recited. The wart would disappear within three months (Burton), which seems an inordinately long time to have to wait. OLEANDER is used for a wart charm on the island of Chios. The practice is to put a leaf on each wart, then, in the wane of the moon, the leaves are put under a stone in a river-bed. the patient has to go away without looking back (Argenti & Rose). SAVIN, bruised, fresh or dried, is still recommended by herbalists for warts (Flück). The root infusion of TREE CELANDINE (*Bocconia frutescens*) has been used in Mexico to treat warts (Usher).

Just using the juice of certain plants on the wart is very widespread and varied, with the favoured plant ranging from FIG (the leaf juice) (V G Hatfield) to GOUTWEED, or HOGWEED (in East Anglia) (V G Hatfield. 1994), or SOW THISTLE, or MILK THISTLE and SCARLET PIMPERNEL (in Somerset) (Tongue. 1965). Another is ST JOHN'S WORT, and ORANGE BALSAM juice is used in America, as is that of TOUCH-ME-NOT in Britain (McLeod). The juice of GREATER CELANDINE is another medicament for the purpose; some of its names proclaim the use – Wart-plant, or Wartweed, for example, or, in French, herbe aux verrues (Schauenberg & Paris). MILKWORT juice can be rubbed on them (Gerard), and so can RHUBARB juice (Stout). The yellow latex of MEXICAN POPPY (*Argemone mexicana*) can be used on them, too (Gooding, Loveless & Proctor). CHICKWEED poultices, or just the fresh juice itself, can be used for warts, as well as most skin complaints. But it seems that the warts had to be pared to the quick first, then they would fall out (Fernie). Culpeper recommended "the juyce of the Leaves and

Flowers [of MULLEIN]" for warts, and the powder of the dried root was also prescribed. CASTOR OIL will do, as well, and SNOWBERRY juice can be used, too, though in Kent, a modern belief is that the juice from the berries actually cause warts (Vickery. 1995). The Physicians of Myddfai prescribed SHEEP'S SORREL juice, "and bay salt, wash your hands and let them dry spontaneously. Do this again and you will see the warts and freckles disappear". DANDELION juice could be used, too. An Irish charm was to give nine dandelion leaves, three leaves to be eaten on three successive mornings. In Britain BASIL, mixed with blacking, has been used (Leyel. 1926). An ONION poultice was used in England to put on a wart (Drury. 1991), and there is a charm from Staffordshire – rub the warts with two halves of an onion, and then bury the onion (Raven). The same charm was used in America (Stout). BUTTERCUPS were also used,and Gerard said "… it is laid upon cragged warts, corrupt nailes, and such like excrescences, to cause them to fall away". Bruised BITING STONECROP has been used (Flück), and so has the juice of most of the SPURGES, though care has to be taken, otherwise blisters on the skin may be the result. A Victorian cure for warts is given as "the bruised leaves [of MARI-GOLDS] mixed with a few drops of vinegar" (Dodson & Davies). The Romans were using the juice for just this purpose in their day, hence the old Latin name of Verrucaria for Marigolds.

CINNAMON has been used for warts. The practice in Illinois was simply to cover the wart with cinnamon powder in order to get rid of it (H M Hyatt). Navajo Indians used to burn some powdered SUNFLOWER pith on the wart (Wyman & Harris). Other groups would use the juice of the fresh fruit of SMOOTH SUMACH, and also for skin diseases like tetter or ringworm (Lloyd).

Getting rid of warts by rubbing a snail on them and then impaling the snail on a BLACKTHORN thorn used to be a common practice; or, from East Anglia, you could rub the wart with a green sloe, and then throw the sloe over your left shoulder (Glyde). They are both transference charms. An Irish cure is to prick them with a GOOSEBERRY thorn passed through a wedding ring (Fernie). The Scottish practice was to lay a wedding ring over the wart, which was then pricked (through the ring) with a gooseberry thorn. Ten thorns were to be picked, the other nine simply being pointed at the wart, then thrown over the left shoulder (Beith).

Thus far, the emphasis has been on getting rid of warts, but one superstition from Cornwall warns us that picking the flowers of SHEEP'S BIT will actually cause warts (Vickery. 1995).

WASSAIL

literally means good health, and is a word of Saxon origin. The legend has it that when Vortigern, prince

of the Silures, fell in love with Rowena, the niece of Hengist, she presented him with a bowl of spiced wine, saying, "Waes Heal, Hlaford Cyning" (Be of health, Lord King). Vortigern married her, and his kingdom was conceded to the Saxons. Since then Waeshael became the name of the drinking cups of the Anglo-Saxons (Howells). The wassail bowl, though, has a more special significance, something like a punchbowl, but the original idea of spiced wine is retained. The proper drink to use as a wassail is Lamb's Wool, spiced ale with roasted apples in it, and the proper time to drink it is the eve of Twelfth Day, or Twelfth Night.

But wassailing is a much more important custom than just drinking lamb's wool. Quite a lot of Hallowe'en (i.e., Celtic new year) customs have been absorbed into those of Twelfth Night. Apple wassailing, or apple howling, as it is sometimes called, is one of them (Hull). Wassailers were known as howlers in Sussex (Sawyer). The purpose of the custom is set out in Tusser's rhyme:

> Wassail the trees, that they may bear
> You many a plum and many a pear;
> For more or less fruit they will bring
> As you do them wassailing.

The whole object of the wassail ceremony is to make the trees bear fruit, and a lot of it. There was nearly always a rhyme or song to be sung, the best known of which is:

> Here's to thee, old apple tree;
> Whence you may bud, and whence you may blow.
> And whence you may bear apples enow.
> Hats full, caps full
> Bushel-bushel sacks full
> And my pockets full, too (Brand).

In Devonshire, if the parson happened to be popular, the line "Old parson's breeches full", was added.

In the area of east Cornwall and west Devonshire, the custom was to take a milkpanful of cider, into which roasted apples had been chopped, usually to pour over the roots of apple trees (Weston), but sometimes taken and put as near as possible in the centre of the orchard. It was important that *everyone* partook. The children were brought out, so were the sick and invalids. If anyone were missing, the charm would not be effective (Whitlock. 1977). Everyone would take a cup of the drink, and each went to a separate tree, saying the ritual:

> Health to the good apple tree,
> Well to bear, pocketsful, hatfuls,
> Peckfuls, bushel-bagfuls.

Part of the cupful of cider was drunk as a health to the tree, but the rest was thrown at it (Hunt). Sometimes cider-soaked bits of toast and sugar were put in the branches (Farrer). Note that the cider is *thrown* at the tree. Guns were actually fired into the branches. It is as if they threaten the tree, a "let-this-be-a-warning" series of shots to bolster up the entreaties. The Sussex custom was apparently for a youth to climb the tree, seemingly the object of the firing, though of course care was taken not to hit him (Weston). Where this occurs the original intention was for him to answer for the tree. But these threats were certainly confined to the English cider producing areas. The Santals, in India, always shot arrows into the "Sal Tree", a sacred tree to them (Biswas). The ordinary way to treat a walnut tree to make it bear a good crop was to beat it. As the outrageously politically incorrect rhyme has it:

> A woman, a spaniel, and a walnut tree,
> The more you whip them, the better they be
> (Halliwell. 1869).

There are very similar practices recorded in both French and Japanese folklore. There is some rationalisation in all this. It is claimed that the shock and smoke of gun-firing had a practical purpose, and tended to dislodge insects, so that the birds could get them. Or, the shot would tear the bark in places, and so quicken budding (Minchinton).

A Somerset variant of the apple wassailing custom is to take the lowest branches of the apple tree and pull them down and dip them in cider. When this is done, everybody bows three times. Then they raise themselves from the third bow as if with great effort, miming the bearing of a heavy load (Wilks). This, of course, is sympathetic magic, like jumping high in the air to show the corn how high you want it to grow.

WATER ARUM

(*Calla palustris*) Like all the arums, this is an unlucky plant to have in the house (Bergen. 1899). The rhizomes yield an edible starch after grinding and washing (Lindley), and they were ground to serve as an ingredient for bread in Sweden in Linnaeus's day. Rotzius found it still used for this purpose in Finland in the 3rd quarter of the 19th century (J G D Clark. 1952). They also contain an acrid compound used as an antidote to snake venom (Schauenberg & Paris).

WATER AVENS

(*Geum rivale*) The roots are aromatic, slightly astringent, and once used to flavour ale, and to keep it from turning sour (Hulme). A decoction of the rootstock was a favourite beverage among the American Indians (Yanovsky), and it is still used, boiled in milk and sweetened, as a beverage not too different from chocolate (Sanford). The plant is actually called Chocolate-root (Sanford), or Indian Chocolate (Leighton). It has its medicinal uses, too, as an aromatic bitter, and the root was chewed to sweeten the breath (Fairweather). The Welsh text known as the Physicians of Myddfai had a remedy for hoarseness:

"Take the water avens, and St John's Wort, boil in pure milk. Mixing butter therewith when boiling. Boil a portion thereof briskly every morning and drink".

WATER BETONY

(*Scrophularia auriculata*) This is a stinking herb, "sympathetically clapped on to stinking sores and ulcers", as Grigson put it. The leaves could be applied to cuts, and are used in the Fen country to treat chapped hands and feet, sore heels, and the like (Marshall). One important usage relies heavily on the very name of the genus: *Scrophularia*, which means the swelling of the neck glands we know as scrofula, once known in England as the King's Evil, for the touch of the anointed monarch was taken as the only cure for the condition. Rose-noble is a name given to this and its close relative, Figwort. It was the name of an old English gold coin, with a figure of a rose on it. It must surely have had the king's head on it, for the name occurs in connection with the King's Evil. There is an Irish remedy for scrofula with a concoction of burdock roots, common dock, bogbean and rose-noble, boiled in water, of which the patient was required to drink three times a day (Wilde. 1890). Another Irish usage, from County Tipperary, was for the leaves to be boiled in water to make a "tonic drink" (Barbour).

Gerard reported that "if the face be washed with juyce thereof, it taketh away the rednesse and deformity of it". Culpeper also reported it, and capped the description by informing his readers that "it is an excellent remedy for sick hogs".

WATER GERMANDER

(*Teucrium scordium*) A French belief was that if a woman wanted to make a man love her, she had to put a piece of Water Germander (if the identification was correct) in his pocket without his knowing it (Sebillot). There have been medicinal uses involving this plant. It was, for instance, once esteemed as a poison antidote (Grieve. 1931). Gerard was careful to point this out, and he also claimed, among others, that it would "mitigate the pain of the gout ...". Later, Hill was recommending it as a remedy "against pestilential fevers". But it seems that the only genuine use was for worms, when the dried leaves, powdered, were employed (C P Johnson).

WATER MELON

(*Citrullus vulgaris*) There have been some strange uses for the plant. The Arabs, for example, used the ripe fruit, charred in the fire and pulverized, to prepare gunpowder, tinder and fuses. Similarly, in Egypt, the green fruit was crushed on a piece of fabric, which absorbed the juice, and when dried it acted like tinder (Dalziel). Equally odd is the practice in Morocco of a young man who wants his beard to grow to rub his skin with a piece of water melon, for the juice was thought to produce the desired effect (Westermarck).

But eat water melon rinds for a smooth complexion is the maxim in Alabama (R B Browne).

Water melons are an ingredient in American domestic medicines, particularly for fevers (Beck). A tea made from the seeds is also said to be good for high blood pressure (R B Browne); so it is for newborn babies whose kidneys have failed to act, according to Alabama belief. In these circumstances, a tea should be made of a handful of seeds and a pint of water. The baby should be fed a spoonful frequently, until results are obtained (R B Browne). The rind and pulp are used in China and Japan to treat jaundice and diabetes. The pulp relieves sore throat or a sore mouth (Perry & Metzger).

There are some pieces of advice about planting water melons from the southern states of America. They should be planted on the 1st of May, before sunrise, for good luck (R B Browne), and by poking the seed in the ground with the fingers (Puckett), or, according to Kentucky wisdom, in your night clothes, before sunrise; then the insects will never attack them. Carry the seed out in a wash tub. Then the melons will grow as large as the tub (Thomas & Thomas).

WATER MINT
(*Mentha aquatica*) "the savor or smell … rejoyceth the heart of man, for which cause they use to strew it in chambers …" (Gerard). It was used in Lincolnshire to make a tea to be taken for heart complaints (Gutch & Peacock). Actually, this is quite a common popular belief, but usually it is said to be best in some alcoholic medium (Schauenberg & Paris). An infusion of the leaves gives a good carminative, like peppermint, but without the menthol. Africans powder the root bark, to be eaten for the relief of diarrhoea (Watt & Breyer-Brandwijk), and herbalists use it for this all over the world. It is "… good against the stinging of bees and wasps, if the place be rubbed therewith" (Gerard). For scrofula, in the Balkans, it is cooked in water, and the hot plant applied externally, and the liquid drunk (Kemp).

WATER PEPPER
(*Polygonum hydropiper*) They say that fleas will not come into a room where this herb is kept (Fernie). Rub it on warts, and throw the plant away (Stout). Herbalists use the tea to treat piles (Thomson. 1978). Tea is also used as a febrifuge in the southern states of America (Puckett), and, taken cold, for colds, in Iowa (Stout). The leaves infused in boiling water, or a strong decoction of them, were applied to bruises and contusions (Barton & Castle), and the juice, they say in Iowa, makes an excellent liniment for sprains (Stout). In Norway, the herb was chewed for toothache (Barton & Castle).

Arsmart is a name given to the plant, concerning which Cockayne says: "it derives its name from its use in that practical education of simple Simons, which village jokers enjoy to impart". It is the internal effect of the leaves that is responsible for the name, which varies into Smartarse (Elworthy. 1888), or Smartweed (Grigson. 1955). PERSICARIA also has the name, but Culpeper called it 'Mild Arsmart', and Gerard 'Dead Arsmart' – "it doth not bite as the other doeth".

WATERCRESS
(*Nasturtium officinale*) or (*Rorippa nasturtium-aquaticum*) A cress with exceptionally high Vitamin C content (Mabey. 1972), which explains why it has always been so widely eaten, especially as it is also iron-rich, so that it has long been used particularly for anaemia (Conway). Although it is possible to pick the wild plant, it is best, given that the water it grows in could be polluted, to buy it from the greengrocer. A Somerset legend has it that watercress will only grow near sewerage. Even if it grows well away from human habitation, they will say that there must be an old cesspit nearby (Storer). There is a belief that it should not be gathered for eating when there is no letter 'R' in the name of the month, perhaps because it is in flower during the summer months, or, as water levels are likely to be low in summer, the watercress might be unclean then (Vickery. 1995).

Watercress soup, made with cream, was a traditional Hertfordshire dish (Jones-Baker. 1974). It was even eaten stewed at one time (C P Johnson), which must have destroyed all that it is genuinely good for.

Most of watercress lore has to do with folk medicine, but there are one or two more beliefs connected with it. A Devonshire saying, of some simple person, is that he "never ate his watercress" (Vickery. 1995), giving the idea that the plant was one that gave intelligence (like fish). It seems also to have magical powers of its own, according to Highland witchcraft lore (and always provided that this plant is what is meant by "watercress" (see Watts. 2000)). It was used to steal milk. The witch cut the tops of the plant with a pair of scissors, while a charm was spoken along with the name of the cow's owner, and ending with the words "the half mine, the other half thine". A handful of grass from the thatch over the byre would apparently do just as well. The counter charm was groundsel put with the milk (Polson. 1932). In the language of flowers, it is the symbol of stability and power (Leyel. 1937).

This is a valuable plant, still in use for some medicinal purposes, for some skin diseases (Schauenberg & Paris), even some cancerous growths (Baïracli-Levy). Just put the juice on the sore place (Page. 1978). Boiled with whisky and sugar, it is an Irish cure for bronchitis (Wood-Martin), and Ô Súilleabháin mentions its use there for asthma. Watercress tea is drunk for a cold in Trinidad (Laguerre), while it is taken in the Highlands for reducing fevers, and also as a "blood purifier" (Beith). Herbalists still use it for treating rheumatism (Conway), and it is even said in Ireland that, eaten raw, it is good for heart disease (Vickery. 1995).

The Anglo-Saxon version of Apuleius claimed, among a long list of illnesses, "in case that a man's hair fall off, take juice; put it on the nose; the hair shall wax" (Cockayne). Watercress actually is a good hair tonic. There is a saying in French that a bald man "n'a pas de cresson sous le caillou" – loosely, has no watercress on his head (Palaiseul).

WAYFARING TREE

(*Viburnum lantana*) "Wayfaring", because of the characteristic mealy appearance, like a dusty wayfarer. Evelyn mentions its use for protecting cattle from witchcraft, and see Aubrey. 1847: "… they used … to make pinnes for the yoakes of their oxen of them, believing it had vertue to preserve them from being forespoken, as they call it; and they use to plant one by the dwelling-houses, believing it to preserve from witches and evil eyes". Another of Evelyn's claims was that "the leaves decocted to a lie, not only colour the hairs black, but fastens their roots", a point that Gerard had already made.

In Devonshire, there was a saying, "as tough as a whitney stick", whitney being one of the names of our tree, and farmers always made whips from this wood, which is very tough (Friend. 1883). "… the branches are long, tough, and easie to be bowed, and hard to be broken …"(Gerard). There are various "whip" names for the tree, "Whipcrop" being the best known. Gerard also had: "it is reported that the barks of the root of the tree buried a certaine time in the earth, and afterwards boyled and stamped according to Art, maketh a good Bird-lime for Fowlers to catch birds with".

Coven-tree is one of the names given to the tree, varied into Coban-, or Cobin-tree, a name that appears in the children's games quoted by Northall:

> Keppy-ball, keppy-ball, Cobin-tree,
> Come down and tell me
> How many years old
> Our (Jenny) shall be …

The number of keps, or catches before the ball falls is the age. It seems that the game was played under one of these trees (the rhyme is said by Northall to be a Tyneside one, so presumably the name was local there, too). But these names were given to any tree that stood before a Scottish mansion house; it was under this tree that the laird always went out to greet his visitors. So the word is covyne, a meeting or trysting place. In these circumstances, "keppy" in the song is likely to be Scottish kep, to meet (Grigson. 1955).

WEATHER LORE

PINE cones stay open when the weather is to turn fine, but as soon as they close, then it is believed

rain is on the way (Waring). The same is believed of SILVER FIR cones (M E S Wright). ASH has its share of weather lore, the best known being the comparison with the OAK to foretell a good or bad summer:

> If the oak before the ash come out,
> There has been, or will be, a drought (Northall).

There are quite a number of jingles of the same import, the most succinct of which is, from Surrey:

> Oak, smoke,
> Ash, squash (Northall),

or sometimes:

> Oak, choke.
> Ash, splash

i.e. if the oak leafed first, there would be dry, dusty weather (Baker). A lengthier version is:

> If the ash before the oak
> We shall have a summer of dust and smoke (Page).

There are many more of these rhymes, but they tend to get rather ambiguous, sometimes completely contradictory.

When an ONION skin is thin and delicate, we can expect a mild winter; if it is thick, it foretells a hard season (Inwards). Put into verse, we have:

> Onions' skin very thin,
> Mild winter's coming in;
> Onion's skin thick and tough,
> Coming winter cold and rough (Krappe).

In a similar way, the thickness of HAZEL nut shells is an indication of the weather to come – the thicker they are, the harder the winter to come; conversely, of course, thin shells, mild winter (Conway). An American version expects a large crop of nuts to be followed by a hard winter (H M Hyatt), and all over Europe a large crop of ACORNS presaged a long, hard winter, pessimistically, "a bad year for everything": 'Viel Eicheln lassen strenger Winter erwarten' (Swainson., 1873), or 'Année de glande, année du cher temps', and so on. BLACKTHORN is an unlucky plant. Even its early blooming brings talk of a "blackthorn winter". There are often some warm days at the end of March and beginning of April, which are enough to bring it into flower, and they are nearly always followed by a cold spell, the Blackthorn Winter. "Beware the Blackthorn Winter", or "blackthorn hatch" as it is sometimes called, is a well known admonitory saying. The north-east winds that seem to prevail in spring, about the time the shrub is in flower, were known as "blackthorn winds". A blackthorn winter means a spoiled summer, they say in Somerset (Tongue. 1965). Sometimes, it seems, there is a second blackthorn winter, which is said to fall in the second week in May. This may be just coincidence, for the festivals of the Ice Saints (Mamertus, etc) fall then (Jones-Baker. 1974). Even when the sloes themselves appear, there is foreboding. Is it not said that:

> Many haws, many sloes,
> Many cold toes? (Denham).

The more berries there are, so the worse the coming winter is said to be. Note the Devonshire rhyme:

> Many nits [nuts]
> Many pits;
> Many slones,
> Many groans (Choape).

A Cheshire belief was that if the weather breaks when the ELDER blossoms were coming out, it would be soaking wet till they fade (Hole. 1937) – or vice versa, for the belief seems to be that the weather never changes while the flowers are in blossom (Baker). Anyway, it all seems safe when the flowers are out:

> You may shear your sheep
> When the elder blossoms peep. (A C Smith),

or, more obscurely:

> When the elder is white, brew and bake a pack;
> When the elder is black, brew and bake a sack (Denham. 1846).

Belgian people once used elder to foretell future weather by putting a branch in a jug of water on 30 December. If buds developed and opened, it would be a sign of a fruitful summer to come; if no buds, then the harvest would be bad (Swainson. 1873). Large buds on a BEECH tree will foretell a wet summer (Addison. 1985).

If the leaves of HORSE CHESTNUT spread like a fan, then warm weather would come; but long before rain arrives the leaves begin to droop and point downward (Trevelyan). Similarly with WHITLOW GRASS – if the leaves droop, it is a sign of rain (Inwards). A Lincolnshire weather pointer is that DUCKWEED rises in a pond when the weather is going to be fine (Rudkin).

As HOPS became a more and more important crop, some weather lore attached itself to the plant:

> Till St James's Day be come and gone,
> There may be hops or there may be none (Dyer).

St James's Day is 25 July, which seems rather a late day to judge the well-being of the crop (perhaps the reference is to St James the Less – 1 May). Another rhyme seems more realistic:

> Rain on Good Friday and Easter Day,
> A good crop of hops, but a bad one of hay.

This is from Herefordshire (Leather). Another saying might apply to any crop:

Plenty of ladybirds, plenty of hops (Dyer).

There was a belief that when a BRAMBLE blooms early in June, an early harvest could be expected (Swainson. 1873). A Yorkshire tradition tells that an abundance of blackberries in autumn foretells a hard winter to come (Gutch. 1901), on the "many haws, many snows" basis. But at least, the weather is usually good when the blackberries ripen, and that period at the end of September and beginning of October is quite often called the blackberry summer (Denham. 1846). They say that SWEET BRIAR has a fresher fragrance before rain (Trevelyan). LADY'S BEDSTRAW, too, has a stronger smell when it is going to rain (Inwards).

SCARLET PIMPERNEL is a noted weather fore-caster, as many of the local names given to it will confirm. Gerard said, "… the husbandmen having occasion to go unto their harvest worke, will first behold the floures of Pimpernel, whereby they know the weather that shall follow the next day after; as, for example, if the floures be shut close up, it betokeneth raine and foule weather; contrariwise, if they be spread abroad, faire weather".

> Pimpernel, pimpernel, tell me true,
> Whether the weather be fine or no;
> No heart can think, no tongue can tell,
> The virtues of the pimpernel (M E S Wright).

This is quite true; the flowers open when it is going to be sunny, and close when it is going to rain (Page. 1977). It will forecast twenty four hours ahead, so it is claimed (Trevelyan). Similarly, GERMANDER SPEEDWELL will forecast rain by closing its petals, and opening them again when the rain has stopped (Inwards). PICNIC THISTLE (*Cirsium acaulon*) is a similar forecaster; if the flower is open, there is good weather to come, and the opposite if they are closed (Gubernatis). And the same applies to MARIGOLD; if it does not open its petals by seven o'clock in the morning, the signs are that it will rain or thunder that day. It is also one of those flowers that closes up before a storm (Swainson. 1873). Children use the seed tufts of DANDELIONS as a barometer. When the down is fluffy, then there will be fine weather, but when it is limp or contracted, then there will be rain (Swainson., 1873). The flower heads will close directly rain falls, or just before, and always before dew-fall (Rohde. 1936). DAISIES, too, will shut when bad weather is coming (Page. 1977), and WOOD ANEMONES close their petals and droop before rain (Inwards). There is, too, an old belief about COLTSFOOT, from Coles, *Knowledge of plants*, 1656, that "if the down flyeth off colts's foot, dandelyon, and thistles, when there is no winde, it is a sign of rain". The TAMARIND tree folds its leaflets at night and in overcast weather. That and the fact that the tree almost always feels damp makes it the ideal abode of the rain god in Burmese thought (Menninger). Indians have a prejudice against sleeping under one of these trees, probably because of this usual dampness. Tents pitched near one will certainly have their canvas affected by it (Leyel. 1937).

If the COWSLIP'S stalks are short, then we are in for a dry summer (Inwards). Some say, too, that we never get warm settled weather till the cowslips are finished (Page. 1977). There is an American saying that if deadnettles are in abundance late in the year, it is a sign of a mild winter to come (Inwards).

WEDDINGS
ALMONDS, in the Mediterranean area, have always been important at weddings, and there is a lot of sexual symbolism associated with them. Greek mythology has the Phrygian tale of Attis. In one version he is castrated by the gods and dies. His testicles fell to the ground and sprouted new life in the form of an almond tree (Edwards). At Spanish gypsy weddings everyone heaps pink almond blossom on the bride's head as she dances. Such throwing is known in Greece as "pouring", as if the blossom were life-giving water. Almond paste, known as 'matrimony' because it blends bitter and sweet flavours, usually appears on wedding cakes. Sugared almonds are used these days at Greek weddings, the sugar adding another dimension – white for purity. So universal are they in Greece that the question "when will we eat sugared almonds?" is asked instead of "when will the wedding be?" They are equally prominent at Indian weddings, and some are put at every table-setting at a reception. They may indicate both prosperity and children, i.e., money fertility and sexual fertility. At wedding ceremonies in Rhodes, the bride's hands were anointed with CINNAMON (Rodd) (as an oil, presumably). At Roman weddings, the bridal wreath was of VERVAIN, gathered by the bride herself, and in Germany until quite recently a wreath of it was presented to the newly married bride (Dyer. 1889). In Greece and Rome young married couples were crowned with MARJORAM (Dyer. 1889).

GARLIC. Swedish bridegrooms used to sew sprigs of garlic, thyme or some other strongly-scented plant, into their clothing to avert the evil eye, and in south Arabia the bridegroom wears it in his turban (M Baker). Among gypsy marriage customs was one that required the bride to hang up bundles of garlic in her house – for luck, and against evil, for the garlic turns black after attracting all the evil to itself, and so protects her (Starkie).

HAZEL was the medieval symbol of fertility. Throwing nuts at the bride and bridegroom is sometimes the practice at Greek weddings, and sugar coated nuts are known to take the place of the better-known sugared almonds (see above). Until quite recently, Devonshire brides were given little bags of hazel nuts as they left the church. These had the same significance as rice and confetti have today (Hole. 1957). Ruth Tongue told the story of a Somerset village girl who returned from London to get married. She openly said that she did not intend to be hampered with babies too soon, and would take steps to ensure this. Such talk outraged village morality, and when she got to her new house, she found among the presents a large bag of nuts, to which most of her neighbours had contributed. She had four children very quickly.

There was an old country custom of putting a spray of GORSE in the wedding bouquet (Grieve. 1931), and a Somerset version of the wedding dress rhyme runs:

> Something old,
> Something new,
> Something borrowed,
> Something blue,
>   And a sprig of vuz

for the belief is that it brings gold to the house (Raymond). The "something blue", at least in Gloucestershire, would be a piece of PERIWINKLE. Some say that it must be worn in the garter for fertility (Vickery. 1995), and in Italy, JASMINE is woven into bridal wreaths. There is a proverb that says that a girl who is worthy of being decorated with jasmine is rich enough for any husband (McDonald). MIGNONETTE was included in the bouquet in France, for it was believed that it would hold a husband's affection (M Baker. 1979).

MYRTLE was said to have the power of creating and perpetuating love (Philpot), and as such was the symbol of married bliss. It was often used on the Continent in the bridal wreath, the forerunner of orange blossom as the bridal emblem. Since the mid-19th century, royal brides have carried in their bouquets a sprig of myrtle from a bush said to have grown from a piece carried by Queen Victoria on her wedding day, growing in the gardens of Windsor Castle (Higgins). The habit spread, until it was quite common for a sprig of myrtle from the bridal wreath to be planted in the bride's garden, but always by a bridesmaid, never by the bride herself (Baker. 1980), for that would be very bad luck. If the sprig did not strike, then the destiny of the planter was to stay an old maid, an unlikely fate in this context, because myrtle roots very easily. On Guernsey, YELLOW IRIS was one of the favourite flowers used for strewing in front of the bride at a wedding (MacCulloch).

WELCOME-HOME-HUSBAND-THOUGH-NEVER-SO-DRUNK
Surely the most picturesque of all vernacular plant names. It is used for HOUSELEEK.

WELD
A name for DYER'S ROCKET (*Reseda luteola*), and probably more important than the usual common

name. It comes from a Germanic source, possibly Wald, forest, but note that Wau is the German for this plant (see Browning).

WELLS

Many holy wells have been marked by a guardian tree, often a HAWTHORN. These sacred thorns are met frequently, particularly in Ireland. There was one near Timshally, County Wicklow, known as Patrick's Bush. Devotees attended on 4 May, rounds were made about the well, and offerings made to the thorn (Wood-Martin), if that is the right way to describe the ritual of hanging rags or trinkets on the tree that is the companion to the holy well. The same inviolability applies to these holy well thorns as to the solitary fairy thorns, equally sacred.

BOX was used to dress wells at Llanishen, Gwent, on New Year's Eve (Baker. 1980).

WENS

A strange old cure was to "annoynt them with oil of snails, oil of swallows, and HOUSELEEK, but do this not on a Friday or the first quarter of the moon" (Jeffrey). Gerard recommended RED DEADNETTLE, as a poultice.

WHERRY

is the name of a drink made in Yorkshire from the pulp of CRABAPPLES after the verjuice had been expressed (Holloway).

WHIN

One of the common names for GORSE (*Ulex europaeus*), used in the Scandinavian areas of England (the O Norse was hvin) (Grigson. 1974). The Yorkshire expression "a whinny road" means a thorny path, or a difficult road in general (Robinson).

WHIPCROP

is a name given to the WAYFARING TREE (*Viburnum lantana*), for this was a favourite for making whips, being very tough, but also very pliant. Twistwood is another relevant name, meaning fit to be twisted into whip handles (Cope).

WHITE ASPHODEL

(*Asphodelus ramosus*) The roots, dried and boiled, yield a mucilaginous substance which is mixed with grain or potato to make Asphodel bread. In Portugal the root is a specific against ringworm and other skin diseases (Gallop). This is apparently genuinely beneficial.

WHITE BROOMWEED

(*Parthenium hysterophorus*) It is called 'balai amer' in Haiti, where it is used for l ove magic. You take seven small bushes, tie them together and throw them into a river, while saying the correct prayer, and this would give you the love of the girl you want (F Huxley). They also use it to find thieves.

WHITE BRYONY

(*Bryonia dioica*) Mandrake, of course, is *Mandragora officinalis*, but in England it was bryony that was taken to be mandrake, and credited with the same powers and attributes. Imitation mandrake puppets used to be made out of the roots, often used by witches in malevolent charms; "they take likewise the roots of mandrake, according to some, or, as I rather suppose, the roots of briony, which simple folk take for the true mandrake, and make thereof an ugly image, by which they represent the person on whom they intend to exercise their witchcraft" (Coles). People could open the earth round a young bryony, taking care not to disturb the lower fibres, and then put a mould round the root, after which it could all be covered up again, and then left. The mould, of course, would have to bear some resemblance to a human figure. Bryonies grow very quickly, and the object was generally accomplished in one summer. The leaves were also sold for those of mandrake, although there is no resemblance (Barton & Castle). There is a story of a French peasant who got a bryony root from a gypsy. He buried it on a lucky conjunction of the moon with Venus in the spring, on a Monday, in a churchyard, and then sprinkled it with milk in which three field mice had been drowned. In a month, it became more human-looking than ever. Then he put it in an oven, and wrapped it in a shroud. As long as he kept it, his luck never failed him at work or at play (Jacob). But its chief use in England was as a fertility stimulant, and in some parts, Lincolnshire for instance, it was reckoned to be a specific for causing women to conceive (Gutch & Peacock). In East Anglia, a childless woman who wanted a baby would drink "mandrake tea" (Porter. 1969), presumably made from the roots, but not necessarily so. They were even given to mares as an aid to conception (Drury. 1985). All this had by 1646 attracted the attention of Sir Thomas Browne, who dismissed it like any other superstition: "… for the roots [of mandrake] which are carried about by impostors to deceive unfruitful women, are made of the roots of canes, briony, and other plants; for in these, yet fresh and virent, they carve out the figures of men and women, first sticking therein the grains of barley or millet where they intend the hair to grow; then bury them in sand until the grains shoot forth their roots which, at the longest, will happen in twenty days; they afterwards clip and trim these tender strings in the fashion of beards and other hairy teguments …". Indeed, bawdy competitions, known as Venus Nights, used to be held in Cambridgeshire village inns, for the most strikingly anthropomorphic bryony root in female form. The prize-winning root hung in the sow's sty, as an encouragement for her to produce large litters, and when withered it was dropped in the household money-stocking as a stimulant there (Porter. 1969).

But bryony roots were also given to horses in their feed, to make them look sleek (Friend. 1883), a practice recorded in Buckinghamshire and East Anglia. The roots were scraped, and moistened, and then added to the feed. In Cambridgeshire, it was the chopped leaves, gathered before the flowers appeared, that were given to the horses (Porter. 1969). But it was always known to be a dangerous practice. See the rhyme:

Bryony, if served too dry,
Blinded horses when they blew (G E Evans. 1966).

People working in the Cambridgeshire harvest fields used to garland themselves with sprays of bryony to keep off flies, and leaves of it were put in the privy pits of Fen cottages as a deodorant in hot summer weather. Another usage was to crush the roots and stuff them into rat holes in barns to drive the vermin away (Porter. 1969). But in Warwickshire, it was said to be bad luck to cut through a root (Palmer. 1976), and bryony's uncanny nature was recognized as far away as Russia. There is a story there that says that if bryony is hung from the girdle, all the dead Cossacks would come to life again (G E Smith. 1919).

The root is poisonous (it contains a resin, bryresin, that is a drastic purgative (Schauenberg & Paris)), and has been known to have caused the poisoning of whole families who have eaten them in mistake for turnips or parsnips (Long. 1924). Nevertheless, it has been used medicinally for a long time. They used to say in Dorset that white bryony was a good substitute for castor oil (Dacombe). In both Germany and Sweden they had a more picturesque way of preparing it – they used to scoop out a portion of the root and fill the cavity with beer, which in the course of a night would become emetic and purgative (Barton & Castle). Indeed, the root was well known as what Thornton described as "a strong rough purgative", and it was equally well known that using the root in this way could be a dangerous practice, because of the powerful and highly irritant nature of the plant (Grieve. 1931).

The roots have been used for a number of other ills, too. In fact, slices of "mandrake" were sold in London at a penny each early in the 20th century, as a cure-all (letter in *Folk-lore. vol 24; 1913*). They used to provide a poultice to relieve lumbago and sciatica, or rheumatism, by making a liniment from the boiled root (V G Hatfield, 1994), and it could act as a painkiller for other conditions, including apparently, painful tumours (Barton & Castle)). It was even used in the 14th century as an antidote to leprosy, under the name of Wild Nep (Grieve. 1931). Herbalists still sometimes prescribe it in small doses for coughs, influenza, bronchitis and pneumonia (Grieve. 1931), but in days gone by even epilepsy was treated with it, for an Irish manuscript of about 1450 gave a

recipe – "for the falling sickness: a plaster made of mandragore and ground ivy, boiled and laid upon the head. If the patient sleeps well he will do well, and if not, he will not" (Wilde. 1902). It could be used for bruises, too, hence the French name 'herbe aux femmes battues' (Cullum). And the ladies of Salerno were reputed to have steeped bryony roots in honey to put on their faces, "which gives them a marvellous blush" (Withington). Round about the same time, a prescription was published "for nits in the head. Make a lye of wild nept [i.e., bryony] and therewith wash thine head, and it will destroy them" (Dawson. 1934), a recipe that might very well have been successful.

Aubrey. 1686/7 quoted an instance of the use of the leaves for gout, and Lupton also recommended it for dropsy, These usages seem to stem from the doctrine of signatures; the root suggested a swollen foot, so it would be used for such complaints as dropsy and gout. Chilblains were treated in Essex by rubbing the crushed berries on them (V G Hatfield. 1994). Of course, there are also examples of what can only be charms, for example, a so-called anodyne necklace used to be made of beads turned out of the root; it was hung around the necks of babies to help teething, and to ward off convulsions. Probably the original use was to distract the evil eye (Maddox). There are also records of the roots being put at the head of bedsteads as charms to assist childbirth (Lovett), an obvious relic of the belief in the fertility powers of the root. Like a good many other things, a piece of bryony root carried in the pocket was reckoned to be a rheumatism cure, at least in Norfolk (M R Taylor. 1929).

WHITE CAMPION

(*Silene latifolia*) As unlucky as its close relative, Red Campion. They would say in Wiltshire that picking white campion brought thunder and lightning, and perhaps death from lightning (Wiltshire), hence names (though not from Wiltshire) like Thunder-flower or Thunder-bolts. Another superstition explaining the name Mother-die (cf **RED CAMPION**) was that if children picked campion it would result in the death of a parent, the father if it were Red Campion, or the mother if it is White (Radford). One oddity concerns the name Cow-mack, used in the north of Scotland – "some husbands (to make the cow take the bull the sooner) do give her of the herb called Cow-mack, which groweth like a white gilly-flower among the corn" (Britten & Holland).

WHITE CLOVER

(*Trifolium repens*) Is this the true shamrock? Many have argued so, but this is a question of some complexity (see **SHAMROCK**). Not surprisingly, with this background, clover had something magical about it. It was a protector, able to drive witches away (Dyer), and anyone carrying it about his person will be able to detect the presence of evil spirits (Wood-Martin). It was gathered with a gloved hand, and brought into the house where there was a lunatic, without anyone knowing; its very presence was enough to cure madness (Wood-Martin). If a farmer carries one, all will be well with his cattle at that most difficult time, May Day. Pliny mentions the ability of trefoil to cure those suffering from the effects of bites of poisonous creatures. So the legend of St Patrick banishing all snakes from Ireland may have derived from some such superstition as this.

How much greater the efficacy if it happens to be a four-leaved clover! There are many beliefs concerning it (see **FOUR-LEAVED CLOVER**). But there can be five-leaved specimens, too, though in America, finding one is believed to bring nothing but misfortune (Bergen. 1896). That is not the Scottish view, though – there is a proverb "He found himself in five-leaved clover", i.e., in very comfortable circumstances. Two-leaved Clover also had some significance:

A clover, a clover of two,
Put it in your right shoe;
The first young man you meet,
In field, street or lane,
You'll get him, or one of his name (Dyer).

Similarly, there is a Quebec superstition that if you put a four-leaved clover in your shoe, you will marry a man having the first name of the man you meet first after doing so (Bergen. 1896), and there are many more with similar import. You could put your four-leaved clover over the door, too. Then the first person to walk under it will be your future mate. A two-leaved specimen would do as well in Wales, provided it was found by accident. Then it should be put under the pillow, so that the finder should dream of the future partner (Howells).

There is one piece of weather lore involving clover: they say that their leaves fold before rain, and expand some hours before a heat wave (Trevelyan). To dream that you are in a field of clover is an omen of health and happiness (Raphael). What other interpretation could there be in popular imagination?

WHITE DEADNETTLE

(*Lamium album*) Boys made whistles from the stalks, Curtis told us, but, as was pointed out by Vickery. 1995, girls made whistles, too, for the stems are easy to cut. "The flowers have been particularly celebrated in uterine fluors, and other female weaknesses …" (Curtis), and it is still being recommended, either in infusion or tincture for these conditions (Schauenberg & Paris). Gerard had earlier advised "that the white floures … stay the whites" (which sounds suspiciously like doctrine of signatures). He went on, "the floures are baked with sugar as Roses are which is called Sugar Roset: as also the distilled water of them, which is used to make the heart merry, to make good colour in the face and to refresh the vitall spirits".

WHITE HEATHER

(*Calluna vulgaris var. alba*, if it is a genuine separate variety). This is the white Scotch heather, said to be the print left by the fairies (Simpson), and it is lucky to find a piece of it. If a sprig of it is given by a man to a woman, it is a declaration of love (Addy), and "happy is the married life of her who wears the white heather at her wedding" (Cheviot). It is a lucky plant in France, too, and in the Liège district of Belgium. The girl who finds it will have good luck in her household management (Sebillot).

WHITE HELLEBORE

(*Veratrum album*) No relation to the true hellebore, but still highly poisonous to man and cattle, producing among other symptoms a fall in blood pressure (the tincture is used medicinally to treat hypertension (Schauenberg & Paris)). The powdered rhizome sprinkled on currant and gooseberry bushes protects them against insect pests; apparently the powder loses its toxic quality after three or four days, so it is safe to apply to ripening fruit (M Baker. 1977).

Medicinally speaking, this is an emetic, and drastic purgative, rarely used internally (Fluckiger & Hanbury). Gerard spoke of "this strong medicine … ought not to be given inwardly into delicate bodies without great correction, but it may be more safely given unto countrey people which feed grossly, and have had tough and strong bodies". Lindley, though, seemed to be recommending it for a variety of ills, "melancholia, mania, epilepsy, herpes, gout, chronic affections of the brain …". It was occasionally used in the form of an ointment to treat scabies (Fluckiger & Hanbury). "The itch … mix Powder of white Hellebore with Cream for three days … It seldom fails" (Wesley). Gerard had already quoted Pliny, "that it is a medicine against the Lowsie evill". They were quite right – the powder is highly toxic to fleas and lice (Flück). In Russian folk medicine, it is said that the infusion would stop hiccuping immediately (Kourennoff). Of course, the doses would have to be very small.

WHITE HICKORY

(*Carya tomentosa*) A very tough timber, used for skis and other sports equipment, such as lacrosse sticks, a game invented by the American Indians (Edlin). This is the best wood for its resistance to impact, so it is ideal for the handles of hammers, axes, picks, etc. Hickory bark has long been known as a dye. The colour is yellow, if set with alum (R B Browne), and the twigs and leaves give a tan colour, with alum (S M Robertson).

Carolina wives used to hold straying husbands by driving a hickory stub into the doorpost. Hickory is very slow to rot, and while the peg is sound, husbands remain faithful (M Baker. 1977). It is said in Alabama that a toothbrush made out of a hickory twig will stop the teeth from falling out (R B Browne). Another prescription from Alabama is to "take the ashes from burnt hickory wood, put them in water, and drink it for fever. Make it very weak, as it will eat the stomach" (R B Browne).

WHITE HOREHOUND

(*Marrubium vulgare*) Evelyn recommended the use of white horehound in beer instead of hops, and horehound beer was an East Anglian specialty (Clair). Randall records how his mother would always put a sprig of horehound in her brew, to improve the flavour, and to improve appetite. In Dorset, horehound and wood sage boiled and mixed with sugar made a cooling drink called woodsage beer, which was drunk at harvest time (Dacombe). Candied horehound was made, too (Grieve. 1933). But the herb is best known in folk medicine. This horehound candy was popular in American medicine before it became popular as a confection (Lloyd). The leaf expectorant has always been used (as an expectorant) to cure coughs and colds (Vesey-Fitzgerald; Dacombe), and it is still used in lozenges to control a cough (Cameron). As long ago as the Anglo-Saxon period it was prescribed for colds in the head (Cockayne), and leechdoms of similar date are recorded for coughs (Dawson. 1934). Gerard, too, recommended the infusion, for it "prevaileth against the old cough", and the syrup made from the leaves "is a most singular remedy against the cough and wheezing of the lungs …".

Similarly, horehound was used for all pains in the chest, and for lung disease. The Lacnunga prescribed a draught for lung diseases: "boil marrubium in wine or ale; sweeten somewhat with honey. Give it warm to drink after the night's fast. And then let him lie on his right arm as much as he can" (Grattan & Singer). A tisane of this herb is often taken for weak stomach, lack of appetite, etc. (Flück), and horehound was remembered in Cheshire as the cure for loss of appetite (Cheshire FWI). Indigestion and dyspepsia, too, were treated with this preparation. Even the Navajo Indians were reported to use this herb for indigestion (Wyman & Harris), and it is certainly an American domestic medicine for dyspepsia still (Henkel). There is nothing new in this. The Anglo-Saxon Apuleius prescribed it "for sore of maw" (Cockayne), and it went on to advise its use for "tapeworms about the navel", a recipe still in use many centuries later, as candy or tea in Alabama (R B Browne), or simply by using the powdered leaves (Macleod).

Horehound has still more uses, one being as an application to wounds (Flück). In Africa, it is used for fevers, especially typhoid (Watt & Breyer-Brandwijk); the Navajo, too, used it to reduce fever (Wyman & Harris). In Wales, the infusion of the chopped herb is used both externally and internally for eczema and shingles (Conway). Oedema is another condition to be remedied with this plant. A 16th century recipe from France reads: "pisser, neuf matins sur le marrube avant

que le soleil l'ait touché; et à mesure que la plante mourra, le ventre se desenflera" (Sebillot) – but that is a simple transference charm. It is a counter-poison, too, or rather it was thought to be one. There are a number of authorities, though, who were sure of it, the Lacnunga for one, and the Anglo-Saxon Apuleius, and into the 14th century, too (Henslow). Perhaps the strangest use was in a sleeping draught, used in the Fen country. It was made of white horehound and rue, followed by a good dose of gin mixed with laudanum. It is quoted as being a last resort means of stopping a mother giving birth on 1 May (an unlucky day). It just put her to sleep for twenty-four hours (Porter. 1969).

WHITE MUGWORT

(*Artemisia gnaphalodes*) A North American species, of some ceremonial importance to some of the native Americans. The Chippewa, for instance, put the dried flowers on live coals, and the resultant fumes were looked upon as an antidote to "bad medicine" (Densmore), particularly at the beginning of any ceremony. Another usage was for ceremonial cleansing. For example, a man who had broken a taboo or who had touched a sacred object had to bathe with it (Youngken).

WHITE POND LILY

(*Nymphaea odorata*) An American waterlily, much used in the past by native American groups for food and medicine (Yarnell). The flower buds were eaten, either cooked or pickled (Coffey), and so were the leaves, boiled as greens. The roots were used as a substitute for soap, and they also yield a brown dye, as well as producing a liquid that, mixed with lemon juice, was supposed to remove freckles (Sanford). The root was used, too, in dysentery and diarrhoea, and the leaves and roots have been used as a poultice to boils, tumours, etc., (Grieve. 1931).

WHITE POPLAR

(*Populus alba*) "White", because, although the bark is black, it becomes grey or even yellowish-white higher up the branches (Leathart). The leaves, too, are white on the undersides, and blackish-green above, a fact that was used to explain the reason for its being the symbol of time, for those leaves show the alternation between night and day. That and the fact that they always seem to be in motion were enough to excite somebody's imagination into inaugurating the symbolism (Dyer. 1889). Greek mythology had the two colours of the leaves as representing the underworld (the dark side) and the world of the living (the light side) (Baumann).

To carry a wand of poplar when walking, to prevent the legs getting chafed (C J S Thompson. 1947) sounds nonsense, and is probably a garbled version of something quite different, perhaps the magical use of ash for a similar problem when on horseback. One superstition has it that poplars (not necessarily just this one) always lean to the east (Nall). There are a few genuine uses in folk medicine: Somerset people used to boil the bark, and drink the infusion for flatulence and fevers (Tongue). Gerard also recommended the bark, but for sciatica "or ache in the huckle bones", and for the strangury. "The same barke is also reported to make a woman barren if it be drunke with the kidney of a Mule, which thing the leaves also are thought to performe …".

WHITE WATERLILY

(*Nymphaea alba*) A symbol of charity, according to the ideas of Raban Maur (Haig). There used to be an idea that waterlilies submerged themselves at night. See Tennyson, *The Princess*:

> Now folds the lily all her sweetness up,
> And slips into the bosom of the lake.

Actually, they are careful to keep their leaves dry, by floating; water runs off the upper surface of the leaves (Young).

It was said once that if you fall while holding the flower in your hand, you would become epileptic (Grimm). The rhizome was used as an anti-aphrodisiac, to suppress sexual excitement, and the plant was nicknamed "destroyer of pleasure" because of its supposed power as a love-killer (Schauenberg & Paris). Gerard knew about this, and recommended the decoction, "or use the seed or root in pouder in his meats; for it drieth up the seed of generation, and so causeth a man to be chaste …". There was a cosmetic use once – the dew gathered from the cups of these waterlilies was used in Hampshire to enhance the appearance of the eyes (Hampshire FWI).

WHITLOW GRASS

(*Erophila verna*) "As touching the qualitie hereof, we have nothing to set downe: only it hath been taken to heale the disease of the nailes called a Whitlow, whereof it tooke his name" (Gerard). It is a sign of rain if the leaves of the Whitlow Grass droop (Inwards).

WHITLOW-WORT

(*Paronychia spp*) The name of the genus, *Paronychia*, is from Greek words meaning 'close to the nail', alluding to the original use of the plant to treat whitlows. That is why the genus is known generally as Whitlow-wort, or sometimes, particularly for *P jamesii*, Nailwort.

P argentea is a plant from the Middle East and North Africa. Palestinian children would eat the tips of the young stems, and because they are red at the joints, they give the name Dove's Foot, the dove who always has henna on her feet. For when Noah sent a raven and a dove from the Ark, the raven never came back, but the dove did, and Noah blessed her (Crowfoot & Baldensperger).

WHITLOWS

The use of WOODY NIGHTSHADE for skin complaints can be confirmed by the name sometimes given to this plant – Felonwort. That is a sure sign that it would have been used in curing whitlows which were called in Latin 'furunculi', little thieves – felons, in other words (Prior). "Country people commonly used to take the berries of it, and having bruised them, they apply them to Felons, and thereby soon rid their fingers of such troublesome Guests" (Culpeper). Irish country people have a "herb poultice" with which to dress a whitlow – YARROW leaves, fresh grass and a herb called finabawn, whatever that is. Equal parts of the herbs are ground up thoroughly, and then beaten up with white of egg. This is put on the inflamed finger, and it must not be changed for 48 hours (Logan). Another Irish charm is to point a GOOSEBERRY thorn at it nine times in the name of the Trinity (Wilde. 1902). A chewed TOBACCO leaf has been used in Scotland to cure a whitlow (Rorie. 1914). But the best known cure is by use of the plant known as WHITLOW-WORT, *Paronychia spp*. The generic name is from Greek words meaning 'close to the nail', an allusion to its original use to treat the condition. See also WHITLOW GRASS (*Erophila verna*).

WHITSUN

Whitsun used to be known in Italy as Pasqua Rosata, or Pasqua della Rosa, because the festival is celebrated at the time when you would expect the roses to be in bloom (Dyer). Sometimes the Whitsuntide association is with flowers in general, as when Yorkshire milkmaids from Hornsea and Southrop at one time collected flowers to make garlands at the cowherd's house, ready for a Monday festival (M Baker. 1974). But, though the feast is a movable one, some flowers seem to be particularly associated with it, and it can only be because it would be reasonable to find them in flower whenever the day happens to fall. Peonies are called Pentecost Roses (Pfingstrose) in Germany, for example. The GUELDER ROSE, which actually is, or was, the ecclesiastical emblem for Whitsuntide, was called Whitsun Boss in Gloucestershire (boss here refers to the shape of the flower buds). WOOD SORREL is Whitsun Flower in Dorset (Grigson), though this would seem to be rather too early flowering to merit the name; but Whitsun Gilliflower, a Somerset and Dorset name for the double variety of Sweet Rocket (Britten & Holland), seems to be safe enough (though Gerard called the same plant Winter Gilliflower, surely a misunderstanding on his part). Greater Stitchwort was known as Whit Sunday, or White Sunday, in Devonshire (Friend. 1882), but the fact that one can find Easter Flower, or Easter Bell, for the same plant shows the length of flowering time to be expected of it. CHICKWEED, its close relative, will be found in bloom virtually the whole year through. A spray of BROOM flowers was a traditional feature of Whitsuntide decorations (A W Hatfield). LILIES-OF-THE-VALLEY are also associated with Whitsuntide, so churches are decorated with them at that time (J Addison).

The town gates of Dunbar were once dressed with flowers at Whitsuntide (Banks). Kentish windmills were also decorated, and a pail with a small tree in it was hung from one of the arms, with a basket of bread and butter from the other (the sails would be standing as a St Andrew's cross; a vertical cross meant the mill was being repaired (Igglesden)). Trees would be used too for decoration on this day. BIRCH branches in particular were used to decorate churches; they were also a favourite church ornament in Germany (Tyack). Burne mentions the custom in Shropshire, and it was known at Raydon St Mary, in Suffolk (Gurdon). From Staffordshire, the church accounts of Bilston provide further proof of the custom in the following entries:

> 1691 "For dressing ye chapel with birch, 6d"
> 1697 "For birch to dress ye chapel at Whitsuntide, 6d"
> 1702 "For dressing ye chapel, and to Ann Knowles for birch, and a rose, 10d" (Tyack),

in spite of Herrick's implication, in:

> When yew is out, then Birch comes in
>   And many flowers besides,
> Both of a fresh and fragrant kinne,
>   To honour Whitsuntide.

YEW was certainly used to decorate the church at this time. At Winterslow, in Wiltshire, for instance, not only was it used on Whit Sunday, but it was kept in the church till the following Christmas (Vaux). Yew was also employed in the same way at many places in the West Midlands, e.g., at Kington, in Herefordshire, where it was fastened to the tops of the pews (Leather). MAPLE boughs were used for the decoration at Heybridge, near Maldon, in Essex (Vaux), where rushes were also strewn on Whit Sunday (Wright), as they were at Turley, and Haw, in Gloucestershire (Hartland). In the latter case, an acre of ground was given over to maintain the custom. At Tatton, in Somerset, John Lane left half an acre of ground, called the Groves, to the poor for ever, reserving a quantity of grass for strewing the church on Whit Sunday (Burton). St Mary Redcliffe, Bristol, is specially decorated with flowers on this day, the custom being to lay posies on every seat (Hole. 1943), and it too was strewn with rushes, in accordance with the will of William Mede, who gave a tenement to defray the expenses, in 1497 (Burton). The graves were covered with rushes at Farndon, in Cheshire (Tyack)

Green GOOSEBERRY pie used to be a traditional Whit Sunday dish (Savage).

WHOOPING COUGH

Sebillot says that children with coqueluche, which must be whooping cough, were passed through split ASH trees (see **HERNIA** also). He quoted an ancient ash in Richmond Park which was visited by mothers "dont les enfants étaient ensorcelés, malade de la coqueluche ou d'autres affections". It had to be done before sunrise, and no stranger could be present. The child was passed nine times under and over. It seems too that whooping cough could be cured by pinning a lock of the patient's hair to an ash tree (Addison & Hillhouse), a typical transference charm. Another way to deal with whooping cough is to crawl under a blackberry arch, with a certain amount of ritual. The Dorset remedy was to pass the child nine times under and over a bramble (Udal). Sometimes a rhyme had to be recited, like this one from Staffordshire (Raven):

> Under the briar, and over the briar
> I wish to leave the chin cough here.

That usage was known to Aubrey, in the 17th century. The cure in Warwickshire involved passing the child three times or perhaps nine times, under a "moocher", as it was called – a bramble that had bent back to root at both ends (Palmer. 1976). The Essex whooping cough remedy was to draw the child under "the wrong way", presumably, that is, by the ankles (Newman & Wilson). In the Midlands, the child had merely to walk under the bramble arch "a certain number of times". The same method was used for hernia, too, or for boils, etc.

Domestic medicine agrees on the efficacy of NETTLE in chest complaints, whether for a cough or something more serious. Gerard had recommended it for "the troublesome cough that children have, called the Chin-cough, [whooping cough in modern parlance], taken as a tea". A MILK THISTLE decoction was used in Ireland, or a CHICKWEED leaf infusion (Maloney; Ô Súilleabháin), and FLAX has been used in America: "take three pounds of flax seed, steep in one quart of water for three hours, mix with two lemons and two cups of sugar or some honey. Give this often as hot as the patient can take it" (R B Browne). A MARJORAM infusion is often given for whooping cough (Flück), and THYME, too, was given for the complaint (Camp) (thymol, a good antiseptic, is still used in gargles and mouthwashes). When the tree was plentiful in the United States, an infusion of the dried leaves of AMERICAN CHESTNUT (*Castanea dentata*) was given there for whooping cough (Weiner), just as in Europe a similar preparation is taken from SWEET CHESTNUT leaves (Page. 1978). A decoction of JUNIPER berries is good "against that which children are now and again extremely troubled, called the Chin-cough" (Gerard). A root decoction of ROSEBAY WILLOWHERB can be taken, too (Leyel. 1937). An infusion of the fresh roots of

ELECAMPANE, often given with honey, was used for the complaint (Thornton).

Advice from the Highlands of Scotland enjoins a decoction of APPLES and ROWAN, sweetened with brown sugar, to be taken for whooping cough (Beith), and in County Down, GORSE juice, taken night and morning was the remedy (St Clair). ONION juice rubbed on the soles of the feet is a folk cure for the condition (Camp), or GARLIC, in a similar way. But the simplest way to take it is to cut a clove in thin slices, put them in a saucer, and pour on just enough golden syrup to cover them. Leave it for two or three hours, and take a teaspoonful when necessary (Quelch).

Jut the smell of BROAD BEAN flowers is enough to cure the complaint, so it is still claimed in Norfolk (V G Hatfield. 1994). SUNFLOWER seeds are a specific for whooping cough, and Sunflower Gin used to be made quite often for the disorder – the boiled seeds in Hollands gin (Leyel. 1937). An infusion of the flowers of MARI-GOLD has long been a country remedy for the complaint (V G Hatfield. 1994). BLACKCURRANT, sovereign for sore throat, or quinsy, was also taken for whooping cough in the form of a tea, in Cumbria (Newman & Wilson). Herbal remedies using the bark and leaves of LESSER EVENING PRIMROSE have been used not only for whooping cough, but for asthma, too (Grieve. 1931). Wesley recommended PENNYROYAL for the complaint: "Chin-cough or Hooping-cough … give a spoonful of Juice … mixt with brown Sugar-candy, twice a day". In France, young babies with whooping cough (coqueluche) were given infusions made from RED POPPY (Loux), and in West Africa the GUINEA-HEN WEED (*Petiveria alliacea*) is used (Dalziel). Tie a bag of ASAFOETIDA round the child's neck and that will cure it (Iowa) (Stout).

WHORTLEBERRY,

or **BILBERRY**, as it is often called (*Vaccinium myrtillus*) is associated in Ireland with the Lughnasa festival of 1 August, particularly with the picking of whortleberries. In fact, Blaeberry Sunday is the name given to the first Sunday in August, and it is known as that, or something like it, in most parts of Ireland. As well as Blaeberry Sunday, there are Heatherberry, Bilberry, Fraughan, Whort or Hurt Sunday. The fruits are called Mulberries in some parts of County Donegal, so the Sunday festival is known as Mulberry Sunday in those parts (Mac Neill). Even if all aspects of the festival have disappeared, there still remain outings to pick bilberries on the mountains. The custom on Carn Trenna was for the boys to make bilberry bracelets for the girls. They were worn on the hilltop, but were carefully left there when it was time to go home (Mac Neill).

Dartmoor people used to chant, when going bilberrying:

The first I pick, I eat;
The second I pick, I throw away;
The third I pick, I put in my can.

If they said that, they would find a lot of berries (Crossing).

A syrup called Rob was made at one time from bilberries – "the juyce of the blacke Soirtleberries is boyled till it become thicke, and is prepared or kept by adding hony or sugar to it : the Apothecaries call it Rob …" (Gerard). The fruit is distinctly astringent, and is still sometimes prescribed for diarrhoea (Schauenberg & Paris), just as it has been for centuries in the Highlands of Scotland (Beith). In parts of France. Bilberries were said to be "souveraines contre la colique des enfants" (Loux). Both the berries and the leaves are used in an Irish cure for kidney trouble (Ô Súilleabháin), In north-east Sutherland, too, they were used for dissolving kidney stones (Beith). Bilberry tea (up to ten cups a day, very hot), was a Russian folk remedy for asthma; diabetes was treated there by strict diet and a prescription of infusion of dried bilberry leaves – a tablespoonful to a cup of boiling water infused for half an hour. Another use for the leaves or young shoots in Russia was for arthritis – one part to one part of alcohol, kept in a warm place for 24 hours, and then strained. A tablespoonful in warm water was to be taken twice a day (Kourennoff).

WILD ANGELICA

(*Angelica sylvestris*) A traditional food on the Faeroes for St John's Day, when the stem is chopped in small pieces, and served with sugar and cream (Williamson). Gypsies used to smoke the stem as a tobacco substitute, as they did also with the stems of hogweed. Of course, wild angelica inherited some of the beliefs invested in its more august relative, Archangel (*Angelica archangelica*). So we find it to have been a remedy against poison or the plague, by chewing the root, apparently (F P Wilson),
(… it has always been famous against pestilential and contagious disease" (Hill. 1754). Wood-Martin recorded its use in Ireland as a cure for hydrophobia, and so on. Even some of Archangel's names attached themselves to this plant, Holy Ghost, for example, the medieval radix Spiriti Sancti, or Holy Plant. Ghost Kex is a descendant of that concept (Grigson. 1955). Kex in one form or another is the name given to any hollow-stemmed umbellifer, and usually refers to the dried stems. Trumpet Keck is another of these names (Britten & Holland), so called because children used to make "trumpets" of the hollow stems, which have other uses, too, for in Scotland they are Ait-skeiters, or Bear-skeiters (Grigson. 1955). Children shoot oats through the stems, like peas through a pea-shooter. The Somerset name Water Squirt (Mabey. 1972) seems to point to yet another use.

WILD CHERRY

(*Prunus avium*) To dream of cherries means misfortune, according to Dyer. 1889, but:

A cherry year,
A merry year (A C Smith; Swainson. 1873).

The wild cherry had some magical uses; to get rid of
a fever all one had to do was to lie naked under the
tree on St John's Day, and to shake the dew on one's
back (Dyer. 1889). This was from Germany, but there
was a very similar usage from the south of France,
the tree being the peach this time. There were genuine
attempts at medicinal usage, though. "The distilled
water of Cherries", according to Gerard, "is good for
those that are troubled with heate and inflammation
in their stomackes, and prevaileth against the falling
sicknesse given mixed with wine". He also noted
that "the gum of the Cherry tree taken with wine and
water, is reported to help the stone …", something
on which Lupton had already reported. Cherry gum
dissolved on wine was a remedy for coughs and colds
(Earwood). Wild Cherry seems to maintain normal
uric acid levels in people suffering from gout, and was
much used for the purpose before synthetic treatment
was available (Lewis & Elvin-Lewis).

WILD GARLIC
See under RAMSONS (*Allium ursinum*)

WILD PLUM
See under BULLACE (*Prunus domestica 'institia'*)

WILD RADISH
(*Raphanus raphanistrum*) Not the ancestor of the
garden radish, but this is a common farm and garden
weed. The seeds, which are toxic to livestock, have
been used as a remedy for haemorrhage and malaria
(Watt & Breyer-Brandwijk). Among Europeans in
South Africa, the plant has been used for gravel,
a remedy listed by Hill a long time ago. An early
17th century auburn hair dye had as its principal
constituents radish and hedge-privet (Wykes-Joyce).

WILD RICE
(*Zizania aquatica*) A North American plant, found in
marshes, shallow ponds, and by lake shores, etc. It was
particularly associated with the Menomini Indians,
whose name seems to mean something like "Wild
Rice Men", because they were so intimately connected
with the harvest and use of the plant. They even
transplanted it into new waters. Since it springs up
from under the earth and water, it was reckoned to be
the gift of the "Underneath" beings. They believed that
the birds on their migrations followed these beings,
and brought rice to them. The usual harvest season is
about the middle of September, when the Menomini
gathered in camps on the shores of the lakes. When the
rice was heavy, the chief of each band made a sacrifice
of tobacco to the "Underneath" beings, the Master of
Rice, and begged permission to harvest it (A Skinner).

WILD SAGE
(*Salvia horminoides*) In the Cotswolds, it is said to
be a legacy of the Roman occupation of Britain. The

soldiers, they say, dropped the seed as they marched
across the country. In proof of this, country people
will point out that it often flourishes along the line of
old Roman roads (Briggs. 1974). It is a true native,
though, even if found only locally, but it can be quite
frequent in grassy places.

The seed drunk with wine was reckoned to be
aphrodisiac, a view to which Culpeper subscribed,
but there were less recondite uses in medicine. A
decoction, for instance, was used in Lincolnshire for
sprains (Gutch & Peacock). But the other prescriptions are much older, and less particular, like this 15th
century remedy: "for botches: Take … oculus Christi
and vervain, and make a plaster of them; and lay it
from the boil two finger-breadths, and again put it as
far further. And so do till it come to the place where
you will break it". (Dawson. 1934). Hardly a model
of clarity. Oculus Christi, is, of course Christ's Eye
("most blasphemously called Christ's Eye, because it
cures Diseases of the Eye" (Culpeper)).

There were some veterinary usages as well, noticed
by Martin in his account of the Western Isles. Horses
were wormed with it, he said, and "a quantity …
chewed between one's teeth, and put into the ears of
cows and sheep that become blind, cures them, and
perfectly restores their sight, of which there are many
fresh instances both in Skye and Harris, by persons of
great integrity".

WILD STRAWBERRY
(*Fragaria vesca*) Often confused with the Barren
Strawberry, which is no relation, and belongs to an
entirely different genus. Nobody seems to agree about
how strawberry got its name. Of course, we put straw
around the plants to protect the fruit, but the name
was streawberige in OE times. The *Oxford Dictionary
of English Etymology* admits it does not know the
derivation. Anyway, the fruit has always been much
appreciated; to quote Aubrey – "strawberries have
moist delicious taste, and are so innocent that a
woman in childbed, or one in a feaver, may safely eat
them".

WILLOW
(*Salix spp*) The Latin word salix, so it is said,
comes from salire, to leap, bestowed because of
the extraordinarily quick growth of the tree. Not
for nothing is there a saying, "The willow will
buy a horse, before the oak will pay for a saddle"
(Denham. 1846). Willows have for a very long time
been symbols of sorrow, and of forsaken love; in the
Scriptures they are generally a symbol of woe and
sadness (Dyer. 1889). Aubrey records the usage in
Oxfordshire, albeit "in a frolique", but Shakespeare
used the symbolism seriously as, for instance, in
Much ado about nothing, when Benedick says:
"I offered him my company to a willow-tree, either to
make him a garland, as being forsaken, or to bind him

up a rod, as being worthy to be whipped". Or, in *III. King Henry VI*, when Bona says:

> Tell him, in hope he'll prove a widower shortly,
> I'll wear the willow garland for his sake.

In Wales, willow caps used to be presented to people disappointed in love (Trevelyan), and willows were the symbols of grief, especially that of the disappointed lover. Weeping willows in particular used to be death and mourning symbols, often used in some stylised form embossed on Victorian mourning cards, though they first appeared as such at the end of the previous century as a tomb decoration, usually with the figure of Hope, or the widow weeping and clinging to the urn beneath its boughs (Burgess).

But according to the dream books, if one has a dream of oneself mourning under a willow over some calamity, it is actually a happy omen, forecasting good news! (Gordon. 1985). Because of their association with water, willows could be symbols of resurrection (Curl), and that may be why branches of willow are carried by mourners at a mason's funeral (Puckle). In China they have from very ancient times been looked on as tokens of immortality (Curl). They are even credited with aphrodisiac qualities – "spring water in which willow seeds have been steeped was strongly recommended in England as an aphrodisiac, but with the caveat that he who drinks it will have no sons, and only barren daughters" (Boland. 1977). The human backbone, according to Ainu tradition, was originally made of a willow branch, and the backbone, they say, is the seat of life (Munro). Willow was a symbol of vitality in China, too, and it is that aspect of it that probably accounts for the wearing of willow wreaths as a protection against scorpion poison (Tun Li-Ch'en).

Irish harps were usually made of willow wood, for these trees have a soul in them which speaks in music. Brian Borohm's ancient harp, still in existence when Lady Wilde was writing in 1890, was made of willow. There was another Irish belief that is related – willow had the gift of inspiring an uncontrollable desire to dance. A willow-wand, pared to a four-sided stick, with the necessary spells cut upon it, would cause all the inmates of a house to dance if it were put over the lintel (Wood-Martin). Perhaps that is why sawn willow was so unlucky in the house; if it were admitted at all, the wood had to be shaped with an adze (Whitlock. 1982). Magical beliefs like this had their sinister side, too. Dyer said that in Hesse, one may kill an enemy at a distance with knots tied in willows, and there was a gypsy magical rite that consisted in watering a branch of weeping willow for nine days and pouring this water in front of the house of the person who was to suffer (Clebert). This is homeopathic magic, of course, for the weeping willow symbolizes tears. And willows were supposed to have a sinister

habit of following a traveller on a dark night. See the Somerset folk song:

> Ellum do grieve,
> Oak he do hate,
> Willow do walk
> If you travel late (Briggs. 1978).

Chinese farmers used to ask for rain by prayers to the Dragon King Lung Wang, who has control over rain, and they would wear willow wreaths during the ceremony (Tun Li-Ch'en), for willows grow in wet country, and homeopathic magic accounts for their use in this context (they were worn for other purposes, as already mentioned). So, too, the doctrine of signatures would ensure that they would be used for diseases like ague, caused by damp. There was a Welsh charm for ague that said: "Go in silence, without crossing water, to a hollow willow tree. Breathe into the hole three times, then stop the aperture as quickly as possible, and go home without looking round, or speaking a word" (Trevelyan). But there was another Welsh usage recorded by the same author, that of treating toothache by picking at the decayed tooth with a sharp twig of willow, until it bled. After that, the twig had to be thrown into a running stream. Simply chewing some willow bark would have been useful, for it contains salicin, from which salicylic acid was obtained. Later, this was compounded into acetyl-salicylic acid – aspirin, in a word (Grigson. 1955). Salicin, or Willow Quinine, as it is sometimes called (Savage), is apparently still used in the treatment of rheumatic fever.

Russian folklore contains a belief that willow branches put under the marriage bed would ensure a pregnancy (Kourennoff). But this is a fruitless tree, and so it would be used for contraception. Even in quite modern times German women believed that drinking willow tea would make them barren (Simons).

Gull is a Sussex word that seems to apply to any willow catkins (Parish); so too does Goslings, at least in Hampshire (Cope). These two names are connected, for in some places (Herefordshire among them) it is unlucky to bring willow catkins (called Gulls) into a house where there are young goslings (also called Gulls).

It was by the rivers of Babylon that the captive Jews hung their harps upon the willows, so the boughs were bent with their weight, and so they have always remained as Weeping Willows (Psalm 137: "We hanged our harps upon the willows in the midst thereof"). One belief connected with the tree is that a branch of it was used as the scourge with which Christ was beaten, since when it has always drooped its boughs and wept (Friend. 1883). There is a saying recorded from Alabama – "plant a weeping willow and by the time it casts a shadow, it will shade your grave" (R B Browne). For quite a different reason it was thought to be bad luck to plant a weeping willow in Kentucky; there the result would be to remain an old maid (Thomas & Thomas). This tradition that it was used as Christ's scourge accounts for the practice in the early Christian church, when the willow was used to punish those who did not go to early mass at Easter (Fogel). There is, too, quite a common belief that animals struck with a willow rod will be seized with internal pains, and children beaten with one will stop growing (Burne. 1914), or that it will cause swellings in both children and animals who are struck with it (Fogel).

WINDFLOWER
A name given to most species of ANEMONE, particularly to the WOOD ANEMONE (*Anemone nemerosa*). It is explained by asserting that some species flourish in open exposed places, or that they would not open till the March winds begin to blow (Friend. 1883). The belief is from Pliny, and the Greek anemos is the word, and the name of the flower means literally "daughter of the wind" But this is folk etymology really, for the true origin is the Semitic word na'aman, which means the one "who was pleasant", or "lovely", with actual reference to the POPPY ANEMONE (*A coronaria*) (Grigson. 1976).

WINTER ACONITE
(*Eranthis hyemalis*) It is toxic, at any rate to animals, but little harm is normally done, for the burning taste will usually stop them eating it (North), and it is not so dire as Gerard's remark ("this herb is counted to be very dangerous and deadly") would suggest. But he went on to describe its uses, one of which involved scorpions: "… it is reported to prevail mightily against the bitings of scorpions". It seems that scorpions had only to touch the plant, and they became "dull, heavy and senceless", but White hellebore was the scorpion's antidote.

WINTER JASMINE
(*Jasminum nudiflorum*) It is reported that in Caistor, in Lincolnshire, yellow jasmine was the gift brought by the first-foot on New Year's Day (Rudkin).

WISHING CAP
The Welsh wishing cap was usually made of the leaves and twigs of HAZEL, though sometimes JUNIPER could be used. The hazel leaves had to be gathered at midnight, at new or full moon, and made up as quickly as possible (Trevelyan). The cap was worn for protection, and also for good luck, particularly by sailors, or anyone connected with the sea and ships (R L Brown). It was also said that, wearing one of these caps, a person could easily remember nursery tales or other stories (Roderick).

WITCH HAZEL
(*Hamamelis virginica*) The twigs were used in America as divining rods, as those of hazel still are in England. As water divining was once looked on as the result of occult power, the name Witch Hazel was given (Weiner). This is a useful tree in medicine, the leaves and bark being used in pharmacy as Pond's Extract, a lotion to be applied for skin inflammation, bruises (Sanecki), or for scalds and burns (Browning). This preparation, properly diluted, can be used as an eye lotion, too (Leyel. 1937).

WITCHES
GARLIC, in older belief, holds witches at bay, by putting some under a child's pillow, as is the Polish custom (Leland), while the Bosnian belief was that everyone should taste garlic before going to bed at the time when witches were traditionally active. Around Sarajevo garlic was rubbed on children's chests, on the soles of the feet and the armpits at Christmas and Easter, which are the times when witches attack people and eat them. While rubbing children with garlic on those days, a formula used to be recited: "When the witch has counted up all the blades of grass on the ground and all the leaves on the trees, then let her kill my child". It was also Serbian practice to put a garlic bulb (or a juniper twig) on the windowsill on the evening of St Thomas's Day (19 October in that calendar), which would keep witches from the house all the year (Vukanovic).

TURNIPS have been used to keep witches away. African Americans in the deep south of the USA used to scatter turnip seed around the house for that purpose (R B Browne).

ASH was regarded as all-powerful against witch-craft. A Greek historian said that the Assyrians wore amulets made from its wood round their necks as a charm against sorcery (Berdoe). Rather more recently, in Lincolnshire, the female ash, called Sheder, would defeat a male witch, while the male tree, Heder, was useful against a female one (Baker). Eating ash buds provided invulnerability to witchcraft (Banks). Ash-wood sticks were preferred to any other, as they would protect the cattle from witchcraft. A beast struck with one could never be harmed, as witchcraft could never then strike a vital part (Wiltshire), while branches of ash were wreathed round a cow's horns, and round a cradle, too (Wilde). English mothers rigged little hammocks to ash trees, where their children might sleep while field work was going on, believing that the wood and leaves were a sure protection against danger. A bunch of the leaves guarded any bed from harm, and a house surrounded by an ash grove would always be secure (Skinner). BIRCH, too, was potent against witchcraft. Over quite a large part of the Continent, it used to be said that if a witch were struck with a birch broom, she would lose all her power (Lea). The Irish believe that the birch is disliked by fairies, and in the west of England, birch crosses were hung over cottage doors to repel enchantment (Grigson), while Herefordshire waggoners would couple birch branches to protect stables on May Eve. WAYFARING TREE wood also protected. Evelyn mentions its use for protecting cattle from witchcraft, and Aubrey. 1847 had "… they used … to make pinnes for the yoakes of their oxen of them, believing it had vertue to preserve them from being forespoken, as they call it; and they use to plant one by the dwelling-houses, believing it to preserve from witches and evil eyes". An aspect of HAWTHORN is as an abode, not just of fairies, but of witches, too. It was an accepted belief in the Channel Islands that witches used to meet under solitary hawthorns (MacCulloch), and there used to be quite a widespread superstition that it was dangerous to sit under a hawthorn on Walpurgis Night, May Eve in our terms, because it was then that a witch was most likely to turn herself into a thorn tree (Jacob). On the other hand, and quite in keeping with accepted belief, hawthorn would also protect against witchcraft. In Gwent tradition, one of the commonest ways of breaking a witch's spell was reckoned to be putting whitethorn (or birch) over the house door (Roderick). Far from there, the Serbs believed that a cradle made from hawthorn wood would be a most powerful protective device (Vukanovic). To "drive witches out of milk" by beating it with hawthorn used to be a Pennsylvanian German saying (Fogel). But it is the thorns themselves that are the deterrent. So, too, with BUCKTHORN, but with a buckthorn stick a man can strike witches and demons. There are examples as far back as Ovid, who described a ceremony for countering a vampire witch. The final act was to put branches of buckthorn in the window (Halliday). CHRIST-THORN is another case. A statement, originally from Dioscorides, runs "it is said that the branches thereof, being layd in gates or windows, doe drive away the enchantments of witches" (Dioscorides *edited by* R T Gunther). The ambivalence noted with Hawthorn is also apparent with HONEYSUCKLE. It is such a familiar and well-loved plant that it is difficult to connect with witches, whether to thwart them or as a plant that they would use for their own ends. Certainly, its evil-averting powers outweigh its evil-working claims. On May Day in particular, when, so it seems, there was always a lot of ill-wishing about, honeysuckle took care of the butter and milk, and the cows (Grigson). It was a favourite in Scotland (pregnant women were advised to wear it (Dempster)), along with the rowan:

> The ran-tree an' the widd-bin
> Haud the witches on come in (McPherson).

Widd-bin, i.e., woodbine, is the honeysuckle. To be mentioned in the same breath as rowan is praise indeed. The widd-bin would be wound round a rantree wand, and then put over the byre door (Milne). Perhaps it was the spiral-growing (and clockwise) habit that gave it such a reputation (D A Mackenzie). The protection, then, lies not in any inherent quality

of the plant itself, but in the way it grows. For there was a firm belief that witches or those who had the evil eye are forced to stop whatever they are doing and to follow out every detail of an involved design that they see. So interlacing and complex interwoven braided cords were deliberately made to distract, delay and confuse the evil eye, and were worn specifically for that purpose (Gifford) (JUNIPER operated in the same way; the deterrent was the leaves, every one of which had to be counted before the witch could proceed (M Baker. 1977)) Even when the honeysuckle wreath was used (by a witch, for often it was the witch who had to use the plant in order that its curative powers could succeed), to cure an ailment, as was certainly done in 18th century Scotland, it did not stop a witch from being charged (see the case of Janet Stewart under **HONEYSUCKLE**). Exactly the opposite can be quoted, as is always the case in magical patterns. The witches themselves will use honeysuckle against victims who will be using the same plant to protect themselves. In the ballad of Willie's Lady, for instance, the witch tries various means of preventing the birth of the Lady's child, including a "bush o' woodbine" planted between her bower and the girl's. Once this restricting, constricting, plant had been removed, the birth proceeds normally (Grigson). WOODY NIGHTSHADE is in the same category as Honeysuckle. Like any poisonous plant, it was said in the Highlands (Kennedy) to have great powers of evil. Such a witch plant would actually give people rheumatism! But it was used in medicine to cure rheumatism, as Lindley confirmed.

VERVAIN is a great protector, either of the home (plant it on the roof and it will guard the house against lightning (Sebillot)), or of the person. At one time, in the Isle of Man, neither the mother nor a new-born baby were let out of the house before christening day, and then both had a piece of vervain sewn into their underclothes for protection (Gill. 1963), while in Sussex the practice was to dry the leaves and put them in a black silk bag, to be worn round the necks of sickly children (Latham), probably rather to avert witchcraft than to effect a cure. ST JOHN'S WORT is just as effective. When hung up on St John's Day together with a cross over a house door, it kept out all evil spirits (Napier). Pennsylvania Germans fasten a sprig to the door to keep out witches (and flies) (Fogel), while in the Western Isles the emphasis was on preventing ordinary folk from seeing the witches (or "grisly visions", as it was described); it had to be sewn into the neck of a coat (Bonser), and left there. If it were interfered with in any way it could lose its power (Spence. 1959). White witches used it, too, in an "unwitching" medicine, which consisted of, among other things, three leaves of sage and three of St John's Wort, steeped in ale, to be taken night and morning (Seth).

RAGWORT is in a different class. It is a fairy plant, dedicated to them in Ireland, and called Fairies' Horse, for it was believed to be a fairy horse in disguise (so is St John's Wort). Folklore equates fairies with witches in many particulars, the belief in their using ragwort to ride on at midnight being very widespread (Henderson; Hunt; Wentz, etc.). "As rank a witch as ever rode on ragwort" was a common saying in those areas where the belief was held (Cromek). CABBAGE stalks served, too (Wood-Martin), and of course so did BROOM. GROUNDSEL owes its very existence to witches, so it was once thought in the Fen country. A small patch growing beside an old trackway showed that a witch had stopped there to urinate; large patches meant that a number of them had met there to plot. Groundsel growing in the thatch was a sign that a witch had landed on the roof during a broomstick flight. It was also believed that witches could never die in winter, but only when the groundsel was in flower (even though it seems to be in flower all the year round). The point was that the witch could then take with her a posy of the flowers, by which the devil could then recognise her as his follower (Porter. 1969). On the other hand, it was quite common once as a counter-charm to witchcraft. In the Western Isles, on St Martin's Day, groundsel was used particularly when milk was being stolen by witchcraft (Polson), and there are records of the use of pieces of the root as amulets against the evil eye (Folkard). WATER-CRESS was used by witches in the Highlands to steal milk, by cutting off the tops of the plant with a pair of scissors. At the same time a charm was spoken along with the name of a cow's owner, and ending with the words "the half mine, the other half thine". A handful of grass from the thatch over the byre would apparently do just as well. The counter charm was groundsel put with the milk (Polson. 1932). In Greece, BASIL was the instrument for finding a witch. While it was burning, a number of names were repeated in succession. A loud pop or crackle tells that the name of the offender has been reached (Lawson).

BROAD BEANS are fairy food, but are associated with the dead, and with witches. Ovid said that a witch put beans in her mouth when she tried to call up spirits. On the other hand, spitting beans at a witch was a Wiltshire way of rendering her spells ineffective (Whitlock, 1992).

ELDER is the most important plant when connections with witchcraft, at least in folklore, are considered. From ancient times, it has been believed that some kind of spiritual being lived in the tree, whether goddess, fairy, witch, or the souls of the dead. That is why, along with hawthorn, there has always been an injunction against cutting elders down, or even lopping off a branch. Permission had to be asked of the being that lived in the tree. But belief went further, for in some cases, the elder actually *was* the witch (see the legend of the **ROLLRIGHT STONES**):

Thou and thy men hoar stones shall be,
And I myself an eldern tree (A J Evans).

The proof that the "elder is a witch is that it bleeds when it is cut" led to the custom, on Midsummer Eve, for people from the area to meet at the stones, and to cut the elder, or at least the tree they thought to be the real one. There is, too, a folk tale from Somerset, called the Elder Tree Witch, where again the witch is the elder, and moves around the farm in the form of the tree (Tongue). At the same time, the tree offered protection. In America, it was said that an elder stick burned on Christmas Eve would somehow or other reveal all the witches in the neighbourhood; another American belief was that, if a small piece of elder pith was cut, dipped in oil, lit, and then floated in water, it would point to any witch present (originally a German belief, carried to America). If a farmer was losing livestock, and wanted to know whether or not the losses were due to witchcraft, he was advised to take six knots of elder wood and to put them "in orderly arrangement" under a new ash bowl or platter. If they were later found "all squandered about", then he would know that the cattle were dying from witchcraft (Brockie). An elder hedge secured many a home in Scotland from "undesirable attention" (McPherson), and at difficult dates in the calendar, May Eve for example, elder leaves were put on doors and windows to keep witches away, and an elder cross was fixed to stables and byres (McNeill). See under **ELDER** for many more examples. JUDAS TREE (*Cercis siliquastrum*) is another example of a thoroughly unlucky tree, for Judas was supposed to have hanged himself on one, being a haunt of witches (Dyer. 1889).

FENNEL was used to ward off evil spirits (Emboden), and it was hung over the door along with other herbs of St John at Midsummer (C P Johnson). The 'bena-ndanti' of 16th century Friuli, who were the "night-walkers" who fought the witches on a psychic level, carried fennel as their weapon, while the witches carried Sorghum as theirs. It was said that these 'benandanti' ate garlic and fennel "because they are a defence against witches" (Ginzburg). A CARNATION was also used in Italy at Midsummer Eve, when the witches were specially active. All one had to do was give them a carnation. For any witch had to stop and count the petals, and long before she had done that, one was well out of their reach (Abbott).

YEW, the "dismal yew" in Shakespeare's words, slips of which "slivered in the moon's eclipse" were among the ingredients of the witches' cauldron in *Macbeth*. But, as usual, yew was a strong anti-witch protector. Any place sheltered by yew trees is safe from witches (Lowe), and it was often planted at the south-west corner of a house for this reason (Jacob). "Vor den Eiben kann kein Zauber bleiben" is still a current proverb in parts of Germany (Elliott), and in the Forest of Dean a yew stick nailed behind the door prevented a witch from entering the house (Elliott). A plant as poisonous as HEMLOCK would naturally have an association with witches. "Root of hemlock digged in the dark" was one of the ingredients in the witches' cauldron in *Macbeth*, and Summers claimed that it was used by them either as a poison or as a drug, favoured mainly because of its soporific effects, and he quoted a formula for witch ointments that had hemlock as one of the ingredients. DEADLY NIGHTSHADE also appeared as an ingredient in these witch ointments as 'Morelle endormante'. SWEET FLAG was another plant used in witch ointments, one formula for which was "De la Bule, de l'Acorum vulgaire, de la Morelle endormante, et de l'huyle" (Summers. 1927, though these formulae were written by the 16th century Italian physician Della Porta). CINQUEFOIL is another ingredient used in these witch ointments, though just what it is doing amid the other poisonous or plain revolting ingredients, is not at all clear. But it was certainly taken as an antidote. Reginald Scot refers to the custom of those "who hang in their entries an herb called Pentaphyllon Cinquefoil", with haw-thorn gathered on May Day, in order "to be delivered from witches". Cinquefoil appears again in a witch philtre for love or hate, composed of adders, spiders, cinquefoil, the brains of an unbaptised baby, and so on (Summers). BRAMBLE offered a physical barrier, as in Guernsey, where wreaths of it were hung from the rafters, the idea being that the witches would get scratched while flying through the air (McCulloch). Sometimes a sprig of bramble was used like rowan to put in or under the milking pail to ensure that the substance of the milk could not be taken by some evil agency (MacGregor), even burning bramble offers protection, for it is said that a bridal bedchamber ought to be fumigated as a safeguard against ill-wish-ing (Boland. 1977). ORPINE had the reputation in the Gironde region of France of being sensitive to the presence of sorcery – if there was a witch around it would wither as soon as it was brought near her (Sebillot).

BLACKTHORN thorns were the ones used to stick into wax images made for black magic, and black-thorn wood was often used for a witch's walking stick (Wiltshire). This "black rod" carried by witches caused miscarriages, and when Major Weir was burned in Edinburgh in 1670, a blackthorn staff was burned with him as the chief instrument of his sorcer-ies (Graves). In Russian tradition, ASPEN was the wood used to make the staff driven through a witch's heart, or simply laid on the grave, to make sure she stayed there (J Mason).

THORN-APPLE (*Datura meteloides*) capsules were used in black magic, and it was certainly considered to be used by witches in their incantations (Trev-elyan). In Puritan times, those who grew it in their gardens were in danger of being burned as witches. Such hallucinogens, particularly Toloache, the cult of many of the Indian tribes of the American south-west, which are powerful enough to kill, or to bring about temporary derangement, or even permanent insanity, must have entered the practice of witchcraft (Furst). Certainly a malicious shaman could use the drug for his own ends (see Gayton). In Africa. a way of detecting witches is by Guinea Pepper , which may be *Capsicum annuum,* or perhaps *C frutescens* [see **TABASCO PEPPER**]. "You light a fire under the tree where witches habitually meet, and you put pepper into the fire. If there is a witch on the tree and the smell entangles her, she can't fly away and will be found on the tree in the morning" (Debrunner). One way of finding a witch in rural areas of India was to write the names of all the women of over twelve years of age on the bark of a branch of the SAL TREE (*Shorea robusta*). The branches were then steeped in water for a given length of time (four and a half hours apparently). If after that time one of them withered, the woman whose name was written on it was deemed a witch (Porteous).

BLACK BRYONY was useful in dealing with a witch-hare, that is, a hare that is actually a witch metamor-phosed into the animal shape. It was well known that such a hare could not be killed with ordinary shot. A piece of silver in the shot would do the trick, and if no silver, a piece of Black Bryony root would serve with the powder (Evans & Thomson).

WITHY
See OSIER (*Salix viminalis*) OE withig, and generally used as frequently as Osier, though they are both applied to any of the basket-making willows. Widdecombe, in Devon, is withycombe, a valley overgrown with willows (Copley).

WOAD
(*Isatis tinctoria*) Its use as a dyeplant is very ancient indeed, for it was known as a dye in Egypt, and later on, in Roman times (Ponting). Pliny described its use. It has to be assumed that the Britons in Caesar's time knew about it, if we are to judge rightly from his description of the way they used it. The original runs "omnes vero se Britanni vitro inficiunt, quod caeruleum efficit colorem, atque hoc horribiliore sunt in pugna adspectur". It was a war paint, then, with blue or green stains on faces or limbs to look more terrible, the idea perhaps being that an enemy could not possibly withstand an army of such grim aspect (Elton). Far from a sign of savagery, Graves pointed out, the use of woad either to stain or tattoo themselves, is a proof of advanced culture, for the extraction of the blue dye is a very complicated chemical process (for descriptions of the technical processes, see Ponting. 1976; Leggett. 1944; and Hurry. 1930). Pliny said the matrons and girls of

Britain stained themselves all over with woad, and Graves said this was probably done in honour of the goddess Anu, as goddess of the dark blue night sky, and the dark blue sea. It was unlucky to dye cloth with woad while a man or boy was in the house, and the legend of St Ciaran was given as an illustration of the superstition. While St Ciaran was still a boy and at home, his mother, who believed in the superstition, told him to leave the house so that she could get on with some dyeing. Ciaran, in a fit of mischief, ill-wished the dye and twice spoiled the process. The third time, his mother begged him to bless the dye, which he did, and it proved to be perfect (Hurry). So male exclusion seems to have been the rule in Ireland, and if this was also the case in Thrace and the northern Aegean, it would account for the nasty smell which, according to Apollodorus, clung to the Lemnian women, and which made the men quit their company. For the extraction and use of the dye is such a smelly business that the woad-dyeing families of Lincolnshire have always been obliged to inter-marry (Graves). It is recorded that Queen Elizabeth I could not stand the smell, and the sowing of woad seed was forbidden within five miles of any of her residences (Leggett). Apparently, these families had traditional chants of their own which they sang during the plucking of the woad leaves. One of them has as its first verse:

Molly of the woad and I fell out,
O, what do you think it was all about?
For she had money and I had none,
And that is how the strife begun (Hurry).

Garments dyed with woad were often worn by special sections of the population – the "blue-coat" was the habit of serving-men in the 16th and 17th centuries, and the same garment formed the dress of a beadle in Shakespeare's time. The blue uniform of Christ's Hospital boys dates from the time of Edward VI, the school's founder, although woad is no longer used in the dyeing of the cloth. Certain colours seem to have been fashionable at different seasons of the year. Blues (woad, that is) predominated from September to December, and they almost disappeared from January to May, when russet or reddish fabrics came in (Hurry).

Cloth-dyeing was not the only thing for which the woad vat was useful in the Middle Ages. They supplied a blue pigment used by artists, especially in the illumination of missals. It was got from the scum that floated on the surface of the vats, known as the "flower" of woad, or "flory". It was used by Italian artists from the 13th to the 16th century. Evidently, there were two shades of blue that could be got, for there was a distinction made between "indigo" and "azure". There are yet other uses, for the seedlings of woad, when about five centimetres high, can be eaten as a salad. They taste rather like mustard and cress. And an oil similar to linseed oil can be extracted from the seeds (Hurry). There were medicinal uses, too. An ointment, for instance, has been made from woad to heal ulcers (Brownlow), while Gerard said that "the decoction of Woad drunken is good for wounds in bodies of a strong constitution, as of countrey people, and such as are accustomed to great labour and hard coarse fare".

Woad is OE wad, which accounts for the pronunciation change in versions like Wad, pronounced, at lkeast in Lincolnshire, to rhyme with 'mad', though the usual name for woadmen, "waddies", is pronounced to rhyme with 'bodies' (Wills). Woad was once called in Latin 'vitrum', which could mean either glass (hence a supposition that the plant was used in glass-making), or the blue colour that woad produced. The late Latin for the plant was 'glastum', (Glastonbury has a first element that seems to mean "place where woad grows" (Cameron)), and glastum gave the French 'guide', Italian 'guado' and Spanish 'gualda', and thus was indirectly responsible for our place-name.

WOOD ANEMONE

(*Anemone nemorosa*) It closes its petals and droops before rain (Inwards), and it is a fairy flower; in wet weather they shelter in them. So it is an unlucky flower to pick, for that would provoke a thunderstorm (it was actually called Thunderbolt in Staffordshire (Vickery. 1995)). To dream of anemones predicts love (Mackay), though in Staffordshire it was reckoned unwise to take one anywhere near a wedding, for it would be bound to cause bad luck (Raven). This ill-luck nature is of some antiquity, for the Egyptians looked upon it as the emblem of sickness, and the Chinese call it the Flower of Death. This reputation is also quite widespread in Europe. The Romans, though, picked the first anemone as a charm against fever, and this idea still prevails in places (Grieve. 1931).

This is a homeopathic doctor's "cure-all", or rather, the tincture is (Schauenberg & Paris), but apart from that, it is not a greatly used plant. Certainly, it served in Ireland as a plaster for wounds (Wood-Martin), and they also used the plant there to relieve a cold in the chest (Moloney). Also, it is "good for the headache, if you put the leaves of it on your head" (Gregory. 1970).

Gerard called it Windflower, though the name is usually applied to the genus as a whole. It is explained by asserting that some species flourish in open exposed places, or that they would not open till the March winds began to blow (Friend. 1883). The belief is from Pliny: "Flos numquam se aperit nisi vento spirante, unde et nomen eius" (see Rambosson). Greek anemos is the word, and the name of the flower means literally "daughter of the wind". But this is a sort of folk etymology, for the true origin is the Semitic na'aman (see POPPY ANEMONE).

WOOD SAGE

(*Teucrium scorodonia*) Inhaling an infusion of Wood Sage (if Wood Sage was actually meant) was a Yorkshire remedy for quinsy (Hartley & Ingilby), and it is still used for the complaint in homeopathic medicine. People in the Dursley district of Gloucestershire used to pick the leaves in spring, and dry them, for a tea against rheumatism (Grigson. 1955). The tops, according to Hill. 1754 "drank for a continuance, is excellent against rheumatic pains". It is used in Ireland for colic (Moloney), and also for colds and even consumption (Ô Súilleabháin). New Forest gypsies combined Wood Sage and Ground Ivy in a tea for treating colds (Boase), and that tea was taken there to cure swellings, and also for biliousness (Hampshire FWI).

The older herbalists used it for several more illnesses – "for him that cannot hold his water", for example (W M Dawson), and from the previous century, there was a remedy "Ad purgandum pectus", with rue and wood sage together (Henslow). Wesley recommended it for something he called "palsy of the mouth", while Gerard prescribed it "against burstings, dry beatings, and wounds", etc, and it has even been used to cure St Vitus' Dance (Vickery. 1995). Wood Sage is, or was, called Gulseck-girse in Orkney (Leask), implying that it was used for jaundice there, for the complaint used to be known as gulsa in the far north.

WOOD SAGE BEER

In Dorset, horehound and wood sage boiled and mixed with sage made a cooling drink called woodsage beer, which was drunk at harvest time (Dacombe).

WOOD SORREL

(*Oxalis acetosella*) It is strange that such a well-known flower, with many vernacular names, should have so little folklore attached to it. The only superstition recorded comes from France, where it used to be said that if two lovers out walking should step on a wood sorrel, their marriage would be delayed (Sebillot). Some American Indians fed the roots to their horses, apparently to increase their speed (Mabey), but whether that is folklore or not is difficult to ascertain.

Oxalic acid is dangerous, but wood sorrel has been used, and is still in use by herbalists, for a variety of ailments (it could cure anything, so it was claimed in Wicklow. "Any badness in the system it would drive it out" (O Cleirigh)). An ointment used to be made in Ireland from the leaves, for cancer (Egan). It is always difficult to know whether 'cancer' or 'canker' is meant in old prescriptions; usually the latter, but the same authority also said that the leaves were eaten for stomach cancer. Interestingly, a leechdom for cancer

also appears in Anglo-Saxon medicine. In Cockayne's translation, the requirement was to boil cuckoosour, singreen and woodruff in butter, in which cuckoosour is wood sorrel and singreen houseleek. This is obviously a preparation to employ on a canker rather than cancer.

WOODBINE

see HONEYSUCKLE A very common name for the plant. It is OE wudubind, wudu being a tree, and binden, to bind. It exists in a number of forms, the most usual being Woodbind. But there are Woodwind, Widewind, Woodvine, and so on, the Scottish version of which is Widbin.

WOODRUFF

(*Asperula odorata*) "*Odorata*" it certainly is, but the fresh leaves are quite scentless, and it is only in the dried state that the familiar hay smell becomes apparent. It is the coumarin content of the plant that is the cause of this fragrance, which transfers well to liquids. Richard Mabey enthused over the result of putting a sprig in a bottle of pure apple juice, and letting it steep for a week or two (Mabey. 1972). The Swiss put it in cognac and benedictine (Painter), while it is customary in Germany to use it to make Maiwein. Sprigs are steeped in Rhine wine, and this is the traditional May Day drink (G B Foster), fine as long as it is not taken in excess, for then it can cause loss of memory (Schauenberg & Paris).

In England, garlands of woodruff were hung in parish churches in the 15th century, particularly on St Barnabas's Day, 11 June (Grigson, 1955), and Gerard wrote of the sweet-smelling bunches of it brought into the house. It was popular, too, for scenting dried linen, and for laying in beds (Mabey. 1972), while in Holland, it has often been used to stuff mattresses, on the soporific hop pillow principle (Thomson. 1976).

Woodruff has been quite important medicinally in its day. To start with, coumarin is an anti-coagulant, so it had been useful for drugs used in heart disease (Mabey. 1972). Put another way, it is a blood-thinner, so woodruff tea was taken as a "spring tonic" (G B Foster), in the days when that was thought to be necessary. Such a tea, taken by itself or with strawberry leaves, helps to relieve headache and depression (Painter), or in Gerard's language, "put into wine, to make a man merry". Herbalists still use it to treat liver infections and jaundice (Grigson. 1955), while in the Highlands of Scotland the liquid in which woodruff had been boiled was given to consumptives (Grant) – indeed, the Gaelic name for it means "wasting plant" (Beith). It was used for the same complaint in Brittany, where minor ailments like colds were likewise treated with woodruff tea (Grigson. 1955). In Anglo-Saxon times, woodruff and brooklime, both rich in tannin, were applied to a burn, in butter, with Madonna Lily. It was said to heal without leaving a scar (Cameron).

WOODY NIGHTSHADE

(*Solanum dulcamara*) Most plants with red berries have some kind of magical association, rowan being the prime example. The connection may be with lightning, and the consequent protective faculty, or it may be with the fairies. Like honeysuckle, woody nightshade used to have names in Germany that fixed it firmly in the latter category – names like Alprauke, Alpkraut, etc., – elfwort, that is (Grimm). In Lincolnshire, collars made from the branches of this plant were hung around pigs if it were reckoned they were "overlooked", and elsewhere, garlands of the flowers, with holly, were put round the necks of horses if they were hag-ridden (Wright. 1913). To quote John Aubrey: "A receipt to cure a horse of being Hag-ridden – take Bittersweet [woody nightshade], and Holly. Twist them together, and hang it about the Horses neck like a Garland; it will certainly cure him". Culpeper thought so, too: "it is excellent good to remove Witchcraft both in Men and Beasts". In the Warwickshire villages of Charlecote and Whitchurch, the collars were even put round babies' necks to prevent convulsions (Palmer. 1976), which presumably were reckoned to be fairy diseases, but the collar or necklace was also put round the babies' necks as an amulet to help teething, and this plant was hung about the neck of cattle that had the staggers (Brand/Hazlitt). Norwegian practice combined woody nightshade, heath spotted orchid and "tree sap" to make a remedy for protecting both people and animals against the demonic (Kvideland & Sehmsdorf).

Like any poisonous plant, woody nightshade was said in the Highlands (Kennedy) to have great powers of evil. Such a witch plant would actually give people rheumatism! But it was used in medicine to cure rheumatism, as Lindley confirmed. The use for skin complaints is shown by a name sometimes given to this plant, Felonwort. That is a sure sign that it was used for whitlows, which were called in Latin 'furunculi', little thieves – felons, in other words. "Country people commonly used to take the Berries of it, and having bruised them, they apply them to Felons, and thereby soon rid their fingers of such troublesome Guests" (Culpeper). Warwickshire people used to do exactly the same for their chilblains, and made a habit of preserving the berries in bottles for just that purpose in winter time (Bloom).

WOOLLY YARROW

(*Achillea lanulosa*) A North American species. Like its European counterpart, it has its magical as well as its medicinal virtues. From the Fire fraternity of the Pueblo Indians, Zuñi men used to chew the blossoms and root, and rub the mixture on their limbs and chest before going through the spectacular performance of passing live coals over their bodies. The same mixture was used for bathing the bodies of those who danced in fire. Not surprisingly, they used the plant medicinally for burns (Stevenson). Navajo Indians say that yarrow acts just like iodine when mixed with water and applied to cuts (Elmore), while some of the Plains Indians are quoted as using the leaves as a poultice to cure the bite of a spider; the dried flower heads are used as tobacco to form part of a ceremonial smoking mixture (H H Smith. 1945). Saddle sores are treated by Navajo Indians by grinding the plant up and applying the solution (Elmore), while the Cheyenne drank an infusion for coughs, and a tea made from the leaves for colds and nausea (Youngken).

WORMS

AMERICAN WORMSEED (*Chenopdoium ambrosioides*), by its common name, proclaims its use for worms, and has long been used for the purpose. Care has to be taken, though, for an overdose could be dangerous, fatal even (Tampion). Another American remedy is the use of JERUSALEM OAK (*Chenopodium botrys*). An Alabama prescription runs: for worms, one teaspoonful of the seed or the stalk tea mixed with syrup, three times a day (R B Browne). GARLIC, "Garlyke … doth kyll all maner of wormes in a man's body" (Boorde), even if its effectiveness operates in strange ways. Louisiana traiteurs give the patient a little ball of garlic to wear round the neck. The worms, they say, are afraid of the smell. ROWAN berries were used in Irish country districts at one time for worms. The prescription suffocates and kills them (Dorson). In Brittany, a necklace of garlic cloves was put round children's necks to keep them free from worms (Sebillot), and some garlic wrapped in a cloth round a child's waist would be the norm in Louisiana, though domestic medicine there sounds more reasonable. There they mash up garlic with sugar and feed that to the patient (R B Browne), while in Trinidad a tea made from the cloves serves the same purpose (Laguerre). CHIVES has also been used for worms, both in Europe and America, and so have TURNIPS. Gerard recommended that the oil pressed from the seeds should be given "to young children against the worms, which it both killeth and driveth forth". Logan records the Irish use of boiled NETTLE roots for worms – it probably is a strong purgative. Even GROUNDSEL has been tried – there are records from Germany of its use as a child's vermifuge (Fernie), and gypsies use an infusion of the leaves and flowers for the purpose (Vesey-Fitzgerald). Of course CASTOR OIL is as efficient a vermifuge as it is a laxative. One of the more notorious cures for worms in children was by the use of STINKING HELLEBORE. Thornton wrote that "a decoction of about a drachm of the green leaves, or 15 grains of the dried, is given to children and repeated three mornings, when it seldom fails expelling the round worms, or a tea-spoonful of the juice, mixed with syrup, may be given for that purpose". Gilbert White mentions the cure, and recognized it as "a violent remedy … to be administered with caution". By the mid-19th century it was condemned, not surprisingly, as all the hellebores are very poisonous, as "far too uncertain in the degree of its action to be safely administered" (C P Johnson). Other hellebores were used in the same way, and are just as dangerous (CHRISTMAS ROSE, for example, or GREEN HELLEBORE). SAVIN (*Juniperus sabina*), too, is another poisonous substance once regularly used for worms; even children were regularly wormed with it, a dangerous practice that survived well into the 19th century. Earlier, Gerard published a receipt that required the physician to "anoint their bellies therewith" – safer, even though ineffectual. MUGWORT too has been used – the dried flower heads used to be sold by herbalists as "wormseed" (Earle). BALSAMINT seed "expelleth all manner of wormes out of the belly" (Gerard), "wormes both small and great", in Langham's words. TANSY, Balsamint's near relative, has always been used for the purpose. The infusion of the young tops and seeds is a gypsy remedy(Vesey-Fitzgerald). Martin mentions this use in the Isle of Skye, and so does Leask in Orkney, and it is certainly used that way in Ireland (Moloney); it is common enough in America, too (Bergen. 1899) (a tansy bag round children's necks used to be quite common in New England (Beck)). ELECAMPANE has been used since ancient times for worms, even, apparently, by laying the preparation on the stomach. It is not very clear what Gerard wanted his patients to do with it – presumably drink "the juyce … boyled", for it "driveth forth all kinde of wormes of the belly …" LEVANT WORMSEED (*Artemisia cina*) proclaims by its name its use in this context. The medicine is obtained from the flower heads (Hutchinson & Melville), tiny as they are, and often called seeds. They are quite often made up into tablets, but large doses have been known to be fatal (Le Strange) – to the patient, that is.

ROWAN berries were used in Irish country districts at one time for worms. The prescription was to eat a few of them before breakfast for a day or two (Egan). Another Irish usage was given to children and horses. In the first case, it was enough to boil a handful of GORSE flowers in milk, and give that to the child to drink (Vickery. 1995). For horses, the tops were cut with a sickle and pounded on a block with a mallet, and this would be given to the horse, often with a pint of linseed oil (Logan). Gerard reckoned that WALNUTS "with a Fig and a little Rue, prevent and preserve the body from the infection of the plague, and being plentifully eaten they drive wormes forth of the belly". Even in recent times, chopped walnut leaves have been used for worming horses. The root of MALE FERN (*Dryopteris filix-mas*) once served as a vermifuge, and in the 19th century it was possible to buy 'Oil of Fern' for this purpose. The root was apparently marketed in the 18th century by a Madame

Niuffleen "as a secret nostrum", for the cure of tapeworm. After he had paid a lot of money to buy it, Louis XV and his physicians discovered that it had been used ever since Galen's time (Paris). PUMPKIN seeds, crushed and made into a paste with milk and honey are efficient for worming, when taken three times before breakfast (Page. 1978), and CUCUMBERS are also useful as an anthelmintic (Watt & Breyer-Brandwijk). PEACH leaves were once recommended for children with worms (Black). Gerard wrote that "the leaves of Peach Trees … being applied plaisterwise unto the navel of young children, they kill the worms, and drive them forth. The same leaves boiled in milke do kill the worms very speedily". the dried and powdered leaves of WATER GERMANDER have been used, too (C P Johnson).

Martin gave an example of a cure in Harris for drawing "worms" out of the flesh. It involved applying a "Plaister of warm BARLEY-dough to the place affected". Eventually the swelling went down, and it drew out "a little Worm, about half an inch in length, and about the bigness of a Goose-quill, and many little feet in each side". They called this creature, whatever it was, a Fillan. BITING STONECROP must have been used for worms, for among its many names are Jack-of-the-buttery, or Jack-in-the-buttery. According to Prior, these are corruptions of Bot-theriacque, to Buttery Jack. He went on to point out that the plant was used as a theriac, or anthelmintic, particularly for Bots and other intestinal worms. SLIPPERY ELM could be taken to deal with worms – "makes the intestines so slippery the worms can't hold on" (H M Hyatt). EDIBLE VALERIAN (*Valeriana edulis*) was used by some native Americans as a tapeworm remedy, the Menomini being one. After the worm was expelled it was washed clean, pulverized, and swallowed again, to make the patient fat and healthy once more! (H H Smith. 1923). Tobacco seeds taken in molasses were recommended in Indiana (Brewster).

WORMWOOD

(*Artemisia absinthium*) Wormwood was cultivated quite extensively in the 18th century around London, for use as an aromatic tonic and vermifuge (Salisbury. 1964). The drink known as absinthe was actually taken as a "tonic drink"; it became very popular by the end of the 19th century. Made from oils of wormwood, combined with anise, coriander and hyssop, absinthe is a narcotic alcoholic drink, banned now that it is realised that it causes permanent neural damage (Emboden. 1979). At one time, wormwood was used in the preparation of all sorts of medicated wines and ale. One of them was called purl, "which hard drinkers are in the habit of taking in the morning to go through their hard day's labours" – this was wormwood mixed with ale (Thornton). Nowadays, extract of aniseed has replaced wormwood in aromatic liqueurs – in Pernod, for example, though small amounts of wormwood are still added to vermouth, which is a fortified white wine (Le Strange).

As with southernwood, wormwood's aroma can be put to use as an insect-repellent. The dried leaves are put among clothes to keep the moth away (Rohde), and it is also used to keep rooms free from fleas. As Tusser said:

> While wormwood hath seed, get a handful or twain,
> To save against March, to make flea to refrain.
>
> Where chamber is sweeped and wormwood is strowne.
> No flea for his life dare abide to be knowne.

"Whoever would destroy fleas, let him steep wormwood in the sea for an hour, and afterwards dry it in the sun. When sufficiently dry …. fleas coming in contact therewith will die" (Physicians of Myddfai). It "keepeth and saveth books and clothes from fretting of mice and of worms, if it be laid therewith in chests or coffers" (Bartholomew Anglicus). Perhaps similar reasoning accounts for the Monmouthshire practice of putting wormwood, together with rue and hyssop, in a coffin (Wherry).

The very strong smell must account for its use in some parts of Europe as a protection against evil spirits. Bunches of it would be hung on doors and windows, or at the belt (Beza). Arnaldo de Vilanova was writing in the early 14th century, and he said that wormwood placed at the door will act as a preventive of sorcery (Lea). So, too, in Somerset, where it was used against the evil eye. The same idea must have been present in France at one time, for it is a St John's herb there (Beza). Another aspect of its preventive role can be seen in manuscripts of Apuleius, where carrying a sprig of wormwood was recommended as a specific against the weariness and danger of travel in wild and mountainous country (Blunt & Raphael), a belief that carried into the 20th century: "as recently as 1925, it was recorded that a driver of the "Auto-Post", on a precipitous road with hairpin bends leading to Maloja, was observed to have a branch of this plant hanging from his windscreen" (Arber). To dream of wormwood is a good sign, they say, and this must be from the general esteem in which it was held.

Thomas Dekker, in *Wonderful Yeare*, 1603, spoke of persons apprehensive of catching the plague, when "they went most bitterly miching and muffled up and downe, with rue and wormwood stuft into their eares and nosthrils, looking like so many bore's heads stuck with branches of rosemary, to be served in for brawne at Christmas". Wormwood steeped in vinegar and kept "in a close-stopped pewter piece" was commonly carried in plague years, to be sniffed in dangerous places (Painter). It used to be said in Alabama that wormwood tea was good for cholera (R B Browne). This of course is a case of the strong smell drowning infection;

southernwood was used for the same purpose in Orkney. But so famous was wormwood as a medicine that it was actually used as a symbol of health (Painter), and it has been used over the centuries for an enormously long list of ills, from a simple cold (a hot water decoction of the leaves) and tonic, to more complicated ailments like jaundice – "the decoction cureth the yellow jaundice, or the infusion. If it be drunke thrice a day some ten or twelve spoonfuls at a time" (Gerard).

The leaf decoction is also used to expel worms, a folk remedy recorded in Orkney (Leask) and in America (Sanford), and in fact well known all over the place. It has been suggested that this use against worms arose from a reading of the name wormwood, and the alteration of the spelling from wermod to the modern name, for it is not worm + wood. The original word is Germanic, and it became weremod in OE, Wermut in German, and vermouth in French (Potter & Sargent). But it is still being used for worms, and is a genuine anthelmintic, even if dangerous because of the large amounts needed (Flück). An overdose can cause heart damage (Mabey). Quite possibly the name wormwood came into being because of the remedy rather than the other way round. Certainly, some of the recipes are old – Cockayne has, from Apuleius, "In case that round worms are troublesome about the navel, take the wort, and horehound, and electre [lupin], alike much of each, seethe in sweetened water or wine, lay it twice or thrice to the navel". A Yorkshire leechdom was more complicated: "take wormwood, rue, bull's gall, and hog's grease; fry all together; apply to the child's navel, and anoint the stomach with the same" (Nicholson). While on the subject, note Wesley's "Earach from Worms … Juice of Wormwood, which kills them", but that is quite another matter.

Wormwood tea is good for indigestion, though the dose should not be continued for more than a day or two (Brownlow). On the island of Chios, wormwood is still the great standby for stomach upsets; there is a saying there that translates "bitter on the lips, sweet to the heart" (Argenti & Rose). In Morocco, we find it being used for heartburn and stomach-ache. Mixed with tea, it is also supposed to promote proper digestion after a meal; but if people have such tea on two or three consecutive days, they will quarrel and separate (Westermarck), surely a recognition of the deleterious effect of the continued dosage.

It is just possible that the bitterness of wormwood may have some genuine effect in an Irish remedy for the falling sickness, in which it was claimed that the juice of wormwood, fennel or sage put into the patient's mouth while he was in the fit, would have an immediate effect (Wilde. 1890), but it sounds just as much from the world of fantasy as Aubrey's story of Sir John Hoskin's wife, "when she lay in of her eldest son, had a swelling on one side of her belly, the third day, when the milk came, and obstructions: she dreamt that syrup of elderberries and distilled water of wormwood would do her good, and it did so …" (Aubrey. 1686). What are "vanities of the head"? Dizziness, perhaps? Anyway there is a recipe for this in the Gentleman's Magazine for a "good oyntment" – "take the juice of wormewood and salte, honye, waxe and incens, and boyle them together over the fire, and therewith anoynte the sick heade and temples". The same source has a receipt for "a man or woman that hath loste their speeche – take wormewood and stampe it, and temper it with water, and strayne it, and with a spoone doe of it into their mouthes". Surely, the most outrageous is the recipe for an elixir, taken from an Irish manuscript of 1770: 1oz cochineal, 1 oz gentian root, 2 drachms each of saffron, snakeroot, salt of wormwood, and the rind of 10 oranges. The whole to be steeped in a quart of brandy, and kept for use" (Wilde. 1890).

WOUND HERBS

GARLIC has always been used as an antiseptic, though its original use relied on the Doctrine of Signatures, its signature being the shape of its leaf. The word garlic is OE garleac, where 'gar' means spear, a recognition of the "taper-leaved" or spear shaped outline. So it soon became used to combat wounds inflicted by spears (Storms). This use as a wound herb, for which there are sound medical reasons, continued into the 20th century. It has always been applied externally as an antiseptic, and during World War 1 the raw juice was put on sterilized swabs to apply to wounds to prevent their turning septic. LEEKS too, surprisingly, enjoyed early reputation as a wound herb. A Middle English medical treatise claimed that, with salt, they "helpe a wounde to close some" (I B Jones), and the Physicians of Myddfai included a prescription "to restrain bleeding from recent wounds". TUTSAN owes its inclusion here to the doctrine of signatures. The dark red juice that exudes from the bruised capsules is taken as representing human blood (they say in Hampshire that it originated from the blood of slaughtered Danes). Actually, the leaves do have antiseptic properties, and were certainly used to cover flesh wounds before bandaging became common (Genders. 1991). GROUNDSEL seems as unlikely as garlic to be a wound herb. But in early times, viz Anglo-Saxon version of Apuleius, it was recommended for wounds, pounded "with old lard, lay it to the wounds" (Cockayne). Gerard repeated the recommendation, this time quoting Dioscorides, who apparently said that "with the fine pouder of Frankincense, it healeth wounds in the sinues". Cockayne also described a wound salve to include groundsel, as well as ribwort, yarrow and githrife. They had all to be pounded together, boiled in butter, and squeezed through a cloth There are other, later, leechdoms, using groundsel for the healing of wounds. RIBWORT

PLANTAIN, mentioned above, is a wound herb in its own right: "Plantain ribbed, that heals the reaper's wounds". The leaves are simply applied to the cut, and are used that way in Ireland (Moloney), Scotland (Beith), and there is a record from India, too (Watt & Breyer-Brandwijk). A refinement to the technique is the traditional Irish remedy for stopping bleeding from a cut, which is to chew the plant before applying it (O'Toole). The earlier herbalists were keen to use RED DEADNETTLE for wounds. From the 15th century: "to heal wounds full of blood. Stamp red nettle in a mortar with red vinegar, and lay on the wound: and it shall do away the blood and cleanse the wound" (Dawson. 1934). This is in fact an old remedy for "stopping the effusion of blood" (Pratt),as Hill, in the 18th century, was still recommending such a cure: "the decoction is good for flooding, bleedings at the nose, spitting of blood, or any kind of haemorrhage. It also stops blood, bruised and applied outwardly".

There was a belief that ASH wood, provided it was cut at certain holy seasons, was incorruptible, and so would heal wounds (Kelly); hence Aubrey, even if the moment does not agree with "holy seasons": "to staunch bleeding, cut an ash of one, two or three years' growth, at the very hour and minute of the sun's entering Taurus: a chip of this applied will stop it". James II's nosebleed was staunched in this way in 1688, so it was claimed.

COMFREY, though its reputation rests on its use for knitting fractures, is good to use on cuts and bruises. The standard Irish method to ease the pain is simply to apply a poultice of the roots. But elsewhere, things are not so simple. The purple-flowered variety is for a man, and the white for a woman. Gypsies too refer to the purple kind as male and the white female, but it has to be male flowers that are good for a woman, and female for a man (Boswell). In the Isle of Man, it was said that the leaves, one side rough and the other smooth, would heal a wound if put on in the right order, by first drawing and cleaning it, and then healing (Killip) Another so-called consound is GOLDEN ROD (Beith), but often used for wounds. Gypsies take an ointment made with the fresh leaves to heal wounds and sores (Vesey-Fitzgerald). Martin also records this use in Lewis, Outer Hebrides, and before that, Gerard, who said that "it is extolled above all other herbes for the stopping of blood in bleeding wounds …". HARTSTONGUE leaves have been used, both in Wales and in Scotland for a wound application (C P Johnson), and TANSY has been used in Ireland. The method was to boil it in unsalted butter, strain and keep for later use (Maloney). ADDER'S TONGUE leaves, when pressed, produce a green oil, sometimes known as Green Oil for Charity, considered by the older herbalists as a balsam for green wounds, and often called Green Adder ointment, still to be bought in the 20th century, but of ancient origin (Savage).

SANICLE has been a wound herb for a very long time (the name is from Latin sanare, to heal.) These days it is usually prescribed as a compress made from the roots (W A R Thomson. 1978), but in earlier times it was taken internally, as a "wound drink". A 15th century wound drink was made from Sanicle, yarrow and bugle, pounded, and given with wine. "This is the vertu of this drynke: bugle holdith the wound open, mylfoyle [yarrow] clenseth the wound, sanycle helith it". But it was emphasised that sanicle must not be given for wounds in the head, or a broken skull, for fear of killing the patient (Grigson. 1955). The tincture of MARIGOLDS is used as a wound application (Flück), though the leaf itself would do just as well; it stops bleeding quickly, and just wrapping a leaf round a cut finger is quite effective for a surface cut, but never for a deep one (Painter & Power). GOOSE-GRASS is traditionally used to soothe wounds and ulcers (Schauenberg & Paris). The juice of OX-EYE boiled with honey was applied externally for wounds in Scotland (Beith), and the bruised plant of BITING STONECROP was also used (Flück), as was PELLITORY-OF-THE-WALL, since ancient Greek times (Baumann). Oil of BALM is useful for drying up sores and wounds (Gordon. 1977). It is certainly a wound herb in the Balkans – balm, the leaves of centaury, and the dust of a live coal, all pounded together (Kemp). The flower or leaf infusion of CENTAURY is a useful wash for wounds and sores, for it is strongly antiseptic (Conway). This is also used as a wound salve in the Balkans (Kemp). AGRIMONY is another plant with a reputation for healing wounds – "… sod in red wine, wherewith if wounds be washed, it cleanseth all filth and corruption from it. And the leaves … beaten or stamped, and tied on wounds that be ill joined, or knit together, by and by doth open them" (Lupton). GREAT BURNET is another plant with a reputation for staunching blood, both internal and external. Its generic name, *Sanguisorba*, Latin sanguis, blood, pronounces its suitability for the task, and it was called Bloodwort (Clair) in English, too. "Burnet is a singular good herb for wounds … it stauncheth bleeding and therefore it was named Sanguisorba, as well inwardly taken as outwardly applied …" (Gerard), in other words it dealt with wounds as well as haemorrhages. It was so used in Chinese medicine, too (Geng Junying). Cornish practice required GROUND IVY leaves as a wound salve – just bind the leaves on to the wound (Deane & Shaw). Another long-standing wound herb is MEADOW CRANESBILL, and it was used in the Highlands of Scotland to stop the bleeding after a tooth had been pulled out (Fairweather). Another Cranesbill, better known as HERB ROBERT, is a wound herb, too. But this may very well be an example of doctrine of signatures, for the whole plant has a red look about it, particularly the stems and the fading leaves.

YARROW occupies a special place among the wound herbs. The very name of the genus, *Achillea*, suggests it, for this refers to Achilles, the first discoverer of the properties of the plant (though there are some who say it was not Achilles at all, but a doctor named Achillo (Le Strange)). Gerard, though, subscribed to the orthodox view. As he wrote, "this plant Achillea is thought to be the very same wherewith Achilles cured the wounds of his soldiers …". Whether it really was the same plant is doubtful, but it is still said that yarrow was always carried by the Greek and Roman armies (A W Hatfield), and the name Soldier's Woundwort is a witness to this. Certainly, the belief in yarrow as a wound herb continued into modern times. The leaves were simply applied, as a kind of poultice. Various North American Indian peoples used it this way. The Anglo-Saxon version of Apuleius has, in Cockayne's translation "for wounds which are made with iron (ad vulnera ferrofacta), pound with grease; lay it to the wounds". Interestingly, the Karok Indians of California also used this plant for arrow or gunshot wounds, in the same way. PRIMROSE leaves were often used on cuts. "Primrose leaves stamped and laid on the place that bleedeth stancheth the blood", said Lupton, and Culpeper agreed "of the leaves of primrose is made as fine a salve to heal wounds as any that I know". The leaves, rubbed on a cut are still being used by men working in the fields (Hampshire FWI), and ointments are made by boiling flowers and young leaves in lard, to heal cuts and chapped hands. Cuts were until quite recently treated both in Ireland and Scotland by binding on a TOBACCO leaf to stop the bleeding and to heal it (Egan).

GOATWEED (*Ageratum conyzoides*) is much used in Africa and Asia to help the healing of wounds. The Mano of Liberia, for example, simply squeeze the juice directly into the wound (Harley). American Indians had a number of resources to stay bleeding, SMOOTH SUMACH (*Rhus glabra*) was one. They made a styptic wash from the boiled fruit to check bleeding (Sanford), especially to stop bleeding after childbirth (Corlet). The powdered seeds would also have been applied to wounds (Lloyd). The Maori used the resin of RIMU (*Dacrydium cupressinum*), a New Zealand conifer, to stop the flow of blood from a wound, and a lotion for bathing wounds was made by cutting the bark into pieces and boiling them in water (C Macdonald).

WOUNDWORT

A name given to various members of the genus *Stachys*, notably *S palustris*, MARSH WOUNDWORT, and *S sylvatica*, HEDGE WOUNDWORT. Gerard spoke of Marsh Woundwort: "the leaves hereof steeped with … Hogs grease, and applied unto green wounds in maner of a pultesse, heale them in a short time". The best known of the genus is BETONY, *S officinalis*.

WYCH ELM

(*Ulmus glabra*) A protective tree, very much in the manner of Rowan. In fact, we find them used in tandem, as it were, in the Cromarty legend of Willie Miller, who went to explore the Dropping Cave – "he sewed sprigs of rowan and wych-elm in the hem of his waistcoat, thrust a Bible into one pocket and a bottle of gin into the other, and providing himself with a torch, and a staff of buckthorn which had been cut at the full of the moon … he set out for the cave …" (H Miller). Smollett knew of Wych Elm's prophylactic qualities, for he has a character in *Humphrey Clinker* say "As for me, I put my trust in the Lord; and I have got a slice of witch elm sewed in the gathers of my under petticoat". Opie & Tatem were able to provide a quotation from a time as recent as 1958: "the butter wunna come in that", she said firmly. "There's no wych elm in it, and anybody in their right senses knows as butter wunna gather unless there's wych elm in the churn". Yorkshire carters, too, put a sprig on their horses, and carried a piece of the wood in their pockets (Nicholson). Another example comes from Somerset, where it used to be said that a few sprigs of it should be put in vases indoors on 15 July, to prevent a curse from St Swithin (Watson).

There are but few usages that require Wych Elm rather than any other kind of elm, but for some peculiar reason, the seat planks of dinghies were always made of it (Wilkinson. 1978). As with the American Slippery Elm, the cambium of Wych Elm has been used until comparatively recently to make a kind of bread, known as bark-bread in Scandinavia, much like Scots pine-bark (Dimbleby). There have been a few records of Wych Elm being used as a shrew-tree (see **ASH**) instead of the more usual ash. There was one apparently made in Somerset in the 20th century. A cottager had a child sick with polio, and the idea was that a shrew-mouse would be allowed to run over the affected limb and then be imprisoned in a hole bored for it in the shrew-tree, in this case a Wych Elm. A decoction of the twigs of the tree that had caused the death of the shrew would act as a remedy for the child's sickness (Whistler).

'Wych' is from a Germanic base meaning 'pliant', or 'bending', much as modern German weich – soft, pliant; switchy is another way of putting it, hence Switch Elm (Grigson. 1955). 'Wych' all too easily becomes 'witch', with consequent confusion in local beliefs, like suggesting that witches dread it, or the opposite. Burning wych elm would bring the malignant power of the witch upon the household (Morley), and so on. So we find Witch Elm, Witchwood, etc., (Nicholson, Elworthy. 1888).

X

Xanthium spinosum > SPINY COCKLEBUR

Xanthium strumarium > COCKLEBUR

Y

YARROW

(*Achillea millefolium*) A common wild flower, but at one time one of the greatest of magical and protective plants. It is one of the seven herbs that, in Irish belief, nothing natural nor supernatural could injure (the others are vervain, St John's Wort, speedwell, eyebright, mallow and self-heal (Wilde. 1902)). This almost automatically makes it one of the herbs of St John; it was recognised as such both in France and in Ireland. The Irish used to hang it up on St John's Eve to turn away illness (Grigson. 1955), and it was also believed to have the power of dispersing evil spirits (Dyer. 1889). In the Fen country it was said to be able to avert evil spells; if it were strewn on the doorstep, no witch would dare enter the house (Porter. 1969). Similarly, elsewhere it was said that yarrow should be kept hanging in the toolshed, "for safety" (it is a wound herb – see below), but later on the protection was reckoned to be in stopping entry by thieves (Boland. 1977). Often a bunch of yarrow was put in the churn if the butter would not come (E E Evans. 1957), and it was tied to the cradle to protect both baby and mother (R L Brown), or to make the babies grow up happy and even-tempered (Porter. 1969). The idea of general well-being seems to lie behind the belief in Cambridgeshire that cattle who grazed in a field where yarrow grew would be more docile than if they were in fields where the plant was absent (Porter. 1969).

When going on a journey, pull ten stalks of yarrow, keep nine, and throw the tenth away (as the spirit's tithe, of course), put the nine under the right heel, and evil spirits will have no power over you (Wilde. 1902). Again, there was a belief that, put under the foot, it allowed the user temporary fluency of speech; in the Hebrides, it was said that a leaf held against the eyes gave second sight (M Baker. 1980).

Such a plant had to be gathered with proper ceremony. Gaelic speakers never pulled yarrow without reciting some formula at the same time. Here is Carmichael's translation of one of these incantations:

> I will pluck the yarrow fair
> That more benign will be my face,
> That more warm shall be my lips,
> That more chaste shall be my speech,
> Be my speech the beams of the sun,
> Be my lips the sap of the strawberry.
> May I be an isle in the sea,
> May I be a hill on the shore,
> May I be a star in the waning of the moon,
> May I be a staff to the weak.
> Wound can I every man,
> Wound can no man me.

The magic of the plant extended into the dream world, too, for it was widely used for love divinations, especially on May Eve and at Hallowe'en (Wood-Martin). Irish girls filled a stocking with it, more specifically, the left stocking, tied with the right garter (Cooke), and put it under their pillow, while saying some recognised rhyme, like the Irish:

> Yarrow, yarrow, yarrow,
> I bid thee good morrow,
> And tell me before tomorrow
> Who my true love shall be (Wilde. 1902).

Aberdeen girls went out to the fields on May morning, always in silence, to gather yarrow. They shut their eyes, and pulled what first came to hand, repeating:

> O it's a bonnie May morning;
> I cam' t' pu' the yarrow;
> I hope before I go
> To see my marrow.

Or perhaps:

> Good morrow, good morrow,
> To thee, braw yarrow,
> And thrice good morrow to thee;
> I pray thee tell me today or tomorrow
> Who is my true love to be.

Then they would open their eyes, and look around in every direction as far as the eye could see. If a man was visible, the girl who spied him would get her mate that year. In some districts, they went out on the first night of May (again in silence), carried the yarrow home, and went to bed without speaking a word. During the night, the future husband would appear in a dream (F M MacNeill. Vol 2), though to be sure of him, he had to be facing the dreamer. If he had his back to her, they would never marry (Beith). Some said the yarrow had to be picked at the new moon (Valiente). This applied in Cornwall, where the rhyme was:

> Good night, fair yarrow,
> Thrice good night to thee;
> I hope before tomorrow's dawn
> My true love I shall see (Courtney).

The yarrow divination travelled to America as well. In Massachusetts, for instance, the formula while walking three times round the yarrow, was:

> Good evening, good evening, Mr Yarrow.
> I hope I see you well tonight,
> And trust I'll see you at a meeting tomorrow.

Then the girls would pluck the head, put it inside their dress, and sleep with it. The first person they met, or spoke to, at church, would be their husband (Bergen. 1896).

In Dorset, they said that if a girl picked yarrow from a young man's grave, and put it under her pillow on

Midsummer Eve, she would see her future husband in a dream (Udal). This churchyard yarrow seems to have had quite a reputation for the magical discovery of witches, etc; see for example the Yorkshire legend "The maid of the golden shoon" (in Blakeborough. 1924), in which "kirkyard yarrow" was one of the things thrown in a burning sheet to force the appearance of witches.

Fenland girls used yarrow for a love charm, not as a divination agent, but by pinning it on their dresses, and then taking every opportunity to get as near as possible to young men, in order to declare their love by means of the flowers. If the girl found that the man she was interested in ignored the hint, then she was likely to wait for a full moon, go to a patch of yarrow, and walk barefoot among them. She would then shut her eyes, bend down and pick a bunch. If she found next morning that the dew was still on the yarrow, then all was not yet lost – it was a sign that he would soon come courting in earnest. If the flowers were quite dry, on the other hand, she would be well advised to wait until the next full moon, and to try again (Porter. 1969), or look elsewhere, of course. In some parts of the country, yarrow was often put in the bridal wreath (McDonald). It was said that this guaranteed seven years of married bliss (Conway), and explains the name 'Seven Years' Love', recorded in Gloucestershire (Fernie). After all this, it seems only natural that dreaming of yarrow itself should be an excellent sign. Irish people thought so – finding yarrow in a dream means good luck in the future (Wood-Martin).

Yarrow had its mundane uses – for instance, they made a herb beer from it in Oxfordshire (Flora Thompson). The name Field Hop points to this former use in beer. The drink made from it is said to be more devastating than the ordinary kind (Skinner); so they said in Sweden, too, and, apparently, in America, (Chandler). Is yarrow so inebriating? Or is it the scent that produces a heady feeling?. There was certainly a belief that working near yarrow on a hot day made people almost delirious with its strong scent – this is why the name "yarrow" was often given to men who talked too much or who were over-given to boasting (Porter. 1969). Perhaps it is no surprise to find that in Orkney, yarrow tea was looked on as a cure for melancholy (Skinner).

Naturally, a plant that has such virtues in the magical sphere would be very popular as a medicinal herb. The very name of the genus, *Achillea*, suggests it, for this refers to Achilles, the discoverer of the properties of the plant (though there are some who say that it was not Achilles at all, but a doctor named Achillo, who first used it as a wound herb (Le Strange)). Gerard, though, subscribed to the orthodox view. As he wrote, "this plant Achillea is thought to be the very same wherewith Achilles cured the wounds of his soldiers …".

Whether it really was the same plant is doubtful, but it is still said that yarrow was always carried by the Greek and Roman armies (A W Hatfield). The leaves were simply applied, as a kind of poultice; the Miwok Indians, too, used to bind the mashed leaves, either green or dried, to a wound to stop pain (Barrett & Gifford), or in the case of the Gosiute Indians, to relieve rheumatic pains (Chamberlin). Gerard talked about this wound use, and went on to say, "it stancheth blood in any part of the body, and it is likewise put into bathes for women to sit in …". An Anglo-Saxon version of Apuleius has, in Cockayne's translation "for wounds which are made with iron (ad vulnera ferrofacta), pound with grease; lay it to the wounds". Interestingly, the Karok Indians of California also used this plant for arrow or gunshot wounds, in the same way (Schenk & Gifford).

Yarrow tea made either from the dried herb or from the fresh plant (a handful of the whole plant to a pint of boiling water) is taken for a bad cold (Jones-Baker. 1974), and for bronchitis or measles (V G Hatfield. 1994), or even a depression (Le Strange). Some of the American Indian groups used it in exactly the same way (Barrett & Gifford). The Ojibwe break up a fever by putting the flowers on a bed of live coals, and then inhaling the smoke (H H Smith. 1945). In Britain, there was an odder way of dealing with the problem – "For an ague … boil Yarrow in new Milk, 'till it is tender enough to spread as a Plaister. An Hour before the cold Fit, apply this to the Wrists, and let it be on till the hot Fit is over …" (Wesley). Equally odd is the relatively recent Alabama superstition – a folk practice to get the bowels moving – you had to boil yarrow and thicken it with meal, and then apply it to the stomach (R B Browne).

"Most men say that the leaves, chewed, and especially greene, are a remedy for the Tooth-ache" (Gerard), something that was known well in Anglo-Saxon times, for Cockayne has, from Apuleius Herbarium, "for toothache, take a root, give it to eat, fasting". An old Irish remedy for toothache advised the patient to chew the leaves (Moloney). The American Indians used a preparation for earache (Sanford); the Winnebago, for example, steeped the whole plant, and poured the resulting liquid into the ear (Weiner).

The use of yarrow for various skin complaints was quite widespread. Its fresh tops were made into a poultice for eczema by some of the American Indian peoples (Corlett). Irish country people have a "herb poultice" with which to dress a whitlow – yarrow leaves, fresh grass and a herb called finabawn, whatever that is. Equal parts of the herbs are ground up thoroughly, and then beaten up with white of egg. This is put on the inflamed finger, and it must not be changed for 48 hours (Logan).

There is still a whole catalogue of ills for which yarrow was recommended. A recipe for indigestion that comes from Alabama requires us to steep a pinch of yarrow blossoms in a cup of water, and to drink a little several times a day for three days (R B Browne). Some four hundred years ago Gerard was able to state that "one dram of pouder of the herbe given in wine, presently taketh away the paines of the colicke". A hundred years before his time, it was being given for jaundice: "take the juice of milfoil [i.e., yarrow] and saffron, and seethe them in sweet barley wort; and give it to the sick to drink" (Dawson). The colour of saffron was recommendation enough for its use in jaundice (see DOCTRINE OF SIGNATURES); perhaps yarrow really had some effect? At any rate, we find it earlier still given "in case that a man may difficulty pass water" (Cockayne). It is noted that the decoction was given for all sorts of internal injuries, and even (in the Western Isles) for consumption (Martin. 1703). There are some real oddities among the list of maladies for which yarrow has at one time or another been prescribed, a "slain" body, for example: Cockayne explains this, not very helpfully, as "stricken". From the same source, there is "if a man's head be burst, or a strange swelling appear on it…". A 15th century prescription seems to regard yarrow as a counter poison. There are, too, a few remedies that can only be magical in nature. Perhaps a lot of the foregoing is, too, but there is always a chance that some of the prescriptions were empirically inspired. But the Somerset belief that putting yarrow in your shoe would stop cramp (Tongue. 1965) has to be pure superstition. So is the Irish idea that by sewing it in the clothes, all disease would be averted (Wilde. 1902). The same idea prevailed in Scotland, where a little yarrow and mistletoe put into a bag and worn upon the stomach was thought to prevent ague and chilblains! (Napier).

"The leaves being put in the nose do cause it to bleed, and easeth the pain of the megrim" (Gerard), hence names like Sneezewort and Sneezings. It is called Nosebleed too, over quite a wide area. The French too have saigne-nez. Prior claims that it got this application by mis-translation, the plant actually referred to being the horsetail. Perhaps so, but it is firmly fixed in yarrow's folklore. The propensity was used to test a lover's fidelity. In East Anglia, for instance, a girl would tickle the inside of a nostril with a yarrow leaf, saying at the same time:

> Yarroway, yarroway, bear a white blow;
> If my love love me, my nose will bleed now.

Bergen also quotes this use in America, where the girl says:

> Yarrow, yarrow, if he loves me and I loves he,
> A drop of blood I'd wish to see.

Another Suffolk thyme is:

> Green arrow, green arrow, you bears a white blow,

If my love love me my nose will bleed now;
If my love don't love me, it 'ont bleed a drop.
If my love do love me, 'twill bleed every drop
(Northall).

YELLOW ARCHANGEL

(*Galeobdalon luteum*) What quality prompted the bestowal of the name Archangel? There is certainly nothing outstandingly obvious about it, and it is probably a case of mistaken identification from a medieval original. Anyway, the name has stuck. The specific name, *galeobdalon*, means weasel's snout (or weasel's stench? (Grigson. 1955)). It has been claimed (by Grindon) that Ophelia's "nettles" were in fact yellow archangel, or deadnettle.

YELLOW BUCKEYE

(*Aesculus octandra*) An American member of the genus that contains Horse Chestnut. Just as the nuts of the latter (conkers) are carried around in the pocket to prevent or cure various ailments, so in America the nuts of Yellow Buckeye are carried for rheumatism (Sackett & Koch), or to prevent cramps, keep off chills, or for kidney troubles (R B Browne), or for piles (Puckett). In New England they are just carried to ward off any disease (Beck). But some people would say it is done to keep away witches (R B Browne), and then again it is just for good luck (R B Browne).

YELLOW GENTIAN

(*Gentiana lutea*) The large rhizomes (they can weigh as much as 25 Kg) were used for the production of alcoholic drinks, mainly in German-speaking Alpine country, where the drink is called Edelanzian (Enzian is German for gentian), but these days the alcohol is usually only flavoured with gentian, which has a characteristic bitter principle (Brouk). This is the species that provides the tincture known as Gentian Violet (A W Hatfield), much used since medieval times, when the plant was in general use in domestic medicine, and as an antidote to poison. Now it is in common use as a bitter tonic, Gentian Bitters, in fact, good in cases of jaundice, and useful for toning up the system generally (Brownlow).

YELLOW IRIS

(*Iris pseudo-acarus*) It was known as gillajeur in the Guernsey dialect, and was one of the favourite flowers used for strewing in front of the bride at a wedding (MacCulloch), and in Ireland, it is put outside doors at Corpus Christi (Ô Súilleabháin). Shetland children used to make boats "seggie boats", of the leaves, seg, or seggie being a "sedge" name given to this iris. Children of Stenness, Orkney, were warned that if they chewed seg leaves, they would become dumb (Marwick), or at least have a stammer (Vickery. 1995). The juice from the roots was thought to cure toothache, but it had to be inhaled through the nose (Leask). In County Cork, the leaves of Yellow Iris, called Flaggers there, were put on the doorstep and window sills, or used to decorate the dresser, on May Day. On Cape Clear Island, branches were put in the fishing boats for luck (Danaher).

Water Iris is claimed to be the origin of the heraldic emblem of the fleur-de-lys. The legend is that Clovis, having to do battle on the banks of the Rhine with an army of Goths that outnumbered his, made a surprise attack by crossing the river at a ford that he had noticed, because of the presence of yellow flags. Out of gratitude he adopted the water flag as his emblem. It was chosen, in stylized form, by Louis VII, as decoration for the royal escutcheon. in the Crusades. In Germany, the root is regarded as a luck-bringer, just as mandrake is (G E Smith).

The roots, boiled in water, with copperas added, were once used on Jura as a substitute for ink (Pratt), and it was also used for dyeing black, "Sabbath black", it was called (Macleod), using iron as a mordant (S M Robertson). The flowers will give a yellow dye, and the root is used for wool dyeing on South Uist. The instructions are to take the root when the flower is past, clean, scrape and break up the root, then boil it in water. Afterwards strain it and boil the wool with the juice for an hour or more, until the desired shade of blue or steel-grey is obtained. A little alum has to be used (Shaw). In Cornwall, the leaves were sometimes used as a substitute for reeds in thatching, especially when the latter were scarce (Barton. 1972). The seeds have been used as a coffee substitute (Ingram), especially so in the Channel Islands during the German occupation (Vickery. 1995). The root used to be dried and ground for snuff (Pratt).

In some parts of Sotland, Mull, for instance (Beith), the root is chopped up and used for the relief of toothache. A 14th century recipe prescribed the leaves, stamped with honey, and applied to the cheek (Henslow). In one case, the instruction was to put the juice in the ear on the same side (Langham). The root was used in Ireland for dressing cuts and wounds (Wood-Martin), and they have been taken for constipation, too. They are certainly laxative, but in large quantities they are toxic (Schauenberg & Paris).

YELLOW WATERLILY
(*Nuphar lutea*) In Holland, it was once said that boys should be extremely careful in handling these waterlilies, for if a boy falls with the flowers on him, he immediately becomes subject to fits (Black). Presumably, this is another way of saying what Sebillot quotes as a French superstition, that they are anti-aphrodisiac. On the other hand, any prohibition on picking the flowers might simply be to discourage children from going into a potentially dangerous places (Vickery. 1995). (see also **BRANDY BOTTLES**)

YEW
(*Taxus baccata*) Probably the oldest living of all trees with the exception of the Bristlecone Pine (*Pinus*

aristata). There is a famous one at Fortingal, in Perthshire, whose age has been has been estimated at 3000 years. According to Duncalf, the second oldest tree is one in Tisbury churchyard, Wiltshire, modestly claiming a thousand years or more. The tree in the churchyard at Alton Priors, also in Wiltshire, is claimed to be 1700 years old (Parker & Chandler). Gaelic computations, always working in multiples of three, were fond of working out the age of a typical oak or yew. This Irish one, from the Book of Lismore, runs:

> A year for the stake
> Three years for the field
> Three lifetimes of the field for the hound
> Three lifetimes of the hound for the horse
> Three lifetimes of the horse for the human being
> Three lifetimes of the human being for the stag
> Three lifetimes of the stag for the ousel
> Three lifetimes of the ousel for the eagle
> Three lifetimes of the eagle for the salmon
> Three lifetimes of the salmon for the yew
> Three lifetimes of the yew for the world from its beginning to its end (Hull).

That would give for human life 81 years and for the yew no less than 19683 years! In another, more reasonable, computation, human life remains at 81 years, but the yew attains 2187 years.

Yew is notoriously poisonous, all parts of it to a greater or lesser extent, with the exception of the red mucilage (hence Snotterberries and similar names) surrounding the ripe seeds, which are themselves the most injurious part. The wood and bark are probably less poisonous than the leaves. Indeed in the Pyrenees, so it is said, water bowls are, or were, made from yew wood, this being the preferred timber; and Evelyn reported that "tankards to drink out of" were made of it. Pliny, though, voiced the opinion that people had died after drinking wine out of casks made of the wood (Dallimore). But cattle seem to eat the fresh leaves without harm, although four cows died in 1994 after eating fresh leaves (Bevan-Jones). Deer seem immune, apparently. According to Linnaeus, the list must include sheep and goats (Ablett). Evelyn noted that fresh leaves have been given to children for the cure of worms – effective, no doubt, except that it often killed the children as well. But once the leaves have been dried, they are certainly poisonous. Old writers warned bee-keepers against putting their hives near yew trees, for the bees would work the flowers, and the resultant honey would be poisonous (Dallimore). Caesar, in *De bello gallico*, said that Cativolcus, a king of the Eburones, poisoned himself with the juice of a yew, but this tale raises all sorts of questions of interest to an anthropologist, for eburos was a Celtic word for a yew tree, and there is every likelihood that Eburacon, the Celtic name for York, is derived from it. There are plenty of place names relying on the Anglo-Saxon derivation from yew – Ifield and Iwade, in Kent, Iridge in Surrey, Ewhurst, Ewshott and Iwode in Hampshire, the Gloucestershire village of Uley, and, more obviously, Yewdale, in Lancashire, are all examples. The name in Irish is eo (though this could also mean a tree in general, as well as yew in particular) (Lucas). But Irish place names like Youghall, in Counties Cork, and Mayo, certainly refer to yews (Kinahan).

Yew is closely associated with rune magic. OE yr, which meant a bow made of yew wood, is etymologically the same as the Germanic rune-name eihwaz, yew. The hunting god Ulh appropriately built his hall in Ydalir, the valley of the yews (Elliott), and the name "yew" was given to one of the commoner German runes. There were plenty of magic and ritual associations attaching to the yew anyway, so combine runes with yew wood, and an extremely efficacious magic potential was realised (Elliott). It is even said that a Druidical cult of the yew existed in Iona (McNeill), probably an attempt to explain the name of the island by reference to the tree. Certainly, seasonal ritual was connected with the tree in Scotland, for apparently the Beltane fire was lit in the hollow inside the famous Fortingal tree (Briggs. 1967). Significantly, too, yew branches were kept in the Hebridean house to preserve it against fire and lightning, and they were hung on balconies in Spain for the same reason (Burne. 1914).

There are cases of fairies becoming visible under yew trees (compare this with similar manifestations connected with ELDER). One of the yew cases quoted (by Sikes) occurs in a wood called Ffridd yr Ywen, which means forest of the yew, in Llanwrin, Wales. The magical yew-tree grows exactly in the middle of the forest. The fairy circle under the yew has the usual legend of a mortal being drawn into the dance, and losing all count of time.

Uncanny it certainly is, and unlucky, too, presumably because of its churchyard associations. though Evelyn's opinion was that it was "doubtless some symbol of immortality, the tree being so lasting, and always green". The custom of planting them in churchyards seems to be very old, for Giraldus Cambrensis noted it in 1184 (Lowe), but many of them that grow in churchyards now are of 19th century origin. Giraldus told the story of soldiers quartered in the town of Finglas, County Dublin, who cut for fuel the trees the monks had planted as holy trees. They were smitten "by a sudden and singular pestilence, so that most of them were dead within a few days" (Lucas). Welsh churchyards, it seems, have a higher proportion of yews than English ones (Wilks), and the custom is known too in Brittany, where it is said that the churchyard yew will spread a root to the mouth of each corpse buried there (Mac-Culloch.1911), and so would stop the mouths of the dead (Curl). Care is taken not to cut these yews, nor to pick their leaves (Elliott. 1957). If yew came into the

house by accident among the Christmas evergreens, it was taken as a sign that a death would occur in the family before a year was out (Glyde). Similarly, dreaming of yew was reckoned to foretell the death of an aged person, who would leave considerable wealth behind him. If you dream you are sitting under a yew, it is a sign that you will not live long, but if you just see and admire the tree, it means long life (Raphael), the tree itself is so long-lived, long enough for it to become a symbol of immortality (see Evelyn above), that its presence in churchyards is appropriate enough. Yew branches were used at funerals, too – branches were carried over the coffin by mourners (Tyack), and in Normandy a branch was sometimes put beside a corpse awaiting burial (Johnson). Shakespeare speaks of a "shroud of white, stuck all with yew" (*Twelfth Night* II, iv). In fact, sprigs of yew were tucked into shrouds at late medieval funerals.

The churchyard yew was probably planted because of its original association with places of assembly, churchyards being the only public meeting places in villages. The choice of yew-trees for open-air gatherings survived well into modern times – at Berkhampstead, in Hertfordshire, the New Year was welcomed in under the churchyard yew, and at Totteridge, in the same county, the Hundred Court of the district met in the churchyard beside the old yew-tree (Elliott). It has also been suggested (by Addison) that an ancient yew might very well mark the site of a fair surviving from a wake, and another opinion is that churches were built close to existing, and ancient, yews (Dallimore).

Nevertheless, it is still the "funeral yew" used sometimes to hang round the church on Good Friday, and, at least in Herefordshire, at Whitsuntide, when branches used to be fastened to the tops of pews (Leather). The practice continued in Ireland from Palm Sunday to Easter Day, when sprigs were worn in the hat or buttonhole (Lowe). Branches were often used to decorate churches, and to represent palm, in Palm Sunday processions, and it was actually called Palm in some places (Greenoak). It was taken indoors as "Palm" in Ireland, to be put beside the crucifix, or at the head of the bed (*Notes and Queries; 1858*). The day was actually called Yew Sunday (Domhnach na iuir) in earlier times (Danaher). But, in Shakespeare's words, it is the "dismal yew" – slips of yew "slivered in the moon's eclipse" were among the ingredients in the witches' cauldron in *Macbeth*. But it was a strong anti-witch protection. Any place sheltered by yews is safe from witches (Tyack). 'Vor den Eiben kann kein Zauber bleiben' is still a current proverb in parts of Germany, and in the Forest of Dean a yew stick nailed behind the door prevented a witch from entering the house (Elliott).

Derbyshire farmers would not cut yews down, and they reckoned it was unlucky to burn the wood (Addy.

1895). So they said in the Balkans, too. But it can be cut and made into all sorts of household implements, as well as crosses and amulets, which are then particularly safe and lucky (Kemp). Individual trees in Britain have distinct fertility legends associated with them – that in the churchyard at Stoke Gabriel, Totnes, is one. If you are male, and walk backwards round it, or female, and walk forwards, fertility is assured. Another superstition says that wishes will come true if you walk round it several times (Wilks), the assumption being that the wishes must be for offspring. But a yew twig was worn by Naga girls as a charm to avoid pregnancy (Pandey). Very occasionally, one finds a story of a yew that is the abode of a fairy. There was one in Scotland, in the Morvern district of Argyllshire, in which a Glaistig, a Highland fairy, often helpful to mortals, lived (Bevan-Jones).

Yew has got magical properties of its own, according to some beliefs. For instance, in Derbyshire it was said that if you had lost anything, the thing to do was to hold a branch of yew straight out before you as you walked. That branch would lead you straight to where the missing article was. You would know because the branch would turn around in your hand when you reached the place (Addy. 1895). Perhaps this is the origin of the "hidden things" that Druids were supposed to have divined by the use of wands of yew – or vice versa, of course. There is a superstition recorded in the north of Scotland that a person who has a branch of churchyard yew may speak to anyone he pleases, but the person spoken to will not be able to hear, though everyone around would. The point about it all is that a clansman could confront his enemy, denounce him, have witnesses, and all without the enemy being aware of the denunciation! (Lowe). This idea of the magical yew having power to deprive someone of some of his senses is met again in Normandy. There was a tree at Tourville-les-Ifs, near Le Havre, that had the power. Anyone walking under it lost all idea of the way he was going or any desire to follow that path. He would have to sit beneath it till the Day of Judgment unless some friend broke the spell by dragging him away, after having taken one of the accepted spell-breaking measures – in this case, turning one of his garments inside out. In any case, yew "slayeth such as sleep thereunder" (Bartholomew Anglicus, quoted in Seager).

Yew wood is the most ancient material for wooden weapons yet discovered. Spears were made of it in the early Palaeolithic. There was one found in an interglacial deposit at Clacton, with the tip fire-hardened, which would be Lower Palaeolithic, and there was also one from the last interglacial found at Lehringen, Lower Saxony, again with fire-hardened tip, and nearly eight feet long, between the ribs of an extinct elephant (Clark & Piggott). Yew was especially grown in England, for it makes the best bows for archers (the yew bow was known in Germany as "englischen Bogen").

There were many enactments both for planting and protecting yew-trees, e.g., Richard III, 1483, ordered a general plantation of yew-trees for the use of archers (Lowe). But it has even been suggested that churchyard yews were actually planted to *prevent* people cutting bow-wood, the idea being that the wood would be employed against authority, and that to cut a tree in a churchyard would be a punishable offence (Chandler, who even found a suggestion that is a mixture of two separate pieces of yew-lore – that they were planted in churchyards to provide bow-wood of extra quality, because it was nourished on corpses!). Perhaps, though, too much has been made of this usage; anyway, the best yew-wood for bows apparently came from foreign trees, for the English ones sold for three-farthings each, while foreign ones fetched double that (Lowe). However, *taxus* is probably connected with the Greek toxon, bow, and also with toxicon, the poison with which the arrows were smeared (Graves). The ancient Irish were said to have used a compound of yew-berry, hellebore and devil's-bit for poisoning their weapons. This double meaning is what Shakespeare had in mind when he said, in *Richard II*:

> Thy very beadsmen learn to bend their bows
> Of double-fated yews against thy state.

This insistence on believing in its ultra-poisonous nature led to exaggerated fear. It was even said that sleeping in the shadow of the tree caused sickness or death (Gerard), or that wine could be poisoned by being kept in barrels made of yew (Leyel. 1926). In fact, wine barrels in Ireland were always made of yew staves (Graves).

Relic-boxes were made of yew-wood, and for the same reason beds were made of it, for it was said that such "will most certainly not be approached by bugs" (Lowe). But old carters would keep a brown paper parcel of clipped "she-yew" under their beds for a year or more before adding a little to their horses' bait, to make their coats shine, and generally improve their appearance for any special occasion (Wiltshire). As mentioned before, animals can eat the fresh leaves without apparent harm, and drying them so thoroughly may have got rid of the toxins, but yew leaves were certainly used medicinally in the 17th century. Used like digitalin, they were given for certain heart conditions (Lowe). Yew twigs steeped in tea were used in Lincolnshire for kidney complaints (V G Hatfield. 1994), and yew bark has been used for tea in India (Dallimore). Yew bark was also used by horsemen to make their horses' coats shine. It was dried, and rubbed to a powder, and a *very* little given in their feed (Kightly. 1984).

YGGDRASIL

the tree of the universe according to Scandinavian mythology, generally supposed to be an ASH, perhaps because of the bunches of "keys" that hang from its branches, "like the bodies of tiny men", recalling the practice of hanging sacrificial victims from trees (Davidson). It is pointed out, too, that the ash has wide-spreading roots, and the roots of Yggdrasil are said to spread to different parts of the underworld. There are three roots to support the tree, one stretching to the world of death, another to the world of frost giants, and the third to the world of men. Three wells lay at the base of the tree, one under each root. Yggdrasil is the centre of the divine world, where the gods sit in council every day. It rises to the sky, and its branches spread over the whole world. It is known too as the Guardian Tree, for it nourishes, and at the same time suffers from, the animals that inhabit it, feed on it, and attack it. The dragon Nidhogg gnaws the roots (Crossley-Holland).

This must be the tree on which Odin hanged himself in his quest for wisdom (Turville-Petrie); in fact Yggr is one the many names of Odin (Davidson). The usual interpretation of Yggdrasil is "horse of Yggr", since in a sense Odin rode the tree when he hung upon it. Old Norse poets often spoke of a gallows tree as a horse (Crossley-Holland).

The Eddas describe the stars as the fruit of Yggdrasil, and also say that all mankind is descended from the ash and the elm (Dyer). According to Hesiod, the men of the third age of the world (the Bronze Age race) grew from the ash tree; Hesychius too said that the Greeks believed that the human race was the fruit of the ash (Philpot). Teutonic mythology also recognised that the first men came from the ash (Rydberg) (see also **HUMAN ORIGINS, CHILDBIRTH**). The cooked fruit of Yggdrasil ensures safe childbirth. When Ragnarok draws near, it is said that the ash tree will tremble, and a man and a woman who hide in it, Lif and Lifthrasir, will survive the ensuing holocaust and flood. They stand alone at the end of one cycle and the beginning of another in the world of time and men. From these two the earth will be re-peopled, and Yggdrasil itself will survive Ragnarok. In other words, Yggdrasil is the source of all new life (Crossley-Holland).

The morning dew from Yggdrasil, according to the mythology, was a sweet and wonderful nourishment, so sweet that the bees use it for the making of honey. The morning dew evaporates from the world tree, which stands over Urd's and Mimer's sacred fountains. The world-tree gets its life-saving sap from these fountains, which contain the elixir of life, wisdom and poetry. By an extension of the myth, the same dew formed the basis of mead, the flowers receiving it, the bees extracting it, and so producing honey. So mead contains some of the strength of Urd's and Mimer's fountains, with the same ability to stimulate the mind and inspire poetry.

In Maori myth, the roots of the POHUTUKAWA tree (*Metrosideros excelsa*) lead down into the underworld (Andersen), just like Yggdrasil.

Yucca glauca > SOAPWEED

YULE LOG

The Yule Log goes by a number of names in different parts of Britain. It is, or was, Yule Clog in Yorkshire, 'clog' being the usual dialect words for log there (Morris), or in Cornwall the Christmas Block, or Mock (Courtney. 1890). It was called Christmas Block in Herefordshire as well (Leather). Yule Mock, Mot, or Stock, were other Cornish names (A R Wright). Evidently the 'tronquet de Noel' of the Channel Islands (MacCulloch) is the same thing. BIRCH or ASH were the preferred trees to use ("as bare as the birch at Yule even" is a North country proverb describing someone in extreme poverty (Denham. 1846, Hazlitt. 1882). Further south, ash was the most popular. Christmas Eve is still called Ashen Faggot Night in parts of Somerset (Rogers), when divinations are made according to the bursting of the willow, hazel or green ash bands round the log as the fire grows. Welsh people, too, played some kind of divination game while the log was burning. They watched the shadows on the wall – those that appeared without heads belonged to persons who were to die within the year.

The log is the Cailleach Nollich, the Christmas old wife, in the Highlands (James. 1962), where it was given "the representation of some woman", who, as it were, "stood in" for the bringer of the log, life for life (Stewart). The Yule Log was "sacrificed to propitiate the angel of death, in the hope that he would refrain from visiting the house during the year …" (Polson), for the vegetation spirit is associated with death and sacrifice as well as the renewal of the crops. In Cornwall the figure of a man was chalked on the log when it was brought in for kindling (Courtney. 1890), which must have been exactly what was meant by giving the cailleach "the representation of some woman". Perhaps this was the origin of the custom whereby each member of the family had to sit on the log in turn, sing a Christmas song, and drink to a merry Christmas and happy New Year (Sandys).

Apart from these deeper sentiments, the use of the log at this time of year always had a bearing on the future prosperity of the family, for this is the sun that is being brought into the house at the season of least daylight, and indeed the new year. Yule Log customs are connected with the new lighting of the house fire, transferred from Samhain (the Celtic new year at the first of November) to the winter solstice (Miles). That is why it is so important that the log is never allowed to burn itself out, for that would be the worst of omens (Gomme. 1883). And parts of the log were kept, perhaps indefinitely, in the house after the festivities were over. At Penistone, for instance, the ashes were collected in the morning and put in the cellar, to keep witches away, so it was said, and to bring luck to the house. They were kept for years, forming a great pile (Addy. 1895). The Somerset custom was to put it in the cow-stall, to bring good luck in rearing calves through the year, while still making sure that it would be used to light next year's log (Watson). So too in France, where the log was called the Calendreau, or Caligneau around Marseilles, and chalendal in Dauphine (Grimm), the belief was that if the householder kept the charred remains under his bed, they would act as a talisman, preserving the house from fire and lightning for the whole year (Salle). The same belief turns up in the north-east of England. Jeffrey, writing of Whitby, reckoned the old ends of Yule logs were kept in the house indefinitely. Any small piece of the log would protect the house from burning, and if thrown on the fire, it would quieten a tempest- an important consideration for sailors' families. There is even a belief somewhere that keeping it under the bed will protect the family from getting chilblains! (Waring).

Sometimes one finds a piece of last year's log being saved to light up the new one at Christmas, while keeping the family from harm in the meantime (Gomme. 1883). That was the custom both in Wales (Trevelyan) and in Cornwall (Courtney. 1890), right into the twentieth century. It was put into the fireplace and burned, but the new log was placed on it before it was burnt through, so that "the old fire and the new burn together" (Trevelyan). The custom is also recorded in east Yorkshire (Nicholson). Herrick knew the custom, too:

> With last year's brand
> Light the new block, and
> For good successe in his spending
> On your psalteries play,
> That sweet luck may
> Come while the log is a tending.

There was plenty of ritual about bringing in the log. It could only be done on Christmas Eve, and it was very unlucky to light it before then; it was equally unlucky to stir the fire during Christmas Eve supper (Denham). Getting the log nearly always involved a great deal of fuss and noise. Sometimes it was called Dun the carthorse, and it was supposed to get stuck in the mud. Drawing Dun out of the mud was a common Christmas pastime. Some of the party would try to pull Dun out, but they would need more help, and so it used to go on, until everyone was involved in the job before success was achieved (Sandys). It took the form of a regular tug-of-war in Stromness, Orkney, each year on Christmas Eve (until 1936) for a tree of some kind (after all, trees are scarce in Orkney). The point about this one was that it had to be taken from someone's garden, without the owner's knowledge or consent, and carried off into the middle of the town. Chains or ropes were attached to the tree, and so the contest began (Northenders versus Southenders) to drag it to a traditional goal somewhere (Marwick).

Often, the log was decorated with evergreens as it was brought home (Hole. 1976), and always it was dragged in with a lot of noise and merrymaking. Sometimes it was well-nigh a tree trunk, for it was supposed to last over the whole Christmas season, until, that is, Old Christmas Eve (5 January). There was even a rather comic reason for getting the biggest log available. On large farms in eighteenth century Norfolk, it was always the custom to make two qualities of cider. During the Christmas season and specifically during the time the Yule log was burning, the whole household, master and servants alike, drank the better quality cider. Obviously it was in the interest of the servants to make sure that the biggest and slowest burning log should be kept for Christmas, in order to make the good stuff last longer (Minchinton). So too with the Devonshire Ashen Faggot. A quart of cider was served each time a hoop round the faggot burst. Sometimes it was the job of a particular person, or member of the family, either to bring the log, or to light it, as in east Yorkshire, where it was the wheelwright's apprentice who brought it round, to be given a few coppers for his Christmas box (Nicholson). Or in Provence, where it was the duty of the grandfather to set the log, from a pear tree preferably there, while the youngest child of the family poured wine over it (Bayley). (see also **ASHEN FAGGOT**)

Z

Zantedeschia aethiopica > ARUM LILY

Zanthoxylum americanum > PRICKLY ASH

Zea mays > MAIZE

ZEDOARY

(*Curcuma zedoaria*) As with Turmeric, its close relative, it yields a yellow dye. Popular at one stage as a spice, it is now only used by Indonesians in curry powder (Clair), and in perfumery (Genders. 1972). Zedoary is actually mentioned in Anglo-Saxon medicine, but being a rather unusual substance, is recommended for magical medicine, as an ingredient in the "holy drink" against "elfin enchantment" (Bonser). As late as the mid-17th century, it was still marvel enough to be recommended against the plague. Lupton has "The root of Zedoary (but be sure it be perfect and good) mixed with raisins, and a little liquorice, champed with the teeth and swallowed, preserves them that do so unhurt, or without danger of the plague, if they go to any that are infected with the plague, or that are constrained to speak with them that have the plague". Pomet, a century or so later, was still quite enthusiastic about it, describing it as "esteem'd a good Cordial, and of great Efficacy against all Venom and Contagion", but in real terms, its reputation was going down by the end of the 17th century, until its use is confined to its own habitat.

Zephyranthes atamasco > ATAMASCO LILY

Zingiber officinale > GINGER

ZINNIA

i.e., *Zinnia elegans*, and *Z grandiflora* In its native Brazil, the curandero, or traditional healer, puts a leaf of *Z elegans* on top of a patient's head to cure madness, and it is used as an ingredient in the ritual bath that forms part of all Brazilian healing ceremonies (P V A Williams). *Z grandiflora* is a North American plant, and the Navajo Indians drank an infusion of this plant, or used it as a fumigant, for sexual diseases attributed to ceremonially improper sexual intercourse, or to intercourse too soon after childbirth (Wyman & Harris).

Zizania aquatica > WILD RICE

Zizyphus abyssinica > CATCHTHORN

Zizyphus jubajuba > JUJUBE

Zizyphus spina-christi > LOTUS TREE

BIBLIOGRAPHY

ABBOTT, G F. *Macedonian folklore*. Cambridge, Univ. Press

ABLETT, William H. *English trees and tree-planting*. Smith, Elder

ACKERMANN, A S E. *Popular fallacies explained and corrected*. 3rd ed. Old Westminster Press, 1923

ADAMS, W H Davenport. *Witch, warlock and magician: historical sketches of magic and witchcraft in England and Scotland*. Chatto & Windus, 1889

ADDISON, Josephine. *The illustrated plant lore*. Sidgwick & Jackson, 1985

ADDISON, Josephine *and* Cherry Hillhouse. *Treasury of tree lore*. Deutsch, 1999

ADDY, Sidney Oldall. *Household tales and traditional remains collected in the counties of York, Derby and Nottingham. David Nutt, 1895*

AITKEN, Hannah *editor. A forgotten heritage: original folktales of Lowland Scotland*. Edinburgh, Scottish Academic Press, 1973

AKERMAN, John Yonge. *A glossary of provincial words and phrases in use in Wiltshire*. John Russell Smith, 1842

ALBERTUS MAGNUS. [Liber aggregationis] *The book of secrets of Albertus Magnus of the virtues of herbs, stones and certain beasts* and *A book of the marvels of the world. edited by* Michael R Best and Frank H Brightman. Oxford, Clarendon Press, 1973

ALFORD, Violet. *The hobby horse and other animal masks*. Merlin Press, 1978

ALLAN, Mea. *The gardener's book of weeds*. Macdonald & Jane's, 1978

ALLEGRO, John M. *The Dead Sea scrolls and the Christian myth*. Newton Abbot, David & Charles, 1979

ALLEN, Andrew. *A dictionary of Sussex folk medicine*. Newbury, Countryside Books, 1995

ALLIES, Jabez. *Antiquities and folk lore of Worcestershire*. John Russell Smith, 1896

ANDERSEN, Johannes C. *Myths and legends of the Polynesians*. Harrap, 1928

ANDERSON, Frank J. *An illustrated history of herbals*. New York, Columbia Univ Press, 1977

ANDERSON, William. *Green man: the archetype of our oneness with the earth*. Harper Collins, 1990

ANDREWS, Elizabeth. *Ulster folklore*. Eliot Stock, 1913

ANDREWS, W. *Curious church customs*. 2nd ed. Andrews, 1898

ANSON, Peter F. *Fisher folk-lore: old customs, taboos and superstitions among fisher folk, especially in Brittany and Normandy and on the east coast of Scotland*. Faith Press, 1965

L APULEIUS MADAURENSIS. *The herbal of Apuleius Barbarus from the early twelfth century manuscript formerly in the Abbey of Bury St Edmunds (MS Bodley 136) described by* Robert T Gunther. Oxford, Roxburghe Club, 1925

ARBER, Agnes. *Herbals: their origin and evolution*. Cambridge, Univ Press, 1912

ARGENTI, Philip P *and* H J Rose. *The folk-lore of Chios*. 2 Vols. Cambridge, Univ Press, 1949

ARMSTRONG, Edward A. Mugwort lore. *Folklore. Vol 55; 1944 22–27*

ASHTON, Hugh. *The Basuto: a social study of traditional and modern Losotho*. 2nd ed. Oxford, Univ Press, 1967

ASHTON, John,. *Chap-books of the eighteenth century*. Chatto & Windus, 1882

ATKINSON, David. "The Broomfield Hill" and the double standard. *Lore and Language. Vol 14; 1996 15–30*

ATKINSON, J C. *Forty years in a moorland parish*. Macmillan, 1886

ATKYNS, *Sir* Robert. *The ancient and present state of Gloucestershire*. 2nd ed. 1768

AUBREY, John. *Remaines of Gentilisme and Judaisme* (1686/7). *edited by* J Britten Folklore Society, 1881. *Miscellanies upon various subjects* (1696). 5th ed. Reeves & Turner, 1890. *The natural history of Wiltshire. edited by* James Britten Devizes, Wiltshire Topographical Society, 1847

AWBERY, G M. Plant names of religious origin in Welsh oral tradition. *in* VICKERY, R *editor Plant-lore Studies*. Folklore Society, 1984

AWOLALU, J Omosade. *Yoruba beliefs and sacrificial rites*. Longman, 1979

BAIRACLI-LEVY, Juliette de. *Wanderers in the New Forest*. Faber, 1958

BAKER, Anne Elizabeth. *Glossary of Northamptonshire words and phrases*. 2 Vols. John Russell Smith, 1854

BAKER, H G. *Plants and civilization*. Macmillan, 1964

BAKER, Margaret. *Folklore and customs of rural England*. Newton Abbot, David & Charles, 1974

BAKER, Margaret. *Wedding customs and folklore*. Newton Abbot, David & Charles, 1977

BAKER, Margaret. *The gardener's folklore*. Newton Abbot, David & Charles, 1977

BAKER, Margaret. *Folklore of the sea*. Newton Abbot, David & Charles, 1979

BAKER, Margaret. *Discovering the folklore of plants*. 2nd ed. Princes Risborough, Shire, 1980

BALAZS, J. The Hungarian shaman's technique of trance induction. *in* DIOSZEGI, V *editor Popular beliefs and folklore tradition in Siberia*. Bloomington, Indiana Univ Press, 1968

BALÉE, William. *Footprints of the forest: Ka'apor ethnobotany the historical ecology of plant utilzation by an Amazonian people*. New York, Columbia Univ Press, 1994

BANKS, *Mrs* M Macleod. *British calendar customs: Scotland*. 3 Vols. Folklore Society, 1937–41

BANKS, *Mrs* M Macleod. *British calendar customs: Orkney and Shetland*. Folklore Society, 1946

BARBER, Paul. *Vampires, burials, and death: folklore and reality*. New Haven, Yale Univ Press, 1988

BARBER. Charles. *Early modern English*. Deutsch, 1976

BARBER, Peter *and* C E Lucas Phillips. *The trees around us*. Weidenfeld & Nicolson, 1975

BARBOUR, John H. Some country remedies and their uses. *Folk-lore. Vol 8; 1897 386–390*

BARDSWELL, Francis A. *The herb-garden*. A & C Black, 1911

BARING, Anne *and* Jules Cashford. *The myth of the goddess: evolution of an image*. Viking, 1991

BARKER, S G. Some indigenous dyestuffs of Travancore. *Government of Travancore. Department of Industries. Bulletin. No 11; 1921*

BARRACLOUGH, Daphne. *A flower lover's miscellany*. Warne, 1961

BARRETT, S A *and* E W Gifford. Miwok material culture. *Milwaukee. Public Museum. Bulletin. Vol 2. No 4; 1933*

BARRETT, W H. *More tales from the Fens*. Routledge, 1964

BARNASCHONE, L P. Manners and customs of the people of Tenby in the eighteenth century. *Cambrian Journal Vol 4; 1857 177–197*

BARTHOLOMEW ANGLICUS i e Bartholomaeus Anglicus. [Liber de proprietatabus rerum …c1472] *Medieval lore. edited by* Robert Steele. Eliot Stock, 1853

BARTON, Benjamin H *and* Thomas Castle. *The British flora medica*. 2 Vols. E Cox, 1837

BARTON, Rita M *editor. Life in Cornwall in the late nineteenth century: being extracts from the* West Briton *newspaper in the two decades from 1855 to 1875*. Truro, D Bradford Barton, 1972

BARTON, Rita M *editor. Life in Cornwall at the end of the nineteenth century: being extracts from the* West Briton *newspaper in the years from 1876 to 1899*. Truro, D Bradford Barton, 1974

BASSETT, Fletcher S. *Legends and superstitions of the sea and of sailors in all lands and at all times*. Sampson, Low, Marston, Searle and Rivington, 1885

BAUMANN, Hellmut. *Greek wild flowers and plant lore in ancient Greece ....translated and augmented by* William T Stearn and Eldwyth Puth Stearn. Herbert Press, 1993

BAYLEY, Harold. *Archaic England*. Chapman & Hall, 1919

BAZIN, Germain. *A gallery of flowers*. Thames & Hudson, 1960

BEARD, Charles. *Lucks and talismans*. Sampson Low, nd

BECK, Horace Palmer. *The folklore of Maine*. Philadelphia, Lippincott, 1957

BECKWITH, Martha Warren. *Black roadways: a study of Jamaican folk life*. Chapel Hill, Univ of North Carolina Press, 1929

BECKWITH, Martha Warren. *Hawaiian mythology*. New Haven, Yale Univ Press, 1940

BECKWITH, Martha Warren. Notes on Jamaican ethnobotany. *in* BECKWITH, Martha Warren *Jamaica folk-lore American Folk-lore Society (reprint, New York, 1969)*

BEGG, E I. Scraps of Highland folklore. *Folk-lore. Vol 62; 1951 326–8*

BEIER, Lucinda McCary. *Sufferers and healers: the experience of illness in seventeenth-century England*. Routledge, 1987

BEITH, Mary. *Healing threads: traditional medicines of the Highlands and Islands*. Edinburgh, Polygon, 1995

BELL, Hesketh J. *Obeah ; witchcraft in the West Indies*. 2nd ed. Sampson, Low, Marston & Co, 1892

BELL, William. *Shakespeare's Puck and his folk-lore*. 1852

BENWELL, Gwen *and Sir* Arthur Waugh. *Sea enchantress*. Hutchinson, 1961

BERDOE, Edward. *The origin and growth of the healing art: a popular history of medicine in all ages and Countries*. Swan, Sonnenschein, 1893

BERGEN, Fanny D. Current superstitions collected from the oral tradition of English speaking folk. *American Folklore Society. Memoirs. Vol 4; 1896*

BERGEN, Fanny D. Animal and plant lore. *American Folklore Society. Memoirs. Vol 7; 1899*

BERKSHIRE FEDERATION OF WOMENS' INSTITUTES. *The Berkshire book edited by* Robert Gathorne-Hardy. 2nd ed. Reading, nd (1951)

BESTERMAN, Theodore. The folklore of dowsing. *Folk-lore. Vol 37; 1926 113–133*

BETT, Henry. *English legends*. Batsford, 1950

BETT, Henry. *English myths and traditions*. Batsford, 1952

BEVAN-JONES, Robert. *The ancient yew: a history of* Taxus baccata. Macclesfield, Windgather Press, 2002

BEZA, Maran. *Paganism in Roumanian folklore*. Dent, 1928

BIANCHINI, Francesco *and* Francesco Corbetta. *The fruits of the earth*. Cassell, 1975

BILLINGTON, Sandra. Butchers and fishmongers: their historical contribution to London's festivity. *Folklore. Vol 191; 1990 97–103*

BILLSON, Charles J. *County folklore, Leicestershire and Rutland*. Folklore Society, 1895

BIRCHER, Ruth. *Eating your way to health: the Bircher-Bennor approach to nutrition.* Faber, *1961*

BISWAS, P C. *Primitive religion, social organization, law and government among the Santals. Univ. of Calcutta. Anthropological papers. New series. 4; 1935*

BLACK, William G. *Folk medicine.* Folklore Society, 1883

BLAKEBOROUGH, J Fairfax. *The hand of glory.* Grant, Richards, 1924

BLOOM, J Harvey. *Folk lore, old customs and superstitions in Shakespeare land.* Mitchell, Hughes & Clark, nd (1930?)

BLOUNT, Thomas. *Fragmenta antiquitatis: or, ancient tenures, and jocular customs of manors* (1679) enlarged and corrected by Josiah Beckwith, with considerable additions from authentic sources by Hercules Malebysshe Beckwith. Butterworth, 1815

BLOXHAM, Christine. *May Day to mummers: folklore and traditional customs in Oxfordshire.* Charlbury, Wychwood Press, 2002

BLUNT, Wilfrid. *The art of botanical illustration.* Collins, 1950 (New Naturalist *series*)

BLUNT, Wilfrid. *Flowers drawn from nature edited by* Gerard von Spoendanck. Leslie Urquhart Press, 1957

BLUNT, Wilfrid *and* Sandra Raphael. *The illustrated herbal.* Frances Lincoln in assoc. with Weidenfeld & Nicolson, 1979

BOAS, Franz. The religion of the Kwakiutl Indians. Part II translations *Columbia University. Contributions to Anthropology. Vol 10; 1930*

BOASE, Wendy. *The folklore of Hampshire and the Isle of Wight.* Batsford, 1976 (The folklore of the British Isles *series*)

BOGATYRËV, Petr. *Vampires in the Carpathians: magical acts, rites, and beliefs in Subcarpathian Rus'.* (1929*)* New York, Carpatho-Rusyn Research Center. East European Monographs, 1998

BOISSEVAIN, Pascall. The Maltese islands. *in* SUTHERLAND, Anne *Face values: some anthropological themes.* BBC, 1978

BOLAND, Maureen *and* Bridget Boland. *Old wives' lore for gardeners.* Bodley Head, 1976

BOLAND, Bridget. *Gardener's magic and other old wives' tales.* Bodley Head, 1977

BOLTON, Brett L. *The secret powers of plants.* Sphere, 1975

BONAR, Ann. *The complete guide to conservatory plants.* Collins & Brown, 1992

BONSER, Wilfrid. *The medical background of Anglo-Saxon England: a study in history, psychology and folklore.* Wellcome Hist. Med. Libr., 1963

BOORDE, Andrew. *A compendious Regyment or dyetary of helth.* (1542) Early English Text Society, 1870

BORGES, Jorge Luis. [El libro de los seres imaginarios] *The book of imaginary being.* Harmondsworth, Penguin, 1974

BOSWELL, Silvester Gordon. *The book of Boswell: the autobiography of a gypsy.* Gollancz, 1970

BOTTRELL, William. *Traditions and hearth-side stories of west Cornwall.* Penzance, 1870

BOUISSON, Maurice. *Magic: its rites and history.* Rider, 1960

BOURKE, John G. The medicine-men of the Apache. *Washington. Smithsonian Institution. Bureau of Ethnology. Annual Report. 9th; 1887–88 451–603*

BOURNE, George (writing as George Sturt). *A small boy in the sixties.* Cambridge, Univ Press, 1927

BOWNAS, Geoffrey. *Japanese rain-making and other folk practices.* Allen & Unwin, 1963

BRAIN, Robert. *The decorated body.* Hutchinson, 1979

BRAND, John. *Observations on popular antiquities. edited by Sir* Henry Ellis Charles Knight, 1841. *Observations on popular antiquities. edited by* W Carew Hazlitt Charles Knight, 1870

BRAY, *Mrs. The borders of the Tamar and the Tavy: their natural history, manners, customs, superstitions, scenery, antiquities, eminent persons, etc.* 2 Vols. W Kent & Co, 1879

BREWSTER, Paul G Folk cures and preventives from southern Indiana. *Southern Folklore Quarterly. Vol 3; 1939 33–43*

BRIGGS, Katharine M. *Pale Hecate's team: an examination of the beliefs on witchcraft and magic among Shakespeare's contemporaries and his immediate successors.* Routledge, 1962

BRIGGS, Katharine M. *The fairies in tradition and literature.* Routledge, 1967

BRIGGS, Katharine M. *The folklore of the Cotswolds.* Batsford, 1974 (The folklore of the British Isles *series*)

BRIGGS, Katharine M. *The vanishing people: a study of traditional fairy beliefs.* Batsford, 1978

BRIGGS, Katharine M. *Nine lives: cats in folklore.* Routledge, 1980

BRILL, Edith. *Life and tradition on the Cotswolds.* Dent, 1973

BRIMBLE, L J F. *Trees in Britain: wild, ornamental and economic, and some relatives in other lands.* Macmillan, 1948

BRITTEN, James. Plant-lore notes to Mrs Latham's West Sussex superstitions. *Folk-lore Record. Vol 1 155–9*

BRITTEN, James. *Old country and farming words: gleaned from agricultural books.* Trübner for English Dialect Society, 1880

BRITTEN, James. The shamrock *The Month. Vol 137; January-June 1921; 192–205*

BRITTEN, James *and* Robert Holland. *A dictionary of English plant-names*. English Dialect Society, 1886

BORADWOOD, Lucy E. A Swiss charm. *Folk-lore. Vol 16; 1903 465–7*

BROCKETT, Jan Trotter. *A glossary of north Country words, with their etymology, and affinity to other languages. 3rd ed*. Simpkin Marshall, 1846

BROCKIE, William. *Legends & superstitions of the county of Durham*. Sunderland, 1886 (reprint Wakefield, EP Publishing, 1970)

BRONGERS, Georg A. *Nicotiana tabacum:the history of tobacco and tobacco smoking in the Netherlands*. Amsterdam, Becht, 1964

BROUK, B. *Plants consumed by man*. Academic Press, 1975

BROWN, O Phelps. *The complete herbalist: or, The people their own physicians, by the use of nature's remedies*. The author, 1872

BROWN, Raymond Lamont. *A book of superstitions*. Newton Abbot, David & Charles, 1970

BROWN, Theo. The folklore of Devon. *Folklore. Vol 75; 1964 145–160*

BROWNE, Ray B. *Popular beliefs and practices from Alabama*. Berkeley, Univ of California, 1958 (Folklore studies series. No 9)

BROWNE, *Sir* Thomas. *Vulgar errors* (1646). *edited by* S Wilkin Bell, 1884

BROWNING, Gareth H. *The naming of wild flowers*. Williams & Norgate, 1952

BROWNLOW, Margaret E. *Herbs and the fragrant garden*. Herb Farm, 1957

BUCHAN, William. *Domestic medicine: or, A treatise on the prevention and cure of diseases by regimen and simple medicines. 15th ed*. 1797

BUCHER, Hubert. *Spirits and power: an anaylsis of Shona cosmology*. Capetown, Oxford Univ Press, 1980

BUCK, Anne. The countryman's smock. *Folk Life. Vol 1; 1963 16–34*

BUCKINGHAMSHIRE FEDERATION OF WOMEN'S INSTITUTES. *A pattern of hundreds*. Chalfont St Giles, Richard Sadler for Bucks FWI, 1975

BUCKLAND, A R. *Anthropological studies*. Ward & Downey, 1891

BUCKLEY, Anthony D. Unofficial healing in Ulster. *Ulster Folklife. Vol 26; 1980 15–34*

BUCKLEY, Anthony D. *Yoruba medicine*. Oxford, Clarendon Press, 1985

BUDGE, *Sir* E A Wallis. *Amulets and superstitions*. Oxford, Univ Press, 1930

BUHLER, A. Geographical extent of the use of bark fibres. *CIBA Review.33; 1940*

BUHLER, A. Primitive dyeing materials. *CIBA Review. 68; 1948*

BUNYARD, Edward A. *Old garden roses*. Country Life, 1936

BURGESS, Frederick. *English churchyard memorials*. Lutterworth, 1963

BURN, Ronald. Folklore from Newmarket, Cambridgeshire. *Folk-lore. Vol 25; 1914 363–366*

BURNE, Charlotte S. *Shropshire folk-lore*. Trubner, 1883

BURNE, Charlotte S. More Staffordshire superstitions. *Folk-lore. Vol 8; 1897 91–92*

BURNE, Charlotte S. *The handbook of folklore*. Folklore Society, 1914

BURTON, Alfred. *Rush-bearing*. Manchester, Brook & Chrystal, 1891

BURTON, Elizabeth. *The Jacobeans at home*. Secker & Warburg, 1962

BUSHAWAY, Bob. *By rite: customs, ceremony and community in England, 1700–1880*. Junction Books, 1982

BYRNE, Hugh James. All Hallows Eve and other festivals in Connaught. *Folk-lore. Vol 18 437–439*

BYRNE, M St Clare. *Elizabethan life in town and country*. Methuen, 1925

CALLENDER, Charles *and* Fadwa El Guindi. *Life-crisis among the Kenuz*. Cleveland, Press of Case Western Reserve University, 1972

CAMDEN, William. *Britannia*. newly translated into English by Edmund Gibson. 1695

CAMERON, Kenneth. *English place-names. 3rd ed*. Batsford, 1977

CAMERON, Malcolm Laurence. *Anglo-Saxon medicine*. Cambridge, Univ Press, 1993 (Cambridge studies in Anglo-Saxon England *Series)*

CAMP, John. *Magic, myth and medicine*. Priory Press, 1973

CAMPBELL, John Gregorson. *Superstitions of the Highlands and Islands of Scotland*. Glasgow, Maclehose, 1900

CAMPBELL, John Gregorson. *Witchcraft and second sight in the Highlands of Scotland*. Glasgow, Maclehose, 1902

CAMPBELL, Joseph. *The masks of God. Vol 1: primitive mythology*. Secker & Warburg, 1960

CAMPBELL-CULVER, Maggie. *The origin of plants: the people and plants that have shaped Britain's garden history since the year 1000*. Headline, 2001

CANZIANI, Estella. Piedmontese folklore. 1 *Folk-lore. Vol 24; 1913 213–218*

CANZIANI, Estella. Abruzzese folklore. *Folk-lore. Vol 39; 1928 209–247*

CAPLAN, Patricia. The Swahili of Chole Island, Tanzania. *in* SUTHERLAND, Anne *editor Face values; some anthropological themes*. BBC, 1978

CAREW, Richard. *Survey of Cornwall.* (1602) Francis Lord de Dunstanville's edition J Faulds, 1811

CARMICHAEL, Alexander. *Carmina gadelica.* 2nd ed. Edinburgh, Oliver & Boyd, 1928

CARR, William. *The dialect of Craven, in the West Riding of the county of York. 2 Vols. 2ⁿᵈ ed. William Crofts, 1828*

CENNINI, Cennino d'Andrea. *The craftsman's handbook. translated by* D V Thompson New York, Dover, 1933

CHAMBERLAIN, Mrs E L. *A glossary of west Worcestershire words.* Trübner for English Dialect Society, 1882

CHAMBERLIN, Ralph V. The ethno-botany of the Gosiute Indians of Utah. *American Anthropological Association Memoirs. Vol 2; 1911 331–405*

CHAMBERS, Robert. *The book of days: a miscellany of popular antiquities.* Edinburgh, Chambers, 1863

CHAMBERS, Robert. *Popular rhymes of Scotland.* Edinburgh, Chambers, 1870

CHANDLER, Jennifer. Old men's fancies: the case of the churchyard yew. *FLS News. No 15; July 1992 3–6*

CHAPLIN, Mary. *Riverside gardening.* Collingridge, 1964

CHESHIRE FEDERATION OF WOMEN'S INSTITUTES. *Cheshire within living memory.* Newbury, Countryside Books, 1994

CHEVIOT, Andrew. *Proverbs, proverbial expressions and popular rhymes of Scotland.* Paisley, Gardner, 1896

CHILD, Heather and Dorothy Colles. *Christian symbols, ancient and modern: a handbook for students.* Bell, 1971

CHINESE medicinal herbs of Hong Kong. 3 Vols. 1983–1987

CHISENDALE-MARSH, U B. Folk-lore in Essex and Herts. *Essex Review. Vol 5; 1896 142–162*

CHOAPE, R Pearse. *The dialect of Hartland, Devonshire.* Kegan Paul, Trench, Trubner for English Dialect Society, 1891

CHOPRA, *Sir* Ram Nath, Rattan Lall Badhwar *and* Sudhamoy Ghosh. *Poisonous plants of India.* Calcutta, Government of India Press for Indian Council of Agricultural Research, 1940 (reprint Jaipur, Academic Publishers 1984)

CIPRIANI, Lidio. *The Andaman Islanders.* Weidenfeld & Nicolson, 1966

CLAIR, Colin. *Of herbs and spices.* Abelard-Schumann, 1961

CLAPHAM, A R, T G Tutin & E F Warburg. *Flora of the British Isles.* Cambridge, Univ Press, 1952

CLARE, John. *The natural history prose writings of John Clare. edited by* Margaret Grainger Oxford, Clarendon Press, 1983

CLARK, A L *and others.* Some Wiltshire folk-lore. *Wiltshire Notes and Queries. Vol 1; 1893–5*

CLARK, H F. The mandrake fiend. *Folklore. Vol 73; 1962 257–269*

CLARK, J Grahame D. Fowling in prehistoric Europe. *Antiquity. Vol 22; 1948 116–130*

CLARK, J Grahame D. *Prehistoric Europe: the economic basis.* Methuen, 1952

CLARK, J *and* Stuart Piggott. *Prehistoric societies.* Hutchinson, 1965

CLASSEN, Constance, David Howes and Anthony Synnott. *Aroma: the cultural history of scent.* Routledge, 1994

CLEBERT, Jean-Paul. *The gypsies. translated by* C Duff Vista, 1963

CLINCH, George *and S W Kershaw. Bygone Surrey.* Simpkin, Marshall, Hamilton, Kent, 1895

COATES, Helen. *Weaving for amateurs.* Studio, 1941

COATS, Alice M. *Flowers and their histories.* A & C Black, 1956

COATS, Alice M. *The treasury of flowers.* Phaidon, 1975

COBBETT, William. *Cottage Economy.* 1822 (reprint of 17th edition, 1850 Oxford, Univ Press, 1979)

COCKAYNE, Oswald. *Leechdoms., wortcunning and starcraft of early England.* 3 vols. Longman, 1864

CODRINGTON, K de B. The use of counter-irritants in the Deccan. *Royal Anthropological Institute of Great Britain and Ireland. Journal. Vol 66; 1936 369–377*

COFFEY, Timothy. *The history and folklore of North American wildflowers.* New York, Facts on File, 1993

COLE, R E G. *A glossary of words used in south-west Lincolnshire.* Trübner for English Dialect Society, 1886

COLE(S), William. *Adam in Eden: or, Nature's paradise.* 1657

COLGAN, N. The shamrock in literature. *Royal Society of Antiquaries of Ireland. Journal. Vol 26; 1896*

CONDRY, W M. *The Snowdonia National Park.* Collins, 1966 (New naturalist *series*)

CONWAY, David. *The magic of herbs.* Cape, 1973

COOKE, L. Notes on Irish folklore from Connaught, collected chiefly in north Donegal. *Folk-lore. Vol 7 ; 1896 299–301*

COON, Carleton S. *The hunting peoples.* Cape, 1972

COPE, *Sir* William H. *A glossary of Hampshire words and phrases.* Trübner for EDS, 1883

COPLEY, Gordon J. *Names and places.* Phoenix House, 1963

CORLETT, William Thomas. *The medicine-man of the American Indian and his cultural background.* Springfield, Thomas, 1935

CORNISH, Vaughan. *Historic thorn trees in the British isles.* Country Life, 1941

COULTON, G C. *Social life in Britain from the Conquest to the Reformation.* Cambridge, Univ Press, 1918

COURTNEY, M A. *Cornish feasts and folk-lore.* Penzance, Beare, 1890

COXHEAD, C J W. *Old Devon customs.* Exmouth, Raleigh Press, 1957

CRAIGIE, William A. *Scandinavian mythology*. Aisley, Gardner, 1896

CRANE, Thomas Frederick. *Italian popular tales*. Macmillan, 1885

CRIPPEN, T G. *Christmas and Christmas lore*. Blackie, 1923

CROMEK, R H. *Remains of Nithsdale and Galloway song*. Cadell & Davies, 1810

CROSBY, Alan G. *Dictionary of Lancashire dialect, tradition and folklore*. Otley, Smith, Settle, 2000

CROSSING, William. *Folk rhymes of Devon*. Exeter, Commin, 1911

CROSSLEY-HOLLAND, Kevin. *The Norse myths*. Deutsch, 1980

CROWFOOT, Grace M *and* Louise Baldensperger. *From cedar to hyssop: a study in the folklore of plants in Palestine*. Sheldon Press, 1932

CULLUM, Elizabeth. *A cottage herbal*. Newton Abbot, David & Charles, 1975

CULPEPER, Nicholas. *The English physician enlarged*. 1775

CUNNINGHAM, John J *and* Rosalie J Côté. *Common plants, botanical and colloquial nomenclature*. New York, Garland, 1977

CURL, James Stevens. *The Victorian celebration of death*. Newton Abbot, David & Charles, 1972

CURTIS, William. *Flora londinensis*. 3 Vols. 1777–1798

DACOMBE, M R. *Dorset up along and down along*. Winchester, DFWI, 1935

DALBY, Andrew. *Dangerous tastes: the story of spices*. British Museum Press, 2000

DALE, Ivan R *and* P J Greenway. *Kenya trees and shrubs*. Nairobi, Buchanan's Kenya Estates, 1961

DALLIMORE. *Holly, yew and box, with notes on other evergreens*. John Lane, The Bodley Head, 1908

DALYELL, John Graham. *The darker superstitions of Scotland*. Glasgow, Richard Griffin, 1835

DALZIEL, J M. *The useful plants of west tropical Africa*. (appendix to HUTCHINSON, J *and* J M Dalziel *Flora of west tropical Africa*. Crown Agents for Overseas Governments and Administrations, 1937)

DANAHER, Kevin. *Folk tales of the Irish countryside*. Cork, Mercier Press, 1967

DANAHER, Kevin. *The year in Ireland*. Cork, Mercier Press, 1972

DARTNELL, George Edward *and* Edward Hungerford Goddard. *Wiltshire words: a glossary of words used in the County of Wiltshire*. English Dialect Society, 1893

DAVENPORT, Gertrude C. Folk-cures from Kansas. *Journal of American Folklore. Vol 11; 1898 129–132*

DAVIDSON, Hilda Ruth Ellis. *Gods and myths of northern Europe*. Penguin, 1964

DAVIDSON, Thomas. The needfire ritual. *Antiquity. Vol 29; 1955 132–136*

DAVIES, J C. *Folk-lore of west and mid-Wales*. Aberystwyth, 1911

DAVIES, Lynn. Aspects of mining folklore in Wales. *Folk Life. Vol 9; 1971 79–107*

DAWSON, Warren M. *Magician and leech: a study in the beginnings of medicine with special reference to ancient Egypt*. Methuen, 1929

DAWSON, Warren M. *A leechbook, or collection of medical recipes, of the 15ᵗʰ century*. Macmillan, 1934

DEANE, Tony *and* Tony Shaw. *The folklore of Cornwall* Batsford 1975 (The folklore of the British Isles *series*)

DEBRUNNER, Hans. *Witchcraft in Ghana: a study on the belief in destructive witches and its effect on the Akan tribes*. Kumasi, Presbyterian Book Depot, 1959

DEMPSTER, *Miss*. Folk-lore of Sutherlandshire. *Folk-lore Journal. Vol 6; 1888 215–252*

DENHAM, M A. *A collection of proverbs and popular sayings relating to the seasons, etc.* Percy Society, 1846

DENHAM, M A. *The Denham tracts. edited by* James Hardie Folklore Society, 1892

DENSMORE, Francis. Uses of plants by the Chippewa Indians. *Washington. Bureau of American Ethnology. Annual Report. 44ᵗʰ; 1926–7 1928*

DE ROPP, Robert. *Drugs and the mind*. Gollancz, 1958

DEVLIN, Judith. *The superstitious mind; French peasants and the supernatural in the nineteenth century*. New Haven and London, Yale Univ Press, 1987

DEW, Walton N. *A dyshe of Norfolk dumplings*. Jarrold, 1898 (reprint Wakefield, EP Publ., 1973)

DIMBLEBY, G W. *Plants and archaeology*. John Baker, 1967

DIOSCORIDES. [Herbal] *The Greek herbal of Dioscorides, illustrated by a Byzantine A D 512, Enloished by John Goodyer A D 1655, edited and first printed A D 1933 by* Robert T Gunther Oxford, Univ Press, 1934

DITCHFIELD, P H. *Old English sports, pastimes and customs*. Methuen, 1891

DOBKIN DE RIOS, Marlene. Banisteriopsis in witchcraft and healing activities in Iquitos, Peru. *Economic Botany. Vol 24; 1970 p296–300*

DOBKIN DE RIOS, Marlene. The *wilderness of mind: sacred plants in cross-cultural perspective*. Beverly Hills and London, Sage Publications, 1976

DODSON, Harry *and* Jennifer Davies. *The Victorian kitchen garden companion*. BBC, 1988

DONOVAN, Frank. *Never on a broomstick*. Allen & Unwin, 1973

DORMAN, Rushton M. *The origin of popular superstitions*. Philadelphia, Lippincott, 1881

DORSON, Richard M. *Buying the wind: regional folklore in the United States*. Chicago, Univ Press, 1964

DOUGLAS, James Sholto. *Alternative foods: a world guide to lesser-known edible plants*. Pelham, 1978

DRAKE, Maurice *and* Wilfred Drake. *Saints and their emblems*. T Werner Laurie, 1916

DRIVER, Harold E. *Indians of North America*. Chicago, Univ Press, 1961

DRURY, Susan M. The use of wild plants as famine foods in eighteenth century Scotland and Ireland. *in* VICKERY, R *editor Plant-lore studies*. Folklore Society, 1984

DRURY, Susan M. Herbal remedies for livestock in seventeenth and eighteenth century England: some examples. *Folklore. Vol 96. Part 2; 1985 243–247*

DRURY, Susan M. English love divinations using plants: an aspect. *Folklore. Vol 97; 1986 210–214*. Customs and beliefs associated with Christmas evergreens. *Folklore. Vol 98; 1987 194–199*

DRURY, Susan M. *Plants and wart cures in England from the seventeenth to the nineteenth century: some examples. Folklore. 102; part 1; 1991 97–100*

DRURY, Susan M. Plants and pest control in England circa 1400–1700. *Folklore. Vol 103; 1992 103–6*

DRURY, Susan M. Funeral plants and flowers in England: some examples. *Folklore. Vol 105; 1994 101–3*

DUNCALF, William G. *The Guinness book of plant facts and feats*. Enfield, Guinness Superlatives, 1976

DUNCAN, Angus. *Hebridean island: memories of Scarp*. East Linton, Tuckwell Press, 1995

DURHAM, M E. Whence comes the dread of ghosts and evil spirits? *Folk-lore. Vol 44; 1933 151–175*

DURRELL, Lawrence. *Prospero's cell: a guide to the landscape and manners of the island of Corcya*. Faber, 1945

DUTTON, Ralph. *The English garden*. 2nd ed. Batsford, 1937

DYER, T F Thiselton. *British popular customs*. Bell, 1876

DYER, T F Thiselton. *Domestic folk*-lore. Cassell, Petter, Galpin & Co, nd (1881?)

DYER, T F Thiselton. *Folk-lore of Shakespeare*. Griffith & Farrar, 1883

DYER, T F Thiselton. *The folk-lore of plants*. Chatto & Windus, 1889

EARLE, John. *English plant names from the 10th to the 15th century*. Oxford, Clarendon Press, 1880

EARWOOD, Caroline. Trees and folk medicine. *Folk Life. Vol 38; 1999–2000 22–31*

EASTHER, Alfred. *A glossary of the dialect of Almondbury and Huddersfield*. Trübner for EDS, 1883

EBERLY, Susan S. A thorn among the lilies: the hawthorn in medieval love allegory. *Folklore. Vol 100; 1989 41–52*

ECKENSTEIN, Lina. *Comparative studies in nursery rhymes*. Duckworth, 1906

EDDRUP, *Rev. Canon*. Notes on some Wiltshire superstitions. *Wiltshire Archaeological Magazine. Vol 22; 1885 330–334*

EDLIN, Herbert L. *The natural history of trees*. Weidenfeld & Nicolson, 1976 (World Naturalist *series*)

EDWARDS, Thornton B. The sugared almond in modern Greek rites of passage. *Folklore. Vol 107; 1996 41–52*

EDWARDS, Thornton B. Greek wedding customs. Part 1 *Folk Life. Vol 41; 2002/3 45–60*

EDWARDS, Thornton B. A letter to *FLS News. 41; November 2003 14*

EGAN, F W. Irish folk-lore: medical plants. *Folk-lore Journal. Vol 5; 1887 49–56*

EINZIG, Paul. *Primitive money in its ethnological, historical and economic aspects*. Oxford, Pergamon Press, 2nd ed. 1966

ELLACOMBE, Henry N. *The plant-lore and garden-craft of Shakespeare*. 2nd ed. W Satchell, 1884

ELLIOTT, Brent. *Treasures of the Royal Horticultural Society*. Herbert Press, 1994

ELLIOTT, Ralph W V. *Runes, yews and magic. Speculum. Vol 32; 1957 250–161*

ELLIOTT, Ralph W V. *Runes: an introduction*. Manchester, Univ Press, 1959

ELLIS, E S. *Ancient anodynes: primitive anaesthesia and allied conditions*. Heinemann, 1946

ELMORE, Francis H. Ethnobotany of the Navajo. *Albuquerque. Univ of New Mexico. School of American Research Monograph series. Vol 1. No 7; 1943*

ELTON, Charles. *Origins of English history*. Quaritch, 1882

ELWIN, Verrier. *The kingdom of the young*. Bombay, Oxford Univ Press, 1968

ELWORTHY, Frederick Thomas. *The West Somerset word-book: a glossary of dialectal and archaic words and phrases used in the west of Somerset and east Devon*. Trubner for English Dialect Society, 1888

ELWORTHY, Frederick Thomas. *The evil eye*. Murray, 1895

EMBODEN, William A. Ritual use of *Cannabis sativa*: a historical-ethnographic survey. *in* FURST, Peter T *editor Flesh of the gods: the ritual use of hallucinogens*. New York, Praeger, 1972. *Bizarre plants: magical, monstrous, mythical*. Studio Vista, 1974. *Narcotic plants*. new ed. Studio Vista, 1979

ENDICOTT, Kirk. *Batek Negrito religion: the world-view and rituals of a hunting and gathering people of Peninsular Malaysia*. Oxford, Clarendon Press, 1979

ENGLERT, Sebastian. *Island at the centre of the world: new light on Easter Island*. Hale, 1972

ERDÖS, Kamill. Notes on pregnancy and birth customs among the gypsies of Hungary. *Gypsy Lore Society. Journal. 3rd series. Vol 37; 1958 50–56*

EVANS, Arthur J. The Rollright Stones and their folk-lore. *Folk-lore. Vol 6; 1895 6–51*

EVANS, E Estyn. *Irish heritage: the landscape, the people and their work*. Dundalk. W Tempest, 1942. *Irish folk ways*. Routledge, 1957

EVANS, George Ewart. *The horse in the furrow*. Faber, 1960. *The pattern under the plough: aspects of folk life in East Anglia*. Faber, 1966. *The farm and the village*. Faber, 1969

EVANS, George Ewart *and* David Thomson. *The leaping hare*. Faber, 1972

EVANS, Mark. *Herbal plants: history and uses*. Studio Editions, 1991

EVELYN, John. *Sylva: or, A discourse of forest trees*. 4th ed. 1678. *Acetaria: a discourse of sallets*. B. Tooke, 1699

EVERETT, Thomas H. *Living trees of the world*. Thames & Hudson, 1969

EWEN, C L'Estrange. *Witch hunting and witch trials: the indictments for witchcraft from the records of 1373 assizes held for the Home Circuit AD 1559–1736*. Kegan Paul, Trench, Trubner, 1929

EYRE, Margaret. Folk-lore of the Wye Valley. *Folk-lore. Vol 16; 1905 162–179*

FAIRWEATHER, Barbara. *Highland plant lore*. Glencoe and North Lorn Folk Museum, nd

FARNSWORTH, Norman R. Hallucinogenic plants. *Science. Vol 162; 1968 1086–92*

FAROOQI, Mohammed Iqtedar Husain. *Plants of the Qur'an*. 5th ed. Lucknow, Sidrah Publishers, 2000

FARRER, James A. *Primitive manners and customs*. Chatto & Windus, 1879

FAULKNER, Christine. Hops and the hop-pickers of the Midlands. *Folk Life. Vol 30; 1991–2 7–16*

FERGUSON, George W. *Signs and symbols in Christian art*. New York, Oxford Univ Press, 1954

FERNANDEZ, James W. *Tabernanthe iboga*: narcotic ecstasis and the work of the ancestors. *in* FURST, Peter C *editor Flesh of the gods: the ritual use of hallucinogens*. New York, Praeger, 1972

FERNIE, W T. *Herbal simples approved for modern uses of cure*. 2nd ed. Bristol, John Wright, 1897

FIRTH, Raymond. *Tikopia ritual and belief*. Allen & Unwin, 1967

FISHER, John. *The origin of garden plants*. Constable, 1982

FISKE, John. *Myths and myth-makers*. Boston, Houghton Mifflin, 1892

FITZGERALD, David. Popular tales of Ireland. *Revue Celtique. Vol 4; 1879–1880*

FLACELIERE, Robert. *Greek oracles*. Elek, 1965

FLETCHER, Linda. Strewings. *Folk Life. Vol 36; 1997–8 66–71*

FLÜCK, Hans. *Medicinal plants and their uses*. Foulsham, 1976

FLUCKIGER, Friedrich *and* Daniel Hanbury. *Pharmacographia*. Macmillan, 1873

FOGEL, Edwain Miller. *Beliefs and superstitions of the Pennsylvania Germans*. Philadelphia, American Germanica Press, 1915

FOLKARD, Richard. *Plant lore, legends and lyrics, embracing the myths, traditions, superstitions and folk-lore of the plant kingdom*. 2nd ed. Dampson, Low & Marston, 1892

FONTENOT, Wonda L. *Secret doctors; ethnomedicine of African Americans*. Westport, Conn., Bergin and Garvey, 2004

FORBES, Alexander Robert. *Gaelic names of beasts (mammalia), birds, fishes, insects, reptiles, etc*. Edinburgh, Oliver & Boyd, 1903

FORBES, Thomas Rogers. *The midwife and the witch*. New Haven, Yale Univ Press, 1966

FORBY, Robert. *A vocabulary of East Anglia*. 2 vols. J B Nichols & Son, 1830

FORDE, C Daryll. *Habitat, economy and society*. 6th ed. Methuen, 1948

FOREY, Pamela. *Wild flowers of the British Isles and northern Europe*. Limpsfield, Dragon's World, 1991

FORSELL, Mary. *Herbs: the complete guide to growing, cooking, healing and pot-pourri*. Amaya, 1990

FORSYTH, A A. *British poisonous plants*. 2nd ed. HMSO, 1968

FOSTER, Gertrude B. *Herbs for every garden*. Rev. ed. Dent, 1973

FOSTER, J J. Dorset folk-lore. *Folk-Lore Journal. Vol 6; 1888 115–119*

FOSTER, Jeanne Cooper. *Ulster folklore*. Belfast, H R Carter, 1951

FRAZER, *Sir* James George. *The golden bough. Part iv Adonis, Attis, Osiris ; Part vi The scapegoat. Macmillan, 1922*. Jacob and the mandrakes. *British Academy. Proceedings. Vol 8; 1917–1918 57–79*

FREETHY, Ron. *From agar to zenry: a book of plant uses, names and folklore*. Ramsbury, Crowood Press, 1985

FREUND, Philip. *Myths of creation*. W H Allen, 1964

FRIEND, Hilderic. *A glossary of Devonshire plant names*. English Dialect Society, 1882. *Flowers and flower lore*. 2 Vols. Allen, 1883

FÜRER-HAIMENDORFF, Christoph von. *The Gonds of Andhra Pradesh*. Allen & Unwin, 1979

FURST, Peter T *editor. Flesh of the gods: the ritual use of hallucinogens*. New York, Praeger, 1972. *Hallucinogens and culture*. San Francisco, 1976

GALLOP, Rodney. *Portugal: a book of folk-ways*. Cambridge, Univ Press, 1936

GALLOTTI, J. Weaving and dyeing in North Africa. *CIBA Review. 21; 1939*

GAMMIE, G A. A note on plants used during famines and seasons of scarcity in the Bombay Presidency. *Botanical Survey of India. Records. Vol 2; 1902–4*

GARIS, Marie de. *Folklore of Guernsey*. The author, 1975

GARRAD, Larch S. Some Manx plant lore. *in* VICKERY, Roy *editor Plant-lore studies*. Folklore Society, 1984

GAYTON, A H. Yokuts-Mono chiefs and shamans. *Berkeley, Univ of California, Publications in American archaeology and ethnology. Vol 24. No 8; 1930 361–420*

GEERTZ, Clifford. *The religion of Java*. Glencoe, Free Press, 1960

GELDART, Ernest. *The art of garnishing churches at Christmas and other times: a manual of directions.* 1882

GELFAND, Michael. *Medicine and magic of the Mashona.* Capetown, Juta, 1956

GENDERS, Roy. *The scented wild flowers of Britain*. Collins, 1971. *A history of scent.* Hamilton, 1972. *The wild-flower garden.* Newton Abbot, David & Charles, 1976

GENG JUNYING *and others. Medicinal herbs: practical traditional Chinese medicine and pharmacology.* Beijing, New World Press, 1991

GENTLEMAN'S MAGAZINE. *The Gentleman's Magazine: being a classified collection of the Gentleman's Magazine from 1731 to 1868 – Popular superstition. edited by* G L Gomme Elliot Stock, 1884

GERARD, John. *Herball.* (Thomas Johnson's edition 1636)

GIBBINGS, Robert. *Coming down the Wye.* Dent, 1942

GIBSON, Colin. *The folklore of Tayside.* Dundee, Museum and Art Gallery, 1959

GIFFORD, Edward S. *The evil eye: studies in the folklore of vision.* New York, Macmillan, 1958

GILL, William Walter. *A Manx scrapbook.* Arrowsmith, 1929. *A second Manx scrapbook.* Arrowsmith, 1932. *A third Manx scrapbook.* Arrowsmith, 1963

GILLOW, John *and* Nicholas Barnard. *Traditional Indian textiles.* Thames & Hudson, 1991

GILMORE, Melvin Randolph. Uses of plants by the Indians of the Missouri River region. *Washington. Bureau of American Ethnology. Annual Report. 33rd; 1919*

GIMBUTAS, Marija. *Ancient symbolism in Lithuanian folk art.* Philadelphia, American Folklore Society, 1958. *The Balts.* Thames & Hudson, 1963

GIMLETTE, John D. *Malay poisons and charm cures.* Oxford, Univ Press, 1915

GINZBURG, Carlo. [*I Benandanti: stregonweria e culti agrari tra Cinquecento e Seicento*] *The night battles: witchcraft and agrarian cults in the 16th and 17th centuries. translated by* John and Anne Tedeschi Routledge, 1983

GLYDE, John. *The Norfolk garland: a collection of the superstitious, beliefs and practices, proverbs, curious customs, ballads and songs of the people of Norfolk.* Jarrold, 1872

GODDARD, C V. Wiltshire folklore jottings. *Wiltshire Archaeological Magazine. Vol 50; 1942–4 24–46*

GODDARD, C V *and* E H Goddard. Wiltshire words: addenda. *Wiltshire Archaeological Magazine. Vol 46; June 1934 478–519*

GOLDIE, W H. Maori medical lore. *New Zealand Institute. Transactions. Vol 37; 1904 1–120*

GOMME, *Lady* Alice Bertha. *The traditional games of England, Scotland and Ireland.* 2 Vols. David Nutt, 1894–8 (reprint New York, Dover, 1964)

GOMME, *Sir* George Laurence. *Folklore relics of early village life.* Methuen, 1908. *editor* (see Gentleman's Magazine)

GOODING, E G B, A R Loveless *and* G R Proctor. *Flora of Barbados.* HMSO, 1965

GORDON, Lesley. *Green magic: flowers, plants & herbs in lore and legend.* Ebury Press, 1977. *The mystery and magic of trees and flowers.* Exeter, Webb & Bower, 1985

GRAHAM, Henry Gray. *The social life of Scotland in the eighteenth century.* 5th ed. Black, 1969

GRANT, Campbell. *The rock paintings of the Chumash.* Berkeley, Univ of California Press, 1965

GRANT, K W. *Myth, tradition and story from western Argyll.* Oban, Oban Times Press, 1925

GRANT, Michael. *Myths of the Greeks and Romans.* Weidenfeld & Nicolson, 1962

GRATTAN, J H *and* Charles Singer. *Anglo-Saxon medicine and magic.* Oxford Univ Press, 1952

GRAVES, Robert. *The white goddess.* Faber, 1948

GREENOAK, Francesca. *God's acre: the flowers and animals of the parish churchyard.* Orbis, 1985

GREGOR, Walter. *Folk-lore of the north-east of Scotland.* Folklore Society, 1881. Some folk-lore from Achterneed. *Folk-lore Journal. Vol 6; 1888 262–265*

GREGORY, Augusta, *Lady. Cuchullain of Muirthemne.* John Murray, 1919. *Visions and beliefs in the west of Ireland.* 2nd ed. Gerrards Cross, Colin Smythe, 1970

GRIAULE, Marcel. *Conversations with Ogotemmeli: an introduction to Dogon religious ideas.* Oxford, Univ Press for International African Institute, 1965

GRICE, F. *Folk tales of the North country, drawn from Northumberland and Durham.* Nelson, 1944

GRIEVE, Maud. *A modern herbal.* Cape, 1931. *Culinary herbs and condiments.* Heinemann, 1933

GRIGSON, Geoffrey. *Gardenage: or, the plants of Ninhursaga.* Routledge, 1952. *The Englishman's flora.* Phoenix House, 1955. *A herbal of all sorts.* Phoenix House, 1959. *A dictionary of English plant names (and some products of plants).* Allen Lane, 1974. *The goddess of love: the birth, triumph, death and return of Aphrodite.* Constable, 1976

GRIMM, Jacob. *Teutonic mythology. translated by* Stallybrass Swan Sonnenschein, 1880

GRINDON, L H. *The Shakspere flora. Manchester, Palmer & Howe, 1883*

GRINSELL, Leslie V. *Folklore of prehistoric sites in Britain.* Newton Abbot, David & Charles, 1976. *The Rollright Stones and their folklore.* St Peter Port, Toucan Press, 1977

GROOME, Francis Hindes. *Gypsy folk-tales.* Hurst & Blackett, 1899

GROSE, Francis. *A glossary of provincial and local words used in England.* John Russell Smith, 1839

GUAZZO, Francesco Maria. *Compendium maleficorum. translated by* E A Ashwin, *edited by* Montague Summers John Rodker, 1929

GUBERNATIS, Angelo de. *Zoological mythology.* Trubner, 1872. *La mythologie des plantes.* Paris, Reinwald, 1878

GUNTHER, Erna. *The ethnobotany of western Washington: the knowledge and use of indigenous plants by native Americans.* rev ed. Seattle, Univ of Washington Press, 1973

GUNTHER, Robert T. The cimaruta: its structure and development. *Folk-lore. Vol 16; 1905 132–161*

GUPTA, Sankar Sen *editor. Tree symbol worship in India.* Calcutta, Indian Publications, 1965

GURDON, Eveline Camilla, *Lady. County folklore, Suffolk.* Folklore Society, 1892

GUTCH, M. *County folklore, North Riding of Yorkshire.* Folklore Society, 1901

GUTCH, M. *County folklore, East Riding of Yorkshire.* Folklore Society, 1911

GUTCH, M *and* M Peacock. *County folklore, Lincolnshire.* Folklore Society, 1908

HACKWOOD, Frederick William. *Staffordshire customs, superstitions and folklore.* Lichfield, Mercury Press, 1924

HADDON, A C. A batch of Irish folk-lore. *Folk-lore. Vol 4; 1893 340–364*

HADFIELD, Miles *and* John Hadfield. *The twelve days of Christmas.* Cassell, 1961

HAIG, Elizabeth. *The floral symbolism of the great masters.* Kegan Paul, Trench, Trubner, 1913

HAINING, Peter. *The warlock's book: secrets of black magic from the ancient grimoires.* W H Allen, 1972

HALD. Margrethe. The nettle as a culture plant. *Folk-lore. 1942; 28–49*

HALLIDAY, William Reginald. *Greek and Roman folk-lore.* Harrap, 1927

HALLIWELL, James Orchard. *Illustrations of the fairy mythology of a* Midsummer Night's Dream Shakespeare Society, 1845

HALLIWELL, James Orchard. *Popular rhymes and nursery tales.* John Russell Smith, 1869

HALLIWELL, James Orchard. *A dictionary of archaic and provincial words, obsolete phrases, proverbs, and ancient customs, from the fourteenth century.* 10th ed. John Russell Smith, 1881

HAMPSHIRE FEDERATION OF WOMEN'S INSTITUTES. *It happened in Hampshire.* 5th ed. Winchester, 1977

HANAUER, J E. *Folk-lore of the Holy Land:Moslem, Christian. and Jewish. edited by* Marmaduke Pickthall Duckworth, 1907

HARDY, James. Wart and wen cures. *Folk-lore Record. Vol 1; 1878 216–228*

HARDY, James. The popular history of the cuckoo. *Folk-lore Record. Vol 2; 1879 47–91*

HARE, C E. *Bird Lore.* Country Life, 1952

HARLAND. John *and* T T Wilkinson. *Lancashire folk-lore.* Warne, 1867

HARLEY, George Way. *Native African medicine, with special reference to its practice in the Mano tribe of Liberia.* Cambridge, Mass., Harvard Univ Press, 1941

HARPER, Edward B. Shamanism in south India. *Southwestern Journal of Anthropology. Vol 13; 1957 267–287*

HARPER, Roland E. Economic botany of Alabama. part 2 *Geological Survey of Alabama, Monograph. 9; 1930*

HARRIS, A. Gorse in the East Riding of Yorkshire. *Folk Life. Vol 30; 1991–2 17.29*

HART, Cyril *and* Charles Raymond *illustrator. British trees in colour.* Joseph, 1973

HARTLAND, Edwin S. *Primitive paternity.* David Nutt, 1909

HARTLAND, Edwin S. *County folklore, Gloucestershire.* Folklore Society, 1892

HARTLAND, Edwin S. *English fairy and other folk tales.* Walter Scott, nd

HARTLEY, Marie *and* Joan Ingilby. *Life and tradition in the Yorkshire Dales.* Dent, 1968

HARVEY, John H. Vegetables in the Middle Ages. *Garden History. Vol 12; 1984 89–99*

HASSIG, D. *Medieval bestiaries.* Cambridge, Univ Press, 1995

HATFIELD, Audrey Wynne. *A herb for every ill.* Dent, 1973

HATFIELD, Vivienne Gabrielle. Herbs in pregnancy, childbirth and breast-feeding. *in* VICKERY, Roy *editor Plant-lore studies.* Folklore Society, 1984

HATFIELD, Vivienne Gabrielle. *Country remedies: traditional East Anglian plant remedies in the twentieth century.* Woodbridge, Boydell Press, 1994

HATT, Gudmund. The corn mother in America and in Indonesia. *Anthropos. Vol 46; 1951 853–914*

HAVARD, V. Food plants of the North American Indian. *Torrey Botanical Club. Bulletin. Vol 22. Part 3; 1895 98–123*

HAVERGAL, Francis T. *Herefordshire words and phrases.* Walsall, W Henry Robinson, 1887

HAWKE, Kathleen. *Cornish sayings, superstitions and remedies.* 3rd ed. Redruth, Dyllansow Truran, 1981

HAWKINS, Desmond. *Avalon and Sedgemoor.* Gloucester, Alan Sutton, 1982

HAYDON, Peter. *The English pub: a history.* Hale, 1994

HAYES, Elizabeth S. *Herbs, flavours and spices.* Faber, 1963

HAZLITT, W Carew. *English proverbs and proverbial phrases.* Reeves & Turner, 1882

HEANLEY, R M. Burning elder-wood. *Folk-lore. Vol 22; 1911 235–6*

HEARN, Karen *editor. Dynasties: painting in Tudor and Jacobean England 1530–1630.* Tate Gallery, 1996

HEATHER, P J. Folklore from Naphill, Bucks. *Folklore. Vol 43; 1932 104–110*

HEIZER, Robert F. Aboriginal fish poisons. *Washington. Smithsonian Institution, Bureau of American Ethnology, Bulletin 151; 1953*

HELM, P J. *Exploring prehistoric England.* Hale, 1971

HELIAS, Pierre-Jakez. *[Le cheval d'orgueil] The horse of pride. translated by* June Guicharnaud New Haven, Yale Univ Press, 1978

HEMPHILL, Rosemary. *Herbs for all seasons.* Sydney, Angus & Robertson, 1972

HENDERSON, George. *Survivals in belief among the Celts.* Glasgow, Maclehose, 1911

HENDERSON, William. *Notes on the folk-lore of the northern counties of England and the Borders..* W Satchell Peyton for Folk-Lore Society, new ed. 1879

HENKEL, Alice. Weeds used in medicine. *US Dept of Agriculture., Farmers' Bulletin. 188; 1904*

HENSLOW, G. *Medical works of the fourteenth century.* Chapman & Hall, 1899

HEPBURN, Ian. *Flowers of the coast.* Collins, 1952 (New Naturalist *series*)

HEWETT, Sarah. *Nummits and crummits: Devonshire customs, characteristics and folk-lore.* Thos. Burleigh, 1900

HEWITT, J C B. The Iroquoian concept of the soul. *Journal of American Folklore. Vol 8; 1895 107–116*

HIBBERT, Samuel. *A description of the Shetland Islands.* Edinburgh, Constable, 1822

HICKEY, Sally. Fatal feeds?: plants, livestock losses and witchcraft accusations in Tudor and Stuart Britain.. *Folklore. Vol 101. No 2; 1990 131–142*

HIGGINS, Rodney. The plant-lore of courtship and marriage. *in* VICKERY, Roy *editor Plant-lore studies.* Folklore Society, 1984

HILL, Jason. *Wild foods of Britain.* A & C Black, 1939

HILL, *Sir* John. *The family herbal.* 1754

HILL, Thomas. *The gardener's labyrinth.* (1577) *edited by* Richard Mabey Oxford, Univ Press, 1987

HILL, W W. Navajo use of jimsonweed. *New Mexico Anthropologist. Vol 3; 1938 19–21*

HILLMAN. *Tusser Redivivus.* 1710

HOBLEY, C W. *Bantu beliefs and magic.* Witherby, 1922

HOFFMAN, W J. Folk-lore of the Pennsylvania Germans. *Journal of American Folklore. Vol 1; 1888 125–135*

HOGARTH, Peter *with* Val Clery. *Dragons.* Allen Lane, 1979

HOHN, Reinhardt. *Curiosities of the plant kingdom.* Cassell, 1980

HOLE, Christina. *Christmas and its customs.* Richard Bell, nd.

HOLE, Christina. *Traditions and customs of Cheshire.* Williams and Norgate, 1937

HOLE, Christina. *English custom and usage.* Batsford, 1941

HOLE, Christina. Notes on some folklore survivals in English domestic life. *Folklore. Vol 68; 1957 411–419*

HOLE, Christina. *English traditional customs.* Batsford, 1975

HOLE, Christina. *British folk customs.* Hutchinson, 1976

HOLE, Christina. Protective symbols in the house. *in* DAVIDSON, H R E *editor Symbols of power.* Cambridge, Brewer for the Folklore Society, 1977

HOLLAND, Robert. *A glossary of words used in the county of Chester.* Trubner for English Dialect Society, 1886

HOMEOPATHIC DEVELOPMENT FOUNDATION. *Homeopathy for the family: an introductory guide to the use of classical homeopathic medicines in the treatment of common ailments and conditions.* 8th ed. Wigmore Publ, 1988

HONE, William. *The table book; or, Daily recreation and information.* William Tegg, nd

HORA, Bayard *editor. The Oxford encyclopedia of trees of the world.* Oxford, Univ Press, 1981

HOSKINS, W G. *The making of the English landscape.* Hodder & Stoughton, 1955

HOSTOS, A de. Antillean stone collars: some suggestions of interpretative value. *Royal Anthropological Institute. Journal. Vol 56; 1926*

HOUGH, Walter. The Hopi in relation to their plant environment. *American Anthropologist. Vol 10; 1897 33–44*

HOUSE, Homer D. Wild flowers of New York. *State Museum of New York. Memoir. 15 (72nd report of Museum); 1918*

HOWELLS, William. *Cambrian superstitions.* 1831

HOWES, F N. *A dictionary of useful and everyday plants and their common names.* Cambridge, Univ Press, 1974

HOWES, H W. Gallegan folklore. III *Folk-lore. Vol 40; 1929 53–61*

HULL, Eleanor. *Folklore of the British Isles.* Methuen, 1928

HULME, F Edward. *Familiar wild flowers.* 2ⁿᵈ series Cassell, nd.

HULME, F Edward. *Natural history lore and legend.* Quaritch, 1895

HUNT, Robert. *Popular romances of the west of England.* Chatto & Windus, *1881*

HUNTER, Joseph. *The Hallamshire glossary.* Pickering, 1829

HURRY, Jamieson B. *The woad plant and its dye*. Oxford, Univ Press, 1930

HURSTON, Zora Neale. *Mules and men*. Philadelphia, Lippincott, 1935 (reprint New York, Harper & Row 1990)

HUTCHINGS, John. Folklore and symbolism of green. *Folklore. Vol 108; 1997 55–63*

HUTCHINSON, John *and* Ronald Melville. *The story of plants and their uses to man*. Gawthorn, 1948

HUXLEY, Anthony. *Plants and planet*. Allen Lane, 1974

HUXLEY, Francis. *The invisibles*. Hart-Davis, 1966

HYAM, Roger *and* Richard Pankhurst. *Plants and their names: a concise dictionary*. Oxford, Univ Press, 1995

HYATT, Harry Middleton. *Folk-lore from Adams County Illinois*. New York, Alma Egan Hyatt Foundation, 1935

HYATT, Richard. *Chinese herbal medicine: an ancient art and modern healing science*. (1978) New York, Thorsons, 1984

HYDE, Douglas. *Beside the fire: a collection of Irish Gaelic folk stories*. David Nutt, 1910

IGGLESDEN, *Sir* Charles. *Those superstitions*. Jarrolds, 1932

INGRAM, John. *Flora symbolica, or, the language and sentiment of flowers*. Warne, 1870

INVERARITY, Robert Bruce. *Art of the Northwest Coast Indians*. Berkeley, Univ of California Press, 1950

INWARDS, Richard. *Weather lore*. Rider, 1950

JACOB, Dorothy. *A witch's guide to gardening*. Elek, 1964

JAEGER, Edmund C. *Desert wild flowers*. rev. ed. Stanford, Univ Press, 1941

JAGO, Frederick W P. *The ancient language, and the dialect of Cornwall*. Truro, Netherton & Worth, 1882

JAMES, Edwin O. *Sacrifice and sacrament*. Thames & Hudson, 1962

JAMES, Edwin O. *The tree of life: an archaeological study*. Leiden, E J Brill, 1966

JAMIESON, John. *An etymological dictionary of the Scottish language*. 2 Vols. Edinburgh, Univ Press, 1808

JEFFERIES, Richard. *Round about a great estate*. 1880 reprint Bradford-on-Avon, Ex Libris Press, 1987

JEFFREY, Percy Shaw. *Whitby lore and legend*. 2nd ed. Whitby, Horne & Son, 1923

JENKINS, D E. *Bedd Gelert: its facts, fairies and folk-lore*. Portmadoc, Llewelyn Jenkins, 1899

JENKINS, J Geraint. Traditional methods of dyeing wool in Wales. *Folk Life. Vol 4; 1966 64–74*

JENKINS, J Geraint. *Life and tradition in rural Wales*. Dent, 1976.

JENNESS, Diamond. The Ojibwa Indians of Parry Island. *National Museum of Canada. Bulletin. Vol 12; 1922*

JENNINGS, James. *The dialect of the west of England, particularly Somersetshire*. 2nd ed. John Russell Smith 1849

JOBSON, Allan. *Under a Suffolk sky*. Hale, 1964

JOHNSON, C Pierpoint. *The useful plants of Great Britain*. Hardwicke, 1862

JOHNSON, W Branch. *Folk tales of Normandy*. Chapman & Hall, 1929

JOHNSTON, Alex. Blackfoot Indian utilization of the flora of the north-western Great Plains. *Economic Botany. Vol 24; 1970 301–324*

JONES, David E. *Sanapia: Comanche medicine woman*. New York, Holt, Rinehart and Winston, 1974

JONES, Francis. *The holy wells of Wales*. Cardiff, Univ Wales Press, 1954

JONES, Ida B. Popular medical knowledge in xivth century England. *Institute of the History of Medicine. Bulletin. Vol 5; 1937 405–451 538–588*

JONES, Lavender M. Some Worcestershire calendar customs. *Folklore. Vol 72; 1961 320–322*

JONES, Malcolm *and* Patrick Dillon. *Dialect in Wiltshire, and its historical, topographical and natural science contexts*. Trowbridge, Wiltshire County Council, 1987

JONES, T Gwynn. *Welsh folklore and folk customs*. Methuen, 1930

JONES, William. *Credulities past and present*. Chatto & Windus, 1880

JONES- BAKER, Doris. *Old Hertfordshire calendar*. Chichester, Phillimore, 1974

JONES- BAKER, Doris. *The folklore of Hertfordshire*. Batsford, 1977

JORDAN, Michael. *A guide to wild plants: the edible and poisonous species*. Millington, 1976

JUDGE, Roy. *The Jack-in-the-green: a May Day custom*. Ipswich, D S Brewer, 1979

KEARNEY, Michael. Spiritualist healing in Mexico. *in* MORLEY, Peter *and* Roy Wallis *editors Culture and curing: anthropological perspectives on traditional medical beliefs and practices*. Peter Owen, 1978

KELLY, Isabel *and* Angel Palerm. *The Tajin Totonac*. Washington, Smithsonian Institution, 1952

KELLY, Walter K. *Curiosities of Indo-European tradition and folk-lore*. Chapman & Hall, 1863

KEMP, P. *Healing ritual: studies in the technique and tradition of the southern Slavs*. Faber, 1935

KENDRICK, T D. *The druids: a study in Keltic prehistory*. Methuen, 1927

KENNEDY, H E. Polish peasant courtship and wedding customs and folk-song. *Folk-lore. Vol 36; 1925 48–68*

KENNEDY, James. *Folklore and reminiscences of Strathspey and Grandtully* Perth. Munro Press nd (1927?)

KENSINGER, Kenneth M. *Banisteriopsis* usage among the Peruvian Cashinhua. *in* HARNER, Michael J *editor Hallucinogens and shamanism*. New York, Oxford Univ Press, 1973

KERENYI, C. *Dionysos: archetypal image of indestructible life*. Routledge, 1976

KIGHTLY, Charles. *Country voices: life and lore in farm and village*. Thames & Hudson, 1984

KILLIP, Margaret. *The folklore of the Isle of Man*. Batsford, 1975 (The folklore of the British Isles *series*)

KILVERT, Francis. *Diary*. illustrated edition Century, 1986

KINAHAN, G H. Irish plant-lore notes. *Folk-lore Journal. Vol 6; 1888 265–267*

KINGSBURY, John M. *Poisonous plants of the United States and Canada*. Englewood Cliffs, Prentice-Hall, 1964

KINGSBURY, John M. *Deadly harvest: a guide to common poisonous plants*. Allen & Unwin, 1967

KLETTER, Christa *and* Monika Kriechbaum *editors. Tibetan medicinal plants*. Stuttgart, Medpharm Scientific Publishing, 2001

KLUCKHOHN, Clyde, W W Hill *and* Lucy Wales Kluckhohn. *Navaho material culture*. Cambridge, Mass., Belknap Press of Harvard Univ Press, 1971

KNIGHT, Katherine. A precious medicine: tradition and magic in some seventeenth-century household remedies. [extract] *Folklore. Vol 113; 2002 237–247*

KNIGHT, W F Jackson. Origins of belief. *Folklore. Vol 74; 1963 289–304*

KORITSCHONER, Hans. Ngoma ya sheitani *Royal Anthropological Institute of Great Britain and Ireland Journal. Vol 66 1936 209–219*

KOURENNOFF, Paul M. *Russian folk medicine*. W H Allen, 1970

KRAPPE, Alexander Heggerty. *The science of folk-lore*. Methuen, 1930

KROEBER, A L. Handbook of the Indians of California. *Washington. Bureau of American ethnology. Bulletin. 78; 1925*

KRYMOW, Vincenzina. *Healing plants of the Bible: history, lore and meditations*. Cincinnatti, St Anthony Messenger Press, 2002

KUNZ, George Frederick. *The curious lore of precious stones*. Philadelphia, Lippincott, 1913

KVIDELAND, Reimund *and* Henning K SEHMSDORF, *editors. Scandinavian folk belief and legend*. Oslo, Norwegian University Press, 1991

LA BARRE, Weston. *The peyote cult*. New Haven, Yale Univ Press, 1938

LA FONTAINE, Jean. *Initiation*. Harmondsworth, Penguin, 1985

LAGUERRE, Michel. *Afro-Caribbean folk medicine*. South Hadley, Mass., Bergin & Harvey Publishers, 1987

L'AMY, John H. *Jersey folk lore*. (1971) limited edition, La Hause, 1983

LANCASHIRE FEDERATION OF WOMEN'S INSTITUTES. *Lancashire lore: a miscellany of country customs, sayings, dialect words, village memories and recipes*. Preston, 1971

LANG, Andrew. *Custom and myth*. new ed. Longmans, 1893

LANGHAM, William. *The garden of health. 1578*

LANNOY, Richard. *The speaking tree: a study of Indian culture and society*. Oxford, Univ Press, 1971

LAROUSSE. *Larousse encyclopedia of mythology*. Hamlyn, 1959

LATHAM, Charlotte. Some west Sussex superstitions lingering in 1868. *Folk-lore Record. Vol 1; 1878 1–67*

LAWRENCE, Berta. *Somerset legends*. Newton Abbot, David & Charles, 1973

LAWSON, John C. *Modern Greek folklore and ancient Greek religion*. Cambridge, Univ Press, 1910

LEA, Henry C. *Materials towards a history of witchcraft*. Yoseloff, 1957

LEASK, J T Smith. *A peculiar people, and other Orkney tales*. Kirkwall, W R Mackintosh, 1931

LEATHART, Scott. *Trees of the world*. Feltham, Hamlyn, 1977

LEATHER, Ella M. *Folk-lore of Herefordshire*. Sidgwick & Jackson, 1912

LE BAS, John. Jersey folklore notes. *Folk-lore. Vol 25; 1914 242–251*

LEETE, J A. *Wiltshire miscellany*. Melksham, Venton, 1976

LEE, Richard Borshay. *The !Kung San: men, women, and work in a foraging society*. Cambridge, Univ Press, 1979

LEGEY, Françoise. *The folklore of Morocco*. translated by Lucy Hotz Allen & Unwin, 1935

LEHNER, Ernst *and* Johanna Lehner. *Folklore and odyssey of food and medicinal plants*. New York, Tudor, 1962

LEIGHTON, Ann. *Early English gardens in New England*. Cassell, 1970

LEIX, Alfred. Ancient Egypt: the land of linen. *CIBA Review. 12; 1958*

LELAND, Charles Godfrey. *Gypsy sorcery and fortune telling*. Unwin, 1891

LELAND, Charles Godfrey. *Etruscan Roman remains in popular tradition*. Unwin, 1898

LELAND, Charles Godfrey. *Aradia: or the gospel of the witches*. David Nutt, 1899

LE STRANGE, Richard. *A history of herbal plants*. Angus & Robertson, 1977

LEWIN, Louis. *Phantastica: narcotic and stimulating drugs*. translated by P H A Wirth. Kegan Paul, Trench, Trübner, 1931

LEWINGTON, Anne. *Plants for people*. Natural History Museum, 1990

LEWIS, Don. *Religious superstition through the ages*. Oxford, Mowbrays, 1975

LEWIS, W H *and* Memory P F Elvin-Lewis. *Medical botany: plants affecting man's health*. New York, Wiley, 1977

LEYEL, *Mrs* C F. *The magic of herbs*. Cape, 1926

LEYEL, *Mrs* C F. *Herbal delights: tisanes, syrups, confections, electuaries, robs, juleps, vinegars and conserves*. Faber, 1937

LEYEL, *Mrs* C F. *Elixirs of life*. Faber, 1948

LINCOLN, Bruce. *Emerging from the chrysalis: studies in rituals of women's initiation*. Cambridge, Mass., Harvard Univ Press, 1981

LINDLEY, John. *Medical and economical botany*. Bradbury & Evans, 1849

LING ROTH, H. *The aborigines of Tasmania*. Halifax, 1899

LLOYD, John Uri. *Origin and history of all the pharmacological vegetable drugs, etc*. Vol 1 Washington, American Drug Manufacturers Association, 1921

LOCKWOOD, W B. *Languages of the British Isles past and present*. Deutsch, 1975

LOEWENFELD, Claire *and* Philippa Back. *Britain's wild larder*. Newton Abbot, David & Charles, nd

LOEWENTHAL, L J A. The palms of Jezebel. *Folklore. Vol 83; 1972 20–40*

LOGAN, Patrick. *Irish country cures*. Dublin, Talbot Press, 1972

LONG, Harold C. *Common weeds of the farm and garden*. Smith, Elder, 1910

LONG, Harold C. *Plants poisonous to livestock*. 2nd ed. Cambridge, Univ Press, 1924

LONG, W H. *A dictionary of the Isle of Wight dialect, and of provincialisms used in the island*. 2nd ed.. Portsmouth, W H Barrell, and London, Simpkin Marshall, 1921

LOOMIS, Roger Sherman. *The Grail from Celtic myth to Christian symbol*. Cardiff, Univ Wales Press, 1963

LOUX, Françoise. *Le jeune enfant et son corps dans la médicine traditionelle*. Paris, Flammarion, 1978

LOVETT, Edward. *Folk-lore and legend of the Surrey hills and of the Sussex downs and forests*. Caterham, 1928

LOWE, John. *The yew-trees of Great Britain and Ireland*. Macmillan, 1897

LOWSLEY, B. *A glossary of Berkshire words and phrases*. Trubner for English Dialect Society, 1888

LUCAS, A T. Furze: a survey and history of its uses in Ireland. *Béaloideas. Vol 16; 1958 1–204*

LUCAS, A T. The sacred trees of Ireland. *Cork. Historical and Archaeological Society. Journal. Vol 68; 1963 16–54*

LUCAS, A T. Irish food before the potato. *Gwerin. Vol 3; 1964 pp 8–43*

LUCAS, Joshua Olumide. *The religion of the Yorubas: being an account of the religious beliefs and practices of the Yoruba peoples of southern Nigeria, especially in relation to the religion of ancient Egypt*. Lagos, C M S Bookshop, 1948

LUPTON. T. *A thousand notable things, on various subjects*. 1660 (reprint Thomas Tegg 1827)

MABBERLEY, David. *Ferdinand Bauer: the nature of discovery*. Merrill Holberton/ Natural History Museum, 1999

MABEY, Richard. *Food for free*. Collins, 1972

MABEY, Richard. *Plants with a purpose: a guide to the everyday uses of wild plants*. Collins, 1977

MABEY, Richard. *Flora Britannica: the concise edition*. Chatto & Windus, 1998

MABUCHI, Toichi. Tales concerning the origin of grains in the insular areas of eastern and south-eastern Asia. *Asia folklore studies. Vol 23. Part 1; 1964 1–92*

McCLINTOCK, David *and* R S R Fitter. *The pocket guide to wild flowers*. Collins, 1956

MacCOLL, Ewan *and* Peggy Seeger. *Till doomsday in the afternoon: the folklore of a family of Scots travellers, the Stewarts of Blairgowrie*. Manchester, Univ Press, 1986

MacCULLOCH, *Sir* Edgar. *Guernsey folklore*. Elliot Stock, 1903

MacCULLOCH, J A. *The misty isle of Skye*. Edinburgh, Oliphant, Anderson & Ferrier, 1905

MacCULLOCH, J A. *The religion of the ancient Celts*. Edinburgh, T & T Clark, 1911

MacDONALD, Christina. *Medicines of the Maori, from their trees, shrubs and other plants, together with foods from the same sources*. Auckland, Collins, 1974

MacDONALD, Donald. *Fragrant leaves and flowers: interesting associations gathered from many sources, with notes on their history and utility*. Warne, nd

MacDONALD, James. *Religion and myth*. David Nutt, 1893

MacGREGOR, Alasdair Alpin. *The peat-fire flame: folk-tales and traditions of the Highlands and Islands*. Edinburgh, Moray Press, 1937

MACKAY, C. *Memoirs of extraordinary popular delusions*. Bentley, 1841

McKAY, John G. *More West Highland tales*. Vol 1 Edinburgh, Oliver & Boyd for the Scottish Anthropological and Folklore Society, 1940

MACKENZIE, Donald A. *Scottish folk-lore and folk life: studies in race, culture and tradition*. Blackie, 1935

MACKENZIE, William. *Gaelic incantations, charms and blessings of the Hebrides*. Inverness, 1895

MacKINNEY, Loran. *Medical illustrations in medieval manuscripts*. Wellcome Historical Medical Library, 1965

MacLAGAN, Robert Craig. *Evil eye in the western Highlands*. David Nutt, 1902

MACLEOD, Dawn. *A book of herbs*. Duckworth, 1968

MACMILLAN, A S *compiler*. *Popular names of flowers, fruits, etc., as used in the county of Somerset and the adjacent parts of Devon, Dorset and Wiltshire*. Yeovil, Western Gazette, 1922

McNEILL, Florence Marian. *The silver bough*. 4 Vols. Glasgow, Maclehose, 1957–1968

MacNEILL, Maire. Wayside death cairns in Ireland. *Béaloideas. Vol 16; 1946*

MacNEILL, Maire. *The festival of Lughnasa: a study of the survival of the Celtic festival of the beginning of harvest.* Oxford, Univ Press, 1962

McNEILL, Murdoch. *Colonsay: one of the Hebrides.* Edinburgh, Douglas, 1910

McPHERSON, J M. *Primitive beliefs in the north-east of Scotland.* Longman, 1929

MADDOX, John Lee. *The medicine man: a sociological study of the character and evolution of the shaman.* New York, Macmillan, 1923

MALINOWSKI, Bronislaw. Myth in primitive psychology. (Frazer lecture. 1925) *in* DAWSON, Warren R *editor The Frazer lectures, 1922–1932.* Macmillan, 1923

MALONEY, Beatrice. Traditional herbal cures in County Cavan part 1. *Ulster Folklife. Vol 18; 1972 66–79*

MANDEVILLE, *Sir* John. *Travels.* many editions

MANKER, Ernst. *People of eight seasons.* Watts, 1965

MANLEY, Victor Strode. *Folklore of the Warminster district.* Warminster, Coates & Parker, 1924 (new edition, Warminster, John Weallens, 1987)

MAPLE, Eric. *The dark world of witches.* Hale, 1962

MARSHALL, Sybil. *Fenland chronicle: recollections of William Henry and Kate Mary Edwards collected by their Daughter.* Cambridge, Univ Press, 1967

MARTIN, Martin. *A description of the western isles of Scotland.* (1703) *edited by* Eneas Mackay Stirling, 1934

MARWICK, Ernest W. *The folklore of Orkney and Shetland.* Batsford, 1975 (The folklore of the British Isles *series*)

MASON, Charlotte Craven. *Essex: its forest, folk and folklore: a footnote to history.* Chelmsford, J H Clarke, 1928

MASON, James. The folk-lore of British plants. *Dublin University Magazine. Vol 82. 1873 313–328; 424–440; 554–570; 668–686*

MASSINGHAM, H J. *Fee fi fo fum: or, The giants in England.* Kegan Paul, 1926

MATHEWS, F W. *Tales of the Blackdown borderland.* Somerset Folk Press, 1923

MAYHEW, Ann. *The rose: myth, folklore and legend.* New English Library, 1979

MEADE, Gordon M. Home remedies in rural America. *New York. Academy of Sciences. Annals. Vol 120; 1965 823–8*

MEANEY, Audrey L. The Anglo-Saxon view of the causes of illness. *in* CAMPBELL, Sheila, Bert Hall *and* David Klausner *editors Health, disease and healing in medieval culture.* Macmillan, 1992

MEGAS, George A. *Greek calendar customs.* 2nd ed. Athens, Philipottis, 1963

MEIGGS, Russell. *Trees and timber in the ancient Mediterranean world.* Oxford, Clarendon Press, 1982

MENNINGER, Edwin A. *Fantastic trees.* New York, Viking, 1967

MEYEROWITZ, Eva L R. *The divine kingship in Ghana and ancient Egypt.* Faber, 1960

MILES, Clement A. *Christmas in ritual and tradition, Christian and pagan.* 2nd ed. Unwin, 1913

MILLER, Hugh. *Scenes and legends of the north of Scotland.* 2nd ed. Johnstone & Hunter, 1850

MILLER, William. *A dictionary of English names of plants.* John Murray, 1884

MILNE, John. *Myths and superstitions of the Buchan district* (1891) 4th ed Maud, Rosalind A Jack 1988

MINCHINTON, Walter. Cider and folklore. *Folk Life. Vol 13; 1975 66–79*

MITTON, F *and* V Mitton. *Mitton's practical modern herbal.* rev ed. Foulsham, 1982

MOLDENKE, Harold N *and* Alma L Moldenke. *Plants of the Bible.* Waltham, Mass., Chronica Botanica, 1952

MOLONEY, Michael F. *Irish ethno-botany and the evolution of medicine in Ireland.* Dublin, Gill, 1919

MOONEY, James. The medical mythology of Ireland. *American Philosophical Society. Proceedings. Vol 24; 1887 136–166*

MOORE, A W. *Folk-lore of the Isle of Man.* David Nutt, 1891

MOORE, A W, Sophia Morrison, *and* Edmund Goodwin. *A vocabulary of the Anglo-Manx dialect.* Oxford, Univ Press, 1924

MORLEY, George. *Shakespeare's greenwood: the customs of the country.* David Nutt, 1900

MORRIS, M C F. *Yorkshire folk-talk, with characteristics of those who speak it in the north and east Ridings..* Henry Frowde, 1892

MORRIS, Richard. *Churches in the landscape.* Dent, 1989

MUNRO, Neil Gordon. *Ainu creed and cult.* Routledge, 1962

MURRAY, Margaret A. *The god of the witches.* Faber, 1931

NALL, John Greaves. *Great Yarmouth and Lowestoft ... and an etymological and comparative glossary of the dialect of East Anglia.* Longmans, Green, Reader & Dyer, 1866

NAPIER, James. *Folklore: or, Superstitious beliefs in the west of Scotland within this century.* Paisley, Gardner, 1879

NASH, Harry *and* Judy Nash. Swans and spear at Abbotsbury. *The Countryman. Vol 96. No 6. Dec/Jan 1991/2 47–52*

NASH, John. *English garden flowers.* Duckworth, 1948

NELSON, E Charles. Shamrock 1988. *Ulster Folklife. Vol 36; 1990 32–42*

NELSON, E Charles. *Shamrock: botany and history of an Irish myth.* Kilkenny, Boethius Press, 1991

NEWALL, Venetia. *An egg at Easter: a folklore study*. Routledge, 1971

NEWALL, Venetia. The Jew as a witch figure. *in* NEWALL, Venetia *editor The witch figure: folklore essays by a group of scholars in England honouring the 75ᵗʰ birthday of Katherine M Briggs*. Routledge & Kegan Paul, 1973

NEWALL, Venetia. West Indian ghosts. *in* DAVIDSON, H R E *and* W M S Russell *editors The folklore of ghosts*. Cambridge. D S Brewer for Folklore Society, 1981 73–93

NEWMAN, L F *and* E M Wilson. Folk-lore survivals in the southern "Lake counties" and in Essex: a comparison and Contrast. *Folk-lore. Vol 62; 1951 252–266*

NICHOLSON, Irene. *Mexican and Central American mythology*. Hamlyn, 1967

NICHOLSON, John. *Folk lore of east Yorkshire*. Simpkin Marshall, 1890

NICOLSON, James R. *Traditional life in Shetland*. Hale, 1978

NORBECK, Edward. *Religion in primitive society*. New York, Harper, 1961

NODAL, John Howard *and* George Milner. *A glossary of the Lancashire dialect*. Manchester, Alexander Ireland & Co, 1875

NORFOLK FEDERATION OF WOMEN'S INSTITUTES. *Within living memory: a collection of Norfolk reminiscences*. King's Lynn, NFWI, 1972

NORTH, Pamela. *Poisonous plants and fungi in colour*. Blandford Press, 1967

NORTHALL, G F. *English folk-rhymes: a collection of traditional verses relating to places and persons, customs, superstitions, etc*. Kegan Paul, 1892

NORTHCOTE, *Lady* Rosalind. *The book of herbs*. Bodley Head, 1912

O CLEIRIGH, Tomas. Gleanings in Wicklow. *Béaloideas. Vol 1; 1928 245–252*

O'FARRELL, Padraic. *Superstitions of the Irish country people*. rev ed. Cork, Mercier Press, 1982

OLIVIER, Edith *and* Margaret K S Edwards *editors. Moonrakings: a little book of Wiltshire stories*. Warminster, Coates & Parker, (1930)

O'NEILL, John. *The night of the gods: an inquiry into cosmic and cosmogonic mythology and symbolism*. 2 Vols. Quaritch, 1893, 1897

OPIE, Iona *and* Peter Opie. *The lore and language of schoolchildren*. Oxford, Clarendon Press, 1959

OPIE, Iona *and* Peter Opie. *The singing game*. Oxford, Univ Press, 1989

OPIE, Iona *and* Moira Tatem *editors. A dictionary of* superstitions. Oxford, Univ Press, 1989

OPIE, Peter. England, the great undiscovered. *Folk-lore. Vol 65; 1954 149–164*

O RIORDAIN, S P. Tara. *in* DANIEL, G E *et al Myth or legend?* Bell, 1955

Ô SÚILLEABHÁIN, Sean. *A handbook of Irish folklore*. Dublin, Folklore of Ireland Society, 1942

OTO, Tohihiko. The taboos of fishermen. *in* DORSON, Richard M *editor Studies in Japanese folklore*. Bloomington, Indiana Univ Press, 1963

O'TOOLE, Edward. A miscellany of North Carlow folklore. *Béaloideas. Vol 1; 1928 316–218*

OTTO, Walter F. *Dionysus: myth and cult*. Bloomington, Indiana Univ Press, 1965

OWEN, Elias. *Welsh folk-lore*. Oswestry, 1896

OXFORDSHIRE AND DISTRICT FOLKLORE SOCIETY. Annual Record. No 2 - no 12; 1950–1960

PAGE, Robin. *Weather forecasting, the country way*. Davis-Poynter, 1977

PAGE, Robin. *Cures and remedies, the country way*. Davis-Poynter, 1978

PAINTER, Gillian *and* Elaine Power *illustrator. The herb garden displayed*. Auckland, Hodder & Stoughton, 1978

PALGRAVE, Olive H Coates *illustrator and* K C Palgrave. *Trees of central Africa*. Salisbury, National Publications Trust, Rhodesia and Nyasaland, 1956

PALAISEUL, Jean. *Grandmother's secrets: her green guide to health from plants*. Barrie & Jenkins, 1973

PALMER, A Smythe. *Folk-etymology: A dictionary of verbal corruptions or words perverted in form or meaning, by false derivation or mistaken analogy*. Bell, 1882

PALMER, Roy. *The folklore of Warwickshire*. Batsford, 1976 (The folklore of the British Isles *series*)

PALMER, Roy. *The folklore of Leicestershire and Rutland*. Wymondham, Sycamore Press, 1985

PALMER, Roy. *The folklore of Radnorshire*. Woonton, Logaston Press, 2001

PANDEY, Brahma Prakash. *Sacred plants of India*. New Delhi, Shree Publishing House, 1989

PARIHAR, Dharam *and* Sikhibhushan Dutt. (on *Zingiber*) *Indian Soap Journal; Nov 1950*

PARIS, J A. *Pharmacologia*. (1812) 9ᵗʰ ed. Samuel Highley, 1843

PARISH, William Douglas. *A dictionary of the Sussex dialect*. expanded edition Chichester, printed for H. Hall by R J Acford, 1957

PARISH, William Douglas *and* W F Shaw. *A dictionary of the Kentish dialect and provincialisms in use in the county of Kent*. Trübner for English Dialect Society, 1887

PARKE, H W. *The oracles of Zeus: Dodona, Olympia, Annon*. Oxford, Blackwell, 1947

PARKER, Derek *and* John Chandler. *Wiltshire churches: an illustrated history*. Stroud, Alan Sutton, 1993

PARKINSON, John. *Paradisi in sole: paradisus terrestris*. reprinted from the 1629 edition Methuen, 1904

PARKINSON, John. *Theatrum botanicum*. 1640

PARMAN, Susan. Curing beliefs and practices in the Outer Hebrides. *Folklore. Vol 88; 1977 107–9*

PARRINDER, Geoffrey. *West African religion: a study of the beliefs and practices of Akan, Ewe, Yoruba, Ibo, and kindred peoples*. 2nd ed. Epworth Press, 1961

PARRINDER, Geoffrey. *Witchcraft: European and African*. 2nd ed. Faber, 1963

PARTRIDGE, J B. Cotswold place-lore and customs. *Folk-lore. Vol 23; 1912 332–342; 443–457*

PATON, C I. *Manx calendar customs*. Glaisher for Folklore Society, 1939

PATON, Lewis Bayles. *Spiritism and the cult of the dead in antiquity*. Hodder & Stoughton, 1921

PAVITT, William Thomas *and* Kate Pavitt. *The book of talismans, amulets and zodiacal gems*. (1914) reprint Bracken Books, 1993

PEACOCK, Edward. *A glossary of words used in the Wapentakes of Manley and Corringham, Lincolnshire*. Trubner for English Dialect Society, 1877

PEACOCK, Robert Backhouse. A glossary of the dialect of the Hundred of Lonsdale, north and south of the sands, in the county of Lancaster. *Philological Society. Transactions; 1867 Supplement 1*

PENNANT, Thomas. *A tour in Scotland and voyage to the Hebrides 1772*. 2nd ed. Benjamin White, 1776 (reprint Edinburgh, Birlinn, 1998)

PERRY, Frances. *Flowers of the world*. Hamlyn, 1972

PERRY, Frances. *Beautiful leaved plants*. Scolar Press, 1979

PERRY, Lily M *and* Judith Metzger. *Medicinal plants of east and southeast Asia: attributes, properties and uses*. Cambridge, Mass., MIT Press, 1980

PETTIGREW, Thomas Joseph. *On superstitions connected with the history and practice of medicine and surgery*. Churchill, 1844

PHILLIPS, George Lewis. The chimney sweepers' assimilation of the milkmaids' garland. *Folklore. Vol 61; 1952 383–387*

PHILPOT, *Mrs* J H. *The sacred tree: or, The tree in religion and myth*. Macmillan, 1897

PHYSICIANS OF MYDDFAI. *Physicians of Myddfai: Meddygon Myddfai. translated by* John Pughe *and edited by* John Williams ab Ithal Llandovery, Roderic for the Welsh MSS Society, 1861

PIGGOTT, Juliet. *Japanese mythology*. Hamlyn, 1969

PLASS, Margaret Webster. *African miniatures: the goldweights of the Ashganti*. Lund Humphries, 1967

PLATT, Colin. *Medieval England: a social history and archaeology, from the Conquest to AD 1600*. Routledge, 1978

POIGNANT, Roslyn. *Oceanic mythology*. Hamlyn, 1967

POLLOCK, Linda. *With faith and physic: the life of a Tudor gentlewoman: Lady Grace Mildmay, 1552–1620*. Collins & Brown, 1993

POLSON, Alexander. *Our Highland folklore heritage*. Dingwall, George Souter, 1926

POLSON, Alexander. *Scottish witchcraft lore*. Inverness, W Alexander & Son, 1932

POMET, Pierre. *A compleat history of druggs – done into English from the originals*. 3rd ed. J & J Bonwicke, 1737

PONTING, K G. Indigo and woad. *Folk Life. Vol 14; 1976 75–88*

POPE, Harrison C. *Tabernanthe iboga*: an African narcotic plant of social importance. *Economic Botany. Vol 23; 1969 174–184*

POPE, Rita Tregellas. Bean lore. *in* VICKERY, Roy *editor Plant-lore studies*. Folklore Society, 1984

POOLE, Charles Henry. *The customs, superstitions and legends of the county of Stafford*. Rowney, 1875

PORTEOUS, Alexander. *Forest folklore, mythology, and romance*. Allen & Unwin, 1928

PORTEOUS, Crichton. *The beauty and mystery of well-dressing*. Derby, Pilgrim Press, 1949

PORTER, Enid. Some folk beliefs of the Fens. *Folklore. Vol 69; 1958 112–122*

PORTER, Enid. *Cambridgeshire customs and folklore*. Routledge, 1969

PORTER, Enid. *The folklore of East Anglia*. Batsford, 1974 (The folklore of the British Isles *series*)

POTTER, Stephen *and* Laurens Sargent. *Pedigree: essays on the etymology of words from nature*. Collins, 1973

POWER, Eileen. *Medieval people*. 10th ed. Methuen, 1963

POWERS, Stephen. Aboriginal botany. *Californian Academy of Sciences. Proceedings. Vol 5; 1873–1875 373–9*

PRANCE, Ghillean Tolmie *and* Anne E Prance. *Bark: the formation, characteristics, and uses of bark around the world*. Portland Or., Timber Press, 1993

PRATT, A. *Wild flowers of the year*. Religious Tract Society, 1913

PREST, John. *The garden of Eden: the botanic garden and the re-creation of paradise*. New Haven, Yale Univ Press, 1981

PRIOR, R C A. *On the popular names of British plants*. Norgate, 1879

PUCKETT, Newbell Niles. *Folk beliefs of the southern negro*. Chapel Hill, Univ. of North Carolina Press, 1926

PUCKLE, Bertram S. *Funeral customs: their origin and development*. T Werner Laurie, 1926

PUTNAM, Clare. *Flowers and trees of Tudor England*. Hugh Evelyn, 1972

QUELCH, Mary Thorne. *Herbs for daily use. In home medicine and cookery*. Faber, 1941

QUENNELL, Peter. *Shakespeare: the poet and his background.* Weidenfeld & Nicolson, 1963

QUILLER-COUCH, T. *Glossary of words in use in Cornwall. 2. East Cornwall.* Trübner for English Dialect Society, 1880

RACKHAM, Oliver. *Trees and woodland in the British landscape.* Dent, 1976

RACKHAM, Oliver. *The history of the countryside.* Dent, 1986

RADFORD, E *and* M A Radford. *Encyclopaedia of superstitions.* Rider, 1946

RAMBOSSON, J. *Histoire et legendes des plantes utiles et curieuses.* Paris, Formin Didot, 1868

RAMSAY*, John (of Ochtertyre). Scotland and Scotsmen in the eighteenth century. edited by* Alexander Allardyce Edinburgh, 1888

RANDELL, Arthur B. *Sixty years a Fenman. edited by* Enid Porter Routledge, 1966

RANDOLPH, Charles Brewster. The mandragora of the ancients in folk-lore and medicine. *American Academy of Arts and Sciences. Proceedings. Vol xl. No 12; 1905 487–537*

RAPER, Elizabeth. *The receipt book of Elizabeth Raper: and a portion of her Cipher Journal ... written 1756–1770 and never before published.* Nonesuch Press, 1924

RAPHAEL *[pseud.]. The book of dreams: being a concise interpretation of dreams.* Foulsham, nd (18--)

RAPPOPORT, Angelo S. *The folklore of the Jews.* Soncino Press, 1937

RAVEN, Jon. *The folklore of Staffordshire.* Batsford, 1978 (The folklore of the British isles *series*)

RAWCLIFFE, D H. *The psychology of the occult.* Derricke Ridgway, 1952

RAYMOND, W D. Native poisons and native medicines in Tanganyika. *Journal of Tropical Medicine and Hygiene. Vol 142; 1939 295 303*

RAYMOND, Walter. *Under the spreading chestnut tree: a volume of rural lore and anecdote.* Folk Press, nd (1920)

READ, D H Moutray. Hampshire folk-lore. *Folk-lore. Vol 22; 1911 292–329.* Scraps of English folk-lore. VI *Folk-lore. Vol 23; 1912 p350*

REDFIELD, Robert *and* Alfonso Villa. *Chan Kom: a Maya village.* Washington, Carnegie Institute. Publications. No 448; 1934

REES, Alwyn D. The measuring rod. *Folk-lore. Vol 65; 1954 30–2*

REEVES, James. *The idiom of the people: English traditional verse.* Heinemann, 1958

REICHEL-DOLMATOFF, Gerardo. The cultural context of an aboriginal hallucinogen. *Washington. Smithsonian Institution. Bureau of American ethnology. Bulletin. 55; 1976*

REYNOLDS, Barrie. *Magic, divination and witchcraft among the Barotse of Northern Rhodesia.* Chatto & Windus, 1963

RHODES, H T F. *The Satanic Mass.* 1854

RHYS, *Sir* John. *Celtic heathendom: the Hibbert lectures for 1886.* Williams & Norgate, 1892

RICKERT, Edith *compiler. Chaucer's world.* New York, Columbia Univ Press, 1948

RIMMEL, Eugene. *Book of perfumes.* Chapman & Hall, 1865

RIVAL, Laura *editor. The social life of trees: anthropological perspectives on tree symbolism.* Oxford, Berg, 1998

RIX, Michael M. More Shropshire folklore. *Folklore. Vol 71; 1960 184–7*

ROBERTS, Peter. *The Cambrian popular antiquities.* E Williams, 1815

ROBERTSON, C M. Folk-lore from the west of Ross-shire. *Inverness. Gaelic Society. Transactions. Vol 26; 1904–7 262–299*

ROBERTSON, J Drummond. *A glossary of dialect and archaic words used in the County of Gloucester.* Kegan Paul, Trench Trübner for English Dialect Society, 1890

ROBERTSON, Ronald Macleod. *More Highland folktales.* Edinburgh, Oliver & Boyd, 1964

ROBERTSON, Seonaid M. *Dyes from plants.* New York, Van Nostrand Reinhold, 1973

ROBIN, P Ansell. *Animal lore in English literature.* John Murray, 1932

ROBINSON, F K. *A glossary of words used in the neighbourhood of Whitby.* Trubner for English Dialect Society, 1876

ROCKWELL, Joan. The ghosts of Evald Tang Kristensen. *in* DAVIDSON, H R E *and* W M S Russell *editors The folklore of ghosts.* Cambridge, D S Brewer for Folklore Society, 1981

RODD, Rennell. *The customs and lore of modern Greece.* David Stott, 1892

RODERICK, Alan. *The folklore of Gwent.* Cwmbran, Village Publishing, 1983

ROGERS, Kenneth H. *Warp and weft: the Somerset and Wiltshire woollen industry.* Buckingham, Barracuda, 1986

ROGERS, Norman. *Wessex dialect.* Bradford-on-Avon, Moonraker Press, 1979

ROHDE, Eleanour Sinclair. *Herbs and herb gardening.* Medici Society, 1936

ROHEIM, Geza. *Animism, magic, and the divine king.* Kegan Paul, Trench, Trubner, 1930

ROLLESTON, J D. The folklore of children's diseases. *Folk-lore. Vol 54; 1943 287–307*

ROLLINSON, William. *Life and tradition in the Lake District.* Dent, 1974

RORIE, David. *Folk tradition and folk medicine in Scotland: the writings of David Rorie. edited by* David Buchan Edinburgh, Canongate Academic, 1994

RORIE, David. The mining folk of Fife. *in* SIMPKINS, J E *County folklore, Fife.* Folklore Society, 1914

ROSE, H J. The folklore of the *Geopontica. Folk-lore. Vol 44; 1933*

ROSS, Anne. *Pagan Celtic Britain: studies in iconography and tradition.* Routledge, 1967

ROUSE, W H D. Folklore first fruits from Lesbos. *Folk-lore. Vol 7; 1896 142–159*

ROWE, W John. Old-world legacies in America. *Folk Life. Vol 6; 1968 68–82*

ROWLAND, Beryl. *Animals with human faces: a guide to animal symbolism.* Allen & Unwin, 1974

ROWLING, Marjorie. *The folklore of the Lake District.* Batsford, 1976 (The folklore of the British Isles *series*)

ROYS, Ralph L. The ethno-botany of the Maya. *New Orleans. Tulane University of Louisiana. Middle American Research Papers. Publications. No 2; 1931*

RUDDOCK, Elizabeth. May-day songs and celebrations in Leicestershire and Rutland. *Leicestershire Archaeological and Historical Society. Transactions. Vol 40; 1964–5 69–84*

RUDKIN, Ethel L. *Lincolnshire folklore.* Gainsborough, Beltons, 1936

RUNEBERG, Arne. *Witches, demons and fertility magic: analysis of their significance and mutual relations in West European folk religion.* Helsingfors, 1947 (*Societas scientiarum fennica. Commentationes Humanarum litterarum. xiv. 4)*

RUSSELL, William Moy Stratton *and* Claire Russell. The social biology of werewolves. *in* PORTER, Joshua Roy *and* WM S Russell *editors Animals in folklore.* Ipswich, Brewer for Folklore Society, 1978

RUTHERFORD, Meg. *A pattern of herbs.* Allen & Unwin, 1975

RUUD, Jorgen. *Taboo: a study of Malagasy customs and beliefs,* Oslo, Univ Press, 1960

RYDBERG, Viktor. *Teutonic mythology. translated by* Rasmus B Anderson Swan Sonnenschein, 1889

RYDÉN, Mats. The contextual significance of Shakespeare's plant names. *Studia Neoplilologica. Vol 56; 1984 155–162*

RYDER, M L. Teasel growing for cloth raising. *Folk Life. Vol 7; 1969 117–119*

SAFFORD, W E. Narcotic plants and stimulants of the ancient Americans. *Washington. Smithsonian Institution. Annual Report; 1916*

ST CLAIR, Sheila. *Folklore of the Ulster people.* Cork, Mercier Press, 1971

ST LEGER-GORDON, Ruth E. *The witchcraft and folklore of Dartmoor.* Hale, 1965

SALAMAN, Redcliffe N. *The history and social influence of the potato.* Cambridge, Univ Press, 1949

SALISBURY, *Sir* Edward. *Weeds and aliens.* 2nd ed. Collins, 1964 (New naturalist *series*)

SALISBURY, Jesse. *A glossary of words and phrases used in SE Worcestershire.* the author, 1893

SALLE, Laisnal de la. *Croyances et legendes du centre de la France.* 2 vols. Paris, 1875

SANDERSON, Stewart F. Gypsy funeral customs. *Folklore. Vol 80; 1969 p184*

SANDYS, William. *Christmastide: its history, festivities and carols.* John Russell Smith, nd (1890?)

SANECKI, Kay N. *The complete book of herbs.* Macdonald, 1974

SANFORD, S N F. New England herbs: their preparation and use. *New England Museum of Natural History. Special publication. 2; 1937*

SARGENT, Gilbert. *A Sussex life: the memories of Gilbert Sargent, countryman. edited by* D Arthur. Barrie & Jenkins, 1989

SAVAGE, F G. *The flora and folklore of Shakespeare.* Cheltenham, 1923

SAVAGE, George. *Porcelain through the ages.* Penguin, 1954

SAVAGE, George. *Chinese jade: a concise introduction.* Cory, Adams & Mackay, 1964

SAWYER, Frederick E. Sussex folk lore and customs connected with the seasons. *Sussex Archaeological Collections. Vol 33; 1883*

SCARBOROUGH, John. Theophrastus on herbals and herbal remedies. *Journal of the History of Biology. Vol 11; 1978 353–385*

SCHAPERA, Isaac. *The Khoisan people of South Africa.* Routledge, 1930

SCHAUENBERG, Paul, *and* Ferdinand Paris. *Guide to medicinal plants.* Guildford, Lutterworth, 1977

SCHENK, Sarah *and* E W Gifford. Karok ethnobotany. *Univ. of California. Anthropological records. Vol 13. No 6; 1952*

SCHERY, Robert W. *Plants for man.* Allen & Unwin, 1954

SCHOFIELD, Eunice M. Working class food and cooking in 1900. *Folk Life. Vol 13; 1975 13–23*

SCOT, Reginald. *The discoverie of witchcraft* 1584. *edited by* Brinsley Nicholson. Elliot Stock, 1886

SCUDDER, Thayer. *The ecology of the Gwembe Tonga.* Manchester, Univ Press, 1962 (Kariba studies. Vol 2)

SEAGER, H W. *Natural history in Shakespeare's time: being extracts illustrative of the subject as he knew it.* Elliot Stock, 1896

SEBILLOT, Paul. *Le folk-lore de France. Vol 3: La faune et la flore.* Paris, Guilmoto, 1906

SEKI, Keigo. The spool of thread: a subtype of the Japanese serpent-bridegroom tale. *in* DORSON, Richard *Editor Studies in Japanese folklore.* Bloomington, Indiana Univ Press, 1963

SETH, Ronald. *Children against witches.* Hale, 1969

SHAW, Margaret Fay. *Folksongs and folklore of South Uist.* 3rd ed. Aberdeen, Univ Press, 1986

SHEEHY. Jeanne *and* George Mott *photographer*. *The rediscovery of Ireland's past: the Celtic Revival 1830–1930.* Thames & Hudson, 1980

SIKES, Wirt. *British goblins: Welsh folk lore, fairy mythology, legends and traditions.* Boston, Osgood, 1881

SIMONS. G L. *Sex and superstition.* Abelard-Schumann, 1973

SIMOONS, Frederick J. *Plants of life, plants of death.* Madison, Wi., Univ of Wisconsin Press, 1998

SIMPKINS, John Ewart. *County folklore, Vol 7: Fife.* Folklore Society, 1914

SIMPSON, Eve. *Folklore in Lowland Scotland.* Dent, 1908

SIMPSON, Jacqueline. *The folklore of Sussex.* Batsford, 1973

SIMPSON, Jacqueline. Witches and witchbusters. *Folklore. Vol 107; 1996 5–18*

SIMPSON, Jacqueline *and* Steve Roud. *A dictionary of English folklore.* Oxford, Univ Press, 2000

SIMPSON, P A. Native poisons of India. *Pharmaceutical Journal and Transactions. 3rd series. Vol 2; 1871–72 604–606; 626–627*

SKEAT, Walter William. *Malay magic: being an introduction to the folklore and popular religion of the Malay peninsula.* Macmillan, 1900

SKINNER, Alanson. Material culture of the Menomini. *New York. Museum of the American Indian. Heye Foundation. Indian Notes and Monographs; 1921*

SKINNER, Charles M. *Myths and legends of flowers, trees, fruits and plants, in all ages and in all climes.* Philadelphia, Lippincott, 1925

SMITH, A C. On Wiltshire weather proverbs and weather fallacies. *Wiltshire Archaeological Magazine. Vol 15; 1875*

SMITH, A W. *A gardener's dictionary of plant names: a handbook on the origin and meaning of some plant Names.* Cassell, (rev ed. by William T Stearn) 1972

SMITH, F Porter. *Chinese materia medica: vegetable kingdoms.* 2nd ed. Shanghai, American Presbyterian Mission Press, 1911

SMITH, Grafton Elliot. *The evolution of the dragon.* Manchester, Univ Press, 1919

SMITH, Huron H. Ethnobotany of the Menomini Indians. *Milwaukee. Public Museum. Bulletin. Vol 4. no 1; 1923*

SMITH, Huron H. Ethnobotany of the Meskwaki Indians. *Milwaukee. Public Museum. Bulletin. Vol 4. no 2; 1928*

SMITH, Huron H. Ethnobotany of the Ojibwe Indians. *Milwaukee. Public Museum. Bulletin. Vol 4. no 3; 1945*

SMITH, J B. Ruth Tongue's 'Legalizing of Bastards'. *Folk Life. Vol 35; 1996–7 91–95*

SMITH, John. *A dictionary of the popular names of the plants.* Macmillan, 1882

SMITH, Julia. *Fairs, feasts and frolics: customs and tradition in Yorkshire.* Otley, Smith Settle, 1989

SOFOROWA, Abayomi. *Medicinal plants and traditional medicine in Africa.* Chichester, Wiley, 1982

SPECK, Frank G. Decorative art and basketry of the Cherokee. *Milwaukee. Public Museum. Bulletin. Vol 2. No 2; 1920*

SPENCE, John. *Shetland folklore.* Lerwick, Johnson & Greig, 1899

SPENCE, Lewis. *The magic arts in Celtic Britain.* Rider, 1945

SPENCE, Lewis. *British fairy origins.* Watts, 1946

SPENCE, Lewis. *Myth and ritual in dance, game and rhyme.* Watts, 1947

SPENCE, Lewis. *The fairy tradition in Britain.* Rider, 1948

SPENCE, Lewis. *The history and origin of Druidism.* Rider, 1949

SPENCE, Lewis. *Second sight: its history and origins.* Rider, 1951

STARKIE, Walter. *Raggle-taggle.* John Murray, 1933

STEFANISZYN, Bronislaw. *Social and ritual life among the Ambo of northern Rhodesia.* Oxford, Univ Press, 1964

STERNBERG, Thomas. *The dialect and folklore of Northamptonshire.* John Russell Smith, 1851

STEVENS, Catrin. *Welsh courting customs.* Llandysul, Gomer Press, 1993

STEVENSON, Matilda Coxe. Ethnobotany of the Zuñi Indians. *Washington. Smithsonian Institution. Bureau of American Ethnology. Annual Report. 30th; 1908–9*

STEWART, W Grant. *The popular superstitions of the Highlanders of Scotland.* Glasgow, Constable, 1823

STORER, Bernard. *Sedgemoor.* Newton Abbot, David & Charles, 1972

STORMS, G. *Anglo-Saxon magic.* The Hague, Martins Nijhoff, 1945

STOUT, Earl J *editor.* Folklore from Iowa. *American Folklore Society. Memoirs. Vol 29; 1936*

STOW, John. *A survey of London in 1598.* edited by Henry Morley Routledge, 1890

STURTEVANT, William C. History and ethnography of some West Indian starches. *in* UCKO, Peter J *and* G W Dimbleby *editors The domestication and exploitation of plants and animals.* Duckworth, 1969

SUGGS, Robert C. *Marquesan sexual behaviour.* Constable, 1966

SUMMERS, Montague. *The geography of witchcraft.* Kegan Paul, 1927

SWAHN, J O. *The lore of spices: their history and uses around the world.* New York, Crescent Books, 1995

SWAINSON, Charles. *A handbook of weather folk-lore.* Blackwood, 1873

SWAINSON, Charles. *The folk-lore and provincial names of British birds.* Folklore Society, 1886

SWIRE, Otta F. *Skye: the island and its legends.* Blackie, 1961

SWIRE, Otta F. *The Highlands and their legends*. Edinburgh, Oliver & Boyd, 1963

TALBOT, P Amaury. *Some Nigerian fertility cults*. Cass, 1927

TAMPION, John. *Dangerous plants*. Newton Abbot, David & Charles, 1977

TANNER, Heather *and* Robin Tanner. *Wiltshire village*. 3rd ed. Robin Garton, 1977

TAYLOR, George M. *British herbs and vegetables*. Collins, 1947

TAYLOR, Joseph. *Arbores mirabiles*. W Darton. 1812

TAYLOR, Mark R. Norfolk folklore. *Folk-lore. Vol 40; 1929 113–133*

TEBBUTT, C F. Huntingdonshire folk and their folklore. *Cambridgeshire and Huntingdonshire Archaeological Society. Transactions. Vol 6. part 26; 1947 119–154*

TEIT, James. The Thompson Indians of British Columbia. *American Museum of Natural History, Memoirs (Jessup North Pacific Expedition. Vol 1; 1898–1900)*

TENNANT, Pamela. *Village notes, and some other papers*. Heinemann, 1900

THOMAS, Daniel Lindsey *and* Lucy Blayney Thomas. *Kentucky superstitions*. Princeton, Univ Press, 1920

THOMPSON, C J S. *The mystery and romance of alchemy and pharmacy*. Scientific Press, 1897

THOMPSON, C J S. *Magic and healing*. Rider, 1947

THOMPSON, Flora. *Lark Rise to Candleford: a trilogy*. Oxford, Univ Press, 1945

THOMSON, William A R. *Herbs that heal*. A & C Black, 1976

THOMSON, William A R. *Healing plants: a modern herbal*. Macmillan, 1978

THONGER, Richard. *A calendar of German customs*. Wolff, 1966

THORNDIKE, Lynn *editor*. *The Herbal of Rufinus*. Univ Chicago Press, 1946

THORNTON, Robert John. *A new family herbal: or, a popular account of the nature and properties of the various plants used in medicine, and the arts*. Phillips, 1810

THORPE, BENJAMIN. *Northern mythology*. Lumley, 1851

TILLE, Alexander. *Yule and Christmas: their place in the Germanic year*. David Nutt, 1899

TILLHAGEN, C-H. Food and drink among the Swedish Kalderasa gypsies. *Gypsy Lore Society. Journal. 3rd series. Vol 36; 1957 25–52*

TONGUE, Ruth L. *Somerset folklore*. Folklore Society, 1965

TONGUE, Ruth L. Folk-song and folklore. *Folklore. Vol 78; 1967 293–303*

TONGUE, Ruth L. *Forgotten folk-tales of the English counties*. Routledge, 1970

TOONE, William. *A glossary and etymological dictionary of obsolete and uncommon words, antiquated Phrases*. 2nd ed. Thomas Bannett, 1834

TOPSELL, Edward. *[The history of four-footed beasts* 1607 *; The history of serpents* 1608 *selections] Topsell's histories of beasts. edited by* Malcolm South Chicago, Nelson-Hall, 1981

TOSCO, Umberto. *The flowering wilderness*. Orbis, 1972

TRACHTENBERG, Joshua. *Jewish magic and superstition: a study in folk religion*. New York, Behrman's Jewish Book House, 1939

TRAIN, Joseph. *An historical and statistical account of the Isle of Man, from the earliest times to the present date*. 2 vols. Douglas, Mary A Quiggin, 1845

TREVELYAN, Marie. *Folk lore and folk stories of Wales*. Elliot Stock, 1909

TUN LI-CH'EN. [Yen-ching Sui-shih-chi]*Annual customs and festivals in Peking. translated and annotated by* Derek Bodde Peking. Henri Vetch, 1936

TURNER, Nancy Chapman *and* Marcus A M Bell. The ethnobotany of the Coast Salish Indians of Vancouver Island. *Economic Botany; 1969*

TURNER, Victor W. *Revelation and divination in Ndembu ritual*. Ithaca, Cornell Univ Press, 1975

TURNER, William. *The names of herbes*. (1548) *edited by* James Britten English Dialect Society, 1881

TURNER, William. *A new Herball*. 1551

TURVILLE-PETRIE, E C G. *Myth and religion of the north: the religion of ancient Scandinavia*. Weidenfeld & Nicolson, 1964

TUSSER, Thomas. *Five hundred points of good husbandry*. (1557) reprint (of 1580 edition) Oxford, Univ Press, 1984

TYACK, G S. *Lore and legend of the English church*. W Andrews, 1899

TYLER, Varro E. *Hoosier home remedies*. West Lafayette, Ind., Purdue Univ Press, 1985

TYNAN, Katharine *and* Frances Maitland. *The book of flowers*. Smith, Elder & Co, 1909

UCHENDU, Victor C. *The Igbo of south-east Nigeria*. New York, Holt, Rinehart & Winston, 1965 (Case studies in cultural anthropology *series*)

UCHIMAYADA, Yasuchi. 'The grove is our temple': contested representations of *Kaavu* in Kerala, South India. *in* RIVAL, Laura *editor The social life of trees: anthropological perspectives on tree symbolism*. Oxford, Berg, 1998

UDAL, John Symonds. *Dorsetshire folk-lore*. Hertford, 1922

UNDERHILL, Ruth M. *Red man's religion: beliefs and practices of the Indians north of Mexico* Chicago, Univ Press, 1965

UPADHYAYA, K D. Indian botanical folklore. *in* GUPTA, Sankar Sen *editor Tree symbol worship in India: a new survey of a pattern of folk-religion.* Calcutta, Indian Publications, 1965

URLIN, Ethel L. *Festivals, holy days, and saints' days.* Simpkin Marshall, 1915

USHER, George. *A dictionary of plants used by man.* Constable, 1974

VALIENTE, Doreen. *An ABC of witchcraft, past and present.* Hale, 1973

VAN ANDEL, M A. Dutch folk-medicine. *Janus. Vol 15; 1910 452–461*

VAUGHAN, J G *and* C Geissler. *The new Oxford book of food plants.* Oxford, Univ Press, 1977

VAUX, J Redward. *Church folklore.* Griffith, Farran, 1894

VERGER, Pierre Fatumbi. *Ewé: the use of plants in Yoruba society.* São Paolo, ed. Schwarz, 1995

VESEY-FITZGERALD, Brian. *Gypsies of Britain: an introduction to their history.* Chapman & Hall, 1944

VESTAL Paul A *and* Richard Evans Schultes. *The economic history of the Kiowa Indians, as it relates to the history of the tribe.* Cambridge, Mass., Botanical Museum, 1939

VICKERY, Roy. Traditional uses and folklore of *Hypericum* in the British Isles. *Economic Botany. Vol 35; 1981 289–295*

VICKERY, Roy. *Lemna minor* and Jenny Greenteeth. *Folklore. Vol 94. No 2; 1983 247–250*

VICKERY, Roy. Plants, death and mourning. *in* VICKERY, Roy *editor Plant-lore studies.* Folklore Society, 1984

VICKERY, Roy. *Unlucky plants.* Folklore Society, 1985

VICKERY, Roy. Plants used for pest control: some twentieth century examples. *Folklore. Vol 103; 1993 170–172*

VICKERY, Roy. *A dictionary of plant lore.* Oxford, Univ Press, 1995

VOIGTS, Linda E *and* Robert P Hudson. A *drynke that men callen dwale to make a man to slepe whyle men kerven him: a surgical anaesthetic from late medieval England in* CAMPBELL, HALL & KLAUSNER *editors Health, disease and healing in medieval culture.* Macmillan, 1992

VUKANOVIC, T P. Witchcraft in the central Balkans 1. Characteristics of witches. *Folklore. Vol 100. Part 1; 1989 9–24.* 2. Protection against witches. *Folklore. Vol 100. Part 2; 1989 221–235*

WALKER, Timothy. These plants could save your life. *Country Life. Vol 192. No 38; 17 Sept 1998 104–5*

WARING, Philippa. *A dictionary of omens and superstitions.* Souvenir Press, 1978

WARNER, Elizabeth A. Russian peasant beliefs and practices concerning death and the supernatural collected in Novosokol'niki region Pskov province, Russia, 1995 part 1: the restless dead, wizards and spirit beings. *Folklore. Vol 111; 2000 67–90*

WARNER, Marina. *From the beast to the blonde: on fairy tales and their tellers.* Chatto & Windus, 1994

WARREN-DAVIS, Ann. Herbal medicine. *in* RUTHERFORD, Meg *A pattern of herbs.* Allen & Unwin, 1975

WASSON, R Gordon, Albert Hofmann *and* Carl A P Ruck. *The road to Eleusis: unveiling the secret of the Mysteries.* New York, Harcourt Brace Jovanovich, 1978

WASSON, Valentina Pavlovna *and* R Gordon Wasson. *Mushrooms, Russia and history.* 2 Vols. New York, Pantheon, 1957

WATERS, Ivor. *Folklore and dialect of the lower Wye Valley.* Chepstow, Moss Rose Press, 1982

WATSON, W G Willis. *Calendar of customs, superstitions, weather-lore, popular sayings, and important events connected with the county of Somerset.* reprinted from the *Somerset County Herald,* 1920

WATT, John Mitchell *and* Maria Gerdina Breyer-Brandwijk. *The medicinal and poisonous plants of southern and eastern Africa.* 2nd ed. Edinburgh, Livingstone, 1962

WATTS, Donald Charles. *Elsevier's dictionary of plant names and their origins.* Amsterdam, Elsevier, 2000

WEBSTER, Helen N. *Herbs: how to grow them, and how to use them.* Massachusetts Horticultural Society, 1935

WEBSTER, Wentworth. *Basque legends.* Griffith & Farran, 1877

WEIL, Andrew T. Nutmeg as a narcotic. *Economic Botany. Vol 19; 1965 194–217*

WEINER, Michael A. *Earth medicine – earth foods: plant remedies, drugs and natural foods of the North American Indians.* New York, Macmillan, 1972

WELLCOME, Henry S. *Hen feddegyaeth Kymrie (ancient Cymric medicine).* Burroughs Wellcome, 1903

WENTZ, Walter Y Evans. *The fairy faith in Celtic countries.* Oxford, Clarendon Press, 1911

WESLEY, John. *Primitive Physick: or, an easy and natural method of curing most diseases.* 1747

WESTERMARCK, Edward. Midsummer customs in Morocco. *Folk-lore. Vol 16; 1905 27–47*

WESTERMARCK, Edward. *Ritual and belief in Morocco.* 2 Vols. Macmillan, 1926

WESTROPP, Thomas J. A folk-lore survey of County Clare. *Folk-lore. Vols 21,22 1910–11*

WHERRY, Beatrix A. Miscellaneous notes from Monmouthshire. *Folk-lore. Vol 16; 1905 63–67*

WHISTLER, C W. Sundry notes from west Somerset and Devon. *Folk-lore. Vol 19; 1908 88–91*

WHITCOMBE, *Mrs* Henry Pennell. *Bygone days in Devonshire and Cornwall.* Bentley, 1874

WHITE, Gilbert. *The natural history of Selborne.* White & Son, 1789

WHITE, Leslie A. Notes on the ethnobotany of the Keres. *Michigan Academy. Papers. Vol 30; 1944 557–568*

WHITING, Alfred F. Ethnobotany of the Hopi. *Flagstaff. Northern Arizona Society of Science and Art. Museum of Northern Arizona. Bulletin. 15; 1939*

WHITLOCK, Ralph. *The folklore of Devon*. Batsford, 1977 (The folklore of the British Isles *series*)

WHITLOCK, Ralph. *The lost village: rural life between the wars*. Hale, 1988

WHITLOCK, Ralph. *Wiltshire folklore and legends*. Hale, 1992

WHITNEY, Annie Weston *and* Caroline Canfield Bullock. Folk-lore from Maryland. *American Folklore Society. Memoirs. Vol 18; 1925*

WHITTLE, Tyler *and* Christopher Cook. *Curtis's flower garden displayed*. Oxford, Univ Press, 1981

WICKHAM, Cynthia. *Common plants as natural remedies*. Muller, 1981

WIDDOWSON, John D A. The things they say about food: a survey of traditional English foodways. *Folk Life. Vol 13; 1975 5–12*

WIDDOWSON, John D A. Plants as elements in systems of traditional verbal social control. *in* VICKERY, R *editor Plant-lore studies*. Folklore Society, 1984 202–236

WILDE, *Sir* William R. *Irish popular superstitions*. Dublin, James McGlashan, 1852

WILDE, *Lady. Ancient cures, charms and usages of Ireland*. Ward & Downey, 1890

WILDE, *Lady. Ancient legends, mystic charms and superstitions of Ireland*. Chatto & Windus, 1902

WILKINSON, Gerald. *Trees in the wild and other trees and shrubs*. Stephen Hope Books, 1973

WILKINSON, Gerald. *Epitaph for the elm*. Hutchinson, 1978

WILKINSON, Gerald. *A history of Britain's trees*. Hutchinson, 1981

WILKS, J H. *Trees of the British Isles in history and legend*. Muller, 1972

WILLETTS, R F. *Cretan cults and festivals*. Routledge, 1962

WILLIAMS, Alfred. *Folk songs of the upper Thames*. Duckworth, 1923

WILLIAMS, Frank R. Some Sussex customs and superstitions. *Sussex Notes and Queries. Vol 10. No 3; August 1944 p 58–*

WILLIAMS, Paul V A. *Primitive religion and healing: a study of folk medicine in north-east Brazil*. Cambridge, Brewer for Folklore Society, 1979

WILLIAMS, Thomas Rhys. *The Dusun: a North Borneo society*. New York, Holt, Rinehart & Winston, 1965

WILLIAMSON, Hugh Ross. *The flowering hawthorn*. P Davies, 1962

WILLIAMSON, Kenneth. *The Atlantic islands: a study of Faeroe life and scene*. Collins, 1948

WILLIS, John C. *A dictionary of the flowering plants and ferns*. 8th rev ed. Cambridge, Univ Press, 1973

WILLS, Norman T. *Woad in the Fens*. 3rd ed. Long Sutton, the author, 1979

WILSON, Frank Percy. *The plague in Shakespeare's London*. Oxford, Clarendon Press, 1927

WILTSHIRE, Kathleen. *Wiltshire folklore*. Compton Chamberlayne, Compton Press, 1975

WIMBERLEY, Lowry Charles. *Folklore in the English and Scottish ballads*. Chicago, Chicago Univ Press, 1928

WINSTANLEY, L *and* H J Rose. Welsh folklore items, III. *Folk-lore. Vol 39; 1928 171–178*

WIT, H C D de. *Plants of the world: the higher plants*. Thames & Hudson, 1966

WITCUTT, W P. Notes in Warwickshire folk-lore. *Folk-lore. Vol 55; 1944 41–42*

WITHINGTON, Edward Theodore. *Medical history from the earliest times: a popular history of the healing art*. Scientific Press, 1894

WOOD-MARTIN, W G. *Traces of the elder faiths of Ireland*. Longman, 1902

WOODCOCK, Hubert B Drysdale *and* William Thomas Stearn. *Lilies of the world: their cultivation and classification*. Country Life, 1950

WOODFORDE, James. *The diary of a country parson, 1758–1802*. [selections] Oxford, Univ Press, 1935

WOOKEY, Barry. *Rushall: the story of an organic farm*. Oxford, Blackwell, 1987

WRIGHT, A R. *British calendar customs. edited by* T E Lones 3 vols. Folklore Society, 1936–1938

WRIGHT, Elizabeth. *Rustic speech and folk-lore*. Oxford, Univ Press, 1913

WRIGHT. M E S. *A medley of weather lore*. Bournemouth, Commin, 1913

WRIGHT, Thomas. *St Patrick's Purgatory*. John Russell Smith, 1844

WYKES-JOYCE, Max. *Cosmetics and adornment: ancient and contemporary usage*. Owen, 1961

WYMAN, Leland C *and* Stuart K Harris. Navajo Indian medical ethnobotany. *Albuquerque. Univ of New Mexico. Bulletin. 366 (Anthr. Series. Vol 4. No 5; 1941)*

YANOVSKY, Elias. *Food plants of the North American Indians*. Washington, US Dept of Agriculture, 1936

YARNELL, Richard Asa. Aboriginal relationships between culture and plant life in the Upper Great Lakes region. *Ann Arbor. Univ of Michigan. Museum of Anthropology. Anthropological papers. 23; 1964*

YEATS, William Butler *editor. Fairy and folk tales of the Irish peasantry*. Walter Scott, nd

YOUNG, Andrew. *A prospect of flowers: a book about wild flowers*. Cape, 1945

YOUNGKEN, Heber W. The drugs of the North American Indian. *American Journal of Pharmacy. Vol 96. No 7; 1924 485–502; Vol 97. No ; 1925 158–185; Vol 97. No 4; 1925 257–271*

ZOHARY, Michael. *Plants of the Bible*. Cambridge, Univ Press, 1982

ERICACEAE

VACCINIUM MYRTILIS L

BILBERRY

PAPILIONACEAE

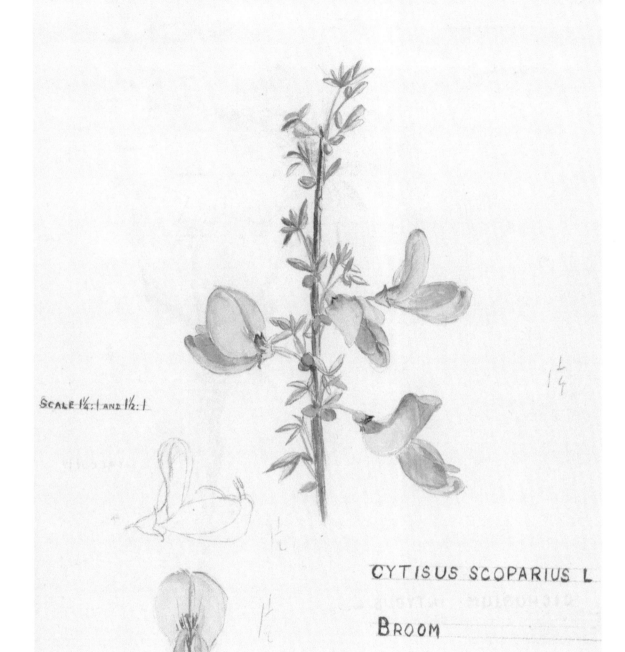

SCALE 1½:1 AND 1½:1

L
¼

CYTISUS SCOPARIUS L

BROOM

SCALE 1½:1

CICHORIUM INTYBUS L.

CHICORY

PRIMULACEAE

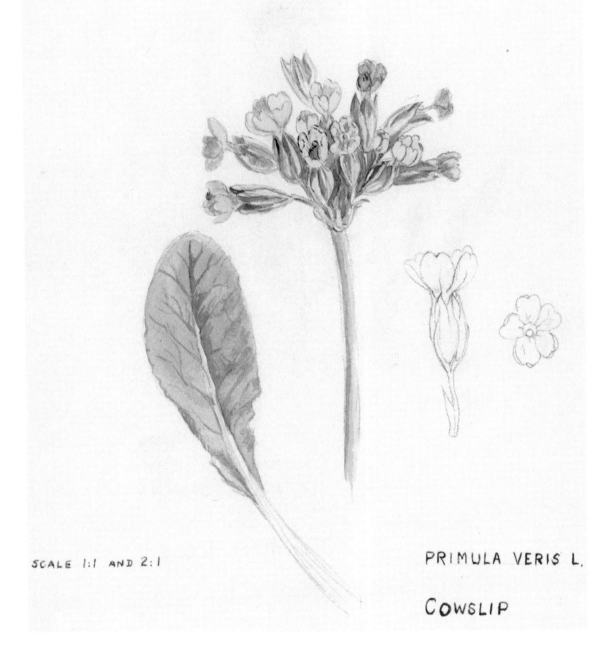

SCALE 1:1 AND 2:1

PRIMULA VERIS L.

COWSLIP

ROSACEAE

SCALE 1:1

ROSA CANINA L.

DOG ROSE

SCROPHULARIACEAE

VERONICA CHAMAEDRYS L.

GERMANDER SPEEDWELL

SCALE 1:1

GERANACEAE

GERANIUM ROBERTIANUM L.

HERB ROBERT

SCALE 1¾ : 1

CRUCIFERAE

SCALE 1:1 AND 3:1

LOWER LEAFLETS

CARDAMINE PRATENSIS L.

LADY'S SMOCK

RANUNCULACEAE

SCALE 1½:1   RANUNCULUS FICARIA L.

                              LESSER CELANDINE

BORAGINACEAE

SCALE 1:1 AND 2:1

PULMONARIA OFFICINALIS L.

LUNGWORT

RANUNCULACEAE

CALTHA PALUSTRIS L.

MARSH MARIGOLD

SCALE 1:1

POLYGALACEAE

POLYGALA VULGARIS L.

MILKWORT

COMPOSITAE

CHRYSANTHEMUM LEUCANTHEMUM

OX-EYE DAISY

PRIMULACEAE

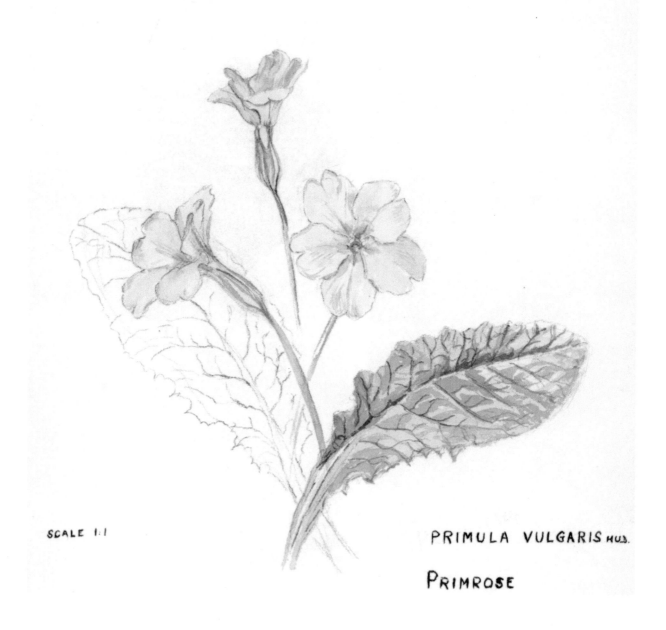

SCALE 1:1

PRIMULA VULGARIS HUD.

PRIMROSE

LILIACEAE

SCALE 1:1

ALLIUM URSINUM L.

RAMSONS

CARYOPHYLLACEAE

LYCHNIS DIOICA L.

RED CAMPION

SCALE 1½:1 AND 1¾:1

COMPOSITAE

SCALE 1:1

ASTER TRIPOLIUM L.

SEA ASTER

LILIACEAE

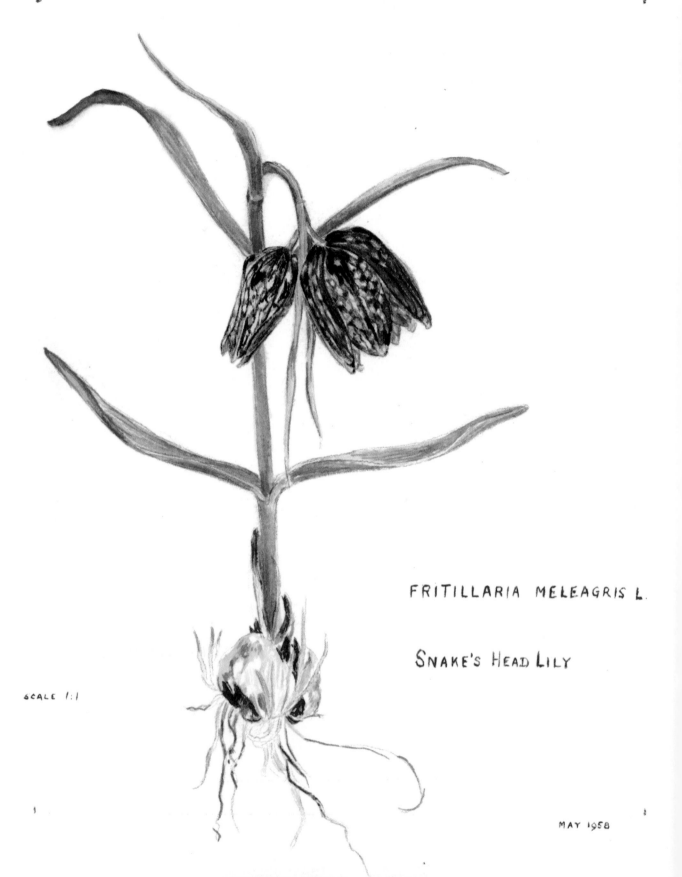

FRITILLARIA MELEAGRIS L.

SNAKE'S HEAD LILY

SCALE 1:1

MAY 1958

VIOLACEAE

SCALE 1:1

VIOLA ODORATA L.

SWEET VIOLET

ROSACEAE

POTENTILLA ERECTA L.

TORMENTIL

SCALE 1¾:1

HYPERICACEAE

SCALE 1:1

HYPERICUM ANDROSAEMUM L.

TUTSAN

RANUNCULACEAE

SCALE 1½ : 1

ANEMONE NEMEROSA L.

WOOD ANEMONE

OXALIDACEAE

OXALIS ACETOSELLA L.

WOOD SORREL

SCALE 1:1

$\frac{1}{4}$

NUPHAR LUTEA.

YELLOW WATER
LILY

Printed and bound by CPI Group (UK) Ltd, Croydon, CR0 4YY

03/10/2024

01040316-0013